McGRAW-HILL YEARBOOK OF
Science & Technology

1993

McGRAW-HILL YEARBOOK OF
Science & Technology

1993

Comprehensive coverage of recent events and research as compiled by
the staff of the McGraw-Hill Encyclopedia of Science & Technology

McGraw-Hill, Inc.
New York St. Louis San Francisco Auckland Bogotá Caracas Lisbon London Madrid Mexico
Milan Montreal New Delhi Paris San Juan São Paulo Singapore Sydney Tokyo Toronto

McGRAW-HILL YEARBOOK OF SCIENCE & TECHNOLOGY
Copyright © 1992 by McGraw-Hill, Inc.
All rights reserved. Printed in the United States of America.
Except as permitted under the United States Copyright Act of 1976,
no part of this publication may be reproduced or distributed in any
form or by any means, or stored in a database or retrieval system,
without prior written permission of the publisher.

1 2 3 4 5 6 7 8 9 0   DOW/DOW   9 8 7 6 5 4 3 2

Library of Congress Cataloging in Publication data

McGraw-Hill yearbook of science and technology.
1962–   . New York, McGraw-Hill Book Co.

    v. illus. 26 cm.
    Vols. for 1962–   compiled by the staff of the
McGraw-Hill encyclopedia of science and technology.
    1. Science—Yearbooks. 2. Technology—
Yearbooks. 1. McGraw-Hill encylopedia of
science and technology.
Q1.M13   505.8   62-12028

ISBN 0-07-046736-6
ISSN 0076-2016

International Editorial Advisory Board

Editorial Staff

Sybil P. Parker, Editor in Chief

Arthur Biderman, Senior Editor
Jonathan Weil, Editor
Betty Richman, Editor
Ginger Berman, Editor
Patricia W. Albers, Editorial Administrator

Ron Lane, Art Director
Vincent Piazza, Assistant Art Director

Joe Faulk, Editing Manager
Ruth W. Mannino, Editing Supervisor

Thomas G. Kowalczyk, Production Manager
Suzanne W. Babeuf, Senior Production Supervisor

Suppliers: Electronic Technical Publishing Services Company, Portland, Oregon, generated the line art, and composed the pages in Times Roman, Helvetica Black, and Helvetica Bold.

The book was printed and bound by R. R. Donnelley & Sons Company, The Lakeside Press at Willard, Ohio.

Consulting Editors

Consulting Editors (continued)

Contributors

A list of contributors, their affiliations, and the titles of the articles they wrote appears in the back of this volume.

Preface

The *1993 McGraw-Hill Yearbook of Science & Technology* continues a long tradition of presenting outstanding recent achievements in science and engineering. Thus it serves both as an annual review of what has occurred and as a supplement to the *McGraw-Hill Encyclopedia of Science & Technology*, updating the basic information in the seventh edition (1992) of the Encyclopedia. It also provides a preview of advances that are in the process of unfolding.

The Yearbook reports on topics that were judged by the consulting editors and the editorial staff as being among the most significant recent developments. Each article is written by one or more authors who are specialists on the subject being discussed.

The *McGraw-Hill Yearbook of Science & Technology* continues to provide librarians, students, teachers, the scientific community, and the general public with information needed to keep pace with scientific and technological progress throughout our rapidly changing world.

Sybil P. Parker
EDITOR IN CHIEF

Aeronautical meteorology

Aviation must cope with the weather, and because of increased aircraft and system capabilities, pilots tend to fly in nearly all weather conditions. Science has advanced understanding of weather hazards, and some of these advances are entering the operational aviation environment.

Today's aviation system, which includes airports, aircraft, pilots, and controllers, is designed to function in all types of weather. However, the safety of aircraft and passengers and the efficiency of airport operations continue to be threatened by potential aviation weather hazards. In response, new technology to sense and quantify aviation weather hazards is being developed in parallel with increased phenomenological understanding.

Aviation and weather. Most aircraft operate in the troposphere, the layer of the Earth's atmosphere that is the site of all important weather processes. New methods for sensing weather that is hazardous to aviation operations, and a better understanding of that weather are critical to efficient and safe operations.

Weather dictates the rules under which aircraft are operated. If the weather is good, airplanes operate under visual flight rules (VFR); if the weather is too bad to allow visual flight rules, pilots may obtain an instrument flight rules (IFR) clearance and fly solely by reference to instruments as opposed to visual references. With the enhanced capabilities of the airplanes, the pilots, and the air-traffic control system in the United States, it is often more expeditious to operate aircraft under instrument flight rules regardless of the weather. Such aircraft can be exposed to more hazardous weather conditions than they would encounter if operating under the far more conservative visual flight rules, where bad weather is avoided.

Weather sensing and depiction. Satellite photographs remain one of the best ways for pilots to obtain an overall perspective of the weather.

Photographs from the *Geostationary Operational Environmental Satellite (GOES)* located 22,300 mi (35,700 km) above the Equator provide the only photographic data input to systems used by pilots and the rest of the aviation weather system. Outlets for these photographs include the Weather Channel (cable television), local television stations, and the aviation weather briefer. When one of these satellites failed in 1986, the United States was left with a single *GOES* system to cover the entire country. The remaining *GOES* is operating well beyond its projected lifetime, and another weather satellite was not launched until 1992. The new satellite, *GOES-Next*, will provide vastly improved weather information in the form of more detailed resolution and more frequently transmitted photographs, plus data on the vertical structure of the atmosphere such as information on winds and water vapor.

Continued improvement in the weather models of the National Weather Service (NWS) provides more accurate forecasts for the aviation users. The National Weather Service runs a specific aviation global-scale model that provides higher-resolution information on winds and temperatures at aircraft cruising altitudes. The computer power available today enables all atmospheric models to evolve continually to finer resolution. The three most commonly used weather models have grid points (points in space where information is calculated) every 48–96 mi (80–160 km) in the horizontal, have 7–20 different pressure levels in the vertical, and are run twice per day. A new model has been designed to run every 6 h with up to 18 mi (30 km) resolution at 30 different altitude (pressure) levels. This model was made possible by improved meteorological understanding, greater computer power, and additional measurements that aided in initialization of the model. SEE WEATHER FORECASTING AND PREDICTION.

The capability to obtain wind and temperature data automatically from air carriers was completed in 1991. These asynchronous weather data are provided to the

National Weather Service to improve the initialization of weather prediction models. The National Weather Service has installed a demonstration network of 30 wind profilers in the central United States (**Fig. 1**); each provides hourly vertical profiles of the wind above the site. While the above capabilities have not yet been totally integrated into the national weather database, they are far superior to the current method of upper-air wind sensing that requires launching weather balloons twice daily at locations across the United States. Expanded use of wind profilers may require the allocation of additional frequencies to transmit the necessary measurement. SEE RADIO SPECTRUM ALLOCATIONS.

Thunderstorms. Thunderstorms initiate a vast array of associated potential weather hazards, and they are the meteorological event most feared by pilots. The general purpose is to avoid thunderstorms by a wide margin, but at times this may not be possible.

Lightning. Historically, it was necessary to hear thunder or visually observe lightning to verify that a convective cloud was indeed a thunderstorm. A major advance in recent years is the national lightning network, consisting of a series of sensors that detect and locate cloud-to-ground lightning strokes. Depiction of lightning locations and stroke frequencies are now available to the aviation weather system and provide vastly improved information for the location of thunderstorms and the consequent lightning hazard.

Hail and heavy rain. These hazards can be detected by ground-based and airborne radars. These radars have improved over the past few years, but they are still unable to discriminate between heavy rain and hail, although this distinction is theoretically possible and has been demonstrated with research radars. When modernization of the entire National Weather Service weather radar system is complete, aviation users will be provided with national weather radar coverage. In 1991, the National Weather Service began installing a new Doppler weather radar (WSR-88D) to provide a complete volumetric scan of all airspace above 10,000 ft (3000 m) over the entire United States, replacing the 1957-vintage radars. The WSR-88D digital radar has a much smaller beam size so that it can detect the high-radar-reflectivity areas associated with heavy rain and hail with more precision and resolution than the older radar systems. SEE METEOROLOGICAL RADAR.

Microburst. The most hazardous of all thunderstorm weather phenomena during aircraft takeoffs and landings is a form of severe windshear known as the microburst. In the United States, this hazard has been associated with 149 incidents that claimed 465 lives over a 10-year period. Microbursts are produced by powerful, small-scale downdrafts of cold, dense air that can occur beneath thunderstorms. The downdrafts produce divergent flow when they impact the surface of the Earth. Aircraft flying through a microburst during takeoffs and landings experience a strong headwind, then a downdraft, and finally a tailwind that produces an immediate, sharp airspeed reduction and

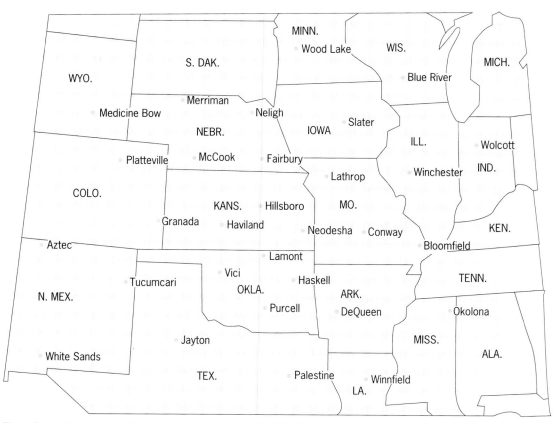

Fig. 1. Sites of the wind profiler demonstration network. (*Profiler Group Systems Laboratory, Boulder, Colorado*)

loss of lift. Because of the lethal nature of this hazard, a multifaceted approach to the problem was instituted in the early 1980s. The Federal Aviation Administration developed the Windshear Training Aid, which is the basis for mandatory windshear training for all air-carrier pilots. The Windshear Training Aid is designed to increase pilot awareness and understanding of severe windshear and microbursts. In-place warning devices have also been developed for installation on aircraft to alert a pilot when the aircraft is actually in a microburst. These devices are now mandatory equipment on all large commercial passenger-carrying aircraft. Infrared sensors are in the final stages of development; they will allow pilots to "see ahead" of their aircraft and detect microbursts before they fly into them.

The development of ground-based sensors to detect microbursts and warn pilots is the third part of this coordinated approach to minimize the microburst hazard. The Terminal Doppler Weather Radar (TDWR) is a ground-based system designed to detect and specifically identify microbursts. The first of these systems was installed in 1992. **Figure 2** shows the Terminal Doppler Weather Radar display depicting microburst alerts. The radar is also capable of detecting a number of additional weather events in the airport terminal area, such as radar reflectivity (associated with rain rate), gust fronts, storm motion, snowfall rate, and wind shift.

Gust fronts. These strong, rapid changes in wind direction and velocity are another form of windshear associated with thunderstorms. Gust fronts represent a potential hazard in the terminal area and often disrupt airport operations. The Low Level Windshear Alert System (LLWAS) was developed by the FAA in the early 1980s to detect windshear associated with gust fronts. This system measures wind speed and direction at several locations on and around the airport with a series of anemometers. Modernization of 110 of these sytems was completed in October 1991. The Low Level Windshear Alert System and the Terminal Doppler Weather Radar can detect gust fronts and alert both pilots and controllers.

Knowledgeable operational planning also requires a future prediction of the location of wind-shift lines (gust fronts) in the airport terminal area. Installation of the Low Level Windshear Alert System and the Terminal Doppler Weather Radar makes it possible to locate gust fronts and predict their future position in greater detail. The Terminal Doppler Weather Radar can locate gust fronts based on Doppler radar shift and track their positions, enabling forecasters to predict their future locations.

Turbulence. Doppler radar technology can also be used with airborne weather radars to measure convective cloud turbulence ahead of aircraft, and a similar technique is being tested for use by ground-based radars. While there are enough particles (insects, raindrops, and so forth) in the boundary layer close to the Earth's surface and in clouds for this technique to work, most commercially available radars are not sensitive enough to detect turbulence in clear air.

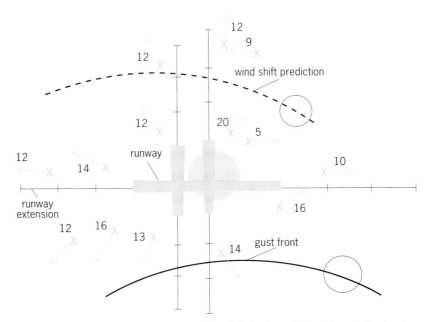

Fig. 2. Geographic situation display. The colored circle shows the location of microbursts. Open circles show locations of windshear. Arrows show sensor locations of the Low Level Windshear Alert System and measured wind direction and velocity in knots (1 knot = 1.85 km/h). Shaded areas along runways indicate where aircraft could be exposed to microburst hazards.

Icing. In-flight icing is a major hazard for small aircraft. Weather research is seeking to transfer some of the new physical understanding of icing conditions into the forecast process, including the mechanisms for the production and depletion of supercooled liquid water, which produces ice on aircraft in flight. Effective forecasts of icing also depend on knowledge of temperature and droplet size. Remote and in-place measurements of temperature are improving, but currently there is no operational technology for remote measurement of droplet size.

The principal problem for large aircraft while waiting in the terminal area is ice accumulation from snowfall and frost. There are deicing methods for the removal of snow and ice from the aircraft prior to take-off, but they depend on the amount of snow, temperature, and water content of the snow and wind. Techniques are being developed to forecast these detailed parameters to enable a more comprehensive approach to deicing.

Ceiling and visibility. The capability of pilots to operate safely and legally is based on their ability to see the runway and surrounding area during takeoff and landing. While technology has been developed to allow aircraft to take off and land automatically, even in weather conditions with zero ceiling and zero visibility, pilots must have a minimum ceiling and visibility to operate. Even pilots of aircraft capable of landing in zero-visibility conditions must be able to see to taxi.

Devices to measure visibility at the touchdown point (runway visual range, RVR) have been widely deployed in recent years. Cloud ceilings, however, have been measured by a variety of techniques, many of which require human input and judgment. Recently deployed laser ceilometers continuously and automatically measure the height of the cloud base.

Automatic surface observations. Traditionally, surface observations of key weather parameters have been taken each hour by certified weather observers. The Automatic Surface Observing System (ASOS) and the Automatic Weather Observing System (AWOS) are being deployed, and within a few years over 1000 of these systems will be available to make automatic observations of key weather parameters on a continuous basis.

For background information SEE AERONAUTICAL METEOROLOGY; APPLICATIONS SATELLITE; CLEAR-AIR TURBULENCE; LIGHTNING; METEOROLOGICAL RADAR; RADAR METEOROLOGY; THUNDERSTORM in the McGraw-Hill Encyclopedia of Science & Technology.

Wayne Sand

Bibliography. J. Cole and W. Sand, Statistical study of aircraft icing accidents, *Proceedings of the 29th Aerospace Sciences Meeting*, Reno, AIAA 91-0558, January 7–10, 1991; M. K. Politovich and R. Olson, An evaluation of aircraft icing forecasts for the continental United States, *4th International Conference on the Aviation Weather System*, Paris, American Meteorological Society, June 24–26, 1991; W. Sand and J. McCarthy, Microburst wind shear: Integration of ground-based sensors to produce effective aircraft avoidance, *Royal Aeronautical Society's One Day Conference on Wind Shear*, London, November 1, 1990; P. Schultz and M. K. Politovich, Automated guidance for forecasting conditions conducive to aircraft icing, *7th International Conference on Interactive Information and Processing Systems for Meteorology, Hydrology and Oceanography*, New Orleans, American Meteorological Society, January 14–18, 1991.

Affective disorders

Affective disorders, also known as mood disorders, are common conditions that frequently require medical treatment. Data from National Institute of Mental Health epidemiological studies indicate that at some point in their lives, about 5% of people will have mood problems serious enough to warrant a diagnosis of major affective disorder. The symptoms of an affective disorder generally involve changes in normal patterns of sleep, appetite, energy level, and interest or enjoyment of daily activities, as well as pervasive changes in mood. A large body of evidence demonstrates alterations in the functioning of the nervous system in patients suffering from affective disorders. Fortunately, these conditions can often be treated effectively; and apparently, many of these nervous system changes tend to normalize when patients recover.

Classification. Although a variety of schemes have been used to classify affective disorders, most scientific research now recognizes two main types, bipolar affective disorder (also known as manic-depressive illness) and unipolar affective disorder (also known as major depression). The difference is determined by whether or not the patient has ever had an episode of mania. Mania is a mood disturbance characterized by an abnormal increase in energy and a decreased need for sleep, accompanied by an elevated (elated or euphoric) or very irritable mood. Depression, on the other hand, is manifested by a decrease in energy and a depressed (dysphoric) mood. In bipolar disorder, there are episodes of mania and usually episodes of depression as well. In unipolar disorder, patients experience only depressive episodes.

This distinction is important in the diagnosis and treatment of these disorders. Although bipolar depression and unipolar depression are often quite similar in appearance and therefore difficult to differentiate, they frequently respond to treatment quite differently. For instance, lithium appears to be better for treating bipolar depression than unipolar depression. There are also differences in the age and gender distribution of patients: on average, bipolar disorder has an earlier age of onset than unipolar disorder (early twenties versus late twenties to thirties), and bipolar disorder occurs about equally in men and women while women outnumber men about two to one in unipolar disorder. Measurable changes in brain function have been demonstrated in both types of affective disorders by using the technology of positron emission tomography. Furthermore, recent studies show that, while there are some similarities in how the brain is affected by bipolar and unipolar disorders, there are also dissimilarities.

Positron emission tomography (PET). PET is an analytic imaging technique that allows the measurement of local rates of biochemical reactions in tissue within an organism. These measurements are made by using very small amounts of biologically active compounds labeled with positron-emitting isotopes. Positron emission from these labeled compounds results in the production of gamma rays, which can be measured in a tomograph—an array of radiation detectors placed around the head. The collected data permit the study of the biochemistry and physiology of localized brain regions without disturbing ongoing physiological processes.

Various biochemical processes can be studied by using PET. The studies described below measure local cerebral metabolic rates for glucose, that is, the rates at which various brain areas utilize glucose for generating energy. A positron-labeled glucose analog, fluorine-18 fluorodeoxyglucose (FDG), is used. Tracer kinetic models developed for this compound make it possible to use PET scans for quantitative studies of the local cerebral metabolic rate for glucose in humans.

The brain is unusual among the body's organs in that, except under starvation conditions, glucose metabolism provides all of its energy needs. Since glucose consumption largely reflects nerve-cell activity (both excitatory and inhibitory), cerebral glucose metabolism provides a sensitive measure of brain function.

Metabolic change. When cerebral glucose metabolism was studied in individual patients with affective disorder, only those patients with bipolar disorder showed a regular and consistent pattern of metabolic change as their mood changed. Bipolar patients showed an increase in cerebral metabolism when they changed from a state of depressed or mixed mood to a more euphoric condition. Thus, the cerebral metabolic rate of a bipolar depressed patient is lower when the patient is depressed than when that patient is euthymic (in a normal mood) or manic. However, unipolar depressed patients show no consistent pattern of change in cerebral metabolic rate when they change from the depressed to the euthymic state. This difference is found even when the severity and appearance of the depressions of the two groups are indistinguishable. This finding is evidence that bipolar and unipolar depressions, while similar in appearance, may have differing underlying biological foundations.

Caudate nucleus. Another difference between bipolar and unipolar depressions is seen in the local cerebral metabolic rate for glucose for a brain structure called the caudate nucleus. This structure is part of the basal ganglia, a group of gray-matter structures located deep within the brain. The caudate nucleus and a structure called the putamen are part of a basal ganglia subunit called the corpus striatum. These structures seem to function as a gating mechanism or "filter" that processes signals sent in to them by the cerebral cortex, the very complex outer surface of the brain.

The caudate/hemisphere ratio (the local cerebral metabolic rate for glucose of the caudate nucleus divided by the value of the metabolic rate for the cerebral hemisphere of which it is part) has been found to be lower in patients with unipolar depression than in patients with bipolar depression. This difference is of even greater magnitude if the ratio for the local cerebral metabolic rate for glucose of the caudate nucleus is calculated with respect to a part of the cerebral hemisphere called the prefrontal cortex, which is the front part of the outer surface of the brain. This finding may be significant since one of the areas from which the caudate nucleus specifically processes or filters messages is the prefrontal cortex. This metabolic difference between bipolar and unipolar depressions may also indicate a difference in the underlying brain mechanisms.

Prefrontal cortex. A change in brain metabolism that is common to both bipolar and unipolar depressions involves the prefrontal cortex. This region is considered to be among the most recently evolved areas of the brain, and presumably serves in "higher" or complex cognitive functions. The metabolic rate for this region relative to the metabolic rate of the hemisphere (the prefrontal cortex/hemisphere ratio) is decreased in both bipolar and unipolar depressed patients compared to a nondepressed control group (see **illus.**). Some evidence indicates that this may be a stronger finding for the left prefrontal cortex since the left prefrontal cortex/hemisphere ratio was also decreased in patients with depression secondary to obsessive-compulsive disorder and in bipolar depressed patients compared to a bipolar manic group. Further, the prefrontal cortex/hemisphere ratio showed a significant correlation with a standard measure of severity for depression, the Hamilton depression rating scale. This correlation indicates that there is a significant trend for patients with a greater decrease in their prefrontal cortex/hemisphere ratio to manifest a more severe de-

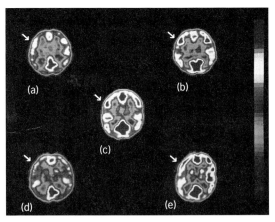

Positron emission tomographic scans demonstrating the glucose metabolic rate of each area of the brain, divided by that of the whole brain (darker areas represent higher relative metabolic rates). Scans are from five middle-aged women and illustrate findings in the left prefrontal cortex (arrows). (*a*) Unipolar depression. (*b*) Obsessive-compulsive disorder without secondary major depression. (*c*) Normal control. (*d*) Bipolar depression. (*e*) Obsessive-compulsive disorder with secondary major depression. The patients with depression show lower relative rates of metabolism in the left prefrontal cortex compared to the normal control individuals and obsessive-compulsive disorder patients without depression. (*From L. R. Baxter et al., Reduction of prefrontal cortex glucose metabolism common to three types of depression, Arch. Gen. Psychiat., 46:243–250, March 1989; copyright 1989 by American Medical Association*)

pression. When a group of patients were studied after their depression improved, there was a significant increase in their left prefrontal cortex/hemisphere ratio, and a significant correlation between the percentage of increase of this measure and the percentage of improvement on their Hamilton depression rating scale.

Prospects. There are numerous studies implicating other aspects of brain function in affective disorders. For instance, promising advances have been made in neuroendocrinology, the study of how the brain regulates the hormonal responses of the body. Interesting changes have been seen in nervous system regulation of hormones such as cortisol, growth hormone, and thyroid hormone in patients with affective disorders. The regulation of neurotransmitters such as norepinephrine and serotonin has also been an area of intensive investigation.

As knowledge and understanding of brain function improves, it may become possible to distinguish those biological changes that are causally linked to disorders in mood from those that are the result of other aspects of the disorder, for example, motivational or cognitive changes. Treatment approaches may thus be enhanced.

For background information SEE AFFECTIVE DISORDERS; MEDICAL IMAGING; NORADRENERGIC SYSTEM in the McGraw-Hill Encyclopedia of Science & Technology.

Jeffrey M. Schwartz

Bibliography. L. R. Baxter, PET studies of cerebral function in major depression and obsessive-compulsive disorder: The emerging prefrontal cortex consensus, *Ann. Clin. Psychiat.*, 3:103–109, 1991; F. K. Goodwin and K. R. Jamison, *Manic-Depressive Illness*, 1990; H. I. Kaplan and B. J. Sadock (eds.),

Comprehensive Textbook of Psychiatry, 5th ed., 1989; J. M. Schwartz et al., The differential diagnosis of depression: Relevance of positron emission tomography studies of cerebral glucose metabolism to the bipolar-unipolar dichotomy, *JAMA,* 258:1368–1374, 1987.

Agriculture

The rising costs of chemical pesticides and fertilizers, the restrictions on use, the reduction in available land and water, and the prospect of increased competition from agricultural imports have increased the pressure on the farmer to produce high-quality food and fiber at the lowest possible cost. Farmers in the United States are under increased pressure, both regulatory and economic, to reduce levels of inputs of chemical pesticides, fertilizers, and water. Agricultural economists have found that chemical pesticides are often applied as a form of "insurance," and not necessarily as a response to real pest pressure. The overuse of other inputs such as fertilizer and water can actually reduce yields. Therefore, it makes both economic and environmental sense for farmers to apply inputs only at a level appropriate to the needs of their crop. To assess these needs, the farmer must treat the crop as a functioning ecosystem and determine how the addition of inputs at a particular level would affect this system. This systems-based approach is sometimes called integral crop management.

Need for expert systems. If farmers wish to reduce their use of chemicals and other inputs, they must be more precise in both the timing and the amount of inputs. Therefore, farmers must have better information about the current and projected status of their crop. With the advent of low-cost personal computers readily available for on-farm use, it is natural for computers to be used to help meet the increased demand for information. As a result, computer-based decision support software (expert systems) for use by farmers in crop management has been developed.

Operation and features. An expert system is a computer program with the following attributes: (1) it solves real-world problems (as opposed to abstract problems in statistics, mathematics, or optimization, for example); (2) it uses heuristic and nonnumeric knowledge in developing a solution; and (3) it is readily adaptable to new information. As a practical matter, most expert systems currently are rule-based. A rule-based expert system consists of four basic components (**Fig. 1**): a set of one or more knowledge bases (or rule bases), an inference engine for processing the knowledge in the rule bases, a dynamic memory for storing information about the problem and its solution, and a user interface. The rule base contains all knowledge specific to the problem domain, that is, to the specific subject of expertise. The inference engine contains all the logical capacity to form chains of reasoning by using the knowledge in the rule base.

Virtually all rule-based expert systems currently employ a software structure based on the program MYCIN, developed at Stanford University in the late 1970s as a medical diagnostic tool. The simple rule-based expert system in Fig. 1 illustrates the basic operation of such systems. Its rule base consists of a set of production rules, a goal, and one or more queries, which are instructions on how to query the user when seeking information. The rule base in **Fig. 2** diagnoses disorders of cotton in California. It chooses among the four primary causes of economic damage: lygus bugs (*Lygus* spp.), which feed on the flower buds, called squares, and on young fruit, called bolls; spider mites (*Tetranychus* spp.), which feed on leaves; Verticillium wilt, a fungus (*Verticillium dahliae*) disease; and Fusarium wilt, involving a fungus (*Fusarium oxysporum*) that forms a disease complex with parasitic nematodes.

Backward chaining. The most common form of inference is backward chaining, in which the system starts from a goal and attempts to link together a chain of reasoning leading to this goal. In the rule base of Fig. 2, the goal is the value of the variable "cause." The inference engine searches the rule base until it finds a rule with this variable in its "then" part; in this case it first finds rule 2. To use this rule, it must first determine the values of the variables "agent" and "galls." It recursively searches the rule base for rules with "agent" in the "then" part and finds rule 1. To use rule 1, the inference engine must determine the values of the variables "external damage" and "stem discoloration." A search for the variable "external damage" produces no rules, so the system queries the user for its value. If the value is "false," then the process is repeated for "stem discoloration." If the value of "external damage" is "true," then rule 1 is abandoned, and a search for other possible rules is mounted. The process continues until either a sequence of rules that leads to the goal is

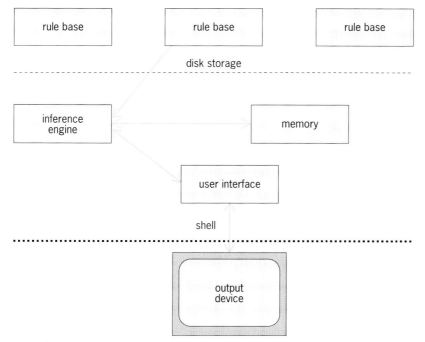

Fig. 1. Schematic diagram of a rule-based expert system showing its four primary components. Each of the three rule bases may be separately accessed by the inference engine.

RULE 1: IF external damage = false
AND stem discoloration = true
THEN agent = pathogen

RULE 2: IF agent = pathogen
AND galls = false
THEN cause = Verticillium wilt

RULE 3: IF agent = pathogen
AND galls = true
THEN cause = Fusarium wilt

RULE 4: IF external damage = true
THEN agent = arthropod

RULE 5: IF agent = arthropod
AND leaf color = red
THEN cause = spider mites

RULE 6: IF agent = arthropod
AND leaf color = green
AND square or boll damage = true
THEN cause = lygus bugs

GOAL: cause

QUERY external damage:
Is external damage visible?

QUERY: galls
Does the plant have galls?

QUERY: stem discoloration
Are the insides of stems discolored?

QUERY: leaf color
What color are the leaves?

QUERY: square or boll damage
Are squares or bolls damaged?

Fig. 2. Schematic of a simple rule base for distinguishing among four possible causes of damage to a cotton crop.

found or the list of possibilities is exhausted. A sample of a full consultation session is shown in **Fig. 3.**

The example shows that rule-based expert systems are readily capable of incorporating heuristic and nonnumeric information into their problem solving. Indeed, the rule base may, in some circumstances, consist primarily or entirely of rules of thumb, that is, of empirical information. Such information is much more difficult to integrate into other types of computer-based decision support software.

Rule-based expert systems have a number of other useful built-in features. One is the ready availability of an automatic explanation system through which the system can disclose its problem-solving procedure to the user. As shown in Fig. 3, the user interface can

provide, in response to the question "Why?", the rule currently under consideration. Moreover, the memory facility can be made to hold the rule used to deduce a particular conclusion together with the conclusion itself. When a particular conclusion is announced, the expert system can respond to the query "How?" by displaying the rule used to deduce this conclusion. The entire logical chain of reasoning can then be unwound in response to repeated "How?" queries referring to specific statements in the rule being displayed.

Incorporation of uncertainty. A second commonly used feature of rule-based expert systems is the incorporation of uncertainty. In many cases, particularly in a field such as agriculture, the conclusions associated with a given rule cannot be made with absolute certainty. For example, while lygus bugs are the most common cause of damage to squares and small bolls of cotton plants in California's San Joaquin Valley, other insects occasionally cause such symptoms as well. Therefore a more appropriate rule for identifying the type of insect might be:

RULE 6: IF agent = arthropod
AND leaf color = green
AND square or boll damage = true
THEN cause = lygus bugs (certainty factor = 0.7)
cause = heliothis (certainty factor = 0.2)

Here the certainty factor is a numerical representation of the degree of belief in the conclusion. A common convention is that a certainty factor of 1.0 indicates absolute certainty. If certainty factors are to be incorporated into an expert system, then a means must be provided for calculating the certainty of a final conclusion based on the certainties associated with

Is external damage visible?
Yes
What color are the leaves?
Why
The program is trying to use the rule
IF agent = arthropod
AND leaf color = red
THEN cause = spider mites
How arthropod
arthropod was verified by the rule
IF external damage = true
THEN agent = arthropod
What color are the leaves?
Green
Are squares or bolls damaged?
Yes
The cause is lygus bugs
How
lygus bugs was verified by the rule
IF agent = arthropod
AND leaf color = green
AND square or boll damage = true
THEN cause = lygus bugs

Fig. 3. Sample consultation session that uses the rule base of Fig. 2. Communications from the expert system are shown in plain text, and user responses are shown in bold.

intermediate conclusions. A number of such formulas are available. One way is to interpret the certainty factors as probabilities and use the rules of probability to do the calculations. Other formulas are available, however, that do not require the certainty factors to be considered as probabilities.

Domain knowledge and inference mechanism. A third important feature of rule-based expert systems is the separation of the domain knowledge from the inference mechanism. All knowledge specific to the problem domain (for example, diagnosis of cotton disorders in the example discussed above) is contained in the rule base in the form of individual rules. The rule base does not, however, have any capability in linking the rules together in a chain of inference. This capability resides in the inference engine and memory, which, together with the user interface, make up the expert system shell. Individual rule bases are generally stored on the computer disk as separate files, similar to the separate files used to store individual documents in a word processor. As shown in Fig. 1, the same shell can be used with any one of the files, so that the expertise of the system is readily changed from one subject to another.

The ability of the expert system shell to sequentially process information from several different knowledge bases is especially useful in crop management. A fundamental tenet of integrated crop management is the incorporation of information from all facets of the agroecosystem into the decision-making process. Decisions on pest management, for example, must take into account plant health, water status, fertilization requirements, and so forth.

CALEX system. While the standard expert system shell does not provide any means of retaining and integrating information from previously processed knowledge bases when it is processing the current one, shells designed specifically for crop management do provide this feature. For example, the CALEX shell retains information on a special structure called the blackboard and makes this information available to other knowledge bases.

The CALEX blackboard is more than a passive slate on which information is stored. The blackboard concept, now widely used in artificial intelligence applications, originated in programs for understanding spoken language. Such programs must use contextual information to make sense of similar-sounding phrases such as "till Bob brings" and "tell Bob rings." This contextual information is provided by phrases recently heard and understood. Similarly, information from previously processed knowledge bases in an integrated crop management system provides the context in which decisions are made. The blackboard provides a control structure that determines which knowledge bases are processed, and in what order.

For example, suppose that a user of CALEX/Cotton, an implementation of the CALEX shell for cotton management in the San Joaquin Valley, California, has identified lygus bugs as a potential problem and calls on the program to determine whether the problem is sufficiently serious to warrant pesticide treatment.

The expert system bases its decision making on the recognition that the cotton plant normally produces many more bolls than it can support to maturity and harvest. A large fraction of the bolls are shed while they are immature. Therefore, the decision of whether or not to recommend a pesticide treatment for lygus bugs is based on a comparison of the number of excess bolls present in the crop with the rate of boll consumption by the current lygus population level. If the number of bolls that would be shed for physiological reasons exceeds the estimated consumption, then a pesticide application is not necessary. In this case a future scouting for lygus bugs is recommended. The rule base for lygus bugs needs information from the rule base for irrigation to know when the field may be wet from irrigation, since scouting is impossible during this period. The irrigation rule base in turn needs information from rule bases determining the crop vigor, the status of the crop disease Verticillium wilt, and the projected date when the crop will "cut out" (cease producing harvestable fruiting structures). Each of these rule bases is therefore processed first, followed by the irrigation rule base and then by the lygus bug rule base.

Problems and prospects. Paradoxically, the ability to use empirical knowledge and rules of thumb, which is one of the great advantages of the expert system approach, is also one of its greatest weaknesses: this ability often leaves the resulting expert system vulnerable to a phenomenon called brittleness. Brittleness refers to the inability of the expert system to deal with situations slightly different from those for which the system was designed and programmed, such as the situation in which the same crop is grown in a different environment. It is caused by the shallow, empirical nature of the knowledge encoded in the system. It can be alleviated by the inclusion of fundamental, mechanistic knowledge about the problem domain. For example, in a crop management expert system a crop growth model can be included. The cotton management system GOSSYM-COMAX, for example, consists of the linkage of the crop simulation model GOSSYM and the expert system COMAX. If such a system is to be used in a new environment, the crop model may need only to be recalibrated for the system to function, rather than completely reworked to reflect different empirical knowledge. A major drawback of this approach is the length of time needed to create and test a crop simulation model; this factor makes the development of such systems difficult for crops for which no such model already exists. Current research is focused on the development of crop management systems that represent a compromise between the pure rule-based system and the system relying primarily on a detailed crop simulation model. It is hoped that such systems will provide some of the robustness associated with the crop model without requiring detailed physiological realism and the associated long development time.

CALEX/Cotton and GOSSYM-COMAX are two of the most fully developed crop management decision-support systems. Both have been implemented for a number of years, and are used by a growing number of farmers. As yet, neither can be said to have really made a substantial difference in the way cotton is grown. Both systems, and others like them, will require some years of further refinement. In addition, many farmers are conservative in their application of new technologies. By the end of the twentieth century, however, it is likely that successful farmers of cotton and other crops will be relying on such programs for advice in crop management.

For background information SEE AGRICULTURAL SCIENCE (PLANT); AGRICULTURE; COTTON; EXPERT SYSTEMS in the McGraw-Hill Encyclopedia of Science & Technology.

Richard E. Plant

Bibliography. J. R. Barrett and D. D. Jones, *Knowledge Engineering in Agriculture*, 1989; B. G. Buchanan and E. H. Shortliffe, *Rule-Based Expert Systems*, 1984; R. E. Plant and N. S. Stone, *Knowledge-Based Systems in Agriculture*, 1991.

Agroforestry

During the 1980s, agroforestry attained prominence as a land-use practice particularly suited to the tropics and subtropics. Agroforestry is a new name for an old set of practices. Essentially, it represents the purposeful growing of trees and crops in interacting combinations for a variety of objectives. This article summarizes the developments leading to the emergence of agroforestry, reviews the range of practices included under the term, and outlines some specific advantages and disadvantages of agroforestry.

History. The cultivation of trees and agricultural crops in intimate combination is an ancient farming practice that has been used throughout the world. In Europe, until the Middle Ages, it was the general custom to clear-cut abandoned or degraded forest, burn the slash, cultivate food crops for varying periods on the cleared area, and plant or sow trees before, along with, or after sowing of the agricultural crops. Though no longer popular in European countries and other industrialized countries, farming systems involving trees continue to be popular in most of the tropics and subtropics. Farmers in these regions have a very long tradition of raising food crops, trees, and animals together, as well as exploiting a multiple range of products from natural woodlots.

In the past few decades, scientific advances in agriculture and forestry in the tropics have proceeded largely along the pattern of the developments in the industrialized nations, with an emphasis on the production of single commodities. This approach resulted in substantial advances in agriculture and forestry in the developing nations. However, the combined integrated production systems, on which a large number of farmers in these nations depended, did not fit into the policy framework for development of monocultural commodity systems. As a result, integrated production systems have been ignored or dismissed by development experts.

At the beginning of the 1970s, serious doubts began to be expressed about the relevance of current development policies and approaches. In particular, there was concern that the basic needs of the poorest farmers, especially those in the rural areas, were neither being considered nor adequately addressed. It was soon realized that many of the technologies that contributed to the Green Revolution of the 1970s, such as irrigation systems, fertilizers, and pesticides, were not affordable to the poor farmer. It also came to be recognized that most tropical soils, which are poorer and more easily degraded than temperate-zone soils, were unable to withstand the impact of high-input technology. At around the same time, the disastrous consequences of deforestation of the world's tropical regions, which increased at an alarming rate, were recognized. It soon became clear that a major cause of deforestation was the search for more land to provide food and fuel wood for rapidly increasing populations.

Faced with these problems, land-use experts and institutions intensified their search for appropriate land-use approaches that would be socially acceptable, enhance the sustainability of the production base, and meet the need for production of multiple outputs. These collective efforts led to studies of age-old practices based on combinations involving trees, crops, and livestock on the same land unit. Recognition of the inherent advantages of traditional land-use practices involving trees—sustained yield, environmental conservation, and multiple outputs—grew quickly, and the decade beginning in the mid-1970s represented the coming of age of agroforestry.

Concepts. Basically, agroforestry involves the deliberate growing of woody perennials on the same unit of land as agricultural crops or animals, either in some form of spatial mixture or in temporal sequence. There is significant interaction (ecological and economical) between the woody and nonwoody components of the system.

It is difficult to encompass all the concepts of agroforestry in a single, precise definition. However, available definitions imply that an agroforestry system normally involves two or more species of plants (or plants and animals), at least one of which is a woody perennial; two or more outputs; a cycle of more than 1 year; and both ecology and economics that are more complex than a monocropping system.

Tropical agroforestry. The term agroforestry as used today represents an interface between agriculture and forestry. It encompasses a wide range of land-use systems and practices, which are most widespread in the developing countries of the tropics where a combination of biophysical and socioeconomic conditions not only favor but also necessitate such integrated land use. The major agroforestry practices in the tropics are shown in the **table**.

The basic components in all agroforestry systems are woody perennials (trees), herbaceous and other agricultural species (crops), and livestock. Based on this composition, agroforestry systems are grouped as agrosilvicultural (crops and trees), silvopastoral (trees and pasture/animals), or agrosilvopastoral (crops, trees, and pasture/animals).

Other specified agroforestry systems can also be defined, for example, apiculture with trees, aquaculture involving trees and shrubs, and woodlots of multipurpose trees.

An agroforestry system is site-specific. It is characterized by certain types of practices that, taken as a whole, form a dominant land-use system in a specified locality. In any one system, there can be more than one agroforestry practice. The distinction here is that an agroforestry practice refers to a specific arrangement of components in a defined land-management unit, such as a farmer's field. Because of the site-specific nature of these systems, several hundred agroforestry systems have been recorded. The number of distinct practices of which these systems are composed is, however, small (see table).

Temperate-zone agroforestry. Compared to tropical agroforestry systems, temperate-zone agroforestry systems are less complex and diverse; efforts toward their development have been rather slow. Basically there are two major types of temperate-zone agroforestry systems: silvopastoral systems involve animal grazing under managed coniferous forests or plantations; and agrosilvicultural systems involve agricultural crops under intensively managed high-value hardwood plantations.

Tree-based agriculture involving a large number of multipurpose trees such as chestnuts (*Castanea* spp.), oaks (*Quercus* spp.), carob (*Ceratonia siliqua*), olive (*Olea europa*), and figs (*Ficus* spp.) is an ancient system of land use in the Mediterranean region. The "dehesa" system of land use, involving grazing under oak trees with strong linkages to recurrent cereal cropping in rangelands, is also a very old system in this region. Interest in the practice of agroforestry in the temperate regions grew with the trend toward wider initial spacing and subsequent thinning of forests, especially pine (*Pinus radiata*) plantations in the 1970s. Efforts to develop these land-use systems in the temperate zones are now being strengthened.

Potential of agroforestry. The advantages of agroforestry relate generally to the woody perennials in the system and can be broadly grouped as productive or protective functions. The productive advantages include tree products such as animal fodder, fuel wood, small timber for construction and other uses, and a myriad of food products. The protective and service functions stem mainly from soil enrichment. Woody perennials add organic matter via leaf litter and root biomass, enhance nutrient cycling, help reduce soil erosion, and improve the physical and chemical properties of soil.

Consequently, specific agroforestry practices have been developed for soil conservation, soil fertility improvement, reclamation of degraded lands, protection of watersheds, and use of trees and shrubs as windbreaks and shelterbelts. These productive and protective advantages can also lead to economic advantages and minimization of the risk associated with monocultural enterprises. Furthermore, low-input, multiple-output agroforestry systems are often socially more acceptable to resource-poor farmers.

Even in the tropics, agroforestry is no panacea for all the problems of land management. The production emphasis of agroforestry is on the optimization of diversified products per unit of land area as opposed to maximization of a single commodity. A larger number of species and products will demand more diversified skills and greater management attention than in monoculture. Moreover, mixed plant communities may pose greater hazards in terms of pest and disease problems; for example, trees on farms may also harbor birds or

Major agroforestry practices in the tropics

Agroforestry practice	Description
Improved fallow	Fast-growing, preferably leguminous woody species are planted and left to grow during the fallow phase of shifting cultivation; the woody species cause site improvement and may yield economic products.
Taungya	Agricultural crops are grown during the early stages of establishment of forestry plantations.
Hedgerow intercropping (alley cropping)	Fast-growing, preferably leguminous woody species are grown in crop production fields; the woody species are periodically pruned to reduce shading of crops; the prunings are applied as mulch into the alleys as a source of organic matter and nutrients, or are removed from the field to be used for other purposes such as animal fodder.
Multilayer tree gardens	Multispecies, multilayer combinations of different woody species of varying forms and growth habits.
Multipurpose trees on farms and rangelands	Fruit trees and other multipurpose trees are scattered in crop or animal production fields; trees provide fruits, fuel wood, fodder, timber, and so on.
Plantation crop combinations	Integrated multistory mixtures of tree crops such as coconut, cacao, coffee, and rubber with other tree crops, shade trees, or herbaceous crops.
Home gardens	Intimate, multistory combinations of various trees and crops in homesteads; livestock may or may not be present.
Trees in soil conservation and reclamation	Trees on bunds, terraces, and other such structures with or without grass strips; use of trees for reclamation of saline, acidic, or otherwise degraded soils.
Shelter belts and windbreaks	Use of trees to protect fields from wind damage, sea encroachment, floods, and so on.
Pasture under plantations	Cattle grazing pasture under widely spaced rows of plantation species.
Protein banks	Block planting of fodder trees on farms or rangelands for cut-and-carry fodder production.

other animals that damage grain crops. Finally, mechanization will be very difficult in multispecies plant communities.

Although the art of agroforestry is old, the science is new. Many of the hypotheses about agroforestry remain to be proven by research. However, the research that has been done indicates the scientific merits of these time-tested systems and points to several possibilities for improvement. Perhaps the greatest opportunity, as well as challenge, in agroforestry lies in exploiting the large number of indigenous, multipurpose trees and shrubs in both the tropics and the temperate zones.

For background information SEE AGRICULTURE; AGROECOSYSTEM; MULTIPLE CROPPING in the McGraw-Hill Encyclopedia of Science & Technology.

P. K. Ramachandran Nair

Bibliography. K. G. MacDicken and N. T. Vergara (eds.), *Agroforestry: Classification and Management*, 1990; P. K. R. Nair (ed.), *Agroforestry Systems in the Tropics*, 1989; P. K. R. Nair, *The Prospects for Agroforestry in the Tropics*, World Bank Tech. Pap. 131, 1990; A. Young, *Agroforestry for Soil Conservation*, 1989.

Air-traffic control

With the advent of aeronautical mobile satellite communication, the nonradar air-traffic control environment is undergoing revolutionary changes. In the near future, satellites will be used to provide communication, navigation, and surveillance on a global basis for aircraft flying over areas without radar and direct radio coverage. This will significantly increase the safety, efficiency, and flexibility of air-traffic control operations in nonradar environments.

Air-traffic control provides an airplane pilot with the safest route to fly from origin to destination. Air-traffic controllers on the ground carry out this service with the help of three supporting functions: communication, navigation, and surveillance. Communication allows the pilot to convey the desired route of flight to the controller and to request any changes to the route while in flight. It provides the controller with a means to give instructions to the pilot in order to ensure the safest route of flight. Navigation allows the pilot to maintain the route of flight and to modify the route if required. Surveillance provides the controller with knowledge of an aircraft's position at any given time. In addition to these supporting functions, controllers can be aided by computer automation, which provides a display of all aircraft under their control and performs monitoring and alerting functions for any situation requiring controller intervention.

Current nonradar operations. Surveillance of aircraft in nonradar areas is currently done using voice position reports transmitted via high-frequency (HF) radio from the flight crew to the air-traffic controller. These reports may be sent as infrequently as once per hour. High-frequency radio is subject to atmospheric disturbances. In oceanic areas where

the United States provides air-traffic control services, communication between pilot and controller requires voice messages to be transmitted from the cockpit to a radio operator via high-frequency radio and then transcribed and sent as data to the controller. Controller-initiated communication also takes place through a radio operator. The use of high-frequency radio can result in communication delays on the order of 10 min.

Aircraft navigation capabilities degrade over areas with no ground navigation stations, most significantly in oceanic airspace. The most common navigation system in use by commercial aircraft flying in oceanic areas is an inertial navigation system. This system is based on accurate gyroscopes and accelerometers that sense aircraft accelerations, which are integrated to yield a continuous output of aircraft position. Two or more units are carried on board for error detection and redundancy. Inertial systems do not depend on any navigation input from outside the aircraft and are characterized by navigation errors that increase with flight time. At the end of an oceanic flight, navigation errors commonly exceed 10 nautical miles (18 km).

Because of the limitations associated with indirect, unreliable communication and surveillance and inac-

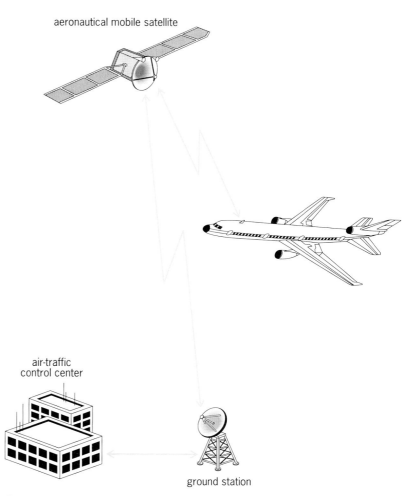

aeronautical mobile satellite

air-traffic control center

ground station

Fig. 1. Operation of automatic dependent surveillance. A satellite data link is used for both direct two-way (pilot–controller) data communication and the transmission of surveillance information.

curate navigation, aircraft in nonradar airspace must adhere to rigid route structures, and relatively large separations between aircraft must be maintained. Typical oceanic rules require that aircraft be separated by one of the following criteria: 2000 ft (600 m) vertical, or 60 nmi (110 km) lateral, or 10 min in trail longitudinal. This is in contrast to areas under radar coverage, which require only 1000 ft (300 m) vertical, or 5 nmi (9 km) laterally or in trail. Additionally, oceanic airspace use is characterized by heavy peak periods, during which demand for specific routes or altitudes exceeds the airspace capacity. Because of the separation requirements and peak periods, aircraft are often required to fly less than optimum routes that use more fuel and time than the preferred routing.

Satellites in air-traffic control. Increasing interest is being shown by the international community in the application of satellites for air-traffic control communication, navigation, and surveillance in nonradar environments. Satellite-based communication and surveillance offers the possibility of high-quality air-traffic control service in areas where line-of-sight communication and surveillance techniques are unfeasible, unavailable, or inadequate. The prin-

cipal advantage of satellites, especially those in high-altitude orbits, is that they provide excellent coverage over very large areas for aircraft flying at low altitudes, over remote land areas, and over oceanic regions. In addition to the communication and surveillance functions offered by satellites, satellite-based navigation can provide appropriately equipped aircraft with extremely accurate navigation data.

The key system components required for a satellite-based air-traffic control system are a satellite, appropriately equipped aircraft, a satellite ground station, a ground distribution network, and an air-traffic control center.

Satellite-based communication. Direct two-way (pilot-controller) data communication can be provided using a satellite data link. The same satellite data link will be used to transmit surveillance information (**Fig. 1**), as discussed below. In the initial implementation of a satellite-based system, pilot-controller data-link communication will be entirely by data messages. In the next few years, digitized voice communication via the satellite will be phased in, primarily for use in nonroutine and emergency situations.

Data communication will continue to be used for routine air-traffic control operations because of its many benefits. A fixed-format message scheme agreed upon between the ground automation and avionics automation systems will permit automatic processing and generation of messages, including automatic record keeping. Safety and efficiency will be enhanced by the use of fixed-format data communication because of the reduced probability of incorrect relaying or misinterpretation of messages.

The same satellite communication system used by air-traffic control can be used to provide data and voice services to passengers in flight. Plans are being made to accommodate individual passenger telephone calls, facsimile, and data-communication service to anywhere in the world via satellite.

Satellite-based navigation. Satellite-based navigation will be provided by the Global Navigation Satellite System (GNSS), a space-based position, navigation, and time-distribution system designed for worldwide use. The GNSS will consist of the U.S. Global Positioning System (GPS) and the Russian Global Orbiting Navigation Satellite System (GLONASS). GPS and GLONASS will each consist of 21 orbiting satellites plus 3 active spares in six orbital planes, ground-control facilities, and lightweight inexpensive receivers. The GNSS satellites in each system will transmit navigation signals at the same time on the same frequency. The aircraft's avionics will determine the aircraft's position by processing these signals.

In the future, GNSS is expected to be a common source of highly accurate navigation data in all areas of the world, including oceanic airspace, where ground navigation stations do not exist. With GNSS an aircraft can establish its position with errors no greater than 330 ft (100 m). The addition of GNSS into the automatic dependent surveillance environment, discussed below, increases the accuracy of the surveillance data

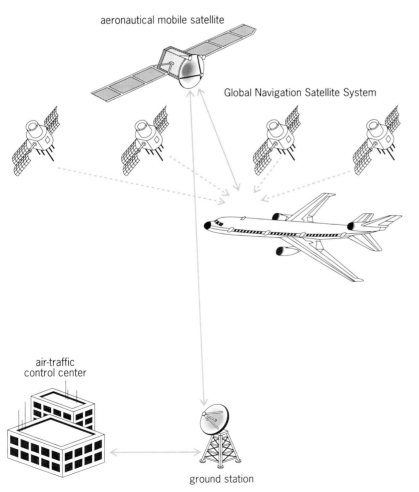

aeronautical mobile satellite

Global Navigation Satellite System

air-traffic control center

ground station

Fig. 2. Operation of automatic dependent surveillance in the environment of the Global Navigation Satellite System (GNSS).

being transmitted, although GNSS is not required for automatic dependent surveillance to function (**Fig. 2**).

Satellite-based surveillance. In a satellite-based air-traffic control environment, the surveillance capability is provided by a function called automatic dependent surveillance, in which aircraft avionics automatically transmit, via a satellite data link (Fig. 1), position data derived from on-board navigation systems. This data link will be provided by geosynchronous commercial communication satellites that are currently in operation, with global coverage between approximately 75° north latitude and 75° south latitude. The position data provided by automatic dependent surveillance is not limited to a particular type of navigation system; the automatic dependent surveillance function will transmit data from any navigation system available on the aircraft. However, in order for an aircraft to provide automatic dependent surveillance data, it must be equipped with the appropriate automatic dependent surveillance automation and satellite communication systems.

An automatic dependent surveillance report contains current and expected position data, aircraft heading information, and wind and temperature data. The position data consist of the latitude, longitude, and altitude of the aircraft as derived from the on-board navigation system, along with the time of the position report. The rate of automatic dependent surveillance reports can range from one report every 10 s to one every 20 min. The standard automatic dependent surveillance reporting rate is expected to be once every 5 min over oceanic airspace.

Automation aids. Use of satellites for surveillance, and two-way data link communication between pilots and airborne automation systems at one end and automation systems in oceanic air-traffic control centers at the other, will provide benefits to both air-traffic controllers and airspace users. Automated report processing and analysis will result in nearly real-time monitoring of aircraft movement. The automation will be able to detect any potential conflicts based on the planned routes of the aircraft in the airspace, and any aircraft deviations off the planned route will be automatically detected based on automatic dependent surveillance position reports. These features, along with the increased accuracy of navigation data provided by GNSS, will support increased safety, eventual reductions in separation minima, and increased accommodation of user-preferred routes and trajectories. Graphic display of aircraft movement and automated processing of data messages, flight plans, and weather data will significantly improve the ability of a controller to manage oceanic air traffic. The ability of a controller to have immediate communication with a pilot will facilitate the incorporation of a number of automation tools, which will eliminate many of the routine tasks currently performed by the controller. To take advantage of the potential benefits of satellite-based communication, navigation, and surveillance, adequate automation support for both the ground air-traffic control and the cockpit elements of the system must be implemented.

For background information *SEE AIR NAVIGATION; AIR-TRAFFIC CONTROL; INERTIAL GUIDANCE SYSTEM; SATELLITE NAVIGATION SYSTEMS* in the McGraw-Hill Encyclopedia of Science & Technology.

Jane Hamelink

Bibliography. International Civil Aviation Organization, Special Committee on Future Air Navigation Systems, *Final Report*, 4th meeting, Montreal, May 2–20, 1988; *Proceedings of the 1st Annual Symposium on Automatic Dependent Surveillance*, Federal Aviation Administration Technical Center, Atlantic City, New Jersey, March 13–15, 1990; *Proceedings of the 1989 Radio Technical Commission for Aeronautics Assembly*, 1989.

Aircraft instrumentation

The electronic flight instrument system (EFIS), commonly called the glass cockpit, provides both flight and systems information to the flight crews of modern commercial airliners. These systems are available on all new airplanes and may be installed on some older models; in the future they will become the standard instrument presentation. Although not every manufacturer has fully exploited the capability of the electronic flight instrument system, all such systems have clearly improved the quality of information over that of the electromechanical indicators.

Display techniques. The term glass cockpit derives from the fact that conventional flight-deck gages and indicators are replaced by cathode-ray tubes and flat-panel displays. These devices have many similarities with computer displays; however, several differences are necessary to accommodate the unique requirements of aircraft instrumentation. Of necessity, aircraft displays are much smaller, on the order of 6–8 in. (15–20 cm) square.

To achieve a much higher resolution, intensity, and clarity, the mechanics of displaying images is also different. On computers and televisions, images are drawn by displaying varying patterns of closely spaced dots (picture elements or pixels). To achieve the higher graphic resolution required in the cockpit, a technique known as ray tracing is used for the flight-deck cathode-ray tubes. Instead of thousands of pixels, a single beam of light, or ray, traces out the desired graphic image on the face of a phosphor-coated screen. The phosphor remains illuminated for a brief period of time after the ray strikes it. The trace is repeated at a high rate. The combination of a high-frequency trace, the characteristic persistence of illuminated phosphor, and persistence of vision yields a stable, high-clarity image.

Flat-panel displays have been used in applications where lower resolution was acceptable, such as alphanumeric engine displays. In the future, higher-grade flat panels will be used for the flight instruments as well. In contrast to the characteristic picture-tube shape of cathode-ray tubes, flat panels are generally less than 1 in. (2.5 cm) thick. Small caplets of material (liquid crystal, gas plasma, and so forth) that illumi-

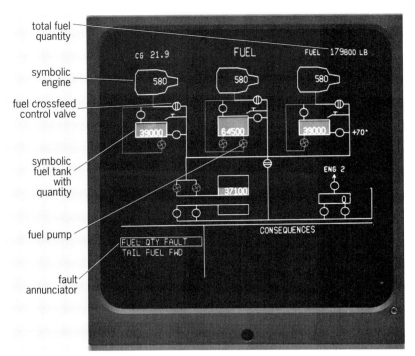

total fuel quantity

symbolic engine

fuel crossfeed control valve

symbolic fuel tank with quantity

fuel pump

fault annunciator

Fig. 1. Schematic cathode-ray-tube display of fuel system.

nate when they are excited by electricity are arranged in a matrixlike pattern on the display. The required characters are formed by selectively energizing the appropriate areas in a manner similar to that used in liquid-crystal displays of wristwatches and hand calculators.

Flexibility. The new electronic systems can provide the crew with all of the information necessary for aircraft operation and, because of their versatility, do

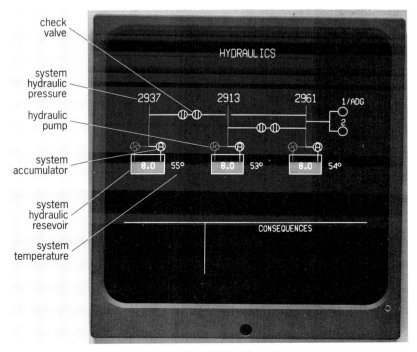

check valve

system hydraulic pressure

hydraulic pump

system accumulator

system hydraulic resevoir

system temperature

Fig. 2. Schematic cathode-ray-tube display of hydraulic system.

so in a much more efficient manner than the "steam gage" instruments they replaced. Information can be presented in the traditional "clock" form, or it can be presented in a graphic or pictorial manner, more closely depicting the bit of information it represents. Fuel indications, for instance, can be shown in their traditional format, or they can appear in the shape of transparent tanks, each an intuitive pictorial display of actual fuel levels (**Fig. 1**).

Reliability. Prior to the introduction of electronic-flight-instrument-system displays into service, manufacturers predicted that these displays would be far more reliable than their predecessors. This claim, contested at the time, has been amply borne out in almost a decade of actual line operations so that improved reliability is no longer seriously questioned.

The electronic displays also allow the most critical display to be retained through selective switching after the failure of any particular cathode-ray tube. For example, the primary flight display (PFD) or attitude detection indicator (ADI) can be switched to a second cathode-ray tube if the one on which it is normally portrayed fails. This switching was not possible with electromechanical instrumentation and is a major advantage for both designers and pilots.

Information overload. Such a vast array of information is available in a modern commercial airliner that designers must guard against overloading the pilots. The question is no longer what information can be displayed but what information should be displayed and in what manner. A delicate balance must be reached between providing all information that the crew members really want or need and overwhelming them with trivia. Manufacturers have been generally successful in achieving this balance with electronic systems, primarily because of the flexibility such systems afford.

Situational awareness. The desirability of improving or enhancing the flight crew's awareness of the flight situation, described by the term situational awareness, is generally acknowledged. That the glass cockpit can be one of the most effective tools in promoting situational awareness among flight crews is a primary reason that operators have been willing to bear the additional cost of the electronic display systems. The payback is a safer operation by a much better informed crew.

Display content and format. The efficiency level that a particular flight crew is able to achieve is dependent on the quality of information that is provided. In the past, information was presented in the cockpit in a bewildering variety of different forms, usually determined by constraints at the source of the particular bit of information. An example is an electromechanical instrument whose physical characters limit it to a simple left or right movement about a central position. Through the use of electronic displays, this information can be changed into a form much more easily understood by, or intuitive to, the crew. This transformation is normally done by presenting the information in pictorial or graphic form, as a picture or schematic of its real-world counterpart if it has one,

ground speed
true airspeed
wind arrow
wind direction and speed
heading rose
way point
control-display-unit message
navaid bearing and distance

heading/track information
clock and stopwatch
navaid (navigational aid)
symbolic aircraft
system annunciator

Fig. 3. Navigation display (ND).

or as an intuitive graph if it does not. Information from diverse instruments or symbols can be integrated into a single display. Pictures can display information in a way that is meaningful and accords with mental models of the world by condensing information into easily recognized formats with clear interrelationships. Again, the flexibility of electronic displays makes this improvement relatively easy to achieve.

Applications. A good example is the standard electronic-flight-information-system format for displaying data about aircraft subsystems, such as the electrical or hydraulic system. The conventional meth-od for displaying this information with a schematic, similar in design to the flow-switch panels that have been used for years, is generally accepted (**Fig. 2**). The objective is to make the display resemble the actual configuration of the subsystem as closely as possible. All major commercial manufacturers follow this convention in one form or another. It is generally agreed that these displays enhance the users' awareness of the state of the subsystems.

Electronic-flight-information-system intuitive pictorials are even more effective in portraying flight-situation information, and greater advantages are pos-

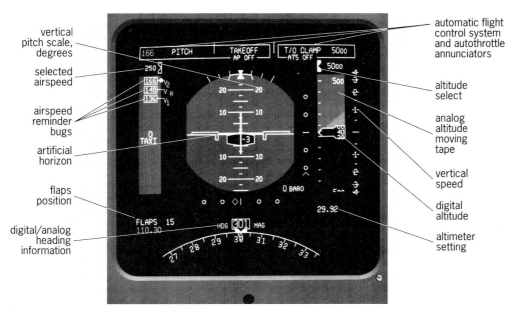

vertical pitch scale, degrees
selected airspeed
airspeed reminder bugs
artificial horizon
flaps position
digital/analog heading information

automatic flight control system and autothrottle annunciators
altitude select
analog altitude moving tape
vertical speed
digital altitude
altimeter setting

Fig. 4. Primary flight display (PFD).

sible because the information is more critical. The instruments or displays that portray the information are generally understood to be the horizontal situation indicator (HSI), which is now being replaced by the navigation display (ND) or map, and the attitude detection indicator (ADI), now being replaced by the primary flight display (PFD).

Navigation display. Replacing the horizontal situation indicator with the navigation display has proven a major breakthrough in improving the crew's position and navigational awareness. Again, a readily accepted and time-proven method of presentation was available to designers in the form of the map or navigational chart, a format familiar to all pilots. The navigation display shows the aircraft's position on a moving map, relative to airports, navigational aids, en route fixes or way points, and the intended course or track (**Fig. 3**).

This display is so intuitive that little training is necessary beyond learning what the different symbols represent. At a glance, the crew knows the aircraft's position relative to all relevant navigational points within the map's range. This display affords a great improvement in overall position awareness over that of a conventional horizontal situation indicator, and such an improvement is especially pronounced when the aircraft is operating in the flight-management-system lateral-navigation (L-NAV) mode. When the crew is using the navigation display, the aircraft position, relative to the real world, is readily apparent to the crew at all times, while a horizontal situation indicator would furnish only heading and course-error information.

Primary flight display. Equivalent gains have not been achieved as easily with the attitude detection indicator or primary flight display. In this case, a readily acceptable method or format for improving vertical-flight-path awareness was not readily available. Manufacturers decided, in their first applications of the electronic flight instrument system, to simply replicate a television picture of a World War II–vintage artificial horizon on the electronic attitude-direction-indicator screens. The capabilities of the new technology were not fully exploited, and the flight crew realized no advantage in either performance or awareness compared with older electromechanical instruments.

Information is being added during the transition to the primary-flight-display format (**Fig. 4**). Airspeed, in moving-tape form, is being added to the left of the artificial-horizon representation, along with airspeed trend; and altitude, in both digital and moving-tape form, is being added to the right of the horizon.

For background information SEE AIRCRAFT INSTRU-MENTATION; ELECTRONIC DISPLAY in the McGraw-Hill Encyclopedia of Science & Technology.

J. Randolph Babbitt

Bibliography. J. G. Oliver, *Improving Situational Awareness Through the Use of Intuitive Pictorial Displays*, Soc. Automot. Eng. Tech. Pap. 901829, 1990; E. L. Wiener and D. C. Nagel (eds.), *Human Factors in Aviation*, 1988.

Allelopathy

Plants resist infection caused by pathogenic microbes by using many different mechanisms, some of which are passive and others of which require active responses. One of the most striking forms of resistance is the hypersensitive response, which can offer complete resistance in the presence of the appropriate pathogen populations. Many different biochemical changes, most of which have yet to be fully characterized, occur in plant cells undergoing the hypersensitive response. Biochemical changes that are similar to those that occur in the hypersensitive response can be induced in plant cells upon treatment with compounds known as elicitors. The term elicitor refers to a diverse group of biotic and abiotic agents that, when applied to plant cells, are capable of inducing many of the same defense responses observed in plant–pathogen interactions. The similarity between the hypersensitive response and many elicitor-induced responses raises the possibility that elicitors may in fact function in the induction of the hypersensitive response. The relationship between the hypersensitive response and the elicitor-induced plant-defense response remains to be defined, but the information gained in the study of both systems has provided valuable insight into how plants resist microbial attack.

Plant defense against pathogens. Plants come into contact with numerous microbial agents during their growth and development. From the time that seeds enter the soil until the developing plants mature and produce seed, plants are placed in an environment teeming with microbial life. The majority of microbial organisms have little detrimental effect on plants; in some instances, they even provide beneficial effects (for example, symbiotic nitrogen fixation by *Rhizobium* species). It is the relatively rare instance where the interaction of plants with microbes results in disease, thus reducing the yield, quality, or esthetic appeal of the plant.

Plants are not defenseless against disease-causing microbes. Indeed, plants have a vast array of highly developed mechanisms to protect themselves against disease-causing microbes. A defense mechanism that humans have exploited in seeking to protect crop plants from pathogenic microbes is the hypersensitive response. In a hypersensitive interaction the plant cells that come into contact with the pathogen undergo catastrophic changes that result in death of the affected cells. Growth and development of the pathogen thus either stops or is greatly inhibited; by sacrificing a few cells the plant prevents disease development that could eventually destroy the entire plant. This form of plant defense is usually controlled by one gene that can be utilized easily by plant breeders in developing disease-resistant plants. Hypersensitive resistance has been used effectively to control many plant pathogens, including fungi, bacteria, viruses, and nematodes. A drawback to the use of hypersensitive resistance is that it is often overcome when the pathogen undergoes genetic changes.

Despite the exploitation of hypersensitive resistance in plant protection, the understanding of the biochemical processes resulting in the hypersensitive response is incomplete. Some of these processes have been elucidated in recent years, and the biochemistry is quite complex, involving numerous active responses by the plant cell. Attempts to study the biochemistry of plant defense responses in plant–pathogen interactions are complicated by the presence of the pathogen. Great care must be taken to separate the responses of the pathogen from that of the plant. Elicitors have provided a means of studying plant defense responses without the contaminating effects of a second organism and have been a valuable tool for understanding the ways in which plants respond to external stimuli.

Another tool that has proven valuable in studying plant defense mechanisms is cell-suspension culture. Cell-suspension cultures consist of loosely aggregated undifferentiated cells and can be developed for many plant species. They allow almost simultaneous treatment of large numbers of cells. The combination of fungal elicitors with cell-suspension cultures has allowed detection of cellular responses with much greater sensitivity than for actual pathogen interactions in whole plant tissues. However, care must be exercised in directly comparing results of elicitor and cell-culture experiments to responses found in microbial infection of whole plant tissues.

Membrane function and integrity.   It seems odd that a major form of resistance to pathogen infection in plants should involve necrosis. Necrosis is itself one of the most obvious signs of disease. Through use of microscopy it has been observed that pathogen growth in hypersensitive reactions is often inhibited well ahead of cell death. This observation has led scientists to look at the biochemical processes leading up to cell death in search of factors to explain this form of resistance. A common character of many elicitors is their ability to kill plant cells. The timing of cell death in response to elicitors is similar to that of the hypersensitive response, taking 8–24 h to be noticed in the absence of other factors. Cell death

in response to elicitors and during hypersensitive response is the end result of a cascade of events in which many different defense responses are simultaneously stimulated.

One of the earliest responses documented in cells undergoing hypersensitive response is alteration of plasma membrane permeability (**Fig. 1**). Potassium ions (K^+) are selectively released from the cell while calcium ions (Ca^{2+}) and hydrogen ions (H^+) are taken up. These changes in permeability lead to extracellular alkalization, intercellular acidification, and eventual disruption of membrane function. The overall reaction appears to be energy-dependent. Similar reactions have been observed in response to fungal elicitors. Changes in plasma membrane permeability in response to elicitors have been observed after time periods of less than 5 min and are typically of short duration, lasting about 60 min. Permeability changes resulting from hypersensitive response–causing pathogens often require several hours to be measured and may continue for more than 8 h. What effect these immediate responses have on microbial growth is unclear, but their effect on subsequent plant cellular responses is thought to be substantial.

The generation of active oxygen has been documented in several hypersensitive response reactions (Fig. 1). The term active oxygen is used to classify highly reactive species of oxygen (for example, $O_2{}^-$, OH^-, and H_2O_2) that are capable of oxidizing other molecules. An increase in production of active oxygen occurs concomitantly with changes in membrane permeability. The generation of active oxygen is a normal process in most biological systems, but the production and conversion of active oxygen to less reactive species is tightly regulated by enzymatic activities. The generation of unregulated active oxygen in plant tissues results in toxic effects on plant tissue as well as on the pathogen.

Increased active-oxygen production has been documented in plant tissues responding to elicitors and enzymes that degrade plant cell walls. Products of cell wall degradation are known to function as elicitors

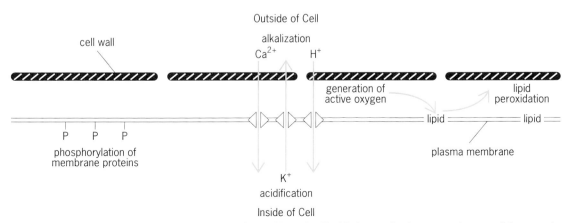

Fig. 1. Defense responses affecting membrane function and integrity. Rapid changes in plasma membrane activity occur in plant cells undergoing the hypersensitive response or responding to treatment with many elicitors.

of plant defense responses. The relationship between elicitors and active-oxygen production warrants further study.

Although active oxygen can react with many different substrates, the initial site of action within the plant cell may be the plasma membrane. Oxidation of the fatty acids of lipids within the plasma membrane can lead to new radical formation, which in turn initiates further lipid breakdown. This type of scenario would lead to extensive membrane damage. Increased lipid peroxidation has been demonstrated both in plant cells undergoing the hypersensitive response and in plant cells responding to elicitors.

The importance of protein phosphorylation in hypersensitive response and elicitor-mediated responses is presently under intensive study. It has been suggested that many of the responses that occur during the hypersensitive response result from recognition events on the plasma membrane. It is thought that these events stimulate phosphorylation of specific membrane proteins by one or more protein kinases, resulting in altered membrane activity. Phosphorylation of membrane proteins in response to elicitors has been demonstrated in plants, and inhibitors of protein kinases have been shown in at least one case to block elicitor-induced ion fluxes.

Enhanced biosynthetic processes. There has been great interest in the enhanced biosynthetic processes that take place immediately after pathogen or elicitor interactions with plant cells. These responses often include enhanced transcriptional and translational activities that result in increased levels of specific enzymes or structural proteins involved in plant defense, including phytoalexins, extensins, callose, and pathogenesis-related protein. Many of these induced responses occur in susceptible interactions in which disease develops as well as in resistant hypersensitive response interactions. The distinguishing factor during the hypersensitive response appears to be the speed with which the responses occur.

Phytoalexins are small molecules such as phenolics and terpenoids that inhibit microbial growth. These molecules are often found in tissues infected by pathogens. Elevated levels of phytoalexin biosynthesis have been observed in plant cells undergoing the hypersensitive response, and the area of enhanced phytoalexin biosynthesis often extends out of the infected cells into neighboring uninfected cells. Elicitors are very effective inducers of phytoalexin biosynthesis in both whole tissues and cell cultures, and have proven particularly useful in studying regulation of phytoalexin biosynthesis.

Plant cells often localize infection sites through the biosynthesis and deposition of compounds that act as structural barriers to further invasion, such as extensins, callose, and lignin. Extensins (hydroxyproline-rich glycoproteins) are thought to function in cell wall organization. They may be synthesized in response to pathogen infection or elicitor treatment. Deposition of callose, a β-1,3 glucan, often occurs in fungal infection sites as well as in response to elicitors. Callose deposition may actually surround infection structures in incompatible plant–pathogen interactions. Enhanced biosynthesis of lignin is also a well-characterized defense response to pathogen attack.

A large class of proteins has been identified in plant tissues undergoing hypersensitive response, and are called pathogenesis-related proteins. Pathogenesis-related proteins, many of which have no known function, are found in widely divergent plant species.

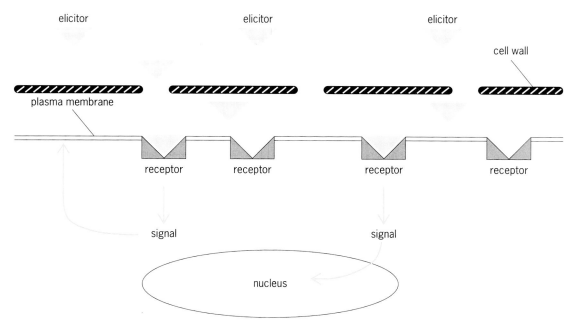

Fig. 2. A model for receptor-mediated induction of plant defense responses by elicitors. The model proposes that the interaction of specific plasma membrane receptors with elicitor molecules produces signals that stimulate altered nuclear and membrane activities.

Two classes of pathogenesis-related proteins have been identified as β-1,3 glucanases and chitinases. The β-1,3 glucans, substrates for β-1,3 glucanases, and chitin form major components of fungal cell walls, and their digestion products may function as elicitors. Some elicitors are known to be potent inducers of pathogenesis-related protein biosynthesis.

Biosynthesis of the gaseous plant hormone ethylene is stimulated by many different types of stress as well as by some normal cellular processes. Enhanced ethylene biosynthesis has been observed in many different plant–pathogen interactions and is often associated with the hypersensitive response and elicitor treatment. Many different plant defense responses are stimulated or enhanced by preexposure of the plant tissue to ethylene. Ethylene biosynthesis in response to pathogen attack or elicitor treatment can be enhanced by first exposing the tissue to ethylene. Many of the effects of ethylene on plant tissues result from enhanced transcription and translational processes. The ethylene-forming enzyme and ACC (1-aminocyclopropane-1-carboxylic acid) synthase have recently been cloned and are the subjects of intensive study. The regulation of ethylene biosynthesis during plant defense responses should soon be elucidated.

There are many other hypersensitivity- and elicitor-associated defense responses. Knowledge in some areas such as induction of systemic resistance and the function of salicylic acid as an inducer of defense responses is in a state of rapid advance. The importance of elicitors in these and other areas is yet to be determined.

Control of defense responses. Although many different plant defense responses have been elucidated in recent years, the actual signaling mechanisms controlling these responses are unknown. The most commonly proposed hypothesis is the presence of specific receptors (**Fig. 2**) on the plasma membrane that directly interact with elicitor molecules or other pathogen-generated signals. Binding sites for glucan elicitors have been identified, and there is evidence to support the existence of receptors for proteinaceous elicitors. However, since these receptors have yet to be characterized, the mechanisms by which receptors signal the cell to induce the multitude of cellular responses remain unknown. A good candidate for mediating receptor signals to the cell is protein phosphorylation in reactions mediated by protein kinase. It is also unknown if these receptors have any direct relationship to the induction of the hypersensitive response in plant–microbe interactions. Future research in isolating and characterizing plant receptors that are important in the induction of plant defenses will undoubtedly rely heavily on the use of elicitors.

The list of elicitors is continually expanding. Plants are able to perceive and respond to many different molecules of widely varying structure. Small changes in an elicitor's structure can render it inactive, thus suggesting that the plant's recognition of many elicitors is highly specific. This type of structural specificity is consistent with the functioning of elicitors in plant defense. The full value of elicitors in elucidating recognition phenomena and rapid cell responses in plants is just beginning to be exploited.

For background information SEE ALLELOPATHY; ETHYLENE; PHYTOALEXINS; PLANT PATHOLOGY in the McGraw-Hill Encyclopedia of Science & Technology.

B. A. Bailey; J. D. Anderson

Bibliography. D. J. Bowles, Defense-related proteins in higher plants, *Annu. Rev. Biochem.*, 59:873–907, 1990; T. Kosuge and E. W. Nester (eds.), *Plant–Microbe Interactions*, 1989.

Alzheimer's disease

Alzheimer's disease is a common, degenerative disorder of the central nervous system characterized by gradual loss of memory, reasoning, and judgment. A true diagnosis of Alzheimer's disease is possible only upon autopsy, when the brain can be examined for the presence of classic hallmarks of the disorder. Two major types of lesions are observed, both of which involve the formation of abnormal proteinaceous aggregates. One type, known as amyloid, appears outside neurons in the form of amyloid plaques, or so-called senile plaques, and blood vessel deposits. The other type, neurofibrillary tangles consisting of twisted fibers derived from the proteins of the cellular skeleton of neurons (cytoskeleton), is observed inside neurons. Research on how these lesions are formed, and in particular on the origin of amyloid, has led to a virtual explosion of new information regarding the cause of Alzheimer's disease. The majority of data has come from either genetic studies aimed at localizing unknown gene defects responsible for familial Alzheimer's disease (FAD) or from work targeted at identifying genes that play a role in the formation of the characteristic lesions of the disease. Although the newest information does not yet afford the possibility of a cure, the light it sheds on the etiology of Alzheimer's disease improves the prospects for future treatments.

Alzheimer's disease is classically considered to be a spontaneously occurring disorder with relatively high prevalence in the United States. In the early 1980s, epidemiological studies in the United States revealed that the number of people with Alzheimer's disease increased dramatically with age. Between 5 and 15% of the population over the age of 65 were initially believed to have Alzheimer's disease, and the disorder was calculated to be the fourth highest cause of death among elderly individuals. In a more recent study carried out on 3000 persons living in the Boston area, 10% of those over 65 years of age and almost 50% of those over 85 were diagnosed with probable Alzheimer's disease. Based on these estimates and the fact that people today are living longer than in the past, Alzheimer's disease is rapidly reaching epidemic proportions. In the absence of treatment, it will clearly become one of the most pressing health problems in the United States in the twenty-first century.

Genetics. The risk of developing Alzheimer's disease appears to be higher among relatives of patients with the disease, especially those with early onset of the disease (younger than 65 years). In the 1980s, a large number of families were identified in which Alzheimer's disease was observed to be inherited in a dominant manner; that is, in every generation, about half get the disease. Although the largest groups of related individuals with familial Alzheimer's disease involve individuals who develop symptoms of the disease at a particularly early age (40–50 years), familial Alzheimer's disease is otherwise neuropathologically and clinically identical with sporadic cases of Alzheimer's disease. Efforts to localize the FAD gene defect were initially concentrated on chromosome 21, based on the observation that the brains of older patients with Down syndrome (trisomy 21) have numerous senile plaques and tangles. In 1987, genetic analysis of four large families containing individuals with early-onset Alzheimer's disease resulted in the localization of an FAD gene defect on chromosome 21. At that same time, the gene encoding amyloid protein precursor (APP), which is the protein that gives rise to Alzheimer's-associated amyloid, was isolated and mapped to the same general region of chromosome 21 as the newly discovered FAD gene defect. These combined results ignited the hope that the primary genetic defect in familial Alzheimer's disease may have been not only genetically localized (by genetic linkage analysis) but also identified. This initial spark of excitement was soon doused, however, when genetic linkage analysis of the same four FAD families previously shown to contain a gene defect on chromosome 21 revealed a lack of tight genetic linkage to the APP gene.

Since then, research groups around the world have endeavored to test additional families with familial Alzheimer's disease for genetic linkage to chromosome 21. While some laboratories confirmed the presence of an FAD gene defect on chromosome 21, others found evidence for the existence of additional FAD defects elsewhere in the human genome. Significant evidence for genetic heterogeneity (the existence of more than one gene defect) in familial Alzheimer's disease was obtained through a massive international collaboration. The results reported from this study showed that only a subset of FAD families were genetically linked to chromosome 21 and these primarily contained individuals who developed the disease at an early age. On the other hand, families with members who developed the disease at a late age did not appear to harbor a gene defect on chromosome 21.

Having established that different FAD families can involve separate and distinct gene defects, the human genome was scanned for additional FAD genes using genetic linkage analysis. Suggestive evidence for a possible gene defect that is linked with late-onset familial Alzheimer's disease was obtained for chromosome 19. However, this finding has not been confirmed. Overall the results indicate that familial Alzheimer's disease is clearly not a genetically homogeneous disorder. Early-onset familial Alzheimer's disease appears to be caused by a different gene defect than late-onset familial Alzheimer's disease. In

addition, the type of gene defects inherited in each FAD family may not necessarily be identical. For example, while some may be directly causative, others may merely predispose the gene carrier to an increased risk of getting Alzheimer's disease in the presence of other genetic and environmental factors. In general, the genetics of familial Alzheimer's disease as depicted by current studies appears to be highly complex. Future studies aimed at identifying possible gene-to-gene and gene-to-environment interactions will likely be necessary to gain a clearer understanding of the etiology of this disorder.

Amyloid protein precursor gene. A major focus in studies of familial Alzheimer's disease has been the APP gene, which encodes the amyloid protein precursor of the principal component of amyloid plaques. This component comprises 39–42 amino acids and is referred to as the βA4 peptide. The location of the APP gene on chromosome 21 suggests that the presence of amyloid deposits in the brains of patients with Down syndrome is most likely the result of having three (rather than the normal two) copies of the gene, and consequent overproduction of the precursor protein. Although the APP gene was originally excluded from being tightly genetically linked to familial Alzheimer's disease, the recent finding that different genes may underlie the disease in different families has raised the possibility that the APP gene may still contain the defect in some FAD families. A recent study has revealed two families with apparent Alzheimer's disease in which affected individuals inherit a mutated APP gene. The mutation involves a single base substitution in the portion of the gene responsible for encoding the βA4 region of amyloid protein precursor (exon 17 of the APP gene). Exon 17 was initially chosen for deoxyribonucleic acid (DNA) sequencing analysis on the basis of a recent discovery that another mutation (also a single base substitution) in this exon represents the gene defect in a rare disorder termed Dutch hereditary cerebral hemorrhage with amyloidosis. In this disorder, Alzheimer's-type amyloid accumulates in large amounts in cerebral vessels and ultimately leads to stroke and death by the fifth or sixth decade of life.

The Alzheimer's-associated APP mutation causes the substitution of the amino acid isoleucine for valine at position 717. This is in a portion of the amyloid protein precursor that spans the plasma membrane when the precursor protein is anchored in the cell surface. This has led to speculation that the substituted amino acid may destabilize the placement of amyloid protein precursor in the membrane and lead to its early release. Premature release could lead to the abnormal breakdown of amyloid protein precursor since it has been shown that under normal circumstances cell membrane–associated amyloid protein precursor is cleaved within the βA4 region. Once cleaved, it is unable to form amyloid. The release of intact, uncleaved amyloid protein precursor from the cell membrane could conceivably result in the subsequent formation of Alzheimer's-associated amyloid.

The APP mutation associated with Alzheimer's disease has more recently been shown to be present in seven families containing individuals with early-onset familial Alzheimer's disease. Meanwhile, it has been shown to be absent in over 300 normal individuals and does not occur in over 100 additional FAD families. These results lend considerable support to the notion that this single base substitution represents a very rare mutation accounting for less than 5% of inherited Alzheimer's disease.

Whether this mutation underlies classic Alzheimer's disease as described by specific neuropathological hallmarks is at present unclear. An examination of the brains of deceased patients in one FAD family reported to contain this mutation reveals a peculiar set of pathological features. This includes so-called Lewy bodies, which normally occur in Parkinson's disease, and abundant amounts of amyloid in blood vessels, leading to stroke in some cases. This latter feature is very similar to that found in patients with Dutch hereditary cerebral hemorrhage with amyloidosis, in which dementia is sometimes observed to follow the onset of cerebral strokes. This then suggests the possibility that the two mutations in exon 17 of the APP gene may actually represent gene defects leading to similar disorders that involve abundant amounts of blood-vessel amyloid and cerebral hemorrhage and stroke. Meanwhile, it appears that the primary gene defects responsible for most of familial Alzheimer's disease have yet to be identified.

The extent to which mutations in the APP gene underlie familial Alzheimer's disease is not yet clear. It is also unclear whether genetic linkage of familial Alzheimer's disease to chromosome 21 is entirely explained by APP mutations. The largest FAD family shown to be linked to chromosome 21 reveals strong evidence for a gene defect on chromosome 21, yet it does not appear to be linked to, or involve a mutation in, the APP gene. These findings, along with other genetic linkage results, continue to support the possibility of a second FAD locus on chromosome 21. Since the chromosome 21–linked pedigree displays classic Alzheimer's disease pathology (devoid of events of cerebral hemorrhage and stroke), further genetic linkage analysis of this and similar kindreds could eventually lead to the identification of more universal FAD gene defects on chromosome 21 or elsewhere in the genome.

For background information SEE ALZHEIMER'S DISEASE; CONGENITAL ANOMALIES; HUMAN GENETICS in the McGraw-Hill Encyclopedia of Science & Technology.

Rudolph E. Tanzi

Bibliography. L. A. Farrer et al., Segregation analysis reveals evidence of a major gene for Alzheimer's disease, *Amer. J. Hum. Genet.,* 48:1026–1033, 1991; A. M. Goate et al., Segregation of a missense mutation in the amyloid precursor protein gene with familial Alzheimer's disease, *Nature,* 349:704–706, 1991; P. H. St. George-Hyslop et al., Familial Alzheimer's disease: Progress and problems, *Neurobiol. Aging,* 10: 417–425, 1989; P. H. St. George-Hyslop et al., The genetic defect causing familial Alzheimer's disease maps on chromosome 21, *Science,* 235:885–889, 1987; P. H. St. George-Hyslop and the FAD Collaborative Group, Genetic linkage studies suggest that Alzheimer's disease is not a single homogeneous disorder, *Nature,* 347:194–197, 1990; R. E. Tanzi et al., Amyloid beta protein gene: cDNA, mRNA distribution, and genetic linkage near the Alzheimer locus, *Science,* 235:880–884, 1987.

Antibody

Recent research has involved the synthesis of enzymelike catalysts known as catalytic antibodies and the production of related species known as metallo-antibodies.

Catalytic antibodies. Monoclonal antibodies have recently been utilized as tailor-made, stereospecific catalysts for many reactions of interest in chemistry and biology. This approach to catalysis complements efforts based on host–guest chemistry and attempts to generate novel catalysts by altering existing enzymes by mutagenesis. While efficiency and specificity are hallmarks of enzymes, these features also limit their use in nonnatural reactions. Often, no known enzyme catalyzes the reaction of interest, or an available enzyme will tolerate only small variations in substrate. However, the diversity of the vertebrate immune system—10^8 to 10^{10} individual binding specificities—virtually guarantees that an antibody can be isolated that binds a given antigen, whether natural or synthetic. Thus, the problem of substrate binding is solved by the immune system. Further, by careful choice of the eliciting antigen, antibodies that not only bind a related substrate but also catalyze a subsequent reaction can be isolated. In addition, the generation and characterization of catalytic antibodies allows basic theories of enzyme action to be tested experimentally in novel systems.

Transition-state stabilization. This provides the theoretical support for antibody catalysis. All chemical reactions proceed through at least one high-energy transition state, and the reaction rate can therefore be increased by stabilizing this fleeting intermediate relative to the starting materials or products. Such stabilization is manifested in several forms, including pi bond–to–pi bond interactions and electrostatic, hydrogen-bonding, or solvent effects. Enzymes and antibodies utilize the same fundamental forces in binding their targets; they differ primarily in that the former bind transition states more tightly than starting materials or products, while the latter bind ground states most tightly. However, an antibody that provided a surface complementary to the transition state of a reaction would act as a catalyst. Experimentally, such antibodies are isolated by raising antibodies against stable compounds that mimic the transition state of a reaction. In general, these compounds were previously identified as potent inhibitors of enzymes that catalyzed similar transformations. Three examples of antibody-catalyzed reactions are discussed below. In the first two, the antibodies apparently act as entropy traps, "freezing out" rotational degrees of freedom and locking the substrates into conformations conducive to the respective reactions. In the third, a large rate acceleration is achieved by a multistep pathway reminiscent of the enzyme chymotrypsin.

Claisen rearrangement. The concerted Claisen rearrangement [reaction (1)] of chorismic acid (I) to

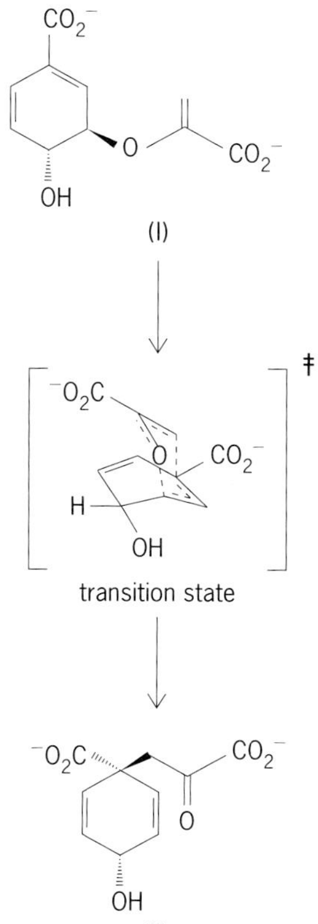

$$
\tag{1}
$$

(I)

transition state

(II)

prephenic acid (II) proceeds via a chairlike transition state and exemplifies many features of pericyclic reactions. This rearrangement has attracted a great deal of interest, both because it is a model for substituent effects on pericyclic reactions and because it remains the only example of an enzyme-catalyzed pericyclic reaction. Several compounds have been synthesized as potential inhibitors of this enzyme with the aim of reproducing the character of the transition state in a stable compound. *Endo*-bicyclic diacid (III) was iden-

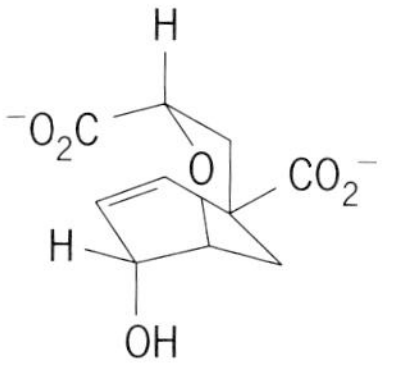

(III)

tified as the most potent inhibitor of the enzyme from the bacterium *Escherichia coli*. A flexible linker was attached to diacid (III) via its secondary alcohol, and this was conjugated to carrier protein for production of monoclonal antibodies.

Antibodies that bound antigen (III) tightly and also accelerated the rearrangement of chorismic acid by factors of 10^2 and 10^4 over the uncatalyzed rate were independently isolated by two research groups. Mechanistic studies indicated that these antibodies produced their rate enhancements by selectively binding the substrate in the conformation required for rearrangement. These results should be contrasted with the 2×10^6 rate acceleration afforded by *E. coli* chorismate mutase, which apparently augmented simple entropy reduction with other techniques to achieve an impressive rate enhancement. The antibodies utilized the two carboxylates of structure (I) for much of the binding affinity, and the dimethyl ester of chorismic acid was not a substrate. In addition, one of these catalytic antibodies was completely specific for the natural isomer of structure (I); its enantiomer was not recognized as a substrate. This illustrates the potential of antibodies as stereospecific catalysts and points to the possibility of isolating pairs of antibodies, each one specific for a single enantiomer of a given substrate.

Cyclization. The cyclization of chiral hydroxy ester (IV) to lactone [V; reaction (2), where the aster-

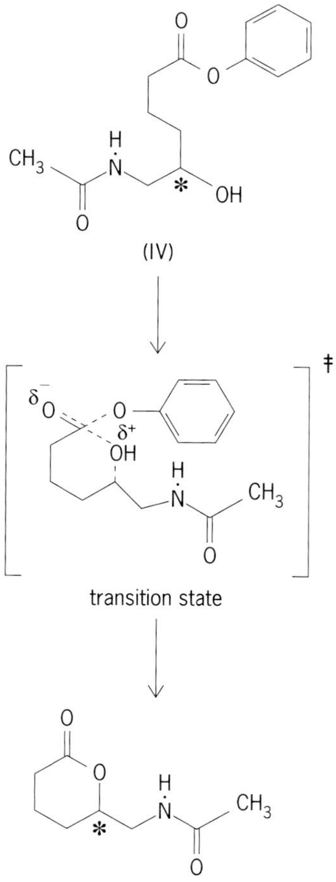

(IV)

$$
\tag{2}
$$

transition state

(V)

isk denotes the chiral center] provided another example of stereospecific antibody catalysis. The hapten (VI), whose phosphonate mimics the lengthened carbon-oxygen (C—O) bonds and electronic polariza-

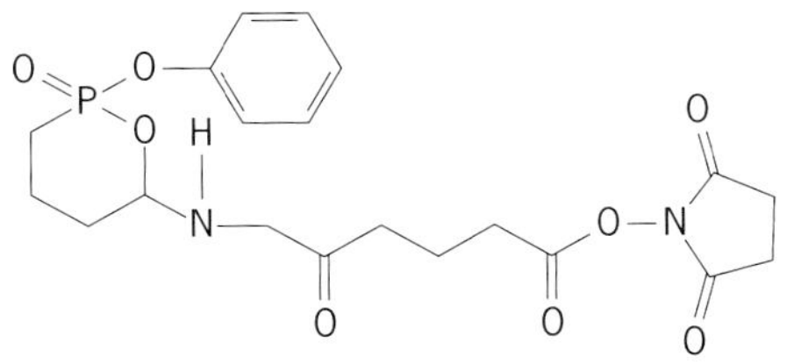

(VI)

tions of the tetrahedral intermediate leading to lactone (V), was synthesized in racemic form. After conjugation to carrier protein, 24 monoclonal antibodies that bound hapten (VI) were isolated by standard means. One of these also catalyzed the cyclization of one enantiomer of the racemic form of ester (IV). After correcting for the background (uncatalyzed) lactonization of structure (IV), lactone R-(V) [R is the designation for the configuration of the chiral center] was produced in $94 \pm 6\%$ enantiomeric excess. The pH rate profiles (which can provide clues to the reaction mechanisms involved) indicated that both the uncatalyzed and antibody-catalyzed versions of this reaction utilized the same mechanism. Like the chorismate mutase catalysts, this antibody also appeared to function as an entropy trap, binding the substrate in a conformation conducive to subsequent reaction.

In addition to lactonization, this antibody catalyzed the stereospecific synthesis of amide (VII) from phenylenediamine and lactone [V; reaction (3)]. As expected, only R-(V) was a substrate for amide synthesis.

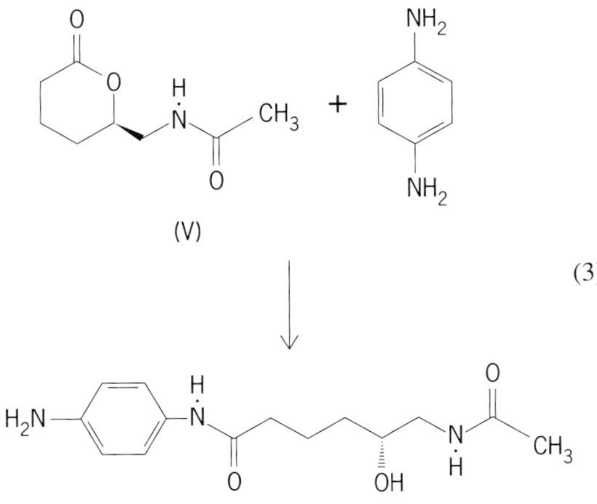

(V)

(3)

(VII)

This versatility resulted from the choice of eliciting hapten, which modeled a tetrahedral intermediate with alkoxy and aryloxy substituents. In the cyclization direction, an alkoxy attacked the carbonyl, and phenoxide was subsequently expelled. In amide synthesis, an aromatic amine functioned as the nucleophile, and the alkoxide was expelled. Thus, the differing nucleophilicities of the reacting groups governed the directions of these reactions; the antibody merely stabilized the intervening transition states.

Amide hydrolysis. Amide hydrolysis (a reaction

in which the amide is divided into its constitutive acid and amine fragments) is an especially important reaction in biological systems; and enzymes that catalyze these energetically demanding reactions have been intensively investigated. Compounds that approximate the key tetrahedral intermediate have proven to be potent inhibitors of these enzymes. In particular, substrate analogs in which the reactive amide has been replaced by a phosphonamidate are among the most potent classes of inhibitors discovered. In an effort to isolate antibodies with amidase activity, phosphonamidate (VIII; broken-line bonds indicate resonance

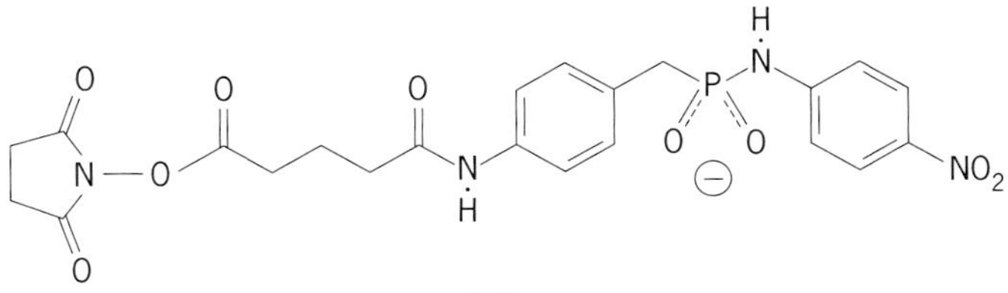

(VIII)

forms) was designed as a mimic for the intermediate produced during the hydrolysis of p-nitroanilide [IX; reaction (4)] to acid (X) and p-nitroaniline (XI). Of the antibodies that tightly bound phosphonamidate (VIII), several catalyzed the hydrolysis of p-nitrophenyl ester [XII; reaction (5)]. In addition, one isolate also hydrolyzed amide (IX). Extensive kinetic studies of this amidase antibody allowed construction of a Gibbs free-energy reaction diagram for the hydrolysis of amide (IX; see **illus.**). In this diagram, Ab·S represents the noncovalent complex of antibody and amide (IX). This reacts to give an intermediate, Ab·I, which decomposes to give a ternary complex of antibody, acid (X; P_1), and p-nitroaniline (XI; P_2). The acid and aniline are then released sequentially to regenerate the catalyst.

Several conclusions emerged from this analysis. Most importantly, the number and organization of catalytic steps bear a remarkable resemblance to those of a highly evolved protease such as chymotrypsin, even though the antibody was generated to a single hapten. Hydrolysis of both ester and amide proceeded by an intermediate, most likely an acyl-antibody, which did not accumulate to significant levels during turnover. The catalytic efficiency of the antibody was lower than that of an enzyme, however, mainly because of the very slow dissociation of products from the ternary complex of antibody [acid (X)] and p-nitrophenol (XIII). Product inhibition has proven to be a general problem in antibody-catalyzed reactions, caused by insufficient discrimination in antibody binding for the transition state relative to substrates or products. This limitation may be overcome by careful design of hapten and substrates.

Prospects. Antibody catalysis will undoubtedly remain an area of active research. Current efforts center on utilizing bacteria, particularly *E. coli*, for the production and selection of catalytic antibodies. Such a system would greatly simplify the engineering of altered antibodies with the aim of improving their

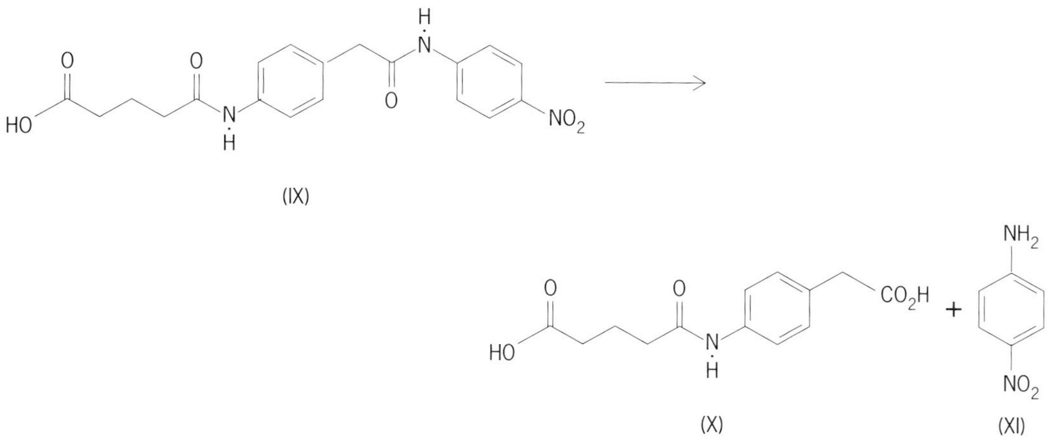

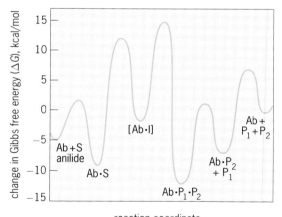

(IX)

(4)

(X)   (XI)

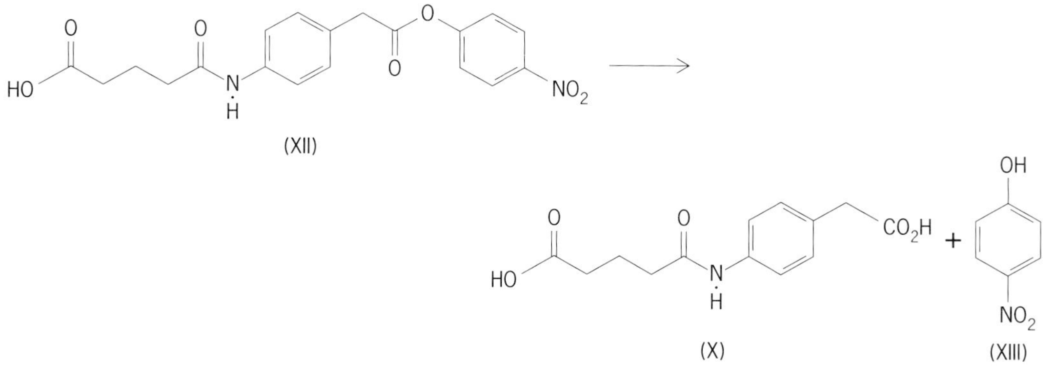

(XII)

(5)

(X)   (XIII)

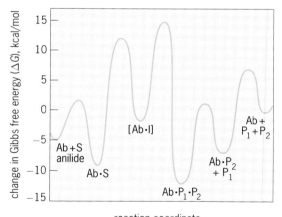

Gibbs free-energy reaction profile for antibody-catalyzed anilide hydrolysis at pH 7. (*After S. J. Benkovic et al., The enzymatic nature of antibody catalysis: Development of multistep kinetic processing, Science, 250:1135–1139, 1990*)

catalytic activities. In addition, simple methods for recruiting cofactors would expand the classes of reactions accessible by these novel catalysts.

Stephen J. Benkovic; Jon D. Stewart

Metalloantibodies. Metalloantibodies are designed to harness the chemical "firepower" of metal ions for use in catalytic antibodies. A metalloantibody is produced by carefully positioning a coordinated metal ion in an antibody binding pocket; the goal is to dramatically expand the repertoire of reactions that can be catalyzed by antibodies.

Binding sites. Antibodies have binding sites that are exactly complementary to the molecule they tend to bind. This molecular complementarity, often described as being analogous to a lock and key, is responsible for the exquisite specificity of antibody catalysis, since only molecules that bind in the antibody binding pocket will undergo a reaction with the catalytic antibody. Best of all, this molecular specificity can be programmed into catalytic antibodies via immunizations with molecules of appropriate structure. The unique features of programmability and specificity will make catalytic antibodies an important technology. The most difficult task in producing catalytic antibodies is deciding how to introduce powerful chemical catalysis into the antibody binding site.

Enzymes may be viewed as molecule-sized chemical reactors; they are responsible for building virtually all of the complex chemicals that make up living things. Enzymes use a number of tricks to help carry out their complex chemistry. Perhaps the most common of these tricks is the use of a metal ion such as zinc $[Zn^{2+}]$, copper(II) $[Cu^{2+}]$, or iron(II) $[Fe^{2+}]$ to assist with difficult chemical reactions. Metal ions are used by a large variety of enzymes to facilitate important reactions, for example, protein cleavage, cleavage of deoxyribonucleic acid (DNA), gene regulation, oxygen transport, or photosynthesis. Enzymes that use metal ions are known as metalloenzymes; the key to their success is that the metal ion is coordinated (bound) by the enzyme in precisely the correct geometry and location to ensure that the desired reaction will be accelerated.

Design. Metalloantibodies are designed to use an appropriately positioned metal ion to increase catalytic activity inside an antibody binding pocket analogous to the metalloenzymes. Computer modeling studies have been carried out to locate an area in antibodies that would be suitable for building a metal-ion coordination site. The chosen location was on the antibody light chain, deep in the antibody binding site. A molecule bound by the antibody is therefore predicted to come in contact with the coordinated metal ion, thus allowing a catalytic chemical reaction to take place in a manner similar to that which occurs in natural metalloenzymes.

The catalytically active metal coordinating site from the metalloenzyme carbonic anhydrase B was used as a prototype for the metalloantibody metal coordinating site. Carbonic anhydrase B is one of the fastest enzyme catalysts known. It is a metalloenzyme that catalyzes the reversible reaction between carbon dioxide (CO_2) and water (H_2O), a reaction important for maintaining proper concentration of CO_2 in the blood and other tissues. The metal coordination site of carbonic anhydrase B is composed of three histidine amino acid residues, and the metal ion is coordinated via nitrogen atoms on the three histidine imidazole rings. The geometry around the metal ion is roughly tetrahedral, with the imidazole nitrogens accounting for three of the four corners of the tetrahedron, the metal ion being at the center. The fourth corner of the tetrahedron is not coordinated by any amino acid residues of the protein, and it usually is occupied by a water molecule when the enzyme is inactive. During the catalytic reaction, the substrates replace this water molecule and interact directly with the metal ion. Although Zn^{2+} is the catalytically active metal ion observed inside living systems, other metal ions such as Cu^{2+} or cadmium (Cd^{2+}) can be placed in the carbonic anhydrase B metal-ion coordination site in a test tube. The carbonic anhydrase B metal-ion coordination site was chosen for use in the metalloantibodies because of its extremely high level of catalytic activity, and there are certain key similarities in overall three-dimensional structure between carbonic anhydrase B and antibodies.

Tests. The metalloantibody design was tested on a certain antibody known as 4-4-20 that specifically binds the highly fluorescent molecule fluorescein. The detailed three-dimensional structure of 4-4-20 is known, and this aided the computer modeling studies. Just as important, the antibody was already available commercially in the so-called single-chain Fv form.

Single-chain Fv molecules are derived from antibodies that have been trimmed down to include only the Fv region. The Fv region of an antibody is the smallest structure that retains full binding properties. Unfortunately, normal Fv structures can be unstable, since they are composed of two different polypeptide chains that are not permanently attached to each other. The single-chain Fv molecules contain a polypeptide linker that permanently attaches the light-chain portion of the Fv to the heavy-chain portion, thus increasing the stability of these constructs.

Site-directed mutagenesis is the technique used to change amino acid residues of a protein into different amino acid residues. Site-directed mutagenesis was carried out on the 4-4-20 single-chain Fv to create a metal coordination site on the chosen region of the light chain. To make the metal coordinating site, two serine residues and one arginine residue on the 4-4-20 protein were replaced by three histidine residues. These replacements were intended to reproduce the three-histidine metal coordinating site of carbonic anhydrase B, since computer modeling indicated that in the antibody the three histidine imidazole groups could adopt the same tetrahedral arrangement required for metal-ion coordination.

Exploiting the fact that metal ions such as Cu^{2+} can quench the fluorescence of nearby fluorescent molecules such as fluorescein, it was convincingly demonstrated that the site-directed mutagenesis had in fact produced a new metal coordinating site in the binding pocket of the 4-4-20 single-chain Fv molecule. This new metal coordinating site bound not only Cu^{2+} but also Zn^{2+} and Cd^{2+}. As expected, these metal ions bind to the metalloantibody somewhat less tightly than to carbonic anhydrase B. Importantly, the relative order of binding for the different metal ions is exactly the same for the two proteins; thus the key features of the carbonic anhydrase B metal coordination site were successfully reproduced in the metalloantibody.

Scientists are currently attempting to use metalloantibodies in catalytic reactions. Studies are also being carried out to place different types of metal coordination sites in metalloantibodies. Although it is still too early to say for sure, based on the importance of natural metalloenzymes in living systems, metalloantibodies should soon become valuable additions to the emerging technology of catalytic antibodies.

For background information SEE ANTIBODY; CATALYSIS; COORDINATION CHEMISTRY; ENZYME; PERICYCLIC REACTION; REACTIVE INTERMEDIATES; STEREOSPECIFIC CATALYST in the McGraw-Hill Encyclopedia of Science & Technology.

Brent Iverson

Bibliography. S. J. Benkovic et al., The enzymatic nature of antibody catalysis: Development of multistep kinetic processing, *Science*, 250:1135–1139, 1990; R. E. Bird et al., Single-chain antigen-binding proteins, *Science*, 242:423–426, 1988; D. Hilvert et al., Catalysis of concerted reactions by antibodies: The Claisen rearrangement, *Proc. Nat. Acad. Sci. USA*, 85:4953–4955, 1988; B. L. Iverson et al., Metalloantibodies, *Science*, 249:659–662, 1990; D. Y. Jackson et al., An antibody-catalyzed Claisen rearrangement, *J. Amer. Chem. Soc.*, 110:4841–4842, 1988; R. A. Lerner, S. J. Benkovic, and P. G. Schultz, At the crossroads of chemistry and immunology: Catalytic antibodies, *Science*, 252:659–667, 1991; V. Roberts et al., Antibody remodeling: A general solution to the design of a metal-coordinating site in an antibody binding pocket, *Proc. Nat. Acad. Sci. USA*, 87:6654–6658, 1990.

Antigen

Recent research in immunology has included antigen processing and presentation, and minor lymphocyte-stimulating antigens.

Antigen Processing and Presentation

T-cell receptors, unlike antibodies, do not recognize free antigen. The antigen has to be processed and is presented to the T cells on the surface of so-called antigen-presenting cells by molecules derived from the major histocompatibility complex (MHC). There are two sets of MHC molecules involved in the antigen-presentation process, class I molecules and class II molecules; and there are two types of T cells. While the T-cell receptors are of the same type on both T-cell subsets, the cells are distinguished by their expression of different accessory proteins. The CD8-bearing T cells selectively recognize antigen in the context of class I MHC molecules, and the CD4-bearing T cells interact with antigen presented by class II molecules. The CD8 and CD4 molecules interact directly with the class I and class II molecules, respectively, so, in molecular terms, the antigen-recognition event involves at least the T-cell receptor, an MHC molecule, and a CD4 or CD8 molecule.

The ability of the T cell to interact with antigen is restricted. In contrast to antibodies, which are able to react with most types of foreign molecules, T-cell receptors preferentially react with peptides. The recognized peptides are limited by what can be bound to the various class I and class II molecules. Although class I and class II molecules are promiscuous in their peptide-binding specificities, any one MHC molecule can bind only a limited number of peptides. However, since an individual has between three and six varieties of class I and class II molecules, respectively, with different peptide-binding specificities, the T-cell receptors are exposed to a vast array of different peptides. The exact nature of the structural requirements for the peptide binding is poorly understood, but certain amino acids in some positions of a peptide appear critical for the binding.

Antigen presentation. Structural analyses of class I and class II molecules have suggested that their three-dimensional structures might be very similar, although their two pairs of nonidentical subunits are of quite different sizes. This similarity has raised the question of why evolution has generated two types of structurally and functionally similar molecules. The answer is now apparent. While class I and class II molecules in many cases can present exactly the same antigenic peptide, they obtain their peptides in different subcellular compartments. Class I molecules primarily bind peptides that are derived from proteins manufactured by the cell presenting them (endogenous antigen presentation), whereas class II molecules bind peptides from proteins that have been imported into the cell (exogenous antigen presentation). Thus, class I molecules mostly present antigenic peptides of newly synthesized proteins of viruses and other intracellular pathogens, while class II molecules preferentially present peptides derived from proteolysis of intact viruses and bacteria that have been taken up by the antigen-presenting cell.

Peptide generation. Since proteins imported into a cell by endocytosis are sequestered from those in the nucleus, the cytoplasm, and the organelles of the exocytotic pathway, it is obvious that at least two different peptide-generating machineries must exist. Proteins accumulating in the endosomal/lysosomal system of the cell, which is rich in proteases that function optimally at low pH, are probably broken down by such enzymes, and the antigenic peptides are subsequently presented by class II molecules. The generator of antigenic peptides destined for class I molecules is less obvious. As several antigenic proteins are known to reside in the cytoplasm, proteolysis probably occurs in the location. Many cellular proteins that normally function in the cytoplasm are very susceptible to proteolysis. Therefore the machinery that generates the antigenic peptides must be controlled so that it does not cause general proteolysis in the cell. Highly regulated proteolytic systems exist in all cells. One such system comprises particles called proteosomes, composed of 10 to 15 different types of subunits, which break down proteins into peptides. Several of the proteosomal subunits are regulated by interferon-gamma, a lymphokine that alerts different defense systems against viral infections. This observation and the fact that two of the proteosomal subunits are encoded by genes situated in the MHC region on the human chromosome 6 implicate the proteosome as one, if not the only, generator of peptides destined for class I molecules.

Class I molecule assemblage. Class I molecules are assembled in the endoplasmic reticulum and are then transported, via the Golgi complex, to the cell surface. Thus, class I molecules are always separated from the cytoplasm by a lipid bilayer. Therefore, peptides that have been generated by the proteosome in the cytoplasm must be transported across a membrane to reach the class I molecules. Two proteins that may accomplish this task have been identified. Their sequences are homologous to those of a number of other transporter proteins that are dependent on adenosine-triphosphate (ATP) and are present in bacteria, yeast, and mammalian cells. The exact mode of operation of the peptide transporters is presently unknown.

The two subunits of the class I molecules, the heavy chain and beta-2-microglobulin (β2m), assemble within a few minutes after synthesis. Whether peptide is bound to the heavy chain prior to the binding of β2m is presently unclear. However, only molecules containing both peptide and β2m are transported out of the endoplasmic reticulum to the cell surface. A number of auxiliary proteins involved in regulating the peptide addition onto class I molecules and the intracellular transport of class I molecules have been identified, but the exact roles of these proteins have yet to be determined. However, the sophisticated control system that selects class I molecules intracellularly complexed with a peptide for expression at the cell surface ensures that class I molecules at the cell surface do not passively pick up peptides released from neighboring cells. This is an important safeguard that prevents T cells from destroying innocent bystander cells that do not actively synthesize antigenic peptides, such as viral proteins.

Class II molecule assemblage. The class II alpha- and beta-chain subunits assemble in the endoplasmic reticulum much like the class I subunits, but they do not obtain antigenic peptides in this organelle. This realization was surprising because peptides that can bind either class I or class II molecules are present in the endoplasmic reticulum. However, class II molecules associate with another polypeptide at the time of assembly. This protein, called the invariant chain, binds to the class II molecules in a manner that renders the peptide-binding groove of the MHC molecule inaccessible to peptides. Thus, the invariant chain may be the most important protein distinguishing the endocytotic from the exocytotic antigen-presentation pathway.

Peptide transport. Peptides presented by class II molecules are generated in the endosomal/lysosomal system in the cell. Therefore, class II molecules have to move from the exocytotic to the endocytotic compartments for contact with the antigenic peptides. Molecules that are destined for the endosomal/lysosomal compartments can use two different routes. Some molecules follow the exocytotic pathway to the cell surface, where they accumulate in coated pits, which subsequently form coated vesicles that are transported to early endosomes. In these organelles, molecules are sorted so that some proteins are recirculated to the cell surface while others move to late endosomes and lysosomes. The second route takes molecules through the exocytotic compartment as far as the trans-Golgi network. In this organelle, some molecules are diverted from transport to the cell surface and are instead routed to the late endosomes and lysosomes. Whether molecules follow one or the other of the various pathways seems to depend on discrete structural groups that are present in the transported proteins and are recognized by the sorting machineries.

Class II molecules reach the endosomal/lysosomal compartments by both routes. Newly synthesized class II molecules bound to the invariant chain go from the trans-Golgi network to the late endosomal/lysosomal compartments. The exact destination is still controversial, but it is clear that class II molecules can bind peptides generated both in late endosomes and in lysosomes. Whether such peptides are transported from the two organelles to a third one, where the class II molecules reside, or whether class II molecules recirculate between these compartments, is still debated. In any case, the invariant chain is released from the class II molecules in the endosomal/lysosomal compartment, and subsequently the MHC molecules become competent to bind peptides. It still remains to be determined whether class II molecules must be complexed with peptides for transport from the endosomal/lysosomal compartment to the cell surface.

Once expressed, class II molecules recirculate between the cell surface and early endosomes. Since the latter compartment is slightly acidic, a condition that favors peptide binding, peptides can be loaded onto class II molecules in this environment. However, little if any proteolysis occurs in early endosomes, so the peptides present in this compartment must have been taken up from the surrounding cytoplasm.

The existence of auxiliary proteins other than the invariant chain, which are essential for the class II peptide-loading process, seems likely. The existence of mutant cell lines, which are unable to present peptides by class II molecules, supports this notion.

Per A. Peterson

Minor Lymphocyte-Stimulating Antigens

The identity of minor lymphocyte-stimulating (Mls) antigens has been an enigma in immunology for two decades. The antigens were detected by the proliferation, in mixed culture, of lymphocytes from animals with identical major histocompatibility genes. It has now become apparent that the minor lymphocyte-stimulating antigens are vital superantigens encoded by endogenous and exogenous mouse mammary tumor viruses. The biological significance of these antigens has not been determined. Further, viral superantigens have not yet been found in other species, including humans. However, *Herpesvirus saimiri*, a herpesvirus of primates, has recently been shown to have a deoxyribonucleic acid (DNA) sequence similar to that of viral superantigens. Although not known, the gene product may be a superantigen that binds to MHC class II molecules and induces the proliferation of lymphocytes observed in early stages of infection.

Viral products. The Mls antigens are expressed in certain strains of mice. B cells from such mice induce a large proportion of T cells from Mls antigen-negative mice to proliferate without prior immunization. By contrast, the frequency of T cells that are capable of responding to foreign antigens without prior immunization (except for those antigens encoded by the MHC) is so low that the proliferation of these T cells is impossible to detect in cell culture. Owing to their ability to stimulate such a relatively large proportion of T cells, these antigens have been called superantigens. Recently, a number of virus coded molecules with functional characteristics similar to Mls antigens have been discovered. They appear to be the products of retroviruses, including endogenous mouse mammary tumor proviruses that are integrated into germ-line DNA but do not produce infectious particles, and infectious mouse mammary tumor viruses that are transmitted in the milk. The genetic code for the superantigens appears to reside in an open reading frame (Orf) in the $3'$ long terminal repeat of the viral sequence. The appropriate genetic sequence has been shown to induce expression of the superantigen in cells, but isolated products of the genes have not yet been demonstrated to act as superantigens.

Superantigens and T cells. Superantigens activate the proliferation of mature T cells through a specific variable element of the beta chain (Vβ) of the T-cell receptor. Twenty-three functional Vβ elements have been discovered in mice. The degree of activation of mature T cells by superantigens differs.

Immature T cells that bind self antigens through the T-cell receptor with high affinity, including superantigens, are eliminated in the thymus. The activation of immature T cells leads to programmed cell death (apoptosis). The outcome is deletion of the affected clones of cells, that is, negative selection of the responders. The deletion may make the animals subsequently unable to respond immunologically to challenge with the initial antigen, or superantigen, in which case they will have become immunologically tolerant to the antigen. The negative selection triggered by particular superantigens has the same specificity as the activation of mature T cells. The extent of deletion of the respective Vβ-bearing T cells varies among superantigens. Positive selection occurs when immature T cells recognize self MHC molecules through the T-cell receptor at lower affinity. Such cells are allowed to differentiate to mature T cells in the thymus. Certain Vβ-bearing T cells are positively selected during development in the thymus, but it has not been shown that superantigens are involved in this selection.

Binding of MHC class II molecules. Immunological presentation of conventional peptide antigens (usually consisting of 9–17 amino acids) to T-cell receptors requires the antigen to be embedded in the groove of MHC class I or class II molecules. Superantigens, which bind MHC class II molecules but not MHC class I molecules, do not interact with the peptide binding groove. However, the precise mechanism of interaction between superantigens and MHC class II molecules is not known. There are two isotypes of mouse MHC class II molecules: A and E. These molecules are very polymorphic; that is, there are many variants with minor differences of amino acid sequences. All superantigens are presented by E molecules, although some are also presented by A molecules. Unlike conventional peptide antigens, superantigens are promiscuously presented by many different MHC class II molecules.

The T-cell receptor binds the MHC–peptide complex through three hypervariable regions: CDR1, CDR2, and CDR3. However, superantigens appear to be bound through CDR4, but not CDR1, CDR2, or CDR3, although the precise interaction between the T-cell receptor and the superantigen–MHC complex has not yet been defined. Further, CD4-bearing T cells and CD8-bearing T cells selectively recognize conventional antigenic peptides embedded in MHC class II molecules or in class I molecules. By contrast, both CD4-bearing T cells and CD8-bearing T cells are activated by superantigens in association with MHC class II molecules.

The biological significance of superantigens is still a mystery. Possibly, viral superantigens have a function that brings some selective advantage to the host and therefore promotes their conservation in the mouse population. However, the selective advantage would have to be quite marked in order to compensate for the presumably substantial disadvantage caused by the loss of a significant percentage of the Vβ repertoire.

For background information SEE ANTIGEN; CELLULAR IMMUNOLOGY; HISTOCOMPATIBILITY in the McGraw-Hill Encyclopedia of Science & Technology.

Kyuhei Tomonari

Bibliography. H. Acha-Orbea et al., Clonal deletion of Vβ14-bearing T cells in mice transgenic for mammary tumor virus, *Nature*, 350:207–211, 1991; P. J. Borkman et al., Structure of the human class I histocompatibility antigen, HLA-A2, *Nature*, 329:506–512, 1987; Y. Choi, J. W. Kappler, and P. Marrack, A superantigen encoded in the open reading frame of the 3′ long terminal repeat of mouse mammary tumor virus, *Nature*, 350:203–207, 1991; P. J. Dyson et al., Genes encoding ligands for deletion of Vβ11 T cells co-segregate with mammary tumor virus genomes, *Nature*, 349:531–532, 1991; C. Harding et al., Liposome-encapsulated antigens are processed in lysosomes, recycled, and presented to T cells, *Cell*, 64:393, 1991; V. Lotteau et al., Intracellular transport of class II MHC molecules directed by invariant chain, *Nature*, 348:600–605, 1990; K. Tomonari and S. Fairchild, The genetic basis of negative selection of Tcrb-V11$^+$ T cells, *Immunogenetics*, 33:157–162, 1991; A. R. Townsend and H. Bodmer, Antigen recognition by class I restricted T lymphocytes, *Annu. Rev. Immunol.*, 7:601–624, 1989.

Aquaculture

Fresh-water crayfishes are the dominant bottom-dwelling invertebrates in many temperate and subtropical aquatic habitats. They are used as food, fish bait, teaching and research subjects, and pets. There are three families of crayfishes: the Astacidae (found in western North America and Europe), Cambaridae (found in North and Central America and eastern Asia), and Parastacidae (found in South America, New Guinea, Australia, New Zealand, and Madagascar). Representatives of all families have been widely introduced around the world, most notably North American cambarids into natural habitats in Europe, Africa, and Asia.

Species of commercial value. There are fewer than 10 recognized species of astacids, over 300 of cambarids, and over 100 of parastacids. Despite this large number, less than 20 have commercial value. Significant species include: in Astacidae, noble crayfish (*Astacus astacus*), Turkish crayfish (*A. leptodactylus*), and signal crayfish (*Pacifastacus leniusculus*); in Cambaridae, papershell crayfish (*Orconectes immunis*), rusty crawfish (*O. rusticus*), northern crayfish (*O. virilis*), red swamp crayfish (*Procambarus clarkii*), and white river crayfishes (*P. zonangulus*, *P. acutus acutus*, and other *Procambarus* species); and in Parastacidae, marron (*Cherax tenuimanus*), red claw (*C. quadricarinatus*), and yabbie (*C. albidus* and *C. destructor*). Other species are exploited when locally abundant, and many have unrealized potential for aquaculture.

With the exception of the papershell crayfish, all the above crayfishes reach more than 4 in. (10 cm) in length and 1 oz (28 g) in weight, a size that is generally considered to be large enough for human consumption. The papershell crayfish is a popular fish bait in the north-central and northeastern United States and rarely exceeds $2^1/2$ in. (6.3 cm) and $^1/4$ oz (7 g). The three *Cherax* species are Australian and are often referred to as giant fresh-water lobsters. They range in weight from $^1/4$ to 4 lb (0.11 to 1.82 kg).

The most commercially and ecologically important crayfish species is the red swamp crayfish, accounting for over 70% of the world's annual crayfish production. It is native only to the south-central United States and northeastern Mexico, but has been successfully introduced around the world since the 1930s. This durable, aggressive crayfish yields annual harvests of 50,000 tons (45,455 metric tons) in the United States, 40,000 tons (36,364 metric tons) in the People's Republic of China, 4000 tons (3637 metric tons) in Europe, and over 100 tons (91 metric tons) in Africa.

Cultural methods. The area of most intense aquaculture is southern Louisiana in the United States with over 125,000 acres (50,600 hectares) of crayfish ponds. Red swamp crayfish are cultured there by establishing perpetuating populations in earthen ponds. These are managed by simulating the natural hydrological cycle to which this species has adapted its life cycle.

Life cycle. Low-lying areas of southern Louisiana are dry during the summer because evaporation exceeds precipitation. During this season, adult crayfish survive in simple burrows that are 2–5 ft (0.6–1.5 m) in depth. Most females lay eggs in these burrows during the autumn. An average $3^1/2$-in. (9-cm) female produces about 300 eggs that are secured to swimmerets, which are appendages located under the abdomen. The young hatch after $2^1/2$–3 weeks and cling to the swimmerets, undergoing two molts over the next 2 weeks.

After the second molt, the young crayfish, about $^1/3$ in. (0.8 cm) in length, are capable of leaving the mother. However, the mother will not emerge from her burrow unless standing water is present at the surface. Young crayfish grow rapidly by consuming plankton, insect larvae, snails, worms, green plants, and detritus (microbially enriched decomposing vegetation), which is their principal food source. Winter rains and spring floods expand the natural habitat. By midspring, most crayfish have reached maturity at sizes around $3^1/2$ in. (9 cm) in length after undergoing another nine molts. The 6–8-month-old adults burrow as water levels begin to decline. Final adult size is dependent on a variety of conditions, including crayfish density, water levels, water quality, food availability, and water temperature. The range in length is from $2^1/2$ to 5 in. (6.3 to 12.7 cm).

Crayfish ponds. Artificial crayfish ponds are shallow (12–18 in. or 30–46 cm deep) and must be built on clay soil. The size of the pond may range from less than 1 acre to over 500 acres (0.4 to 202 ha), although most ponds are 20–40 acres (8–16 ha). Ponds are generally divided into several sections by low, open-ended levees so that water may be thoroughly circulated through them.

New crayfish ponds are stocked with about 50 lb of adult crayfish per acre (57 kg/ha) in mid to late spring. Water is slowly removed over a 2–4-week period to encourage the crayfish to burrow. A forage crop such as rice is often planted, although some farmers still depend on the growth of natural wetland vegetation on the exposed pond bottom. Ponds are refilled in mid-autumn. Harvesting with baited traps begins when most of the young reach usable size, as early as November under ideal conditions, or as late as March. Harvesting is discontinued in mid to late spring, and the pond is once again slowly drained to encourage burrowing of the unharvested crayfish. Restocking of artificial ponds is rarely needed.

Ponds permanently committed to crayfish production may be completely or partially cleared of trees. Removal of trees is advisable because they interfere with water circulation and growth of wetland vegetation or planting of forage crops. Louisiana's average crayfish production is 500–600 lb/acre (568–682 kg/ha). Wooded ponds rarely produce half this amount, while ponds cleared of trees and planted with rice regularly produce twice this amount.

Ponds that are rotated with agricultural crops are called rice-field ponds. Common rotations are rice–crayfish–rice, rice–crayfish–soybean, and rice–crayfish–grain sorghum–crayfish–rice. In the first option, rice is planted in March or April. Ponds are flooded for weed control when the rice is 4–6 in. (10–15 cm) tall. Crayfish are then stocked. Because the level of the water is kept low, the pond becomes very hot during the summer. The crayfish therefore burrow into levees during this time. The rice is harvested in August or September after the water is drained. The pond is then refilled for completion of the crayfish life cycle. Crayfish harvesting usually continues into April, when the pond is once again drained for rice planting. Surviving crayfish burrow so that restocking is not needed.

The rice–crayfish–rice cycle may be repeated continuously, but good rice husbandry requires that the ponds be drained every 2–4 years and left fallow or planted with a dryland crop such as soybeans or grain sorghum. If soybeans or grain sorghum are planted, the rice–crayfish sequence described above is followed for the first year. However, in the summer of the second year, if soybeans are planted, the pond is left dry the following fall, winter, and spring and no crayfish are produced. This is because there is not enough foliage from the soybeans to support crayfish growth, and the sequence must start again with restocking of crayfish in the following spring. If grain sorghum is planted, there is enough forage to support crayish production so the ponds are flooded and restocking is not required.

Problems. Dissolved oxygen in pond water is critical because the decomposing vegetation in the ponds may use up all of the oxygen. Therefore, properly constructed ponds have pumps, aeration towers, and inner levees that permit recirculation of oxygenated water through the entire pond every 3–4 days.

Hardy fishes, such as bullhead catfish and green sunfish, may prey so efficiently on small crayfish that entire crops can be lost. Ponds must be thoroughly dried each year to prevent the establishment of such predators.

Farmers have no real control over the numbers of young crayfish that hatch because they have little control of the numbers of burrowed females. Stunting is a serious problem when the population of juveniles becomes too large. No effective management strategies have been developed other than to remove as many small crayfish as possible.

Soft-shell crayfish. Crayfish must molt their old exoskeletons in order to grow. Crayfish that are used before the new exoskeleton hardens are called soft-shell crayfish and are excellent fish bait and gourmet food. Soft-shell crayfish are produced by transferring desirable-sized crayfish from ponds into shallow tanks at 20–30 crayfish per square foot (210–310 per square meter). They are fed and moved to separate tanks when ready to molt. Biofilters remove wastes and conserve water and energy.

Other cultured species. The white river crayfishes and the papershell crayfish are cultured in the same way as the red swamp crayfish. The white river crayfishes often appear in red swamp crayfish ponds, although they are rarely dominant. Other North American crayfishes are residents of ponds or rivers that are permanently filled, and are very poor burrowers. As a result, they must be cultured in permanent ponds, or the young must be hatched separately and stocked into separate ponds. The same is generally true for European and Australian crayfishes. While the technology for constructing and operating crayfish hatcheries is readily available, they are not widely used because of the expenses involved.

Native European crayfishes have no known resistance to a disease called the crayfish fungus plague, caused by *Aphanomyces astaci*, to which North American crayfishes have high resistance. This disease has devastated European crayfish populations for over 100 years. The situation is so critical that European crayfish farms specializing in native crayfishes often suffer continuous outbreaks of the disease. Few of the European crayfish farms produce food-sized crayfish; rather, most produce young crayfish for restocking plague-stricken natural waters.

There are conflicting data about the resistance of parastacid crayfishes to the crayfish fungus plague. While hatchery and pond management systems have been developed in Australia for the three important species of *Cherax*, it seems inadvisable to pursue cultural endeavors in Europe and North America, where the plague is known or presumed to be present.

For background information SEE *AQUACULTURE; DECAPODA (CRUSTACEA)* in the McGraw-Hill Encyclopedia of Science & Technology.

Jay V. Huner

Bibliography. D. Culley and L. Duobinis-Gray, *Culture of the Louisiana Soft Crawfish,* 1991; H. H. Hobbs III and J. P. Jass, *The Crayfishes and Shrimp of Wisconsin,* Milwaukee Public Museum, 1988; D. M. Holdich and R. S. Lowery (eds.), *Freshwater Crayfish: Biology, Management, and Exploitation,* 1988; J. V. Huner and J. E. Barr, *Red Swamp Crawfish: Biology and Exploitation,* 3d ed., 1991.

Asthma

In the United States, asthma is the most common chronic respiratory disease among children and young adults. An estimated 5–6% of the American population suffers from this disease, and asthma annually is the principal cause of nearly 500,000 hospitalizations, approximately 5000 deaths, and uncounted days lost from school, work, or routine activities. Despite major advances in the understanding of the pathogenesis of asthma and the widespread availability of potent therapies, the prevalence and severity of this disease appear to be increasing.

Asthma frequently has its onset in early childhood, most often before the age of 2 years. However, asthmatic symptoms may develop at any age, and it is not uncommon for asthma to emerge in early or late adulthood. In at least half the cases, childhood asthma, especially if mild, goes into remission as the child grows older; adult asthma rarely goes into remission entirely.

Asthma is a chronic inflammatory disease of the air passageways (bronchi and bronchioles) of the lungs. It is distinct from other inflammatory conditions involving the airways, such as chronic bronchitis, in that it confers on the airways a susceptibility to muscular constriction, with consequent abrupt airway narrowing. Smooth muscle (muscle without voluntary control) rings the airways of the lungs. In asthmatic persons, stimuli that have no effect on airways of nonasthmatic persons trigger widespread contraction of this smooth muscle and often elicit additional airway inflammation. Most triggering stimuli reach the airways by inhalation. They include aeroallergens (such as pollens or animal danders), irritant gases and fumes, viral infections involving the bronchi, and large volumes of inhaled air (such as occur during exercise). Other stimuli reach the airways via the bloodstream after ingestion of substances such as certain food preservatives and certain medications (for example, aspirin and beta-adrenergic blocking agents). In severe disease, airway inflammation itself may contribute importantly to narrowing of airways by two general mechanisms: encroachment of the lumen by a thickened airway wall, an effect that is magnified in the presence of muscular contraction; and obstruction of airway lumens by mucus, sloughed airway lining cells, and other inflammatory debris.

Mechanisms. The precise etiology and pathogenesis of asthma are unknown, but two potentially important mechanisms are the allergic or immunologic model and the neurogenic model involving sensory nerves.

Allergic model. In patients with a tendency toward allergic diseases, asthma appears to be the expression of allergy involving the airways. Related allergic manifestations may include ocular and nasal symptoms and rashes. Common to these conditions, and to many patients with asthma, is the excess production of immunoglobulin (IgE). This class of antibody has receptors for one or more antigens to which the patient is sensitized, and receptors for molecules found on the surface of mast cells. IgE-bearing mast cells are present in excess quantities both superficially and deep within the airway walls of persons with asthma. Mast cells appear to be of central importance in the development of asthma. They contain potent chemical mediators capable of promoting inflammation that are packaged within granules. Upon stimulation, the mast cells release the contents of these granules into the surrounding tissue environment and also synthesize additional mediators of inflammation. One way in which mast cells can be activated is by the interactions of cell-surface-bound IgE with the appropriate antigen recognized by the IgE receptor. A series of enzymatic reactions ensues and results in the explosive release of the chemical mediators, including histamine, platelet-activating factor, leukotrienes, prostaglandins, proteases, and chemoattractant substances. Together, these chemicals can induce the following effects, all of which are observed in asthma: direct smooth muscle constriction as well as reflex smooth muscle constriction mediated by stimulation of sensory nerve endings; increased permeability of small veins and capillaries that leads to edema formation and facilitates migration of cells from blood to bronchi; stimulation of mucus production; and chemical attraction of other inflammatory cells, especially the subtype of white blood cells called eosinophils. Eosinophils are present in increased numbers in the airways of asthmatic persons, and they are thought to amplify the inflammatory response by releasing additional chemical mediators into the local environment. One important product of the eosinophil is major basic protein, which is toxic to epithelial lining cells and promotes their sloughing from the surface of airways into the lumens.

Neurogenic model. Some patients with asthma have no history of allergic diseases and have normal levels of IgE antibody in their blood; they are not sensitive to any known antigens. It is suspected that alternative, nonallergic pathways may promote airway inflammation in these persons. One mechanism currently receiving intensive study is referred to as neurogenic inflammation. It involves sensory nerves that arborize through the superficial layer (mucosa) of the airways. Sensory nerves generally function by transmitting signals centrally to the spinal cord or brain, but retrograde conduction of nerve impulses can result in the local release of neuropeptides from sensory nerve endings. In the airways, end terminals of these sensory nerves (called C-fibers) innervate smooth muscle, blood vessels, and mucous glands. The peptides released from these nerve endings trigger coughing and promote smooth muscle contraction, increased permeability of small blood vessels, and increased secretion of mucus, and so create many of the features seen in asthma. In asthma an imbalance perhaps exists between the production and degradation of these neuropeptides and favors increased local concentrations. A viral tracheobronchitis, reported as the initial event in many patients with nonallergic asthma, may create or exacerbate such an imbalance by disrupting the surface epithelium or by depleting important regulatory enzymes.

Symptoms. Persons with asthma may be entirely well for days and weeks at a time and may develop symptoms only after exposure to relevant stimuli. The cardinal symptoms of asthma are coughing, shortness of breath, and wheezing. The cough may be dry or may raise some of the thick secretions present within airways. Wheezing is the musical, whistling sound made as gas passes through narrowed airways and creates vibrations, like air passing over the reed of a wind instrument. Wheezes may be audible to the patient and are detectable by a physician listening to the chest through a stethoscope. A sensation of chest tightness and chest congestion may also accompany active asthma.

Severe asthmatic symptoms may develop relatively abruptly, over a few minutes to hours, and constitute an asthmatic attack. Patients may develop severe distress with a sensation of suffocation. Heart rate and breathing rate increase, and accessory breathing muscles, including some of the neck muscles, are recruited to help expand the chest. Although there is increased resistance to inspiring during active asthma, the dominant physiologic derangement is the limitation to exhalation. During inspiration, forces surrounding the lung act to expand the chest and, in parallel, to widen airways, whereas during exhalation these forces tend to compress the chest and to narrow airways.

The severity of the obstruction to expiratory airflow can be quantified by having the patient inspire fully and then exhale as rapidly and forcefully as possible into a peak-flow meter, an instrument that records the maximal rate of exhalation achieved, expressed in liters per minute (called the peak expiratory flow rate). In general, asthmatic persons become symptomatic when their expiratory flows fall to less than 70–80% of normal; severe attacks may cause reductions in expiratory flow to less than 30–40% of normal, and urgent medical attention is usually necessary. In recent years, small, relatively inexpensive, and simple-to-use peak-flow meters have become available for home use, so asthmatic patients can regularly monitor their expiratory airflow. Then, patients can initiate or increase their antiasthmatic medications in accordance with previously specified plans of action.

Therapy. Asthma therapy begins with preventive measures to avoid exposure to known triggers of asthmatic symptoms. Allergy testing may be performed to detect the presence of an unsuspected allergic precipitant or to confirm or refute the existence of a potential one. These tests can be performed in two ways: skin tests involve the introduction of a small amount of antigen into the skin and observation for an immediate cutaneous reaction (formation of a hive); blood tests use a sensitive radioallergosorbent assay to detect minute amounts of IgE antibody directed at specific antigens. Positive skin-test responses or antibody levels in the blood correlate with the existence of airway sensitivity to the particular antigen. Immunotherapy is an approach to asthma management that attempts to reduce bronchial sensitivity to a specific allergen by repeatedly injecting small but incremental concentrations of the antigen into the skin over a period of several months. Results with immunotherapy have been variable, and its role in the management of this disease remains controversial.

The pharmacologic treatment of asthma can be divided into two major categories, bronchodilators and anti-inflammatory therapies. Bronchodilators are medicines that prevent or reverse contraction of airway smooth muscle; and anti-inflammatory therapies involve medicines that prevent or reverse airway inflammation.

Bronchodilators. Adrenaline and its derivatives and methylxanthine compounds are the two major groups of bronchodilators used in the treatment of asthma. Adrenaline, also known as epinephrine, is an effective therapy for severe attacks. It binds to beta-type adrenergic receptors on the surface of airway smooth muscle cells and stimulates a sequence of enzymatic reactions that increases intracellular concentrations of cyclic adenosinemonophosphate and ultimately inhibits the functioning of the contractile apparatus, so that muscles relax. Adrenaline has a number of disadvantages, including a relatively short duration of action (1–2 h); stimulation of other adrenergic receptors, including those on heart muscle and blood vessels; and rapid metabolism by enzymes that render it inactive following oral administration. Chemical modifications of the adrenaline molecule have led to the development of improved beta-adrenergic receptor stimulants, or agonists. These agents have a longer duration of action, relative specificity for the particular beta-adrenergic receptors on the smooth muscle of airways, and, in many instances, resistance to degradative enzymes in the gastrointestinal tract, so that oral administration is permitted. For the treatment of asthma, the preferred route of administration of beta-adrenergic agonists is inhalation.

The other major group of bronchodilators, methylxanthines, includes theophylline and caffeine. Theophylline and its salts are dimethylxanthines with moderate bronchodilator potency; caffeine is a trimethylxanthine with very weak bronchodilator activity. Specially compounded tablets and capsules of theophylline allow slow, uniform release of the medication following ingestion, and a long duration of action. This long duration of action is particularly advantageous to persons suffering from interruption of sleep because of asthmatic symptoms that occur during the early morning.

Anti-inflammatory therapy. The most effective treatments for the reduction of airway inflammation are the glucocorticosteroids, a class of medications whose anti-inflammatory activity is utilized in the treatment of numerous medical conditions, from skin rashes to arthritis. The precise mechanism of action of these steroid medications in asthma is unknown, but important effects include an inhibition of the synthesis of inflammatory mediators, a reduction in the permeability of blood vessels, and a decrease in the concentrations of blood and airway eosinophils. When administered orally or intravenously to patients whose asthma has not responded to bronchodilators, corticosteroids generally effect improvement within several hours, with improved airflow and diminished cough and sputum production. Systemically administered corticosteroids are the cornerstone of therapy for severe asthmatic attacks. However, side effects are common and, whenever possible, systemically administered corticosteroids are given for a short period of time (for example, 1–2 weeks) and then withdrawn as the asthmatic attack ameliorates.

In 1977 corticosteroid preparations that can be delivered in aerosol form and that are active locally, with minimal absorption into the bloodstream, became available in the United States. Inhaled corticosteroids are not as potent as systemically administered corticosteroids, but are important in chronic maintenance therapy of asthma and in the prevention of severe attacks. When inhaled on a regular basis, these drugs reduce airway hyperresponsiveness significantly, and patients often report fewer symptoms of asthma and less frequent need for bronchodilator medications.

Cromolyn sodium is an anti-inflammatory therapy for prevention of asthmatic symptoms. Delivered by inhalation, it is thought to block the release of inflammatory mediators from mast cells, even in the presence of antigen–IgE antibody interaction on the cell surface. It is particularly useful for maintenance therapy of childhood asthma and for preventive use prior to exposure to a known antigen. Also, a single dose inhaled prior to exercise effectively blunts exercise-induced constriction of bronchial smooth muscle. Inhaled cromolyn is virtually free of side effects.

Thus, the modern medical treatment of many asthmatic patients relies on inhaled medications alone to achieve maximized lung function and minimized airway hyperresponsiveness with the fewest possible side effects. The regular use of inhaled corticosteroids or cromolyn and the episodic use of inhaled beta-adrenergic agonists provide good control in a large percentage of patients. For asthma that worsens despite these measures, the dose and frequency of administration of the inhaled beta-agonists can be increased, the number of inhalations of corticosteroids taken daily can be increased, and corticosteroids can be given orally or intravenously. Well-informed patients collaborating with health-care providers can utilize these treatments to reduce significantly asthma morbidity and mortality.

For background information SEE ALLERGY; ASTHMA in the McGraw-Hill Encyclopedia of Science & Technology.

Christopher H. Fanta; Albert L. Sheffer; Elliot Israel

Bibliography. P. J. Barnes, A new approach to the treatment of asthma, *N. Engl. J. Med.*, 321:1517–1527, 1989; Expert Panel Report, National Asthma Education Program, National Heart, Lung, and Blood Institute, Guidelines for the diagnosis and management of asthma, *J. Allergy Clin. Immunol.*, 88(3, pt. 2):425–534, 1991; A. L. James et al., The mechanics of airway narrowing in asthma, *Amer. Rev. Respir. Dis.*, 139:242–246, 1989; E. R. McFadden, Jr., Pathogenesis of asthma, *J. Allergy Clin. Immunol.*, 73:413–428, 1984; J. M. O'Byrne (ed.), *Asthma as an Inflammatory Disease*, 1990.

Asymmetric synthesis

For many reasons, organic chemists want to make only one enantiomer of a chiral molecule. Notably, pharmaceutical drugs are often chiral; biological receptors with which drugs interact are chiral. In this situation, the interaction of one enantiomer of a drug with the biological receptor will be different in energy from the interaction of the other enantiomer with the same receptor. An apt analogy is that two right hands shaking each other is quite different from a left hand trying to shake a right hand. One enantiomer of the drug is effective; the other at best is useless, but at worst it may actually be harmful, as was the case with thalidomide. The problem then for an organic chemist is how to go about making a single enantiomer.

Chirality. The source of chirality is always the store of naturally occurring compounds, most of which are available as single enantiomers, and many of which are abundant and cheap. These molecules, if they are suitably chosen and, if necessary, modified, can set up an environment that is no longer achiral—the enantiomers are no longer equal in energy, nor are the pathways to them. Opportunities for applying an asymmetric environment are present in three places. One or both of the starting materials can be made to carry chiral information into the reaction, usually by modifying one of them so that it is a single enantiomer instead of an achiral form. The second opportunity involves incorporation of chirality into the catalyst or into the environment in which the reaction is carried out. The third involves leaving chirality until the end and simply separating the enantiomeric products. The last of these, although the least satisfying intellectually, still represents the simplest solution, the oldest, and the most widely applied in practice if not in research.

Resolution. An asymmetric environment can be achieved after the reaction has been completed. The separation of one enantiomer from the other is known as resolution, and needs something with which to create an achiral environment in which the two enantiomers will differ in energy. The most recent addition to this technology is the chiral chromatography column, where the stationary phase on the column is chiral in only one sense. The enantiomers moving through the column interact differently with the molecules of the chiral stationary phase (as in the analogy of a right hand shaking a right or a left hand) and therefore flow through the column at different rates, one eluting from the end before the other. Enormous progress has been made in the last few years in designing chiral stationary phases that are effective in separations of molecules containing functional groups such as the hydroxyl ($C-OH$) and the amide [$-C(=O)NH-$] groups, where hydrogen bonding can be important, and even more recently in finding stationary phases that can interact with molecules lacking these binding sites. One of these stationary phases (cyclodextrin) is derived from a ring of six sugar molecules; the nonpolar cavity inside the ring is chiral in one sense and able to accept chiral nonpolar molecules to a greater or lesser extent, depending upon the sense of their

chirality and the ease with which each enantiomer can fit inside the ring without bumping into the atoms of the ring itself.

Synthesis from the chiral pool. This approach to the problem involves the starting materials. It does not seek to create the chirality selectively in one sense, but it starts with a natural product and carves the asymmetric part of the target molecule out of something that is already available in nature. The source of compounds coming from the stock of natural materials is called the chiral pool of molecules. Amino acids, sugars, and terpenoids such as pinene and camphor have been used, but the penalty is that much of the original molecule may have to be thrown away, and many steps have to be carried out in selectively removing the parts of the original molecule that are not needed in the final product. An example of this approach is the synthesis shown in reaction (1),

(1)

where mannose (I) is converted to a sugar derivative (II), which after a number of steps, is converted to shikimic acid (III). In these reactions the three chiral centers marked with asterisks in the starting material are preserved throughout the sequence of 10 steps and appear unchanged in the final product. The first step, giving the sugar derivative (II), illustrates how the functional groups can be changed without changing the sense of chirality at the carbon atoms marked with the asterisks.

Attaching a chiral auxiliary. If the starting material is not inherently chiral, it can be made chiral by attaching a chiral auxiliary derived from a natural product. Whereas the methylation of achiral enolate ions inherently gives an equal amount of the enantiomeric products, the enolate ion (IV) in reaction (2) gives unequal amounts of the two products (V) and (VI), the former being 89% of the mixture. Note that structures (V) and (VI) are not enantiomers—they differ in one chiral center but not in both, and the properties of each can be quite different, although the difference can sometimes be inconveniently small. The reason for the high selectivity in favor of the formation of structure (V) is that the chiral auxiliary, derived from the amino acid valine, makes the two faces of the enolate double bond different—one is hindered by

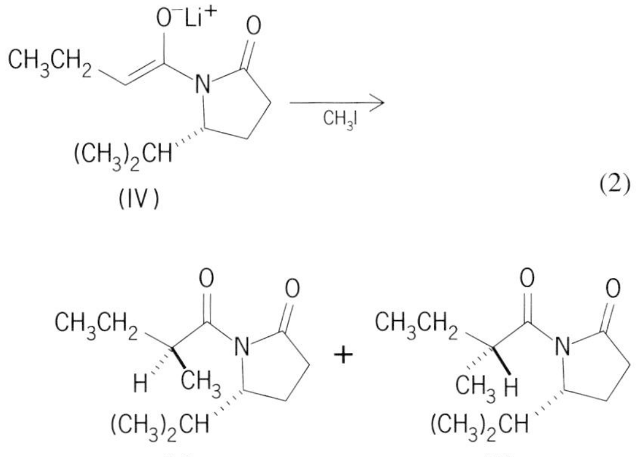

(2)

the bulk of the isopropyl [$(CH_3)_2CH$] substituent and the other is not, as can be seen in the projection [VII, in reaction (3)] of the three-dimensional structure of

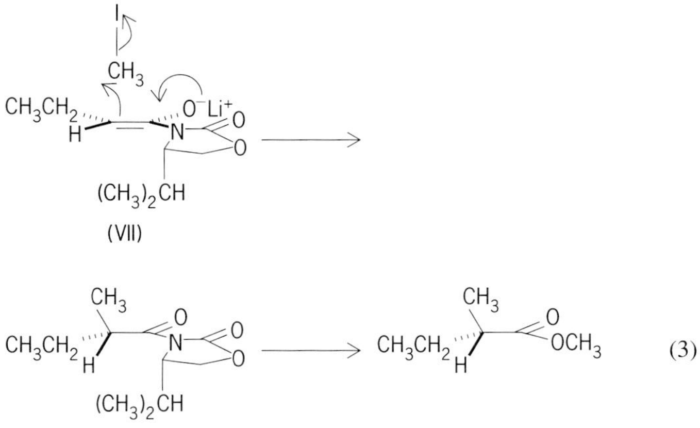

(3)

(IV). Attack by methyl iodide (CH_3I) on the upper surface of structure (VII) leads to the major product (V). Attack on the lower surface is impeded by the isopropyl group. After the chiral auxiliary has performed its function, it can easily be removed, since it is attached to the rest of the molecule merely by an amide [$C(\!=\!O)\!-\!N$] linkage. Thus the product (V) can be converted into the single enantiomer (VIII), which was not available directly.

Chiral catalysts. The method of attaching a chiral auxiliary has two disadvantages: the chiral auxiliary has to be attached and then removed, so that two extra steps enter the total synthesis; and for every mole of product, one mole of chiral auxiliary has to be used. In most cases, the chiral auxiliary can be recovered from the reaction after it has been removed, and hence can be recycled. Nevertheless, some losses are always entailed, and the extra steps are undesirable for a large-scale process. Thus a better place from which to induce the formation of a chiral product in only one sense is in the reagent used to speed up the reaction (catalyst). The best situation is to have the chiral product present in a catalytic quantity, that is, much less than 1 mole of reagent for every mole of product, with the chiral reagent functioning over and

over by regenerating itself as the reaction proceeds.

Highly efficient chiral catalysts are available for reactions such as the hydrogenation of carbon-to-carbon ($C\!=\!C$) double bonds, the reduction of the carbon-to-oxygen ($C\!=\!O$) groups of ketones, the epoxidation of alkenes, and, by way of illustration, dihydroxylation [reaction (4)], where an alkene (IX) is converted to

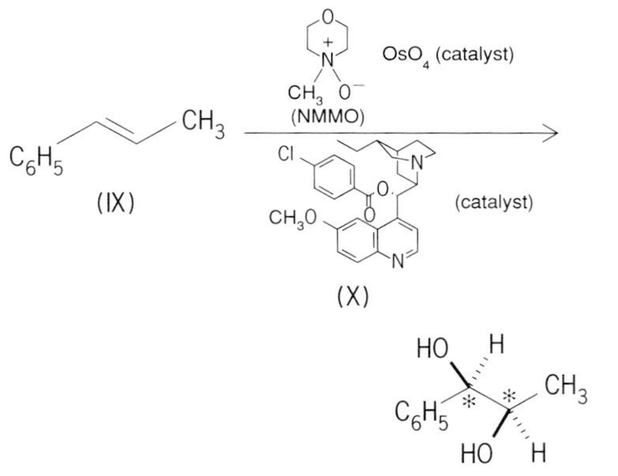

(4)

a diol (XI). The catalyst is a simple derivative (X) of a quinidine alkaloid, but the shape of the reaction complex is not yet understood. The catalyst works well to enable the reagent, osmium tetroxide (OsO_4), to deliver the two hydroxyl groups to the top surface of the alkene (IX) to give the diol (XI). Even the toxic osmium tetroxide is needed only in catalytic quantities, because it is regenerated as the reaction proceeds by oxidation with N-methylmorpholine oxide (NMMO).

Following this line of reasoning, the ultimate in selectivity is often achieved by using an enzyme, taken directly from a naturally occurring compound. Enzymes react selectively to catalyze very particular functional-group interconversions, such as the hydrolysis of esters, which are catalyzed by enzymes called esterases. Pig liver esterase in particular catalyzes the hydrolysis of the diester (XII) in reaction (5) [$Z\!=\!COOCH_2C_6H_5$; $Me\!=\!CH_3$], and selectively

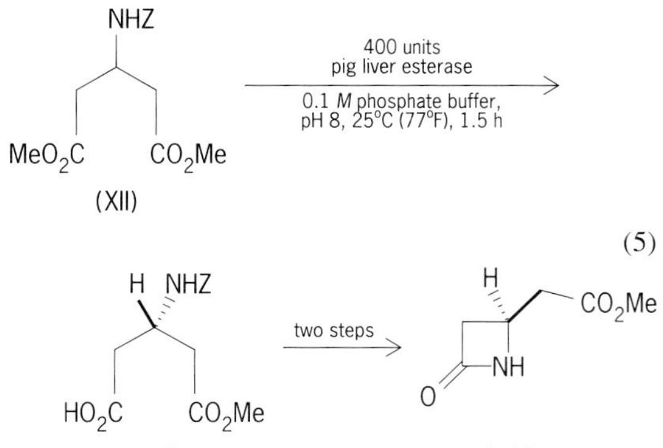

(5)

interacts with only one of the ester groups. An achiral molecule (XII) is thus converted into a chiral molecule (XIII) entirely in one sense (a single chiral form), usually with very high efficiency. The product (XIII) in this case can be converted in two steps into a very simple β-lactam (XIV), ready for making penicillin analogs. The problem in this general area is the finite number of enzyme-catalyzed reactions that can be manipulated, and the even smaller number of enzymes that are promiscuous enough to accept a range of unnatural substrates such as the diester (XII).

For background information *SEE ASYMMETRIC SYNTHESIS; ENANTIOMER; ENZYME; STEREOCHEMISTRY; STEREOSPECIFIC CATALYST* in the McGraw-Hill Encyclopedia of Science & Technology.

Ian Fleming

Bibliography. G. W. J. Fleet, T. K. M. Shing, and S. M. Warr, Enantiospecific synthesis of shikimic acid from D-mannose, *J. Chem. Soc., Perkin Trans. 1*, pp. 905–908, 1984; J. B. Jones, Enzymes in organic synthesis, *Tetrahedron*, 42:3351–3403, 1986; J. D. Morrison (ed.), *Asymmetric Synthesis*, vol. 3, 1984; J. S. M. Wai et al., A mechanistic insight leads to a greatly improved osmium-catalyzed asymmetric dihydroxylation process, *J. Amer. Chem. Soc.*, 111:1123–1125, 1989.

Atomic physics

The process of electron ejection from matter following the absorption of electromagnetic radiation has been under investigation for just over a century. The earliest measurements involved electron emission from metal surfaces irradiated by ultraviolet radiation. The theoretical interpretation of this phenomenon, known as the photoelectric effect, played an important role in establishing quantum theory.

In a low-density environment the photoeffect is called either photoionization (in the case of atoms and positive ions) or photodetachment (in the case of negative ions). In general, these bound-to-free transitions can involve multiphoton absorption and multielectron ejection, but most studies have been limited by the intensity and energy of the photons to the most probable process, namely, single-photon–single-electron ejection. Studies of the photodetachment of atomic ions are discussed in this article.

Theory of photodetachment. In the initial state of photodetachment, a photon is absorbed by a negative ion (an atom with one or more extra electrons). In the final state, the detachment continuum electron moves in the weak and relatively short-range field of the residual atom. This final-state interaction arises from polarization and exchange effects, which are manifestations of electron correlation. The enhanced role of correlation in photodetachment is due to the absence of the Coulomb interaction, which, in contrast, dominates the photoionization process.

Each mode of fragmentation defines a final-state exit channel that is characterized by the energy and angular momentum of the detached electron as well as the excitation state of the residual atom. Since photodetachment is an endoergic process (the reaction needs external energy to proceed), each channel has a well-defined threshold photon energy below which the channel is energetically closed. The energy carried off by the outgoing electron represents the balance between the energy supplied by the photon and the binding energy of the electron plus the excitation energy, if any, of the residual atom. The orbital angular momentum of the detached electron is $l = l_0 \pm 1$, where l_0 is the angular momentum of the electron prior to detachment (since the photon carries one unit of angular momentum). The outgoing electron is, in general, represented by two partial waves ($l_0 \pm 1$) that interfere as a function of photon energy. This interference determines the angular distribution of the photoelectrons.

To understand the detailed manner in which a negative ion interacts with radiation requires an investigation of the cross section for photodetachment as a function of the photon energy (the cross-section curve). Generally, such curves start at zero at threshold, rise to a maximum just beyond threshold, and then decrease monotonically. Occasionally, resonant structure is superimposed on the cross-section curve over a narrow range of photon energies. Resonances result from an interference between resonant and nonresonant pathways to the same final state. The resonant process arises from the transient photoexcitation of a doubly excited state of the negative ion in an intermediate step of the photodetachment process. Photodetachment cross sections have been measured for only about 20 species of atomic negative ions, and only a few cross sections are known to within 10%.

Due to their enhanced sensitivity to electron correlation effects, photodetachment cross sections pose a challenge to theory. Calculations based upon the independent-electron model generally fail to agree satisfactorily with experimental data. Correlated wave functions are needed to represent both the bound and continuum states. Photodetachment cross sections are not only of practical and intrinsic interest. Frequently they form the most accessible path to determining, via the principle of detailed balance, cross sections for the far less probable, inverse process of radiative attachment, an exoergic process whereby an electron attaches itself to an atom, releasing the excess energy radiatively.

Experimental methods. Negative ions are fragile quantum systems and are thus easily destroyed in interparticle collisions. This is a consequence of the weak and relatively short-range nature of the interaction between the outermost electron and the atom to which it binds. The binding energy of the extra electron in a negative ion (the electron affinity) is typically an order of magnitude smaller than the corresponding quantity in an atom. The production of negative ions in sufficient quantities for spectroscopic studies is challenging, since processes involved in their creation must compete with more probable destruction processes. A low-density and low-temperature environment is essential for the survival of negative ions. For photodetachment studies, the concentration of ions into a small source volume permits an efficient overlap with a spatially small but intense laser beam. Recently, photodetachment measurements have been made on negative ions stored in electromagnetic traps, but the most commonly used source has been a tenuous beam produced by an accelerator. After production, the fragile ions travel unidirectionally through a high-vacuum environment to the photon–ion interaction region. Both crossed and merged laser-beam and negative-ion-beam geometries have been used in photodetachment studies. **Figure 1** shows a crossed-beam apparatus designed for energy- and angle-resolved spectroscopic measurements of the energies, yields, and angular distributions of photoelectrons. A fast-moving and tenuous beam of negative ions is prepared, in this case, by charge transfer when a beam of positive ions from an accelerator passes through an alkali vapor cell. The negative-ion beam is intersected perpendicularly by a plane-polarized beam of photons from a pulsed dye laser. Following photodetachment, electrons, ejected from the moving ions in the direction of their motions, are collected and energy-analyzed by means of an electron spectrometer placed in the path of the ion beam. The angular distribution of the detached photoelectrons can be determined by measuring their yield as a function of the angle between their fixed collection direction and the polarization vector of the laser beam, which can be rotated by the use of a half-wavelength phase retarder.

Cross sections. A cross section may be determined using a crossed-beam apparatus by measuring the photoelectron yield and angular distribution, the photon flux, and the ion-beam intensity and velocity. In addition, to make an absolute measurement requires knowledge of the overlap of the two beams in the interaction region and the efficiency for collection and detection of the photoelectrons. One way of circum-

venting these difficult measurements is to determine, under identical conditions, the ratio of the cross sections for photodetaching an electron from a beam of the ions of interest and a beam of reference ions whose cross section is well known. A novel feature of the apparatus is the way that kinematic effects, associated with detachment from moving ions, are exploited in order to eliminate the need to know the relative efficiency for collecting and detecting electrons from the two beams. The combination of fast, unidirectional ion beams and forward-directed electron collection provides a wide latitude for shifting electron energies from their value in the ion frame to a greater value in the laboratory frame. This so-called kinematic amplification of peak energies is illustrated in **Fig. 2**. The spectra in Fig. 2a and b show peaks associated with the photodetachment of the metastable negative helium (He$^-$) ion via the $2^3S\,\varepsilon s,d$ and $2^3P\,\varepsilon p$ channels, respectively. The spectrum peak in Fig. 2c is associated with photodetachment of the negative hydrogen (H$^-$) [in this case, negative deuterium (D$^-$)] reference ion via the $^2S\,\varepsilon p$ channel. [Here, channels are labeled by quantum numbers describing the states of the residual atom and the photodetached electron. For example, in the $^3P\,\varepsilon p$ channel of Fig. 2b, the atom is left in the $1s2p\;^3P$ state and the outgoing electron has an orbital angular momentum of 1 (p).] The energy coincidence of the three peaks in the laboratory frame, brought about by appropriate choices of the ion-beam energies E_i, ensures equal efficiencies for electrons associated with each peak and therefore the cancellation of these efficiencies in the cross-section ratios. The elimination

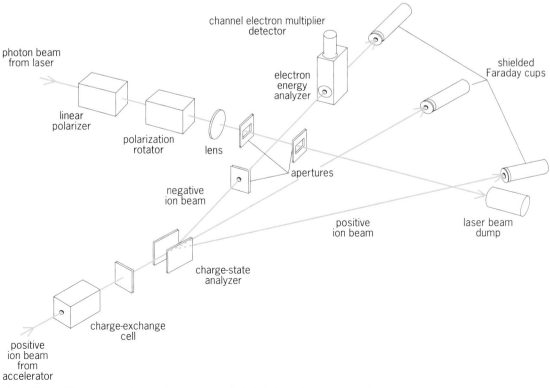

Fig. 1. Crossed-beam apparatus used to measure photodetachment cross sections.

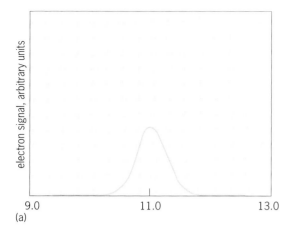

(a)

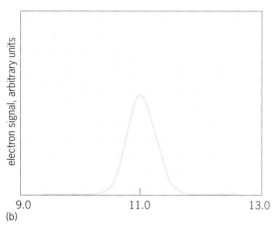

(b)

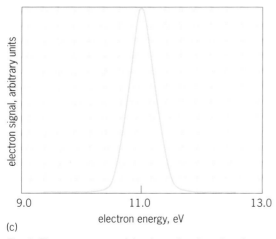

(c)

Fig. 2. Electron spectra arising from the photodetachment of negative helium (He⁻) ions, (a) via the $2^3S\varepsilon s,d$ channels for ion-beam energy E_i = 29.19 keV and (b) via the $2^3P\varepsilon p$ channel for E_i = 50.47 keV, and (c) negative deuterium (D⁻) ions via the $2S\varepsilon p$ channel for E_i = 19.17 keV.

of the dependence of the ratios on efficiency and the use of theoretical H⁻ photodetachment cross sections (known to be better than 3%) for normalization have resulted in the most accurate cross-section measurements so far (approximately 5–10%).

Thresholds. It was demonstrated by E. Wigner that the behavior of a photodetachment cross section

σ in the close vicinity of a channel opening depends solely on the final-state interaction between the residual atom and the slowly moving detached electron. As a consequence of the enhanced importance of electron correlation in photodetachment, the threshold law exhibits a dependence on the orbital angular momentum l of the detached electron. The Wigner threshold law for photodetachment takes the form $\sigma \sim (\Delta E)^{l+1/2}$, where ΔE is the excess electron energy beyond threshold. This threshold behavior is illustrated in the bottom half of **Fig. 3**, where the change in shape of cross section for photodetaching the negative lithium (Li⁻) ion near the opening of the $^2P\varepsilon s$ channel (corresponding to the lithium atom being left in the 2^2P excited state) is shown. The data follow the expected threshold dependence ($\Delta E^{1/2}$ for s-wave detachment) for approximately 6 meV beyond threshold. The top half of Fig. 3 illustrates anomalous behavior in the $^2S\varepsilon p$ channel cross section at the threshold for the opening of the $^2P\varepsilon s$ channel. The structure, which takes the form of a cusp in this case, arises from the strong coupling between the two channels brought about by polarization of the excited Li atom in the field of the detached electron. Due to the coupling, the sharp onset of photoelectron production in the $^2P\varepsilon s$ channel (where the cross-section curve has infinite slope at the threshold for s-wave detachment) produces an equally sharp decrease in the $^2S\varepsilon p$ channel in the postthreshold region. The sharp increase in the prethreshold region is due to the occurrence of virtual transitions. The electron affinity of Li can be accurately determined

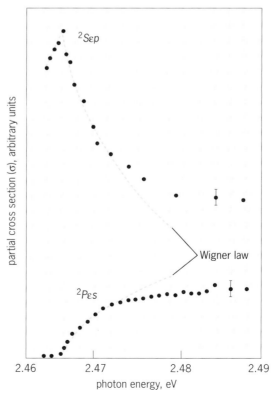

Fig. 3. Photodetachment of negative lithium (Li⁻) ions in the vicinity of the 2^2P threshold.

from a measurement of the photon energy corresponding to the onset of photoelectron production in the $^2P\varepsilon s$ channel.

For background information SEE ATOMIC STRUCTURE AND SPECTRA; ELECTRON AFFINITY; SCATTERING EXPERIMENTS (ATOMS AND MOLECULES) in the McGraw-Hill Encyclopedia of Science & Technology.

David J. Pegg

Bibliography. D. Bates (ed.), *Atomic and Molecular Processes*, 1962; R. R. Corderman and W. C. Lineberger, Negative ion spectroscopy, *Annu. Rev. Phys. Chem.*, 30:347–376, 1979; A. Dalgarno et al. (eds.), *The Physics of Electronic and Atomic Collisions*, 1990; D. J. Pegg et al., Partial cross sections for the photodetachment of metastable He^-, *Phys. Rev. Lett.*, 64:278–281, 1990.

Aves

The earliest bird, *Archaeopteryx lithographica*, was a pigeon-sized animal, known from six fossilized specimens unearthed in limestone quarries of Bavaria. Although *Archaeopteryx* lived almost 140 million years (m.y.) ago (Late Jurassic Period), its avian nature is immediately evident—several specimens are at least partially cloaked in flight and contour feathers that are surprisingly similar to those of many living birds. However, another attribute of *Archaeopteryx* is equally obvious: much of its skeleton is reptilian and bears striking resemblance to a number of late Mesozoic small bipedal dinosaurs, the coelurosaurid theropods. Indeed, a skeleton of *Archaeopteryx*, preserved without feather impressions (the Eichstatt specimen), was long misidentified as that of an immature dinosaur.

Interpretation of *Archaeopteryx*'s general mode of life and capacity for powered flight has historically been the source of continuous, sharp debate. Most

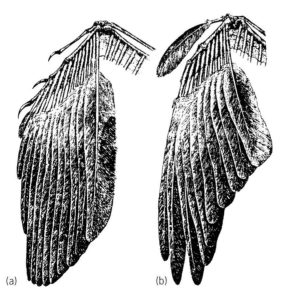

Fig. 1. Similarity of the wings and flight feathers of (*a*) *Archaeopteryx* and (*b*) the pigeon. (*After G. Heilman, The Origin of Birds, Appleton, 1926*)

paleontologists believe that *Archaeopteryx* was an active, ground-dwelling predator that was also able to move about in the trees. It is frequently described as a capable glider and a feebly powered, or flapping, flier; many paleontologists assume it was incapable of take-off and flight upward from the ground. However, other scientists believe that *Archaeopteryx* was an adept flier, as most birds are today.

Flight-related structures. Although its skeleton remained distinctly dinosaurlike, *Archaeopteryx* was aerodynamically advanced. Forelimb proportions and arrangement, as well as anatomy of primary and secondary flight feathers, were generally similar to those of modern birds (**Fig. 1**). The wing shape was like that of birds that fly easily through broken vegetation (for example, woodcocks and woodpeckers), and the asymmetry of the primary flight feathers indicates that they probably functioned as airfoils during flapping (powered) flight, just as they do in living birds. This combination of features in *Archaeopteryx* is strongly indicative of flapping flight.

Interpretation of flight in *Archaeopteryx* is complicated by the apparent disparity between internal and external flight-support structures: the pectoral region appears too poorly developed to have accommodated the volume of flight musculature commensurate with powered, flapping flight. Modern birds possess an extensive pectoral region (from the furcula, or "wishbone," to the posterior extension of the keeled sternum, or "breastbone") and enlarged crests on the humerus, or arm bone, to accommodate flight muscles that constitute about 15–25% of total body mass. In *Archaeopteryx*, development of the humeral crest is comparable to that of many modern birds. However, lack of an ossified keeled sternum, and particularly, the likely presence of gastralia ("belly" ribs) in the region that a well-developed, keeled sternum would have occupied, are indicative of a comparatively small flight-muscle mass (**Fig. 2**). Therefore, several authors suggest that *Archaeopteryx* lacked sufficient muscle power to take off upward from the ground or to even sustain horizontal powered flight.

Thus, *Archaeopteryx* possessed highly specialized fore and hind appendages, but conventional thought holds that each pair functioned optimally in very different habitats: its hindquarters were adapted primarily to a ground-dwelling, cursorial existence, but the highly aerodynamic wings could generate powered flight only if *Archaeopteryx* first managed to scramble up a tree. These contradictory aspects of the scenario are resolvable if *Archaeopteryx* was capable of flight upward from the ground, as well as down from the trees. A more reasonable alternative would then emerge, in which *Archaeopteryx* could have ranged easily through terrestrial or arboreal habitats without having to forego its apparent and highly specialized capacity for powered flight.

"Flying-ectotherm" hypothesis. Recently, it was suggested that if *Archaeopteryx* retained a cold-blooded, or ectothermic, reptilelike physiology rather than a warm-blooded, or endothermic, birdlike physiology, the flight-support capacities of its internal and external structures would appear more evenly matched. Surprisingly, if *Archaeopteryx* was ectothermic, it may well have achieved powered, flapping flight, as well as takeoff upward from the ground, with less than one-half the flight-muscle volume of modern birds.

This possibility is related to a largely unrecognized attribute of reptilian muscle physiology: during burst-level activity, major locomotory muscles of a number of active terrestrial reptiles generate at least twice the power per mass of muscle tissue as do such muscles of birds and mammals. Even though lizards and mammals attain similar sprint speeds, the locomotory muscle mass in reptiles is commonly less than in mammals of equivalent size.

Utilization of high-power, reptile-type flight muscle to support powered flight seems consistent with *Archaeopteryx*'s relatively reduced pectoral surface area. Muscle tendons often insert on bony processes, or crests, the size of which reflects the magnitude of muscular force to which the bony processes are subjected. *Archaeopteryx*'s humeral crest was somewhat larger than that of many modern birds of equivalent body size. These observations are strongly indicative that *Archaeopteryx*'s capacity for generation of flight-muscle power was probably at least equal to that of many living birds. If *Archaeopteryx* possessed pectoral muscle that was only 7% of total muscle mass (rather than the 15–25% of many extant birds) but its skeletal muscle was physiologically reptilian, then its flight muscles would have generated power comparable to that of many extant birds (150–225 watts per kilogram of skeletal muscle), but these muscles would have required far less skeletal area for their origin. This argument would resolve much of the apparent discrepancy between external and internal flight apparatus in *Archaeopteryx*. Thus, if *Archaeopteryx* was physiologically reptilian, it probably could have achieved takeoff upward from the ground, as well as horizontal flapping flight.

Powered flight in an ectothermic, reptilelike *Archaeopteryx* would have had certain disadvantages as well. During intense exercise, reptiles typically rely on low-endurance, anaerobic metabolism to support muscular activity; the muscles of birds and mammals utilize primarily high-endurance, aerobic metabolism. Consequently, following bouts of intense exercise, reptiles must deal with marked muscle fatigue and accumulation of high levels of intramuscular lactic acid, a by-product of anaerobic metabolism.

Anaerobic muscle power. A question exists as to whether a physiologically reptilian *Archaeopteryx* could have relied on anaerobic muscle power for flapping flight without chronic danger of exhaustion-induced crash landings or prolonged periods of post-flight fatigue. Based on the metabolic physiology of a number of active modern reptiles, an ectothermic, mostly anaerobically powered *Archaeopteryx* would have been capable of nonstop flapping flight for at least 0.9 mi (1.5 km). This capacity is not remarkably different from the locomotor capacity of some large lizards (such as the Komodo monitor, *Varanus komodoensis*), which are capable of nonstop 0.6-mi (1-km)

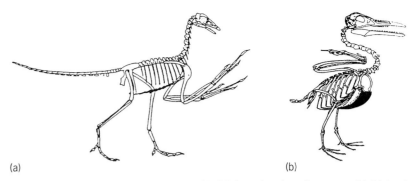

(a)   (b)

Fig. 2. Skeletons of (*a*) *Archaeopteryx* and (*b*) *Ichthyornis*, a Late Cretaceous bird (about 75 m.y. old) that was fairly similar to birds living today. Compared to *Archaeopteryx*, *Ichthyornis* possessed a well-developed, keeled sternum that probably accommodated enlarged flight muscles. (*Part a after J. A. Ostrom, Archaeopteryx and the origin of birds, Biol. J. Linn. Soc., 8:91–182, 1976; part b after O. C. Marsh, Odontornithes: A Monograph of the Extinct Toothed Birds of North America, 1880*)

sprints at speeds approaching 19 mi/h (30 km/h).

Shorter flights could have been taken with relative impunity: flights of 11–22 yd (10–20 m) might have been followed by approximately 60-s periods of recovery from very low fatigue. Thus, normal daily activity patterns of an ectothermic *Archaeopteryx* could readily have included repeated short-distance flights upward from the ground to tree limbs, or from tree to tree.

Evolution of modern bird flight. This Jurassic flying-ectotherm scenario helps account for some subsequent events in avian flight evolution. By the Early Cretaceous Period (about 15 m.y. subsequent to *Archaeopteryx*), birds had developed relatively advanced flight-support structures, including a keeled sternum and a fairly typical avian furcula. These are also features of modern birds and reflect Early Cretaceous evolution of greatly enlarged flight muscles. There is, however, no reason to assume that these modifications made birds more aerodynamically advanced than *Archaeopteryx*. Perhaps, following attainment of fully powered flight in Jurassic birds, the advances were more closely allied with Cretaceous Period development of avian endothermy.

High rates of aerobic metabolism are tightly linked to endothermy in modern birds and are fundamental to long-duration, aerobically powered flight. Attainment of efficient, short-distance powered flight in *Archaeopteryx* might have been followed by selection for expanded capacity for aerobic metabolism to support increased stamina and flight duration. Additionally, flight-muscle capacity for anaerobiosis probably declined significantly, because high rates of lactic acid formation are potentially disruptive to aerobiosis. Consequently, the switchover to endothermy, and to an aerobically based avian activity physiology, was probably inextricably linked to reduced reliance on reptile-type physiology and diminished capacity for flight-muscle power output. The evolution of avian endothermy likely was accompanied by expansion of flight-muscle volume because, like the flight muscle of modern birds, the muscles would necessarily have been composed of high-endurance but lower-power aerobic fibers. Thus, the keeled sternum and furcula of Early Cretaceous birds may signal avian attainment of endothermy and expanded flight stamina in birds subsequent to *Archaeopteryx*.

If the flying-ectotherm scenario is correct, *Archaeopteryx* was already an adept flier, aerodynamically similar in many respects to extant birds. In that case, flight in *Archaeopteryx* may reveal little more about the origins of avian flight than does flight in modern birds.

For background information SEE ARCHAEORNITHES; AVES in the McGraw-Hill Encyclopedia of Science & Technology.

John Ruben

Bibliography. M. K. Hecht et al. (eds.), *The Beginnings of Birds*, 1985; J. A. Ostrom, *Archaeopteryx and the origin of birds*, *Biol. J. Linn. Soc.*, 8:91–182, 1976; J. A. Ruben, Reptilian physiology and the flight capacity of *Archaeopteryx*, *Evolution*, 45(1):1–17, 1991.

Bioinorganic chemistry

Recent advances in bioinorganic chemistry include discoveries concerning the mode of action of the antitumor drug cisplatin and development of new inorganic reagents for site-specific cleavage of proteins.

Key: ○ nitrogen
○ platinum
○ oxygen
◉ phosphorus
○ carbon

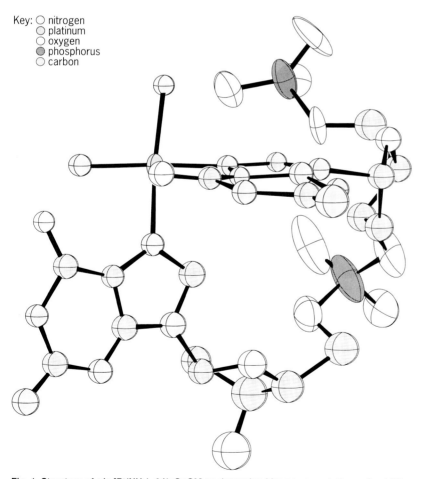

Fig. 1. Structure of *cis*-[Pt(NH₃)₂{d(pGpG)}] as determined by x-ray crystallography. (*After S. E. Sherman et al., Crystal and molecular structure of cis-[Pt(NH₃)₂{d(pGpG)}], the principal adduct formed by cis-diamminedichloroplatinum(II) with DNA, J. Amer. Chem. Soc., 110:7368–7381, 1988*)

Antitumor Drug Cisplatin

The simple inorganic compound *cis*-diamminedichloroplatinum(II), or cisplatin, remains one of the most effective anticancer agents currently approved for clinical use. Its major impact is in testicular cancer; more than 90% of the patients treated with the drug can expect long-term survival. Although cisplatin has been approved for treatment in ovarian and head and neck cancers, it is less effective against these tumors; and it is even less effective for the most common cancers, those of the lung, colon, and breast. Once the molecular mechanism of action of the drug in cancer cells is understood, it should be possible to design and implement more effective chemotherapeutic protocols, and to synthesize better compounds for treating tumors that are or have become resistant to chemotherapy based on platinum-containing compounds.

Mode of action. An understanding of the mode of action of an anticancer drug must derive from elucidating the details of its binding to biological targets that lead ultimately to the selective toxicity for tumor tissue. In the case of cisplatin, there is general agreement that it damages deoxyribonucleic acid (DNA) and that this damage leads ultimately to the desired selective toxicity. Much is known about the nature of cisplatin adducts on DNA. In the cytoplasm, the release of chloride ions from the platinum opens sites for binding to the nucleophilic bases of DNA, especially the nitrogen atoms of the purines guanine and adenine. The major adducts are intrastrand crosslinks formed between two adjacent guanine bases and, to a lesser extent, a guanine and an adenine base. The structure of one such cross-link, as determined by x-ray crystallography, is shown in **Fig. 1**. Recent evidence has revealed that this intrastrand cross-link results in unwinding of the double helix by 13° and bending by 34° toward the major groove. The resulting structure is postulated to serve as a recognition signal for cellular factors that might bind specifically to DNA damaged by cisplatin and in some manner potentiate its mode of action.

Accordingly, several strategies were employed to identify such cisplatin–DNA structure specific recognition proteins (SSRPs) with considerable success. By using platinated DNA as a probe, several SSRPs in the 25,000–30,000 and 90,000–100,000 molecular weight ranges were identified in human HeLa cells, a cervical cancer cell line. In this research, the probe was found to undergo a shift in its gel electrophoretic mobility following incubation with cell extracts. The gel shift arises from tight association of proteins with the platinum-induced distortions, specifically the d(GpG) cross-links, shown in Fig. 1, or the d(ApG) cross-links. In a separate set of experiments, a gene encoding a cisplatin-damaged DNA binding protein, SSRP1, of approximate molecular weight 81,000, was cloned and sequenced. The sequence revealed homology to a pair of nuclear proteins from the high-mobility-group (HMG) protein class, HMG1 and HMG2. Subsequently, HMG1 obtained from a clone was found to bind specifically to DNA containing the same major cisplatin intrastrand d(GpG) or d(ApG) cross-links.

Hypotheses. Although the true functions of the HMG1 and HMG2 proteins and of the 90,000-molecular-weight damage recognition proteins are unknown, the former have been associated with critical cellular processes such as replication, mitosis, recombination, and transcription. Various hypotheses have been set forth to account for how the proteins might contribute to the mode of action of cisplatin (**Fig. 2**). One is that the proteins are part of the cellular repair apparatus, with the particular function of recognizing and signaling the presence of damaged DNA. If there is less of such a protein in tumor cells than is required to process platinum damage, then inhibition of DNA synthesis by the drug in that population of cells might lead to the anticancer effect. In the case of the cloned SSRP1 and the protein HMG1, however, the viability of this hypothesis is questionable, given the high levels of both in a variety of tissues and species, the failure to induce SSRP1 by exposing cells to cisplatin, and the fact that no cross-reactivity of the proteins with DNA damaged by any other agent has been discovered.

A second hypothesis is that these structure specific recognition proteins have functions critical to the survival of tumor cells, for example, mediating the transcription of a specific protein involved in the regulation of cell growth. Removal of such proteins from the cellular pool by binding to regions of DNA that contain cisplatin adducts could result in loss of function and either reversion of the tumor to the normal cell phenotype or inability to control cellular homeostasis, leading to cell death. A related, third hypothesis is that elevated levels of the protein in tumor cells, perhaps associated with their rapid division, actually serve to mask cisplatin adducts from recognition by the normal repair proteins. At the onset of cell division, the masked platinum might be sufficient to block replication, so that anticancer activity is produced. Recent work using DNA containing a single intrastrand d(GpG) or d(ApG) cross-link has shown that these major cisplatin adducts are efficient in blocking replication.

It has been known since the work of B. Rosenberg, who first reported the anticancer activity of cis-

platin, *cis*-[Pt(NH$_3$)$_2$Cl$_2$] (I), that the trans isomer, *trans*-[Pt(NH$_3$)$_2$Cl$_2$] (II), is inactive. When DNA is al-

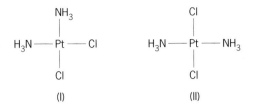

lowed to react with the trans isomer, adducts that consist mainly of long-range intrastrand and interstrand cross-links are formed. This isomer is stereochemically incapable of forming 1,2-intrastrand cross-links characteristic of the major cisplatin adducts. Although replication is efficiently inhibited in DNA containing *trans*-[Pt(NH$_3$)$_2$Cl$_2$]-induced cross-links, there is no binding of the SSRPs to DNA that has been damaged by this and other inactive compounds. These results reinforce the notion that the SSRPs are involved in the mode of action of cisplatin.

Other platinum compounds. Carboplatin, [Pt(NH$_3$)$_2$(1,1-{O$_2$C}$_2$C$_4$H$_6$)], a cisplatin analog, has been approved for anticancer therapy. This compound is generally less toxic than the parent drug and can be used with fewer side effects. Its mode of action involves hydrolysis of the relatively inert cyclobutane dicarboxylate carrier ligand, freeing the same {Pt(NH$_3$)$_2$}$^{2+}$ moiety that evolves from hydrolysis of the chloride ligands of cisplatin. Carboplatin and cisplatin thus both release, at different rates, the critical platinum unit that damages DNA and leads to antitumor properties.

Under development is a class of platinum compounds that might be suitable for oral administration. These compounds are derivatives of platinum in the 4+ oxidation state. They contain one ammonia, one cyclohexylamine, two chloride, and two carboxylate ligands. They probably function by crossing the gastrointestinal tract, being reduced with loss of their carrier carboxylate ligands, and undergoing subsequent pharmacology and biochemistry analogous to that of cisplatin and carboplatin. Platinum(IV) carboxylates are scheduled to enter clinical trials in mid-1992. Preliminary studies of their DNA binding properties indicate that they form the same recognition complex for the SSRPs as do cisplatin and carboplatin. Thus, the above hypotheses concerning the role of these proteins in the mode of action of platinum anticancer drugs apply in a unified way to the therapeutically active compounds.

If the role of the structure specific recognition proteins in the molecular mechanism can be established, it might be possible to extend the chemotherapeutic potential of the drug to more common tumor types. Perhaps by altering the levels of these proteins, possibly through gene therapy, these more refractory tumors might become treatable at very low doses of the platinum drug.   *Stephen J. Lippard*

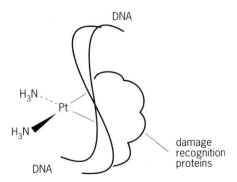

Fig. 2. Cisplatin-damaged DNA serves as a signal for structure specific recognition proteins (SSRPs) in the cell that may facilitate the antitumor mechanism of the drug.

Site-Specific Cleavage of Proteins

Major building blocks of living cells, proteins play central roles in all biological processes, such as transport and storage, mechanical support, immune protection, molecular recognition, information transfer, regulation, and catalysis of biochemical reactions with remarkable specificity. These functions depend on the molecular structure at particular sites on the proteins. Physical methods such as x-ray crystallography, nuclear magnetic resonance spectroscopy, electron microscopy, and circular dichroism studies have been employed to elucidate the structures of these biologically significant macromolecules. The proteolytic enzymes and chemically synthesized reagents that react with specific amino acids of proteins and cleave the polypeptide backbone at selective sequences provide useful information about protein structure and function. Because of the limitations of naturally occurring peptidases and synthetic peptide cleaving agents, new reagents are being developed that are directed by proximity (cleavage of neighboring bonds in folded proteins) rather than residue type. This new strategy can be employed for sequence analysis, in which the spatial arrangement of subunits within a supramolecular structure is determined, and for the design of new targeted therapeutic agents.

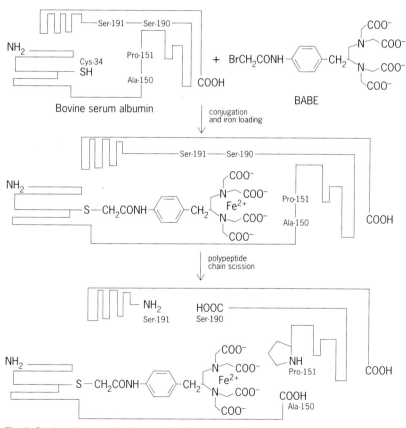

Fig. 3. Conjugation reaction between bovine serum albumin and (*p*-bromo-acetamidobenzyl)-EDTA (BABE), showing iron loading on the chelate and polypeptide-chain scission at 25°C (77°F) for 10 s in the presence of sodium ascorbate and hydrogen peroxide. Cys = cysteine, Ser = serine, Pro = proline, Ala = alanine. (*After T. M. Rana and C. F. Meares, Specific cleavage of protein by an attached iron chelate, J. Amer. Chem. Soc., 112:2457–2458, 1990*)

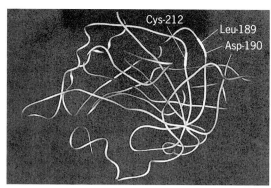

Fig. 4. Human carbonic anhydrase I showing residues of cysteine (Cys-212), leucine (Leu-189), and aspartic acid (Asp-190) in the polypeptide backbone. Cys-212 is modified with iron-(*p*-bromo-acetamidobenzyl)-EDTA (Fe-BABE), and the peptide bond between Leu-189 and Asp-190 is hydrolyzed. The sulfur of Cys-212 is 0.53 nanometer from the carbonyl carbon of Leu-189, and 0.51 nm from the carbonyl carbon of glycine (Gly-145). (*From T. M. Rana and C. F. Meares, Iron chelate mediated proteolysis: Protein structure dependence, J. Amer. Chem. Soc., 113:1859–1861, 1991*)

Proteins are susceptible to damage by reactive oxygen species generated by the oxidation of reduced transition metals such as ferrous iron (Fe^{2+}) and cuprous copper (Cu^+) by oxygen (O) or hydrogen peroxide (H_2O_2). The reaction of generated reactive oxygen species with neighboring bonds causes destruction of amino acid side chains and fragmentation of proteins. These reactions are catalytic in the presence of reducing agents such as ascorbate or dithiothreitol. New inorganic reagents that cleave proteins site-specifically are developed by introducing artificial metal-binding sites in the protein sequences. A specifically bound metal ion serves as a site for generation of reactive species to cleave the peptide backbone. Separation and identification of the cleavage products provides useful information about the three-dimensional structure of the protein, revealing, for example, which segments of the polypeptide chain are close to the metal site.

Experimental strategy. In general, artificial metal-binding sites are introduced into the sequence of the proteins by chemical modification of the protein side chains with bifunctional chelates. A bifunctional chelating agent is a reagent whose molecular structure contains a strong metal-chelating group and a chemically reactive functional group. In the case of bovine serum albumin, a unique cysteine residue is modified with an iron-containing bifunctional chelate (**Fig. 3**). Protein cleavage is initiated by treating the protein-chelate conjugate with H_2O_2 and ascorbate, and the production of three peptide fragments that together account for the entire polypeptide chain is observed. Protein fragments are isolated by polyacrylamide gel electrophoresis and are transferred to polyvinylidene membranes, and the cleavage sites are identified by sequencing. Amino termini of the fragments are sequenced by automated sequence analysis using a special technique known as Edman degradation chemistry. Unfortunately, there is no automated technique to sequence the carboxyl termini of the proteins. Carboxypeptidase Y enzyme is used to release amino acids sequentially from the carboxyl terminus of the peptide chain. Sequential release of the amino acids at different time points gives the sequence of amino acids from the carboxyl terminus to the amino terminus. The amino acids at the sites of cleavage are detected unaltered; the sequences of the fragments exactly match that of the parent protein, with no gaps. For standard sequencing procedures, the cleavage has the same result as cleavage by a proteolytic enzyme.

Protein structure dependence. The observed cleavage at both alanine-proline (Ala-Pro) and serine-serine (Ser-Ser) shows that this process does not depend on the chemical reactivity of the amino

acid residue that is to be cleaved (Fig. 3). Since the peptide bond between the other Ser-Ser sequence in the protein (Ser-270 and Ser-271) is not affected, cleavage appears to depend on the proximity of the reagent to the polypeptide chain. However, because no crystal structure is available for bovine serum albumin, no deductions concerning the steric requirement or selectivity of the cleavage reaction can be made. This question has been addressed by attaching the iron chelate to cysteine-212 (Cys-212) of human carbonic anhydrase I (HCAI), whose structure has been determined at high resolution (**Fig. 4**). In the presence of ascorbate and hydrogen peroxide, the site-specifically-attached iron chelate quickly hydrolyzes the human

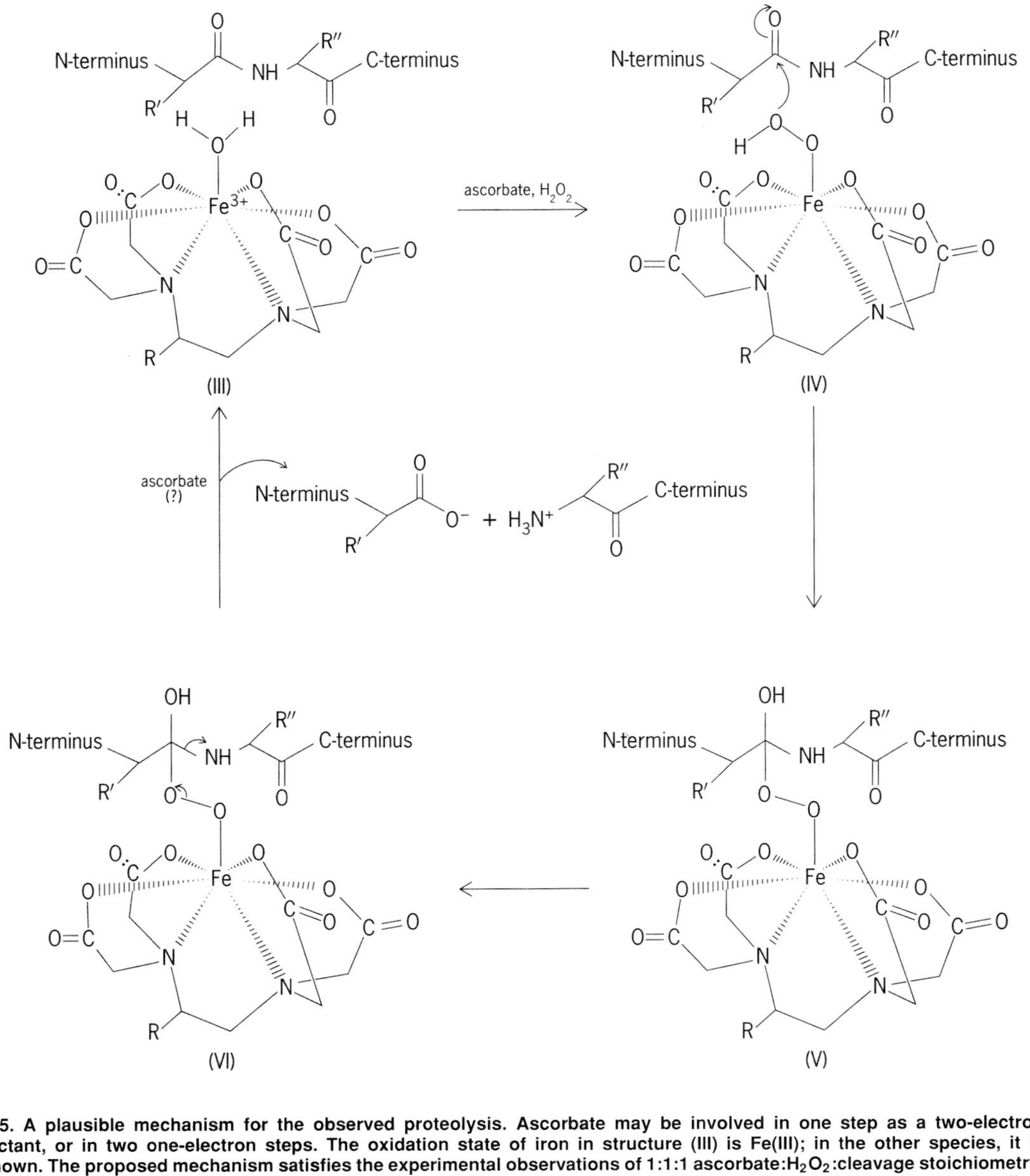

Fig. 5. A plausible mechanism for the observed proteolysis. Ascorbate may be involved in one step as a two-electron reductant, or in two one-electron steps. The oxidation state of iron in structure (III) is Fe(III); in the other species, it is unknown. The proposed mechanism satisfies the experimental observations of 1:1:1 ascorbate:H_2O_2:cleavage stoichiometry, oxygen atom transfer to the new carboxyl group, extremely narrow geometric constraints, insensitivity to hydroxyl radical scavengers, and apparent regeneration of the starting metal species. The R terms represent functional groups. (*After T. M. Rana and C. F. Meares, Transfer of oxygen from an artificial protease to peptide carbon during proteolysis, Proc. Nat. Acad. Sci. USA, 88(23):10578–10582,1991*)

carbonic anhydrase I at a single site to produce two discrete fragments that account for the entire protein. To demonstrate the specificity of the observed cleavage and its dependence on protein tertiary structure, control experiments are performed. In contrast to cleavage by the Cys-212 conjugated chelate, free Fe^{2+} in the reaction medium does not afford any cleavage, nor does chelate-conjugated unfolded protein.

A comparison of this result to expectations from the crystal structure of human carbonic anhydrase I can be made. As shown in Fig. 4, Cys-212 lies between two polypeptide loops, near the glycine residue (Gly-145) and the leucine residue (Leu-189). Since only one peptide bond [between Leu-189 and the aspartic acid residue (Asp-190)] is hydrolyzed, the observed cleavage process depends exquisitely on the three-dimensional structure of the protein-chelate conjugate.

Cleavage efficiency and mechanisms. The cleavage of human carbonic anhydrase I is very efficient; with a concentration of H_2O_2 equal to the protein concentration, all of the molecules of human carbonic anhydrase I that bear chelates are cleaved in 10 s. The stoichiometry of the protein cleavage reaction is determined by either titrating with H_2O_2 in the presence of excess ascorbate or titrating with ascorbate in the presence of excess H_2O_2. These results indicate that one molecule of H_2O_2 and one molecule of ascorbate afford the hydrolysis of one peptide bond (1:1:1 stoichiometry) and that both H_2O_2 and ascorbate are required for cleavage.

It is commonly thought that protein damage is caused by hydroxyl radicals that are generated during oxidation of iron (Fenton-type reaction). This perception can be explored by carrying out the cleavage reactions in the presence of high concentrations of radical scavengers such as thiourea, mannitol, and tertiary butyl alcohol, which react with hydroxyl radicals at very high rates. These experiments show that the cleavage is independent of scavenger concentration in the reaction medium. Further, the fact that resultant cleavage fragments contain free amino and carboxyl termini suggests that the cleavage is not initiated by hydroxyl radicals. Thus, possibly the peptide bond is cleaved through the direct attack of an oxygen-activated complex of iron and ethylenediaminetetraacetic acid (EDTA).

To investigate further mechanistic details, the cleavage reactions of chelate-conjugated human carbonic anhydrase I are carried out in the presence of $[^{18}O]H_2O_2$. The larger fragment (containing residues 1–189, including the newly formed C-terminus) is digested with carboxypeptidase Y, and the amino acids initially released are analyzed by fast atom bombardment mass spectrometry (FAB-MS). This analysis reveals that one atom of ^{18}O is incorporated into the carboxyl group of the C-terminal Leu-189 with about 93% efficiency. The cleavage reaction is also carried out under 1 atmosphere of $^{18}O_2$ for 30 h; mass-spectrometric analyses show that ^{18}O is incorporated into the carboxyl group of Leu-189 with about 92% efficiency. The carboxyl group of Leu-189 does not incorporate detectable amounts of ^{18}O when the reaction is carried out in the presence of $[^{18}O]H_2O$.

All of the above observations are consistent with the proposal that in the presence of O_2 (or H_2O_2) and ascorbate, iron-EDTA may form an intermediate oxygen-activated complex that leads to nucleophilic attack by oxygen on the carbonyl carbon of the peptide bond. Such a proposed reaction mechanism is shown in **Fig. 5**. In the first step, structure (III) is converted to intermediate (IV) by the direct binding of H_2O_2 to the metal chelate. Intermediate (IV) may also be formed after O_2 has been reduced to H_2O_2 by ascorbate in the presence of the iron chelate. In the next step, coordinated peroxide acts as a nucleophile to attack the peptide bond, so that the key intermediate (V) is formed. Possibly the efficiency of this nucleophilic attack is such that the attack occurs before formation of the hydroxyl radical can take place, in which case the proximity and orientation of the reacting groups would be crucial to the observed selectivity of the reaction. The final step is heterolytic cleavage of the carbon-nitrogen (C—N) bond and of the peroxide to yield a new carboxyl terminus, or a more complex series of steps leading to incorporation of an ^{18}O atom into Leu-189 [structure (VI)]. Subsequently the metal complex is presumed to be converted back to its original form (III), since clear evidence for multiple cycles is observed in the case of bovine serum albumin.

Applications. Further developments of this technology could allow the mapping of ligand binding sites on proteins, analogous to currently available methods for mapping binding sites on nucleic acids.

For background information SEE *AMINO ACIDS; BIOINORGANIC CHEMISTRY; CHELATION; CHEMOTHERAPY; ONCOLOGY; OXYGEN TOXICITY; PROTEIN; REACTIVE INTERMEDIATES* in the McGraw-Hill Encyclopedia of Science & Technology.

Tariq M. Rana

Bibliography. S. F. Bellon, J. H. Coleman, and S. J. Lippard, DNA unwinding produced by site-specific intrastrand cross-links of the antitumor drug *cis*-diamminedichloroplatinum(II), *Biochemistry*, 30: 8026–8035, 1991; S. L. Bruhn et al., Isolation and characterization of human cDNA clones encoding an HMG-Box protein that recognizes structural distortions in DNA caused by binding of the anticancer agent cisplatin, *Proc. Nat. Acad. Sci. USA*, 89:2307–2311, 1992; S. L. Bruhn, J. H. Toney, and S. J. Lippard, Biological processing of DNA modified by platinum compounds, *Prog. Inorg. Chem.*, 38:477–516, 1990; B. A. Donahue et al., Characterization of a DNA damage-recognition protein from mammalian cells that binds specifically to intrastrand d(GpG) and d(ApG) DNA adducts of the anticancer drug cisplatin, *Biochemistry*, 29:5872–5880, 1990; C. A. Lepre and S. J. Lippard, Interaction of platinum anticancer compounds, *Nucl. Acids Mol. Biol.*, 4:9–38, 1990; T. M. Rana and C. F. Meares, Iron chelate mediated proteolysis: Protein structure dependence, *J. Amer. Chem. Soc.*, 113:1859–1861, 1991; T. M. Rana and C. F. Meares, Specific cleavage of a protein by an attached iron chelate, *J. Amer. Chem. Soc.*, 112:2457–2458, 1990; T. M. Rana and C. F. Meares, Transfer of oxygen from an artificial protease to peptide carbon during proteolysis, *Proc. Nat. Acad. Sci. USA*, 88(23):10578–10582, 1991.

Bioremediation

Bioremediation of soils contaminated with organic chemicals is a type of treatment that can often clean up hazardous waste sites. Biological processes have been identified that can serve as mechanisms for attenuating contaminants during transit through the vadose zone to the groundwater.

In bioremediation, specific organic constituents, that is, parent compounds, are degraded. The term degradation may refer to complete mineralization of the constituents to carbon dioxide, water, and inorganic compounds, and incorporation of constituents into microbial cell mass. The ultimate products of aerobic metabolism are carbon dioxide and water, while under anaerobic conditions the products of metabolism are simple organic substances that are incompletely oxidized, such as organic acids or other substances such as methane or hydrogen.

Biodegradation. Biological degradation of a compound is frequently a stepwise process involving many enzymes and many species of organisms. Therefore, in the environment, rather than being completely degraded a constitutent may be transformed to intermediate products; these may be less, equally, or more hazardous than the parent compound, as well as more or less mobile in the soil environment. The goal of on-site bioremediation is degradation that results in detoxification of the parent compounds into products that are no longer hazardous to human health and the environment.

Microbial ecologists have identified ranges of critical environmental conditions that affect the activity of soil microorganisms (see **table**). Many of the conditions can be managed and can be changed to enhance biodegradation of organic constituents.

Systems. On-site bioremediation of contaminated soils is generally accomplished by using one of the three types of systems: in-place, prepared-bed, or in-tank bioreactor.

In-place. In this type of system, contaminated soil is not moved from the site; it is treated in place, generally by naturally occurring microorganisms. The process may be enhanced by a variety of physical and chemical methods, such as fertilization, tilling, adjustment of soil pH, and moisture control. In some instances, addition of supplemental populations of adapted organisms may serve to enhance treatment.

Prepared-bed. In a prepared-bed system, the contaminated soil is handled in either of two ways. It may be physically moved from its original site to a newly prepared area, which has been designed to enhance bioremediation and to prevent transport of contaminants from the site. In the other method, the contaminated soil is removed from the site to a storage area while the original area is prepared for use; it is then returned to the bed, where treatment is accomplished. Preparation of the bed may consist of installing a clay or plastic liner to retard transport of contaminants from the site, or the addition of uncontaminated soils to provide treatment medium. Treatment may also be enhanced with physical and chemical methods, as with in-place systems.

In-tank. For in-tank treatment systems, the contaminated soil is physically removed and placed in an enclosed reactor that uses the batch, complete-mix, or plug-flow systems, which are commonly used in chemical and environmental engineering systems. The soil may be in an unsaturated (soil pores not completely filled with water) or saturated (soil pores completely filled with water) physical form. Tank treatment systems usually include composting, slurry-phase treatment, and solid-phase treatment.

Treatability studies. To determine whether a specific site is suitable for bioremediation, information derived from research known as treatability studies is combined with information concerning the characteristics of the site and the waste in order to determine potential applications and limitations of the technology at the site. Ultimate limitations to the use of bioremediation at a site are usually related to the time required for cleanup, the level of cleanup attainable, and the cost of cleanup using bioremediation. Treatability studies for soil contaminated with organic materials can be conducted in the laboratory, in pilot-scale facilities, or in the field. They provide data concerning the fate and behavior of the organic contaminants at specific sites, permitting an estimate of the rate and extent of the bioremediation that would be possible for the surficial soil and the deeper soils of the vadose zone.

Information from treatability studies is also used to design and implement the engineering of a bioremediation system at a specific site. Engineering design elements include (1) determination of what type of containment is required to prevent contamination of

Environmental factors that are critical for microbial activity

Environmental factor	Optimum levels
Available soil water	25–85% of water-holding capacity; −0.01 MPa (−145 lb/in.^{2})*
Oxygen (O_2)	Aerobic metabolism: greater than 0.2 mg per liter of dissolved oxygen, minimum air-filled pore space of 10% Anaerobic metabolism: O_2 concentrations less than 1%
Redox potential	Aerobes and facultative anaerobes: greater than 50 mV Anaerobes: less than 50 mV
pH	5.5–8.5
Nutrients	Sufficient carbon (C), nitrogen (N), phosphorus (P), and other nutrients so that microbial growth is not limited (suggested C:N:P ratio of 120:10:1)
Temperature	59–113°F (15–45°C; mesophiles)

*Water is held in unsaturated soils under negative potential (suction).

off-site receiver systems; (2) development of techniques to maximize mass transfer of chemicals affecting microorganism activity (such as the addition of mineral nutrients, oxygen, energy sources, products for controlling pH, and removal of toxic chemicals or products of degradation) in order to enhance bioremediation; and (3) design of a monitoring program to evaluate effectiveness of treatment.

To assess the potential for biological degradation at a specific contaminated site, treatability studies that incorporate concepts of materials balance and mineralization to determine the environmental fate and behavior of the constituents in the specific soil are often used. Materials balance involves measurement of extractable material that can be recovered, either the parent compound or the transformation products. Mineralization involves measurement of either the production of gas (carbon dioxide or methane) from the parent compound or the release of substituent groups, for example, chloride or bromide ions.

The rate of degradation is calculated by measuring the loss of the parent compound and the production of carbon dioxide during a given time of treatment. Degradation rate is often reported as half-life, which represents the time required for 50% of the compound to disappear, based upon a first-order kinetic model.

Calculation of the rate of decrease of the parent compound by itself does not provide complete information concerning mechanisms and pathways by which organic constituents are interacting with the soil environment. Further information is necessary in order to understand whether a constituent is simply transferred from one phase to another, for example, from solid to air, or is chemically altered so that the properties of the parent compound are destroyed. Therefore, evaluation of the fate of a constituent in a soil also requires identification and measurement of the distribution of the constituent among the physical phases that make up the system, as well as differentiation of the mechanisms by which the constituent may be chemically altered in a soil system.

Bioassays may be used in treatability studies to demonstrate detoxification of parent compounds. They measure the effect of a chemical on a test species under specified test conditions. The toxicity of a chemical is proportional to the severity of the monitored response of the test organism to that chemical. Often a battery of bioassays is utilized, including measurements of the effects of the substances on general microbial activity as well as assays relating to activity of subgroups of the microbial community. Bioassays to evaluate effects on human health may also be included in the battery of tests.

The results of treatability studies provide information relating to rates and extent of treatment of hazardous organic constituents when mass-transfer rates of the potentially limiting substances are not limiting the treatment. Treatability studies usually represent optimum conditions with respect to mixing, contact of the solid materials of the soil with waste constituents and with microorganisms, and homogeneous conditions throughout the test area. Thus they permit assessment of the levels of treatment that may be achievable at a specific site.

Applications and limitations. Many organic compounds from a wide range of chemical classes have been shown to be amenable to biodegradation in laboratory studies (both in aqueous cultures and in soil microcosms), using both single strains of microbial species or consortia of microbial populations. Examples of specific chemical classes shown to be biodegradable include amines and alcohols, polycyclic aromatic hydrocarbons (PAHs), chlorinated aromatic hydrocarbons, chlorinated and nonchlorinated phenols, halogenated aliphatic compounds, and pesticides. Removal efficiencies generally are greater for nonhalogenated compounds than for halogenated compounds.

Even though a specific organic constituent has been shown to undergo biological degradation under laboratory conditions, actual degradation in a specific soil–site system is dependent on many factors. Potential degradability requires investigation in site-specific treatability studies. Availability of sufficient oxygen may be a limiting factor in some cases, while in others, anaerobic conditions are necessary. Other environmental conditions that may place restrictions on biological activity include pH, temperature, and moisture. Upon exposure to the soil environment, the constituent may be biologically or chemically altered so as to be rendered persistent and toxic in the environment. The system may lack nutrients required for microbial activity. Other chemicals present may serve as preferred substrates, or they may act to repress required enzyme activities. High concentrations of metal salts may be inhibitory or toxic to many microorganisms.

Most chemicals require the presence of consortia of microbial species for mineralization, some of which may not be present at the specific site. Also, most organisms require a period of acclimation to the constituent before metabolism occurs. During this period, the level of constituent must be high enough to promote acclimation without being toxic or inhibitory. Prior exposure to the constituent or similar constituents may help to shorten the acclimation period.

Bioremediation of sites contaminated with organic chemicals is a promising technology, especially if it is incorporated in a remediation plan that uses an integrated approach to the cleanup of a complete site, that is, a plan that involves the concept of a so-called treatment train of physical, chemical, and biological processes to address remediation of all sources of contaminants at the site.

For background information SEE BIOASSAY; HAZARDOUS WASTE; SOIL CHEMISTRY in the McGraw-Hill Encyclopedia of Science & Technology.

Judith L. Sims

Bibliography. R. C. Loehr and J. F. Malina (eds.), *Land Treatment: A Hazardous Waste Management Alternative*, Water Resources Symposium 13, Center for Research in Water Resources, University of Texas, Austin, 1986; K. S. Park et al., Fate of PAH compounds in two soil types: Influence of volatilization, abiotic loss, and biological activity, *Environ. Toxicol. Chem.*, 9:187–195, 1990; J. L. Sims, R. C. Sims, and

J. E. Matthews, Approach to bioremediation of contaminated soil, *Hazard. Waste Hazard. Mater.*, 7:117–149, 1990; R. C. Sims, Soil remediation techniques at uncontrolled hazardous waste sites: A critical review, *J. Air Waste Manag. Ass.*, 40:703–732, 1990.

Birth control

The development of sustained-release contraception began in 1966 at the Population Council's Center for Biomedical Research. By 1969 the Center had developed a long-acting, sustained-release system and had reported the initial clinical experience with it. Subsequently, various synthetic progestins and lengths and numbers of implants were investigated, and in 1974 the use of six capsules containing the synthetic progestin levonorgestrel was identified as best for sustained-release contraception. The capsules, when placed under the skin, provide very effective contraception for up to 5 years but can be removed at any time. The slow, steady release of the hormone from the capsules over the 5-year period results in much lower concentrations of the hormone in the blood than would occur with birth control pills. By 1976 clinical trials of this system were under way in six countries, including the United States.

Advantages. Clinical trials have shown levonorgestrel to be one of the most effective reversible contraceptives ever developed. During the effective 5 years, fewer than 1 woman out of every 100 users will become pregnant. The capsules can be removed at any time should a woman want to become pregnant or want to terminate use for other reasons.

An important advantage over oral contraceptives is the absence of estrogen. The low, constant-dose, progestin-only method is safe for some women who should not use contraceptive pills, for example, those with elevated blood pressure or a history of thromboembolic disease. However, some women should not use levonorgestrel, for example, women who have acute thrombophlebitis or acute thromboembolic disease, benign or malignant liver tumors, or known or suspected breast cancer, or women who are pregnant. There are no significant effects on the growth or health of infants whose mothers used levonorgestrel while breastfeeding.

Mode of action. The contraceptive effects of levonorgestrel are threefold. First, it acts on the hypothalamus and pituitary to suppress the luteinizing-hormone (LH) surge responsible for ovulation. Both LH and follicle-stimulating hormone (FSH) production are decreased, but the sharp LH peak is eliminated completely, while the FSH level is reduced to about half the normal levels. Estradiol levels fluctuate and, in the absence of progesterone elevations, indicate follicular activity without ovulation. The continued secretion of FSH by the pituitary means that production of estradiol by the ovary stays within normal limits. This is important because low levels of estradiol can lead to decreased bone density and other abnormalities. Second, the levonorgestrel has a marked effect on

cervical mucus: the mucus thickens and decreases in amount, so that sperm penetration is barred. Third, the low steady concentration of levonorgestrel in the blood also affects the lining of the uterus so that a fertilized egg would have difficulty growing there. This effect is probably not very important in the contraceptive action because the eggs of levonorgestrel users are rarely fertilized. The antiovulatory and cervical mucous actions of the levonorgestrel are, therefore, the primary mechanisms of contraceptive action.

Metabolic effects. Exposure to the sustained, low dose of levonorgestrel results in few metabolic changes. Only minor changes occur for carbohydrate metabolism, liver function, blood coagulation, immunoglobulin levels, serum cortisol levels, and blood chemistry. Cholesterol/high-density lipoprotein ratios either improve or are unaltered during levonorgestrel use. Thus, long-term exposure to low concentrations of levonorgestrel is unlikely to influence the deposition of cholesterol in the arteries of users, an earlier, but unsubstantiated, concern about birth control pills.

Side effects and terminations. Most users of this method of contraception will experience one or more side effects during 5 years of use, but serious side effects are very rare. The most common side effect of levonorgestrel is menstrual irregularity, including prolonged menstrual bleeding, spotting between periods, amenorrhea, or a combination of these patterns. Most women can expect an altered bleeding pattern to become more regular after 6–12 months of use. Despite the increased frequency of bleeding in some women, monthly blood loss is usually less than normal menses; hemoglobin levels thus tend to rise. Because ovulation is suppressed in only about two-thirds of the cycles and is related to serum levonorgestrel levels, functional ovarian cysts sometimes occur, but they usually disappear spontaneously so that surgery is rarely necessary.

Other side effects include headaches, acne, weight change, breast pain, hyperpigmentation over the implants, hirsutism, depression, mood changes, anxiety, nervousness, unexpected milk flow, and ectopic pregnancy. Many of these side effects are also encountered among users of oral contraceptives. Although the common side effects occurring with levonorgestrel are not threats to health, they sometimes cause users to discontinue the method.

Effects on future fertility. Within 48 hours of removal of the capsules, plasma levonorgestrel concentrations drop below contraceptive levels, and the return of fertility is prompt. Most women resume normal ovulatory cycles during the first month after removal. Rates for women attempting pregnancy after removal are higher than for intrauterine devices or birth control pills.

Biodegradable implants. The cost of implant contraception could be reduced by simpler systems with fewer implants or biodegradable implants that would eliminate the need for surgical removal. In addition, an implant contraceptive that prevents conception for less than 5 years could be an attractive and less expensive alternative for women who may want pregnancy within a year or two. A levonorgestrel system using two capsules may soon be available.

Biodegradable implants deliver sustained levels of progestin for variable periods of time from a vehicle that dissolves in body tissues. Preliminary results indicate that biodegradable implants are safe and effective. Two types of biodegradable implants are under evaluation: a polymer capsule containing levonorgestrel, and pellets consisting of the progestin norethindrone mixed with a cholesterol base.

For background information *SEE BIRTH CONTROL* in the McGraw-Hill Encyclopedia of Science & Technology.

Philip D. Darney

Bibliography. V. Brache et al., Anovulation, inadequate luteal phase, and poor sperm penetration in cervical mucus during prolonged use of Norplant[R] implants, *Contraception*, 31(3):261–278, 1985; P. D. Darney et al., Sustained release contraceptives, *Curr. Prob. Obstet. Gynecol. Fertil.*, 13:87–125, 1990; S. Roy et al., Long-term reversible contraception with levonorgestrel-releasing Silastic rods, *Amer. J. Obstet. Gynecol.*, 148(7):1006–1013, 1984; S. Segal, A new delivery system for contraceptive steroids, *Amer. J. Obstet. Gynecol.*, 157(4):1090–1092, 1987; I. Sivin, International experience with Norplant[R] and Norplant[R]-2 contraceptives, *Stud. Fam. Plan.*, 19(2):81–94, 1988.

Blood

Platelet-activating factor (PAF) is the common name for a phospholipid (1-*O*-alkyl-2-acetyl-*sn*-glycero-3-phosphocholine) with the structure below. PAF con-

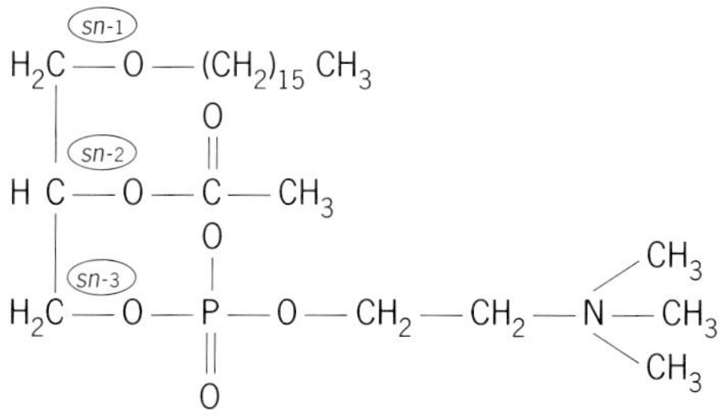

sists of a glycerol backbone with substituents at each of the three carbons. One of these is a phosphocholine group. An acetate group is attached stereospecifically to the middle carbon via an ester linkage. The other carbon contains an ether-linked fatty alcohol that is usually 16 carbons long.

The name platelet-activating factor is misleading because it implies a single role for the molecule. However, since its discovery 20 years ago, the molecule has been implicated in many biological processes, including activation of platelets and white blood cells, mediation of extreme allergic reactions such as anaphylactic shock, and participation in shock due to bacterial infections and infarction of the bowel. It also is important in reproduction, because it participates in

the implantation of the embryo and in stimulation of the onset of labor.

PAF serves as a messenger between cells. Whether it serves a beneficial or detrimental physiological function probably depends largely on whether the biochemical regulation of its synthesis and degradation operates normally. That is, PAF could be helpful when appropriately regulated, but damaging when too much is made or when it is made at the wrong time or in the wrong place.

Synthesis. PAF is synthesized by many types of cells, including platelets, leukocytes, endothelial cells, and neuronal cells, and has been found in almost all tissues of mammals. Two routes of synthesis have been described. One pathway is thought to produce a small amount of PAF continuously in the kidney and perhaps other organs. A second pathway is used by leukocytes, endothelial cells, and platelets. These cells do not make PAF all the time, nor do they store it. In response to the appropriate signal, the enzymes of this second pathway are activated and synthesis begins. An enzyme, phospholipase A_2, catalyzes the hydrolysis of phospholipids in the cell membranes, preferring those that contain an arachidonic acid. Thus, this enzyme produces both an intermediate in PAF synthesis and free arachidonic acid. The arachidonic acid can then undergo the addition of oxygen to yield one or more members of another large family of lipid messengers, the prostaglandins and leukotrienes. Many of the stimuli for PAF production are the same as those for the production of prostaglandins and leukotrienes.

Activation and inactivation. The actions of PAF are mediated by receptors on the outer membrane of susceptible cells. Drugs that block the action of PAF have been shown to act by preventing its binding to the receptor. These and related experiments have suggested that there are two or more types of receptors, which are very specific for PAF or closely related molecules. For example, the receptor does not recognize the optical isomer of PAF, and a compound with an ester linkage at the *sn*-1 position (see **illus.**) is 100–300 times less potent than a compound with an ether linkage. Potency markedly diminishes as the carbon chain length of the *sn*-2 position is lengthened. Once PAF binds to its receptor, a subsequent series of reactions causes an increase in the amount of calcium in the cell and the activation of several enzymes, including protein kinases, which regulate other proteins by adding phosphate to them. These processes then cause cellular responses such as adhesion, secretion of granules, and synthesis of other messengers that result in the physiological and pathological action of PAF.

The effects of PAF are turned off when it is degraded. This turnoff is accomplished by a group of enzymes named PAF acetylhydrolases, which cleave the acetyl group at the *sn*-2 position. Several forms of this enzyme are found inside cells. One form is found in the plasma of blood, and is associated with lipoproteins. The PAF acetylhydrolases have different specificities than the PAF receptor: they share the requirement for a short chain at the *sn*-2 position, but have little specificity for substituents at the other positions. Abnormal levels of PAF acetylhydrolase activity in plasma have been found in several human diseases, but whether this is an important factor in causing any of the diseases has not been determined. The activity of the enzyme in plasma falls markedly at the end of pregnancy. This drop in activity is thought to be a signal to begin labor since PAF will consequently accumulate and cause contraction of uterine muscle.

Effects. PAF was originally described as a soluble factor that caused platelet aggregation in the blood of rabbits undergoing anaphylaxis. It is now clear that PAF has many more functions and acts on many cell types. Acting through its high-affinity receptor, PAF is active at nanomolar concentrations. Additionally, much of the PAF synthesized by some cells is not secreted; instead it is either expressed on the surface or kept intracellularly, so that it serves as a signal for cell-to-cell interactions or as an intracellular mediator.

Platelets. Platelets from most of the species tested are activated by PAF. Human platelets require the presence of fibrinogen to be activated by PAF. Platelets exposed to thrombin or collagen (which are important signals for arrest of bleeding) are activated and secrete PAF. However, the responses of platelets to thrombin or collagen do not seem dependent on the PAF that is made, since they are not inhibited by drugs that block the PAF receptor. Thus normal platelet function apparently does not require PAF. However, there is evidence that PAF may be one of the components leading to excessive production of blood clots (thrombosis), which is the final step in heart attack, strokes, and other blockages of blood vessels.

White blood cells. Neutrophils, the white blood cells that cause acute inflammation, both synthesize PAF and are activated by it. Most of the PAF synthesized by neutrophils remains associated with the cell, and its role is not known. The PAF secreted by neutrophils is thought to activate nearby cells. This effect would result in an amplification of the inflammatory response, since PAF released by one neutrophil would activate neighboring platelets, neutrophils, and other cells, which would in turn synthesize more PAF and other messengers. The actions of PAF in such a situation are likely to occur over only a short distance, because the PAF acetylhydrolase in plasma would degrade PAF before it could travel through the circulatory system.

PAF regulates neutrophil adhesion to endothelial cells. These cells, which line all blood vessels, make PAF in response to oxidants, bacterial toxins, thrombin, bradykinin, and histamine. The majority of the PAF is expressed on the surface of the endothelial cells. This expression occurs within minutes. Under the same conditions, a protein known as P-selectin is transferred from a granule inside the cell to the surface. This protein and PAF act together to cause neutrophils to bind to endothelial cells and then to leave the blood vessel for the tissues. This is an important step in normal inflammatory responses, such as fighting a bacterial infection. PAF-mediated binding of neutrophils and their subsequent emigration is likely the initial event in many inflammatory processes.

PAF is a potent activator of eosinophils, a type of white blood cell involved in allergic reactions. It attracts eosinophils and induces them to secrete superoxide anion, which is important in killing pathogenic microorganisms, and leukotriene C_4, which is another mediator of anaphylaxis. Interestingly, injection of PAF into the skin causes an accumulation of eosinophils in the skin of allergic patients, and inhalation of PAF causes the accumulation of eosinophils in the lungs. When the experiment is repeated in normal patients, neutrophils are the predominant leukocytes that accumulate. Eosinophils also make PAF, and the eosinophils from patients with allergic diseases, such as asthma, have an increased capacity to synthesize and secrete PAF.

The effects of PAF on eosinophils, and the observation that it causes contraction of the muscles in the airways, suggest that it plays a major role in asthma. In fact, several of the antagonists of the PAF receptor were found during screening of Chinese herbal medicines that have been used to treat asthma. In both animals and humans, PAF inhalation causes a short-lived bronchoconstriction directly, and also a prolonged period of hyperactivity to other agents. This response is analogous to the nonspecific airway hyperreactivity seen in patients with asthma.

Another white blood cell that responds to PAF is the monocyte and its tissue form, the macrophage, which are involved in chronic inflammation and in atherosclerosis. PAF attracts monocytes and induces them to secrete prostaglandins and other messengers derived from arachidonic acid. In response to endotoxins released by bacteria, monocytes and macrophages secrete tumor necrosis factor and interleukin 1, two proteins that are important for chronic inflammation. Preexposure of the cells to PAF augments the production of these proteins. One of the major physiological functions of monocytes is the ingestion of particles, including bacteria and dead cells. This process is a strong stimulus for them to synthesize PAF, and they subsequently secrete most of it. As monocytes differentiate into macrophages, their secretion of PAF diminishes, because they begin to secrete the enzyme that degrades PAF. This response suggests a complex role for these cells in the initiation and regulation of inflammation.

Animal models. In animal models, the intravenous administration of endotoxin or PAF leads to rapid fall in blood pressure and death. Both effects can be attenuated by antagonists of the PAF receptor. PAF decreases the force of contraction of heart muscle, and this effect is blocked by PAF receptor antagonists. In some species, this action of PAF is also blocked by antagonists of the receptors for thromboxane and leukotrienes, products of arachidonic acid; this effect suggests that the actions of PAF on the heart muscle go through these intermediates. PAF causes constriction of coronary arteries, which supply blood to the heart, and the decreased blood flow further impairs heart function. A similar effect has been described in other vascular beds, and part of the effect is blocked by thromboxane antagonists.

For background information SEE BLOOD; PHOSPHA-TIDE in the McGraw-Hill Encyclopedia of Science & Technology.

Mark Kozak; Guy A. Zimmerman;
Thomas M. McIntyre; Stephen M. Prescott

Bibliography. P. Braquet et al., Perspectives in platelet-activating factor research, *Pharmacol. Rev.*, 39(2):97–145, 1987; S. M. Prescott, G. A. Zimmerman, and T. M. McIntyre, Platelet-activating factor, *J. Biol. Chem.*, 265(29):17381–17384, 1990; K. Saito and D. Hanahan, *Platelet-Activating Factor and Diseases*, 1989; F. Snyder, *Platelet-Activating Factor and Related Lipid Mediators*, 1987.

Book preservation

Research libraries and archives throughout the world have vast numbers of old books and documents that have become too fragile for circulation to readers. It is estimated that there are more than 75 million books, published since the midnineteenth century, on now-brittle chemical wood pulp paper in American research libraries.

Embrittlement. About 50% of the volumes in the New York Public Library and 25% in the Library of Congress are considered brittle (**Table 1**). Among the causes of embrittlement are deterioration of acid paper used in manufacture, air pollution, poor conditions of storage, and heavy use. Libraries have sought methods for treating large numbers of books, with emphasis on the neutralization of acids in the paper. If they prove successful and cost-effective, such treatments will significantly reduce the rate of embrittlement for volumes that are acid but not yet unusable. Considerable progress has been made in this area, with at least 15 processes in use or in development; though a substantial literature has developed on individual processes, few independent comparative studies have been undertaken. The more urgent task of strengthening a large number of already-brittle books by impregnation with polymers or by other chemical intervention is further from cost-effective solutions.

Some reduction in the rate of production of acid-paper books has been observed as economic forces and technical developments have begun to make alkaline-paper manufacture more attractive. In alkaline papers, alkyl-ketene-dimer or alkenyl succinic anhydride with calcium carbonate as a filler is used as sizing to prevent ink penetration, make the paper smooth, and render it opaque. It is estimated that less than 25% of the book paper produced in the United States may be classified as alkaline. The calcium carbonate introduced as a sizing material provides the additional benefit of a substantial alkaline reserve. An expansion of the use of alkaline papers, often referred to as permanent papers, will reduce the rate of introduction of potentially brittle books to collections.

Mass deacidification. Mass treatments are generally defined as those that permit the simultaneous treatment of substantial numbers of books (minimum of around 50) without removal of the bindings. A mass deacidification process is one that neutralizes acid in the paper and provides a residual alkaline buffer, that is, an alkaline reserve. The general goal for a full-scale plant is the treatment of over 1 million books per year at a cost of around $5.00 per volume. The processes that are under active development or in use are diethyl zinc, Wei T'o, Archival Aids, Bookkeeper, Book Preservation Associates, FMC, and vapor-phase deacidification (interleaf).

Diethyl zinc process. Research on mass deacidification that began at the Library of Congress in 1973 led to the development of a vapor-phase process using the organometallic compound diethyl zinc. The process sequence is given in the **illustration**. The process is currently being carried out in a $1/25$-scale pilot plant.

In the reactions with acids in the paper and other book materials, zinc salts are formed [reactions (1) and (2)]. Reaction with residual moisture (H_2O) in the

$$H_2SO_4 + (C_2H_5)_2Zn \rightarrow ZnSO_4 + 2C_2H_6 \quad (1)$$
Sulfuric Diethyl   Zinc   Ethane
acid   zinc   sulfate

$$2CH_3COOH + (C_2H_5)_2Zn \rightarrow Zn(CH_3COO)_2 + 2C_2H_6 \quad (2)$$
Acetic acid Diethyl   Zinc acetate   Ethane
            zinc

book deposits zinc oxide [ZnO; reaction (3)] to yield an alkaline reserve that may, subsequent to treatment, neutralize intrusive or unreacted acids [reaction (4)].

$$H_2O + (C_2H_5)_2Zn \rightarrow ZnO + 2C_2H_6 \quad (3)$$

$$ZnO + H_2SO_4 \rightarrow ZnSO_4 + H_2O \quad (4)$$

Essential to the process is vacuum operation, safe handling, and control of paper moisture content prior

Table 1. Brittle books in major research libraries*

Institution	Number of volumes, millions	Number of embrittled volumes, millions	Percent of embrittled volumes
Harvard University	12	4.8	40
Library of Congress	20	5	25
New York Public Library	11	5.5	50
Stanford University	5.5	1.4–2.2	24–40
Yale University	8	3	37

*After B.E. Frye, Toward a collaborative national preservation program, in F. Pflieger (ed.), *Preservation of Library and Archival Materials*, Association of Higher Education Facilities Officers, 1991.

to treatment with diethyl zinc, a hazardous material that will ignite spontaneously upon exposure to air. Diethyl zinc also reacts vigorously with water. In the early development of the process at the NASA-Goddard Space Flight Center, two fires occurred, the second accompanied by a small explosion. One conclusion of the Accident Investigating Board was that the pilot plant was designed with inadequate instrumentation to monitor the inventory of diethyl zinc. Thus, some in the condenser reacted with brine that backstreamed from a seal tank. The second pilot plant, in Houston, Texas, was designed with careful attention to management of engineering and chemical safety. However, development of a full-scale plant will require substantial work involving selection of site and plant management.

Wei T'o process. This mass deacidification process is in regular use. The Wei T'o liquefied gas process uses methoxy methyl magnesium carbonate ($CH_3OMgOCOOCH_3$) in a nonaqueous solvent system comprising methanol, trichlorotrifluoroethane (Freon 12), and dichlorodifluoromethane (Freon 113). The solution of deacidifying compound in solvent is passed through a filter to remove particulate impurities, for example, paper fragments from books previously treated, and then introduced to the process tank. Vapors from the processing tank are recovered and returned to the supplier for reprocessing. Books selected for treatment are dried to reduce moisture content to 0.5%, the total drying procedure taking about 24 h. Moisture control is important since the treatment solution will form a gel on reaction with excess water.

The methoxy methyl magnesium carbonate reacts with water to form magnesium compounds [reactions (5)–(8)]. The magnesium hydroxide and magnesium

$$CH_3OMgOCOOCH_3 + 2H_2O \rightarrow$$

$$\underset{\substack{\text{Magnesium} \\ \text{hydroxide}}}{Mg(OH)_2} + 2CH_3OH + CO_2 \quad (5)$$

$$H_2O + CO_2 \rightarrow \underset{\substack{\text{Carbonic} \\ \text{acid}}}{H_2CO_3} \quad (6)$$

$$Mg(OH)_2 + H_2CO_3 \rightarrow \underset{\substack{\text{Magnesium} \\ \text{carbonate}}}{MgCO_3} + 2H_2O \quad (7)$$

$$MgCO_3 \rightarrow \underset{\substack{\text{Magnesium} \\ \text{oxide}}}{MgO} + CO_2 \quad (8)$$

carbonate react with acids in the paper to form neutral salts [for example, reaction (9)]. Excess magnesium

$$Mg(OH)_2 + H_2SO_4 \rightarrow \underset{\substack{\text{Magnesium} \\ \text{sulfate}}}{MgSO_4} + 2H_2O \quad (9)$$

oxide and magnesium carbonate form the alkaline reserve. Though the process has been used since 1979 by the Canadian National Library and Archives, the use of chlorofluorocarbons in the process and the need for preselection of books to avoid undesirable solvent-system effects on the inks and cover materials have led

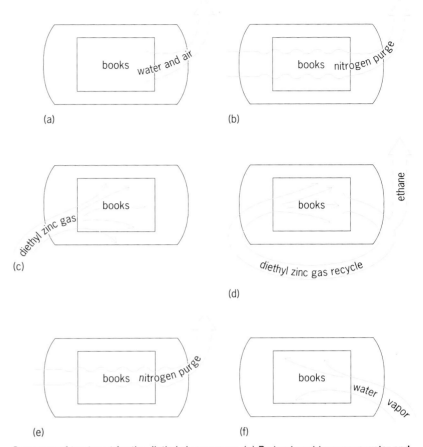

Sequence of treatment for the diethyl zinc process. (*a*) Pretreatment to remove water and air. (*b*) Purging with nitrogen. (*c*) Treatment with gaseous diethyl zinc. (*d*) Removal of gaseous products. (*e*) Second purging with nitrogen. (*f*) Reintroduction of water vapor to attain normal equilibrium moisture content. (*After Office of Technology Assessment, U.S. Congress, Book Preservation Technologies, 1988*)

to the search for process modifications and alternative methods.

Archival Aids process. This process is similar to the Wei T'o method and follows the same basic chemistry. It was developed by the Centre de Recherches pour la Conservation des Documents Graphiques in Paris and is used in the Bibliothèque National de France. The process deposits carbonated magnesium methoxide ethoxide by using a chlorofluorocarbon/methanol solvent.

Bookkeeper process. In this method the books are treated with particles of basic metal oxides less than 1 micrometer in diameter suspended in a mixture of chlorofluorocarbons and a surfactant. It has also been proposed that the basic metal oxide particles could be introduced directly by electrostatic transfer as in electrostatic photocopying. Deacidification occurs over time, though there is question as to the degree to which penetration of the paper structure is achieved.

Book Preservation Associates process. In this method, deacidification is achieved from ethanolamines produced in the vapor-phase treatment with ethylene oxide and ammonia. Some concern has been expressed over residual ammonia and ethylene oxide in

Table 2. Comparison of mass deacidification processes*

Process	Composition	Process medium	Use of chlorofluoro-carbons	Preselection for books	Dehydration	Maximum temperature and time	pH of treated paper	Alkaline reserve
			Criteria and characteristics†					
Ideal process	—	—	None	None	None	Room temperature	7.0–8.5	~2%
Archival Aids	Carbonated magnesium methoxide ethoxide	Nonaqueous solution	Yes	Yes	0.5%, 32 h	60°C (140°F), 32 h	5.8–10.5	0.22–1.5%
Bookkeeper	Magnesium oxide particles	Nonaqueous solution	None	Yes	None	Room temperature, 3–5 h	8.0–9.0	2%
Book Preservation Associates	Ethylene oxide + ammonia	Gas	None	Yes	None	37.7°C (99.9°F), 18 h	7.0–9.0	0.85–2.14%
Diethyl zinc	Diethyl zinc	Gas	None	Yes	0.5%, 18–30 h	65°C (149°F), 6–9 h	7.0–8.5	1–2%
FMC	Magnesium butoxy triglycolate	Nonaqueous solution, impregnation	Yes	None	2%, 3 h	50°C (122°F), 2 periods of 3 h each	7.0–9.5	1.5–2.3%
Vapor-phase deacidification	Cyclohex-ylamine carbo-nate	Interleaf	None	Yes	None	Room temperature	5.0–8.7	None
Wei T'o	Carbonated magnesium methoxide	Nonaqueous solution	Yes	Yes	0.5%, 26–36 h	58°C (136°F), 26 h	7.5–9.5	0.7–1.2%

*D. N.-S. Hon, Critical evaluation of mass deacidification processes for book preservation, in S. H. Zeronian and H. L. Needles (eds.), *Historic Textile and Paper Materials II*, 1989; and A. Lienardy, A bibliographical survey of mass deacidification methods, *Restaurator*, 12(2):75–103, 1991.
†The data are generally based on publications supplied by the manufacturer. Few independent comparative evaluations have been conducted.

the treated books and also a tendency toward yellowing of papers containing lignin. Possibly the chlorofluoro-carbon carrier gas in the process will be replaced by carbon dioxide. The current version of the process requires preselection to remove books with leather bindings, which may gelatinize on reaction with ethylene oxide.

FMC process. This method has been used since 1990 in a demonstration plant capable of treating 300,000 books per year in Bessemer City, North Carolina. The process is claimed both to neutralize acid in paper and to strengthen damaged papers by the introduction of magnesium butoxy triglycolate, whose structure is shown below. It is believed that this compound forms hydrogen bonds between cellulose polymers cleaved at the hemiacetal linkage. The magnesium carbonate portion of the molecule neutralizes the acid and provides an alkaline reserve.

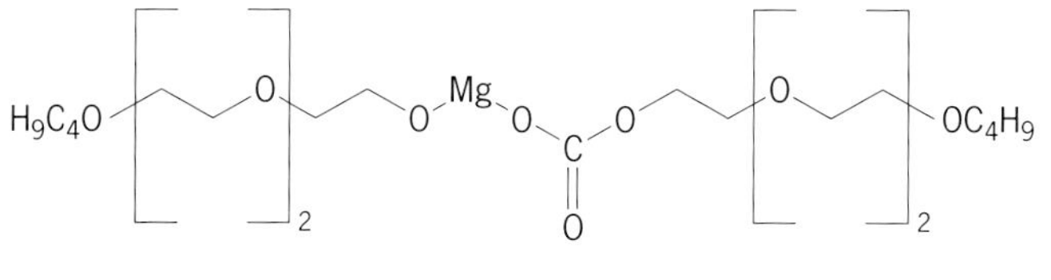

Vapor-phase deacidification (interleaf). In this process, one sheet of paper treated with cyclohexylamine carbonate per each 50–100 book pages is interleaved in the book. In another form of the process, books are exposed to pellets or granules of cyclohexylamine carbonate, which migrates through the book and neutralizes the acids. The comparatively inexpensive method has been used in the United States and in the United Nations Collections in Geneva. The degree of neutralization and the alkaline reserve are deemed inadequate; in addition, there is concern over residual odor and possible toxic effects, in part due to the formation of cyclohexyl amine under the normal conditions of use of cyclohexylamine carbonate.

Comparison of processes. Many factors enter into the design and selection of a mass deacidification treatment for books. In **Table 2** the characteristics of various processes are compared to those for an ideal process. Other factors to be considered are unit cost, effects on nonpaper book materials (adhesives, dyes, photographs, leather, cloth, and so forth), the mechanical properties on treatment and after aging, uniformity of results, and residual odor.

For background information SEE BOOK MANUFAC-TURE; PAPER; PILOT PRODUCTION in the McGraw-Hill Encyclopedia of Science & Technology.

Norbert S. Baer

Bibliography. A. Lienardy, A bibliographical survey of mass deacidification methods, *Restaurator*, 12(2):75–103, 1991; Office of Technology Assessment, U.S. Congress, *Book Preservation Technologies*, 1988; G. Petherbridge (ed.), *Conservation of Library and Archival Materials and Graphic Arts*, 1987; C. J. Shahani and W. K. Wilson, Preservation of libraries and archives, *Amer. Sci.*, 75:240–251, 1987; S. H. Zeronian and H. L. Needles (eds.), *Historic Textile and Paper Materials II*, 1989; C. Zimmermann, *Bibliography on Mass Deacidification of Books*, 1991.

Borehole logging

For many years geophysicists have sought to investigate the internal structure and composition of the Earth by making measurements at its surface. Borehole logging is essentially the practice of geophysics in drill holes. A downhole probe (sonde) is lowered into the hole and then raised at constant speed. The sonde measures one or more physical characteristics of the surrounding rocks, recorded as a continuous depth record or log. These measurements have been essentially physical in nature, and have been directed at thermal, mechanical, magnetic, nuclear, and electrical properties of rocks. Objectives have included the delineation of petroleum reservoirs, fresh-water aquifers, orebodies, and sources of geothermal energy; pollution prevention; earthquake prediction; and engineering foundation studies. The philosophy of interpretation is one of indirectness. Logs rarely measure the desired quantity directly, but rather some physical indicator of the target characteristic.

Because of the proximity of the sensor to the sensed material, logging tools do not require a substantial depth of investigation into the surrounding formations; therefore they can be designed to be compact. Logs are less affected by geophysical equivalence and suppression than their surface counterparts. They can be used to calibrate the interpretation of surface geophysical data.

Since the first documented open-hole log was run in France in 1927, borehole logging has grown into a technical discipline in its own right. This growth has largely been driven by commercial forces within the petroleum industry. Therefore most logging tools are designed for oil-field use. Borehole logs constitute the basis for evaluating hydrocarbons in place within petroleum reservoirs. Traditionally, well-log evaluation has been practiced in virtual isolation from surface geophysics, but the contemporary drive in industry is toward integrated reservoir description in which geophysical, geological, logging, and laboratory data are merged into a unified reservoir model. This drive is becoming an integral part of scientific programs. It is increasingly recognized that borehole logging is intermediate within a family of measurements that range from the megascale of surface geophysics to the microscale of pore studies in the laboratory. Logging can play a unique pivotal role relative to surface geophysical surveys and laboratory analysis.

Conventional logs. With certain exceptions, conventional logs are those developed for and used routinely by the oil industry. A conventional open-hole oil-field logging survey comprises three basic logging runs. The first run contains deep- and shallow-sensing electrical resistivity tools, designed for distinguishing at an early stage between water and hydrocarbons on the basis of conductivity; a sonic tool, for measuring acoustic (compressional) velocity, which indicates changes in porosity and provides a link to surface seismics; and a natural gamma-ray tool, for recording the natural radioactivity of the rocks penetrated by the borehole as an indicator of lithology. The first logging run is intended to provide sufficient information for a basic evaluation of the formations if the hole should then be lost, for example, through collapse. The second run, directed at the evaluation of porosity, contains a density tool, which indicates porosity from the measured bulk density of a rock for which the matrix density is known, and a neutron tool, which monitors neutron capture after bombarding neutrons are slowed down through collisions with hydrogen nuclei, the measured counts being interpreted empirically in terms of porosity. The third run comprises deep-, medium-, and micro-sensing resistivity tools for more precise evaluation of fluids. The natural gamma-ray tool is run as part of every tool string to facilitate depth merging of the different logs. Several other tools are also run as part of the three primary tool combinations. These provide measurements of borehole diameter, temperature, borehole fluid resistivity, and electrochemical potential in the wellbore; electrochemical potential is a lithology indicator, and it can also furnish useful information about pore water resistivity.

Although the suite of basic logs has remained essentially unchanged in recent years, tool technology and measurement capability have advanced greatly as a result of developments in digital electronics. Digital data transmission has meant that spectral and multichannel tools can be deployed downhole, and a number of useful tools have emerged. These include sonic waveform tools capable of determining compressional and shear velocities as well as attenuation parameters; nuclear spectral tools with a greater diagnostic capability for lithology; and digital induction tools, which use both the in-phase and phase-quadrature components of the measured signal for data enhancement. Advances in data processing have sharpened effective tool resolutions and improved so-called quick-look interpretations at the well site. Log quality control is practiced in real time.

Other tools have been developed by the oil industry, but these are not run as standard logs. Examples are the geochemical logging tool, which provides elemental concentrations of up to 12 elements, and the formation microscanner, which provides a detailed electrical image of the borehole wall based on electrical contrasts. Both these tools emphasize the links with geology. Geochemists have been cautious in their approach to geochemical logging because of doubts concerning the interpretative model. Sedimentologists have generally welcomed the microscanner, with its high spatial resolution and potential role in core orientation and core-log correlation.

Scientific logs. Borehole logging carried out for scientific purposes draws upon commercial technology, but also includes some logging tools that are not run routinely in oil fields. In some cases the tools may have been developed specifically for sci-

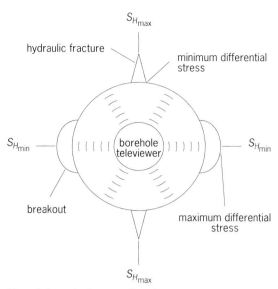

Fig. 1. Schematic diagram of breakout and hydraulic-fracture orientation relative to maximum horizontal principal compressional stress (S_{Hmax}). S_{Hmin} = minimum horizontal principal compressional stress.

entific logging purposes, although the high costs associated with the engineering development of downhole tools does not allow all scientific opportunities to be grasped. Scientific logging can be carried out on land or at sea. Two current scientific programs in which borehole logging has a full and integral role are the German Continental Deep Drilling Project (KTB) and the multinational Ocean Drilling Program (ODP). The Ocean Drilling Program is of particular interest because it runs a standard logging suite, augmented by specialist tools, and therefore its data can be compared with the oil industry's data.

The Ocean Drilling Program is concerned with exploring the structure and history of the Earth as revealed beneath the oceans through shipborne drilling at target locations around the globe. It has four technical thrusts for the future: structure and composition of the crust and upper mantle; dynamics, kinematics, and deformation of the lithosphere; fluid circulation in the lithosphere; and cause and effect of oceanic and climatic variability. Logging can contribute to these efforts by providing information on lithology, structure, mineralogy, sedimentology, porosity, permeability, elastic properties, stress regime, and stratigraphy. The standard logging suite and additional specialist tools used in the Ocean Drilling Program have been chosen because they provide substantial input to these subject areas. Thus geochemical and microscanner logs are part of the standard suite. In this respect the standard logging suite of the Ocean Drilling Program is more comprehensive than that used routinely in the oil industry, where the containment of operating costs and commercial risks is a strong governing factor. Supporting tools include the borehole televiewer, an acoustic camera that can provide an ultrasonic image of the borehole wall and a 360° record of borehole diameter for recognition of breakout (failure or collapse of the borehole wall); a downhole magnetometer

and susceptibility tool for magnetostratigraphy; packer tools (devices that contain packers or seals) for determination of field permeability; and downhole water samplers.

The set of borehole logging tools available to the Ocean Drilling Program provides the most technically advanced capability for downhole measurements that is being deployed routinely in the world today. A comprehensive logging program has allowed the interpretation of drilling results from Ocean Drilling Program holes to be viewed from a fuller quantitative perspective than would otherwise have been possible.

Scientific applications. Borehole logging carried out as part of the Ocean Drilling Program has made several noteworthy contributions to geoscience. In order to understand the way in which oceanic crust forms and evolves, more must be learned about crustal composition and structure. Acoustic logs have contributed to a better understanding of the seismic layering of oceanic crust. Log-derived variations in chemical composition have suggested different magma sources or zones of alteration. The borehole televiewer and formation microscanner have delineated fracture zones. The integrated use of diverse logs contains the key to deriving maximum information from downhole measurements.

Measurement of the stress distributions around the world is directed at learning more about the mechanism of plate tectonics. Where the stress distribution is anisotropic, failure of the borehole wall can occur in the form of breakouts that are oriented perpendicular to the directions of the maximum horizontal principal compressional stress (S_{Hmax}). Breakouts have been identified with the borehole televiewer (**Fig. 1**), which in the caliper mode provides better resolution of breakouts than mechanical calipers, which might miss

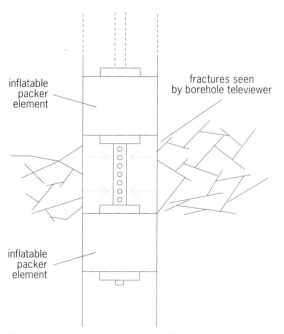

Fig. 2. Schematic diagram of packer deployment in oceanic basement rock. Arrows indicate direction of flow.

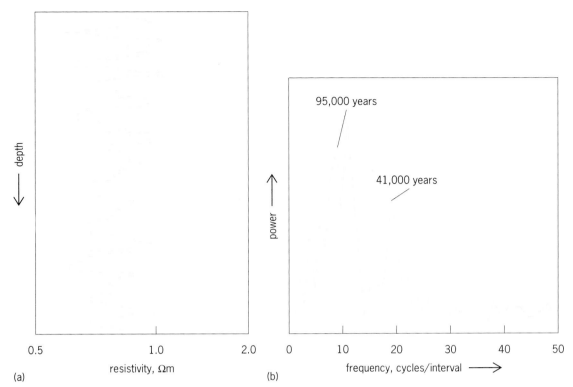

Fig. 3. Example of a downhole resistivity log in sedimentary rocks showing its cyclic nature: (*a*) the log indicates alternating higher and lower resistivity; (*b*) the Fourier power spectrum of the log indicates energy peaks with periods that correspond to those of astronomically driven climatic cycles, that is, 41,000 and 95,000 years.

a void during deployment. This recognition of breakouts has allowed the direction of stress to be inferred. Some investigators have claimed that the magnitude of the stress can be deduced from the dimenions of the breakouts, but this is not a widely held view. Another approach is controlled hydraulic fracturing. One possible goal is to create a global atlas of horizontal stress azimuths.

The exchange of matter between seawater and subsea pore waters takes place through fluid circulation as one of the mechanisms that define the global geochemical budget. This mechanism depends on the porosity and permeability of the subsea sediments and basement rock. Logs have enabled porosity to be evaluated with a fair degree of confidence. Permeability cannot be obtained from logs directly, but it has been determined in basement rocks through packer tests (pressure tests of sealed zones; **Fig. 2**). For example, if a borehole televiewer senses fractures but no alteration products are seen by the geochemical log and a locally high permeability is interpreted from the packer test, the fractures are open and active circulation is indicated. If, however, the geochemical log identifies alteration products associated with the fracture occurrence and a low permeability is interpreted from the packer test, the fractures have been sealed and paleocirculation is indicated. The porosity and permeability data, used in conjunction with other logs, have allowed zones of active and paleo-hydrothermal circulation to be identified.

Environmental changes that have caused variations in lithology and composition can be recognized in the sedimentary succession. For example, some clay minerals might indicate arid conditions, while others indicate humid conditions. The continuous nature of borehole logs makes them ideal for studying paleoenvironmental changes, provided that these changes have discernible physical manifestations. This condition essentially reduces to one of significant parametric contrasts between layers that are sufficiently thick to be resolved by logging tools. In some cases, the pattern of change has been found to be cyclic, with the periodicities corresponding to those of known cyclic variations in the Earth's orbit (**Fig. 3**). Thus the continuity of borehole logs, which is not always achievable in core data, has contributed to identifying astronomically driven climatic variations preserved in the sedimentary record. This application suggests a possible dating facility for sedimentary rocks that might complement and provide finer resolution than micropaleontology.

Technological advances. Several new logging tools have been recently developed or are at the development stage. Examples are a magnetic-resonance-imaging tool for movable fluid identification and possibly an empirical indication of permeability, a geochemical tool with a high spectral resolution that is capable of providing the concentrations of more elements for improved geochemical characterization, and a high-magnetic-resolution magnetometer/susceptibility tool for seeing reversals. Such developments suggest that the application of borehole logs in commercial and scientific areas will become even more sophisticated over the next few years.

For background information *SEE BOREHOLE LOGGING; PETROLEUM ENGINEERING; WELL LOGGING* in the McGraw-Hill Encyclopedia of Science & Technology.

Paul F. Worthington

Bibliography. R. Desbrandes, *Encyclopedia of Well Logging*, 1985; Lamont-Doherty Geological Observatory, Borehole Research Group, *Wireline Logging Manual*, 1990; P. F. Worthington, The direction of petrophysics: A five-year perspective, *The Log Analyst*, 32(2):57–62, 1991; P. F. Worthington et al., Scientific applications of downhole measurements in the ocean basins, *Basin Res.*, 1:223–236, 1989.

Borehole mining

Borehole mining, also known as slurry mining, is a process in which a tool incorporating a water-jet cutting system and a downhole slurry pumping system is used to mine minerals through a borehole drilled from the ground surface to the buried mineralized rock. Water jets from the mining tool erode the ore to form a slurry, which flows into the inlet of a slurry pump at the base of the tool. The slurry is then pumped to the surface in a form suitable for transfer to a processing plant by pipeline (**Fig. 1**).

Borehole mining offers important advantages over conventional open-pit and underground mining methods, and it can access mineral deposits that presently are not mined because of technical or economic difficulties. Borehole mining can achieve immediate production, because there is no need to drive openings to and in an ore body to admit workers; such conventional mining methods require 3–5 years of development before production and return on investment can be expected.

The systems for transportation and fragmentation of ore are incorporated into a single machine that is operated remotely from the surface by a two- or three-person crew, thus eliminating health and safety problems inherent in underground mining. Disturbance to the environment is minimal and short-term; no overburden is removed and subsidence can be avoided by backfilling. Ore fragmented by the water jet is brought to the surface in a slurry; thus it is ideally suited to economical transport by pipeline (**Fig. 2**).

Borehole mining is selective and can extract deposits that are small or erratically mineralized, thereby

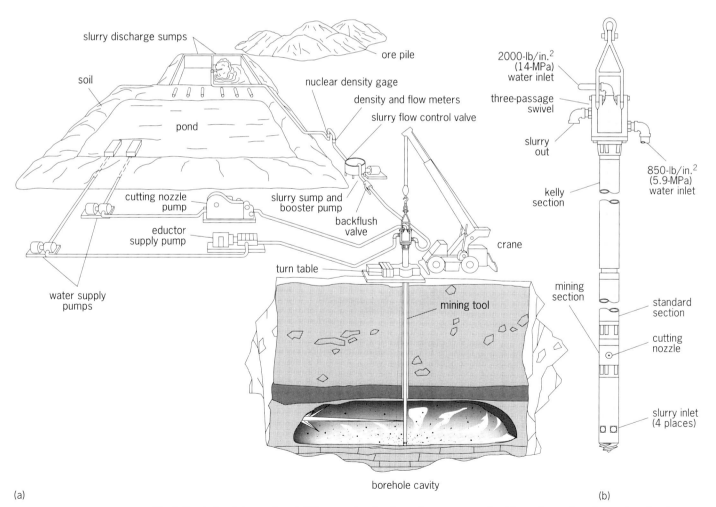

Fig. 1. Borehole mining system. (*a*) Diagram showing the various components of the entire system; the nuclear density gage measures the proportion of water slurry in the total slurry discharge. (*b*) Mining tool details.

Fig. 2. Phosphate slurry at the outlet of a borehole mining tool.

broadening the resource base. This selectivity allows the ore to be extracted without disturbing the surrounding rock, thereby avoiding dilution. Crushing and grinding costs are minimal, since the ore is reduced to grain size by jet impact.

Rock fragmentation. The water-jet cutter in the borehole miner must be capable of producing enough broken ore to pay for the cost of the mining operation and to yield a profit. In order to maximize profit, the largest possible volume must be cut; thus the effective reach of the jet must be as long as possible.

Water jets in air dissipate by entrainment of air at the water–air interface; thus the ratio of surface area to volume of a jet influences the rate at which a jet will dissipate. Since jet volume increases as the square of the jet radius while surface area increases linearly with increasing radius, smaller jets dissipate in a shorter distance than larger jets. As jet volume increases, a smaller portion of the mass of the jet resides near the surface, where it can be dissipated by air entrainment. Thus, in order to maximize the effective range of a jet, its diameter should be as large as possible.

Rock fragmentation, however, requires that a high energy density be deposited on the rock surface. Maximizing the density of the kinetic energy on the rock surface requires that the jet be as small as possible. Because pump power is finite and the power delivered by a jet is a product of jet pressure and flow rate, it is necessary to find an appropriate compromise between pressure and flow rate that will yield the maximum volume of slurry ore per borehole. The jet must have an energy density high enough to disaggregate the rock and a diameter large enough to enable the jet to stay coherent over long distances. Given constant pump power, high jet pressures give small jets with high velocities, while lower pressures give longer, slower jets.

The particular pressure and flow rate chosen depend upon the power available and the hardness of the ore. The energy required to break hard rock at long distances is so large as to make borehole mining of hard rock impractical; thus borehole mining is limited to softer rocks. Some jetting parameters that have been employed in borehole mining are as follows: 4500 lb/in.^{2} (31 megapascals), 200 gal/min (760 liters/min) for coal in Wilkeson, Washington; 1500 lb/in.^{2} (10 MPa), 350 gal/min (1330 liters/min) for sandstone in Natrona County, Wyoming; 1000 lb/in.^{2} (6.9 MPa), 300 gal/min (1200 liters/min) for oil sands at Taft, California; and 1800 lb/in.^{2} (12 MPa), 420 gal/min (1600 liters/min) for phosphate ore in St. Johns County, Florida.

Borehole-mining cavity radii have been measured at 26 ft (7.9 m) in sandstone in Natrona County, Wyoming, and 20 ft (6.1 m) in phosphate ore in St. Johns County, Florida. The Wyoming cavity was created by mining with the jet in air, while the Florida cavity was mined underwater. Water jets can be effective underwater at long distances if the jet is shrouded by compressed air.

Water jets for borehole mining are most effective if the nozzle can be rotated 360° and translated along the axis of the borehole. Translating the jet permits cutting a large vertical interval while the slurry pump stays at the base of the borehole where the slurry density is the highest.

Borehole mining is most successful in granular rocks, especially sandstones, which are amenable to disaggregation with water jets; the jet pressurizes the grain boundary pores and liberates the individual grains. Nonporous ores such as kaolin or bauxite have not proved to be amenable to borehole mining with water jets. Work is in progress to develop a method of borehole mining whereby bauxite is fragmented with explosives prior to being pumped out of the borehole.

Slurry pumping. Centrifugal pumps, Moyno pumps (which operate on a progressing cavity principle), jet pumps, and air lifts have been used as downhole slurry pumps in borehole mining equipment. Moyno and staged centrifugal pumps are not used as frequently as jet pumps or air lifts. Centrifugal pumps require a larger-diameter borehole than jet pumps, and Moyno pumps are not capable of pumping abrasive slurries. Jet pumps and air lifts are more reliable because they have no moving parts and can pump abrasive slurries with minimal wear.

Applications. Borehole mining has been attempted in the United States and the former Soviet Union. This mining has demonstrated the technical feasibility of the remote extraction of coal, oil sands, uranium ore, phosphates, iron ore, sand, gravel, and amber as a slurry through a borehole.

Borehole mining of deep phosphate ore, ore too deep for economical open-pit mining, provided the most successful of the field trials in the United States. Productivity was higher than that of other commodities because of the lack of induration of the phosphate ore and because the mining took place under 250 ft (76 m) of groundwater. The ability to mine underwater increased the efficiency of the slurry pump by placing 250 ft (76 m) of positive suction head on the inlet of the pump. Mining underwater also has the advantage of providing roof support during mining. Tests in Florida indicated that the roofs of phosphate borehole mining cavities will collapse if the mining takes place in air.

Borehole mining is selective; only ore is extracted, leaving the adjacent barren rock intact. In order to exploit this advantage, either the rock must be strong enough to remain standing when the ore is removed, or the ore must be overlain by a cap rock strong enough to support the overburden and form the roof of the borehole-mined cavity. Otherwise, the barren rock above the ore will fall into the cavity and dilute the ore. This uncontrolled caving could propagate to the surface and endanger the borehole mining tool. Many sandstones, such as the uraniferous Teapot sandstone of Wyoming, are self-supporting; and a cylindrical cavity with a diameter in excess of 50 ft (15 m) will stand unsupported. The Hawthorne Formation in St. Johns County, Florida, contains layers of weak phosphate ore overlain by a cap rock of dolomite that keeps the barren overburden from falling into the ore.

Borehole mining fulfills the need for a method to mine incremental uranium ore. Incremental ore comprises small, irregular, high-grade uranium ore bodies that are adjacent to working open pits but cannot be included in the pit because of engineering limitations. The small sizes and irregularities of these deposits make them ideal candidates for borehole mining because of the high areal selectivity of the method.

Field tests of borehole mining of oil sands and coal have demonstrated the technical feasibility of the remote extraction of these commodities through boreholes, but the mining rate was too low for commercial viability because of the low unit value of the ore. Generally, ores with high unit value are better candidates for borehole mining, because the value of the ore mined is the product of the production rate and the unit value of the ore. SEE OIL MINING.

Environmental impact. Surface subsidence and the presence of piles of tailings represent potentially adverse environmental impacts resulting from borehole mining operations. Both impacts can be mitigated by backfilling the borehole-mined cavities with slurry jets as shown in **Fig. 3**. Mill tailings can be used as backfill.

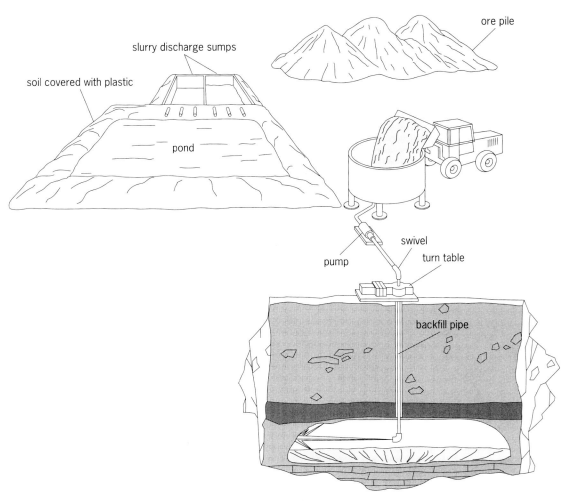

Fig. 3. Diagram of the method of backfilling a borehole-mined cavity.

However, borehole mining is one of the mining methods that are least disruptive to the environment. Ground subsidence is eliminated by backfilling, no foreign substances are introduced into the groundwater, and no overburden is removed. Thus borehole mining seems ideally suited for mining in environmentally sensitive areas.

For background information, SEE MINING in the McGraw-Hill Encyclopedia of Science & Technology.

George A. Savanick

Bibliography. H. Hartman (ed.), Hydraulic mining: Borehole slurry, *SME Mining Handbook*, 2d ed., 1992; G. A. Savanick, Borehole mining of deep phosphate ore in St. Johns County, Florida, *Min. Eng. J.*, 37(2):144–148, 1985; G. A. Savanick, *Borehole (Slurry) Mining of Coal, Uraniferous Sandstone, Oil Sands, and Phosphate Ore*, U.S. Bureau of Mines, Rep. Investig. 9101, 1987.

Carbon

A fullerene is a molecule containing an even number of carbon atoms arranged in a closed hollow cage. The fullerenes were discovered as a consequence of astrophysically motivated chemical physics experiments that were interpreted by using geodesic architectural concepts. Fullerene chemistry, a new field that appears to hold much promise for materials development and other applied areas, was born from pure fundamental science.

In 1985, fifteen years after it was conceived theoretically, the molecule buckminsterfullerene (C_{60} or fullerene-60) was discovered serendipitously. Fullerene-60 (**Fig. 1**) is the archetypal member of the fullerenes, a set of hollow, closed-cage molecules consisting purely of carbon. Some representative structures are shown in **Fig. 2**. The fullerenes can be considered, after graphite and diamond, to be the third well-defined allotrope of carbon. Macroscopic amounts of various

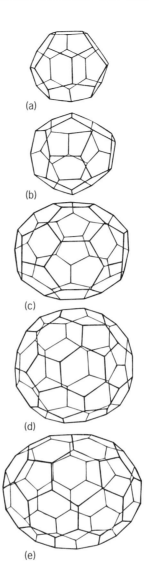

Fig. 2. Some of the more stable members of the fullerene family. (*a*) C_{28}. (*b*) C_{32}. (*c*) C_{50}. (*d*) C_{60}. (*e*) C_{70}.

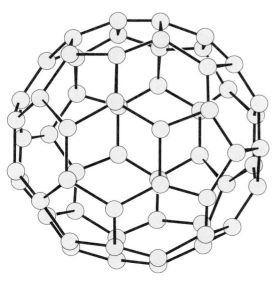

Fig. 1. Structure of fullerene-60 (C_{60}).

fullerenes were first isolated in 1990, and since then they have been studied intensively. It has been discovered that members of this new class of spheroidal aromatic organic molecules have numerous novel physical and chemical properties. The fullerenes promise to have synthetic, pharmaceutical, and industrial applications. Already, derivatives have been found to exhibit fascinating electrical and magnetic behavior, in particular superconductivity and ferromagnetism.

Structures. In the fullerene molecule an even number of carbon atoms are arrayed over the surface of a closed hollow cage. Each atom is trigonally linked to its three near neighbors by bonds that delineate a polyhedral network, consisting of 12 pentagons and n hexagons. (Such structures conform to Euler's theorem for polyhedrons in that n may be any number other than one, including zero.) All 60 atoms in fullerene-60 are equivalent and lie on the surface of a sphere distributed with the symmetry of a truncated icosahedron. The 12 pentagons are isolated and interspersed symmetrically among 20 linked hexagons; that is, the symmetry is that of a modern soccerball. The molecule was named after R. Buckminster Fuller, the inventor of geodesic domes, which conform to the same underlying structural formula. Three of the four valence electrons of each carbon atom are involved in the sp^2 sigma-bonding skeleton, and the fourth p electron is one of 60 involved in a pi-delocalized molecular-orbital electron sea that covers the outside (exo) and inside (endo) surface of the molecule. The resulting cloud of pi-electron density is similar to that which covers the surface of graphite; indeed, the molecule can be considered a round form of graphite. The smallest member of the family, fullerene-20, is dodecahedrene, whose fully hydrogenated analog ($C_{20}H_{20}$, dodecahedrane) has been created by rational synthesis (production of a compound by using a sequence of chemical reaction steps strategically chosen for the purpose). Several members of the family have now been isolated, in particular C_{70}, C_{76}, and C_{84}. Giant fullerenes with at least 600 atoms also seem to form, and appear to have interesting polyhedral shapes. The highly symmetric icosahedral species C_{60}, C_{240}, and C_{540} are shown in **Fig. 3**.

Discovery and preparation. In 1985 laboratory experiments were initiated that aimed to simulate the physicochemical conditions in a cool red giant star by using a pulsed laser focused on a graphite target to generate a plasma. As the resulting helium-entrained plasma expanded supersonically and cooled, the carbon atoms aggregated in the vapor phase, and an exceptionally strong signal was detected for the C_{60}^+ ion by mass spectrometry. The signal for C_{70}^+ was also prominent. It was proposed that these observations indicated that the C_{60} and C_{70} species were long-

lived, and that on the basis of geodesic and chemical arguments this unexpected stability was commensurate with closed-cage fullerene structures.

Subsequent experiment and theory yielded significant support for the cage proposal. For instance, the formation of endohedral metallofullerene complexes, in which a metal atom is encapsulated in the cage, was demonstrated. Fullerenes with nonabutting pentagons should be especially stable (the pentagon isolation rule), a conclusion that readily explained the observation that both C_{60} and C_{70} are prominent. Although there was initially some skepticism about the fullerene proposal, a great deal of supporting data was subsequently amassed.

In 1990 the dust produced by a carbon arc struck under argon was found to contain around 10% fullerenes, which could be extracted by using solvents or by sublimation. Individual fullerenes were found to be separable by standard chromatographic techniques. The structures of fullerenes-60 and -70 have been confirmed unequivocally by x-ray, nuclear magnetic resonance (NMR), and other spectroscopic measurements. The simplicity of the preparative procedure has enabled the chemistry and physics of the fullerenes to advance very rapidly. It has also been found that a benzene/oxygen combustion flame can be adjusted so that the resulting soot contains up to 4% fullerenes. The implications for a more detailed understanding of soot formation are interesting.

Theory. Soon after fullerene-60 was conceived, a preliminary theoretical study (a Hückel calculation, which is a procedure for estimating bond energy and bonding electron distribution in a molecule) showed that it would be a closed-shell molecule. The discovery that fullerene-60 formed in a carbon plasma triggered intense theoretical study of the likely properties of the

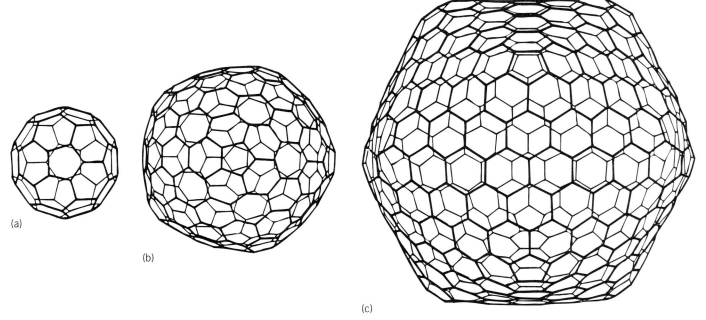

Fig. 3. Comparison of the structure of fullerene-60 with that of two giant fullerenes. (*a*) C_{60}. (*b*) C_{240}. (*c*) C_{540}.

fullerenes, including their possible aromaticity. The importance of pentagon isolation was recognized as the key general fullerene stability criterion. Spectroscopic properties were calculated, and it was recognized that the molecule should possess only four characteristic infrared vibrational frequencies. These frequencies provided crucial clues during the investigations that led to the successful isolation of the fullerenes. A general electronic stability rule has been developed that suggests that fullerenes with $60 + 6k$ atoms (where $k = 0, 2, 3 \ldots$) should possess closed electronic shells.

Properties. In benzene solution, fullerene-60 is magenta and fullerene-70 red. Fullerene-60 forms translucent magenta face-centered cubic (fcc) crystals that sublime. The ionization energy is 7.61 eV and the electron affinity is 2.6–2.8 eV. The strongest absorption bands lie at 213, 257, and 329 nanometers. Studies with NMR spectroscopy yield a chemical shift of 142.7 ppm; this result is commensurate with an aromatic system. Other NMR measurements indicate that the molecule is rotating in the lattice at very low temperatures.

Fullerene-60 behaves as a soft electrophile, a molecule that readily accepts electrons during a primary reaction step. It can accept three electrons readily and perhaps even more. The molecule can be multiply hydrogenated, methylated, ammonated, and fluorinated. It forms exohedral complexes in which an atom (or group) is attached to the outside of the cage, as well as endohedral complexes in which an atom [for example, lanthanum (La), potassium (K), or calcium (Ca)] is trapped inside the cage. Exohedral transition-metal complexes with ligands containing osmium and platinum have been made. Evidence for oxygen addition compounds, such as $C_{60}O$ and $C_{70}O$, has also been reported. A stable endohedral metallofullerene, $C_{82}La$, has been isolated. Fullerene-60 appears to catalyze the formation of singlet oxygen in the presence of light. This ability may be related to the observation that under certain conditions solutions of fullerene-60 decompose in the presence of oxygen and light.

The possible hazards associated with working with fullerene-60 have not yet been determined. However, common sense and the compound's ability to catalyze formation of singlet oxygen suggest that some care should be exercised when working with it until more is known about possible toxicity.

Metal-doped crystals such as $C_{60}K_3$ have been found to be isotropically superconducting at 30 K ($-406°F$), and charge-transfer complexes have been found to be ferromagnetic.

Applications. Fullerene materials have been available for such a short time that applications are yet to be established. However, the properties already discovered suggest that there is likely to be a wide range of areas in which the fullerenes or their derivatives will have uses. The facility for acceptance and release of electrons suggests a possible role as a charge carrier in batteries. The properties of graphite suggest that lubricative as well as tensile and other mechanical properties of the fullerenes are worthy of investigation. Liquid solutions exhibit excellent properties of optical

harmonic generation. The high temperature at which superconducting behavior is observed is particularly exciting. This observation suggests possible applications in microelectronics devices, as does the detection of ferromagnetism in other fullerene derivatives.

Implications of fullerene formation. Perhaps the most important aspect of the fullerene discovery is that the molecule forms spontaneously. This fact has important implications for understanding the way in which extended carbon materials form, and in particular the mechanism of graphite growth and the synthesis of large polycyclic aromatic molecules. It is now clear that as far as pure carbon aggregates of around 60–1000 atoms are concerned, the most stable species are closed-cage fullerenes. A flat, pure graphitic carbon sheet has dangling bonds at the edge and these can be satisfied by closure, so that the energy of the system is reduced. In the case of bulk graphite, which consists of stacks of hexagonally arrayed carbon sheets, the interlayer interactions and size effects appear to stabilize the flat sheets. Also, it may be possible to isolate some of the more stable smaller fullerenes, those with fewer than 60 carbon atoms, such as fullerene-50.

Astrophysical implications. Fullerene-60 was discovered as a direct result of physicochemical investigations that simulated processes occurring in stars and in space. Consequently the likelihood that fullerenes, in particular fullerene-60, and analogs are present in space is a fascinating conjecture. Particularly likely are the exohedral protonated form, $C_{60}H^+$, and exohedral metallated species such as $C_{60}M^+$, where M is an abundant interstellar atom such as sodium (Na), K, Ca, or O.

For background information SEE CARBON; ELECTRON CONFIGURATION in the McGraw-Hill Encyclopedia of Science & Technology.

Harold W. Kroto

Bibliography. R. F. Curl and R. E. Smalley, Probing C_{60}, *Science*, 242:1017–1022, 1988; W. Kraetschmer et al., Solid C_{60}: A new form of carbon, *Nature*, 347:354–358, 1990; H. W. Kroto, Space, stars, C_{60}, and soot, *Science*, 242:1139–1145, 1988; H. W. Kroto et al., C_{60}: Buckminsterfullerene, *Nature*, 318:162–163, 1985; H. W. Kroto, W. Allaf, and S. C. Balm, C_{60}: Buckminsterfullerene, *Chem. Rev.*, 91:1213–1235, 1991.

Cat

The living cats constitute an extremely uniform group of about 36 species that is easily recognized and delimited from other groups of living Carnivora. They are short-faced carnivorous animals with prominent conical canines and elongated carnassials (shearing cheek teeth) that are modified into simple blades. There are no crushing or grinding molars, and the anterior premolars are reduced or absent. An inflated, ossified auditory bulla encloses the middle-ear cavity; the cavity is bisected by a bilaminar septum that is formed by the ectotympanic (tympanic ring) and caudal entotympanic

(**Fig. 1***a*). The limbs are slender and the feet are digitigrade with retractile claws; the limbs are less modified for running than those of some other Carnivora, such as dogs (Canidae), and the forelimbs retain the dexterity required for taking and handling prey.

Fossil deposits of late Eocene through Pleistocene age yield the remains of a variety of catlike animals that until recently had been considered extremely closely related, or ancestral, to living cats. These catlike animals had usually been included in the cat family (Felidae). They share the short faces, retractile claws, and hypercarnivorous dentition of living cats; many species were saber-toothed carnivores, in which the upper canines were elongated into laterally compressed sabers (**Fig. 2**). Recent study has demonstrated, however, that some of these catlike animals lack the diagnostic features of the Felidae; these animals are now referred to a separate family, the Nimravidae, which is not the family of Carnivora most closely related to the Felidae. The implication is that cats, as an adaptive form, evolved twice within the Carnivora.

Historical record. The nimravids appeared in the late Eocene in North America; the first radiation [the Nimravinae; for example, *Nimravus, Dinictis* (Fig. 2*c*), *Hoplophoneus*] persisted through the

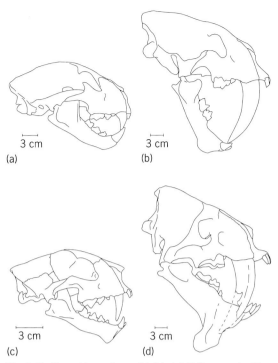

Fig. 2. Skulls and lower jaws of felids (*a*) *Felis concolor* (the cougar) and (*b*) *Smilodon*, and of nimravids (*c*) *Dinictis* and (*d*) *Barbourofelis fricki*. (*Parts c and d after H. N. Bryant, Phylogenetic relationships and systematics of the Nimravidae (Carnivora), J. Mammal., 72:56–78, 1991*)

Oligocene in both North America and Eurasia. A second radiation [the Barbourofelinae; for example, *Sansanosmilus, Barbourofelis* (Fig. 2*d*)] appeared in the middle Miocene of Europe; they later reached Africa, Asia, and North America and became extinct in the late Miocene. The nimravids varied considerably in morphology, but most species demonstrated at least some degree of saber-toothed adaptation. Their skeletons retain many features of early carnivorans; the limbs were short and robust and the digits were spreading. The manus was at least partially digitigrade, and included a fused scapholunar, as in the extant Carnivora.

The earliest known felids (for example, *Proailurus*) appeared in the latter half of the Oligocene of Europe, and the later *Pseudaelurus* reached North America in the middle Miocene. The major adaptive radiation of the Felidae occurred in the late Miocene, and by the Pleistocene the group had spread to all continents except Australia and Antarctica. Even the earliest felids had more slender limbs and more digitigrade feet than did the nimravids. Felids include several lineages that developed saber-toothed adaptations and were highly convergent on their nimravid counterparts (Fig. 2*b*).

Changing views. Soon after the discovery in the nineteenth century of various North American Eocene and Oligocene cats, the significant differences in the arrangement of their basicranial foramina from that of extant felids led to the naming of a separate family, the Nimravidae, for these early cats. The Nimravidae was initially considered a horizontal evolutionary category or grade that was ancestral to the Feli-

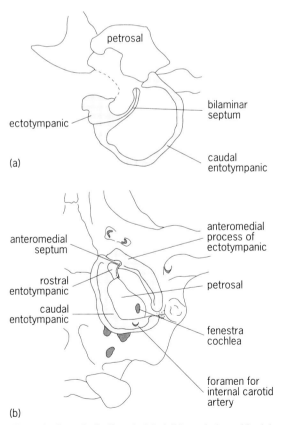

Fig. 1. Auditory bulla (in color) in felids and nimravids. (*a*) Transverse section of the middle-ear region and auditory bulla of a felid, dorsal toward the top; the eardrum is indicated by the broken line (*after R. J. G. Savage, Evolution in carnivorous mammals, Palaeontology, 20:237–271, 1977*). (*b*) Ventral view of the left portion of the basicranium in *Dinictis*, anterior toward the top.

dae. However, by 1900 most classifications included the nimravids in the Felidae. Early phylogenies were based primarily on dental morphology and included separate lineages of cats with saber-toothed and normal canines that extended throughout the later Tertiary. A later view held that although nimravids and felids were extremely closely related, one group was not ancestral to the other. This view was based on the lack of intermediates between the two groups, and is more consistent with some aspects of the fossil record and geographic distribution.

The morphology of the basicranium and auditory bulla has been used extensively in the establishment of relationships among the extant families of the Carnivora. The Felidae are members of the Aeluroidea (or Feloidea), which also includes the Viverridae, Herpestidae, and Hyaenidae. This superfamily is characterized by a unique double-chambered, completely ossified auditory bulla with a bilaminar septum (Fig. 1a), and a unique arrangement of basicranial foramina. The early recognition of the nonaeluroid arrangement of the basicranial foramina in nimravids was not considered a preclusion to an ancestral relationship to felids. The failure to identify a bilaminar septum or completely ossified bulla in Eocene-to-Oligocene nimravids was considered an artifact of poor preservation. However, the continued failure to identify an aeluroid bulla in the early nimravids and the clearly demonstrated absence of the bilaminar septum in the bulla of the Miocene nimravid *Barbourofelis* increased doubts regarding the referral of these cats to the Felidae. Indications of fundamental differences in the basicranium and auditory bulla between the two groups of cats led to the renewed recognition of the Nimravidae as a separate family. The felidlike dentitions and retractile claws have led to some recent arguments for a closest relationship between felids and nimravids, but this view is contradicted by the absence in nimravids of the basicranial features that characterize the entire Aeluroidea. The lack of strong evidence for some relationship to aeluroids prompted suggestions of even wider phylogenetic separation between nimravids and felids.

Recent analyses. Detailed anatomical study of the auditory bulla and basicranium of nimravids confirmed the marked differences, including the absence of the bilaminar septum, from those of all aeluroids. In addition, significant differences were found between the bulla of nimravines and barbourofelines. As in other Carnivora, the nimravine bulla is composed of an ectotympanic and caudal and rostral entotympanics (Fig. 1b). The medial and posterior walls of the nimravine bulla are formed by the caudal entotympanic, which has a unique sandwichlike construction with thin outer sheets of bone and thicker inner cancellous bone. The ectotympanic has a unique anteromedially projecting process that overlaps, but does not fuse with, the caudal entotympanic. The ventral floor of the bulla is not ossified, and was probably cartilaginous, so that the petrosal is visible in ventral view (Fig. 1b). In contrast, the barbourofeline bulla is at least superficially more like those of living Carnivora. It is

completely ossified, and the probable entotympanic is unilaminar. The absence of sutures suggests fusion between constituent elements early in development. Because of the significance of the basicranium and auditory bulla for taxonomic groupings within the Carnivora, the marked differences in this morphology between nimravines and barbourofelines suggested that even these two groups might not be closest relatives (sister groups). A different study identified additional similarities between the nimravine and barbourofeline bullae, and suggested a scenario for the derivation of the nimravid basicranium and auditory bulla from the proposed primitive aeluroid morphology.

In contrast, the earliest felids had a demonstrably aeluroid bulla. *Proailurus* had at least a rudimentary bilaminar septum, and *Pseudaelurus* of the North American middle Miocene had a virtually modern felid auditory bulla.

Cladistic analysis of the nimravines, barbourofelines, and the major carnivoran clades based on characters taken from the entire skeleton and dentition confirmed the monophyly (exclusive common ancestry) of the Nimravidae (**Fig. 3**). Nimravids have a number of synapomorphies (shared derived features), including a short palate, a continuous ectosylvian sulcus on the cerebral hemispheres, enlarged upper canines, propor-

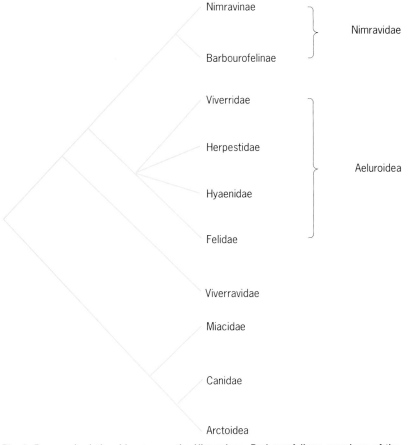

Fig. 3. Proposed relationships among the Nimravinae, Barbourofelinae, members of the Aeluroidea, and other major carnivoran clades. The Viverravidae and Miacidae are early Cenozoic carnivorans.

tionally large deciduous upper canines, late eruption of the permanent upper canines, and characteristic dental serrations. Although the auditory regions of nimravines and barbourofelines differ considerably, they share a small vertical septum in the anteromedial corner of the middle-ear cavity (Fig. 1*b*) and a similarly oriented ectotympanic. The internal carotid artery enters the auditory region in a more lateral position than in other Carnivora (Fig. 1b); however, this similarity may be a primitive feature. The barbourofeline bulla may have originated from the fusion and complete ossification of the components found in the nimravine bulla.

The most parsimonious cladogram places the Nimravidae as the sister group of the Aeluroidea (Fig. 3). Synapomorphies for the Aeluroidea and Nimravidae include the anterior opening of the palatine canal through the maxilla, highly developed retractile claws, and particular modifications to the upper dentition. Although some nimravine features, such as the prominent teres major process on the scapula, suggest a relationship to arctoids (bearlike Carnivora), these similarities conflict with the strong evidence for the monophyly of the Nimravidae and are interpreted as instances of convergent evolution. Nonetheless, the small number of synapomorphies supporting higher-level relationships among the Carnivora indicates that these conclusions require further corroboration.

It is now evident that nimravids are not involved in the ancestry of felids and that the two groups of cats evolved independently. Similarities in their cranial morphology and dentition, including the tendency to develop saber-toothed upper canines, reflect convergent adaptations to a hypercarnivorous life style rather than common ancestry. The extant felids are an extremely homogeneous group and represent only a small portion of the diverse array of cats that has existed over the past 35 million years.

For background information *SEE CARNIVORA; CAT* in the McGraw-Hill Encyclopedia of Science & Technology.

Harold N. Bryant

Bibliography. M. J. Benton (ed.), *The Phylogeny and Classification of the Tetrapods*, vol. 2: *Mammals*, Systematics Association Special Volume 35B, 1988; H. N. Bryant, Phylogenetic relationships and systematics of the Nimravidae (Carnivora), *J. Mammal.*, 72:56–78, 1991; R. M. Hunt, Jr., Evolution of the aeluroid Carnivora: Significance of auditory structure in the nimravid cat *Dinictis, Amer. Mus. Novit.*, 2886:1–74, 1987.

Cell senescence

The life-span and characteristics of aging in different organisms is assumed to reflect the coevolution of thousands of genes. These genes control biochemical systems and structures that are optimized for survival and for reproduction early in life. However, many of these systems have side effects late in life that cause senescence (that is, the deleterious changes associated with aging). Genes controlling cell replication in somatic (nonreproductive) cells are among those that contribute to aging. Cellular senescence is the process in which normal somatic cells lose their ability to divide after a critical number of divisions. During development, this process may help define mature body size and reduce the probability of cancer, but late in life the loss of proliferative potential in cells reduces the capacity for tissue regeneration and wound repair.

The finite replicative capacity of normal somatic cells has been called the Hayflick limit after Leonard Hayflick, who with his coworkers first associated the limited growth potential of cells in the laboratory with aging in the organism. The Hayflick limit reflects mitotic events, that is, cell divisions, and not just chronological time. Thus, cells taken from older humans divide significantly fewer times in the laboratory than cells from young humans; presumably the fewer divisions reflect the greater number of cellular doublings that have already occurred in the bodies of older people. Moreover, the life-span of cultured cells from different species is roughly proportional to the life-span of the species. These observations support the idea that cellular senescence, as observed in the laboratory, relates to aging of organisms.

Telomere hypothesis. The mitotic clock that counts cell divisions and ultimately causes senescence in somatic cells has not been conclusively identified. However, the telomere hypothesis, proposed about 1970, explains many characteristics of cellular aging and has recently been supported by several lines of experimental evidence. The two key assumptions of this hypothesis are that a small amount of deoxyribonucleic acid (DNA) from chromosomal ends, or telomeres, is lost with each round of DNA replication, and that critical deletions are eventually made, leading to cellular senescence. The telomere hypothesis allows for immortality of germ-line (reproductive) cells and cancer cells by assuming that special mechanisms circumvent the deletion process in these cells.

Telomeres have long been known to play a vital role in chromosome structure and function; they help to maintain chromosomal integrity by preventing aberrant recombination, or joining, between chromosomes. They may also function to protect the ends of the chromosome from degradation and to help organize chromosomes during meiotic and mitotic cell divisions. Thus, it is quite plausible that loss of telomeres with age could lead to chromosomal abnormalities and the inability of cells to divide further.

There are logical reasons to suspect that telomeric DNA might be lost at a fixed rate during successive cell divisions. Three biochemical characteristics of standard DNA replication lead to incomplete copying of DNA at the ends of linear chromosomes (**Fig. 1**). First, the two strands of the DNA double helix always run in opposite directions; that is, they are antiparallel. Second, DNA polymerase, the enzyme that synthesizes DNA by using one strand of the helix at a time as the template, can replicate DNA in only one direction. Third, DNA polymerase requires a ribonucleic acid

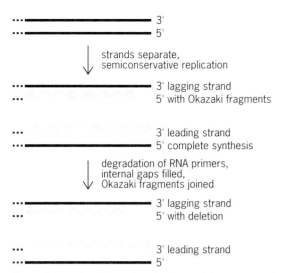

Fig. 1. Schematic of the end-replication problem at one end of a linear DNA double helix. Strand separation and unidirectional (5′→3′) synthesis of new DNA lead to discontinuous replication of the lagging strand in DNA segments called Okazaki fragments. These segments are initiated with an RNA primer (□) that is later degraded. Since only the internal gaps can be filled, the 3′ end of the parental strand is left incompletely copied, so that there is a terminal gap. The size of this terminal gap should be between the sizes of an RNA primer and an Okazaki fragment.

(RNA) primer to initiate DNA synthesis, with this primer later being removed. As a consequence of these characteristics of DNA and its polymerase, the two strands are copied in a slightly different manner, and only one is fully replicated at each telomere (Fig. 1).

Various immortal cells and viral genomes have evolved mechanisms to overcome the end-replication problem. The mechanism used by immortal eukaryotic cells, including human tumor cell lines and virally transformed cells, involves an enzyme called telomerase. Telomerase is a special DNA polymerase that elongates telomeres, so that cells can balance incomplete replication with new synthesis of telomeric DNA. In humans, telomeres contain many repeats of the base sequence TTAGGG (T = thymine, A = adenine, G = guanine). Thus, human telomerase adds TTAGGG repeats to the ends of linear DNA. The telomere hypothesis of cellular aging assumes that normal somatic cells lack telomerase and, as a consequence, lose some telomeric DNA at every round of cell division.

Evidence for telomere hypothesis. Studies in simple eukaryotes provided a precedent for cell death associated with gradual telomere loss. Mutants in yeast and protozoa with altered telomere maintenance grow poorly and eventually die. However, the first direct evidence for telomere loss during aging of higher organisms was obtained by analysis of cultured connective tissue cells (fibroblasts) from humans at various ages. In all cases, the amount and length of telomeric DNA decreased as cells doubled in culture. Moreover, fibroblasts from older individuals had shorter telomeres and were unable to divide as many times as cells from younger individuals. Cells from other somatic tissues (such as blood and intestinal

lining) also showed an inverse relationship between telomere length and donor age. In control experiments, nontelomeric sequences did not decrease in size or abundance with replicative age in cell culture, and when cells were maintained for long periods of time without cell division, there was no loss of telomeres. Together, these observations showed that the terminal TTAGGG repeats of human chromosome ends were specifically lost during aging in cell culture in a replication-dependent manner.

When cells become senescent, the number of chromosomal aberrations involving telomere fusions (resulting in dicentric chromosomes) increases dramatically. This increase is exactly what would be predicted if some chromosomes had lost most of their terminal TTAGGG repeats. In this sense, telomere loss is both a mitotic clock and a genetic time bomb: average telomere length reflects the prior replicative history and remaining proliferative potential of a cell, and when the length becomes critically short, it may induce chromosomal abnormalities that lead to cell senescence.

Further support for the telomere hypothesis of aging has come from the observation that telomeres from at least one type of germ-line cell (sperm) are longer than those in somatic cells, and that these telomeres do not decrease in size with donor age. These data suggest that telomerase is active at some stage of gametogenesis so that telomere length is maintained in germ cells between generations, and that it becomes repressed early in embryonic development so that a finite life-span is conferred on somatic cells. As expected, telomerase activity has not been detected in normal somatic cells. The presence of telomerase in tumor cell lines demonstrates that human cells have the genetic information for telomerase, and that reactivation of telomerase during tumorigenesis may be an obligatory event for cell immortalization.

A scheme of the telomere hypothesis for cell aging and immortalization is shown in **Fig. 2**. The hypothesis begins with the assumpiton that telomere length in germ-line cells is maintained by telomerase, but as yet there is no direct experimental evidence. Nor is it known whether telomerase is active continuously in stem cells of early embryonic development and becomes repressed in specialized somatic cells at a later stage, or whether it is inactive at conception and becomes activated only at a later time in germ-line tissue. In somatic cells, telomerase cannot be detected, and telomeres clearly shorten with age. For fibroblasts from human connective tissue, the rate of telomere loss is about 50 base pairs per cell division in cell culture and about 15 base pairs per year in the body. It is assumed that when nearly all of the terminal TTAGGG repeats are lost on one or perhaps a few chromosomes, cells stop dividing (the Hayflick limit). Certain agents such as viral oncogenes (cancer genes) can bypass the Hayflick limit, so that cells can divide further while telomeres become even shorter. Gross chromosomal aberrations, particularly telomere fusions, are prominent at the Hayflick limit; these aberrations may become lethal at a point in the culture called crisis, when most cells die, unless further telomere loss is

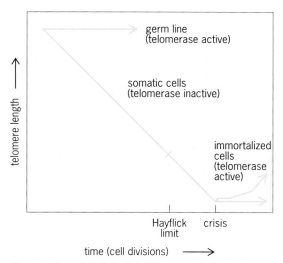

Fig. 2. Schematic of the telomere hypothesis for human cell aging and immortalization. (*After C. B. Harley, Telomere loss: Mitotic clock or genetic time bomb?, Mutat. Res., 256:271–282, 1991*)

prevented. Thus, only immortal cells that have reactivated telomerase survive crisis. In these cells, telomere length can either remain short and stable or increase in length (Fig. 2).

Prospects. Direct evidence that telomeres and telomerase play a causal role in cell senescence and immortalization requires experimental manipulation of telomerase expression. For example, if telomere length could be artificially increased or decreased in normal somatic cells, replicative life-span should be affected accordingly. Alternatively, inactivation of telomerase should convert immortal cells to mortal cells. Thus, antitelomerase drugs might prove effective against cancer cells and some pathogenic microorganisms. Finally, if manipulation of cellular life-span were technically possible through regulation of telomerase, it might be possible to decrease morbidity or increase life-span of the organism.

For background information *SEE CELL SENESCENCE AND DEATH; DEOXYRIBONUCLEIC ACID (DNA); NUCLEIC ACID; ONCOLOGY* in the McGraw-Hill Encyclopedia of Science & Technology.

Calvin B. Harley

Bibliography. S. Goldstein, Replicative senescence: The human fibroblast comes of age, *Science*, 249:1129–1133, 1990; C. B. Harley, A. B. Futcher, and C. W. Greider, Telomeres shorten during aging of human fibroblasts, *Nature*, 345:458–460, 1990; R. K. Moyzis, The human telomere, *Sci. Amer.*, 265(2):44–55, August 1991; A. M. Olovnikov, A theory of marginotomy, *J. Theoret. Biol.*, 41:181–190, 1973.

Cellular immunology

Topics of current research in cellular immunology include adhesion molecules in the immune system and the mechanisms by which Epstein-Barr virus affects B lymphocytes.

Adhesion Molecules

The fundamental task of the immune system is to detect potentially threatening changes in the body and to eliminate the threats before they cause further damage. To carry out this function, a multitude of different cells, such as lymphocytes and monocytes, continuously circulate in the body and search for antigens, substances that signify body damage and injury. In order to mediate an immune response, the various cells in the immune system must physically interact in a carefully orchestrated series of events. Since such physical interaction between cells is an energetically unfavorable event, there are molecules expressed on the surface of immune cells that facilitate it. Often referred to as adhesion molecules, they play an important role in initiating and maintaining intercellular interactions. The identification of lymphocyte adhesion molecules in the early 1980s has led to an explosion of information on the structure, expression, and function of adhesion molecules. These studies have shown that many cell surface molecules mediate adhesion, and that these molecules also transduce intracellular signals on the cell on which they are expressed.

Role in immune system. There are two distinct but functionally related roles that adhesion molecules play in the immune system. The first role involves the intercellular interactions that must occur for the immune system to recognize and respond to a foreign antigen. For instance, a T lymphocyte must interact with an endless array of other cells in order to detect the rare cell that expresses the appropriate combination of foreign antigen and major histocompatibility complex (MHC) protein recognized by the T cell's antigen-specific T-cell receptor (TCR). Other immune responses, such as production of antibody by B lymphocytes, also require cell–cell contact. The second functional role for adhesion molecules is in directing the specific movement or migration of lymphocytes to specific anatomic sites. Lymphocytes are constantly circulating throughout their lifetime, but their circulation is highly specific rather than random. Thus, lymphocytes that have been specifically activated by exposure to foreign antigen at one site, such as in a peripheral lymph node, acquire a tendency to return to that site. The nonrandom migration of lymphocytes is mediated by a diverse and still incompletely defined group of adhesion molecules, termed homing receptors, that mediate adhesion to high endothelial venules, which are specialized endothelial cells found in lymphoid organs.

Identification and analysis. Initial studies on the function of adhesion molecules documented the ability of monoclonal antibodies directed against surface molecules other than the antigen-specific T-cell receptor to inhibit T-cell-mediated killing of a target cell expressing the specific foreign antigen. Subsequent studies demonstrated that T cells could interact with cells that did not express the appropriate foreign antigen, and that this antigen-independent adhesion could be blocked by the same monoclonal antibodies that inhibited a T-cell functional response such as cytotoxicity or proliferation. Various assays have now been developed for assessing the molecules involved in the adhesion between two cell types of interest. Adhesion is mediated by the physical interaction of two molecules on distinct surfaces; thus, every adhesion molecule has at least one corresponding ligand, a molecule to which it binds. Definitive identification of an adhesion molecule on a T cell and its ability to bind to a ligand on another cell requires the physical purification of the adhesion molecule or ligand and the demonstration of specific adhesion of cells to the purified protein.

Families of molecules. The immune system utilizes a large array of cell surface molecules to mediate adhesion. Three protein families play particularly important roles as adhesion molecules. The first includes members of the immunoglobulin superfamily, a large group of molecules with structural homology to the extracellular domains of immunoglobulin. Immunoglobulin superfamily members involved in adhesion include the CD2 molecule and its ligand, leukocyte-function-associated antigen 3 (LFA-3); the LFA-1 ligands, intercellular-adhesion molecule 1 (ICAM-1) and ICAM-2; the very-late-antigen-4 molecule (VLA-4) ligand, vascular cell adhesion molecule (VCAM-1); and the CD28 molecule and its ligand, B7.

Integrins are a second, large family of cell surface proteins that are expressed on many different cells. They were first shown to mediate adhesion to components of the extracellular matrix (ECM), the complex network of macromolecules secreted by cells that fill the extracellular space within tissue. However, integrins can also mediate adhesion to other cells. Furthermore, a bacterial cell surface protein, invasin, binds to several integrins and may be responsible for the ability of these bacteria to invade human cells. In addition to LFA-1 and macrophage antigen 1 (Mac-1), which play major roles in adhesion in the immune system, lymphocytes express members of the VLA integrin subfamily. There is considerable flexibility in ligand recognition within the integrin superfamily. For example, one adhesion molecule can have more than one ligand. Conversely, one ligand can have more than one adhesion molecule.

The LECAMs or selectins represent a third family of adhesion molecules in the immune system. These molecules have an extracellular structure consisting of a lectinlike domain, a domain with homology to epidermal growth factor, and a variable number of complement regulatory protein repeat sequences (**Fig. 1**). The ability of simple polysaccharides to inhibit LECAM-dependent adhesion, coupled with the presence of the lectin domain in the extracellular region, has suggested that these molecules mediate adhesion by binding to carbohydrates. Recent studies have indeed identified putative carbohydrate ligands for each of the three LECAM molecules. The LECAM-1 molecule is expressed on lymphocytes and neutrophils and has been implicated as a peripheral-lymph-node homing receptor. The endothelial–leukocyte adhesion molecule 1 (ELAM-1) is expressed on activated endothelium and mediates the adhesion of neutrophils, monocytes, and a subset of T lymphocytes to activated endothelium. The CD62 molecule (also known as GMP-140 or PADGEM) is expressed on activated platelets and activated endothelial cells. The CD62 molecule appears to be involved in the adhesion of activated platelets to certain leukocytes and may also be involved in cell adhesion to endothelium.

Besides the immunoglobulin, integrin, and LECAM families, other structurally unrelated molecules mediate adhesion in the immune system. One example is the CD44 molecule, which mediates adhesion to

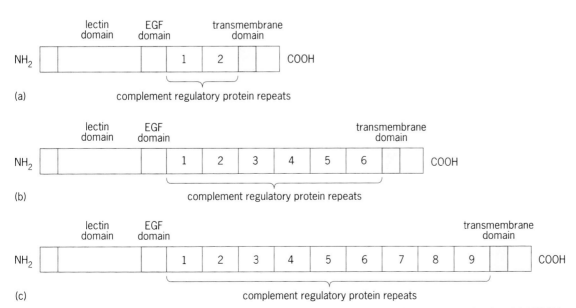

Fig. 1. Schematic diagram of the three members of the LECAM–selectin family of cell adhesion molecules. (*a*) LECAM; (*b*) ELAM-1; (*c*) CD62. The extracellular portion of each molecule consists of a lectin domain, a domain with homology to epidermal growth factor (EGF domain), and a variable number of complement regulatory protein repeats.

the extracellular-matrix glycosaminoglycan hyaluronic acid and has also been implicated as a lymphocyte homing receptor.

The immune system utilizes this array of adhesion molecules in various combinations to mediate important cell–cell and cell–matrix interactions. For example, the interaction of a cytotoxic T cell with a target cell is mediated by the LFA-1 molecule expressed on the T cell binding to its ligand ICAM-1 on the target cell, and the CD2 molecule on the T cell binding to its ligand LFA-3 on the target cell. Lymphocyte adhesion to activated endothelial cells, which is critical to lymphocyte migration and the inflammatory response, involves at a minimum four different adhesion molecule–ligand interactions: the LFA-1 molecule on the T cell interacting with both its ligands ICAM-1 and ICAM-2 on the endothelial cell; the VLA-4 molecule on the T cell interacting with its ligand VCAM-1; and an unidentified carbohydrate molecule on the T cell interacting with its ligand ELAM-1 on the endothelial cell (**Fig. 2**).

Regulation of adhesion. Although lymphocytes generally circulate in a continuous search for antigen, the actual encounter with antigen requires that lymphocytes stop moving in order to carry out a specific functional response. Since lymphocytes must be able to alternate between adhesive and nonadhesive states, the function of adhesion molecules is tightly regulated in various ways. Changes in the activation state or the differentiation state of a cell can result in changes in the level of expression of adhesion molecules. For example, an inflammatory response is marked by an increased expression of ICAM-1, VCAM-1, and ELAM-1 on endothelial cells at the site of inflammation (Fig. 2). This expression results in the increased adhesion of various cell types, such as neutrophils and lymphocytes, to endothelium and a rapid influx of these cells into the site of tissue injury. Changes in adhesion-molecule expression also occur during the process of T-cell differentiation. Previous

encounter with antigen distinguishes two major subsets of peripheral T cells designated as naive and memory T cells; naive T cells have yet to encounter foreign antigen, while memory T cells have been previously activated. A key feature that distinguishes naive and memory T cells is an increased expression on memory cells of a number of adhesion molecules, including LFA-1 and VLA integrins, CD2, and LFA-3. Greater expression of adhesion molecules on memory T cells is associated with greater adhesion of memory cells to other cells and to the extracellular matrix.

The functional activity of adhesion molecules can also be dramatically regulated by activation of the cell. Resting human T cells adhere poorly to integrin ligands such as ICAM-1 and fibronectin. However, cross-linking of the antigen-specific T-cell receptor with antibody, which is thought to mimic the encounter of a T cell with foreign antigen, within minutes results in an increase in integrin-mediated adhesion to ligands such as ICAM-1 (via LFA-1), VCAM-1 (via VLA-4), and fibronectin (via VLA-4 and VLA-5). The exact mechanism by which activation regulates integrin function remains unknown, since no change occurs in the levels of cell surface expression of these molecules in the time during which these changes in adhesion occur. These initial findings have been extended in two important ways. First, activation of other lymphoid cell types, such as B cells and monocytes, also results in rapid changes in LFA-1 function. Second, the adhesive potential of other adhesion molecules, such as LECAM-1, is also regulated by activation. The increase in adhesion as a consequence of activation is thought to result in the arrest of lymphocyte migration when a lymphocyte encounters foreign antigen, and a corresponding increase in the strength of adhesion between a lymphocyte and a cell bearing the foreign antigen.

Transduction of signals. The function of adhesion molecules in the immune system is not limited to mediating adhesion. In fact, interaction of an

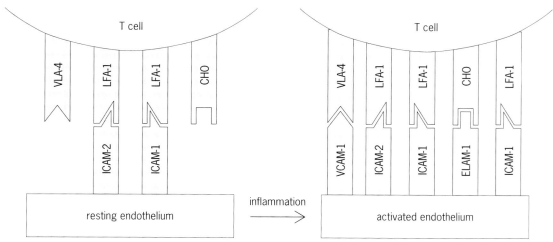

Fig. 2. Diagrammatic representation of the multiple adhesion molecules mediating T-cell adhesion to endothelial cells. Weak T-cell adhesion occurs with resting endothelium, but adhesion is considerably strengthened at a site of inflammation owing to increased expression of ICAM-1 and induced expression of VCAM-1 and ELAM-1 on activated endothelial cells. CHO = carbohydrate.

adhesion molecule with its ligand often results in the generation of intracellular signals that have an important impact on the functional responses of the interacting cells. There are numerous examples of such signal transduction by adhesion molecules. For example, specific pairs of monoclonal antibodies recognizing the CD2 molecule are able to activate human T cells. Engagement of the LFA-3 molecule on human monocytes with its ligand CD2 results in the production of interleukin 1 (IL-1) and tumor necrosis factor, two important cytokines involved in the immune response. Such studies provide evidence that adhesion molecules are multifunctional in nature, and that intracellular signals can be generated in both interacting cells once adhesion has occurred. Conceptually, adhesion molecules can be thought of as environmental sensors for a roving lymphocyte that provide critical information regarding the extracellular environment and the appropriate response that the lymphocyte must generate in that environment.

Clinical relevance. Adhesion molecules play an important role in disease processes. A hallmark of inflammation is an increased expression of adhesion molecules on endothelial cells at the site of injury. Consequently, immune cells in the bloodstream bind strongly to the inflamed endothelium and rush into the surrounding injured tissue. Further understanding of the expression and function of adhesion molecules will be critical in elucidating the pathogenesis of the inflammatory response.

Alterations in adhesion-molecule expression and function can also lead to disease. Leukocyte adhesion deficiency is a genetic disorder in which expression of three integrins, LFA-1, MAC-1, and p150/95, is lacking in affected individuals. Consequently, patients with leukocyte adhesion deficiency suffer from recurrent bacterial infections and impaired wound healing. Alterations in the expression of VLA integrins may play a role in the development of cancer; changes in the adhesion of normal cells to the surrounding extracellular matrix may be a critical step in the development and metastasis of tumor cells.

It may be possible to modulate adhesion-molecule function and manipulate the immune response by using specific drugs or other therapeutic agents. Current research is directed at inhibiting or enhancing the function of specific adhesion molecules to control inflammation, to prevent the rejection of transplanted organs, and to treat autoimmune diseases such as arthritis. Thus, not only is an analysis of adhesion molecules fundamental to the understanding of the normal immune response, but it may also lead to important clinical breakthroughs in the treatment of a variety of disorders.   *Yoji Shimizu*

Epstein-Barr Virus

Epstein-Barr virus is a human herpesvirus that infects the majority of people worldwide. The virus persists for the lifetime of the host, with a stable virus–host balance normally being established. Epstein-Barr virus is renowned for its effects on the antibody-producing cells of the immune system, the B lymphocytes, whose destiny it can fundamentally alter by guaranteeing their survival when they might otherwise be selected to die, and by locking them into a program of continuous cell division. These immortalizing properties of Epstein-Barr virus have led to its acceptance as an oncogenic agent—not surprising in view of its clear association with malignant B-cell tumors, notably African Burkitt lymphoma and lymphomas of immunosuppressed individuals.

Infection of B lymphocytes. Epstein-Barr virus gains access to the B-cell interior through a cell surface receptor, CD21; this receptor is present on the vast majority of mature B cells, not for the convenience of the virus but to act as a receptor for a variety of host regulatory proteins. Once internalized, Epstein-Barr virus exerts its powers of immortalization through the activity of only a small portion of its large deoxyribonucleic acid (DNA) genome. Accordingly, alteration of a B lymphocyte from the resting state to a state of continuous proliferation results from the coordinate action of only 8 of the 50–100 proteins that the virus can encode. These proteins, known as the Epstein-Barr virus latent proteins, comprise the six Epstein-Barr nuclear antigens and the two latent membrane proteins. They take over the normal growth-controlling mechanisms of the B cell, so that there is a state of continuous activation such that, as observed in the laboratory, the cell replicates successively as a so-called lymphoblastoid cell line.

Features of B-cell proliferation. The mechanisms by which the Epstein-Barr virus latent proteins interfere with normal cellular control systems are yet to be determined. However, the individual cells of an Epstein-Barr virus–immortalized lymphoblastoid cell line show many features in common with B cells that are activated in the course of their normal program of development.

B cells that are activated to respond to foreign molecules (antigens) produce a cohort of cells that synthesize highly specific antibodies, and express a variety of molecules at high levels on their surface. Some of these molecules have been well characterized and serve to mediate intercellular adhesion (for example, LFA-1, ICAM-1, and LFA-3) or are involved in B-cell growth-signaling processes (for example, the CD23 molecule). The genes encoding these molecules are activated either de novo or to higher levels of expression only transiently during the normal response of B cells to antigen, but expression of Epstein-Barr virus latent proteins maintains these genes in a state of continued activation.

As in activated B cells, close intercellular contact in lymphoblastoid cell lines is achieved through interaction of the LFA-1 molecule with its major ligand, ICAM-1, on adjacent cells. Cleavage of the CD23 molecule into smaller, soluble fragments provides the lymphoblastoid cell line with potent B-cell growth-factor activity. At least two growth factor/receptor pathways operate in lymphoblastoid cell lines (**Fig. 3**). In the autocrine pathway, individual B cells produce both soluble fragments with growth-factor activity and a specific receptor for the fragment; in the paracrine

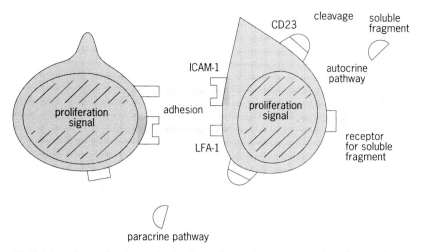

Fig. 3. Intercellular adhesion and autocrine and paracrine growth are key elements in the maintenance of Epstein-Barr virus–induced B-cell proliferation.

pathway, B cells produce soluble fragments to which closely adjacent B cells can respond. Intercellular adhesion facilitates growth factor/receptor interactions between adjacent Epstein-Barr virus–activated B cells. It is likely that the two processes, intercellular adhesion and growth-factor production, are central to the continuity of Epstein-Barr virus–driven cell division observed in lymphoblastoid cell lines.

Programmed cell death. In addition to their role as inducers of continuous proliferation in resting B cells, Epstein-Barr virus latent proteins serve a function that ensures that the virus gains access to the long-lived memory-B-cell pool. In order to reach this pool, B cells must undergo a stage of differentiation in germinal centers of lymphoid tissue (lymph nodes, spleen, and gut-associated lymphoid tissue). Here, the genes that encode the B cells' receptors for antigen, the immunoglobulin molecules that determine the antibody specificity of the cell, are subject to mutations that can change their antigen-binding properties. The purpose of these genetic changes is to provide memory B cells with receptors of improved affinity for their specified antigens, so that subsequent exposure results in improved responsiveness. Successful mutation produces receptors with improved affinity for their antigens. Following interaction with antigen on neighboring antigen-presenting cells (follicular dendritic cells), the B cells with successful mutations are given a survival signal. However, if the resulting mutant receptors do not bind antigen efficiently, the cell receives no survival signal and enters a predetermined program of self-destruction, known as apoptosis. Probably, Epstein-Barr virus provides infected cells with protection from apoptosis by overriding the normal process of antigen selection, so that the need for successful receptor gene mutations is obviated.

Programmed cell death, which is an active form of cell suicide, must be distinguished from death resulting from accidental injury. The latter process, known as necrosis, is caused by noxious stimuli and environmental extremes that result in irreversible cell-membrane damage, leading to swelling and bursting of the cell.

The degradative contents that are spilled by the cell in turn lead to inflammation in the surrounding tissue. By contrast, apoptosis is controlled from within the cell and is characterized by cell shrinkage, condensation of chromatin, and activation of an enzyme capable of cleaving the DNA of the cell into regularly sized fragments. Cells undergoing apoptosis are rapidly recognized and phagocytosed by tissue macrophages; thus inflammation is prevented. Programmed cell death is a widespread phenomenon, occurring, for example, during the development of embryonic limbs and organs as well as in the control of white blood cell output.

The key to the question of how Epstein-Barr virus protects B cells from death by apoptosis lies with the latent membrane protein, LMP-1, and its link with a cellular gene, *bcl*-2. The *bcl*-2 gene was originally discovered through its association with a chromosomal translocation present in one of the most common human lymphomas, follicular lymphoma. Introduction and expression of the *bcl*-2 gene in a variety of cell types has shown that its product serves to enhance cell survival by blocking programmed cell death, although the gene is probably not the sole mediator of this function. Epstein-Barr virus takes advantage of this property by using its LMP-1 gene to activate *bcl*-2, so that apoptosis in susceptible cells is prevented. Therefore, the virus harnesses the power of a cellular gene by again using a normal cellular regulator mechanism to its own advantage.

B-cell malignancy. Entry into the long-lived memory-B-cell pool is probably an important strategy for persistence of Epstein-Barr virus. The oncogenic potential of the virus in B cells leads to many questions relating both to the mechanisms controlling infection and persistence and to the role it plays in the evolution of B-cell malignancies that are positive for Epstein-Barr virus. In the healthy individual, proliferation of B cells infected with Epstein-Barr virus is probably controlled by virus-specific cytotoxic T lymphocytes that can be identified in all virus carriers. Such cytotoxic T lymphocytes recognize peptides from Epstein-Barr virus latent proteins expressed by lymphoblastoid cell line, and are further facilitated in recognizing infected B cells by the Epstein-Barr virus–induced up-regulation of intercellular adhesion molecules on the B-cell targets. The emergence of malignant B cells with a striking resemblance to lymphoblastoid cell lines in patients on immunosuppressive therapy bears testimony to this control by cytotoxic T lymphocytes. Furthermore, the regression of the tumors upon withdrawal of immunosuppression adds weight to the importance of cytotoxic T lymphocytes in the prevention of Epstein-Barr virus–infected B-cell proliferation in the body. In the immunocompromised environment, therefore, the virus probably plays a direct role in the malignant process. By contrast, the Epstein-Barr virus–positive malignant B cells of African Burkitt lymphoma emerge in the face of active responses of cytotoxic T lymphocytes. In these cells, however, the interaction with the virus is different from that of lymphoblastoid cell lines in that Epstein-Barr virus–latent protein expression is restricted to only one of the nuclear antigens. Other factors are known to contribute to the evolution of Burkitt lymphoma.

For background information SEE CELLULAR ADHE-SION; CELLULAR IMMUNOLOGY; EPSTEIN-BARR VIRUS; IM-MUNOGLOBULIN; LYMPHOMA in the McGraw-Hill Encyclopedia of Science & Technology.

Christopher D. Gregory

Bibliography. S. M. Albelda and C. A. Buck, Integrins and other cell adhesion molecules, *FASEB J.*, 4:2868–2880, 1990; B. K. Brandley, S. J. Swiedler, and P. W. Robbins, Carbohydrate ligands of the LEC cell adhesion molecules, *Cell*, 63:861–863, 1990; J. J. Cohen, Programmed cell death in the immune system, *Advances in Immunology*, 50:55–85, 1991; M. L. Dustin and T. A. Springer, T-cell receptor crosslinking transiently stimulates adhesiveness through LFA-1, *Nature*, 341:619–624, 1989; J. Gordon and J. A. Cairns, Autocrine regulation of normal and malignant B lymphocytes, *Adv. Canc. Res.*, 56:313–334, 1991; I. C. M. MacLennan and D. Gray, Antigen-driven selection of virgin and memory B cells, *Immunol. Rev.*, 91:61–85, 1986; L. Osborn, Leukocyte adhesion to endothelium in inflammation, *Cell*, 62:3–6, 1990; A. B. Rickinson and C. D. Gregory, Immunology of Epstein-Barr virus infection, in P. J. Lachman et al. (eds.), *Clinical Aspects of Immunology*, 5th ed., 1992; Y. Shimizu and S. Shaw, Lymphocyte interactions with extracellular matrix, *FASEB J.*, 5:2292-2299, 1991; T. A. Springer, Adhesion receptors of the immune system, *Nature*, 346:425–434, 1990.

Chaos

Recent developments involving chaos theory include applications to the fields of meteorology and climatology and of biopsychology.

Atmospheric Chaos

Weather may be defined as the complete state of the Earth's atmosphere at a particular instant in time. The weather patterns in the atmosphere change constantly and, within the scope of current observations, have never exactly repeated themselves. Although the weather patterns are observed to be similar from time to time, all of these approximate repetitions have been of brief duration. This lack of regularly repeated behavior indicates that the atmosphere is in a chaotic state. An important implication of this atmospheric chaos is that the weather and climate are unpredictable in detail into the distant future.

Aperiodicity and unpredictability. Any meteorological time series, such as the temperature at a particular place or an average for a particular region, can be expressed as the sum of periodic and aperiodic parts. The periodic components include the annual and diurnal cycles induced by direct solar forcing and are characterized by regularly repeated behavior. However, even if all of the known or suspected periodic components are removed from a time series, a very large aperiodic component will remain. For example, migratory cyclones do not pass over a given location at regular intervals. This aperiodic behavior is a characteristic of all chaotic systems. Without the presence of this aperiodicity, weather prediction would be trivial. Simple extrapolation into the future of the periodic components of the weather would suffice.

Although aperiodicity is an excellent indicator of unpredictability, it is not the cause. Unpredictability is caused by instability with respect to small perturbations. Mathematically, this sensitive dependence on initial conditions means that the solution to a set of equations can change dramatically in response to very small changes in the initial conditions. As a result, two or more nearly identical initial states, governed by the same laws of motion, will eventually evolve into widely differing states with the advance of time.

Predictability of weather. In principle, weather prediction is a mathematical initial-value problem. Given the present state of the atmosphere (the current weather), and the set of equations that embody the physical laws governing the atmosphere's motion, it would seem possible to solve the equations subject to the initial conditions in order to determine the state of the atmosphere at any future time.

In the early 1960s, E. N. Lorenz performed experiments with a very simple numerical model of the atmosphere's general circulation. As he repeated a portion of one experiment, he noticed that the solution in his second run was quite different from the original solution (**Fig. 1**). Although the solutions agreed initially, after a period of time they bore no resemblance. In his second simulation, Lorenz used as initial conditions the solution of the model at a particular time accurate to only three decimal places. The very small differences between the assumed initial condition and the "true" initial condition ultimately led to the divergence of the solutions.

This simple model can be used as an analogy to the much more complex numerical models used in weather prediction. The initial conditions used in these models are determined from measurements taken at various sites over the Earth by weather balloons and satellites; such atmospheric parameters as wind speed and direction, temperature, and humidity are included. These measurements contain average errors of roughly a degree in temperature, a few meters per second in wind, and 10% in relative humidity. The small-scale motions where these initial errors are reflected are dynamically coupled to the large-scale motion. Eventually, the small-scale errors induce errors in the large scales of motion through nonlinear interaction, rendering the predicted state of the atmosphere as different from the actual state as a randomly chosen state would be. Uncertainty in the initial conditions due to measurement errors causes the model solution (even if the model were perfect) to diverge from the time behavior of the atmosphere itself.

Generally, the shortest range of predictability occurs for small-scale motions, and the largest predictability exists for the largest-scale motions (**Fig. 2**). For example, the predictability limit of thunderstorms, whose spatial scale is on the order of tens of kilometers, may be measured in hours. At midlatitude, migratory

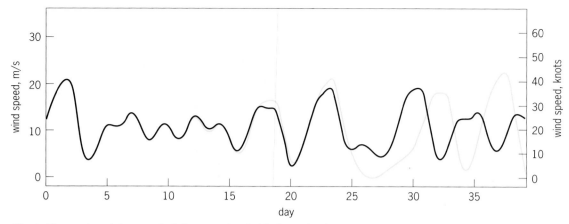

Fig. 1. Time series of the speed of the westerly wind in Lorenz's simple general-circulation model. Two solutions with very slightly different initial conditions are nearly identical for 2 weeks but then begin to diverge dramatically. (*After R. Monastersky, Forecasting into chaos, Sci. News, 137(18):280–283, 1990*)

cyclonic storms with spatial scales on the order of a few thousand kilometers may be predictable a few days ahead, and the predictability limit for the features comparable in size to the continents and oceans in midlatitudes (the planetary-scale waves) is roughly 2 weeks. These predictability limits were obtained from simple mathematical models. Though crude in many respects, these models contain good representations of the physical process that is the principal source of the unpredictability, namely the transport of atmospheric properties (such as heat, momentum, and moisture) by the atmosphere itself. This transport is represented by quadratic nonlinearities in the atmosphere's equations of motion. Additional research with more sophisticated models has supported these results.

Efforts to reduce the magnitude of errors in the initial conditions have led to more accurate weather predictions. However, even if the measurements that make up the initial conditions of the prediction models are reduced to their smallest conceivable values, the fundamental predictability limits dictated by the chaotic nature of the atmospheric circulation cannot be overcome. Unavoidable errors in the initial conditions, no matter how small, lead to uncertainty about the present state of the atmosphere. Since the exact present state of the atmosphere cannot be determined from the multiplicity of nearly identical possible states, there is no basis for judging which of the myriad of widely differing possible states might occur at some sufficiently distant point in the future.

Climatic predictability. The climate of the Earth may be defined by an average over many instances of the weather. (This average should also include the state of the world's oceans, ice sheets, and land surfaces.) Ideally, the average should be taken over an infinite time span so that all possible states of the atmosphere are included. However, such an average is impossible to attain; therefore, climatological statistics must be taken over long but finite time intervals, typically 30 years. The question of climatic prediction arises because statistics taken over such time intervals are found to differ considerably from those from subsequent intervals. For example, within recorded history it is known that regions that were once characterized by agriculture for periods of centuries are now deserts. Climatic prediction involves determining how these climatic statistics change as the beginning and ending of the averaging time-span advance.

The theory of chaos, which indicates that the day-to-day variability of the weather cannot be predicted in detail a month in advance, much less a century in advance, also indicates that the climate cannot be predicted indefinitely into the future. Even if all the relevant physical processes governing the evolution of the climate were known, the present condition of the climate, including the atmosphere, oceans, cryosphere, and lithosphere, cannot be determined without some error. The error in these initial conditions will lead inevitably to a loss of predictability after some finite time. However, there may not be a uniform limit on the range of climatic predictions. For example, annual

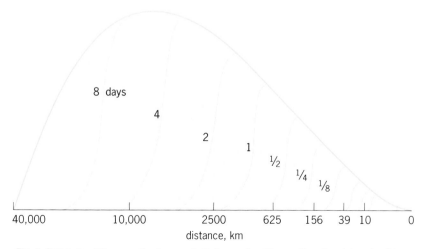

Fig. 2. Estimate of the growth of errors initially confined to small scales determined from a theoretical model. The set of curves indicates the time taken for the initial errors to propagate into the large-scale motion. The upper curve is the assumed atmospheric motion spectrum; the lower (horizontal) curve is the zero line. The vertical curves connecting these are spectra of errors at the indicated times, and each curve coincides with the lower curve to the left and the upper curve to the right. Areas under the curves are proportional to kinetic energy. 1 km = 0.6 mi. (*After E. N. Lorenz, Estimates of atmospheric predictability at medium range, Predictability of Fluid Motions, AIP Proc. 106, 1984*)

means may be predictable 2 years in advance but not 10. Ten-year means may be predictable 20 years in advance but not 50, and so forth, regardless of the lack of periodicity in the system.

The time limits on predictability may vary for different parts of the climate system. The possibility for climatic prediction resides in the more slowly evolving parts of the system. For example, the sea surface temperature certainly changes more slowly than the weather. The pattern of the sea surface temperature is influenced by the weather, and in turn it influences the weather. If the ensemble of weather patterns associated with one set of sea surface temperatures differs significantly from an ensemble associated with a different set, it may be possible to predict the time-averaged weather pattern up to the range at which the sea-surface-temperature pattern itself is predictable. Intensive current research is focused on whether the variations in tropical Pacific sea surface temperatures associated with the El Niño/Southern Oscillation phenomenon may lead to predictability of monthly or seasonal mean weather patterns.

Another potential source of weather or climate predictability could arise if some portion of the atmosphere or climate system adopts a configuration that is persistent in time and that eventually changes to a configuration that is distinguishable from its predecessor. An example may be the alternation between glacial and interglacial epochs in the past million years of the Earth's climate. In weather, an example may be the alternation between the strong midtropospheric westerly flow and the so-called blocked flow patterns, in which the westerly flow is impeded by an amplified high-pressure ridge. If a set of such configurations or regimes exists in the atmosphere or climate system, the range of predictability may be increased by the higher probability that the coming week, month, year, or century would depart from normal in a known way depending on the initial climatic regime.

Anthony R. Hansen

Chaos Theory in Biopsychology

Psychological sciences are undergoing a major transformation, due in part to the contribution of chaos (dynamical) theory. This revolution is changing the way that theories are being proposed and the design and analysis of the experiments by which they are tested. It is anticipated that soon all psychological theories will be dynamical.

Every facet of psychology deals with the complexity of the interaction of features evolving over time. Dynamical systems theory provides the mathematical tools for theory and research design and analysis of such complexity. In this complexity, parameters will usually be such that chaos prevails, and the chaos will represent normal healthy functioning, not just abnormal conditions. An understanding of bifurcation theory and chaos theory will help in the improvement of personal and social systems.

Dynamics and chaos. Strictly speaking, chaos theory is a branch of dynamical systems theory, a field of mathematics. However, the term chaos is often used to refer to the whole of dynamical systems theory, probably to emphasize the accessibility and generality of chaos theory compared to earlier forms of systems theory, and its important role in current developments in science and in understanding the ubiquity and nature of chaotic features of normal processes.

A dynamical system is a set of interacting variables. Dynamical systems theory examines how those variables change together over time. The rules governing these changes are usually expressed as differential or difference equations. These equations describe a vectorfield of forces that act upon each state of the system. Their solution by integration yields a unique trajectory (path of successive states) for each starting point in the state space. The collection of all the trajectories is known as a phase portrait. Advantages lie in (1) the ability to deal with the complexity of interacting

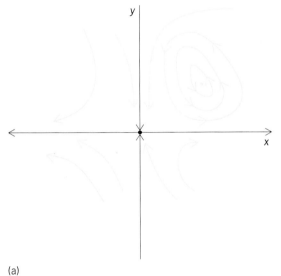

(a)

(b)

Fig. 3. Phase portraits. (*a*) Two interacting responses whether from two people (a mother's tension and a son's anger), one person (singing and eating), or two neurons (firing rates). (*b*) A three-dimensional chaotic attractor for the hypothalamo-pituitary-gonadal neuroendocrine system with gonadotropic releasing hormone (GnRH), luteinizing hormone, and estrogen (EST) is shown in the GnRH-EST plane; omitted is the third variable, luteinizing hormone (LH), which can be considered perpendicular to this plane. A bifurcation from point to periodic attractor occurs at puberty; chaos may occur with greater GnRH output sensitivity to EST. (*After F. D. Abraham, R. H. Abraham, and C. D. Shaw, A Visual Introduction to Dynamical Systems Theory for Psychology, Aerial Press, 1990*)

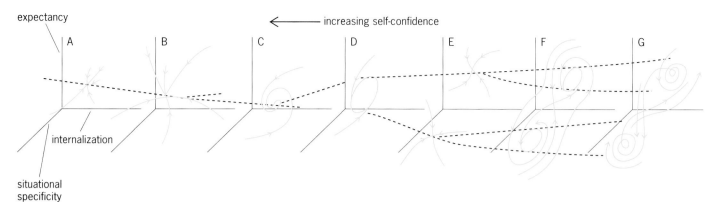

expectancy

←———— increasing self-confidence

internalization

situational
specificity

Fig. 4. Response diagram showing a simplified attribution model of expectancy, situational specificity, and internalization with a very conjectural and atypical bifurcation sequence. The control parameter is self-confidence. The labels on the axes for A apply to all sets of axes. A, B = radial-point attractors; C = focal-point attractor; D = cyclic attractor; E = two radial-point attractors; F = two cyclic attractors; G = chaotic attractors.

variables, (2) the parsimonious explanation of bifurcations (transformations from one phase portrait to one quite different even though the equations remain the same, while only the value of some parameter changes by only a very small quantity), (3) the capability of extrapolation to more complex networks of systems (when control parameters may be under the control of other systems), (4) the incorporation of self-organization and development, and (5) the intuitive ease of geometric representation.

Features of phase portraits include local limit sets such as repellors (trajectories depart), attractors (trajectories approach), and saddles (some trajectories approach, some depart). Basins of trajectories headed for different attractors in the state space are separated by trajectories (also known as actual separatrices) that do not approach any attractor. Attractors may be fixed points (homeostatic), periodic, or neither, in which case they are called chaotic. The latter term covers everything from nearly periodic to very random.

Phase portrait. Typical systems include a family system involving a mother's tension (x) and a son's anger (y); or an animal with two competing operant responses (x,y are measures of their strength); or two neurons reciprocally innervating each other, one normally quiet with an excitatory input from the other, the other normally active but with an inhibitory input from the former (x and y are their firing rates). For each of these three synergistic/competitive systems, the tendency of each variable to change depends not only on its present value but also on the value of the other variable. One such model is shown by the mass action equations (1) and (2), where $a, b, c,$ and d

$$dx/dt = ax - bxy \qquad (1)$$
$$dy/dt = cxy - dy \qquad (2)$$

are constants. Each equation depends upon the cross-product of both variables, a nonlinearity that is necessary in order for bifurcations to occur.

The phase portrait in **Fig.** 3*a* indicates that the mother's behavior and son's behavior cycle about, with the mother's tension triggering the son's anger, which symbiotically diverts attention from and re-

lieves the mother's tension; that two responses such as singing and eating alternately wax and wane; and that two neurons' activity levels are similarly reciprocally periodic. This portrait has a center in the positive quadrant, and a saddle at the origin. A center in a phase portrait is a nest of periodic trajectories, each cycling endlessly through its starting point. The center is actually quite unstable and occurs right at the bifurcation point, with a point attractor and a periodic attractor to either side of the bifurcation point. Such theoretical systems can represent competitive and synergistic neural, behavioral, and societal behavior.

Behavioral, mental, neural, physiological, and environmental factors continually interact. For example, mood, emotions, and reproductive behavior interact with social environment and with neuroendocrine feedback loops involving hypothalamic gonadotropic releasing hormone (GnRH), pituitary luteinizing hormone (LH), and gonadal hormone (estrogen and testosterone). There is usually a painful Hoph bifurcation from a point attractor to a circalunar (almost monthly) periodic attractor at puberty (in some cultures the age of this bifurcation may be affected by nutrition and social work habits). Bifurcations to other phase portraits may occur, such as to a chaotic attractor (Fig. 3*b*). Several abnormal conditions, such as infertility from hypothalamic amenorrhea (low GnRH), have been corrected by restoring pulsatile hormonal activity. A chaotic attractor does not imply abnormality.

Response diagram. There are several dynamical systems whose phase portraits have two attractive areas. It is tempting to use them as starting points to model psychological systems that may have distinctly different attractive areas, such as manic-depressive conditions, and social systems that involve strong ideological conflict. Sometimes these systems have a single cyclic or chaotic attractive region; and then some control parameter representing stress on the system is responsible for a bifurcation to a phase portrait with dual attractive regions, where some trajectories may end up in a normal healthy area and some may end up in a very undesirable area of the state (psychological) space. One example can be taken from cognitive attri-

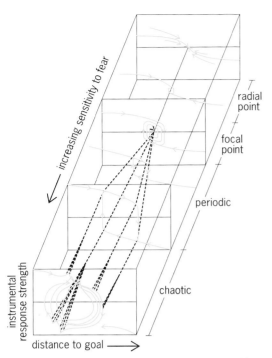

Fig. 5. Response diagram showing the approach–avoidance conflict. The vertical dimension, instrumental response strength, is approach to the goal when above the midline, and avoidance when below it. The nonlinearity creating the hysteresis effect is due to a lag in the stimulus sampling procedure that is direction- and rate-sensitive. The sequence proceeds from a point, to a periodic, to a chaotic attractor as the control parameter, sensitivity to fear, increases. (*After F. D. Abraham, R. H. Abraham, and C. D. Shaw, A Visual Introduction to Dynamical Systems Theory for Psychology, Aerial Press, 1990*)

bution theory, a major area of personality theory. It is necessary to reduce the rich multivariate space of this framework here for simple three-dimensional viewing to expectancy of success, situational specificity, and the extent of internalization of the locus of control over a given outcome. The speculation in this case involves a particular sequence of phase portraits as some control parameter, such as self-confidence, changes. The response diagram reveals this bifurcation sequence (**Fig. 4**). Bifurcations are of three types. Subtle bifurcations are those where an attractor changes type, as with the Hoph bifurcation from a point to a periodic attractor (C to D in Fig. 4). Catastrophic bifurcations

are those in which an attractor appears or disappears, as with the appearance of a second point attractor (D to E). An explosive bifurcation, not shown, is a sudden change in the magnitude of an attractor.

Approach–avoidance conflict provides another example. This involves the interaction of conditioned appetitive and aversive gradients for some goal. Miller's original models assumed that these gradients were linear functions of the distances from the goal; and that instrumental approach–avoidance tendencies were the additive result of these gradients, each of which had different slopes and strengths. In addition, they assumed that under typical conditions the gradients added to zero at some distance from the goal where the gradients crossed, and the animal was rendered immobilized at that point. However, some observations indicated that the behavior could be nonlinear. If there are nonlinear effects in stimulus sampling as the animal moves in this space, and if this cognitive distortion is sensitive to the direction and rate of movement, then this nonlinear generalization could include bifurcations. These bifurcations could be modeled by any number of known equation systems. The Van der Pol system of equations has been suggested, using elements of both statistical learning theory and drive reduction learning theory. The response diagram (**Fig. 5**) for the bifurcation sequence might have a control parameter such as sensitivity to fear.

Self-organization and networks. The original linear approach–avoidance model was generalized to social psychological systems in which the gradients became a function of other individuals' behavior. A similar extension of the nonlinear model can be made. If Jack and Jill went to the playground where they had had many pleasant experiences but were once frightened by a bully, each would have a specific response diagram, as in Fig. 5. If they go to the playground again with their friend, Miss Piggy, who is fearless, the behavior of any one of them might influence the sensitivity-to-fearfulness parameter of the other two (**Fig. 6a**). The relative strengths of these couplings might give rise to different social and individual dynamics. If Miss Piggy was tough and confident and strongly influenced one or both of the others, there might be bifurcations to a point attractor located at the playground itself; that is, all three would end up back at the playground instead of the road-

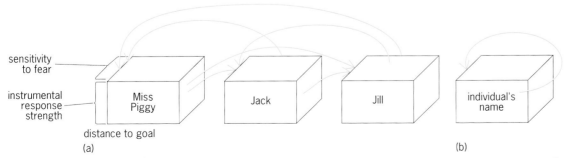

Fig. 6. Networks and self-organization. (*a*) Network of three individuals, each having a specific dynamical system of approach–avoidance conflict as shown in Fig. 5. (*b*) A diagram showing that any given individual's control parameters may be directly under the influence of the individual's dynamic system, which is direct self-control or self-organization.

side between home and playground. If all six lines of influence are nonzero, there is self-organization. Self-organization occurs when the control parameter is under the influence of the same system. Learning to discriminate these control parameters and their influence can be made part of such a complex system. Here learning is mediated by feedback loops. Miss Piggy exerts influence on Jack and Jill, who influence each other and Miss Piggy, so the changes she exerts on them come back to change her own behavior. She may display even more bravado or less, depending on all these strengths and the nature of each coupling. There can be self-control feedback without loops mediated by the participation of other individuals. Jill may learn to control her own fear, or a person may learn to control his or her life's control parameters without the use of a therapist or support group (Fig. 6*b*).

A final note might be made about growth and irreversibility. An individual or society usually undergoes irreversible growth and individuation or complexification. This process may be viewed as a progressive bifurcation sequence of a complex dynamical system along some set of control parameters. Maturity of an individual or social system implies ability to plan for the future along with an ability to discriminate the control parameters, the setting of growth goals, and the exercise of self-control (self-organization).

For background information *SEE ATMOSPHERIC GENERAL CIRCULATION; CATASTROPHE THEORY; CHAOS; PERIOD DOUBLING; SIMULATION; WEATHER FORECASTING AND PREDICTION* in the McGraw-Hill Encyclopedia of Science & Technology.

Frederick David Abraham

Bibliography. F. D. Abraham, R. H. Abraham, and C. D. Shaw, *A Visual Introduction to Dynamical Systems Theory for Psychology*, 1990; R. H. Abraham and C. D. Shaw, *The Dynamics of Behavior*, part I: *Periodic Behavior*, part II: *Chaotic Behavior*, part III: *Global Behavior*, part IV: *Bifurcation Behavior*, 1982–1988; E. Basar (ed.), *Chaos in Brain Function*, 1990; J. Gleick, *Chaos: Making a New Science*, 1987; R. Levine and H. Fitzgerald (eds.), *Analysis of Dynamic Psychological Systems*, 1992; E. N. Lorenz, Atmospheric predictability experiments with a large numerical model, *Tellus*, 34:505–513, 1982; E. N. Lorenz, Deterministic nonperiodic flow, *J. Atm. Sci.*, 20:130–141, 1963; E. N. Lorenz, Irregularity: A fundamental property of the atmosphere, *Tellus*, 36A:98–110, 1984; A. Skarda and W. J. Freeman, How brains make chaos in order to make sense of the world, and commentary by others, *Behav. Brain Sci.*, 10:161–195, 1987; I. Stewart, *Does God Play Dice? The Mathematics of Chaos*, 1988; J. M. T. Thompson and H. B. Stewart, *Nonlinear Dynamics and Chaos*, 1986.

Climatology

Methane (CH_4) is an important trace gas in the atmosphere. It controls various chemical processes and species in both the troposphere and the stratosphere. It is a strong infrared absorber (greenhouse gas), as

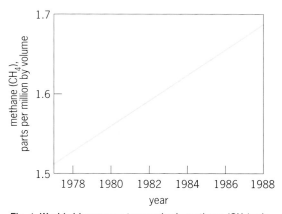

Fig. 1. Worldwide average tropospheric methane (CH_4) mixing ratio versus time. The apparent rate of increase is about 1% per year. (*After D. R. Blake and F. S. Rowland, Continuing worldwide increase in tropospheric methane, Science, 239:1129–1131, 1988*)

are carbon dioxide (CO_2), nitrous oxide (N_2O), ozone (O_3), and the chlorofluorocarbon compounds; it affects atmospheric temperature directly and indirectly. Methane influences tropospheric and stratospheric levels of ozone, and it is a major source for stratospheric water vapor. The reconstruction of atmospheric concentrations of CH_4 through the last 160,000 years from ice cores, and the observation of the present-day rise in concentration, have stimulated research into the global budget of this trace gas, its sources, and sinks.

Atmospheric methane concentration. The modern worldwide average tropospheric mixing ratio is about 1730 parts per billion by volume (ppbv; end of 1990). There has been a steady linear increase of the concentration of CH_4 at a rate of about 16 ppbv per year, or a concentration change of 1% per year, since 1978 when comprehensive atmospheric measurements were begun (**Fig. 1**). This percent increase, considerably larger than that for CO_2, is believed to be caused by anthropogenic influence. Global measurements show an interhemispheric concentration difference of about 90 ppbv, higher in the Northern than the Southern Hemisphere. The indication is that the strengths of Northern Hemisphere sources are larger than those of the Southern Hemisphere.

The relative importance of CH_4 as a greenhouse gas, as compared to the other radiatively important trace gases, is explored in climate models. The expected global surface-temperature warming from doubling today's atmospheric CH_4 concentration is about 15–25% of that warming expected from doubling the concentration of all the radiatively important species (CO_2, N_2O, CH_4, O_3, and chlorofluorocarbon compounds). The latter increase is estimated to range between about 2 and 4°C (3.5 and 7.2°F).

The history of the atmospheric CH_4 concentration has been reconstructed from measurements of air occluded in bubbles from deep ice cores, which are thought to contain air samples from the past atmosphere. Large natural and human-made variations in concentration have been observed. These ice cores, retrieved from both Antarctica and Greenland, cover

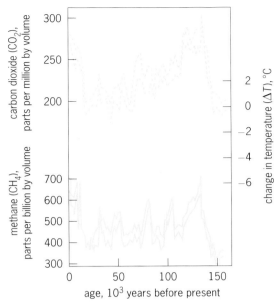

Fig. 2. Mixing ratios of carbon dioxide (CO_2) and methane (CH_4) in air bubbles from the 2083-m (6832-ft) Vostok (Antarctica) ice core versus years before present. Top curves: observed range for CO_2; bottom curves: observed range for CH_4; middle curve: temperature difference (°C) from current air temperature over Antarctica as inferred from deuterium/hydrogen (D/H) measurements in ice. (*After C. Lorius et al., The ice core record: Climate sensitivity and future greenhouse warming, Nature, 347:139–145, 1990*)

the period of the last climatic cycle and extend into the penultimate glacial time. During the last glacial epoch [about 110,000–20,000 years before present (BP)] and at the end of the previous glacial epoch (about 160,000 years BP), the atmospheric CH_4 concentration was between 350 and 400 ppbv (**Fig. 2**). In the present warm period, during the period from about 15,000 years BP until preindustrial times and also during the previous interglacial epoch between about 120,000 and 140,000 years BP, the atmospheric CH_4 concentration was higher by about a factor of 2, between 650 and 700 ppbv.

Shorter ice cores from regions with high snow accumulations have revealed the atmospheric CH_4 concentration record for the last two centuries (**Fig. 3**). It was found that during 150–200 years BP the atmospheric concentration started to rise progressively from the previous 650–700 ppbv to the level of today, a concentration increase by a factor of about 2.5. Thus the atmospheric CH_4 concentration over the last 160,000 years has never been higher than it is today.

The mixing ratios of both CO_2 and CH_4 during the climatic cycle are correlated with temperature, which is derived from the isotopic ratio of deuterium (D) to hydrogen (H) in the ice (Fig. 2). Climatic change is thought to be triggered by periodic changes in the orbital parameters of the Earth (Milankovitch theory). These changes are not large enough to fully explain the observed temperature changes. Analysis of the Vostok ice core data shows that 40–65% of climate forcing is due to changes in the concentration of the radiatively important trace gases CO_2 and CH_4.

The Vostok ice core record (Fig. 2) illustrates the

basic differences between the global CO_2 cycle and the global CH_4 cycle. The atmospheric CH_4 concentration seems to follow more closely the temperature variations throughout the glacial cycle and within the last glacial epoch than does CO_2, in both amplitude and time. Atmospheric CO_2 exchanges with the ocean and the biosphere (both large reservoirs of carbon), while CH_4 does not. Thus the atmospheric CH_4 concentration is more directly influenced by climatic change. In preindustrial times, including the glacial epoch, natural wetlands were the main sources of CH_4, and their extent and production are influenced by changing precipitation.

Sources of methane. There are various sources for atmospheric CH_4, of both biogenic and abiogenic origin. Biogenic CH_4 is produced by bacteria (methanogens) under anaerobic conditions via two major pathways: bacterial production of CH_4 from acetate fermentation, and reduction of CO_2 by hydrogen (H_2), which is also of bacterial origin. The necessary anaerobic conditions prevail in inundated environments, such as wetlands, swamps, peat bogs and fens, and tundra, where methane is produced in substantial amounts. Rice paddies, which are inundated during part of the growing season, are another source of CH_4. Methane produced bacterially from these environments originates mainly from the acetate fermentation pathway. It involves major parts of the carbon cycle, uptake of atmospheric CO_2, decomposition of organic carbon, and bacterial formation of CH_4. Methane escapes from these environments by diffusion or by ebullition of bubbles. Methane is also produced from landfills, which constitute a smaller source. Another important source for CH_4 is ruminants. In the rumen, a complex foregut, food is fermented by symbiotic microorganisms. Gases produced in this process are CO_2 and CH_4. Most of the CH_4 is formed from reduction of CO_2 by H_2 and is emitted through belching. A typical 500-kg (1100-lb) domestic cow produces about 200 liters (7 ft^3) of CH_4 per day. Termites also produce CH_4, but their role as a source is still debated.

Abiogenic CH_4 originates from fossil sources and biomass burning. Fossil CH_4 occurs from coal mining

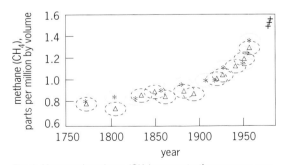

Fig. 3. Measured methane (CH_4) concentrations versus age, from air bubbles occluded in an ice core from Siple Station, Antarctica. The results are obtained from melt extraction (⋆) and from dry extraction (△); ellipses represent experimental uncertainties in the measurements and in the gas enclosure process. The most recent data (+) indicate direct measurements in the atmosphere. (*After B. Stauffer et al., Increase of atmospheric methane recorded in Antarctic ice core, Science, 229:1386–1388, 1985*)

and from losses during drilling and distribution of natural gas. Methane from biomass burning is due to incomplete combustion.

Sinks. The major sink for tropospheric CH_4 is a series of destruction reactions in which CH_4 is ultimately oxidized to CO_2. The oxidation process starts with the reaction of CH_4 with the hydroxyl (OH) radical. The lifetime of atmospheric CH_4 is 8–12 years. In the lower stratosphere also, its oxidation is initiated by OH; at higher altitudes, reactions with excited oxygen atoms (O^1D) and with chlorine (Cl) are important. A minor sink is consumption of CH_4 by bacteria on dry soils.

It has been suggested that part of the observed increase in the atmospheric mixing ratio of CH_4 since preindustrial times might be caused by a decrease in the atmospheric mixing ratio of the OH radical. Anthropogenic addition of CH_4 and of CO, which is also oxidized by OH, may have decreased the mixing ratio of OH by as much as 30%, so that the oxidative capacity of the atmosphere is reduced.

Methane production rate estimates. The annual production rate of CH_4 and the magnitude of the individual source terms are not well known. Various estimates have come from considerations of atmospheric chemistry, flux measurements, biostatistical surveys, modeling of atmospheric concentration data, and investigations of the isotopic composition of CH_4 [the ratios of carbon-13 to carbon-12 ($^{13}C/^{12}C$) and deuterium to hydrogen (D/H) and the carbon-14 (^{14}C) content].

The steady-state sink for the destruction reaction for CH_4 is about 500 teragrams per year. The atmospheric inventory of CH_4 is about 4900 Tg; thus with a 1% per year increase, the annual input is about 550 Tg per year, with a range between 400 and 640 Tg per year.

Numerous flux measurements for emanation of CH_4 from wet, inundated ecosystems, such as wetlands, swamps, peat bogs, tundra, and rice cultivation, have been made; and they have been used, together with the areal extent of these systems, to extrapolate to global source strengths. The observed fluxes vary over orders of magnitude; they are influenced by many factors, such as temperature, water content, length of growing season, and extent of fertilization, so that global extrapolations are uncertain. The contributions by domestic and wild ruminants were estimated, considering

differences in diet type, food amounts, and age; however, uncertainties in the strength of this source remain. It is estimated that the amount of CH_4 produced by domestic ruminants has increased by about a factor of 4 since 1890. Natural wetlands, rice production, and ruminants are the three major biogenic sources of methane. Each source contributes about 15–20% to the global annual production (see **table**). Smaller contributions, about 10% from each, come from landfills (biogenic) and from biomass burning (abiogenic). The remainder (about 20%) is essentially fossil CH_4. Overall, about 60–80% of the modern global annual production of CH_4 is of biogenic origin, and about 40–70% of the modern global annual production is influenced by human activities.

Additional information on the relative strengths of different CH_4 sources comes from the analysis of the stable carbon and hydrogen isotopes, and of the radionuclide ^{14}C in methane. It has been found that the isotopic composition of CH_4 from each different source is rather distinct and characteristic; this is due to the different isotopic compositions of the various substrate materials and the different kinetic isotope fractionations for the various methane production pathways. Thus it is possible to derive constraints on the strengths of the individual sources by comparing their isotopic composition to that of atmospheric CH_4. The isotopic composition is usually expressed as deviation δ [per mille (‰)] of an isotope ratio (R_{sample}) from that of an adopted standard ($R_{standard}$), as shown in the equation below. For $\delta^{13}C$, R is equal to $^{13}C/^{12}C$ and

$$\delta = \left(\frac{R_{sample}}{R_{standard}} - 1 \right) \times 1000$$

the standard is Pee Dee Belemnite (PDB); and for δD, R equals D/H and the standard is standard mean ocean water (SMOW). Atmospheric CH_4 has a $\delta^{13}C$ of about −47‰ PDB and a δD of about −83‰ SMOW. In contrast, CH_4 from biomass burning has a $\delta^{13}C$ of about −27‰, and the biogenic CH_4 ranges −50 to −80‰, and δD −250 to −400‰. The isotopic composition of fossil methane is intermediate between these values.

The radionuclide ^{14}C, produced by cosmic-ray interaction in the atmosphere, is particularly useful in deriving an estimate of the strength of the source of fossil CH_4. All biogenic CH_4 contains ^{14}C at contemporary concentrations, as atmospheric $^{14}CO_2$ is

Current budget of methane

Source	Annual release, Tg	Annual release, %	Range, Tg
Natural wetlands and tundra	120	22	100–200
Rice paddies	100	18	60–140
Ruminants	100	18	70–120
Termites	30	6	10–100
Landfills	40	7	30–70
Biomass burning	50	9	50–100
Fossil	110	20	80–130
Total	550	100	400–860

incorporated into organic matter. Fossil CH_4 contains no ^{14}C and thus tends to dilute the atmospheric ^{14}C concentration of methane. The magnitude of the fossil source determined via ^{14}C is about 20% of the global annual CH_4 production. The current status on CH_4 production is summarized in the table.

Atmospheric CH_4 is likely to continue to increase, since major sources such as rice production, domestic ruminants, and the burning of biomass and agricultural wastes are growing. No measures are currently being applied to reduce the CH_4 input into the atmosphere. Some slight reduction might result from the declining extent of natural wetlands. In principle, reduction of the CH_4 production from beef cattle can be achieved by application of certain antibiotics. Careful control of losses of natural gas in exploration and distribution could stabilize the fossil source. On the other hand, it was found that agricultural cultivation and nitrogen fertilization decreases CH_4 oxidation on soils and increases the emission of N_2O. The tundra is an example for positive feedback upon global warming. Large amounts of organic carbon are stored in the permafrost. If the permafrost level were lowered, additional substrate would become available for bacterial CH_4 production.

For background information *SEE ATMOSPHERIC CHEMISTRY; BIOGEOCHEMISTRY; CLIMATE MODELING; CLIMATOLOGY; GREENHOUSE EFFECT; METHANE; METHANOGENESIS (BACTERIA)* in the McGraw-Hill Encyclopedia of Science & Technology.

Martin Wahlen

Bibliography. D. R. Blake and F. S. Rowland, Continuing worldwide increase in tropospheric methane, *Science*, 239:1129–1131, 1988; R. J. Cicerone and R. S. Oremland, Biogeochemical aspects of atmospheric methane, *Global Biogeochem. Cycles*, 2:299–327, 1988; C. Lorius et al., The ice core record: Climate sensitivity and future greenhouse warming, *Nature*, 347:139–145, 1990; B. Stauffer et al., Increase in atmospheric methane recorded in Antarctic ice core, *Science*, 229:1386–1388, 1985; M. Wahlen et al., Carbon-14 in methane sources and in atmospheric methane: The contribution from fossil carbon, *Science*, 245:286–290, 1989.

Cold hardiness (plant)

The responses of plants to low temperatures have been of interest to plant physiologists because the survival, growth, and reproduction of many economically and ecologically important species are affected by subzero temperatures. Subzero temperatures cause ice nucleation in extracellular air spaces. Because the chemical potential of water (that is, the energy of water, or its ability to participate in chemical reactions) in solid ice is lower than that for liquid water in cells, water flows out of cells, down its chemical potential gradient (a process known as distillation), and freezes onto extracellular ice crystals. The rate and amount of cell water efflux are dependent upon the rate of cooling, the absolute amount of cooling, and the duration of the subzero temperature episode. If the distillation of water from cells is excessive, cell death due to dehydration occurs.

Cytological damage leading to cell death can occur in many ways. For example, rapid water efflux across cell membranes can cause irreversible mechanical lesions to membranes. Likewise, the reduction in water available for hydration can change the conformation and function of macromolecules, thus affecting the rate of biochemical reactions and the integrity of structural components in the cell.

Many plant species have evolved adaptations that reduce the damage caused by massive cellular water loss and allow increased survival during exposure to subzero temperatures. One of the ways that plants survive freezing temperatures is by the accumulation of sugars and related compounds. The type, rate, and amount of sugar accumulation, and the means by which such accumulation is advantageous, are highly species-specific. The role of sugars in freezing tolerance has been characterized on the basis of metabolic, osmotic, and cryoprotective effects.

Metabolic effects. In the autumn, low temperatures and short photoperiods cause a relative reduction in the amount of photosynthesis that occurs. This reduction is likely due to a combination of Q_{10} effects (the effect of a $10°C$ shift in temperature on the rate of a reaction), changes in the biophysical characteristics of the photosynthetic apparatus, alterations to the biochemistry of carbon fixation, and ultimately, changes in the amounts and utilization of photosynthetically derived sugars. Respiration is also altered during low-temperature acclimation (again, due to Q_{10} effects), with a general decrease observed with decreasing air temperatures. The changes in the photosynthetic production of sugars and starch, and in the respiratory use of sugars during low-temperature acclimation, may feed back upon these processes, thus complicating the overall regulation of sugar metabolism.

As ambient air temperatures decrease toward the end of the year, starch and lipid reserves increase for many plant species. The accumulation of reserves is essential as a source of energy and carbon skeletons; plants depleted of photosynthates do not acclimate to low temperatures. At the same time, the enzymes of glycolysis shift to the pentose phosphate pathway, and reducing reactions that utilize energy are favored. The resultant high levels of adenosinetriphosphate (ATP) and the reduced form of nicotinamide adenine dinucleotide phosphate ($NADPH_2$) allow for the synthesis of nucleic acids, proteins, and phospholipids. These substances can be used for structural, metabolic, and cryoprotective alterations to cells. For example, wheat seedlings that are subjected to a chilling stress exhibit an increase in sucrose synthase activity within 1 h of the stress. Indeed, the activity of this enzyme, which is involved in sucrose metabolism, continues to increase throughout the acclimation period, and is due to the novel synthesis of polypeptide rather than a low-temperature-induced change in enzyme activity. The changes in carbohydrate metabolism are reversed during spring growth or when environmentally controlled plants are returned to warmer temperatures. For instance, during the autumn, alfalfa exhibits an increase in amylase activity and a decrease in starch content in taproots. This is accompanied by an increase in the soluble sugar content and freezing tolerance of taproots. In the spring, enzyme activity and the soluble sugar content in taproots decrease and growth resumes. Although this pattern is common for many species, the precise role of soluble sugar accumulation in alfalfa taproots is not entirely clear.

The metabolic shifts in sugar chemistry during low-temperature acclimation are highly species-specific. For example, herbaceous species such as chicory, dandelion, and artichoke metabolize high-molecular-weight polysaccharides into smaller sugars, while winter-hardy cacti, such as *Opuntia humifusa* and *O. polyacantha*, accumulate extracellular mucopolysaccharide. Also, conifers accumulate raffinose and stachyose, and grasses produce fructosans and arabinoxylans in response to low temperatures. Thus, the type of sugar or related compound that accumulates differs widely, suggesting that sugar accumulation can be advantageous in many different ways.

Water relations. The chemical potential of a cellular solution is lowered by the addition of osmotically active solutes. Thus, an increase in cellular sugars can cause the potential difference between the liquid water in cells and the solid water in ice to approach zero, lowering the drive for water efflux from cells. Several species appear to modify the colligative properties of cell sap in response to low temperatures by accumulating nonreducing sugars such as sorbitol and mannitol. This presumably confers an advantage by lowering the freezing point of cell sap. However, the reduction in water efflux that occurs as a result of sugar accumulation is limited because the colligative properties of water dictate that a 1 molal solution will experience only $3.35°F$ ($1.86°C$) of freezing-point depression. Using the van't Hoff relation at $69°F$ ($20°C$) and considering a simple reducing sugar such as glucose, a 1070 mol/liter increase in concentration would be required for a $3.6°F$ ($2°C$) freezing-point depression. Therefore, reduction in freeze-induced dehydration by accumulation of sugars and freezing-point depression is a strategy available only to species capable of mobilizing large amounts of low-molecular-weight sugars, or those that experience a reliably narrow threshold of extremes of low temperatures, such as the tropical high-elevation species of the genera *Lobelia, Espletia,* and *Dendrosenecio.*

Some species reduce the potential for mechanical damage to membranes and the loss of water available for hydration by accumulating large-molecular-weight mucopolysaccharides in extracellular spaces. For example, the winter-hardy prickly pear cactus, *O. humifusa*, exhibits a doubling of extracellular mucopolysaccharide content over 7 weeks at day/night air temperatures of $41/23°F$ ($5/-5°C$), compared to stems maintained at $77/59°F$ ($25/15°C$). Alcian blue staining indicates that the mucopolysaccharide accumulates on the outside of cell walls, and water potential isotherms of extracted mucopolysaccharide show that it is an excellent water-storage molecule. Thus, the mucopolysaccharide is situated between liquid water in cells and solid water in extracellular ice, and acts as a passive storage molecule that modifies the rate or extent of water efflux from cells to ice during extracellular ice formation. Indeed, thermal analysis of stems acclimated to day/night air temperatures of $77/59°F$ ($25/15°C$) indicates that ice nucleation is followed by a rapid increase in stem temperature due to the release of the heat of fusion as water is distilled from cells to growing ice crystals. However, when stems are acclimated to $41/23°F$ ($5/-5°C$) for 14 days, there is no rapid rise in stem temperature, consistent with the proposed role of mucopolysaccharide as a passive water capacitor that slows the rate of water efflux during extracellular ice formation. Polysaccharides can also modify the freezing of water by interfering with the bond angle of adjacent water molecules, and may also provide a protective hydration layer adjacent to membranes.

Cryoprotective effects. The advantage to cells that accumulate sugars as cryoprotectants is not clearly understood at the physicochemical level. It is possible that sugars provide a molecular hydration layer for membranes and other macromolecules during freeze-induced dehydration. It is also possible that sugars and related compounds increase the viscosity of cell sap and thus increase the amount of unfreezable water. It has also been suggested that sugars reduce the damage due to freeze-induced concentrating of ions during dehydration because they do not readily form crystalline structures. Hopefully, recent advances in imaging techniques and isolation and quantification procedures will yield further information on the cryoprotective nature of sugars and related compounds that plants accumulate in response to low temperatures.

For background information SEE COLD HARDINESS (PLANT); OSMOSIS; PLANT METABOLISM; PLANT PHYSIOLOGY; PLANT-WATER RELATIONS in the McGraw-Hill Encyclopedia of Science & Technology.

Michael E. Loik

Communications satellite

The Advanced Communications Technology Satellite (ACTS) System, sponsored by the National Aeronautics and Space Administration (NASA), is an experimental satellite operating in the Ka frequency band (30 and 20 GHz) and scheduled for launch in the first quarter of 1993. ACTS will provide an on-orbit testbed for experimental, advanced communications technologies with the potential to dramatically enhance the capabilities and reduce the user service costs of satellite communications. ACTS technology will allow more cost-effective delivery of data, video, and voice services, as well as provide a more efficient use of both the radio-frequency spectrum and the geostationary orbit. Using multiple, dynamically hopping spot beams and advanced on-board switching and processing hardware, ACTS will open new vistas in communications satellite technology important to the future communications needs of the United States.

The satellite will be launched by the space shuttle and placed in the geostationary orbit at 100° West longitude on the equatorial plane at an altitude of 22,300 mi (36,000 km), where it will have the same period of rotation as the Earth. Satellites in this orbit appear stationary to an observer on Earth, and the directions in which the Earth-station antennas point at these satellites remain fixed. A number of satellites, spread out over the geostationary orbit, can provide complete coverage of the Earth except for the high-latitude north and south polar regions. By assigning a different part of the frequency spectrum to each signal, many signals may be transmitted simultaneously. Placing satellites 1000 mi (1600 km) apart will preclude signals from one satellite system (spacecraft and Earth stations) interfering with another, so the same frequency spectrum can be reused at each orbital position.

Frequency reuse techniques. The ACTS System will employ another technique of frequency reuse. Unlike present domestic communications satellites, which cover the contiguous United States with a single, stationary antenna coverage pattern, ACTS will concentrate its communication signals into narrow spot coverage over an area approximately 130 mi (209 km) in diameter. Using narrow, multiple, in-

dependent beams to cover different geographic areas will permit the same frequency to be reused from a given spacecraft without causing signal interference. The polarization properties of radio waves also allow signals transmitted with orthogonal polarization to be electronically separated at the receiving antenna, even though the signals occupy the same frequency spectrum. Such frequency-reuse technologies allow a given satellite to handle a large increase in system communications capacity, which in turn provides for more efficient use of the geosynchronous-orbital-arc resource and more cost-effective services to the communication end user. Another key advantage of spot beams is increased strength of signals as the beams are narrowed, so that smaller antennas can be used in the Earth stations.

Fixed and scanning beams. The narrow beams, however, illuminate only a small portion of the Earth's surface. As an experimental satellite, ACTS will cover only approximately 20% of the United States, a sufficient test to prove out its advanced concepts. Within even this limited area, 56 individual beam spots are required. The ACTS multibeam antenna will provide a number of isolated fixed beams as well as a series of overlapping beams in the midwestern and eastern parts of the United States. The latter beams will be scanned or illuminated in a stepped or hopped fashion according to the demands of the users in these different beam "footprints." During the time the beam dwells on an area, information will be

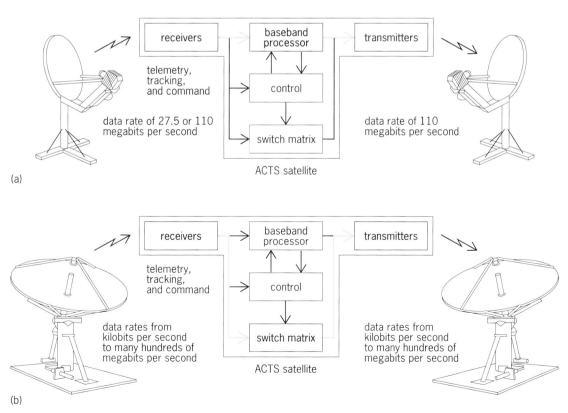

Fig. 1. ACTS dataflow. (*a*) Baseband processor mode using two scanning beams and time-division multiple access. (*b*) Microwave switch matrix mode, using three active fixed beams and either time-, frequency-, or code-division multiple access.

transmitted and received in short bursts. Separate frequencies will be used for the uplink and the downlink transmissions. Each user in a dwell area will transmit in a different portion of the uplink dwell time and use only the received signal from a different portion of the downlink dwell time, a mode of operation referred to as time-division multiple access (TDMA). The time slot allocated to a given user can be lengthened or shortened to accommodate a varying amount of communication traffic. These time slots will be adjusted and synchronized by a master control station, located at the NASA Lewis Research Center in Cleveland, Ohio, which will oversee the satellite network operations.

On-board switching and processing. Use of multiple spot beams requires a switching system on board the satellite to interconnect the beams and route signals or messages to their appropriate destinations. Two distinct types of switching systems will be tested on ACTS: microwave switch matrix and baseband processing.

The baseband processor (**Fig. 1***a*) will be the heart of the scanning spot-beam system. Acting as a switchboard in the sky, it will provide efficient routing of low-volume communications from small-antenna-diameter (4-ft or 1.2-m) Earth stations, such as those located directly on the customer's premises. The baseband processor will reduce the radio-frequency signals received by the satellite down to their baseband content: digital data consisting of binary ones and zeros. These data will be stored briefly in memory and then routed to their destination. In addition to permitting signal error detection and correction, the baseband processor will provide highly dynamic routing of individual telephone calls and, in essence, move the switchboard from the ground to the sky.

The switch matrix mode (Fig. 1*b*) will not store information, but will allow three different beams to interconnect to each other in a very fast dynamic way. This switch can be used to interconnect users with high-volume (multimegabit) data requirements. In addition, this switch can be used to establish a long-term connection between two ground terminals, as in current frequency-division multiple-access practices; or it can, at a very high rate, alter sequentially the interconnections so that each terminal can communicate with each of the others in successive time periods by using time-division multiple-access techniques.

Adaptive techniques. Signals in the 30-GHz uplink and 20-GHz downlink (Ka-band) frequency bands being used by ACTS are degraded more by rain and other atmospheric phenomena than the more commonly used C-band (6-GHz uplink and 4-GHz downlink) and Ku-band (14-GHz uplink and 12-GHz downlink) allocations. ACTS will employ techniques that allow the system to adapt to the degradations. Adaptability will be achieved by applying forward-error-correction coding to the signals and by reducing transmission rates when signals are degraded. In addition, a ground transmitter capable of operating over a range of power levels so as to overcome signal fading will be tested.

Antenna coverage. As an experimental satellite, ACTS is not intended to provide full United States coverage. **Figure 2** portrays the ACTS antenna cov-

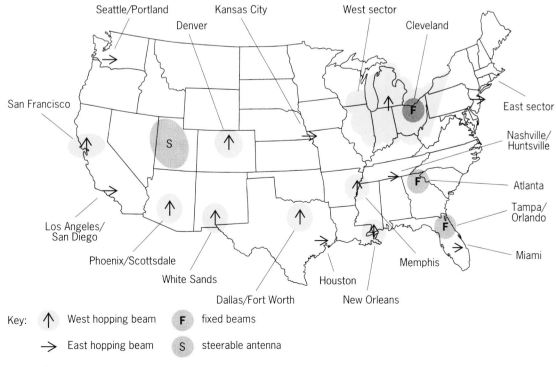

Fig. 2. ACTS multibeam antenna coverage.

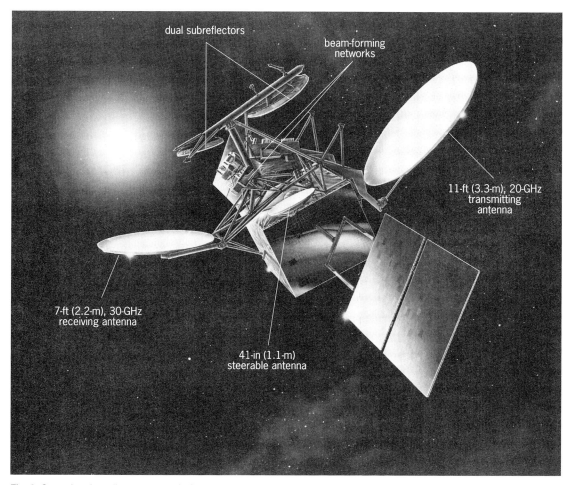

Fig. 3. Operational configuration of ACTS flight system. (*GE Astro Space*)

erage, an area sufficient to demonstrate the advanced multibeam concepts. The Ka frequencies, higher than the C or Ku bands currently in use by most communications satellites, will allow narrow spot-beam patterns to be formed by antennas capable of fitting inside the shuttle's cargo bay, 15 ft (4.5 m) in diameter. The ACTS 20-GHz downlink transmitting antenna will be 11 ft (3.3 m) in diameter, while the 30-GHz receiving antenna will be 7 ft (2.2 m) in diameter. The antenna beam-forming network will have five basic input and output ports: two scanning beams, designated East and West, and three fixed beams centered on Cleveland, Atlanta, and Tampa. The East and West scanning beams will be made up of numerous individual feed horns and can be switched very rapidly (in 100 nanoseconds) to hop between the locations shown in Fig. 2. The locations include a contiguous area in the northeastern and midwestern parts of the United States as well as individual isolated regions. A separate mechanically steerable antenna on the satellite, 41 in. (1.1 m) in diameter, will provide coverage to any area within the satellite's hemispherical field of view, including Alaska, Hawaii, and South America. Future operational satellites will incorporate additional scanning beams to provide full United States coverage or coverage of the particular geographic regions where they are providing communications services.

Spacecraft configuration. **Figure 3** depicts ACTS with solar panels and antennas deployed in orbit. During launch the antennas will be folded in a so-called ear-muff configuration, and the solar panels will be folded against the sides of the spacecraft so as to fit inside the shuttle's cargo bay. The spacecraft measures 30 ft (9 m) across the antennas and 48 ft (14.5 m) across the solar panels, which generate 1770 watts of power. The rectangular body of ACTS will measure 80 in. × 84 in. × 75 in. (2.0 m × 2.1 m × 1.9 m). ACTS will weigh 3250 lb (1472 kg) when it arrives on-station above the Equator.

For background information SEE COMMUNICATIONS SATELLITE in the McGraw-Hill Encyclopedia of Science & Technology.

Ronald J. Schertler

Bibliography. American Institute of Aeronautics and Astronautics, *13th International Communications Satellite Conference*, AIAA 90-0835, 1990; R. A. Bauer and T. C. von Deak, *Advanced Communications Technology Satellite Program and Experiments: Program Descriptive Overview*, NASA Lewis Research Center, 1991; *IEEE International Communications Conference*, Boston, CH 2655-9/89/0000-1566, September 1989.

Concurrent processing

Parallel processing (also known as concurrent processing) is gaining widespread use for two reasons: performance and price performance. A typical calculation needing 10^{15} arithmetic operations takes about 3 days on a fast uniprocessor; it takes under an hour if the work is divided among 100 of them. Further, a parallel machine consisting of hundreds of microprocessors can achieve supercomputer performance at a fraction of the cost of a conventional supercomputer.

Classification. A widely used classification divides parallel machines into three classes. Conventional computers are called SISD because they have a *single* *instruction* stream working on a *single* *data* element. A typical operation is adding the contents of one memory location to the contents of another.

One class of parallel machines is called SIMD because there is a *single* *instruction* stream working on *multiple* *data* elements. A typical operation is adding 50 numbers starting at one memory location to 50 numbers starting at another location.

The third class is called MIMD machines because there are *multiple* *instruction* streams working on *multiple* *data* elements. Each processor might do a different operation on distinct data, one doing a multiplication and another an addition.

Performance. A program distributed across p processors will not run p times faster than it does on one processor. It is unlikely that the entire problem can be run in parallel; some parts of the job must be done one at a time.

Amdahl's law. If the fraction of the work that must be done in serial mode is f, then the time $T(p)$ required for a run on p processors is subject to inequality (1), where $T(1)$ is the time required on one

$$T(p) \geq f T(1) + \frac{1-f}{p} T(1) \qquad (1)$$

processor. Inequality (1) can be rewritten in terms of the speed-up S, giving Amdahl's law, inequality (2).

$$S \leq \frac{T(1)}{T(p)} = \frac{1}{f + [(1-f)/p]} \qquad (2)$$

This inequality shows that even with infinitely many processors the application will never speed up by more than a factor of $1/f$. A program that runs 99% of its operations in parallel mode can never run more than 100 times faster than its serial version. Analysis similar to this led many people to believe that parallel processors with hundreds or thousands of processors would never be cost-effective.

Problem scaling. Amdahl's law neglects one important point. A person with a faster computer is likely to run a larger problem, one that takes about the same time to complete. Thus, Amdahl's law makes the incorrect assumption that the problem size stays fixed as the number of processors increases.

The serial part of the code is often a fixed number of calculations or a slowly growing function of problem size. If the simplifying assumption is made that the serial time is independent of problem size, and the parallelizable time is proportional to the number of processors, a version of Amdahl's law is obtained, inequality (3), that is suitable for scaled-size problems.

$$S \leq f + (1-f)p \qquad (3)$$

This inequality shows that the maximum speed-up is proportional to the number of processors. If the parallel fraction is 99%, the maximum speed-up on 1000 processors will be 990, compared to 91 for the fixed-size analysis. Real programs perform somewhere between these two extremes.

Overhead. An important factor has been neglected: there is work associated with a parallel run that is not needed when only one processor is used. The time to do this work, called overhead, limits the performance of the parallel processor even for scaled problems. No matter how slowly the overhead increases as processors are added, a point is reached where adding more processors increases the run time. More time is spent coordinating activities than performing them.

The overhead takes several forms. Each processor must decide what part of the work it will do. During the run, processors must exchange data, which takes time away from computation. Frequently, an algorithm that runs well on one processor cannot be made to run in parallel. In this case, the parallel run must use an algorithm that requires more arithmetic than the sequential program. Finally, the work done by the various processors may not be the same. In extreme cases, one processor works for a long time while all

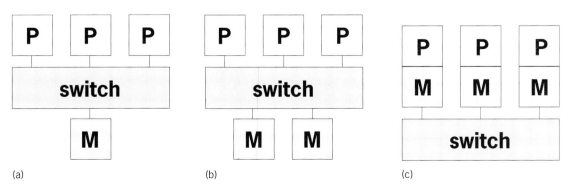

(a)   (b)   (c)

Fig. 1. Various ways of building shared-memory machines. (*a*) Scheme in which all processors (P) can access a single memory (M) through a switch. (*b*) Dance-hall machine, in which memory is distributed over separate modules. (*c*) Nonuniform memory access (NUMA) machine, in which the memory modules are attached to the processors.

the others wait for it to finish. This problem is called load imbalance.

Hardware. A parallel processor is designed for high performance. The machine should consist of as many processors as practical. Further, the overhead associated with coordinating them should be as low as possible.

Memory sharing. One way to keep the overhead low is to have all processors share a common memory. The most straightforward approach is to connect several processors to the system memory with a switch (**Fig.** 1*a*). Since any processor can read (or write) any word of memory, one processor can write data into memory to be read by another processor. This exchange takes only a few machine cycles. Unfortunately, the memory has a limited bandwidth. If too many processors are added, say more than 20, the memory will not be able to keep up with the requests and the processors will have to wait for data.

One way around the problem is to spread the memory out over several independent modules (Fig. 1*b*). Now it is possible for several processors to use the memory simultaneously. The key in this configuration is the switch. Since it can make many connections between memory modules and processors, its total bandwidth can be substantially higher than a single memory. Such systems are called dance-hall machines because the modules resemble men and women facing each other at the start of a square dance.

Another variant is to attach the memory modules to the processors (Fig. 1*c*). Any processor accessing data out of its own memory will get the data faster than if its data are in a remote memory. Such a system is called a nonuniform memory access (NUMA) machine.

Switching. The switch is key to the performance of the system. A full cross-bar switch allows any processor to reach any memory module very quickly. Such a switch is impractical for a large number of processors since the number of connections is proportional to the square of the number of nodes.

Many other connection schemes are possible. A butterfly network (**Fig. 2**) allows any node to connect to any other node. The switch needs exactly n stages to connect 2^n processors. Each of the circles in Fig. 2 is a switch. Depending on a particular bit in the difference of the addresses of the modules wishing to communicate, the switch will send the data horizontally (if the bit is 0) or along the diagonal (if the bit is 1).

Dance-hall and NUMA systems perform well if memory access patterns are random. Unfortunately, often they are not. It is common for all processors to need the same word from memory, which creates a memory hot spot. Even if only some of the processors are trying to use the same word, the switching network can become clogged with requests. Then other processors cannot get data from the rest of the memory modules.

Building some intelligence into the network alleviates this problem somewhat. In a combining network, each switch remembers the memory address. It can then satisfy several requests with a single memory access.

Multicomputers. Each node in Fig. 1*c* contains

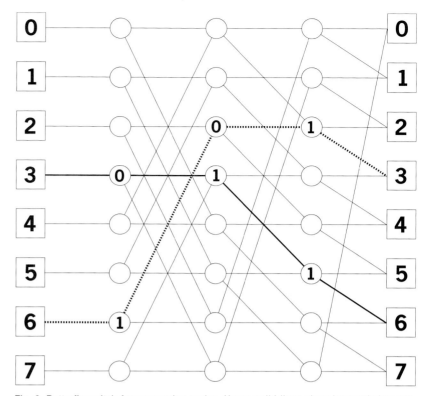

Fig. 2. Butterfly switch for connecting nodes. Heavy solid lines show how switches are connected for node 3 to connect to node 6, a difference of 3 (011 in binary), and heavy broken lines show how switches are connected for node 6 to connect to node 3. For the connection of node 6 to node 3, 3-6 modulo the number of nodes (8 in this case) is 5, which is 101 in binary.

a processor and some memory; it is a complete computer. If a shared memory can be sacrificed, then independent computers can simply be lashed together over a network. Data are shared by sending data as a message from one processor to another. Such systems are called message-passing machines or multicomputers.

There are many ways to interconnect the processors, some of which are shown in **Fig. 3**. Processors with only two communications ports can be configured only as a ring of processors (Fig. 3*a*). The disadvantage of a ring is the diameter of the parallel processor; it takes the time of $p/2$ messages to spread data to p nodes, where a node is a processor and its memory. A two-dimensional mesh that needs only four communications ports per node has a diameter of $\sqrt{p}$ (Fig. 3*b*).

If each processor has $\log_2 p$ communication ports, it is possible to build a hypercube such as the four-dimensional one shown in Fig. 3*c*. Such a system is built recursively by taking two hypercubes of one lower dimension and connecting each corner of one to the corresponding corner in the other. A two-dimensional hypercube is a square. It is a simple matter to simulate a ring or a grid with a hypercube. Hypercubes have a diameter of $\log_2 p$.

Parallel processing can also be done with computers connected over a general-purpose network. Such systems typically have slower communications than multicomputers. Also, the network bandwidth limits the number of processors that can be used effectively.

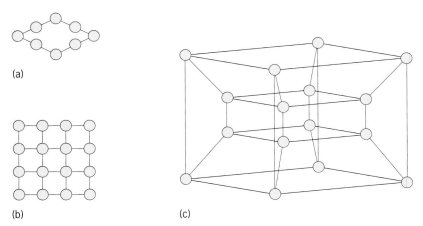

Fig. 3. Interconnection schemes for multicomputers. (*a*) Ring of processors. (*b*) Two-dimensional mesh. (*c*) Four-dimensional hypercube.

One disadvantage of multicomputers is the rather long time it takes to get data from one machine to another. This time has two parts, the startup time t_s and the transmission time. If the data move at a rate r, and the message is of length L, the time to send a message is given by Eq. (4).

$$t = t_s + \frac{L}{r} \qquad (4)$$

Typically, t_s is between 50 microseconds and 1 millisecond, while r is around 1 or 2 megabytes per second. A long message will move data more efficiently than many short ones. Also, a typical multicomputer processor can do about 500 arithmetic operations in 50 μs. Thus, it is advantageous to design algorithms that do a lot of computing between communications. Unfortunately, there are some algorithms that cannot be parallelized with sufficiently large granularity to perform well on these machines.

Software. There are several programming styles used for parallel machines. One of the first used was the fork-join model. A program starts running. When it is time to do some work in parallel, this root task forks one or more child processes. These tasks run in parallel with the root. They can either share memory with each other and their parent or pass messages. When the parallel work is done, the child processes end. This is the join step.

Each child process can run a completely different subroutine; in this case, a functional decomposition is said to have been carried out. It is also possible to have all the children running the same subroutine but operating on different data; this scheme is called single-program multiple-data (SPMD).

There are significant differences between programming shared-memory machines on the one hand and message-passing machines on the other.

Message-passing systems. Message-passing machines need only a way to send a message and continue processing and a way to wait for a message to arrive. Some systems support more advanced functions. A nonblocking receive allows the program to do other work if the message is not yet available. A blocking send does not return control to the sending

process until it is safe to reuse the area of memory that contained the message. It is also useful to be able to set priorities on messages so that an important message can be handled before a less important one that arrived earlier.

Other functions are useful, too. In the case of summing a list of numbers, each processor forms a partial sum from its part of the data. Sending all the partial results to a single processor leads to a serial bottleneck. A better approach is a spanning tree, in which each node collects partial results from its nearest neighbor. The processors combine them with results from those one hop away, then two hops, and so forth. If done properly, all p processors will have the result after $\log_2 p$ steps.

Shared-memory systems. Shared-memory systems typically need more software support. The case of summing a list of numbers again provides an example. To form the global sum on a shared-memory computer, each processor must fetch the old value, add its contribution, and store the result. If two processors fetch the old value simultaneously, the final result will be incorrect. These so-called race conditions are avoided by controlling access to shared data.

Access to shared data can be limited by using a critical section, a piece of code all processors execute one at a time. Any process trying to enter the critical section while another is there waits. Other constructs can be used to keep the processors in step. For example, all processors will wait at a barrier for all the others to arrive. In addition to fork-join parallelism, a shared-memory system is often programmed with a parallel loop. The system gives different iterations of the loop to different processors. Parallel cases are constructs that distribute different blocks of code to different processors. SEE SUPERCOMPUTER.

The processors of an SIMD machine operate in lock step; they all execute the exact same instructions at the exact same time. These machines use a data-parallel style of programming. Data parallelism may not be suitable for some problems, but many applications have been run on SIMD machines.

For background information SEE CONCURRENT PROCESSING; MULTIPROCESSING; SUPERCOMPUTER in the McGraw-Hill Encyclopedia of Science & Technology.

Alan H. Karp

Bibliography. J. Gustafson, Reevaluating Amdahl's law, *Comm. ACM*, 31:532–533, 1988; R. W. Hockney and C. R. Jesshope, *Parallel Computers 2*, 1988; K. Hwang and F. Briggs, *Computer Architecture and Parallel Processing*, 1984; A. H. Karp. Programming for parallelism, *IEEE Computer*, pp. 43–57, May 1987.

Control systems

The ability to actively or passively manipulate a flow field to effect a desired change is of immense technological importance. In its broadest sense, the art of flow control probably has its roots in prehistoric times, when streamlined spears, sickle-shaped boomerangs, and fin-stabilized arrows evolved empirically. For centuries, farmers have known the value of windbreaks to keep topsoil in place and to protect fragile crops.

The term boundary-layer control includes any mechanism or process through which the boundary layer of a fluid flow (a low-velocity viscous region near a solid surface) is caused to behave differently than it normally would were the flow developing naturally along a smooth, straight surface. An external boundary-layer flow, such as that developing on the exterior surface of an aircraft or a submarine, can be manipulated to achieve delay of the transition from laminar to turbulent flow, separation postponement, lift enhancement, drag reduction, turbulence augmentation, or noise suppression. These objectives are not necessarily mutually exclusive. For example, by maintaining as much of a boundary layer in the laminar state as possible, the skin-friction drag and the flow-generated noise are reduced. However, a turbulent boundary layer is, in general, more resistant to separation than a laminar one. By preventing separation, lift is enhanced and the form drag is reduced. An ideal method of control that is simple, inexpensive to build and operate, and does not have any trade-offs does not exist, and the skilled engineer has to make compromises to achieve a particular goal. *SEE DRAG REDUCTION.*

Early in the twentieth century, L. Prandtl used artificial control of the boundary layer to show the great influence such control exerted on the flow pattern. He used suction to delay boundary-layer separation from the surface of a cylinder. Notwithstanding Prandtl's success, aircraft designers in the following three decades accepted lift and drag of airfoils as predetermined characteristics. This situation changed chiefly because of German research in boundary-layer control pursued vigorously shortly before and during World War II. In the two decades following the war, extensive research was conducted on laminar flow control, where the boundary layer formed along the external surfaces of an aircraft is kept in the low-drag laminar state, culminating in the successful flight test program of the X-21 where suction was used to delay transition. The oil crisis of the early 1970s brought renewed interest in novel methods of flow control to reduce skin-friction drag even in turbulent boundary layers. In the 1990s, the need for supermaneuverable fighter planes, faster and quieter underwater vehicles, and hypersonic transport (for example, the proposed U.S. National Aerospace Plane) are providing new challenges in the field of flow control.

Transition delay. Both the science and technology to maintain a laminar boundary layer up to a Reynolds number (Re) of about 4×10^7 are well established, although some details remain to be worked out. The linear stability theory provides a solid analytical framework, at least for the important first stage of the transition to turbulent flow. Barring large disturbances in a conventional boundary-layer flow, the linear amplification of Tollmien-Schlichting waves is the slowest of the successive multiple steps in the transition process. Stability modifiers inhibit this linear amplification and, therefore, determine the magnitude of the transition Reynolds number. Shaping to pro-

vide extended regions of favorable pressure gradient is the simplest method of control and is well suited for small underwater vehicles or for the wings of low- or moderate-speed aircraft. Flight tests have demonstrated the feasibility of using suction to maintain a laminar flow on a swept wing up to Re $\simeq 4.7 \times 10^7$. The required suction rate is very modest, and 20% net drag reduction is possible. The remaining problems are technological in nature and include the maintainability and reliability of suction surfaces and further optimizaton of the suction rate and its distribution. Suction is less suited for underwater vehicles because of the abundance of particulate matter that can clog the suction surface and destabilize the boundary layer. For water applications, compliant coatings that increase the transitional Reynolds number by a factor of 5–10 are available in the laboratory, but their performance in the field is still unknown. This technique is very appealing because of its simplicity and the absence of energy requirements. Moderate surface heating also increases the transition Reynolds number for water flows by an order of magnitude, but a source of rejected heat must be available to achieve net drag reduction. To ensure successful implementation of any of the transition-delay methods in the ocean, a particle-defense mechanism is needed to prevent suspended particulates from entering the boundary layer and causing a premature transition. For futuristic aircraft using cryogenic fuels, surface cooling may be a feasible method to delay transition.

Separation control. Of all the various types of shear flow control now extant, control of flow separation is probably the oldest and most economically important. Separation control is of immense importance to the performance of air, land, or sea vehicles, turbomachines, diffusers, and a variety of other technologically important systems employing fluid flow. Generally it is desired to postpone separation so that form drag is reduced, stall is delayed, lift is enhanced, and pressure recovery is improved. However, in some instances it may be beneficial to provoke separation.

Flow-separation control is currently employed, for example, via vortex generators on the wings of most Boeing aircraft; via blown flaps on older-generation supersonic fighters or leading-edge extensions and strakes on newer generations; and via passive bleed in the inlets of supersonic engines on, for example, the SR-71 and Concorde. Future possibilities for aeronautical applications of flow-separation control include providing structurally efficient alternatives to flaps or slats; cruise applications on conventional takeoff and landing aircraft, including thick spanloader wings (where most of the payload is distributed along the wing's span); as well as cruise applications on high-speed civil transports, for wave-drag reduction through favorable interference effects, increased leading-edge thrust, and enhanced fuselage and upper-surface lift. In fact, much of the remaining gains to be made in aerodynamics appear to involve various types of flow control, including separated flow control. *SEE VORTEX.*

The tremendous increases in the capabilities of computational fluid dynamics, which have occurred as a direct result of increases in computer storage capacity and speed, are transforming flow-separation control from an empirical art to a predictive science. Control techniques such as mitigation of imposed pressure gradients, blowing, and suction are all readily parametrized via viscous computational fluid dynamics. Current inaccuracies in turbulence modeling can severely degrade predictions of computational fluid dynamics once separation has occurred; however, the essence of flow-separation control is the calculation of attached flows, estimation of separation location, and indeed whether or not separation will occur, tasks which computational fluid dynamics can perform reasonably well within the uncertainties of the transition-location estimation.

Particularly intriguing in the future are possible applications at cruise, especially at high speeds, as well as standby techniques that are meant to be used in situations other than those for which the aircraft was designed, thereby allowing extremely tight designs. In addition, increased knowledge and control innovations for vortical flows should allow solutions to the problems of maneuvering at high angles of attack and stall/spin prevention or recovery. Alternate techniques should also be sought for either supplementation or replacement of conventional high-lift flap systems. Candidate approaches include the use of leading-edge suction systems for laminar flow control, along with air-jet vortex generators and perhaps either rotating wings or airport ski jumps (where the short runway consists of a mild-downhill portion followed by a cliff). Also, there exists tremendous energy-saving potential in further separation control or form-drag reduction for vehicles that have notoriously high drag, such as automobiles, helicopters, and tractor-trailer trucks.

Drag reduction. Techniques to reduce pressure drag are better established than turbulent skin-friction reduction techniques. Streamlining and other methods to postpone separation can eliminate most of the pressure drag. The wave- and induced-drag contributions to the pressure drag can also be reduced by geometric design. Skin friction constitutes about 50%, 90%, and 100% of the total drag on commercial aircraft, underwater vehicles, and pipelines, respectively. Most of the current research effort in drag reduction concerns the reduction of skin-friction drag for turbulent boundary layers.

Three flow regimes are identified. For $Re < 10^6$, the flow is laminar, and skin friction may be lowered by reducing the near-wall momentum. Adverse pressure gradients, blowing, and surface heating or cooling lower the skin friction but increase the risk of transition and separation. For $10^6 < Re < 4 \times 10^7$, active and passive methods to delay the transition to turbulent flow can be used, thus avoiding turbulent skin friction, which is much larger than laminar drag. Finally, at the high Reynolds numbers encountered after the first few meters of a fuselage or a submarine, methods to reduce the large skin friction associated with turbulent flows are sought. These methods fall in the categories of reduction of near-wall momentum, introduction of foreign substances, geometrical modification, relaminarization, and synergism.

The introduction of foreign substances leads to the most impressive results. The introduction of small concentrations of polymers, surfactants, particles, or fibers into a turbulent boundary layer leads to a reduction in the skin friction coefficient of as much as 80%. Among the practical considerations requiring further study are the cost of the additive, methods of delivering it to the boundary layer, the potential for recovering and recycling, degradation, and the portion of the payload that has to be displaced to make room for the additive.

Recently introduced techniques fall mostly under the category of geometrical modification and seem to offer more modest net drag reduction. These methods are, however, still in the research stage and include riblets ($\sim 8\%$), large-eddy breakup devices ($\sim 20\%$), and convex surfaces ($\sim 20\%$). Potential improvements in these and other methods will perhaps involve combining more than one technique with the aim of achieving a favorable effect that is greater than the sum. Due to its obvious difficulties, this area of research has not been widely pursued but deserves attention.

Along these lines, the selective suction technique combines suction to achieve an asymptotic turbulent boundary layer and longitudinal riblets to fix the location of low-speed streaks. Although they are still far from indicating net drag reduction, the available results are encouraging, and further optimization is needed. Potentially the selective suction method is capable of skin friction reduction that approaches 60%.

Practical applications of methods targeted at controlling a particular turbulent structure to achieve a prescribed goal would perhaps require implementing a large number of surface sensors and actuators together with appropriate feedback loops and control algorithms. The emerging field of microfabrication, where microscopic machinery is fashioned on a single silicon chip, may in the future be utilized to realize such smart arrays at reasonable cost. Additionally, newly formulated theories for controlling nonlinear dynamical systems could be used to optimize the relevant processes. SEE MECHANISM.

Noise control. Noise, or undesired sound, is generated by machines, vehicles moving in a fluid, or even natural phenomena. Noise control, a relatively young field of research, is required in order to maintain an acceptable (even pleasant) environment, avoid detection of certain vehicles, and allow the proper operation of sonars on underwater vehicles. Active noise suppression systems are based on the generation of sound by an auxiliary source with such an amplitude and phase that in the region of interest the sound-wave interference from the original and auxiliary sources results in considerable reduction of the noise levels. The linearity of the governing equations makes these systems possible. The older systems with fixed gain in the feedback loop could not achieve high noise reduction. However, with the recent availability of adaptive filter systems, much more impressive suppression of noise is now feasible. These systems are capable of adjusting the feedback loop for the magnitude and phase relationship of the spectral components and can quickly compensate for sound-path changes. These active control devices seem to work best for low-frequency sound, where passive silencers are relatively ineffective, and when the source is localized and accessible. Local control is obviously easier than a global system. Moreover, effective control is achieved when the system response within the frequency band of interest is dominated by relatively few modes. Active noise control using antisound is a new research area that, remarkably, seems to have gone into applications in just a few years.

For background information SEE ADAPTIVE SOUND CONTROL; AERODYNAMIC WAVE DRAG; AIRFOIL; BOUNDARY-LAYER FLOW; FLUID FLOW; SKIN FRICTION; STRAKE; STREAMLINING in the McGraw-Hill Encyclopedia of Science & Technology.

Mohamed Gad-el-Hak

Bibliography. M. Gad-el-Hak, Control of low-speed airfoil aerodynamics, *AIAA J.*, 28:1537–1552, 1990; M. Gad-el-Hak, Flow control, *Appl. Mech. Rev.*, 42:261–293, 1989; M. Gad-el-Hak and D. M. Bushnell, Separation control: Review, *J. Fluids Eng.*, 113:5–30, 1991.

Dairy cattle production

Several new technologies are likely to have a major impact on the dairy industry in the United States. These technologies include embryo transfer, cloning by nuclear transplantation, freezing of embryos, sexing of embryos, in vitro fertilization, use of bovine growth hormone, and production of transgenic animals. Successful implementation of these technologies could substantially increase milk production.

Embryo transfer. This technology is a method of expanding the reproductive capabilities of genetically superior cows. A cow is capable of producing up to 75,000 fertilizable eggs in her lifetime, although usually only one egg matures in each normal estrous cycle. Embryo transfer technology allows a superior cow to generate up to 50 offspring per year. The steps comprise superovulation, insemination, recovery of embryos, storage, and transfer of embryos to recipients.

Superovulation is a procedure in which a donor cow is given intramuscular injections of hormones to stimulate maturation and ovulation of more than one egg. Individual donors can produce up to 50 eggs. At the proper time of estrus, the donor is inseminated, usually twice in 24 h to assure that as many eggs are fertilized as possible.

Embryos are commonly recovered by nonsurgical methods, which cause less trauma for the donor. On day 6 or 7 after the first insemination, both horns of the donor's uterus are flushed with a buffered saline solution. An average of five or six transferable embryos is produced by each flush. Embryos can be stored for up to 24 h in a solution of buffered saline supplemented with serum or serum albumin. More elaborate procedures, such as freezing, are required for long-term storage.

Recipient cows are injected with specific hormones that synchronize their estrus with that of the donors and ensure that the uteri will be receptive to the transferred embryos. Nonsurgical implantation and flank surgery are two methods used to transfer embryos to recipients. The most common nonsurgical method is to place embryos into the uterine horn via the cervix. In flank surgery an incision is made in the flank wall, and the embryo is transferred through the uterine wall. Success rates of 50–70% have been obtained for both methods.

Cloning by nuclear transplantation. Cloning is a procedure by which exact genetic replicates are made. Early cloning work involved splitting fertilized embryos at the two-cell stage. The separate cells were then transferred to the reproductive tracts of animals for development into identical individuals. More recent work involves splitting multicellular embryos, after which individual embryonic cells are transferred to recipients by nonsurgical procedures.

Most cloning research has involved the use of individual embryonic cells, but it is not possible to accurately identify from embryos the specific desirable production traits, such as high levels of milk production, milk protein content, and body conformation. Such traits have only low-to-medium frequency of transmission in conventional breeding. Potential benefits from cloning by nuclear transplantation from an adult cell are substantial. Ideally, the major economic benefits would result from cloning adult nuclei obtained from genetically superior adults that have demonstrated traits of economic importance. The procedure involves transplanting the nucleus of an adult donor into an egg that has been enucleated. The traits that have been demonstrated in the adult donor could thus be cloned. Cloning from adult cells is not possible with current technology because embryos will not develop to maturity. Apparently the nuclei and genes of adult cells have gone through a biological process that prevents the expression of all genes making up individuals.

Most of the current benefits from cloning are in research applications. Procedures to perform nuclear transplantation are expensive and require much skill, so that the present procedure is impractical for commercial use. Most nuclear transplantation carried out so far has utilized amphibians, mice, and rabbits.

Freezing embryos. The commercial application of embryo technology is largely dependent on the ability to freeze embryos for long periods. In embryo transfer procedures, coordination of the number of recipients and embryos is difficult because there is significant variability in the number of embryos that are flushed from a donor. Thus, an advantage of freezing is that embryos are available when recipients are ready for them. A longer storage life also enables more extensive use of embryos produced by select animals.

Freezing of embryos is an industry practice. Embryos undergo cryopreservation, rapid freezing at extremely low temperatures, in a special medium. To maintain viability, embryos are usually stored at $-321°F$ ($-196°C$) in liquid nitrogen.

Technical problems related to temperature and time cycles for freezing and thawing embryos will limit the use of freezing unless there are some technological breakthroughs. Currently, conception rates are considerably lower with frozen embryos.

Sexing of embryos. Each of the cells of male embryos has an X and a Y chromosome, whereas each of the cells of female embryos contains two X chromosomes. Two methods for sexing sperm and embryos rely on identification of Y chromosomes. One method is karyotyping, in which embryo cells are opened to identify and stain Y chromosomes. Another method, carried out on sperm, uses a probe that binds to Y chromosomes, so that Y-chromosome-bearing sperm can be identified. An alternative approach is to separate the semen into X- and Y-chromosome-bearing sperm by techniques that utilize differences in characteristics such as density, electrical charge, or swimming patterns. Another procedure for sexing embryos involves an identifying H-Y antigen, which is produced on the outside of male cells but not female cells.

There is considerable interest in sexing of sperm and embryos. A principal advantage of sexing would be reduction of the cost of replacement heifers since unwanted bull calves would not be produced. If accurate sexing procedures are developed, most commercial dairies would generally produce heifer calves. Sperm sexing would speed adoption of embryo transfer technology since costs of producing heifer calves would be drastically reduced, depending upon the level of sexing success.

Although these techniques hold great promise, technological hurdles must be overcome before they become economical. Current laboratory methods for sexing sperm and embryos have not proven to be cost-effective and trouble-free.

In vitro fertilization. In vitro fertilization of eggs occurs in an artificial environment outside the body. Cultured embryos are then transferred to cows for continued development. In vitro fertilization procedures have also been studied to develop primary oocytes removed from cows at the time of slaughter into fertilizable secondary eggs. This technique may lead to the use of "manufactured" embryos in research and on farms. In vitro techniques have shown promise as a means of controlling bovine diseases such as bovine leukemia virus and infectious bovine rhinotracheitis virus.

Before the technology of in vitro fertilization can be applied to commercial livestock, several technical problems must be resolved. More information on processes of micromanipulation of embryos to obtain twins and quadruplets and on clonal reproduction is needed.

Even though in vitro techniques have found little application so far in commercial dairies, the technology does have much promise. The greatest limiting factor is cost effectiveness since in vitro techniques require expensive equipment and high levels of expertise.

Bovine growth hormone. Somatotropin, a growth hormone for dairy animals, has been studied for many years. Research was initiated over 50 years ago, when it was demonstrated that cows that were injected with extract from the pituitary gland had increased milk production. Somatotropin can now be produced synthetically by genetic engineering, so that commercial quantities at lower prices are obtainable.

Of all the technologies discussed in this article, growth hormone use will have the most immediate impact on dairies since the financial incentives are high enough to warrant commercial adoption. Research has demonstrated that the hormone increases gross efficiency of dairy cows. Gross efficiency is defined as the ratio of milk energy (calories per unit of milk) to consumed energy. However, biologic and economic risks are associated with the use of somatotropin. Risks include mastitis and fertility problems. Use of somatotropin requires intensive management procedures. Currently, the hormone must be injected in order to generate the production response.

Transgenic animals. The injection of foreign genetic material into fertilized eggs produces transgenic animals. Genes of a foreign species attach to the chromosomes of the embryo. The altered embryos are then transferred into surrogate mothers. If gene transfer is effective, the transgenic animal will be able to transmit its new genetic characteristics to its offspring.

Three methods are used to introduce new genetic material into animals. The first is the microinjection of genetic material into the pronucleus of a fertilized egg. This method has the advantage of generating transgenic animals that express the genes in a predictable manner. The major disadvantage is that new genetic material cannot be introduced at later developmental stages. The second method is retrovirus injection. A single virus is introduced into an embryo at a given chromosomal site. This technique has the advantage that the virus can be inserted at various developmental stages. A problem is the reproduceability of the transduced gene in later generations. The third method uses embryonic stem cells. These cells are established in vitro from explanted blastocysts. After being injected into host blastocysts, they contribute to the genetic structure of the new animal.

Anticipated benefits from transgenic procedures are numerous. For example, medically important proteins could be extracted from the milk of transgenic cows. Milk with differing nutritional values or properties suitable for the dairy industry could be produced. More importantly, transgenic procedures provide cows that produce larger quantities of milk. These effects would be realized by introducing genes that are expressed in the mammary glands. Desirable characteristics could be transferred from breed to breed more rapidly than by conventional breeding. Genetic material could also be transferred between species, and genetic defects could be corrected. Other benefits include enhanced growth, greater ovulation, and improved disease resistance.

Increases in gross efficiency could have an impact on the demand for traditional forages and grains.

Transgenic technologies have the potential of reducing feed requirements per unit of food produced. For example, bacteria in the rumens of animals could be genetically altered so that the nutritional value of feeds such as saltbush, tropical grasses, and shrub browse could be enhanced.

Transgenic technology is still in its infancy. Most transgenic research has been conducted on mice, and use of this technology for large agricultural animals appears to be in the future.

For background information *SEE BREEDING (ANIMAL); DAIRY CATTLE PRODUCTION; GENETIC ENGINEERING* in the McGraw-Hill Encyclopedia of Science & Technology.

Farrell E. Jensen

Bibliography. F. E. Jensen et al., Policy implication of new technologies in the U.S. dairy industry, *J. Prod. Agr.*, 3:13–20, 1990.

Data compression

Data compression is the process of transforming information from one representation to a smaller representation from which the original, or a close approximation to it, can be recovered. The compression and decompression processes are often referred to as encoding and decoding.

Data compression has important applications in the areas of data storage and data transmission. Many data-processing applications require storage of large volumes of data. Despite the fact that the costs of data storage have been declining and are expected to continue to decline, the expanding volume of data to be stored makes the economics of data compression self-evident. Simultaneously, there is an increasing requirement of data transfer over communication lines. The ability to transfer the same information using a smaller number of bits not only enables a more rapid information transfer but also mitigates the need to expand the capacity of communication channels.

The decision as to which of the many data compression-decompression techniques is the best to be used for a particular application is determined by the needs dictated by that application. Besides compression savings, other parameters of concern include encoding and decoding speeds and workspace requirements, the ability to access and decode partial files, and error generation and propagation.

The data compression process is said to be lossless if the recovered data are assured to be identical to the source; otherwise the compression process is said to be lossy. Lossless compression techniques are requisite for applications involving textual data and typically produce compression savings of 40–50%. Other applications, such as those involving voice and image data, may be sufficiently flexible to allow controlled degradation in the data. Compression savings are typically 70–90% and can be as high as 98%. Some lossy compression techniques gracefully accept increased data degradation for increased compression savings. Alternatively, some compression-transmission techniques enable rapid recovery of a highly degraded version of the source and, at the expense of additional time, allow recovery of a version with less degradation. This ability to browse and optionally gain further detail is essential for some applications such as satellite–Earth transmission.

Data compression techniques are characterized by the use of an appropriate data model, which determines the selection of the elements of the source on which to focus; data coding, which determines the mapping of source elements to output elements; and data structures, which are responsible for the efficiency of implementation.

Source data. The source data typically are presented in a format in which each of several possible values of a datum is represented by a fixed number of bits. For textual data, each character is typically represented in eight bits using, for example, the extended ASCII mapping. For speech, each datum consists of a real-valued sample of the waveform expressed to a set precision. For images, each picture element or pixel typically is represented by 1 bit (for black-white), 8 bits (for gray scale), or 24 bits (for color images, 8 bits intensity for each of three primary colors). For speech and images, the source data presented for compression are themselves an approximation to the original.

Models. Information theory dictates that, for efficiency, fewer bits be used for common events than for rare events. Compression techniques are based on using an appropriate model for the source data in which defined elements (source values in general or within a particular context) are not all equally likely. The encoder and the decoder must use an identical model. The encoder determines the relevant elements under the agreed model and outputs sufficient information to enable the decoder to determine these elements.

A static model is one in which the choice of elements and their assumed distribution is invariant. For example, the letter "e" might always be assumed to be the most likely character to occur. A static model can be predetermined with resulting unpredictable compression effect, or it can be built by the encoder by previewing the entire source data and determining element frequencies. This latter, off-line, process suffers from the necessity of the encoder making two passes over the data, one pass for determining frequencies and another for encoding, and also from the delay imposed on producing any output until after the entire source has been scanned. In addition, the relevant aspects of the determined frequency distribution must be sent to the decoder, which would otherwise have no means of determining it. The benefits of using a static model include the ability to decode without necessarily starting at the beginning of the compressed data. This can be important for database applications in which it may be necessary to access only a small part in the middle of a compressed file.

An alternative dynamic or adaptive model assumes an initial choice of elements and distribution and, based on the beginning part of the source stream that has been processed prior to the datum presently under consideration, progressively modifies the model so that the encoding is optimal for data distributed similarly to recent observations. Some techniques may weight recently encountered data more heavily. Dynamic algorithms have the benefit of being able to adapt to changes in the ensemble characteristics. Most important, however, is the fact that the encoding process is on-line; the source is considered serially and output is produced directly without the necessity of previewing the entire source.

In a simple statistical model, frequencies of values (characters, strings, or pixels) determine the mapping. In the more general context model, the mapping is determined by the occurrence of elements, each consisting of a value which has other particular adjacent values. For example, in English text, although generally "u" is only moderately likely to appear as the "next" character, if the immediately preceding character is a "q" then "u" would be overwhelmingly likely to appear next.

In a sequential model, the source is considered in order of its (one-dimensional) presentation. In a hierarchical model (typically used for image compression), the source is recursively partitioned into smaller subparts. The quadtree representation, which recursively splits a square into four subsquares, exemplifies the use of this model.

Coding techniques. The use of a model determines the intended sequence of values. (For example, if a dictionary model is used for text compression, the output may be indices into a hash table. If a gray image is compressed, the output may correspond to gray-scale values, typically in the range 0–255.) An additional mapping via one or a combination of coding techniques is used to determine the actual output.

Either a variable-length mapping is defined, in which relatively frequent elements require fewer bits and relatively infrequent elements require a greater number of bits, or a fixed-length mapping is defined, in which a relatively long sequence of frequently occurring elements or a relatively short sequence of infrequent elements will generate the same number of output bits. The net effect is similar.

Several commonly used data-coding techniques will be described. The decoder is made aware of which coding technique has been employed either by prearrangement or by an inserted codeword.

Dictionary substitution. A repeated sequence of characters is replaced with a reference to an earlier occurrence. A table of possible values for reference is maintained; a value from this table is presented as an index into the table. In ordered table indexing, the table of possible values is ordered by some predetermined criterion. Actual frequency can be used, necessitating either a prescan or the maintenance of a corresponding dynamic frequency table. Alternatively, a strategy for self-organizing a list, such as transpose or move-to-front, may be used. The method of static dictionary substitution is useful for representing alphanumeric representations of information, such as gender or month. In a Lempel-Ziv (or dynamic dictionary) approach, as additional source characters are scanned, some table entries may be deleted and

appropriate candidates may be considered for addition to the table. The variations on this method differ in terms of how the references to earlier occurrences are presented and how the deletions and additions to the table are selected.

A typical Lempel-Ziv procedure is as follows. The table is initialized to contain the set of singleton characters. Iteratively the longest string s in the table that matches the source is determined, and the corresponding table index is output. If the following nonmatched source character is c, then the concatenation of s followed by c is added to the table if space remains. When the table becomes full, this method reverts to static dictionary substitution. Alternatively, after the table becomes full and the saving from compression is observed to degrade below a trigger level, the table is reinitialized. Another approach maintains a sliding window in which string references may occur. For example, a longest match of the to-be-compressed source is sought within the immediately preceding 4095 characters of the already-compressed source.

Run-length encoding. A sequence of values is presented as a sequence of pairs (c, v), where c is the count of the number of values in a row that have the same value v or are within a predefined margin of allowable error.

Predictive coding. A sample value is presented as the error term formed by the difference between the sample and its prediction. In difference mapping, it is expected that the next value is close to the current value, and so the predicted value is defined to be equal to the previous sample. In a linear predictive model, the expectation of the next value is the extrapolation of the present and previous values. Thus, if the previous value was 200 and the current value is 220, the next value will be 240. There are many other predictive models, both one- and two-dimensional, in use.

Huffman coding. Associated with the set of candidate values is a frequency or probability distribution. A code tree is constructed in such a manner as to minimize the expected weighted path length of the tree. A value is presented as a path from the root of the code tree to a leaf, expressed as a sequence of 0's and 1's denoting left and right tree branches.

Arithmetic coding. Associated with the set of candidate values is a frequency or probability distribution from which can be determined a cumulative frequency distribution and, ultimately, associated subintervals of the unit interval. For example, if the only characters are X with probability 20%, Y with probability 70%, and end-of-string (EOS) with probability 10%, then X occupies the interval 0 to 0.2, Y occupies 0.2 to 0.9, and EOS occupies 0.9 to 1. Accordingly, associated with a sequence of values will be an increasingly smaller subinterval of the unit interval. A sequence of values is presented as a real number within that ultimate small subinterval. Continuing with the earlier example, the string YX would be associated with the interval 0.326 to 0.34 because the first character (Y) focuses on the interval 0.2 to 0.9, the second character (X) focuses on the first 20% of that interval (0.2 to 0.34), and the string end focuses on the last 10% of that. Similarly, the string *XY* would be associated with the interval 0.166 to 0.18. The code is the final real number to whatever precision is required to distinguish it from other contenders.

Variable-length representation of integers. A number of different coding techniques have been developed that, depending on the relative expected frequency of integers, can present the values of the integers in compact form. Such codes include Elias codes, Fibonacci codes, and start-step-stop codes.

Scalar quantization. A value is presented (in approximation) by the closest, in some mathematical sense, of a predefined set of allowable values. In some cases, the allowable values may be nonuniformly spaced. For example, a real number may be presented rounded to the nearest three decimal places or truncated to five significant digits.

Vector quantization. A sequence of *k* values is presented as resembling the template (from among the choices available in a given codebook) that minimizes a distortion measure. A typically used distortion measure is that of square-error distortion, which is evaluated as the sum of the squares of the difference between the source and presented values. Under this approach, the encoder must output information sufficient for the decoder to be aware of the contents of the codebook as well as the selection.

Transform coding. A block transform (such as discrete cosine transform) is applied to a vector (or a subimage) of sampled values. The transformed coefficients are then scalar-quantized.

Finally, a number of data structures can be used that provide a trade-off between the parameters of time and space. For example, the use of a hash table instead of an ordered table (especially for Lempel-Ziv and context models) enables searching to be performed very rapidly on average at the cost of an unlikely worst-case slow search.

For background information SEE INFORMATION THE- ORY in the McGraw-Hill Encyclopedia of Science & Technology.

Daniel S. Hirschberg

Bibliography. T. Bell, J. G. Cleary, and I. H. Witten, *Text Compression*, 1990; D. A. Lelewer and D. S. Hirschberg, Data compression, *Comput. Surv.*, 19(3): 261–297, 1987; N. M. Nasrabadi and R. A. King, Image coding using vector quantization: A review, *IEEE Trans. Commun.*, 36(8):957–971, 1988; J. A. Storer, *Data Compression: Methods and Theory*, 1988.

Dating methods

Recent advances in dating methods include new techniques for determining concentrations of isotopes of rhenium and osmium, which have applications in dating various rock deposits, and the use of new methods of thorium-230 dating of corals to determine the timing of events in Earth history that might be related to changes in climate.

Rhenium-osmium geochronometry. Of the natural radionuclides used in radioactive geochron-

ometry, rhenium-187 (^{187}Re), with a half-life of around 4.5×10^{10} years decaying to osmium-187 (^{187}Os), has been the least exploited until recently. The first attempts in the early 1960s showed the difficulty of obtaining a significant signal by the thermal ionization of Os in normal positive-ion mass spectrometry. The low concentrations of the elements Re and Os consequently retarded the development of suitable techniques of dating.

The discovery of the utility of surface ionization mass spectrometry (ion microprobe) by J.-M. Luck and C. J. Allegre at the University of Paris to determine the concentrations of Re and Os and the isotopic composition of Os paved the way for a serious return to the Re-Os system for geochronometry. The isotopic composition of laboratory-grade Os and Re is shown in the **table**, which gives the relative abundances of the seven stable isotopes of Os and the relative abundances of radioactive ^{187}Re and stable ^{185}Re. The protocol of using ^{186}Os, a nonradiogenic, nonradioactive isotope, as the normalization isotope permits the construction of the usual isochron plots in which the ratio ^{187}Os/^{186}Os is plotted against the ratio ^{187}Re/^{186}Os to yield an age for a system with components having the same initial ^{187}Os/^{186}Os ratio but a different Re/Os ratio. Following the success of the ion microprobe technique of analysis, other methods of mass spectrometry have been developed for the determination of these isotopes. One method uses normal positive-ion thermal ionization with laser resonance enhancement to optimize the signal of the Re and Os isotopes being emitted. The method of choice at the present time, however, for Os isotopic measurements at least, is negative-ion thermal ionization mass spectrometry, which has the advantage of improving both sensitivity and precision over the other methods.

Methods of chemical isolation of pure Os and Re in nanogram quantities suitable for measurement require elaborate procedures. These include the usual method of chemical dissolution of the rock or the so-called fire assay technique of isolating Os and Re in a nickel sulfide bead after adding the appropriate fluxes. The extraction of Os, after the initial preparation, is based on the strong volatility of osmium tetroxide. Isotope dilution techniques commonly used in radioactive geochronometry allow for the accurate measurement of the ^{187}Re/^{186}Os ratio.

Abundances of Os and Re isotopes

Isotopes	Abundances, %
^{184}Os	0.023
^{186}Os	1.600
^{187}Os	1.510
^{188}Os	13.286
^{189}Os	16.251
^{190}Os	26.369
^{192}Os	40.957
^{185}Re	37.390
^{187}Re	62.602

Aside from the normal requirement of a closed system in radioactive geochronometry, the accurate knowledge of the decay constant (λ) is required so that ages obtained using the Re-Os system can be compared with ages obtained by other radionuclide systems. Because of the long half-life of ^{187}Re (and therefore small λ) and the difficulty of assessing the absolute decay rate of beta-emitting nuclides, the methods of determining λ depend either on analyzing systems for Re and Os isotope concentrations that have accurately known ages determined by other radioactive geochronometric techniques, or on actually measuring the increase in ^{187}Os in a pure Re solution set aside for a known period of time in the laboratory.

Luck and Allegre used an isochron plot of the ratio ^{187}Os/^{186}Os versus the ratio ^{187}Re/^{186}Os for a suite of iron meteorites presumed to be formed at the same time as the independently well-dated type of stony meteorites called chondrites. The resulting value of λ was 1.52 ± 0.08 $(2\sigma) \times 10^{-11}$ y^{-1}, where σ represents the standard deviation. M. Lindner and colleagues set aside a solution of a purified rhenium salt for several years and then measured the amount of ^{187}Os produced by the decay of ^{187}Re. This direct method of determining the decay constant is subject to the problems of measurement involving spike calibration and equilibration with the carrier. Therefore the method of estimating the error is complex. The resulting value for λ obtained by this procedure is reported as 1.64 ± 0.05 $(2\sigma) \times 10^{-11}$ y^{-1}. The value for λ based on meteorite analyses and that based on the ^{187}Os ingrowth experiment are just within the stated analytical errors of each method. The choice of λ becomes important if comparisons are made with rock systems that are dated by other techniques such as rubidium-87–strontium-87 (^{87}Rb-^{87}Sr), uranium-lead (U-Pb), or samarium-147–neodymium-143 (^{147}Sm-^{143}Nd). In the following sections the value for $\lambda = 1.52 \times 10^{-11}$ y^{-1} will be used, based on the iron meteorite data, which corresponds to a half-life of 4.56×10^{10} years.

The Re-Os method has been used in dating meteorites, mantle-derived igneous rocks, molybdenite deposits, and black shales. The technique is still so new that many of the problems of establishing criteria for closed-system behavior and the degree of constancy of the initial ^{187}Os/^{186}Os ratios remain to be established with the same degree of certainty now expected for other dating techniques.

Meteorites. Iron meteorites are presumed to be derived from the metallic cores of parent bodies that had undergone internal differentiation before disaggregation. Due to the siderophile behavior of both Re and Os, the greater part of these two elements is partitioned into the metallic cores during planetary differentiation. Thus, irons have the very high concentrations of Re and Os that were necessary for successful analysis in the early 1980s. Small differences in the Re/Os ratios among the iron meteorites result in the array of points shown in **Fig. 1**. Luck and Allegre used the slope of the line and the assumed age of irons to arrive at a value of λ as discussed above. If, however, the laboratory value of λ is used, iron meteorites are decidedly younger

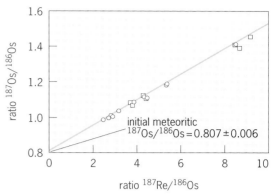

Fig. 1. Re-Os isochron diagram with data for iron meteorites from two laboratories.

(by about 300×10^6 years) than the chondritic (stony) meteorites dated by other techniques. This implies that the metallic cores of meteorite parent bodies may take longer to cool down to the temperature at which Os and Re no longer diffuse significantly (the so-called closure temperature) than the associated silicate systems. Support for this possibility may be found in the observation that many chondritic meteorites plot slightly above the iron isochron on a Re-Os isochron diagram and lie along a slightly steeper line, implying a greater age than the irons. This possibility, however, cannot be reconciled with the observed existence of excess silver-107 (^{107}Ag) seen by analysis of separate phases in iron meteorites. The observed differences in ^{107}Ag could have been produced only by the decay of different amounts of palladium-107 (^{107}Pd; half-life = 6.5×10^6 years) within a core that had cooled to the closure temperature for Pd less than 50×10^6 years after the formation of the parent bodies. Since Os behaves like Pd, closure should be established at an early time for Os as well. It is conceivable that chondritic and iron meteorites plot disparately on a Re-Os isochron because of terrestrial or possibly preterrestrial loss of Re from chondritic meteorites. This could result in the array observed on the isochron diagram.

Komatiites and other mantle-derived igneous rocks. The first geochronologic work on mantle-derived igneous rocks was on komatiites. These highly magnesian volcanics [magnesium oxide (MgO) = 15–30 wt %], largely confined to the Precambrian, have elevated concentrations of Os relative to modern basalts [for example, Os = 1.0–2.2 parts per billion (ppb) in komatiites while basalts have 0.001–0.4 ppb]. In the late 1980s R. Walker and colleagues analyzed komatiite samples from the Canadian Shield. A Re-Os isochron for these samples yielded an age of 2726 ± 93 $(2\sigma) \times 10^6$ years, in agreement with whole rock ages determined using Sm-Nd isotopic systematics. This result was similar to an earlier preliminary study by the Paris group. Obviously Re and Os were not differentially mobilized during the chemical alteration and the low-to-moderate degrees of metamorphism that the komatiite bodies had experienced. Other studies of igneous and metamorphic rocks indicate that differential mobility of Re and Os is possible under certain conditions, however.

Molybdenite. Molybdenite, or molybdenum sulfide (MoS_2), commonly contains Re at approximately the level of 100 ppm while being virtually devoid of Os. For this reason the dating of molybdenite mineralization is feasible by measuring the ingrowth of ^{187}Os from ^{187}Re decay. An early attempt at dating in the early 1960s was followed 20 years later by an extensive study made by other workers. Although partial success was evident in both studies, the possibility of postformational migration of Re was also evident. Future developments may elucidate criteria for closed systems. One potential prospect is dating the time of deformation of granitic gneiss where molybdenite, a platy-cleavage mineral, has been reported to occur along foliation planes much like mica, another platy-cleavage mineral, indicating that crystallization of these minerals is contemporaneous with deformation of the gneiss.

Black shales. Under reducing conditions in the oceanic water column, sediments are deposited that are enriched in metals such as U, Mo, and Re. Although the platinum-group elements, including Os, may also be enriched there, the concentrations of these elements in seawater are so low compared to Re that a strong enrichment of Re over Os occurs in the sediments rich in organic matter. A detailed study of such an environment, the Black Sea, confirmed the feasibility of dating of sedimentary rocks rich in organic matter in the geologic record. These types of sediments are found in the geologic record associated with characteristic fossils that provide prospects of worldwide correlations. For this reason the dating of black shales, as these deposits are called, by the Re-Os method has excellent prospects and is valuable for dating major parts of the geologic time chart of the last 500×10^6 years. G. Ravizza and K. K. Turekian obtained results for black shale horizons on the Devonian-Mississippian boundary found in deep drill holes in North Dakota, and shallower deposits in Indiana and Tennessee have been studied. These fossil-bearing rocks gave an age for the New Albany Shale in Indiana (the best record of the three sites), based on the Re-Os system, of $354 \pm 14\ (2\sigma) \times 10^6$ years. This age is as precise as the best ages obtained from sedimentary deposits using potassium-40–argon-40 (^{40}K-^{40}Ar) or U-Pb dating of volcanic minerals.

Karl K. Turekian; William J. Pegram

Thorium-230 geochronometry. Thorium-230 (^{230}Th) dating is a method for determining the age of a variety of natural materials, including fossil coral skeletons, sediments, soils, peats, cave deposits, and volcanic rocks. It is unique among the various dating methods because it is appropriate for materials that formed during the past 500,000 years. The carbon-14 (^{14}C) method is appropriate only for quite young materials, formed in the past 50,000 years. Most other dating methods are appropriate for older materials, formed millions to billions of years ago. Thorium-230 dating is the only method that permits examination in detail of the timing of events that occurred in the last 500,000 years. This is a very important and exciting portion of Earth history, because it is characterized by climatic conditions commonly referred to as the ice ages. Using the ^{230}Th method, scientists have obtained records of these past climatic conditions. Recently, the ^{230}Th method has attracted renewed attention because of the development of precise measurement techniques that permit determination of ^{230}Th ages with much higher precision than was previously possible. The resulting high-resolution records of past climatic conditions provide scientists with new insight into the causes of the ice ages.

Ice ages. The last 2.4×10^6 years of Earth history (the Quaternary Period) are particularly interesting because the Earth's climate shifted dramatically during this period. Over much of this time, the Earth experienced climatic conditions known as ice ages or glacial conditions. These are characterized by the buildup of huge glaciers, similar to the ones currently covering Antarctica and Greenland. At their furthest extent, the glaciers covered much of northern Eurasia and North America, as far south as the Ohio River Valley. Over this time, the Earth's climate cycled more than 20 times between interglacial conditions (similar to today's climate) and glacial conditions. Ever since the cycles were discovered in the midnineteenth century, scientists have wondered what caused these great shifts in climate.

Of the many hypotheses put forward to explain these cycles, one of them can be tested, if the timing of these past climatic shifts can be determined. In 1941, M. M. Milankovitch, using fundamental physical principles, calculated that the geometry of the Earth's orbit changed with time. This change in geometry caused the intensity of the Earth's seasons to change with time. He showed that changes in the Earth's orbit had resulted in changes of up to 8% in the amount of sunlight reaching the high latitudes of the Northern Hemisphere during the summer months. He hypothesized that cycling between glacial and interglacial conditions was caused by these slight changes in the intensity of Northern Hemisphere summers. He reasoned that cold conditions would allow snow cover to last through the summer and favor the buildup of glaciers. Snow cover would not last through a warm summer, precluding the growth of glaciers and favoring interglacial conditions. His calculations also indicated the times in the past when relatively high and relatively low amounts of sunlight reached the high northern latitudes during the summer months. If Milankovitch's hypothesis is correct, the timing of past glacial periods should correspond to those times when the northern latitudes received relatively small amounts of sunlight in the summer, and the timing of past interglacial periods should correspond to those times when large amounts of sunlight reached the northern latitudes during the summer. Because it is appropriate for the latter part of the Quaternary Period, ^{230}Th dating is one of the few methods that can be used to answer this question and test Milankovitch's hypothesis.

Past climatic conditions. Fossil coral skeletons can be considered as recorders of past climatic condi-

Fig. 2. Horizontal terraces along the north shore of the Huon Peninsula, Papua New Guinea.

tions. Efforts to reconstruct the timing of past climatic shifts (during the Quaternary) have focused on the record contained in fossil coral skeletons. The ^{230}Th method has been tested extensively on this material, and there is reason to believe that under favorable conditions ^{230}Th ages accurately reflect the true age of a coral. Furthermore, certain fossil corals contain important information about the Earth's climate at the time the corals grew. These are the coral species that currently grow close to the sea surface. Fossils of these species record past sea-level elevations. The age of such a coral is the time sea level was at that elevation. By analyzing a number of such corals, of various ages, it is possible in principle to obtain a record of past changes in sea level. For example, numerous horizontal terraces occur along the north shore of the Huon Peninsula, Papua New Guinea (**Fig. 2**). Each terrace is dominantly made up of fossil coral skeletons. Much like a bathtub ring reveals an earlier bath water level, each terrace records a past position of sea level. By dating the fossil corals contained in the terraces, geologists can reconstruct a history of sea-level changes. Such a record is a proxy for the total volume of ice stored in continental glaciers. When continental glaciers melt (interglacial conditions), sea level rises; when glaciers grow (glacial conditions), sea level falls. Thus, if a sequence of corals that record past sea level can be dated, the timing of sea-level rises and falls can be compared to Milankovitch's prediction as a test of his theory. Such a test is possible only if the coral samples can be dated.

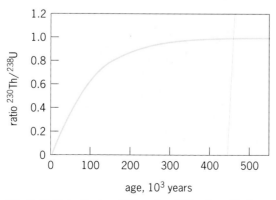

Fig. 3. Shift in thorium-230/uranium-238 ratio with time, which can be used to calculate an age.

Principles of thorium-230 dating. Thorium-230 dating is based on the uranium-238 (^{238}U) decay scheme. Uranium-238 (half-life of 4.47×10^9 years) is a naturally occurring radioactive isotope that decays to produce ^{234}U (half-life of 245,000 years), which in turn decays to produce ^{230}Th (half-life of 75,400 years). When corals grow, small amounts of U are incorporated into their skeletons, but essentially no Th is incorporated. Thus, when a coral first grows, the ratio ^{230}Th/^{238}U is zero. As time passes, ^{238}U and ^{234}U decay to produce ^{230}Th, and the ^{230}Th/^{238}U ratio increases. Because the laws that govern radioactive decay are well known, the rate at which the ^{230}Th/^{238}U ratio changes with time is easily calculated (**Fig. 3**). If the ^{230}Th/^{238}U ratio of a fossil coral can be measured, it is possible to use the relationship between the ^{230}Th/^{238}U ratio and time to calculate an age.

Measurement of thorium-230, uranium-238, and uranium-234. The ^{230}Th method has been in use since the 1950s. However, before 1987 it was plagued by the inability to measure the ^{230}Th/^{238}U ratio precisely, and by the corresponding imprecision in ^{230}Th age. By the traditional method (alpha counting), Th and U isotopes are determined by detecting alpha particles that are given off when an atom decays. Because all three isotopes have long half-lives, the analyst must wait long periods of time to detect small numbers of alpha particles. Because only small numbers of particles are detected, the analyses are not particularly precise. In 1987, a new technique was developed for the measurement of the three isotopes in question. A thermal ionization mass spectrometer (**Fig. 4**) was used to ionize Th and U atoms, sort the atoms with a magnetic field, and count the sorted atoms directly. The mass spectrometric technique resulted in large increases in the precision of measurement and corresponding increases in the precision of a ^{230}Th age. For example, it was demonstrated that the age of a coral that was 100 years old could be measured to within ±3 years; the age of a 10,000-year-old coral could be determined to ±50 years; and the age of a 100,000-year-old coral could be determined to within ±1000 years. With this new tool in hand, scientists set out to put together a high-resolution sea-level curve and test Milankovitch's hypothesis.

Sea-level curve and Milankovitch's hypothesis. Using the mass spectrometric method, scientists have measured ^{230}Th ages of corals that grew during the transition between the last glacial period and the present interglacial period (20,000–6000 years ago), corals that grew during relatively ice-free times during the last glacial period (112,000–88,000 years ago), and some that grew during the last interglacial period (131,000–119,000 years ago). In general, the timing of climatic changes, determined from these measurements, agrees with the timing of climatic changes predicted by Milankovitch. In particular, the timing of interglacial periods corresponds to the times when the high latitudes of the Northern Hemisphere received the largest amounts of summer sunlight. Thus it appears that changes in the geometry of the Earth's orbit are largely responsible for the great shifts in climate that

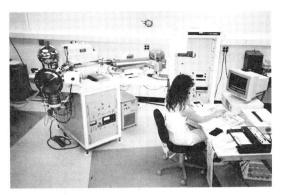

Fig. 4. Thermal ionization mass spectrometer, at the University of Minnesota, which is used to sort and count uranium and thorium atoms.

the Earth experienced over the last 2.4×10^6 years. However, the puzzle is not yet complete. Some of the climate shifts do not correspond in an obvious manner to shifts in northern latitude summer sunlight and may be the result of other mechanisms. Furthermore, the mechanism whereby changes in the Earth's orbit affect climate is not well understood. Scientists are actively working to answer these questions.

Future applications. The field of mass spectrometric measurements of ^{230}Th is in its infancy. Because of the resolution with which ages can be measured over a very critical period of time, scientists will continue to make important contributions using this method. The main areas where contributions will be made are in understanding climate change, in understanding tectonics (a field concerned with movement of material within and on the surface of the Earth, for example, during earthquakes), and in understanding how magma is generated within the Earth.

For background information SEE CLIMATIC CHANGE; DATING METHODS; GEOCHRONOMETRY; GLACIAL EPOCH; SEA-LEVEL FLUCTUATIONS; SECONDARY ION MASS SPECTROMETRY (SIMS) in the McGraw-Hill Encyclopedia of Science & Technology.

R. Lawrence Edwards

Bibliography. E. Bard et al., Calibration of the C-14 timescale over the past 30,000 years using mass spectrometric U-Th ages from Barbados Corals, *Nature*, 345:405–410, 1990; R. L. Edwards et al., Precise timing of the last interglacial period from mass spectrometric determination of thorium-230 in corals, *Science*, 236:1547–1553, 1987; R. L. Edwards, J. H. Chen, and G. J. Wasserburg, U-238–U-234–Th-230–Th-232 systematics and the precise measurement of time over the past 500,000 years, *Earth Planet. Sci. Lett.*, 81:175–192, 1987; M. Lindner et al., Direct determination of the half-life of ^{187}Re, *Geochim. Cosmochim. Acta*, 53:1597–1606, 1989; J.-M. Luck and C. J. Allegre, ^{187}Re-^{187}Os systematics in meteorites and cosmochemical consequences, *Nature*, 302:130–132, 1983; G. E. Ravizza and K. K. Turekian, Application of the ^{187}Re-^{187}Os system to black shale chronology, *Geochim. Cosmochim. Acta*, 53:3257–3262, 1989; R. J. Walker, S. B. Shirey, and O. Stecher, Comparative Re-Os, Sm-Nd, and Rb-Sr isotope and trace element systematics for Archean komatiite flow from Munro Township, Abitibi Belt, Ontario, *Earth Planet. Sci. Lett.*, 87:1–12, 1988.

Decision theory

Decision support is the process of providing data and models to decision makers, usually within a computerized environment. Decision makers, such as managers in a business organization, operate the decision support system (DSS) as they analyze complex problems. The problems are often of a technical nature and can be analyzed mathematically. The decision support system does not solve the problem; rather it enables systematic analysis of the problem and possible solutions. Decision support systems came into widespread use in the late 1970s and blossomed with the advent of the personal computer and easy-to-use software interfaces.

In the 1990s, as computing expands from the desktop to the conference room and from stand-alone computers to computers linked via local-area and wide-area networks, group decision support has become of interest. Group decision support expands the notion of providing data and models to individuals to include the realm of multiparty work, such as meetings and conferences. Members of a work group are able to analyze problems, discuss issues, develop plans, and evaluate alternative courses of action via specialized software that is simultaneously available to everyone in the group. Group members may meet face to face in a conference room (see **illus.**) or via a computer network. So far, most implementations have been in the conference-room setting.

Group decision support sytems (GDSSs) are distinguished from what computer vendors call groupware products. Groupware products operate exclusively on computer networks and support functions such as sending messages, sharing documents, and organizing mail and other material in the office. Group decision support systems do all this and more. They operate in face-to-face meetings as well as on networks and, most importantly, they provide group decision models and specialized techniques for decision making. Whereas groupware products have been available since the late 1980s, the first commercially available group decision support system was introduced in 1990.

Components. The term group decision support system refers to software that is specially designed to aid in group decision activities. Use of the software requires computer hardware that can accommodate multiparty input and retrieval of information. In some cases, specialized furnishings also may be required. For example, in a conference room (see illus.), participants may sit at a U-shaped table with recessed computer workstations. The furnishings are designed to make the technology as unobtrusive as possible and to accommodate eye contact among all participants. A projection device attached to a separate workstation is used to display computer-generated lists, graphs, or other output on a large common viewing screen.

Each participant has a workstation consisting of a

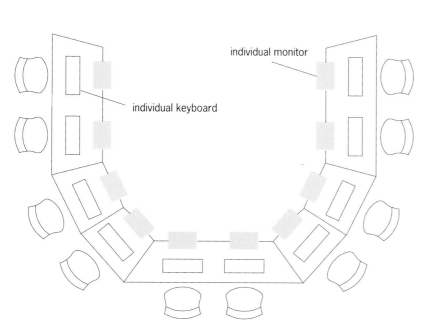

public screen

individual monitor

individual keyboard

Conference room furnished for a group decision support system. Computer support is provided to each participant, and results of group work are displayed on the large screen at the front of the room.

computer terminal and keyboard through which ideas, votes, comments, and so on can be entered. Typically, such information is entered anonymously. Simultaneous and anonymous entry of information speeds up the data-gathering process and encourages group members to be creative and uninhibited in self-expression. Once entered, information can be publicly shared on the common viewing screen. In a face-to-face meeting, a common, large screen is at the front of the room. If group members are dispersed across a network, public information is viewed in a computer window or other designated screen at the individual's private workstation. Computer cables or a local-area network is needed to connect all of the workstations together into a common workstation, or server, where the software for the group decision support system resides.

Software designs for group decision support systems vary, but most include at least two types of computer modules or programs: (1) a so-called private module, which collects information from individual members and responds to whatever commands the group member enters at his or her workstation, and (2) a so-called public module, which aggregates individual information for display on the common viewing screen. In some implementations, only a group leader, technician, or other facilitator can operate the public module. In others, each group member has direct access to the public module. In all systems, at least some of the public functions operate automatically.

Group decision support systems aim to improve decision making by encouraging full participation of all members and by providing structured decision tech-

niques for the group. The storage, retrieval, and manipulation of data facilitates group "memory" and makes the group decision support system much more powerful than a desktop computer or electronic blackboard. Information generated during a group meeting can be used in subsequent meetings or to monitor group work over time. Typically, groups do not rely entirely upon a group decision support system during a meeting but rather use a subset of the features at various points during their discussions. Standard verbal discussion, or telephone or electronic messaging in the case of dispersed groups, supplements use of the group decision support system.

Functions. Group decision support systems support seven major aspects of group decision making: (1) idea generation, (2) idea evaluation, (3) idea organization, (4) planning, (5) problem formulation, (6) resource allocation, and (7) open, structured discussion. These are supported using functions of the sort given in the **table**. For example, to facilitate creative thinking in the group, each participant enters ideas at an individual workstation. Lists of ideas are displayed on the common viewing screen. The group then discusses the ideas, often without identifying who entered particular ideas. Ideas are praised or criticized based on their soundness rather than on who entered them. Ideas can be systematically evaluated using weighting, rating, ranking, or voting routines. Members are anonymously polled by having them enter evaluations privately at their own terminals. The common viewing screen then shows aggregated summaries of the polls, such as averages, ranges, or a frequency distribution. Ideas can also be aggregated into clusters or categories.

Planning, problem formulation, and resource allocation are supported via step-by-step procedures that

Support functions for group decision-making activities

Group decision-making activities	Support functions of group decision support system
Idea generation	Anonymous, simultaneous entry of ideas followed by display of idea lists on the common viewing screen
Idea evaluation	Weighting ideas (weights of all ideas sum to 100 points) Ranking, or ordering, ideas Rating, or scaling, ideas (such as on a 1-to-7 scale) Voting yes/no on each idea
Idea organization	Clustering ideas into common categories or themes, outlining
Planning	Planning models, such as forecasting, stakeholder analysis, or contingency analysis
Problem formulation	Case-based reasoning, causal analysis, developing analogies to similar problems
Resource allocation	Budget models, analyzing alternative budget categories against preset criteria
Open, structured discussion	Agenda management, group note taking, storage and retrieval of meeting minutes, commenting (generally or by topic)

require individuals to enter information about the problem at their workstations; the information is then aggregated in a form consistent with the particular decision technique and displayed on the common viewing screen. Many of the techniques are mathematically based, combining numeric evaluations of text-based ideas according to a preset model. Results do not provide the group with a final decision but rather with feedback for decision analysis. The group is responsible for interpreting the information generated by the technique or model and for developing the final plan, budget, or decision.

Information generated during a meeting with a group decision support system can be printed or stored for later retrieval. At any time during the meeting, participants can enter comments or take notes. Sometimes commenting may be done by topic or category, whereas at other times it is completely open. The group decision support system also can be used to set and implement a meeting agenda.

Implementations. Group decision support system technology is used most frequently for three types of applications: general planning, quality team meetings, and group-information-needs determination and requirements specification on such important tasks as software development.

General planning meetings include such activities as issues identification, resource management, personnel evaluation, and strategic planning. When groups face complex problems like these that include many issues and are in need of innovative solutions, the group decision support system can help in getting the group to quickly surface and "put on the table" all the issues at hand. Idea generation and evaluation functions, as well as specialized decision models, can be helpful in generating and evaluating creative ways to address the issues.

Quality teams are small groups of workers in a business who meet to develop ways of improving organizational efficiency and effectiveness. Quality teams vary in their design and style of implementation, but most emphasize common principles: union–management cooperation, participative decision making, and systematic problem solving. Using a group decision support system in quality team meetings enables realization of these principles and augments the use of quality manuals, flip charts, and trained facilitators.

Development of software within a business typically involves meetings between technical staff and staff who will actually use the system once it is completed. Group decision support systems are being used in these meetings to enhance communication and enable identification of the information requirements of the software and design objectives.

Group decision support systems are not widely used at this point; there were fewer than 100 installations, worldwide, within government and private organizations in mid-1991. But interest is growing rapidly, and several major computer vendors now offer commercial group decision support systems. In time, these will blend with groupware products that are already available.

The viability of group decision support systems for improving organizational meetings has been demonstrated. However, dramatic improvements in group decision making are not guaranteed with this technology. Groups that have a practical grasp of the use of group decision techniques and that confront problems for which the group decision support system is particularly suited are more likely to benefit. Researchers are currently exploring how the range of facilities and applications of group decision support systems might be expanded.

For background information SEE DECISION THEORY; LOCAL-AREA NETWORKS in the McGraw-Hill Encyclopedia of Science & Technology.

Gerardine DeSanctis

Bibliography. L. M. Jessup and J. Valacich (eds.), *Group Support Systems: New Perspectives*, 1993; S. Kinney, R. Bostrom, and R. Watson (eds.), *Computer Augmented Teamwork: A Guided Tour*, 1991; A. P. Sage, *Decision Support Systems Engineering*, 1991.

Diamond

The metastable synthesis of diamond from hydrocarbon gases has been achieved by using numerous techniques, the most common being various glow-discharge plasma methods (microwave, radio-frequency, and direct-current) and thermally activated, or hot-wire, methods. These coatings can be differentiated from so-called diamondlike films, which are plasma-deposited noncrystalline carbon and hydrocarbon films, by Raman spectroscopy and x-ray diffraction, which demonstrate that the material is indeed well-crystallized diamond. Whereas in the early to mid 1980s the early reports met with much skepticism, and some workers doubted whether the quality of the material could be high enough to be useful, since 1989 most researchers have become convinced that material rivaling the quality of natural and high-pressure synthetic diamond can be grown by a vapor-phase process. Many important issues remain to be resolved before a reliable assessment can be made of the impact this development will have, but numerous industrial, academic, and government laboratories have already begun to develop and exploit this technology for a large number of applications. Engineering, scientific, and commercial interest has continued to grow as both the commercial possibilities and recent research achievements become more widely recognized and better appreciated. A wide variety of objectives exist because of diamond's extreme and unusual combination of properties. Major objectives include layers and coatings for infrared and visible optical systems; the preparation of semiconductor-quality material for radiation-hard, high-temperature semiconductor electronics; the packaging of high-speed digital electronics; and the coating of many different industrial tools.

Growth process. Even before the pioneering work in the 1950s and early 1960s, there were sporadic reports that diamond could be synthesized by decomposing gaseous hydrocarbons. In the early 1980s Japanese scientists confirmed earlier Soviet experiments and showed that stable diamond growth was possible if certain conditions were established and maintained during growth. Paramount among these conditions is a steady supply to the solid surface of atomic hydrogen, which is now believed by almost all researchers to be the central reason for the growth of diamond instead of graphite.

Although the kinetically stable growth of noncrystalline or poorly crystallized forms of carbon, boron nitride, boron carbide, and many other compositions had been long known, the growth of well-crystallized metastable forms is rare, and much research continues to focus on understanding the diamond growth process. Two fundamentally different points of view have emerged to explain why well-crystallized diamond, and not graphite or vitreous carbon, is observed in many experiments. The first argues that graphite is gasified by atomic hydrogen at a rate higher than diamond and hence diamond is kinetically stable with respect to graphite. If diamond formation is kinetically controlled, then the deposition mechanism is critical, and much debate still centers on the growth mechanism. Other researchers suspect that at the growth interface the diamond surface is preferentially stabilized by hydrogen terminations that are maintained by the presence of atomic hydrogen. If this viewpoint is correct and bulk reorganization can be ignored, then there may emerge general principles leading to the growth of other well-crystallized metastable phases, notably cubic boron nitride.

Numerous chemical models, some differing with respect to the carbon species believed important, have been proposed in the effort to understand the process. While they all attribute a major role to the presence of atomic hydrogen, the role assigned varies between the different models, with a higher gasification rate for graphite than for diamond being emphasized in some, and stabilization of sp^3-hybridized carbon at the solid surface being emphasized in others. Depending in part on which of these models proves most accurate, the stable growth of well-crystallized diamond may herald a new age of metastable materials synthesis or may represent only a peculiarity of carbon chemistry. However, even in the latter case, the development of vapor-phase diamond and diamond-composite coatings has had, and will continue to have, a profound effect on many important industries worldwide.

Diamond versus diamondlike carbon. Many of the techniques of physical vapor deposition, such as the various ion-beam and sputtering techniques, that have been used to prepare diamondlike carbon films have also been used to prepare films containing nanocrystalline diamond in addition to various disordered carbon structures. Among those reports of films containing essentially no hydrogen (approximately 1 atomic percent or less), several have shown clear evidence for the presence of diamond. However, most such films additionally show evidence for the presence of various noncrystalline carbons and poorly

crystallized forms including phases that have yet to be identified. Hence, differentiation between what is diamond and what has become known as diamondlike carbon remains a frequently debated issue. Because of the difficulty of accurately characterizing very thin films, there have been recent errors in which researchers believed that they had found a new method for diamond synthesis, only to discover later that the material produced was something else. Nevertheless, the list of reproducible methods has grown dramatically with recent developments, which include the growth of diamond from oxyacetylene flames and the pyrolysis of halohydrocarbons such as methyl fluoride.

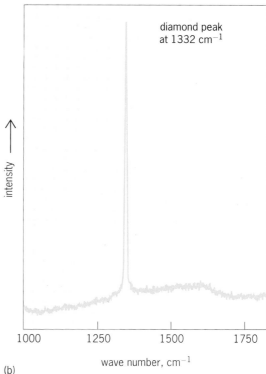

Fig. 1. Diamond crystals grown on silicon with activation by a hot tantalum wire of a dilute mixture of methane in hydrogen. (*a*) High-magnification scanning electron photomicrograph. (*b*) Raman signature obtained with a microfocus Raman spectrophotometer. The sharp peak at 1332 cm⁻¹ corresponds to the optical phonon in diamond permitted by crystal symmetry.

Characterization methods. Two means of characterization are commonly used to differentiate these coatings. The first is x-ray or electron diffraction. X-ray diffraction is generally preferred because a greater level of crystalline order is needed for a positive result and the likelihood of a spurious conclusion is reduced. This very feature, however, renders the technique of limited value in the study of very thin films or of materials that are available in only small amounts. Electron diffraction (particularly, selected-area diffraction from transmission electron microscopy) is a much more sensitive technique and has been used to demonstrate the presence of diamond. However, errors and uncertainties can arise as a result of the extremely small volume of material being sampled and the strong possibility of texture or preferred orientation in vapor-phase-deposited thin films. Indeed, it has been shown that highly oriented thin graphitic films can give electron diffraction patterns very much like that of diamond if the diffraction pattern is taken with the electron beam perpendicular to the plane of the layer. Various electron spectroscopies (Auger electron spectroscopy, electron energy loss spectroscopy, and so forth) have also been used for the characterization of carbon coatings, and the detection of sp^3-bonded carbon has been used to demonstrate the presence of diamond. However, these techniques lead to some uncertainty, as they probe only approximately the top nanometer of the surface; hence a possibility exists that the data are not representative of the bulk solid.

The characterization technique that has proved most useful and powerful is Raman spectroscopy. Crystalline diamond has a single symmetry allowed and a characteristic Raman peak at 1332 cm^{-1}, and this peak has become the single most accepted and definitive test for the presence of diamond. **Figure 1** shows the Raman signature and a scanning electron photomicrograph of a diamond film deposited on silicon; the method involved a tantalum wire at approximately 2200°C (4000°F), located 0.5 cm (0.2 in.) from a silicon substrate at 930°C (1700°F), and a dilute (0.5%) mixture of methane in hydrogen gas at a pressure of 3.3 kilopascals (25 torr). Although other Raman peaks may be present occasionally, the position and breadth of this peak has become a major test and is also extensively used as a measure of crystalline perfection. As disorder or defects are introduced, this peak broadens, and thus the peak width at half maximum is often cited as a measure of crystal quality. The best vapor-phase-deposited diamond will generally give a peak-width value of approximately 1.7–3.0 cm^{-1}, depending upon the resolution of the instrument used. Those carbon coatings and layers that do not exhibit the characteristic diffraction pattern of diamond, or its characteristic Raman shift, but that do have many of its physical properties, including hardness, chemical resistance, and optical transparency, are those that have become known as diamondlike carbon.

Growth methods. The methods employed for diamond growth typically combine a dilute hydrocarbon-hydrogen mixture with a means of dissociating molecular hydrogen and the hydrocarbon precursor to yield a mixture of free-radical species, including atomic hydrogen. Present methods include not only glow-discharge plasmas and hot refractory metals (commonly tungsten, tantalum, and rhenium) but also the combustion of acetylene in oxygen. In the latter method, diamond is easily grown from an ordinary brazing or welding torch, and this approach is now reported to be a high school chemistry experiment. A mixture of oxygen and acetylene is used, at a ratio close to unity or slightly rich in acetylene, and the incandescent initial-combustion portion of the flame, close to the torch nozzle, is made to impinge on a water-cooled substrate, typically silicon or molybdenum. Microscopic crystals of diamond are seen to grow on the substrate, usually within approximately the first 10 min (**Fig. 2**). The substrate temperature range at which diamond has been deposited is quite broad, approximately 350–1100°C (650–2000°F). However, the optimum surface-temperature range, one balancing the requirements of both acceptable growth rate and crystal quality, is believed to be 800–1000°C (1500–1800°F).

Substrates. The range of substrates that have been successfully coated by various methods is also very broad and includes such materials as silicon, silicon dioxide, aluminum oxide, silicon nitride, cemented tungsten carbide, and other ceramics, as well as many metals, including molybdenum, tantalum, tungsten, and copper. Diamond has also been grown on graphite, cubic boron nitride, and diamond itself. The growth of diamond on cubic boron nitride, which has a lattice constant (0.362 nm) nearly the same as diamond (0.357 nm), has been reported to be heteroepitaxial.

Diamond polishing. Early in the work, it was found that to achieve rapid and high-density diamond nucleation on many substrates, the substrate surface should be polished with a micrometer-to-submicrometer diamond abrasive. Whether the resulting enhanced nucleation is a consequence of some specific tribological interaction or is simply the result of carbonaceous

Fig. 2. Scanning electron photomicrograph of a polycrystalline diamond film grown by using an oxyacetylene brazing (welding) torch with a mixture of oxygen and acetylene, slightly rich in acetylene. Note the (100) facets of crystalline diamond, a commonly seen crystal morphology in flame-grown diamond films and layers.

material, including possibly diamond, left adherent to the substrate remains an issue of considerable interest. However, the efficacy of diamond polishing to enhance nucleation has been observed in most experiments.

Goals and challenges. Much effort has been focused on understanding the nature, distribution, and types of defects found in diamond synthesized by chemical vapor deposition, and their causes. The resistivity of diamond is extremely sensitive to its crystalline perfection and the presence of minor impurities, with values for natural diamond that range from approximately 10^4 ohm-cm to approximately 10^{20} ohm-cm (in the absence of light) for the highest-purity gem-quality natural stones. A major goal is a material with a resistivity similar to the best natural diamond. The better polycrystalline vapor-deposited diamond films have values in the range 10^{10}–10^{12} ohm-cm, with some values reported as high as 10^{16} ohm-cm. Although cutting tools and other nonelectronic products have been commercially manufactured, numerous scientific and technological challenges remain. These include the fabrication of adherent diamond coatings on substrates that can be degraded or destroyed by the conditions normally used for the growth of diamond. Many such materials are technologically important, such as the steels for high-speed tools, plastics, and materials used in infrared or optical windows [such as silicon dioxide (SiO_2) and zinc sulfide (ZnS)]. For such substrates, deposition temperatures will have to be lowered substantially while adequate growth rate is maintained.

Applications. Many diamond-coated products have been introduced, ranging from speaker diaphragms and laser-diode heat sinks to diamond-coated turning, boring, and drilling tools. New, pioneering methods of manufacture, which include the use of high-speed plasma torches, give linear growth rates as high as approximately 300 micrometers/h over areas as large as several square inches. Low-temperature deposition offers the greatest hope for fabricating adherent films on thermally sensitive substrates, such as hardened steels and plastics, and methods for this purpose continue to be pursued. The technological applications of both diamond and diamondlike carbon are numerous and are growing as larger-scale deposition technologies become commercially available. The use of diamondlike carbon for the coating of plastics and other thermally sensitive surfaces (such as magnetic media) for hardness (scratch resistance) and optical transparency is the major application to date. Currently available or developing products range from eyeglasses and supermarket checkout countertops to magnetic recording media, both tape and disk.

For background information *SEE CRYSTAL GROWTH; DIAMOND; RAMAN EFFECT* in the McGraw-Hill Encyclopedia of Science & Technology.

Walter A. Yarbrough

Bibliography. J. C. Angus and C. C. Hayman, Low pressure, metastable growth of diamond and "diamondlike" phases, *Science*, 241:913–921, 1988; W. A. Yarbrough, Current research problems and opportunities in the vapor phase synthesis of diamond and cubic boron nitride, *J. Vac. Sci. Technol.*, A9(3):1145–1152, 1991; W. A. Yarbrough and R. Messier, Critical issues and problems in the chemical vapor deposition of diamond, *Science*, 247:688–696, 1990.

Dinosaur

Since a dinosaur was first described more than 150 years ago, more than 300 valid species of dinosaurs have been named. Forty percent of these species were established over the past 25 years. Specialists are currently refining knowledge of the morphology, functional anatomy, variation, and interrelationships of dinosaurs and are delving into the areas of dinosaur physiology, behavior, and extinction. Dinosaur research has become multidisciplinary and multinational.

Types of dinosaur resources. A variety of dinosaur resources, including skeletons, footprints, skin impressions, and eggs, have been discovered. Whole skeletons are common at some sites. Sometimes skeletons were completely dismembered before burial, and the bones were mixed with the remains of other individuals and concentrated in bonebeds. Concentrations of more than 200 bones per square meter have occurred, and some of these bonebeds can stretch for several kilometers along the bottoms of ancient river channels.

During its lifetime a single dinosaur could make millions of footprints, and under the right conditions some of these footprints were preserved. The study of dinosaur footprints and trackways provides information on the activities of dinosaurs. Impressions of the pads and even of the skin under the feet may be preserved in footprints. From trackways, it is evident that almost all dinosaurs held their tails off the ground when they walked and that carnivorous dinosaurs held their legs directly under their bodies when they walked and ran. A formula, based on a constant relationship between speed, animal sizes, and stride length, has been developed to determine from trackways how fast a dinosaur was walking or running. So far, the fastest dinosaur recorded was a carnivorous dinosaur that left its trackway in the Early Cretaceous muds of Texas, and was calculated to have been running at about 25 mi (40 km) per hour. Trackway sites often show how animals interacted. For example, several localities around the world show numerous individuals of the same species walking in the same direction as if they were moving together in a herd. Trackway sites in Texas suggest that young sauropods walked in the center of the herd.

Skin impressions have been found for every major type of dinosaur. In tyrannosaurids and ceratopsians from Alberta, Canada; sauropods from Wyoming; hadrosaurs from Alberta, Mongolia, New Mexico, and Wyoming; and a stegosaur from southern China, the skin has a lightly pebbled texture reminiscent of the patterns seen in some modern thick-skinned mammals. Ankylosaurs from Alberta and a large theropod, *Carnotaurus*, from Argentina have much coarser tubercles that impart the scaly appearance often seen in reconstructions of dinosaurs.

Dinosaur eggs in nests were discovered by the Central Asiatic Expeditions of the American Museum of Natural History in the 1920s. Since that time, thousands of dinosaur eggs have been found. But only since 1972 have embryos of theropods, sauropods, hadrosaurs, ankylosaurs, and ceratopsians been found within eggs. The largest dinosaurs laid eggs not much larger than those of an ostrich. Even sauropod species, some of which exceeded 100 ft (30 m) in length, would have had babies less than 3 ft (1 m) long. It has been suggested on the basis of bone growth that some dinosaurs were able to fend for themselves as soon as they were born but others would have required parental assistance for feeding. For example, the ends of the bones of hypsilophodonts are well formed at birth, whereas those of hadrosaur hatchlings are still largely cartilaginous. The suggestion is that young hadrosaurs could not walk efficiently or far, and would be unable to fend for themselves.

Anatomy. There is still much to be learned about the anatomy of dinosaurs. Some species have been established solely on diagnostic skeleton parts that cannot be assigned to any other known species. For example, *Troodon* was based on a very distinctive tooth described in 1856. No other known animal had such teeth, and different workers referred to the individual as a lizard, theropod, hypsilophodont, or pachycephalosaurid. A positive identification was not made until 130 years later, when a jaw with teeth was finally discovered; it showed that *Troodon* was a small carnivorous dinosaur. Nearly complete skeletons of this animal and its closest relatives have been found only recently in Montana and China.

Microscopic and chemical examinations of bone have revealed further information on dinosaur biology. Study of the microstructure of the bones of young animals suggests that dinosaurs had extremely rapid growth rates, comparable to those of modern birds and mammals (which are significantly higher than those of modern reptiles and amphibians). Several laboratories are now attempting to liberate genetic material from dinosaur bone. The most promising results have come from 65-million-year-old dinosaur bones from Alaska, which for some reason were not permineralized (fossilized).

Reconstructions of muscles have been made by observing muscle attachment points on well-preserved bones and by making comparisons with a range of modern animals. Knowledge of skeletal structure and reconstructions of the musculature are used to study the functional morphology of dinosaurs. For example, study of the jaws and teeth of hadrosaurs has revealed grinding mechanisms as sophisticated and efficient as those of modern herbivorous mammals.

Size. The Dinosauria included the longest, tallest, and heaviest land animals known. For example, the poorly known *Therizinosaurus* from central Asia had claws that were more than 2 ft (60 cm) long. Current fieldwork continues to reveal larger individuals and species. For example, a multiyear excavation in New Mexico is slowly revealing the gigantic proportions of a sauropod known as *Seismosaurus*. This animal is closely related to the well-known *Diplodocus* but is estimated to be more than 140 ft (43 m) long.

Small dinosaurs, such as the chicken-sized *Compsognathus*, also existed. However, the remains of smaller dinosaurs tend to be rare because these bodies were more susceptible to destruction by scavengers and the elements. From recent recovery of smaller dinosaurs, paleontologists now know that small species were common and even dominated some habitats. Juvenile dinosaurs are also being recovered with greater regularity, and include a 9-in.-long (22-cm) skeleton of *Psittacosaurus* from Mongolia, a 1-ft (30-cm) skeleton of *Pinacosaurus* from China, and a 1-ft (30-cm) *Mussasaurus* from Argentina.

Classification. The classification of dinosaurs is being refined by using cladistic methods, which emphasize the derived characters shared by related taxa. Cladistics not only has improved the understanding of the interrelationships of dinosaurs but has also shown to the satisfaction of the majority of specialists that birds are descended from theropods, which were carnivorous dinosaurs. More than a hundred unique characteristics found in theropods and birds are not found in any other major group of animals. Under a cladistic classification, birds are a subset of the Dinosauria, and in that sense dinosaurs have not become extinct. *SEE AVES.*

Diversity. Although several hundred species of dinosaurs have been identified, many thousands of species must have evolved over their 140-million-year history. Each ecosystem around the world would have had its own unique fauna. Each of two Cretaceous sites, Dinosaur Provincial Park in Canada and the Nemegt Valley of Mongolia, has produced more than 35 species of dinosaurs, totaling almost a quarter of the known dinosaur species. The high diversity of dinosaurs at these sites can be correlated with the richness of the food available to them. Not all ecosystems supported this kind of diversity, however. The Cretaceous Djadokhta Formation of central Asia represents an arid or semiarid environment. Although the Djadokhta has produced the remains of several thousand dinosaurs, all can be assigned to only a half dozen genera that were specialized for living in deserts.

Variation. For some species of dinosaurs, sufficient numbers of fossils have been collected to show individual variation, changes during growth, and sexual variation. Monospecific bonebeds often represent the mass death sites of herding dinosaurs. These sites, known for some species of theropods, ornithopods, ceratopsians, and ankylosaurs, represent single gene pools and thus provide information on the variation possible in single biological species of dinosaurs.

Among ceratopsians, the study of suites of specimens from monospecific bonebeds of *Pachyrhinosaurus* and *Centrosaurus* from Alberta and *Styracosaurus* from Montana has been particularly interesting. The adults of these genera are easily identified by distinctive horns on the face and frill, but the juveniles have only small horns over the nose and eyes and no horns on their short frills. It would be difficult to distinguish the juveniles of these three genera unless they were found associated with the adults in bonebeds. The horns developed as the animals approached sexual maturity, when differences in appearance between the males and females also became noticeable. During the 1970s it was shown that hadrosaurs did not develop a crest on the top of the head until around the time of sexual maturity. The same had long been suspected to be the case with the horns and frills of ceratopsians, but proof had to wait for the study of juveniles in bonebeds.

Behavior. There can be little doubt that many species of dinosaurs were gregarious and spent at least some of the year in huge herds just as many species of mammals do today. Nests of hadrosaur and hypsilophodont eggs in Montana suggest that some species returned to the same communal nesting sites every year. When babies of the hadrosaur *Maiasaura* hatched, they probably remained in the nest for several months, during which time food must have been brought to them by their parents.

The monospecific bonebeds that provide evidence of herding behavior may have been formed by migrating species. Dinosaurs inhabited the polar regions during Cretaceous times and took advantage of the high plant productivity stimulated by up to 24 h of daily sunlight during the polar summer. During the winter, food would have been more difficult to obtain. Evidence from Australia and Antarctica suggests that smaller dinosaur species could overwinter in polar regions either by scrounging enough food from dormant plants or by hibernating. Some species of hadrosaurs and ceratopsians in the Northern Hempisphere may have collected into large herds as winter approached, and then have pushed south to the midlatitudes. This behavior would explain the presence of large herds in Alberta and Montana and the high dinosaur diversity of midlatitude sites.

Extinction. A growing body of evidence suggests that the Earth may have been hit by an asteroid at the end of the Cretaceous Period. Sites around the world where the Cretaceous-Tertiary boundary has been identified by floral and faunal changes have produced higher-than-normal levels of iridium (an element normally found in higher concentrations only in meteorites), often associated with shocked quartz and micro diamonds. An asteroid hitting the Earth would have put so much dust into the air that sunlight might not have reached the ground for months. Plants would have either died or become dormant, and food chains would have collapsed. The majority of dinosaurs were herbivores and would have died out quickly. Large carnivores thereby would have been deprived of their food and would have become extinct as well. Small animals, such as insects, frogs, lizards, birds, and mammals, would have been able to find enough food throughout this period and could have survived by eating dead plants and seeds.

Although the asteroid theory provides a very elegant and simple model to explain the great extinctions at the end of the Cretaceous, it does not provide all of the answers and has not been universally accepted. The smaller theropods could have survived, for example, by continuing to eat lizards, mammals, and other small animals. The fact that dinosaur teeth are abundant in early Paleocene rocks in Montana and China suggests that some dinosaurs may have survived the extinction event. However, the teeth might also have been washed out of older sediments by Paleocene rivers.

Although the extinctions at the end of the Cretaceous continue to receive much attention, paleontologists still do not know if dinosaurs died out rapidly or slowly. Evidence indicates that dinosaur diversity peaked 10 million years before the final disappearance of dinosaurs; after that peak, diversity dropped steadily. The record of dinosaur life at the end of the Cretaceous is well documented in North America, but evidence to show that dinosaurs died out at the same time on other continents is not strong.

For background information SEE DINOSAUR; PALEO-ECOLOGY; TRACE FOSSILS in the McGraw-Hill Encyclopedia of Science & Technology.

Philip J. Currie

Bibliography. K. Carpenter and P. J. Currie (eds.), *Dinosaur Systematics: Approaches and Perspectives*, 1990; D. A. Russell, *An Odyssey in Time: The Dinosaurs of North America*, 1989; D. B. Weishampel, P. Dodson, and H. Osmólska (eds.), *The Dinosauria*, 1990.

Direct broadcast radio satellites

Several projects are under way to broadcast radio programming directly from satellites in high orbit to inexpensive home, car-mounted, and portable radio receivers. These projects share a common emphasis on the use of digital signal modulation, the compression of audio source information into a reduced bandwidth, and transmission in select portions of the microwave region at frequencies of 1400–2600 MHz. The satellite sound broadcasting projects are scheduled to culminate with worldwide service capability by the year 2000. Service is planned to start in various parts of the world in 1994.

The radio receivers in a satellite sound broadcasting system employ either patch antennas of only 10 in.^{2} (64 cm^2) or 3-in.-tall (8-cm) helixes, although these antennas must receive signals relayed from 22,300 mi (35,800 km) above the Equator. These remarkably sensitive portable satellite radio receiving terminals evolved from digital satellite navigation receivers that are now commercially available. A dramatic example of this portable satellite terminal technology is its use by hikers in wilderness areas. A handheld digital satellite navigation receiver can direct a lost hiker to safety. The new satellite sound broadcasting projects rely on digital satellite receivers that are functionally similar but are tuned to radio broadcasting rather than navigation frequencies and operate at much faster data rates.

Satellite sound broadcasting. Satellites have been used for many years to transmit radio programming across large areas for reception by large, directive ground antennas and subsequent retransmission by ground transmitters. However, the satellite power received on the ground was never strong enough to be received by a portable or mobile radio, with its necessarily near-omnidirectional antenna. Consequently, satellite radio broadcasting to a general mobile audience was not possible.

Satellite sound broadcasting represents the culmination of 30 years of effort to reduce the size of satellite receiving terminals (see **illus.**). Spread-spectrum satellite data networks in the 1980s paved the way for mobile satellite navigation and portable satellite sound broadcasting systems in the 1990s. By the early 1990s it became economically practical to provide much of the signal gain (about 20 decibels) generally offered by satellite ground antennas via the use of spread-spectrum coding techniques. Also, the ratio of satellite power to downlink frequency increased due to higher-power satellite transmitters and new frequency allocations in the lower microwave region between 1400 and 2600 MHz. Increasing the ratio of satellite power to downlink frequency reduces the amount of antenna gain needed to receive a satellite radio signal, as compared with earlier point-to-multipoint satellite radio redistribution systems. This is due to the inverse-square relationship between wavelength and the necessary transmission power for a given received power flux density on the surface of the Earth. Finally, the need for directive satellite ground antennas was also reduced by sophisticated audio source coding techniques, which demonstrated bandwidth compression ratios of up to 6:1 through reliance on psychoacoustical factors. This factor minimizes the necessary data rate that must be decoded by the satellite radio receiver. *SEE* DATA COMPRESSION.

Satellite sound broadcasting techniques also have been developed to deal with the variable fading characteristics of a mobile channel. In traditional satellite communications systems, variable fading arises only from rain, and Earth stations are sized appropriately to compensate for rain losses with built-in margin (greater gain). But in satellite sound broadcasting systems, the quality of the free-space channel between a radio receiver and the satellite can change almost constantly and unpredictably, as in the case, for example, of a satellite radio-equipped car passing by buildings, trees, and other line-of-sight obstructions.

Satellite sound broadcasting techniques to mitigate the variable fading characteristics of a mobile channel are called space diversity, time diversity, and frequency diversity. The common element in each of these techniques is a reliance on the fact that mobile-link fades occur differentially in space, time, and frequency. For example, by spreading a source signal among several frequencies, a signal loss at any particular frequency need not be perceptible to the listener because the radio still receives the major portion of the signal at the other frequencies. Similarly, a source signal can be divided randomly in many microsecond-level time blocks. Fades that affect some time blocks need not be perceptible to the listener so long as most of the time blocks are not affected. The most effective satellite sound broadcasting antifading techniques

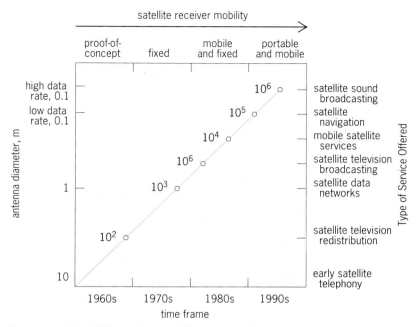

Development of satellite receiver capability and reduction in antenna size over time since the 1960s. 1 m = 3.3 ft. The numbers 10^n are the approximate numbers of satellite receivers in the network. (The exponent n generally increases by 1 after 10 years of operation.)

combine frequency and time diversity. Space diversity requires multiple antennas, a feature that to date has retarded its attractiveness among system developers.

Application to radio networks. Satellite sound broadcasting is a replacement technology for terrestrial radio broadcasting systems: shortwave radio (high-frequency band, 3–30 MHz), long-wave radio (AM band, 0.5–1.5 MHz), and medium-wave radio (FM band, 88–108 MHz). The key differences in technology among radio broadcasting systems relate to coverage, fidelity, and convenience.

Satellite sound broadcasting is functionally similar to shortwave and AM radio in that it is designed to cover very large geographical areas, such as countries or continents. However, a satellite sound broadcasting system accomplishes its coverage with direct line-of-sight connections between user radios and the satellite, whereas a shortwave radio broadcasting system achieves coverage via ground-wave and sky-wave propagation. A satellite transmitter requires only about 50 W of solar-panel power to achieve the same radio coverage as a standard 500 kW ground-based shortwave transmitter. Indeed, the largest commercial shortwave radio network in sub-Saharan Africa is also the third largest user of electricity in West Africa. A single radio broadcasting satellite called *AfriStar 1*, approved by the U.S. government in 1991 and scheduled for launch in 1994, will provide commercial audio broadcasting to all of Africa with insignificant ground-based transmitter electricity requirements.

Satellite sound broadcasting differs from shortwave and AM systems in its development primarily as a higher-quality audio medium. Although satellite sound broadcasting need not be digitally modulated, all current projects anticipate digital modulation to enable the sophisticated coding schemes needed to permit low-gain, mobile receiving antennas. Also, digital modulation need not provide an improvement in audio fidelity over shortwave and AM radio, but all current satellite sound broadcasting projects are planning for the audio fidelity associated with either FM radio or compact-disk recordings. Improved audio fidelity is planned for satellite sound broadcasting to encourage consumers to purchase the new satellite radios, in addition to providing more reliable signal reception across large geographical distances.

Current audio source coding technology allows the following correlation to be made between satellite sound broadcasting data rates and audio fidelity—256 kilobits per second: stereo compact-disk quality, CCIR (International Radio Consultative Committee) level 4.5 (on a scale whose maximum value is 5.0); 64 kb/s: monophonic FM-radio quality, CCIR level 4.0; for FM stereophonic broadcasting, combine two channels; 32 kb/s: monophonic AM-radio quality, CCIR level 3.0. Generally, satellite sound broadcasting projects targeted at Europe, North America, and Japan are being designed for compact-disk quality. Developing-country satellite sound broadcasting systems are being designed to operate at 64 kb/s to enable a qualitative fidelity upgrade from shortwave radio, while reducing by a factor of 4 satellite power requirements (and hence cost) as compared to a 256-kb/s compact-disk-oriented system.

Satellite sound broadcasting radio receivers are designed to be much more convenient to use than the shortwave systems they replace. Tuning of the new satellite digital radios will be by arbitrary channel numbers (1, 2, ...) rather than by kilohertz or megahertz. This is because the digital radio systems will be able to reassign radio frequencies to different channel numbers, and command all radios to display the currently appropriate channel number for each frequency. Also, each broadcast will include digital information regarding the program's name and format (for example, NEWS). The new digital radios will be able to display these names, thereby enabling listeners to also tune by this textual information. Finally, the new radios will be individually addressable via a built-in unique identification code. This feature enables each digital radio to also serve as an alphanumeric paging terminal and as a receiver for facsimile messages. The digital paging and facsimile message information is interspersed with the digital audio information. The *AfriStar 1* system, for example, uses 56 kb/s for digital audio and 8 kb/s for digital messaging. The addressability feature of digital radio also enables the creation of software-defined radio subaudiences based on national markets or monthly subscription fees. For national markets, certain broadcasts could be restricted to those identification numbers reserved by a particular country for the radios it imported or manufactured.

Interoperability. There are three concepts for interoperability of satellite and terrestrial digital audio radio systems. The satellite sound broadcasting system could be used to simply uplink, rebroadcast, and downconvert terrestrially broadcast digital signals. This is similar to the way home satellite dishes receive "superstation" television programming, that is, uplinked and rebroadcast local television signals. A second concept is that the terrestrial digital radio transmitters can simply rebroadcast the satellite sound broadcasting signals. The purpose of this scheme is to permit continuous, reliable reception of digital audio signals even when buildings or tunnels block the line of sight to the satellite. Experiments by the European Broadcasting Union in 1990–1991 proved that digital signal processing techniques permit cochannel transmission of satellite and terrestrial digital audio signals, with the radio receivers constructing a coherent signal based on received signal strength. A third concept for interoperable satellite and terrestrial digital audio radio systems is for independently operated systems to be receivable on a single radio receiver. For example, a terrestrial digital radio service might operate in frequencies adjacent to a satellite digital radio service. A single radio receiver would tune across both bands, picking up local (terrestrial) broadcasts on one band and national (satellite) broadcasts on the other.

The benefits of satellite-terrestrial radio interoperability create substantial pressure for standardization. Digital radio standards must be agreed upon at a detailed technical level to ensure that a radio can operate with both a satellite system and dozens of local digital broadcasting stations. Much of the work toward digital radio standardization is being carried out in the International Telecommunications Union. However, it appears that different standards will be adopted for Europe, the United States, and other countries. Europe is moving toward an integrated (cochannel) terrestrial and satellite system. The U.S. Federal Communications Commission announced in 1991 its preference for a complementary (adjacent-channel) local and satellite digital radio system. In developing countries, the geography of high satellite elevation angles (due to equatorial proximity) minimizes the need for terrestrial repeaters. Hence, a simple downconverting and demodulation of satellite signals for listening over existing radio sets may prove to be the most sensible approach.

Economic considerations. It is generally accepted that satellite sound broadcasting provides an economic means of covering a large area with a radio signal. Current digital satellite radio projects have channel costs that compare favorably to those experienced by major international shortwave broadcasters. However, the latter group accesses millions of existing shortwave radios, while the satellite radio broadcaster must face a dearth of radios at the inception of service.

A digital radio must include, at minimum, an antenna, downconverter, demodulator, and audio decoder. An audio processing board, speaker, and housing will generally also be needed. Satellite radio prices are now comparable to those of other high-fidelity audio equipment. However, there are over 2 billion radios in use, more than the number of telephones and televisions combined, and successful penetration of so large a potential customer base will ensure the economies of scale needed to reduce the costs of digital radios to the levels of analog radios.

In order to achieve a large market share, there must first be attractive radio programming that encourages radio sales. However, radio programmers cannot be expected to pay for satellite sound broadcasting channels in the absence of a large audience. This "chicken-or-egg" dilemma is being addressed in three ways. Satellite digital radios are being designed into multi-band radios for gradual market growth. Satellite digital radios are being sold to developmental aid organizations for free distribution in poorer countries. Finally, programmer fees are being calibrated to the level of radio set distribution. Since there is little operating cash flow from which satellite sound broadcasting systems can be economically self-sufficient in their early years, such systems generally are proposed as experimental programs until they achieve a commercial threshold of radio receiver set distribution.

For background information *SEE COMMUNICATIONS SATELLITE; DIRECT BROADCASTING SATELLITE SYSTEMS; RADIO BROADCASTING; SPREAD SPECTRUM COMMUNICATION* in the McGraw-Hill Encyclopedia of Science & Technology.

Martin A. Rothblatt

Bibliography. *1990 National Association of Broadcasters Engineering Conference Proceedings*; Reaching for a starman, *Broadcasting*, 120:73, July 1, 1991; *Satellite Sound Broadcasting with Portable Receivers and Receivers in Automobiles*, CCIR Rep. 955-1 (MOD-F), International Telecommunication Union, 1991; G. M. Stephens, The changing face of audio, *Sat. Commun.*, September 16–18, 1990.

Drag reduction

The energy crisis of the 1970s and recent renewed concern regarding the environment, energy efficiency, and industrial competitiveness have refocused attention upon drag reduction technology as a means of reducing energy usage and cost. Drag reduction is traditionally discussed in terms of the various drag forces a body experiences. Form or pressure drag usually produces the highest levels of drag and is caused by flow separation (that is, flow which no longer adheres to the surface, **Fig. 1**) and resulting large deviations from inviscid flow pressure recovery. Friction drag is the result of satisfying the no-slip condition on the body surface and is usually much lower than the pressure drag associated with separated flows. Drag due to lift, induced by the three-dimensional pressure field of lifting bodies of finite aspect ratio, occurs on surfaces having a normal force component, whether utilized for lift, propulsion, or control. For aircraft, drag due to lift is of the order of the attached-flow friction drag. Bodies traveling at supersonic speeds or at the air–water interface experience various forms of wave drag, which, depending upon body design and speed, can also result in sizable drag levels.

For nonlifting bodies, minimum drag corresponds to attached viscous friction drag along with some residual but relatively low-level pressure drag associated with attached viscous flow-induced decambering of the surface in external flow cases. These drag levels are approached in long-distance pipelines and on neutrally buoyant, fully submerged streamlined bodies. Subsonic aircraft experience friction drag plus drag due to lift, while supersonic aircraft are subjected to friction, lift, and wave drag. Bluff bodies, such as automobiles, trucks, and helicopters, typically exhibit large regions of flow separation (Fig. 1*b*) and drag levels an order of magnitude larger than those of more streamlined shapes. A 20% aircraft drag reduction, possible with recently developed technology, corresponds to a substantial fuel saving for airlines. Fuel savings would also be sizable if tractor trailer trucks utilized flow separation control devices.

Form-drag reduction. The most obvious approach to form-drag reduction is to reduce the severity of the causative adverse pressure gradients, thereby obviating the occurrence of flow separation. Advances over the past several decades in computer speed and memory along with corresponding improvements in computational fluid dynamics allow the design of bodies with extremely low pressure drag. This technique is routinely used in many applications, notably aircraft and underwater vehicles. Design considerations other than energy expenditure, such as styling, volumetric efficiency, and operational requirements, have thus far limited this approach for most ground vehicles (automobiles and trucks) and helicopters. If flow separation cannot be reduced or eliminated by improved body design, then a large number of other approaches are available to reduce pressure drag. In general, these techniques increase the longitudinal momentum in the

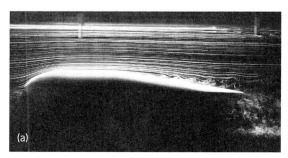
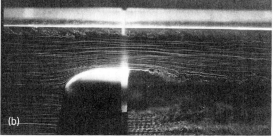

Fig. 1. Flow over bodies. (*a*) Flow over streamlined body with attached flow and low drag, and (*b*) flow over bluff-based body with separated flow and high drag.

near-wall region of the flow. Viable approaches include suction, wall injection in the downstream direction at high velocity, wall motion (for example, rotating cylinders), turbulence amplification for enhanced cross-stream momentum transfer, and probably most widely used, longitudinal vortex generators. The last have been routinely employed on the wings of commercial aircraft and typically consist of small upright vanes attached to the surface (**Fig. 2**). Vortex generators also occur naturally on both swimming and flying animals, for example, so-called pop-up feathers on bird wings and shark scales that deform differentially under load. Recent research indicates that discrete fluid jets can also act as vortex generators. This approach reduces the parasitic drag of the vortex-generating device but incurs system penalties associated with pumping and injecting fluids.

Since form drag can produce such large drag forces, the first priority in drag reduction, whether for human technology or analysis of natural systems, should be attention to, and minimization of, flow separation. The currently preferred techniques include suction, injection, and vortex generators. In addition to drag reduction, flow-separation control can provide lift enhancement, wing-stall delay, and pressure recovery enhancement in diffusers. SEE CONTROL SYSTEMS.

Friction-drag reduction. The basic question regarding friction-drag reduction is whether the unmodified flow is laminar or turbulent. For Reynolds numbers typical of human technology, laminar viscous flows exhibit a lower drag level than the turbulent case. Turbulent flows are composed of, and result from, nonlinear interactive instabilities that transfer momentum via complex dynamic motions, significantly increasing friction drag levels. If the unmodified flow is laminar, then viscous drag reduction approaches are limited to such techniques as reduced wetted area and careful (to avoid producing turbulent flow) use of convex longitudinal curvature, adverse pressure gradients, and normal fluid injection. If the unmodified flow is turbulent, viscous drag can be reduced via either laminar flow control or turbulent flow alteration.

Laminar flow control was first investigated in the 1940s and 1950s but was essentially abandoned in

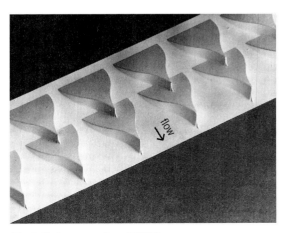

Fig. 2. Vortex generators. (*NASA*)

the 1960s due to the availability of inexpensive fuels and difficulties with the establishment and maintenance of the requisite surface smoothness. Laminar flow control research was restarted in the late 1970s by the National Aeronautics and Space Administration (NASA) as a result of the energy crisis. This ongoing program has resulted in several successful flight tests that have proven that, with contemporary technology, laminar flow control will work on transonic swept-wing aircraft in simulated airline service, that is, flights over airline routes, on airline schedules, and in the environments (weather, maintenance) encountered by aircraft in airline service. This success is due, in a major way, to improvements since the 1950s in materials and structural technology, which provide the required surface smoothness, as well as advances in viscous-flow stability theory and computational fluid dynamics, which enable efficient control techniques to be designed. Laminar flow control approaches for air include favorable pressure gradients for nearly two-dimensional flows and suction for three-dimensional flows. The currently favored technique is a combination termed the hybrid approach, wherein suction is utilized near swept-wing leading edges where the flow is highly three-dimensional and a moderate favorable pressure gradient is used farther back on the wing where the flow is more nearly two-dimensional. Laminar flow control on aircraft wings can provide very significant aircraft drag reductions, on the order of 25% if the benefits of aircraft resizing are included.

Turbulent drag reduction must be resorted to in cases where laminar flow control cannot be utilized and viscous drag reduction is desired. Such cases include flows involving sizable freestream disturbances or particulate content, innate surface roughness, or inordinately high Reynolds numbers. For water, the addition of certain dilute long-chain polymer solutions of 10–100 parts per million can provide large viscous drag reductions, in excess of 50%. The mechanisms responsible for this effect are not yet clarified, but apparently a non-newtonian effect for small-scale helical vortex motions is involved. Such motions are an intrinsic part of near-wall turbulence. Studies indicate that the outer so-called slime covering on several species of fish exhibits similar drag-reducing behavior. Polymer drag reduction has been employed to increase flow rate and reduce pressure drop in such applications as the Alaskan oil pipeline, storm drains, and fire hoses. The injection of microbubbles of air into water can also reduce turbulent viscous drag on the order of 50% or more, with much of the effect due to a lowering of the fluid density near the wall. Both polymer and microbubble techniques involve the injection of an additive incurring, in the process, various system penalties that must be evaluated for the application of interest.

Other turbulent drag reduction techniques, which can be utilized in either air or water, produce much lower levels of drag reduction, on the order of 10%. These techniques include riblets and normal or tangential injection of fluid from a low-drag source such as laminar-flow-control suction air or inlet-compressor

Fig. 3. Riblets molded into surface of adhesive film. (*NASA*)

bleed utilized for separation control. Riblets (**Fig. 3**) are small longitudinal surface striations (on the order of 0.002 in. or 50 micrometers for many applications) that produce a slow, creeping flow with low viscous drag within the groove. The 8% net viscous drag reduction is the relatively small difference between two large numbers, a large increase in wetted area and an even larger decrease in viscous drag per unit area. Riblets violate the conventional wisdom that a smooth surface always exhibits the lowest drag. Studies of the dermal denticles on shark skin indicate that, on the fast deep-water species, the denticles' size and configuration correspond to optimal riblet dimensions. Riblets achieved a modicum of notoriety as a result of their use on the successful America's Cup challenge by D. Conner on the craft *Stars and Stripes* in Australia in the late 1980s. An adhesive film with the riblets molded into the film surface has been developed (Fig. 3), which can be used to "wallpaper" surfaces to obtain drag reduction. The combination of laminar flow control on the wings and riblets on the fuselage may enable the development of an economically viable supersonic commercial transport for overwater routes.

Drag-due-to-lift reduction. Aerodynamic theory provides the most powerful and straightforward approach to drag-due-to-lift reduction, that is, increased span and aspect ratio to reduce the severity of the three-dimensional pressure field induced on finite-span lifting surfaces. Increasing the span creates a stress problem at the wing-fuselage junction that is addressable by an old-fashioned remedy, strut bracing. The development of computational fluid dynamics should allow the design of efficient, large-span, strut-braced wings, including minimization of strut interference drag. Alternate drag-due-to-lift reduction techniques include tip-mounted engines, nonplanar lifting surfaces, and thrust extraction from the angled tip flow. The last approach functions in a similar fashion to a boat sail tacking upwind, and was incorporated into the Whitcomb winglet drag-reduction device. Tip devices can comprise either vertical airfoils or several angled bodies, the latter directly analogous to the tip feathers on soaring birds. Nonplanar lifting surfaces can be realized via swept-back tapered wing tips and wing upsweep, both commonly observed on natural fliers and swimmers. Drag-due-to-lift reduction can

also ameliorate the so-called wake vortex hazard induced on following aircraft, inasmuch as the tip vortex flow is altered in the process. The reduction of drag due to lift is obviously fostered by avoiding lift on bodies having small aspect ratio.

Supersonic wave-drag reduction. There are essentially two ways to reduce shock-wave drag: reduce shock strength or employ favorable interference. Shock strength is reduced typically via employment of long slender bodies and wings and subsonic leading edges (swept behind the Mach line). Favorable interference involves impinging the forebody shock onto the wing to provide interference lift and subsequent reflection of the shock onto the after section of the body to provide thrust and mitigate drag. Classical examples of shock-wave drag reduction via favorable interference include ring wings and the Busemann biplane, which, for the inviscid nonlifting case, internally cancels the shock system. In transonic flow, shock waves that form due to viscous-induced interaction can be considerably weakened by means of a porous surface-subsurface plenum combination, which allows the viscous flow to passively bleed, thereby mitigating extreme growth of displacement thickness and producing a train of weaker waves.

For background information *SEE AERODYNAMIC FORCE; AERODYNAMIC WAVE DRAG; AIRFOIL; AIRPLANE; BOUNDARY-LAYER FLOW; SKIN FRICTION* in the McGraw-Hill Encyclopedia of Science & Technology.

Dennis M. Bushnell

Bibliography. D. M. Bushnell, *Supersonic Aircraft Drag Reduction*, AIAA Pap. 90-1596, 1990; D. M. Bushnell and J. N. Hefner (eds.), *Viscous Drag Reduction in Boundary Layers*, 1990; D. M. Bushnell and K. J. Moore, Drag reduction in nature, *Annu. Rev. Fluid Mech.*, 23:65–79, 1991; M. Gad-El-Hak and D. M. Bushnell, Separation control: Review, *J. Fluids Eng.*, 113:5–30, 1991.

Electric power systems

Since 1979, questions have emerged regarding the possibility that magnetic fields from power lines might increase the risk of cancer, particularly in children. This issue was first raised in a study of the effects of electric power lines near the homes of children. The study found that children who had developed cancer were more likely than healthy children to have lived in homes with nearby wiring that was thought to produce higher-than-average levels of magnetic fields. Homes located near power lines carrying large amounts of current were presumed to have elevated magnetic fields, since such fields are not shielded by trees or building materials. In contrast, electric fields, also produced by power lines, are effectively shielded so that levels in the home are virtually unaffected by nearby power lines.

The concern raised in the original study and those that followed is not just (or even primarily) with high-tension power lines but rather with neighborhood distribution lines and even electrical appliances that produce elevated and prolonged exposures, such as electric blankets and heated water beds. Overhead transmission lines do produce magnetic fields, but it is relatively rare for people to live near enough (within a few hundred feet) to receive increased exposure from this source. In contrast, many homes (about 20% in the Denver, Colorado, area, for example) are near enough to distribution lines that carry large amounts of current to be classified as having high exposure. The potential public-health implications from even a small adverse effect are profound, given the extensiveness of such exposures in modern society.

Epidemiologic studies. Further studies have provided some support for the hypothesis that prolonged exposure to elevated levels of 60-Hz magnetic fields may increase the risk of cancer in children. Several epidemiologic studies have addressed residential exposure from nearby power lines in relation to childhood cancer subsequent to the original report. A study in Stockholm, Sweden, found that children who were diagnosed with cancer were more likely to live near certain electrical constructions (transmission lines, substations) and more likely to have high fields (above 0.3 microtesla) measured near their homes. A childhood cancer study in Denver identified all childhood cancer cases from 1976 to 1983 and found that these children were more likely to live in homes near outside power lines that produce elevated magnetic fields compared to healthy children. The pattern was recently confirmed in a study of childhood leukemia in Los Angeles, California.

Several studies have included interviews with the parents of cancer cases and controls in order to address possible correlates of exposure that might distort the results. The associations with magnetic field indicators have been found to be present independent of such factors as the parents' social class, smoking habits, and childhood x-ray use; not all studies, however, have found this pattern. Negative results have been reported in studies conducted in Rhode Island and in England. Nonetheless, the body of research generally supports the presence of an association between certain types of power lines and childhood cancer. The increases in cancer risk for persons living in higher-exposure homes have been on the order of 1.5–3.0 times the background levels. This excess risk is consistent across the different types of cancers. Adult cancers have been examined less extensively but, with one exception, have not been linked to fields from power lines.

Research methods. The challenges and uncertainties in this research should be appreciated. Childhood cancer, which consists of a number of diseases such as leukemia, brain tumors, and lymphomas, is quite rare (about 1 in 10,000 children per year develops cancer), so that research on the topic is difficult to conduct. It is too expensive and burdensome to monitor magnetic field exposures to hundreds of thousands of children over the extensive time periods in which cancer develops. Instead, cancer cases that occur within a defined community are identified, and their exposure histories must be reconstructed along with those of unaffected, healthy persons. Then, investigators try to determine what exposures the individuals with cancer received in the past, typically starting with a history of the homes that they occupied.

Within a given home, there is the challenge of estimating what magnetic fields have been present up to 10–15 years previously. Wiring configurations provide a marker that is indirect but stable over time, since it is very unusual to modify distribution or transmission lines once they are installed. Equipment is also available to go into homes and take direct measurements of electric and magnetic fields, with the value of such measurements depending on the assumption that present levels provide a reflection of past exposures. Studies that have included both types of indicators (wire codes and measured fields) have tended to show stronger associations for the indirect measure, wire codes, as compared to the more direct marker, measured fields. If there really is an effect of magnetic fields on cancer, this would suggest that wire codes, though indirect, are a superior long-term exposure indicator and that measured fields are too noisy over long periods of time to be as reliable. Alternatively, it is possible that the magnetic fields, which the measurements more correctly indicate, really have no effect but that some other aspect of the wire code (for example, as a marker of housing density) confers increased cancer risk. The very limited understanding of the causes of childhood cancer makes such speculation difficult to resolve.

Occupational studies. In a separate set of studies, workers in certain electrical occupations (power-line repair, radio and television repair, power-station operation) have been found to be at increased risk of developing leukemia and brain cancer. This does not prove that electric and magnetic fields cause cancer, because it has not been proven that such workers really have elevated levels of exposure, and because such workers may be exposed to other agents in their jobs that are the actual cause of their increased cancer risk. Nonetheless, studies of an entirely different exposure source (workplace equipment) and a different population (adult males) point in the same general direction as studies of power lines and childhood cancer.

Laboratory studies and theories. Laboratory studies indicate that these very low energy fields are biologically active, but there are no studies directly examining whether such fields are capable of causing cancer in laboratory animals. Leading theories about mechanisms of action focus on changes in calcium metabolism in cells and suppression of melatonin, a hormone produced by the pineal gland. There is a great deal of controversy over how such fields could be amplified to ultimately produce such a profound biological effect as cancer.

Evaluation of evidence. Most reviewers of this evidence conclude that it falls short of demonstrating a causal relationship between exposure and disease when scrutinized from the perspective of scientific cautiousness. The epidemiologic studies have shortcomings, both individually and collectively, and direct experimental evidence for carcinogenicity has not been produced. Even a well-accepted theoretical mechanism for cancer production has not been developed. In direct answer to the question of whether a hazard has been proven, there is currently no convincing evidence that fields from power lines and related sources cause cancer.

On the other hand, there are a number of unrefuted studies that suggest that magnetic fields do produce a threat to health. The replication of associations between residential exposures and childhood cancer and between occupational exposures and adult cancers suggests that some association is present. These studies have been criticized, and there are a number of strategies for improving future research. Nonetheless, from a public-health perspective, there are clear suggestions of potential adverse effects on health that should be taken seriously.

Prudent avoidance. For advising the public or policy makers in the face of this uncertainty, the notion of prudent avoidance is very useful. Interpreted broadly, this encourages incorporating the present level of evidence into personal decision making, including the choice of home and job. Studies of residential exposure and childhood cancer suggest a 1.5–3-fold increase in risk from living in a higher-exposure home. Given that cancer occurs in roughly 1 in 10,000 children per year, living in a higher-exposure home might raise that incidence to 1.5–3.0 in 10,000 per year. This possible increase in a rare but dreaded disease should be incorporated into the array of factors impacting on decisions regarding health risks in daily living.

The principle of prudent avoidance was developed to be applied to power-line siting decision. It is suggested that alternative routings be examined for cost and public exposure, with a formal examination of the extra expense per averted exposure. Based on current evidence, it is clearly worth some expenditure to reduce exposures (even though it may eventually be determined that such fields are without harm), but it is not worth great expense to society. Unfortunately, there are no inexpensive methods for reducing magnetic-field exposures from existing power lines.

The public interest makes the present level of knowledge a sufficient basis for some concern (which is often grossly exaggerated in the lay press) but not necessarily sufficient for action. Epidemiologic and laboratory studies are in progress that should, within the next several years, lead to substantial progress in understanding whether or not magnetic-field exposures truly pose a health threat and, if so, how they operate.

For background information SEE CALCIUM METABOLISM; ELECTRIC POWER SYSTEMS; EPIDEMIOLOGY; ONCOLOGY; PINEAL BODY; RISK ANALYSIS in the McGraw-Hill Encyclopedia of Science & Technology.

David A. Savitz

Bibliography. A. Ahlbom, A review of the epidemiologic literature on magnetic fields and cancer, *Scan. J. Work Environ. Health*, 14:337–343, 1988; Office of Technology Assessment, *Biological Effects of Power Frequency Electric and Magnetic Fields: Background Paper*, OTA-BP-E-53, 1989; D. A. Savitz, N. E. Pearce, and C. Poole, Methodological issues in the epidemiology of electromagnetic fields and cancer, *Epidemiol. Rev.*, 11:59–78, 1989; N. Wertheimer and E. Leeper, Electrical wiring configurations and childhood cancer, *Amer. J. Epidemiol.*, 109:273–284, 1979.

Electrical utility industry

The year 1991 saw a continuation of the merger activity that is new to the United States electrical utility industry. In May, Northeast Utilities completed its takeover of the Public Service Company of New Hampshire, a formerly independent company. Public Service of New Hampshire had been forced to declare bankruptcy in 1990. Legislation had prohibited rate collection for construction work in progress. The several billions of dollars that Public Service of New Hampshire had expended in building the delayed Seabrook nuclear unit could not then be rate-based, and bankruptcy eventually resulted.

State regulatory and legislative officials in California ruled against the proposed merger of Southern California Edison Company and the San Diego Gas and Electric Company, which would have created the largest utility in the United States. Kansas Power and Light Company agreed to merge with Kansas Gas and Electric Company; Iowa Southern Utilities Company approved a merger with Iowa Electric Light and Power Company; Fitchburg (Massachusetts) Gas and Electric Company agreed to a merger with Unitil, which owns Concord (Massachusetts) Electric Company and Exeter & Hampton (New Hampshire) Electric Company.

Yankee plant controversy. A development of particular import to the nuclear power industry was the controversy surrounding Yankee Atomic Electric Company's plant in Rowe, Massachusetts. Commercial operation began in 1961, with an operating license for 40 years. Special-interest groups have called into question the safety of continued operation at this time on the basis that neutron bombardment of the main reactor vessel has made the steel so brittle that sudden injection of cool water, a standard procedure during certain types of emergencies, would cause the vessel to fail, with a major release of radioactivity. The reactor is currently shut down for maintenance, and the Nuclear Regulatory Commission has entered the case. This is an especially important matter for utilities because 10 major units are scheduled for relicensing within the next 20 years, and a shortening of operating life by a decade would be a severe blow to the individual utilities and to nuclear power in general. The utility is exploring methods of heat-annealing the reactor vessel to restore its ductility.

Clear Air Act compliance. Since the passage of the Clean Air Act Amendments late in 1990, utilities have been struggling to assess the impact of the new regulations. One feature of the act is a system whereby utilities operating clean plants, that is, with pollutant emissions below those specified by the regulations, can trade or sell those "credits" to other utilities, which can use them to avoid expensive pollution control measures on any of their units that exceed the specified limits. About 10% of the 80 GW of capacity directly affected by the legislation can be brought into compliance by applications of such credit. Another 65% of this capacity can be brought into compliance by switching to cleaner-burning fuels. The remaining capacity must be equipped with flue-gas scrubbers, retired, or treated in some yet-undetermined fashion.

Technological developments. Two technological developments during 1991 were noteworthy. Alabama Electric Cooperative brought into service the first compressed-air energy storage installation in the United States, and only the second in the world. This 110-MW unit uses off-peak power during the nighttime hours to compress air to 1100 lb/in.^{2} (7.6 megapascals), and stores it in an underground salt-dome reservoir. During the hours of peak demand, the stored air is withdrawn, heated with oil or gas, and used to drive a combustion turbine-generator. The success of the plant has led more than 80 other utilities under the guidance of the Electric Power Research Institute to form a working group to explore the possibilities for their own systems.

In the second development, three California utilities—Southern California Edison, Los Angeles Department of Water and Power, and Sacramento Municipal Utilities District—joined with the U.S. Department of Energy to design and build a 10-MW solar plant. This plant will use computer-controlled mirrors to focus the sun on a reservoir of nitrate salt in a 300-ft (90-m) tower. The heat captured by the molten salt will be used to generate steam to drive a conventional turbine-generator.

Capacity additions. Capacity additions continue to lag the increase in system demand nationwide. During 1989, electrical utilities in the United States added a total of 7905 MW to their systems. This was followed by 6118 MW in 1990, and an estimated

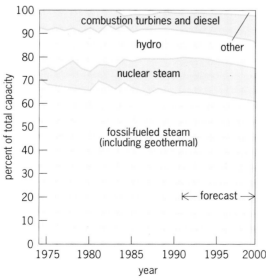

Probable mix of net generating capacity. "Other" includes solar, wind, and other nonconventional generating technologies. (*After North American Electric Reliability Council, Electricity Supply and Demand 1991–2000, 1991*)

increment of 4821 MW in 1991, the lowest increment in the last 41 years.

Reasons for low growth. The extremely low figure for annual capacity additions, compared with the average annual load growth of about 12,000–15,000 MW, can be attributed to several factors. First, there was the abrupt shift from long-term growth in the 7–8% range that obtained prior to the energy crisis of 1973, to the current consensus long-term growth forecast of about 2.5–3.0%. This shift caused utilities to cancel construction of those plants that, considering a normal construction lead time of roughly 8–10 years, would have entered service in the post-1981 period.

Second, regulatory discouragement of new construction through the disallowance of any return or only partial return on capital expenditures for new capacity ruled to be in excess of immediate need, coupled with intense environmental opposition, has caused utilities to sustain the hiatus in ordering new units. Consequently, annual capacity additions may continue to decline until the 1995–1997 period. There is increasing concern by utilities and regulators that insufficient capacity will be available to meet the demands of the 1990s, and a renewal of orders over the next few years should increase capacity additions beginning late in the 1990s. Regional economic growth patterns, however, coupled with the general hiatus in construction of new, large, base-load units, have already produced periods of capacity shortages in New England, Florida, and the Middle Atlantic states, so that forced brownouts have occurred. Utilities are now planning to meet future load growth through heavy reliance on capacity provided by nonutility installations, demand-side management of load, and extensive use of combustion turbines, rather than on the traditional fossil-fired, steam-driven turbine-generators.

Aging generating base. A major effect of the sharply reduced construction program is a rapidly aging generating base. In past decades, utilities generally used 30 years as the economic life of a generating unit, after which the accelerating cost of operating and maintenance, falling availability, and the development of new technology made it more economical to replace the unit with a new one. An analysis of the existing population of units shows that, by the year 2000, about 70,000 MW of existing fossil-fired and nuclear generating capacity will be more than 40 years of age. This will be roughly 10% of total installed capacity at that time. Further, more than 100,000 MW of capacity will be between 30 and 40 years of age by that date. This aging could pose serious problems in unit reliability, and in the reliability of customer service.

Division by plant type and utility type. The division of generating capacity by plant type in 1990, along with other United States electric power industry statistics, is given in **Table 1**. The historical breakdown and a forecast of these figures are shown in the **illustration**. Capacity additions in 1991 and a comparison of capacity in 1990 and 1991 are given in **Table 2**.

The electrical utility industry in the United States is pluralistic, divided among investor-owned utilities, cooperatives, municipal utilities, federal organizations

Table 1. United States electric power industry statistics for 1990*

Parameter	Amount	Change compared to 1989, %
Generating capacity, MW		
Hydroelectric	87,535 (11.9%)	0.35
Fossil-fueled steam	476,686 (64.9%)	0.23
Nuclear steam	110,207 (15.0%)	3.26
Combustion turbines, internal combustion	60,451 (8.2%)	2.10
TOTAL[†]	734,880	0.84
Noncoincident demand,[‡] MW	545,500	4.22
Energy production, TWh[δ]	2,807.1	0.82
Energy sales, TWh		
Residential	910.3	1.01
Commercial	734.6	2.61
Industrial	926.2	1.47
Miscellaneous	96.3	2.97
TOTAL	2,667.4	1.77
Revenues, total, 10^9 dollars	175.5	3.46
Capital expenditures, total, 10^9 dollars[¶]	24,289	−1.90
Customers, 10^3		
Residential	96,771	1.20
TOTAL	109,766	1.19
Residential usage, (kWh/customer)/year	9,472	0.00
Residential bill, cents/kWh (average)	7.82	2.22

*After 42d annual electric utility industry forecast, *Elec. World*, 205(10):11–22, October 1991; 1991 annual statistical report, *Elec. World*, 205(4):9–14, April 1991; and Edison Electric Institute, *Statistical Yearbook of the Electric Utility Industry, 1990.*

[†] Does not include nonutility capacity available to utilities.

[‡] Noncoincident demand is the sum of the peak demands of all individual utilities, regardless of the day and time at which they occurred.

[δ] TWh (terawatt-hour) = 10^{12} watt-hours.

[¶] Investor-owned utilities only. Does not include cooperatives or public power utilities, estimated at about $3.6 billion.

such as the Tennessee Valley Authority, and state or public power districts. The division of capacity among these various entities at the end of 1991 was as follows: investor-owned, 570,904 MW (77.2% of total); cooperatives, 26,629 MW (3.6%); federal, 67,313 MW (9.1%); municipals, 39,944 MW (5.4%); and state and public districts, 34,766 MW (4.7%).

Alternative measures of capacity. All capacity figures above represent the nameplate rating of the units, that is, the rating assigned by the manufacturer at the time of installation. However, some units may not be capable of achieving their nameplate rating for various reasons, such as low water levels in hydroelectric reservoirs, boiler-tube leaks, component degradation, or partial derating caused by operating problems. Therefore, the North American Electrical Reliability Council prefers to assess the ability of individual units or entire utility systems in terms of capability, that is, the ability to perform to a given level at periods of

annual maximum demand, rather than by capacity. The actual capability of the United States systems during the summer peak-demand period in August 1991 was only 687,341 MW, considerably below the aggregate nameplate capacity of those same systems.

The aggregate nameplate capacity or capability of the United States somewhat understates the capacity available to serve the maximum demand. The reluctance of utilities to construct new base-load generating units in the present regulatory climate, together with the provisions of the Public Utility Regulatory Policy Act and state regulation, both of which provide incentives for nonutility entities to build generating facilities and sell power to utilities, has spawned a vigorous sector of nonutility generators of electricity. In 1991, a total of 20,853 MW of this non-utility-owned capacity was available to utilities at the time of summer peak demand. Almost a third of that national nonutility capacity, 6700 MW, resides in California. Estimates by

Table 2. Generating capacity added in 1991, and comparison of capacity in 1990 and 1991

| Plant type | Capacity added in 1991 | | Capacity, MW | | |
	Power, MW	Number of units	1990	1991	1991 total capacity, %
Fossil-fueled steam	2,907	10	476,686	479,593	64.8
Nuclear steam	0	0	110,207	110,207	14.9
Combustion turbines	1,187*	18	55,433	56,620	7.7
Hydroelectric	706	8	87,535	88,241	11.9
Diesel engine or generators	20	2	5,019	5,039	0.7
Solar	1	1	3	4	
TOTAL	4,821	39	734,883	739,704	

*Includes 110-MW compressed-air-storage installation.

the North American Electric Reliability Council show nonutility capacity additions of 17,200 MW over the period 1991–2001.

Energy conservation. Pressure from regulators has also focused attention on the conservation of electric power as an alternative to adding new capacity. Utilities have embarked on two programs: direct control of customer load (demand-side management) and interruptible load. Firm estimates of the full effect of the current programs are difficult to obtain, but best assessments place the reduction in summer peak demand at 14,000 MW in 1991, rising to 20,000 MW in 2000.

Plant retirement. Utilities expect to retire 4644 MW during the period 1990–2000. This figure, however, either could decrease under pressure of growth in demand unsatisfied by new construction or could increase as utilities decide upon strategies to comply with the Clean Air Act Amendments of 1990. Older plants that might otherwise be upgraded through plant rebuilding programs to provide capacity in lieu of building new units could be retired if the new provisions require the installation of flue-gas scrubbers. Preliminary analysis indicates that retirements due to the cost of pollution abatement should be relatively few, amounting to perhaps less than 500 MW.

Demand. The demand in summer is, on a national basis, the critical measure of load on the utility system. The high ambient temperature results in less efficient cooling of electrical equipment, and warmer water in the steam condensers reduces the efficiency of the thermal cycle in which steam-turbine units operate. Preliminary estimates are that the total national summer peak demand in 1991 was 550.4 GW, a rise of 0.89% over the summer peak of 545.5 GW in 1990. This peak is on a noncoincident basis; that is, it is the sum of the peak demands reached by individual utilities but not necessarily at the same exact time.

One major measure of the reliability of the electric system is the reserve margin, that is, the surplus of capacity installed over maximum peak demand. On a national basis, a margin of 25% is generally accepted as adequate. Reserve margin in 1990 was 34.7%, and in 1991 was 32.1%. This figure may be deceptive because it does not fully account for capacity that may be unavailable for various reasons at the time of maximum annual peak. The North American Electric Reliability Council prefers a slightly different measure of adequacy, the capability margin. This is a measure of the surplus of capacity actually available at the time of peak over actual maximum demand. In 1990 capability margin was 22.0%, and only 21.8% in 1991. The danger in using national figures, however, is that while an adequate reserve margin or capability margin may exist at that level, some individual utilities or regions may actually have severe shortages of capacity or capability, and others have a substantial surplus.

Usage. In 1990, the last year for which figures are available, energy sales to the ultimate customer rose 1.8% over those in 1989, and reached a total of 2667.4×10^9 kWh. For 1991, preliminary figures indicate a further rise of about 2% to 2736.7×10^9 kWh.

Residential sales, which constitute about 34% of the total, in 1990 rose by a bare 1% over 1989 to $910,296 \times 10^6$ kWh. Commercial sales, which constitute about 28% of the total, rose substantially more, 2.6%, to $734,584 \times 10^6$ kWh. Industrial sales, which are roughly 35% of the total, ended up only 1.5% higher at $926,161 \times 10^6$ kWh, reflecting the depressed economy. The remaining percentage is made up of street lighting, railroads, public authorities, and interdepartmental usage.

Use per customer. Annual residential use per customer in 1990 was 9472 kWh, close to the all-time high of 9470 kWh in 1989, reflecting the depressed economy, a mild summer, and a mild winter. Individual residential customers paid an average of 7.82 cents/kWh, which translates to an average annual bill of approximately $741. This is an increase of about 2% over the 1989 figure of $725 and, in real terms, reflects a continuation of the downward trend in real costs to the customer that has obtained since 1983. In those regions where oil makes a major contribution to total generation, specifically the Northeast, the run-up in oil prices after the Iraqi invasion of Kuwait in 1990 boosted prices substantially, since regulators permit fuel-cost adjustments in rates. New England, for example, experienced a rise from 9.13 cents/kWh in 1989 to 9.71 cents/kWh in 1990.

On the basis of total ultimate customers rather than just residential customers, the annual average use was 24,462 kWh in 1990, a bare rise of only 0.01% over 1989. Average cost on this same basis was 6.58 cents/kWh, or an annual average bill of $1609 per customer of all classes.

Number of customers. The number of ultimate customers served by electrical utilities rose 1.2% in 1990 to 109,766,389, comprising residential, 96,771,096 (88.2%); commercial, 12,105,163 (11.0%); industrial, 513,228 (0.47%); street and highway lighting, 179,728 (0.16%); other public authorities, 195,164 (0.18%); and railroads and interdepartmental use, 2010 (0.002%).

Canadian imports. A portion of the total annual consumption of the United States is met by importation of electricity from Canada. In the New England region, such imports constitute slightly more than 10% of total consumption; but on a national level, imports make up only about 2% of the total. Imports from Canada and from Mexico totaled 22.5×10^9 kWh in 1990. United States exports, which occur at different times of the year from imports, amounted in 1990 to 20.5×10^9 kWh.

Revenues. Total revenues of the electrical utility industry in 1990 amounted to $175.5 billion, an increase of 3.46% over 1989. Residential revenues made up 40.6%, or $71.17 billion, a 3.5% increase over 1989; commercial customers contributed 30.7%, or $53.88 billion, a 4.5% increase over 1989; and industrial customers paid 25.3% of the total, or $44.78 billion, a 2.4% jump from 1989. The remaining was paid by street lighting, railroads, and interdepartmental customers. Revenues from off-system sales to Canada and Mexico amounted to $460.8 million.

Fuels. The costs of fuel and of capital are the two largest expenses of utilities. The relatively low price of natural gas is exerting pressure for switching to it where possible. Additional pressure is growing from the increasng use of combustion turbines fired with gas by utilities in order to avoid construction of coal-fired, base-load capacity. However, use of gas actually decreased slightly in 1990 compared to 1989, primarily because total generation by all fuels decreased. Gas consumption in 1990 was 2786×10^{12} ft^3 (78.8×10^{12} m^3) compared to 2787×10^{12} ft^3 (78.9×10^{12} m^3) in 1989. Total energy generated by gas as a fuel in 1990 amounted to 264×10^9 kWh, or 9.4% of the total, not significantly changed from 1989. Oil dropped from 267.5×10^6 barrels (42.5×10^6 m^3) burned in 1989 to only 196.2×10^6 barrels (31.2×10^6 m^3) in 1990, a decrease of 27%. Oil-generated energy represented just 4.2% of total generation during 1990. Coal continues its dominance as the fuel of choice. In 1990, utilities fired 771.7×10^6 short tons (698.0×10^6 metric tons) compared to 1989 quantities of 766.9×10^6 tons (693.7×10^6 metric tons). Coal supplied 55.5% of total energy generated in 1990, only slightly down from 55.8% in 1989. Nuclear fuel contributed 576.9×10^9 kWh of the total electric energy generated in 1990, or 20.6% of the total. This compares to the 1989 total of 529.4×10^9 kWh, which was 19% of the total generated by all fuels and sources. Hydroelectric generators produced 279.8×10^6 kWh during 1990, which was 10.0% of total generation, compared to 265.1×10^9 kWh in 1989, or 9.5% of total national generation. All other sources, including wind, solar, geothermal, and other nonconventional sources, contributed only 0.4% of the total, about the same as during 1989.

Expenditures. Total expenditures for capital accounts during 1990 amounted to $28.0 billion, of which $24.3 billion was by the investor-owned utilities. Expenditures for generating equipment constituted roughly 40% of the capital expenditures, about $8.9 billion for the investor-owned sector. This amount, however, included $1.7 billion in allowance for funds used during construction. This is a noncash item of current income that is added to the capital plant accounts for future collection when facilities under construction are completed and enter the rate base. Nuclear fuel made up $1.7 billion of the total. Total capital budgets for 1991 should rise to $29.0 billion.

Capital expenditures for new and reinforced distribution facilities for the investor-owned sector of the industry amounted to $9.1 billion in 1990. Rural cooperatives and municipals spent an additional $2.4 billion for distribution lines and substations during 1990. In 1991, capital spending on distribution facilities by the investor-owned utilities was estimated to be up about 11%, or $10.1 billion. Rural cooperatives and public power entities also planned an increase to roughly $2.9 billion.

Capital expenditures for transmission facilities in 1990 was $2.1 billion in the investor-owned sector, and $1.3 billion for the public-power and cooperative utilities. During 1991, investor-owned utilities raised their expenditures to $2.5 billion, while the cooperatives and

public power sector saw only a mild increase to $1.35 billion. Utilities installed a total of 3800 circuit miles (6116 km) of overhead lines operating at 22 kV and above in 1990, bringing the total installed to 631,893 circuit miles (1,016,933 km).

Overall electric operating expenses of the investor-owned sector, including items such as taxes, depreciation, property losses, and so forth, rose 5.0% in 1990 to $116.9 billion. Operating and maintenance expenses (those expenses associated directly with the operation and maintenance of physical plant only) were $81.6 billion, consisting of production, $59.2 billion; transmission and distribution expense, $7.2 billion; customer accounts, $3.2 billion; customer service and information, $1.1 billion; sales, $0.2 billion; and administration and general expenses, $10.6 billion.

Total assets of the investor-owned utilities as of the end of 1990 were $471.9 billion.

Number of employees. The average number of employees in investor-owned utility electric departments was 511,650 in 1990, a substantial decrease from the 1986 figure of 529,665.

For background information *SEE ELECTRIC POWER GENERATION; ELECTRIC POWER SYSTEMS; ENERGY SOURCES; ENERGY STORAGE; SOLAR ENERGY* in the McGraw-Hill Encyclopedia of Science & Technology.

William C. Hayes

Bibliography. Edison Electric Institute, *1990 Post-Summer Electric Power Survey*, 1990; Edison Electric Institute, *1990 Statistical Yearbook of the Electric Utility Industry*, 1991; 42d annual electric utility industry forecast, *Elec. World*, 205(10):11–22, 1991; 1991 annual statistical report, *Elec. World*, 205(4):9–14, 1991; North American Electric Reliability Council, *1991 Electricity Supply and Demand*, 1991.

Energy sources

Since the oil shortages of the 1970s, there has been concern that the world's resources of energy are limited and could be exhausted in the foreseeable future. Interest has therefore been focused upon renewable sources of energy, which could never be used up because they would immediately be replaced.

Renewable energy sources have frequently been tapped in the past, and a fair number of such systems are now in use, but modern technology has opened up many new possibilities for tapping old sources. Those being actively considered include hydroelectricity, wind energy, biofuels, wave power, geothermal power, solar power, tidal power, and ocean thermal energy.

Environmental hazards. In recent years, the rate of increase in energy consumption has slowed considerably, and renewable energy sources are now being considered more for their environmental friendliness than for the unlimited resources they represent.

It used to be thought that nuclear reactors would provide virtually limitless amounts of inexpensive energy. However, public concern over releases of radioactivity in accidents such as the Chernobyl explosion is augmented by fears of low-level radiation from nuclear fuel reprocessing and nuclear waste disposal, which may have long-term consequences.

Fossil fuels are used to provide most of the world's energy, but they can also create very unpleasant types of pollution. Acid rain is caused by the release of oxides of sulfur and nitrogen into the atmosphere, while smoke and carbon monoxide can be created if combustion is not under careful control at all times. Such pollutants can be removed at a cost, but all fossil fuels contain carbon, so the production of carbon dioxide is inevitable when they are burned.

Carbon dioxide is one of the main atmospheric gases responsible for the global greenhouse effect, retaining heat that would otherwise radiate to outer space. Atmospheric carbon dioxide levels have been increasing steadily, causing concern that retaining more of the Earth's heat could lead to global warming, which could melt the polar ice caps with disastrous consequences.

While renewable technologies also have environmental consequences, in general these consequences are not widespread. They affect only the local population who will benefit from the energy that is produced.

Hydroelectricity. Water-wheel technology has developed from the overshot, undershot, and cross-flow wooden wheels used in past centuries to modern hydroelectric turbines (Francis, Kaplan, Pelton, cross-flow, and Turgo) with erosion-resistant, high-performance metal blades that can be more than 80% efficient. In many countries, hydroelectricity is the primary source of power. The great dams that collect water for hydroelectricity bring the benefits of flood control at the same time. Run-of-river schemes working on lower heads of water generate much less power, but novel techniques are being devised for mini and micro hydro installations of a few megawatts or considerably less. These small-scale applications are promising economically, but the total resource is quite limited.

Geothermal power. The same underground heat that is responsible for volcanoes can produce hot aquifers under suitable conditions, and these can be tapped to provide hot water or steam. Considerable geothermal power from such sources is produced in Iceland, Italy, New Zealand, and other countries with a history of volcanic action.

Hot rocks can be found beneath every part of the Earth's surface if holes are drilled deeply enough, and methods of heat mining are being developed for situations where the hot rocks are dry. Many techniques established by the oil industry are being developed to improve heat extraction. A pair of holes can be drilled close together so that cold water can be pumped down one to collect heat from a fractured volume of rock and be returned as steam out of the other. This technique should be economical if multiple doublets are drilled with centralized generation.

Wind energy. The old style of windmill with four blades was widely used for land drainage as well as corn milling, while millions of multiblade wind pumps were used in North America before the great hydroelectric dams were built. Modern designs use two or three (or occasionally one) tapered blades with advanced airfoil cross sections. Up to 59% of the wind energy over the area of a rotor can theoretically be captured, and modern wind turbines achieve efficiencies much closer to the theoretical limit than was previously possible.

A rotor 100 ft (30 m) in diameter, the most economical size of wind turbine at present, will generate a fraction of a megawatt. Many machines are needed to produce a significant amount of power, and wind farms consisting of groups of wind turbines about 10 diameters apart operate in the United States, Denmark, the Netherlands, and other countries. Modern fatigue-resistant materials and designs have reduced construction costs, but tax credits and subsidies are still needed to make the economics of wind energy attractive. However, wind energy is seen as one of the most promising of the renewable technologies.

Solar power. Solar cells use the photovoltaic effect to convert light energy from the Sun directly into electrical energy. Thin-film, silicon-based devices have proved most successful, with efficiencies of 10–20%, but the search for higher efficiencies continues, for example, by using multilayer devices.

A considerable amount of mechanical and civil engineering work is needed to provide support for the thin films of silicon and protect them against the wind and the rain as well as the force of gravity. The costs of doing this dominate those of solar power. (Studies have been made of solar power stations on orbiting satellites, which would eliminate these requirements.)

A less direct approach is for heliostat mirrors to focus the Sun's rays upon a central boiler at the top of a so-called power tower. Since such a focusing system can be used only in direct sunlight and the mirrors must track the Sun, this method is also expensive.

Simpler and more economical technology can be used when rows of heat-absorbing tubes are filled with circulating water to form a solar panel that can be used for space or water heating. Vacuum and other techniques can be applied to this basic concept to achieve improved performance.

Biofuels. Biofuels are different from fossil fuels. Since the carbon dioxide that they release when they are burned must have been absorbed very recently from the atmosphere or would be returned there in due course by biodegradation, they do not add to the burden of greenhouse gases. Biofuels represent a form of solar power in that sunlight provides the energy needed for photosynthesis, the process by which carbon dioxide and water are combined to form carbohydrates and other organic compounds.

Many developing countries get up to 90% of their primary energy from firewood and animal dung, particularly cattle and camel dung. Cultivation of plants is possible in equatorial forests, but most advanced economies have industrial and commercial residues, such as wood chips and food-processing wastes, which can be used as biofuels. Agricultural and horticultural residues, including cereal straw, vegetable tops, and livestock manure, can also be used, as can substantial amounts of domestic waste and refuse from individual households, including sewage sludge. Avoiding the costs of waste disposal, which can be very large, is a considerable economic benefit.

The simplest approach is to use such waste materials directly as combustible fuels. Alternatively, more sophisticated fuels and other valuable products such as charcoal can be produced by pyrolysis or gasification. Fermentation or anaerobic digestion can produce alcohol or methane, both of which are valuable clean fuels. At refuse landfill sites and sewage plants, these processes tend to occur naturally, and they can be enhanced when conditions are properly controlled.

Tidal power. Energy from the rise and fall of tides can be trapped by an enclosed basin created by a barrage across a bay or estuary. A barrage across the Rance estuary between St. Malo and Dinard in Brittany on the northern cost of France has been generating up to 240 MW of electrical power since 1966.

The world's highest tides, with a range of 50 ft (16 m), are in the Bay of Fundy in Nova Scotia, Canada, and a 20-MW pilot scheme was commissioned at Annapolis Royal in 1985. Many studies have been made of projects that would generate up to 12,000 MW from the River Severn between England and Wales.

Wave power. In the middle of the Atlantic Ocean, a typical average wave height is 13 ft (4 m), and huge volumes of seawater rising and falling every 10 s or so represent a considerable amount of energy. This can be tapped with efficiencies as high as 80–90% by a number of different devices, including a so-called nodding duck, a hinged raft, a flexible tube, and oscillating water columns.

Mooring and anchoring, power transmission, offshore maintenance, and other problems of marine technology would be costly to overcome. Shore-based systems have achieved some success, for example, in Norway, where natural fiords channel the waves that reach the coast into energy-capturing devices with considerable cost savings; but it is difficult to design systems that can withstand the enormous power of storm-force waves.

Ocean thermal energy. Ocean thermal energy conversion uses the relatively small temperature range between water at the bottom of the sea, perhaps $39°F$ ($4°C$) at a depth of 3300 ft (1000 m), and that at the surface, which in tropical regions can average up to $72°F$ ($22°C$) or more. A steam plant can operate on a conventional thermodynamic cycle between such low temperatures, provided it is at greatly reduced pressure, but that requires a correspondingly large and costly volume, particularly with regard to the heat-transfer surface in the evaporator or boiler. Better performance would be achievable using a working fluid such as ammonia. Research on this alternative is being pursued, but costs appear to be prohibitive.

Economics. However environmentally friendly a supply of energy may be, it will not be accepted unless it is also economic compared with the alternatives that could be used. Most of the cost of energy

from any renewable source is in the initial cost of the plant that must be built to capture the energy: the dam and turbogenerators for hydroelectricity, the wind turbines for wind energy, the boreholes for geothermal power, and so forth. This is in contrast to conventional methods of energy production, where it is the running costs that tend to dominate, principally the costs of fuel.

To find the cost per unit of energy, the initial capital costs must be spread over the generating units and over a period of time, typically the life of the plant. Since interest must be paid on the capital, the cost of energy depends upon the rate of interest to be charged. Interest rates can vary widely according to numerous nontechnical factors, so the cost of energy from a given technology of fixed capital cost and known performance can also vary widely.

Another important aspect of the economics of environmentally friendly technologies is the method used to reflect their environmental advantage. A centrally planned economy will simply legislate that particular technologies will be used or avoided according to careful analysis and evaluation to balance environmental and other consequences. A market economy must ensure that the price mechanism properly reflects this balance. However, it is difficult to determine the total financial benefit of environmental friendliness to various people and even more difficult to persuade them to pay for it. Considerable thought is being given to ways of bringing external environmental costs into the prices that are charged for energy by tax allowances, subsidies, or other means.

Intermittent sources. Many of the environmentally friendly and renewable sources of energy are intermittent, notably wind, wave, tidal, and solar. This means that they cannot be used on their own to provide a steady or reliable power supply. Battery or other storage systems can be used to improve the availability of energy, but the only effective way to ensure a very reliable supply is to have a system of independent units capable of working together with spare capacity.

If a moderate amount of any intermittent source of energy is connected to a power system, up to say 10–20% of the installed capacity, reliability is provided by the spare capacity of the system. The intermittent source contributes its energy and therefore its average power to the system capacity without the need for any special provisions, such as additional control or storage.

For background information SEE BIOMASS; ELECTRIC POWER GENERATION; ENERGY SOURCES; ENERGY STORAGE; GEOTHERMAL POWER; SOLAR ENERGY; TIDAL POWER; WATERPOWER; WIND POWER in the McGraw-Hill Encyclopedia of Science & Technology.

Donald T. Swift-Hook

Bibliography. P. D. Dunn, *Renewable Energies*, 1986; M. A. Laughton (ed.), *Renewable Energy Sources*, 1990; A. A. M. Sayigh (ed.), *Energy and the Environment: Proceedings of the World Renewable Energy Congress*, 1990; J. W. Twidell and A. D. Weir, *Renewable Energy Resources*, 1986.

Engineering

The European Community is one of the world's three major trading blocks, and engineering is a mainstay of its prosperity. The 12 member states of the European Community (Belgium, Denmark, France, Germany, Greece, Ireland, Italy, Luxembourg, the Netherlands, Portugal, Spain, and the United Kingdom) have a population of 345,000,000 with a gross domestic product per capita approaching those of the United States and Japan. Europe's engineering strength lies in many areas of chemical, aeronautical, and mechanical technology; for example, it produces more machine tools than the United States and Japan combined. But it lags in electronics and computers.

Engineering production, infrastructure, and services in the European Community require the attention of more than 1,000,000 engineers, qualified to professional level. Because of an increasingly rapid rate of Europeanization of industry, there is an urgent need for these engineers to be able to practice across Europe. Various steps have been taken by the European Community governments and by the profession itself to abolish the restrictive effects of national frontiers on the mobility of engineering professionals. To understand the situation, it is necessary to distinguish between governmental measures and those undertaken by the profession itself.

Governmental measures. The idea of creating a single European economy based on a common market originated in the 1957 Treaty of Rome. It is the policy of the European Community to remove all obstacles in order to maximize trade and opportunities and to allow Europe to operate as an entity; hence, the often-used phrase "completing the internal market." The Single European Act provided that the internal market should be completed by December 31, 1992. Its implications are immense, and it emphasizes commitment to create an area without internal frontiers, in which the free movement of goods, persons, services, and capital is assured.

To ensure the freedom of movement of professionals across national frontiers, one of the 300 directives adopted by the European Community to complete the internal market deals with the mutual acceptance of higher-education qualifications. It establishes the right of those possessing an appropriate qualification and migrating to another member state to practice on equal terms with those obtaining their qualification in the host country.

The minimum academic qualification stipulated by the Council directive of November 21, 1988, is a 3-year university-level qualification, usually a degree, awarded by a competent authority under the law of a member state. While there is no specific period of training needed in addition to the degree, the directive includes a provision for successful completion of professional training where such is required by national custom. The directive also provides for those whose total education and training falls significantly short of that required in the host country to undertake either a period of supervised practice or an aptitude test before being fully recognized.

The general directive on higher-education diplomas covers all professions unless, like the architects, they already have their own separate directive. Engineering does not yet have a separate directive, and therefore comes within the scope of the general directive.

Usually, in Europe anyone can practice as an "engineer" even without any qualifications. What is protected are the national titles, for example, Diploma Ingenieur (Dipl Ing) in Germany or Chartered Engineer (CEng) in the United Kingdom. While these are esteemed titles, they are not a license to practice, although in each member state there are some occupations and duties for which a person holding the national title must be employed. One effect of the directive on higher-education diplomas is that nationals of a member state who satisfy the requirements of the directive can migrate to any of the other member states and use the professional title of the host state. This is necessary so that they can compete for work on an equal basis with those who gained their qualifications in the host country.

Unlike the professional measures discussed below, the European Community proposed no new pan-European title for engineers.

FEANI and the European title. On October 28, 1987, a historic ceremony took place in Paris, when the first 60 European Engineers (Eur Ing) received the title. This ceremony was the culmination of 4 years of intensive negotiation and international collaboration within the European Federation of National Engineering Associations (abbreviated FEANI, after its French title, Fédération Européenne d'Associations Nationales d'Ingénieurs).

FEANI was established in 1951 and has its headquarters in Paris. It has the following aims: (1) to secure the recognition of European engineering titles and to protect those titles in order to facilitate the freedom of engineers to move and practice within and outside Europe, (2) to safeguard and promote the professional interests of engineers, (3) to foster high standards of education and professional practice and regularly review them, and (4) to promote cultural and professional links within the engineering profession, especially in Europe.

FEANI brings together national engineering associations from 21 countries as national members. Each country has established a National Committee for FEANI with a secretary-general and a membership that is representative of the profession as a whole in the country concerned.

The 21 national members include all the countries of the European Community plus those of the European Free Trade Area (Austria, Finland, Iceland, Norway, Sweden, and Switzerland) and some others (Cyprus, Hungary, and Malta). Hungary was the first eastern European country to join. It is expected that others will follow soon. (East Germany is automatically included following German reunification.)

From its beginnings, FEANI operated a register, but

until the impetus provided by the completion of the internal market there had been no specifically agreed pan-European basis for registration. The reason was the great variation of the educational and professional systems throughout Europe, which made comparisons difficult. For example, the lengths of engineering degree courses vary from 3 years to more than 5 years. Also, professional titles are conferred differently and acquired by different means. In the United Kingdom and Ireland the Chartered Engineer (CEng) title is not awarded by the government but by a chartered professional institution on the basis of approved training and experience, in addition to the possession of an accredited engineering degree. For other countries in Europe, the titles, such as Ingénieur Diplômé in France, are given for successful completion of a recognized academic course without necessarily any associated training in a job. The title is awarded by an educational institution recognized by the government for that purpose, so in effect it is governmentally conferred.

FEANI Register. The purpose of the FEANI Register is as follows: (1) to facilitate the movement of practicing engineers inside and outside the FEANI ambit and to establish a framework of mutual recognition of qualifications in order that engineers who wish to practice outside their country can carry with them a hallmark of competence; (2) to give sufficient data about the formation (education, training, and experience) of the individual engineer for the benefit of a prospective employer; (3) to encourage a continuous updating of the quality of engineers by setting, monitoring, and reviewing standards; and (4) to provide a source of information about the great variety of formation systems in member countries. Negotiation among the professional bodies represented in FEANI has resulted in minimum standards both for registration simply on the basis of education but without a European title, and for registration on the basis of formation as a European Engineer (Eur Ing).

The minimum standard for registration on the basis of education is given by expression (1), where B repre-

$$B + 3U \qquad (1)$$

sents a high level of secondary education, validated by one or more official certificates awarded at about the age of 18 years; and U represents a year (full-time or equivalent) of approved engineering education given by either a university or other recognized body at the university level that is included in the FEANI Index, a list of schools and courses approved by FEANI for access to the Eur Ing title.

The minimum standard for registration on the basis of formation as a European Engineer is a total of 7 years, as given by expression (2). Here, T represents

$$B + 3U + 2(U \text{ and/or } T \text{ and/or } E) + 2E \qquad (2)$$

a year (full-time or equivalent) of training through a program whose aim is to increase knowledge through work within technical fields, as part of engineering formation. The work can be done in a construction site, factory, laboratory, office, or other working environment that is defined, supervised, and approved by a university or by a body accepted by FEANI. E represents a year (full-time or equivalent) of relevant engineering experience assessed and approved by a body accepted by FEANI.

Registration as a European Engineer confers the right to use this title in the language of the national member and to use the designatory letters Eur Ing, which is invariable in all the member countries, subject to national law. Persons registered as Eur Ing must abide by the FEANI code of professional conduct.

National Monitoring Committee. The mere fulfillment of the above formulas by an individual does not necessarily mean that the knowledge and skills appropriate to registration have been acquired. FEANI has therefore developed a system to ensure that its minimum standards are reached by all those approved for registration.

Each of the 21 FEANI National Committees has established a National Monitoring Committee composed of representatives from engineering associations, industry, and education. This committee is responsible for checking the qualifications and experience of individual applicants. The basic philosophy is that different systems can coexist, the focus being on attained competence. Thus, for Eur Ing applicants, in addition to checking academic qualifications, the National Monitoring Committee looks at engineering experience gained according to the following guide: (1) the solution of problems requiring the application of engineering science, in fields such as research, development, design, production, construction, installation, maintenance, engineering sales, and marketing; and (2) management or guiding of technical staff; or (3) the financial, economic, statutory, or legal aspects of the engineering task; or (4) industrial or environmental problems.

The National Monitoring Committees make recommendations on individual applications to the central European Monitoring Committee, whose task is to maintain a European-wide standard. It is also the responsibility of each National Monitoring Committee to keep the European Monitoring Committee well informed on the structure of engineering education and the standard of the individual schools and courses in its own country.

Applications for registration. Application is open to individuals only if they are members of an engineering association represented in FEANI. Applications must be made to national members, not directly to FEANI. The appropriate form in one of the three official FEANI languages (English, French, or German) must be completed, the required documentation must be attached, and a fee determined by the national member must be paid. (A second language is among the listed requirements to become a European Engineer, but this requirement is not being enforced at present.) A national member is not obliged to support an application involving a school or course from another member country.

Applicants whose formation took place outside the

FEANI area cannot be considered for admission to the Register or for the award of the Eur Ing title unless they are sponsored by a FEANI national member and have undergone a formation that meets FEANI criteria. FEANI does not give general decisions about whether a particular local diploma or degree is regarded as equivalent to those accepted in FEANI countries. In cases where the applicant is not a citizen of a FEANI country and neither education nor experience required for Eug Ing title took place within the FEANI area, at least 2 years of valid experience as a resident in the area is required.

However, the Eur Ing title, although desirable, is not necessary for the practice of engineering in any of the FEANI countries. Subject to any overriding work-permit regulations, which apply to all occupations, non-European engineers can practice engineering anywhere in Europe, although in some countries they may not be able to use the title Engineer. Multinational companies operating in Europe customarily employ engineers from inside and outside Europe on an equal basis, and this practice will no doubt continue. With regard to wider qualifications, the World Federation of Engineering Organizations (WFEO) is becoming interested in promoting the mutual recognition of qualifications to promote freedom of movement in the global economy, although its deliberations are still in the early stages.

A further recent development is that the U.S. Accreditation Board for Engineering and Technology (ABET) has opened discussions with FEANI on the possible mutual recognition of accredited degrees in the United States and Europe.

Progress. Since the launch of the European Engineer scheme in 1987, some 10,000 Eur Ing titles have been conferred, and the rate of application is steadily increasing. The greatest numbers (in terms of relative populations) come from Ireland, the United Kingdom, Finland, and France.

The rate across Europe is far from uniform, partly because some countries found that their existing internal systems allowed the establishment of their National Monitoring Committee more readily than others. It is to be expected, however, that over the next 5–10 years Eur Ing will become well established and better known everywhere as the pan-European title. Possibly, FEANI will convince the European Commission that the higher standard of Eur Ing compared with the standard required by the European Community's own Higher Education Directive will result in a separate directive for engineers, based upon the Eur Ing formula.

There are already some signs that employers welcome the Eur Ing qualification. For one thing, it will enable them to advertise for staff with a single qualification throughout Europe. At present, mention of any of the national titles appears to exclude all titles from other countries. In time it will be natural for engineers to move from London to Madrid or from Paris to Rome as easily as they do from New York to Chicago. This new mobility will enhance individual prospects and undoubtedly promote engineering development across the whole of Europe.

For background information SEE ENGINEERING in the McGraw-Hill Encyclopedia of Science & Technology.

J. C. Levy

Bibliography. European Commission, *Panorama of EC Industry*, annually; Fédération Européenne d'Associations Nationales d'Ingénieurs, *Formation System of Engineers and Technicians in the Countries Represented by FEANI*, 1990.

Environmental health

Smoking of tobacco is the leading cause of preventable deaths in the world today. In the United States alone, smoking kills over 400,000 people every year, more than the total of premature deaths due to alcohol, accidents, suicide, homicide, acquired immune deficiency syndrome (AIDS), and illegal drugs combined (see **illus.**). Smoking kills because cigarette smoke contains a wide variety of carcinogens and other toxic substances. Since the first Surgeon General's Report on the heath effects of smoking in 1964, it has been widely accepted that smoking causes disease and death in smokers. Scientists, however, have only recently turned their attention to the effects of cigarette smoke on nonsmokers, so-called passive smoking. Although the smoke that pollutes the air from a burning cigarette is more diluted than that a smoker inhales, it contains the same toxins. When nonsmokers inhale this secondhand smoke, it produces many of the same effects as are produced in smokers, including lung cancer, heart disease, and pulmonary disorders. When viewed as an air pollutant, secondhand smoke is much more serious than commonly regulated outdoor pollutants. In fact, following active smoking and alcohol, secondhand smoke is the third leading preventable cause of death in the United States today.

Constituents of secondhand smoke. A burning cigarette produces three different kinds of smoke: mainstream smoke, inhaled by the smoker; sidestream smoke, emitted by the lit end of the cigarette when it is not being smoked; and exhaled mainstream smoke, exhaled after a smoker has drawn on the cigarette. Secondhand smoke is a mixture of sidestream smoke and exhaled mainstream smoke. When an individual smoking a cigarette inhales, he or she draws oxygen through the lit end of the cigarette, which increases the temperature of the combustion, producing more complete, and hence less dirty, combustion. The combustion products of the tobacco are then filtered somewhat by the cigarette itself before being inhaled by the smoker. This situation contrasts with that which exists when the cigarette is simply idling, during which the cigarette burns at a relatively low temperature and hence produces less complete and dirtier combustion. Sidestream smoke actually contains higher concentrations of toxins than mainstream smoke does (see **table**). Secondhand smoke contains 43 known or suspected human carcinogens, as well as a variety of irritants and poisons such as cyanide and arsenic. Many of the compounds in secondhand smoke, such as benzene, vinyl chloride, arsenic, and cyanide, are regulated substances when they appear as air pollutants from industrial processes or automobiles; the concentrations that nonsmokers are exposed to, while much lower than those that smokers are exposed to, are still hundreds, if not thousands, of times above levels that would be considered acceptable in pollution from any other source.

Effects on children. Secondhand smoke has a wide variety of adverse effects on children, particularly infants. Children of parents who smoke develop more asthma and bronchitis, are hospitalized more often, and have slower development of pulmonary function than children of nonsmoking parents. Children of parents who smoke also appear to be at an increased risk of lung cancer and heart disease as adults compared to children of nonsmoking parents.

Lung cancer. The first studies linking passive smoking with lung cancer appeared in 1981. Since then the number of studies on passive smoking and lung cancer has increased rapidly. By 1990, there were 22 studies that were deemed of sufficient quality to warrant consideration by the U.S. Environmental Protection Agency (EPA) in a formal risk assessment of lung cancer produced by passive smoking. After reviewing these studies, the EPA concluded that passive smoking caused lung cancer in healthy nonsmokers and accounted for approximately 3700 deaths annually in the United States. Nonsmoking wives have a 30–40% elevation in their risk of dying from lung cancer if their husbands are smokers rather than nonsmokers.

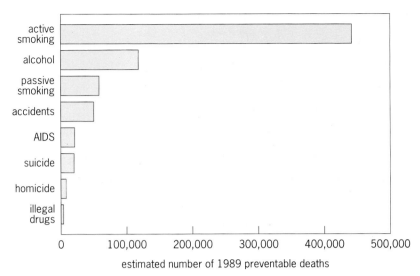

Preventable deaths from various sources.

In reaching this conclusion, the EPA was reinforcing and extending earlier conclusions that passive smoking causes lung cancer.

Heart disease. As of 1991, there were 13 epidemiological studies on the relationship between exposure to secondhand smoke in the home and the risk of heart disease death in the nonsmoking spouse. All but one of these studies showed an elevated risk of heart disease death for nonsmoking spouses of

Some toxic and tumorigenic agents in undiluted cigarette sidestream smoke*

Compound	Type of toxicity[†]	Amount in sidestream smoke per cigarette	Ratio of amount in sidestream smoke to amount in mainstream smoke
Vapor phase			
Carbon monoxide	T	26.8–61 mg	2.5–14.9
Carbonyl sulfide	T	2–3 μg	0.03–0.13
Benzene	C	400 μg	8–10
Formaldehyde	C	1500 μg	50
3-Vinylpyridine	SC	300–450 μg	24–34
Hydrogen cyanide	T	14–110 μg	0.06–0.4
Hydrazine	C	90 ng	3
Nitrogen oxides (NO_x)	T	500–2000 μg	3.7–12.8
N-Nitrosodimethylamine	C	200–1040 ng	20–130
N-Nitrosopyrrolidine	C	30–390 ng	6–120
Particulate phase			
Tar	C	14–30 mg	1.1–15.7
Nicotine	T	2.1–46 mg	1.3–21
Phenol	TP	70–250 μg	1.3–3.0
Catechol	CoC	58–290 μg	0.67–12.8
o-Toluidine	C	3 μg	18.7
2-Naphtylamine	C	70 ng	39
4-Aminobiphenyl	C	140 ng	31
Benz(a)anthracene	C	40–200 ng	2–4
Benzo(a)pyrene	C	40–70 ng	2.5–20
Quinoline	C	15–20 μg	8–11
N'-Nitrosonornicotine	C	0.15–1.7 μg	0.5–5.0
4-(Methylnitrosamino)-(3-pyridyl)-1-butanone	C	0.2–1.4 μg	1.0–22
N-Nitrosodiethanolamine	C	43 ng	1.2
Cadmium	C	0.72 μg	7.2
Nickel	C	0.2–2.5 μg	13–30
Polonium-210	C	0.5–1.6 pCi	1.06–37

*From D. Hoffman and S. S. Hecht, Advances in tobacco carcinogenesis, *Handbook of Experimental Pharmacology*, Springer-Verlag, 1988.

[†]C, carcinogenic; CoC, cocarcinogenic; SC, suspected carcinogen; T, toxic; TP, tumor promoter.

smokers compared to nonsmoking spouses of nonsmokers. There are several lines of biological evidence that make this association plausible. Firstly, exposure to secondhand smoke reduces exercise tolerance of healthy individuals and people with existing coronary artery disease. This reduced exercise capability is one of the compromises to the coronary circulation associated with heart disease. Secondly, there is good evidence from both human and animal studies that exposure to tobacco smoke, including passive smoking, increases the aggregation of blood platelets. Such increases in platelet aggregation are an important step in the genesis of arteriosclerosis. In addition, increasing platelet aggregation contributes to the risk of coronary thrombosis, a cause of acute myocardial infarction. Lastly, carcinogenic agents in secondhand smoke have been shown to injure the cells that line the arteries. Such injuries are the first step in the development of arteriosclerosis. Thus exposure to secondhand smoke can contribute to both short- and long-term insults to the cardiovascular system and heart. These physiologic studies help explain the elevated risk of approximately 30% that has been demonstrated in epidemiological studies.

The 30% increase in risk of death due to secondhand smoke for heart disease is comparable to that observed for lung cancer. This increase translates into about 10 times as many deaths for secondhand smoke-induced heart disease as lung cancer because heart disease is a much more common problem than lung cancer. Thus, while passive smoking accounts for about 3700 excess or premature deaths annually in the United States to lung cancer, it accounts for about 37,000 excess deaths or premature deaths due to passive-smoking-induced heart disease.

Combining the deaths from passive-smoking-induced lung cancer and heart disease with other cancers associated with passive smoking (such as cervical cancer) leads to an estimated 53,000 deaths annually from passive smoking in the United States.

Control measures. The growing evidence that passive smoke is dangerous to nonsmokers has led to widespread adoption of legislation that restricts smoking in the workplace and public places. In addition to legislative steps to protect nonsmokers from secondhand smoke, some businesses are voluntarily enacting policies to restrict or end smoking entirely. The trend to smoke-free workplaces is also being accelerated by a series of legal decisions holding employers liable for damage done to nonsmoking employees by secondhand smoke. In 1991, the Federal Court of Appeals in San Francisco held that the Eighth Amendment to the Constitution (which prohibits cruel and unusual punishment) protected a prisoner from being incarcerated in an environment polluted by secondhand smoke. The presence of secondhand smoke in households has also been a factor in deciding several child custody cases in which one parent was a smoker and the other a nonsmoker.

The scientific evidence that passive smoking is dangerous, together with the growing social unacceptability of smoking, is a powerful force to reduce smoking in society today.

For background information SEE CANCER MEDICINE; CIRCULATION DISORDERS; ENVIRONMENTAL PATHOLOGY; RESPIRATORY SYSTEM DISORDERS in the McGraw-Hill Encyclopedia of Science & Technology.

Stanton A. Glantz

Bibliography. S. A. Glantz and W. W. Parmley, Passive smoking and heart disease: Epidemiology, physiology, and biochemistry, *Circulation*, 83:1–12, 1991; U.S. Environmental Protection Agency, *Health Effects of Passive Smoking: Assessment of Lung Cancer in Adults and Respiratory Disorders in Children*, Doc. EPA/600/6-90/O06A, 1990; U.S. Public Health Service, *The Health Consequences of Involuntary Smoking: A Report of the Surgeon General*, DHS(CDC) 87-8398, 1986; A. Wells, An estimate of adult mortality in the United States from passive smoking, *Environ. Int.*, 14:249–265, 1988.

Enzyme

Recent research in biochemistry has involved the study of a new class of enzymes known as free-radical metalloenzymes and the development of artificial enzymes.

Free-Radical Metalloenzymes

Biochemical reactions are catalyzed by enzymes, adapted over several billion years of biological evolution to the wide range of reactions required for the existence of life. Studies probing the structures and mechanisms of enzymes at the molecular level with advanced methods of spectroscopy and x-ray crystallography are revealing the fundamental principles of enzyme catalysis. Data derived from these studies are beginning to elucidate the molecular origins of catalytic reactivity for a large number of enzyme types. Recently, a new class of redox enzymes in which an organic free radical is associated with metal ions in the protein structure has been recognized. These free-radical metalloenzymes are part of an emerging field, free-radical enzymology, focusing on the structure and reactivity of radicals in biological catalysis.

Free radicals. Any molecule containing unpaired electrons may be called a free radical. An unpaired electron leaves a molecular orbital half-filled; since the Pauli exclusion principle permits a second electron with opposite spin to be added at relatively low energy, this incompletely filled valence shell will normally represent an unstable configuration. Because of this chemical instability, free radicals are often unusually reactive and short-lived.

Free radicals may be formed by removing or adding one electron to a closed-shell molecule through oxidation or reduction steps. The presence in free radicals of an unpaired electron spin with its associated magnetic moment leads to a paramagnetism that dramatically distinguishes them from other molecules, which are typically diamagnetic. This difference can be used to detect free radicals with electron paramagnetic resonance (EPR) spectroscopy, which can detect as few as

10^{-10} mole of unpaired electrons in a sample. Used quantitatively, this technique provides an analytical tool for the detection, identification, and characterization of free radicals.

Free radicals have long been relegated to either obscure or deleterious roles in biology, associated with aging, oxygen toxicity, and radiation damage. However, this view is rapidly giving way to an interest in a wider range of radical reactions in biochemistry, with a growing recognition that radical mechanisms occur in a broad spectrum of enzymes and biological processes. Free radicals have been identified as components of the charge accumulation complex that is involved in the water oxidation chemistry of photosynthesis; and they appear to participate in the biosynthesis of essential biomolecules, including deoxyribonucleotides [the building blocks of deoxyribonucleic acid (DNA)] and prostaglandins. The cleavage of an unactivated carbon-hydrogen (C—H) or C—C bond in the substrate appears to be a common feature in enzyme mechanisms involving free radicals. Two well-characterized free-radical metalloenzymes, ribonucleotide diphosphate reductase and galactose oxidase, are of particular interest; crystal structures have recently been reported that reveal the molecular structures at nearly atomic resolution. These structures, combined with extensive spectroscopic characterization of the enzymes that addresses the electronic structural details, are providing insight into the catalytic mechanisms of free-radical enzymes and the stabilization and reactivity of the radical sites.

Free-radical sites in proteins. Free radicals in proteins generally occur as the products of one-electron oxidation of a nonradical precursor. The driving force for this oxidation reaction may be described quantitatively by the redox potential, a thermodynamic quantity relating to relative stabilities of initial and final molecular species. Most amino acid side chains forming the structure of proteins have a large positive redox potential (are relatively difficult to oxidize) and do not appear to be involved in radical reactions. Others, more susceptible to oxidation, can form fairly stable electron-deficient radical products. Tyrosine, tryptophan, and cysteine side chains, in particular, have been implicated in free-radical reactions of proteins. For the first two, the radical is delocalized in the π-electron system of the aromatic ring, while in the third a sulfur-centered radical is present. These radicals may be stabilized by sharing electrons from neighboring groups to form electron donor-acceptor or charge-transfer complexes. Isolation of the biological radicals in the protein core may protect them from reduction and introduce a degree of selectivity in their reactivity.

The single-electron redox chemistry involved in generating a free radical in a protein generally appears to require the participation of metal ions. In ribonucleotide reductase, a binuclear ferrous complex reacting with dioxygen (O_2) is responsible for the formation of tyrosine free radical; in photosystem II of green plants, tyrosine free radicals appear to be associated with the manganese ions of the water-splitting active site; in cytochrome c peroxidase, a tryptophan radical is formed on oxidation of the heme prosthetic group of the protein; in galactose oxidase, a fungal redox enzyme, the free-radical site is formed in association with a cupric ion (Cu^{2+}) bound in the protein. In the latter case, ligation of the radical by a metal ion may also play a role in the control of free-radical reactivity in the active site. In all of these cases, the detection of the biological radicals depends on the application of sensitive spectroscopic methods. Crystallography effectively complements spectroscopy in these studies by revealing additional structural details.

Ribonucleotide diphosphate reductase. The conversion of ribonucleotide diphosphates (NDP) to deoxyribonucleotides (dNDP) is essential for replication of DNA and therefore the reproduction of living cells [reaction (1)].

$$\text{(1)}$$

Three varieties of ribonucleotide reductase have been identified, containing either iron (Fe), manganese (Mn), or cobalt (Co) as an essential metal cofactor. The Fe- and Mn-containing enzymes appear to be similar, and quite distinct from the Co-containing enzyme, which is found only in bacteria and requires an adenosyl cobalamin (B_{12}) cofactor. In spite of differences in structure and metal content, all three types of enzyme appear to use free radicals in their active sites to accomplish the C—OH to C—H reduction chemistry [reaction (1)]. For the Fe-containing enzyme, the radical is stabilized as an intrinsic part of the enzyme structure; thus it is available for detailed characterization. EPR signals from this radical provided the basic evidence for the free-radical nature of this enzyme and the assignment to a specific tyrosyl residue (tyrosine-122). Knowing the identity of the radical has led to questions of how the radical is generated and stabilized in the protein structure and of how it is involved in catalysis.

The crystal structure solved for the radical-containing subunit of ribonucleotide reductase (**Fig. 1**) is interesting. The tyrosine residue known from spectroscopic experiments to form the radical site in the active enzyme lies buried deeply in the interior core of the protein, remote from the protein surface and surrounded by relatively inert amino acid side chains. This inaccessibility probably contributes to the stability of the radical. The nearest Fe atom lies at least 0.5 nanometer away, in a specific metal-binding site; it is likely that in the course of oxidation of the ferrous iron (Fe^{2+}) by O_2, a high-potential oxidant is generated that abstracts an electron from the nearby tyrosine. While resolving certain questions, the crystal structure has raised new ones—for example, the question of how a radical that is localized deep in the protein structure can participate in catalysis in an active site on the protein interface.

Galactose oxidase. Wood rot fungi produce an armamentarium of redox enzymes for degradation of lignin polymers, including a variety of peroxidases and galactose oxidase. The function of galactose oxidase appears to be the in-place formation of the hydrogen peroxidase that is required as a cosubstrate by the peroxidases from O_2 and reductants as shown in reaction (2).

$$RCH_2OH + O_2 \longrightarrow RCHO + H_2O_2 \qquad (2)$$

Primary alcohol — Aldehyde — Hydrogen peroxide

Galactose oxidase catalyzes the two-electron redox chemistry of reaction (2), where oxidation of an alcohol (RCH_2OH) to an aldehyde ($RCHO$) is coupled to the reduction of dioxygen to hydrogen peroxide. This reaction occurs in the enzyme at a monomeric copper complex (**Fig. 2**) for which single-electron reactivity would normally be expected. A systematic and quantitative characterization of the enzyme in each accessible redox form with a combination of powerful spectroscopic approaches has solved this puzzle by showing that the enzyme contains a free radical that ligates the cupric ion and participates in the redox reaction as a free radical–coupled copper complex. Removal of the metal ion and chemical reoxidation of the metal-free protein permits the radical to be generated in a form accessible to EPR spectroscopy, leading to its identification as a tyrosine-derived radical.

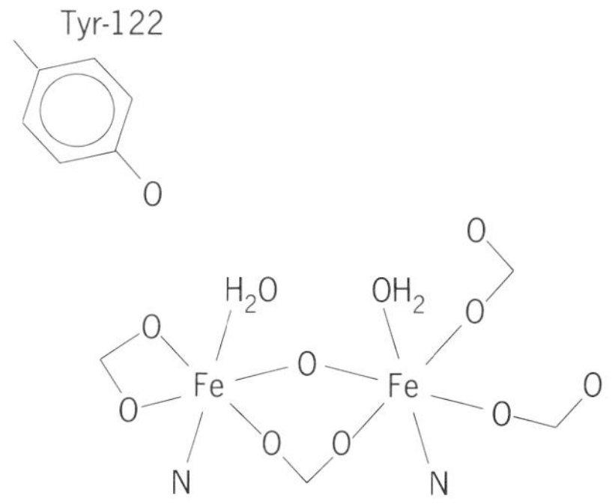

Fig. 1. Free radical and iron (Fe) centers in ribonucleotide reductase; tyrosine-122 (Tyr-122) is close to the iron center but is not directly coordinated to it. (*After P. Norlund, B.-M. Sjöberg, and H. Ecklund, Three dimensional structure of the free radical protein of ribonucleotide reductase, Nature, 345:593–598, 1990*)

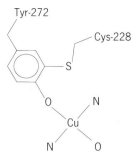

Fig. 2. Active-site copper complex of galactose oxidase, showing the tyrosine (Tyr-272) and cysteine (Cys-228) cross-link. (*After N. Ito et al., Novel thioether bond revealed by a 1.7Å crystal structure of galactose oxidase, Nature, 350:87–90, 1991*)

Crystallographic studies inspired by these spectroscopic insights have recently yielded a high-resolution crystal structure for galactose oxidase (Fig. 2). The radical is lost during crystallization by reduction, so the crystal structure relates to a catalytically inactive form of the enzyme. However, the structure reveals a novel modification of one of the copper ligands, a covalent cysteine-tyrosine cross-link. Spectroscopic data support an identification of this modified amino acid as the radical site in galactose oxidase. Thiosubstitution of the tyrosine ring (that is, introduction of sulfur atoms) very likely contributes to the substantial lowering of the redox potential of the radical site in galactose oxidase relative to an unperturbed tyrosine side chain.

Mechanisms. These examples show how enzymes have evolved special structures for the generation and stabilization of radicals essential for a catalytic mechanism. Simple principles of protection and isolation appear to be effective in stabilizing free radicals in the biological structures. Metal ions participate in both the generation of radicals and the control of radical reactivity in active sites. Covalent modification of amino acids in the protein structure allows a tailoring of radical redox potential to a specific reaction. The diversity of the biochemical reactions that are now suspected of involving free radicals is continuing to increase, and further studies can be expected not only to shed light on the significance of free radicals in biology but also to lead to more general insights into potential applications of free radicals in chemistry.

James W. Whittaker

Artificial Enzymes

Artificial enzymes are small synthetic models or mimics of naturally occurring enzymes. One application of artificial enzymes is modeling of the key aspects of the catalytic machinery of a natural enzyme. The goal is to learn about the natural enzyme by comparing its mechanism with a simpler, better-understood model. Another application is the development of catalysts for reactions that have no natural enzymatic counterpart. In this case, the goal is to mimic the general approach used by enzymes to perform their reactions, and then to apply this approach to the development of a catalyst with commercial applications.

This concept has great potential for the production of catalysts for almost any desired reaction, such as those reactions used in the petrochemical and specialty chemical industries, and in the mass production of pharmaceuticals.

Catalyst design. In the typical approach to mimicking a natural enzyme, functional groups are synthetically built into the artificial enzyme in an array similar to that found in the enzyme active site. Functional groups (imidazoles, carboxylates, or even cofactors such as pyridoxamine or thiazole) are often appended to a binding cavity. Binding of the substrate in the synthetic receptor can be driven by binding forces such as hydrogen bonding, coulombic attraction, or the hydrophobic effect. The enzyme analog, which is usually smaller than the natural protein, is then studied in terms of its kinetics and catalytic mechanism in order to draw parallels to the kinetics and mechanism of the natural enzyme. The technique has yielded insight into the catalytic mechanisms of many hydrolytic enzymes such as lysozyme, ribonuclease A, chymotrypsin, carboxypeptidase A, alkaline phosphatase, and carbonic anhydrase.

More recently, the development of artificial enzymes has been oriented toward a new goal. This goal is to take the general lessons learned from natural enzymes and not just mimic these essential features in a catalyst that performs the same reaction as its natural counterpart, but to make a catalyst for a reaction with practical applications. Three essential factors in catalyst design have emerged through studying natural enzymes: (1) preassociation of the substrate to the catalyst; (2) proximity and preorganization of functional groups; and (3) increased binding of the transition state over the substrate. Each of these factors plays a different but related role. The first two factors are the simplest to understand. The catalyst must recognize and bind to the substrate in order to act upon it. In addition, the extent to which functional groups are in proximity and correctly preorganized to bind the substrate and perform catalysis, controls the amount of entropy that must be decreased to reach the transition state.

The increased binding of the transition state is more complex. Increased binding of the transition state is the most well accepted theory of how enzymes impart large rate enhancements. The idea is that the enzyme is more complementary to the transition state of the reaction than the substrate, so that forming the transition state requires less endothermic free energy than if the enzyme were not present. The lowering of the transition-state energy arises from energetic interactions at the active site that release free energy as the transition state is formed. This release of free energy partially counteracts the endothermic free energy required to reach the transition state. The energetic interactions that partly decrease the activation energy are increased strength of hydrogen bonds; the release of steric, bond, or torsional strains; and increased ion pairing, or in general just a better fit of the transition state to the enzyme active site. The key premise is that this better fit is relative to the fit of the substrate. This is demonstrated in **Fig. 3**. If the transition state is bound to the same extent as the ground state, the change in free energy (ΔG) for the reaction with or without the catalyst is the same, and thus the rates of the reactions are the same. If, however, a greater amount of free energy is released upon formation of the transition state than upon binding the substrate, the barrier to the reaction is lower than without the catalyst, and a rate enhancement is observed.

The three factors (substrate binding, proximity and preorganization, and increased transition-state binding) are related. In fact, the first two are prerequisites for the last. In order to bind and stabilize a transition state of a reaction, the catalyst must first bind the substrate, since the chances of encountering a transition state in solution are quite low. In addition, the preorganization and

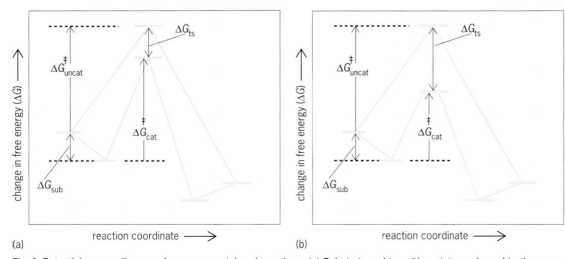

Fig. 3. Potential-energy diagrams for enzyme-catalyzed reactions. (*a*) Substrate and transition state are bound to the same extent, and the uncatalyzed rate is the same as the catalyzed rate. (*b*) Transition state is bound better than the substrate, and the catalyzed rate is faster than the uncatalyzed rate. uncat = uncatalyzed, cat = catalyzed, ts = transition state, sub = substrate; ‡ denotes an activated complex.

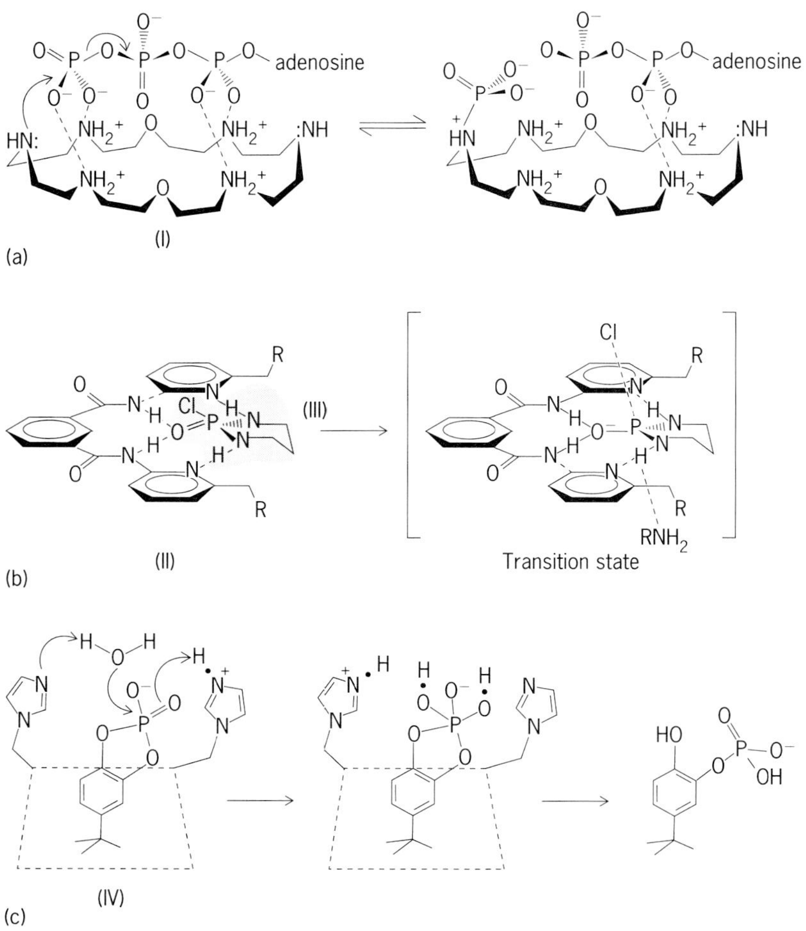

Fig. 4. Phosphoryl transfer catalysts. (a) ATP-to-ADP interconversion (after M. W. Hosseini, A. J. Blacker, and J.-M. Lehn, Multiple recognition and catalysis: A multifunctional anion receptor bearing an anion binding site, an intercalating group, and a catalytic site for nucleotide binding and hydrolysis, J. Amer. Chem. Soc., 112:3896–3904, 1990). (b) Phosphoryl chloride–to–amine transfer; highlighted area shows the phosphoryl chloride (after P. Tecilla, S.-K. Chang, and A. D. Hamilton, Transition-state stabilization and molecular recognition: Acceleration of phosphoryl-transfer reactions by an artificial receptor, J. Amer. Chem. Soc., 112:9586–9590, 1990). (c) Phosphodiester hydrolysis; broken lines denote the cyclodextrin cavity (after E. Anslyn and R. Breslow, Proton inventory of a bifunctional ribonuclease model, J. Amer. Chem. Soc., 111:8931-8932, 1989).

proximity of functional groups in the enzyme active site that are complementary to the transition state will also inherently bind the substrate, since the substrate and transition state have similar structures.

Phosphoryl transfer reactions. Phosphoryl transfer reactions are key natural metabolic steps in energy release and storage, and they are used to construct the genetic code DNA and ribonucleic acid (RNA). The development of synthetic catalysts to manipulate phosphoryl groups could lead to improved methods of genetic engineering, and new methods for storage of chemical energy. **Figure 4** shows several synthetic phosphoryl transfer catalysts.

Compound (I) in Fig. 4a catalytically assembles and hydrolyzes pyrophosphates. Substrate binding is achieved with complexation of the anionic pyrophosphate to the cationic azacyclam, a type of macrocycle that has a crown ether structure in which some of the oxygen atoms have been replaced with nitrogen atoms. The essential functional group of catalysis is a

nucleophilic amine that is part of the cyclam unit. The substrate binding places this amine functional group in proximity to the pyrophosphate. The amine attacks the pyrophosphate adenosinetriphosphate (ATP) and becomes phosphorylated. The artificial enzyme forms a covalent intermediate, which can now associate a different adenosinediphosphate (ADP) and thus reform the pyrophosphate by phosphate nucleophilic attack and expulsion of the protonated amine leaving group. Catalytic interconversion between ATP and ADP was observed.

Compound (II) in Fig. 4b has also been developed as an artificial enzyme for phosphoryl transfers. In this case, the reaction catalyzed does not possess a direct natural counterpart. The design concentrates upon increased binding in the transition state. Compounds such as structure (II) are known to complex planar barbiturate derivatives strongly; and these derivatives were used as transition-state analogs for the equatorial substituents of the trigonal bipyramidal transition-state phosphorane formed upon addition of an amine to the phosphoryl chloride (III). Indeed, prior coordination of structure (III) to structure (II) does lead to rate enhancements for the reaction with an amine.

Another approach involves the hydrophobic binding of catechol-derived cyclic phosphates with bis-imidazole-derivatized β-cyclodextrins (IV). This arrangement is a mimic of the functional groups present in ribonuclease A, and it has provided a key to understanding this enzyme. When the imidazoles are placed approximately 90° apart along the cyclodextrin primary rim (the rim of the cyclodextrin that is all primary hydroxyl groups), a rate for hydrolysis that is only 250 times slower than the natural enzyme ribonuclease A was found; this represents a significant rate enhancement. A mechanism in which an imidazole acts as a general base to deliver water to the phosphate while an imidazolium protonates the forming phosphoranelike transition state was revealed from regiochemical and proton inventory studies, which were done with different placements of the imidazoles around the cyclodextrin rim. The success of this artificial enzyme demonstrates the three design principles discussed above. First, binding of the substrate into the cavity places the imidazoliums in direct proximity to the phosphate group. The approximate 90° placement of the imidazoles yields an optimum preorganization compared to other arrangements. Finally, increased transition-state binding and stabilization is achieved with proton transfer from the imidazolium to the trigonal bipyramidal phosphorane intermediate.

Structures (V) and (VI) are artificial enzymes that incorporate features from both ribonuclease A and staphylococcal nuclease. These features include binding of the substrate with hydrogen bonds to acidic protons, which can be transferred during the hydrolysis reaction. A design for transition-state binding is included in each.

Compound (VII) is a control that possesses no acidic protons. The essential idea is that a monoanionic substrate is bound in a dicationic artificial enzyme in such a way that, upon formation of the dianionic trigonal

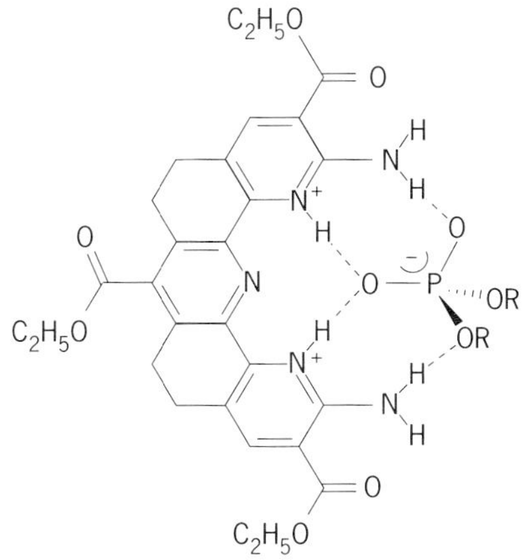

(V)

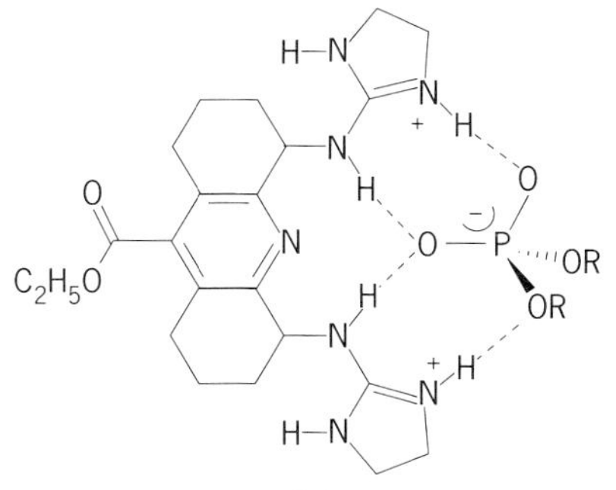

(VI)

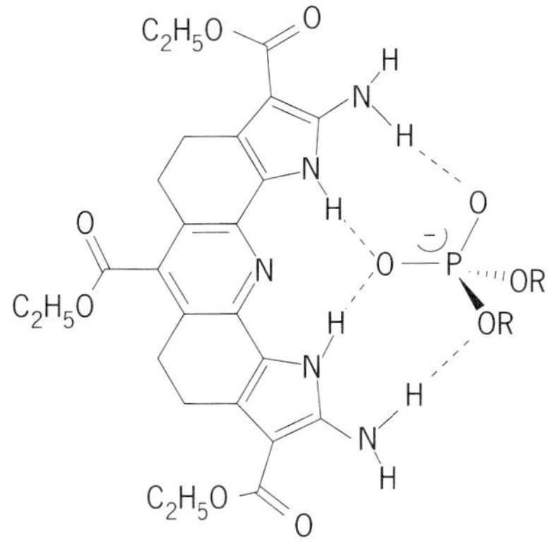

(VII)

The binding constants range from 10^2 to 10^5. The kinetics of hydrolysis and transesterification reactions are currently under study.

For background information SEE CATALYSIS; DELOCALIZATION; ELECTRON PARAMAGNETIC RESONANCE (EPR) SPECTROSCOPY; ELECTRON TRANSFER REACTION; ENZYME; EXCLUSION PRINCIPLE; FREE RADICAL; HYDROGEN BOND in the McGraw-Hill Encyclopedia of Science & Technology.

Eric Anslyn

Bibliography. E. Anslyn and R. Breslow, Proton inventory of a bifunctional ribonuclease model, *J. Amer. Chem. Soc.*, 111:8931–8932, 1989; A. Fersht, *Enzyme Structure and Mechanism*, 2d ed., 1985; P. A. Frey, Importance of organic radicals in enzymatic cleavage of unactivated C-H bonds, *Chem. Rev.*, 90:1343–1357, 1990; M. W. Hosseini, A. J. Blacker, and J.-M. Lehn, Multiple recognition and catalysis: A multifunctional anion receptor bearing an anion binding site, an intercalating group, and a catalytic site for nucleotide binding and hydrolysis, *J. Amer. Chem. Soc.*, 112:3896–3904, 1990; J. Stubbe, Protein involvement in biological catalysis?, *Annu. Rev. Biochem.*, 58:257–285, 1989; J. Stubbe, Radicals in biological catalysis, *Biochemistry*, 27:3893–3900, 1988; P. Tecilla, S.-K. Chang, and A. D. Hamilton, Transition-state stabilization and molecular recognition: Acceleration of phosphoryl-transfer reactions by an artificial receptor, *J. Amer. Chem. Soc.*, 112:9586-9590, 1990.

Extinction (biology)

While hundreds of terrestrial and fresh-water species of animals and plants are known to have become extinct as a result of human activity in the past four centuries, understanding of the extinction of recent marine invertebrates is extraordinarily limited despite the widespread perception that shallow-water marine organisms have been extensively decimated by humans. So limited is this understanding that the only well-documented case of the extinction of a marine invertebrate in historical time is that of a small snail that once lived upon eelgrass in the North Atlantic Ocean. The case history of the eelgrass limpet provides insight into the processes of modern-day extinction in the world's oceans.

Attributes of marine invertebrates. There are at least three major attributes that would render a marine invertebrate prone to extinction: a restricted geographic range, a restricted habitat, and restricted dispersal powers or dispersal potential. A species does not have to possess all three characteristics in order to become extinct.

Limited geographic distributions may be created by the sequential elimination of peripheral populations through climatic change or by the loss of habitat through natural or human processes. If these processes are ongoing, the distribution of a species could continue to be reduced.

A restricted habitat may be caused by the requirement of a specialized food resource, by physiological

bipyramidal phosphoranelike transition state, a second ion pair will be formed, reinforcing the binding. The proposed mechanism would proceed as follows. As the nucleophile attacks a phosphodiester bound in the cavity of structure (VI), a proton from one of the guanidinium groups (a type of amine group) can transfer to the phosphate to stabilize the newly developing negative charge. In a second step, the intermediate trigonal bipyramidal phosphorane intermediate breaks down by proton transfer from the other guanidinium to assist departure of the leaving group. Preliminary studies toward this goal are quite encouraging. Compounds (V), (VI), and (VII) have been found to bind phosphodiesters in acetonitrile, chloroform, and mixtures of dimethyl sulfoxide and water, respectively.

limitations, or by a combination of the two. Both specialization and physiological limitations result from the evolutionary consequences of natural selection, often occurring over millions of years. A severe alteration of the availability of a unique food resource would put a specialized species at risk. A narrow habitat range may be linked to narrow physiological tolerances of environmental variables (such as temperature or salinity extremes). An added complexity would occur when the habitat range of the predator is narrower than the habitat range of the prey, a pattern that could arise from different physiological tolerances to the same environmental gradients. In this case, localized extinction of the prey could result in complete extinction of the predator.

Limited dispersal abilities may arise from having preadult (larval) stages that cannot be dispersed widely within a suitable geographic region. Most marine invertebrates have planktonic larval stages that are transported for varying distances by tidal or ocean currents. The other major dispersal strategy in marine invertebrates involves production of larvae that exit as miniature adults from the parent or from an egg case. While having widely dispersing planktonic larvae would appear to provide immunity from extinction, this may not be the case if the larvae disperse to a potentially extinguishable habitat.

Atlantic eelgrass limpet. Most marine invertebrates have broad geographic ranges and broad physiological tolerances within their habitats. Although many thousands of invertebrates are reported as having extremely limited ranges, these are mostly poorly known species perhaps represented by only a few specimens, whose distributions are probably actually larger. A known exception is the small (up to 10 mm in length) limpet *Lottia alveus* (also known as *Acmaea alveus*). This limpet lived only upon the long, narrow blades of the eelgrass *Zostera marina* along the shallow shores of the northwestern Atlantic Ocean, from Labrador in the north to Long Island Sound in the south. Similar eelgrass limpets, subspecies of *L. alveus* or distinct species in their own right, occur in restricted regions of the North Pacific Ocean (Alaska to British Columbia and Sakhalin Island).

Nineteenth- and early-twentieth-century accounts record the Atlantic eelgrass limpet as abundant. Canadian naturalists noted that it was "very abundant on eel-grass at low water" on Grand Manan Island in the Bay of Fundy in 1890. Other biologists recorded that in 1910 in the Boston region "in certain places hundreds may be collected in a short time." And in 1929 "thousands of individuals were readily accessible" at low tide on Mount Desert Island in Maine. However, since 1929, the eelgrass limpet has not been reported alive. Field work and a thorough search of reports and museums have revealed no further records, and in 1991 a group of marine biologists concluded that the eelgrass limpet had become extinct. It had disappeared before even its most basic biology had been studied.

Eelgrass limpet extinction. As do most snails, limpets feed by scraping their food with a ribbonlike structure called a radula. The radula, upon which are rows of tiny "teeth," protrudes from the mouth. The shape of the teeth is correlated with the food consumed. The radula of the eelgrass limpet is adapted for feeding upon the surface cells of the eelgrass itself (as opposed, for example, to scraping microscopic algae off the surface of the grass, which many other snails do). The eelgrass limpet was thus a trophic specialist, relying solely upon the eelgrass for its food supply.

Locations from which the eelgrass limpet was collected, and naturalists' records of other marine organisms found with the limpet, indicate that it occurred in fully marine water rather than in brackish water. However, eelgrass commonly lives in both marine and brackish water. The eelgrass limpet thus had a narrower physiological range than its host plant. It is assumed that *L. alveus* had planktonic larvae (as do all members of the genus *Lottia*), and that the larvae possibly had specialized chemical mechanisms to locate eelgrass blades.

Between 1930 and 1933, eelgrass precipitously disappeared from the North Atlantic Ocean. The primary cause of this decline was a "wasting disease" caused by the slime mold *Labyrinthula*. More than 90% of all eelgrass populations were eliminated, with concomitant and often striking changes in associated plants and animals. The eelgrass decline led to extensive reductions in migratory waterfowl populations and loss of commercial bay scallop fisheries. Populations of eelgrass did, however, survive in brackish water, which is beyond the salinity tolerance range of the slime mold. Similarly, the narrow salinity range of the eelgrass limpet prevented it from surviving on refugial eelgrass populations in lower-salinity waters.

Another eelgrass specialist of the western Atlantic Ocean, the tiny seaslug *Elysia catulus*, did not become extinct. The eelgrass slug ranges from Massachusetts to Virginia. Unlike the eelgrass limpet, *Elysia* can live in both marine and brackish water and thus was able to survive on low-salinity populations of eelgrass.

Thus, the refugia (brackish water) of the eelgrass limpet's sole food source during a period of catastrophic decline were outside the limpet's physiological tolerances, and the limpet became extinct. *Lottia alveus* has the dubious distinction of being the first marine invertebrate species to be documented as going extinct in historical time. The evidence indicates that it was a natural extinction, not related to any human activities.

Other possible extinctions. Other marine invertebrates have been reported as possibly extinct, but each of these invertebrates has been subsequently shown either to be alive, to be encumbered with taxonomic problems, or to be insufficiently documented. The lack of records, however, does not mean that no species have become extinct.

A primary habitat in which evidence for extinctions may have been sought are the now-destroyed estuaries and marshes of many coastlines. Scores and perhaps hundreds of species of invertebrates from coastal waters described from the nineteenth century have never been subsequently reported. Many of these are from coastal localities that have been obliterated by human population expansion and littoral urbanization. Some of these species may simply not have been recollected by specialists from remaining comparable habitats. For others, extinction may have occurred long before the first formal twentieth-century biological surveys.

Invertebrate extinctions may also have occurred in association with well-known vertebrate extinctions. For example, extinct marine mammals (such as Steller's sea cow, last seen alive in 1768) may have had species-specific parasites or symbionts. Thus, when the mammal became extinct, so too would its parasites or symbionts.

Other cryptic extinctions may also have occurred, even among well-known groups of marine organisms. Many local populations of marine invertebrates (and algae and fish) are known to have been destroyed by human activity during the last 150 years, particularly in Europe and North America. "Once found locally" is a common annotation in many faunal and floral lists. It has been long presumed that because populations of "the same species" survive elsewhere the eliminated populations do not represent extinctions. In the 1970s and 1980s, advances in biochemical systematics revealed that a range of common animals (including, for example, the polychaete worm *Capitella capitata* and the marine mussel *Mytilus edulis*) once thought to belong to a single species (based upon external morphology) are in fact complexes of a number of separate species [based upon differences in proteins or deoxyribonucleic acid (DNA), for example]. Thus, the worm *Capitella* may represent a dozen or more species, and the mussel *Mytilus* three or more species. It is possible that a number of what have been thought of as population extinctions actually represent extinction of species that were eliminated long before techniques were available to distinguish them from similar-looking taxa.

While evidence indicates that marine invertebrates have in general escaped historical extinctions, human activities have been, and clearly continue to be, capable of severely reducing and eliminating entire populations of marine organisms. A striking and perhaps increasing potential exists for such activities to involve species that possess the appropriate biological and ecological characteristics that would make them susceptible to extinction.

For background information SEE EXTINCTION (BIOLOGY); MARINE ECOLOGY in the McGraw-Hill Encyclopedia of Science & Technology.

James T. Carlton

Bibliography. J. T. Carlton et al., The first historical extinction of a marine invertebrate in an ocean basin: The demise of the eelgrass limpet *Lottia alveus, Biol. Bull.*, 180:72–80, 1991; S. J. Gould, On the loss of a limpet, *Nat. Hist.*, 6/91:22–27, 1991; L. K. Muehlstein, Perspectives on the wasting disease of eelgrass *Zostera marina, Dis. Aquatic Organ.*, 7:211–221, 1989; L. K. Muehlstein et al., *Labyrinthula zosterae* sp. nov., the causative agent of wasting disease of eelgrass, *Zostera marina, Mycologia*, 83:180–191, 1991.

Food

Food is fuel for the most important system of any arsenal: the military personnel. Changes in the world situation, with ever-increasing complexity of battlefield conditions coupled with advances in food and packaging technology, have led to corresponding changes in the way the U.S. Armed Forces are fed.

Evolution of military field feeding. At the time of the American Revolution, combat field feeding was an unsophisticated matter. Fresh, dried, or salt-cured food items were the only products available; little was known about nutrition or the effects of diet on performance; and combatants had simple tastes and limited expectations. The pace of warfare was such that it was routine to allow time for soldiers to return to their farms to plant or harvest crops. Herds of livestock trailed with the army as a convenient source of meat and milk. During the course of a military campaign, soldiers were expected to forage for a significant portion of their subsistence needs.

Emergence of modern warfare negated this approach to combat field feeding. With the modern potential for hostilities to occur anywhere in the world and with highly mobile forces, competition for available storage and transportation space has become acute. In addition, the tastes and expectations of American service personnel reflect their culture, where a wide variety of excellent food is readily available. In concert with these changes, a better understanding of human nutrition has led to the development of performance-oriented operational rations with increased emphasis not only on acceptability but also on high levels of consumption. In keeping with advances in food technology, increasingly sophisticated consumer demands, and better understanding of nutrition and metabolism, the U.S. Armed Forces have developed and fielded a wide range of military operational rations designed to provide the best possible food to personnel, with primary consideration to their physical well being and performance across the entire spectrum of battlefield conditions.

Standard individual combat ration. The standard individual operational ration for the U.S. military forces in the field is known as the Meal, Ready-to-Eat (MRE). It is a thermally processed ration using a triply laminated retort pouch as the primary package. The MRE was introduced in the early 1980s and has since undergone major revisions. These include replacing 9 of the 12 entrees with more highly acceptable items, increasing serving size of the entree from 5 to 8 oz (142 to 227 grams), adding a cold beverage mix, eliminating dehydrated entrees, replacing the "military" candy components with commercial candies, and adding hot pepper sauce to several menus. The MRE (**Fig. 1**), like all operational rations, is designed to meet or exceed the Office of The Surgeon General's Military Recommended Dietary Allowances. There are 12 different menus, each of which provides approximately 1300 calories and contains up to nine components. This variety includes an entree, fruit, cookie bar, jelly, a spread of cheese or peanut butter, cracker, cocoa,

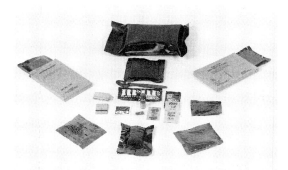

Fig. 1. Contents of one of the 12 Meal, Ready-to-Eat (MRE) menus.

beverage base, candy, plus an accessory packet and spoon. Further improvements are under development or have recently been introduced.

Baked goods. Until recently a bread that could be stored up to 3 years at 80°F (27°C) or 6 months at 100°F (38°C) and still retain its flavor, appearance, and high acceptability was unattainable. A packaged bread that meets these criteria has been developed, and it was used during Operation Desert Shield/Storm. Staling, which occurs in baked items having a wheat flour base, is inhibited by incorporation of sucrose fatty acid esters, emulsifiers that interact with the amylose starches. Furthermore, these emulsifiers condition the bread structure so that it is readily returned to an essentially fresh-baked state by mild warming, even after prolonged storage over a wide range of temperatures.

For this bread, commonly used mold inhibitors in the maximum quantities allowed were not sufficient to assure prevention of mold growth during long-term storage. Molds, which require oxygen for growth, may be totally inhibited by use of vacuum or nitrogen packaging, but the former would crush the flexibly packaged breads and the latter is too slow and costly for the high-speed methods used in commercial baking production. An oxygen scavenger or absorber is a substance, generally an innocuous iron compound, that chemically combines with the residual oxygen in a sealed container of food. As a result of that reaction, iron oxide is formed and little or no oxygen remains to react with food components such as fats that are oxygen-sensitive. An oxygen scavenger packet that reduces intrapackage oxygen levels to 1% or less was developed; this provides absolute mold inhibition. In addition, the low residual oxygen resulting from use of the oxygen scavenger prevents degradation of flavor and aroma in the stored bread. Water activity is a measure of the unbound (free) water present, expressed as the ratio of the degree of moisture saturation of air over the head space of foods to the degree of saturation over the head space of pure water. Water activity is controlled to 0.87 or lower, well below the level necessary to prevent growth of organisms.

This technology has also permitted the successful development of a wide range of dessert baked goods, such as pound cakes, as well as combination items containing meats and cheeses, such as pizza and burritos

(**Fig. 2**). Because of the meat and cheese components, techniques to reduce water activity, pH, and head space oxygen are employed to ensure that the unrefrigerated shelf stability meets the long-term storage required for military food. Further, the use of flexible, high-barrier pouches in combination with the oxygen scavenger has permitted the packaging of a broad range of snack items that attain the requisite 3-year shelf life.

Entrees. A family of new entrees has been developed to expand the variety available in the MRE. These include a smoky (link) frankfurter with the firm texture of a familiar American frankfurter rather than the more mushy texture of a Vienna sausage; a hamburger and cheeseburger; and various ethnic dishes, such as chicken chow mein.

The production of thermally stabilized breaded-and-fried meat items in rigid containers is not feasible, because products processed in these containers must either be packed solidly or have the head space volume filled with a sauce or brine to prevent stress damage to the container, owing to changes in pressure during retort processing. Breadings do not survive thermal processing in a fluid medium. However, breaded-and-fried chicken, fish, beef, and pork that are vacuum-packaged in flexible retort pouches do not require fluids for heat transfer during sterilization. Recent laboratory research has demonstrated that such processing is feasible, since heat is transferred directly over essentially the entire surface of the food item, which is in intimate contact with the package.

Sweets. Food researchers have developed a line of fig-based chew bars in seven flavors. Each bar provides not only necessary fiber but also 25% of the minimum daily requirement of calcium. This is important because this mineral is generally in short supply in military rations due to the difficulty of providing in the field fresh dairy products, the usual source. Furthermore, these bars can provide calcium to personnel who suffer from lactose intolerance and as a result cannot consume dairy products.

Milk products. Dehydrated milk products are unstable. Nonfat dry milk, for example, has a useful commercial shelf life of about 6 months at 70°F (21°C) and is not widely accepted as a beverage because it tastes different from whole milk. An acceptable, sta-

Fig. 2. Shelf-stable burrito containing ground beef, cheese, and spices encased in a dough stabilized for water activity.

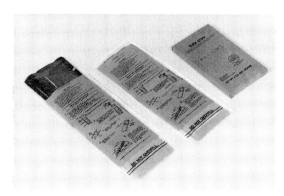

Fig 3. Flameless ration heater.

ble, nonfat milk item is needed to assure dietary calcium. Encapsulation technology, employing maltodextrins and hydrocolloids, a technique used only since the 1980s for spray-dried dairy products, has been effective in preventing some of the oxidative flavor degradation that typically occurs in conventional spray-dried milk. Incorporation of an oxygen scavenger package in the container of dried milk further enhances stability. Encapsulation and the addition of cream flavor provide a taste and mouth feel that enhance acceptability. This product has been used in powdered milkshake mixes that can be reconstituted with potable water in the field.

Fruits. The fruit component in the MRE, with the exception of applesauce, is freeze-dehydrated. Acceptance studies have shown that people have a significant preference for a canned-type wet-packed fruit compared with dehydrated products. Since rigid metal cans cannot be used in the MRE, retort pouches are required and, as with any operational ration component, a 3-year shelf life is necessary. Standard commercially available canned fruits derive their extended shelf life from the tin in the composition of the metal can; the tin helps protect the fruits from changes in color and texture. The flexible packaging materials do not contain tin; therefore, other mechanisms had to be developed to maintain acceptable characteristics of color, flavor, and texture during prolonged storage.

Four steps were taken to maximize the shelf life: (1) sucrose was used instead of high-fructose corn syrup, the sweetener typically used in the canned fruit industry; (2) the head space gas was kept below 0.6 in.^{3} (10 cm^3); (3) the pH was adjusted to 4.0; and (4) ascorbic acid was added to bring its concentration up to 500 parts per million. These changes minimized natural oxidation processes that occur in wet-pack fruit and in large part replaced the essential protection previously provided by tin. These steps, along with careful control of the maturity of the fruit at the time of packing, appear to be adequate to provide the required shelf life. After being stored for more than 3 years, test packs are still receiving high acceptance. It is anticipated that these same results will be achieved in field use.

Flameless heating. One shortcoming of operational rations has been the fact that in general they are consumed cold. In many tactical situations a flame would reveal field position and invite enemy fire; in addition, the time required to heat the meal is often unavailable. What seemed to be required was a flameless means of heating the ration—if necessary, while the soldier is on the move and unable to consume the meal immediately. This requirement has been met by the development of the flameless ration heater (**Fig. 3**), an electrochemical device that generates heat at a controlled rate. The reaction mechanism is incorporated into a porous pad measuring $4^1/2 \times 3^1/2 \times ^1/8$ in. (11.4 × 8.9 × 0.3 cm) and weighing 0.7 oz (20 g); it is capable of raising the temperature of an 8-oz (226-g) retort-pouched entree approximately 100°F (60°C) in 12 min. The heating pad is a composite material, consisting of a highly reactive metallic alloy (magnesium with 5 atomic percent iron, that is, 5 iron atoms for every 95 magnesium atoms, produced by high-energy powder metallurgical milling techniques) dispersed throughout a porous polyethylene matrix. Pressureless sintering of a mixture of the alloy powders and a powder of ultrahigh-molecular-weight polyethylene provides the porous matrix material. Sodium chloride is added (3% by weight); when dissolved it activates and sustains the reaction. A small amount of hydrogen gas is released; the other products of the reaction are water vapor and magnesium hydroxide. The MRE entree is heated by placing the entree pouch inside a polyethylene bag that contains the heating pad. Water is added (1.5 oz or 42 cm^3 by volume) and wets the pad. The polyethylene bag is then placed inside the paperboard carton of the entree; the carton acts as an insulating sleeve while the heating takes place. Should a tactical situation preclude immediate consumption of the heated item, the pad continues to provide heat to maintain an acceptable temperature for approximately 1 h. The ensemble can be inserted in the utility pocket of the battle dress uniform to afford the soldier heat-on-the-move capability. The U.S. Army procured a large quantity of these items for use in Operation Desert Shield/Storm during the colder months.

Continuing research. The correlation between a combatant's nutritional intake and maintaining or enhancing physical and cognitive performance on the battlefield cannot be overstated. Significant technological strides have been made to improve the variety, acceptability, and consumption rate of the MRE. Efforts to overcome additional technical barriers are under way to enable the introduction of even more highly desirable food components. The goal of recent and planned developments of new MRE menus and components is to ensure that the U.S. military will continue to receive the best type of food possible under combat conditions.

For background information SEE FOOD; FOOD MAN-UFACTURING in the McGraw-Hill Encyclopedia of Science & Technology.

Philip Brandler; Gerald A. Darsch

Bibliography. G. L. Schulz, E. Hirsch, and A. Salant, *New Operational Ration Development Issues*, Commonwealth Defense Service Organization, Malaysia, 1988; U.S. Department of the Army, *Army Technology Base Master Plan*, November 1990.

Food engineering

Recent advances in food engineering include improvements in controlled atmosphere storage for produce, and the development of high-speed optical sorting systems for removing defects and foreign matter from raw food products prior to processing.

Controlled Atmosphere Storage

Controlled atmosphere (CA) storage is a process whereby fresh produce is placed in a sealed refrigerated chamber immediately after harvest and held in an atmosphere that has a lower content of oxygen (O_2) and a higher content of carbon dioxide (CO_2) than does air. The temperature and atmospheric composition are carefully controlled to meet the optimum metabolic needs of the stored produce. The reduced level of O_2 and elevated level of CO_2 reduces the metabolism of the produce below that achieved by refrigeration alone. By minimizing metabolism, the rate of produce deterioration is slowed, so that postharvest quality is extended and the marketing season lengthened. Controlled atmosphere storage is used when the economic benefits of extended quality of produce create increased market value sufficient to offset the higher costs.

Principles. Typical ranges of steady-state O_2 and CO_2 concentrations in controlled atmosphere storage are approximately 1–3% and 1–5%, respectively, although each cultivar has its unique optimum requirement. Produce will experience oxygen starvation and become anaerobic if O_2 levels are maintained too low. Likewise, the produce will be irreversibly injured if CO_2 levels become too high. Temperatures from -1 to $3°C$ (30 to $37°F$) are used, depending upon the chilling sensitivity of the crop being stored. Relative humidity must be maintained at the optimum level for the stored product, and is usually controlled indirectly through the operation of the refrigeration system.

The control aspect of this technology is achieved by the continuous regulation of the steady-state gas concentrations inside the storage chamber. Chambers designed for controlled atmosphere storage are connected to an O_2/CO_2 analyzer that monitors the concentration of both gases. Levels are checked at least daily, and adjustments in gas concentration are made to maintain the desired range.

When the chamber is first filled, cooled, and sealed, the void volume is filled with air containing 79% nitrogen (N_2), 21% O_2, and a negligible amount of CO_2. An atmosphere generator is usually employed to remove the excess O_2 and create the desired atmosphere. Instead of using a generator, the chamber can be purged with a gas mixture rich in N_2 to displace unwanted O_2. Purging or generator operation is terminated when the desired O_2 concentration is achieved. Alternatively, the respiration of the produce may be allowed to consume the oxygen slowly, and to create a self-generated atmosphere. Once the atmosphere is established, a small quantity of air is added daily to provide the O_2 necessary for continued product respiration. The respired CO_2 must be scrubbed from the atmosphere to prevent it from reaching levels that would be toxic to the produce. Scrubbers containing solutions of sodium hydroxide (NaOH), dry hydrated lime [$Ca(OH)_2$], or an activated carbon molecular sieve material are used. Under steady-state conditions, air infiltrates the chamber to replace the removed CO_2. Since the chamber is nearly airtight, a variable-volume expansion bag is often connected to it to prevent small pressure differentials between the chamber and the outside. A safety vent is always provided to relieve large pressure differentials, which could damage the chamber structure. The components of the controlled atmosphere chamber are shown schematically in **Fig. 1**.

Recent developments. Recent developments in the practice and technology of controlled atmosphere storage, coupled with new applications, have led to several significant advances worldwide. For example, rapid and low-ethylene controlled atmosphere storage technologies have found commercial application. New and modified technology includes automatic computer-based analysis and control systems, cost-efficient ethylene scrubbers, and commercial use of cryogenic nitrogen and on-site air separation systems. Controlled atmosphere systems for container shipping are being introduced, and modified-atmosphere produce packaging is becoming widespread; this storage method does not include external monitoring or adjustment of gas concentrations. Use of controlled atmosphere technology is increasing in many developing countries to support production of high-value, perishable produce.

Rapid processing. Rapid controlled atmosphere storage is being adopted internationally by the commercial apple storage industry. The term rapid is relative, meaning that the entire process of harvesting, loading the chamber, cooling the produce, and establishing the controlled atmosphere is significantly faster than with traditional procedures. By shortening the waiting time for the harvested produce to be cooled and to have the atmosphere established, a longer period for storage in a controlled atmosphere can be realized. Thus the benefits of rapid controlled atmosphere storage are either longer-term storage or improved poststorage quality as compared to the traditional controlled atmosphere technology.

Several technological advances make the rapid system commercially practical. One is improved cultural practices and better postharvest handling systems. These improvements have helped reduce the time and labor required for delivering harvested product to the controlled atmosphere storage facility, so that less time is needed to fill the chamber initially. Present trends in construction are to build smaller, better-sealed, and more numerous chambers in the storage complex to enable faster filling and more rapid establishment of controlled atmospheres in individual chambers.

Chamber purging with nitrogen (N_2) from cryogenic sources is the fastest, most positive method of rapid establishment of a controlled atmosphere. By using a properly designed liquid-nitrogen system, chamber oxygen levels can be reduced from 21% to 1% in as little as 6 h. The main drawbacks are the high cost of

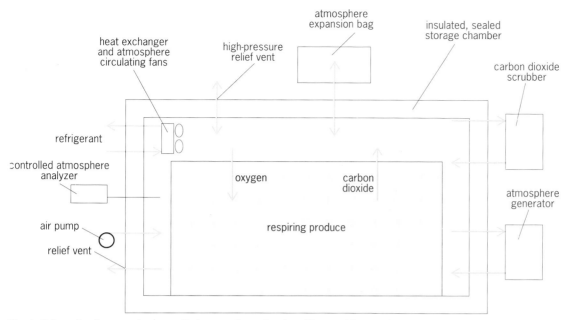

Fig. 1. Schematic diagram of a controlled atmosphere chamber. The respiring produce consumes oxygen and liberates carbon dioxide.

the N_2 and the need to maintain a cryogenic storage facility on the premises.

Air separation systems. There is much interest in generating N_2 on site for atmosphere establishment and, to a lesser extent, for long-term atmospheric control. These generating systems separate compressed air into a fraction rich in N_2 and a fraction rich in O_2. Thus N_2 is available on demand without the need for a compressed-gas or cryogenic-liquid storage system.

Two methods of air separation are being used. One utilizes polymer membranes that retain N_2 but allow O_2 to permeate, with the pressure differential across the membrane providing the driving force for permeation. The membrane is configured as a series of hollow fibers (hollow-fiber permeation), and it separates N_2 continuously from a compressed air supply. A second method employs adsorption under pressure by a carbon molecular sieve to trap O_2 while N_2 passes through. In this batch process the O_2 is desorbed periodically by relieving the pressure in the adsorber bed. The fluctuating pressures that are required for the alternate adsorption and desorption cycles give rise to the process name, pressure swing adsorption.

Automatic atmosphere control. Computer use for controlled atmosphere storage has expanded as a result of the development and use of automatic electronic atmosphere analyzers, followed by the introduction of computer-managed atmosphere control systems. This technology permits the reliable, precise, and consistent maintenance of very low concentrations of O_2 (as low as 0.6% in some applications) and a reduced dependence on human intervention in regulating the atmosphere in the storage chamber.

Traditionally, the atmosphere in each chamber was analyzed manually once or twice each day. Control actions such as addition of air to raise the level of O_2 or operation of the scrubbers to remove excess

CO_2 were also done manually. Usually each chamber required slightly different adjustments. The procedure demanded diligent observation by a trained operator. The amount of effort depended upon the number of chambers being operated and the precision of control desired. Operation of chambers with low levels of O_2 near the produce starvation threshold required extreme vigilance to ensure that the produce was not being damaged. Adjustments in gas concentrations were made as step changes once or twice each day, and required overshooting the O_2 set point and undershooting the CO_2 set point to compensate for the continuous effects of respiration.

Automatic analysis and control systems, on the other hand, are programmed to sample each chamber sequentially and to adjust the atmosphere as needed, several times each hour. A printed record of this activity is prepared each day. Through frequent but small adjustments, the atmosphere control is extremely precise, and safe use of low O_2 concentrations is possible. State-of-the-art control systems are now capable of monitoring and regulating atmosphere concentrations, storeroom temperature, and relative humidity, as well as alerting the operator if a refrigerant leak occurs inside the chamber. These systems still require the attention of a well-trained operator, but they enable one person to efficiently manage a large number of chambers from a central location and to program each chamber precisely to meet the requirements of the stored produce. Research and development is proceeding on the next generation of control systems, which will be able to monitor the metabolism of the produce and adjust the chamber environment, according to the instantaneous metabolic needs of the stored product.

Reduction of ethylene level. Most fruits and vegetables produce ethylene (C_2H_4) as part of the complex respiration process, and over time ethylene levels inside sealed controlled atmosphere chambers tend to increase. Elevated ethylene concentrations stimulate ripening and senescence of sensitive crops, which is detrimental to long-term postharvest quality. In certain situations it is economically worthwhile to employ some means of removing ethylene so as to maintain the concentration below the threshold level that stimulates ripening or senescence.

Various techniques are used for removing ethylene, including reaction with potassium permanganate ($KMnO_4$) at normal temperatures and direct oxidation through the use of catalytic scrubbers. Special shipping and packaging systems often utilize a premeasured, expendable quantity of proprietary ethylene-absorbing material containing $KMnO_4$ as the active ingredient. Catalytic scrubbers are gaining popularity for low-ethylene controlled atmosphere storage of apples. Recirculating scrubbers are being tested that use ultraviolet radiation to generate ozone (O_3), which in turn oxidizes the ethylene contained in the storage atmosphere. Successful storage with a controlled atmosphere low in ethylene requires extremely careful preharvest selection of produce and at least weekly postharvest monitoring of the ethylene in the chamber with a gas chromatograph to ensure that sufficiently low levels are maintained.

Transport. The use of refrigerated controlled atmosphere containers for ocean transport of produce is being explored. One promising system incorporates a small hollow-fiber permeation unit on the refrigerated container that constantly maintains an atmosphere having the desired N_2/O_2 ratio. The potential advantage is that surface shipment under controlled atmosphere may be used for commodities that cannot be shipped economically by air, so that the export potential and worldwide availability of many perishable products will be increased.

Developing countries. Controlled atmosphere storage systems based on indigenous, appropriate technology are being utilized throughout the world. In many locations, controlled atmosphere and modified atmosphere storage is being applied in the absence of refrigeration, scrubbing machinery, or even electricity. One ingenious method in China utilizes enclosures made of polymer film to achieve commercial controlled atmosphere storage in cool underground caves. Tents made of polyvinyl chloride film containing a silicone membrane window are filled with apples, sealed, and placed in underground caverns. The silicone membrane passively regulates the influx of O_2 and the diffusion of CO_2 from the tent. Oxygen levels are monitored with a manual wet chemical analyzer, and air is added to the tents as needed to sustain respiration. Although the method is labor-intensive, it is simple, and the positive benefits of controlled atmosphere storage are realized at a low technological cost.

Modified atmosphere packaging. Polymer films are being widely used for modified atmosphere packaging of fruits and vegetables. Presently it is possible to commercially produce films with precise O_2 and CO_2 permeability. By matching the properties of the film to the respiration characteristics of the produce

and taking into account the package surface area and volume, it is possible to develop and maintain the desired atmosphere inside the package. It is not yet possible to independently control the O_2 and CO_2 levels inside such packages, and this is an area of current research. There is also much interest in developing modified atmosphere packages with permeability characteristics that parallel the rates at which the respiration characteristics of specific produce vary with temperature. Modified atmosphere packaging may someday be used to offset the need for refrigeration during shipment of produce, retail display, and home storage prior to consumption. Modified atmosphere packaging of single apples has been demonstrated, but the unknown and variable respiration rate of each apple requires the film permeability to be custom-matched to the individual piece of fruit for optimum long-term benefits.

James A. Bartsch

High-Speed Optical Sorting

Processors of potatoes, fruits, vegetables, and other foods are increasing their use of automated sorting systems to remove defects and foreign matter from raw products. Growing demand for higher-quality product, coupled with increasing shortages of manual labor for defect sorting, has led to the development of high-speed/high-volume optical sorting systems. Typical volumes sorted by such systems range from 10,000 to 40,000 lb/h (4500 to 18,000 kg/h). For a commodity such as peas, this means rates approaching a half million peas per minute. The original optical sorters used in the food-processing industry channeled a single stream of product through a sensing area. The capacity required in today's vegetable and fruit industries demands a system that can sense and analyze a broad, two-dimensional stream of product, and remove both defects and foreign matter. Modern imaging techniques, high-speed digital processing, and sound mechanical systems for product handling must be used.

Components. The basic elements of a high-speed/high-volume optical sorter are material handling for infeed to the system, a mechanism that presents the product to an imaging system, the vision system (sensors, electronics, and a sophisticated computer system), a reject removal system, and conveyors to handle

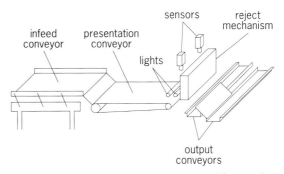

Fig. 2. Diagram showing the components and layout of a high-speed/high-volume optical sorter. The sensors are part of a vision system that includes a computer and interface electronics.

the sorted product. A typical layout of such systems is shown in **Fig. 2.**

Most high-volume sorters have the general mechanical elements in common, with differences only in execution and performance. Incoming product is generally fed to a vibratory conveyor, which tends to spread the product uniformly across its surface and accelerates product to a speed of 50–100 ft/min (15–30 m/min). The conveyor's surface may be made of a screening material to help remove water from the product if desired. The product typically discharges from the vibratory conveyor onto a belt conveyor, which accelerates it and separates and spreads it in the direction of flow, before it is viewed by the vision system.

Belt conveyor speeds generally range from 500 to 700 ft/min (150 to 215 m/min). Since the speed of the belt conveyor is much faster than that of the vibratory conveyor that feeds it, the transition between the two conveyors is critical. If the product is dropped from one to the other, it will bounce, perhaps even off the system, and not become stabilized before analysis by the machine vision system. The product being viewed must be stable to prevent data aliasing and to maintain good registration between imaging and defect rejection. Counterrotating stabilizers and other techniques are used to match the speed of the product and defect to the speed of the belt. Then, product and defect are viewed by the vision system, either while still on the belt or immediately after being discharged from the end of the belt.

Representative system. Figure 3 shows a representative sorting system introduced in 1991. The angle of view shows the infeed vibratory conveyor, which feeds product onto a belt conveyor (out of view behind the vision and reject systems). Also out of view are the bottom cameras, but the top cameras are visible. They are mounted so as to have an off-angle view of 50% of the line of sight. The usable line of sight for this machine is 54 in. (137 cm), and when run at 500 ft/min (150 m/min) belt speed the unit can sort more than 30,000 lb/h (13,500 kg/h) of most small vegetables, such as peas and diced carrots. Stretched across the center of the machine is the air rejector system used to separate defective from good product. The unit in Fig. 3 has 144 rejectors spaced at 0.375-in. (0.953-cm) intervals. The total height is just over 8 ft (2.4 m), and the maximum width is just under 9 ft (2.7 m). The length is governed by the conveyors leading to and away from the system.

Not shown in Fig. 3 is the enclosed operator control interface, which has a 19-in. (48-cm) touch screen with menu-driven controls. As product (and defect) are viewed by the vision system, data either on groups lined up with the separation system or on discrete pieces are taken. Individual picture elements (pixels) are evaluated as good or defective, and counts of defective pixels are accumulated. If the count exceeds a given level, a command is sent to the separation system to activate specific air valves when the defect is directly below those valves. The product is discharged from the end of the belt and passes through the sepa-

Fig. 3. ColorSort II high-speed/high-volume optical sorting system. (*Key Technology, Inc.*)

ration system. The individual good pieces of product follow a natural trajectory, landing on a take-away conveyor, while foreign matter or defective pieces of product are diverted by a puff of air to a different conveyor.

Unfortunately, the speeds of the valves used to separate the defective from the good product cause a certain amount of good product to be rejected. If the commodity is of sufficiently high value, the rejected portion may be recycled onto a portion of the incoming conveyor to allow the system a second chance to sort out defects, usually at a lower infeed rate; the disadvantage of such a feature is that it complicates the conveying system used for taking away the sorted product.

There are other, product-specific variations in the infeed systems. French fry strips, for example, are elongated in one dimension, and can be aligned with their axes along the direction of flow to enhance sorting. This aligning is usually done on an infeed vibratory conveyor; the structured surface, or troughs, on the bed of the conveyor aligns individual strips along the direction of motion.

Vision systems. While most of the general mechanical features of a sorter are common among various manufacturers, significant differences exist among the various vision systems. All vision systems include sensors, lights, and digital processing. In many systems, no matter whether sorting is done on defect size and gray scale or on color differences, lighting is provided by fluorescent bulbs over or under the path of the moving product stream. These bulbs may be specially designed to highlight specific spectral characteristics of the product or defect. One manufacturer of a bichromatic (two-color) sorter illuminates the product stream with two lasers, which create apparent red and green lines on the product as light output from each laser is swept by a high-speed rotating polygonal mirror. The lighting system must provide uniform illumination over the full width of the product stream in the area where the sensors are taking data.

High-speed/high-volume sorters use either a linear array of charge-coupled image sensors or small, two-dimensional sensors. As the stream of product and defect passes by the sensors, images are created and processed by the supporting computer and electronics. **Figure 4** shows a typical arrangement for image collection and the resulting image stored in memory. While the data are collected one scan at a time, they are processed in groups of lines to assess sizes of defects or foreign material. Each point or picture element (pixel) in the image represents a level of contrast, dark to light. In a typical monochromatic machine, each pixel is categorized as background, product, or defect, based on gray-scale contrast. The elements are then grouped or counted, and a machine decision that is based on the number of picture elements classified as defect is made. In a simple monochromatic machine, defects are either lighter than product or darker than product. These gray-scale differences can disappear on colored products, that is, any product that is not white or black, for example, peas or carrots. In this case, defects can be lighter than, darker than, "redder" than, "greener" than, or "bluer" than the product (or vice versa), so that the decision is even more complex. Color cameras typically produce two to three times the data produced by their monochrome counterparts. Because of the nature of the color differences between product and defect, compromises are used in some units that provide a two-color solution rather than a standard three-color solution. The data, in any case, are collected as two or three standard gray-scale images. Once in the computer, they may be processed together as RGB (red, green, blue) images or as HSL (hue, saturation, luminance) images, or possibly in some less conventional manner. Decisions are then made for each element to define whether it represents good product or some class of defect. Because of the digital nature of the data sampling, some pixels lie on boundaries between product and background, product and defect, or defect and background. This problem is dealt with either by electronic filtering or by providing a background that is the same color as the product. Unless the data are processed in discrete two-dimensional blocks, care must be taken to know the position of the piece of product or defect relative to the mechanism that separates good product from defect. Actual decision times vary from system to system, but typical in-air sorters make up to 140 parallel decisions in 0.02 s.

Utility. In understanding the value of the high-speed/high-volume optical sorting machines to the user, it is necessary to consider the difficulties involved in obtaining consistent sorting with manual labor, the cost and availability of such labor, and the increasing demand for higher-quality food products. High-volume sorters first made a significant impact when the potato (french fry) industry needed consistent product that met the demands of the fast-food industry. In this application, sorters are coupled with automatic defect removal systems that employ optical and computing systems similar to these used with reject devices, but that utilize a cutting system to remove the defects from

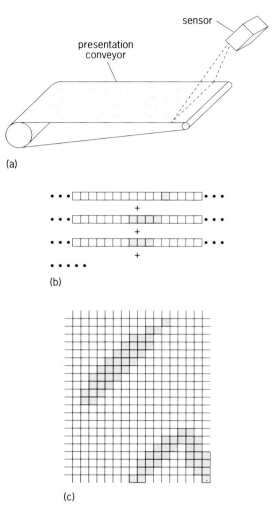

(a)

(b)

(c)

Fig. 4. Imaging for sorting. (a) Diagram of a sensor (camera) collecting image. (b) Camera data processed line by line. (c) Lines accumulated and the data processed as two-dimensional images.

individual pieces of the fried potatoes while removing a minimum amount of the product. Such a machine can do the work of 50–100 people; and since the machine is consistent and never needs a rest period, it maintains a higher average grade of product.

The newest machines are being used in vegetable and fruit processing plants and utilize color to search for a wider variety of defects. The next generation of machines, currently in development, will provide even higher levels of performance by removing defects related to shape and texture, as well as improved performance on color defects. Eventually, automated machines should replace most, if not all, of the human inspectors on the food processor's inspection line.

For background information SEE COMPUTER VISION; FOOD ENGINEERING; FOOD MANUFACTURING; OPTICAL INFORMATION SYSTEMS in the McGraw-Hill Encyclopedia of Science & Technology.

William E. Shaw

Bibliography. W. E. Shaw, *The Food Processors' Guide to Optical Defect Detection*, 1988; R. M. Sperber, Laser beans!, *Food Process.*, 52(7):60–62, 1991.

Food manufacturing

Recent advances in food manufacturing have involved development of insights into the mechanistic basis of puffing a number of porous food products, new methods for production of gum-based texturized products, clarification of fruit juices, and liquefaction of solid constituents of fruit to increase fruit juice yields.

Puffing

Puffed cereals, popcorn, bread, biscuits, cookies, and meringues are porous, expanded foods. Pore volume, size, and structure, as well as the thickness of the solid between pores, affect their texture, deformation properties, breaking and cutting strength, heat- and mass-transfer resistance, appearance, optical reflectance, and the sounds they produce when they are being consumed.

Product behavior. White bread is a typical plastic and porous food. **Figure 1** shows a stress (S) versus strain (ϵ) curve for compression of bread. When pore walls buckle, yield, or fracture, ϵ increases markedly but S changes very little. For crisp, porous foods, this region is very jagged; and if they are not confined, they crumble away and densification does not occur. The porosity (ψ) of an expanded material is given by the equation below, where ρ is the den-

$$\psi = 1 - \frac{\rho}{\rho_s}$$

sity of the expanded material, and ρ_s is the density of the solid matrix. Stress-strain behavior depends on the ratio ρ/ρ_s. Strength attributes are proportional to $(\rho/\rho_s)^n$, where $n = 1.5$ or 2 for open pores, and $n = 2$ or 3 for closed pores. Product properties are strongly affected by those of the solid matrix. Strength decreases and crisp foods become more plastic as moisture content and temperature increase.

Methods of generating pores. Pores in foods can be created or enlarged by generating gas [usually carbon dioxide (CO_2)], heat-induced expul-

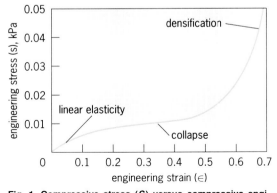

Fig. 1. Compressive stress (S) versus compressive engineering strain (ϵ) for white bread. In the region of linear elasticity, cell walls bend; in the region of collapse, cell walls buckle, yield, or fracture; in the region of densification, cell walls crush together. Stress varies linearly with ϵ (deformation is elastic when the value for ϵ is close to 0). (*After M. Peleg et al., Mathematical characterization of the compressive stress-strain relationship of spongy food products, J. Food Sci., 54(4):947–949, 1989*)

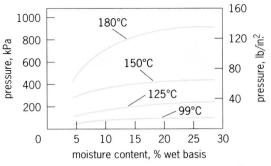

Fig. 2. Equilibrium water-vapor pressure versus moisture content for corn starch at various temperatures. °F = (°C × 1.8) + 32°.

sion of dissolved gas, admixing or beating gases into liquids or doughs, and heat-induced formation and expansion of bubbles of water vapor or of water vapor and gas.

Pores expand when stresses generated by excess pressure in bubble nuclei or existing pores exceed the strength of the surrounding matrix; and the matrix yields, stretches, unfolds, or flows. During expansion, pore walls become thinner, gas and vapor diffuse out and escape, and evaporative cooling occurs. If walls rupture, gas and vapor escape rapidly; pores depressurize; and if walls do not harden, contraction occurs. Wall thinning reduces wall strength, but moisture loss and evaporative cooling facilitated by thinning increase strength and viscous resistance to expansion.

Baked goods. During leavening and baking of bread, CO_2 diffuses into pores previously created by folding air into dough, and these pores expand. Hydrated gluten in wheat dough forms extensible fibers that accommodate expansion at low excess pressure. Dough is "punched down" and rekneaded to subdivide pores. Pores expand as temperature rises during baking; but at around 80°C (176°F), CO_2 starts to escape and expansion stops. Gluten concurrently sets, probably because water transfers to starch, and the starch and gluten interact. This setting prevents contraction. Cookies and cakes have less gluten and also contain sugar, which acts as a plasticizer. Consequently, they shrink when internal pressure falls during cooling. The ratio ρ/ρ_s for breads and cakes ranges from 0.12 to 0.20.

Explosive decompression. Between 160 and 215°C (320 and 419°F), starchy feedstocks containing 5–20% water partially melt and generate water-vapor pressures, for example, 475–910 kilopascals (69–132 lb/in.^{2}) at 180°C (356°F) [**Fig. 2**]. Rapid puffing can be obtained by confining such feeds or maintaining a high external pressure, heating them to 170–200°C (338–392°F), and then suddenly ending the confinement or releasing the external pressure. Extruders, rice-cake machines, and batch and continuous puffing guns are used for this purpose.

Puffing guns. Grain-based particles are processed in puffing guns with walls heated to 205–370°C (401–698°F) and with superheated steam under 0.65–1.76 megapascals (94–255 lb/in.^{2}) pressure. The heated, pressurized particles discharge through venturi tubes in continuous guns or through suddenly opened ports in batch guns. Residence times are roughly 8 min for batch guns and 90 s for continuous guns. Products with ρ/ρ_s as low as 0.066 are obtained. Partly dried fruits and vegetables containing 15–26% moisture are puffed in steam-pressurized systems fitted with star valves (thick wheels with vanes that rotate in tight-fitting cylindrical casings) or double-valve locks; both devices provide explosive decompression and permit continuous operation. Superheated steam under 83–410 kPa (12–60 lb/in.^{2}) pressure and 0.5–2 min holdup times are used.

Extrusion puffing. Starchy feeds containing 8–35% moisture are puffed in cooker extruders operating at 110–180°C (230–356°F). Twin-screw extruders with kneading blocks or single-screw extruders with shallow channels and rifled barrels generate intense shear, specific mechanical energy inputs of 0.10–0.16 kWh/kg (0.22–0.35 kWh/lb) of feed and 4–8 MPa (580–1160 lb/in.^{2}) discharge pressure; they can handle feeds containing 8–15% moisture and provide products with ρ/ρ_s of 0.021–0.13. Medium-shear extruders provide specific mechanical energy inputs of 0.02–0.08 kWh/kg (0.04–0.18 kWh/lb), discharge pressures of 2–4 MPa (290–580 lb/in.^{2}), and barrel temperatures up to 145°C (293°F); they are used for feeds with 20–35% moisture and generate products with ρ/ρ_s of 0.10–0.33.

Expansion increases as specific mechanical energy input increases at constant temperature. It also increases as moisture content decreases and screw speed increases, probably because these changes increase the specific mechanical energy input. For starches, expansion goes through a maximum at a temperature, T_{max}, which increases as amylose content increases. For mixtures of starch and protein, expansion decreases as protein content increases. Unpuffed extrudates are sometimes produced by using die temperatures lower than 100°C (212°F), are then dried, and subsequently are puffed by heating in hot oil or air. Extruders produce amorphous doughs, completely destroy original particle structure, and cause extensive breakdown of starch.

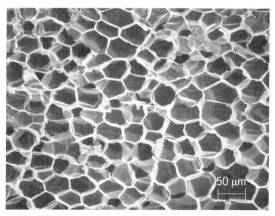

Fig. 3. Scanning electron microscope image of cells in popped corn.

Popcorn popping. When popcorn is heated, its hull acts like a pressure vessel. Little popping occurs if the hull is damaged. Around 185°C (365°F), when internal water-vapor pressure is 760 kPa (110 lb/in.^{2}), the hull ruptures and starch granules in the translucent outer endosperm puff; expansion proper takes roughly $1/30$ s. Temperature drops slightly, then rises again; internal pressure drops to 300–400 kPa (44–58 lb/in.^{2}) or lower, because of vapor release and cooling. **Figure 3** is a scanning electron micrograph of cells in popped corn.

Hot-air puffing. Grains containing 7–9% moisture have been puffed in air-heated fluidized beds and jet-zone dryers by air heated to 205–290°C (401–554°F) with velocities of 5 m/s (20 ft/s). Partially dried, diced potatoes have been puffed in high-velocity air heated to 150°C (302°F). They rehydrate well, but if air at 175°C (347°F) is used they fall apart. Partially dried fruits and imitation fruit pieces made from gels puff at much lower temperatures, but tend to caramelize.

Vacuum puffing. Caramelization can be prevented by puffing under vacuum. Temperatures as low as 30°C (86°F) can be used initially; 80°C (176°F) is used to achieve the desired final moisture level of 2%. Higher final-moisture contents can be used if products are cooled while under vacuum to prevent collapse upon discharge to atmospheric pressure.

Pore characteristics. Pores in popcorn form from individual grains of starch and are polygonal, 30–50 micrometers wide, and fairly uniform in size. The pores are mainly closed when initial moisture contents are lower than 12%, but openness increases as initial moisture content increases. At initial moistures greater than 24%, pores coalesce, pore size increases, and overall expansion decreases. Pores in puffed wheat and puffed rice are also polygonal, 100–200 μm in diameter, and mostly open. They probably are produced by coalescence of smaller pores.

Products made in puffing guns from predried pellets of corn- or oat-based dough have highly irregular open pores that are 100–200 μm wide, but they contain blowholes that are roughly 1 mm (0.04 in.) in diameter. Extruded starch-based products sometimes have closed pores with mean diameters of 2–3 mm (0.08–0.12 in.), but pore openness increases as expansion increases. Pores in extrudates are probably produced by nucleation and growth of vapor bubbles in an amorphous fluidlike or viscoelastic mass.   *Henry G. Schwartzberg*

Gum-Based Texturized Products

Many processed foods contain hydrocolloids (gums) that are responsible for the functionality and organoleptic acceptability. Hydrocolloids are natural, modified natural, synthetic, or biosynthetic polymers that dissolve in water and have the ability to thicken or gel aqueous systems. Examples include algin, starch, modified starches, agar, carrageenan, gelatin, xanthan, and gum arabic, as well as some proprietary formulations. Hydrocolloids are frequently used in formulating and preparing gum-based foods, and they significantly affect the appearance, physical properties, and shelf life of such foods. Hydrocolloids have been selected to fit conditions of manufacture, and specific manufacturing processes have been invented to take advantage of the functional properties of the various gums.

Restructured foods. Many fabricated foods contain hydrocolloid gelling agents. When an alginate solution is dropped into a soluble calcium salt, an insoluble calcium alginate "skin" forms almost immediately. This simple reaction provides a basis for manufacturing many restructured foods. The first patent based on this reaction was granted in 1946 for a process for making artificial cherries (U.S. Patent 2,403,547), which did not contain any fruit. Since the 1970s, many alternative processes for preparing fabricated fruits have been described in patents and technical publications. Fruit pulp is frequently used in creating these fruit analogs. However, pulps with high acidity and pH of 3–3.5 tend to weaken gels. Products with more than 10% of highly acid pulp are so weak that they collapse under their own weight.

Recently a procedure for preparing products containing up to 90% high-acidity pulp (**Fig. 4**) was developed. The pulp is neutralized, so that addition of a high concentration of pulp is possible. Later, a glucono-δ-lactone is added, and an acid/calcium bath is used to acidify the system and to strengthen the final product.

Types. Many different types of fabricated foods can be created by using different main ingredients (such as fish, meat, fruit, and vegetables), simply changing the concentration of alginate, and selecting the source of soluble calcium. The variety of final products is huge.

Fabricated onion rings were introduced in the 1970s. A slurry mainly of sodium alginate and diced onions is extruded into a calcium chloride bath. After the skin forms, the rings are dusted with breading batter, fried in deep fat, blast-frozen, and packed. Cellulose ethers, which form lipophobic protective films, are used to prevent oil absorption during deep fat frying.

Other hydrocolloids—carrageenan, gelatin (and recently gellan), and combinations of gums such as carrageenan and locust bean gum—have been used in fabricating products. Examples include reconstituted pimento strips (based on alginate and gum arabic), imitation caviar, and restructured fish and shellfish.

Extrusion processes for manufacturing snacks use a flour or starch matrix in combination with flavors, colors, and spices. The starch component is transformed by heat and pressure to yield desirable functional properties in the finished product. Hydrocolloids, such as cellulose ethers, are used to improve thermoplasticity and provide lubrication during extrusion. Starches are also widely used in conventional batch canning operations to maintain the consistency of the product after heat treatment. Such starches should be stable over a wide range of storage conditions and must retain their physical and sensory qualities. This requirement is especially important for creamed vegetable soups, gravy-based products, and pie fillings. Modified starches that do not thin out under continuous heating and do not gel when stored at low temperature are used for this purpose.

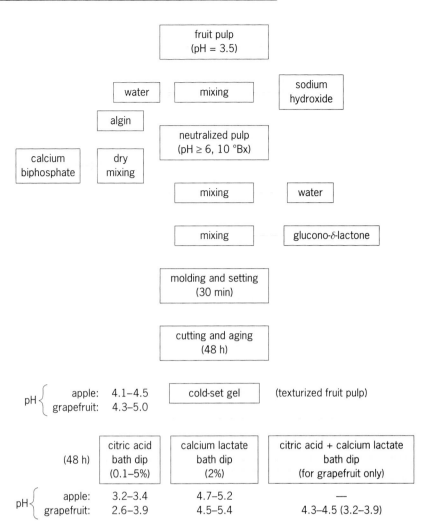

Fig. 4. Block diagram of a process preparing fruit products with high-acidity apple pulp or grapefruit juice; the products are texturized with alginate. °Brix is a unit that represents percentage of sugar content on the Brix scale. Values in parentheses refer to a process where the product was first dipped in calcium lactate and subsequently in an acid bath. Broken lines indicate that cold-set gel (texturized fruit pulp) can be treated in any or all of the proposed methods.

Molded products. A common method of manufacturing gum-based texturized product is to pour a hot or cold solution of the gum, mixed with other ingredients, into a mold and let it set. Agar forms a firm gel at a low concentration and is used to prepare several Japanese dessert products. These traditional desserts are molded into ingot-shaped bars. Generally, they are packed and sealed in a laminated-film container. These jellylike confections are based on agar, sugar, and mashed Azuki beans. Instead of beans, other ingredients such as sugars, powders, and fruit pulp may be used.

A modernized version of the dessert, "sweet agar jelly," is a highly sweetened, mildly flavored candy of low acidity. Processing includes dissolving the agar in a boiling solution of sucrose and invert sugar, followed by brief cooling and addition of juice, acidulant, flavoring, and coloring. Moisture content is reduced by drying in an oven. Since this product is sticky, it is coated with an edible film of agar and starch. The film is produced by brushing a hot solution of the agar–starch mixture onto a polished metal surface.

The process of hardening of slabs can also be used to produce canned, multilayered dessert gels. Such a product can be fashioned from a commercially produced hydrocolloid. A so-called simultaneous method for preparing layered desserts from solutions of blended xanthan gum/locust bean gum and carrageenan was reported recently. Solutions of the hydrocolloids (the blend and carrageenan) are poured into a mold at $158°F$ ($70°C$) and cooled, and two distinct layers separate. These products can contain juices, coffee extracts, and dietary fibers as well as different colorings, flavorings, and texturizing materials.

Drying. In food drying, the goal is to remove water at the lowest possible temperature to optimize product quality. For fruits and vegetables, this sometimes results in a tough, leathery surface on the dried particle that can adversely affect rehydration and eating qualities. The problem can be reduced by wetting the fruit or vegetable particles with solutions of cold, water-soluble gums such as guar or carboxymethylcellulose prior to drying. Several methods are used, for example, soaking under vacuum to improve penetration of the hydrocolloid into the particles.

Drying techniques, such as spray drying, are used to encapsulate food components such as flavorings and protect them from environmental influences. The mixtures or emulsions of the flavoring and a gum in water must be dried rapidly. During spray drying, water is removed and the flavored particle becomes coated with a gum film. Gum arabic is superior for this purpose, but gum ghatti, another plant exudate, is also used. Gum arabic reduces surface tension and allows emulsions to form more easily. It also coats dispersed oil droplets so that they are surrounded with an electrically charged surface. Since all the droplets then have the same charge, they repel each other, and collision and coalescence is inhibited. Gum arabic is used widely for microencapsulation. Modified starches have been used for the same purpose when gum arabic has been scarce, but with less success.

Stabilization. Hydrocolloids are used as emulsion stabilizers in salad dressings and ice cream. Salad dressings are oil-in-water emulsions prepared by mechanical homogenization. The dispersed phase is stabilized by the increased viscosity and yield strength provided by the hydrocolloid. Gum tragacanth, propylene glycol alginate, and xanthan can be used.

In ice cream and similar frozen desserts, stabilizers serve to promote smoothness, reduce drip, improve body, and increase resistance to heat shock. The hydrocolloids limit incorporation rates in crystals and thus minimize growth of ice crystals and sugar crystals, which adversely affect texture.

Freezing often causes undesirable changes in foods. Hydrocolloids are used to improve the quality of such foods. In order to produce an ice cream of high quality, a blend of guar gum or carboxymethylcellulose with a smaller amount of carrageenan may be used. If xanthan and guar gum are used instead, viscosity is lower and faster processing is obtained. Gum karaya was used

in the past as a stabilizer for frozen desserts, but has been replaced almost completely by other gums. Carrageenan, guar gum, and carboxymethylcellulose have also been used as stabilizers in other frozen products. In foods containing starch as the main ingredient, there is a tendency for water to exude from the gel. Thus starch-based products curdle and undergo syneresis (loss of water) after freezing and thawing. Modified starches were developed to deal with the problem.

Many milk and juice products are processed aseptically. Special carrageenans are used in such milk products. They are soluble at ultrahigh temperatures, stabilize the milk protein, suspend fat globules, and preserve the product taste; they also permit hot or cold packaging of the finished product. Carrageenan is used widely in other milk products, such as processed beverages, pasteurized milk, dry-mix beverage powders, milk puddings, and cheese products.

Formation of stable foams is essential in preparing many food products ranging from bread to whipped cream and mousses. Incorporation of air is needed to produce the desired textures in the final product. Gums help stabilize these foams. Mixtures of gums such as carrageenan, sodium alginate, guar, and carboxymethylcellulose are frequently used. Cellulose ethers are used to stabilize whipped toppings. In foams based on proteins, the foam can be stabilized by physical means (for example, heating or cooling), but gums such as carrageenan, alginate, and locust bean gum are often added to react with the protein to form a stable foam.   *Amos Nussinovitch*

Clarification of Fruit Juices

Some fruit juices, such as apple and grape, are clarified because of consumer preference. Other juices are clarified to facilitate further treatment such as concentration by reverse osmosis and, as in some citrus juices, debittering.

Right after extraction, fruit juices are turbid because of suspended solids. The amount of suspended matter depends on the method of extraction. Extraction processes that involve milling and high-pressure expression usually result in high contents of suspended solids. Diffusion extraction, involving a counterflow of hot water through fruit slices, is used to produce juice with a lower content of suspended solids, such as pectin and hemicellulose, but with a higher content of polyphenols and acids in the final product.

Causes of turbidity. Juice from an extractor contains unstable components that can coagulate during treatment with enzymes and heat; as a result, sedimentation occurs during storage. Another common cause of haze in fruit juices is microbial growth in juice that has not been properly heat-treated or stored. Haze can develop in apple juice from aggregation of the tannins, proteins, and starches. Eventually the association of these molecules causes visible haze when the aggregates reach 0.1 micrometer. As the particles continue to grow, they settle to the bottom of the container.

Tannins and protein complexes. Tannins are a common source of haze in some juices, particularly apple, pear, and grape. They are produced by the polymerization of phenolic compounds. Even though polyphenols are slow to form in juice, after long storage periods they tend to form a brown sediment. Polyphenol formation is accelerated by low pH and high temperature. Polyphenols or tannins give a desirable astringent taste to the juice, and are usually suitable only in moderate amounts.

Fruit juices usually do not have appreciable amounts of indigenous protein. However, some protein is introduced during treatment with enzymes and gelatin. These proteins readily associate with tannins to form complexes that result in dark sediments that settle in the bottom of the container.

Gums. During the extraction of fruit juices such as apple and pear, components from the cell walls, such as gums normally associated with the hemicellulose, are released into the juice. This problem is exacerbated when apples that have been stored for long periods of time are used in juice production. Other gums in the fruit such as arabinans tend to remain in apple juice even after conventional processes of clarification and fining are applied, and, after long storage, result in haze formation.

Starches and dextrose. Starch is naturally present in some fruits, up to 8% in apples. Agglomeration of short starch molecules in the juice results in haze formation after the particles have become large enough to scatter light. Starches can also form complexes with proteins and tannins that result in haze.

Methods of clarification. These include standard filtration and membrane processes.

Standard filtration. Diatomaceous earth is commonly used as a filter aid for clarifying fruit juices. It is used in conjunction with a mechanical filtration system such as filter presses and rotary filters. However, this system allows particles smaller than 1 μm to pass through the filter along with the clear juice, and so the particles must be eliminated by a fining process. In some instances the amount of suspended solids is reduced by centrifugation to facilitate filtering. Gelatin is commonly used as a fining agent to eliminate haze produced by long-chain tannins. It is also used in combination with bentonite and silica sol to remove protein and tannin–protein complexes in the juice. The polymer polyvinylpyrrolidone is also used to remove polyphenols in juices and beverages, particularly short-chain tannins. Heating the juice prior to filtration helps to clarify fruit juices by coagulating some of the suspended solids, such as proteins, which can then be easily separated during filtration.

Membrane processes. More recently, membrane processes have been introduced as a means of clarifying fruit juices. Membrane filtration can be divided into ultrafiltration and microfiltration. In general, microfiltration screens out suspended particles larger than about 0.1 μm and thus allows the passage of smaller particles that can produce haze. Therefore, the utility of microfiltration is as a pretreatment step for a fining process. Ultrafiltration, however, screens out suspended and dissolved particles as small as 0.001 μm.

Ultrafiltration is now widely used for clarification and fining of fruit juices, and it is beginning to replace conventional methods. Ultrafiltration makes it possible to clarify fruit juices in a single step after extraction and heat treatment. Membranes used for ultrafiltration can remove polyphenols, starch, protein, and gums. An added advantage is the minimizing of heat treatment, because ultrafiltration membranes are able to remove bacteria from the raw material.

Prior to ultrafiltration, the juice is usually treated with enzymes to lower the viscosity and facilitate passage through the membrane. Suspended pectin and hemicellulose in the juice can foul the ultrafiltration membrane, so that it is necessary to use pectinases and hemicellulases right after the extraction stage. Starch-digesting enzymes can also be added to the juice to convert starch into sugar, whose molecules will not form haze in the final product. Since the polysaccharide arabinan is a common cause of haze development during storage of the final clear product, enzymes that hydrolyze arabinans (arabinases) are now added to the commercial enzyme formulations for pretreatment of fruit juice.

Tubular ultrafiltration systems can handle juices with higher contents of suspended solids. In this system the feedstock flows inside the tubes, and the permeate (the juice product) is collected in a housing outside the tubes. These systems operate with turbulent flow and at transmembrane pressures of 30–40 lb/in.^{2} (207–276 kPa). The membranes are easy to clean and to replace. However, this ultrafiltration system has a low area-to-volume ratio: 46 ft^2/ft^3 (150 m^2/m^3).

One of the most widely utilized ultrafiltration systems in the juice industry is the hollow-fiber cross-flow system. It is commonly used for the clarification of apple juice. This system is similar to the tubular membrane system, except that the diameter of the hollow fibers is much smaller and shear rates are usually high (4000–14,000 s^{-1}); this design helps prevent fouling of the membrane by the suspended solids. Operating pressures are 10–25 lb/in.^{2} (69–172 kPa). This system can handle only those juices with limited content of suspended solids.

Plate-and-frame ultrafiltration design utilizes a flat-sheet configuration in which the membrane and drainage screen are placed between plates. A large number of plates are placed on top of or next to each other. This system can operate at higher pressures (150 lb/in.^{2} or 1030 kPa), and replacement of membranes is relatively easy. Energy consumption is intermediate between spiral-wound and tubular systems.

The spiral-wound systems are the least expensive and the most compact of the available ultrafiltration systems. The membrane is wound around a central permeate-collection tube with a meshlike spacer sheet for the feed and another spacer for the permeate. The feed flows lengthwise, and the permeate is forced through the membrane and spirals toward the central pipe of the cartridge. This ultrafiltration system is very susceptible to plugging and difficult to clean, so it is not suitable for juices with high content of suspended solids. The surface-to-volume ratio is 200–

300 ft^2/ft^3 (660–990 m^2/m^3), and the pressure drop is 15–20 lb/in.^{2} (103–138 kPa). This system consumes less power, and the cost of membrane replacement is relatively low.

Rotary ultrafiltration systems were introduced recently for the clarification of fruit juices. They operate by using both shear and centrifugal forces so that there is efficient clarification of fruit juices with high pulp content at lower temperatures and transmembrane pressures. The membrane cartridge rotates at high speed, up to 6000 revolutions per minute, and the resultant vortices at the walls of the membrane help prevent fouling. The feedstock flows inside the rotating cartridge, and the permeate is forced axially toward the outside. This ultrafiltration system is reported to provide better permeate flow rates per membrane area than conventional cross-flow ultrafiltration systems (3–15 times higher) with minimum shear imparted to the product.   *Ernesto Hernandez*

Liquefaction of Fruit

Exogenous enzymes are used in processing fruits to improve yield and quality of the product. Generally these enzymes belong to the group of pectinases, although for some applications cellulases also appear to be necessary. Fruit juice manufacturers can add pectinases at several stages of production, depending on the technologies used.

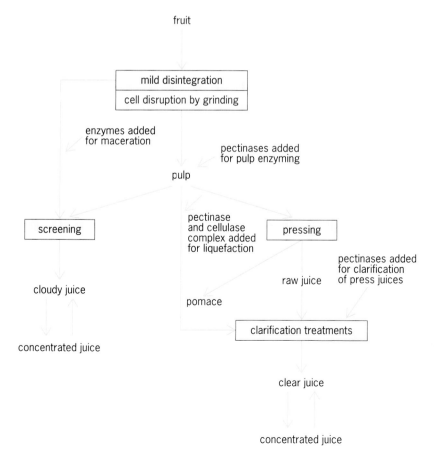

Fig 5. Flow diagram of fruit juice manufacture; arrows indicate eventual enzyme treatments.

There are four main processes: juice clarification, pulp enzyming, liquefaction, and maceration (**Fig. 5**). These processes differ with respect to type of enzyme used, the stage at which the enzyme is added, and the extent of cell wall degradation. Fungal pectinases have been used for juice clarification for more than 50 years. The enzyme preparation is added to the press juice (juice obtained from straight pressing of fruit) to degrade pectic polysaccharides in order to remove cloud particles, as well as to decrease the viscosity of the raw juice by depolymerizing dissolved pectin before final filtration (thus the term filtration enzymes). The pulp enzyming process was originally developed to produce juice from soft fruits such as black currants, which are very difficult to press; however, it was later also applied to producing juice from apples and grapes. Cloudy juices and nectars or other pulpy (viscous) products can be produced when the more specific polygalacturonase, pectin lyase, or pectate lyases are applied for maceration. The mechanism of enzymatic cloud stabilization is well documented, while the use of specific pectinases for maceration is still under investigation.

A new development is the use of a preparation in which pectinase and cellulase are combined for the complete liquefaction of fruit tissue. In that case, pectin, hemicellulose, and cellulose are degraded to such an extent that the cell walls almost completely disintegrate and are solubilized. The breakdown products increase the amount of soluble solids in the juice, so that yields of more than 100% can be obtained on a soluble-solid basis.

Enzymes and substrates. Commercial pectinase preparations are generally derived from *Aspergillus niger*; they are mixtures of enzymes, particularly polysaccharidases that act on pectin and pectin-related hemicellulosic polysaccharides. Endocellulase (the endoenzyme carboxymethylcellulase) also is commonly present. Degradation occurs differently in the homogalacturonan and the rhamnogalacturonan part of the pectin molecule. The homogalacturonan is built up by 1,4-linked α-D-galacturonic acid residues. Part of the galacturonic acid residues is esterified with methanol. The homogalacturonan chain can be degraded by a mixture of pectin esterase and polygalacturonase or pectate lyase (**Fig. 6**). The esterified homogalacturonan is an ideal substrate for pectin lyase. The splitting of the glycosidic bond by polygalacturonase is a hydrolytic reaction, while the lyases split the glycosidic linkage by β-elimination to give unsaturated galacturonides.

The rhamnogalacturonan part of the pectin molecule consists of alternating sequences of galacturonic acid and L-rhamnose residues. These structures are called ramified hairy regions because oligomeric and polymeric side chains, consisting of the neutral sugars L-arabinose and D-galactose, are linked to the main chain at this location. D-Xylose occurs as short side chains. An enzyme that can degrade the rhamnogalacturonan chain to branched and unbranched oligomeric products has recently been identified, purified, and characterized. How this enzyme affects liquefaction is unknown

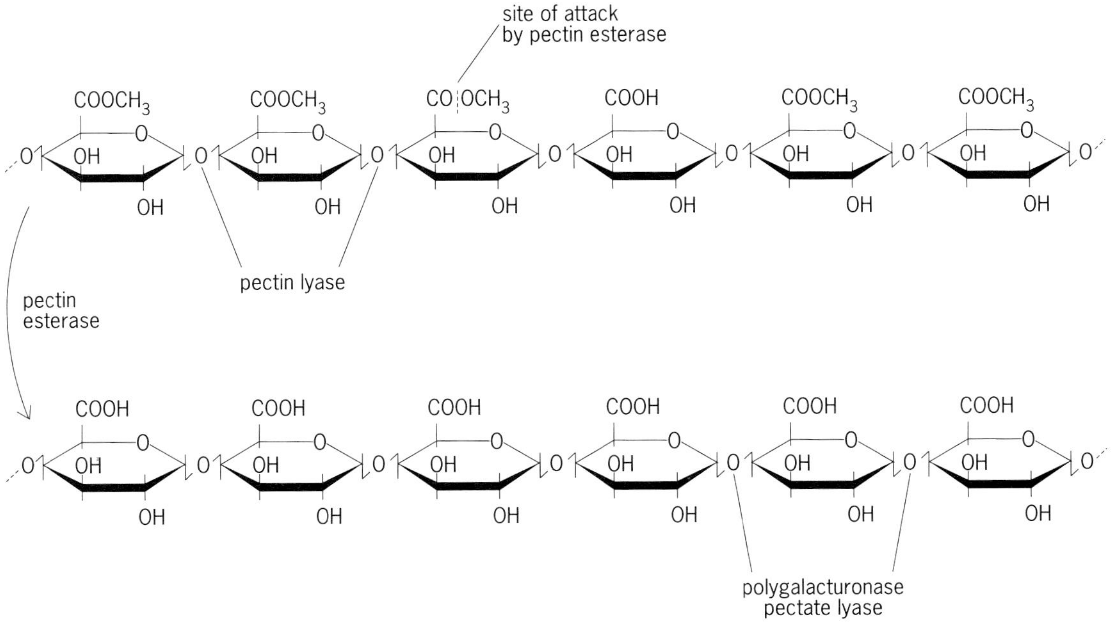

site of attack
by pectin esterase

COOCH$_3$ COOCH$_3$ CO¦OCH$_3$ COOH   COOCH$_3$ COOCH$_3$

pectin
esterase

pectin lyase

COOH   COOH   COOH   COOH   COOH   COOH

polygalacturonase
pectate lyase

Fig. 6. Fragments of a pectin molecule and points of attack by pectic enzymes.

as yet, but the commercial preparation from which the enzyme has been isolated is known to have a mode of action that is different from that of another preparation in which the enzyme is missing.

Enzymatic degradation of pectic polysaccharides by commercially produced pectinases will result in the appearance of methanol, saturated monomers and oligomers of galacturonic acid, unsaturated oligomers of galacturonic acid, and monomers and oligomers of arabinose, galactose, and xylose in the raw juice. In liquefaction processes, glucose is formed from cellulose by the action of cellulolytic enzymes. A prerequisite for liquefaction is the the use of cellulases that can degrade crystalline cellulose, for instance, cellulases derived from *Trichoderma reesei*. These preparations contain the whole complex of cellulases, including several endo-1,4-β-glucanases and cellobiohydrolases. These two types of enzymes act in concert on the crystalline regions of the cellulose fibril, so that mainly cellobiose and some cello-oligosaccharides are released. β-Glucosidases, present in the *T. reesei* preparation and especially in the pectinase enzyme from *A. niger*, degrade these oligomers further to glucose, so that there is less product inhibition of the glucanases.

Process. Treatment of pulp with the combination of pectinase and cellulase results in a synergistic action, which leads to almost complete liquefaction. **Figure 7** shows the decrease in viscosity of apple pulp during incubation with pectinase, cellulase, and a mixture thereof. Separation of the liquefied mash into juice and pomace can easily be carried out on an industrial scale with a decanter centrifuge, although even a horizontal press, rotary filter, or vibrating screen can be applied, depending on the desired clarity of the juice. Rotary filters that are operated under vacuum give rise to considerable aroma loss. Use of a vibrating screen provides a pulpy juice with a stable cloud. Enzymatic

liquefaction can also be used with press residue after straight pressing of fruits to increase the yields. Yields obtained with straight pressing and with liquefaction of apples are compared in **Table 1**. It can be seen that for the production of 1000 kg (2200 lb) of concentrated juice [72 °Bx (Brix degrees)] almost 2000 kg (4400 lb) of raw material can be saved. The technology of liquefaction is also suitable for the production of juices from nonpressable fruits and vegetables as well as from fruits for which no presses have been developed, for example, mangos, guavas, and bananas. Because it is a simple, low-capital technology that is easily maintained, it is appropriate for use in developing countries. Another advantage is the almost total elimination of costly waste disposal.

Quality aspects. These involve haze, color, and aroma.

Haze. The primary quality requirement for many juices, namely sparkling clarity, can be achieved with

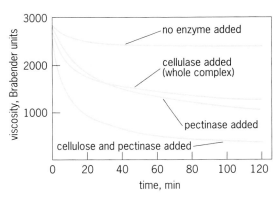

Fig. 7. Graph showing decrease in viscosity of stirred apple pulp at 25°C (77°F) in Brabender units as a function of time, as measured with a Brabender amylograph.

the liquefaction process as well as with the other enzymatic treatments mentioned in Fig. 5, provided that pectins are sufficiently degraded. Nevertheless, haze has appeared in concentrated apple and pear liquefaction juices and sometimes in concentrated juices from pulp enzyming. This haze material consists of 90% arabinose, of which 88% is α-1,5-linked. The haze-forming mechanism can be explained on the basis of the mode of action of *A. niger* arabinanases (enzymes) on the branched apple-juice polysaccharide arabinan, which is released after enzyme treatment of the pulp. These branched arabinans originate from the ramified hairy regions of the pectin molecule, and they consist of a main chain of α-1,5-linked L-arabinose residues and short branches of α-1,3-linked units. One of the enzymes present in the commercial pectinase preparation, an exo-α-L-arabinofuranosidase, preferentially splits the α-1,3-linkages of the side chains, with resultant debranching. The remaining α-1,5-chains can then retrograde and crystallize to generate haze. Addition of sufficient endoarabinanase activity to depolymerize the α-1,5-chain inhibits haze formation.

Color. Apple juices obtained by the liquefaction process have a yellowish color, caused by the flavonol glucoside quercitine; quercitine is retained in the press case in the traditional technology, but is solubilized in the enzymatic process. Aeration of the pulp cannot prevent this, as is the case with other phenolic compounds, which precipitate as a polyphenol–protein complex by oxidation.

Enzyme-treated juices undergo more rapid browning than traditional juices. This can be explained by the presence of unsaturated oligomer uronides, which have been shown to be powerful browning precursors. Another consequence of the higher uronide content in enzyme-treated juice is the enhanced acid taste; this may be a desirable characteristic.

Table 1. Comparison of yields from apples with good pressing characteristics*

Method	Yield
Pressing	786 kg (1729 lb; 750 liters; 198 gal) of juice, 12 °Bx from 1000 kg (2200 lb) of apples
	131 kg (228 lb) of concentrate, 72°Bx, from 1000 kg (2200 lb) of apples
	1000 kg (2200 lb) of concentrated juice, 72 °Bx, from 7633 kg (16,793 lb) of apples
Enzymatic liquefaction	950 kg (2090 lb; 900 liters; 238 gal) of juice, 13.5 °Bx, from 1000 kg (2200 lb) of apples
	178 kg (392 lb) of concentrate, 72°Bx, from 1000 kg (2200 lb) of apples
	1000 kg (2200 lb) of concentrated juice, 72 °Bx, from 5618 kg (12,360 lb) of apples

* °Bx represents Brix degrees, a unit indicating percentage of sugar content on the Brix scale.

Table 2. Aroma compounds of apple juice obtained from pulp (not heat-treated) by pressing and by enzymatic liquefaction

Compound	Pressing, parts per million	Enzymatic liquefaction, parts per million
2-Butylacetate	0.15	0.05
Butylacetate	28	0.3
2-Pentylacetate	0.6	0.25
Butylbutyrate	0.3	0.2
Hexylacetate	5.1	0.05
Hexanal	1.3	1.4
3-Hexanal	0.3	0.15
2-Hexanal	18	15
Benzaldehyde	0.05	0.2
Butanol	14	26
2-Methylbutanol	0.1	0.65
Pentanol	0.05	0.2
Hexanol	3.7	14
2-Hexanol	0.6	1.9

Aroma. Table 2 compares the volatile components of apple juices obtained by pressing and those obtained by liquefaction. Differences result from the saponification of esters and the reduction of aldehydes to alcohols. An increased endogenous lipoxygenase activity is probably responsible for the relatively high levels of C_6 alcohols, which in turn are derived from a higher initial concentration of C_6 aldehydes. In general, aroma changes occur more rapidly in the liquefied system because of better access of enzymes to their substrates. Appropriate technology can minimize these changes in flavor.

Other applications. Although the liquefaction process was originally developed for the processing of fruits to juices, it can be applied more generally to the conversion of plant biomass; examples include the production of fermentable sugars from waste material, the extraction of oils from oleaginous fruits and seeds, and the extraction of natural pigments and other constituents from plants.

For background information SEE ALGINATE; CARRAGEENAN; ENZYME; FILTRATION; FOOD MANUFACTURING; GUM; ULTRAFILTRATION in the McGraw-Hill Encyclopedia of Science & Technology.

Gerrit Beldman; Fons Voragen

Bibliography. M. F. Ashby, The mechanical properties of cellular solids, *Metallurg. Trans. A AIME*, 14A:1755–1769, 1983; J. M. V. Blanshard and J. R. Mitchell, *Polysaccharides in Food*, 1979; M. Cheryan, *Ultrafiltration Handbook*, 1986; D. L. Downing (ed.), *Processed Apple Products*, 1989; M. Glicksman, *Food Hydrocolloids*, vol. 1, 1982, vol. 2, 1983, vol. 3, 1986; R. C. Hoseney, *Principles of Cereal Chemistry*, 1986; W. Janda, Total liquefaction of apples: Technological and economic aspects, *Flüssiges Obst*, 50:312–313, 1983; J. J. Jen (ed.), *Quality Factors of Fruits and Vegetables: Chemistry and Technology*, ACS Symp. Ser. 405, 1989; G. Kaletunc, A. Nussinovitch, and M. Peleg, Alginate texturization of highly acid fruit pulps and juices, *J. Food Sci.*, 55(6):1759–1761, 1991; A. Nussinovitch and M. Peleg, Mechanical properties of a raspberry product texturized with alginate, *J.*

Food Process. Preserv., 14:267–278, 1990; M. Peleg et al., Mathematical characterization of the compressive stress-strain relationship of spongy food products, *J. Food Sci.*, 54(4):947–949, 1988; A. H. Rose (ed.), *Microbial Enzymes and Bioconversions*, 1980; G. Siemoneit, Description of different ways for clarification and preservation of fruit juices, *Confructa*, 29:44–94, 1985; J. F. Sullivan and J. C. Craig, Jr., The development of explosion puffing, *Food Technol.*, 38:52–55, 131, February 1984; J. R. Whitaker and P. E. Sonnet (eds.), *Biocatalysis in Agricultural Biotechnology*, ACS Symp. Ser. 389, 1989; J. R. Whitaker, Pectic substances, pectic enzymes and haze formation in fruit juices, *Enzyme Microb. Technol.*, 6:341–349, 1984; P. Zeuthen et al. (eds.), *Thermal Processing and Quality of Foods*, 1984.

Food spoilage

The role of yeasts in food spoilage, although frequently unrecognized, is a significant economic problem in industrial food and beverage production and in the consumer market with regard to the quality, acceptability, and shelf life of many food products. While food industries seldom report the frequency and costs of problems caused by yeast spoilage, such problems commonly occur in the production and distribution of certain foods. Economically the losses can be severe, encompassing costs of raw materials, production, packaging, warehousing, transportation, and distribution, as well as the costs of recall and disposal of the spoiled product, loss of future sales due to negative responses by consumers, and insurance and legal expenses for determining responsibility and compensation.

Consumers normally do not detect spoilage until yeast populations have reached approximately a million cells per gram of product. However, unless packages swell due to gas production or product appearance grossly changes, the spoilage is likely to be ignored, because many consumers have favorable associations of yeast-caused odors and tastes in normally fermented foods and do not associate them with spoilage. In fact, since the major products of yeast growth, other than carbon dioxide gas, are normally alcohols, organic acids, and ester (aromatic) compounds, many consumers readily consume, and may even prefer, the spoiled product.

Spoilage yeasts. Ecological studies reveal that yeasts are associated with various parts of plants and animals, soil, and water, and in fact, with most of the equipment encountered in a food-processing plant. However, yeasts that are normal in one product may be considered a spoilage organism in another product. A yeast that is responsible for undesirable changes in foods or beverages, either during processing or subsequently, may be considered to be a spoilage yeast. Spoilage may be expressed as an alteration in the appearance of the product. The physical presence of the yeast may result in a slimy or powdery film or an unnatural turbidity in liquid products. Metabolic substances may cause a change in the flavor or odor of a particular product, an excessive gaseousness of the product, or swelling of the package. Swelling, however, may also be caused by electrolysis of tin cans, by some bacteria, and by a few molds. The ability of certain yeasts to utilize organic acids (such as lactic acid) that are added to control bacterial growth can enable spoilage bacteria to develop. Some spoilage yeasts can even utilize certain acids (such as benzoic and sorbic acids) that are commonly used as preservatives to control yeast growth.

There is no record of food poisoning being caused by the presence of spoilage yeasts, since their metabolic products are not considered toxic at the levels found in spoiled foods. Possibly, certain individuals are allergic to specific yeast proteins, and in high concentrations yeast cells have been suspected of causing some cases of gastroenteritis. However, people have been consuming large numbers of yeast cells in fermented foods and beverages since the start of civilization without obvious ill effects. The few species of yeasts known to be pathogenic are not known to be transmitted via food products.

Yeast selectivity and activity. One compilation from published studies of yeasts associated with foods comprises about 120 species representing 30 genera. Most investigators would agree that the majority of these are incidental contaminants from natural sources and normally are not capable of food spoilage. Under normal manufacturing conditions, about 20 species account for more than 90% of incidences of food spoilage caused by yeast. The initial yeast flora of a food often contains many species, introduced during harvest of the raw material, transportation, processing in inadequately cleansed equipment, addition of condiments and other ingredients in processing, or storage. However, the spoiled food product ultimately has but a few, or even a single, species present. This selectivity is strongly influenced by the composition of the food as well as temperature, method of processing, type of packaging, amount of oxygen present, and the presence of various inhibitory substances.

Yeast activity depends upon injury that may occur during harvest or transportation and that permits access to the moist tissues and juices of the products. At this point, the selectivity process initiating spoilage begins. For example, the tissues of sound fruits and vegetables are inherently sterile, but become contaminated when the protective skin is broken or when the processing equipment and added ingredients introduce yeasts. Alcoholic fermentation of packaged fruit juices is common, even at refrigerating temperatures in the store or, more often, after opening and refrigerating at home. The sugary composition and the acidity of juices promote the growth of yeasts that are capable of fermentation. Initially, species of *Hanseniaspora* (and its related genus *Kloeckera*) and other yeast species start the fermentation in juice, but are soon overgrown by strains of *Saccharomyces* that are more alcohol-tolerant. If preservatives are added to control spoilage, the shelf life of the juice may be extended. However, preservative-resistant yeasts, often a *Zygosaccharomyces* species, will continue to grow and will eventually spoil the juice. Even fruit concentrates, if stored at temperatures above 41°F (5°C), will eventually spoil. In this case the species of spoilage yeasts will be selected by ability to grow at low temperatures and high sugar concentrations, such as some species of *Zygosaccharomyces, Candida*, and *Hanseniaspora*. Heating is commonly used to control spoilage of such juices and concentrates, but the heat treatment is usually minimal so that the fresh-fruit characterists are preserved. Some spoilage yeast species form ascospores that have greater heat resistance than the vegetative cells and may survive the heat treatment. Some strains of spoilage yeasts are more tolerant to the higher temperatures of pasteurization than other strains of the same species.

Spoilage of dairy products. Liquid milk, either raw or pasteurized, contains few yeasts under normal conditions. Yeast spoilage of milk is limited to those species that can ferment lactose. Storage at refrigerating temperatures and the inability to compete with the rapid growth of cold-tolerant bacterial species present in the milk further restrict yeast growth. However, "acidophilus" milks, which contain lactic acid–producing bacteria, may be spoiled by yeasts capable of utilizing the lactic acid. Condensed milks (sweetened with sucrose) can undergo fermentation by certain yeasts that can outcompete bacterial species. Bacterial growth is deterred by the high sugar concentration of condensed milks. Flavored, sweetened, fruit-containing varieties of yogurt can be spoiled by yeasts that are introduced by the added ingredients or by contaminated equipment. Cheeses support yeast growth, although spoilage is relatively insignificant in the early stages of cheese production. During maturation, however, the cheese has a low pH, relatively low moisture, and high salt content, and is stored at low temperatures—all conditions selective for yeast growth. Certain varieties of cheese do require yeast activities to contribute to the development of normal texture and flavor, but yeasts can also cause surface films, abnormal texture, and off-flavors. Selectivity for yeasts is influenced by the ability to enzymatically split milk lipids and proteins, ferment lactose, utilize lactic acid, and in the case of mold-ripened cheeses, to grow in the presence of the molds.

For background information SEE *FOOD ENGINEERING; FOOD MANUFACTURING; FOOD MICROBIOLOGY* in the McGraw-Hill Encyclopedia of Science & Technology.

Martin W. Miller

Bibliography. L. R. Beuchol (ed.), *Food and Beverage Mycology*, 2d ed., 1987; G. H. Fleet, Yeasts in dairy products: A review, *J. Appl. Bacteriol.*, 68:199–211, 1990; D. Y. C. Fung and Chao-Liang, Critical review of isolation, detection and identification of yeasts from meat products, *Crit. Rev. Food Sci. Nutr.*, 29:341–379, 1990; M. E. Rhodes (ed.), *Food Mycology*, 1979; A. H. Rose and J. S. Harrison (eds.), *The Yeasts*, vol. 3: *Yeast Technology*, 1970; J. F. T. Spencer and D. M. Spencer (eds.), *Yeast Technology*, 1990.

Food web

From zooplankton through fish to whales, it is essential to study the food chain in the ocean. By the use of new acoustic methods, animal populations can be accurately surveyed at relatively high speed and resolution. Size, spatial and temporal distributions, and sometimes species-level information are obtainable through acoustic or echo-sounding techniques.

Regardless of what method is used, the need to obtain these data can be explained as follows: If most of the fish in the ocean were caught in one year, there would not be enough fish left to reproduce and supply people with food in the following years. Thus it is important to make accurate estimates of the abundance and distribution of fish every year so that the fishing industry can catch the optimum amount of fish, leaving a sustainable population. Understanding the distribution and behavior of the zooplankton upon which fish feed and the impact of environmental hazards on the zooplankton will help to quantify this supply of food for the fish. This information will make it possible to estimate the distribution and growth rates of the fish. This type of argument holds true up and down the food chain. A change in one element of the food chain affects the rest of the chain.

Marine sampling methods. Because of the importance of understanding the population dynamics of marine animals, a variety of techniques have been applied to measure marine populations at sites around the world. The traditional method has been to tow nets of various sizes and meshes behind a ship. By deploying the nets at a fixed depth or range of depths, the animals in a small volume of ocean water can be collected and brought back to the laboratory for counting and analysis. There are, however, major drawbacks to this method. The smaller the animal, the greater the chance it will escape through the holes in the nets. The larger the animal, the greater the chance that it will avoid being caught by simply swimming around the nets. A major practical problem is that the size of the nets is limited. Thus, even if all the animals were caught in the path of the nets, the catch would represent only an extremely small fraction of the animals in that ocean area. The problem is aggravated by the slow speed of the boat, which limits the total number of net samples that can be taken in a day. Thus, when using trawls, it is necessary to intelligently sample in many locations in order to produce a representative picture of the distribution of animals over a large region in the ocean.

Acoustics offers a partial solution to this need, since sound can travel through water with great speed. Acoustic methods have been developed that can rapidly and remotely sense the motion and contents of large volumes of water. The variety of acoustic techniques available include active sonars that transmit their own signals and listen for the backscattered echoes. These devices can help direct the use of sampling methods, such as towing nets to capture organisms, and they can help provide for continuous interpolation of data between samples collected by conventional means. Investigations using sonars along with other techniques can produce large-scale acoustic maps of the area of interest as well as samples containing high-quality morphological information at various locations.

Principles of sonar systems. The critical element in a sonar system (**Fig. 1***a*) is the transducer, which can convert electricity to mechanical motion and mechanical motion to electricity with equal efficiency. This reciprocal device can come in many forms, but is typically a piezoelectric ceramic in the shape of a flat disk (or an array of flat disks) for the echo sounders used for detection of marine life.

The transmitter in the system applies a short (electrical) tone burst at high power to the transducer. The transducer converts that energy into a short acoustical tone burst in the water. Most of the acoustic energy then travels rapidly away from the transducer, mostly in a narrow range of directions. Diffraction effects determine how narrow is the range of angles in which most of the signal is confined. Diffraction also causes sidelobes in undesired directions. (A typical diffraction pattern appears below the transducer in Fig. 1*a*.)

The tone burst travels quickly through the water, scattering off various inhomogeneities such as fish. The sound waves are scattered into many directions by each fish, and the portion of the scattered signal, or echo, that travels back toward the sonar is ultimately detected by the transducer. The acoustic echo is converted back into electrical energy. A typical display of the echo envelope from a collection of fish is given in Fig. 1*b*. The peak on the left is a portion of the transmit signal that feeds through from the transmit

electronics to the receive electronics, while the time-varying signal to the right of that corresponds to the echoes, delayed in time, from the various animals. The delay corresponds to the distance between the sonar and the animals.

The system discussed above is very simple in that it transmits an acoustic signal at one frequency with a single beam. In order to extract more information on the organisms, it is beneficial to use more sophisticated systems with multiple narrow beams, so that imaging of the scattering volume is possible, or with multiple frequencies, to take advantage of the frequency dependence of the scattering in the interpretation of the data.

Interpretation techniques. Converting a pattern on an oscilloscope to a quantity that is useful in biological or ecological studies is a tremendous challenge that involves sophisticated mathematical descriptions of the scattering processes of the organisms, laboratory and controlled field data for verification of the models, and finally inversion methods that will use this information to convert the echo voltages into quantities such as a size distribution of the animals.

The scattering depends upon the acoustic wavelength, and size, shape, orientation, and material properties of the animals. In order to describe the scattering of sound by the animals, simplifications must be made in order to make the problem mathematically tractable. Studies have shown that when the scattering models are reduced to the dominant scattering mechanisms, the animals can be modeled by much simpler objects. Zooplankton such as shrimplike euphausiids can be modeled as a bent and tapered cylinder of finite length, composed of a fluid material, while fish can be modeled, to first order, by ignoring the flesh and bones, leaving only the swim bladder. This great simplification of the fish is due to the fact that the bladders contain gas, which has a much higher acoustic-impedance contrast than the fluidlike flesh and thin bones. In spite of the simplifications in these models, the mathematical solutions are not exact, and research is being conducted to best estimate the scattering by the various animals.

Field results. In spite of the approximations and assumptions in the scattering models, acoustic data collected in the ocean can be interpreted with a fair degree of success. If a multiple-frequency sonar is used, the data can be converted into a size distribution of the animals by use of a scattering model and matrix inversion. (An inversion, in this case, refers to inferring the distribution from the data and mathematics.) Typically, one size range can be resolved for every different acoustic frequency. From the size distribution, the volume of animals per volume of water (biovolume), which is a measure of animal abundance, can be estimated.

Figure 2 illustrates an inversion of data from a 21-frequency sonar (100 kHz–10 MHz) that was used to detect microzooplankton (animals ranging in length from tens of micrometers to several millimeters) in the ocean off southern California. There is good agreement between data collected by a pump that sucked the

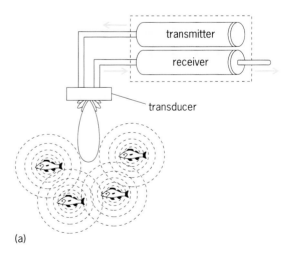

(a)

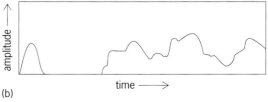

(b)

Fig 1. Simple sonar system. (*a***) Diagram of system. A cloud of scatterers (fish) is shown, whose individual sonar echoes overlap. (***b***) Display of echo envelope.**

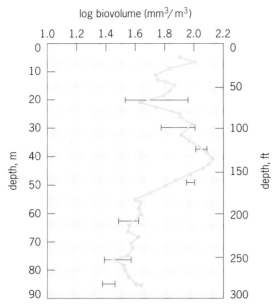

log biovolume (mm³/m³)

Fig. 2. Comparison of the logarithm of the biovolume estimated from multifrequency acoustic data (circles) and high-volume pump samples (bars). The limits of each bar are the biovolumes calculated from each of two samples taken sequentially at the specified depths. (*After R. E. Pieper, D. V. Holliday, and G. S. Kleppel, Quantitative zooplankton distributions from multifrequency acoustics, J. Plank. Res., 12:433–441, 1990*)

animals through a tube to the ship and the biovolume inferred from the sonar inversion. The comparison is not perfect, of course, since there are imperfections in the scattering model that was assumed, there is noise in the sonar system (as in any electrical system), and there was a finite number of frequencies (the higher the number of frequencies, the better the resolution of size distribution). Furthermore, the pump system is not perfect. Some animals can escape before being collected by the pump, and others may be damaged in the pump, hence giving erroneous results.

In spite of these practical difficulties, Fig. 2 illustrates the usefulness of acoustics in providing a remote-sensing tool in the ocean. The ability to make rapid estimates of size distributions of animals over a large region in the ocean is invaluable for studies of the food chain. Research is under way to make the acoustic techniques more accurate and robust.

For background information SEE FOOD WEB; MARINE ECOLOGY; SONAR; UNDERWATER SOUND; ZOOPLANKTON in the McGraw-Hill Encyclopedia of Science & Technology.

Timothy K. Stanton

Bibliography. C. S. Clay and H. Medwin, *Acoustical Oceanography: Principles and Application*, 1977; K. G. Foote, On representing the length dependence of acoustic target strengths of fish, *J. Fish. Res. Board Can.*, 36:1490–1496, 1979; D. N. MacLennan, Acoustical measurement of fish abundance, *J. Acous. Soc. Amer.*, 87:1–15, 1990; R. E. Pieper, D. V. Holliday, and G. S. Kleppel, Quantitative zooplankton distributions from multifrequency acoustics, *J. Plank. Res.*, 12:433–441, 1990; T. K. Stanton, Sound scattering by zooplankton, *Rapp. P.-v. Reun. Cons. Int. Explor. Mer.*, 189:353–362, 1990.

Forensic anthropology

Forensic anthropology has long been associated with helping law enforcers solve crime. Now some of its techniques are being applied to open windows to the past. The ability of the forensic anthropologist to recover and identify human skeletal remains has been at the core of the usefulness of this field in crime solving. Going beyond the work of the traditional physical anthropologist, who studies bones, the forensic anthropologist collects evidence that can be used in a courtroom to convict a murderer. Especially in sensational cases such as those involving serial killers, the public can perceive the importance of forensic anthropology in the investigation and prosecution.

In June 1991, the exhumation of President Zachary Taylor, who perhaps had been assassinated by arsenic poisoning, illustrated new attempts to apply forensic anthropology to investigation of the cause of death of a historical figure.

Identification of the skeleton. To support law enforcement, the forensic anthropologist recovers and attempts to identify human skeletal remains. In homicide cases, by the time a body is discovered, it may have decomposed and only bone is left. Depending upon the climate and season, a human body may completely skeletonize within a few weeks. While the police generally look for a missing person of a particular sex and population group, the anthropologist must deal with skeletal remains often lacking such basic characteristics. Associated evidence such as headgear, clothing, and jewelry may sometimes help in the identification process. But admixture of genetic traits among different populations may sometimes make it difficult to match the race of the missing person to the race shown by the skeleton. Since there is also as much variability within a population as between populations for many genetic traits, assignment of race may not always be meaningful. Transvestism may occasionally confuse the match of gender between a missing person and a skeleton. Estimates of the age of the skeleton may also vary widely.

Skeletal characteristics. The anthropologist uses a systematic, ordered approach to identify skeletal characteristics such as species, age, sex, race, and stature. First, the anthropologist must correctly identify the species of skeletal remains. If the remains are human, the law enforcement community becomes involved.

The next task is determining the historical age of human skeletal remains. If the skeleton dates back far enough, it is not of interest to law enforcers. If the skeleton can be identified as of current date (generally less than 30–50 years since death, or of interest as part of an open criminal investigation), the important process of individual identification can begin. The sex of the skeleton can usually be determined by analyzing specific characteristics, primarily in the skull and the pelvis. If these skeletal elements are unavailable, incomplete, or damaged beyond recognition or repair, measurements can be taken on other skeletal parts to determine sex based upon the ranges of measurement differences between males and females.

The skull and the pelvis are also the focus for determination of age in adults. The closure of the cranial sutures (where the bones of the skull meet) along the internal and external surfaces of the skull provides landmarks in adults. The pubic bones at the symphysis (where the two halves of the pelvis meet in front) show patterns of grooving and pitting that are correlated with age in adults. In women, these bones and others may also show evidence of childbirth. The ribs and other bones may be used to help determine age in adults. In microscopic technique, a sample from the outer portion of a long bone can be used to count the number of osteons (bone cells) within a given field. This osteon count can be correlated to age. The age of a child's skeleton is determined by the growth plates on the ends of bones, which become fused at a rate and in a predictable sequence associated with age. In ascending order, the sequence is elbow, hip, ankle, knee, wrist, and shoulder.

The assignment of race to the skeleton based upon features of the skull may be the most problematic aspect of the identification, since the anthropologist can measure and observe only the features that are present. Nonetheless, the correct race is assigned in a significant majority of cases. Stature (height) is determined by measuring various long bones, such as the femur (thigh bone), and using equations for predicting the total skeletal height. The species, age, sex, race, and stature can usually be assigned to skeletal remains that are relatively intact. Additional distinguishing features such as congenital anomalies and evidence of injury or medical therapy are helpful. These features are used to compare information about missing persons to make a probable identification.

Positive identification and DNA testing. A positive identification can be made with unique criteria, such as matching recovered dental work to individual dental records. The teeth are among the most survivable and recognizable parts of the skeleton, and identification may rest on them in badly damaged, burned, or decomposed skeletons. It is now becoming possible to use deoxyribonucleic acid (DNA) "fingerprints" to achieve a positive identification by matching unique genetic patterns in the DNA in bone to a known sample from the individual in life, or by matching to living blood relatives.

A remarkable proposal for DNA testing was announced in early 1991 by the National Museum of Health and Medicine in Washington, D.C., which holds autopsy samples of bone from the gunshot wound of the assassinated president Abraham Lincoln. It has long been debated whether President Lincoln suffered from Marfan's syndrome, the most common inherited connective tissue disorder, accounting for his striking tall, gaunt appearance and possibly placing him

at risk of imminent death. The proposal to obtain and amplify DNA from the bone samples by using polymerase chain reaction (PCR) led to a historical symposium on Lincoln's health as well as the formation of a national panel of experts to study the ethical, legal, and social implications of such a proposal. The panel approved the proposal to examine the technical and scientific aspects of DNA testing on Lincoln's bone samples. The research on Lincoln's bone samples will test the ability to obtain technically adequate DNA from samples that are many decades old.

Postmortem changes. Many bodies are discovered after decomposition has set in but before skeletonization is complete. The traditional recommendation had been for the forensic anthropologist to completely skeletonize such remains so that analysis could proceed on dry bone. However, there has been an increasing role for the anthropologist in the identification of fleshed or partially fleshed remains. An important category of evidence is the postmortem interval, the period of time between death and the discovery of the body. Such evidence can be used to help establish identity, interpret the crime scene, and investigate and prosecute a homicide case. There is a sequence of changes and losses in soft tissue, of invasion of microorganisms and insects, and of disarticulation of the skeleton, that is affected by certain physical and biological agents. Anthropologists have begun to systematically describe and document such changes to contribute to investigation and to develop the science of postmortem change.

Reconstructions. Facial reconstruction involves using clay sculpture, based upon average tissue thicknesses upon the face of the skull to recreate the appearance of the living individual. To attempt to identify such a reconstruction, the media may disseminate the image. Video cameras may also be used to superimpose a skull on a photograph of a missing person to determine whether a match can be made.

The anthropologist may sometimes be able to move beyond determination of postmortem interval and identify more subtle evidence of posture, life-style, stress, occupation, environment, and culture. For example, certain changes in the bones of the shoulder may indicate that the individual had been a longbow archer; a forward-thrusting jaw may indicate a woodwind player; and an extra bone in the left hand may indicate a violin player. The reconstruction of lifeways from skeletons is most tantalizing. Discovery of the activities, diet, environmental exposures, and other aspects of bygone peoples from skeletal populations holds great promise for understanding the biology and culture of humans.

For background information SEE ANTHROPOMETRY; FORENSIC MEDICINE; PHYSICAL ANTHROPOLOGY in the McGraw-Hill Encyclopedia of Science & Technology.

Marc S. Micozzi

Bibliography. M. Y. Iscan and K. A. R. Kennedy, *Reconstruction of Life from the Skeleton*, 1989; M. S. Micozzi, *Postmortem Change in Human and Animal Remains*, 1991.

Forest and forestry

Since about 1970, formal forest plans have been used in the management of national forests. Much of the impetus for these plans has been in response to the National Environmental Policy Act (NEPA) of 1969 and the National Forest Management Act (NFMA) of 1976. These legislative acts included requirements for documentation of how forest actions were determined and how plans were developed. NFMA also emphasizes and codifies the policy that national forests are used for multiple rather than single uses.

Systems analysis for forest planning came into wide use concurrently with these legislative mandates. Forest planning includes assembling information for the variety of land resources found in a planning area, calculating the effect of various alternative treatments or practices on these land resources, and determining whether combination of treatments could meet various output goals. An objective of forest planning is to meet the output goals at least cost. The allocation of land for various treatments becomes a major part of the forest plan. These problem aspects are readily stated as a linear programming problem in which multiple outputs are listed as constraints, and in fact linear programming is the most commonly used systems analysis technique for forest planning.

Typical output goals include determination of recreation visitor days, water volume, acres of habitat for wildlife, and timber harvest volume.

Linear programming approach. The linear programming problem is usually stated with the following parts:

1. A numerical list (that is, a vector) **b** of constraints that include the number of acres or hectares of each different land unit, and the desired output goals or production constraints for various outputs.

2. A list of possible land management practices or prescriptions that can be used on the different land units, and kinds of output products that can be produced. A numerical vector **x** associated with this list is the solution of the linear programming problem, and shows the number of land units to be managed with each of the listed practices and the number of units of each product.

3. A unit cost and benefits list (or vector) **c** for each of the items in the list of practices and products. The first part of the vector **x** is the number of land units used with each practice, and each item in this part is multiplied by the appropriate monetary costs in vector **c**. The second part of the vector **x** is the number of units of each product, and each item in this part is multiplied by the appropriate monetary benefits in vector **c**. Thus, the sum of products when the vector **x** is multiplied by the vector **c** is the net value for the problem statement. The values of **x** are computed so that the net value is maximized or the net cost is minimized.

4. A matrix of coefficients **A** that links the land units and other constraints of the vector **b** with the quantities selected in the solution of the vector **x**. Generally, these coefficients are determined by independent research

results or by other studies, and include numbers such as the acre-feet (or cubic meters) of water that will result if a given land unit is managed according to a prescribed management practice. Other coefficients could represent the volume of timber that could be harvested per acre with specified harvest regimes, or the visitor days that could be serviced if a given campground were constructed.

Proponents of the linear programming approach have felt that this problem statement, and completion of the various matrix and vector elements, is a useful way to bring together data from a variety of disciplines into a common framework and a common analytical procedure. Most practitioners of linear programming recognize that the various mathematical assumptions (additivity, divisibility, proportionality, and certainty) do not strictly hold in forest management situations. However, to these practitioners, the advantages for interdisciplinary discussion, and explicit expression of public interests in the goal statements, are so great that concerns about these assumptions are not seen as a major difficulty. Other difficulties include lack of complete information for such analyses, and complaints that many of the desired outputs from a forest area (for example, scenic beauty) cannot be expressed quantitatively. However, surrogate numerical statements have been developed to include consideration of esthetic and other goals that are difficult to quantify. In applications to simpler and smaller land management problems or in developing conceptual and strategic plans, the linear programming approach does provide many benefits.

Time and space dimension. Even in more comprehensive applications to larger problems, the classical mathematical assumptions of linear programming usually have not presented serious difficulties. However, issues of time and space dimensions of the planning problem have arisen to create very serious shortcomings of forest plans developed from a linear programming base.

In timber harvest scheduling, the time dimension has long been recognized as difficult to handle. The nature of the issue lies not so much in the analytical techniques as in the forest management problem of conversion from an older, less productive forest to a more intensively managed forest. Typically, for a Pacific Coast forest in the western United States and Canada, an average stand age of 150 years would be converted to an average stand age of about 80 years in order to increase the stand growth rate and thus the economic output. However, such conversions invariably result in a calculated temporary harvest shortfall until higher production rates can be achieved. Foresters have lived with this classic conversion problem for years, and usually have filled the shortfall gap with old-growth harvests and other manipulations to reduce the likelihood of a real reduction in timber harvest. Better stand-growth models, inventories, and problem formulations may allow management with greater detail and accuracy.

Another major problem in forest plans based on linear programming is space. In small and simple land planning problems, the total land area is divided into contiguous land units, each of which is described as being homogeneous. These land units may be called ecological land units, ecological response units, or capability areas. The ability of each capability area to produce outputs under a variety of land treatments gives rise to the data for the elements of the linear programming matrix.

When these planning techniques were being developed, computer limitations were such that the number of contiguous capability areas was too great for any reasonable analysis. Noncontiguous land areas (analysis areas), comprising similar capability areas, were formed to alleviate the computational burden. With this approach, the spatial information content, if any, of the original capability areas is entirely lost in the aggregation. There may be a constraint, for example, against clear-cutting next to a recreational campground. However, the analysis areas, made up of several capability areas, would have no way to express this constraint. Many land managers have found that these spatial constraints are more serious than the temporal difficulties of classic timber harvest scheduling.

Prospects. At present, for the general class of spatially distributed allocation problems there is no optimization algorithm (a procedure that has a definite solution) that is amenable to the overall forest planning process. However, there are certain specific spatial considerations that lead to improved analysis. For example, scheduling of forest road construction requires a rather lengthy iterative network analysis, because roads developed to reach a given timber stand can be used for access to additional stands in future periods. Logging-systems design is another area where analytical improvements can be made. Often the design of a specific logging operation will show that the proposed harvest quantity based on the stand-growth model cannot be accomplished because of engineering problems, even though the timber volume may be present.

Current developments in geographic information systems (GIS) and computer graphics offer considerable hope that improvements are possible for more general spatial problems through heuristic rather than algorithmic methods. (A heuristic method has no definite solution, even though it may offer an improvement.) Spatial disaggregation of proposed harvests assigned to analysis areas is a good example of the potential application of heuristic methods. To use these techniques, the geographic locations of capability areas that were combined to form each analysis area would have to be retained on file. Following the linear programming solution and allocation of harvests to an analysis area, this harvest would be disaggregated on a per-acre basis to each capability area within the analysis area. If the spatial locations of capability areas are retained, the amount of harvest on each capability area can be displayed graphically in the proper location on a computer map. Using the techniques of geographic information systems, a manager could point to a capability area displayed on a computer screen and move some of the proposed harvest amount to another area.

Linear programming is still a valuable tool in initial development of strategic plans and economic analyses, but its dominance as a comprehensive and rational analytical technique has passed. As forest planners learn that they cannot solve all of their problems with rational-comprehensive plans developed with linear programming, other analytical procedures, such as improved user interactions through spatial modeling and graphic displays, will have increasing importance. However, the requirement for clear analytical and well-documented planning procedures will remain.

For background information *SEE FOREST AND FOREST-RY; LINEAR PROGRAMMING; SYSTEMS ANALYSIS* in the McGraw-Hill Encyclopedia of Science & Technology.

Donald A. Jameson

Bibliography. T. W. Hoekstra et al., *Critique of Land Management Planning [CLMP]*, vol. 4: *Analytical Tools and Information*, USDA Forest Service Policy Analysis Staff, FS-455, 1990; G. Larsen et al., *CLMP*, vol. 1: *Synthesis of the Critique of Land Management Planning*, FS-452, 1990; D. E. Teeguarden, *CLMP*, vol. 11: *National Forest Planning Under RPA/NFMA: What Needs Fixing?*, FS-462, 1990.

Forest ecosystem

The buildup of greenhouse gases in the atmosphere threatens to precipitate a global warming via the greenhouse effect. The principal greenhouse gas is carbon dioxide. Although this gas is a natural constituent of the Earth's atmosphere, it has been building up there especially as the result of human activities. The principal human source of carbon dioxide is the burning of fossil fuels; a secondary source is tropical deforestation. There are two major approaches to controlling this carbon dioxide. The first is to reduce total carbon emissions, thereby reducing the carbon flow into the atmosphere. The second is to increase the carbon flow out of the atmosphere by fostering carbon sink enhancement. Although a number of potential carbon sinks exist, to date most discussion of sink enhancement has been focused upon forests, and specifically upon the effects of large-scale reforestation and creation of new forests.

Forests and the carbon cycle. Forests dominate the dynamics of the terrestrial carbon cycle. They contain 86% of the world's aboveground carbon and an estimated 73% of the belowground carbon. In 1985 it was estimated that 31% of the world's land area was covered with forests.

The process of photosynthesis captures carbon in plants, thereby reducing atmospheric carbon, while releasing oxygen into the atmosphere. Forests are particularly effective at capturing carbon since, unlike vegetative systems that operate on an annual cycle such as grasslands, the tree biomass of forests continues to accumulate for decades or centuries. Throughout the life of a forest, carbon is continually sequestered as a building material and stored in the woody biomass of trees. Furthermore, litter and detrital material that build up on the forest floor also hold carbon captive in their cells and thus contribute to carbon sequestering. Finally, forest soils continually capture carbon

Table 1. Carbon and specific growth of selected tree species

Species or species group	Proportion of carbon	Specific gravity	Kilograms of carbon/m³
Southern pines	.531	.510	271
Spruce-fir	.521	.369	192
Douglas-fir	.512	.473	242
Southern hardwoods	.530	.639	339
Western hardwoods	.496	.380	189

through interactions with tree root systems. Thus, forest ecosystems serve as a storage facility for carbon that would otherwise be freed into the atmosphere, and forest growth serves as the mechanism to add carbon to that already held in the forest ecosystem.

Differences in carbon sequestration. Not all forest systems sequester the same amounts of carbon. Standing forests hold carbon, even if they are not experiencing net growth. However, to sequester additional carbon, the forest ecosystem must be experiencing a net growth in volume. The volume of forest biomass is an indicator of the amount of carbon that is being stored, a stock; the rapidity of net forest growth is an indicator of the rate at which additional net carbon is being captured. Thus, high-biomass forests such as most tropical forests or Pacific Northwest old-growth forests will hold captive large volumes of carbon, while rapidly growing forests will capture large amounts of additional carbon from the atmosphere.

In addition, the species of tree is important in determining the amounts of carbon held or sequestered. Different species have different rates of growth and hold different amounts of carbon per wood volume (trees with high specific gravity hold more carbon than trees with a low specific gravity). **Table 1** provides values for some major species. From this table it can be determined that a mature southern hardwood forest would sequester more carbon than a western hardwood forest with the same tree volume.

Increasing the biomass of forests. The world's forest biomass currently constitutes a major global carbon sink. An expanding forest stock will sequester additional carbon, while a reduction in that stock, as occurs with deforestation, typically releases carbon. For the world's forests to increase the carbon held captive and thus reduce atmospheric carbon, the global stock of forest biomass must increase. This can be accomplished through an increase in the biomass of existing forests, the creation of new forests, or the regeneration of previously cleared forests.

Worldwide, at any point in time, parts of the global forest system are experiencing growth while other parts are experiencing decline. Forest decline, which is occurring at a significant rate in the tropics, is contributing to emission of carbon into the atmosphere. However, creation of new forests and reforestation in the temperate and the boreal areas of the Northern Hemisphere are sequestering substantial volumes of atmospheric carbon. Data indicate that the area of temperate and boreal forest has been expanding in much of the world (**Table 2**). Much of this expansion

reflects the reforestation of areas deforested in the past, particularly the latter half of the nineteenth century. Additional expansion is due to the establishment of artificially established forest plantations, some on recently logged forest sites but others on sites that have not been in forest for decades. In the United States, for example, well over 2.5×10^6 acres (1×10^6 hectares) of artificially established tree plantations are established annually. The widespread control of forest fires in much of the temperate world has also contributed to the buildup of forest biomass. Over the past several decades, Europe, the Soviet Union (now Commonwealth of Independent States), and North America have all experienced substantial expansion of their forest biomass, and estimates suggest that the carbon sequestered annually by these regenerated forests is substantial.

Plantation forests. Annually about 2.9×10^9 tons more carbon are being released into the atmosphere than are recaptured in the terrestrial system or the oceans. The human influences on the global carbon budget are presented in **Table 3**. It has been estimated that this annual "excess" of atmospheric carbon could be sequestered by approximately 1275×10^6 acres (465×10^6 ha) of rapidly growing plantation forests—an area almost the size of the United States west of the Mississippi. This much new forest would increase the world's total forest land area by about 16%. Once the plantations have been established, they could be expected to continue to sequester carbon for several decades or until the forest reaches maturity and the growth dramatically diminishes. The precise amount of plantation forests required depends upon the rate at which the new forests grow, since carbon sequestration depends largely on the volume of biomass growth. If plantations are established on poor soils or in areas

Table 2. Change in the volume of timber in temperate and boreal forests*

Country or continent	Timber volume increase, 10⁹ m³/yr	Representative period
Soviet Union	0.400	1973–1984
Europe	0.100	1985
United States	0.168	1986
Canada	0.025	1985
TOTAL	0.693	

* After ECE/FAO: *European Timber Trends and Prospects to the Year 2000 and Beyond*, 1986; *Outlook for the Forest and Forest Products Sector of the USSR*, 1989; *Timber Trends and Prospects for North America*, 1990.

Table 3. Global carbon budget*

Carbon	Change (median estimate), $\times 10^5$ g/yr
Released	
Fossil-fuel combustion and cement production	5.3
Tropical forest clearing	0.7
Accounted for	
Nontropical forest accumulation	−0.7
Atmospheric increase	−2.9
Ocean uptake	−2.2
Unaccounted for	0.2

* After R. P. Detwiler and C. A. S. Hall, Tropical forest and the carbon cycle, *Science*, 239:42–47, 1988.

of limited moisture, tree growth is much slower and so too is the corresponding buildup of biomass and carbon. Thus, the poorer the growing conditions, the larger the area of plantations that is required to capture a given amount of carbon. Again, the tree species that are planted must be considered. While rapidly growing tree species capture carbon more rapidly, they often have a relatively short period to maturity. Thus, the total biomass and the total carbon accumulation potential of this forest may be less than that of a forest containing slower-growing species. For example, forests in the southern United States grow more rapidly in the early years after planting than do the forests of the Pacific Northwest. However, after about 30 years the growth of the southern forests declines, while that of the western forests continues at a high rate for several more decades. In addition, the amount of tree planting required would depend upon the amount of carbon that is captured in the entire forest ecosystem, including that in the buildup of soil and litter.

A massive amount of land worldwide would be required to have a substantial effect on the level of carbon in the atmosphere. Large areas of tropical wastelands as well as marginal agricultural sites in the temperate world could be used. It appears unlikely that creation of forests and reforestation would totally offset the buildup in atmospheric carbon. More realistically, forest plantations would probably be used as a component of a multifaceted strategy that would involve reducing carbon emissions and increasing carbon sequestering. Nevertheless, the role of creating new forests and reforesting recently cleared ones could well be substantial. A number of studies have suggested that the tree-planting approach may be one of the most cost-effective ways to address the carbon issue.

Although reforestation and creation of new forests have considerable appeal as a means to reduce levels of atmospheric carbon dioxide, this approach cannot be a permanent answer to carbon emissions. While there is some latitude for increasing the world's stock of forest, there are limits to the extent to which the forests can be expanded. Continued emissions from fossil-fuel use have the capacity to overwhelm any conceivable expansion of forests. Thus, forest expansion can at best buy time, perhaps three to five decades during which period the newly established forests are growing toward maturity, until a non-carbon-emitting energy source becomes a cost-competitive alternative. However, creating forests to postpone an atmospheric carbon buildup should not be belittled, since it is recognized that "quick" technological fixes to problems are typically very high cost and often generate unforeseen problems. Plantation forests have the capacity to provide additional time for other long-term technology to be developed.

For background information SEE FOREST AND FORESTRY; FOREST ECOSYSTEM; FOREST MANAGEMENT; FOREST RESOURCES; GREENHOUSE EFFECT in the McGraw-Hill Encyclopedia of Science & Technology.

Roger A. Sedjo

Bibliography. R. A. Birdsey, *Carbon Budget Realities at the Stand and Forest Level*, paper presented at the National Convention of the Society of American Foresters, Washington, D.C., July 31, 1990; R. J. Moulton and K. R. Richards, *Costs of Sequestering Carbon Through Tree Planting and Forest Management in the United States*, USDA Forest Service, GTR WO-58, December 1990; W. D. Nordhaus, The cost of slowing climate change: A survey, *Energy J.*, 12(1):37–65, 1991; N. Rosenberg et al. (eds.), *Greenhouse Warming: Abatement and Adaptation*, 1988.

Forest resources

Forests have an enormous capacity to assimilate carbon dioxide through the ecosystem process of net primary production. Knowledge of the rate and distribution of this productivity is as vital to earth science as it is to commercial interests. Worldwide the condition of many forests has changed, often in dramatic ways such as tropical deforestation. Remote sensing, a tool to measure ecosystems without coming into contact with them, is especially appropriate for assessing the productivity and condition of forests.

Remote sensing of productivity. Productivity (net primary production) is the actual amount of energy that is used to produce new tissue per year, less than 1% of the total solar radiation received on Earth per year. By using remote sensing, production can be estimated through multistage statistical inventories and through computer simulation of mechanistic ecosystem processes.

Statistical methods. Traditional aerial photography is widely used as a first-stage estimator in forest inventories of timber volume and growth through mapping of stand properties and boundaries. A second stage involves detailed ground measurements at a few field plots that have been selected from the mapping. The measurements from these areas are then extrapolated to estimate the mean value and variance of timber volume and growth for the entire study region.

Digital imagery has been substituted for traditional aerial photographs in such inventories with mixed success. The problem is that even with better spectral information, the analysis of the coarser imagery data rarely extracts the detailed information expected from photography. Yet satellite data are now in more common use in forest inventories. The reason is that multispectral classification methods can map very large regions into meaningful classes of cover type, including species assemblages, density or crown closure, and size class. In recent years, techniques to retrieve tree density, average crown diameter, and cover type from the spatial and directional variances in multispectral data have been perfected. These variances are related to patterns of sunlit and shaded tree crowns and background.

Simulation of ecosystem processes. This approach takes two related but distinctly different lines. One line has been to couple hydrologic and ecophysiological principles into models to simulate the ecosystem processes of photosynthesis, respiration, and decomposition. These models use parameters derived mainly from remote-sensing data. The second line, again using remotely sensed data, focuses on the slower ecosystem processes of plant community development with population models that simulate the life histories, growth, and competition of plants. Research is under way to integrate both type of models. This integration promises to produce simulation of processes, including productivity, across a wide range of temporal and spatial scales. Models already conceptualized at a global scale, such as the global carbon models, are helping scientists to understand the aggregate behavior of forests in Earth system processes. While useful at the global scale, these models are currently less useful for regional studies.

Perhaps the key ecosystem variable required for the hydroecological models is leaf area index (LAI). Research using data from the *Landsat* satellites and a National Oceanographic and Atmospheric Administration (NOAA) satellite has shown that leaf area index can reasonably be estimated over large forested regions. These data have also been used to estimate both the absorption of photosynthetically active radiation (APAR) and the stomatal resistance of leaves. Supplementing these broadbanded satellite data (and spatially coarse NOAA data) are new high-spectral-resolution imaging sensors that operate from aircraft. Early evidence suggests that these data may be used to estimate the biochemical composition of forest canopies, such as nitrogen and lignin contents. These estimations can then be used to elucidate ecosystem productivity by relating nutrients to nitrogen turnover and leaf decomposition. These variables as well as meteorologic ones are needed to drive hydroecologic models to predict productivity.

Remote sensing of forest condition. The comprehensive coverage and the uniform, repeatable measurements of remote sensing are useful in studies that monitor the condition of forests. For example, remote sensing can be used to estimate the extent of defoliation, monitor the effects of acid precipitation, and monitor abrupt changes in forest condition.

Defoliation. Droughts can often lead to large areas of dieback and insect infestation. Insects such as gypsy moths can also produce drastic losses of foliage over large regions. Since remote sensing has been used to estimate foliar properties such as leaf area index, monitoring of forest condition by sensing changes in the index or surrogate variables has been attempted a number of times. The results have been only partly successful. In general, conditions of severe defoliation can be accurately discriminated, whereas moderate levels of foliar loss cannot.

Acid precipitation. Chronic acid precipitation can lead to changes in ecosystem functioning and condition. Scientists have recently shown that chronic inputs of nitrogen from acid precipitation into northeastern forests of the United States lead eventually to nitrogen availability in excess of biological demand, causing at first a net increase in production, a rise in foliar nitrogen concentration, and a decrease in foliar lignin content with greater susceptibility to insect attack. Some of the excess nitrogen is leaked from the ecosystem as nitrous oxide, a greenhouse and ozone-scavenging gas, through microbial processes. Field samples of foliage have been chemically analyzed and verify the predicted changes in biochemistry. *Landsat* data have been used successfully to detect and to map complete dieback of spruce-fir forests in the United States and in Europe.

Abrupt changes. Remotely sensed data have long been used to monitor abrupt changes in forest condition. These changes include disruptions due to clearcut harvesting, conversion of forestland to agricultural purposes, and wildfires.

In conifer forests, a clear-cut operation of virtually any size is easily detectable because the spectral brightness of the colonizing plants and bare soil differs substantially from the previous dark appearance of the forest. Once the radiometric and geometric differences between two scenes of imagery are completed, the spectral shifts in disturbed areas are fairly obvious and can be accurately mapped.

The conversion of forestland to agricultural purposes is particularly acute in the Amazon Basin. Slash-and-burn agricultural practices have kept these tropical forests cycling from forestland to cropland for centuries. In recent years, road-building into states such as Rondonia have greatly expanded the conversion of forest to pasture. This tropical deforestation, both in the Amazon and in virtually every other tropical-forest region worldwide, has become a major environmental concern because of the resultant increases in carbon dioxide and other greenhouse gases in the atmosphere. An effort is now being made to estimate the extent of tropical deforestation in tropical countries by using satellite remote sensing. Photographic products of *Landsat* data are being interpreted, the boundaries transferred to a map base, and the boundaries digitized in a geographic information system. When the processing is completed for two different years, the rate, location, and changes will be accurately estimated for the first time.

In many regions of the world, reforestation and ecological restorations are being pursued. The remote-sensing data can be used to map the condition of the remaining forests and disturbed areas, and these results can be used to develop a plan of restoration. This approach is particularly useful in remote, inaccessible regions. Remotely sensed data can also be used to monitor the reestablishment of forests.

To understand the consequences of fire on ecosystems and the atmosphere, one needs to determine the thermal energy being released from active fire fronts, the rate of spread and duration of the fire front, the residual temperature after the fire, and the amount and moisture content of fuel before the fire. These conditions call for sensors that can measure temperatures from ambient to 2370°F (1300°C) at reasonably high spatial resolution for repeated and frequent observations, as well as rapid analysis of imagery.

The smoke from forest fires completely obscures visible and near-infrared sensing of the actual fire conditions. However, observations in the longer infrared wavelengths can sense directly through the smoke plume to monitor the emitted thermal radiation of the fire front and the hot smoldering areas. In the past several years, airborne sensors that do not saturate at low thermal energy levels have been developed to make these measurements.

For background information *SEE AERIAL PHOTOGRAPH; APPLICATIONS SATELLITES; FOREST RESOURCES; REMOTE SENSING* in the McGraw-Hill Encyclopedia of Science & Technology.

David L. Peterson

Bibliography. G. Asrar (ed.), *Theory and Applications of Optical Remote Sensing*, 1989; L. E. Band et al., Forest ecosystem and processes at the watershed scale: Basis for distributed simulation, *Ecol. Model.*, 56:171–196, 1991; P. J. Curran et al., *Remote Sensing of Soils and Vegetation in the U.S.S.R.*, 1990; F. G. Hall et al., Large-scale patterns of forest succession as determined by remote sensing, *Ecology*, 72(2):628–640, 1991; R. B. Myneni and J. Ross (eds.), *Photon-Vegetation Interactions: Applications in Optical Remote Sensing and Plant Ecology*, 1991.

Gaia hypothesis

The Gaia hypothesis holds that the surface of the Earth behaves as a physiological system. The chemistry of a physiological system differs from that of a geological or geochemical system, in which the chemical reactions are not under active biological control. The Gaia hypothesis asserts that aspects of the living surface of the Earth (the biosphere) are regulated and modulated at a planetary level. These aspects (or variables) include the mean global temperature, the gaseous composition of the atmosphere, and the salinity and alkalinity of the oceans. In the absence of a global physiology postulated by Gaia, these variables would be highly predictable, because they would respond to changes in the Sun's output of energy, and would conform to the relatively simple rules of chemistry and geology. However, an examination of the Earth's surface shows that such variables vary widely from predictions based on physics, chemistry, and other, nonbiological sciences. This disparity led to the Gaia hypothesis. This hypothesis has been criticized because of its controversial claim that the Earth is more

like a living thing than an inert environment. In its extreme form, Gaia lends credence to the idea that the Earth is a giant organism. Since this idea resonates with ancient beliefs and brings about a radically different way of looking at the world and life, it has come under suspicion in some scientific circles. Nonetheless, evidence for organismlike monitoring of the planetary environment is wide-ranging.

Evidence for Gaia. At least three main bodies of evidence support the Gaia hypothesis. They are from the areas of atmospheric chemistry, astronomy, and oceanography.

Atmospheric chemistry. The strongest evidence comes from atmospheric chemistry. The Earth's atmosphere, which is approximately one-fifth oxygen, differs radically from that of the Earth's nearest planetary neighbors, Mars and Venus. The atmospheres of Mars and Venus are over 90% carbon dioxide; on Earth, however, carbon dioxide is only 0.03%. Many gases present in relatively stable concentrations in the Earth's atmosphere react quickly with oxygen. According to chemical calculations, reactive gases such as hydrogen, methane, ammonia, methyl chloride, methyl iodide, and various sulfur gases should be present in the Earth's atmosphere in concentrations too minute to be detectable. However, nitrogen, carbon monoxide, and nitrous oxide are respectively 10^{10}, 10, and 10^{13} times more abundant than they should be according to chemical calculations alone. The continued presence of oxygen and the hydrogen-rich gases that react with it provide ready evidence that the atmosphere is being regulated. The atmosphere is an extension of the biosphere. If the Earth's surface were not covered with oxygen-emitting plants, methane-producing bacteria, and countless other organisms, the atmosphere would long ago have reached chemical equilibrium, as have the atmospheres of Mars and Venus.

Astronomy. Another strong argument for Gaia comes from astronomy. According to accepted astrophysical models of the evolution of stars, the Sun used to be up to 40% cooler than at present. However, fossil evidence shows that life has existed almost since the Earth's formation some 4.6 billion years ago. The presence of life on Earth for the last 3-billion-plus years can be verified by fossil evidence of early bacteria. Since organisms survive within the limited temperature range within which water is a liquid (32–212°F or 0–100°C), the fossil life suggests that the global mean temperature has not varied dramatically since life's inception. However, given the major increaase in luminosity of the Sun, the surface temperature of the Earth should also have dramatically increased. The fact that it has not suggests that as the Sun grows more luminous the Earth's biosphere cools to compensate. Although it is not known precisely how the temperature of the planetary surface has been regulated, scientists strongly suspect that greenhouse gases play a role. The most likely candidate for a greenhouse gas that was crucial in cooling the Earth over geological time is carbon dioxide. In photosynthesis, carbon dioxide reacts with hydrogen or its compounds to make the carbon- and hydrogen-based tissue of organisms. Although carbon dioxide lately has been accumulating in the atmosphere because of the burning of fossil fuels (resulting in the greenhouse effect), over geological time the amount of carbon dioxide in the Earth's atmosphere has been dramatically reduced, so that the planet cooled. This temperature regulation may be a geochemical coincidence, but it seems more likely that global life, Gaia, has played a role in keeping the Earth's temperature down. Thermostasis, or temperature regulation, is a typical physiological trait of many animals, including mammals, and some plants, such as skunk cabbage. The details of global temperature control are not yet well known. Nonetheless, global thermostasis may be a property of organisms interacting at the Earth's surface, just as the autonomic temperature regulation of an individual human being is the property of cells interacting in the body.

Oceanography. Oceanic salt and acidity levels may also be actively sensed and stabilized by the biota. Chemical calculations suggest that salts would accumulate in the oceans to levels that are dangerous to life. World oceans, however, have remained hospitable to life for hundreds of millions of years. The suggestion is that the oceans are continuously undergoing some form of desalination. Although precise mechanisms are not known, concentrations of marine salt that are too high for most life may be relieved during the formation of evaporite flats, which are hypersaline expanses rich in halophilic (salt-loving) microbes.

Search for Gaian mechanisms. Some biologists initially objected to the Gaia hypothesis on the grounds that physiological entities can evolve only by natural selection. They argued that since there is ony one Earth, it is not possible that the Earth evolved a physiology; for the Earth to be considered a living, physiological being, space would have to be littered with dead planets that did not survive.

This objection to Gaia has been countered by the development of computer models, the most famous of which is Daisyworld. The Daisyworld model is a theoretical view of a planet in homeostasis. Daisyworld is a planet that is about the same size as the Earth and spins on its axis and orbits, at the same distance as the Earth, a star of the same mass and luminosity as the Sun. The model is simplified so that the environment is reduced to a single variable, temperature, and the biota to a single species, daisies. The mean temperature of Daisyworld is determined by the average shade of color of the planet (the albedo). If the planet is a dark shade (because of the presence of more dark daisies than light), it absorbs more heat and its surface is warmed. If the planet is a light shade (more light daisies than dark), it refects more heat and its surface is cooled. In this model, temperature is shown to be maintained within a narrow range of values, that is, to be regulated effectively over a wide range of solar luminosities, by the simple exponential growth of dark and light daisies.

Such computer models demonstrate in principle how Gaian mechanisms work: the biological regulation of planetary variables is the cumulative result of the incessant growth and activity of billions of organisms on the surface of the Earth. Each one exchanges gases, liquids, and other fluids with other organisms and with its environment. In principle, there is nothing mystical or non-Darwinian about the concept of geophysiology, which views the biosphere as a Gaian, or living, phenomenon. In practice, however, the search for Gaian mechanisms has just begun. For example, it was recently suggested that dimethyl sulfide, a gas floating over the ocean surface and produced by plankton, could be a candidate in the global regulation of temperature. Marine algae grow more precipitously when there is less cloud cover and the ocean receives more sunlight. The growing algae emit dimethyl sulfide. This gas, it was further postulated, provides condensation nuclei, sulfate particles around which raindrops form. In this sequence, absorption of sunlight leads to local higher temperatures and greater ambient blooms of planktonic algae, which produce more dimethyl sulfide and attendant rainclouds. Rainclouds, in turn, lower ambient light levels and lead to cooler surfaces. Thus ends the cycle, in which higher temperatures lead through greater microbe-generated cloud cover to lower temperatures, resulting in rain and stabilizing temperatures over the oceans.

For background information SEE ATMOSPHERIC CHEMISTRY; EARTH; GREENHOUSE EFFECT; PLANETARY PHYSICS in the McGraw-Hill Encyclopedia of Science & Technology.

Dorion Sagan; Lynn Margulis

Bibliography. L. Joseph, *Gaia: The Growth of an Idea*, 1990; J. E. Lovelock, *The Ages of Gaia*, 1989; L. Margulis and L. Olendzenski, *Environmental Evolution*, 1992; D. Sagan, *Biospheres: Metamorphosis of Planet Earth*, 1990; S. Schneider and P. Boston (eds.), *Scientists on Gaia*, 1992.

Gastrointestinal tract disorders

Although gastritis and peptic ulcers were among the oldest and most prevalent gastrointestinal diseases, no specific etiology had been demonstrated. Similarly, spiral-shaped organisms had been observed in the gastric mucosa of animals and humans for over 75 years with little attention to their potential role as agents of disease. Not until 1982, when a gram-negative, microaerophilic, urease-producing, curved or spiral-shaped bacterium was successfully cultured from gastric biopsy specimens from patients with histologic gastritis, was it realized that these diseases could be infectious in nature. This discovery has prompted an unprecedented amount of investigation, with results suggesting that the organism *Helicobacter pylori* may be responsible for most cases of gastritis, and that it is probably a major factor in the pathogenesis of peptic ulcer disease.

Classification. Initially, *Campylobacter pyloridis* was the name given to the bacterium recovered from patients with gastritis. This classification was based on its morphology, growth requirements, and guanine/cytosine ratio, which were similar to the campylobacters, an already-delineated group of bacteria associated with gastrointestinal disease in humans and animals. Subsequent investigation, following a nomenclature change to *C. pylori*, demonstrated that the organism is taxonomically not related to the campylobacters. Based on differences in ultrastructure, cellular fatty acid composition, respiratory quinones, enzyme content, and ribosomal ribonucleic acid (RNA) sequencing, the organism was assigned to a newly established genus, *Helicobacter*. Other closely related organisms recovered from the stomachs of ferrets, *H. mustelae*, and cats, *H. felis*, have subsequently been assigned to this genus.

Morphology and growth. *Helicobacter pylori* is a curved or spiral-shaped, gram-negative bacterium with four to six lophotrichous sheathed flagella. Structurally, the flagellum consists of a filament composed of flagellin (a protein monomer), a hook, a cell-anchored basal plate, and, unlike the majority of flagellated bacteria, a sheath that is continuous with the outer cell-wall membrane. The outer cell wall of *H. pylori* is smooth, unlike that of campylobacters, and is surrounded by a glycocalyx up to 40 micrometers thick, which is thought to play a primary role in the adherence of the organism to gastric epithelial membranes. *Helicobacter pylori* can express two morphological forms, depending on environmental conditions. In the body and in young cell cultures the organism expresses a typical curved bacillary morphology, but prolonged culture results in the emergence of coccoid forms that usually cannot be cultured. This transformation, in response to unfavorable conditions, potentially represents the transmissible form of the organism.

Although the specific nutritional growth requirements of *H. pylori* have not been determined, several important factors required for growth in cell culture have become evident. Primary isolation from gastric biopsy specimens requires a complex basal medium supplemented with either 7% whole horse blood or 20% serum; however, charcoal, cornstarch, egg yolk, or casein can be substituted for whole blood or serum with successful results. *Helicobacter pylori* requires a moist microaerophilic (7–10% CO_2, 5% O_2) environment for growth with an optimum temperature of $98.6°F$ ($37°C$) and a range of $91.4–104°F$ ($33–40°C$), although some strains will grow at $108°F$ ($42°C$). Surprisingly, due to its presence in the human stomach where the pH is as low as 1.5–2.0, *H. pylori* can grow in cell culture only in the pH range of 6.6–8.4, although it can remain alive at pH 1.5–2.0 in the presence of 5-*mM* urea.

Metabolically, *H. pylori* produces a number of preformed enzymes that include short-chain fatty acid esterases, arylamidases, aminopeptidases, alkaline phosphatase, catalase, oxidase, deoxyribonuclease, and most notably a potent urease. The last enzyme is the most distinguishing metabolic characteristic of *H. pylori*. The urease enzyme is located on the cell surface and is composed of two subunits of approximately 30 and 60 kilodaltons. The average rate of hydrolysis of urea by this enzyme in cell lysates is

$36 ± 26$ μmol NH_3/(min)(mg urease), which is more than twice that of *Proteus mirabilis*, another potent urease-producing bacterium. Despite the obvious importance of this enzyme, its specific function remains speculative. However, the following three functions have been proposed: provision for nutritional nitrogen in the form of ammonia, production against gastric acid, and involvement in the pathogenesis of gastric ulcer.

Perhaps the most interesting aspect of *H. pylori* is its ability to thrive in the hostile environment of the human stomach. This ability is due to the development of several specialized characteristics that have permitted a high level of adaptation common among organisms that inhabit gastrointestinal mucus. The organism's spiral morphology in conjunction with motility facilitates movement through viscous environments such as gastric mucin. Another adaptive mechanism for survival in the gastric environment is the ability to thrive in environments with very low oxygen concentration. The gastric epithelial surface and mucus layer, where the organism is found, have low oxygen concentrations but are not totally anaerobic, which is similar to the optimum for growth of *H. pylori*. In conjunction with a potent urease, these characteristics are indicative of an organism that is highly adapted for its environment, as evidenced by *H. pylori*'s apparent organotropism for gastric mucosa.

Pathogenesis and potential virulence. In addition to specialized adaptive mechanisms, many bacteria possess factors that when produced under appropriate circumstances in a susceptible host result in pathogenesis and disease. Prior to infection, all pathogenic bacteria must be capable of colonization. For *H. pylori* this involves initial penetration of the gastric mucus layer in order to reach the preferred neutral environment of the gastric mucosa and epithelium. This requirement emphasizes the importance of *H. pylori*'s motility in viscous environments via its flagella. At the epithelium, *H. pylori* is found in close association with gastric mucosa cells, particularly at intracellular junctions. This association is suggestive of specific binding of gastric epithelial cell surfaces with *H. pylori* adhesions or adherence factors. A fibrillar hemagglutinin that binds specifically to N-acetyl neuraminyl-lactose, which is a type of glycopeptide found in gastric mucosa, has been proposed as the primary adhesion of *H. pylori* that mediates gastric mucosal colonization.

Once established, *H. pylori* possesses several potential virulence factors that could account for the production of gastritis with subsequent ulcer formation. Many strains of *H. pylori* produce a heat-labile protein cytotoxin that causes vacuolization and cytopathic effects in a variety of mammalian cell lines. *Helicobacter pylori* also produces an extracellular protease that degrades gastric mucin. The degradation of mucin could result in a loss of its barrier function, with the subsequent exposure of the gastric epithelium to the back diffusion of hydrogen ion leading to cell damage. Similarly, the potent urease of *H. pylori* has been proposed as a virulence factor through a mechanism analogous to that of its protease. Production of ammonia from urea hydrolysis could cause a loss of ionic integrity in the protective gastric mucin layer, resulting again in the back diffusion of caustic hydrogen ions. Urease, as well as a specific protein acid secretion inhibitor, may be responsible for the abnormally small concentration of hydrochloric acid in the stomach seen during acute *H. pylori* infection. Although any or all of these potential virulence factors could result in gastritis and ulcer production, further investigation is required to elucidate their specific role in pathogenesis.

Prevalence and transmission. The prevalence of *H. pylori* is in large part age-related, with healthy persons less than 30 years old having prevalence rates of approximately 10% and those over 60 years of age having rates approaching 60%. Other factors associated with higher prevalence rates include race, ethnic background, and socioeconomic condition. As suggested initially, *H. pylori* is highly associated with gastritis, gastric ulcer, and duodenal ulcer. There is evidence of *H. pylori* infection in essentially all patients with duodenal ulcer and in approximately 80% of patients with gastric ulcer. The mode of transmission of *H. pylori* is not known, although person-to-person spread is suggested from currently available evidence. The organism has been cultured only from the stomach, with the exception of a single isolate recovered from human saliva.

Helicobacter and disease. Although a substantial body of evidence has accumulated concerning the relationship between *H. pylori* and gastritis and ulcerative disease, the proof that it is the primary etiological agent is largely indirect. This evidence includes the intentional ingestion of *H. pylori* by two human volunteers, with a resulting clinical course that was consistent with that of patients infected with *H. pylori*. Although a good practical animal model for *H. pylori* infection remains to be developed, some success has been obtained in neonatal gnotobiotic piglets with a host response slightly different than that observed in humans. In addition, as in all true infections, there is a strong specific systemic IgG and IgA antibody response as well as the inflammatory response to *H. pylori*. Finally, the most convincing indirect evidence that the organism causes disease comes from antimicrobial treatment studies. Although limited, these studies have demonstrated a clear relationship between suppression or elimination of the organism and resolution of disease.

Helicobacter pylori may be the most common cause of gastrointestinal infection in humans, although the majority of people do not have specific health problems despite its presence in the stomach. The specific factors that result in disease are not known, although the relationship between increasing age and disease does suggest that immunologic host factors may play a role in the transition from colonization to disease. Similarly, it could be argued that gastritis and peptic ulcers are associated with the presence of *H. pylori* but the relationship is not one of cause and effect. This issue cannot be resolved until more data are available in terms of well-controlled, randomized trials.

For background information *SEE ILEITIS; MEDICAL BACTERIOLOGY* in the McGraw-Hill Encyclopedia of Science & Technology.

James D. Dick

Bibliography. M. J. Blaser, *Campylobacter pylori in Gastritis and Peptic Ulcer Disease*, 1989; J. D. Dick, *Helicobacter (Campylobacter) pylori*: A new twist to an old disease, *Annu. Rev. Microbiol.*, 44:249–269, 1990; W. L. Peterson, *Helicobacter pylori* and peptic ulcer disease, *N. Engl. J. Med.*, 324:1043–1047, 1991.

Geological engineering

Geological engineering is a branch of science that combines petroleum geology with petroleum engineering. Professional geological engineers are employed in both the hard-rock mining industry and the petroleum industry. Within the petroleum industry, they are involved in the highly technical business of drilling and producing accumulations of subsurface oil and gas. Their primary objective is to find and recover efficiently and profitably the petroleum and natural gas in porous, heterogeneous, subsurface rock strata known as reservoirs. Geological engineering encompasses the acquisition and integration of key information about reservoir rock pore systems, their fluid content, and reservoir fluid-flow behavior during production. New technological developments in each of these subjects are appearing throughout the industry.

Drilling for oil and gas is an expensive, high-risk industry, even when the drilling is carried out in a proven field. Petroleum development and production must be sufficiently profitable over the long term to withstand a variety of economic uncertainties. These uncertainties include formation complexities within the reservoir, dry holes, widely fluctuating market conditions, rising operating and capital costs, competition for drilling leases, taxes, royalties, and world politics. By working closely with other professionals (such as geologists, geophysicists, reservoir engineers, drilling engineers, and log analysts), geological engineers integrate their expertise to help reduce the technical risks and thereby to contribute to profitable operations.

Geological engineers are trained to interpret and integrate technical data acquired from a wide variety of sources, including seismic surveys, wellbore electric logs, records of production history, and cores representing the reservoir rock. Correct synthesis of all valid information is essential to derive a geological model of the most likely subsurface reservoir conditions. The time and effort spent to create a realistic model of a particular reservoir that fits all the valid data are worthwhile. The model may reveal unexpected investment opportunities by identifying bypassed reserves in existing wells or may point to possible pool extensions for future development. A good model also helps to avoid wrong decisions or false leads and thereby minimizes costly dry holes. The petroleum scientist must follow a systematic approach to data evaluation and technical problem solving, based only on sound physical evidence.

Besides the usual geological and engineering working tools, a variety of new developments continue to enhance geological engineering practice.

Horizontal drilling. Most petroleum reservoirs are very complex because of their long geological history of deposition, burial, and subsequent alteration by the physical process of compaction and the chemical process of mineral dissolution. There are many low-quality reservoirs containing enormous volumes of hydrocarbons that are not economically recoverable by conventional drilling methods. Such reservoirs will be increasingly important sources of hydrocarbons as the conventional pools, developed with vertical wells, ultimately become depleted and replacements for them become more difficult and costly to find. Although more costly, a horizontal well can be used to search laterally within a hydrocarbon-bearing reservoir having predominantly poor porosity and permeability properties, in an attempt to locate thick, permeable, and productive areas or so-called sweet spots (see **illus**.).

Improvements to horizontal drilling methods and specialized drilling tools, coupled with growing operating experience, are reducing drilling costs. Meanwhile, the number of benefits and applications (compared to vertical wells) is growing. Among the more important are (1) higher gas and oil production rates, (2) better, more widespread drainage of the reservoir's hydrocarbons, and (3) successful production of oil and gas from unconventional reservoirs such as low-permeability sandstones or fractured shales.

The drill bit produces small rock chips (drill cuttings) as it progresses horizontally through the formation. These cuttings, representative of the rock present in the horizontal hole, are screened from the circulating drilling mud. Samples of the cuttings are collected at regular depth intervals, cleaned, and then prepared for study. At the wellsite, a geological engineer examines the samples with a microscope for signs of

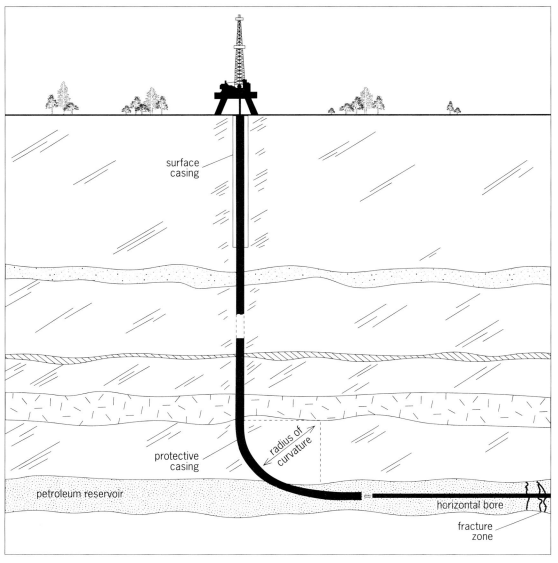

Diagram showing positions of the components of a horizontal well.

hydrocarbons and for an appraisal of changes in the quality of the reservoir rock along the horizontal hole. Continuous wellsite monitoring of cuttings and other data generated during the drilling allows prompt revisions to the operating plan in response to any new information that affects the original geological model or drilling objectives.

It is now possible to drill great distances horizontally within a prospective oil- or gas-bearing reservoir, while steering control of the bit's trajectory and position are maintained. Remote control of the drill bit's precise position, relative to the drilling rig located at the surface (often several kilometers from the bit), is in itself a remarkable achievement. The opportunities offered by these advances in horizontal-drilling technology depend on the ability of the individual engineer to recognize reservoir situations where significant benefits could be gained. For example, the geological engineer may decide to drill perpendicular to natural, open, and very permeable vertical fracture planes that could exist in some low-permeability rock strata. An extensive, slightly open fracture system that is intersected by a horizontal borehole can assist drainage of reservoir hydrocarbons from low-permeability, porous rock at favorable production rates. Two outstanding cases are the Pearsall and Giddings oil fields in southern Texas, where a very large amount of horizontal drilling is taking place in the generally low-permeability, fractured, oil-bearing Austin Chalk formation. An estimated 4000 horizontal wells will be drilled in the Austin Chalk of Texas by 1995.

Major improvements to the specialized technology of drilling horizontal oil and gas wells have been made in the past several years. Numerous horizontal wells have been drilled worldwide as companies seek to test this still-evolving technology in a variety of oil and gas reservoir situations. Many of these somewhat experimental attempts were successful, and some very creative applications, based on geological engineering approaches, have yielded good production results in unconventional reservoir situations. Increased production rates from horizontal wells enhance profitability relative to vertical wells drilled in the same reservoirs. Since the late 1980s, horizontal drilling has become a distinct growth industry with its own special technical and economic requirements. SEE OIL AND GAS WELL DRILLING; OIL MINING.

Reservoir fluid flow. In the past, petroleum reservoirs were considered to be more or less homogeneous entities in the subsurface. Supplemental recovery operations, horizontal drilling, and modern sedimentological/diagenetic knowledge have shown that this is seldom the case. The task of the geological engineer is to clarify the internal complexities of reservoirs in order to effectively plan, carry out good reservoir management, and extract valuable hydrocarbons.

Backed by a wide array of powerful well-logging devices that are supported by comprehensive petrographic analyses of samples representing the reservoir itself, the geological engineer can begin to unravel complex petroleum reservoirs and to plan effective pool development. Fluid-flow paths in the reservoir are strongly influenced by the continuity and permeability of the rock pore systems. Expert analysis of rock and log data, supported by performance histories of well production, can be evaluated to help sort out these important reservoir trends. The geological engineer is in the best position to predict fluid-flow patterns in the reservoir and to plan development drilling to achieve optimum hydrocarbon recovery.

A similar approach is used in carefully engineered operations involving maintenance of reservoir pressure. These are intended to increase the total amount of recovered oil by providing supplemental energy to flush crude oil to the producing wells. Often, the incremental oil produced represents a large proportion of the original volume of oil that would be left behind if wells were simply allowed to produce until reservoir pressure became too low to maintain economic production rates. Such plans for pressure maintenance or enhanced recovery are collectively known as supplemental recovery operations.

In a supplemental recovery operation, fluids, such as water, steam, carbon dioxide, condensate, or other chemical combinations, are pumped into injection wells to sustain or raise reservoir pressure and, at the same time, displace oil through the reservoir rock pore system toward producing wells. In most cases this type of operation is costly, with high initial capital investment. The array of equipment for fluid-injection and related oil-producing wells must be positioned to conform to those natural-continuity properties of a reservoir that influence fluid-flow patterns. If condensate, which is similar to gasoline, is used (in the dual role of solvent and displacing fluid), it also must be recovered later by using a less valuable fluid such as water.

Rapid acquisition and processing of geological and engineering data are important parts of designing a supplemental recovery operation. Data, whether for a single well, a group of wells, a large field, or even an entire sedimentary basin, are generally stored electronically in large petroleum database systems. Enormous amounts of diverse information can be retrieved and evaluated by using computer workstations that are linked to the databases. The computer workstations ensure that less time is spent acquiring and assembling technical data than in the past, with more time allowed for interpretation.

Petrography. Petrography involves more than just the systematic classification and description of rocks. Unlike most methods for evaluating petroleum reservoirs, petrography utilizes actual reservoir rock material. Thus, rock samples, such as cores and drill cuttings, provide information derived directly from the reservoir. They are used to calibrate a variety of electronic well-logging tools whose responses to rocks and entrained fluids in the subsurface are subject to interpretation.

Reservoir rock material is obtained either as expensive cores that are 7–10 cm (3–4 in.) in diameter and up to 18 m (60 ft) long, or as inexpensive drill cuttings. The value of collecting and examining these rock samples cannot be exaggerated. Experience has shown that the amount of vital reservoir information that can be obtained, even from cuttings samples, is virtually unlimited. The value of the information depends mainly upon the training, knowledge, and experience of the examiner, and the kinds of equipment used for rock examinations.

Rapid advances in electronics have made available many techniques not previously used for petrographic evaluations; many techniques were developed for other applications such as the computer industry and medical diagnosis.

Important improvements in the scanning electron microscope (SEM) allow detailed evaluation of rock samples. Sophisticated electronic equipment used for x-ray, quantitative image analysis, and image archiving purposes can be linked to the scanning electron microscope. These combined devices yield quantitative measurements of the rock-forming minerals and their associated petrophysical properties such as porosity and permeability. The information is used to make day-to-day decisions regarding the costly operations of well drilling and completion. Similarly, previously stored images and data can be retrieved for comparative analysis. Such an integrated system designed around a scanning electron microscope has capabilities far beyond those of the conventional optical microscopes that were formerly relied upon for rock examinations by many geologists.

Even the optical microscope has been modified by using laser-beam technology. This development, the scanning confocal microscope, provides high resolution and exceptional depth of field. However, these devices complement rather than replace optical microscopes. Each specialized instrument provides some additional information not obtainable from the others.

Petrophysics. Petrophysics is a specialized field of technical expertise that has evolved along with the development of numerous types of electronic well-logging devices and the interpretation of measurements they produce.

Important advances in well-logging technology are continuing to yield better characterization of reservoir heterogeneities. This information, in turn, aids in the prediction of reservoir fluid flow. The most effective plan for oil and gas field development takes into account the heterogeneous distribution of those reservoir properties that influence subsurface fluid flow.

Computers housed in modern logging units may be used to integrate and interpret an array of well-log measurements. The result is a quick analysis of subsurface rock strata and their hydrocarbon fluid content. Logging information can also be transmitted by satellite from remote wellsites as it is being acquired. The signals are received at a workstation located near company offices, where the logs are reproduced for quick examination by professionals. In fact, wellsite personnel can communicate with other company engineers at the home office to jointly examine and discuss logging results and to make any program or tool changes required during a logging operation.

Cross-well imaging. In a continuing effort to learn more details about the geology of a reservoir in the areas between wells, new approaches to seismic signal generation and detection are being developed. One of the more promising is cross-well imaging, also known as cross-well tomography.

Conventional surface seismic methods detect energy waves reflected from rock layers that are often 2 mi (3 km) or more below the surface. In the process the high-frequency sound waves, which carry details about changes in subsurface rock character, become weakened and lost. Better resolution of the geology can be achieved by placing the seismic source and detectors across zones of interest in separate wells in a field. Signal travel paths are shortened, and better coupling of the receivers to solid rock means that higher-frequency energy waves are detected. The source can be moved up and down a well, so that a particular layer of interest can be examined from different angles, as reflected sound waves are reaching the receiver in an opposite well. The benefit of this technology is that it is possible to obtain more detailed information about reservoir properties between the wells in a field. This information in turn can be used to improve the geological engineering design of hydrocarbon recovery operations.

For background information SEE CONFOCAL MICROSCOPY; ENGINEERING GEOLOGY; OIL AND GAS WELL DRILLING; PETROLEUM ENGINEERING; PETROLEUM ENHANCED RECOVERY; PETROLEUM GEOLOGY; PETROLEUM PROSPECTING; PETROLEUM RESERVOIR ENGINEERING; SCANNING ELECTRON MICROSCOPE; SEISMIC EXPLORATION FOR OIL AND GAS; WELL LOGGING in the McGraw-Hill Encyclopedia of Science & Technology.

Robert M. Gies

Bibliography. P. C. Crouse & Associates and World Oil Magazine, *2d International Conference on Horizontal Well Technology* (special conference volume of 40 technical papers compiled by Gulf Publishing Company, Houston, Texas), October 1990; J. H. Doveton, *Log Analysis of Subsurface Geology: Concepts and Computer Methods*, 1986; W. J. Lang and M. B. Jett, High expectations for horizontal drilling becoming reality, *Oil Gas J.*, pp. 70–79, September 24, 1990; J. C. Russ, *Computer Assisted Microscopy: The Measurement and Analysis of Images*, 1990.

Grassland ecosystem

Grasslands occupy 7.4×10^9 acres (3×10^9 hectares) of the land surface of the Earth and are one of the largest renewable resources. They are found in all major climate zones capable of supporting plant growth. Annual primary production ranges from less than 0.22 ton/acre (0.5 metric ton/ha) in grasslands of the sub-Sahara (perhaps zero in drought years) to over 58 tons/acre (130 metric tons/ha) in intensively managed napiergrass (*Pennisetum purpureum*). Maximum annual primary productivity is about 36 tons/acre (80 metric tons/ha) in tropical grasslands and 11 tons/acre (25 metric tons/ha) in temperate grasslands. However, productivity of many grasslands is restricted by various factors, including temperatures that are below or above optimum, inadequate or excess water, plant nutrient deficiency and toxicity, plant disease, physical and chemical properties of soil, erosion, fire, and overgrazing by livestock and other herbivores.

Progenitors of modern herbivores moved into the primitive grasslands as they evolved mechanisms to extract energy from the complex carbohydrates of the plant cell walls. From that time on, the grasslands and the herbivores coevolved to form the complex grassland ecosystems that humans have recently learned to exploit.

Foraging theory. It is probable that grassland herbivores, including livestock on intensively managed grasslands and on range, behave according to foraging theory. Grassland researchers agree that energy is the primary limiting nutrient of herbivores of most grassland ecosystems. Foraging theory suggests that herbivores attempt to maximize energy intake with the minimum expenditure of energy in the work of grazing.

Water may also limit the productivity of some dry rangeland, but some herbivores, for example, sheep and camels, have minimized their water needs by metabolic and behavioral mechanisms.

Herbage protein may also limit grassland herbivores. For proper functioning of digestion, ruminants need herbage with about 6–8% crude protein. Herbage produced by tropical and subtropical pastures dominated by grasses with the C4 system of photosynthesis is often very low in protein, and the extraction of energy may be impaired.

Plant–animal interactions. The interactions between the animals and the plants upon which they graze are extremely complex, and are increasingly a focus of research on grassland systems.

Herbage intake. In livestock operations, as in systems analysis of grasslands, it is of primary importance to be able to estimate the amount and composition of herbage ingested over time. Yet it is very difficult and expensive to measure the amount, quality, and species of herbage grazed each day by, for example, free-ranging cattle.

Herbage dry-matter intake is the product of grazing time and the rate of herbage dry-matter intake per day. Cattle that are grazing under ideal conditions ingest herbage dry matter at about 0.5% of body weight per hour of grazing. Typically, cattle require 2–3% of their body weight as herbage dry matter each day to meet their nutritional requirements, and thus they must graze a minimum of 4–6 h each day. Pasture or range conditions rarely allow cattle to graze at their maximum rate, and they may compensate by grazing for up to 14 h per day. Grazing time cannot fully compensate for sparse pastures and low rates of intake because time is needed for other activities such as resting, watering, suckling, rumination (about 1 h of rumination is needed per hour of active grazing), and herd interactions.

Cattle typically have two main grazing meals each day, one beginning near dawn and one beginning in the late afternoon. They do not graze at night unless days are uncomfortably hot or the animals have very high energy requirements. Hunger initiates grazing, and satiety causes grazing to cease. When consuming low-quality herbage that passes slowly through the gastrointestinal tract, cattle have fewer and longer grazing meals each day than when they graze on high-quality herbage that has high rates of passage. Many ruminant-nutrition researchers believe that satiety is triggered by the level of food and other materials in the gastrointestinal tract of grazing herbivores, but the mechanisms involved are not known. In a herd, grazing time per day, as well as the number and length of grazing meals, may also be moderated by the grazing and social activities of the herd.

Rate. The rate of herbage intake is the product of rate of biting and bite size. Bite size is more important than rate of biting in determining rate of intake. Maximum bite size is a function of the dimensions and mechanics of the ingestive apparatus, including the muzzle, the lips, the distance between incisors, the area of grinding surfaces of the teeth, and the volume of the buccal cavity. Maximum bite size is closely related to body mass: the larger the animal, the larger the bite. Actual bite size, however, is determined by the properties of the herbage and is moderated by hunger and satiety. Animals may compensate for small bites by accelerating the biting rate, but the compensation is not complete. Cattle may take up to 80 bites per minute, sheep up to 120/min, and horses up to 25/min.

Deterrents. Physical properties of herbage such as surface height and herbage mass density are important in establishing bite size and, therefore, rate of biting and rate of intake. Recently, the pseudostem (a false stem composed of concentric rolled or folded blades and sheaths that surround the growing point) of tillers of vegetative cool-season grasses has been identified as a physical barrier to biting for sheep and cattle.

Feeding deterrents may limit intake of some grassland species. Tall fescue (*Festuca arundinacea*), which occupies about 50×10^6 acres (20×10^6 ha) of the mid-south in the United States, contains alkaloids that act as feeding deterrents for most, if not all, of the herbivorous species, including nematodes, insects, rabbits, sheep, cattle, and horses.

Selectivity. Herbivores attempt to maximize the amount of nutrients ingested per unit of time and effort spent grazing. They prefer some species over others because of differences in taste, texture, and odor. Cattle and horses, for example, prefer forage legumes to improved grasses. In grasslands where some species are grazed in preference to others, drastic changes may occur in the species composition and in the species contribution to plant biomass.

Grazing animals usually prefer young tissue over old, leaf tissue over stem, and green tissue over brown. Thus, herbivores preferentially eat the young leaves, stems, and growing points. Consequently, grazing severely reduces the assimilatory surfaces and generally stresses plants. Fortunately, coevolutionary processes have ensured that many grassland plants can withstand the stresses of grazing.

Systems analysis. Natural and managed grasslands are both extensive and complex, so research and analysis are limited. Grassland systems must be understood in order for them to be managed and their stability and productivity to be maintained. Simulation models of grassland ecosystems are being used to help analyze reactions of grasslands to various management regimens and to answer questions of a global nature. For example, models are being used to analyze the effect of global warming on the productivity and stability of the grasslands. Other models describe smaller grassland ecosystems or parts of ecosystems.

Many early grassland simulation models were detailed and very complex, and required large computers. These simulation models were not used extensively, but they became the logical and algorithmic base for current desk-top computer models and many expert-system models. These computer models address specific problems of grassland systems and are being used in an increasing number of applications.

For background information *SEE AGRICULTURAL SCIENCE (ANIMAL); AGROECOSYSTEM; GRASS CROPS; GRASSLAND ECOSYSTEM* in the McGraw-Hill Encyclopedia of Science & Technology.

C. T. Dougherty

Bibliography. R. Dennis and E. K. Byington, *Potential of the World's Forages for Ruminant Animal Production*, 1981; J. Hodgson, *Grazing Management: Science into Practice*, 1990; F. P. Horn et al., *Grazing-Lands Research at the Plant-Animal Interface*, 1987; D. W. Stephens and J. R. Krebs, *Foraging Theory*, 1986; K. Wisiol and J. D. Hesketh, *Plant Growth Modeling for Resource Management*, 2 vols., 1987.

Growth factor

The hematopoietic growth factors are glycoprotein hormones that are instrumental in regulating normal hematopoiesis. These factors are produced by a variety of tissues, including the peritubular cells of the kidney, T lymphocytes, fibroblasts, and endothelial cells. The principal role of the hematopoietic growth factors is the control of the process of proliferation and differentiation as the immature hematopoietic cells in the bone marrow become the mature cells found in the peripheral circulation. Six hematopoietic growth factors— erythropoietin, granulocyte macrophage-colony stimulating factor (GM-CSF), granulocyte-colony stimulating factor (G-CSF), macrophage-colony stimulating factor (M-CSF), interleukin 3 (IL-3), and stem cell factor (SCF)—have been purified, cloned, and produced on a large scale through recombinant deoxyribonucleic acid (DNA) technology. All six entered clinical trials and have proven to be beneficial for patients with a variety of diseases ranging from the anemia of renal failure to the neutropenia associated with cancer chemotherapy.

Basic hematopoiesis. The mature red cells, white cells, and platelets that circulate in the peripheral blood are short-lived and must be replaced continuously throughout life. These cells are derived from pluripotent hematopoietic stem cells, which are located mainly within the bone marrow. Stem cells must give rise to a pool of cells that are committed to proliferate and differentiate to become functionally mature peripheral blood cells. These mature cells replace those that are lost because of senescence, damage, and other causes. Stem cells must also be capable of self-renewal to prevent their depletion (**Fig. 1**). Both self-renewal and committed differentiation must be closely regulated. If the process of self-renewal were not closely regulated, the stem cell pool either could become depleted and eventually exhausted, or could fail to differentiate so that committed progenitors would be lacking. The process of committed differentiation is likewise under tight control; the daily consistency of the number of mature cells in the peripheral blood is an obvious demonstration of this fact. Much of the control of hematopoiesis is mediated by hematopoietic growth factors that act directly on marrow cells. The **table** shows the factors that have been isolated. Each factor reacts with specific receptors on hematopoietic cells. The response seen when these molecules react with their target depends in part on the particular target. The usual response for immature cells is to proliferate and begin to differentiate, while for mature cells the response is to become functionally activated.

Hematopoietic growth factors being used in clinical trials

Factor	Abbreviation	Source	Progenitor targets	Mature targets
Erythropoietin	EPO	Peritubular cells of kidney, Kupffer cells	Erythroid progenitors	None
Granulocyte macrophage-colony stimulating factor	GM-CSF	T cells, monocytes, fibroblasts, endothelial cells	Early hematopoietic progenitors, including myeloid, erythroid, and megakaryocytic precursors	Granulocytes, eosinophils, monocytes
Granulocyte-colony stimulating factor	G-CSF	Monocytes, fibroblasts, endothelial cells	Myeloid precursors	Granulocytes
Macrophage-colony stimulating factor	M-CSF	Monocytes, fibroblasts, endothelial cells	Monocytic precursors	Monocytes
Interleukin 3	IL-3	T cells	Early hematopoietic progenitors, including myeloid, erythroid, and megakaryocytic precursors	Eosinophils, monocytes
Stem cell factor	SCF	Marrow stromal cells, T cells, hepatocytes, fibroblasts	Very early hematopoietic progenitors, mast cells, melanoblasts	Mast cells

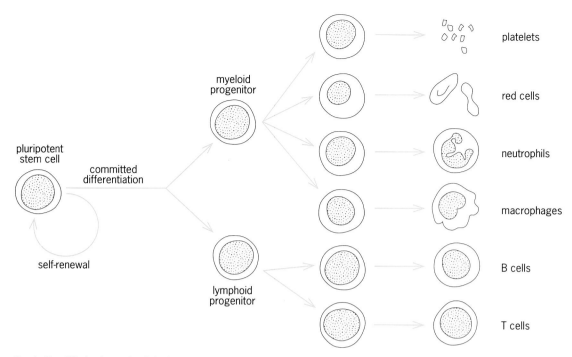

Fig. 1. Simplified schematic of the hematopoietic system.

Biology. Bone marrow cells placed in a semisolid media such as agar will not proliferate to form colonies unless a growth factor is added (**Fig. 2**). Over the last several decades, materials have been identified that are relatively rich in growth factors. For example, urine from patients with severe anemia can stimulate erythroid colony formation in cell cultures. Cell lines produced from tumors of patients with unexplained elevated neutrophil counts will in many cases release substances that increase granulocyte colony formation in cell cultures. Eventually, the distinct glycoproteins responsible for these colony-stimulating activities were entirely characterized, and on the basis of their amino acid structure the DNA encoding these proteins was isolated, cloned, and introduced into yeast, bacteria, or mammalian cells. Recombinant DNA technology allows for the production of large quantities of purified hematopoietic growth factors for use in clinical investigation.

A single gene encodes each hematopoietic growth factor. The genes for three of the growth factors (GM-CSF, M-CSF, and interleukin 3) are all localized to a small region of the short arm of chromosome 5; G-CSF is localized to chromosome 17; and erythropoietin is on chromosome 7. The genes encoding for these products become active and produce their respective growth factor only in certain cells. Erythropoietin is produced predominantly by the peritubular cells in the kidney, although 10–15% of erythropoietin synthesis can take place in the liver. GM-CSF, G-CSF, and M-CSF are produced by multiple cell types, including monocytes, fibroblasts, endothelial cells, T lymphocytes, and marrow stromal cells. Interleukin 3 is made principally by T cells. Stem cell factor is produced principally by marrow stromal cells, but other cells probably contribute to production as well.

Regulation. The regulation of production of the hematopoietic growth factors is incompletely understood. The best-understood case is that for the production of erythropoietin. Hypoxia, as sensed in the kidney, is a major stimulus for increased production of this factor. Production of some of the other factors, including GM-CSF, G-CSF, M-CSF, and interleukin 3, is increased after cells that produce these factors are exposed to inflammatory stimuli, including bacterial endotoxin, tumor necrosis factor, and interleukin 1 (IL-1). The increased production of these hematopoietic growth factors after exposure to inflammatory mediators is important in a patient's response to infection or acute inflammation, but probably is not the way in which hematopoiesis is controlled during normal states of health.

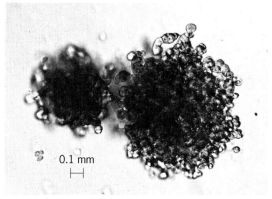

0.1 mm

Fig. 2. Photomicrograph of erythroid colonies. These cultured colonies were grown with erythropoietin as a stimulus.

Once released, hematopoietic growth factors achieve their biologic activity by binding to specific high-affinity receptors on their target cells. Much of this binding occurs locally, and for some of the hematopoietic growth factors, circulating levels are never achieved normally. For others, including G-CSF, M-CSF, and erythropoietin, low levels circulate in the blood in states of normal health.

Activities of factors. The hematopoietic growth factors have two major roles, regulating the proliferation and differentiation of hematopoietic progenitors and enhancing the function of mature blood cells. The hematopoietic growth factors can induce nondividing precursors to start dividing within several hours. If growth factor levels are maintained in cell culture, the cell cycle time (that is, the time between cell divisions) will be shortened and maturation seems delayed. Thus, a larger number of precursors will remain in the proliferating pool. The result is that, with constant presence of growth factor, more cells can be produced and they can be produced faster. Several of the hematopoietic growth factors also act on mature cells, and promote the survival of these cells and stimulate their functional activities.

By analyzing the effects of the hematopoietic growth factors in cell culture, it is possible to have some indication of how these agents act in the body (see table). Erythropoietin activity is limited in cell culture to red cell progenitors. Likewise, in the body, erythropoietin affects only red cell production. The activity of G-CSF in cell culture is also relatively limited, and is restricted to granulocyte precursors and mature granulocytes. In the body, this factor increases granulocyte counts and increases the functional activity of mature granulocytes. Likewise, the activity of M-CSF is relatively restricted both in cell culture and in the body to monocyte precursors and mature monocytes. The activities of GM-CSF and interleukin 3 in cell culture are somewhat broader, with activities seen on myeloid, erythroid, and megakaryocytic precursors. Studies using GM-CSF in the body confirm this broader range of activity. Stem cell factor has the broadest activity on hematopoietic progenitors and appears to act on the earliest progenitors of any of the factors. Additional data from cell cultures demonstrate that for many hematopoietic growth factors, when they are used in combination, their activities are additive and often synergistic. For example, more vigorous responses are seen with combinations of GM-CSF and interleukin 3 than with either agent alone.

Clinical studies. The following describes recent studies on the six hematopoietic growth factors.

Erythropoietin. Recombinant human erythropoietin was the first recombinant hematopoietic growth factor to enter clinical trials in humans. Since most erythropoietin in the body is produced by the kidneys, patients with damaged or absent kidneys have extremely low erythropoietin levels, resulting in severe anemia. In clinical studies, recombinant human erythropoietin was shown to be remarkably effective in stimulating red blood cell production in transfusion-dependent patients with kidney failure. When it was

given intravenously three times per week over a broad dose range (15, 50, 150, and 500 units/kg), a dose-dependent increase in red cell production was seen. Human recombinant erythropoietin has recently been approved by the Food and Drug Administration (FDA) as a safe and effective therapy for the anemia of renal failure.

The success of recombinant human erythropoietin in the anemia of renal failure led to its use in other anemias, including those associated with acquired immune deficiency syndrome (AIDS), rheumatoid arthritis, and cancer. Positive results have been seen in several of these studies, although none as striking as the response in patients with renal failure.

GM-CSF and G-CSF. When given to normal individuals, both GM-CSF and G-CSF result in an increase in circulating neutrophils in the peripheral blood to levels that are more than 10 times the normal level. GM-CSF, but not G-CSF, also increases numbers of circulating monocytes and eosinophils. Given these effects, both factors have been studied in the treatment of patients with neutropenia, a condition in which there is an abnormally small number of neutrophils in the circulating blood and which is associated with an increased risk of severe or fatal infections. By far the most common form of neutropenia is that induced by cancer chemotherapy. Approximately 50% of patients receiving chemotherapy for disseminated cancers experience at least one neutropenic episode leading to a significant infection, and at least a 5% death rate accompanies such infection. If either G-CSF or GM-CSF is given to patients shortly after the administration of chemotherapy, both the degree and duration of neutropenia can be lessened. The result is fewer episodes of fever and a reduction in the duration of each episode. The benefits of G-CSF and GM-CSF are most apparent following very high dose chemotherapy, and are less obvious when more easily tolerated, lower-dose chemotherapy is administered.

Bone marrow transplantation is a similar situation in that intensive chemotherapy or chemoradiotherapy is given prior to the transplant and a prolonged period of severe neutropenia is seen after the transplant until the new marrow begins to function. As with cancer chemotherapy, severe and sometimes fatal infections are often seen. However, if either GM-CSF or G-CSF is given after the marrow transplant, the period of severe neutropenia can be lessened, with fewer days of fever and fewer infections. Given these results, both factors have been approved by the FDA as safe and effective therapy.

Other forms of neutropenia have been successfully treated with GM-CSF or G-CSF. These include congenital neutropenias, acquired idiopathic neutropenia, neutropenia associated with viral infections such as AIDS, and neutropenias that are the direct result of malignant invasion of the bone marrow. Occasionally, patients are born with severe neutropenia (Kostmann's syndrome) and suffer a clinical course of recurrent infections with considerable morbidity and mortality. Administration of G-CSF leads to a marked increase in neutrophil counts in these patients, which in turn results in substantial clinical benefit. Another congenital neutropenia successfully treated with G-CSF is human cyclic neutropenia, a disorder characterized by periodic drops in the numbers of neutrophils in bone marrow and blood. G-CSF reduces the extent of the drop and essentially eliminates the complications normally associated with the disease. Occasionally, patients develop severe neutropenia for no apparent reason, as can occur in acquired aplastic anemia. Both GM-CSF and G-CSF have been used in aplastic anemia with mixed results. Patients with milder forms of this disorder often respond to treatment with these factors, but those with the most severe forms do not. Patients with AIDS, particularly those on zidovidine antiviral therapy, often develop neutropenia. GM-CSF can increase the neutrophil counts in many such patients, so that long-term therapy with zidovidine is possible.

M-CSF, interleukin 3, and stem cell factor.
M-CSF, interleukin 3, and stem cell factor have yet to be approved by the FDA (as of late 1991), but all three are under study. M-CSF increases both monocyte counts and function. Interleukin 3 causes a modest increase in neutrophils and lymphocytes. Stem cell factor has been isolated only recently and is the only hematopoietic growth factor that causes an increase in all hematopoietic cell types, including red cells, platelets, neutrophils, and lymphocytes. Stem cell factor has now entered clinical trials.

For background information *SEE HEMATOLOGIC DISORDERS; HEMATOPOIESIS; MOLECULAR BIOLOGY* in the McGraw-Hill Encyclopedia of Science & Technology.

Frederick R. Appelbaum

Bibliography. J. E. Groopman et al., Hematopoietic growth factors: Biology and clinical applications, *N. Engl. J. Med.*, 321:1449–1459, 1989; D. Metcalf, The molecular control of cell division, differentiation commitment and maturation in haemopoietic cells, *Nature*, 339:27–30, 1989; J. Nemunaitis et al., Recombinant granulocyte-macrophage colony-stimulating factor after autologous bone marrow transplantation for lymphoid cancer, *N. Engl. J. Med.*, 324:1773–1778, 1991; C. L. Sawyers et al., Leukemia and the disruption of normal hematopoiesis, *Cell*, 64:337–350, 1991.

Guidance systems

A new radar system has made possible a substantial increase in the landing capacity of airports with parallel runways. With improved surveillance of approaching aircraft, 30–40% more landings can be authorized during poor weather conditions than previously. For the largest airports, limitations in capacity lead to delays and restrict growth. The new radar displays aircraft positions more accurately and more frequently, and it includes an automated alert for the controller if an aircraft deviates from its course.

Avoiding collision with other aircraft is essential. When weather conditions prevent pilots from seeing other aircraft, air-traffic controllers assign nonconflicting flight paths. How close the flight paths can be to one another depends on the precision with which

the pilot can keep the aircraft on an assigned path, on the quality of the surveillance of aircraft position presented to the controller, and on the response time of the controller to resolve a conflict. These factors are most critical when airplanes must converge to land at a single airport.

Need for increased capacity. Airlines have made extensive use of hub airports. Aircraft from a large circle of cities fly to an airport near the center of the circle, exchange passengers, and return to the cities. In this way, passengers can fly from any city in the system to any other with only a single change of planes. This practice demands more of air-traffic control at the hub because the planes need to arrive in a narrow time span, an hour or less, so that the connections can be made with minimal layover.

Most arrival schedules at a hub airport are set so that when operating in visual conditions the airport operates near saturation. If the runway configuration necessitates reduction in capacity when the airport is operating in instrument conditions, aircraft are delayed because there is not enough time to get them all on the ground in the period of the arrival window.

Use of multiple runways. Airports have responded by building multiple runways so that a number of aircraft can land at the same time. Runway configuration is not particularly critical for simultaneous use of runways when pilots can see. When aircraft separation depends on ground air-traffic control, a complicated set of restrictions arises. For parallel runways spaced at least 4300 ft or 1310 m apart (in the United States; 5000 ft or 1525 m internationally), controllers can control operations on one of the runways independent of the others. For closer spacings, aircraft cannot land simultaneously. Two runways can be used for landings during the same time period, but landings must be alternated between them (**Fig. 1**).

The minimum spacing allowing independent use of the runways is determined by the need for sufficient time to respond should an aircraft blunder from its approach and fly toward another aircraft on the parallel approach path. Blunders, although very rare, can be traced to confusion about the runway assignment, faulty aircraft or navigation equipment, or an inattentive pilot. For a blunder to be corrected, it must first be observed by the controller, who must then radio an evasive maneuver to the pilot of the endangered aircraft, who in turn must move that aircraft out of the way. These steps are diagrammed in **Fig. 2**.

The 4300-ft (1310-m) minimum was determined from the judgment of experienced pilots and controllers who were familiar with the time required for each blunder-correction action according to communications, navigation, and radar surveillance equipment widely available in the 1970s. Now, improvements in air-traffic control radar and the procedures for pilots and controllers make lower minima possible.

Improved radar and displays. The new radar provides improvements for three fundamental quantities compared with the radar when the old minimum was established: accuracy, update rate, and display quality.

Accuracy. The accuracy of airport surveillance radars in the 1970s allowed them to report position to an azimuth accuracy of 5–10 milliradians, or about 0.5°. The new radar improves on it tenfold, to about 1 milliradian. This improvement allows the controller to clearly see a blunder from among the lower random errors of the radar and to detect it sooner. The improved accuracy also makes possible a computer prediction of the aircraft's track and an automated caution alarm when a blunder is occurring. The alarm helps to ensure a quick response from a controller who may see blunders only rarely.

The accuracy improvements are due to the monopulse feature of the radar. The accuracy of conventional air-traffic control radar is limited by the width of the antenna beam that receives replies from the aircraft. Aircraft position is determined after the antenna receives several replies from an aircraft being scanned by the rotating beam. The accuracy improvement of monopulse results from the ability to sense the population within the beam on the basis of one reply.

Update rate. Conventional airport surveillance radars update at intervals of 4–5 s. Radars for closely spaced runways update at intervals of 0.5–2.4 s with two antenna configurations. One configuration consists of two flat arrays mounted back to back on a pedestal rotating at a conventional 4.8-s rate, with 2.4 s between updates of the aircraft position. The other is a stationary, circular array antenna with electronic beam steering, which updates as often as 0.5 s. The more rapid the update, the sooner the controller can detect a blundering aircraft.

Display quality. A high-resolution, color, cathode-ray-tube display measuring 20 × 20 in. (51 × 51 cm) shows the aircraft radar position. The high resolution and large screen make it possible for controllers to view and operate with the higher detection accuracy achieved by the radar itself. The display logic makes use of color to distinguish between aircraft deviating from course and others operating normally. The processor also includes the automated alarm.

Pilot response. Procedural improvements are necessary to ensure a quick response from the pilot. Tests in flight simulators showed that most pilots responded quickly when directed to leave their approach path to avoid a blundering aircraft, but a few failed to understand the urgency of the instruction and responded less quickly. Special phraseology was developed, and it was decided that all pilots would be trained in the avoidance maneuver. Steps were taken to ensure that a clear communications channel was

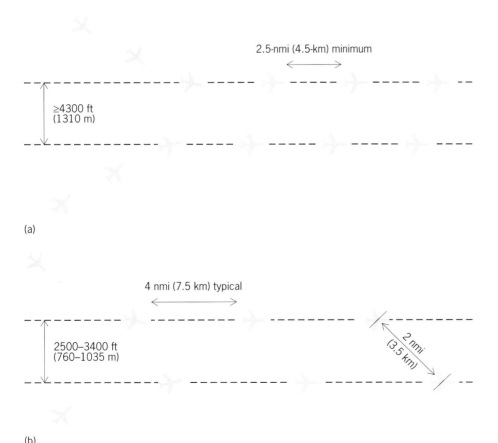

(a)

(b)

Fig. 1. Independent and dependent parallel approaches. (*a*) Independent case. Aircraft need be spaced only with regard to aircraft destined for the same runway. (*b*) Dependent case. No aircraft can be closer than 2 nautical miles (3.5 km) to an aircraft approaching the adjacent runway.

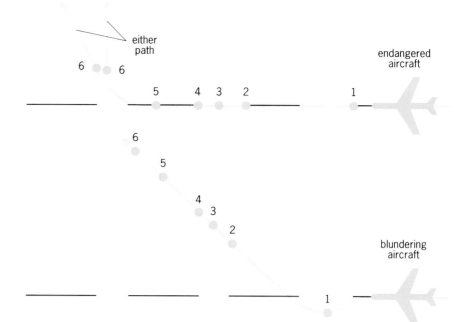

Fig. 2. Sequence of events during a blunder. The distance traveled by the aircraft during the time required for each of the steps limits the safe spacing between runways. Numbers indicate events: (1) blunder begins; (2) caution alarm sounds; (3) breakout decision is made; (4) command is received by endangered aircrew; (5) maneuver acceleration begins; (6) increasing separation is achieved.

available to the controller for delivery of the avoidance instruction.

Minimum runway spacing. The U.S. Federal Aviation Administration announced in 1991 its intention to reduce the minimum separation between parallel runways used for independent approaches to 3400 ft (1035 m), down from the 4300-ft (1310-m) minimum that has been in effect for several years. The reduced minimum would require a 2.4-s (or faster) update interval and 1-milliradian accuracy, with high-resolution, color, predictive displays. Tests are under way to determine if the separation could be reduced further.

The limiting factor appears to be the inability to distinguish between small navigational errors—which are normal to aircraft, particularly in strong crosswinds or turbulent air—and the beginning of a blunder requiring controller intervention. As the runways are moved closer together, there is less time for the controller to respond to a blunder. If the small errors that occur on almost every flight would prompt the controller to issue an avoidance maneuver to the adjacent aircraft, the system would be removing aircraft from the approach stream too frequently. Only small improvements in the 3400-ft (1035-m) separation are anticipated.

Alternative systems. A more promising approach may be a system that could make the pilot aware of the position of the other aircraft solely by electronic means, without assistance from a controller. Closely spaced approaches are already possible if one pilot can see the aircraft on the adjacent approach. Should that aircraft blunder, the pilot can see the threat and maneuver clear without ground instruction. If an

electronic system could replace that visual link, similar minima could be expected. One possibility is the traffic alert and collision avoidance system (TCAS), which will be used by all air-carrier aircraft in the United States by 1993. Although the system was designed for detecting and resolving potential collisions during flight between airports, its capabilities could be modified to show aircraft on a parallel approach. Another possibility is to use the radar monitor described above but to transmit the radar picture to the pilot.

For background information SEE AIR-TRAFFIC CONTROL; AIR TRANSPORTATION; AIRCRAFT COLLISION AVOIDANCE SYSTEM; MONOPULSE RADAR; RADAR in the McGraw-Hill Encyclopedia of Science & Technology.

Kenneth V. Byram

Bibliography. U.S. Department of Transportation, Federal Aviation Administration, Research and Development Service, *Precision Runway Monitor Demonstration Report*, DOT/FAA/RD-91/5, 1991.

Heat transfer

Boiling is a process in which a liquid phase is converted into a vapor phase. The energy for phase change is generally supplied by the surface on which boiling occurs. Boiling differs from evaporation at predetermined vapor/gas–liquid interfaces because it also involves creation of these interfaces at discrete sites on the heated surface. Boiling is an extremely efficient process for heat removal and is utilized in various energy-conversion and heat-exchange systems and in the cooling of high-energy-density components.

Boiling process. **Figure 1** shows qualitatively the dependence of the wall heat flux on the wall superheat. The wall superheat is defined as the difference between the wall temperature and the saturation temperature (boiling point at a given pressure) of a liquid. The plotted curve is for a horizontal surface underlying a pool of liquid at its saturation temperature.

Heterogeneous nucleation. As the heat input to the surface is increased, the first mode of heat transfer to appear in a gravitational field is natural convection (region I in Fig. 1). At a certain value of superheat (point A), vapor bubbles appear on the heater surface. The bubbles form on cavities or scratches on the surface because less surface energy is required there to form the vapor/gas–liquid interface. This type of nucleation is called heterogeneous nucleation. Homogeneous nucleation, on the other hand, is the process of creation of a vapor bubble nucleus in the host superheated liquid away from the bounding walls and in the absence of any foreign material. Homogeneous nucleation occurs at 80–90% of the critical temperature. This temperature is much higher than the wall temperature required for heterogeneous nucleation at low pressures.

For liquids that wet the surface, the size, D_c, of the cavities that nucleate at low wall heat fluxes, q, can be related to the wall superheat, ΔT, by Eq. (1), where σ

$$D_c \simeq \frac{4\sigma T_{\text{sat}}}{\rho_v h_{fg} \Delta T} \qquad (1)$$

is the surface tension, T_{sat} is the saturation temperature of the liquid at the system pressure, ρ_v is the vapor density, and h_{fg} is the latent heat of vaporization. After inception, a bubble continues to grow until forces causing it to detach from the surface exceed those pushing it against the wall. In a simplistic approach, bubble diameter at departure, D_d, is obtained by balancing

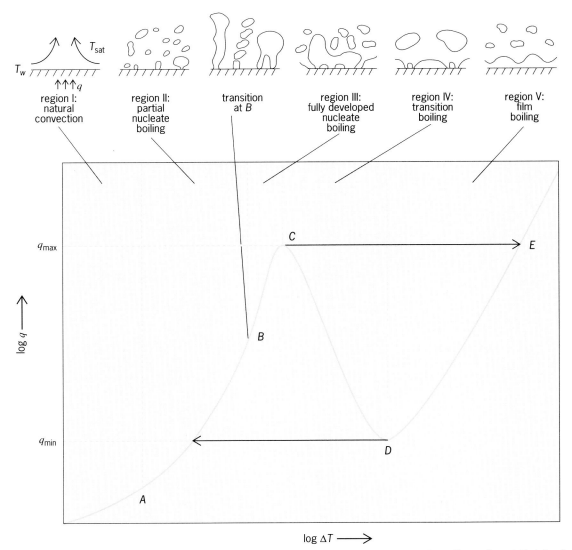

Fig. 1. Typical boiling curve, showing qualitatively the dependence of the wall heat flux q on the wall superheat ΔT, defined as the difference between the wall temperature T_w and the saturation temperature T_{sat} of the liquid. Schematic drawings show the boiling process in regions I–V. These regions and the transition points A–E are discussed in the text.

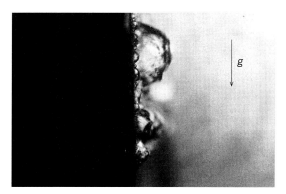

Fig. 2. Fully developed nucleate boiling on a vertical surface.

the surface tension and the buoyancy force; the result is Eq. (2), where ρ_l is the density of the liquid and

$$D_d \sim \sqrt{\frac{\sigma}{g(\rho_l - \rho_v)}} \qquad (2)$$

g is the acceleration of gravity. A complete analysis should also include liquid inertia, vapor inertia, and liquid drag. At present there is conflicting evidence with respect to the effect of wall superheat on bubble departure diameter.

Partial nucleate boiling. In liquids that wet the surface well, the onset of nucleation may be delayed. The sudden inception of a large number of cavities at a certain wall superheat causes a reduction in the surface temperature, while the heat flux remains constant. This behavior is not observed when the boiling curve is obtained by reducing the heat flux, and hysteresis results. After inception a dramatic increase in the slope of the boiling curve is observed. In partial nucleate boiling, corresponding to region II (curve AB) in Fig. 1, discrete bubbles are released from randomly located active sites on the heater surface. The density of active bubble sites and the frequency of bubble release increase with wall superheat. Several parameters, such as the procedure used to prepare a surface, the wettability of the surface, the interaction between neighboring sites, and oscillations in temperature, affect the dependence of nucleation-site density on wall superheat.

The mechanisms contributing to the total heat flux under pool boiling conditions are transient conduction at the area of influence of a bubble growing on a

nucleation site, evaporation (a fraction of which may be included in the transient conduction) at the vapor–liquid interface, enhanced natural convection on the region in the immediate vicinity of a growing bubble, and natural convection over the area that has no active nucleation sites and is totally free of the influence of the former three mechanisms. However, the importance of these mechanisms depends strongly on the magnitude of the wall superheat and other system variables, such as the orientation of the heater geometry with respect to gravitational acceleration, and the magnitude of gravitational accleration.

Fully developed nucleate boiling. The transition (point B in Fig. 1) from isolated bubbles to fully developed nucleate boiling (region III) occurs when bubbles at a given site begin to merge in the vertical direction. Vapor appears to leave the heater in the form of jets. The condition of formation of jets also approximately coincides with the merger of vapor bubbles at the neighboring sites. Thus, vapor structures appear like mushrooms with several stems (**Fig. 2**). Now most of the heat transfer occurs because of evaporation at the vapor–liquid interface of the vapor stems implanted in the thermal layer adjacent to the heater surface. At present no totally theoretical model exists for the prediction of nucleate boiling heat fluxes. However, the dependence of the heat flux on the wall superheat is given roughly by Eq. (3).

$$q \sim \Delta T^{3 \text{ or } 4} \qquad (3)$$

Maximum heat flux. The peak heat flux sets the upper limit on fully developed nucleate boiling or safe operation of equipment. For well-wetted surfaces the upper limit on the nucleate-boiling heat flux is set by the vapor removal rate. Hydrodynamic theory based on vapor jet instability away from the heater has been successful in predicting the maximum heat flux on well-wetted surfaces, given by Eq. (4). This equation

$$q_{max} = C_1 \rho_v h_{fg} \sqrt[4]{\frac{\sigma g(\rho_l - \rho_v)}{\rho_v^2}} \qquad (4)$$

is valid at low pressures, and the constant C_1 depends on the heater geometry. For a large horizontal plate it has a value of 0.15. However, the maximum heat flux on surfaces that are not well wetted is probably set by the upper limit of the evaporation rate near the surface. The maximum heat fluxes on partially wetted surfaces are lower than those on well-wetted surfaces.

Film boiling. After the occurrence of the maximum heat flux condition, most of the surface is very rapidly covered by vapor. This condition nearly insulates the surface, and the surface temperature rises very rapidly. When heat flux is controlled, the surface will very quickly pass through regions IV and V in Fig. 1 and stabilize at point E. If the temperature at E exceeds the melting temperatures of the heater material, the heater will fail (burn out). The curve ED (region V) represents stable film boiling, and the system can be made to follow this curve by reducing the heat flux. In stable film boiling, the surface is totally covered by vapor film and liquid does not contact the solid.

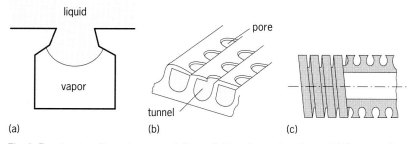

Fig. 3. Reentrant cavity and commercially available enhanced surfaces. (a) Cross section of reentrant-type cavity. (b) Thermoexcel-E surface. (c) Gewa-T surface.

On a horizontal surface the vapor release pattern is governed by the Rayleigh-Taylor instability. At low wall superheats where the radiation contribution is small, the film-boiling heat flux is related to the wall superheat by Eq. (5).

$$q \sim \Delta T^{5/4} \tag{5}$$

Minimum heat flux. The heat flux at which the stable film breaks down is called the minimum heat flux. Heater surface properties and geometrical parameters affect the minimum heat flux. From purely hydrodynamic considerations involving the instability of the interface, the minimum heat flux is predicted by Eq. (6).

$$q_{min} = C_2 \rho_v h_{fg} \sqrt[4]{\frac{\sigma g (\rho_l - \rho_v)}{(\rho_l + \rho_v)^2}} \tag{6}$$

For a horizontal surface the constant C_2 has a value of 0.09. However, depending on experimental conditions, large variations in the value of C_2 are observed.

Transition boiling. Region IV in Fig. 1 (the curve CD), falling between nucleate and film boiling, is called transition boiling. It is a mixed mode of boiling that has features of both nucleate and film boiling. Transition boiling is very unstable, since it is accompanied by a reduction in the heat flux with an increase in the wall superheat. As a result, it is extremely difficult to obtain steady-state data on transition boiling. In general, quenching or temperature-controlled systems are used to obtain transition-boiling data. Transition boiling is the least understood mode of boiling.

Boiling enhancement. The boiling process is influenced by several variables, such as surface finish, wettability, and cleanliness; heater geometry, size, material, and thickness; system pressure; gravitational acceleration; imposed time variation of heat flux or wall superheat; liquid subcooling (the difference between the boiling temperature and the liquid temperature); and flow velocity. Liquid subcooling and flow velocity have little effect on fully developed nucleate boiling. However, these parameters significantly enhance the maximum heat flux by increasing the upper limit of the vapor removal rate from the heated surface, and they enhance the transition-boiling, film-boiling, and minimum heat fluxes as well. At very high subcoolings, vapor bubbles may not leave the heater surface, and bubbles sitting on the surface act as miniature heat pipes. Under subcooled, forced-flow conditions, heat flux as high as 17 kW/cm^2 has been observed. This heat flux is about three times the heat flux at the surface of the Sun but is obtained at about one-fifth as large a temperature difference between the source and the sink.

Recent advances in enhancing nucleate-boiling heat flux have included the development of heater surfaces with high-density interconnected artificial cavities of the reentrant type. **Figure 3** shows a reentrant-type cavity and the structures of two of the commercially available surfaces. The cavities on these surfaces nucleate at very low superheats. The nucleate-boiling heat fluxes are enhanced not only by the high density of nucleation sites at low superheats but also by the

evaporation of a thin liquid film formed on the cavity walls. This film results from the liquid that is sucked into the cavity after a bubble leaves. The enhanced surfaces have led to an order-of-magnitude increase in already-high nucleate-boiling heat-transfer coefficients; however, enhancement is much less at high wall superheats or near the maximum heat flux.

For background information SEE BOILING POINT; HEAT TRANSFER; SURFACE TENSION in the McGraw-Hill Encyclopedia of Science & Technology.

V. K. Dhir

Bibliography. V. K. Dhir, Nucleate and transition boiling heat transfer under pool and external flow conditions, *Int. J. Heat Fluid Flow*, 12(4):290–314, 1991; D. M. Fontana, Simultaneous measurement of bubble growth rate and thermal flux from the heating wall to the boiling fluid near the nucleation site, *Int. J. Heat Mass Transfer*, 15:707–719, 1972; W. R. Gambill and N. D. Greene, Boiling burnout with water in vortex flow, *Chem. Eng. Prog.*, 54(10):68–76, 1958; R. L. Webb, The evolution of enhanced surface geometries for nucleate boiling, *Heat Transfer Eng.*, 2(3–4):46–69, 1981.

Hormone

Models of how hormones influence behavior have been generated by investigators examining the actions of three types of chemical compounds: steroids, amines, and peptides. A diversity of animals, including representative primates, rodents, birds, annelids, mollusks, insects, and crustaceans, have been used in these studies. In general, the models take the form illustrated in **Fig. 1**. In response to environmental cues, such as the availability of food, the approach of a potential mate, the approach of a predator, or even the time of day, information is transferred via the central nervous system (CNS) to tissues capable of releasing hormonal substances. These tissues can be specialized glandular structures located outside the nervous system, such as the adrenal glands or the gonads, or they can be

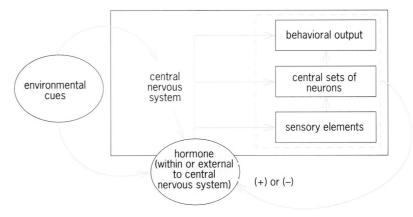

Fig. 1. Diagrammatic representation of hormonal sensitization of the response properties of sets of neurons. (*From E. A. Kravitz, Hormonal control of behavior: Amines and the biasing of behavioral output in lobsters, Science, 241:1775–1781, September 30, 1988; copyright 1988 by the American Association for the Advancement of Science***)**

neurons within the peripheral or central nervous systems. Upon release, the hormonal substances find their target sites within the central nervous system either via the general circulation or via a local circulation, act on those targets, and change the way the targets function for prolonged (minutes to hours to days) but finite periods of time. The targets are arrays of neurons that contain sensory elements, central sets of neurons that read out programs of behavior, and motor and other elements that, when activated, result in behavioral output. The result is to bias behavior in highly stereotypical directions in response to appropriate sensory stimuli for the duration of the period of hormone action.

Aggressive behavior in lobsters. Studies exploring the biological basis of aggressive behavior in lobsters illustrate such a model. When they are placed together in a confined space, lobsters invariably engage in fights called agonistic encounters. The fights are complex, involving a series of patterns of social interaction, which include threats and displays, limited contacts, and ultimately, violent contact in which animals try to damage each other (**Fig. 2**). In most cases, one to several such encounters, each of which can last from minutes to hours, occur over several days; at the end of the final encounter, a hierarchy is established, with one animal assuming a dominant role and the other a subordinate role. Thereafter, the pattern of behavior between the two animals becomes completely different, with the dominant animal continually advancing in an elevated posture and the subordinate

animal continually backing away in a lowered posture. The remarkable thing is that social encounter has completely changed the way the two animals interact, and that such changes in behavior must be accompanied by changes within their nervous systems.

In about 1981, an experiment opened the way toward understanding the changes in the lobster nervous system that might underlie such behavioral changes. After one or the other of two different amines were injected into lobsters and the animals were permitted to move about, the lobsters exhibited opposing postures resembling those seen in dominant and subordinate animals. The amines were serotonin, found in the human nervous system, and octopamine, a structural analog of norepinephrine, which is also found in the human nervous system. The dominant-looking posture (triggered by the serotonin injection) results from the contraction of all postural flexors in the lobster, while the subordinate-looking posture (triggered by the octopamine injection) results from contraction of the postural extensors. Further analysis using nerve and muscle preparations showed that amines triggered opposing postural changes by actions at several sites. Amines primed exoskeletal muscles to respond more vigorously to nerve stimulation and acted on the central nervous system to trigger the readout of opposing motor programs (octopamine caused the readout of a program for extension, and serotonin caused the readout of a program for flexion). Within the central nervous system of invertebrates have been found neurons that, when activated, cause the readout of similar motor

Fig. 2. Agonistic encounters, showing some of the stereotypical aspects of lobster "fights." Threats and displays are exhibited by (*a*) meral spread, which usually begins encounters, and (*b*) stretching and reaching. Limited contact is exhibited by (*c*) lock and push-and-pull. Establishment of a hierarchy is exhibited by (*d*) the dominant animal, on the right, and the subordinate. (*Photos by R. Huber; from E. A. Kravitz, Hormonal orchestration of behavior: Amines and the biasing of behavioral output in lobsters, in N. Elsner and H. Penzlin, Synapse-Transmission-Modulation, Georg Thieme Verlag, 1991*)

programs. Such neurons are called command neurons, and it was proposed that amines interact with command neuron circuitries in unspecified ways to activate flexor or extensor commands. These results offered an explanation of how injected amines generated opposing postures, and prompted a search within the lobster nervous system for individual amine-containing neurons concerned with postural regulation. The question was whether such neurons functioned in the same way as did injected amines, and, more importantly, whether changes could be seen in any aspect of the functioning of these cells as a consequence of social interaction between lobsters.

Amine-containing neurons. With a combination of microbiochemical assays and immunocytochemical methods, the lobster nervous system was surveyed for neurons containing either octopamine or serotonin and for the sites of their synthesis, storage, and release. The central nervous system of the lobster consists of a ventral nerve cord, which is a chain of ganglia that includes a brain (the supraesophageal ganglion) and 13 other ganglia. The ganglia are linked by connectives that carry information from one ganglion to the next, so that information can be conveyed from one end of the animal to the other. Each ganglion is more or less concerned with events happening in one segment of the animal, and the ganglia connect with the periphery via nerve trunks carrying information to and from the ventral nerve cord. The exoskeletal muscles are innervated by motor neurons whose cell bodies lie within the ventral nerve cord, and the sensory neurons are scattered in peripheral tissues or in specialized structures such as antennae, antennules, or eyes. There is no direct innervation of exoskeletal muscles by serotonergic or octopaminergic neurons. Instead, amines reach these muscles via the circulation, into which they are released from neurosecretory endings along peripheral nerve trunks associated with each ganglion in the thoracic part of the animal.

The origins of these neurosecretory endings prove to be specific amine-containing neurons located within the ventral nerve cord. By using antibodies to serotonin and octopamine, the locations of the cell bodies of the amine-containing neurons have been mapped, and the candidate neurons that supply peripheral neurosecretory endings have been identified. There are approximately 120 serotonin-containing and 85 octopamine-containing neurons in the central nervous system of the lobster. For serotonin, the focus of the studies has been on only two pairs of the 120 serotonin-containing cells. These two pairs are found in the 5th thoracic and the 1st abdominal ganglia. Like vertebrate amine-containing neurons, serotonin-containing neurons in the 5th thoracic and 1st abdominal ganglia have enormous arbors of endings with processes in every anterior thoracic ganglion, where they also send branches to the peripheral neurosecretory regions. These two pairs of cells appear to be the origin of circulating serotonin in lobsters and therefore are likely to be of prime importance in postural regulation. Physiological studies showed that the cells are spontaneously active and are under constant bombardment by inhibitory

synaptic contacts, which regulate the cells' rate of firing.

However, unlike injected or externally applied serotonin, serotonin-containing neurons upon activation do not direct the readout of motor programs from the central nervous system. Instead they apparently amplify the output of command neurons concerned with postural regulation (**Fig. 3**). Thus the serotonin-containing neurons are "gain-setters," enhancing or suppressing the output of circuitries with which they interact. When command neurons concerned with postural flexion are activated, the serotonin-containing cells increase their rate of firing; when command neurons concerned with extension are activated, the serotonin-containing neurons decrease their rate of firing. Serotonin-containing cells, therefore, are wired into circuitries in ways appropriate to involvement with postural flexion. As part of a flexion command circuit, the cells enhance the output of the circuit. However, if serotonin-containing cells are activated by microelectrodes and forced to fire, they also can enhance the output of extensor circuitries. Thus the serotonin-containing cells are general gain-setters in the lobster nervous system, and their specificity is determined by their pattern of wiring. Less is known about the lobster's octopamine neurons, which are under active investigation.

Gain-setters are logical points of change to produce long-term alterations in behavior. No circuitry need be modified to bias the behavior of an animal in one direction or another if the activity of the amplifier that acts on that circuitry can be enhanced or suppressed. This idea is being investigated by examining the function of serotonin-containing cells in lobsters with different hierarchical status. In this regard, some preliminary developmental studies are of interest. Lobsters grow by a process called molting, in which they regularly shed their external cuticle (exoskeleton) and replace it with a larger underlying one. Adult animals molt about once a year, while larval and juvenile animals

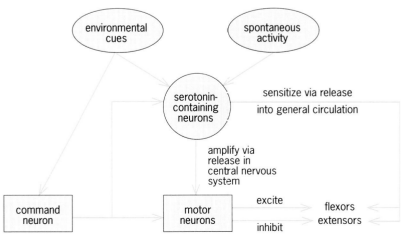

Fig. 3. Schematic representation of serotonin circuitry involving a flexor command stimulation. (*From E. A. Kravitz, Hormonal orchestration of behavior: Amines and the biasing of behavioral output in lobsters, in N. Elsner and H. Penzlin, Synapse-Transmission-Modulation, Georg Thieme Verlag, 1991*)

molt more frequently. Immediately before a molt, lobsters become highly aggressive, presumably to clear their territory of other, possibly dangerous animals. After a molt, when the exoskeleton is soft for several days, lobsters are highly submissive and hide. In certain serotonin-containing cells, amine levels seem to cycle during the molt stage, with an elevated peak immediately before a molt and a drop shortly after a molt. While such changes could be coincidental with the behavioral changes, their temporal correlation is intriguing.

Prospects. In the past, the use of invertebrate models has led to major contributions to knowledge of fundamental cellular processes, such as the ionic mechanisms underlying the generation of action potentials, or the role of calcium in synaptic function, or the identity of the chemicals used for signaling in synapses. The large size of invertebrate neurons and the ability to find precisely the same cells in different animals have been the keys to the many discoveries. Now, higher-level processes, such as studies of the development of the nervous system, and studies of behavior at levels ranging from whole animals to individual neurons, have yielded to analysis with invertebrate models and have generated exciting results. Such studies would be extraordinarily difficult or impractical at the present time with vertebrate systems.

Aggressive interactions between lobsters are interesting to examine for several reasons. First, there is the inherent interest in the fundamental mechanisms underlying a behavioral difference between two animals that is triggered by a social interaction. In lobsters, that study may be extended to the level of the single nerve cells that may be changed by the social interaction. Second, while there is little information on how violent behaviors are triggered in people, model systems such as the lobster should at least offer ideas with which to work. Finally, good systems with which to learn how amines function should be of great human import because many drugs that influence mood and emotional state, such as cocaine, amphetamines, and the major depressants and antidepressants, appear to act by blocking or potentiating the actions of amines. Cocaine, for example, blocks the uptake of catecholamines (whose analog in lobsters is octopamine) by nervous tissues, so that the actions of the amines are greatly potentiated. The lobster model may yield valuable information on all these important topics.

For background information SEE EFFECTOR SYSTEMS; HORMONE in the McGraw-Hill Encyclopedia of Science & Technology.

Edward A. Kravitz

Bibliography. A. P. Arnold, Logical levels of steroid hormone action in the control of vertebrate behavior, *Amer. Zool.*, 21:233–242, 1981; J. P. Changeux and M. Konishi, *The Neural and Molecular Basis of Learning*, 1987; E. A. Kravitz, Hormonal control of behavior: Amines and the biasing of behavioral output in lobsters, *Science*, 241:1775–1781, 1988; J. W. Truman, Hormonal control of invertebrate behavior, *Horm. Behav.*, 10:214–234, 1978.

Human genetics

The field of human genetics is advancing rapidly. In particular, the techniques and tools of molecular biology have allowed faster and more accurate methods of genetic disease diagnosis and a better understanding of inherited disease, and offer the potential for new therapeutic options.

The first stage in unraveling the mechanism of an inherited disorder is cloning (or isolation) of the appropriate gene. Then disease-causing mutations (or alterations) within the gene can be detected, and diagnostic procedures based on deoxyribonucleic acid (DNA) can be developed. Isolation of a disease gene permits study of the structure and function of its normal protein product, and leads to further understanding of the disease that results from a missing or altered protein. Knowledge of the gene product and disease mechanism may lead to better medical strategies or drug development. Recombinant proteins (that is, those synthesized from cloned genes) can be used as pharmaceutical agents, usually with decreased cost and risk of contamination compared with proteins isolated from animal or human tissue. An alternative to providing the protein missing in a genetic disease is the delivery of a functional gene to the appropriate tissue so that the genetic defect itself is corrected; that is known as gene therapy. The Human Genome Project is a coordinated worldwide effort to define the genetic makeup of humans and other organisms. Information and technology from this ongoing project have already led to rapid advances in medical genetics.

Cystic fibrosis. Cystic fibrosis occurs in approximately 1 in 2000 Caucasians and affects primarily the lungs and pancreas. In 1989 the cystic fibrosis gene was cloned and the most common mutation identified (deletion of a phenylalanine residue at amino acid position 508). Since that time, over 125 different cystic fibrosis mutations have been characterized, and rapid DNA diagnostic tests can detect 85% of these mutations.

The protein of the cystic fibrosis gene has been named CFTR (cystic fibrosis transmembrane conductance regulator) because the flow of the chloride across the cell membrane is abnormal in cells from cystic fibrosis patients. CFTR is a large protein that spans the cell membrane and forms a channel through which chloride ions can pass; alterations in this protein can disrupt its function and therefore result in cystic fibrosis. Comparison of CFTR with similar proteins has assisted in defining different functional domains within the protein. The fact that many mutations within the cystic fibrosis gene have been described is a disadvantage for diagnostic purposes because a simple test to detect all mutations is not possible. However, the variety of observed alterations and the clinical heterogeneity of the disorder have permitted some speculation as to the relative importance of various domains of the protein. For instance, that a mutation near one end of the protein (position 1255 of 1480) leads to pancreatic insufficiency but only mild pulmonary disease suggests that this region of the protein has a more important function in the pancreas than in the lung.

Neurofibromatosis. Neurofibromatosis type 1 (NF1, also known as von Recklinghausen disease) affects an estimated 1 in 3500 of the general population and is the autosomal dominant disorder frequently (although erroneously) associated with Joseph Merrick, the "Elephant Man." Symptoms of the disease are variable in nature and severity, even within the same family, and include café-au-lait spots, neurofibromas, optic and bone lesions, and learning disabilities.

In 1987 the NF1 gene was mapped to the long arm of chromosome 17. In 1989 chromosomal translocations involving chromosome 17 in two unrelated NF1 patients were shown to be less than 600 kilobases apart; this evidence localized the gene more precisely and led to intensive efforts to search for the NF1 gene in this region. In order to demonstrate that a candidate was in fact the NF1 gene, some association of the proposed NF1 gene with the disease had to be demonstrated. Several genes expressed in nervous tissue were quickly isolated from the appropriate region, but all were smaller than expected (because the gene was expected to span at least the distance between the two translocation breakpoints), and none was altered in NF1 patients. Finally, in 1990, groups in Michigan and Utah independently discovered the NF1 gene with several convincing lines of evidence: in hybrid cell lines containing each of the translocation breakpoints, it was shown that both the gene and the ribonucleic acid (RNA) produced from it were disrupted; one NF1 patient had within the candidate gene an insertion that had not been inherited from either parent; another patient had a deletion within the gene; and several other patients had small DNA variations that were not found in control individuals.

The NF1 gene is large (300 kilobases) and has the interesting feature of containing several entire genes within its own noncoding regions. The protein product of the gene appears to function as a negative regulator of normal cell growth, that is, a tumor suppressor gene.

Fragile X syndrome. Fragile X syndrome is the most common form of inherited mental retardation in humans, with a frequency estimated to be approximately 1 in 1250 males. The inheritance of this disorder is basically X-linked dominant, although 30% of carrier females display some symptoms, and 20% of males carrying the chromosomal abnormality are apparently normal. The association of this syndrome with a visible constriction or breakpoint at the end of the long arm of the X-chromosome was observed in 1977, but it was not until 1991 that the gene at the fragile X site was identified. This so-called FMR-1 gene is unusual in that there is a stretch of about 30 repeats of the three nucleotides cytosine, guanine, guanine (CGG). It appears that the CGG repeat region in the DNA is actually the fragile site itself, since individuals with fragile X syndrome have hundreds of copies of the repeat and no detectable FMR-1 protein. The expanded repeat region presumably renders that part of the DNA particularly unstable and prone to breakage under certain conditions, and results in the reduction or elimination of FMR-1 protein production.

Cancer genetics. The controlled growth of a cell within an organism is a finely balanced process. Cell proliferation is stimulated by the products of proto-oncogenes and is under the negative control of tumor suppressor genes that prevent overgrowth. Several gene products that act as negative regulators of growth in the normal situation have now been identified; when these genes are disrupted, uncontrollable cell growth (cancer) results. For example, interruption of the retinoblastoma gene leads not only to ocular retinoblastomas but also to bone, breast, and lung cancers; mutations within the NF1 gene lead to a variety of tumors; inactivation of the Wilms tumor gene is important in the pathogenesis of this childhood kidney tumor; and the p53 gene encodes a protein essential to controlled cellular growth in numerous tissues. It has been proposed that colon cancer, a common malignancy, evolves in multiple stages from the normal phenotype through adenoma and carcinoma to metastasis, each of these stages being associated with the loss of a different tumor suppressor gene. Because tumor suppressor genes and their products offer protection against cancer, it is hoped that better understanding of them will eventually lead to improved methods of cancer prevention and therapy.

Recombinant drug products. An obvious treatment of deficiency disorders is replacement of the missing protein. Unfortunately, chemical synthesis or extraction from human or animal tissue of the protein can be costly, impractical, or hazardous. Recombinant proteins can be generated from cloned genes carried in cultured bacteria, yeast, or mammalian cells; two examples of recombinant proteins currently marketed or under trial are combinant coagulation factors and recombinant erythropoietin.

In receiving coagulation factors such as factors VIII and IX extracted from human plasma, hemophiliacs have been exposed to various blood-borne viruses, in particular human immunodeficiency virus type 1 (HIV-1) and hepatitis C virus. As many as 75–90% of people with severe factor VIII deficiency have serological evidence of infection with one of these viruses; HIV-1 infection is now the leading cause of death in hemophilia patients, and chronic infection by hepatitis C virus is the most common cause of liver dysfunction in these patients. Recombinant coagulation factors, such as factor VIII, provide an alternative source of therapeutic material without the hazards of viral contamination.

Erythropoietin, generated in the kidney, is the primary regulator of red blood cell synthesis; thus anemia is one of the main complications of chronic renal failure. It is not practical to consider the isolation of erythropoietin from tissue; however, isolation of the erythropoietin gene has led to the marketing of recombinant erythropoietin. The drug has been shown to increase hemoglobin levels and reduce the need for blood transfusions in patients with end-stage renal disease, and further studies are being conducted. Other possible uses of recombinant erythropoietin include treating chronic anemias and increasing hemoglobin levels before autologous blood donations.

Gene therapy. Although gene therapy that alters the entire genetic makeup of an organism, so that the genetic change is passed on to its offspring (germ-line gene therapy) is possible, this is considered controversial in humans. Somatic-cell gene therapy, on the other hand, is already undergoing clinical trials because this is the treatment of the affected tissue alone and is analogous to an organ transplant. Before gene therapy can be considered, several requirements must be met: the disorder should result from disruption of a single gene, and that gene must have been cloned; sufficient knowledge of the disease should indicate that supplying the missing gene will fully or partially correct the phenotype; there should be an appropriate target cell and delivery system for introduction of the therapeutic gene; adequate data from cultured cell and animal studies should suggest that the vector, gene construct, and target cell are suitable; and the risk–benefit ratio should weigh in favor of experimental treatment, usually because little or no alternative treatment is available. Some diseases considered as candidates for gene therapy are adenosine deaminase deficiency (which results in a severe immune deficiency), cystic fibrosis, and Duchenne muscular dystrophy.

The first clinical trial of gene therapy for adenosine deaminase deficiency began in 1990 with the treatment of a 4-year-old girl. A functional gene carried in a retroviral vector was delivered to cultured T cells isolated from the blood of the patient, and the altered T cells were then returned to the girl in several doses. Initial results have been encouraging, and other children have begun the experimental therapy, but it is still too early to tell if this will be an effective general approach to the treatment of adenosine deaminase deficiency.

Cystic fibrosis and Duchenne muscular dystrophy are more common single-gene disorders than adenosine deaminase deficiency, but each presents particular problems in consideration of gene therapy. Viral delivery of the cystic fibrosis gene has successfully corrected the defect in cultured cells, but a major problem for therapy in individuals is delivery of the gene to the lining of the lung, the tissue which determines the mortality of the disease. The Duchenne muscular dystrophy cDNA is too large to be accommodated by the retroviral vectors used in other gene transfer protocols. Although direct injection of the Duchenne muscular dystrophy gene into muscle is feasible, this presents the difficulty of treating a significant fraction of the skeletal muscle mass.

Human genome project. The human genome contains 50,000–100,000 genes within approximately 3 billion base pairs of DNA. The goal of the Human Genome Project is to generate genetic and physical maps of the entire sequence of the human genome and of other organisms. A genetic map is an ordered array of inherited markers along a chromosome, and a physical map measures distance between DNA markers; both kinds of map are useful to geneticists in characterizing the genome and locating new genes. The ultimate physical map is the nucleotide sequence, and one of the objectives of the Human Genome Project is the improvement of

sequencing technology so that the entire human nucleotide sequence can be determined at a reasonable cost. Achievement of project goals will require new improved genetic technologies and the development and use of large computer databases.

For background information *SEE GENE ACTION; GENETIC ENGINEERING; GENETIC MAPPING; HUMAN GENETICS; ONCOLOGY; SEX-LINKED INHERITANCE* in the McGraw-Hill Encyclopedia of Science & Technology.

C. Thomas Caskey; Belinda J. F. Rossiter

Bibliography. M. Pieretti et al., Absence of expression of the FMR-1 gene in fragile X syndrome, *Cell*, 66:1–20, 1991; I. M. Verma, Gene therapy, *Sci. Amer.*, 263(5):68–84, 1990, J. D. Watson, The Human Genome Project: Past, present and future, *Science*, 248:44–49, 1990; G. Xu et al., The neurofibromatosis type 1 gene encodes a protein related to GAP, *Cell*, 62:599–608, 1990.

Hydrodynamics

Fluid dynamics and quantitative predictions of fluid motions have been a continuing scientific endeavor for centuries. The Navier-Stokes equations state the law of conservation of momentum. These equations, together with the continuity equation and the energy equation for the flow of a gas in thermodynamic equilibrium, are the most complete set of equations for fluid flow. Less complete versions of the Navier-Stokes equations have been developed for different physical situations: the Euler equations for zero viscosity, the potential (Laplace) equations for constant density with zero viscosity and vorticity, thin-layer Navier-Stokes equations for zero streamwise viscous terms, and boundary-layer equations for a zero normal-pressure gradient. In most potential flows, the partial differential equations can be formulated as boundary integral equations. Numerical hydrodynamics, an amalgamation of fluid dynamics, applied mathematics, and numerical analysis, develops numerical algorithms to solve equations of incompressible fluid flows about or within bodies, regardless of the governing equations or solution method used.

RANS analysis. The Navier-Stokes equations are nonlinear partial differential equations that are extremely difficult to solve. Only subsets of flows about complex geometries with simple physics or about simple geometries with more complex physics have been solved by computers. An indication of the relative magnitude of the inertial and viscous forces is given by the Reynolds number. High Reynolds numbers (of the order of 10^6–10^9) occur in most hydrodynamic applications. The flows over most ships and airplanes in this range of Reynolds number are turbulent. If a solution were attempted to account for all of the small details (eddies) in a turbulent flow at high Reynolds number, the world's entire computation capacity now, and in the foreseeable future, would not be able to complete the solution. Therefore, all useful Navier-Stokes implementations use time averaging of flow variables and modeling of turbulent stresses as a simplification. Programs that embody such simplifications are usually referred to as Reynolds-averaged Navier-Stokes (RANS) formulations. Turbulence modeling is still in its infancy and will continue to be a major limitation in the application of numerical hydrodynamics.

Use of potential theory. This is not to say, however, that greatly simplified analytical versions are not important. Even before digital computers were available, potential-flow and boundary-layer solutions were a valuable adjunct to model experiments. Practical predictions of ship and airplane performance, however, relied mostly on model experimental results in the design of a wide range of vehicles such as ships, submarines, missiles, subsonic aircraft, and helicopters. For all practical purposes, access to programmable digital computers in the early 1960s marked the beginning of numerical hydrodynamic applications.

The programmable digital computer made possible the application of potential theory to more complex three-dimensional configurations. Many aircraft and ship research organizations immediately developed practical computer codes based on potential theory for use in the design process. The surfaces of complex bodies, airplane fuselages, wings, control surfaces, and engine nacelles, and of such marine structures as ship hulls, rudders, and propellers were divided into several hundred to several thousand small adjoined areas or panels on which potential singularities (sources, sinks, doublets, or vortices) were distributed. This arrangement enabled the designer to predict inviscid velocity fields around a body. Pressure distributions, lift forces, and moments on a body were then predicted with a precision not previously achievable. Many new applications such as the free-surface problems associated with a ship moving in an air–water interface are now being addressed with considerable success.

Panel codes are now the principal computational fluid dynamics tool used for ship, propeller, aircraft, and missile design. However, it must be recognized that the assumption of inviscid flow precludes the prediction of drag, flow separation, and other important viscous effects. Many shortcomings have been addressed, to a limited extent, by including the effects of boundary-layer development and small-flow separation interactively in the panel codes. The lift and drag of an entire airplane and some flow details of a submersible are predicted by this viscous-inviscid interactive approach in the current design process. *SEE DRAG REDUCTION.*

Application to yacht racing. Numerical hydrodynamics has increasingly been used in the comparative evaluation of the America's Cup 12-m (39-ft) yacht design and design changes. America's Cup yachts are unusual in that 0.1-knot (0.05-m/s) speed improvements can mean the difference between a winner and early elimination. A keel is often used to provide lateral force to counter the lateral force of the sail. A small horizontal surface mounted at the after portion of the keel tip (similar to a winglet) is used to decrease induced drag associated with the lateral force

of the keel. A panel code was used to increase the maximum usable sail lift by designing keel-mounted winglets with reduced induced drag and improved lateral stability. Slender-body analysis was used to calculate the differences in wave-making resistance and the added resistance from waves as the hull geometry was changed. While available panel codes may not predict the induced resistance of the winged keels correctly, they can predict resistance changes related to geometry changes. This simple numerical aproach, employed with game theory, probability theory, and real-time simulations, played a major role in the design of the 1987 America's Cup winner. Every serious designer now uses a similar approach.

Analysis of viscous flows. When viscous forces dominate, such as when flow separation and vortex generation occur, the designer's only alternatives are to use model experiments or the Navier-Stokes equations in which viscosity effects are retained. The long-term future of numerical hydrodynamics therefore lies in the development of efficient algorithms that take advantage of the processing speed, memory storage, and parallel processing capability of supercomputers to solve the incompressible Navier-Stokes equations. Analysis of the viscous flow field around a complex body will provide solutions to many of the most vexing fluid-flow problems. A viscous Navier-Stokes analysis can provide accurate predictions not only of the overall forces but also of the flow physics such as vorticity strengths and tracks, boundary-layer details, and flow separations. *SEE SUPERCOMPUTER.*

Numerical hydrodynamics research encompassing several decades has only recently provided several computer programs that are efficient and robust enough for practical applications. Efficient programs can provide a useful solution to a problem with a few million grid cells in a few hours on a supercomputer. A robust program is one that converges rapidly to a solution in the iterative process used almost universally for Navier-Stokes analysis. Performance achievements are the result of contributions made by thousands of researchers, each of whom provided improvements in some aspect of the process.

A RANS analysis is initiated by first creating a grid cell structure that encompasses the entire flow field from the body surface to the outermost boundary of the analysis area or volume. The number and distribution of the grid cells and the precision with which body geometry is replicated fix an upper limit on the solu-

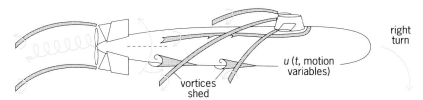

Fig. 1. Maneuvering submerged body, which is shedding vortices. Velocity u depends on time t and motion variables. This example involves a fluid flow whose accurate analysis exceeds the capacities of current supercomputers.

tion accuracy and resolution achievable. A reasonably accurate analysis of an entire submersible, such as that shown in **Fig. 1**, would require about 10^7 grid cells, which exceeds the physical memory capacities of current supercomputers. Most current analyses have been limited by the size of available computer memory to a few million grid cells. **Figure 2** depicts the results of a RANS analysis of the flow field created at the junction of a simple wing with a flat plate. For this junction, 1.8×10^6 grid cells were used, and remarkably detailed flow structures were computed. An efficient and robust grid generation process is a prerequisite to the use of RANS computation as a design evaluation tool. The effort required to create a grid structure for a RANS computation usually ranges from worker-weeks for a relatively simple body to worker-months for a more complex one. Some complicated grids, such as those for curved rotating blades, have not been satisfactorily created.

Validation. RANS applications are further restricted by absence of accuracy assessments. Wind tunnels and model towing tanks, the traditional sources of validation data, are not well suited for flow-field assessments since they provide data that are global in nature, that is, total body forces and pressures and wall shear stresses on the hull surface, and few pro-

vide the flow-field details needed. Furthermore, model test facilities have interference from walls and mountings (strings, struts, and wires), which can severely degrade measurement accuracy, especially at nonzero angles of attack. This difficulty is being rapidly rectified by new flow-field measurement systems that are now producing quantitative results. Measurements are now made by using particle displacement velocimetry. This method employs a pulsed laser to illuminate a thin two-dimensional sheet of a flow field seeded with microscopic fluorescent particles, and uses a camera to record the displacements of particles between pulses emitted at known intervals (say, a few milliseconds). A number of particle displacements recorded on high-resolution film can be tracked and converted into two-dimensional components of fluid-field velocities. Three-dimensional fields can be reconstructed if a number of laser illuminators displaced in space and time are used. Supercomputers and digital image recognition techniques are used for data analysis. Particle displacement velocimetry used with a free-running, self-propelled model that avoids wall and mounting effects is providing flow-field measurement data with sufficient accuracy for code validation. **Figure 3** is an example of a measurement by particle displacement velocimetry of a vortex structure generated by a fin

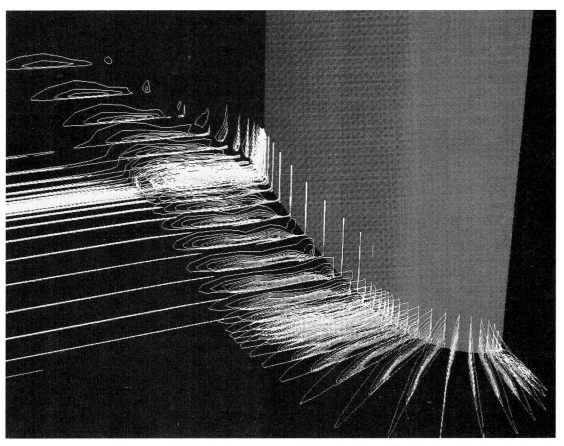

Fig. 2. Results of Reynolds-averaged Navier-Stokes (RANS) analysis, using 1.8 × 10⁶ grid cells, of the flow field at the junction of an airfoil with a flat plate. Contour lines represent constant values of the longitudinal vorticity (the component of vorticity parallel to the mean flow direction). Successive contours are separated by an increment of U_∞ / C in the longitudinal vorticity, where U_∞ is the ambient velocity and C is the chord length. (*Computed by C. H. Sung and M. J. Griffin*)

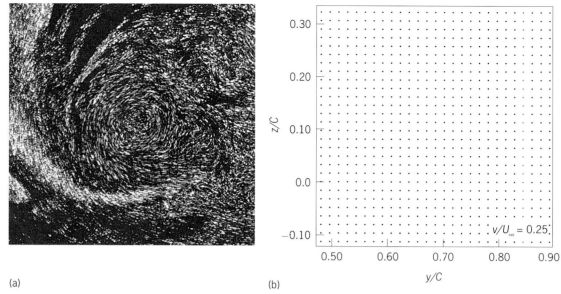

(a)

(b)

Fig. 3. Quantitative flow visualization of a tip vortex shed from a fin attached to an axisymmetric body. (*a*) Double-exposed image of the tip vortex. (*b*) Velocity-vector map of 60 mm × 60 mm (2.4 in. × 2.4 in.) section of this image. The length is indicated for a velocity vector that corresponds to a velocity whose magnitude *v* is such that $v/U_\infty = 0.25$, where U_∞ is the ambient velocity. The horizontal and vertical coordinates, *y* and *z*, are scaled to the chord length C. The origin of coordinates, $y/C = z/C = 0.0$, is chosen at the tip of the fin.

on an axisymmetric body. Hundreds to thousands of individual velocity vectors can be extracted from a photograph of each laser-illuminated sheet. Multiple light sheets are used to map unsteady vortex flows.

Prospects. With continued improvements in numerical analysis and grid generation methods and the availability of accurate particle displacement velocimetry and other types of flow-field measurements, a validated RANS code may become the principal evaluation tool in the hydrodynamic design process. As these improvements occur, the viscous free-surface problems associated with a ship moving in an air–water interface will be one of the many important new application areas.

For background information SEE FLUID FLOW; FLUID-FLOW PRINCIPLES; LAPLACE'S IRROTATIONAL MOTION; NAVIER-STOKES EQUATIONS; NUMERICAL ANALYSIS; REYNOLDS NUMBER; SUPERCOMPUTER; TURBULENT FLOW in the McGraw-Hill Encyclopedia of Science & Technology.

Thomas T. Huang; William E. Smith

Bibliography. American Institute of Aeronautics and Astronautics, *9th Applied Aerodynamics Conference*, AIAA 91-3307, September 1991; American Institute of Aeronautics and Astronautics, *29th Aerospace Science Meeting*, AIAA 91-0022, January 1991; J. S. Letcher, Jr., et al., Stars and stripes, *Sci. Amer.*, 257(2):34–40, August 1987; *19th International Towing Tank Conference*, September 1990.

Immunoglobulin

Antibodies are immunoglobulins (Ig's) consisting of heavy chains and light chains, each of which is composed of a variable and a constant region. The variable regions of both chains determine the specificity of the antibody, and the constant regions of the heavy and light chains define the class and type, respectively, of the antibody molecule. There are two types of antibody light chains: kappa and lambda; the number of classes varies with the organism. Human beings have nine classes of antibodies: IgM, IgD, IgG1, IgG2, IgG3, IgG4, IgE, IgA1, and IgA2. These classes contain the heavy chains μ, δ, $\gamma1$, $\gamma2$, $\gamma3$, $\gamma4$, ϵ, $\alpha1$, and $\alpha2$, respectively. However, IgG1–IgG4, and IgA1 and IgA2 are often considered to be subclasses of the traditional classes IgG and IgA, respectively.

Function of the various classes. The function of antibodies of all classes is to bind antigen. The class determines where the antibody is available for binding as well as the physiological consequences of binding. Most resting small B lymphocytes have IgM and IgD on their surface. The binding of antigen triggers a chain of events that results in these B lymphocytes differentiating into plasma cells that secrete a pentameric form of IgM. IgD is rarely secreted, and its function is unknown. After binding antigen, IgM and some IgG subclasses fix complement, which destroys bacteria. IgM is too large to cross the placenta. However, IgG does so in humans and thereby confers the mother's immunity onto the fetus, so that, in case of Rh incompatibility, erythroblastosis fetalis may result. All immunoglobulin classes are present in serum, but IgA is dominant in secreted fluids (tears, saliva, milk).

IgE is of great clinical interest. It is commonly believed that IgE is synthesized in order to fight larger parasites, such as worms; however, it is not very effective. IgE is responsible for immediate-type hypersensitivity; perhaps 10% of all people suffer from an allergy of this type. If an antigen cross-links at least two IgE molecules bound to the surface of a mast cell,

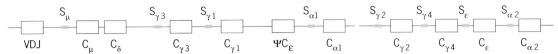

Fig. 1. Organization of the human immunoglobulin heavy-chain locus in a newly formed B cell. The VDJ gene segment encodes the variable region. The constant regions of the various immunoglobulin heavy chains are encoded by C_μ, C_δ, $C_{\gamma3}$, $C_{\gamma1}$, $C_{\alpha1}$, $C_{\gamma2}$, $C_{\gamma4}$, C_ϵ, and $C_{\alpha2}$; ΨC_ϵ is a "pseudogene" segment that cannot be expressed. Switch regions (S) are located 5′ to the constant gene segments.

degranulation is triggered, and histamine and other mediators of allergic reactions are released.

Heavy-chain class switch. All B lymphocytes start off with IgM on the surface, and the early immune response is dominated by IgM. The later immune response is dominated by IgG, IgA, and (rarely) IgE. This so-called heavy-chain class switch occurs within a clone, that is, any of the cells that are the progeny of a single antibody-producing lymphocyte. The antigen specificity of all the clone cells remains the same; thus the variable region remains the same but the constant region changes.

Genetic basis for the switch. In the germline of vertebrates, there are no functional genes encoding the polypeptide chains of immunoglobulins; there are only gene segments. During B-cell differentiation, selected segments are joined to yield functional heavy- and light-chain genes. Two types of deoxyribonucleic

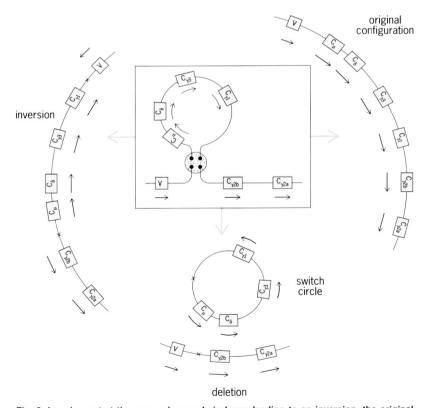

Fig. 2. Looping-out at the mouse heavy-chain locus leading to an inversion, the original configuration, or deletion and formation of a switch circle. The stippled circle represents the putative enzyme complex holding together the four free DNA ends, which can be ligated in three different ways to create the chromosomal configurations shown on the sides and below. X marks the recombination breakpoint. Arrows under the gene segments define the direction of transcription. (*From U. v. Schwedler, H. M. Jäck, and M. Wabl, Beswitched: The looping out model for immunoglobulin class switching, New Biol., 2(8):657–662,1990*)

acid (DNA) rearrangements, resulting in deletions or inversions, join gene segments that may be far away from each other in the germline. The first type, used for the initial construction of the immunoglobulin heavy- and light-chain genes, is quite precise; it is site-specific and governed by a recombinase (a putative enzyme complex that performs DNA rearrangement) that recognizes known signal sequences flanking the exon segments to be joined. Exons encoding the variable part of the immunoglobulin chains, and thus the antibody specificity, are generated by such a process.

The second type of DNA rearrangement, used to reconstruct the heavy-chain gene during further development of the B cell, is rather imprecise; the site of recombination (breakpoint) varies widely within an intron. Since the intron will be deleted by ribonucleic acid (RNA) splicing, this imprecision does not impair the outcome of the process. As a result of this reconstruction, the gene segment encoding a given constant region is replaced by another. The constant region C_μ is replaced by $C_{\gamma3}$, $C_{\gamma1}$, $C_{\alpha1}$, $C_{\gamma2}$, $C_{\gamma4}$, C_ϵ, or $C_{\alpha2}$, which are closely linked to C_μ in that order. Thus, this rearrangement is the genetic basis of the immunoglobulin heavy-chain class switch.

The switch rearrangement generally results in deletion of the DNA sequences between a site within the intron 5′ to C_μ and a site within the intron 5′ to the particular constant-region gene segment that is to be expressed. The breakpoints usually fall within regions that contain repetitive sequences, the switch regions, but sometimes they do not. The organization of the constant-region gene segments at the heavy-chain locus is shown in **Fig. 1**. Except for C_δ, each such segment is preceded by a switch region of several kilobases.

Looping-out and deletion of DNA. Three mechanisms were proposed to account for the DNA deletion resulting in the immunoglobulin class switch: (1) unequal recombination between homologs; (2) unequal recombination between sister chromatids; and (3) looping-out and deletion. These are the simplest mechanisms that can account for any deletion of DNA in diploid cells. Although a priori all three mechanisms must result in switch deletions at some frequency, the genomic configurations predicted by recombination between homologs or sister chromatids are too infrequent to be demonstrated. On the other hand, the products predicted to be produced by looping-out—inversions and circular DNA—have been isolated.

The way that a deletion is produced by looping-out is shown in **Fig. 2**. The heavy-chain locus of the mouse is given, for this organism was used to study the genetic mechanism of the switch; the heavy-chain loci of human beings and mice do not differ substantially. Looping-out and deletion are also involved in the rearrangement of the variable segments at the initial construction of the T-cell receptor and immunoglobulin genes. After looping-out and cutting, the four free DNA ends thus created can be re-ligated in three different ways to produce the original configuration (no switch), an inversion of the looped-out sequences (loss of heavy-chain expression), or deletion of the looped-out sequences from the chromosome (switch). Although they are of no apparent use to the organism, many cells with an inversion between two switch regions are found in cell cultures.

The end of the excised DNA is ligated to form a circle containing C_μ, including the $3'$ part of S_μ, and all of the constant regions between C_μ and the constant region of the heavy chain to which the cell switches. The breakpoint is located somewhere between the $3'$ part of S_μ and the $5'$ part of the switch region of the heavy chain; these two switch regions, characteristic for the switch circle, are joined in an order that is reciprocal to the configuration remaining on the chromosome. Switch circles with precisely these characteristics are present in cells that have undergone immunoglobulin class switching. This finding is good evidence that the switch rearrangement involves looping-out.

The origins of replication on animal chromosomes are estimated to be 50 to 330 kilobases (kb) apart, so it would not be surprising if there were an origin somewhere within the approximately 200 kb of DNA encompassing the constant-region gene segments of the immunoglobulin heavy chain; thus, switch circles might well contain an origin of replication that also functions extrachromosomally. But so far, there is no evidence that switch circles replicate, at least not at the same rate as the chromosomes. In non-B cells, the immunoglobulin heavy-chain locus functions as a single replicon with its origin downstream of C_α; however, additional origins of replication might be activated in B cells.

The intervening sequences between the constant-region exons of the heavy chain have been sequenced, but it is not known whether any additional genes occur in these regions. Deletion of these putative genes after switching might give the cell a growth advantage or put the cell into the category of a memory cell.

Directed class switching. The switch regions contain multiple repeats of both short sequences–pentamers and decamers–and long sequences. Essentially, the switch-region sequences are sufficiently different from one another that the eye or the computer can readily identify a particular switch region from its sequence, and it is likely that enzymes would be at least as facile in making such distinctions. Thus, switch recombinases may prefer one switch region over another and thereby influence the class to which the cell will switch. Indeed, the following observations can be explained by directed class switching: many immunoglobulin-secreting cells have both heavy-chain

alleles switched to the same class; pre-B cells of the mouse transformed by virus switch mainly to the $\gamma 2b$ heavy chain; and B cells stimulated by mitogen or by interleukin 4 express certain isotypes but not others. It is thought that this directed class switch is due to induction or activation of a specific switch recombinase—or of a specific factor associated with it—that directs the action of the switch recombinase to a particular isotype.

For background information SEE ANTIBODY; ANTIGEN; EXON; HYPERSENSITIVITY; IMMUNOGLOBULIN; INTRON in the McGraw-Hill Encyclopedia of Science & Technology.

Matthias Wabl

Bibliography. F. D. Finkelman et al., Lymphokine control of in vivo immunoglobulin isotype selection, *Annu. Rev. Immunol.*, 8:303–333, 1990; T. Honjo and T. Kataoka, Organization of immunoglobulin heavy chain genes and allelic deletion model, *Proc. Nat. Acad. Sci. USA*, 75:2140–2144, 1978; U. von Schwedler, H. M. Jäck, and M. Wabl, Circular DNA is a product of the immunoglobulin class switch, *Nature*, 345:452–456, 1990.

Immunotherapy

Medical science has often sought to make "magic bullets" that could seek out and destroy disease without affecting the normal functioning of the body. In practice, this means delivering a therapeutic agent to infected or malignant cells either to cure them or to destroy them. It is difficult to realize such an ideal therapy, but immunotoxins are quite promising magic bullets. Immunotoxins are conjugates of antibodies and toxic proteins and have been at least partially successful in treating cancers and other diseases.

Immunotoxins. In the original conception, immunotoxins used an antibody to recognize target cells, and a chemically linked enzyme toxin of plant, fungal, or bacterial origin to kill the targeted cell. Modern biochemistry and molecular biology have allowed this concept to be expanded greatly to include other targeting molecules and modes of linkage.

Antibodies. Produced by the B cells of all vertebrates, antibodies are protein molecules that can be easily isolated by modern biochemical methods. Antibodies are composed of light and heavy chains; they assume a Y shape and have two sites (variable regions) that bind foreign molecules (antigens) like a lock accepting a key. It is through these molecular interactions that antibodies recognize specific targets.

In principle, antibodies that are complementary in shape to any other macromolecule can be produced. The goal in constructing an immunotoxin is to produce antibodies against an antigenic marker that is unique to the cell type to be destroyed. For example, if tumor cells had on their surfaces a protein or other molecular marker that existed on no other cell type, it would be possible to produce antibodies that bind only to them. Biochemical methods could be used to isolate the protein from cancerous tissue. The purified protein could then be injected into an animal to elicit antibody formation. Once isolated, the antibodies could be conjugated to produce an exquisitely specific magic bullet. Unfortunately, this ideal does not often occur; tumor cells arise from normal progenitor cells and strongly resemble them. In general, the best that can be done is production of an antibody that takes advantage of slight chance differences between a given tumor cell and normal tissue.

Once an antibody binds to a target cell surface, it will be taken into the cell by endocytosis. Passenger proteins, such as toxins attached to the antibody, will also be taken into endosomes.

Toxins. Extremely potent protein toxins can be isolated from various plants, fungi, and bacteria. All of the toxins used for immunotoxin construction act by disrupting protein synthesis in the target cell, although the modes of toxicity differ in detail. Probably the most commonly used toxin in immunotoxin design is ricin toxin A chain (RTA), a protein isolated from the seeds of castor plants (*Ricinus communis*); pokeweed antiviral protein (PAP) and gelonin are also commonly used. These toxins act by attacking the ribosomes of eukaryotic cells. They remove a single vital adenine base from the ribosome and thereby inactivate it. Because the toxins are enzymes, a single toxin molecule can travel throughout the cell and disrupt thousands of ribosomes, inhibit cellular protein synthesis, and thereby kill the cell. Bacterial toxins such as diphtheria toxin (DT) from *Corynebacterium diphtheriae* and exotoxin A from *Pseudomonas aeruginosa* (PEA) can also be used in immunotoxin construction. These enzymes also disrupt protein synthesis, but rather than attacking ribosomes directly they attack elongation factor 2 (EF-2), a protein necessary for protein synthesis.

Linkers. Antibodies and toxic proteins are isolated separately and must be joined to form an active conjugate agent. This joining is commonly done by using bivalent reagents called linkers that have different chemically reactive groups on each end. One reaction condition allows the linker to attach to the antibody while the other end remains free. The toxin is then added, and new reaction conditions allow the linker to attach to the toxin. The most useful linkers are cleavable; that is, although stable in blood circulation, they have a disulfide bond in the middle that will break under the reducing conditions found inside the cell. When an immunotoxin binds to the surface of a cell carrying the marker antigen, it is taken into the cell by endocytosis. The linker is then reduced, and the toxin can cross the endosomal membrane to reach the ribosomes in the cytoplasm. The free toxin is much more effective than when it is conjugated to the bulky antibody.

Genetically engineered immunotoxins. Recombinant deoxyribonucleic acid (DNA) technology can now be used to design immunotoxins in which the cell recognition protein and toxin are fused into a single unit. In practice, light- and heavy-chain gene fragments coding for the antigen binding site are linked by genetic engineering methods; this unit is then fused

to the gene or gene fragment coding for a toxic protein. The protein expressed from this genetically engineered construct has several therapeutic advantages: it is homogeneous and well defined with fewer side-product containments. It is smaller, since large portions of the antibody not required for target recognition have been eliminated; therefore, it can penetrate more deeply into tumor tissue. Initial studies on human cell lines using this new technology suggest that in certain cases the recombinant immunotoxins are at least as effective as analogs that were chemically linked.

Other targeting proteins. The concept of immunotoxins can be extended to other targeting molecules. Because antibodies can be produced against virtually any target, they are very commonly used, but other molecules can be very effective in specific cases. For example, protein hormones such as human gonadotropin have receptors on a subset of cells. Conjugates made with the hormone and a given toxin are able to pick out those cells and kill them selectively. This same approach has been employed in the design of compounds used against acquired immune deficiency syndrome (AIDS).

Testing. Before an immunotoxin can be used on human patients, it must be thoroughly tested. An immunotoxin is purified in order to remove any free toxin or antibody and to isolate conjugates of a given size. The purification and conjugation may alter the antibody affinity or the toxin enzymatic activity, and this possibility must be assessed.

Initially, potential immunotoxins are incubated in culture medium with cells expressing the target antigen and also with cells lacking the antigen; the immunotoxin should kill only the antigen-bearing target cells. The concentration of immunotoxin that is lethal to 50% of the tumor cells (lethal dose 50, or LD_{50}) and control cells is measured. LD_{50}'s for target cells are typically 1–1000 picomolar; for nontarget cells the value should be 100- to 10,000-fold higher.

Once the immunotoxin is seen to be effective against cultured target cells, it is generally tested in animal models. The test animal may be infected with the tumor cell line to be attacked, or the immunotoxin may simply be checked in healthy animals for unexpected toxicities. An immunotoxin that caused severe damage to healthy monkey tissue could be expected to cause similar problems in humans. After intravenous administration the immunotoxin circulates in the bloodstream, with a half-life of minutes to hours. This half-life is significantly shorter than the half-life of unmodified antibody. Increasing the effective lifetime of immunotoxins is a major area of research. Antitumor studies in mice suggest that local exposure of tumor cells to immunotoxins often appears more efficacious than systemic treatment. Another problem encountered with immunotoxin treatment is nonspecific toxicity in which nontargeted cells are destroyed. Liver damage is commonly seen, and neuronal damage has also been reported.

A few immunotoxins have proven efficacious in animal models and are now undergoing human clinical trials. For example, immunotoxins have been used to retard and treat graft-versus-host disease. In these cases, patients typically receive radiation or chemotherapy to destroy malignant cells, but the treatment also destroys vital stem cells. Stem cells can be supplied from a matched donor (an identical twin is the best). The donor cells, however, also contain T cells that will reject the transplant. It is these T cells that are killed by immunotoxins to prevent tissue rejection. In addition, clinical trials have begun to test immunotoxins against both T-cell and B-cell lymphomas and leukemias. Immunotoxins have also been tested against a number of solid tumors, including metastatic melanoma, colorectal carcinoma, breast carcinoma, and ovarian carcinoma. To date, the results have been somewhat disappointing. Poor tissue penetration and a number of nonspecific toxic reactions have been observed.

Therapy against AIDS. Acquired immune deficiency syndrome is caused by an attack on human T-helper cells by the human immunodeficiency virus (HIV). A viral surface protein, gp120, binds to a T-cell protein, CD4, to initiate viral uptake. The destruction of T cells weakens the immune defense system and leaves the body susceptible to other infectious agents that cause wasting and eventual death. HIV-infected cells often fill with virus, and certain viral proteins such as gp120 are exposed on the surface. The display of various AIDS-related surface markers has served as the basis for the design of a number of immunotoxins and related reagents.

In one experiment, pokeweed antiviral protein has been conjugated to antibodies against CD4. It might be anticipated that this action would aggravate HIV infection by killing even more T cells, but in fact researchers have found immunotoxin levels that are high enough to inhibit HIV replication almost completely but show no deleterious effect on proliferation of uninfected T cells. Apparently, synthesis of viral proteins is more susceptible to toxin action than is normal synthesis. In another experiment, ricin toxin A chain was conjugated to soluble CD4. This complex could recognize cells displaying gp120 on the surface and killed HIV-infected cells 1000 times faster than uninfected cells of the same line. In a similar strategy, genetic engineering was used to couple a portion of CD4 to exotoxin A. Again, this compact single protein could recognize and selectively kill HIV-infected cells displaying the gp120 protein. These reagents are not yet in clinical use, but it is hoped that one or more will soon be approved for clinical use in the treatment of AIDS.

For background information SEE GENETIC ENGINEERING; IMMUNOLOGY; IMMUNOTHERAPY in the McGraw-Hill Encyclopedia of Science & Technology.

Jon D. Robertus

Bibliography. A. E. Frankel (ed.), *Immunotoxins*, 1988; J. M. Lord (ed.), Redirecting nature's toxins, *Sem. Cell Biol.*, vol. 2, no. 1, 1991; I. Pastan, M. C. Willingham, and D. J. P. FitzGerald, Immunotoxins, *Cell*, 47:641–648, 1986; E. S. Vitetta et al., Redesigning nature's poisons to create anti-tumor reagents, *Science*, 238:1098–1104, 1987.

Infant respiratory distress syndrome

The respiratory distress syndrome (RDS), also known as hyaline membrane disease, occurs in premature infants who have a deficiency or absence of endogenous pulmonary surfactant. Surfactant functions physiologically to reduce the surface tension forces that tend to collapse the lung's tiny air sacs (alveoli). During the past decade, replacement therapy with exogenous surfactants has significantly reduced the morbidity and mortality of this disease.

Biochemistry and biophysics. Pulmonary surfactants are complex mixtures of phospholipids and proteins that are synthesized and secreted by the type II alveolar epithelial cells. The phospholipid fraction is largely lecithin (also known as dipalmitoylphosphatidylcholine or DPPC) and contains lesser amounts of other phospholipids, including phosphatidylglycerol and sphingomyelin. The protein fraction, which is less than 10% by weight, consists of four surfactant specific apoproteins known as SP-A, SP-B, SP-C, and SP-D.

The pulmonary alveoli are coated with a thin film of water that generates surface-tension forces that tend to collapse the alveoli. Pulmonary surfactant adsorbs to the air–liquid interface, reduces the surface-tension forces, and stabilizes the alveoli to permit oxygenation and ventilation. The hydrophobic surfactant apoproteins SP-B and SP-C facilitate the adsorption of the phospholipids at the alveolar air–water interface; SP-A facilitates the recycling of phospholipids by the alveolar type II cells; and SP-D has recently been described, but its biological role is still unclear.

The genes for the surfactant apoproteins SP-A, SP-B, and SP-C have been isolated and sequenced. The developmental expression of these genes is regulated by glucocorticoids, thyroid hormone, beta-adrenergic agents, and multiple growth factors.

Clinical features. The respiratory distress syndrome causes progressive respiratory insufficiency during the first few hours following birth in premature infants with surfactant deficiency. Severity varies depending on the relative degree of developmental surfactant deficiency. Before the use of ventilators, mortality during the first few days was very high in severe cases of respiratory distress syndrome. Milder cases reached a peak of severity by 3–4 days of age with a gradual resolution over the next 7–10 days. The unopposed surface-tension forces cause alveolar collapse and decreased lung compliance. Infants develop rapid breathing, inspiratory nasal flaring, and retraction of the respiratory muscles during inspiration, and require increasing concentrations of oxygen to prevent cyanosis. Typical roentgenographic features include poor lung expansion, diffusely granular lungs that are whiter than normal, and air-filled dilated bronchi that are darker. Therapy prior to surfactant replacement consisted of oxygen administration, assisted ventilation, and maintenance of continuous positive airway pressure. In spite of these measures, mortality rates remained high (particularly in infants of less than 30 weeks' gestation), and morbidity was common. Acute and chronic forms of lung injury result from the oxygen toxicity and pressure-related trauma (barotrauma) of assisted ventilation. Acute barotrauma may cause the formation of tiny air bubbles trapped throughout the lungs (interstitial emphysema) and the collapse of one or both lungs (pneumothorax). One long-term complication of respiratory distress syndrome is a chronic form of lung disease known as bronchopulmonary dysplasia.

Surfactant replacement therapy for respiratory distress syndrome was first attempted in the 1960s using pure lecithin administered via an aerosol mist. These early attempts were unsuccessful, and further trials were delayed until additional knowledge of pulmonary surfactant was available. In the late 1970s and early 1980s, two exogenous surfactant preparations derived from bovine lungs were studied in animal models of respiratory distress syndrome and in human newborns. These preparations were administered by direct instillation via an endotracheal tube. The efficacy of these bovine surfactant preparations was attributed to the presence of the two surfactant apoproteins, SP-B and SP-C. Multiple randomized clinical trials, involving several thousand premature newborns, have confirmed the safety and efficacy of these as well as other exogenous surfactant preparations.

Exogenous surfactants for therapy. Human surfactant has been isolated from human amniotic fluid obtained at cesarean section. The purification process kills viruses that may contaminate amniotic fluid, such as hepatitis and human immunodeficiency viruses. Although the safety and efficacy of human surfactant has been demonstrated in several large clinical trials, the difficulty of collecting large volumes of human amniotic fluid prohibits large-scale commercial production.

Both bovine and porcine surfactants were used extensively for replacement therapy during the 1980s. Lungs obtained from meat-packing houses are a source of multiple commercial pulmonary surfactants. These surfactants are extracted from minced bovine and porcine lungs and from the lavage fluid used to irrigate bovine lungs. They are purified by an organic extraction process that preserves the apoproteins SP-B and SP-C. Different bovine and porcine surfactants have been used to treat several thousand newborn infants and, to date, no adverse immunological reactions have been reported.

Extensive clinical trials have also been conducted with two synthetic surfactant preparations that are free of any animal proteins. One preparation is a specially formulated mixture of lecithin and phosphatidylglycerol, and the second is a mixture of lecithin and two alcohols (hexadecanol and tyloxapol) that promote the spreading of lecithin.

Current work is focused on the development of a surfactant composed of lecithin and the human recombinant apoproteins SP-A, SP-B, and SP-C produced by genetic engineering.

Laboratory evaluation. A variety of laboratory methods have been used to evaluate the biophysical properties of surfactants from bovine, porcine, and synthetic origins. Dynamic surface activity measurements of surfactants have been made using the pulsating bubble technique. The bubble apparatus is based on the law of Laplace given by the equation below,

$$\Delta P = \frac{2y}{R}$$

where ΔP is the pressure difference across the bubble, R is the bubble radius, and y is the surface tension. The bubble apparatus is useful for the evaluation of potential new surfactant preparations as well as for the quality-control monitoring of batch-to-batch variations in commercial surfactants.

The bubble technique has also been used to demonstrate that albumin, hemoglobin, and meconium can inhibit the biophysical activity of pulmonary surfactant. This finding suggests that surfactant replacement therapy may be useful in full-term newborns with meconium and blood aspiration syndromes. An excised rat lung model that has been made deficient in surfactant by multiple lavage is useful for evaluating the physiological properties of surfactant preparations. This model is used to evaluate the ability of instilled surfactants to restore normal pressure–volume curves. Surfactant replacement preparations have also been studied extensively in prematurely delivered lambs with surfactant deficiency.

Strategies for administration. Both prophylactic and rescue approaches have been used to administer surfactant preparations. In prophylactic therapy, the first dose of surfactant is administered via the endotracheal tube immediately or very soon after birth. The prophylactic approach has been used primarily in those infants thought to be less than 30 weeks' gestation at birth, based on obstetrical and ultrasonographic data. With the prophylactic approach, the endogenous surfactant mixes with the fetal lung fluid that is still present at birth. As breathing is established, this fluid is reabsorbed by the alveoli and the surfactant is distributed throughout the lungs. This approach also permits delivery of surfactant to the alveoli before the onset of lung injury due to the oxygen toxicity and barotrauma that results from assisted ventilation. A disadvantage of the prophylactic approach is that 30–40% of infants born at less than 30 weeks' gestation do not develop moderate to severe respiratory distress syndrome and are, therefore, treated unnecessarily.

With a rescue strategy, surfactant therapy is administered after the development of established respiratory distress syndrome. This strategy allows the surfactant to be given under controlled conditions following the roentgenographic confirmation of endotracheal tube position. It also permits the treatment of only the 60–70% of infants of less than 30 weeks' gestation who develop respiratory distress syndrome. A major disadvantage to the rescue strategy is that lung injury is already present when the surfactant is administered. When these two strategies were compared in a large multicenter randomized trial, the group receiving prophylactic therapy had a significantly better survival rate.

Evaluation of surfactant replacement. Over 25 large randomized trials of surfactant replacement therapy have been reported. It is difficult to compare the results since these trials used different surfactant preparations, different patient populations, and different administration strategies. However, many trials have reported an improved survival rate and a decrease in the incidence of pneumothorax. Unfortunately, most of the trials have not shown a decrease in the incidence of chronic lung disease. Trials comparing different surfactant preparations are under way. The long-term follow-up of these patients is currently an area of intense interest.

Prospects. The efficacy and safety of surfactant replacement therapy have been verified in clinical trials involving thousands of premature infants. Additional studies are needed to optimize surfactant therapy and to examine the safety and efficacy of surfactant replacement therapy in full-term infants with acute respiratory failure, secondary to pneumonia and meconium aspiration syndrome.

For background information SEE INFANT RESPIRATORY DISTRESS SYNDROME; RESPIRATORY SYSTEM DISORDERS in the McGraw-Hill Encyclopedia of Science & Technology.

James W. Kendig

Bibliography. M. E. Avery and T. A. Merritt, Surfactant-replacement therapy, *N. Engl. J. Med.*, 324: 910–911, 1991; B. Lachmann (ed.), *Surfactant Replacement Therapy in Neonatal and Adult Respiratory Distress Syndrome*, 1988; R. J. Martin, Neonatal surfactant therapy—Where do we go from here?, *J. Pediat.*, 118:555–556, 1991; D. L. Shapiro and R. H. Notter (eds.), *Surfactant Replacement Therapy*, 1989.

Information technology

Information technology is concerned with improvements in a variety of human problem-solving endeavors through the design, development, and use of technologically based systems and processes that enhance the efficiency and effectiveness of information in a variety of strategic, tactical, and operational situations. Ideally, this is accomplished through critical attention to the information needs of humans in problem-solving tasks and in the provision of technological aids, including computer-based systems of hardware and software and associated processes, that assist in these tasks. Information technology activities complement and enhance, as well as transcend, the boundaries of traditional engineering through emphasis on the information basis for engineering.

The information and systems approaches to engineering science and technology complement and enhance the more traditional approaches based on physical and material science. A systems-based approach is a top-down approach to systems design that, when associated with information technology products and services, provides a conceptual basis for the integration of the information basis for engineering with the traditional physical- and material-science basis.

Information technology and information systems engineering are professional areas that encompass hardware and software and that involve and enable the acquisition, representation, storage, transmission, and use of information. Success in information technology and engineering-based efforts is dependent upon being able to broadly understand the interactions and interrelations that occur among the components of large systems, as well as being able to cope with the overall architecture of systems, their interfaces with humans and organizations, and their relations with external environments.

Activities. The knowledge and skills required in information technology and engineering come from the applied engineering sciences, especially those of the computer and systems engineering sciences, and from the world of professional practice. Professional activities in information technology and information systems engineering vary from requirements definition or specification to conceptual and functional design and development of computer-based systems for information support. They are much concerned with such topics as architectural definition and evaluation. These activities include integration of new systems into functionally operational existing systems and maintenance of the resulting systems as user needs change over time. The activities occur at a variety of points in the system life cycle, one version of which is illustrated in **Fig. 1**, and are needed for functional integration, maintainability, reliability, and the interfaces that ensure that the system is designed for human interaction. This human interaction with systems and processes, and the associated information-processing activities, may take any of several diverse forms. It may involve human supervisory control of physical processes, such as the robots that are used in automated manufacturing. It may involve typically cognitive tasks of situation assessment and issue resolution at the operational levels of fault diagnosis, detection, and correction, or at the level of strategic planning. It is especially important that corrective actions be coordinated with, and implemented at, the levels of institutions, organizations, and values. Much experience has shown, in information technology as well as in many other areas, that solutions that can only resolve symptoms are not solutions at all. SEE SYSTEMS ENGINEERING.

Efforts in information technology and information systems engineering have a problem orientation or issue orientation in that they are concerned with formulation, analysis, and interpretation of issues and organization of information to support these ends. They have a design or architecture orientation, and the technical direction and systems management efforts associated with this orientation.

Systems design methodologies. The hardware and software of computing, communications, and control form the basic tools for information technology. These are implemented as an information technology system through use of systems design methodologies. Such methodologies can provide the top-down perspectives that support future needs and successful contemporary professional practice. For this

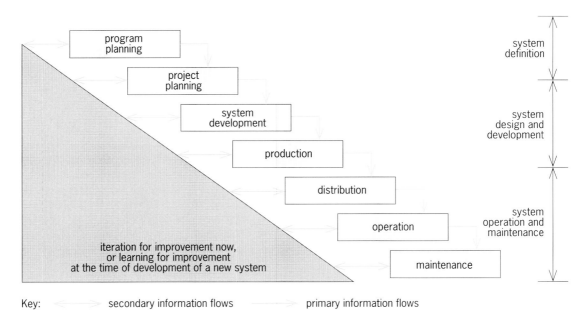

Key: secondary information flows ——→ primary information flows

Fig. 1. Seven phases in the systems engineering life cycle for information technology development.

reason, there is a need for a thorough understanding of the capabilities and limitations of system designs such that meaningful and appropriate operational implementations of information technology–based systems and services are possible over a broad range of applications.

Design of new systems. While information technology and information systems engineering do indeed enable better designs of systems and existing organizations, they also enable the design of fundamentally new organizations and systems. Thus, efforts in this area not only include interactivity in working with clients to satisfy present needs but also include proactivity in the sense of being aware of future technological, organizational, and human concerns such as to support graceful evolution over time to new information technology–based services.

Evolution of systems. The initial efforts at provision of information technology–based systems concerned implementation and use of new technologies to support office functions. These have evolved from electric typewriters and electronic accounting systems to include very advanced technological hardware, such as facsimile (FAX) machines and personal computers, to perform such needed functions as electronic file processing, accounting, and word processing. This represents one form of information technology–based support, support through more advanced information technology–based hardware and software.

Among the many potentially critical information technology–based tools are database machines, electronic mail, artificial-intelligence tools, facsimile transmission devices, fourth-generation programming languages, local-area networks (LANs), integrated service digital networks (ISDNs), optical disk storage (CD-ROM, or compact-disk read-only memory) devices, personal computers, parallel-processing algorithms, word-processing software, computer-aided

software engineering packages, word-processing and accounting software, and a variety of algorithm-based software packages. There are many others. *SEE CONCURRENT PROCESSING; OPTICAL INFORMATION SYSTEMS.*

Management information systems. These can often be used as isolated technologies. For example, a single person might utilize a word-processing software package on a personal computer to prepare a hard-copy, paper report. Here, a relatively simple information technology has replaced an older technology, possibly that of the same person using a mechanical typewriter. It was recognized not only that support could be provided to individuals in accomplishing such tasks as report preparation, but that the ubiquitous computer could provide support for groups in answering queries of a what-if nature with an if-then response. This led, in the 1960s, to the development of support through management information systems (MIS). These systems have become quite powerful, and are used for a variety of purposes, such as scheduling airplane flights and booking passenger seats on them and registering university students in classes.

Decision support systems. As management information systems began to proliferate, it soon was recognized that at least two difficulties remained. While the management information system was very capable of providing support for organizing data and information, it did not necessarily provide much support for human judgment and choice activities. Many such activities need support. They range from assessing situations in order to better detect issues or faults, to diagnosis in order to enable the identification of likely causative or influencing factors. Nor did the classical management information system provide support for decision-related issues that involve selection of alternatives that have multiple and noncommensurate attributes. This capability was provided by support through judgment and decision support systems. These systems involved linking the database management systems (DBMS), so common in the era of management information systems, with the capability of model-base management systems (MBMS), made possible through advances in operations research and artificial intelligence, and with the visualization and interactive-presentation capability made possible through dialog generation and management systems (DGMS). The resulting systems are generally known as decision support systems (DSS). These systems provide needed support for information processing by individuals and organizations. *SEE DECISION THEORY.*

Information systems integration. An additional difficulty is that it has become essentially impossible to cope with the plethora of new information technology–based support systems. The major reason is the lack of systems integration across the large variety of such products and services. This has led to the identification of an additional role for information technology professionals, one involving support through information systems integration engineering. An information systems integration engineer is responsible for overall systems management, including configuration management, to ensure that diverse products and services are identified and assembled into total and integrated solutions to information systems issues of large scale and scope. There are many contemporary technological issues here. There is a need for what are often called open-systems architectures, or open-systems environments, that provide for such needs as interoperability of applications software across a variety of heterogeneous hardware and software platforms. The key idea here is the notion of open, or public, a notion that is intended to produce consensus-based developments that will ameliorate difficulties associated with the lack of standards and the presence of proprietary interfaces, services, and protocols.

When brought to fruition, these information-systems integration developments, and the associated open-systems environments, will enable more efficient and effective configuration management. This will result in the development of information technology solutions in enabling existing organizations to function more effectively. However, there is also a need for better organizational designs. These needs exist in a variety of areas that range from more efficient and effective enterprise management, through more efficient and effective education of students in universities, to more efficient and effective manufacturing processes. It is in this area that contemporary proactive information technology developments promise the greatest payoff. **Figure 2** illustrates the evolution of information technology–based systems over time and the increased dependence of contemporary systems upon proper design and management approaches.

Developments. It is not possible to provide a detailed discussion of the present and many possible future developments in the field of information technology. The use of computers by top management, as decision support systems or executive support systems, has been the subject of much study. In the decision and control trilogy of strategic planning, management

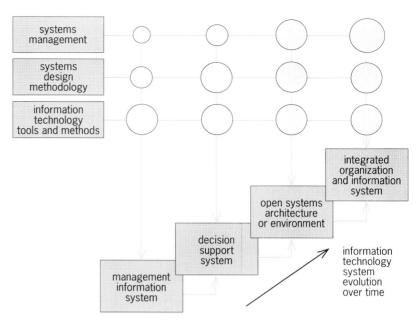

Fig. 2. Evolution of information technology systems over time and development effort. Increasing size of circles indicates increasing frequency of interaction with the information technology system, and increasing depth of shading indicates increasing intensity or depth of interaction with the information technology system.

control, and task control, computer-based information systems are particularly useful in the management control function. It is especially necessary to be able to valuate potential investments in information technology, and much work has been done on this subject. There are many legal implications to implementations of information technology, especially when considerations of systems integration and systems management are involved, as is invariably the case. The impact of information technology innovations on human performance is a subject of much current interest. Information technology has the potential for support in a variety of organizations and for a variety of purposes. To achieve effective support, it will be essential to manage information technology developments, to integrate information technology and institutions and organizations, and thereby to enable appropriate design through information technology.

Impacts on environment. It is important to also consider the impacts that decision support systems and other forms of contemporary information technology have on the environment. G. P. Huber has identified a number of propositions that relate to the effect of information technologies on organizations and on associated organizational situation assessment and decision making.

With regard to subunit structure and processes, the use of information technology can lead to a larger number and variety of people participating as information sources in the making of decisions; to decreases in the number and variety of people making up the traditional face-to-face decision unit; and to less organizational time being absorbed by decision-related meetings. With regard to the organization as a whole, the use of information technology can lead

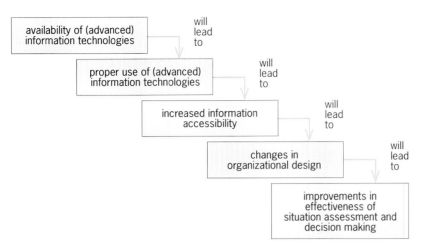

Fig. 3. Causal model of the effects of information technology on situation assessment and decision making in organizations.

Bibliography. G. P. Huber, A theory of the effects of advanced information technologies on organizational design, intelligence, and decision making, *Acad. Manag. Rev.*, 15(1):47–71, 1990; P. G. W. Keen, *Shaping the Future: Business Design through Information Technology*, 1991; J. F. Rockart and D. W. DeLong, *Executive Support Systems: The Emergence of Top Management Computer Use*, 1988; A. P. Sage (ed.), *Concise Encyclopedia of Information Processing in Systems and Organizations*, 1990; A. P. Sage, *Decision Support Systems Engineering*, 1991.

to a more uniform distribution, across organizational levels, of the probability that a specific organizational level will make a particular decision; to a greater variation across organizations in the levels at which a particular type of decision is made; to a reduction in the number of organizational levels involved in authorizing proposed organizational actions; to fewer intermediate nodes within organizational information-processing networks; to more frequent development and use of computerized databases as components of organizational memory; and to more frequent development and use of in-house expert systems as components of organizational memories. With regard to situation assessment, the use of information technology can lead to more rapid and more accurate identification of problems and opportunities, and to organizational situation assessment that is more accurate, comprehensive, timely, and available. Finally, with regard to the effects of information technology on decision making, the use of information technology can lead to decisions of higher quality, to a reduction in the time required to authorize proposed organizational actions, and to a reduction in the time required to make decisions.

On the basis of these propositions, the causal structural model shown in **Fig. 3** can be obtained. This model indicates the considerable role that modern information and decision support technology can be expected to have in present and future organizational environments. This will involve generating the context for information technology through integrating and aligning the strategies for the organization and for information technology developments; engineering the design of information technology–based systems through involvement of users in all life-cycle phases of systems design; and fielding the system in a manner that provides maximum benefits to organizations and humans.

For background information SEE DATABASE MANAGEMENT SYSTEMS; INFORMATION SYSTEMS ENGINEERING; SOFTWARE ENGINEERING; SYSTEMS ENGINEERING in the McGraw-Hill Encyclopedia of Science & Technology.

Andrew P. Sage

Instrumentation

An electronic instrumentation system is an integrated combination of sensors, data-processing modules, communication links, controllers, and user interfaces. The important trends are lower cost per function, improved performance and automation, and simplification of the user interface. Modern instrument systems are called upon to solve increasingly complex problems, so a large fraction of the cost of instrumentation is related to software, maintenance, and the often-neglected costs of installation and user training. In most cases the costs of instrumentation are a small fraction of the cost of the system it monitors or controls, and a good instrument system is frequently an excellent investment. The topics of sensors, networks, user interfaces, and data processing will be considered, and examples will be given to indicate how a complete system might operate.

Sensors. A sensor is a component that converts a physically observable phenomenon into an electrical signal that can be interpreted by data-processing modules. Although there are thousands of physical measurements that can be made, sensors are usually available for converting the physical phenomena into a voltage, current, impedance, frequency, or time interval. Many modern sensors have electronic conversion integrated into the sensor, as with semiconductor pressure and gas sensors used in automotive engine control. A recent trend is toward the use of so-called smart sensors, wherein properties of the sensor, such as built-in sensor identification codes, calibration constants, and historical records of use, are communicated directly to the electronic modules. The data-processing module allows a wide variety of different sensors to be plugged into a standard connector, and the processor interrogates the sensor to find its type and hence the type of measurement and computation that is required. The use of calibration constants stored in the sensor allows lower tolerances in manufacturing sensors and higher accuracy in interpreting measurements. The use of historical records allows automatic determination of maintenance needs and possible anticipation of degradation or failure.

Networks. The key to a large system is the interconnection network that allows a multiplicity of sensors, data-processing modules, and interfaces to share data. A networked instrumentation system can be likened to the human body with a central computer providing the overall intelligence, but with remote processors providing most of the low-level input and output and performing most of the routine control independent of the central processor. There are many types of networks in use, but most instruments now use 4–20-milliampere analog signals, and EIA 232 (the modern form of RS 232C, which is also widely used for computer terminals) and the IEEE 488 General Purpose Interface Bus (GPIB) as interfaces to connect sensors and processors. Although these older networks seem likely to last indefinitely, newer designs tend to use standard computer network technologies such as Ethernet and Token Ring. These networks operate without a master control and with vastly higher bandwidth, so they can be expanded as needed to add new functions. Other examples of emerging standards for networks are TBase-10, a new standard for high-speed data transfer over twisted-pair wires; the Advanced Radio Data Information Service (ARDIS); and new forms of cellular telephones, such as ones using code-division multiplexing, which allow inexpensive and robust radio communication with minimum interference. *SEE SPREAD SPECTRUM COMMUNICATION.*

A key problem is the communication between instruments from a variety of manufacturers. Many industries have or are creating standards. For example, marine instrument manufacturers have created the NMEA 0183 standard, which allows a variety of instruments to communicate using a standard format with operation similar to EIA 232. As an example, a Global Positioning System receiver can send information to remote displays and computer controls. In the future, all vehicles of a certain class may be required to have Global Positioning System receivers and accompanying transmitters that send this position information to any nearby vehicle. In this case the instrumentation network could include thousands of vehicles using radio as well as wire links. Such large systems must be be carefully designed, analyzed, and simulated if they are to fulfill their mission without early obsolescence.

Another example is the seven-level Manufacturing Automation Protocol (MAP), which has been developed to be used in factory automation systems so that equipment from many vendors can share common information. The problem with the more general systems is the multiplicity of levels and standards that makes truly universal interconnection quite difficult. The best hope is probably the networked instrumentation systems that will evolve from the huge networking systems being developed for computers.

User interface. The user interface for most early instruments consisted of a few knobs or buttons and a meter, indicator lights, and possibly a relay activated by an alarm condition. Display technology has gone through many stages of evolution, but because of the widely varying demands of different classes of users, there is no clearly superior display. For low-cost and low-power applications the liquid-crystal display, such as is used in most wristwatches, is the most widely employed and can also provide both alphanumeric and graphical output. The semiconductor light-emitting diode is also widely used for alphanumeric displays, and more recently the vacuum fluorescent display has found use because of its good visibility under poor light and its reasonable cost. The best display would be the cathode-ray tube used in television receivers, but for most instrumentation this is too large, costly, and fragile except for use with a central computer. The display of the future may be a flat-panel display that is derived from displays now being developed for television and computers, such as the active matrix display. *SEE AIRCRAFT INSTRUMENTATION.*

With the advent of multiline displays and keypads for data entry, the modern user interface resembles a computer with menus, graphics, user-controlled display format, and audio and visual alarms. A beginning trend is the use of voice input and output and easily programmable data processing.

An increasingly important output device is an electronically controlled mechanical actuator that automatically carries out the function that used to be performed by a human operator in response to an instrumentation output. Thus an instrumentation system will increasingly begin to resemble a specialized robot.

Data processing. In older instruments the sensor usually produced an analog signal that was amplified and modified, but was often used almost directly to drive a display or to create an analog output signal, as with the 4–20-mA instrumentation standard. In a modern design a microprocessor acts as an intermediary between various analog-to-digital and digital-to-analog converters and frequently combines the output of several sensors to create a single measurement output; an example is the use of a standard pH probe and a temperature sensor to determine the ion concentration of a solution.

Although the electronic circuitry is the key to modern instrumentation, it is usually not the most expensive part. Even within the data-processing modules, only a fraction of the cost is associated with the electronics, so it is usually wise to use more sophisticated electronics to allow use of less expensive sensors, networks, and user interfaces. Modern electronic instruments all use a microprocessor for data processing and display, and the trend is for 16-bit and 32-bit embedded processors to replace older 4- and 8-bit models. Many instruments now use very sophisticated digital signal processing in order to improve performance.

Examples. Four examples of instrumentation systems that are either in operation or planned will be discussed. Each system involves over 100 instrumentation components, customized software, expensive installation, and considerable user training.

Water purification. A water purification control system has sensors, local controllers, and displays, scattered over a large area. The local controllers operate a multitude of pumps, valves, and alarms, while the central computer monitors the operation and can be used to change the mode of operation, set points, and so forth. Any downtime or unexpected contamination is so expensive that it is desirable to make many measurements of conductivity, pressure, pH, temperature, and flow, and to automatically take action to replace faulty parts of a system with redundant components installed for just that purpose.

Marine applications. The nerve center of an America's Cup sailing "machine" is an instrumentation system with sensors and displays throughout the boat, all linked to a computer that provides advice for steering, navigation, and performance evaluation and enhancement (see **illus**.). For general marine use, there are proposals to use the Global Positioning System in conjunction with an electronically stored chart to automatically warn of impending grounding.

Electric power. An electrical utility is installing a monitoring system using remote sensors and sophisticated signal processing to identify potential failures in transformers, circuit breakers, pumps, and valves. There is increasing evidence that expensive failures can be prevented and unnecessary maintenance avoided by an instrumentation system that is programmed to sense the early indicators of the most common types of failure. Impending failure can be communicated over standard telephone lines so that no on-site operator is necessary. *SEE OPTICAL FIBERS.*

Transportation. A transportation control system uses distributed sensors and controls for automated

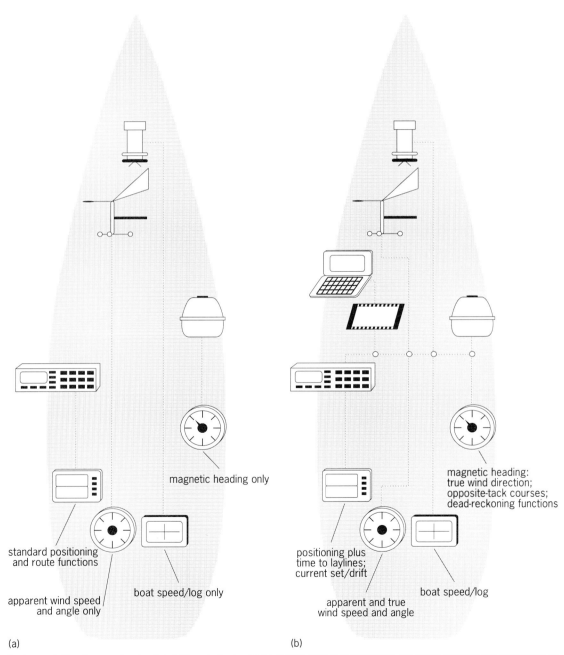

(a)   (b)

Instrumentation for a racing boat. (*a*) **Discrete instruments, which read independently.** (*b*) **Same instruments integrated in a system, offering the benefits of a comparative analysis of the independent measurements.** (*After J. Marshall, Circuits in synergy, Sailing World, pp. 57–60, April 1991*) monitoring and control of a large number of vehicles sharing a common highway, airport, or guideway. Modern guided, high-speed surface transportation is among the safest because automation can eliminate accidents due to the most common driver failings.

For background information SEE DISTRIBUTED SYS- TEMS (CONTROL SYSTEMS); ELECTRONIC DISPLAY; INSTRU- MENTATION; LOCAL-AREA NETWORKS; MICROPROCESSOR; TRAFFIC-CONTROL SYSTEMS; TRANSDUCER in the McGraw-Hill Encyclopedia of Science & Technology.

Richard D. Thornton

Bibliography. J. Marshall, Circuits in synergy, *Sailing World*, pp. 57–60, April 1991; G. Stix, Data communication, *IEEE Spectrum*, 26(1):43–46, January 1989; R. D. Thornton and T. S. Light, A new approach to accurate resistivity measurement of high purity water, *Ultrapure Water*, 5(5):14–26, July/August 1989.

Intelligent weapons

In 1991, the Persian Gulf war between Iraq and a coalition of United Nations forces ushered in a new era in warfare. The information technology revolution arrived on the battlefield as coalition forces, led by the United States, applied information-processing technology to weapons, electronic warfare, reconnaissance, battle management, communications, navigation, and data processing.

Smart weapons. Allied forces daily showed dramatic videotapes of bombs being flown precisely to the targets. So-called smart bombs have sensors that can be programmed to detect or recognize distinctive features of their targets, and data transfer systems that allow information to flow between the bomb and a guidance unit for course direction and changes in flight. Some systems, such as cruise missiles, are au- tonomous and require no in-flight inputs from outside sources. Others require some form of target designa- tion such as radar or laser-beam energy.

Cruise missiles fly at about the speed of sound. On-board computers store information used for iner- tial guidance, position location, and target recognition. The missile flies to its target by using computerized maps and terrain recognition sensors along with iner- tial navigation systems (**Fig. 1**). Once it arrives in the target area, optical sensors in the missile search for the precise point of impact on the target, and the com- puterized steering system alters the missile's course to hit that point. Over 250 cruise missiles were launched into Iraq during the 1991 war—some from submarines, most from large surface ships.

Guided bombs (**Fig. 2**) have an electronic or optical sensor in the nose that can receive a beam reflected off the target. The reflection is created by transmitting a beam of energy from a designator, which can be the same craft delivering the bomb or another source in the air or on the ground. The bomb has a special circuit that guides it to the source of the reflection. The designator simply keeps the beam focused on the target, and the bomb's information-processing technol- ogy does the rest.

Electronic warfare. In the Persian Gulf war, electronic warfare reached a new height as Ameri- can electronics, processed and directed by advanced information-processing technology, overwhelmed Iraqi electronics.

A number of aircraft carried special pods under their wings, such as the ALQ-135, or in their fuse- lage, full of electronic gear for jamming or deceiv- ing Iraqi receivers. In some cases, jammers simply overwhelmed Iraqi air-defense radar and communica- tions devices with powerful beams of energy. In oth- ers, specially designed black boxes, whose operating

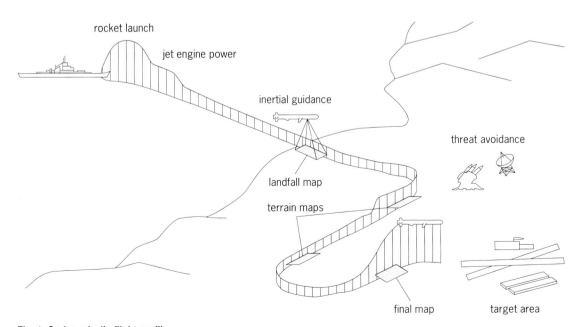

Fig. 1. Cruise-missile flight profile.

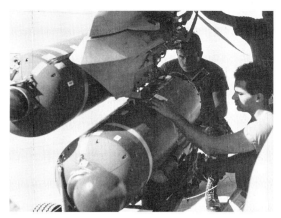

Fig. 2. Ordnance technicians mounting smart bombs under the wing of a tactical fighter during Operation Desert Storm.

principles and physics remain classified, created on Iraqi radar screens false but believable images showing many more targets than were actually there or showing allied aircraft in places where there were none. The Iraqis also used some older-technology jammers, which the allies defeated with even more advanced electronic counter-countermeasures devices.

One of the more deadly electronic warfare systems is the High-Speed Anti-Radiation Missile (HARM). This missile is carried under a fighter. A sophisticated radar detector aboard the fighter picks up the hostile air-defense radar, and a computer determines the location and likely identity of the source. These data are transferred to a computer on the missile which, when launched, is guided into the source of the radiation, usually a radar transmitter, by the computer.

The hearts of these electronic warfare and weapons systems were programmable read-only memories (PROMs), which handle data requirements such as weapon emission characteristics, electronic order of battle, target data, and geographic information. Until recently, PROMs had to be reprogrammed by physically replacing the chips or by burning new data in. This procedure requires the removal of the processor from its aircraft, missile, or bomb. A recent improvement permits reprogramming of these chips while they are still on board their host system, and the time required to make changes is greatly reduced.

Command and control. Among the first aircraft to arrive in Saudi Arabia when the crisis unfolded in August 1990 were advanced models of the U.S. Airborne Warning and Control System (AWACS) aircraft. Saudi Arabia had purchased several models of this airplane, based on the 707 airframe, for its own use. By the time the fighting broke out, the planes were providing continuous coverage for the airspace over the theater of war. Inside the AWACS, Air Force officers serving as battle controllers monitored computerized radar screens (**Fig. 3**) and operated radios linked directly to fighters in the air. As enemy fighters took to the skies, these battle controllers used sophisticated airborne computers programmed to decide rapidly which coalition aircraft could best respond. The controllers then directed those pilots to their tar-

gets. A related system, the Airborne Communications, Command, and Control System (ABCCCS), which is a specially equipped version of the C-130 military transport, controlled the use of aircraft employed directly in the support of ground troops.

The planning and execution of the air campaign was managed by the allied air command using mission planning computers and software that had been developed during the 1980s for just this kind of conflict. Every day nearly 2000 aircraft required that precise mission assignment instructions be sent to the pilots, meteorologists, ordnance loaders, and intelligence analysts. The document making those assignments was the Air Tasking Order (ATO) and typically was over 600 pages. The specially assembled information-processing equipment was used to automate the process, and the Air Tasking Order was downloaded onto optical disks for transmission manually to each airbase carrier. In many cases the instructions were sent digitally over the airwaves, so that the time needed for mission planning by aircrews and pilots was dramatically reduced.

The latest technology to be deployed to the Persian Gulf was the still-experimental Joint Surveillance Target Attack Radar System (JSTARS). Like AWACS, JSTARS is flown aboard a modified Boeing 707 that houses a dozen computer and communications consoles linked to an advanced radar system mounted under the fuselage. This radar detects, locates, identifies, classifies, tracks, and targets enemy ground forces through any kind of weather. The data it collects are transmitted instantly to commanders, ground troops, and airbases. Based on information provided by JSTARS on the ground situation, commanders can decide what kind of force to use against the enemy. The radar is so sensitive that it can tell the difference between a truck and a tank, and it provides mapping data on topographical features. It proved to be crucial in discovering that the Iraqis had failed to achieve their objectives in the battle of Khafji in late January and in detecting the retreat in February when Iraqi forces were finally ordered out of Kuwait.

Reconnaissance. The ability to "see" the modern battlefield far behind enemy lines became key to the combined United Nations air and land campaign

Fig. 3. Battle captain monitoring his assigned sector aboard an Airborne Warning and Control System (AWACS) aircraft during Operation Desert Storm.

that dramatically defeated the Iraqi army occupying Kuwait. Information-processing technology gave the coalition a decisive advantage over Iraqi forces in reconnaissance capabilities.

Sophisticated surveillance assets were applied to the problems of Operation Desert Storm, and virtually all of them were computer-controlled or -aided. The most important systems were the space-based spy satellites that were tuned to watch for Iraqi Scud missile launches. Using these space-based sensors, computers were able to relay data rapidly to ground-based Patriot ballistic missile defense systems. As a result, coalition forces had up to 6 min warning time to intercept Iraq's Scud missile attacks on Israel and Saudi Arabia. Other space-based sensors provided optical, infrared, and radar images to intelligence and operations personnel.

Even commercial satellites were used by the coalition during Desert Storm. France's SPOT system is used in peacetime to produce images for any paying customer in business or government in a so-called open-skies, open-access policy. During the war the French suspended this policy and provided digital photographs transmitted in near-real time for a number of high-priority missions undertaken by the coalition. With most sophisticated satellite imagery capabilities turned to the war effort, there was, in fact, too much information coming too fast for processors to convert it to usable products for the combat forces; many military commanders criticized the overload. United States intelligence agencies intend to apply information-processing technology to make the processing, analysis, and dissemination system more effective for the future.

Communications. Modern military organizations require information-processing technology to manage communications networks from tactical to strategic levels. At the upper levels of command for Operation Desert Storm, sophisticated computerized communications networks such as the Defense Switched Voice and Message Net handled the routine and secure communications needs of more than 500,000 deployed U.S. forces. For highly sensitive needs, special satellites known as MACSAT provided so-called store-and-dump ultrahigh-frequency communications capabilities across the theater.

On the ground, the Army deployed its Multiple Subscriber Equipment (MSE) for the first time in combat. Multiple Subscriber Equipment uses sophisticated network control software and minicomputers to control radio and microwave communications across the depth and width of a battlefield that a typical army division or corps would have to occupy. In Desert Storm, this area was often stretched to the limits of the system's capabilities, which is about 90 mi (150 km) wide by 150 mi (250 km) deep. The system comprises 42 communications nodes providing service to 1900 mobile and 8500 static subscribers. It allows users to keep the same telephone number as they move around the battlefield, and automatically routes calls around inoperable nodes. It is similar in operation to a commercial telephone system with mobile-radio telephone service and data transmission capability.

Navigation. The NAVSTAR Global Positioning System became a key to the allies' ability to outmaneuver and outshoot the Iraqis. NAVSTAR is a joint Army, Air Force, and Navy system that provides accurate, continuous, all-weather, common-grid, worldwide navigation, positioning, and timing information to land, sea, air, and space-based users. Its space segment will eventually have 24 satellites; the ground control segment manages the entire system; and the user segment consists of a receiver that provides usable data.

The receivers can be hand-held or vehicle-mounted and can also be mounted aboard fast-moving ships and aircraft. Aircraft such as the JSTARS used information from the Global Positioning System to track themselves in flight and to stabilize their high-technology radar. Thousands of commercial versions of the hand-held receivers, known as SLUGGERs, for Small Lightweight GPS Receivers, were bought on short notice by the U.S. Department of Defense for issue to troops in the field. They were invaluable to ground forces in navigating across the trackless expanses of the Iraqi and Kuwaiti deserts and in calling in artillery or coordinating logistics support. This navigation capability left the Iraqi forces incredulous that any military force could make its way across those deserts. The U.S. Navy used the Global Positioning System not only to supplant the traditional loran-based maritime navigation system but also to provide flight navigation for several of its experimental Stand-off Land Attack Missiles (SLAMs).

Data processing. Thousands of computers supported U.S. forces in the field during Operation Desert Storm. Most logistics and administrative systems are fully automated, and they employ ruggedized field computers and software. These systems often require special security specifications, known as Tempest qualifications, which make the computers resistant to enemy jamming or interception and provide some protection from the effects of radiation from either nuclear weapons or directed-energy weapons such as lasers or high-power microwaves. Advanced research is being carried out in order to attain high-throughput computers capable of performing over 4×10^{10} operations per second in a system occupying only 5 in.^{2} (32 cm^2).

Air defense. The best-known allied weapon of the Persian Gulf war was undoubtedly the Patriot missile. The Patriot's fast reaction capability, high firepower, and ability to operate in a severe electronic countermeasures environment are highly dependent on its advanced information-processing technology. The combat element of the system is the fire unit, which consists of a radar set, an engagement control station, a power plant, an antenna mast group, and eight remotely located launchers. The highly automated system combines high-speed digital processing with various software routines to effectively control surveillance, target detection and tracking, and support of missile guidance. The only crewed element of fire during air battle, the engagement control station, provides the human interface for control of automated operations.

On the second night of the war, Patriots began to intercept Scuds with remarkable consistency. Not every Scud was intercepted because the Patriot was programmed to allow nonthreatening Scuds to fall harmlessly, without wasting interceptors. The Patriot was originally designed to protect a single point on the ground, so any incoming aircraft not recognized by the computer program as threatening to the site was permitted to pass. For Operation Desert Shield this programming had been modified to instruct the system to protect three points at once, so that area coverage could be approximated. The modified programming was not foolproof, since there were some places between the center and the flank that were not completely protected, but it was as near to area coverage as could be achieved by the system.

Prospects. Advanced research into the application of information technology for military uses is acquiring new frontiers. Future information-processing capabilities, using artificial intelligence techniques, will provide commanders and staffs with battle-planning aids that will be able to anticipate enemy actions even before the enemy formulates them. Weapons systems will have embedded information technologies that will permit artillery and air strikes to be launched virtually autonomously. Data distribution techniques will allow every commander to have a real-time picture of the entire battlefield at extremely detailed levels of specificity. Ballistic missile defenses will soon field information-processing capabilities that will permit limited protection from strategic nuclear strikes in the 1990s. Finally, simulation technologies will permit commanders to rehearse entire campaigns to provide training opportunities for individual soldiers, sailors, aircraft crew members, and marines prior to battle, and to give staffs the chance to work out the minutest details of planning.

For background information SEE AIR ARMAMENT; ARMY ARMAMENT; ELECTRONIC WARFARE; GUIDANCE SYSTEMS; GUIDED MISSILE; MILITARY AIRCRAFT; MILITARY SATELLITES; NAVAL ARMAMENT; SATELLITE NAVIGATION SYSTEMS; SEMICONDUCTOR MEMORIES in the McGraw-Hill Encyclopedia of Science & Technology.

James A. Blackwell, Jr.

Bibliography. J. Blackwell, *Thunder in the Desert: The Strategy and Tactics of the Persian Gulf War*, 1991; Jane's Information Group, *Jane's C3I Systems*, 3d ed., 1991–1992; U.S. Air Force, *Quick Reference Guide: Aircraft and Weapons*, 1990; U.S. Army, *Weapon Systems, 1991*, 1991.

Land reclamation

Mining is one of the fundamental building blocks of any industrialized economy; without mining, most of the needs of an industrialized society could not be met in an economic fashion. Fortunately, the United States has access to many of the minerals required to satisfy these needs, at reasonable cost. While these minerals are available from various sources, a number of additional costs are associated with the mining operations. These costs include social and environmental trade-offs in the extraction (removal from the ground), beneficiation (concentrating the mineral values), and processing of the minerals to produce marketable products or feedstocks for other processes.

Mining waste. To accomplish the goals of bringing minerals to the market, great volumes and tonnages of material often must be excavated to reach the ore body. Once reached, the ore is excavated, and it is usually hauled to a mill where up to 99% of the excavated material is rejected in the production of an ore concentrate. This concentrate is upgraded by various processing technologies, depending on end use. As an example, copper, a mineral essential in electrical and many other applications, is often mined at an ore grade of 0.7% by weight. Prior to extraction of the ore, often two to three times as much as the amount of material actually mined must be removed to uncover the deposit. Once the ore is removed, it is concentrated to a grade of approximately 27–30% by weight. The remaining material is disposed of as tailings. The concentrate is smelted and refined to produce metallic copper. For every pound of copper produced, 435 lb of waste, including waste rock, overburden, tailings and slag, is generated; this waste must be disposed of properly.

In the United States the minerals industry typically generates $1.5-3.0 \times 10^9$ tons ($1.4-2.7 \times 10^9$ metric tons) of waste materials per year from non-coal mining. The coal industry generates an equivalent amount of waste materials. These wastes contain very low levels of metal-bearing minerals, and they pose only a marginal potential threat to the environment. Most of these wastes are classified as high volume and low toxicity by the U.S. Environmental Protection Agency (EPA). However, by their sheer volume and tonnage, these wastes may have significant impacts on the environment and can lead to deterioration of water quality, land erosion, and contamination of the soils. The impact would be even greater except for the fact that coal mines in active operation since 1978 treat their effluent water chemically.

Contamination of soils and water supplies by metals is a major environmental problem associated with wastes from mining and mineral processing, and research directed at solving these cleanup problems is a priority. The detoxification of metal-bearing wastes is nearly analogous to the major objective of most metal mining operations: the economic extraction and recovery of metal values from low-grade ores. Thus the research directed at detoxification often employs the techniques of extractive metallurgy to remove toxic metals from mineral wastes and from contaminated soils and waters.

Rejects from mining and milling are considered as wastes, because economical processing to recover residual metals could not be accomplished. Therefore, existing technology must be modified and extended to treat wastes containing relatively low but potentially toxic levels of contaminants; alternatively, entirely new technology must be developed to detoxify these wastes. In some cases, these modified or new technologies could be introduced as unit processes into current process streams, which would greatly reduce the volume and toxicity of any wastes generated in the future.

While the minerals industry currently suffers from a negative image, most of the negative environmental impacts result from historic mining and minerals-processing activities that are no longer practiced. This legacy of perceived lack of concern for the environment is an impediment to productive and environmentally sound recovery of minerals. In addition, while the minerals industry is working with the regulatory authorities to comply with myriad state and federal regulations, the current marginal profitability in the metals markets often precludes any significant expenditures for research with the exception of site-specific problems. Some of the major problems include the production of acidic waters from mines and waste piles or acid mine drainage, the fate of cyanide and its compounds used for gold and silver recovery, the contamination of soils and neighboring areas by windblown dust containing metallic minerals, and the degradation of water quality by contaminants such as arsenic and selenium. While other minerals-related problems exist, these encompass the majority of classifications for environmental impacts.

In the United States most of the research targeted at environmental problems within the mining and minerals industry is conducted by the federal government. Research efforts related to solving problems of mineral waste disposal have received increased emphasis. One of the principal agencies involved is the Department of the Interior's Bureau of Mines. Recent efforts related to mitigating the effects of mining-waste streams on the environment have investigated both biochemical and chemical methods.

Biochemical methods. It is well known that bacteria can change the form and mobility of metals. Bacteria operate on metals through the mechanisms of bioleaching, biological adsorption (biosorption), and biological reduction (bioreduction). Biosorption and bioreduction are most applicable to contaminated waters. A promising technique for removing metal contaminants from wastewaters involves immobilizing thermally killed microorganisms and plant material in porous polymeric beads. The 1.5–3.0-mm-diameter (0.059–0.18-in.) beads, designated BIO-FIX (an acronym for biomass-foam immobilized extractant), are prepared from readily available raw materials, including algae, duckweed, sphagnum moss, and yeast.

Laboratory tests have demonstrated that BIO-FIX beads selectively remove toxic metal contaminants from a variety of wastewaters. These contaminants include cadmium, chromium, mercury, copper, lead, manganese, and zinc. The beads are readily regenerated for reuse, and they also produce a concentrated effluent stream that can be treated to recover the metals. Effluents from treated waters often meet or exceed drinking-water standards and water-quality criteria for aquatic life. The beads can remove metal contaminants from very dilute solutions and may be used in secondary treatment of wastewaters.

In 1983, the U.S. Fish and Wildlife Service determined that fish and waterfowl inhabiting the Kesterson National Wildlife Refuge, San Joaquin Valley, California, were being adversely affected by ingested selenium. Selenium is a necessary nutrient as a trace element, but in the concentrations found at Kesterson it can cause malformation and death in wildlife. A novel approach to solving this complex problem is the utilization of natural biological systems. Research has focused on bacterial reduction of soluble selenium to the elemental state in agricultural drainage waters and in tailings pond water containing precious metals. Early microbial research that involved removal of selenium from a tailings pond water was unsuccessful because of the solution's high alkalinity (pH 11) and concentration of cyanide. Removing the cyanide by means of chemical oxidation produced a solution suitable for subsequent bacterial reduction of selenium, which was reduced 71% in 5 h by these bacteria.

Contaminated mine drainage is one of the most persistent mining-related problems in the United States. According to a recent inventory, over 7000 mi (11,200 km) of rivers and streams are adversely affected by acidic drainage, mostly from mines that have been abandoned for many years. Therefore much of the research in this area is directed toward the development of alternatives to chemical treatment that are less expensive, as well as control methods appropriate for abandoned mines.

One approach that has proved useful at both active and abandoned mines is biological water treatment, using wetlands constructed for that purpose on the mine property. Metals, such as iron and zinc, are precipitated, and acidity is neutralized by biological activity. Constructed wetlands have been built at a number of mine sites to assist in the treatment of acid mine drainage. An experimental wetland, built in cooperation with the National Park Service, has demonstrated that processes of chemical and bacterial iron oxidation and anaerobic bacterial sulfate reduction play an integral part in improving overall water quality.

Chemical methods. Research involving detoxification by chemical methods has been directed at treatment of tailings and removal of cyanide.

Tailings treatment. Tailings piles and ponds in the new and old lead belts of Missouri contain as much as 300×10^6 tons (270×10^6 metric tons) of material. Present operations are increasing this total by nearly 9×10^6 tons (8×10^6 metric tons) per year. The presence of minerals bearing heavy metals in these tailings may harbor the risk of environmental degradation. While this could result in environmental problems, the tailings also represent a potential resource for heavy metals. Copper, lead, silver, and zinc are all present in quantities sufficient to warrant recovery efforts, provided a suitable method can be developed. Recovery of the heavy metals would also render the tailings less hazardous, and they could then be used as agricultural limestone or returned to the mined-out workings as fill. The host rock material of these tailings is a dolomitic limestone, that is, a magnesium-enriched calcium carbonate. By removing the residual sulfides to very low levels, the material is suitable for use as agricultural limestone.

A recently developed process utilizing froth flotation to remove heavy-metal-bearing minerals from the tailings is being tested. By using a combination of conventional and nonconventional flotation reagents, the heavy-metal content of the tailings can be reduced by as much as 95%. The first step in the flotation process is to add the sulfur ion as sodium sulfide. This coats the mineral surface of oxidized sulfide minerals (usually as sulfates) to make them appear as sulfide minerals to the chemical collector. The minerals, both sulfides and altered sulfides, are collected as a mineralized froth. This is followed by the addition of a primary amine collector to scavenge most of the remaining heavy-metal sulfides from the tailings. The bulk concentrate can then be treated to recover the metal values.

Development of this flotation process includes the development of the air-sparged hydrocyclone—an advance in flotation technology that combines compact size with high throughput. This device operates with retention times of less than a second, compared to retention times of tens of minutes for conventional flotation machines. It could lead to the development of a highly portable, compact, high-capacity plant to treat tailings on site.

Cyanide. Ores containing finely disseminated gold are primarily leached with a carefully controlled sodium cyanide solution. High-grade ores may be leached in vats, whereas the lower-grade ores are more commonly leached in heaps placed above an impermeable liner for collection of the solution. Several questions exist about the short- and long-term fate and the environmental effect of the cyanide-bearing wastes that result from these processes. These questions include considerations such as the long-term persistence in the environment of cyanide and cyanide compounds; the mass balance of cyanide in the leaching process; the mobility of metals in the heap; seepage and leaks through liners; and the protection of wildlife.

Scientists are investigating new methods of rinsing heaps to remove cyanide. Rinsing parameters for such removal have been investigated in the laboratory for a variety of precious-metal ores. Researchers have determined that interrupted (pulsed) water rinses cyanide from heaps more efficiently and produces less liquid waste needing to be chemically neutralized or destroyed.

The use of bacteria to assist in the mitigation of cyanide and cyanide species has provided some encouraging results. A family of bacteria that can survive in cyanide solution containing 250–280 mg cyanide/liter (0.033–0.038 oz/gal) has been isolated and cultured. These bacteria reduce the cyanide to a level of 10–15 mg/liter (0.0014–0.0020 oz/gal) by biological oxidation into benign compounds. Efforts are under way to test the feasibility of this method in the field. A commercial operation currently uses a biological reactor to reduce cyanide in the processing stream from

a range of 12–15 mg/liter (0.0019–0.0020 oz/gal) to below the discharge limit of 0.5 mg/liter (0.000067 oz/gal). Combining these approaches may lead to natural decontamination of heap leach pads and mill tailings.

Chemical neutralization methods are being studied for a number of cyanide complexes that are typically found in mining wastes. Chemists are working to identify and quantify intermediate compounds in the destruction reaction pathways to better understand the long-term stability of the compounds. This information will enable the development of alternative destruction or sequestering methods, which may be even more effective in removing cyanide from the environment.

Prospects. As the previous examples indicate, environmental research, conducted by private industry and federal agencies, is solving many of the environmental difficulties associated with current mining operations and helping to repair the damage caused by earlier practices. These efforts will help make society's mineral requirements more compatible with maintaining a sound environment and will help ensure that the mineral needs are met in an acceptable manner.

For background information *SEE BIOLEACHING; FLOTATION; LAND RECLAMATION; SOLUTION MINING* in the McGraw-Hill Encyclopedia of Science & Technology.

Benjamin W. Haynes

Bibliography. F. M. Doyle (ed.), *Mining and Mineral Processing Wastes*, Proceedings of the Western Regional Symposium on Mining and Mineral Processing Wastes, Berkeley, California, May 30–June 1, 1990; D. J. Lootens, W. M. Greenslade, and J. M. Barker (eds.), *Environmental Management for the 1990's*, Proceedings of the Symposium on Environmental Management for the 1990's, Denver, Colorado, February 25–28, 1991; G. M. Ritcey (ed.), *Tailings Management: Problems and Solutions in the Mining Industry*, 1989.

Laser

Semiconductor diode lasers have revolutionized the way that information is stored and is transmitted over long distances, as exemplified by compact disks and fiber-optic long-distance communication systems. The development of these technologies was made possible by the relatively small size, low cost, and high reliability of diode lasers compared to other lasers. A new form of diode laser, called the vertical-cavity surface-emitting laser (VCSEL), is now emerging from laboratory development. The new structures can be scaled to sizes much smaller than those of conventional diode lasers, and their beam quality is superior. Their light beams emit perpendicularly from the wafer, and they can be fabricated in one- or two-dimensional arrays with conventional integrated-circuit processing techniques. It is likely that these new diode lasers will lead to many applications not yet imagined.

Operation of diode lasers. The present edge-emitting diode laser (EEDL) is simply a forward-biased *pn* junction (hence the term diode laser) with the active material in the center. Its optical cavity (the space between two mirrors where the optical field can resonate) is formed by cleaving the semiconductor crystal at both ends of the device (**Fig. 1***a*). Because of the high gain in semiconductors and the length of the device, equal to or greater than 200 micrometers, the approximately 30% reflectivity from each cleaved facet is sufficient to permit lasing. Electric current flows perpendicular to the cavity (vertically in Fig. 1*a*) and fills the active region with electrons and holes, which recombine to emit photons. When the current is sufficiently high, the photons multiply exponentially along the cavity axis until an equilibrium is reached. This condition is called lasing. The light beam emitted from the cleaved facet is generally about $1 \times 3 \ \mu m$ in cross section, so that it diverges in an approximately $30 \times 10°$ ellipse. To convert this beam to one having a symmetric cross section, a pair of prisms is often used.

The principle of the VCSEL is the same. However, the optical cavity, and therefore the output beam, is oriented vertically (Fig. 1*b*). No cleaving is necessary to form the cavity. The electrical contacts to the laser can be printed on the chip, just as they are printed in electronic circuits. The laser cavity can be arbitrarily shaped by the semiconductor fabrication process, and is generally circular with a diameter of 10–20 μm. This shape produces circularly symmetric beams with output diameters of 5 to 10 μm, corresponding to divergences of about 6 and 3° respectively.

Applications of VCSELs. Since they can be scaled to very small sizes and fabricated in two-dimensional arrays, VCSELs address a broad spectrum of diode-laser applications and requirements, from the low-power, high-speed lasers for optical computing and photonic interchip connections; through medium-power, individual addressable, two-dimensional arrays for scanning and displays; to high-power pumps for solid-state lasers. For many applications they may prove superior to their horizontal counterparts. The unique configurations made possible by VCSELs pose challenges in design, fabrication, and optical engineering, as well as in finding a (probably huge) number of applications as yet unknown.

Microlasers. For example, very small VCSELs called microlasers have been fabricated at densities of over $2 \times 10^6/cm^2$ (**Fig. 2**). The diameters of the microlasers in Fig. 2 are 1, 1.5, 2, 3, 4, and 5 μm. Electrically pumped lasing was achieved in sizes down to 1.5 μm. To achieve lasing in such small devices, highly anisotropic etching to depths greater than $5\mu m$ was required. More recently, optically pumped lasing has been obtained from devices less than 0.5 μm in diameter. The eventual practicality of these ultrasmall microlasers depends on reducing the large electrical resistance, removal of damaged material from the sidewalls, and filling in the space between them to allow electrical contacting and heatsinking. The chief applications foreseen for them are photonic information processing or optical communications between electronic computer chips. SEE OPTICAL INFORMATION SYSTEMS.

An important parameter used to characterize diode

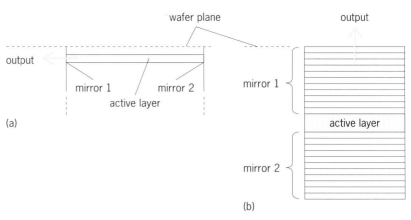

Fig. 1. Diagrams of diode lasers, showing their geometries relative to the wafer plane, mirrors 1 and 2, and their output beams. (*a*) Edge-emitting diode laser (EEDL). Typical sizes are ∼200 μm long by several micrometers wide. The active layer is ∼5–100 nm thick. (*b*) Vertical-cavity surface-emitting laser (VCSEL). Typical dimensions are ∼3–20 μm diameter with active-layer thickness ranging from 8 nm to ∼1 μm.

lasers is the current required to reach the lasing threshold. Although still in their infancy, VCSELs have attained thresholds almost as low as EEDLs, about 0.5 milliampere. A requirement for achieving ultralow thresholds is to have a very small volume of active material, and VCSELs can be scaled to sizes much smaller than EEDLs. The lengths of the microlasers in Fig. 2 are about 5.5 μm, and more than 99.5% of that is passive material, such as mirrors and spacers. The active-material length is only 24 nanometers, and the diameters of working lasers are as small as 1.5 μm. Thus the active-material volume (a critical parameter limiting the minimum threshold) is less than 0.05 μm^3, more than an order of magnitude smaller than that of edge emitters. Such tiny lasers have the potential for

Fig. 2. Portion of an array that contains over 10^6 electrically pumped VCSELs of diameters 1, 1.5, 2, 3, 4, and 5 μm. All sizes are lased except the smallest one.

thresholds in the 10-microampere range. The results of optically pumped lasing experiments indicate that the lasers can be made smaller still, perhaps to volumes of 0.001 μm^3 and thresholds less than 1 μA.

High-power arrays. At the opposite end of the application spectrum are high-power arrays. The ability to fabricate VCSELs by lithographically printing them over large two-dimensional areas opens the possibility of wafer-size laser arrays emitting kilowatts of power. Efficiency and cooling capability will place practical limits on the power, with reasonable levels for improved devices being tens of watts in a square centimeter. This level is not, however, the equivalent of a single laser putting out the tens of watts. A large laser such as a neodymium:yttrium-aluminum-garnet (Nd:YAG) laser puts out a beam that can be focused to a diffraction-limited point. Any such light source is called a point source, and it requires that the beam be spatially coherent over its cross section. Each VCSEL will generally lase independently from all the others, and the resultant spatially incoherent source cannot be focused to a point and will not propagate over large distances without significant diffractive spreading. On the other hand, a partially coherent source can read holograms without the bothersome speckle that stems from spatial coherence, and coherence is not desired for other applications, such as pumping solid-state lasers (yttrium-aluminum-garnet, for example) in a slab geometry. Coupling of lasers into a single coherent in-phase mode has never been easy, and formidable difficulties await attempts to couple large arrays by any known techniques.

Medium-power arrays. Some of the most interesting applications of VCSELs lie in the medium-power range with matrix-addressable surface-emitting laser arrays (MASELAs), wherein each laser in a two-dimensional array is controlled individually. Such an array is scalable to large sizes because, rather than requiring an individual contact and wire for each laser (N^2 wires for an $N \times N$ array, or 4096 for a 64 × 64 array), it needs contacts only for each row and each column ($2N$ wires, or 128 for the 64 × 64 array). A large VCSEL array with indium-gallium-aluminum-phosphorus (InGaAlP) active material, emitting at wavelengths of about 670 nm, that is, in the red part of the spectrum, could be used to display in a "black-and-red television" manner. Infrared MASELAs up to 32 × 32 in size have been used to address a hologram containing 16 distinct images, each laser reading out a separate image. This configuration is useful for pattern recognition and optical neural networks. Related MASELA applications include laser microscopy, confocal microscopy, and short-range laser radar. The two-dimensional MASELA is a technology to which edge-emitting diode lasers have no practical counterpart.

Prospects. The past few years have seen burgeoning activity in VCSEL research. There are technological achievements that must still be accomplished for VCSELs to make their impact. The next few years should determine not only what research can accomplish but how soon VCSELs will reach commercial production and what innovative products they will make available.

For background information SEE LASER in the McGraw-Hill Encyclopedia of Science & Technology.

Jack L. Jewell

Bibliography. J. L. Jewell et al., Surface emitting microlasers for photonic switching and interchip connections, *Opt. Eng.*, 29:210–214, March 1990; J. L. Jewell et al., Vertical-cavity surface-emitting lasers: Design, growth, fabrication, characterization, *IEEE J. Quant. Electr.*, 27:1332–1346, 1991.

Laser deposition

Laser deposition refers to the process of growing films on substrates from a material that is evaporated by the laser radiation. This process is usually carried out in vacuum or in a partial atmosphere of a reactive gas (see **illus.**). Radiation from a laser is introduced through a transparent window into a deposition chamber, and the beam is directed to the target material to be evaporated. The laser power heats the irradiated region of the target above its melting temperature, so that the material evaporates; the material subsequently condenses on the substrate and forms a thin film.

Thermal processes. Laser deposition is a physical-vapor-deposition technique that can be described as an equilibrium process, a nonequilibrium process, or a combination of both, depending on the density of the laser power and the properties of the materials.

Equilibrium processes. Examples of the conventional equilibrium processes of physical vapor deposition are resistive evaporation, electron-beam (e-beam) evaporation, and molecular beam epitaxy. For laser power densities less than 10^6 W·cm^{-2}, laser deposition is usually an equilibrium process. Continuous-wave (cw) lasers (such as a continuous-wave carbon dioxide), or pulsed lasers [such as a loosely focused neodymium:ytttrium-aluminum-garnet (Nd:YAG) operated in a free oscillation mode] with relatively long pulse durations are used in this power regime. Pulsed lasers operating in the low-power-density regime can generate temperatures high enough to dissociate the target material, so that a stream of neutral atomic species is produced, similar to molecular beam epitaxy. The kinetic energies of these atoms are a few electronvolts; they obey a Maxwell-Boltzmann energy distribution characteristic of an equilibrium process. A requirement for laser deposition in this regime is that the target be optically absorbing at the wavelength of the evaporating laser. Optical absorption can occur through the interaction of the laser radiation with free carriers, optically active lattice vibrations (phonons), electronic intraband or interband transitions, and grain-size effects. The type of optical absorption determines the efficiency with which heat is coupled into the target.

Nonequilibrium processes. For laser power densities exceeding 5×10^8 W·cm^{-2}, laser deposition is usually a nonequilibrium process. Examples of conventional nonequilibrium physical-vapor-deposition processes are sputtering and ion-beam deposition. In this regime, a focused pulsed laser with a relatively short pulse duration must be used. Lasers that satisfy this condition are Q-switched ruby, Nd:YAG, and Nd:glass, transverse electroacoustic (TEA) CO_2, nitrogen (N_2) and ultraviolet excimer lasers. The evaporated species is supersonic; it can contain electronically excited neutral particles, ions, electrons, ultraviolet photons, and soft x-rays. In this regime, a high-temperature plasma is created, and its evolution is governed by gas dynamics.

In a phenomenological description of the evaporation process, the initial portion of the laser pulse rapidly heats a thin surface layer and produces a plasma cloud. The time required to produce the plasma may be considerably shorter than the duration of the laser pulse. As the plasma cloud becomes more dense, it absorbs more laser radiation and allows less of it to penetrate to the target surface. During this period, the target surface is also heated by the plasma cloud. As more ions from the target join the cloud, the plasma becomes dense enough to be opaque to the laser radiation. At this point, all the laser radiation is absorbed, and the plasma reaches temperatures on the order of 10^4 K. At these temperatures, the plasma rapidly expands so that the cloud becomes transparent, and the cycle starts again.

Plasmas with heavier atoms expand more slowly and are able to absorb more energy, so that their temperatures are higher. The plasma evolves from a collisional and thermalizing mode to a free molecular flow over a period of a few microseconds. Ions generated in this regime can constitute 10–100% of the evaporants, depending on the laser conditions. Generally, the higher the laser power density, the higher the ionic percentage of the evaporants. The kinetic energy of the ions can be as high as 10,000 eV. All of the evaporated species in the high-power-density regime take part in the growth process of thin films, although it is not always possible to separate the individual contributions.

Mixed processes. In the intermediate-power-density regime, with power densities between 1×10^6 and 5×10^8 W·cm^{-2}, there is a mixture of equilibrium and nonequilibrium processes. This range can be achieved by using either a loosely focused Q-switched Nd:YAG laser or a tightly focused acoustooptical Q-switched Nd:YAG laser. Although some degree of plasma formation occurs, the plasma is not dense enough to absorb a significant portion of the radiation. The evaporants contain a mixture of neutral particles and ions, with the ionic percentage typically less than 1%.

Advantages. Laser deposition has many advantages over the other processes of physical vapor deposition. Congruent evaporation, the tendency of the laser-deposited film to have the same composition or stoichiometry as that of the target material, is considered the most advantageous characteristic of laser deposition. Congruent evaporation has been used extensively in the growth of copper oxide superconductors. In conventional physical-vapor-deposition

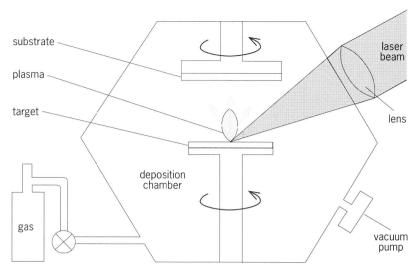

Schematic diagram of a typical laser deposition process.

techniques, the oxygen from the source tends to combine to form molecular oxygen, which results in films that are severely oxygen-deficient. However, the congruent evaporation of laser deposition minimizes the formation of O_2, and stoichiometric (or nearly stoichiometric) superconducting films can be produced, if the deposition is carried out in an oxygen-containing atmosphere.

Laser heating is more efficient than conventional resistive or e-beam heating, since the laser power is coupled to a localized area, and a much higher temperature can be attained. Thus the number of materials that can be deposited is much greater than that possible with conventional physical vapor deposition. In addition, thermal radiation from the source material is substantially reduced, so that the ambient temperature of the reactor is lowered, outgassing of the chamber is lessened, and higher-purity films are produced.

In laser deposition the heating is localized; thus the target forms its own crucible. This advantage over the other processes of physical vapor deposition means that the introduction of impurities from crucibles is negated. For example, rare-earth metals tend to react with their crucibles. Because of the localized heating of laser deposition, the consumption of source material is less than that occurring in other physical-vapor-deposition processes. Reactive gases such as O_2 or hydrogen (H_2), commonly used in physical vapor deposition, tend to react with the heating elements used in the conventional processes. Since no heating elements are required in laser deposition, this problem is avoided.

Multiple sources can be evaporated simultaneously from a single laser by using simple optics to split the radiation and direct it to multiple targets. Multiple sources can be evaporated sequentially from a single laser by rotating segments of the sources in and out of the laser radiation. This process has been used very successfully in the growth of superlattices and heterostructures. The laser radiation can also be used effectively to clean (remove oxide layers) and anneal substrates in the deposition chamber.

Particulates. The most significant disadvantage of laser deposition is the generation of particulates from the laser–target interaction. These particulates can lead to a deterioration of the quality and the surface morphology of the deposited films. The formation of the particulates depends on the target material as well as the laser conditions. At low power densities, the particulates usually are irregularly shaped, microsized solid pieces of the target. Spherical or disk-shaped pieces (ranging in size from less than a micrometer to tens of micrometers) that condense from molten globules are observed at higher power densities.

Particulates can arise from a rapid expansion of trapped gas within the target. This phenomenon can rupture the target and send fragments onto the film; this effect is small and can be avoided by degassing the target prior to the actual laser deposition. A more serious source of particulates is the generation of too much heat by the laser pulse. In this case, a subsurface layer is superheated before the surface has vaporized. This effect can be decreased by lowering the laser power density. Unfortunately, reducing the power density also causes a reduction in the ionic percentage of the evaporants, which is seldom desirable. Another source of particulates is a rough target morphology. After the laser radiation has evaporated a smooth target surface, the new surface can be cracked and scattered with micrometer-sized spheres that have condensed from molten globules. The next time that the laser radiation strikes this region, the spheres are ejected from the surface, and the cracks allow target pieces to be jettisoned from the surface without experiencing a thermal evaporation. These fragments strike the substrate, and those that adhere become particulates.

The problem of particulates in the films is the most serious impediment to wider use of laser deposition. Consequently, much work has been applied to this problem, and several interesting solutions have been reported; however, no current single solution eliminates the problem of particulates universally for all materials.

Materials. Although first demonstrated in the 1960s, laser deposition has become popular only recently because of the high-quality copper oxide superconducting films that can be grown. The ability to tune the deposition process from an equilibrium to a nonequilibrium technique with a single system enables the growth of a thin film of virtually any solid. Metallic, semiconducting, insulating, polymeric, and ferroelectric thin films have been grown by laser deposition. For instance, heterostructures of semiconductors (used for band-gap engineering purposes) and superconductor–insulator–superconductor heterostructures (useful as a Josephson device) have been fabricated by laser deposition and found to function as well as heterostructures grown by molecular beam epitaxy. Some systems, such as epitaxial cubic boron nitride on silicon, can be grown only by laser deposition. When a film grows with its crystalline axes in a particular orientation to a substrate, the film is in epitaxy with the substrate.

Boron nitride (BN) is an interesting compound because of its similarity to carbon compounds. For instance, the hexagonal form of BN is structurally similar to graphite, while the cubic form is similar to diamond. Because the physical properties of the cubic phase of BN are also similar to diamond, there is great interest in growing cubic BN epitaxially on substrates such as silicon. Although cubic BN can be made by several physical-vapor-deposition techniques, none of these were able to make expitaxial films. The laser deposition of BN usually involves a hexagonal BN target and a pulsed laser. If the laser wavelength has an energy much less than the band gap of the target (E_g = 5.2 eV), the evaporated species is a large molecular aggregate of BN. On the other hand, if the laser wavelength has an energy near or above the band gap, the evaporated species is atomic. The dissociation of the molecular bonds of the hexagonal BN target is important to the growth of the cubic BN phase, since the bonding arrangement for cubic BN is an sp^3 type, while hexagonal BN has an sp^2 bonding configuration. In addition, the presence of ions is required to provide energy to form the cubic BN phase. Nonepitaxial cubic BN films can also be grown by an equilibrium evaporation of hexagonal BN with a continuous-wave CO_2 laser, while a stream of N_2 ions from a secondary ion source is directed at the substrate. The ions produced by the pulsed laser evaporation of the hexagonal BN target have kinetic energies on the order of tens of electronvolts and do not adhere to a Maxwell-Boltzmann energy distribution. If the appropriate conditions are satisfied, laser-deposited cubic BN can be grown epitaxially on (001) faces of silicon. The epitaxy is established through a 3:2 coincidence ratio between the lattice constants of cubic BN silicon.

Prospects. Laser deposition has many advantages over conventional physical-vapor-deposition processes. The versatility of the process and its tendency to evaporate materials congruently permit the growth of a thin film of virtually any solid. The rapidly increasing popularity of laser deposition will lead to advances in thin-film technology and perhaps to a solution to the problem of generation of particulates that is more independent of the characteristics of individual materials.

For background information *SEE CHEMICAL BONDING; CRYSTALLOGRAPHY; LASER; VAPOR DEPOSITION* in the McGraw-Hill Encyclopedia of Science & Technology.

Gary L. Doll

Bibliography. J. T. Cheung and H. Sankur, Growth of thin films by laser-induced evaporation, *CRC Crit. Rev. Sol. State Mater. Sci.*, 15:63–109, 1988.

Lateral meristem

The stems of many plants increase in girth by means of lateral meristems. In most plants, the vascular cambium, one of the lateral meristems, forms a continuous ring of vascular tissue. However, some woody plants have so-called anomalous secondary growth. These plants have several structural patterns that relate to vascular cambium origin and mature structure. One of these variants is the presence of successive cambia.

Formation of successive cambia. Successive cambia sequentially form either complete rings or bundles of vascular tissue arranged concentrically (see **illus.**). Whether the vascular cylinders are in a continuous ring or in separate bundles depends on the structure of the young stem. The successive vascular rings form between the original vascular bundles and the peripheral cork cambium. Each new cambium that is added produces secondary xylem internally and secondary phloem externally. The thickness of the secondary xylem varies from several centimeters to a few millimeters. The secondary phloem often appears as strands rather than as a continuous cylinder. In many species, areas of conjunctive parenchyma form between adjacent secondary vascular cylinders.

Successive cambia form in gymnosperms and angiosperms, and probably arose many separate times in the evolutionary history of seed plants. Notable gymnosperms with successive cambia are *Cycas*, *Encephalartos*, *Lepidozamia*, and *Macrozamia* in the order Cycadales; and *Gnetum* in the order Gnetales. Among angiosperms, several dicotyledonous families form successive cambia in their stems and roots. Included are many tropical and subtropical vine-forming taxa and members of the order Caryophyllales (Amaranthaceae, pigweed family; Caryophyllaceae, pink family; Chenopodiaceae, goosefoot family; Nyctaginaceae, four-o'clock family; and Phytolaccaceae, pokeweed family).

There are important differences in the origin, development, and activity of the successive cambia in stems of various plants. The differences are probably associated with the site of insertion of lateral appendages and with changes that efficiently move water or photosynthates along the stem.

Growth and origin of successive cambia. Secondary growth in plants having successive cambia occurs in at least two patterns. In sago palm (*Cycas*), *Gnetum*, black mangrove (*Avicennia*), and amole (*Stegnosperma*), the first vascular cambium originates from procambium, and secondary growth is established before the formation of the first of the successive cambia. Alternatively, in bougainvillea (*Bougainvillea*), bird-catching tree (*Heimerliodendron*), pokeweed (*Phytolacca*), and jojoba (*Simmondsia*), the initial cambium originates from procambium but forms little or no secondary tissue within the primary bundles. Thus, the first complete secondary vascular cylinder develops in the first of the successive cambia and occurs early in development.

Dedifferentiation of cells may occur in different areas of the stem. In members of the Caryophyllales, for example, each new cambium may originate by tangential divisions of primary phloem parenchyma just outside the oldest recognizable sieve tubes of the last ring of bundles. Tangential divisions in this zone produce radial files of parenchyma cells. The first vascular elements to differentiate within these files are sieve-tube elements. In other plants, such as sago palm, the new cambium arises from tangential divisions of inner cortical cells. After a file of new cells is established, the differentiation of sieve cells and tracheids begins toward both sides of the zone of division. Each new cambium is initially discontinuous and later forms a complete ring of secondary vascular tissue.

A complex pattern of differentiation of successive cambia occurs in black mangrove. The second cambium differentiates within the inner cortical zone, and the third originates partially from the cortex and partially from parenchyma derivatives of the second cambium. Subsequent cambia develop from the outermost parenchyma derivatives produced by the previous cambium. After the zone of division is established, xylem differentiation immediately follows, along with the production of 6–10 files of parenchyma cells toward the periphery. Phloem differentiates from the parenchyma cells after the cambium has stopped functioning. Despite these differences relating to the establishment of secondary growth and the origin of successive cambia, the resultant mature structures are similar with respect to their internal organization of secondary vascular tissue.

Induction of successive cambia. In sugarbeet (*Beta*), the formation of the first three leaves after the cotyledons induces the development of a second cambium. The new cambium is initiated from the procambial remnants of each of the three large leaf

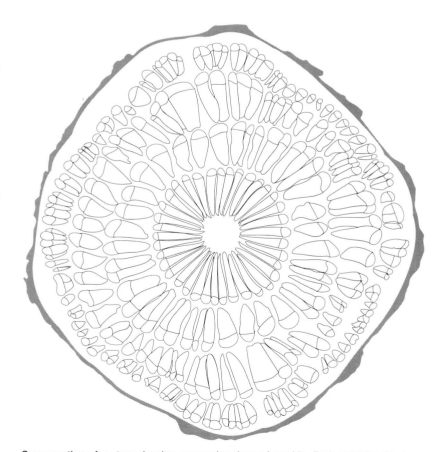

Cross section of a stem showing successive rings of cambia. Each vascular ring has bundles with secondary xylem internally (color) and secondary phloem externally (white).

traces and from interfascicular parenchyma. The third cambium is initiated by the small leaf traces associated with the first four leaves. As growth continues, there is no correlation between numbers of leaves and the initiation of additional cambia. A clear correlation between phyllotaxis and successive cambia occurs in *Phytolacca dioica*. The vascular interconnections that occur in several plants having successive cambia apparently connect the cambia to the point of leaf insertion. The stimuli that induce the cambia are probably made in the leaves and transported basipetally in the stem. Although in black mangrove stems there is no correlation between the number of successive vascular cylinders and leaf or lateral branch formation, radial growth is controlled by the tree crown, since the number of vascular cylinders in a stem is related to the overall vigor of the plant.

Cambial activity. A characteristic feature of some plants with successive cambia is that the original cambium is short-lived. When the initial cambial activity ceases, the first of the successive cambia develops outside the original vascular cylinder. This behavior is repeated as many times as new vascular cylinders develop in the stem. Thus, there is a continuous regeneration of cambial activity. However, in some species the initiation of the second and all subsequent cambia is gradual. As the new cambium forms, the older cambium may not cease functioning immediately. The reason is that certain cambial zones remain active longer because of the unequal development of the bundles.

All cambial layers in sago palm and sugarbeet are active simultaneously. In both species, the number and diameter of cells that extend radially with each cylinder increase basipetally and radially in each vascular cylinder. The radial increment in cell number and diameter results directly from cell division and differentiation of derivatives within each cambial layer. Thus, rings grow simultaneously and not by substitution of cambial activity.

Vascular connections. To study the three-dimensional structure of the conducting system in plants, structural botanists have used direct surface cinematography. In this procedure, the plant axis is perfused with a dye and placed in the microtome. A movie camera is stationed above the cut surface of the plant axis. As the specimen is advanced along its axis through the microtome, a photograph of each successive newly exposed surface is taken. The photographs are then used to determine the three-dimensional vasculature of the stem. This technique has shown that connections exist between and within vascular cylinders. The connections usually occur below the node. There are two types of connections. Radial connections extend through conjunctive tissue between successive vascular rings, whereas tangential connections occur within the same vascular ring between bundles. The interconnection of numerous bundles within and between the vascular cylinders suggests that several rings function as a unit, and it is likely that simultaneous cambial activity is more common.

Each new cambial ring directly descends from the original cortical cells or from parenchyma cells formed by the previous cambium. Many questions remain, however, with respect to the factors that induce cambial initiation and function.

For background information SEE LATERAL MERISTEM; STEM in the McGraw-Hill Encyclopedia of Science & Technology.

Teresa Terrazas

Bibliography. M. Iqbal (ed.), *The Vascular Cambium*, 1990; B. K. Kirchoff and A. Fahn, Initiation and structure of the secondary vascular system in *Phytolacca dioica* (Phytolaccaceae), *Can. J. Bot.*, 62(12): 2580–2586, 1984; T. Terrazas, Origin and activity of successive cambia in *Cycas* (Cycadales), *Amer. J. Bot.*, 78(10):1335–1344, 1991; E. Zamski, Sugar beet vasculature, I. Cambial development and the three-dimensional structure of the vascular system, *Bot. Gaz.*, 142(3):334–343, 1981.

Leaf

Recent research on plant leaves has included studies on the distribution of leaf water with nuclear magnetic resonance techniques, the structure and function of hydathodes, and leaf development in window-leaved plants.

Nuclear Magnetic Resonance

Much of the knowledge about plant leaves comes from optical or electron microscopy. These techniques give good images of membranes and particles, but they cannot detect the presence of water, the most abundant leaf-cell component. A more complete view of leaf structure is provided by using nuclear magnetic resonance (NMR) imaging. NMR detects water, and so it provides a different kind of window through which to view leaf structures. NMR also can distinguish water in the chloroplasts from water in other parts of the leaf.

Certain atomic nuclei, including those of hydrogen, have energy levels that diverge in a magnetic field. NMR measures the energy difference; an NMR spectrum shows how the levels separate when a sample is placed in a uniform magnetic field. All nuclei of a given isotope are identical. Therefore, when all nuclei in a sample are exposed to the same magnetic field strength, their levels diverge by the same amount, and their NMR spectrum shows only one energy difference. For example, the hydrogen NMR spectrum of pure water contains just one signal (represented as a single, narrow peak), indicating that all hydrogen nuclei in the water experience the same average magnetic field strength. Peaks from hydrogen in other molecules may be shifted (relative to the water peak) by chemical effects that shield nuclei from the full strength of the applied magnetic field.

Living cells are mostly water. Therefore, water dominates the hydrogen NMR spectra of biological samples. Signals from the other molecules are relatively unimportant. For this reason, recording of similar spectra from an onion, a human body, or almost any other

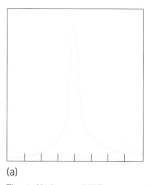
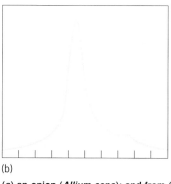
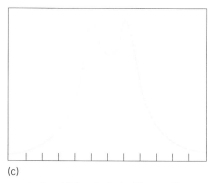

(a)   (b)   (c)

Fig. 1. Hydrogen NMR spectra from (*a*) an onion (*Allium cepa*); and from (*b*) a sun leaf and (*c*) a shade leaf from a Norway maple tree (*Acer platanoides*). In *b* and *c*, the right peak has been assigned to chloroplast water; the left peak relates to all the other water in the leaf. The horizontal scale is graduated in parts per million of magnetic field; it is not numbered because the samples provided no reference signals.

living specimen may be expected. **Figure 1***a* shows a typical example; water in an onion gives a narrow, almost symmetrical peak.

Leaf spectra. Leaves show different spectra than other biological specimens. Water is the dominant source of leaf NMR signals, just as it is in other biological specimens, but internal structures in leaves distort the applied magnetic field, so that water molecules experience different average magnetic field strengths in different parts of the leaf. Most leaves have broad, asymmetrical NMR spectra, and leaves of some plant species have two or even three resolved peaks. Each species has a characteristic spectrum, distinguished from those of other plants by different shapes, positions, widths, and relative intensities of the peaks. Leaf spectra are orientation-dependent; they vary as the angle between the leaf surface and the applied field changes. This orientation dependence demonstrates that some of the water is located in an ordered environment.

Leaf water is confined to a variety of compartments, the most important being the cytoplasm, vac-

uoles, chloroplasts, and extracellular space. Chloroplasts are subcellular organelles containing thylakoid membranes. Often, the chloroplasts are ordered so that all their thylakoids are oriented perpendicular to the leaf surface. Small magnetic fields from paramagnetic ions in these thylakoids add to produce a field offset. With the offset, the NMR signal from chloroplast water shifts to form a peak different from that of water in all the other compartments. Peak displacements depend on the degree of order in the thylakoid membranes, the concentration of manganese in the chloroplasts, the quantity of water in the chloroplasts, and the angle between the leaf surface and the applied magnetic field. In most plant species the peaks are poorly resolved, but in a few the chloroplast and nonchloroplast water signals are well separated. Figure 1*b* and *c* show spectra from Norway maple, one of the best species for NMR study.

Chloroplast and nonchloroplast water. The unique advantage of NMR for leaf studies is that it can distinguish water in the chloroplasts from water in all the other compartments. Therefore, NMR can be

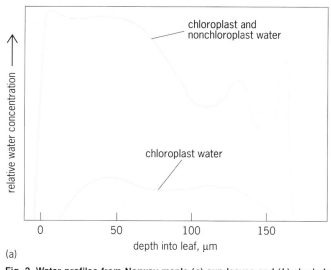
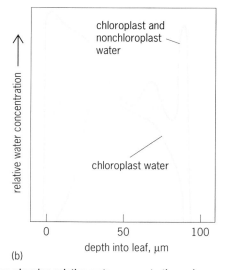

Fig. 2. Water profiles from Norway maple (*a*) sun leaves and (*b*) shade leaves showing relative water concentrations along a vertical cross section through the leaves. Upper surfaces of the leaves are at 0 μm. Both profiles are composites from a number of leaf samples.

used to measure the amount and location of chloroplast water in living leaves. Leaf spectra can be recorded less than 2 min after a small sample is cut out of the leaf. The short time interval ensures that the sample is healthy, living tissue, and that edge effects from the cut are negligible.

One simple use of leaf NMR is to measure relative amounts of chloroplast and nonchloroplast water. Figure 1*b* and *c* compare the spectra of two leaves from the same tree: the sun leaf is growing in full sunlight, and the shade leaf is growing in reduced light. These results are characteristic; sun leaves and shade leaves have different spectra, but all sun leaves from the same tree have similar spectra, as do all shade leaves. Areas under the peaks are proportional to relative amounts of water. Chloroplast water represents a much larger fraction of the total in shade leaves than in sun leaves. Shade leaf chloroplasts contain about 47% of the total leaf water, while sun leaf chloroplasts contain about 17% of the water. Sun leaves typically are thicker; they dissipate heat more effectively, and are capable of greater rates of photosynthesis than are shade leaves. These features are adaptations that improve the efficiency of sun leaves in hot, bright environments. The different relative proportions of chloroplast water in sun leaves and shade leaves may be related to greater need for water reserves in sun leaves.

Nonuniform magnetic fields distort an NMR spectrum. This effect can be used to create images that show how water is distributed through the sample. To produce an image, the applied magnetic field is deliberately distorted. An image can be reconstructed from the shape of the spectrum if the field geometry is known. By using special NMR techniques, it is possible to focus on individual peaks in a complex spectrum and to produce separate images of each. Thus the chloroplast water and nonchloroplast water in a leaf can be separately imaged.

Water profile. Although NMR can be used to make three-dimensional images, it is useful sometimes to reduce the number of dimensions. A magnet with a linear field gradient perpendicular to the leaf surface generates a one-dimensional image called a water profile. The advantage of a water profile over three-dimensional imaging is that it represents the structure of an entire sample; it provides information about the average distribution of water and chloroplasts as a function of depth into the leaf. Conventional microscopy focuses on such small regions that statistical information is almost lost in the mass of detail.

Leaves have layered structures; these are apparent in **Fig. 2**, where the water profiles reveal regions of different water density at different depths into the leaves. The topmost layer (on the left, about 15 micrometers thick) is the upper epidermis; the mesophyll (including vascular tissue) lies in the center; and the lower epidermis (about 10 μm thick) appears on the right. The bump of greater water density that lies to the right of center in both profiles is from water in small veins; the NMR samples were cut so as to avoid large veins. Sun leaves are thicker than shade leaves (165 μm versus 94 μm), mostly owing to the difference

in the thickness of the mesophyll. The mesophyll has lower water density than epidermal layers because it contains air space, and water density is lowest near the bottom of the mesophyll. Chloroplasts are restricted to the mesophyll; in shade leaves, at one small range of depths nearly all of the water is located in the chloroplasts. These water profiles are in excellent agreement with data from conventional microscopy, but NMR provides new quantitative information about leaf structures.

Douglas C. McCain

Hydathodes

Hydathodes are external secretory structures in plants functioning in guttation, the discharge of liquid water (**Fig. 3**). When stomata are closed and transpiration is depressed, for example, at night, the water of guttation may be the primary means of water and mineral transport from the roots into the leaves. Hydathodes occur most often on the upper (adaxial) side of the leaf near the edge (margin). Marginal hydathodes are the basic or primitive condition. Hydathodes are widely distributed in the plant kingdom, and have been reported in desert xerophytes, tropical and temperate mesophytes (including parasitic plants), and hydrophytes.

Structure. Although hydathode structure varies in different plants, three basic components of the gland are normally recognized: the water pore, the subtending epithem often surrounded by a sheath, and vascular tissue that moves water and dissolved substances to the epithem (**Fig. 4**).

Water pores resemble stomatal complexes; however, unlike stomatal guard cells, the guard cells surrounding the water pores usually cannot respond to changes in internal water pressure (turgor pressure). Thus, unlike stomatal guard cells, water pores remain permanently open. Sometimes, structural differences (for example,

Fig. 3. *Monstera* leaf showing guttation near the leaf margins. The photograph was taken at sunrise.

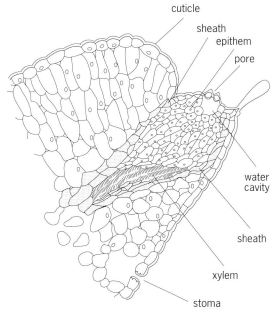

Fig. 4. Typical hydathode structure as exemplified by *Saxifraga*. (After K. Esau, *Plant Anatomy*, 2d ed., John Wiley and Sons, 1965)

size, shape, or wall ornamentation) can be detected between stomatal guard cells and those surrounding the water pores. The water pores may occur in clusters of 10 to 50 or they may occur alone, depending on the plant.

Epithem, a specialized tissue that conducts water from the vascular tissue to the pore, underlies the water pores. Typically, epithem cells contain few or no chloroplasts, have few intercellular spaces, are smaller than other mesophyll cells, and have thin walls. Transfer cells (specialized cells with distinctive wall ingrowths that function in movement of metabolites) occur in the epithem of some species. Epithem in aquatic plants lacks intercellular spaces and may release a brown gummy substance that blocks intercellular spaces between the cells, including the aperture of the guard cell. The cytoplasm of epithem cells is rich in organelles that are normally associated with high metabolic activity such as mitochondria, ribosomes and polysomes, and extensive endoplasmic reticulum. Typically, the epithem is surrounded by sheath cells that have a thin strip of suberin (a waterproofing compound) within their radial and transverse cell walls (Casparian strips) or that have walls completely encircled by a layer of suberin. Sheath cells sometimes have a high tannin or phenolic content. Minerals are probably removed from the guttation water in the epithem tissue and moved to other parts of the leaf. In some plants, water moves from the xylem to the water pore through mesophyll with no anatomically defined specializations.

The xylem of the vascular tissue (vein) transporting water and dissolved substances to the epithem may either underlie the gland (type A hydathode) or end blindly in it (type B hydathode). In both cases, the sheath of the vascular bundle may be continuous with

the sheath of the epithem. Type A hydathodes occur in the leaves of *Monstera* (split-leaf philodendron), *Equisetum*, and members of the rose family. Type B hydathodes occur in succulents such as the jade plant and other dicots such as mustard (*Brassica*) and currant (*Ribes*). In *Ficus*, groupings of water pores occur at the convergence of four to six vascular bundles.

Composition of guttation fluid. The guttation liquid may range from pure water to a dilute solution of inorganic and organic solutes. Various minerals occur in the fluid, with calcium and potassium usually in highest concentration. Some workers report nitrogen-containing compounds, phosphates, sugars, organic acids, vitamins, and nucleotides in the guttation fluid. The pH of guttation fluid ranges from 5.0 to 6.7, depending on the species.

Specialization. Hydathode structure varies greatly among the many groups of higher plants. Some hydathodes are well defined and are considered specialized or advanced phylogenetically. Others show little differentiation from normal leaf tissue and are considered unspecialized or primitive.

Hydathodes of *Monstera* are unspecialized. In fact, one must observe the position of guttation water on top of the leaf to find them. The water pores occur on the upper side of the leaf near the margin or edge (**Fig. 5**). These pores closely resemble stomatal complexes, but do not close as stomata do. Few stomata occur on the upper surface of the leaf, but they are very numerous on the lower surface. Beneath the pores is a large water cavity surrounded by mesophyll cells. The xylem of the vein nearest the margin supplies the water that

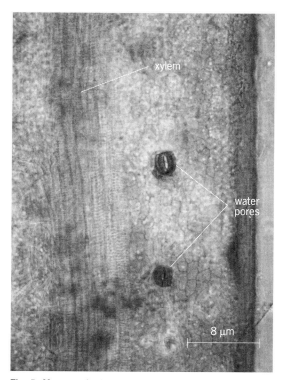

Fig. 5. *Monstera* leaf that has been cleared of chlorophyll, showing adaxial water pores and underlying xylem.

moves through the undifferentiated mesophyll to the water pore and onto the surface of the leaf. Guttation occurs only at night, when the abaxial stomata are closed. The water of guttation in *Monstera* contains a high concentration of calcium relative to other minerals. It also contains small amounts of sugars, amino acids, and phenolic compounds.

The hydathodes of roses and strawberries (both of which are members of the rose family) are considerably more specialized than those of *Monstera*. Water pores occur on the upper leaf surface near the margin in groups of 3 to 38. Inside the leaf, a region of loosely organized epithem occurs between the xylem and the water pore. Other species, such as *Crassula*, have even more highly specialized hydathodes. In *Crassula* the xylem ends in the epithem (type B hydathodes) rather than simply passing below it. Also, the epithem is often compact and surrounded by a water-impermeable sheath that contains tannin.

Mechanism of functioning. The mechanism of hydathode functioning is controversial. Many researchers suggest that guttation is controlled by passive root pressure, and so imply an osmotic gradient from the roots up to the leaf, initially created by active ion movement into the xylem in the roots. Others suggest that guttation may involve an expenditure of energy and thus is an active process. This view is supported by the presence of transfer cells and numerous mitochondria in the epithem cells. However, guttation occurs in plants with poisoned epithem cells, so that a passive mechanism is implied. The fact that hydathodes may retain their activity for only a short period of time in many plants complicates the interpretation of these data. Prevailing opinion seems to favor the hypothesis of passive root pressure; however, the extraction of metabolites from the guttation flow by adjacent cells probably involves an active or energy-requiring mechanism.

Role. The exact adaptive advantage of hydathodes is not known. Perhaps the most widely held view is that hydathodes provide a mechanism of water and mineral movement during times when transpiration is depressed or inhibited, such as at night or when the relative humidity is very high. In some tropical species, ants feed on the guttation fluid and in turn protect the plant from herbivorous insects. In agriculture, insecticides and fungicides sometimes enter the plant through the open water pores and damage (burn) the more delicate internal tissues. Also, bacteria and fungal hyphae can enter leaves through the water pores. However, as is the case with many plant structures, the adaptive advantages of hydathodes appear to outweigh any detrimental effects.

William M. Harris; Elizabeth A. Armstrong

Window-leaved Plants

Window-leaved plants are small succulents native to the deserts of Namibia and South Africa. Structurally diverse, they occur in five families: Piperaceae, Urticaceae, Compositae, Liliaceae, and Mesembryanthemaceae. Leaves of these unusual plants have one or more areas of transparent tissue known as windows.

Window tissue stores water, enhances photosynthesis by permitting light to penetrate to the photosynthetic tissue, and minimizes the surface area of the leaf to reduce water loss.

Kinds of window leaves. There are two types of window-leaved plants: those whose leaves grow entirely above ground (*Peperomia columella*) and those whose leaves grow partly below ground (such as *Frithia pulchra*). *Peperomia columella*, a member of the family Piperaceae, has small (less than 0.2 in. or 5 mm across), fleshy, paraboloid leaves that at maturity weigh about 0.0035 oz (0.1 g), have a surface area of about 1.5 in.^{2} (1440 mm^2), and have a volume of about 0.55 in.^{3} (1730 mm^3). The ratio of leaf surface area to volume is approximately 1.0, which is 25 to 30 times less than that of plants such as *Pothos scindapsus*, with flat, ivylike leaves. Mature leaves of *Peperomia columella* contain about 60% window tissue, accounting for their succulence that helps them store and conserve water.

Frithia pulchra, a member of the family Mesembryanthemaceae, is a small, slow-growing, rosette plant having 6 to 20 conical leaves each up to 2.34 in. (60 mm) long. Each leaf has a dome-shaped, transparent tip made primarily of transparent window tissue that extends to near the base of the leaf. Window tissue constitutes almost 70% of the volume of mature leaves. A cylinder of photosynthetic chlorenchyma tissue surrounds the window tissue. Since only the transparent tip of a leaf extends above the soil in its native environment, most photosynthesis in these window leaves occurs underground. This underground growth of the window leaves reduces the plant's heat load and transpiration.

Development. Young window leaves have approximately 50% chlorenchyma tissue by volume, 30% window tissue, and 20% protective epidermis. The large volume of chlorenchyma in young leaves suggests that they are structurally specialized for photosynthesis rather than for water storage. This pattern of resource allocation typifies other plants, including those that are neither window plants nor succulents.

The relative volumes of window, epidermal, and chlorenchyma tissues change as window leaves grow. For example, young leaves of *F. pulchra*, about 0.2 in. (5 mm) long, have approximately 20% epidermis by volume, 50% chlorenchyma, and 30% window tissue. However, by the time leaves are 1 in. (25 mm) long, the volume of window tissue increases from about 30% to 70%, and the volumes of epidermis and chlorenchyma decrease to approximately 5% and 25%, respectively (**Fig. 6**). Therefore, the structural specialization of leaves of *F. pulchra* for water storage (as exemplified by the increase in window tissue) occurs late in leaf development. Moreover, the relative volumes of epidermal, window, and chlorenchyma tissue do not change as leaf lengths exceed 1 in. (25 mm). Window leaves of *Peperomia columella* develop similarly; the relative volumes of window and chlorenchyma tissues do not change significantly after leaves reach a weight of approximately 0.0014 oz (0.04 g). These results suggest that early stages of window leaf development

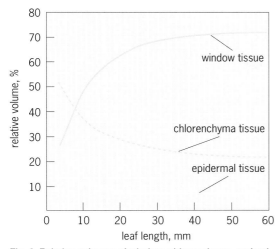

Fig. 6. Relative volumes of window, chlorenchyma, and epidermal tissues during leaf development in *Frithia pulchra*. 1 mm = 0.04 in.

involve preferential reallocations of volume to different tissues, whereas later stages of leaf development involve uniform expansion of all of the leaf's tissues.

Structural specialization for function. Window leaves have two primary functions: photosynthesis and water storage. Photosynthesis occurs in individual chlorenchyma cells and is a function of gas exchange, which in turn is a function of area. Leaves of *F. pulchra* maximize the area available for gas exchange (per unit volume of leaf) by producing many small chlorenchyma cells, each having an absolute volume of approximately 35,000 μm^3. Correspondingly, only the chlorenchyma tissue contains a significant amount of intercellular spaces for gas exchange (approximately 17%). In comparison, window tissue contains about 0.4% intercellular space.

Unlike photosynthesis, water storage is light-independent and is not limited by tissue position. Moreover, water storage requires negligible amounts of gas exchange. Leaves of *F. pulchra* maximize the efficiency of water storage by producing a small number of large, tightly packed window cells. Indeed, the absolute volume of individual window cells is about 30 times larger than that of individual chlorenchyma cells. Similarly, leaves of *F. pulchra* contain 8 to 10 times more chlorenchyma cells than window cells. Thus, the differing functions of chlorenchyma and window tissues partly explain differences in cellular size, number, and tissue compactness characteristic of window leaves in *F. pulchra*.

Significance. Desert plants are evolutionarily specialized to cope with extreme drought and intense light. For example, the small, thin leaves of some desert plants convect heat rapidly, while their reflective surfaces reduce light absorption. Succulence is a common adaptation of desert plants to increase water storage and thereby improve water use efficiency. Above-ground window leaves, such as those of *Peperomia columella*, maximize water storage and the photosynthetic surface area and simultaneously minimize the surface area for transpiration. However, in many desert succulents the advantages of improved water availability (made possible by water-storing tissue) are counteracted by an increased heat load due to increased absorption of infrared light. Window leaves of plants such as *F. pulchra* minimize the chances of overheating by exposing only the transparent leaf tips above the soil. This growth strategy also reduces transpiration and the chances of being eaten by grazing animals.

For background information SEE LEAF; NUCLEAR MAGNETIC RESONANCE (NMR); PLANT-WATER RELATIONS in the McGraw-Hill Encyclopedia of Science & Technology.

Randy Moore

Bibliography. H. Dieffenbach, D. Kramer, and U. Lüttge, Release of guttation fluid from passive hydathodes of intact barley plants, I. Structural and cytological aspects, *Ann. Bot.*, 45:397–401, 1980; D. J. Donnelly and F. E. Skelton, Comparison of hydathode structure in micropropagated plantlets and greenhouse-grown Queen Elizabeth rose plants, *J. Amer. Soc. Hort. Sci.*, 114:841–846, 1989; A. Fahn, *Secretory Tissues in Plants*, 1979; G. A. Krulik, Light transmission in window-leaved plants, *Can. J. Bot.*, 58:1591–1600, 1980; D. C. McCain, J. Croxdale, and J. L. Markley, Water is allocated differently to chloroplasts in sun and shade leaves, *Plant Physiol.*, 86:16–18, 1988; D. C. McCain and J. L. Markley, A theory and model for interpreting the proton NMR spectra of water in plant leaves, *Biophys. J.*, 48:687–694, 1986; D. C. McCain, W. M. Westler, and J. L. Markley, A two-dimensional technique for chemical-shift resolution in one-dimensional NMR images, *J. Magnet. Reson.*, 93:181–183, 1991; R. Moore, Nodes from the underground, *Nat. Hist.*, 95:64–67, 1985; R. Moore and M. Langenkamp, Tissue partitioning during leaf development in ornamentally-grown *Frithia pulchra* (Mesembryanthemaceae), a window plant, *Ann. Bot.*, 67:279–283, 1991; W. Rauh, Window-leaved succulents, *Cacti Succul. J.*, 46:12–25, 1971; T. L. Rost, Vascular pattern and hydathodes in leaves of *Crassula argentea* (Crassulaceae), *Bot. Gaz.*, 130:267–270, 1969.

Magnetic materials

Magnets are indispensable components of a vast range of equipment, including consumer electronics (such as personal stereos and compact disk players), energy converters (such as brushless dc motors for robots and aerospace applications), and specialized instrumentation (such as wigglers and undulators for synchrotron sources). The bulky bars and horseshoes of the past have long since been superseded by compact, powerful rare-earth magnets and inexpensive ferrite magnets that can remain permanently magnetized whatever their shape. The amount of energy that a magnet can store in its magnetic field is proportional to its energy product, a figure of merit that has doubled roughly every 12 years since 1900. This record of improvement is hard to match in any other material property. This article discusses magnets that are based on a new group of interstitial compounds, in particular, $Sm_2Fe_{17}N_3$.

Rare-earth magnets. One landmark was the development of rare-earth cobalt magnets based on the intermetallic samarium-cobalt compounds $SmCo_5$ and Sm_2Co_{17} in the late 1960s and 1970s. Another was the discovery in 1983 of the first iron-based rare-earth magnet, $Nd_2Fe_{14}B$ (neodymium iron boron). These alloys combine a ferromagnetic transition element, iron or cobalt, with a lesser amount of a magnetic rare-earth element R (R = Nd, Sm, . . .), which provides the magnetocrystalline anisotropy needed for coercivity. The magnetization is tied to a unique axis in the crystal structure to ensure that the material, once magnetized, does not relapse to a lower-energy state with many randomly magnetized domains that give no net magnetization and no external magnetic field. Iron is the preferred ferromagnetic transition element because it is 50 times less expensive and more abundant than cobalt. Also, the iron atomic magnetic moment is 30% larger.

Unfortunately, no suitable binary rare-earth iron compounds exist. Any permanent magnet must operate well below its Curie temperature T_c, where ferromagnetism is destroyed by thermal fluctuations, in order to avoid gradual deterioration of the magnetic properties due to thermal effects. The R_2Fe_{17} compounds have Curie temperatures that are little greater than room temperature, and they lack the necessary anisotropy. The RFe_5 phase does not exist.

In $Nd_2Fe_{14}B$, the nonmagnetic boron serves to stabilize a new ternary crystal structure that is not found in the binary neodymium-iron phase diagram. Despite the commercial success of neodymium iron boron, which now makes up a quarter of the permanent magnets manufactured worldwide, the Curie temperature ($320°C$ or $608°F$) is uncomfortably low for certain applications. An intense search for other ternary phases with better intrinsic properties (magnetization, Curie temperature, and anisotropy) has recently led to the discovery of some new intermetallic compounds with very attractive magnetic properties. These materials are made by a simple process called gas-phase interstitial modification. One of the new metals, $Sm_2Fe_{17}N_3$, is especially promising.

Formation and properties of $Sm_2Fe_{17}N_3$. The process involves exposing a finely ground powder of the parent compound with a particle size of about 20 micrometers or less to a gas at about $500°C$ ($930°F$). The gas must be composed of small atoms that can burrow into the crystal structure of the intermetallic compound to occupy interstitial sites next to the rare-earth atoms. The interstitial atoms are electronegative, so they form bonds with the electropositive rare-earth atoms. First results were obtained by treating Sm_2Fe_{17} powder in nitrogen gas. The nitrogen (N_2) molecules dissociate on the surface of the powder and diffuse into the structure to form $Sm_2Fe_{17}N_3$. The temperature must be carefully controlled since diffusion rates increase exponentially with increasing temperature. There is a window near $500°C$ ($930°F$) where the rate for nitrogen is appreciable but that for iron is negligible, so that the 2:17 nitride can form

[reaction (1)] but not break up [reaction (2)]. The

$$R_2Fe_{17} + {}^3/_2N_2 \xrightarrow[\text{nitrogenation}]{} R_2Fe_{17}N_3 \qquad (1)$$

$$R_2Fe_{17}N_3 \xrightarrow[\text{disproportionation}]{} 2RN + 17Fe + N \qquad (2)$$

crystal structure of the metastable ternary nitride is shown in the **illustration**, where the three nitrogen sites lie in a plane around the rare-earth atoms.

Introduction of nonmagnetic nitrogen influences the magnetic properties of Sm_2Fe_{17} in two ways. First, it dilates the structure so that the volume of the unit cell in the illustration expands by 6%. This expansion has a drastic effect on the Curie temperature, which increases from 116 to $476°C$ (from 241 to $889°F$). The extreme sensitivity of the strength of iron-iron coupling to interatomic distances and nearest-neighbor configurations is well documented, but this degree of lattice expansion and the enormous increase in the Curie temperature are quite unusual. Second, the nitrogen alters the environment of the rare-earth atoms. The triangle of nitrogen atoms concentrates electron density in the plane around the samarium atom, so that the strong electric field gradient created couples with the quadrupole moment of the magnetic $4f$ electrons of the rare-earth atom to produce a strong uniaxial anisotropy.

Other interstitial compounds. Applications of the gas-phase interstitial modification process

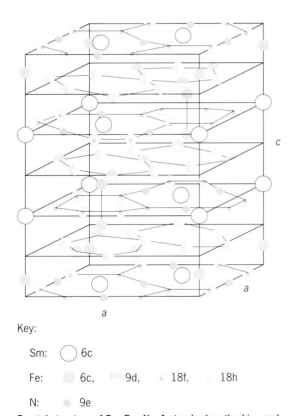

Key:

Sm: ◯ 6c

Fe: ● 6c,   ● 9d,   · 18f,   18h

N: ● 9e

Crystal structure of $Sm_2Fe_{17}N_3$. A standard method is used to designate sites with different crystallographic surroundings: the number of such sites in the unit cell is followed by a letter determined by the crystallographic group.

Intrinsic magnetic properties of rare-earth intermetallic compounds for permanent magnets

Compound	Curie temperature (T_c), °C (°F)	Magnetization (M), teslas	Anisotropy (B_a), teslas	E_{theor},* kJ·m^{-3}
$SmCo_5$	747 (1377)	1.14	28	259
$Nd_2Fe_{14}B$	320 (608)	1.60	9	509
$Sm_2Fe_{17}N_3$	476 (889)	1.53	22	448
$Nd(Fe_{11}Ti)N$	467 (873)	1.37	8	373

* E_{theor} = theoretical upper limit on energy product.

are not restricted to nitrogen or to the 2:17 structure. Other small interstitial atoms such as carbon have been successfully introduced by using hydrocarbon gases such as butane (C_4H_{10}) or acetylene (C_2H_2). Some hydrogen then enters the structure, but it can be removed by pumping at about 500°C (930°F), at which temperature its solubility in the intermetallic compound is small. Other intermetallics that have been interstitially modified include the pseudobinary $R(Fe_{11}Ti)$, with $ThMn_{12}$ structure, which takes up one atom of nitrogen to form $R(Fe_{11}Ti)N$. The volume change (approximately 3%) and the increase in Curie temperature (approximately 150°C or 270°F) are less than for R_2Fe_{17}. Here, the nitrogen atoms form a dumbbell along the c axis on either side of the rare-earth atom, so that an electric field gradient opposite in sign to that of $R_2Fe_{17}N_3$ is created. Therefore, neodymium, which has the opposite sign of quadrupole moment to samarium, is chosen in order to ensure uniaxial anisotropy. Thus, another possible new magnet material is $Nd(Fe_{11}Ti)N$. Intrinsic magnetic properties of the new interstitial compounds are compared with those of established rare-earth materials in the **table**.

Magnet fabrication. However, the table does not provide all the relevant information. A compound with good intrinsic magnetic properties still has to be fabricated into a practical magnet showing hysteresis and coercivity. This process involves creating a metallurgical microstructure that impedes the formation of reversed magnetic domains. Every new magnetic material with favorable intrinsic properties requires considerable development work aimed at optimizing the characteristics of the material before commercial magnets can be produced. A severe processing constraint for the interstitial compounds is imposed by their tendency to decompose when heated much above 500°C (930°F). One possible process that has been demonstrated for $Sm_2Fe_{17}N_3$ magnets is mechanical alloying and heat treatment to generate Sm_2Fe_{17} with a nanocrystalline microstructure that firmly resists demagnetization after the nitriding step. This process yields magnets with excellent coercivity, but their effective magnetization is only about half that shown in the table because the individual nanocrystallites are oriented at random. Further progress toward developing a microstructure where the axes of the crystallites are all oriented in the same direction is needed to arrive at a fully dense oriented magnet whose energy product might approach the theoretical limits listed in the table. Equally desirable is a good coercive powder that can

be directly bonded in polymer and molded to give a plastic magnet. What is needed here is a powder in which each grain is a single crystallite of $Sm_2Fe_{17}N_3$, uncontaminated by any trace of ferromagnetic impurity, such as the iron produced in reaction (2).

Prospects. The new interstitial nitrides and carbides have greatly extended the range of iron-based alloys from which it is possible to create permanent magnets. Their improved thermal stability and high anisotropy promise to increase the applications of rare-earth permanent magnets. However, further doublings of the energy product are rather unlikely, since the inherent limits imposed by the magnetic properties of the elements in the periodic table are now being approached.

For background information *SEE CRYSTAL STRUCTURE; MAGNETIC MATERIALS* in the McGraw-Hill Encyclopedia of Science & Technology.

J. M. D. Coey

Bibliography. J. M. D. Coey et al., Gas-phase carbonation of R_2Fe_{17}, R = Y, Sm, *J. Magnetism Magnetic Mat.*, 98:76–79, 1991; J. M. D. Coey and H. Sun, Improved magnetic properties by treatment of iron-based rare earth intermetallics in ammonia, *J. Magnetism Magnetic Mat.*, 87:L251–253, 1990; K. Schnitzke et al., High coercivity in $Sm_2Fe_{17}N_x$ magnets, *Appl. Phys. Lett.*, 57:2853–2857, 1990.

Magnetic recording

The technology of magnetic recording continues to advance at a rapid rate. Devices that rely on magnetic recording or storage have become an essential part of everyday life. Examples include the magnetic stripes on credit cards, the stereo cassette recorder, the videocassette recorder (VCR), and the disk drive in the personal computer. The pinnacle of technical achievement in consumer electronics is the 8-mm camcorder; and in computer data storage, the high-capacity, small-diameter rigid-disk file.

The rate of advance of magnetic recording technology is just as rapid as that in the semiconductor industry; in fact, it can be argued that the two are necessarily linked. The capacity available in 5.25-in. (130-mm) disk files has increased from 5 megabytes when they were introduced in 1980 to 2 gigabytes in 1991. This represents a capacity growth rate of over 70% per year, equal to the historical growth rates of dynamic random-access memory (DRAM) capacity.

This growth is likely to continue in the foreseeable future. Modern disk drives record approximately 10^5 bits per square millimeter of recording surface. Recent experiments have demonstrated around 2×10^6 bits per square millimeter.

Basic principles. The basic principles of magnetic recording are illustrated in **Fig. 1**. Several turns of wire are wound around a ring of magnetic material in which there is a very narrow gap. This structure is called the head. To record or write information, a current is passed through the wire to generate a very intense field in the proximity of the gap. If a recording medium is now passed very close to the gap, it will become permanently magnetized in response to the current. The recording is played back by again moving the recorded medium past the head and recovering the tiny voltage induced in the wire. The voltage is induced by the changing flux linked into the ring as the magnetic information passes by the gap. In practice, either the head or the medium or both may move.

Although some revolutionary advances are occurring (discussed below), the tremendous progress in the technology has resulted from evolutionary refinements. The density with which information can be recorded (written) and reproduced (read) on a tape or disk surface is determined by the dimensions of the head and the accuracy with which the head can be positioned, by the magnetization level that can be achieved in the recording medium, and by the accuracy and sophistication of the reproduction electronics.

Head size and positioning. To achieve the highest resolution in writing and reading, it is necessary to construct a very small head with a tiny gap and a very narrow contact or track width. Most important, however, an extremely small separation between the head surface and the recording medium must be achieved. Both the field intensity during writing and the sensitivity during reading drop dramatically as that separation increases beyond about one-third the length of the smallest piece of recorded information. Since modern recorders write over 2000 bits per millimeter along a

track, the surface roughness of the tapes or disks is extremely critical. Indeed, the surfaces must be optically smooth since the bit lengths are of the same order of magnitude as the wavelength of light. The width of the recorded track is limited primarily by the accuracy with which the head can be positioned on the medium. Following written tracks as narrow as 10 micrometers requires not only excellent mechanical tolerances but also additional tracking information closely associated with the data in the form of either reference tracks or short bursts (sectors) of positioning signals.

Energy product. The magnetization level that can be achieved in the recording medium is determined by the energy product of the recording medium, given approximately by the product of the coercivity (the magnetic field required to magnetize the medium) and the remanence (the flux density that remains after the magnetic field is removed). At high recording densities the closely spaced magnetic patterns generate strong fields, and can demagnetize themselves if the coercivity is too low. Many modern tapes and disks now use metal films or particles, with energy products an order of magnitude higher than those of the traditional metal oxides common before the 1980s.

Electronics. The third important factor in advancing the technology is the increasingly sophisticated electronics and signal processing. As recording densities increase, the available signal levels decrease, and extremely sensitive, low-noise preamplifiers are required. More sophisticated detection techniques, such as maximum-likelihood detection, are being adapted from communications technology. Advances in silicon technology enable these techniques to be implemented compactly and inexpensively at high data rates.

Camcorders. Camcorders embody the most advanced analog magnetic recording technology in consumer electronics. The signals from a miniature color charge-coupled device (CCD) camera are recorded by using analog frequency modulation (FM) and a miniature helical-scan format. (This technique is adopted from videocassette recorder technology. The tape is wrapped helically around a drum that contains a rapidly rotating head.) The new Hi-Band 8-mm standard combines the helical-scan format with two innovations from the earlier formats, color-under and slant-azimuth. Color-under is a technique in which the chroma signals are recorded without requiring extra bandwidth by using the high-frequency FM luminance carrier as an alternating-current bias signal. Slant-azimuth allows tracks to be more closely spaced since the interference between tracks is dramatically reduced if two heads record alternate tracks and the angle of the gap across the track is, for example, $90 - 20°$ for one head and $90 + 20°$ for the other head.

In the Hi-Band 8-mm format, track densities are increased from 50 tracks per millimeter (in the VHS or Video Home System format) to 100 tracks per millimeter by using pilot tracking tones multiplexed into each track. The linear density of the FM recording is raised from 1000 cycles per millimeter to 2000 cycles per millimeter. The increase in density is supported by the use of metal tapes and metal heads. The transmission

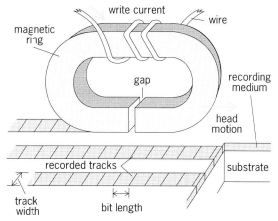

Fig. 1. Gapped-ring head for recording and playback on a magnetic recording medium, illustrating the principles of magnetic recording.

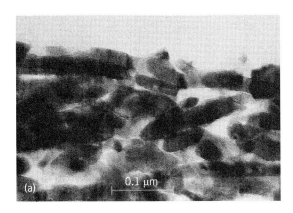

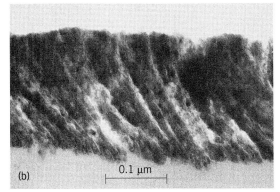

Fig. 2. Transmission electron micrographs of magnetic tapes, showing differences in microstructure. (a) Standard VHS tape. (b) New metal-evaporated tape. (Dexler Corp., Mountain View, California)

electron micrographs in **Fig. 2** contrast the new Hi-8 metal-evaporated tape with a VHS tape (consisting of ferric oxide particles, surface-modified with cobalt). The new tape has a smoother surface, much smaller particles (reducing the tape noise level), and an energy product four times as large (raising the signal level more above the noise of the preamplifier electronics). The increases in recording density are accompanied by thinner tapes (13 μm), leading to longer playing times in a small cassette.

Rigid-disk drives. Small-format rigid-disk drives represent the most advanced technology in recording digital computer data. Data are recorded on the surfaces of a set of rapidly spinning disks. One head serves each surface. To avoid contact and wear, each head is designed into a catamaranlike slider that glides just above the disk surface, supported on a self-generated air bearing. The array of heads can be rapidly moved by a linear or rotary voice-coil motor to access different recorded tracks on the disks.

Progress in rigid-disk drives has followed a path similar to that of consumer videocassette recorders. Again, all three dimensions have been attacked: bits per millimeter, tracks per millimeter, and surfaces per millimeter. Very high linear densities have been achieved by scaling down the head structure and by reducing the flying height from approximately 1 μm in 1980 to almost 0.1 μm in 1991. Track densities have increased by about a factor of 4 over the same time

period because of improvements in servomechanism technology and the use of position information in close association with the data tracks. The most widely used drives come in a standard form, with either 5.25- or 3.5-in. (130- or 95-mm) disks. Ten or more disks can be packed, very closely spaced, into a drive. Metal-film disks have replaced particulate oxide disks in almost all applications. The use of zone recording or banded recording allows higher data rates on the outer tracks of the disk, where the velocity is higher. With a combination of such techniques, as much as 2 gigabytes of data can be stored in the standard 5.25-in. (130-mm) form. Such drives are used increasingly in workstation environments. Even smaller forms based on 2.5- or 1.8-in. diameter (65- or 48-mm) disks are being used in applications such as laptop and notebook personal computers.

New techniques. As indicated above, most of the improvements in magnetic recording can be related to evolutionary improvements in the basic technology. However, a number of revolutionary techniques are also being introduced. One of the most promising is the magnetoresistive head, which relies not on induced voltages but on the property that the resistivity of some materials changes with their magnetic state. Although magnetoresistive heads have been widely employed in computer tape recorders since 1984, their application to rigid-disk drives has only recently occurred.

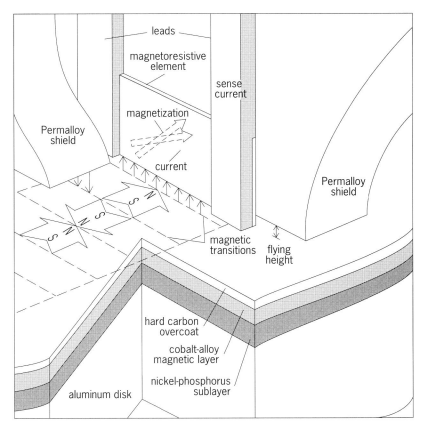

Fig. 3. Cutaway view of a magnetoresistive head gliding a few tenths of a micrometer over a thin-film metal disk. (After R. Wood, Magnetic megabits, IEEE Spectrum, 27(5):32–38, May 1990)

Such heads offer much higher sensitivity at high densities. Magnetoresistive heads have been used in laboratory demonstrations in which recording densities of $1\text{--}2\times10^6$ bits per square millimeter have been achieved on rigid disks, over an order of magnitude higher than is used in current drives. **Figure 3** illustrates the operation of a magnetoresistive head. The direction of magnetization in the thin magnetoresistive element varies with the recorded patterns in the disk. The information is thus recovered by monitoring the resistance of the element, which depends on the angle between the electric current and the magnetization.

Prospects. The applications will continue to motivate advances in the technology of magnetic recording. Consumer digital audio recorders (RDAT) were introduced in the 1980s, but cost and copyright laws inhibited their widespread acceptance. A new digital audio format, DCC, promises to be less expensive and will be compatible with existing analog audio cassette. Development continues on consumer digital video recording, but perhaps the most exciting application is a high-definition consumer video-cassette recorder. Such a device requires aggressive digital image compression as well as very high recording densities. One megabit per square millimeter has been set as the goal, and this density has indeed been demonstrated on an experimental digital videocassette recorder. *SEE DATA COMPRESSION.*

Similarly, the applications will continue to motivate advances in computer data storage technology. High-resolution graphics and image storage are increasingly important in modern computing. Such features require orders-of-magnitude more storage than simple text. Progress in magnetic recording continues to be uninhibited by any fundamental limits and promises another two-orders-of-magnitude improvement in capacity

For background information *SEE COMPUTER STORAGE TECHNOLOGY; MAGNETIC MATERIALS; MAGNETIC RECORDING; MAGNETORESISTANCE; SEMICONDUCTOR MEMORIES; TELEVISION; TELEVISION CAMERA* in the McGraw-Hill Encyclopedia of Science & Technology.

John Mallinson; Roger Wood

Bibliography. F. Jorgenson, *The Complete Handbook of Magnetic Recording*, 3d ed., 1988; J. C. Mallinson. *Foundations of Magnetic Recording*, 1987; C. D. Mee and E. D. Daniel, *Magnetic Recording*, 1987; Special section on magnetic recording, *Proc. IEEE*, vol. 74, no. 11, November 1986.

Mammalia

The reconstruction of mammalian history is much more than the ordering of extinct species into ancestor–descendant trees. For many paleontologists, determining the behaviors of extinct species, tracking evolutionary changes in adaptation, and understanding something of the ecological relationships of a species within its ancient community are the more interesting puzzles of this reconstruction. The history of horses (family Equidae) has been heavily investigated from

this point of view and presents a fascinating example of adaptive evolution as horses shifted gradually, and somewhat sporadically, from a predominantly browsing to grazing way of life. The analysis of locomotory and feeding habits of extinct horses and other species has usually relied almost entirely on observations and measurements of the outer shape and size of bones and teeth. However, the study of the behavior of extinct mammals has been advanced in recent years by the innovative application of both old and new technologies, such as x-ray imaging and electron microscopy, to questions of animal design and function. Paleoecological studies of community evolution have also gained in importance, not so much from new technologies as from recent syntheses of biostratigraphic data on both continental and global scales.

Scanning electron microscopy. The scanning electron microscope has made a major contribution to studies of tooth function and dietary habits in a variety of fossil mammals, ranging from sabertooth cats (*Smilodon* species) to ancient hominids. Photographs taken with the scanning electron microscope of the wear surfaces of teeth reveal patterns of scratches and pits created as food and associated grit are crushed or sliced between opposing teeth. Foods that differ in texture produce distinct wear patterns. Consequently, analyses of these dental microwear patterns can provide additional evidence of the diets of extinct animals and are particularly useful when studying species whose teeth are without modern analogs. For example, sabertooth cats evolved a number of times in the past and clearly represented a successful morphology, and yet they have no living counterparts. As a result, their diets and killing behavior have been the subject of much controversy. Although it is generally agreed that the elongate, knifelike canines were used in killing, there is less concordance concerning subsequent feeding behavior. Some have argued that the unusual teeth of a sabertooth cat such as the North American *S. fatalis* (**Fig. 1**) would have made it diffi-

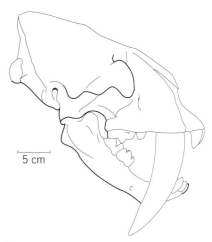

5 cm

Fig. 1. Skull of the North American sabertooth cat, *Smilodon fatalis*. (*After S. B. Emerson and L. Radinsky, Functional analysis of sabertooth cranial morphology, Paleobiology, 6:295–312, 1980*)

cult for them to gnaw on bones and efficiently clean a carcass, and thus these predators would have provided scavengers, such as wolves, hyenas, or hominids, with many leftover meals. A scanning-electron-microscope study comparing the microwear patterns on the lower first molars of *Smilodon* and living cheetahs, lions, leopards, hyenas, wolves, and wild dogs suggested that, in fact, this was the case. The lower molars of *Smilodon* exhibited relatively little wear, consisting mostly of parallel fine scratches and very few pits, a pattern most similar to, but more extreme than, that of the cheetah, a cat known to consume few bones (**Fig.** 2*a*). By contrast, more frequent bone crushers, such as the spotted hyena and the lion, display a distinct microwear pattern dominated by pits and short, broad scratches that criss-cross the wear surface (Fig. 2*b*). Similar comparisons of dental microwear in living primates have distinguished leaf eaters from hard-object (tough fruits and nuts) feeders; as is true of the carnivores, hard-object feeders display many more pits and fewer scratches than leaf eaters. Based on these comparative data, microwear analysis of robust and gracile australopithecine hominids suggests that the two differed in dietary behavior, with the gracile species being greater leaf eaters than the robust species.

The scanning electron microscope has also been applied to studies of evolution of tooth microstructure. Mammalian teeth are capped by enamel, an extremely durable material composed of calcium phosphate crystals aligned as rods or prisms. Scanning-electron-microscope photographs of prism orientation in the enamel of many living mammals reveal a common pattern of crossed sets, known as Hunter-Schreger bands, of enamel prisms that produce a banded appearance. It has been suggested that the crossed sets increase the strength of the tooth in a manner analogous to the structure of plywood, making it more difficult for serious cracks and splits to propagate. Interestingly, the banded arrangement of prisms first became common among mammals in the middle Paleocene (about 62×10^6 years ago), as mammals began to evolve larger body sizes and more varied diets. Prior to that time, prisms tended to be oriented in parallel and no

layering was visible. It appears that the new, layered arrangement was favored by natural selection because it allowed an increase in bite force and therefore a wider variety of food items.

X-ray imaging. Standard x-ray imaging is a relatively old technology that has been increasingly utilized by paleontologists. Recently, it has been applied with some success to the problem of estimating body size in extinct species. Body size is an extremely important characteristic of animals, affecting countless aspects of their lives, including diet, locomotory habits, life history features, and anatomy. Because bone is a living tissue that responds to the loads it must carry, the thickness of compact bone in an animal's limbs reflects its body mass. Using x-rays of long bones, the distribution of compact bone can be quantified and the relationship between bone dimensions and body weight determined for living species. Compact bone of the fossilized skeletal elements of extinct species can be similarly quantified and then used to estimate body weight. Weight estimates based on compact bone thickness are significantly more accurate than previous estimates based on more indirect measures of mass such as tooth size or linear dimensions of the skeleton. In addition to its utility for weight estimation, x-ray imaging of compact bone distribution holds great potential for studies of feeding behavior in extinct mammals. Recent radiological studies of the lower jaws of large predatory mammals indicate that cats (family Felidae) exhibit a different pattern of compact bone distribution than dogs (family Canidae) and that these differences reflect distinct killing behaviors. Felids tend to kill with a single, strong crushing bite and have more strongly buttressed mandibles than dogs, which tend to use multiple, weaker bites to subdue their prey. Preliminary analysis of *S. fatalis* mandibles indicates that the sabertooths had an extremely powerful catlike bite.

Syntheses of biostratigraphic data. Studies of fossil mammals within an ecological framework have benefitted from the technologies of x-ray imaging and electron microscopy inasmuch as community reconstructions are dependent on the quality of the inferences concerning the component species. Any

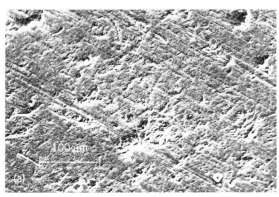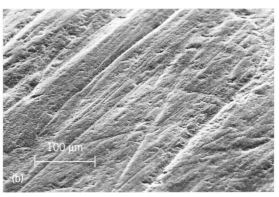

Fig. 2. Scanning-electron-microscope photographs of the wear facet of the lower first molars of (*a*) a cheetah, *Acinonyx jubatus*, and (*b*) a lion, *Panthera leo*. (After B. Van Valkenburgh, M. F. Teaford, and A. Walker, Molar microwear and diet in large carnivores: Inferences concerning diet in the sabretooth cat, Smilodon fatalis, J. Zool., 222:319–340, 1990)

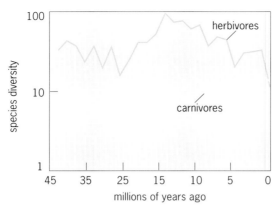

Fig. 3. Species diversity of large mammalian carnivores and their presumed prey, large herbivores, over the last 40×10^6 years in North America. Time scale is approximate. (*After R. E. Ricklefs and D. Schluter, eds., Species Diversity: Historical and Geographical Perspectives, University of Chicago Press, 1993*)

advance that can provide more detailed information about the paleobiology of species within a fossil community assists the paleoecologist in interpreting past environments. However, a less tangible but equally important advance for mammalian paleoecologists has been the recent publication of important syntheses of biostratigraphic data. Such data are essential for the study of large-scale patterns of evolution at the community, continental, and global levels. For example, to answer the question of whether mammalian predator and prey diversity levels have tracked each other over the last 65×10^6 years in North America, it was necessary to compile continent-wide lists of contemporaneous predator and prey species for a number of time intervals. The plot of predator versus prey diversity (**Fig. 3**) shows that the two groups were roughly correlated, with the number of large carnivore species rising and falling in concordance with those of their presumed prey, large herbivores. However, the match is not exact; herbivore numbers fluctuated more widely than those of the carnivores, suggesting that predator diversity was limited by factors other than numbers of herbivore species. Without good biostratigraphic data on which fossil deposits are coeval with others, this would be an extremely difficult and probably impossible task. All questions of changing diversity patterns and their causes, whether between predator and prey or potential competitors (for example, rodents and multituberculates in the early Cenozoic), require evidence of temporal and spatial overlap and thus are unanswerable outside a biostratigraphic framework. Such a framework must be built on decades of work by individual paleontologists, each of whom contributes relatively detailed data concerning one or more fossil deposits from a given geographic area. After more than a century of paleontological work in North America, there are now enough data available to produce comprehensive surveys of mammalian evolution on a continent-wide level. Although details of these surveys are continually being updated as new material arises, they remain stable in broad outline and serve as the starting point

for many potentially profound studies of ecosystem evolution.

For background information *SEE FOSSIL; MAMMALIA; PALEOECOLOGY; PALEONTOLOGY* in the McGraw-Hill Encyclopedia of Science & Technology.

Blaire Van Valkenburgh

Bibliography. J. Damuth and B. J. MacFadden (eds.), *Body Size in Mammalian Paleobiology*, 1990; D. E. Savage and D. E. Russell, *Mammalian Paleofaunas of the World*, 1983; M. F. Teaford, A review of dental microwear and diet in modern mammals, *Scan. Microsc.*, 2:1149–1166; M. O. Woodburne, *Cenozoic Mammals of North America*, 1987.

Marine archeology

The relatively new discipline of marine archeology, made possible by the development of self-contained underwater breathing apparatus (scuba), is devoted to the study of the history of seafaring. Marine archeology provides unique glimpses of early maritime commerce, naval warfare, ship construction, and shipboard life. But because virtually all human products, from tiny bits of jewelry to the huge stone architectural members of temples and churches, have at one time or another been transported by sea, marine archeology is also a study of human technical and artistic development.

Importance of seafaring. Watercraft have allowed the exploration, colonization, supply, and defense of much of the Earth's surface. Further, seafaring is one of the oldest of human activities, preceding farming, the domestication of animals for food, or the production of pottery and metals. The earliest evidence is indirect, going back more than 30,000 years, when humans first reached and settled Australia. Ten thousand years ago, in the Mediterranean, seafarers are known to have made round trips to the island of Melos to obtain obsidian for manufacture into the tools that are found in cave dwellings on the Greek mainland. The island of Crete seems to have been colonized deliberately about 6000 B.C.

Since those early times, ships such as the *Santa Maria* and the *Mayflower*, sailors like Columbus and Magellan, and naval battles from Salamis to Midway have had a profound impact on the course of history. The modern world would be very different without the tankers that carry oil across the oceans.

Preservation on shipwrecks. From the beginning of seafaring, ships of all types have been wrecked. Shipwrecks are especially valuable to archeologists as excavation sites because the artifacts in them have been well protected from later human destruction. Most of the monumental Greek bronze statues now known, for example, have come from the sea. The earliest known glass and tin ingots, raw materials that would have been fashioned into more complex artifacts if they had arrived at their intended destinations, were found on a fourteenth-century B.C. shipwreck off Uluburun, Turkey (**Fig. 1**), that also yielded the oldest known wooden writing tablet and the first known royal gold scarab of Nefertiti, a queen of Egypt. Other examples of unique finds are complete sets of balance-pan weights from all periods of antiquity, and a mechanical astronomical computer from ancient Greece that would have been thought impossible by modern scholars.

Organic materials are also better preserved on shipwrecks than on land in most instances, providing information on trade in foodstuffs, spices, and perfumes. The *Mary Rose*, flagship of Henry VIII, held hundreds of rare examples of the longbows for which English soldiers were famous. Even the clothes of drowned sailors were well preserved on the seventeenth-century Swedish warship *Vasa*.

Dating. Another important feature of shipwreck excavation is the precise dating often provided. If a wreck can be closely dated by coins or written documents found in it, all of the wreck's other contents can be said to have been in use at that time. The best-dated collection of seventh-century Byzantine pottery comes from a wreck excavated off the Turkish coast, and an eleventh-century wreck excavated off Turkey is providing the best chronological guide for medieval Islamic pottery and glass with its uniquely large collections of both. Land-based archeologists use such collections as comparative material to date the ceramics and other finds they excavate, thereby often dating their terrestrial strata with greater accuracy than would otherwise be possible.

Fig. 1. Excavation of four-handled copper ingots still lying in overlapping rows in a ship that sank off Uluburun, Turkey, in the fourteenth century B.C. The excavation is at a depth of about 165 ft (50 m). (*Institute of Nautical Archaeology; photograph by D. A. Frey*)

Development of field. Marine archeology has developed through several phases. The earliest, a phase that began in antiquity, embraced the chance finds of artifacts by fishermen and sponge divers. The development of scuba in the 1940s allowed divers to work carefully in fragile ships' remains, and pioneering research was conducted on Roman wrecks then and in the 1950s, especially along the French and Italian coasts. Various types of suction hoses and pipes for the removal of sediment, and balloons for raising heavy objects, were adapted for archeological use.

The third phase of marine archeology began when archeologists began diving and directing underwater excavations. The first complete excavation of an ancient ship on the seabed was conducted in 1960 on a wreck of about 1200 B.C. off Cape Gelidonya, Turkey. This was also the first wreck excavated to acceptable archeological standards.

Because an excavator's time is limited under water, often to less than an hour a day in two dives, depending on the depth, marine archeologists devoted much of the 1960s to the development of more efficient methods of excavation. Mapping techniques, including grids, stereophotography, and sonic triangulation systems, now allow plans to be made more than 150 ft (45 m) deep that place each bead and ceramic fragment to within 0.4 in. (1 cm) of accuracy.

Shipwreck location. At the same time, remote-sensing devices were adopted to aid the search for sunken ships. Sonar, by bouncing sound waves off the ocean floor, can detect ancient cargoes that rise up above the sand or mud. Because it cannot differentiate between a mound of pottery and a natural rock outcrop on the seabed, however, each anomaly detected must be examined by divers or by underwater television. Magnetometers detect iron and are thus of more limited use in the search for Greek and Roman ships than for later ships, which carried iron anchors and cannons. Techniques differ for other reasons between the search for ancient wrecks and more modern ones. Because no accurate records exist to lead archeologists to ancient ships, ancient ships are usually discovered by accident, often by local divers who report them to archeologists. During and after excavation, the archeologist tries to identify the ship, at least as to date and nationality. For ships of more recent times, archival research can direct the archeologist to a specific area for a specific ship. In that case, the archeologist must determine if the discovered ship is the one sought.

Excavation. During excavation, a wreck must be carefully mapped to show the exact location of every artifact and hull remnant on the seabed. Accurate plans later allow the archeologist to determine which items were cargo and which were for personal shipboard use, and this often helps to identify the nationality of a ship and its route. Plans of the wooden remains are essential for the reconstruction of the hull on paper, and sometimes for the complete restoration of the hull. After being mapped, all remains are brought to the surface, where different materials require individual conservation treatments. Although diving plays a major role in marine archeology, 2 years of subsequent laboratory conservation and recording are normally required for every month of diving. The distinction between treasure hunting and marine archeology, not always understood by the public, is that salvors intent on making a profit through the sale of finds cannot spend up to 20 years conserving, studying, interpreting, and publishing their sites, and thus much valuable information is always lost.

Study of wooden hulls. As mapping and conservation techniques were refined during the 1960s and 1970s, an ancient ship's wooden hull often became the artifact of primary importance to the marine archeologist. A fourth-century B.C. ship excavated off Kyrenia, Cyprus, provided the first ancient hull ever restored. Its thousands of wood fragments were conserved in polyethylene glycol to prevent their warping and shrinking out of recognition, and then were reassembled for display in Kyrenia. Only about 75% of the hull was found because any wood that was not covered by seabed sediment was quickly devoured by marine borers such as shipworms. The Swedish warship *Vasa* was also treated with polyethylene glycol, but in that case the ship was raised nearly intact from Stockholm harbor (**Fig. 2**), whose low salinity does not support marine borers.

This interest in the ships themselves led to a branch of marine archeology known as nautical archeology. Nautical archeology, however, may be practiced on dry land as well, as with the study of hundreds of ships found beneath land that has been reclaimed from the sea in the Netherlands and elsewhere.

Academic discipline. In its most recent phase, marine archeology has become an academic discipline. Universities in several countries grant advanced degrees to students who, in addition to studying the courses required of all archeologists, take specialized courses in the history of seafaring (including maritime commerce and naval warfare), and the conservation of finds from the sea.

Fig. 3. Archeologist using white thumbtacks to mark the ends of wooden pegs of a late-fourth-century A.D. Byzantine ship excavated off Yassiada, Turkey, at a depth of 130 ft (40 m). The tenons within the mortise-and-tenon joints holding together the planks of the ship were fixed permanently in place by a pair of such pegs that ran through the thickness of a joint, one to each side of the planking seam. (*Photograph by G. F. Bass*)

Evolution of ship construction. Although the types of watercraft used by the very earliest seafarers remain unknown, remarkable progress has been made in tracing the evolution of wooden hull construction from the Bronze Age until modern times. From at least the fourteenth century B.C. until the fifth century A.D., European ships were built in a manner called shell-first. Unlike modern ships, their hulls were built up one plank at a time from the keel, without any interior framework (**Fig. 3**). In the Mediterranean the planks were held together either by mortise-and-tenon joints along their edges or by rope lacing; in northern Europe, as was the case with Viking ships, each plank usually overlapped the plank below and was nailed to it, although lacing was also used there in early times, as on boats from about 1500 B.C. found at North Ferriby in England, or a boat of about 350–300 B.C. found in southern Denmark. Frames (commonly called ribs) were installed later, but they provided less strength to the hull than in modern ships.

Most wooden hulls are now built in the frame-first manner, in which a skeleton of frames is first attached to the keel, and the planks are then fastened to it. The slow evolution of the ancient manner into the modern has been traced through a seventh-century A.D. ship excavated at Yassiada, Turkey, whose hull was built shell-first below the waterline and in the frame-first method above. The earliest example of a seagoing vessel built entirely in the modern manner is an eleventh-century A.D. ship excavated at Serçe Limani, Turkey.

Although the transition from shell-first to frame-first construction represented a technical improvement, allowing the construction of the large seagoing vessels that first rounded Africa for the East or crossed the Atlantic, the causes of the transition are still sought. The reasons may have been economic, for frame-first construction is less labor-intensive and may reflect a decline in slave labor; at the same time, it uses timber more sparingly, and thus deforestation may have played a role.

Terrestrial sites. Although marine archeology is primarily devoted to the study of seafaring, it in-

Fig. 2. Raised hull of the Swedish warship *Vasa*, which sank in Stockholm harbor in 1628, mounted on a concrete pontoon in the framework of an aluminum building that housed it during conservation and restoration. (*Swedish Information Service*)

cludes the study of terrestrial sites, such as Port Royal, Jamaica, which sank during an earthquake in 1692, as well as early harbor installations, such as at Kenchreai, an early port of Corinth in Greece, and Roman Caesarea in Israel.

Noteworthy marine excavations. Noteworthy marine excavations, besides those mentioned above, include those of an Early Bronze Age ship of about 2500 B.C. off the Greek island of Dokos; a fishing boat in the Sea of Galilee from the time of Christ; an Etruscan wreck off Giglio in Italy; a classical Greek ship off Porticello, Italy; a Roman wreck at Madrague de Giens in France; five Viking ships of about A.D. 1000 in Roskilde Fiord, Denmark; a fourteenth-century Chinese trader found off Korea; an early-sixteenth-century wreck in the Turks and Caicos Islands in the Caribbean; a sixteenth-century Basque whaler in Red Bay, Labrador; warships from the Spanish Armada lost off Scotland and Ireland; Dutch East Indiamen in Australian waters; and the seventeenth-century Portuguese *Santo Antonio de Tanna* in Mombasa Harbor, Kenya. Marine archeology is spreading around the world, with current or planned excavations in lands including Thailand, China, Bulgaria, Ukraine, Mexico, Jamaica, Brazil, and Syria.

For background information SEE ARCHEOLOGY; DIVING; SHIPBUILDING; SHIPWORM in the McGraw-Hill Encyclopedia of Science & Technology.

George F. Bass

Bibliography. G. F. Bass (ed.), *A History of Seafaring Based on Underwater Archaeology*, 1972; G. F. Bass (ed.), *Ships and Shipwrecks of the Americas*, 1988; K. Muckelroy (ed.), *Archeology Underwater: An Atlas of the World's Submerged Sites*, 1980; P. Throckmorton, *The Sea Remembers*, 1987.

Marine ecosystem

Marine ecosystems are highly varied, depending on water depth and latitude. Around the ocean margins, large seaweeds, vascular plants, and their detritus form the base of food webs. In the open sea, microscopic plants predominate. In polar regions, ice plays key roles in ecosystem structure. In all marine ecosystems, microorganisms are major movers of energy and materials. They are essential to the stability of the total food web, because they rapidly recycle elements such as nitrogen and phosphorus, the supply of which is essential to continued photosynthesis.

Except for those at the ocean margins, marine food webs begin with the production of organic carbon compounds by microscopic plants and bacteria. These compounds are mostly synthesized from carbon dioxide and water during the process of photosynthesis. Because the ocean is on average 15,740 ft (4800 m) in depth, and the water quickly absorbs the entering sunlight, only the uppermost 2% of the ocean is illuminated and can thus be a habitat for green plants and photosynthetic bacteria (which are collectively called photoautotrophs). Except for inputs of organic materials from rivers on the ocean's periphery and fallout

of organic matter blown into the atmosphere from the continents, both of which are relatively small inputs, all of the organic matter to supply the ocean's food web must originate in that illuminated 2%. The ocean's great depth also dictates that there can be no attached photoautotrophs except in the shallow waters around the edges of continents and around islands. As a result, with the exception of two species of *Sargassum* floating in the North Atlantic and the Gulf of Mexico, all photoautotrophs of the open ocean are microscopic. Microscopic plants are not damaged by waves and can more easily remain afloat than can larger plants such as *Sargassum*.

Microscopic photoautotrophs. The smallest photoautotrophs, *Synechococcus*, are blue-green bacteria less than 1 micrometer in diameter. The largest of the microscopic photoautotrophs are colonial blue-green bacteria (cyanobacteria) of the genus *Oscillatoria*, found in the tropical and subtropical oceans. An *Oscillatoria* colony is about 0.04 in. (1 mm) in diameter. When these bacteria are abundant, the colonies can be seen as tiny specks in the water. While photosynthetic bacteria are important contributors to marine food chains, the microscopic plants ranging in size from 0.4 μm to 1 mm, collectively called phytoplankton, are the major producers of organic matter in most food webs.

The distribution of photoautotrophs in the sea seems to be regulated largely by the supply of essential nutrient elements, such as nitrogen, phosphorus, or iron, all of which are depleted in ocean surface waters. Therefore, the largest concentrations of photoautotrophs occur around the margins of ocean basins or in locations where deeper water is welling up and bringing nutrients into the lighted surface waters. The distribution of photoautotrophs is seen dramatically in satellite imagery that shows the distribution of chlorophyll at the ocean's surface.

Consumers. Since most consumer organisms eat other organisms that are about 10% of their body size, most of the consumers of the microscopic photoautotrophs in the ocean are also microscopic. The smallest photoautotrophs are eaten largely by ciliated and flagellated protozoa (see **illus.**). The largest are eaten by zooplankton or larval fishes, which are themselves mostly microscopic or nearly so. However, many of the photoautotrophs in the sea are never encountered and therefore never eaten by zooplankton.

Autotrophic production is driven by the supply of nutrients; since the supply is often erratic, bursts of production of photoautotrophs tend to be followed by their death from nutrient depletion. The photoautotrophs are then consumed by heterotrophic (nonphotosynthetic) bacteria, of which there are approximately 10^8 in every quart (liter) of ocean water. Most of these bacteria are extremely small, 0.2–0.4 μm in diameter, and can be seen only with an electron microscope. The heterotrophic bacteria also utilize dissolved organic materials released by phytoplankton and zooplankton.

By responding rapidly to changing conditions and utilizing dissolved materials in extremely dilute concentration, marine heterotrophic bacteria often utilize 25–50% of the organic matter produced by photoautotrophs in the sea. The bacteria become part of the food web, but because of their extremely small size they are eaten principally by flagellates that are 5–10 μm in diameter and some small ciliates. Most of these protozoa are eaten by larger protozoa, and then the larger protozoa are consumed by zooplankton (see illus.). Various lines of evidence suggest that five microscopic food-web stages typically separate photoautotrophs from organisms having the size of juvenile fishes. When organisms consume other organisms, their assimilation efficiency is rarely more than 50%. So, after five such energy conversions, not more than 5% of the organic matter produced by photoautotrophs in the sea reaches juvenile fishes, and less than 1% reaches larger fishes. This distribution certainly affects the abundance of fishes.

Food webs in deeper water. Most of the organic matter produced by phytoplankton and blue-green bacteria in the ocean is utilized quickly by the heterotrophic consumers in the upper ocean water. Probably less than 5% of the production of phytoplankton reaches the deep waters and the sea floor through the rapid sedimentation of relatively large particles such as dead phytoplankton and zooplankton, and fecal pellets of zooplankton. Distributed down through the water column are some larger planktonic organisms that intercept falling bits of organic matter by casting mucus nets or by pumping water through filtering organs. The organisms living on the sea floor and buried in the sediments of the deep sea tend to be small and very long lived, and adapted to a meager and uncertain food supply. It is not unusual to find mollusks in the deep sediments that are less than the size of a fingernail but over 25 years old. For example, shells of the abyssal clam *Tindaria callistiformis*, found at a depth of 12,350 ft (3800 m) in the North Atlantic, had a mean length of 0.33 in. (8.4 mm) and a mean age of 100 years. Heterotrophic bacteria are even more abundant in bottom sediments than in the water column; here too they utilize a major share of the settling phytoplankton and other organic particles. Bacteria in sediments are utilized as food by a wide variety of micro- and macroorganisms. The paucity of organic matter settling into the ocean depths is in one sense a good thing—if the amount were much greater, bacteria and other organisms would deplete all of the oxygen dissolved in the water as they converted the falling organic matter to carbon dioxide and water, and the deep sea would become uninhabitable for all except anaerobic bacteria—those that can survive without oxygen. This situation exists in the Black Sea, and in the geologically distant past it probably happened in some major ocean basins.

Polar marine food webs. The marine food web in polar regions is influenced by the seasonality of sunlight, and by snow and ice cover, which can limit light penetration. If little snow covers annual sea ice, enough sunlight can penetrate 3–6 ft (1–2 m) of ice to support photosynthesis. Microbial communities in sea ice develop in early spring, when there is 24 h of sunlight per day but the ice has not yet melted enough to promote development of phytoplankton in ice-free water. A complete food web develops in sea ice, including diatoms, heterotrophic bacteria, and protozoa within the ice, and zooplankton that graze along the underside of the ice. In late spring and summer, as sea ice and icebergs melt, luxuriant blooms of phytoplankton develop around the melting ice edges. Because of this two-stage development of photoautotrophs, polar regions have a relatively long period of productivity in spite of the short summer season. The long, largely sunless winter, however, accounts for polar regions not being exceptionally productive on an annual basis.

Chemosynthetic webs. A specialized marine food web occurs on the deep sea floor along rifts and mid-ocean ridges, where seawater has come in contact with very hot volcanic rocks so that the abundant sulfate in seawater is converted to hydrogen sulfide. Chemosynthetic bacteria, some capable of living at 212°F (100°C) and at pressure of 500 atm (50 megapascals), use the energy they acquire

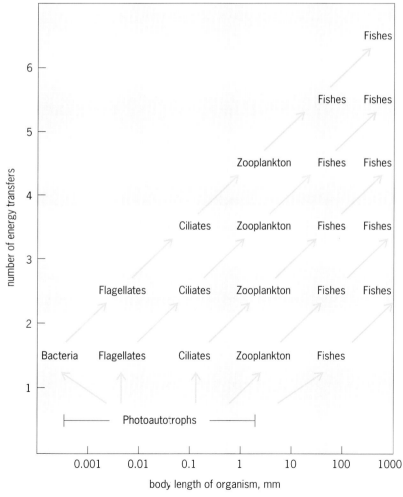

Simplified conceptual model of the marine food web, showing the relation of body size of consumers to the number of energy transfers between consumers and photoautotrophs. Arrows lead from prey to predator. 1 mm = 0.04 in. (*After F. Azam et al., The Ecological Role of Water-Column Microbes in the Sea, Mar. Ecol. Prog. Ser., vol. 10, pp. 257–263, Springer-Verlag, 1983; and C. Moloney, A Sized-Based Model of Carbon and Nitrogen Flows in Plankton Communities, Doctoral Dissertation, University of Cape Town, 1988*)

from oxidizing the sulfide back to sulfate to synthesize organic matter. The bacteria are the base of a food chain that includes filter-feeding clams and scavenging crustaceans. The bacteria are also present as symbionts inside the bodies of highly specialized worms (Vestimentifera) living in the sulfide plumes. The worms lack a functional gut and subsist entirely on organic matter supplied by the symbiotic bacteria.

Tidal mud flats. Shallow tidal mud flats combine aspects of all of the above food webs, but the major producers of organic matter are macroscopic algae or, often, higher plants: eel grass (*Zostera*) in temperate and subarctic waters or turtle grass (several genera) in the tropics. While some sea grasses are consumed directly by turtles, manatees, or geese, much of the grasses is worn away by storms and becomes detritus. The detritus from sea grasses and seaweeds is a significant food source for many invertebrates and fishes in coastal waters, but, as in the food web of the open sea, a large fraction of it is utilized directly by bacteria before the higher organisms have an opportunity to use it. As a result of the very high productivity of sea grasses, excess organic matter accumulates in the sediments of most mud flats. Bacteria utilizing some of the organic matter deplete the oxygen in the sediments, and anaerobic bacteria then convert some of the sulfate dissolved in seawater to sulfide. A clam, *Solemya*, that lives in temperate mud flats has been found to contain symbiotic sulfide-oxidizing bacteria, and thus has a food source just like that of the vestimentiferan worms of the deep sea.

For background information *SEE FOOD WEB; MARINE ECOSYSTEM; MARINE MICROBIOLOGY; PHYTOPLANKTON; ZOOPLANKTON* in the McGraw-Hill Encyclopedia of Science & Technology.

Lawrence R. Pomeroy

Bibliography. M. J. R. Fasham (ed.), *Flows of Energy and Materials in Marine Ecosystems*, 1984; L. R. Pomeroy and W. J. Wiebe, Energetics of microbial food webs, *Hydrobiologia*, 159:7–18, 1988; P. C. Reid, C. M. Turley, and P H. Burkhill (eds.), *Protozoa and Their Role in Marine Processes*, 1991.

Mechanism

Mechanisms are mechanical devices such as gears, shafts, pulleys, rods, and joints that perform some useful function. Such mechanisms are normally assembled to form a system, such as an automobile transmission, which involves not only mechanical components but also electronic control components for sensing the speed and acceleration. Micromechanisms typically are small versions of these complex systems. The construction of these small systems is motivated by applications that cannot accommodate large dimensions. Tools for surgery on the inside of the eye or precision drug-dispensing systems are good examples. The major dimensions of micromechanisms are measured in micrometers. A "large" system may involve sizes that are one or two human-hair diameters (typically about 75 μm). A "small" system may have dimensions in

the submicrometer range, measured in terms of wavelengths for photons in the visible light spectrum.

Micromechanics, the technology of micromechanisms, is now passing the feasibility stage and is advancing rapidly. The areas of application are very large, including high-speed mechanical control systems, microsurgery tools, and many unusual devices yet to be discovered and identified.

Attributes of micromechanisms. Micromechanisms have some unusual attributes. Because the systems are very small and weigh very little, speed control by a large mass, such as the flywheel in an automobile engine, is not available. Instead, high-speed electronic circuits are used to sense and adjust speeds virtually instantaneously. In an airplane engine, such instantaneous adjustments would lead to catastrophic failure. In micromechanics, this type of adjustment is acceptable because weights, or more precisely moments of inertia, are low. In micromechanisms, material properties for metals and ceramics with high surface area–to–volume ratios and nearly perfect surface finishes change significantly from the bulk behavior. Higher tensile strengths and less fatigue- and crack-induced failures are typical.

These properties are, of course, beneficial, but there are also some negative aspects. In a 50-hp electric motor, for example, air friction is a nearly insignificant problem. In a micromotor, air friction becomes a large issue because the available power output is low. Furthermore, since moving surfaces may be within a few atomic distances of each other, additional drag forces that are not accounted for in classical friction theories may be produced. The same comment applies to the behavior of gases in small volumes. Since classical physics deals with the average behavior of gases, a question arises as to what happens if the total number of gas molecules is very small and average behavior is replaced by individual behavior. The positive aspect of this question hints at the possibility of constructing micromechanisms that work only because they are small and therefore have no larger counterpart.

Construction techniques. Micromechanisms, as stated above, are assembled mechanical components that use electronic controls. These systems are normally three-dimensional and must exploit properties such as strength, hardness, and frictional behavior of many different materials. Because of the small size, standard milling machines and lathes are not applicable. New, different construction techniques must be used.

Integrated-circuit fabrication. The basis for these fabrication procedures is found in the integrated-circuit industry, where the key process for local machining involves chemical machining or removal of material by chemical etching via optical pattern definition. A typical task in integrated-circuit engineering is that of cutting a very precisely defined pattern with submicrometer dimensions and tolerances into a 1-μm-thick silicon dioxide layer covering a rigid silicon substrate uniformly. The process is completed in three steps. In the first step, the desired pattern is drawn on a computer at large size. The pattern is transferred

Fig. 1. Exposed and developed photoresist pattern, displaying vertical photoresist flank. The polymer thickness is 18 μm.

to a mask, which is essentially a high-contrast black-and-white photograph of the desired image. Accuracy is obtained by optical reduction of the drawing to image size. The second step involves the silicon wafer itself. The silicon substrate is converted to an optical receiver by coating the oxide with a uniform layer of photoresist. Photoresist is a polymer such as Plexiglas that changes its molecular weight and chemical behavior when it is exposed to photon fluxes on the blue side of the visible spectrum. Exposure through the mask of the photoresist layer transfers the mask geometry to the photoresist. Immersion into the developer, a liquid that dissolves the modified-molecular-weight material but does not attack the unexposed polymer, transfers the mask pattern into photoresist-free regions on the wafer. The third step uses the chemical resistance of the remaining polymer. Immersion of the processed wafer into hydrofluoric acid causes oxide removal from the regions of the wafer not protected by photoresist. The end result is accurately machined, complex, geometrical patterns in the oxide layer that correspond exactly to the mask geometry.

Modification for micromechanisms. This basic process must be modified for the construction of micromechanisms. The idea of a mask and pattern transfer into photoresist is maintained. However, since assembly is desired and requires structures that are not attached to a substrate, much thicker films than those used in integrated-circuit manufacturing must be used. Thicker films are required to develop the intended end product, consisting of mechanical components; also, because very thin films are nonuniformly strained, when freed from their supports they deform like a spring that is initially compressed and then allowed to relax. Patterns in these thicker films must be produced with very high precision. Thus, a round shaft that is to be used as a wheel axle must maintain the same diameter throughout its length to support the wheel without wobble or undesirable friction. This requirement, the run-out problem, eliminates nearly all those processes that chemically or physically remove material (subtractive processes). Additive processes, or procedures in which material is added locally to the wafer, are much preferred, and can be implemented by adding material in the photoresist-free areas of the wafer. The amount of material that can be added into the recess depends on the photoresist thickness, which increases from 1 μm in integrated-circuit applications to 500 μm in micromechanics. The run-out problem can be solved if photoresist patterns with vertical flanks can be produced and if the added material conforms well to the substrate and photoresist flanks.

X-ray lithography. All of these conditions are met by pattern transfer through x-ray lithography, and subsequent electroplating of various low-stress metals. This process begins with a rigid substrate that is typically silicon, but could also be glass, ceramic, or metal. The substrate is covered with a thin release layer if unattached structures are to be produced. This release or sacrificial layer is a polymer that is a special form of polyimide. The unpatterned polyimide is covered with a thin film of titanium that provides adhesion for a thin nickel layer deposited on the titanium. These two metal layers make the substrate surface a conductor, a step necessary for subsequent processing. The substrate, which is still unpatterned, is next covered with a very thick film of photoresist. The film, which is typically thicker than 100 μm, is essentially Plexiglas, and is applied by casting techniques. The processed substrate is exposed by using the appropriate mask. This exposure must be strong enough to change the molecular weight throughout the film thickness. This exposure requires very energetic photons or, since

Fig. 2. Mechanical nickel gears on the working substrate prior to assembly.

Fig. 3. Five-component micromechanism, with two shafts, two pulleys, and a nickel band.

photon energy is inversely proportional to wavelength, very short wavelength light.

X-rays with wavelengths near 0.5 nanometer, the distance between atoms in single-crystal silicon, are ideal. These x-rays penetrate deeply, a fact well known in medical applications; and they do not reflect from surfaces, so high-intensity nonuniformities are prevented. X-rays also replicate the mask geometries nearly perfectly if they are well collimated, because high-fidelity transfer suffers from diffraction effects at roughly two times the exposing photon wavelength. Since 1 μm contains 2000 x-ray wavelengths of 0.5 nm, feature sizes near 1 μm do not suffer from diffraction-induced distortions. There is, however, a problem with x-ray exposure. The necessary flux densities for exposure are in the range 1–2 W/cm^2. This flux density requires a very bright x-ray source, and the only such

source available is a synchrotron or electron storage ring. In this type of machine, electrons with velocities near the speed of light emit x-rays when the electrons are forced to change direction. After exposure, the next step is that of developing. **Figure 1** illustrates the excellent flank acuity that this type of processing can produce.

Electroplating. The photoresist recess exposes the underlying nickel layer. Immersion of the entire substrate into an electroplating bath allows current conduction into the plating bath electrolyte only in the photoresist-free regions. The deposit can be one of many possible metals or alloys. Nickel and its alloys are particularly desirable because they are magnetic substances. Such materials can also be applied without large internal stresses, an important requirement if the coating is to be conformal with the resist flanks. The deposition rates, several micrometers per minute, lead to relatively short processing times for this attractive additive process.

Final processing steps. Chemical removal of all photoresist, unused plating base regions, and the entire sacrificial layer completes the process for parts that are not attached to the substrate. **Figure 2** illustrates the results. If it is desired to produce nickel structures that are rigidly affixed to the substrate, the process can be modified by simply not using the polyimide sacrificial layer. The fixed and free nickel parts can be assembled to produce micromechanisms. **Figure 3** uses five parts: two shafts, two pulleys, and a nickel band that is 4 μm wide and 100 μm high. This band acts as a flexible drive from one pulley to the next. The mechanism illustrates that a great deal of motion can be achieved with small pulleys.

Magnetic micromotor. A key component in micromechanical systems is a device that converts electrical input power to mechanical output power. This device, a motor, can be constructed by using the magnetic properties of the electroplated metal films. **Figure 4a** shows the stator design for a two-pole machine. The stator contains the rotor support shaft

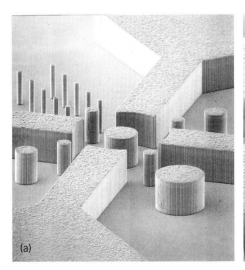

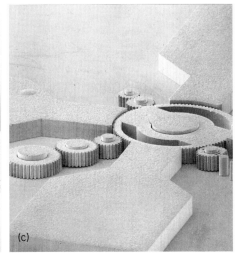

(a)   (b)   (c)

Fig. 4. Magnetic micromotor. (*a*) Stator. (*b*) Assembled micromotor. (*c*) Micromotor with external gears and gear train.

and the four pole pieces that supply the rotating magnetic field to the rotor. Figure 4b is a closeup of the stator-rotor configuration. The rotor design is that of a reluctance motor, a device that develops torque because the magnetic path from pole to rotor to opposing pole is a function of rotor position. This device was operated to speeds of 8000 revolutions per minute with external magnetic flux densities that are roughly 10 times the Earth's maximum magnetic flux density. This surprisingly low flux density is consistent with very low frictional losses, and led to an attempt to load the device with external gears. The resulting structure is shown in Fig. 4c. A field of less than 100 times the limiting saturation field in the magnetic material produced satisfactory rotation. These data further confirm the low-friction behavior and reveal the fact that the motor can produce very significant output torque.

For background information SEE INTEGRATED CIRCUITS; MECHANISM; MOTOR; RELUCTANCE MOTOR; SYNCHROTRON RADIATION in the McGraw-Hill Encyclopedia of Science & Technology.

Henry Guckel

Bibliography. H. Guckel et al., Fabrication and testing of the planar magnetic micromotor, *J. Micromech. Microeng.*, vol. 1, 1991; G. Stix, Golden screws: Micromechanical devices edge toward commercial uses, *Sci. Amer.*, 265(3):166–169, September 1991; *Technical Digest of the 1988 IEEE Solid-State Sensor and Actuator Workshop*, 1988.

Medical imaging

Three-dimensional or 3-D reconstruction in ultrasonic imaging includes (1) the collection of sequential two-dimensional or 2-D ultrasound images of an object, (2) the entry of these data in a computer memory, and (3) the creation of a three-dimensional viewing environment. Such a viewing environment results in either a true three-dimensional display, by the use of stereo images viewed through polarized glasses, or a pseudo three-dimensional display, which presents three-dimensional objects in a two-dimensional fashion on a computer screen but enhances visualization of the spatial relationships and shape by surface shading and depth perspective. SEE MEDICAL ULTRASONIC TOMOGRAPHY.

Three-dimensional ultrasound imaging may be useful for ophthalmological, abdominal, and vascular medicine (through the use of recently developed miniature intravascular transducers). The method has a promising role in echocardiography, because it improves the assessment of structural and functional heart diseases. In comparison with multislice x-ray computer tomography or magnetic resonance imaging (which also enable three-dimensional reconstruction), ultrasound provides studies that are relatively inexpensive, safe (because infusion of a contrast agent and x-ray exposure are not required), and dynamic (with an exposure rate of 30 images per second). In addition, ultrasound machines are portable and can be transported to an operating room or to a patient's bed.

The earliest methods of three-dimensional echocardiography were described in the 1970s, and since that time these techniques have developed rapidly. Originally, three-dimensional imaging was used to reconstruct the left ventricular chamber of the heart, with the assumption that the left ventricle has a simple ovoid shape. Despite this simplification, the method produced accurate volumetric measurements and aided the evaluation of left ventricular function. The reconstruction of several three-dimensional images throughout the heart cycle (that is, a dynamic three-dimensional study, also called four-dimensional reconstruction) provides information about the motion of the left ventricular wall and its abnormalities. Advanced ultrasound machines, produced since the late 1980s, feature transducers with increased resolution. Such transducers enable the reconstruction of three-dimensional images that are more anatomically detailed and can, for example, depict the three-dimensional shape of the mitral valve.

Scanning methods. The collection of the data that specify a three-dimensional object begins by scanning a sequence of two-dimensional images throughout the object. The motion of the transducer has to be exactly defined during the scan in order to obtain the correct orientation of the data in the computer memory. There are several methods.

Compound B-scan. In this method, a transducer producing a single ultrasound ray is positioned on a linearly movable holder in a water bath. One sweep along a line collects backscattered energy in a plane through the object, so that one two-dimensional image is created. Then the holder with the transducer turns horizontally about a predefined angle, and a new two-dimensional image is obtained in the same plane but from a different angle. This procedure is carried out several times. A compound scan is constructed as the sum of individual two-dimensional scans oriented at their respective angles of view. After the compound scan is completed, the holder moves vertically to the next level, and the whole process is repeated until the entire volume is scanned at many planar levels. This method provides high-quality images, but unlike other modalities described below it is limited by the necessity of placing the object in a water bath.

Mechanical articulated arm. In this method, a hand-held transducer is attached to an articulated arm. Potentiometers measure the position of each joint as the transducer is manually scanned. The position of the transducer is continuously detected and stored parallel with the ultrasound images, so that the final three-dimensional image can be correctly reconstructed. The position error of this method, however, can be several millimeters.

Transit-time measuring method. In this method, the location of the transducer is calculated by measuring the sound transit time between three spark gaps (functioning as acoustic emitters) mounted on the transducer and three receivers fixed in defined positions in the space (usually a triangular holder). The transducer is manipulated by hand, but without

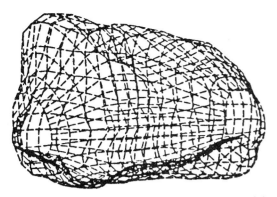

Fig. 1. Three-dimensional wire-frame reconstruction of the left ventricular chamber of the human heart.

any mechanical restraint of its motion. The accuracy of this approach is about 2–3 mm.

Pullback scan. In this method, a scanning transducer moves linearly in regular steps along a side of an object. The plane of scanning is perpendicular to the long axis of the linear motion; therefore, this method is called breadloafing.

Rotational scan. In this method, the scanning transducer is rotated so that the scanning plane performs a fanlike or propellerlike motion. The transducer movement in both the pullback scan and rotational scan methods is usually controlled by a stepper motor with an accuracy of approximately 1 mm and 1°, respectively.

Volumetric scanner. This device represents a different approach to three-dimensional imaging. The goal is real-time acquisition of three-dimensional ultrasonic data. This technique uses a hand-held transducer and control electronics capable of rapid steering of the beam in both up-and-down and sideways directions with simultaneous dynamic focusing in depth. This electronically regulated beam-forming results in a pyramid-shaped scanned volume having its apex at the transducer. Volumetric data from the region of interest can be presented as projected images with depth perspective, as stereoscopic pairs, or as multiple tomographic images.

Image processing. Images acquired by any nonparallel scanning method (fanlike or propellerlike, for example) must be converted into a three-dimensional rectangular volume of parallel slices. This representation simplifies further computer processing.

Interpolation is usually the next step. It ensures that the resolution in all three perpendicular planes of the three-dimensional rectangular volume is identical, so that cubical volume-oriented image elements, called voxels, are created.

Because ultrasound images usually contain many artifacts and noise, image enhancement becomes necessary. It utilizes special mathematical procedures, including contrast enhancement, to give better definition of contours, and image filtering, to remove noise spikes and to smooth edges.

Segmentation is a step in which one object is separated from other objects or the background according to the selected border line. The simplest segmentation

method is to detect edges by manually drawing a sequence of points along the object borders with help of a digitizing table, joystick, mouse, or light pen that controls the position of a cursor on the screen. The computer then matches this operator-defined tracing with corresponding image points, constructs the border line, and sets image elements outside or inside (according to the specification) the outlined object to the predefined values. For example, all picture elements outside the object can be set to zero (which usually means "erase"); hence, only the separated object remains in the image. Because of necessary operator interaction, this method is tedious and substantially subjective.

Computer-assisted border detection represents another solution. It economizes computer time and provides more reproducible results by reducing operator interaction. An example is searching the edge of a cardiac chamber (usually the endocardium of the left ventricle) in radial directions from the starting central point (determined either manually or automatically). Each individual radial search continues outward until a predefined value of a gradient between adjacent image elements is reached. This process is repeated along equally spaced radial lines until the entire endocardial circumference is delineated. The sample points are then connected by a computer algorithm to create a continuous border line representing the segmented ventricular chamber.

Image display. Three-dimensional display has several modalities in echocardiography. So-called wire-frame displays (**Fig. 1**) represent traced edges in a three-dimensional perspective view, providing an approximate definition of the left ventricular endocardial

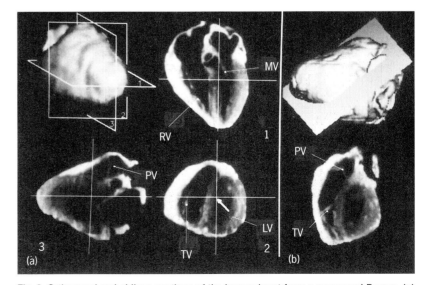

Fig. 2. Orthogonal and oblique sections of the human heart from a compound B-scan. (*a*) Three-dimensional rendered image and three orthogonal sections derived from it. Numbers on the orthogonal planes in the three-dimensional image correspond to numbers on the orthogonal sections. The level of each section can be modified interactively by moving an arrow (here shown in section 2) to reposition any depicted axis. (*b*) Three-dimensional image and oblique section of the same heart. The section is an unusual slice through the right ventricle and the right ventricular outflow tract, showing the tricuspid valve and the pulmonary artery valve in one plane with the left ventricle. LV = left ventricle; MV = mitral valve; PV = pulmonary artery valve; RV = right ventricle; TV = tricuspid valve.

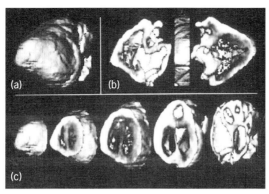

Fig. 3. Volume rendering of the human heart (the same heart as in Fig. 2). (*a*) View toward the heart apex. Shading and surface depiction are demonstrated. (*b*) View in which both left-hand side chambers and right-hand side chambers are cut and opened along the sagittal axis, demonstrating the capability of the technique to reveal the inner structures. (*c*) Coronal slices. Note the preserved backscatter texture in all slices.

and epicardial borders. This method makes it possible to evaluate a general shape and calculate the volume of the left ventricle.

Orthogonal-sections displays (**Fig. 2***a*) enable interactive generation of two-dimensional cross sections in the orthogonal planes. Interior sections can be revealed by slicing exterior planes away.

Two-dimensional oblique sections (Fig. 2*b*) that cut through a three-dimensional volume at any arbitrary angle can also be displayed. In addition to the usual echocardiographic projections, this method displays novel sections.

Surface rendering is a procedure that displays surface voxels as faces of small triangles. The rendering program rotates surface elements to the desired point of view and then processes visibility, shadowing, and depth queues to perform the final depiction or visualization. Surface rendering provides a complex view of detailed surfaces.

Volume rendering (**Fig. 3**) is based on ray tracing and displays surfaces with shading and a volume with voxel density values. The brightness or color of every voxel is selected according to its gradient value, spatial coordinate, and transparency. The gradient value determines surface segmentation, and the spatial coordinate of the voxel can be used for cutting the volume. Any desired angle of view and cut of the object are possible. In Fig. 3, the sectioned surfaces reveal the internal myocardial ultrasound backscatter texture of the original data. Surface rendering is a subset of volume rendering.

Prospects. Rapid progress in computer technology and ultrasonics is making three-dimensional imaging increasingly available. Highly sophisticated imaging systems based on either volumetric scanning or three-dimensional reconstruction of two-dimensional processed images are expected. Three-dimensional ultrasound imaging is a rapidly developing area with potential benefits for clinical practice, research, and education. In clinical practice, three-dimensional imaging can facilitate the exact localization of structures and

the determination of their size and morphology, which might be especially useful for surgeons. In research, the depiction of multiple slices throughout an object can be used to study its shape and backscatter texture in the course of experiments, without any physical cutting. In education, three-dimensional imaging reveals space relationships of anatomic structures.

For background information *SEE BIOACOUSTICS; COMPUTER GRAPHICS; ECHOCARDIOGRAPHY; IMAGE PROCESSING; MEDICAL IMAGING; MEDICAL ULTRASONIC TOMOGRAPHY* in the McGraw-Hill Encyclopedia of Science & Technology.

Marek Belohlavek; James F. Greenleaf

Bibliography. H. A. McCann et al., Multidimensional ultrasonic imaging for cardiology, *Proc. IEEE,* 76:1063–1072, 1988; R. Raqueno et al., Four-dimensional reconstruction of two-dimensional echocardiographic images, *Echocardiography,* 6:323–337, 1989; J. K. Udupa and G. T. Herman, *3D Imaging in Medicine,* 1991.

Medical ultrasonic tomography

A new instrument and methodology that provides a painless noninvasive determination of burn depth has been developed to assess burn injury. Burn depth is an important parameter, since it dictates the clinical regimen to follow in treatment. Full-thickness (third-degree) and some deep-dermal (severe second-degree) burns are excised and replaced with skin grafts, while partial-thickness (first- and second-degree) burns are generally left to natural healing processes. The natural healing processes normally take 2–3 weeks for completion. Until now, the assessment of burn depth has largely been a qualitative, visual inspection procedure, highly subject to the experience and skill of the burn surgeon making the diagnosis. In a recent study designed to assess the accuracy of visual diagnosis, experienced burn surgeons were able to provide the correct prognosis in only 50% of wounds that were neither superficial nor obviously full-thickness injuries. The need for an accurate and rapid diagnostic aid is apparent from this study.

The new instrument allows quantitative measurements of the depth of burn wounds in humans with a resolution of better than 60 micrometers (0.002 in.) by using an ultrasonic technique. Since total skin thickness (epidermis plus dermis) typically varies over the range 1–2 mm (0.04–0.08 in.) over different areas of the body, such resolution provides a measurement accuracy of approximately 3–6%. The accuracy is well within that needed for the most demanding surgery, where necrotic tissue thickness should be known to an accuracy of 10–15%.

Principle of operation. The quantitative assessment of burn depth by using ultrasound is highly dependent on the acoustic properties of the medium through which the sound wave propagates. Of particular importance to the present technology is the acoustic impedance. The acoustic impedance Z is defined as the product of the mass density ρ of the propagation medium and the sound speed v in that medium, Eq. (1). If the acoustic impedance remains constant along

$$Z = \rho v \qquad (1)$$

the propagation path of the ultrasonic wave, the wave will propagate unimpeded through the medium. If, however, the acoustic impedance changes abruptly or discontinuously along the propagation path, the situation shown in **Fig. 1** occurs.

A wave incident from the left in Fig. 1 is shown propagating through medium 1, which is characterized by a constant value of the acoustic impedance. The wave propagates unimpeded through medium 1 until it reaches the interface between medium 1 and medium 2, which is characterized by a different value of the acoustic impedance. The abrupt change in the acoustic impedance at the interface causes part of the incident wave to be reflected through medium 1 to the ultrasonic source. If the incident wave is a pulse, the reflected pulse is called a pulse echo. The remaining part of the wave is transmitted through the interface into medium 2. The amount R of the incident wave reflected from the interface (that is, the ratio of the intensity of the reflected wave to that of the incident wave) is given by Eq. (2). For the case of burned skin

$$R = \left(\frac{Z_1 - Z_2}{Z_1 + Z_2} \right)^2 \qquad (2)$$

tissue, medium 1 represents the necrotic tissue and medium 2 the underlying viable tissue.

The difference in the acoustic impedances of the necrotic tissue and viable tissue is found to be large and produces a large pulse-echo signal from the interface between the necrotic tissue and the viable tissue.

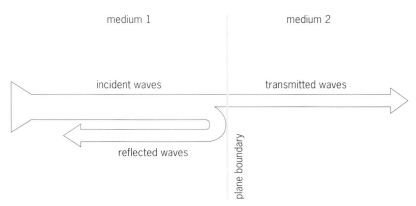

Fig. 1. Reflection and transmission of sound waves incident at a boundary between media of different acoustic impedances.

In **Fig. 2**, a histological section from burned skin tissue is compared with a typical ultrasonic pulse-echo pattern (line-of-sight time-domain spectrum or A-mode) obtained from the same burn site. In addition to the strong pulse reflection at the necrotic tissue–viable tissue interface, there are significant pulse reflections from the surface of the skin (epidermis) and from the dermis–subcutaneous fat interface, which serve as significant anatomical landmarks in assessing the degree of injury. The reflection occurring between the necrotic tissue–viable tissue interface and the dermis–fat interface is from an incidental cutaneous structure. The depth of the burn can be calculated from a knowledge of the sound velocity in necrotic tissue (1540 m/s or 60,630 in./s) and the time-calibrated ultrasonic pulse-echo pattern. The degree of burn can be assessed by comparing the relative positions of the

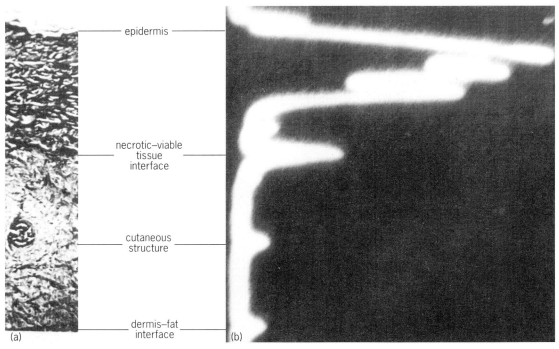

Fig. 2. Comparison of (*a*) a histological section taken immediately postburn with (*b*) the corresponding ultrasonic pulse-echo spectrum.

pulse reflections from the skin surface, the necrotic tissue–viable tissue interface, and the dermis–fat interface.

Physicochemical basis of necrosis. Thermal necrosis has been studied extensively since the 1940s. One of the most significant findings is the dependence of necrosis on the heating temperature and the length of time of heat application. At a temperature of 44°C (111°F) approximately 5 h are required for skin tissue to necrotize, while at a temperature of 65°C (149°F) skin tissue necrotizes for practical purposes almost instantly.

Conductive heat transfer in skin tissue is governed by the Fourier law of heat conduction. Thus, when heat is applied to the surface of skin tissue, the temperature rise at a given point in the tissue is dependent upon the distance from the surface and the time of heat exposure. For a given time of heat exposure, the temperature rise decreases as the distance from the surface increases. Hence, in order to reach a specific temperature at a specific point in the tissue, the time of heat exposure must increase as the distance of the specified point from the surface increases.

The specific temperature at which skin tissue changes from a state of viability to a state of necrosis is called the critical temperature. The physical properties of the tissue change at the critical temperature, and an interface between the viable tissue and the necrotic tissue develops. Since the critical temperature is reached at increasing distances from the surface as the heat exposure time increases, the necrotic tissue–viable tissue interface moves deeper into the skin tissue as a function of the heat exposure time.

The critical temperature T_c at which skin tissue necrotizes is given by Eq. (3). Here T_0 is the normal temperature of the skin, T_s is the temperature

$$T_c = T_s + (T_0 - T_s) \, \text{erf} \, \left[\beta (4a_1)^{-1/2} \right] \quad (3)$$

at the skin surface during the burning processes, a_1 is the thermal diffusion constant of necrotic tissue, erf is the symbol for the error function, and β is a constant relating the depth of burn, ξ, to the heat exposure time t according to Eq. (4). The error function is

$$\beta = \frac{\xi}{\sqrt{t}} \quad (4)$$

defined by Eq. (5), and its graph is shown in **Fig. 3**.

$$\text{erf} \, (x) = \frac{2}{\sqrt{\pi}} \int_0^x e^{-u^2} du \quad (5)$$

The parameter β can be determined directly from the ultrasonic pulse reflections and knowledge of the heat exposure time. From experimentally determined values of the parameters in Eq. (3), the critical temperature for necrosis, when the heating temperature is higher than the critical temperature, is calculated to be 65.3°C (149.5°F). This value of the critical temperature determined from ultrasonic measurements is in agreement with the high-temperature limit for skin necrosis found in other studies.

The ultrasonic measurements also predict that the energy of transition (enthalpy change) associated with thermal burn necrosis is 11.7 calories (49.0 joules) per gram of collagen in the skin tissue. The enthalpy change is the energy or heat that must be added at the critical temperature to produce the change from viable tissue to necrotic tissue. It is similar to the latent heat of melting (or freezing) that is required to transform ice to water at the melting temperature. The value of the enthalpy change and the critical temperature of 65.3°C (149.5°F) deduced from ultrasonic measurements are in agreement with the enthalpy change and critical temperature associated with collagen denaturation. Thus, the data suggest that collagen denaturation is the origin of skin tissue necrosis.

Collagen is a prominent constituent of skin and accounts for as much as 40%, by weight, of some dermal tissue. Studies show that collagen fibers destroyed by heat can no longer fulfill their physiological function and they lead directly to tissue necrosis. At temperatures below the collagen heat denaturation temperature, studies indicate the occurrence of a series of kinetic processes leading to the chemical denaturation of the collagen. When the temperature drops toward 44°C (111°F, the minimum threshold temperature for thermal necrosis), the time required for thermal necrosis becomes longer as an equilibrium between collagen damage and repair mechanisms becomes established. Chemical denaturation of skin collagen is also implicated in nonthermal insults leading to skin necrosis, such as electrical, ionizing radiation, and chemical exposures.

Collagen denaturation. The denaturation of skin collagen is manifested as a change in the collagen structure from a crystalline phase to an amorphous phase, and leads to a shrinkage of the collagen network in the tissue. Collagen shrinkage produces an abrupt and relatively large increase in the mass density of the necrotized skin tissue and hence causes a correspondingly large increase in the acoustic impedance at the necrotic tissue–viable tissue interface. This property accounts for the large reflection of the ultrasonic pulse at the tissue interface.

Skin structure. Although each of the necrotic and viable tissue components of burned skin is modeled thermodynamically and acoustically as being homogeneous and isotropic, skin tissue is actually highly textured and heterogeneous. The small-scale and some-

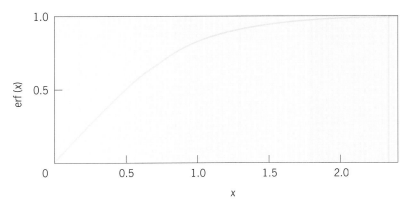

Fig. 3. Graph of the error function, erf (x).

what random nature of the heterogeneity, together with the small acoustic impedance variations associated with the skin texture, gives rise to a low-amplitude background noise in the pulse-reflection spectrum. Other anatomical structures such as hair follicles and pores in the skin tissue (Fig. 2) also serve as sites for pulse reflections, but the reflections are generally incidental and highly localized, in contrast to the globally extended reflections from the burn wound. The occurrence of such incidental structures can be assessed by moving the instrument transducer to an adjacent A-mode position on the skin surface or by employing the instrument B-mode capability, whereby the localized nature of the structure is directly displayed. In B-mode operation, the instrument transducer head rotates in a sweep motion to produce an area-scanned video display of a cross section of skin tissue sliced perpendicular to the skin surface.

Instrumentation. To achieve the resolution necesssary for burn-depth diagnosis, the instrument utilizes broadband, 60-MHz, pulse-echo electronics with a single 15–30-MHz broadband transducer for generating and receiving the ultrasonic pulses. An important feature is the use of a specially developed time-gain compensation network by which the ultrasonic reflections from the skin surface, the necrotic tissue–viable tissue interface, and the dermis–fat interface are brought to approximately equal amplitude in the instrument signal processor. The low-amplitude background noise and reflections from incidental structures are then eliminated by a discriminator or threshold rejection circuit in both the A-mode and B-mode operations of the instrument. This procedure leaves only anatomically significant reflections to consider in the interpretation of the displayed image.

For background information SEE BURN; COLLAGEN; CONDUCTION (HEAT); INTEGUMENT; MEDICAL ULTRASONIC TOMOGRAPHY; SOUND in the McGraw-Hill Encyclopedia of Science & Technology.

John H. Cantrell

Bibliography. C. P. Artz and J. A. Moncrief, *The Treatment of Burns*, 1969; J. H. Cantrell and W. T. Yost, Can ultrasound assist an experienced surgeon in estimating burn depth?, *J. Trauma*, 24(suppl. 9):64–70, 1984; M. E. Nimni, *Seminars in Arthritis and Rheumatism* (proceedings of a symposium sponsored by the Pan American Congress of Rheumatology, Washington, D.C., 1982), 1983.

Meteorite

Recent research has focused on using the isotopic compositions of interstellar grains of silicon carbide (SiC) in meteorites to infer the stellar sources of the grains and to model the evolution of carbon-rich stars.

Different kinds of isotopic anomalies indicating that not all presolar material was homogenized in the early solar system have been known in carbonaceous chondrite meteorites for some time. The recent discovery that microscopic SiC grains are carriers of a whole series of isotope anomalies has heightened interest.

These grains, amounting to only a few parts per million (ppm) of any particular meteorite, are embedded in the matrix of pristine chondrites, such as the Murchison meteorite, and are separated through a very complex microchemical procedure as acid-resistant residues. Because they were produced in the circumstellar envelopes of carbon stars and have survived both residence in the interstellar medium and formation of the solar system, they can be considered cosmic records of events that occurred outside the solar system prior to its birth.

The formation of the SiC grains seems possible in the relatively cool atmospheres of only those stars having a carbon-to-oxygen (C/O) ratio greater than 0.8, that is, double the solar C/O ratio. Among the candidates are red giant carbon stars (C stars), massive Wolf-Rayet stars of type C, novae, and classical type Ia supernovae resulting from low-mass stars in binary systems. The most interesting ones are the C red giants that experience thermal pulses on the asymptotic giant branch, constituting the so-called TP-AGB phase.

Stellar evolution. According to stellar evolution theory, low-mass stars in the range 1 to 3 solar masses, having consumed hydrogen (H) in their cores on the main sequence, rapidly evolve toward the red giant phase with an H-shell-energy-supported structure (**Fig. 1**). Expansion of the stellar radius induces formation of a deep convective envelope that penetrates inward into the zone that previously experienced partial hydrogen burning through the carbon-nitrogen (C-N) chain of nuclear reactions. This penetration, followed

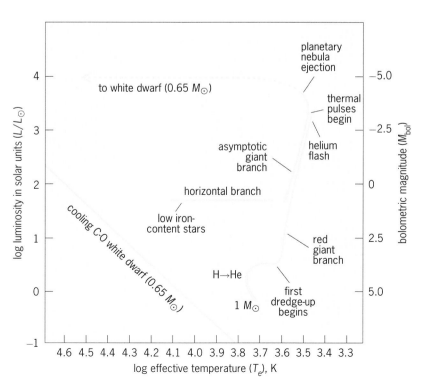

Fig. 1. Evolutionary track of a 1-solar-mass (1 $M_\odot$) star in the Hertzsprung-Russell diagram, from the main sequence to the white dwarf stage. (*After I. Iben, Jr., On the interior properties of red giants, in I. Iben, Jr., and A. Renzini, eds., Physical Processes in Red Giants, Reidel, 1981*)

by movement of processed matter to the surface of the star, is called I dredge-up. Consequently, the surface composition changes: while N is increased by a factor of 2, C is depleted by roughly one-third, O being left unchanged. Helium-3 (^{3}He) and carbon-13 (^{13}C) are also enhanced relative to the initial composition. During this stage, the stellar luminosity strongly increases and mass loss occurs in the form of stellar winds, as shown by infrared observations of red giants with circumstellar envelopes. Such observations were made mainly by the *Infrared Astronomical Satellite* (*IRAS*).

After He is ignited in the center (the He flash), the star leaves the red giant branch and settles on the horizontal branch, burning He quietly in a convective core. Subsequently, a double source of H- and He-shell energy is established while the star comes back to the red, climbing along the asymptotic giant branch (AGB). Late in the asymptotic giant branch, the star undergoes recurrent convective instabilities of the He-shell, known as thermal pulses. It is there that matter is irradiated by neutron fluxes released by captures of alpha particles (^{4}He nuclei) mostly on ^{13}C and marginally on neon-22 (^{22}Ne). Neutron captures on long time scales proceed from the iron (Fe) group nuclei, the most abundant heavy nuclei (**Fig. 2***a*), so that stable isotopes up to lead (Pb) are produced. The path of this so-called s-process, characterized by neutron densities lower than 10^9 cm^{-3}, is schematically represented in the chart of the nuclides (Fig. 2*b*). A second neutron capture mechanism also exists, the r-process, for which much higher neutron densities ($> 10^{20}$ cm^{-3}) are needed. The r-process is complementary to the s-process in nucleosynthesis of heavy elements beyond iron, and it is expected to occur during supernova explosions of massive stars. As shown in Fig. 2, some isotopes are made by only one of the two mechanisms, while most receive contributions from both. Mixing episodes of a small amount of shell material with the envelope (known as III dredge-up) bring ^{12}C and s-processed matter to the surface during the AGB phase.

Isotopic anomalies. The SiC grains exhibit a whole series of isotopic anomalies of noble gases and other elements that must be interpreted by the theory of stellar nucleosynthesis. The measurements are made by using special noble-gas mass spectrometers or ion microprobes. Noble gases contain information about the various mixing phases in red giants. Indeed, the observed large ^{3}He/^{4}He is a direct consequence of the I dredge-up, while the high ^{22}Ne/^{20}Ne is a signature of the III dredge-up. At high temperature, the SiC

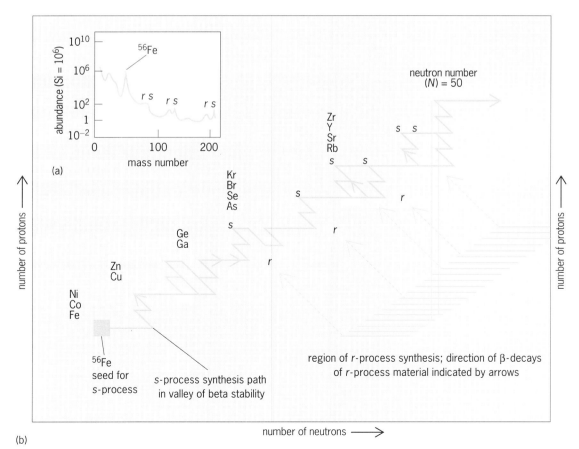

Fig. 2. Isotope graphs. (*a*) Graph of abundances of the nuclides in the solar system versus their mass numbers. (*b*) Section of the chart of the nuclides showing the main features of neutron capture nucleosynthesis. (*After F. Käppeler, H. Beer, and K. Wisshak, s-Process nucleosynthesis: Nuclear physics and the classical model, Rep. Prog. Phys., 52:945–1013, 1989*)

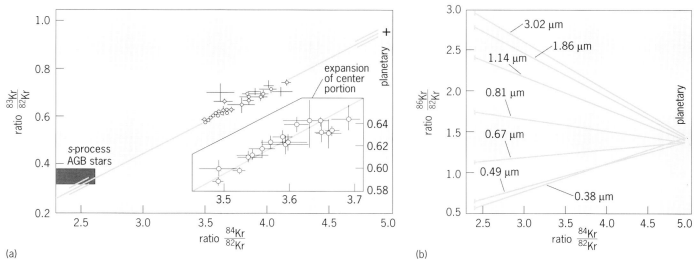

Fig. 3. Three-isotope plots for silicon carbide grains from the Murchison meteorite. (*a*) Plot for ^{82}Kr, ^{83}Kr, and ^{84}Kr showing a binary mixture of a normal component close to normal planetary Kr and an s-process exotic component from asymptotic giant branch (AGB) stars. (*b*) Plot for ^{82}Kr, ^{84}Kr, and ^{86}Kr, showing several linear arrays for SiC grains of different size. (*After R. S. Lewis, S. Amari, and E. Anders, Meteoritic silicon carbide: Pristine material from carbon stars, Nature, 348:293–298, 1990*)

grains release Ne that is almost pure ^{22}Ne. The excess ^{22}Ne originates from captures of alpha particles in the He-shell by the ^{14}N left behind by H burning. The alpha captures first transmute all ^{14}N into ^{18}O and then into ^{22}Ne, the maximum temperature reached in the He-shell being insufficient to induce further alpha captures on ^{22}Ne. Furthermore, the heavy noble gases krypton (Kr) and xenon (Xe) in the SiC grains have a clear s-process signature. For instance, the s-only nuclides ^{128}Xe and ^{130}Xe are strongly enhanced relative to ^{134}Xe and ^{136}Xe. The latter isotopes are not found on the s-path, being mainly ascribable to the r-process.

In order to perform a detailed study of isotopic anomalies, it is useful to display isotopic compositions on a three-isotope graph in which the ratios of two isotopes of the same element with respect to a third are plotted. In such a diagram, a linear correlation among measured compositions means that they result from a mixture of two components. A variation in the measured isotopic compositions of noble gases is obtained from the SiC grains in a stepwise heating procedure in which the gases embedded in the grains are liberated at successively higher temperatures. Two three-isotope graphs for krypton are shown in **Fig. 3**. In Fig. 3*a*, a linear correlation between the ^{83}Kr/^{82}Kr and ^{84}Kr/^{82}Kr ratios is observed in the SiC grains. The straight line extrapolates in one direction through the Kr isotopic composition calculated to be made by the s-process in AGB stars, and in the other direction through the normal meteoritic Kr isotopic composition; the indication is that the SiC grains contain a mixture of these two kinds of Kr.

In Fig. 3*b*, the ^{86}Kr/^{82}Kr and ^{80}Kr/^{82}Kr ratios are plotted against one another. In contrast to Fig. 3*b*, a series of linear correlations is found, each one joining the isotopic composition of normal meteoritic Kr to a different exotic ^{86}Kr/^{82}Kr ratio. The different exotic

compositions can be interpreted as a consequence of branchings in the s-process path at selenium-79 (^{79}Se) and ^{85}Kr that cause the production rates of ^{80}Kr and ^{86}Kr to depend not only on the mean neutron exposure but also on the temperature and neutron density. The inferred variation in the ^{86}Kr/^{82}Kr ratios can be understood if SiC grains of different sizes originated in envelopes of stars of different initial mass and/or initial iron content.

Dating the grains. It is also possible to determine the age of the SiC grains, that is, how long they remained in the interstellar medium before being incorporated into the meteorite. This age determination can be done by a detailed analysis of the ^{21}Ne/^{20}Ne ratio. Indeed, the samples contain an excess of cosmogenic ^{21}Ne that derives from the exposure of the grains to cosmic rays. The estimated cosmic-ray exposure age is up to 130×10^6 years. Actually, this age is obtained from a bulk sample of the SiC grains. However, two populations can be distinguished within this sample: The first population, containing the majority of the grains, is much younger (up to 20×10^6 years) and poor in noble gases. The second population is older (up to 1–2×10^9 years), rich in noble gases, and more reactive chemically.

Interpretations. The most reliable interpretation of the presence of these isotopic anomalies in noble gases is that the gases became associated with the SiC grains by ion implantation. This process can give rise to important elemental fractionations, induced by differences in ionization potentials. In this way, it is possible to explain why more Xe in comparison to Kr is observed, contrary to the predictions of stellar nucleosynthesis.

In addition to the noble gases, isotopic anomalies in C, N, Si, magnesium (Mg), calcium (Ca), titanium (Ti), strontium (Sr), rubidium (Rb), barium (Ba), neodymium (Nd), and samarium (Sm) have been discovered. The heavy elements Ba, Nd, and Sm show an s-signature that is in fair agreement with that shown by heavy noble gases. Moreover, the ratios $^{12}C/^{13}C$ and $^{14}N/^{15}N$ (in many cases lower than solar for $^{12}C/^{13}C$ and higher than solar for $^{14}N/^{15}N$) confirm the formation of SiC grains in carbon star envelopes, even though the finding of a large scatter in isotope ratios in the coarsest single grain indicates the need for a more refined theoretical analysis. The Si anomalies reveal roughly equal excesses of ^{29}Si and ^{30}Si with respect to ^{28}Si. This finding, in turn, can be explained, not in terms of He-shell s-processed material, but by assuming an initial composition for the parent star that is slightly different from solar.

A special problem involves a few coarse SiC grains that possess very different isotopic anomalies than those described above. In particular, their strong enrichment in ^{26}Mg, from the decay of aluminum-26 (^{26}Al), and the excess in ^{44}Ca suggest for them the possibility of a different astrophysical scenario.

Prospects. The analysis of SiC grains is far from complete, and it will certainly provide numerous hints on the way that the solar system formed, on the chemical evolution of the Galaxy, and on the theory of stellar evolution.

Meteoritic graphite or diamond grains have also been discovered, exhibiting their own isotopic anomalies. Interstellar graphite grains are much rarer than microdiamonds (2 ppm versus 400 ppm), or even SiC (~ 6 ppm), while the reverse would be expected. Large fractions of both the graphite and SiC should have been destroyed preferentially. Diamonds are found to have normal $^{12}C/^{13}C$ ratios and a low content of ^{15}N with respect to ^{14}N, together with an enrichment of light and heavy Xe isotopes (the Xe-HL anomaly). The last feature suggests an ion implantation of material that has been present during a supernova explosion. As for the graphite grains, the ratio $^{12}C/^{13}C$ shows a large spread, whereas $^{14}N/^{15}N$ is nearly solar. Moreover, two types of graphite grains have been discovered, both carrying high $^{22}Ne/^{20}Ne$ ratios.

Apparently, a new branch of astrophysical research is being developed based on stellar dust ejected into interstellar space and finally trapped within pristine meteorites at the epoch of formation of the solar system.

For background information SEE ASTROPHYSICS; COSMOCHEMISTRY; METEORITE; NUCLEOSYNTHESIS; STAR; STELLAR EVOLUTION in the McGraw-Hill Encyclopedia of Science & Technology.

R. Gallino; C. M. Raiteri

Bibliography. M. E. Bailey and D. A. Williams (eds.), *Dust in the Universe*, 1988; R. Gallino et al., On the astrophysical interpretation of isotope anomalies in meteoritic SiC grains, *Nature*, 348:298–302, 1990; I. Iben, Jr., and A. Renzini, Asymptotic giant branch evolution and beyond, *Annu. Rev. Astron. Astrophys.*, 21:271–342, 1983; J. F. Kerridge and M. Shapeley Mathews (eds.), *Meteorites and the Early Solar System*, 1988; R. S. Lewis, S. Amari, and E. Anders, Meteoritic silicon carbide: Pristine material from carbon stars, *Nature*, 348:293–298, 1990.

Meteorological radar

The application of radar to meteorology began shortly after World War II. Initial observations were descriptive. Only recently has it become possible to directly infer the dynamical and microphysical properties of convection from radar, and thus open a new era of investigation in convection dynamics.

The special utility of radar in meteorology lies in the large area of coverage [on the order of 500,000 km^2 (200,000 mi^2) for an operational weather radar, or nearly the size of an average-sized state of the United States]; its ability to distinguish between different types of precipitation and to quantify their amounts; and its ability to provide improved warnings of hazards, such as tornadoes.

The radar transmitter broadcasts a focused beam of electromagnetic waves through an antenna. Electromagnetic waves can be described by their amplitude, wavelength, phase, and polarization. Weather radar is based upon the interaction of electromagnetic waves with hydrometeors or other atmospheric inhomogeneities. A small fraction of energy that is incident on a particle is scattered in the backward direction, toward the single antenna that normally is used for both transmitting and receiving. Backscattered energy is used to identify discrete targets, such as aircraft, or distributed targets, such as rain. Position is determined by the direction of the transmitting and receiving antenna, range by the time delay between the transmitted pulse and the received echo, and rain intensity by the amplitude of the received power. The principal difference between conventional (incoherent) radar and Doppler radar is that the conventional radar does not provide phase information; it measures only the amplitude of the received power. Doppler radar uses the Doppler frequency shift to assess the scatterer's component of motion toward or away from the radar. It is important to realize that only the radial component is measured and that three-dimensional flow must be inferred from this one perspective. For weather, most radars transmit in the 3-cm (X-band) to 10-cm (S-band) wavelength band. The letter codes were developed in response to wartime security needs some 50 years ago.

National network. For many years, the National Weather Service network of incoherent weather radars (the WSR-57 and -74 series) has provided information on storm location and characteristics. However, these aging radars (57 and 74 refer to the years of their technology and WSR to Weather Surveillance Radar) are expensive to repair and do not provide the Doppler information, which is useful for better storm warnings and other weather descriptions. In the 1990s, the United States will be largely covered with a network (known as WSR-88D) of over 100 NEXRAD (acronym for *next*-generation weather *rad*ars) radars. These S-band radars will have a beamwidth of less than $1°$ and algorithms to automatically quantify storm characteristics, such as echo top, significant rotation, and the likely presence of hail or damaging winds, just to name a few automatically generated weather products. Thus, the radar will not just gather information; it will process the information and monitor the weather as well.

In preliminary tests conducted in Oklahoma in 1991, the probability of detection of severe weather improved 10% compared to the previous 3 years with the use of Doppler radar. Equally important, the warning lead time for verified significant tornadoes, when a warning was issued, was nearly doubled from an average of 9 min in 1987–1990 to 17 min in 1991, when the Doppler radar was used.

The Federal Aviation Administration (FAA) has a complementary program to provide 60 or so of the busiest airports in the United States with C-band (5-cm wavelength) radars to help protect against aviation weather hazards. An important hazard is the downburst, a small, frequently intense, low-level downdraft that can critically alter an aircraft's flight characteristics at takeoff and landing. The radar detects the small divergence pattern near the Earth's surface from the spreading outflow after the downburst impacts the ground. When an aircraft encounters a downburst, the hazard is due as much to the change in wind speed from the horizontal flow as to the downdraft itself. These hazards are common with thunderstorms and obviously pose a critical threat near airports.

Multiple Doppler networks. A powerful application of Doppler radars is to use two or more radars in combination, or in a network. If u, v, and w represent the east-west, north-south, and vertical wind components, respectively, then the radial component V_r that a radar would measure is a combination of these, depending upon the direction in which the radar is pointing. Mass continuity is usually expressed in the form shown in Eq. (1), where the x, y, and z also

$$\frac{\partial u}{\partial x} + \frac{\partial v}{\partial y} + \frac{\partial \rho w}{\partial z} = 0 \qquad (1)$$

represent the east-west, north-south, and vertical di-

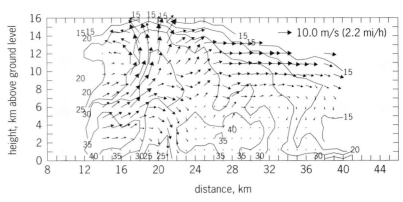

Fig. 1. Vertical cross section of a thunderstorm with winds represented by arrows, which are proportional to the arrow shown at upper right, and rainfall amount (decibels of reflectivity factor) represented by contours. 1 km = 0.62 mi.

rections, respectively. The ρ represents the air density, which decreases with height.

An example of the wind field in a vertical slice through a thunderstorm is shown in **Fig. 1**. Rainfall amount is indicated by the value of its reflectivity factor; larger numbers correspond to larger rainfall amounts. These approaches enable new insights into the processes that govern storm development. Three-dimensional views, such as shown in Fig. 1, have aided in recent understanding of storm processes such as development of tornadoes, storm splitting, storm propagation, and storm rotation. Ultimately, understanding the coevolving wind, water, and electric fields will rest on verification that will include descriptions of three-dimensional wind fields from radar networks. Limitations include the cost of having enough radars to observe storms moving over large distances, and the inability to observe storms in remote locations and over water (except when the storm is close to shore). These gaps are filled by airborne radar.

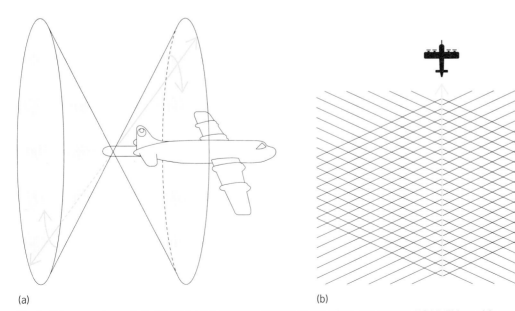

(a)   (b)

Fig. 2. Airborne radar system. (a) Dual-cone scanning system of the NOAA P-3 and the NCAR Eldora airborne radars. (b) Flight level view of data distribution.

Airborne Doppler radar. Airborne Doppler radar systems allow interrogation of storms where they occur, and for longer periods of time as the storms advect to new locations. Although restricted in flight time by fuel limits, aircraft can go over water to investigate hurricanes, over rough terrain, and, in general, over any locations where ground-based radars are unlikely. Use of aircraft expands the geographical diversity of storm types to be studied. Size and weight considerations commend X-band systems for aircraft. **Figure 2** shows the design of two similar systems: the National Oceanic and Atmospheric Administration (NOAA) P-3, and the National Center for Atmospheric Research (NCAR) Eldora radar. Both use a radar with two antennas and two receivers, one antenna pointed about 70° forward and the other 70° backward from the flight direction. As the airplane flies past the area of interest, it views from angles that are separated by 40°. Then the dual-Doppler methodology outlined above can be used. The synthesis approaches are largely similar in concept to multiple Doppler approaches from the ground.

The NOAA airborne Doppler uses a single antenna in which the pointing is alternated fore and aft in each revolution. In an earlier version of the NOAA radar, the antenna scanned only in a direction normal to the flight path, and a helical sampling pattern was formed as the aircraft moved through the air. Pseudo dual-Doppler data can be obtained by flying L-shaped patterns.

Multiparameter techniques. For drops considerably smaller than the radar wavelength, the power backscattered to the radar is related to approximately the sixth power of the drop diameter. The signal is biased to the contribution of the largest drops, which also contribute the most to the mass of rainwater.

While reflectivity alone cannot uniquely determine two parameters of a rainfall distribution, additional information is available through multiparameter techniques. For example, if the values of reflectivity from two orthogonal polarizations are compared, additional

information is gained. Since large drops are distorted (they are "flatter"), they scatter more energy backward when they are illuminated with a horizontally polarized wave than when illuminated with a vertically polarized wave. The difference indicates their size. A spherical drop will have a differential reflectivity (Z_{DR}) value of zero, expressed in decibels. A common measure of this effect is given by Eq. (2), where Z_H represents the

$$Z_{DR} = 10 \log \left(\frac{Z_H}{Z_V} \right) \qquad (2)$$

horizontal reflectivity and Z_V represents the vertical reflectivity; Z_{DR} is proportional to raindrop size, so that a unique solution to an estimation of the drop-size distribution and therefore the rainfall rate is possible.

Perhaps the most useful application of Z_{DR} is in phase discrimination. High values of reflectivity can come from either heavy rain or hail. In middle latitudes, reflectivities in excess of 55 dB usually are associated wth hail. But, whereas rain associated with high reflectivities will have a large (say, 3.5 dB) positive value of Z_{DR}, hail will have values near zero dB, since hail is more nearly spherical. Thus, use of differential reflectivity measurements offers a way of discriminating between hail and heavy rain. In the case of heavy rain, the knowledge of rainfall shape (difference from spherical) gives more information to the spectrum of drop sizes present, and an improved estimate of rainfall rate.

An example of phase discrimination is shown in **Fig. 3.** Figure 3*a* shows the reflectivity (which is proportional to the power received by the radar) in a vertical slice through a thunderstorm near Huntsville, Alabama. The differential reflectivity is shown in Fig. 3*b*. The positive values of Z_{DR} below 4 km (2.4 mi) height show the onset and progression of melting ice into raindrops. The value of zero Z_{DR} above 4 km (2.4 mi) shows the presence of ice.

During the 1990s the operational national network will be implemented. Radar systems will automatically detect and alert meteorologists of possible weather

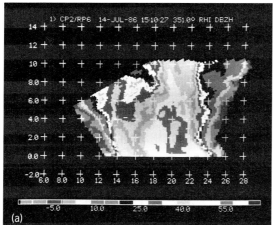
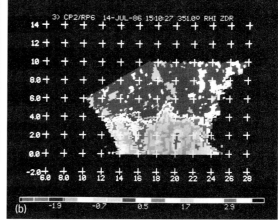

Fig. 3. Vertical slice through a thunderstorm showing (*a*) the reflectivity and (*b*) the differential reflectivity. (*From J. Vivekanandan et al., Microwave radiative transfer studies using combined multiparameter radar and radiometer measurements during COHMEX, J. Atm. Sci., 47(5):549–564, 1990*)

hazards, so that the accuracy and lead time for public response are increased. Research into the processes that cause storms to form and the mechanisms that dictate their evolution will be further clarified through field programs using radar networks with Doppler radars, multiparameter radars, and airborne radars.

For background information SEE DOPPLER RADAR; METEOROLOGICAL RADAR; RADAR; RADAR METEOROLOGY; WEATHER FORECASTING AND PREDICTION in the McGraw-Hill Encyclopedia of Science & Technology.

Peter S. Ray

Bibliography. D. Atlas (ed.), *Radar in Meteorology*, 1990; L. Battan, *Radar Observations of the Atmosphere*, 1973; R. J. Doviak and D. S. Zrnic, *Doppler Radar and Weather Observations*, 1984; P. S. Ray (ed.), *Mesoscale Meteorology and Forecasting*, 1986; M. Skolnik (ed.), *Radar Handbook*, 1990.

Methane

Methane (CH_4; I) is a colorless, flammable gas with a

(I)

boiling point of $-164°C$ ($-263°F$). The most abundant and the least reactive member of the hydrocarbon family, it is the principal constituent of natural gas, with known reserves approaching those of petroleum. A significant amount of the methane produced in the United States is used as fuel. However, methane that is obtained in the more distant areas, such as the Middle East, is generally not utilized elsewhere because of the difficulties associated with transporting a flammable, low-boiling gas. The possible use of methane as an automobile fuel is severely limited by the intrinsic disadvantages of gaseous fuels: low energy content per unit volume, and the hazards associated with handling and distribution.

Conversion to methanol. Methanol (CH_3-OH; II), the product of the partial oxidation of methane,

(II)

is a potentially attractive component of automobile fuels since it is a liquid (boiling point $65°C$ or $149°F$) and it burns more cleanly than gasoline. Use of methanol as an alternative to gasoline has been suggested, because of the expected significant impact on air quality. In addition, methanol is an attractive chemical feedstock. If it were available at a low price, it could become an important building block for the chemical industry. Finally, an on-site process for the conversion of gaseous methane to liquid methanol would alleviate the transportation problem and thus allow better utilization of the world methane output.

An indirect, two-step method for converting methane to methanol has been known for a long time. First, the catalytic reaction of methane with steam (H_2O) forms synthesis gas [reaction (1)], which consists of

$$CH_4 + H_2O \xrightarrow{\text{catalyst}} CO + 3H_2 \qquad (1)$$

carbon monoxide (CO) and hydrogen (H_2) in the ratio of 2:3. Then, this gas mixture, in different proportions, is converted catalytically to methanol [reaction (2)].

$$CO + 2H_2 \xrightarrow{\text{catalyst}} CH_3OH \qquad (2)$$

Both steps involve high temperatures and pressures. For example, reaction (2) is carried out at $250–300°C$ ($482–572°F$) at a combined pressure of 50–100 atm (5–10 megapascals) by using a combination of copper, zinc oxide, and aluminum oxide as catalyst. Thus, a procedure for the direct conversion of methane to methanol at lower temperatures and pressures would be of great practical importance.

Recent advances. Two important problems are associated with the direct oxidation of methane to methanol. First, in order for the process to be economically viable, an inexpensive oxidant is required. Oxygen gas derived from air is ideal. However, methane does not react with oxygen except under combustion conditions. Therefore, a catalyst is required to facilitate the reaction. Second, while methane is relatively inert, methanol is easily oxidized further, eventually to carbon dioxide and water. If the oxidation of methane is not carefully controlled, the methane will undergo complete oxidation to carbon dioxide and water. The same products are also formed when methane burns in air.

In recent years, some progress has been made in the area of selective, partial oxidation of methane to methanol under low temperatures, with metal compounds as catalysts. A suitable metal species has been used to break one of the carbon–hydrogen bonds of methane and to form the corresponding carbon–metal bond. Since carbon–metal bonds tend to be fairly weak, they can be cleaved by a variety of oxidants (including the metal itself) to yield methanol or its derivatives. When the metal itself acts as oxidant, it is reduced. The addition of a second oxidant (cooxidant) that is capable of reoxidizing the metal makes the system catalytic in the metal species. The overall reaction then becomes a metal-catalyzed oxidation of methane by the cooxidant. A typical catalytic cycle is shown in reaction (3), where M^{n+} represents the metal catalyst

$$\begin{array}{c} M^{n+} + CH_3\text{---}H \rightleftharpoons M^{n+}\text{---}CH_3{}^- + H^+ \\ \uparrow \text{Ox} \qquad\qquad \downarrow H_2O \\ M^{(n-2)+} + CH_3\text{---}OH + H^+ \end{array} \qquad (3)$$

and Ox represents the cooxidant.

In one recent discovery, methane has been oxidized to a methanol derivative by hydrogen peroxide (H_2O_2) with the divalent palladium ion (Pd^{2+}) as catalyst [reaction (4)]. In this catalytic process, the divalent

$$CH_4 + H_2O_2 \xrightarrow[80^\circ C\ (176^\circ F)]{Pd^{2+}} CH_3OH + H_2O \qquad (4)$$

palladium ion serves to oxidize methane and is itself reduced to metallic palladium. The role of hydrogen peroxide is to reoxidize the metal to the divalent ion, so that the overall reaction is catalytic in palladium.

However, several problems are associated with the catalytic process in reaction (4). First, the reaction rate is too slow to be practical. Second, hydrogen peroxide is significantly more expensive than oxygen gas, the most desirable oxidant. Finally, the question of selectivity has not been fully addressed. In order to prevent further ready oxidation of methanol, the reaction is carried out in an organic acid. The result is the formation of a methyl ester, a more oxidation-resistant derivative of methanol, rather than methanol itself. While this approach works to a certain extent, further oxidation, eventually to carbon dioxide and water, does occur at longer reaction times.

A second recent report claims that methane can be oxidized to a methanol derivative by oxygen gas, with trivalent cobalt compounds as catalysts. The reaction temperature in this case is much higher ($180^\circ C$ or $356^\circ F$), and significant quantities of carbon dioxide are also formed. Although the mechanism for this reaction has not been elucidated, steps similar to those shown in catalytic cycle (3) are probably involved. Thus, while a few methods for the direct oxidation of methane to methanol have been discovered, they are far from having any practical applications. Nevertheless, the above results should be regarded as encouraging first steps in the design of an economical low-temperature process for the direct oxidation of methane.

Biological model. A class of bacteria called methanotrophs uses methane as the source of carbon and energy for growth. Methanotrophs possess the enzyme, methyl monooxygenase, that catalyzes the oxidation of methane by oxygen gas, initially to methanol. The reaction occurs in water at ambient temperature and pressure. The actual catalyst is believed to be a site containing two iron atoms linked by one or more oxygen atoms. While the sequence of steps involved in the enzymatic oxidation of methane remains poorly understood, a methyl radical, $CH_3\cdot$, formed by hydrogen abstraction from methane by a high-valent iron-oxo species, has been proposed as the reactive intermediate, as in other enzyme systems. If so, the mechanism of biological oxidation of methane differs significantly from the laboratory procedures described above. Although the biological oxidation of methane to methanol occurs under favorable conditions, it is not a practical oxidation procedure. The actual concentration of methanol formed is very low, because at higher concentrations methanol is toxic to the organisms. Nevertheless, the biological model serves to indicate that the direct, selective oxidation of methane to methanol under mild conditions is certainly a realistic goal.

For background information SEE ALCOHOL FUEL; METHANE; METHANOGENESIS (BACTERIA); METHANOL in the McGraw-Hill Encyclopedia of Science & Technology.

Ayusman Sen

Bibliography. N. R. Foster, Direct catalytic oxidation of methane to methanol, *Appl. Catal.*, 19:1–11, 1985; L.-C. Kao, A. C. Hutson, and A. Sen, Low-temperature, palladium(II)-catalyzed solution phase oxidation of methane to a methanol derivative, *J. Amer. Chem. Soc.*, 113:700–701, 1991; M. N. Vargaftik, I. P. Stolarov, and I. I. Moiseev, Highly selective partial oxidation of methane to methyl trifluoroacetate, *J. Chem. Soc., Chem. Commun.*, pp. 1049–1050, 1990.

Microtubules

Microtubules are long, thin, hollow fibers found in almost all eukaryotic cells. They serve as mechanical supports, defining the shape of the cell and the distribution of its contents; and they are responsible for diverse forms of intracellular movement, including the separation of chromosomes during cell division, the beating of cilia and flagella, and the directed transport of membrane-bound organelles and their contents within the cell.

Composition and assembly. The microtubule polymer is composed of subunits of the protein tubulin plus a number of microtubule-associated proteins (MAPs) that bind to the microtubule surface. Tubulin is a dimer of alpha and beta polypeptides. Microtubules can be induced to assemble from tubulin dimers alone, but the microtubule-associated proteins greatly stimulate assembly. The arrangement of microtubules changes when the cell is about to divide or differentiate. This rearrangement is accomplished primarily by reversible disassembly of the microtubules and reassembly in a new pattern as required by the cell.

Disassembly can be induced by a number of drugs such as colchicine, vinblastine, and nocodazole. It can be blocked by the drug taxol, which also potently stimulates the assembly of free tubulin subunits normally present in the cell. Both the destabilizing and stabilizing drugs interfere with microtubule function, and are used as anticancer agents because of their ability to block cell division.

Two general classes of microtubule-associated proteins have been defined, the force-producing MAPs involved in intracellular movement and the structural MAPs. The latter are fibrous proteins that appear as elongated projecting arms on the microtubule surface. In addition to their role in promoting microtubule assembly, they may serve both to link microtubules to other cellular structures and to maintain a uniform distance between microtubules in the cell.

Microtubule-associated proteins have received attention in recent years for their potential role in neurodegenerative diseases. In Alzheimer's disease, for example, abnormal neuronal inclusions known as paired helical filaments are observed which are composed in part of the MAP known as tau. While a role for microtubules in the development of Alzheimer's disease is suggested, the possibility that the inclusions are merely by-products of degeneration cannot be ruled out. Recently, spinal muscular atrophy, a heritable disease, has been genetically mapped to a locus close to the MAP 1B gene. Further work will be needed to determine whether mutations in this gene have a causative role in this disease.

Distribution and function. A typical undifferentiated cell, such as a cultured fibroblast, has from several hundred to a few thousand microtubules emanating radially from a structure called the centrosome near the surface of the nucleus (**illus.** *a*). Microtubules are polar structures: by convention, the end associated with the centrosome is referred to as minus, while the end near the cell membrane is referred to as plus. The radial arrangement of microtubules provides order to the cytoplasm. Many of the membrane-bound organelles of the cell associate with the microtubule array, including the Golgi apparatus, mitochondria, lysosomes, the endoplasmic reticulum, and both secretory and endocytic vesicles. Most, and possibly all, of these organelles move along the microtubules. This movement provides for the directed transport of some organelles, such as secretory vesicles and endosomes. Other organelles use the microtubules as a scaffolding for their own assembly. One example is the Golgi apparatus, a collection of membranes surrounding the centrosome. The Golgi apparatus is dispersed by microtubule-disrupting drugs into small aggregates of membranes, which, however, regroup about the centrosome if the microtubules are allowed to reassemble.

In a dividing cell the centrosome splits, and there appear two radial arrays of microtubules that overlap and interact to form the mitotic spindle (or mitotic apparatus; illus. *b*). The chromosome pairs attach to the plus ends of the microtubules. Once the chromosomes have split, they migrate to opposite ends of the mitotic spindle. This movement is mediated by shortening of the microtubules to which the chromosomes are attached and elongation of the entire mitotic spindle. The elongation results from the sliding of overlapping microtubules in the central region of the spindle.

During cell differentiation, microtubules play an important role in shape determination. The long processes of nerve cells, for example, are very rich in microtubules. If the microtubules are caused to disassemble, the processes retract. In axonal processes, the microtubules are arrayed in parallel, with their minus ends toward the cell body and the plus ends toward the nerve terminal (illus. *c*). This arrangement allows for the orderly transport of organelles and their contents from the nerve cell body to the synaptic terminal and back, a distance of up to several meters in the longest nerve cells of large animals. Blood platelets exhibit a very different arrangement of microtubules. These cells rely on a marginal band of microtubules to

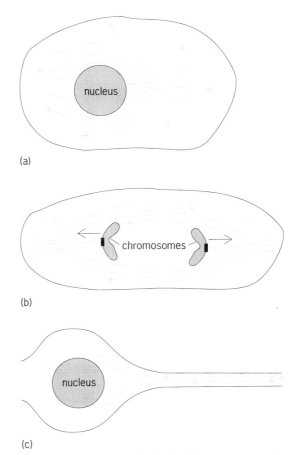

(a)

(b)

(c)

Arrangement of microtubules in cells. Microtubules are indicated by lines with plus and minus ends. (*a*) An undifferentiated cell, showing microtubules radiating from the surface of the nucleus toward the cell margin. (*b*) A dividing cell, showing the microtubules of the mitotic spindle involved in chromosome separation. (*c*) A nerve cell, showing microtubules arrayed parallel to the long axis of the axon.

maintain their disk shape. Extremely complex arrays of microtubules with a range of functions are found in the cytoplasm of many lower eukaryotes. An example is the extensively cross-linked parallel bundles of microtubules in the long cytoplasmic processes of radiolarians.

Cilia and flagella are highly specialized microtubule-containing structures. Both serve as propulsive organelles for a wide variety of single-celled organisms. In vertebrates, cilia are present on the surface of a number of organs, such as the lung and brain, in which case they serve to move mucus or cerebrospinal fluid along the tissue surface. Flagella are responsible for sperm motility. Both cilia and flagella contain a highly conserved bundle of 20 microtubules, two in the center surrounded by nine fused pairs.

Force-producing motor proteins. The bending movement of cilia and flagella is produced by the action of the enzyme dynein. Dynein is distributed between the outer microtubule pairs. It uses adenosinetriphosphate (ATP) as its energy source and causes sliding between adjacent microtubule pairs, much as the enzyme myosin causes sliding between the thin and thick filaments of muscle. Proteins other than dynein link the ciliary and flagellar microtubules, so that sliding is resisted and bending of the entire microtubule bundle occurs. How sliding between the many microtubule pairs is coordinated to produce orderly movement is not yet understood.

Dynein has recently been discovered in a generally distributed cytoplasmic form and is now thought to have multiple functions in the cell. Cytoplasmic dynein, like the ciliary and flagellar forms of the enzyme, uses the chemical energy of ATP to generate force along microtubules. Structures coated with the enzyme move toward the microtubule minus end. Thus, cytoplasmic dynein is thought to be responsible for the movement of organelles such as endosomes and the individual membranous elements of the Golgi apparatus from the cell margin toward the nucleus. A continuous minus-end-directed flow of organelles occurs in neuronal axons. This phenomenon, which is referred to as retrograde axonal transport, delivers signals to the cell body as part of the mechanism of action of neuronal growth factors and in response to injury.

Cytoplasmic dynein may also be important in cell division. The protein has been found at the kinetochore, which is the attachment site for microtubules on the chromosome. It may serve to shorten the kinetochore-associated microtubules by pulling on their ends. To do so, the enzyme would need to remain attached to the microtubules as they shrink.

Kinesin is a protein that produces force along microtubules in the direction opposite to cytoplasmic dynein. Molecular cloning of kinesin has revealed an extensive family of related proteins, also involved in force production. Kinesin and its homologs are thought to be responsible for the transport of organelles toward the plus ends of microtubules, a likely step in secretion. In neurons, this direction of organelle movement is referred to as anterograde axonal transport and is responsible for carrying neurotransmitters to the nerve ending.

Recent evidence indicates that members of the kinesin protein family are involved in chromosome separation. In contrast to cytoplasmic dynein, such proteins would be unlikely to act at the kinetochore, where, rather than separating the chromosomes, they would serve to bring them closer together. However, they could serve to elongate the mitotic spindle by mediating sliding between overlapping microtubules. Thus, both cytoplasmic dynein and members of the kinesin family may act together to carry out mitosis.

For background information SEE CELL MOTILITY; CENTRIOLE: CILIA AND FLAGELLA; CYTOSKELETON; MITOSIS in the McGraw-Hill Encyclopedia of Science & Technology.

Richard B. Vallee

Bibliography. B. Alberts et al., *Molecular Biology of the Cell*, 1989; P. Dustin, *Microtubules*, 1984; R. B. Vallee and G. S. Bloom, Mechanisms of fast and slow axonal transport, *Annu. Rev. Neurosci.*, 14:59–92, 1991; R. B. Vallee and H. S. Shpetner, Motor proteins of cytoplasmic microtubules, *Annu. Rev. Biochem.*, 59:909–932, 1990.

Mid-Oceanic Ridge

The Mid-Oceanic Ridge System is a 50,000-km-long (30,000-mi) submarine volcanic mountain chain formed where the Earth's oceanic tectonic plates spread apart (**Fig. 1**). As the plates separate, new lithosphere is formed (accreted) by solidification of molten mantle material. Formation and cooling of oceanic lithosphere account for two-thirds of the heat lost from the Earth's interior, and constitute a fundamental part of the physical and chemical evolution of the planet.

Along the global Mid-Oceanic Ridge, complex interactions occur among a number of processes (**Fig. 2**). Melting of the Earth's mantle delivers molten rock to ridge-axis magma reservoirs. The magma is cooled to form new oceanic lithosphere, both through circulation of seawater through the lithosphere and through volcanic eruptions onto the sea floor. The circulating seawater, guided by fractures, interacts chemically with the lithosphere and supports sea-floor hot-spring (hydrothermal) activity. Biological communities based on chemical energy rather than photosynthesis thrive on the hydrothermally introduced nutrient fluxes. Hydrothermal plumes rise into the water above the ridge and disperse heat, chemical species, and biota into the overlying ocean and adjacent sediments.

Viewed in a global context, the ridge system can be treated as a single, dynamic system of focused energy flow from the Earth's interior to crustal, oceanic, and biological reservoirs; thus, integrated, coordinated studies are likely to be most valuable in improving understanding of the system. The primary goal of the program known as the Ridge Inter-Disciplinary Global Experiments (RIDGE) is to understand the geophysical, geochemical, and geobiological causes and consequences of that energy transfer within the global ridge system through time. RIDGE will reach this goal through a combination of field and theoretical research projects that will achieve sufficient spatial and temporal definition of the global ridge system to allow development and extensive testing of numerical models that investigate major aspects of energy flow through the system.

Development of RIDGE program. Participants in a 1987 workshop sponsored by the National Academy of Sciences agreed that a major interdisciplinary investigation of the Mid-Oceanic Ridge System would be feasible and have outstanding scientific merit. A steering committee and working groups were established to design a comprehensive, interdisciplinary program incorporating three fundamental approaches to characterizing the ridge system: mapping and sampling, time-series studies, and theoretical modeling of systemic interactions. Each working group held an open workshop in the spring of 1988. On the basis of these discussions, five basic components were proposed as the core of an integrated program to address a wide range of temporal and spatial scales. These components form the basis of the RIDGE program.

RIDGE field efforts. The RIDGE program complements other ridge studies by focusing on key facets of research that are not likely to be accomplished in a timely way without a coordinated program. Ideally, RIDGE studies would be conducted in a series based on progressively smaller areas. However, many areas have already undergone varying degrees of characterization at a range of scales, and components of the RIDGE program were selected to take maximum advantage of work already accomplished. These components are global structure and fluxes, crustal accretion variables, mantle flow and melt generation, event detection and response, and observation of temporal variability.

Global structure and fluxes. Global-scale studies are essential to understanding ridge processes operating on scales of hundreds to thousands of kilometers; for example, long-wavelength geochemical variations that may be related to global-scale mantle flow and structure, and bathymetric variations over 3000 km (1800 mi). Because only a very small percentage of the world's ridges have been sampled, focused global-scale research is also essential to place in context the interpretation of data gathered on smaller scales. Global research will help RIDGE researchers choose appropriate sites for intensive local-scale studies.

The RIDGE global program involves characterization of large parts of the ridge system with respect to the shape of the sea floor, rock and water chemistry, and biological systems. Initial emphasis of this broad-scale mapping and sampling will be on the Pacific-Antarctic Ridge.

Crustal accretion variables. The ridge system is the product of many interacting processes and factors, including the spreading rate between the plates, magma supply, and heat flow. An efficient approach to understanding the relative roles of these factors is an in-depth comparison between two parts of the ridge system that differ significantly in one important variable. RIDGE has chosen spreading rate as a key factor for the first comparison. Intensive studies will focus on the East Pacific Rise (fast-spreading, more than 100 mm or 4 in. per year) and the Mid-Atlantic Ridge (slow-spreading, less than 30 mm or 1.2 in. per year).

The RIDGE Crustal Accretion Variables research program will include five main components on each ridge: (1) characterization of the shape, subsurface structure, and chemical composition of each ridge; (2) geophysical studies to constrain the size and geometry of crustal magma reservoirs; (3) geochemical and geological observations of hydrothermal fluids; (4) studies of biological populations and ecology at hydrothermal vents; and (5) theoretical studies.

Mantle flow and melt generation. Most magma that erupts at the Mid-Oceanic Ridge is generated when solid material from the Earth's mantle rises (flows) to zones of lower pressure, so that melting can occur. The geometry of the melt generation region is controlled by patterns of mantle flow and the temperature and chemical composition of the upwelling mantle. Neither the shape of the melting region nor the

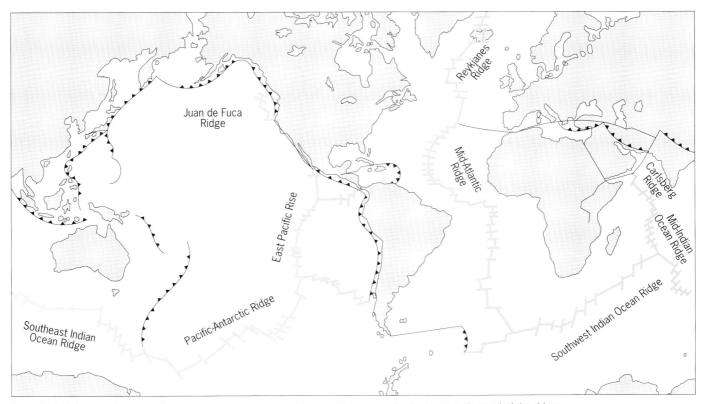

Fig. 1. Global ridge system. Transform faults are shown as short lines cutting perpendicular to the main trend of the ridge; subduction boundaries are shown as lines with solid triangles on the side of the overriding plate.

pattern of mantle upwelling has been definitely established. The objective of the RIDGE Mantle Flow/Melt Generation program is to provide observational constraints on the structure underneath ridges and the processes influencing mantle and melt flow patterns.

By deploying ocean-bottom seismometers on the sea floor to make precise measurements of arrival times of waves that have traveled through the crust and mantle, RIDGE seismologists can draw some conclusions about the temperature, amount of melt, and flow patterns of these regions. As an initial experiment, a large number of ocean-bottom seismometers will be positioned in an array on the East Pacific Rise between 15 and 20°S.

Event detection and response. Of the total transfer of matter and energy from the Earth's interior to its exterior through the ridge system, an unknown but possibly significant proportion occurs during events lasting hours to weeks. Such short-term events may include sea-floor volcanic eruptions, large bursts of heated water (megaplumes), and groups of relatively small earthquake swarms. RIDGE's overall goal of understanding energy transfer through the ridges cannot be realized until some analysis can be made of these portions of the total transfer.

Characterization of short-term events is particularly challenging, because at this stage it is not possible to predict when an event will occur, or precisely where. The first known megaplume—20 km (12 mi) in diameter—was discovered by chance in 1986 during an extensive survey of smaller hydrothermal plumes over the Juan de Fuca Ridge. No one has yet witnessed a submarine volcanic eruption on the ridge.

Because earthquake swarms on the ridge may pre-

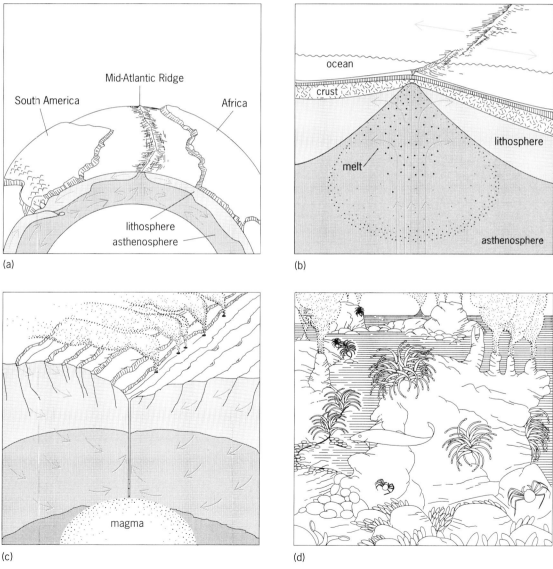

Fig. 2. Diagrams showing the range of study scales in ridge systems. (*a*) Stylized planetary view. (*b*) Magma formation; movement of magma and crust. (*c*) Upper crust. (*d*) Hydrothermal vent field; microscopic bacterial mats on rock surfaces (*courtesy of Sandra Noel*).

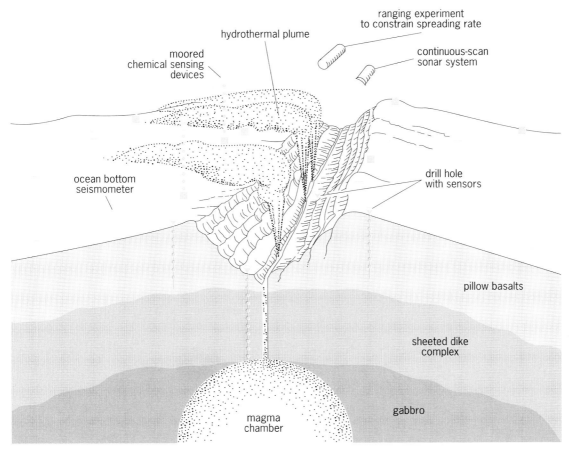

Fig. 3. Schematic diagram of a remote sea-floor observatory along a representative portion of the Mid-Oceanic Ridge System.

cede or accompany volcanic eruptions or formation of large hydrothermal plumes, RIDGE has initiated a coordinated effort to respond rapidly to such swarms. When an earthquake swarm or other event is reported, the RIDGE Office acts quickly to transport equipment and scientists to conduct studies at the site with aircraft or ships. The RIDGE Event Detection and Response program is also studying ways to detect earthquake swarms at distances of thousands of kilometers.

Observing temporal variability. Although the dynamic nature of ridge systems is well established, there have been no long-term interdisciplinary efforts to document rates of change in system components or evaluate links among the complex and interrelated physical, chemical, and biological processes involved. Diverse, coordinated measurements will be essential in studying these relationships and developing improved theoretical models. These measurements will be made through one or more assemblages of sea-floor instrumentation (observatories) placed at particularly active sites on the ridge system. Results from event detection and other observations will be essential in determining which sites are good candidates for an observatory. Instrumentation may include subsurface or sea-floor seismometers, water flow meters, water and rock samplers, tiltmeters to measure ground deformation, video cameras to estimate hydrothermal temperatures and flow rates, and devices to measure physical properties of the water moving above and through the ridge (**Fig. 3**). Initial testing and development has taken place on the Juan de Fuca Ridge.

Theoretical integration. To understand fully the results of the RIDGE field programs, theoretical models must be developed to integrate various hypotheses about ridge processes and make predictions that can be tested through further field work. RIDGE is promoting theoretical research both through providing funding support to the theoreticians' projects and through the RIDGE Theoretical Institute, where researchers and students gather for 2-week programs that include lectures by theoreticians and work on interdisciplinary ridge problems.

For background information SEE EARTH CRUST; EARTH INTERIOR; HYDROTHERMAL VENT; LITHOSPHERE; MAGMA; MID-OCEANIC RIDGE; PLATE TECTONICS in the McGraw-Hill Encyclopedia of Science & Technology.

Patricia Stroh

Bibliography. National Research Council Ocean Studies Board, *The Mid-Oceanic Ridge: A Dynamic Global System*, 1988; E M. Parmentier and C. H. Langmuir (convenors), *RIDGE Theoretical, Experimental, and Analytical Workshop Final Report*, 1988; G. M. Purdy and P. J. Fox (convenors), *RIDGE Mapping and Sampling Workshop Final Report*, 1988; RIDGE Steering Committee, *RIDGE Initial Science Plan*, 1989;

RIDGE Steering Committee, *RIDGE 1989 Working Group Report Volume*, 1989; F. N. Spiess and R. E. McDuff (convenors), *RIDGE Sea-Going Experiments Workshop Final Report*, 1988.

Mobile radio

In June 1990, Motorola announced the development of its Iridium satellite system, which envisions the use of many small satellites in low Earth orbit to provide worldwide cellular personal communications services. The system objective is to be able to communicate from a handheld terminal located anywhere to anywhere else in the world. Subscribers will use portable or mobile transceivers with low-profile antennas to reach a constellation of 77 satellites. These satellites will be interconnected by radio communications as they traverse the globe approximately 413 nautical miles (765 km) above the Earth in seven polar orbital planes. Principles of cellular diversity will be used to provide continuous line-of-sight coverage to and from virtually any point on the Earth's surface, with spot beams providing substantial and unprecedented frequency reuse (more than five times in the continental United States alone).

Each satellite in the Iridium system will be small and have sophisticated electronics capable of communicating with mobile subscriber units, with Earth-station gateways on the ground, and with other satellites

Fig. 1. Iridium constellation of 77 satellites in low Earth orbit.

in the constellation. Iridium's digital cellular design and spot-beam technology are somewhat analogous to present-day cellular telephone systems, except in reverse. In the case of cellular telephones, a static set of cells serves a large number of mobile units, whereas Iridium's cells will move about 24,300 ft/s (7400 m/s) while mobile units remain relatively fixed with respect to the transmitting-receiving cell.

However, the Iridium system, because of its limited capacity and cost structure, is not designed to compete with or replace existing land-line and cellular systems. Instead, the Iridium system will target markets not currently served by mobile communications services, such as (1) sparsely populated locations where demand is insufficient to justify constructing terrestrial telephone systems, (2) areas in developing countries without existing telephone service, and (3) small urban areas presently without a terrestrial mobile communications structure. Iridium will offer to these markets a full range of mobile services, including radio determination satellite service (RDSS), paging, messaging, voice, facsimile, and data services. Iridium is expected to support and serve millions of users throughout the world.

The Iridium system is composed of a space segment comprising a continuously replenished constellation of satellites, a gateway segment consisting of Earth stations and associated facilities distributed throughout the world to support call-processing operations and interconnection with the public switched telephone network, a system control segment for monitoring and controlling the system, a launch segment to transfer the satellites into orbit, and an Iridium subscriber unit segment.

Space segment. The space segment includes a constellation of 77 small, smart satellites in low Earth polar orbit, which are networked as a switched digital communications system (**Fig. 1**). Each satellite will form 37 separate spot beams or cells on the surface of the Earth (**Fig. 2**) in the L-band. The spatial separation of the cells permits increased spectral efficiency through time, frequency, and spatial reuse over multiple cells.

The constellation of satellites and its projection of cells is somewhat analogous to cellular telephone systems. In the case of Earth-based cellular systems, a static set of cells serves a large number of users; in the case of the Iridium system, virtually all users appear static with respect to the velocity of the cells. (That is, the cells traverse the face of the Earth at approximately 24,300 ft/s or 7400 m/s; thus, even high-speed jet aircraft have a velocity vector considerably smaller than that of a cell.) This system poses the dual of a terrestrial cellular system in that the cells move through the users rather than the users moving through the cells. As the satellites move from the Equator toward the poles, the appropriate cells are disabled to eliminate unnecessary overlap.

Each satellite employs intersatellite links to permit cell-to-cell handoff between satellites and to support internetting. These intersatellite links are designed to operate in the Ka-band, and include both forward- and

Fig. 2. Iridium cell pattern.

backward-looking links to the adjacent satellites in the same orbital plane.

Each satellite can also communicate with Earth-based gateways either directly or through the inter-satellite links. The Earth–space gateway links are supported at 20 GHz for the downlink and 30 GHz for the uplink.

Satellite constellation. The satellite constellation (Fig. 1) is composed of seven evenly spaced circular polar planes with each plane containing 11 satellites. Satellites within each plane are spaced $32.7°$ apart, and travel in the same direction at approximately 12,960 knots (24,000 km/h) in a north-south direction and 900 knots (1670 km/h) westward over the Equator. Each satellite circles the Earth every 100 min. This satellite constellation provides coverage over the entire face of the Earth, with single coverage provided at the Equator and increasing levels of redundancy at northern and southern latitudes.

Satellite configuration. Each Iridium satellite (**Fig. 3**) will carry a deployable solar array of approximately 100 ft^2 (10 m^2). The L-band antenna, or main-mission antenna subsystem, includes seven phased-array antennas, each containing an array of transmit-receive modules. Six of the main-mission antennas are identical and form the sides of a hexagonal cylinder that makes up the body of the space vehicle. Each of these six antennas creates 6 of the 37 beams projected to the Earth by the satellite. The seventh antenna creates the single nadir-pointing beam and is located on the side of the space vehicle facing the Earth.

Communication system. The Iridium system is

designed to operate with a seven-cell channel-reuse pattern. The cells are scanned by the satellite antenna arrays in accordance with a specified timing pattern and sequence. During the time slot that the antenna is pointing at a cell, satellite transmissions may be made and receptions of transmissions from Iridium subscriber units may occur during the respective transmit and receive intervals.

Each satellite has scanning-beam antennas that are programmed to point at the correct cell on the Earth at the right time (Fig. 2). The satellites are positioned so that the cells just merge at the Equator, but as the satellites move toward a pole, the cell patterns begin to overlap and thus selected spot-beam antennas are deactivated to permit an orderly reconstitution of the frequency-reuse pattern.

On a global basis, there are 1628 nonredundant cells, and the seven-cell pattern reuses each frequency over 200 times. Within the contiguous United States alone, the Iridium system will achieve more than a five-times frequency reuse.

The multiple-access format for Iridium uses both time-division multiple access (TDMA) and frequency-division multiple access (FDMA); the result is a very efficient use of the spectrum. A 14-time-slot TDMA format allows each cell to be assigned on the average two time slots.

Intersatellite links. The satellites of the Iridium constellation must be interconnected to form a transmission network that can transport subscriber communications around the entire Earth. Each satellite will communicate and route network traffic to the satellites

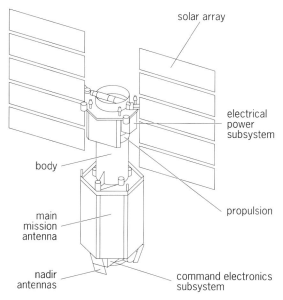

Fig. 3. Iridium space vehicle.

that are fore and aft of the vehicle in the same orbital plane. Two separate fixed-waveguide slot arrays pointing in the forward and aft directions are used for the in-plane links. Similarly, the two movable antennas are used to route network traffic to adjacent satellites in the orbital planes next to the one containing the subject satellite. The four intersatellite links provide direct routing between users throughout the world, as well as substantial redundancy so that traffic flow can be supported in the event of satellite degradation.

Gateway segment. The Iridium gateway requires a minimum of two Earth terminals in order to provide continuous contact with the Iridium constellation of satellites. A typical gateway will contain three Earth terminals, geographically dispersed, to ensure high availability under various environmental conditions. Each Earth terminal contains a satellite-tracking parabolic antenna and the necessary radio-frequency and processing components to provide full-duplex communications with an Iridium satellite. As many as four gateways can be serviced by a satellite, each maintaining a 6.25-megabit-per-second link with one of the four space-vehicle feeder-link antennas. A higher gateway capacity may be achieved by simultaneously processing two of the feeder links at the gateway, thus yielding a total capacity of 12.5 megabits per second from the servicing space vehicle. This configuration will be capable of providing up to 1300 subscriber channels per gateway. The Iridium gateway will provide a billing interface that supplies information to the home billing office.

System control segment. The Iridium system control segment provides monitoring, management, and control of the Iridium network and the individual space vehicles composing the Iridium constellation. It interfaces with the gateways to provide data needed by the gateways to perform their functions, and performs similar functions with respect to

the gateway billing offices and the Iridium engineering facility. Constant communications with the Iridium satellites and gateways will be provided through three geographically diverse Earth-terminal locations.

Subscriber units. Iridium subscriber equipment must function as voice telephones providing full-duplex, high-quality audio. To support this need, the equipment will utilize a 4800-bit-per-second voice coder specifically developed to deliver superior audio quality over typically corrupted radio channels. The technique used is called vector-sum excited linear predictor (VSELP). Testing has shown that voice digitized with this coder sounds equivalent to high-quality analog voice used in present land-mobile radio equipment. In addition to the voice coder, the Iridium subscriber equipment will employ noise-canceling microphones to greatly reduce interference from nearby sounds other than the speaker's voice.

In addition to voice, Iridium customers will find it important to transmit and receive a wide variety of data worldwide. Data could be in the form of text or files from a personal portable computer or images from a facsimile machine. The Iridium subscriber equipment will be available with an industry-standard data port, such as RS-232, to transport data from an external device. Several levels of data service will be available, from basic 2400-baud asynchronous circuits without error checking to higher-speed packet transport sold on demand.

For background information *SEE COMMUNICATIONS SATELLITE; MOBILE RADIO* in the McGraw-Hill Encyclopedia of Science & Technology.

James M. Foley

Bibliography. W. S. Adams and L. Rider, Circular polar constellations providing continuous single or multiple coverage above a specific latitude, *J. Astronaut. Sci.*, 35(2):155–191, 1987; R. J. Leopold, Global cellular communications network, *Proceedings of Mobile Satellite Conference*, Australia, 1991.

Molecular mimicry

Molecular mimicry is the phenomenon in which two organisms that are closely related ecologically, but not phylogenetically, share common macromolecular structures that are not attributable to evolutionary conservation of these structures. In molecular mimicry, some part of an organism's molecular recognition system is deceived by a false molecular signal from another organism. Molecular mimicry has been observed most frequently in antigenic systems, where the possession of shared epitopes (antigenic determinants) between hosts and their infectious agents, including viruses, bacteria, and parasites, has been strongly implicated both in the evasion of immunity by the pathogens and in the generation of immunopathology (autoimmunity) in the host.

Antigenic mimicry as an example of molecular mimicry emphasizes that possession of similar molecular structures by two separate species must occur within a close ecological association, often parasitism but also including certain predator–prey and herbivore–plant associations. Molecular mimicry involves an element of deception in common with other forms of biological mimicry such as classical Batesian visual mimicry, in which a predator is deceived into avoiding the palatable mimic of a noxious or unpalatable model.

An animal's immune system normally functions by discriminating between self and non-self to develop a specific immune response only to the latter. In antigenic molecular mimicry, similarity or identity between regions of certain host and parasite macromolecules may deceive the host's immune system. If the deception results in failure to stimulate the immune response, the parasite will go unrecognized (or more likely, less strongly recognized) as a foreign invader, so that there is a compromised immune response that gives the parasite an advantage in establishing or maintaining itself within the body of the host. If, on the other hand, the deception does trigger the host immune response, the result could be autoimmunity, a reaction that may or may not benefit the parasite but certainly harms the host.

Mechanisms. Sharing of macromolecular structures by disparate organisms can result from several mechanisms. For example, similarities at the gene level can lead to peptide sequence or conformational mimicry, and similarities in biosynthetic pathways can lead to oligosaccharide mimicry in glycoconjugates. These mechanisms in most cases must ultimately stem from preadaptation and natural selection operating over long periods of time. Protoparasites that fortuitously bear hostlike epitopes would be favored, that is, preadapted, to develop the parasitic mode of life, and natural selection would then operate to improve molecular mimicry.

However, a much more rapid process could also be operative. Recent implication of a semiparasitic mite in the introduction of mobile genetic elements (P elements) from one *Drosophila* species into another supports another mechanism for the development of molecular mimicry. Infectious or parasitic agents themselves—through processes such as viral gene capture and integration (that is, incorporation of host deoxyribonucleic acid into the viral genome), or even physical microinjection into embryos of foreign genes, as is thought to have occurred in the mite–*Drosophila* combination—may serve as vectors for the introduction of transposable genes from one species to another, with molecular mimicry of gene products as a consequence.

Molecular mimicry as a means by which parasites evade the host immune response was originally thought to be a passive type of adaptation—a matter of nonrecognition of foreign invaders by the immune system. Passive antigenic molecular mimicry may occur at several points in the immune response train and thus may utilize several mechanisms: at the earliest stage of immune recognition because of holes in the antigen-recognition repertoire of the T cell, after immune recognition through the processes of T-cell regulation, or even at the effector level. More recently, this concept of molecular mimicry in immune evasion

has been modified to include more aggressive forms of mimicry. It is now clear that several intracellular parasites, including certain viruses and leishmanial and malarial protozoa, employ mimicry of ligands for expressed receptors of the host cell in order to gain entry into the cells, where they not only reproduce but also find refuge from the immune response of the host.

Another way that aggressive molecular mimicry can operate to favor parasitism is for host immunoregulatory molecules themselves to be mimicked. An excellent example comes from the recent discovery that a protein of the Epstein-Barr virus, BCRF1, has extensive sequence homology with interleukin 10 (IL-10), a cytokine produced by the TH2 subclass of helper T cells that inhibits the synthesis of a whole set of inflammatory cytokines produced by the TH1 subclass of helper T cells. Significantly, the fact that the BCRF1 protein also exhibits IL-10 activity suggests that the virus may be able to intervene with or manipulate the host immune response to favor its survival and replication.

The human multicellular parasitic worm *Schistosoma mansoni* antigenically mimics a host plasma protein, α_2-macroglobulin, that has been implicated as an immunoregulatory protein. *Schistosoma mansoni* uses another variant of molecular mimicry in its arsenal of evasive tactics. By mimicking certain host receptors, including immunoglobulin and complement receptors, this intravascular parasite is able to partially cover itself with a mask of host proteins. This form of mimicry, called antigenic masquerade, may act aggressively as well as passively. The schistosome worm binds both antibodies and complement proteins in ways that prevent them from performing their normal host protective function of foreign-cell lysis. These and other mechanisms may explain the schistosome's ability to survive for many years in the bloodstream of its host, even in the presence of a stimulated immune response. Obviously, understanding the details of molecular mimicry in these cases of parasite evasion could yield great dividends for antiparasite drug and vaccine development. Much research is being done in this field.

Role in autoimmunity. Molecular mimicry between hosts and pathogens probably plays a role in autoimmunity. This role has been hotly debated for years, but recent evidence favors it. One strong argument for it has been the recognition of the sheer magnitude of instances of molecular mimicry. Extensive comparisons of peptide sequences and the use of monoclonal antibodies to probe host–pathogen cross-reactivity have made this recognition possible. However, the progression from shared molecular structures to the disease state of autoimmunity is thought to be rare relative to the frequency of molecular mimicry. But when autoimmunity does occur, it is usually a serious, even life-threatening condition. Therefore, this is an active field of research.

Link with pathogenesis of disease. Molecular mimicry between various microorganisms and host tissues has been suggested as a factor in the pathogenesis of a variety of diseases, including cardiac diseases (implicated organisms are *Streptococcus*, Coxsackie B virus, and *Trypanosoma cruzi*), demyelinating diseases (implicated viruses include HTLV-1, visna, vaccinia, caprine arthritis, encephalitis, and influenza), ulcerative colitis (*Escherichia* bacteria are implicated), ankylosing spondylitis and Reiter's syndrome (*Klebsiella* bacteria are implicated), myasthenia gravis (herpes simplex virus is implicated), and rheumatoid arthritis (*Mycobacterium* is implicated).

Shared molecular structures have recently been shown to cause disease in several model systems. It was found that injury to the central nervous system could be induced in rabbits by immunizing them with a viral peptide identical to a self peptide, namely, a short peptide sequence of hepatitis B virus polymerase protein also found in the encephalitogenic site of rabbit myelin basic protein. Another useful model is the adjuvant arthritis model, whereby rats develop arthritic joints after immunization with the "adjuvant" bacterium *Mycobacterium tuberculosis*. In this case, the mimicry is between a heat shock protein of the bacterium and a rat cartilage proteoglycan.

The most recent variant of molecular mimicry to be implicated in pathogenesis has been termed messenger mimicry. This applies to retroviruses that, by means of their reverse transcriptase enzyme, can make an excess of single-stranded complementary deoxyribonucleic acid (cDNA) copies of their ribonucleic acid (RNA) genome. These cDNA molecules could theoretically block synthesis of the host protein and thus contribute to, for example, neuron demyelination.

A different type of molecular mimicry in parasitism is the synthesis of host hormone analogs by parasites. This is exemplified by the larva of the tapeworm, *Spirometra mansonoides*, which occurs in a variety of prey animals and matures in carnivores. When the laboratory mice are experimentally infected with the tapeworm, they become very obese. It is now known that the tapeworm produces a growth-hormone-like factor with somatotropic activity. Production of the factor may favor the parasite in nature by making the potential prey host more conspicuous and easier to catch by the predaceous final host of the tapeworm. Other examples of parasite alteration of intermediate-host behavior are being discovered, and some of them are thought to be based on molecular mimicry of host hormones.

Chemical mimicry. Most of the examples of molecular mimicry in parasitism given above have obvious medical significance. Examples of molecular mimicry in other kinds of ecological associations may have other important biological roles. The term chemical mimicry has been used to refer to small-molecular-weight volatile compounds that may be used to deceive olfactory communications in a wide variety of organisms, including insects. This form of molecular mimicry has been implicated in defense by inquiline species against their social insect hosts and in luring prey or pollinator species. In these cases, the mimicry is not due to shared macromolecular structures but rather to duplication of blends of low-molecular-weight odorants. Additionally, some plants

synthesize mimetic hormone molecules that are active in phytophagous insects. The hormonelike molecules could function to desynchronize insect life cycles and plant life cycles and thereby to protect the plants from excessive grazing.

For background information SEE AUTOIMMUNITY; CELLULAR IMMUNOLOGY; MEDICAL PARASITOLOGY in the McGraw-Hill Encyclopedia of Science & Technology.

Raymond T. Damian

Bibliography. P. R. Carnegie and M. A. Lawson, Viral mimicry and disease, *Today's Life Science*, pp. 14–20, February 1991; I. R. Cohen, The self, the world, and autoimmunity, *Sci. Amer.*, 255:34–42, 1988; R. T. Damian, Molecular mimicry revisited, *Parasitol. Today*, 3:263–266, 1987; M. B. A. Oldstone (ed.), Molecular mimicry: Cross-reactivity between microbes and host proteins as a cause of autoimmunity, *Curr. Top. Microbiol. Immunol.*, 145:1–145, 1989.

Molecular recognition

The interaction of organic chemical species with inorganic ions is central to life processes. The inorganic aspect of molecular recognition seeks to emulate this interaction. Many inorganic or organometallic structural units are finding use as receptors for other inorganic ions or organic molecules. Ultimately, it may become difficult to assess the difference between "organic" and "inorganic" molecular recognition.

All chemical reactions occur by collision. The reaction rates are determined by a variety of factors, but selectivity in reactions is determined by whether or not the reacting partners "recognize" structural and electronic features in each other. Recognition includes the formation of hydrogen bonds such as are involved in base-pairing of deoxyribonucleic acid (DNA). Electron-deficient molecules are attracted to and therefore recognize electron-rich compounds.

A molecular receptor is a chemical species that can select one of many possible binding partners and form a complex that is stabilized by interactions such as hydrogen bonding, hydrophobic-hydrophobic forces, or changes in solvation that stabilize the complex relative to the two partners. The goal usually is that a chemical receptor should not differ in function from a biological one. Although structural mimicry has occasionally been the goal, the expectation has generally been that form would follow function. Indeed, the aim is to bind the same molecules recognized in nature, but usually with smaller molecular apparatus than is used naturally.

Since the early 1970s, chemists have attempted both to understand and to emulate recognition properties. Compounds have been designed that can recognize and bind or complex metallic and organic cations, small molecules, and even each other. Macroscopic materials such as clays have been explored so that their ability to complex and recognize other molecular species can be understood and enhanced by alteration. These systems are discussed below, with examples of the interactions and conceptual approaches.

Siderophores. Many natural proteins can recognize and bind alkali, alkaline-earth, and transition metals, which are critical to life functions. One important system involves the recognition and transport of iron (Fe). The family of compounds known as siderophores comprises molecular receptors that bind and transport iron. These compounds must selectively recognize the cations that are required for biological function, binding and transport to the area of biological function must occur, and finally the cation must be released. While this sequence can be accomplished by naturally occurring materials, simpler synthetic structures have also been designed to do so. The synthetic compounds usually involve structural features that are obvious in the natural materials, but the synthetic materials rarely function in just the same way as do the naturally occurring ones. The differences in behavior point to chemical requirements of the systems and help the chemist to understand how the natural systems function. Further, a novel combination of chemical properties may permit the development of drugs to cure certain chemical deficiencies. The natural siderophore enterobactin [structure (I)] is capable of complexing Fe(II); a synthetic mimic is structure (II).

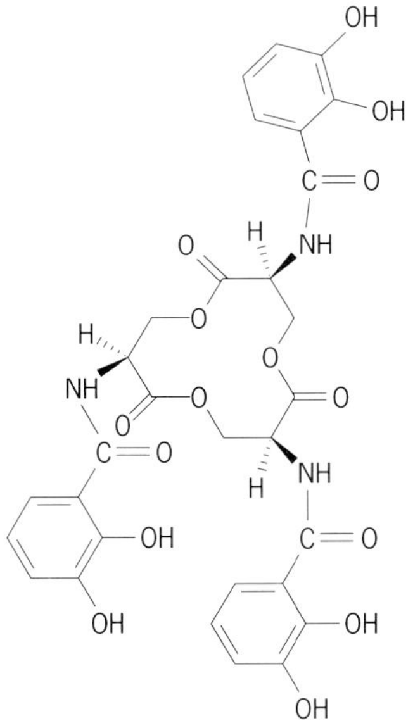

(I)

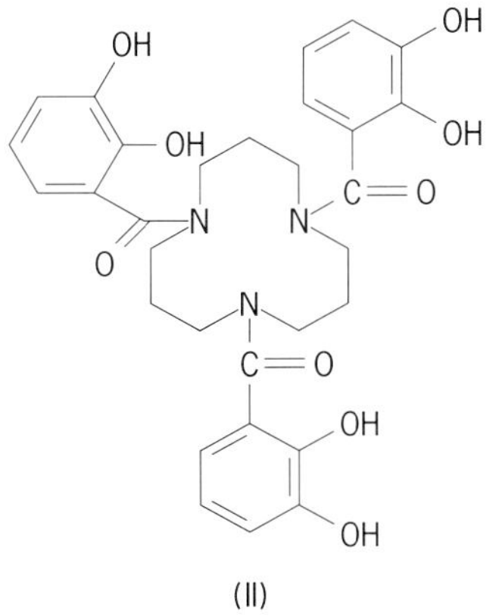

(II)

Porphyrins. The naturally occurring porphyrins containing magnesium cation and iron cation are, respectively, chlorophyll and heme [structure (III); a part

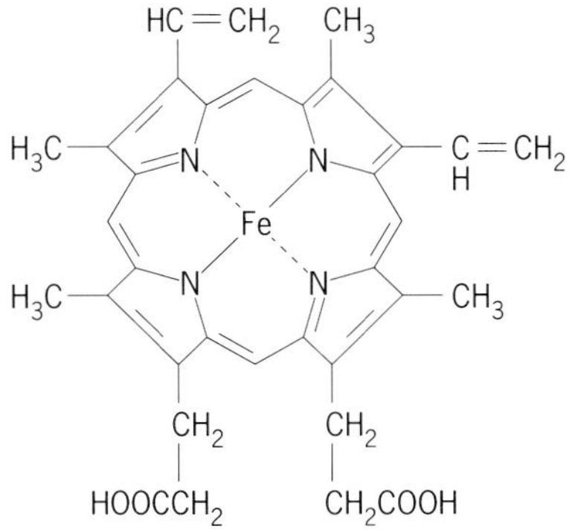

(III)

of the natural protein hemoglobin]. Vitamin B_{12} has a similar structure and binds cobalt. These compounds are flat structures made of four nitrogen-containing, five-membered rings. The nitrogen (N) atoms are focused to the center of the ring and bind the cation. In heme the cation, in turn, binds oxygen (O_2). Extensive studies seeking to mimic the oxygen-binding properties have involved preparation of synthetic structures with both electronic and steric features understood to be present in the natural systems. In addition, completely new porphyrin structures have been devised to bind atoms or molecules that are not known to be bound or transported by porphyrins in naturally occurring systems. These systems may involve caps on the top or bottom of the planar porphyrin structure, steric barriers at the sides, or substituents attached to the ring that alter the electronic properties.

Phenanthrolines and helicates. Chemists have long been fascinated by interlocked molecular systems. In recent years, it has been shown that phenanthroline (IV), a three-ring structure contain-

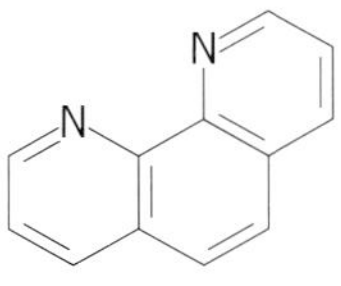

(IV)

ing two coordinating nitrogen atoms, can bind various cations. When the cation is coordinated to two phenanthroline rings simultaneously, it organizes the two organic species and permits them to be connected and interlocked. When the cation is removed, the rings remain interlocked, so that a structure known as a catenane is formed. A helicate is a helical array of molecules formed by the chemical recognition and organization of metals and organic bases.

The concept may be extended into a system that mimics the naturally occurring molecules DNA and

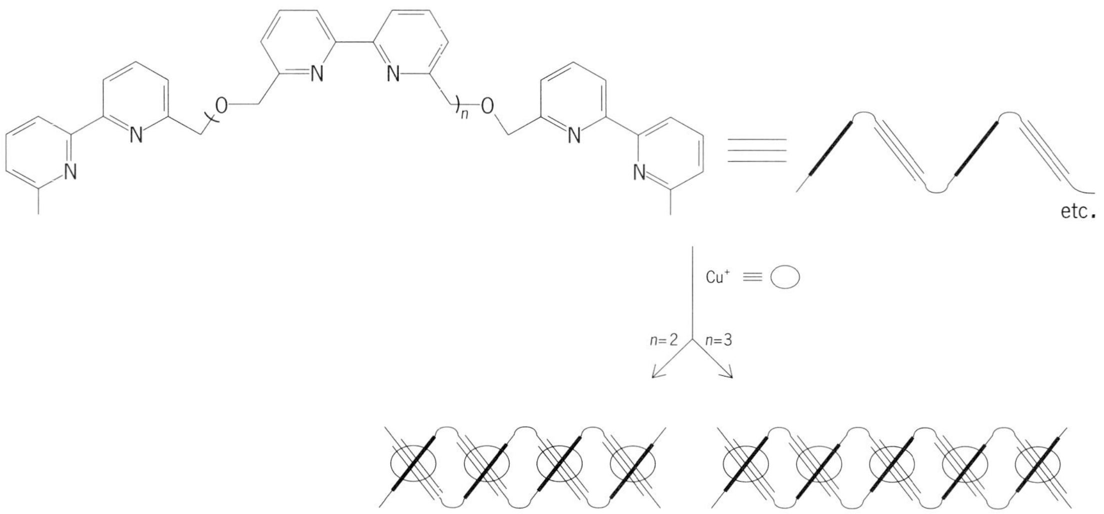

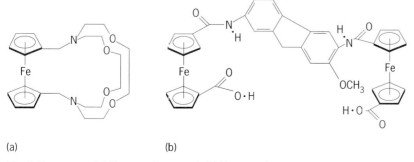

Fig. 1. Mechanism of formation of helicates from bipyridyls.

ribonucleic acid (RNA). These systems are based upon bipyridyl (V) rather than phenanthroline molecules, so

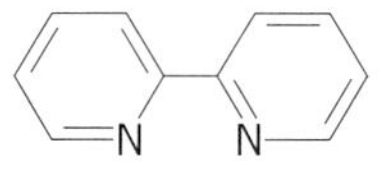

(V)

that extra flexibility is possible. Attachment of the bipyridyls in sequential reactions organized by the cations [copper (Cu) binding to two adjacent nitrogen atoms] leads to a helical array in which two chains wrap about each other (**Fig. 1**). These systems require that the copper cation be recognized in forming the helices and that the chains have opposite twists in order to pair.

Ferrocenes. Ferrocene is the prototype of a family of molecules known as metallocenes or sandwich compounds in which two five-membered aromatic (cyclopentadienide) rings bind to the top and bottom of an Fe atom to sandwich the cation. A number of other cations and other aromatic rings have been found to afford such sandwich structures. Interest in the chemistry of ferrocene has been renewed by recent applications to molecular recognition. For example,

ferrocene has been integrally bound into a cryptand structure (**Fig. 2a**). Because the cation may undergo oxidation-reduction (redox) chemistry, the effective charge inside the cryptand may be altered. Thus, a one-electron oxidation [Fe(II) → Fe(III)] introduces a positive charge. When a metallic cation is bound in the ferrocenyl cryptand (Fig. 2b), this oxidation creates a charge repulsion that causes the bound cation to be released more readily than if the system remained neutral. The cryptand recognizes the cation by size and charge, binds it accordingly, and then is switched to release the cation.

The two aromatic rings of the ferrocene system rotate freely with respect to each other. This property has been used to advantage in the development of a family of small molecule receptor systems that also possess redox-switching properties. The receptors are formed by connecting the top rings of two ferrocenes and then placing carboxyl (COOH) groups on the bottom rings. The carboxyls may rotate freely to create an open cavity into which, for example, a diamine (H_2N—R—NH_2) may fit selectively. The carboxyls then rotate inward to trap and bind the molecule.

Calixarenes. Calixarenes are cyclic structures of the type (—Ar—CH_2—)$_n$, where Ar represents an aryl group. They arise from the condensation of equivalent numbers of phenol (C_6H_5OH) and formaldehyde (HCHO) molecules that form a chalice-shaped molecular receptor. Calixarenes, then, are organic receptors: they are purely synthetic, cyclic oligomers of phenol and formaldehyde. Common ring sizes involve four, six, and eight alternating phenol and methylene (CH_2) units. All the hydroxyl (OH) groups may be directed to the same side of the macrocyclic structure (cone conformation), although other conformations are possible. In the cone, a metal cation may be bound by the hydroxyl groups or by polar residues attached to the hydroxyl groups. In addition, functional groups at the hydroxyl rim may recognize and bind metal ions, and metal-ion complexes may be bound by hydrophobic forces within the cavity. The presence of

Fig. 2. Ferrocenes. (a) Ferrocenyl cryptand. (b) Ferrocenyl receptor.

(a)

(b)

metal-ion-binding sites in proximity to the hydrophobic cavity offers the possibility of mimicking enzyme function. Indeed, more complex systems containing sidearms that serve the function of portions of natural protein chains have been appended to calixarenes, and enzymelike function has been demonstrated.

Natural membranes are formed from molecules that have a polar (hydrophilic) head and a nonpolar (hydrophobic or lipophilic) tail. Many molecules orient in water so that the nonpolar residues interact to form a bilayer in which the polar residues on both sides face outward. The calixarenes fulfill this hydrophobic-hydrophilic requirement and have been shown to form bilayers in the solid state (**Fig. 3***a*). In these systems, the polar surfaces of calixarene bilayers face and interact with water layers that are rich in alkali metal cations. The thickness of the repeating bilayer–polar-layer unit is about 1.4 nanometers, the same dimension found in natural clays.

Clays and other layered materials. Many naturally occurring materials form intricate mineral matrices comprising layers of aluminum oxide–silicon dioxide ($AlO_2^- \cdot SiO_2$) separated by charge-balancing cations. Cavities, holes, channels, or compartments within these structures permit binding of a wide variety of cationic, anionic, or molecular species within the framework. Catalysis of reactions involving these bound substrates is general and has been implicated in the origins of life. In the laboratory, these materials are used in the catalytic syntheses of organic compounds.

The clays are essentially two-dimensional sheets that have recognition properties based upon layer stoichiometry and separation. Alterations in the Al:Si ratios or the spacing between the layers change the recognition properties of the whole system.

A three-dimensional variation is found in the zeolites or molecular sieves. Within the alumino-silicate framework, there exist one-dimensional channels, two-dimensional sheets, and three-dimensional void spaces. Molecular recognition based upon shape, size, and polarity are possible with these versatile systems, and all these facets are currently under exploration. The zeolites that are classified as molecular sieves are used to separate molecules of different sizes, shapes, and polarity from each other. Examples are separation of water from ethanol (drying) and separation of *meta*-xylene from *para*-xylene. The latter separation has been thought to be due to shape differences, but it may actually involve differences in polarity. Catalytic reformation (catalytic cracking) of hydrocarbons is also catalyzed by zeolites in long distillation columns maintained at high temperatures. This technology permits conversion of materials with little economic potential into forms possessing greater commercial value.

For background information SEE CROWN ETHERS; HYDROGEN BOND; METALLOCENES; MOLECULAR RECOGNITION; MOLECULAR SIEVE; ZEOLITE in the McGraw-Hill Encyclopedia of Science & Technology.

Jerry L. Atwood; George W. Gokel

Bibliography. J. L. Atwood, *Inclusion Phenomena and Molecular Recognition*, 1990; J. L. Atwood et al., Second-sphere coordination of transition-metal complexes by calix[4]arenes, *J. Amer. Chem. Soc.*, 113:2760–2761, 1991; J. L. Atwood and J. E. D. Davies (eds.), *Inclusion Phenomena in Inorganic, Organic, and Organometallic Hosts*, 1987; J. L. Atwood, J. E. D. Davies, and T. Osa (eds.), *Clathrate Compounds, Molecular Inclusion Phenomena, and Cyclodextrins*, 1984; Y. Inoue and G. Gokel, *Cation Binding by Macrocycles*, 1990.

Mollusca

With 60,000 existing species, mollusks form the second largest phylum within the animal kingdom. Although two mollusk species may live in proximity in the same environment, feed in a similar way, and be preyed upon by similar animals, one may live for only a year or two while the other may live for 20. For example, a terrestrial slug matures within the first year, breeds in the second, and then dies; whereas a neighboring land snail may mature within 2 years, breed, and then live for another 10 years. Recently, data were collated to examine whether general patterns of longevity can be traced throughout the mollusk phylum. It was found that a mollusk's longevity can be related to its shell morphology, habitat, and size.

Life-spans of mollusks range from 2 months to 200 years. Short-lived mollusks are defined as those living up to 2 years, or those living longer but reproducing during only one season. Long-lived mollusks are defined as those species that live for more than 2 years and breed over at least two seasons. The **illustration** shows the frequency of life-spans in the major mollusk groups. Although many mollusks are long-lived, almost half of existing records are of short-lived mollusks (**Table 1**).

Longevity and shell morphology. Both in terrestrial and marine mollusks, a short-lived mode of life is often correlated with either lack of an external shell or possession of an external shell that is semitransparent. Shell-less mollusks (prosobranchs, opisthobranchs, pulmonates, or cephalopods) are all short-lived (**Table 2**).

This correlation may be explained in adaptive terms. The absence of a shell enables high growth rates; juveniles of shell-less mollusks grow to adult size quicker than shelled ones, and therefore speed through the vulnerable juvenile phase. For example, common squid

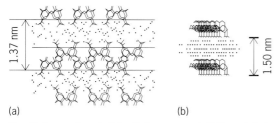

Fig. 3. Layered materials. Dots represent H₂O and Na⁺. (*a*) Bilayer structure of calixarenesulfonates with sodium cations (Na⁺) and water molecules (H₂O). (*b*) Layer structure of hydrated sodium vermiculite.

species can mature at the age of 6 months, whereas *Nautilus* matures only after several years. The consistently faster growth rates of the shell-less mollusks may indeed be an advantageous trait. However, this argument would imply ubiquity in the ecology of entire groups of shell-less mollusks. Such ubiquity is difficult to accept, as there are over 2500 shell-less mollusk species, differing in taxonomic origins, environments, reproductive strategies, methods of moving, and feeding habits, and most of these species have a short life history. The advantages to be gained from a short life-span would have to be overwhelming if such a general correlation were to be explained on this selective basis.

Another adaptive approach could be that when extrinsic mortality risks (such as starvation, accident, disease, or predation) are higher for parents than for offspring, the parent increases its energy expenditure in the reproduction of many offspring. This increase in reproductive energy eventually leads to a short life-span because so much energy is going into reproduction that not enough is left for maintenance of the individual. This age-specific-mortality argument could be applied to many opisthobranchs in which the adults feed upon a transient source and the juveniles feed

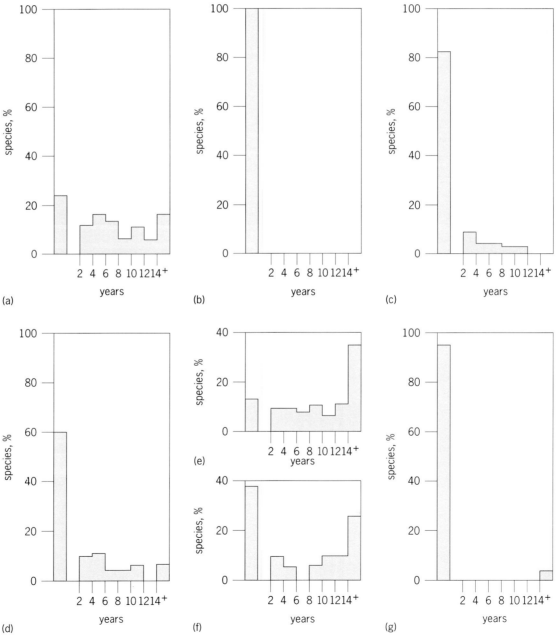

Life-span frequencies in various mollusk groups. (*a*) Marine prosobranchs and pulmonates. (*b*) Opisthobranchs. (*c*) Fresh-water snails. (*d*) Land snails. (*e*) Marine bivalves. (*f*) Fresh-water bivalves. (*g*) Cephalopods. The leftmost bar indicates the short-lived type.

Table 1. Number of short- and long-lived mollusks

Group	Number of mollusks	
	Short-lived	Long-lived
Chitons	—	2 genera, 2 species
Marine snails		
Prosobranchs	16 genera, 25 species	36 genera, 73 species
Opisthobranchs	37 genera, 63 species	—
Pulmonates	—	1 genus, 2 species
Fresh-water snails	23 genera, 48 species	3 genera, 6 species
Land snails	30 genera, 44 species	22 genera, 26 species
Marine bivalves	11 genera, 21 species	73 genera, 116 species
Fresh-water bivalves	4 genera, 19 species	12 genera, 28 species
Cephalopods	16 genera, 26 species	1 genus, 1 species
TOTAL	137 genera, 246 species	150 genera, 254 species

upon a nontransient source (for example, in *Aplysia* the adult feeds upon seasonally abundant green seaweed, whereas the juvenile is a planktonic veliger that feeds upon ubiquitous unicellular algae). However, shell-less mollusks do not consistently feed upon a source that is transient in availability. Some opisthobranchs feed upon a wide variety of prey species, many of which are not of a transient nature. Octopuses and cuttlefish are opportunistic carnivores whose food resources are stable. Slugs feed on a diet that is stable and similar to that of shelled land snails.

However, the correlation between longevity and shell morphology may also be explained in nonadaptive terms. An initial change in the shell that is adaptive to the environment may engender a secondary change in the life-span, irrelevant to adaptation and not under immediate control of the environment. If this nonadaptive explanation is indeed valid, then the short life-span of many mollusks may be a by-product of selection on the shell rather than an independently selected trait.

Adaptive loss of the shell has occurred independently in several molluscan lineages. Selective pressure on marine cephalopods for active swimming has led to the persistence of those with a reduced internal shell or with no shell at all. The ability of early opisthobranchs to burrow and exploit the infaunal environment, combined with their development of chemical defense (rather than the mechanical defense of the shell), led to the reduction of the shell and its eventual loss in many opisthobranchs. Deeper penetration into the ground along with the invasion of calcium-deficient, moisture-rich environments by terrestrial mollusks has led to development of a shell-less slug form in several unrelated taxonomic families of land snails. In any of these mollusk lineages and for any of these selective reasons, once the shell had been lost, the mollusk was automatically compelled to become short-lived.

Within the severe limits of an imposed short life-span, life strategies may vary in evolutionary response to different environmental conditions. Thus, shell-less nudibranch species that have a stable source of food all year have annual life cycles; in contrast, those that have a seasonal transitory food source have short life cycles and pass through numerous generations in a year. However, when considered together in comparison to the short-lived shelled prosobranch, the small

Table 2. Relation between shell and life-span in mollusk groups

Group	Type of shell	Number of genera	
		Short-lived	Long-lived
Marine snails	Opaque	12	36
(prosobranchs)	Semitransparent	2	—
	No external shell	2	—
Marine snails	Opaque	1	—
(opisthobranchs)	Semitransparent	6	—
	No external shell	30	—
Land snails	Opaque	12	21
	Semitransparent	4	—
	No external shell	12	1
Marine bivalves	Opaque	9	74
	Semitransparent	—	1
	No external shell	2	—
Cephalopods	Opaque	—	1
	Semitransparent	—	—
	No external shell	16	—

differences between shell-less species that have a stable food source and those that do not are trivial. Shell absence is the overriding factor in determining whether a mollusk will be short- or long-lived. Once this major factor is set and the mollusk becomes short-lived, environmental factors determine only the fine adjustments.

The advantage of this nonadaptive hypothesis is that it explains the ubiquity of the relationship between shell absence and short life-span. One of its weaknesses is that it requires a nearly single-gene linkage between shell-lessness and longevity, and there is as yet no direct evidence for any such link. A further weakness is that it cannot explain exceptional records of shell-less mollusks that are not short-lived.

Life-span and habitat. Both on land and in the sea, a short life for mollusks with shells is often correlated to a microenvironment exposed to high solar radiation and to high temperatures. The rate of gamete development depends directly on temperature, and high temperatures increase the rate of gonad maturation.

In land snails, the relation between longevity and exposure may be explained in adaptive terms in that ionizing radiation could increase the rate of aging and reduce the average life-span of animals. Large doses of hard radiation (gamma and fast neutrons) shorten life considerably. Also, ultraviolet radiation can cause a breakdown of molecules and thereby alter vital biochemical processes. A short life-span could thus be enforced in land snails dwelling on the tips of vegetation, where they are subject to heavier ultraviolet radiation than snails dwelling underneath stones. Even visual daylight radiation may be an important environmental factor that influences the development of gonads of land snails.

Some land snails dwell in deserts that, in addition to being hot and strongly radiated, are extremely unpredictable. In such deserts, where the proper conditions for growth and reproduction occur infrequently and unpredictably, annual populations consisting of short-lived individuals would quickly become extinct. Mollusks of these habitats are more long-lived than their close relatives in more favorable conditions. Short-lived mollusks are restricted accordingly to environments in which weather is predictable.

Life-span and shell size. A third pattern is found among the shelled gastropods in which longevity is related to size. Short life-spans appear to occur more frequently among very minute gastropods (in which the adult is less than 0.15 in. or 4 mm) than among larger ones. Land snails of the genera *Vertigo* and *Punctum* and marine prosobranchs of the genera *Rissoa* and *Rissoella* are both minute and short-lived. They mature within several weeks early in the season, lay several eggs throughout the remainder of the season, and gradually die off by the end of the season.

Bivalves. A fourth pattern shows that the bivalve group is more long-lived than other groups. The bivalves constitute 88% of the species that live over 25 years and 92% of the species that live over 50 years, and are the only mollusks that live over a century. Short life-spans are not a common strategy among bivalves, and only 15% of the bivalve genera are short-lived (as compared to short life-spans in 63% of the gastropods). A sedentary mode of life apparently bears the potential for a long life-span.

For background information SEE ADAPTATION (BIOLOGY); BIVALVIA; CEPHALOPODA; GASTROPODA; MOLLUSCA in the McGraw-Hill Encyclopedia of Science & Technology.

J. Heller

Bibliography. J. Heller, Longevity in molluscs, *Malacologia*, 31:259–295, 1990.

Mysidacea

The small, shrimplike mysids are a group of crustaceans that show a great diversity in marine environments and are of enormous importance in many marine food chains. They also occur infrequently in fresh-water localities. There are about 800 species of mysids, 24 of which have been recorded from fresh-water habitats. Most species of mysids known from European and Asian fresh-water localities also occur in brackish waters, and some are clearly glacial relicts. Recently, the discovery of a number of fresh-water species in South America and Caribbean coastal waters has provided insight into the origins and adaptations of at least one group of fresh-water mysids.

Fresh-water mysinid species. All but one of the world's fresh-water mysid species (see **table**) are from the tribe Mysini (subfamily Mysinae, family Mysidae) and appear to be closely related. While the suggestion is that these genera of the Mysini have a propensity to adapt to fresh water, none of them is limited to the fresh-water habitat. Indeed, only a few of these species are fresh-water endemics.

The Southeast Asian/Pacific and circumpolar groups represent species that have invaded fresh water from the sea via brackish systems, and these species are still tolerant of brackish conditions; *Mysis relicta* is a relict from lake isolation following periods of glaciation, but can also be found in brackish water in the Baltic Sea.

The central Eurasian group of 10 species is well represented in fresh-water habitats. They represent another relict system, and are presumed to have evolved from a basal group inhabiting the coasts of the large lake filling the Sarmatic Basin in the late Miocene. The area of this basin included what are now the Black and Caspian seas, and these mysid species inhabit the mesohaline waters of these seas and the river systems draining into them.

The Caribbean–South American species form two groups. *Taphromysis* species inhabit brackish-to-marine coastal waters of the Gulf states of the United States, where they have invaded fresh-water systems directly from the sea. The remaining group, the *Antromysis* group, includes recently discovered species exclusive to fresh-water habitats.

Antromysis group. The relationships of the *Antromysis* group were first analyzed in 1977, and the members were amalgamated into a single genus with four subgenera. The discovery of subsequent species has justified the erection of *Parvimysis* and *Surinamysis* to full generic rank, and it may prove logical to raise *Anophelina* to generic rank also. The distribution

Mysid species showing mainly fresh-water distributions

Group	Species
Circumpolar Arctic	*Mysis relicta* *M. litoralis*
Central Eurasian	*Diamysis bahirensis* *D. pengoi* *Limnomysis benedini* *Katamysis warpachowskyi* *Paramysis baeri* *P. intermedia* *P. kessleri* *P. lacustris* *P. ullskyi* *Hemimysis anomala*
Indian Ocean–East Indies–North Pacific	*Gangemysis assimilis* *Nanomysis siamensis* *Neomysis awatschensis* *N. intermedia* *N. mercedis*
Caribbean–South American *Taphromysis*	*Taphromysis bowmani* *T. louisianae*
Caribbean–South American *Antromysis*	*Antromysis cenotensis* *A. cubanica* *A. peckorum* *A. reddelli* *Antromysis (Anophelina) anophelinae* *Parvimysis almyra* *P. bahamensis* *P. pisciscibus* *Surinamysis americana* *S. merista* *S. robertsonae* *?Surinamysis* sp. (only a damaged female is known)

of these forms is shown in **Fig. 1**. All the species are small and range from 0.12 to 0.16 in. (3 to 4 mm) in length. They produce only two to four young at a time, and the frequency of breeding is unknown.

The species of *Antromysis sensu stricto* are adapted to living underground. *Antromysis cenotensis* has been found in fresh waters in caves in Yucatán, Mexico; *A. cubanica* in a cave south of Havana, Cuba; *A. reddelli* in standing pools in a cave south of Acatlán, Mexico; and *A. peckorum* in moderately brackish water in Jackson Bay Cave, Jamaica. All are blind and unpigmented species, as is characteristic of hypogean species living in the dark; ommatidia are absent, although reduced eyestalks are present and separate. The telson is typically short and quadrangular, and tapers to a truncate posterior edge armed with a pair of apical spines, none to two marginal spines, and lateral spines restricted to the distal third or absent entirely (**Fig. 2a**).

Antromysis (Anophelina) anophelinae also has a hypogean habit, and has been recorded from slightly brackish water to fresh water found in the holes of the land crab *Cardisoma crassum* in mangrove inlets at Puntarenas, Costa Rica. This species has reduced but functional eyes (Fig. 2b); the eyestalks are fused proximally, the eyes being present as an anterior ocular band, with ommatidia and black pigment. The telson is similar to that of *Antromysis sensu stricto*, but with two pairs of apical spines, three marginal spines, and lateral spines at its distal and proximal extremes. There is no male process on the antennule.

Parvimysis species (Fig. 2c) have normal, fully de-veloped eyes and eyestalks with pigment and ommatidia and live in open waters ranging from marine to totally fresh. The body has some pigmentation. *Parvimysis bahamensis* occurs in euhaline Caribbean waters in the Bahamas and Puerto Rico at depths of 3–49 ft (1–15 m) over sandy bottoms; *P. almyra* is found in the mixo-oligohaline habitat of coastal brackish-water ditches and a fresh-water creek in Surinam; and *P. pisciscibus* occurs in leaf-litter banks in a "black-water" side stream of the Rio Negro some 1600 mi (2500 km) from the sea (the blackwater systems are humic acid–bearing waters so fresh as to be close to distilled water).

Surinamysis species (Fig. 2d) have been recorded only from fresh and slightly brackish water. They live in open waters and have fully developed eyestalks with pigment and ommatidia, the cornea occupying one-third to one-half of the eyestalk. The body has some pigmentation. *Surinamysis americana* was recorded from fresh-water and brackish ditches at Paramaribo, Surinam; *S. merista* from a fresh-water gully off the River Orinoco near Barrancas, Venezuela; and *S. robertsonae* from a "whitewater" lake near the confluence of the Rio Negro and the Amazon River some 1600 mi (2500 km) from the sea.

Origins of the fauna. Discussion on the origins of the fresh-water mysid fauna of this north-central area of South America must be subject to the caveat that the area is undersampled and more species may remain to be discovered: the first fresh-water mysid from the Amazon was not discovered until

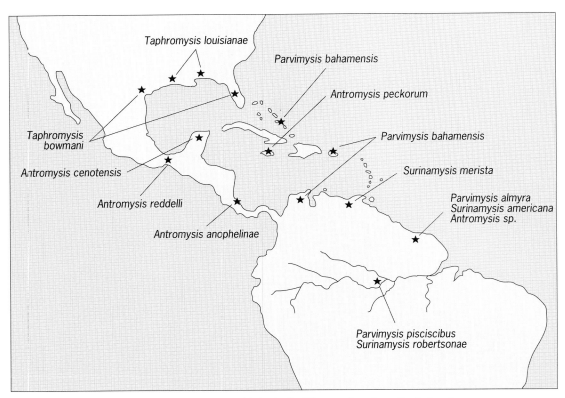

Fig. 1. Distribution of brackish- and fresh-water mysids in the Caribbean and northern South America.

1981. However, from current knowledge the indication is that it is the *Antromysis* group that has extended from Caribbean and Atlantic waters into the fresh-water habitat.

The Central American isthmus has changed through time; it was a zoogeographic barrier in the Paleocene, reflooded in the Miocene, and reformed as an isthmus by the Pliocene (5×10^6 years before present). Thus it may be postulated that this group of mysids evolved and colonized coastal and subsequent inland habitats within the last 5×10^6 years.

How do the most fresh-water of species, *P. pisciscibus* and *S. robertsonae*, found in the center of the South American continent, accord with this hypothesis? In the early Miocene, 15×10^6 years ago, the rise of the Andes closed off the connection to the Pacific of what had been the westerly flowing Amazon River; the Brazilian and Guianan Highlands were still connected in the east, so the Amazon basin became a vast lake or swamp system. The result was the isolation of some species that were once marine in the Pacific, for example, the 20 species of stingrays now living in the central fresh waters of the Amazon. Toward the end of the Miocene 10×10^6 years ago, the Amazon broke through to the Atlantic. Subsequent glaciations caused marked changes in the sea level; 18,000 years ago the sea level was 430 ft (130 m) lower than it is today. The lower levels would have reduced inland transgression of salt water and might have caused faster flows in the river. Conversely, 600 years ago the sea level was 430 ft (130 m) higher than at present, and the Atlantic would have pushed the fresh waters back inland. At present the estuary of the Amazon extends some 200 mi (320 km) offshore.

The basin-lake period is thought to have been a time of speciation and adaptive radiation of other aquatic fauna. The opportunity for colonization from both coasts with an intervening period of radiation has led to the support of the world's richest fish fauna in the Amazon system. Blackwater systems are considered to be closest to the original westward-flowing Amazon; whitewaters (more turbid waters such as those of the Solimões) represent the Andes drainage.

Although the fresh-water antromysine mysids have been found in both types of water, they appear neither common nor diverse in the Amazon system, so are probably of recent origin. Further, they show clear interrelationships with the Atlantic-Caribbean fauna, to which their origin can be attributed.

Despite the lack of knowledge of their biology, it can be hypothesized that animals of such small body size and small brood size must be reproducing quickly; evidence from other groups suggests that the generation time may be as short as 10 days or less. In an ecosystem supporting a great diversity of predators, these mysids will suffer a high attrition rate unless they can find a habitat refuge. The crab holes and small caves inhabited by *Antromysis sensu strictu* species may be such a refuge. However, the species of *Parvimysis* and *Surinamysis* that live in the plankton of lakes or demersally in small streams are more susceptible to predation. A successful reproductive strategy in response to high adult mortality is direction of more energy to reproduction than to somatic growth, so

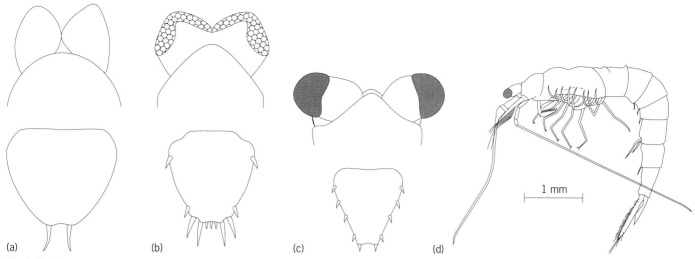

Fig. 2. Fresh-water mysids. (*a*) Eyes and telson of *Antromysis sensu stricto*; (*b*) eyes and telson of *Antromysis (Anophelina)* (*after W. M. Tattersall, A review of the Mysidacea of the United States National Museum, Bull. U. S. Nat. Mus., 201:1–292, 1951*); (*c*) eyes and telson of *Parvimysis*; (*d*) lateral view of *Surinamysis*.

that small size and rapid generation time result. Since mysids have no resting stage in their life history by which to resist adverse conditions, a life-style with brief longevity is likely to be successful only in the nonseasonal environment of the tropics.

Rapid generation should also offer more opportunity for speciation. At this time, more field collecting is under way to target the planktonic and small crustacean fauna of the Amazon system. It is expected that further species of the *Antromysis* group, of similar size and fecundity, will be discovered.

For background information SEE MYSIDACEA; ZOO-GEOGRAPHY in the McGraw-Hill Encyclopedia of Science & Technology.

Roger N. Bamber

Bibliography. R. N. Bamber and P. A. Henderson, A new freshwater mysid from the Amazon, with a reassessment of *Surinamysis* Bowman (Crustacea; Mysidacea), *Zool. J. Linn. Soc.*, 100:393–401, 1990; T. E. Bowman, A review of the genus *Antromysis* (Crustacea: Mysidacea), including new species from Jamaica and Oaxaca, Mexico, and a redescription and new records for *A. cenotensis, Ass. Mex. Cave Stud.*, no. 6, pp. 27–38, 1977; J. Mauchline, The biology of mysids, *Adv. Mar. Biol.*, 18:1–369, 1980; V. P. Weish and M. Türkay, *Limnomysis benedeni*, in Österreich mit Betrachtungen zur Besiedlungsgeschichte (Crustacea: Mysidacea), *Archiv. Hydrobiol.*, suppl. 44, 4: 480–491, 1975.

Neutrino

Results of recent experiments suggest that the neutrino may be more than a thousand times more massive than was previously believed. Studies at four laboratories on the decays of five different isotopes indicate that a neutrino with a mass of 17 keV is emitted approximately 1% of the time in nuclear beta decay. Such a "heavy" neutrino would have profound implications in both particle physics and astrophysics.

Development of neutrino research. Early in the twentieth century, physicists discovered a type of radioactivity that is now known as beta decay. In this process, a neutron inside an atomic nucleus converts into a proton and an electron (beta particle). The proton remains bound inside the nucleus by the strong nuclear force, while the beta particle is ejected. The amount of energy available in such a decay, Q_β, is equal to the difference in rest-mass energy between the initial and final nuclear states. Until this time, every physical process studied had obeyed the law of energy conservation, which states that the total energy of an isolated system does not change. Thus, all beta decays of a given isotope were expected to produce electrons with a unique energy, Q_β. However, when the energy spectrum of the emitted electrons was first measured in the 1920s, the result was surprising. Instead of a single electron energy, a continuous spectrum extending from zero energy up to the full decay energy, Q_β, was observed. At this point, there were two choices: either to give up belief in energy conservation or to hypothesize the existence of an additional unseen particle that is emitted in beta decay. This particle would share the available energy with the electron so that the sum of the two energies would equal Q_β. The particle that accompanies the electron emitted in nuclear beta decay is now known as the electron-type antineutrino. Related processes that have also been observed are (1) positron decay, in which a proton inside a nucleus is converted into a neutron, a positron, and an electron-type neutrino, and (2) electron capture, in which a proton inside a nucleus captures a bound atomic electron and is converted into a neutron plus an electron-type neutrino.

Neutrinos interact with matter only through the weak and gravitational forces. Thus, the direct detection of neutrinos is a formidable experimental challenge. Nearly 30 years passed before the electron-type neutrino was detected in the laboratory. Since then, two additional types of neutrinos have been discovered: one associated with the muon, ν_μ, and one associated with the tau lepton, ν_τ. Neutrino beams are now routinely available at a number of accelerators, and neutrino detectors play a major role in particle physics and astrophysics.

Possibility of massive neutrinos. The fact that the spectrum of electrons emitted in nuclear beta decay appears to extend all the way up to the maximum energy allowed by energy conservation suggests that the neutrino is either massless or very low in mass. Studies of the high-energy portion of the beta spectrum of tritium (^{3}H) have established that the mass of the electron-type antineutrino is less than 10 eV. It was suggested around 1980, however, that the object emitted in beta decay might not be a simple particle with a well-defined mass. The object might instead be a composite made up of two particles with quite different masses. If it is assumed that the major constituent is massless and that the other, smaller component has a mass on the order of kiloelectronvolts, then beta-decay spectra would be more complicated than was previously thought. Each spectrum (**Fig. 1**) would actually consist of two components. The one associated with the emission of the massless neutrino would extend all the way up to Q_β, and the second, weaker component associated with the emission of the massive neutrino would end at an energy of $Q_\beta - m_\nu c^2$, where m_ν is the mass of the heavy neutrino and c is the speed of light. In beta-decay studies, only the electron is detected, not the neutrino. Thus, the sum of these two spectra would be observed. At the energy where the spectrum associated with the emission of the heavy neutrino ends, there would be a change in the slope or a kink in the observed beta spectrum. This spectral feature is the signature of the emission of a heavy neutrino.

First reported observation. The first reported observation of such a kink came in 1985, when J. Simpson claimed to observe an excess number of

events at low energy in the beta spectrum of tritium. He interpreted this excess as being due to the emission of a heavy neutrino. Most beta-decay studies suffer from the problem of the electrons scattering within the radioactive source before reaching the detector, and the problem of the electrons backscattering out of the detector. Both processes mimic the effect of the massive neutrino in that they result in an excess of low-energy events appearing in the beta energy spectrum. In order to observe the tritium spectrum ($Q_\beta = 18.6$ keV) with a minimum of distortion, Simpson implanted the tritium into a silicon semiconductor detector. After the implantation, the beta particles from the tritium decays were detected via the ionization they produced in the silicon. A high voltage applied across the silicon crystal caused negatively charged electrons to drift to one electrode whiile positively charged ions drifted to the other electrode. The amount of ionization produced in such a device is proportional to the energy deposited by the beta particle. From a detailed study of the tritium beta-spectral shape, Simpson inferred that a neutrino with a mass of 17 keV was occasionally emitted instead of the massless neutrino.

This unexpected result aroused considerable interest. Between 1985 and 1988, eight groups of workers attempted to duplicate Simpson's results by studying the beta decays of several different isotopes with a variety of techniques. All of these experiments found no evidence of such a kink, and each claimed to rule out the existence of a neutrino with a mass of 17 keV.

New experiments. In 1989, however, Simpson presented results of two new experiments that again showed evidence for the emission of the heavy neutrino. The first experiment again involved tritium implanted in a semiconductor detector, but germanium was used instead of silicon. The point was to demonstrate that the kink was not an artifact peculiar to the use of a silicon detector. The second experiment was a more conventional beta-decay study utilizing a radioactive source external to the detector. In this experiment, a source of ^{35}S (sulfur-35) was put on a thin backing, and the beta particles were observed in a cooled silicon detector. Both experiments provided further evidence that a neutrino with a mass of 17 keV is mixed with about 1% probability with the massless neutrino.

Supporting results. These new claims have prompted another round of experiments, but the results have been different. New studies of the beta decays of ^{14}C (carbon-14) and ^{35}S and a study of the electron-capture decay of ^{71}Ge (germanium-71) indicate that a neutrino with a mass of 17 keV is emitted approximately 1% of the time when these nuclei decay.

The recent ^{14}C experiment was similar to Simpson's tritium experiments in that the radioactive source was inside the detector. In this case, however, the detector was made from a crystal of germanium grown with ^{14}C inside it. Thus the problem of the radiation damage produced by an implantation process was avoided. The results of this study are shown in **Fig. 2**. The spectral distortion that appears 17 keV below the ^{14}C end-point energy is the signature of the heavy neutrino.

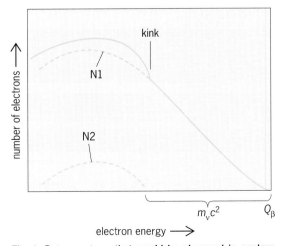

Fig. 1. Beta spectrum that would be observed in nuclear beta-decay studies if a massless neutrino (N1) were emitted 99% of the time and a neutrino (N2) of mass m_ν were emitted 1% of the time. The kink occurs at the energy where the spectrum associated with the emission of the heavy neutrino ends.

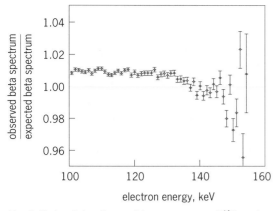

Fig. 2. Ratio of the observed beta spectrum of ^{14}C to that expected if the neutrino were massless. The shape of the spectrum expected for massless neutrinos was determined from the observed data above 140 keV. The change in this ratio from 1.0 that begins to occur at 138 keV is well fitted by the smooth curve through the data points, which shows the deviation expected if a 17-keV neutrino is emitted 1% of the time in the beta decay of ^{14}C. (*After B. Sur et al., Evidence for the emission of a 17-keV neutrino in the beta decay of ^{14}C, Phys. Rev. Lett., 66:2444-2447, 1991*)

The new ^{35}S experiment improved on Simpson's version by utilizing collimators and baffles to prevent scattered electrons from reaching the silicon detector. The results also showed a clear distortion of the ^{35}S beta spectrum characteristic of a neutrino with a mass of 17 keV and an emission probability near 1%.

The ^{71}Ge experiment differs from others in what was detected: photons given off in radiative electron capture (inner bremsstrahlung). These photons share the available decay energy with the neutrino, much as the electron and antineutrino do in ordinary beta decay. Thus, if a massive neutrino is sometimes emitted in electron capture, a kink should appear in the inner bremsstrahlung spectrum. Such a feature, 17 keV below the end point, is just what was observed in this recent study. Thus, three additional groups using quite different techniques have now reported evidence for the existence of the 17-keV neutrino.

Implications of a 17-keV neutrino. If these new results prove to be true, they will have profound implications for both particle physics and astrophysics.

Particle physics. In the so-called standard model of elementary particles, neutrinos are taken to be massless. While some extensions of the standard model allow for neutrino masses on the order of an electron-volt or so, a mass of 17 keV is far greater than had been expected. Experiments performed at high-energy accelerators now confirm that there are only the three types of neutrinos mentioned above, and thus arises the question as to which one has a mass of 17 keV. The limit on the mass of the electron-type neutrino previously mentioned eliminates the possibility that it is the 17-keV neutrino. From so-called neutrino oscillation experiments, in which the conversion of neutrinos of one type into another is sought, the muon-type neutrino is also ruled out. Thus the possibility that the tau neutrino is the object with a mass of 17 keV

that is being observed in these beta-decay experiments remains open.

Astrophysics. Such a heavy neutrino might be expected to play a major role in a number of astrophysical problems, such as the solar neutrino problem (that is, the deficit of neutrinos reaching the Earth from the Sun) and the dark-matter problem (the fact that the universe seems to contain far more matter than can be directly observed). The 17-keV neutrino appears to be too heavy to help resolve the solar neutrino problem. Its effect would be to decrease the flux of electron-type neutrinos reaching the Earth by only about 2%; this decrease hardly makes a dent in the factor-of-2–3 discrepancy between theory and experiment. *SEE SOLAR NEUTRINOS.*

Regarding the dark-matter problem, the situation is not so clear. Neutrinos were produced in huge numbers in the big bang. If these particles are stable, then there are now hundreds of neutrinos per cubic centimeter everywhere in the universe. If neutrinos have a rest-mass energy of approximately 100 eV, they would provide a matter density high enough to eventually halt the universal expansion and close the universe. A 17-keV neutrino is a problem in this regard. If it were stable, it would provide about 200 times the required closure density. Thus, the universe should have long ago collapsed under the weight of its neutrinos. Since this collapse did not happen, the conclusion is that the 17-keV neutrino must be unstable and have a lifetime of less than a million years or so. If its lifetime is near this upper limit, the decay products of the 17-keV neutrino could be the sought-for dark matter. Just what these decay products might be is an open question.

Further experiments. It is quite suggestive that a number of very different recent experiments have observed kinks in beta-decay spectra 17 keV below whatever the end-point energy is for the nucleus being studied. Furthermore, the fact that all of these experiments yield consistent mixing probabilities of approximately 1% is significant. Finding a common error that might be plaguing all of the experiments is difficult. However, so far, all positive reports of such kinks come from experiments performed with solid-state ionization detectors of either silicon or germanium, and magnetic spectrometer or gas detector experiments have not observed a kink. Clearly, if the object that is responsible for these observations is a neutrino, it must show up in every beta spectrum. Thus, many very careful experiments are under way at a number of different laboratories to help clarify the intriguing puzzle of the 17-keV neutrino.

For background information *SEE COSMOLOGY; NEUTRINO; RADIOACTIVITY; SOLAR NEUTRINOS; STANDARD MODEL* in the McGraw-Hill Encyclopedia of Science & Technology.

Eric B. Norman

Bibliography. A. Hime and N. A. Jelley, New evidence for the 17 keV neutrino, *Phys. Lett. B*, 257:441–449, 1991; R. Shrock, New tests for and bounds on neutrino masses and lepton mixing, *Phys. Lett. B*, 96:159–164, 1980; J. J. Simpson, Evidence of heavy-neutrino emission in beta decay, *Phys. Rev. Lett.*, 54:1891–1893, 1985; B. Sur et al., Evidence for the emission of a 17-keV neutrino in the beta decay of ^{14}C, *Phys. Rev. Lett.*, 66:2444–2447, 1991; I. Žlimen et al., Evidence for a 17-keV neutrino, *Phys. Rev. Lett.*, 67:560–563, 1991.

Nobel prizes

The Nobel prizes for 1991 included the following awards for scientific disciplines.

Physiology or medicine. Two Germans, biophysicist Erwin Neher of the Max Planck Institute for Biophysical Chemistry and physiologist Bert Sakmann of the Max Planck Institute for Medical Research, were awarded the prize for their discoveries on basic cell function that spawned a whole generation of research into the way that cells communicate. Using their "patch clamp" technique, Neher and Sakmann showed how individual ion channels, each consisting of a single protein molecule or a complex of molecules, control the passage of positively or negatively charged ions into and out of cells. The technique records how a single channel molecule alters its shape to control the flow of current within a few millionths of a second. The technique also showed that channels are specific to particular ions, such as calcium or chloride. Every cell has 20 to 40 types of ion channels.

The patch clamp technique made it possible to record the electrical activity of very small areas of membrane, so that the effects of opening and closing single ion channels could be observed. In patch clamping, suction is used to clamp a tiny pipette against a patch of cell membrane. An electronic amplifier is connected to the inside of the pipette. As ions traverse porelike channels on the membrane, they create a measurable electric current. Each pulse of current shows that the channel has opened.

This technique quickly evolved into one of the most widely used tools of cellular physiology. Researchers began using it to study cellular mechanisms underlying several diseases. For example, researchers discovered that a defective chloride channel is responsible for cystic fibrosis. The research of Neher and Sakmann also has contributed to understanding processes such as the passage of nerve impulses, the fertilization of eggs, and the regulation of the heartbeat.

Physics. Pierre-Gilles de Gennes of the Collège de France received the prize for his work on liquid crystals and polymers.

De Gennes is known for applying physical principles to problems of seemingly intractable complexity. In the 1960s, following early work on phase transitions in magnetic materials and superconductors, he began studying liquid crystals, whose molecular structure enables them to behave as ordered solids in some respects and as disordered liquids in others. De Gennes and his research group were able to understand and predict the electrically and mechanically induced phase changes in liquid crystals (which provide the basis for their use in electronic displays) by applying principles that have been used to explain phase changes in magnets and superconductors.

In the 1970s, de Gennes turned to the study of polymers, complex substances that have been likened to tangled masses of spaghetti on the molecular level. Again, he was able to gain insight into such systems by comparing them to simpler systems that had already been studied (magnets, superconductors, and liquid crystals) and by discovering mathematical relationships common to all of them. In particular, he showed that polymeric chains obeyed so-called scaling laws, so that their length could be calculated in terms of their thickness. This understanding of small-scale polymer properties makes possible the control of bulk properties such as viscosity. De Gennes and S. Edwards laid the basis for the theory of polymer viscosity and elasticity by showing that the motion of molecules in a flowing, molten polymer resembles that of snakes sliding past each other.

De Gennes has continued studying polymers, colloidal suspensions, and other complex systems, and has explored such varied problems as polymer welding, adhesion, and surface wetting and drying.

Chemistry. Richard R. Ernst of the Federal Institute of Technology (Eidgenössische Technische Hochschule) of Zurich, Switzerland, received the prize for work that helped transform nuclear magnetic resonance (NMR) spectroscopy from a tool with a narrow application to a key analytical technique in chemistry, biochemistry, biology, medicine, and materials science.

Nuclear magnetic resonance spectroscopy is based on the existence of magnetic forces in a large number of atoms. When these atoms are placed in a magnetic field, their nuclei align in the direction of the field. In NMR spectroscopy, the atoms are subjected first to an intense magnetic field, forcing them into alignment with the field. They are then subjected to a radiofrequency field, from which they absorb energy at frequencies that are characteristic for each species of atom. When the radio-frequency field is turned off, the atoms return to their alignment with the magnetic field, and in the process they emit pulses of energy that can be detected and measured. This information is characteristic for each of the species of atoms being measured and the molecular structures in which they are bonded.

Ernst transformed the earlier type of NMR, in which the chemist had to sweep a sample slowly with a spectrum of radio frequencies, into a technique in which a sample is subjected to a single, high-energy pulse of radio waves that contains all frequencies to which the atoms react. The resulting signal is converted into a recognizable spectrum by using a Fourier transform.

His later work provided an improved resolution of the spectrum by using two-dimensional NMR. This technique uses sequences of pairs of pulses in close succession, the interval between each pair being varied. The resulting signals can then be measured both as a function of time during each detection and as a function of the interval between the two pulses. This information provides two-dimensional spectra that identify which nuclei lie close together in a molecule, so that NMR can be applied to the study of molecules that were too large to be studied previously. Two-dimensional techniques also became the basis for a powerful variation of NMR known as magnetic resonance imaging (MRI), used in medicine for noninvasive examination of internal tissues without radiation.

Ernst later expanded the two-dimensional technique to three- and four-dimensional techniques. Such multidimensional techniques have been applied as complements to x-ray crystallography to determine the structures of biological macromolecules in solution.

Yearbook Editors

For background information SEE BIOPOTENTIALS AND IONIC CURRENTS; CELL PERMEABILITY; LIQUID CRYSTALS; MEDICAL IMAGING; NUCLEAR MAGNETIC RESONANCE (NMR) in the McGraw-Hill Encyclopedia of Science & Technology.

Nuclear fusion

On March 23, 1989, two chemists, M. Fleischmann and S. Pons, at the University of Utah announced that they had produced fusion power in a test tube at room temperature. They reported that they had passed electric current into a small cell containing heavy water (deuterium oxide, D_2O) and palladium metal. The current split the water into its constituents, the nuclei of the deuterium atoms being forced into the palladium where, according to Fleischmann and Pons, the nuclei combined (fused), producing new nuclear species and vast amounts of heat. The original claims were that 4 watts of power were generated for every watt expended by the battery. Within a few days the claims had increased by a factor of 10. The possibility of scaling the process up to fuel a fusion reactor with abundant cheap fuel (heavy water occurs naturally in seawater) and negligible pollution (fusion produces no long-lived radioactive waste products, in contrast to conventional nuclear fission plants, nor noxious emissions, which are the bane of chemical processes) generated worldwide interest. However, the majority of the world's scientists now dismiss the claims that cold fusion is a source of useful large-scale nuclear power and deny that the phenomenon is real.

Cold fusion claims. The experience of decades of study in nuclear physics seemed to imply that what Fleischmann and Pons were claiming to have achieved was, in fact, impossible. Indeed, the chance of two deuterium nuclei overcoming their mutual electrical repulsion and fusing at room temperature (as distinct from the temperature of $10^{7\circ}C$ more typical of the center of the Sun) is calculated to be on the order of 10^{50} times smaller than what would be required to explain the claimed heat production. That the claims were not immediately dismissed was in part because a rival group, led by S. Jones at Brigham Young University nearby, had independently been pursuing a similar idea and, seemingly, had come up with similar results, though detecting neutrons rather than heat. However, it subsequently became clear that the scale of the phenomenon claimed by the two groups differed substantially. (Jones detected neutrons, and the amount of heat that would be released during production of his rather feeble neutron signal was calculated to be at least 10^{12} times less than that which Fleischmann and Pons were claiming to have measured.) It also transpired that potential competition for patents with Jones's group had caused Fleischmann and Pons to make a premature announcement before all the necessary tests had been made—for example, replacing the heavy water (D_2O) with ordinary water (H_2O) to see if the effect vanished, since the fusion process requires deuterium. It soon became clear that much of the evidence of Fleischmann and Pons claiming to have detected nuclear fusion products (specifically neutrons and gamma rays) was flawed.

Replication attempts. However, by the time that these shortcomings were becoming generally known, thousands of scientists had begun tests to see if they could replicate the described phenomenon. Early reports appeared in the media, though not in the peer-reviewed scientific literature, that seemed to confirm it.

It turned out that palladium in the presence of hydrogen or deuterium behaves as a complicated chemical system that undergoes sharp changes in temperature as the deuterium is absorbed into the metal. Unaware of these subtleties, as were many in those early days, several small groups assumed that they were observing fusion heat being produced in their experiments, and announced "confirmation" of the Fleischmann and Pons effect.

Meanwhile larger teams, comprising experts in nuclear physics, chemistry, and materials science, were carrying out experiments at government laboratories and leading institutions throughout North America and Europe. A team at Harwell, a major British atomic and nuclear physics laboratory, reported that they had performed over 100 experiments with the most advanced available equipment and had seen no evidence for the claimed cold fusion. The quality of apparatus available at Harwell and other government laboratories was far superior to that available in many other smaller institutions, and these multidisciplinary teams also benefited from the availability of broad expertise that made them alert to the several pitfalls into which the inexperienced might slip (such as the difficulty of measuring neutrons at low levels or of measuring heat balances accurately in the palladium–deuterium system). By May 1989, the negative results from institutions and government laboratories had begun to confirm many scientists' suspicions that the extreme claims made by and on behalf of Fleischmann and Pons were incorrect.

Evaluation of claims. A panel of experts was commissioned by the U.S. Department of Energy to evaluate the claims and the many other experiments in progress. This committee reported in the fall of 1989 that it had not been presented with any convincing evidence for the occurrence of cold nuclear fusion, and that it had identified possible sources of error in several experiments. The committee recommended that any scientists wishing to pursue cold fusion should compete for funds as is the normal practice and that no special funds should be allocated to cold fusion. The panel also recommended that if any support were given, it should be to collaborations between scientists with wide expertise, and in particular that groups claiming contrary results should pool their expertise in order to bring a disciplined focus to the research effort. Similar conclusions were reached by a European study.

Current status. By mid-1991, a National Cold Fusion Institute in Salt Lake City, supported by the State of Utah, had produced little of significance. A few papers were published and one in particular, by a team including Fleischmann and Pons, made some claims that contradicted those that the two had made in 1989. Manifest inconsistencies and errors in their original claims to have measured gamma rays as evidence for fusion were noted.

By 1991, some research continued, and the possibility remained that low-level production of neutrons, related to those claimed by Jones's team, was real, although many physicists dismiss even this as insignificant background rather than a genuine signal. Sporadic reports appeared claiming that one or another group had observed some positive effect, even though, frequently, well-documented and more complete experiments had shown no effects with greater precision than those claiming a positive result. In even the most optimistic reports claiming to have detected fusion products (neutrons, tritium, or helium), not one experiment claimed to find heat and products that correlate as necessary to establish a nuclear reaction. Fleischmann and Pons still claimed to see occasional bursts of heat emerge from their test tubes, but evidence for a nuclear process was not forthcoming.

A conference on cold fusion, attended mostly by committed supporters, was held in Italy during 1991 and claimed that cold fusion was being pursued with success. However, the supporters were only a tiny minority of scientists, and the almost universal reaction of major laboratories in North America and Europe, based on their own experiments and many other careful null experiments worldwide, is that the original cold fusion claims are a chimera.

For background information SEE NUCLEAR FUSION in the McGraw-Hill Encyclopedia of Science & Technology.

Frank Close

Bibliography. F. Close, *Too Hot To Handle*, 1991; M. Fleischmann, S. Pons, and M. Hawkins, Electrochemically induced nuclear fusion of deuterium, *J. Electroanal. Chem.*, 261:301–308, 1989, errata, 263: 187–188, 1989; S. Jones et al., Observation of cold nuclear fusion in condensed matter, *Nature*, 338:737–740, 1989; D. Williams et al., Upper bounds on "cold fusion" in electrolytic cells, *Nature*, 342:375–384, 1989.

Nuclear power

Atomic Energy of Canada Limited (AECL), developer of the CANDU heavy-water-moderated natural-uranium-fueled reactor, has an ongoing effort to ex-

tend the CANDU technology to a system especially designed for future electrical utility markets. This new system under development, designated CANDU 3, is a 450-megawatt electric plant that incorporates modular construction, utilization of natural uranium fuel, a 37-month construction schedule, computer-controlled operation, high availability through use of on-line refueling, a 100-year lifetime design with replaceable components, passive safety features, and other features that will improve safety and reliability. This design is based on the proven, reliable, heavy-water natural uranium technology that characterizes all the previously built CANDU systems. There are currently 26 operating CANDU reactors, which generated 94×10^6 megawatt-hours of electricity in 1990.

Because relatively few CANDU plants operate outside Canada, the technology of heavy-water reactors is not as widely known as that of light-water reactors developed in the United States. A convenient way to describe CANDU technology is to compare it to that of light-water reactors, and, in particular, the pressurized-water reactor. The CANDU and the pressurized-water-reactor plants have many similarities. However, the most obvious difference is the configuration of the CANDU core, which does not employ a large, pressurized reactor vessel. It utilizes a series of fuel channels arranged in a square lattice within a horizontal, cylindrical tank. The structure, called a calandria, is filled with heavy water, which serves as the moderator and operates at a low temperature, near 70°C (158°F), and near atmospheric pressure (**Fig. 1**). The basic core design of CANDU 3 is smaller than earlier CANDU reactors, but features the same fuel and lattice design. Thus, the reactor-physics, fuel, fuel-channel, and thermal-hydraulics aspects of CANDU 3 can be considered to be well proven.

The CANDU 3 design uses natural uranium as fuel, and heavy water as coolant and moderator. The coolant and moderator are separated from each other by regularly and loosely spaced fuel channels. The coolant, which flows inside the channels, is hot and pressurized, and the moderator, which fills the large calandria, is cool and at essentially atmospheric pressure. Thus, the CANDU 3 core has very little excess reactivity and all reactivity coefficients are very small.

Fuel design. The CANDU fuel design is relatively simple in comparison to the light-water reactor fuel (**Fig. 2**). It consists of 37 elements of uranium dioxide (UO_2) sheathed in zircaloy and held together as a bundle by end plates. There are 12 fuel bundles in each fuel channel. Fuel and heavy-water coolant are enclosed in standard zirconium-niobium pressure tubes, which in turn are surrounded by zircaloy calandria tubes. High-purity heavy water contained in the calandria vessel at low pressure and low temperature serves as the moderator. The heavy-water moderator is completely independent of the pressurized heavy-water coolant.

CANDU reactor core design, using natural uranium dioxide as fuel and deuterium oxide (D_2O) as moderator, is dedicated to maximum neutron economy and

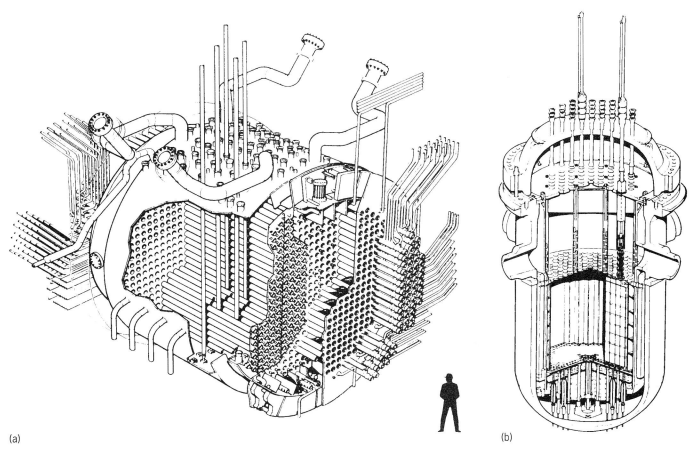

(a)

(b)

Fig. 1. Comparison of (*a*) CANDU and (*b*) pressurized-water reactor structures.

fuel utilization. As the fuel is utilized and uranium-235 is depleted, the buildup of plutonium provides additional reactivity and eventually contributes substantially to the energy production. Saturated and unsaturated fission products account for a very small fraction of the total neutron absorption. The heavy-water moderator and high-purity zircaloy used for fuel sheaths and structural components also have very low neutron absorption. Defective fuel bundles can be located and removed while the reactor is operating at full power, and the result is a relatively clean and uncontaminated reactor system.

Fueling cycle. Because of the very low excess reactivity of a natural uranium core, the CANDU reactor has to be fueled semicontinuously and while operating on power. Thus, the fuel management scheme and the fuel-handling equipment are unique features. A sophisticated fueling cycle, including robot-type machines, has been thoroughly tested both in multiunit stations and in stand-alone plants.

The CANDU system produces more spent fuel by volume than the pressurized-water reactor. However, because of the use of natural uranium, there is no possibility of any criticality occurring with the spent fuel, and spent fuel elements can be stacked in almost any configuration and stored in pools on-site. Also, because of the lower reactivity, the heat generated from these elements is far less on a volume

basis and requires less sophisticated storage shielding.

Refueling operations are carried out on a routine basis with the reactor at full power. The number of fresh fuel bundles introduced into a channel is small, and the shuffling pattern of the bundles along a channel is flexible. By adjusting the fueling rate in various regions of the core, the power distribution in these regions is effectively controlled on a long-term basis. Since power is not changed or interrupted for refueling, it is not necessary to tailor the refueling schedule to the system load requirements of the utilities that are supplied by the plant.

Control and shutdown systems. In CANDU 3, all reactivity devices are inserted in the cool, low-pressure moderator. The reactivity devices, used to control and shut down the reactor, are physically and functionally separated from each other. In addition, for emergency shutdown purposes, there are two independent, diverse, fast-acting systems, capable of achieving and maintaining cold shutdown under all conditions. Since defective bundles can be located and removed on power, there is no long-term buildup of fission products or other contaminants in the entire reactor system. This feature and the high-velocity flow of the coolant lead to a relatively clean reactor coolant system and a very long life for the internal compo-

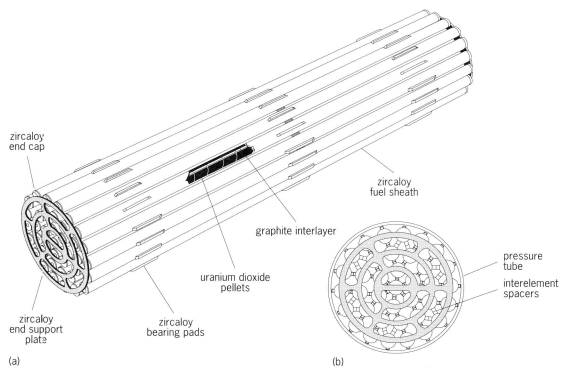

zircaloy end cap

zircaloy fuel sheath

graphite interlayer

pressure tube

interelement spacers

uranium dioxide pellets

zircaloy end support plate

zircaloy bearing pads

(a)

(b)

Fig. 2. CANDU 3 37-element fuel bundle as seen in (a) oblique view and (b) end view inside the pressure tube.

nents. This feature allows the reactor coolant system to be fabricated of carbon steel, which is ductile and immune to stress corrosion cracking. An additional feature is the absence of any reactivity control chemicals.

Pressure tubes. The most obvious feature that differentiates CANDU from pressurized-water reactors is the use of pressure tubes instead of a massive, thick-walled pressure vessel (**Fig. 3**). These tubes are the only CANDU components subject to a combination of high stress and radiation. In the very rare event that a failure occurs in the pressure tubes, they can be easily removed and replaced. It is this feature that permits the projection of a 100-year lifetime, since the only major component that cannot be removed and replaced is the calandria vessel.

Calandria vessel. The calandria vessel is fabricated from carbon steel, and since it is not subject to high pressure, temperature, or radiation levels, it is expected to last indefinitely. The calandria vessel, in conjunction with the integral end shields, supports the horizontal fuel-channel assembly and the vertical and horizontal in-core reactivity control unit component. The entire assembly is integrally supported by the end walls of the shield tank. The shield tank and the upper module (an assembly connected to the top of the calandria) are separate closed vessels that contain light-water shielding. The shield tank, end shields, and upper module protect adjacent areas against radiation from the reactor. As a result, nuclear heat is generated within the shields and transferred to the heat-transport system and the shield-cooling systems.

Heat-transport system. The heat-transport system circulates the pressurized heavy water through the reactor fuel channel, so that the heat produced by the fissioning of the natural uranium fuel is removed. The heat is carried by the reactor coolant to the steam generators, where it is transferred to light water to produce steam. The steam subsequently drives the turbine generator or, alternatively, could be provided in part to users of process heat.

Circulation of the reactor coolant through the reactor fuel channels is maintained at all times during reactor operations, shutdown, and maintenance. A shutdown cooling system, capable of operation at the full temperature and pressure of the heat-transfer system, is used to remove the reactor's shutdown heat. This shutdown cooling system also permits the pumps and steam generators to be drained into the heat-transfer system for maintenance. An emergency core coolant system supplies light water to the heat-transfer system if the reactor coolant is lost from the heat-transfer system because of a pipe rupture. Heavy-water leak sources are minimized in CANDU reactors by using welded construction and bellow-sealed valves where practical. Leaks, pipe ruptures, and joint failures are less likely to occur in modern CANDU reactors than in light-water reactors.

Fuel-handling facilities. The fuel-handling facilities include equipment for the handling and storage of both new fuel for reactor refueling and irradiated fuel. Space for handling and transferring irradiated fuel to transport equipment for shipping or storage in other facilities (for example, on-site dry fuel storage facilities) is provided in the irradiated-fuel bay area. The fueling machine, which loads and unloads fuel at the reactor, and the fuel-transfer system, which transfers new fuel into the reactor building and spent fuel out of the building, are fully automated and operated from the fuel-handling control center. All of the fuel-handling equipment can be serviced in the maintenance lock, an air lock used for the maintenance of the fueling machine, while the reactor operates at full power. The function of the fueling machine is to pick up new fuel from the new-fuel transfer port, load it into the reactor while also removing irradiated fuel from the reactor, and load the fuel removed into the irradiated-fuel storage bay via the irradiated-fuel port. These operations are performed under fully automatic and remote-control conditions.

Computer control. CANDU reactors use highly reliable computers for the control of the process systems such as the reactor regulating system, boiler level and pressure control, pressurizer level, and pressure control. In the CANDU 3 design, the two large central computers used in earlier CANDU reactors have been integrated with a distributed control system consisting of electronic modules distributed throughout the plant and linked by coaxial-cable data highways. Digital control is used in the CANDU design for the actuation of safety systems, including the two reactor shutdown systems. This approach is in line with trends toward increased use of digital protection systems.

Design and construction. The CANDU 3 will be a highly modularized design with a minimum number of modules capable of either shipyard or on-site fabrication. Open-top construction, a very heavy lift crane, and externally assembled modules help to shorten the predicted construction schedule to 37 months or less. Vertical installation of components, such as steam generators, through the open roof of the reactor shortens installation time compared to horizontal access methods. Other systems are assembled either off-site or on-site into modules weighing 10–500 tons, which can then be lifted into place. About 30 heavy lifts in the reactor building will be divided equally between the modules and major components. The design of internal concrete structures has been greatly simplified by using more steel platforms and supports. It is expected that the CANDU 3 will meet the U.S. Nuclear Regulatory Commission's design certification requirements under Regulation 10-CFR Part 52, so that the time period for the overall licensing process will be significantly reduced.

Containment structure. The majority of operating CANDU plants, in particular those owned and

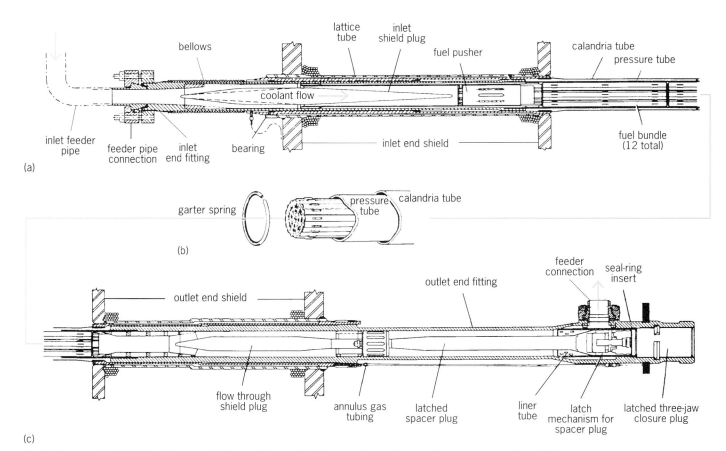

Fig. 3. Diagram of CANDU 3 fuel channel. (*a*) Channel inlet end. (*b*) Fuel bundle and surrounding structures in the middle of the channel. (*c*) Channel outlet end.

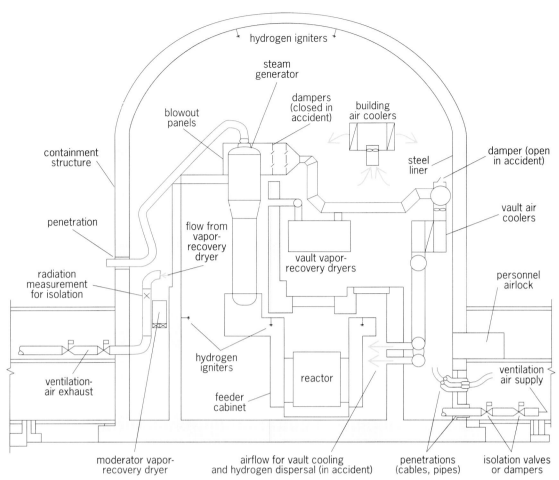

Fig. 4. CANDU 3 reactor containment system. The ventilation system and deuterium oxide vapor recovery are normal operating systems and not part of the containment system. Hydrogen igniters immediately ignite any hydrogen released as a result of an accident, before larger amounts can be accumulated. The radiation measurement unit determines the level of activity of the air inside the containment structure, in the event of an accident, before the air is exhausted or collected.

operated by Ontario Hydro in Canada, are multiunit stations where the plants share common facilities. CANDU 3, however, will be designed as a stand-alone, single plant unit and will have its own containment structure (**Fig. 4**). This containment structure is composed of a reinforced concrete base, cylindrical perimeter walls, and a domed top with a steel liner to provide leak tightness. The design requirements for the containment structure include the consideration of all processing and environmental loading conditions for normal operation, accident conditions, and extreme environmental conditions. A horizontal air lock is provided as a primary entrance and egress route, however; as discussed above, initial construction of the plant will use an open top and major components will be lowered vertically into place.

Application. Worldwide, CANDU plants have traditionally occupied the highest ranks in terms of plant availability, safety, and on-line performance. The CANDU 6, in operation at the LePreau Station for New Brunswick Power, has the highest availability of any power plant in the world at this time. The CANDU 3 represents a nuclear option for United States utilities for the late 1990s and into the twenty-first century. An active program is under way to obtain design certification from the Nuclear Regulatory Commission and to establish utility requirements as part of the CANDU 3 design. Also being explored is the possibility of establishing an infrastructure that would make it possible to deliver a complete system using the CANDU reactor as an American-made product based on Canadian technology for use in the 2000-to-2010 time period.

For background information SEE DIGITAL CONTROL; DISTRIBUTED SYSTEMS (CONTROL SYSTEMS); NUCLEAR FUELS; NUCLEAR POWER; NUCLEAR REACTOR in the McGraw-Hill Encyclopedia of Science & Technology.

Raymond W. Durante

Bibliography. R. J. Atchison, F. C. Boyd, and Z. Domaratzki, *The Canadian Approach to Nuclear Power Safety*, Atomic Energy of Canada Limited, 1983; Atomic Energy of Canada Limited, CANDU Operations, *CANDU: The Advanced PWR with Proven Performance*, Doc. AECL 9963, 1989; Atomic Energy of Canada Limited, CANDU Operations, *CANDU 3 Technical Outline*, vol. 1, Doc. 74-01010-TED-001 (Rev. 9), 1989; K. R. Hedges and E. M. Hinchley, *Nuclear Engineering International*, 1990.

Nuclear reactor

The first section of this article describes the actions and initiatives being taken by the United States electrical utilities, nuclear reactor manufacturers, nuclear plant designers and constructors, and the other organizations supporting the nuclear energy industry in establishing a new generation of advanced light-water reactors (ALWRs). The second section discusses a particular advanced light-water reactor, the AP600, in more detail.

Industry Initiatives on Advanced Light-Water Reactors

The commercial nuclear energy industry has initiated a concerted effort to design, license, construct, and operate a new generation of advanced light-water reactors. The commitment of the industry and the plan for fulfilling it were announced in November 1990, when the Nuclear Power Oversight Committee (NPOC) released *A Strategic Plan for Building New Nuclear Energy Plants*. The Nuclear Power Oversight Committee, which is composed of senior executives responsible for coordinating and directing policy development for the nuclear energy industry, stated that the plan was developed because (1) it is becoming more certain that baseload electric generation capacity will fall short of demand in the next decade; (2) new requirements for air-pollution controls on coal burning, such as those specified in new amendments to the Clean Air Act, will increase the cost and regulatory uncertainty of generating electric power from coal; (3) although the contribution of operating nuclear plants has made it possible to greatly reduce the use of imported oil in electric power generation, such usage is growing again in the electric sector, and new nuclear plants will be required to reverse this trend; and (4) with the increased concern about the possible long-term effects of greenhouse gas emissions, there is a need to develop and utilize electric generation processes in which greenhouse gases are not produced.

The *Strategic Plan* establishes a goal of having an order for a new advanced light-water reactor by the mid-1990s, with the reactor becoming operational around the year 2000. It identifies the conditions and associated actions required to achieve the goal, and assigns specific responsibilities to industry organizations for implementing actions and achieving specific milestones. **Table 1** lists the areas of activity or building blocks addressed in the *Strategic Plan* and the organizations having lead responsibility for each building block.

As is evident from Table 1, the industry has defined a comprehensive program. The program includes actions to assure that the operation of existing light-water reactor plants continues to improve; that meaningful progress is made in the identification and development of disposal sites for both low-level and high-level radioactive waste; and that an adequate fuel supply is available for operation of both existing and future light-water reactors.

Standardization. The electrical utility industry, which will be the purchaser of the next generation of advanced light-water reactors, has specified that those reactors should be designed, constructed, and ultimately operated as standardized plants. Standardization in its simplest terms means that only a small number of different plant designs will exist, and that all units of a particular standardized design would be identical, except for a very limited number of site-specific differences. While most of the existing reactors in the United States are not of standardized designs, the very successful French nuclear program, which provides more than 75% of France's electricity, is based on standardized nuclear plant designs. By using standardized designs, the French were able to shorten construction times from 7 years to 5 years, order and construct a large number of plants in a short period, and operate the plants economically and with high reliability.

Table 1. Building-block summary of NPOC strategic plan

Building blocks	Responsible organizations
Prerequisites from ongoing programs	
Current nuclear plant performance	Utilities
Low-level radioactive waste	Edison Electric Institute, American Committee on Radwaste Disposal
High-level radioactive waste	Edison Electric Institute, American Committee on Radwaste Disposal
Adequate, economic fuel supply	Edison Electric Institute
Generic safety/environmental regulation and industry standards	
Predictable licensing and stable regulation	Nuclear Management and Resources Council
Advanced light-water reactor utility requirements	Electric Power Research Institute, Utility Steering Committee
Project-specific activities	
Nuclear Regulatory Commission design certification	Plant designers
Siting	Electric Power Research Institute, Utility Steering Committee; Nuclear Management and Resources Council
First-of-a-kind engineering	Electric Power Research Institute, Utility Steering Committee
Enhanced standardization beyond design	Nuclear Management and Resources Council
Institutional steps	
Enhanced public acceptance	U.S. Council for Energy Awareness
Clarification of ownership and financing	Edison Electric Institute
State economic regulatory issues	Edison Electric Institute
Enhanced governmental support	American Nuclear Energy Council

The industry commitment to standardization was considered so important that in April 1991 the Nuclear Power Oversight Committee issued a position paper on standardization. This position paper articulates a life-cycle commitment to standardization, which starts with the design and continues through enhanced standardizations beyond design. This enhanced standardization involves standardizing approaches to construction practices, operating practices, maintenance, training, and procurement activities. The position paper predicts that rigorous implementation of standardization can achieve the gains in efficiency and economy normally associated with increases in scale or breakthroughs in technology.

The *Strategic Plan* addresses the actions required for establishing the specifications for such standardized designs as well as appropriate changes to the regulatory review process that would allow the U.S. Nuclear Regulatory Commission (NRC) to more effectively, efficiently, and fairly review such standardized plants. The *Strategic Plan* specifies the actions required to design, license, and site an advanced light-water reactor. Finally, the *Strategic Plan* specifically recognizes the financial and public-acceptance issues that must be addressed as part of establishing the institutional framework required to support a new generation of advanced light-water reactors.

ALWR options. The electrical utility industry, through the Electric Power Research Institute (EPRI), has taken the lead in specifying requirements in the advanced light-water reactors. The Electric Power Research Institute has issued a requirements document that contains a comprehensive set of design requirements for future advanced light-water reactors. This requirements document, whose review has been undertaken by the Nuclear Regulatory Commission, provides the basis for the development of advanced light-water reactor designs by reactor manufacturers, and provides a partial technical basis for licensing these designs by the Nuclear Regulatory Commission.

Two types of advanced light-water reactors are under development, a large, 1300-MWe reactor (meaning that it would generate 1300 megawatts of electric power), and a midsize, 600-MWe reactor.

Large reactors. The large advanced light-water reactor is commonly called an evolutionary plant design because it is based upon the lessons learned and the technology improvements that have been incorporated into the design of the most recently constructed large light-water reactors now operating in the United States. The evolutionary plants include all of the appropriate and necessary changes that evolved during the design and initial operation of existing reactors. In addition, this advanced light-water reactor design incorporates technological improvements such as instrumentation and control systems that use advanced multiplexed fiber optics, improved fuel and reactor core designs, safety systems that improve overall safety by a factor of 10, and many other design changes that improve the availability, maintenance, and operation of these reactors.

Two evolutionary-plant standardized designs are currently being reviewed by the Nuclear Regulatory Commission, the advanced boiling-water reactor (ABWR) and the advanced pressurized-water reactor (APWR). Both designs are expected to complete the Nuclear Regulatory Commission's design certification review process by the mid-1900s and would be available for purchase by utilities at that time.

Midsize reactors. There are also two designs for midsize 600-MWe advanced light-water reactors going through the Nuclear Regulatory Commission design certification process: the simplified boiling-water reactor (SBWR), and the advanced passive pressurized-water reactor (AP600), discussed in more detail below. The standardized designs for midsize advanced light-water reactors differ from the evolutionary plants both in their smaller size and in their use of more passive safety features (such as water flowing because of gravity) in place of the active safety systems (such as water being pumped) that are prevalent in the evolutionary plants. The reliance on passive safety features results in significant decreases in the amount of pipe, control cable, valves, and pumps required in these plants. Because of this decrease, the construction, maintenance, and operation of the midsize advanced light-water reactor are expected to be simpler and easier (for example, in employing modular construction techniques) than for current-generation light-water reactors. Finally, like the evolutionary advanced light-

Table 2. Costs of alternative electric generating plants commencing operation in the year 2000*

Type	30-year-averaged generating costs, ¢/kWh (constant 1991 dollars)	
	600-MWe plant	1200-MWe plant
Nuclear: advanced light-water reactor	4.5	3.8[†]
		4.1[‡]
Pulverized coal	4.8	4.6
Integrated gasification combined cycle (coal)	5.0	4.8
Combined-cycle combustion turbine (gas)	4.6	4.5

*From U.S. Council for Energy Awareness, *Advanced Design Nuclear Energy Plants: Competitive, Economical Electricity*, 1992. Costs of fossil-fired alternatives are provided by the Electric Power Research Institute (EPRI). Costs of advanced light-water reactor alternatives are provided by combination of EPRI, reactor manufacturers, and United Engineers and Constructors.
[†]One-unit evolutionary plant.
[‡]Two-unit midsize plant.

water reactor, the midsize plant incorporates fuel and reactor core improvements, safety improvements, and advanced technologies (such as fiber optics) that will enhance its reliability and overall level of safety. These designs are also expected to receive certification from the Nuclear Regulatory Commission by the early to mid 1990s.

Economic competitiveness. Ultimately, the commercial viability of the advanced light-water reactors will be based on their economic competitiveness with other available electric generating alternatives. Both the midsize advanced light-water reactor and the larger evolutionary advanced light-water reactor are expected to be quite competitive with the alternative generating sources that will be available in the year 2000. **Table 2** shows that the 30-year-averaged generating cost for a midsize advanced light-water reactor commencing operation in the year 2000 is expected to be 4.5¢/kWh (in 1991 dollars) as compared to a range of 4.6 to 5.0¢/kWh for the gas and coal-fired electric generating alternatives. Likewise, at the 1200-MWe size, Table 2 indicates that the advanced light-water reactor's 30-year-averaged generating costs for both a single large plant and two of the midsize plants are significantly less than the viable fossil-fueled alternatives.

The results shown in Table 2 are based upon successful completion of the required actions contained in the Nuclear Power Oversight Committee's *Strategic Plan*. Such completion means, for example, that standardized designs certified by the Nuclear Regulatory Commission exist, and that the Commission licensing process is reformed to recognize the new type of reviews required. Table 2 includes the best available information on capital costs, operating and maintenance costs, fuel costs, and, in the case of the nuclear plant, decommissioning and nuclear waste disposal costs. The results should not be construed to mean that nuclear plants are the only economical option. Rather, they show that, if the industry plan is successful, the advanced light-water reactors will be very cost competitive with their fossil-fueled alternatives, and that, given the need for additional baseload generating capacity, utilities could look to the advanced light-water reactor as a very viable and desirable option to provide that capacity.   *Marvin S. Fertel*

AP600 Reactor

In response to an initiative established by the U.S. Department of Energy and the Electric Power Research Institute for a new generation of smaller, safer, and more economical nuclear reactors, the AP600, an advanced passive pressurized-water reactor that generates 600 MWe, has been designed (**Fig. 1**). Simplification is central to the development of the AP600, the main features being a simplified reactor coolant system, simplified plant systems (including passive safety systems), and a simplified plant arrangement (including a modular construction approach). It has been designed to meet the safety, operational, and financial criteria set by the government, the utility industry, and the public.

Government and industry projections indicate that United States energy demands in the 1990s will grow steadily and will create the need for approximately 115,000 MWe of additional capacity by the year 2000. Although this growth in electricity demand continues to be strong, orders for new nuclear power plants have not kept pace, in part because of licensing delays, prohibitive construction costs, and public uncertainty about safety. In a joint effort to keep nuclear power as an attractive energy option, the Department of Energy and the Electric Power Research Institute initiated a program to develop the conceptual design of a new reactor that would be economical to build, would be simple to operate and maintain, and would have increased safety margins. To this end, Westinghouse Electric Corporation designed the AP600, and has been awarded a 5-year contract by the Department of Energy, the Electric Power Research Institute, and various utilities that should result in plant certification by 1995. The AP600 represents the optimized progression from the licensed and successful two-loop, 600-MWe pressurized-water reactor, which consistently achieved availability factors 10% above the national average during more than 30 years of operation.

Simplified design. The AP600 nuclear plant has a simpler, pared-down design compared to conventional nuclear plants. Its nuclear island uses 60% fewer valves, 75% less pipe, 80% less control cable, 35% fewer pumps, 50% less seismic building volume, and 80% less heating, ventilating, cooling, and ducting than conventional reactors. The AP600 features a low-power-density core with a greater safety margin, proven high-reliability components, and an 18- or 24-month refueling cycle. The plant arrangement allows access to all buildings and systems that are important to safety, and enables all safety-related systems to be located in a single, compact area. Together, these factors result in a reactor that is less expensive to operate and maintain, and easier and less expensive to build.

The low-power-density core consists of 145 fuel rod assemblies arranged in a standard 17×17 rod array, with an active fuel rod length of 12 ft (3.65 m). These fuel assemblies are surrounded by stainless-steel reflector assemblies that reduce neutron leakage. The reduction in neutron leakage improves reactor efficiency, so that a lower level of fuel enrichment and a reduction in fuel cycle cost are possible. The reactor vessel and internal components are of essentially conventional design, so no new manufacturing development is necessary.

Because the reactor produces less power than some other reactors (600 MWe as compared to up to 1300 MWe) and operates at a lower temperature (600°F or 315°C as compared to up to 630°F or 331°C), less residual heat must be dissipated in the event of an accident; thus the safety systems needed to cool the reactor can be simplified. These characteristics also contribute toward an increase in operator reaction time (should it prove necessary) in the unlikely event of an accident.

Safety. AP600 active plant systems are designed to protect the plant and the public from undesirable consequences, such as overheating, which could result from infrequent but anticipated disruptions of normal plant operation. The design minimized the impact of disruptions, such as the disconnection of the turbine and the condenser that remove heat by converting it to electricity and condensing steam, or the rupture of a pipe, resulting in a diversion of coolant water flowing to the core (known as a loss-of-coolant accident). The AP600 maintains safety in three ways: (1) design conservatism that minimizes the chance of such disruptive events, (2) safety systems that mitigate the effects of any such disruptions, thereby precluding severe results such as core melt, and (3) the facilitation of correct human operator responses, even to the extent of allowing the operator not to act.

The AP600 has a simplified loop arrangement that features the combination of the steam generator and hermetically sealed reactor coolant pumps into one structure. This arrangement eliminates the connecting pipe, a potential cause for loss-of-coolant accidents.

Should this first line of defense be disrupted, passive safety systems of the AP600 would automatically provide the ultimate degree of protection. These systems provide long-term supplies of cooling water and rely predominantly on natural forces such as gravity, circulation, convection, and evaporation. Human error has been implicated in all major nuclear reactor mishaps so far. The AP600 plant, especially the control and safety systems, is designed to facilitate human interaction by providing more response time for operators than conventional nuclear plant designs. In fact, no human action is needed to activate the passive safety systems. The use of natural passive safety systems is also a major contributor to the reduction in the quantities of equipment and materials; thus is eliminated the need for, and cost of, large amounts of piping, safety-related pumps, emergency diesel generators, and other complex components and systems, as well as all the related control and monitoring equipment.

The AP600 safety cooling systems are self-contained; that is, reserve reactor coolant tanks are located inside the containment vessel itself (**Fig. 2**), in contrast to conventional reactors, where only the accumulator tank is located inside the containment vessel. Therefore, there is a major reduction in the amount of shielding needed outside the vessel. Core makeup tanks containing borated water are located above and to the side of the reactor vessel to provide water for emergency core cooling injection and decay heat removal. This water is kept at the same pressure as that in the reactor vessel, so that any substantial loss of coolant from the system would trigger air-operated valves automatically and allow water to flow down into the core. The water would continue to cool the flooded core through natural circulation and condensation, as it carried heat from the core to the surrounding steel containment vessel.

Should the pressure keep falling, an additional

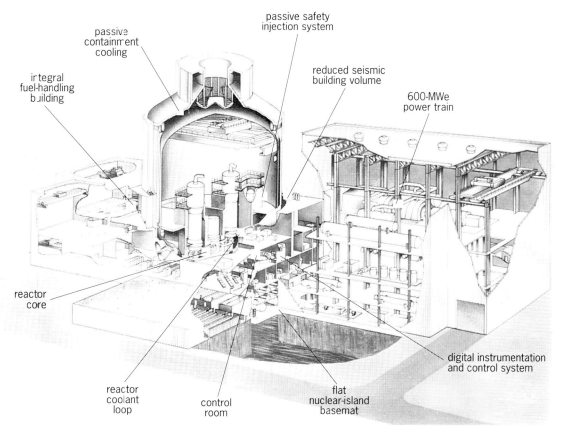

Fig. 1. Cutaway view of the AP600 nuclear power plant.

500,000 gallons (1900 m^3) of gravity-driven water, stored in an in-containment refueling water storage tank, would flood both the inside and the outside of the reactor vessel to above its nozzles. (In normal operation, the water in this tank is used to flood the refueling pit above the reactor vessel after the reactor is shut down for refueling and the reactor vessel head is removed, in order to provide shielding and cooling for the transfer of spent fuel.) The water would continue to cool the reactor indefinitely by transferring heat from the core to the containment vessel. A concrete shield building surrounds the steel containment vessel, and between the building and vessel there is an airflow space. When outside air enters the building, a natural draft transfers containment heat to the airflow space and up through a chimneylike opening in the roof. The vessel is further cooled by convection of air at its surface, with the aid of evaporation of water sprayed on the surface. While these safety systems are functioning, no operator action is necessary for at least 3 days. Other advantages of including all safety systems inside the containment vessel are a reduction in pipes, enhancement of operation to allow quicker operator response in the event of an emergency, and added protection against external damage.

Instrumentation and control. Advanced microprocessor-based instrumentation and control systems also simplify and improve plant operation and maintenance. A digital, multiplexed control system takes the place of hard-wired analog controls and cable-spreading rooms and accounts for the 80% decrease in control cable compared with current nuclear plants. This microprocessor-based system also permits continuous on-line monitoring of plant performance and operating conditions. Multiple rows of switches and more than 1000 alarm windows have been eliminated from the control room design as a result of human-factors studies, and have been replaced with controls that are easier to monitor so that operator action is facilitated. *SEE* INSTRUMENTATION.

Construction and standardization. The AP600 can be constructed and operating within 5 years of a placed order. A modular approach to construction accelerates the process and allows close supervision of the assembly of each module at the factory. Complete, prefabricated sections of the reactor can be shipped to the plant site and assembled. The modular construction approach also facilitates standardization, a key issue with the Department of Energy, the Electric Power Research Institute, and utility companies, as discussed above. Standardization is important because assembly-line manufacturing allows greater quality control, fewer construction delays, shorter on-site construction periods, and lower costs. It would also enable any nuclear engineer or operator to more easily transfer experience and skills acquired in working on one AP600 nuclear plant to any other.

At 600 MWe, the midsize AP600 allows utilities to build a conveniently sized power plant around a single reactor, or a larger one with two or more units. Utilities can add capacity in increments that more closely match their demand growth.

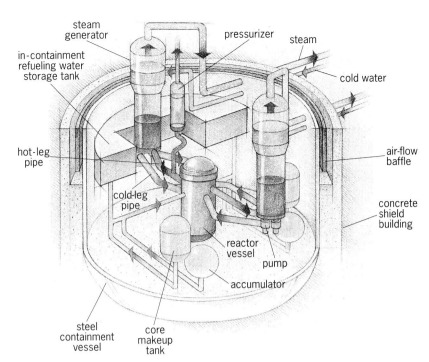

Fig. 2. AP600 natural passive cooling system.

Status. The conceptual design phase of the AP600, which included feasibility and design studies and tests of major components, has been completed. The next phase of the program is certification by the Nuclear Regulatory Commission, followed by detailed design and by orders placed by electrical utilities and consortia.

For background information *SEE* DISTRIBUTED SYSTEMS (CONTROL SYSTEMS); NUCLEAR POWER; NUCLEAR REACTOR in the McGraw-Hill Encyclopedia of Science & Technology.

Howard J. Bruschi

Bibliography. ANS/ASME *Nuclear Engineering Conference*, Newport, Rhode Island, 1990; Electric Power Research Institute, *Advanced Light Water Reactor Utility Requirements Document*, 1990; Nuclear Power Oversight Committee, *Position Paper on Standardization: Building New Nuclear Power Plants*, 1991; Nuclear Power Oversight Committee, *A Strategic Plan for Building New Nuclear Energy Plants*, 1990; U.S. Council for Energy Awareness, *Advanced Design Nuclear Energy Plants: Competitive, Economical Electricity*, 1991.

Nuclear structure

In 1986, discrete states corresponding to extremely elongated or superdeformed nuclear shapes at very high angular momentum or spin were observed for the first time. Since then, much progress has been made in the finding and understanding of these exotic states. Such rapidly rotating nuclei possess some remarkable properties. Three major islands of nuclei where this phenomenon occurs have been found.

Nuclear shapes. Nuclei are now known to display various shapes, for example, spherical, prolate (football shape), oblate (discus or doorknob shape), octupole (pear shape), and triaxial. The short-range attractive interaction among the constituent nucleons (protons and neutrons) is sufficiently strong to allow a small number of valence nucleons moving in anisotropic orbits to polarize and deform the whole nucleus. This situation means that neighboring nuclei may exhibit quite different shape behavior, and indeed that different excitations within a single nucleus may lead to various shapes. Large regions of weakly deformed nuclei, shaped as ellipsoids with major-to-minor axis ratios of 1.3:1, are known to exist throughout the chart of the nuclides. One of the most significant discoveries in recent years has been the observation of superdeformed nuclear shapes, with major-to-minor axis ratios of around 2:1 when the nucleus is spinning very rapidly.

Origin of superdeformation. One successful theoretical approach has been to view the final shape of a nucleus as being governed by the interplay of two major contributions: a macroscopic (liquid-drop) component describing the bulk properties of the nucleus, and a microscopic (shell-correction) component reflecting the quantal character of the orbitals occupied by the valence nucleons. The liquid-drop concept works because the nucleons tend to interact only with their nearest neighbors, rather like molecules in a liquid drop. **Figure 1** shows the calculated energy levels of a single particle in the field of an axially symmetric harmonic oscillator as a function of deformation of the system. This simple model approximates the microscopic component of the energy of the nucleus. The large shell gaps at the particle numbers 2, 8, 20, and so forth are responsible for the existence of spherical nuclei with these numbers of neutrons and protons, often referred to as magic numbers; helium-4 (^{4}He; 2 protons and 2 neutrons), oxygen-16 (^{16}O; 8 of each), calcium-40 (^{40}Ca; 20 of each), and so forth are especially stable. This aspect is analogous to the atomic shell closures in the noble gases. It is interesting to see that large shell gaps away from sphericity, corresponding to deformed magic numbers, are also present in Fig. 1, for deformations corresponding to situations where the lengths of the principal axes form integral ratios such as 1:1, 2:1, and 3:1. (In more realistic potentials, with spin-orbit effects included, this pattern is diluted, but a fairly regular scheme of gaps still persists.) These quantal shell corrections, when added to the liquid-drop contributions, can be sufficient under certain circumstances to lead to the stabilization of a highly deformed nuclear shape, corresponding to a second minimum in the potential-energy landscape.

In the 1960s, physicists studying the spontaneous fission of the heaviest nuclei, with mass numbers A around 240, discovered some unexpected properties that could be explained only if such a 2:1 highly deformed second minimum existed in these nuclei. The presence of this deformed minimum was aided as the strong Coulomb repulsion (among the positively charged protons) in these heavy nuclei helped balance the surface energy, which favored a more spherical shape. These nuclei are known as the fission isomers and thus correspond to superdeformed nuclei. However, their spectroscopic study is extremely difficult and is restricted to very low values of nuclear spin.

In the mid-1970s, several theoretical groups realized that not only do favored shell corrections occur at a 2:1 prolate shape for lighter nuclei (for example, around $A = 150$) but also in these cases this exotic shape can be stabilized and lowered in energy by the effects of rotational Coriolis and centrifugal forces. These calculations indicated that pronounced highly deformed minima occur when favorable proton and neutron shell corrections coincide. As a result, at high angular momentum, islands of superdeformation were predicted in the periodic chart.

Discovery of superdeformed shapes. Nuclei at high spin or angular momentum may be produced by using heavy-ion fusion evaporation reactions induced by projectiles from particle accelerators bombarding target nuclei. The projectile and target fuse to form a compound nucleus at high excitation energy (for example, 70–90 MeV) with angular momentum values up to the limit set by fission instability (for example, 70 $\hbar$, where $\hbar$ is Planck's constant divided by 2π). The compound system, which may be rotating at $\sim 2 \times 10^{20}$ rotations per second, first cools by evap-

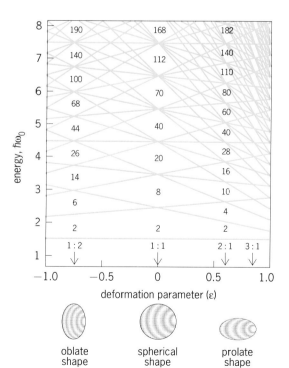

Fig. 1. Single-particle energy levels of the axially symmetric harmonic oscillator potential as a function of the deformation parameter ϵ. Major-to-minor axis ratios are also plotted along the horizontal axis. The large shell gaps and corresponding magic numbers are indicated for spherical (1:1), superdeformed prolate (2:1), and superdeformed oblate (1:2) shapes. Energy is given in units of $\hbar\omega_0$, where $\hbar$ is Planck's constant divided by 2π, and ω_0 is the angular frequency of the oscillator.

oration of particles, usually neutrons, until it is below the nucleon binding energy. It then loses the rest of its excitation energy and almost all of its initial angular momentum by the emission of gamma rays. The total decay process is completed in about 10^{-9} s.

The gamma-ray signals from superdeformed states have very low intensity, and it was not until the advent of special arrays of Compton-suppressed gamma-ray detectors in the 1980s that the first positive results were reported in observing them. The major breakthrough came in 1986 at Daresbury Laboratory, England, when a remarkable rotational spectrum dominated by a sequence (or band) of 19 gamma-ray transitions was observed in dysprosium-152 (^{152}Dy; **Fig. 2**a). From the energy separation between transitions, the nuclear moment of inertia could be determined. This was consistent with that expected for a superdeformed 2:1 shape. Conclusive evidence was obtained later when the intrinsic quadrupole moment of the band was measured. A similar, although slightly less deformed, rotational structure in cerium-132 (^{132}Ce) was also reported.

These first results possessed some startling features that have since been identified as general characteristics of superdeformation: (1) The energy spacing between transitions is usually remarkably constant, a feature that has led some physicists to refer to these objects as nuclear pulsars. (2) No link between the superdeformed states and the low-spin weakly deformed

levels could be identified. (3) There is a sudden and dramatic depopulation of intensity at the bottom of the band over only one or two transitions (Fig. 2b).

Although the exact excitation energy or spin of the superdeformed states has not been determined, owing to the absence of an identifiable link between these states and the low-deformation levels, it is possible to make some reasonable guesses. A spin of 24 was estimated for the bottom of the dysprosium-152 band, meaning that for the first time transitions up to 60 $\hbar$ were observed. The energies of the three known coexisting shapes in dysprosium-152 are plotted as a function of spin in Fig. 2b, with the assumption that the superdeformed band becomes yrast (lowest in energy) around 55 $\hbar$. This figure illustrates the remarkable fact that not only do superdeformed bands allow the limit of spin to increase dramatically but also the excitation at which discrete nuclear states can be observed has been doubled.

Great progress has been made since these initial discoveries. Superdeformed bands have now been observed in seven other nuclei around cerium-132 and nine other nuclei near dysprosium-152. An interesting recent (1991) result has been the observation of a superdeformed band in europium-143, lying between these two regions. It is not known if this case represents simply a stepping stone or the beginning of a bridge between the $A = 130$ and $A = 150$ regions.

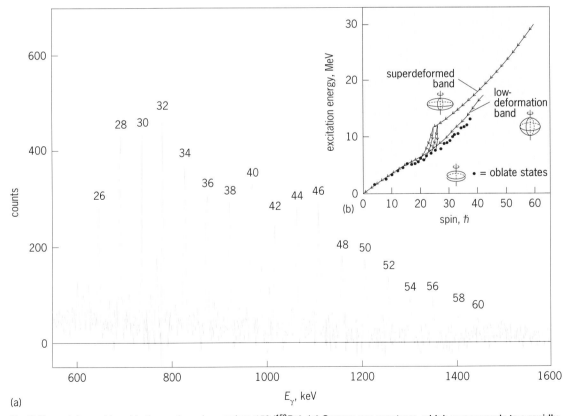

Fig. 2. Superdeformed band in the nucleus dysprosium-152 (^{152}Dy). (a) Gamma-ray spectrum, which corresponds to a rapidly rotating nucleus with a major-to-minor axis ratio of around 2:1. The gamma rays are marked with the estimated spins of the emitting states. (b) Schematic plot of excitation energy versus spin for states belonging to the three different coexisting shapes in ^{152}Dy.

A new region of superdeformation around mercury-192 (^{192}Hg) was also discovered (1989), with bands observed in 14 nuclei so far. In total, about 50 superdeformed bands are presently known to exist to high spin. No evidence for superdeformed oblate shapes (1:2) has so far been observed.

Moments of inertia. From the observed gamma-ray energies, important information can be extracted on the behavior of the moment of inertia as a function of spin or rotational frequency. Thus, the response of the superdeformed nucleus to the stresses of rapid rotation can be studied. For example, as the number of superdeformed bands observed in the region around $A = 150$ increased, a pattern began to emerge. It was realized that the moment-of-inertia curve for a particular nucleus contained a fingerprint of the configuration or number of intruder orbitals that the valence particles occupied. These special orbitals originate from high-lying oscillator shells and drop greatly in excitation energy at large deformations. This effect can be seen in Fig. 1. Because of their high intrinsic angular momentum, these orbitals are strongly affected by the Coriolis force and the rotation of the deformed nucleus. Thus, superdeformed spectroscopy offers the opportunity of obtaining information about the behavior and placement of such exotic orbitals that may not normally be accessible otherwise. In addition, the behavior of the moment of inertia provides information about the properties and stability of pairing or superfluid correlations in rotating, highly elongated nuclear systems.

Identical superdeformed bands. A major step forward in superdeformed spectroscopy occurred in 1989, when excited or non-yrast bands were observed in a single nucleus. Such bands are particularly difficult to observe since they are usually more weakly populated by at least a factor of 2 than the yrast superdeformed band. However, in some special cases as many as three or more superdeformed bands have been observed in certain nuclei. It was found that the moment of inertia (which is partially a measure of the mass distribution and thus the deformation of the nucleus) of the excited band in terbium-151 (^{151}Tb, with one proton less than dysprosium-152) followed extremely closely that of dysprosium-152. This would imply the same high-oscillator-shell occupations for these two bands, which can be understood in terms of particle-hole excitations. This is not very surprising. What was surprising, and indeed unprecedented in nuclear physics, was the fact that the gamma-ray transition energies of these two bands were identical to within a few parts per thousand over such a large spin and energy range (**Fig. 3***a*). Other examples have followed in the regions around $A = 150$ and $A = 190$. The latter region, in fact, has turned out to be the most prolific in this phenomenon and displays some important additional features. For example, it was observed that the superdeformed band in mercury-192 was identical to an excited band in mercury-194 (^{194}Hg; Fig. 3*b*), but only at high rotational frequencies (that is, gamma-ray energies). These results indicate that, under certain circumstances, a superdeformed nucleus can be com-

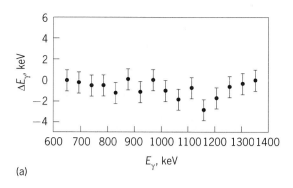

(a)

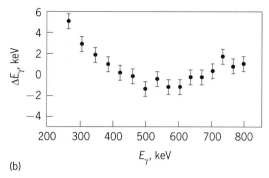

(b)

Fig. 3. Difference in gamma-ray energy (ΔE_γ) between transitions in (*a*) the excited superdeformed band in terbium-151 (^{151}Tb) and those in the yrast superdeformed band in dysprosium-152 (^{152}Dy), and (*b*) the yrast superdeformed band in mercury-192 (^{192}Hg) and an excited superdeformed band in ^{194}Hg, as a function of gamma-ray energy (E_γ).

pletely unaffected by the removal or addition of one or two nucleons. This is very surprising, since there are a number of factors (such as mass, deformation, pairing, and orbital alignments) that can, and usually do in normally deformed nuclei, alter the moment of inertia of neighboring systems and hence their relative transitions energies by tens of kiloelectronvolts.

Apparently, some subtle cancellations are taking place, or more interestingly, these degeneracies are the result of some underlying symmetry that is not yet recognized. There have been numerous speculations involving symmetries and modes of excitation not previously observed in nuclei. However, the major drawback to progress on this question is that the exact spins of the superdeformed bands are not known. Some attempts have been made to ascertain these from the moment-of-inertia values, particularly in the $A = 190$ region where the lowest gamma-ray energies and presumably spins are observed. However, these attempts have been fraught with controversy.

Population and decay. Experimentally, only the energy and intensity of each gamma-ray can be measured. From the energy, information on the behavior of the moment of inertia can be obtained. From the intensity, information can be extracted on the population of superdeformed states at spins close to the fission limit, the flow of gamma-ray intensity through the band as a function of spin, and, at the bottom of the band, the depopulation mechanism out of the superdeformed minimum into the normally deformed minimum.

The observation of discrete transitions up to 60 $\hbar$, so close to the fission limit, was initially surprising. Their intensities exceeded those expected from extrapolation of intensities in normally deformed nuclei by more than an order of magnitude. Much effort has been expended in attempting to understand the mechanism behind this enhanced feeding, since from this knowledge it may be possible to learn the optimal conditions for spectroscopic studies of superdeformation. Progress has been made, but much remains to be done.

An interesting recent (1991) result was the observation that the intensity of the superdeformed states can be enhanced if a more symmetric beam and target combination is used to produce the original compound system. Thus, the final nucleus must have some memory of its origins, a result contradictory to N. Bohr's compound-nucleus hypothesis.

At the low-spin end of a superdeformed band, the fact that the in-band intensity is usually lost very suddenly over a range of one or two transitions is again rather surprising and not understood. So far, none of the decay paths linking the superdeformed bands with normal states has been definitely established, so that a highly fragmented mode is suggested, as illustrated in Fig. 2b. This sudden depopulation is surprising because at the point of decay the superdeformed levels have an excitation energy of 3–6 MeV above yrast. Hence, the superdeformed states are immersed in an increasingly high density of normal states. Thus, a gradual mixing between superdeformed and normal states is expected, and hence a gradual decay. One suggestion has been that the sudden onset of pairing or superfluid correlations in the second minimum triggers the decay. Another model suggests that the depopulation point marks a sharp crossing in the superdeformed minimum between the high-spin band and a low-spin excitation mode, not previously observed in nuclear physics.

Prospects. The observed islands of superdeformed heavy nuclei correspond well to theoretical predictions. However, the detailed spectroscopic study of these exotic shapes has led to some remarkable and unexpected results. While much progress in understanding their behavior has been made, many questions remain and indeed continue to be generated. Probably, some new basic nuclear physics or some new nuclear symmetries are about to be discovered. The future is very promising, with the development and construction of the next generation of gamma-ray detection arrays (GAMMASPHERE in the United States and EUROGAM in Europe). These arrays should provide answers to the above questions, particularly with regard to excitation energies and exact spins, and no doubt some more surprises too.

For background information *SEE DEEP INELASTIC COLLISIONS; GAMMA-RAY DETECTORS; NUCLEAR REACTION; NUCLEAR STRUCTURE* in the McGraw-Hill Encyclopedia of Science & Technology.

Mark A. Riley

Bibliography. S. Åberg, H. Flocard, and W. Nazarewicz, Nuclear shapes in mean field theory, *Annu. Rev. Nucl. Part. Sci.*, 40:439–527, 1990; R. V. F. Janssens and T. L. Khoo, Superdeformed nuclei, *Annu. Rev. Nucl. Part. Sci.*, 41:321–355, 1991; P. J. Nolan and P. J. Twin, Superdeformed shapes at high angular momentum, *Annu. Rev. Nucl. Part. Sci.*, 38:533–562, 1986; J. F. Sharpey-Schafer and J. Simpson, Escape suppressed spectrometer arrays: A revolution in γ-ray spectroscopy, *Prog. Part. Nucl. Phys.*, 21:293–400, 1988.

Ocean circulation

One of the great unknowns of Earth's climate system is how climate is controlled by the circulation of the world ocean. Exchanges of heat, fresh water, and carbon between the ocean and atmosphere, between the upper and deeper layers of the ocean, and between the equatorial and the polar ocean regions greatly affect climate and climate change. Recognizing the importance of understanding the role of ocean circulation in improving the prediction of climate, the World Meteorological Organization, International Council of Scientific Unions, and Intergovernmental Oceanographic Commission established the World Ocean Circulation Experiment (WOCE) as part of the World Climate Research Programme. Of global scale, WOCE is the most ambitious physical oceanographic research experiment ever undertaken and involves scientists and government planners from some 40 nations. It is designed to develop models useful for predicting climatic change and to collect observations necessary to test and verify them.

WOCE is made possible because of scientific insights and technology developed in the 1970s and 1980s. New instruments provide data on a global scale that could not be achieved by earlier techniques. WOCE planners have used these developments to design a field program comprising three core projects; the program is expected to continue through 1997. An innovative data management program will assure that data of high quality are collected and properly archived for future usefulness. The resulting global, integrated ocean data set will improve scientific understanding of ocean circulation and will support the extensive WOCE program to advance the models of ocean circulation used in climate prediction. After 1997, it is expected that a more limited set of observations will be kept operational for long-term measurements.

Core Project 1. Known as the Global Description, this core project obtains global data sets needed to describe the ocean circulation of heat, fresh water, and chemicals; the formation and modification of water masses; and the statistics of ocean variability. It employs a wide range of new and traditional instruments that measure the physical properties of the world ocean throughout its full depth (**Fig. 1**). These instruments include surface drifters; moored current meters; expendable bathythermographs; water samplers; sofar and RAFOS subsurface floats that use sound sources or receivers for tracking; ALACE (Autonomous Lagrangian Circulation Explorer) subsurface floats that periodically cycle to the surface to transmit their locations via satellites; acoustic Doppler current profilers that measure current velocity in the upper 400 m (1312 ft) by using backscattered sound; shore-based stations; communications satellites that relay data from ships, drifters, floats, and instrumental buoys to shore-based stations; and satellites with remote-sensing instruments for global ocean measurements.

Among the measurement techniques in hydrography, which studies vertical and horizontal property distributions, including density, temperature, salinity, dissolved oxygen, nutrients (nitrate, nitrite, phosphate, and silicate), and other geochemical tracers (chlorofluorocarbons, tritium/helium, radiocarbon, and argon/krypton). The WOCE Hydrographic Program measures water samples with high accuracy through a series of research cruises conducted along a global

network of latitudinal and longitudinal transects in the Pacific, Indian, Atlantic, and Southern oceans (**Fig. 2**). Station spacing averages 30 nautical miles (55 km), with much closer spacing across regions with strong horizontal gradients such as boundary currents and large topographic features, and with wider spacing over abyssal plains. Data are used to infer the global distribution, sources, and flow velocities and patterns of water masses. Zonal transects, coupled with measurements of current velocity, provide accurate estimates of the oceanic flux of heat between the Equator and the poles. Meridional transects reveal the relationship of subpolar, subtropical, and tropical circulations. Transects and land form closed boxes that are used to evaluate water flow characteristics that can be used as physical constraints. Hydrographic samples collected

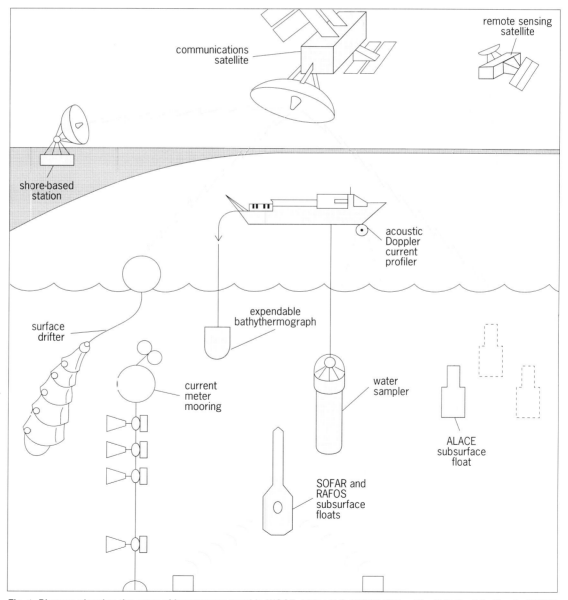

Fig. 1. Diagram showing the array of instruments used in WOCE. (*After U.S. WOCE Office, The U.S. Contribution to WOCE: World Ocean Circulation Experiment Understanding Global Change, 1990*)

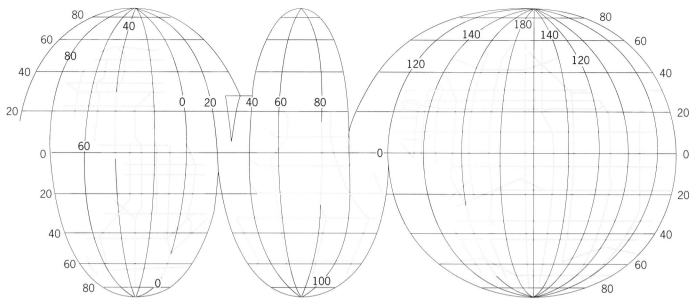

Fig. 2. Map showing locations of the global network of ship transects for the WOCE Hydrographic Program. (*After WOCE International Project Office, Summary of Resource Commitments, WOCE Rep. 64/91, March 1991*)

along repeated transects and at long-term stations help describe temporal variability. Merchant vessels, traveling along regular transoceanic routes several times each year, daily deploy expendable bathythermographs to measure temperature and pressure within the upper 800 m (2625 ft) for analysis of the global thermal structure of the upper ocean and its variability.

Direct measurements of current velocity are obtained from moored current-measuring instruments tethered at various depths, subsurface floats, and surface drifters. Moored measurements provide direct estimates of current structure and transport for study of boundary currents, interbasin exchanges, meridional fluxes, and strength and spatial structure of eddy fields. Floats provide subsurface velocity observations at selected depths for global mapping of absolute velocities and for process studies. Global mapping of subsurface velocities (at about 1000 m or 3280 ft) provides the reference level of known motion needed to compute the field of horizontal velocity through the entire water column and the associated transports of heat and tracers from physical dynamics and hydrographic data. Drifters measure current velocities in the mixed layer at depths of 10–15 m (33–49 ft), surface temperature, and atmospheric pressure, so that tests of modeled surface currents and study of advection of ocean surface properties can be made.

Sea-level gages, located near shores of islands and continents, measure changes in the shape of the ocean surface that are caused by rapidly varying waves and tides, slowly varying ocean currents, and the rise and fall of the ocean as it warms and cools. The WOCE global sea-level network is built upon a selected subset of existing gages and new gages added at key locations. Data are used to investigate large-scale changes in the strength of ocean circulation and, by comparison to historical sea-level data from the past century, long-

term temporal changes in ocean surface topography. The horizontal transports of major currents, such as the Antarctic Circumpolar Current that flows completely around Antarctica, are monitored by measuring changes in sea-level differences with pairs of gages, one on each side of the current. Several gages will provide data for the calibration of high-accuracy measurements of ocean surface topography that will be made from satellite altimeters. TOPEX/POSEIDON, a joint United States–French satellite mission with a precision altimeter, will provide on a global scale sea-level height to within a few centimeters.

Air–sea exchanges of momentum, heat, and fresh water force the global circulation and are needed as ocean boundary conditions for ocean–atmosphere models. Basic measurements that form boundary conditions are sea surface temperature, air temperature, sea-level pressure, humidity, wind direction and speed, solar and long-wave radiation, and precipitation. Satellites measure sea surface temperature, global surface-wind velocity, and shortwave radiation. WOCE-related satellite missions will improve measurements of wind velocity. In-place measurements (for example, from drifters and meteorological buoys) provide spot estimates of all basic variables and also verification and calibration of satellite sensors and algorithms. Fluxes of momentum, heat, and fresh water at spot locations are derived from the basic measurements. The global estimates of air–sea fluxes needed for ocean modeling must be computed as a by-product of atmospheric general-circulation models used for weather prediction. These flux computations will be improved by studies of ocean–atmosphere interactions that use WOCE measurements and model improvements.

Core Project 2. Known as the Southern Ocean, this project evaluates the Southern Ocean and its interactions with the Pacific, Atlantic, and Indian oceans.

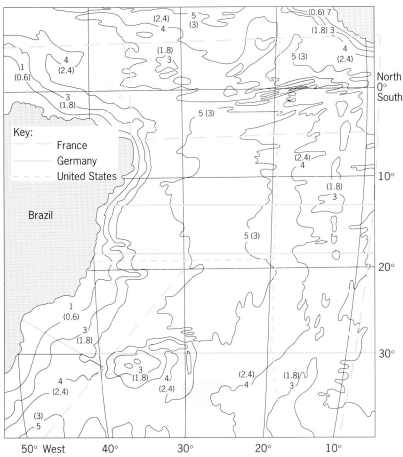

Fig. 3. Hydrographic transects (heavy solid and broken lines) and moored current-meter-array locations (colored rectangles), illustrating the basin-wide coverage of measurements for the Deep Basin Experiment. Contours show depth in kilometers (miles). (*After U.S. WOCE Office, U.S. WOCE Implementation Plan 1991, U.S. WOCE Implement. Rep. 3, May 1991*)

Focus is on the formation and transport of water masses and on the transport and variability of the Antarctic Circumpolar Current. This current connects all the oceans into a single world ocean and transforms regional oceanic transports into a global phenomenon.

Core Project 3. This project, the Gyre Dynamics Experiment, is designed to improve basin-scale ocean models of circulation through a series of ocean process studies conducted in the Atlantic Ocean. The modeling techniques developed in this program will be extended eventually to global ocean models.

Using hydrographic cruises, current-meter moorings, and floats, the Deep Basin Experiment studies the abyssal circulation in the Brazil Basin to investigate deep western boundary currents, upwelling in the interior of basins, effects of topographic passages (passes or gaps through ridges) on water flow, and the ways in which deep water crosses the Equator (**Fig. 3**). The moored arrays are deployed for 2 years and overlap each other in time. Multiple sound sources and floats are deployed at two or three separate depth levels throughout the basin to measure the flows on each level. They operate concurrently with the hydrographic surveys and moored measurements. Advanced models

of deep circulation and its interactions with bottom topography will be developed.

The structure and dynamics of the wind-driven upper ocean layer and the process of subduction, by which water masses formed near the ocean surface through air–sea interactions sink deeper into the ocean, are studied in the Subduction Experiment in the eastern North Atlantic near the Azores. Hydrographic transects linked into triangular or quadrilateral patterns provide data to examine the large-scale response of subpolar and subtropical gyres and tropical circulation to atmospheric forcing. Areas of study include the response of the Gulf Stream system, recirculation of subtropical gyres, dynamics of the North Atlantic current, circulation adjacent to eastern boundaries, and the tropical western boundary circulation.

The Tracer Release Experiment, focused near the Azores, will inject and track a nontoxic tracer, sulfur hexafluoride, at a depth of about 350 m (1148 ft) to study mixing processes in the ocean interior away from continental boundaries. It will place the tracer in a specific density layer and then will observe how the tracer mixes vertically and horizontally off that layer.

Ocean modeling. Major advances in ocean modeling are required to develop models capable of predicting climate change. To provide reliable pictures of how the ocean influences climate, ocean models must be compatible with observed global circulation, property distributions, and surface fluxes. New predictive and inverse or data-assimilative modeling techniques are being developed by using WOCE observations. Predictive models provide a complete description of the ocean circulation by numerically solving the physical equations that govern the phenomena being studied, including equations of continuity and of motion for a rotating system. Models are constrained by boundary conditions, such as observed surface flux fields. All internal fields, such as temperature, salinity, and tracers, and their evolution through time are computed. The WOCE observations of these internal fields test the success of models and measure model reliability. Inverse or data-assimilative models determine the actual state of the circulation from equations similar to those of predictive models, but use observations from the interior and at the boundary to define the model state. The model then is used to determine properties that cannot be observed directly, test the dynamical elements of the model against measurements not used in the model, and determine model parameters needed to make the results compatible with observations. Future studies will utilize the resulting improved WOCE models and data to couple more effectively ocean and atmosphere models, so that climate prediction will be improved.

For background information *SEE ATMOSPHERIC GENERAL CIRCULATION; CLIMATIC PREDICTION; MARITIME METEOROLOGY; MODEL THEORY; OCEAN CIRCULATION; OCEANOGRAPHY; SOFAR* in the McGraw-Hill Encyclopedia of Science & Technology.

Ann E. Jochens

Bibliography. D. Mayes, U.S. WOCE program ready to begin, *Eos, Trans. Amer. Geophys. Union,*

70(40):874, 884–885, October 1989; U.S. WOCE Office, *U.S. WOCE Implementation Plan 1991*, U.S. WOCE Implement. Rep. 3, May 1991; WOCE International Planning Office, *WOCE Implementation Plan*, World Climate Research Programme Series, vols. 1 and 2, WCRP-11 and WCRP-12, WMO/TD 242 and 243, July 1988; WOCE International Project Office, *Workshop on Global Ocean Modelling, Institute of Ocean Sciences, British Columbia, Canada, 5–7 September 1990*, WOCE Rep. 60/91, February 1991.

Oil and gas well drilling

Recent advances in drilling for oil and gas wells include increasing development and application of horizontal drilling, and the integration of surface and downhole measurements made during drilling.

Horizontal Drilling

Methods used for drilling horizontal wells include a combination of tools and techniques that first appeared in the oil industry in the early 1900s; they evolved particularly during the 1980s. Horizontal drilling finds application in all phases of oil recovery as well as in gas reservoirs. Horizontal wells have been commonly used to handle fluid-flow problems and heterogeneous reservoirs.

Fluid-flow problems. These problems are associated mainly with gas and water coning. When a vertical oil well is produced, the pressure around the wellbore is drawdown. This condition elevates the level of the water–oil contact so that a water cone is formed in the vicinity of the wellbore. If there is a gas cap, the pressure drawdown lowers the gas–oil contact so that an inverted gas cone is formed around the wellbore. These are serious problems that lower oil recovery: water cones reduce the permeability to oil in the vicinity of the wellbore, and gas cones lead to large gas/oil ratios and a reduction in reservoir energy.

Horizontal wells help to avoid these problems. For example, if an aquifer is present, the horizontal well can be placed near the top of the reservoir and as far as possible from the water–oil contact in order to delay or decrease water production. If there is a gas cap, the horizontal well can be placed near the bottom of the reservoir and as far as possible from the gas–oil contact in order to delay gas production and help to maintain the energy of the reservoir. If there are both an aquifer and gas cap, the horizontal well should be placed, in principle, somewhere above the middle point of the reservoir.

Heterogeneous reservoirs. These reservoirs are characterized by high anisotropy, that is, variations in mechanical properties of the rock, such as permeability, in different directions of the reservoir. The key to success is identifying the azimuth of the highest permeability and, in most cases, drilling the horizontal well perpendicular to the major permeability. Examples of systems where horizontal wells have proved successful include naturally fractured reservoirs, carbonate rock, channel point bars, and braided

stream systems. Channel-point-bar sands are formed by lateral accretion within fluvial streams. In some instances, these bars contain nonpermeable clay layers deposited during the waning flood stage of river flow.

Naturally fractured reservoirs represent excellent targets for horizontal wells. This fact has been recognized by the oil and gas industry, as more than 70% of horizontal wells drilled to date have been in naturally fractured reservoirs.

If the direction of the horizontal well is parallel to the major permeability, there might be a small increase in productivity with respect to a vertical well. If the horizontal well is perpendicular to the major permeability, the productivity increase can be easily five or more times the productivity of a vertical well.

Drilling methods. Horizontal drilling and completions require much more preplanning than do conventional vertical wells. Communication between all levels of personnel is extremely important to drill and complete a horizontal well successfully. The completion engineer must be involved from the early planning stage, because a large number of completion subtleties can be addressed rather easily in the preplanning stage but may be very difficult to correct once the horizontal well is drilled. Horizontal drilling techniques can be classified according to their angle of buildup rate: the conventional (or low) curvature method, the high-curvature method, and the medium-curvature method (**Fig. 1**).

Conventional (or low) curvature method. The approximate radius of curvature ranges between 1000 and 3000 ft (366 and 714 m) in the conventional curvature method (Fig. 1a). The rate of buildup is 0.02–0.06°/ft (0.07–0.21°/m). Horizontal wells drilled with the conventional curvature technique can be logged and cased and thus can be properly evaluated. Rotary equipment for this type of drilling is usually

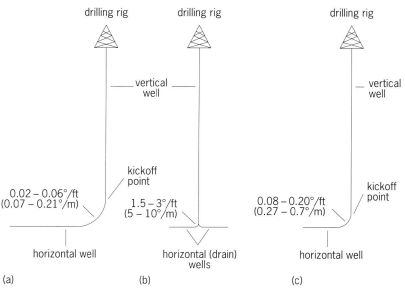

Fig. 1. Horizontal drilling comparison. (*a*) Conventional or low-curvature; radius of curvature ranges 1000–3000 ft (366–714 m). (*b*) High-curvature; reaches 90° in 20–60 ft (6.1–18.3 m). (*c*) Medium-curvature; radius of curvature ranges 285–700 ft (87–213 m).

larger than the equipment for a vertical well at a similar true vertical depth, that is, the depth reached by the horizontal well in a vertical plane.

High-curvature method. In this case the horizontal well builds inclination very rapidly at 1.5–3°/ft (5–10°/m); it can reach 90° in 20–60 ft (6.1–18.3 m; Fig. 1*b*). Special downhole tools are required to drill this type of hole. The horizontal lengths are limited to a few hundred feet. Conventional evaluation tools, downhole tubulars (casing and tubing), and completion tools usually cannot pass the tight radius section. Consequently, these wells can be neither logged nor cased.

Medium-curvature method. This technique is the most favored. In this case the buildup rate is 0.08–0.20°/ft (0.27–0.7°/m), defining a radius of curvature that ranges between 285 and 700 ft (87 and 213 m; Fig. 1*c*). Horizontal wells using this technique have been extended more than 3200 ft (975 m). An important advantage is that, in spite of the relatively small radius of curvature, the wells can be logged and cased. Also, this method usually uses the same rig size, hole size, and drill string as a vertical well drilled to the same true vertical depth. Most conventional evaluation and completion tools used in vertical wells can be used with minor modifications in medium-curvature wells.

Drilling fluids. The average drilling mud used in horizontal drilling operations has specific gravities in the range 0.9–1.5. Drill cuttings tend to slip down through the mud since they have a higher specific gravity. The slip velocity is dependent on drilling-fluid rheology, particle size, and particle shape.

In a horizontal well the slip direction is perpendicular to the mud flow (**Fig. 2***a*). On the other hand, slip direction is parallel to the mud flow in vertical wells (Fig. 2*b*) and diagonal to mud flow in inclined boreholes (Fig. 2*c*).

In a horizontal well the cuttings will slip to the low side of the hole; they can be transported only if the drilling fluid can lift and move them along the borehole axis prior to their settling in the low side of the hole. This transport mechanism is known as saltation. Cuttings can also be transported in horizontal wells via a heterogeneous suspension maintained by turbulence.

Saltation and heterogeneous suspensions can be optimized by increasing fluid viscosity and flow rates while turbulent flow is maintained. An increase in viscosity decreases the slip velocity and augments the distance traveled during one saltation cycle. An increase in the flow rate maintains turbulent crossflow and augments the energy available for saltation. Rotation of the pipe also helps to clean the horizontal well by mechanically crushing the cuttings, thus reducing their size.

Care must be exercised with selection of drilling fluids, because they may induce reservoir damage, which reduces hydrocarbon production and can lead to costly well stimulations.

Downhole motors. Current steerable drilling systems use positive-displacement mud motors, which extend the life of the bit to 50 or more hours at a relatively low angular velocity (number of revolutions per minute). These motors operate on the principle of the Moineau pump. As fluid is pumped into the motor, the rotor is forced to turn so that a fluid can pass. This rotation is transferred to the drilling bit. In the recent past, downhole motors turned the drilling bit at high speeds, and the life of the bit was limited to a maximum of 8–10 h.

The current motors operating at a low number of revolutions per minute can be oriented to make corrections of borehole trajectory, or they can be rotated to drill straight ahead. Advances in the power of portable computers permit rapid processing of data transmitted from the bottom of the hole to the surface via pressure pulses in the mud system; such a survey is known as measurement while drilling. This technique eliminates the need for a wireline connection and permits efficient steering of downhole motors.

Torque and drag. Continuous monitoring of torque and drag during drilling operations is very important. Continuous recording can provide key infor-

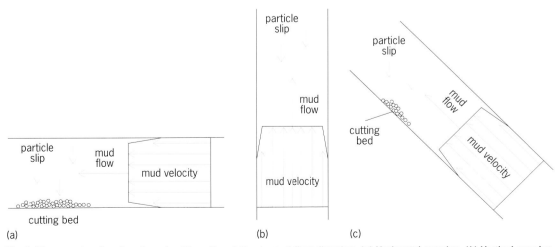

Fig. 2. Diagram showing direction of cutting slip relative to mud flow direction. (*a*) Horizontal annulus. (*b*) Vertical annulus. (*c*) Inclined annulus.

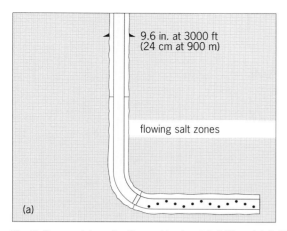 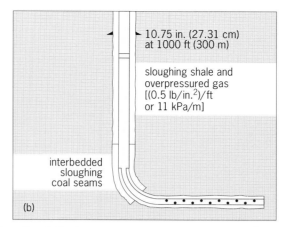

Fig. 3. Successful applications of horizontal drilling. (*a*) Bakken oil well, North Dakota; pressure gradient 0.62 lb/(in.^{2})(ft) [14 kPa/m]; curvature 14°/100 ft (30 m); 5-in. (13-cm) drillpipe throughout. (*b*) Cadomin gas well, Alberta, Canada; pressure gradient 0.35 lb/(in.^{2})(ft) [8 kPa/m]; curvature 20°/100 ft (30 m); 4.5-in. (11-cm) drillpipe throughout.

mation related to hole-cleaning problems and stuck pipe. Hole drag results from contact between the hole wall or casing and the drill string.

In medium-curvature horizontal wells, torque/drag design is required in three sections: the vertical hole, the angle-build hole, and the horizontal hole. The vertical hole down to the kickoff point is treated in the conventional form followed for vertical wells. In this case, drag is the result of minor doglegs and tension in the drill string. The angle-build hole is the section where the hole deviates from vertical and approaches horizontal. It is influenced by inclined hole drag and dogleg drag. The drag in the horizontal sections is dominated by the weight of the drill string and bottom-hole assembly drag. Dogleg in the horizontal section does not contribute significantly to drag, because the drilling pipe is on the low side of the hole and the pipe tension is low.

The sliding-friction drag force in a horizontal hole is approximated by the product of a friction factor (usually ranging 0.2–0.4) and the weight of tubulars in the horizontal hole. For example, in a 2000-ft (610-m) lateral section of hole, using a 41 lb/ft (62 kg/m) drill string and a 0.3 friction factor, the sliding friction drag would be approximately 24,600 lb (11,100 kg). One approach to reduce drag is to reduce the weight of tubulars in the horizontal hole. Another approach is reduction of the friction factor, although this might lead to hole-cleaning problems.

Applications. To be successful in the application of this new technology, horizontal drilling should also be considered a completion operation. Horizontal wells have been completed successfully in North and South America, Europe, Asia, Australia, and the Far East. Applications include one in an overpressured reservoir, the Bakken oil well, in the United States and one in an underpressured reservoir, the Cadomin gas well, in Canada (**Fig. 3**). The Bakken oil well and the Cadomin gas well are medium-radius horizontal wells drilled to maximize economic production. The wells are substantially different, yet both designs meet specific well requirements in cost-effective ways.

Overpressured reservoir. The reservoir in Williston Basin, North Dakota, is an oil-wet, overpressured, fractured shale (known as the Bakken shale) at a true vertical depth of about 10,500 ft (3200 m). Thick salt zones above the kickoff point can cause problems during drilling, and plastic flow has resulted in collapsed production casing in numerous wells. To prevent collapse, operators have concentrated on limiting hole enlargement, achieving good annular cementing, and running heavy wall casing designed for pressure gradients of 1.2 lb/(in.^{2})(ft) [27.6 kPa/m] or more.

A weighted inverted oil mud (an emulsion where water is the dispersed phase) has proven successful for drilling the lateral hole without significant damage. This mud system also controls uphole salts and helps achieve good cement jobs. Although mud costs are rather high because of significant seepage losses, this system allows drilling to total depth (the bottom of the hole) without intermediate casing. Running intermediate casing is an expensive alternative in this case, because the design must include consideration of high collapse gradients and the limitations on the drill string and hydraulics in drilling a small hole at this depth.

The well is completed with 5.5-in. (14-cm) production casing run to total depth, with the lateral section left uncemented below an annular packer and stage collar.

Underpressured reservoir. The Cadomin gas well in Deep Basin, Alberta, Canada, is an underpressured reservoir. Invert oil mud is required for shale control in the vertical well and sloughing coal beds in the angle-build section. Overpressured gas is often encountered high in the vertical well and requires increased mud density. However, the producing zone is an underpressured gas reservoir, which can be damaged by using invert mud and so requires control of fluid loss. The mud system selected for this portion of the well was a water–potassium chloride–polymer fluid, using calcium carbonate as an acid-soluble bridging agent for control of fluid loss.

The deepest coal beds are located immediately above the producing zone. Casing is set at 85° to permit the mud system change from invert oil to the water–potassium chloride–polymer mixture. The 7.63-in. (19.3-cm) intermediate casing is set through the 20°/100 ft (6.6°/10 m) radius and cemented at 85°.

The overall well design requires a rotary rig equipped with 4.5-in. (11-cm) drillpipe. The rheology of the invert oil mud, and the density and circulating pressures of the polymer fluid system, allow the operator to run 4.5-in. (11-cm) drillpipe inside the 7.63-in. (19.3-cm) intermediate casing; thus the solids control is minimized. The high annular velocities reduce rheology requirements for hole cleaning. The overall results are good hole cleaning and control of solids, and the cost of changing to a 3.5-in. (8.9-cm) drill string is saved.

The well was completed by placing an uncemented 4.5-in. (11-cm) production liner in the lateral hole below the intermediate casing. A light acid wash (an operation using a low volume of acid) and squeeze was successful in cleaning up the well to meet economic production rates. Stimulation costs to fracture past mechanical and fluid-induced damage by weighted invert mud would have cost more than intermediate casing. *SEE* OIL MINING. *Roberto Aguilera*

Measurement While Drilling

In general, the term measurement while drilling applies to all measurements made while drilling. However, it particularly refers to downhole measurements that are transmitted to the surface within seconds of being made; they are either recorded in downhole memory or transmitted to the surface. Downhole memory is random access memory in a computer that stores data for later retrieval. Recently there has been a trend toward integration of surface and downhole measurements to provide a more complete analysis of the drilling process.

Research for measurement while drilling expanded in the latter 1970s, with the first commercial systems becoming available in the late 1970s and early 1980s. The first systems provided only directional measurements (inclination, azimuth, and tool-face orientation). Research and development continued and resulted in a variety of downhole measurements that improve drilling safety, directional control, drilling efficiency, and formation evaluation.

Telemetry methods. Today two methods are being used commercially to transmit data from the downhole tool to the surface. The most common is known as mud pulse telemetry. Measurements made by downhole sensors are sent to the surface by encoding the data in either pressure increases (positive pressure pulses) or pressure decreases (negative pressure pulses). The pressure pulses are generated downhole through the operation of a valve (pulser) that is part of the tool designed for measurement while drilling. Positive pressure pulses are generated by restricting the flow of the mud inside the drill string. Negative pressure pulses are generated by momentarily opening a valve that vents some mud from inside the pipe into the annulus. The resulting increase or decrease in pressure is measured at the surface with a pressure transducer.

The pressure data are filtered to remove system noise and then decoded by computer. Many different techniques are used to encode the data, including binary code and measurement of the time between pulses. Mud pulse telemetry is subject to signal attenuation with increasing depth, particularly if air is entrained in the drilling fluid. Since data are transmitted as pressure pulses, information can be sent to the surface only while fluid is being pumped.

The second method is electromagnetic telemetry. This utilizes current injection to create electromagnetic waves that transmit data through the earth in either direction, downhole to surface or surface to downhole. Electromagnetic telemetry has proven effective in shallow applications, but has limitations in deeper applications, depending upon the resistivity of the overlying strata. Current research is directed at developing techniques to extend the depth at which it can be used effectively. An advantage of electromagnetic telemetry is that it does not require circulation of fluid to transmit data, permitting transmission while tripping the drill string.

When tools for measurement while drilling were introduced, the highest rate of data transmission was approximately 1–2 bits/s. Today, data transmission rates are on the order of 2–4 bits/s. Data transmission rates of 8–10 bits/s are anticipated by 1994. Transmission rates for actual information are generally higher than for the data that are transmitted by using enhanced data coding methods.

The data transmission rate required is dependent upon the drilling rate and the amount of information that is to be transmitted to the surface. During fast drilling, fewer measurements per depth increment can be made and transmitted. Providing sufficient data per depth increment for detection of thin formation layers requires either controlling drilling rate or increasing data transmission rate.

Measurements and applications. Measurements made while drilling are classified as directional, formation evaluation, or drilling mechanics. Directional measurements include inclination (relative to vertical), azimuth (direction relative to north), and tool-face orientation. Formation-evaluation measurements involve logging of gamma rays (total and spectral), resistivity (electromagnetic or short normal), neutron porosity, neutron density, and caliper. Drilling-mechanics measurements include weight on the bit, torque on the bit, and downhole vibrations. These three classes of measurements are used to improve directional control, enhance drilling safety, correlate lithology with offset wells, perform reservoir evaluation while drilling, and optimize drilling performance.

Directional measurements. The most common application for measurement while drilling is in directional drilling. All the tools used in this method provide directional measurements, and they can be used for both directional surveying and directional steering. These tools use electronic survey instruments to measure wellbore inclination, direction, and tool-face orientation. The measurements are generally made every 30 ft (9 m), but can be made as often as required.

(Generally, a survey is obtained whenever the pumps are turned on.) If the tool is being used for directional steering, tool-face orientation is updated every 4–20 s, depending upon the tool being used. Many tools for special applications provide only directional drilling measurements. These specialty tools are less expensive to operate and often are more reliable. The design of some directional-only tools allows retrieval of the electronics module with the wireline. The ability to retrieve the tool adds an economic advantage, eliminating the expense of a delay if the drill string becomes stuck. The tool is also easily replaced in the event of a failure.

Formation evaluation. The tools for formation evaluation while drilling are much more complex than the directional-only tools; they are more expensive to use and, because of their complexity, tend to be less reliable. However, this method has a number of advantages. Formation-evaluation logs are obtained in real time while the well is being drilled; thus the completions and testing programs can be optimized. Formation evaluation while drilling reduces the risk of failing to obtain an open-hole log on any given well. It is commonly used on exploration wells or wells where hole problems are expected.

Formation-evaluation-while-drilling data are collected very shortly after the interval is drilled. In most instances, little if any fluid invasion has occurred. The logs are often better than wireline logs, and can be used to provide an indication of formation permeability. Analysis of repeat logging runs obtained over a period of time can provide an indication of fluid invasion, which can be correlated to formation permeability.

Drilling safety has been enhanced by the measurement-while-drilling technique through the use of logs for prediction of pore pressure and selection of depths for casing setting. In some geographic areas, resistivity logs are used to identify increases in pore pressure in shales. Mud weight is adjusted to maintain the hydrostatic pressure higher than the pore pressure and minimize the risk of an influx of fluid into the wellbore. Casing is often set just before or just after a significant change in pore pressure.

Measurement-while-drilling logs are often used for geological correlation to offset wells. The logs are used for selection of casing set depths, core points, and testing intervals. As with wireline logs, the gamma-ray and resistivity logs are commonly used for geological correlation.

With the recent development of neutron density and neutron porosity logs, measurement-while-drilling logs are being used as for petrophysical analysis. These tools now provide the full suite of measurements required for reservoir evaluation. In many development drilling applications and high-angle wells, measurement-while-drilling logs are replacing wireline logs. Since the logs are available during well drilling, important decisions relative to well completions and testing can be made sooner and so reduce well costs.

There are some disadvantages. The tools are complex and expensive to replace if lost in the hole. In steering applications, the sensors are 30–60 ft (9–18 m) behind the bit; this arrangement makes it difficult to use the technique of formation evaluation while drilling for steering. However, the measurements can confirm wellbore location relative to geological markers. The tools for this application are not as stiff as conventional drill collars. The reduced stiffness must be considered in the design of assemblies for controlling directional tendencies.

Drilling mechanics. Some tools for measurement while drilling provide downhole measurements of the weight on the bit and the bit torque. These measurements can be used to improve drilling efficiency, identify bit failure, and reduce the risk of stuck pipe. Downhole vibration measurement can be used to identify harmful operating conditions, which could result in tool and drill-string failures. The use of these applications has been limited but is increasing.

Prospects. Logging of measurements made while drilling is becoming accepted as a replacement for wireline logging in many applications. Initially, logs made during drilling replaced wireline logs only in high-angle wells, where wireline logs would not free-fall to the bottom, and exploration wells, where there was a high risk of the wireline tools sticking. This trend will continue, with measurement-while-drilling logs becoming routine on many conventional wells.

Development of new and improved sensors for measurement while drilling is anticipated. In the near future, inclination measurements will be made near the bit to improve directional control. Logging measurements are being improved to correlate more closely with wireline logs. Work is being done to develop improved borehole calipers and possibly a sonic tool.

In addition, new applications for measurement while drilling are being developed. Horizontal drilling and extended-reach drilling have demonstrated the importance of geological steering (the use of geological measurements to accurately place the wellbore in the reservoir relative to a fluid contact, a geological marker, or a reservoir structural feature). Extended-reach drilling produces high-angle, long-reach wells that are considered to represent the limits of current technology. New sensors with greater depths of investigation and the ability to see ahead of the bit will be required to make geological steering possible.

Directional drilling in the future will involve the use of smart drilling systems with closed-loop control provided downhole. The tool designed for measurement while drilling will be a necessary component of the closed-loop directional bottom-hole assembly. Successful implementation of this technology will require the ability to communicate from the surface to the tool downhole. SEE GEOLOGICAL ENGINEERING.

For background information SEE OIL AND GAS WELL DRILLING; PETROLEUM ENGINEERING; WELL LOGGING in the McGraw-Hill Encyclopedia of Science & Technology.

Douglas A. Gust

Bibliography. R. Aguilera, *Naturally Fractured Reservoirs*, 1980; R. Aguilera et al., *Horizontal Wells*, vol. 9 of *Contributions in Petroleum Geology and Engineering*, 1991; A. B. Cunningham, K. L. Jay, and E. Opstad, Applications of MWD technology in nonconventional well, Prudhoe Bay, North Slope Alaska, *SPWLA 31st Annual Logging Symposium*, June 24–27, 1991; S. D. Joshi, *Horizontal Well Technology*, 1991; C. A. Koopersmith and W. C. Barnett, Environmental parameters affecting neutron porosity, gamma ray, and resistivity measurements made while drilling, *62d Annual Technical Conference of the Society of Petroleum Engineers*, Dallas, SPE 16758, September 27–30, 1987.

Oil-field fires

Near the end of the Gulf War in February 1991, retreating Iraqi soldiers detonated hundreds of explosive charges at the oil wells of Kuwait. These explosives, installed during the 7-month occupation by Iraq, effectively destroyed Kuwait's capacity to produce oil. In all, 732 wellheads were damaged or destroyed. While most of the wells immediately burst into flames, about 10–15% did not ignite. Instead, gushers spewed thousands of barrels of crude onto the desert sands and formed lakes of oil around the wellheads.

Initial planning. As early as September 1990, intelligence reports alerted the government of Kuwait in exile to the mining of the oil-field wellheads, and the Kuwait Oil Company began developing a contingency plan. Four companies were selected to prepare well-control plans and to implement them, if needed, when war ended. Three companies were based in Texas, and the fourth in Alberta, Canada.

Once the war ended in February 1991, each of the four companies was asked to prepare two full crews and fire-fighting and well-control equipment for shipment to Kuwait. The Texas crews began work in Kuwait in March, and the Canadian crews in early April. By mid-April eight full crews were at work there.

The first priority was to control wells close to populated areas like Kuwait City, including the wells that were ablaze and those that were gushing oil to the surface. The most challenging problem was not putting out the fires, but bringing the wells under control and capping them so that the flow of oil to the surface could be stopped. In order to fully appreciate the magnitude and complexity of the task, it is necessary to understand the procedures. Various methods were used, because the crews from the four companies had different options available for fire fighting and well control.

Cooling. The first step was to cool the well site with water sprays, remove surface debris, and establish access to the wellhead. The water was stored either in large lagoons excavated near the wellheads or in several large portable water tanks.

Water was a particular problem in the early weeks of the fire-fighting effort, because it had to be trucked from the ocean or pumped from a limited number of inland wells. By the end of the third month, pipelines that had earlier transported oil to seaside terminals had been reversed to allow seawater to flow inland to support the fire fighting. Water trucks or newly laid

surface pipelines could then distribute the water to the well sites. Thousands of gallons of water were used initially just to cool most of the well sites before debris removal or other well-control work could begin.

Removing debris. Debris around the damaged wellheads was dragged away by cranes, bulldozers, or athey wagons. Unique to oil-field fire fighting, an athey wagon consists of a 50–60-ft-long (15–18-m) track-mounted boom positioned by a bulldozer. A hook at the end of the boom is used to remove debris, deliver explosives, or position a new wellhead.

When the areas around the wellhead were covered with a pool or small lake of crude oil, construction crews had to build causeways of fresh sand and rock to the wellhead. Then a dike was created to hold back the oil and provide a work area. Many of the wellheads were set into deep cellars 10–20 ft (3–6 m) below the ground so that the process of removing debris and clearing the wellhead area was further complicated.

Extinguishing fires. Once the area was cleared of debris, preparations to extinguish the fire could begin. The area was continuously sprayed with water by high-volume pumps. Workers and equipment were also continuously sprayed to make it possible to operate in an environment in which ground temperatures often reached 800°F (427°C).

Crews then determined the appropriate method to bring the well under control. In most cases, the next step was to put the fire out. Two methods were used to deprive the fire of oxygen. In one method, crews mounted an explosive charge on the boom of the athey wagon, and then the charge was positioned directly above the wellhead. The charge was detonated and, if the procedure was successful, the flame was extinguished. Crews then continued to spray the area with water to prevent reignition (**Fig. 1**). Alternatively, the

Fig. 1. After extinguishing a fire, crews protected in monitor shacks train water sprays on an oil well in the Greater Burgan field. (*Safety Boss, Ltd.*)

Fig. 2. Blowout control specialists working on a damaged wellhead in the Greater Burgan field. (*Safety Boss, Ltd.*)

fire could be extinguished by the use of dry chemicals. The crews covered the wellhead area with a high-pressure spray of potassium bicarbonate ($KHCO_3$) in siliconized power form, which does not lump together and behaves like a liquid spray. This dry chemical is propelled by pressurized nitrogen and discharges at a rate of 200 lb (90 kg) per second. Even the largest hydrocarbon fires can be extinguished by this method.

Achieving well control. With the fire out, the next step was the removal of the badly damaged wellheads. **Figure 2** shows two blowout control specialists working on a damaged wellhead in the Greater Burgan field. During this operation, oil continued to escape from the wellhead and coated the area, including the personnel, with a thick film.

Among the various options available to initiate well control, one was to unbolt the damaged wellhead and install a kill spool. Fitted to the wellhead, this device allows crews to connect a high-pressure line through which heavy drilling fluids can be pumped into the well to offset the natural pressure. Once the weight of the column of drilling fluid pumped into the wellbore exceeded the pressure of the well, the well was considered to be controlled, and a new control device was bolted onto the damaged well.

In most cases, enough of the lower flange on the wellhead remained to allow a new wellhead to be bolted in place. Sometimes the explosion that had damaged the wellhead had distorted the lower flange to such a degree that an assembly that would mold to the shape of the deformed flange was required.

Another option extensively used in Kuwait is stinging. A hollow tapered tube, about 3–4 ft (1–1.2 m) long and up to 4–5 in. (10–13 cm) in diameter, was forced into the stream of oil gushing from the well. The stinger, which had a kill line (a high-pressure swivel pipe) attached, was held in place with cables. After closing a valve on top of the stinger, heavy drilling fluid was pumped into the well through the kill line to offset the natural pressure (**Fig. 3**). Once the well was under control, a new wellhead was installed to complete the capping process.

In cases where the entire wellhead assembly had been damaged, crews excavated down to a level where they could cut off all of the damaged assembly and

leave a solid piece of pipe. A metal ring known as a casing bowl was then lowered over the exposed pipe and welded in place. A new blowout preventer was lowered through the flow of oil, aligned with the casing bowl, and bolted in place. Then the valves in the blowout preventer were activated remotely, and the well was shut in and brought under control. **Figure 4** shows well-control personnel engaged in the final stage of shutting in a large well in the South Burgan field.

Project challenges. The work in Kuwait created a number of challenges for fire-fighting and well-control crews that had not been encountered before in an oil-field emergency. The first was the sheer magnitude of the task. The oil wells in Kuwait are among the largest producers in the world, yielding 10,000–60,000 barrels (1600–9600 m^3) per day, which is 50–500 times as much oil as is produced by a typical North American well. Over 700 Kuwait oil wells had been damaged, a disaster on an enormous scale.

The second consideration was the fact that Kuwait had been stripped of heavy equipment such as cranes and bulldozers in the course of the Gulf War. The fire-fighting crews found very little heavy equipment, limited supplies of water, and a countryside littered with the machinery of war, mines, cluster bombs, and other ordnance.

Despite these obstacles, the fire-fighting crews began work, and the Kuwaitis mobilized the heavy equipment needed. Oil-gathering lines that had transported crude oil to seaside terminals and loading docks were reversed, and seawater was pumped to the oil fields to support the fire-fighting crews. Hundreds of loads of equipment and supplies were flown into Kuwait by U.S. Air Force C-5 Galaxy transport aircraft to support the efforts.

A third consideration, particularly in the early months of the fire fighting, was the impaired visibility and extremely polluted air resulting from the hundreds of fires burning within a small area. When the wind dropped, visibility in the center of some oil fields was limited to a few feet, even at midday. In addition, oil continued to fall in a fine rain of particles that covered everything—plants, animals, soil, sand, buildings, and people.

Fig. 3. A stinger, with kill line attached, being directed into position in the Greater Burgan field by well-control personnel. (*Safety Boss, Ltd.*)

Fig. 4. Well-control personnel handling the final stages of shutting in a large well. (*Safety Boss, Ltd.*)

A fourth consideration was the effect of having some wells burn for months before they could be brought under control. Large accumulations of partially burned hydrocarbons built up huge piles of coke or heavy ends around many wellheads. As the months wore on, some of these became mini-volcano formations that had to be removed before crews could gain access to the wellheads.

All 732 damaged wells were capped by November 1991, some 9 months after the damage was inflicted. The work was accomplished by a multinational force. Americans and Canadians were joined by teams from several other countries in the largest emergency response action ever mounted to deal with an oil-field disaster.

For background information SEE *FIRE TECHNOLOGY*; *PETROLEUM ENGINEERING* in the McGraw-Hill Encyclopedia of Science & Technology.

Michael Miller; Richard E. Wyman

Oil mining

Very large reserves of oil exist worldwide in reservoirs from which recoveries by conventional surface wells are extremely low, even with the application of various specialized technologies. Typical examples of such difficult reservoirs include oil sands, heavy oil, and some conventional oil reservoirs. A system of oil mining using horizontal wells, drilled and completed from underground tunnels, shows great potential for considerably improving recovery and economics for many of these difficult reservoirs. The system has been piloted at an underground test facility constructed for this purpose near Fort McMurray in northern Alberta, Canada. The steam-assisted gravity

drainage processes applied from tunnels may have wide application in difficult oil reservoirs in many countries.

Oil sands of Alberta. The oil sands carbonate formations of Alberta are estimated to contain total bitumen (heavy oil) in-place of 2.664×10^{11} m^3 (1.67×10^{12} bbl), approximately four times the producible reserves of conventional oil in all of the Middle East oilfields combined. Unfortunately, production of oil from this vast resource at an acceptable recovery rate and with attractive economics is very difficult. About 5% of the reserve is shallow enough to be surface-mined. The remainder requires in-place recovery operation. Attempts to exploit the reserve using in-place methods applied through wells drilled from the surface have not generally been successful. Consequently, a system using underground access has been examined.

Oil recovery from underground tunnels. The use of tunnels placed in or below an oil reservoir to drain liquid hydrocarbons from the pay zone has been practiced in several countries. Early attempts utilized gravity alone to drain mobile oil to the tunnels, where it was pumped to the surface. Historically, this system was applicable to only a few conventional oil reservoirs, and therefore it was in direct, and generally unsuccessful, competition with standard vertical surface wells.

However, in some countries, such as Canada and the United States, conventional oil reserves were limited. In both countries the reserves of conventional oil at present rates of consumption are in the 7–10-year

Fig. 1. Head frame and surface facilities of the underground test facility.

range. This situation, and the increasing dependence on foreign oil from imports, has naturally focused attention on heavy oil and oil sands, neither of which are amenable to conventional recovery from surface wells. Limited success has been achieved by utilizing in-place methods through wells drilled from the surface, steam injection often being employed to recover petroleum from heavy oil and oil sand reserves by reducing the viscosity of the liquid hydrocarbon. This system is usually characterized by low recoveries and high energy costs.

This drawback has led to the development of a unique system that applies a steam-assisted gravity drainage system through semihorizontal wells, drilled, completed, and operated from tunnels constructed in the rock below the reservoir. The use of mining technology to build and operate tunnels as a means of placing personnel and equipment below the reservoir is known as the shaft and tunnel access concept. The use of steam injection into semihorizontal wells from which the heated oil drains into pipelines in the tunnels below the reservoir is known as steam-assisted gravity drainage.

Underground test facility. Pilot operations for testing the shaft and tunnel access concept and the steam-assisted gravity drainage concept have been undertaken at an underground test facility near Fort McMurray. The facility was designed and built by the Alberta Oil Sands Technology and Research Authority (AOSTRA), a Canadian provincial government organization responsible for developing in-place technologies and strategies for this huge resource.

The facility (**Fig. 1**) consists of two vertical shafts, each 3 m (10 ft) in diameter and 213 m (700 ft) deep, which provide access to tunnels constructed in limestone 15 m (50 ft) below the base of the oil sands. The tunnels are generally 5 m (16 ft) wide and 4 m (13 ft) high and supported by rock bolts. Well chambers have been constructed to accommodate the prototype drilling machine (**Fig. 2**). The well chamber also accommodates well heads and control equipment for the steam injection and production facilities. The two-shaft facility is ventilated by two large surface fans on the top of one shaft, which is equipped with an emergency elevator. The other shaft is equipped with a hoist for transportation of personnel and waste rock from the tunnels.

Surface facilities include offices, workshops, warehouses, and a steam-generating plant.

Steam-assisted gravity drainage. Attempts to develop an effective method to recover bitumen from deeply buried Athabasca oil sands of Alberta have been ongoing for about 50 years. The technology tested in the past had many variations, but it was basically aimed at heating the oil sands with steam injected through vertical wells to reduce the viscosity of the asphaltlike bitumen so that it could flow, then driving the bitumen through the reservoir with the high-pressure steam to an offset vertical production well, where it could be pumped to the surface. The tested methods were not effective, mainly because it is very difficult to push the thick bitumen long distances

Fig. 2. Underground drilling machine.

through the fine sand, which holds the bitumen, and at the same time maintain a heated path from the injector to the producer.

The steam-assisted gravity drainage process is effective because it is based on the simple physical concepts that heat rises and liquids flow downhill. The process is made possible through the use of horizontal wells, drilled from underground tunnels at the underground test facility. As shown in **Fig. 3**, a pair of horizontal wells are placed in the oil sands, with the producer near the base of the oil sands and the injector about 5 m (16 ft) above. Steam is injected through the upper well, where it condenses on the cold oil sands to heat bitumen. The mobilized bitumen and condensed steam (hot water) drain by

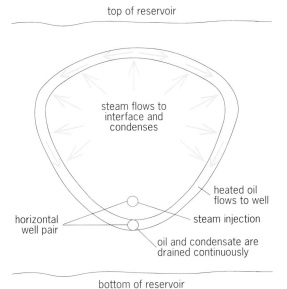

Fig. 3. Diagram showing the configuration of the steam chamber of the steam-assisted gravity process.

gravity to the lower production well. The method is very effective because the injected steam continuously rises to cold sections of the oil sands to heat bitumen and the heated bitumen drains by gravity to the lower production well with a minimum distance to flow in the reservoir, which is constantly heated.

The generation of steam in boilers is the largest single operational cost for producing deeply buried bitumen. The steam-assisted gravity drainage process uses less steam per barrel of bitumen produced than any comparable process developed to date and thus has very favorable bitumen production economics.

Drilling horizontal wells. Horizontal wells proceed a significant distance in a horizontal direction in an oil reservoir. In recent years, horizontal wells have become increasingly common in drilling from the surface, starting with a vertical well that is then deviated to horizontal in the reservoir. Horizontal wells are attractive because the long horizontal section in the reservoir provides a pipeline for oil to flow for production over a greater distance than with a vertical well.

As shown in **Fig. 4**, the horizontal wells at the underground test facility are drilled from tunnels below the oil sands rather than from the surface. This allows control valves on the horizontal producer to be as close as possible to the oil sands reservoir so that a minimum of uncondensed steam escapes from the reservoir through the producer. This close and continuous balanced control between steam injection and bitumen-water production contributes to the efficient usage of steam achieved for bitumen production.

Also, the shorter horizontal wells drilled from tunnels can be more economic than longer wells from the surface, which must pass through the overburden.

A special compact drilling rig has been developed for use in drilling from the underground tunnels. The wells are commenced by drilling up at an inclination of about 16°, with casing installed and cemented to approximately 60 m (200 ft). A casing bowl with a blowout preventer is then installed, and the drilling proceeds through the casing out of the remaining limestone to become horizontal in the oil sands. The underground test facility project has completed the first phase (phase A), in which three sets of twin horizontal wells were drilled with 60-m-long (200-ft) horizontal sections. *SEE OIL AND GAS WELL DRILLING.*

Phase A and phase B. Phase A of the underground test facility project was more successful than anticipated; it achieved a steam/oil ratio of about 2.6 bbl of steam as cold water equivalent per barrel of oil, whereas the rule-of-thumb economic cutoff level is a steam/oil ratio of up to 5.0. The recovery for bitumen in place was excellent, with over 60% recovered. Comparable processes recover only about 20% of oil in place.

The next phase (phase B) is under way with 500-m (1600-ft) horizontal section wells; it is designed as a precommercial demonstration phase with bitumen production of 320 m³ (2000 bbl) per day. As of mid-1991, the wells were successfully drilled, probably the most accurate horizontal wells ever drilled, and steam operations were beginning. The project is operated by AOSTRA with seven industry participants, and with

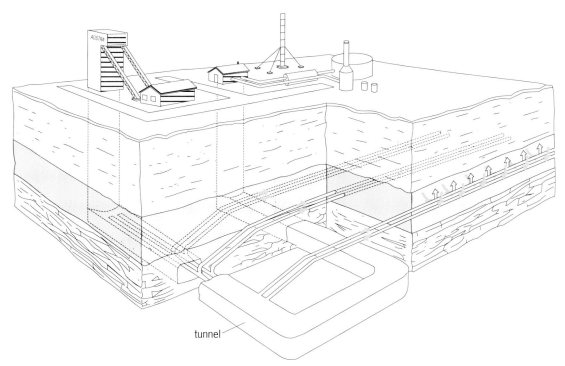

Fig. 4. Diagram showing the placement of horizontal wells and their relationship to the surface facilities. The tunnel is in the limestone below the oil sand. Arrows show the steam-injection (white) and production (color) flows.

support from the CANMET department of the Canadian federal government.

Plant design and economics. A commercial operation based on the results achieved at the underground test facility and utilizing the existing facilities as a base for expansion has been studied. Assuming a bitumen production level of 1600 m^3/day (10,000 bbl/day), the operation would have a life of 16 years and would utilize 70 pairs of wells. Studies have shown that a single commercial operation producing as much as 8000 m^3/day (50,000 bbl/day) of bitumen could be developed with subsequent economies of scale.

Implications for the oil industry. The underground oil mining technology developed in Alberta has application in many oil reservoirs previously considered too difficult to develop. As development proceeds, it is quite likely that reserves now considered uneconomical for conventional oil recovery may become viable by using the technology developed at the underground test facility. Thus, recoverable reserves of oil in several countries, notably Canada and the United States, would be considerably increased.

For background information SEE OIL AND GAS WELL DRILLING; OIL SAND in the McGraw-Hill Encyclopedia of Science & Technology.

H. G. Stephenson

Bibliography. N. R. Edmunds, The use of gravity drainage processes in the U.T.F., *AOSTRA Conference on Advances in Petroleum Recovery and Upgrading Technology*, Edmonton, Alberta, June 1987; J. A. Haston, R. W. Luhning, and S. D. Gittins, AOSTRA underground test facility progress and potential, *CIM Bull.*, 83(938):53–58, Canadian Institute of Mining and Metallurgy, October 1989; H. G. Stephenson and R. W. Luhning, Oil at the end of the tunnel, *Annual Meeting of the Society for Mining, Metallurgy, and Exploration*, Denver, Colorado, Preprint 91-67, February 1991; H. G. Stephenson, H. D. Owen, and L. R. Turner, The AOSTRA underground test facility: An update, *87th Annual Conference of the Canadian Institute of Mining and Metallurgy*, Vancouver, British Columbia, April 1985.

Optical fibers

The introduction of optical communications in instrumentation, communications, and control systems for the electric power industry is making a tremendous impact. The use of optical components in the electric power environment is ideal. Applications that were previously thought to be impossible or impractical are being reconsidered.

When the possibility of using fiber optics in the electric power industry became apparent in the mid-1960s, it was met with some reservations. The electric power industry is very conservative, being committed to the extremely important task of providing a power supply to a set of users who require high reliability. New technology has always been difficult to integrate into this field until proof of supreme reliability is obtained.

Dealing with the engineering problems associated with generating, transmitting, and distributing electrical power is a formidable task. The advantages of optical coupling were recognized from the beginning by those who understood the need to provide reliable, accurate, and less costly control, protection, and revenue-metering devices. For most conventional technologies, limits had been reached in areas such as insulation, saturation, information management, accuracy, speed, and reliability. It was clear that this new technology, which was maturing very rapidly, had few of these limitations.

Power-system applications using optical coupling can be classified into three major categories: communications, measurement, and control. Each can be treated distinctly, but all involve the movement or transformation of data or information.

Communications applications. In this category there are classically three divisions: those applications in which communications are needed from one station to another (interstation links), where path lengths are typically greater than 5 km (3 mi); links within the station itself (intrastation), with path lengths under 1 km (0.6 mi); and special situations, where the station is linked to a nearby facility, such as a radio tower (entrance links), with path lengths under 2 km (1.2 mi).

An important reason for considering fiber optics for interstation links is the very high information transfer rate possible with this technology. In densely populated centers, the use of radio communications requires frequency licensing, which may not be possible because of congestion of the available radio-frequency spectrum. Using satellite communications may not be appropriate, since there are bandwidth and inherent delay problems. (The quarter-second inherent delay in satellite communications would be intolerable, since the signal needed to switch out a power line must arrive at the breaker in far less time.) Finally, there is a need to electrically isolate stations from one another.

The primary reason for selecting fibers for intrastation and entrance connections is the point-to-point isolation, which is critical for reliability and safety concerns. In the past, electronic terminations in power stations had to be designed with very strict attention to input and output protection and personal safety.

Designed correctly, fiber optics (namely, single-mode products operating in the 1300- and 1500-nanometer bands) can be installed with little worry of future obsolescence. Not only can existing systems and designs serve the communications needs of the utilities themselves, but some utilities have also leased out extra capacity to common carriers. Many utilities worldwide have installed fiber-optic cables in the power-line corridors to replace or complement existing wire-line or radio communications.

There are many methods of installing fiber-optic cables along the power-line corridor, including burying the cable beneath the right-of-way; hanging it below the high-voltage conductors (in which case the fiber-optic cable must be completely dielectric); designing it to be an integral part of the high-voltage phase conductor; attaching it directly to the conductor; or making it part of, or attaching it directly to, the overhead ground or static wire, which is usually located above the phase wires (used to protect the power line from lightning). The last method is the most common at voltage ranges greater than 230 kV. Making the cable part of, or attaching it to, the phase conductor is more popular at lower voltages.

Another important advantage of this technology is the cost savings to be expected in storing, maintaining, and installing fiber cables. This reason is that just a few fibers can take the place of many pairs of copper wires, and thus the weight and the size of the cable are much less, and fewer splices are required in installing the cables.

Measurement applications. The ability to communicate directly with power lines, buses, transformers, circuit breakers, and other components used in applications involving high-voltage management requires the features of the uncommonly pure fiber-optic waveguide glass. With this characteristic comes total freedom from noise coupling, ground-mat potential swings, and other sources of dynamic and static signal distortion. This characteristic alone allows for applications that simply could not have been realized with other technologies. The use of optical measurement devices also provides opportunities to increase the quality of the information presented to the station control and protection system. In the past it was very difficult and expensive to obtain high-quality information directly because of the noisy environment.

The measurement of voltage and current is of primary importance in the proper control, protection, and metering of a power station. In fact, the proper operation of a power station is impossible without such information. Thus, up to now, the primary emphasis has been on devising optical systems to perform precision measurements of these two quantities.

There are two basic approaches to designing optical sensors: active designs, where there are electronics at the measurement site; and passive systems, where there are no electronics at the site. Active measurement systems require power for coding and transmitting signals, and this power must be derived from that available at the site. This power can come from batteries, transformers that tap power from the line or bus under measurement, or ground-based light beams coupled into uplink fibers that bring optical energy to a photodiode, which can then be converted to electrical energy. Any of these approaches still requires the use of conventional transducers (resistors, capacitors, transformers, and so forth), but without the bulky and costly insulation materials needed for nonoptical systems.

The use of passive optical schemes offers a potentially more advanced approach than active sensors, since passive units bypass the need to use conventional inputs, with their limitations. A passive transducer is a completely dielectric unit that resides at the measurement site and requires no on-site electric power for operation. The advantages of such devices include being fully dielectric and having inherently low weight, which allows for designs that not only are less expensive but also have very wide bandwidths, high accuracies, and virtually no saturation limits. The low-weight feature makes passive sensors ideal for power systems aboard spacecraft.

The principle behind passive transducers lies in the fact that certain materials show some detectable optical change under the influence of electric or magnetic fields, temperature, pressure, or other static or dynamic changes. In designing an optical passive sensor, selection is made of a material that responds to only the parameter to be measured. Then an optical interrogation beam is used to monitor that optical activity. The light source and the signal process electronics are located in the control house and connected to the sensor module with fiber optics.

For example, the Pockels effect can be used to measure voltage, or the Faraday effect can be used to measure current. The Pockels effect, which is displayed by certain crystals, is a change in birefringence (the difference in the velocity of light traveling down the vertical and horizontal planes of the crystal) due to a change in the impinging electric field. The Faraday effect, which can be exhibited in a glass element, causes the polarization plane of light passing through the glass to rotate (twist) by an amount proportional to the magnetic field in the glass. These effects are also known as electrooptic and magnetooptic effects.

Development of this technology continues, and there has been widespread investigation and construction of prototype passive measurement devices. Several systems are now commercially available. The growth of this technology should be aided by the rapid advance in the manufacture of digital control, protection, and metering systems. Optical transducers interface well with these digital systems, a feature not available with conventional systems.

Not only can voltage and current measurements be performed by using novel optical methods, but virtually any other parameter can also be measured by using similar methods. Examples include temperature, pressure, gas density, stress, strain, positional status, vibration, and gas or oil composition analysis. Moreover, all these measurements can be made in previously unattainable locations. Advanced fiber-optic security systems to sense intrusion into sensitive areas have also been built. In addition, the accurate mapping of electric fields on and around power apparatus has always been difficult, if not impossible, with conventional equipment. This mapping is of particular importance in the design stages of such devices, since the modeling of field-stress concentrations is extremely difficult to accomplish on some of the more complex designs.

To provide for more compact substations, there has been a great deal of research into compressed-gas coaxial power systems. These installations will require novel instrumentation systems that would be particularly dependent on advanced devices, since low cost, small size, and increased accuracy and bandwidth are necessary specifications.

Control applications. Direct control of devices in the power station has been performed with optics in a variety of situations. Most of these applications involve using light beams to turn solid-state power switches at dc converter stations on and off. Also, dedicated communication-link schemes that route optical trigger signals to power apparatus have great promise, since the copper cabling now used to do this electrically is very bulky and expensive. Optical methods also allow a more direct connection to the driver electronics, so that constant checking of the integrity of the link can be made to ensure that the link will be available when it is needed. Earlier wire-line systems were overdesigned in order to achieve the required reliability.

There is even speculation that the conduction of certain gases could be turned on and off with special laser beams, to provide the switching of large voltages and currents that now require large, expensive circuit breakers. Also, with the extremely-low-loss (that is, 0.001 dB/km or 0.0016 dB/mi), far-infrared fibers now envisioned, combined with high-efficiency laser power sources, blocks of power could be transmitted optically to advanced light-triggered power systems. Because of the extremely low weights involved, perhaps future space-station, lunar, and planetary power networks will use fiber-optic power transfer control exclusively.

For background information SEE ELECTROOPTICS; FARADAY EFFECT; FIBER-OPTIC CIRCUIT; FIBER-OPTIC SENSOR; OPTICAL COMMUNICATIONS; OPTICAL FIBERS; TRANSMISSION LINES in the McGraw-Hill Encyclopedia of Science & Technology.

Dennis C. Erickson

Bibliography. Electric Power Research Institute, *Optical Power Line Voltage and Current Measurement Systems*, EPRI Doc. EL-5431, 1987; Electric Power Research Institute and National Institute of Standards and Technology, *The Role of Optical Sensors in Power Systems Voltage and Current Measurements* (workshop), Gaithersburg, Maryland, 1987; Institute of Electrical and Electronics Engineers, *Fiber Optic Applications in Electrical Power Systems* (tutorial course), IEEE Spec. Pub. 84 EH0225-3-PWR, 1984; Institute of Electrical and Electronics Engineers, *Fiber Optics Applications in Electrical Substations*, IEEE Spec. Pub. 83-WM-025-4, 1983.

Optical information systems

The industrialized world has entered the information age, in which production, distribution, and use of information constitute a key industrial sector vital to economic growth. An important component of the information industry is the technology for efficient production, distribution, and use of information. Since about 1950, microelectronics technology has been the primary platform on which information-related products (such as computers, televisions, and telephones) have been developed. Since the early 1980s, however, a new generation of products based on light (photons instead of electrons) has started to appear. This technology has been termed photonics, and perhaps by the year 2000 the photonics industry will rival the electronics industry in importance and dollar volume. In this article the key physical features of photonics are defined, and their implications for information systems are discussed. Following a description of some new developments in photonics technology, a few speculations on futuristic optical information systems will be presented. SEE INFORMATION SYSTEMS ENGINEERING.

Photonics. Photonics currently incorporates three technologies: optics (including lenses, fibers, prisms, and holograms), electrooptics (including scanners and liquid-crystal displays), and optoelectronics (including lasers and charge-coupled-device detectors). In all cases, a variety of physical mechanisms are used in generation, transmission, spatial transformation, spatiotemporal modulation, and detection of optical waves. The term photonics highlights this common aspect of optics, electrooptics, and optoelectronics, and denotes their application to information systems.

Photons are fundamentally different from electrons in that they are chargeless. This characteristic allows them to propagate through free space or clean air without strong absorption or scattering. Propagation of unguided electron beams requires a vacuum and cumbersome electrostatic and magnetic lenses. Photons do not interact with each other, and two beams of light can cross through each other in free space. In contrast, electric currents in two adjacent copper wires will interact with each other through capacitive and inductive effects unless precautions are taken to shield the two wires. Another advantage is the high carrier frequency (10^{14}–10^{15} Hz) of the light waves compared with the frequencies (10^{9}–10^{11} Hz) of microwaves used for communications. Therefore, light waves can be modulated at much higher rates to carry more information. Light waves also have much smaller wavelengths than microwaves, resulting in smaller systems (such as receivers, transmitters, and antennas). These features of light are grounded in physics and are not technology-dependent. Therefore, any advantages of optical information systems based on these features will also be technology-independent and hence lasting.

Characteristics of systems. An information system consists of subsystems for generation, transmission, storage, display, and processing of information. The information generation is performed by sensors that measure various physical parameters (such as temperature, optical images, and sound signals). The transmission can be done through some guiding medium or through free space. Storage, display, and processing operations are necessary for using information. In most applications of information systems, all these operations take place in a tightly coupled environment. For example, a department store cashier uses an optical or magnetic scanner to sense the price and inventory information on the purchase item. The tax and total bill are calculated by the associated computer. The inventory updating is performed by transmitting the sale information along a communication channel (a coaxial or fiber-optic cable) to a central location containing a mass-data-storage system (optical or magnetic disks). The same scenario is repeated with different details in numerous scientific, biomedical, aerospace, and military information systems.

Improvements in sensor technology have led to an explosion in the total amount of information that is generated. Concurrently, the increasing sophistication in the computations to be performed overwhelms the ability of even the latest generation of supercomputers. Thus, the pressure for improved performance in information generating, carrying, storing, processing, and displaying capacity is being exerted by applications ranging from weather modeling and prediction to banking and the retail industry.

Development of optical systems. The use of light in information systems predates the modern information age. Most common among the early optical displays were (and, to some extent, still are) printed paper and film. Paper and film also served the purpose of information storage. Indeed, until the advent of high-density magnetic disks and semiconductor memory chips, film was the preferred medium for storing large amounts of imagery information.

Modern optical information systems began with the invention of the laser in 1960. The highly parallel output of a laser has a very narrow (monochromatic) frequency bandwidth, and the power output is large. These properties made lasers a very sensitive tool for use in information systems. The active media in early lasers were mostly gases, which made the lasers bulky and fragile. Semiconductor lasers (also known as laser diodes) invented in the late 1960s removed these limitations and provided a very compact source of coherent radiation that can be modulated directly at very high rates (greater than 10^9 Hz).

Detection of optical radiation in photography, astronomy, and other scientific applications had been performed almost exclusively with film. Indeed, the resolution and sensitivity of film were unsurpassed. The drawbacks were messy chemical processing and the inability to reuse the film. Optical detectors based on semiconductor devices have now replaced photographic film in several of the applications mentioned above. These semiconductor detectors range from single-element, high-bandwidth (greater than 10^9 Hz) devices to large two-dimensional arrays (4096 × 4096 elements) of charge-coupled devices for imagery applications. The advances in charge-coupled-device technology have made camcorders possible.

Laser diodes and photodetectors have benefitted from advances in thin-film fabrication technology developed for semiconductor electronics. The main material for electronic devices and photodetector arrays is silicon, while for laser diodes it has been gallium arsenide, indium phosphide, and variants of these compounds doped with aluminum and antimony. Advances in growing ultrahigh-purity optical materials have made low-loss fibers possible. As a combined effect of these developments in materials and device-fabrication technology, photonics is providing significant performance improvements over electronics in information systems.

Communications and interconnects. The idea of confining light to a glass strand has been known since ancient times. But very high losses in glass fibers made long-distance links impractical until very pure silica glass fibers with an attenuation of 20 dB/km (a factor of 100 power attenuation over 1 km or 0.6 mi) were fabricated. Further improvements in material purification and selection of the correct wavelength (1550 nanometers) at which absorption losses are minimum have improved this figure to 0.14 dB/km (a power loss of 4% over 1 km or 0.6 mi). The bandwidth limitations on fibers are due to the spreading of optical pulses as they propagate along the fiber (dispersion). Dispersion can be reduced by controlling the size and refractive-index profile of the fiber and by employing laser diodes with a narrow wavelength spread. These two developments have led to optical links that can carry several gigabits of data per second over 100 km (60 mi) without using repeaters. This performance combined with small size, low weight, security against tampering, and immunity to electromagnetic interference has made fiber optics the primary choice, over coaxial or microwave communication links, for trunk lines between cities or across the oceans.

Figure 1 shows several levels of hierarchy in data communications. The length of the links varies from several thousand kilometers for the highest level to a few micrometers for dense integrated circuits. The number of links and bandwidth requirements for communications at these different levels are application-

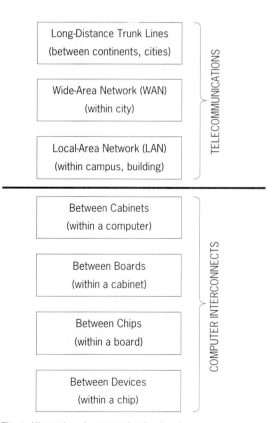

Fig. 1. Hierarchy of communication levels.

driven and vary significantly. Generally, the lower levels require lower bandwidth but greater density. Both bandwidth and density requirements have, however, been increasing with the sophistication of integrated-circuit technology and with the complexity of parallel computer architectures. The bandwidth of conventional electronic interconnects is limited by crosstalk, interference, and power dissipation. The density is limited by topological considerations, which force the devices and interconnects to share the same two-dimensional area in an integrated circuit or a board.

Use of optics for implementing interconnects within high-speed electronic systems has been an active area of research since 1985. In other words, optics has been penetrating deeper into the communication hierarchy shown in Fig. 1. Using fibers instead of multiconductor ribbon cables in board-to-board interconnects is the first step, and commercial products based on this technology are expected to appear soon. Optical thin-film waveguides for realizing high-density, high-bandwidth interconnects within a board have been demonstrated. Within-chip communication with optics is less promising because of the high level of integration required and the large sizes of optical devices (hundreds of square micrometers for optics versus tens of square micrometers for electronics).

A more advanced approach to optical interconnects, now under study, involves free-space optical interconnects at the chip-to-chip and board-to-board level. The distances are small enough and the environment is controlled enough that the scattering and absorption in the intervening air is not a significant factor. Electronic interconnects can make connections between chips or boards only along the edges, so that the total number of data channels is limited to several hundred. On the other hand, a simple lens that can image a 100×100 array of points in a square centimeter is effectively implementing 10,000 high-bandwidth data links between chips or boards. This topological difference between electronic interconnections along the edges and free-space optical interconnects using the entire surface area is depicted in **Fig. 2** for board-to-board interconnects.

The main impediment to the wide-scale use of optical interconnects is the electronic-to-optical and optical-to-electronic conversion devices. While laser diode and detector technologies have made significant progress in the past two decades, the integration of these components with complex electronic circuits (optoelectronic integrated circuits) is still in its infancy. The recent development of surface-emitting microlaser arrays shows promise. These lasers have areas of a few square micrometers and emit light perpendicular to the surface of the chip. An array containing 100,000 such lasers has been demonstrated. Integration of these devices with high-speed gallium arsenide integrated circuits may overcome the obstacles in realizing the system shown in Fig. 2b. SEE LASER.

Optical storage. Compact disks (CDs) have been the most successful consumer product based on photonic technology. Originally developed for recording and playing back video programs, optical disk technology found its chief application in prerecorded digital music. The digital recording gives large dynamic range, and the noncontact operation provided by a laser beam implies no hiss or disk wear, which were common with piezoelectric pickups of record players.

The technology of compact disks is now being applied to storing digital data for computers. Such optical disks are called CD-ROMs (read-only memories), and are being used to distribute software, encyclopedias, catalogs, and training manuals. The disks are produced by stamping into plastic from a metal master and hence can be replicated inexpensively. Magnetic disks, on the other hand, must be serially copied.

A second type of optical disk is called write-once-read-many (WORM), and works by burning tiny holes in optically reflective thin films. Once written, data cannot be erased but can be read many times, so this is an ideal medium for backing up data in large computers. The excellent stability of these media also makes them ideal candidates for applications that require maintaining an audit trail and archival storage.

A third type of optical disk is capable of read, write, and erase operations, similar to magnetic disks. These media are based on several effects, such as magnetooptic effects and phase changes. Optical disk drives for personal computers have recently entered commercial

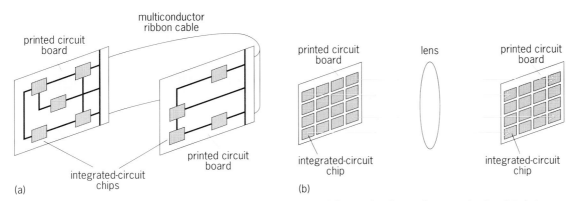

Fig. 2. Comparison of connections between printed circuit boards. (*a*) Conventional way of communicating data between boards via ribbon cables. (*b*) Free-space optical system for implementing board-to-board interconnects. The electronic integrated circuit chips have light detectors and light source-modulators as optical input-output ports.

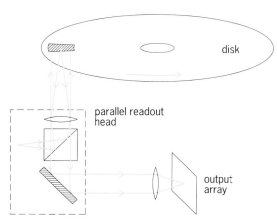

Fig. 3. Parallel-readout optical system for an optical disk. Information over an entire area of the disk is read out simultaneously.

production with a 5¹/₄-in. (133-mm) optical disk having a formatted capacity of 640 megabytes, almost five times that of the latest magnetic disks. These optical disks are removable and therefore can be transported from one machine to another with the ease of a magnetic floppy disk. This feature is unique to optical disks and is due to the large gap between the optical head and the disk (several millimeters versus a few micrometers for fixed magnetic disks). This feature also makes the optical disk drives more rugged, since head crashes are virtually impossible.

Research in optical storage is leading to systems that store and retrieve data in parallel two-dimensional bit patterns simultaneously (**Fig. 3**). This feature will improve the data access time and throughput by several orders of magnitude over the conventional serial mode. New approaches for storing data holographically in ferroelectric crystals to improve storage density are being investigated. These advances combined with optical communication and interconnections can lead to the next generation of optical information systems.

For background information SEE CHARGE-COUPLED DEVICES; COMPUTER STORAGE TECHNOLOGY; LASER; OPTICAL COMMUNICATIONS; OPTICAL DETECTORS; OPTICAL FIBERS; OPTICAL INFORMATION SYSTEMS; OPTICAL RECORDING in the McGraw-Hill Encyclopedia of Science & Technology.

Ravindra A. Athale

Bibliography. R. A. Athale, *Digital Optical Computing: A Critical Review of Technology*, 1990; Optical Society of America, *Technical Digest of Optical Computing Topical Meeting*, March 1991.

Organometallic compounds

The study of anionic complexes of transition metals in negative oxidation states is a rapidly growing segment of inorganic chemistry. Advances have already stretched the conceptions of what is possible in inorganic synthesis, but the development of useful applications for the extraordinary materials made available by this synthetic effort is just beginning.

This area exists because π-acceptor ligands such

as carbon monoxide (CO) can form stable complexes with transition metals in low oxidation states, because of their ability to participate in π-acceptor bonding interactions. In these synergistic bonds, σ-donation of electrons from the ligand is counterbalanced by π-back-donation of electron density from a metal d-orbital into an empty, low-lying orbital on the ligand. In the case of CO, it has long been recognized that this π-acceptor interaction withdraws enough electron density from the metal to stabilize complexes with metals in negative oxidation states, and these are known as carbonylmetalates.

More recently it has been recognized that many other classes of π-acceptor ligands can stabilize complexes of metals in negative oxidation states. These include alkenes, arenes, isonitriles (isocyanides), and carbenes. The importance of this area is threefold: these species test the limits of inorganic synthesis; explaining the stability and reactivity of these complexes advances understanding of bonding in inorganic complexes; and the complexes exhibit unprecedented patterns of reactivity that can be valuable in both inorganic and organic synthesis.

The stoichiometries of highly reduced complexes almost always conform to the 18-electron rule, which states that d-block transition metals tend to form complexes in which their valence electron shells [$1 \times (n + 1)$ s-orbital, $3 \times (n + 1)$ p-orbitals, and $5 \times n$ d-orbitals] are filled. This is a powerful rule of valency for low-valent complexes because π-interaction with acceptor orbitals on the ligand stabilizes filled d-orbitals.

Carbonylmetalates. The first carbonylmetalate, [Fe(CO)$_4^{2-}$] [structure (I)], was reported in the 1930s. It is best prepared by sodium (Na) reduction with a benzophenone catalyst, as shown in reaction (1). The gradual development of general synthetic

$$[\text{Fe(CO)}_5] \xrightarrow[\substack{\text{benzophenone} \\ \text{catalyst}}]{\text{Na}} \text{Na}_2[\text{Fe(CO)}_4] \qquad (1)$$
$$(\text{I})$$

routes to carbonylmetalates has recently culminated in the preparation of the titanium group carbonylmetalates [Ti(CO)$_6$]$^{2-}$, [Zr(CO)$_6$]$^{2-}$, and [Hf(CO)$_6$]$^{2-}$. Monoanionic or dianionic carbonylmetalates are now known for all of the d-block transition metals from the titanium group through the cobalt group.

Substituted carbonylmetalates. A few classes of substituted carbonylmetalates have been known for some time, and 18-electron cyclopentadienyl carbonylmetalates of the general formula [M(π-C$_5$H$_5$)(CO)$_n$]$^{m-}$ (m = 1 or 2) are known for all of the 3d metals. The iron (Fe) complex is the best known and most extensively studied of these systems. It is produced from a precursor containing an iron-iron (Fe—Fe) bond by reduction with sodium amalgam (Na/Hg), as shown in reaction (2).

More recently, synthetic strategies have been developed to allow access to phosphane (PR$_3$) substituted carbonylmetalates of iron and tungsten (W) [such as in reaction (3), where Me = CH$_3$], to arene substituted carbonylmetalates [reaction (4), where pyr = pyridine], and to carbene substituted carbonylmetalates [reaction (5), where NaC$_{10}$H$_8$ = sodium naphthalenide].

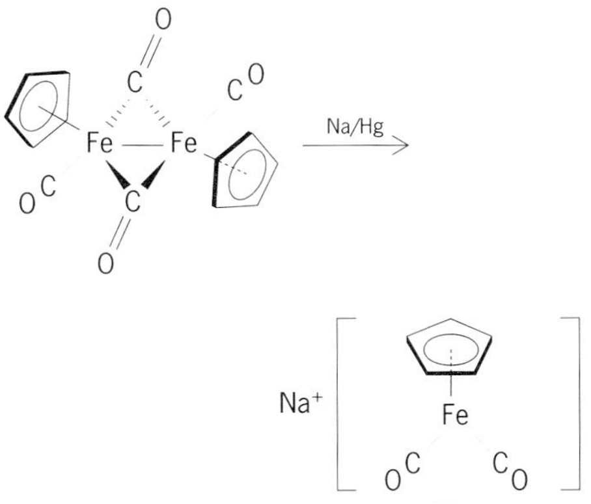

synthesis of complexes such as $[Fe(PF_3)_4]^{2-}$ and $[Co\{P(OCH_3)_3\}_4]^-$ did little to change this, since trifluorophosphine (PF_3) and phosphites were recognized to be powerful π-acceptors comparable with CO. It is, however, well understood that other unsaturated ligands are π-acceptors, and in the early 1980s it was established that alkenes in particular are good enough π-acceptors to stabilize complexes of metals in negative oxidation states, such as the tetraethylene complex of iron (V), produced from ferrocene (VI), lithium (Li), and ethylene (C_2H_4), shown in reaction (7). An adduct of this complex of iron has been struc-

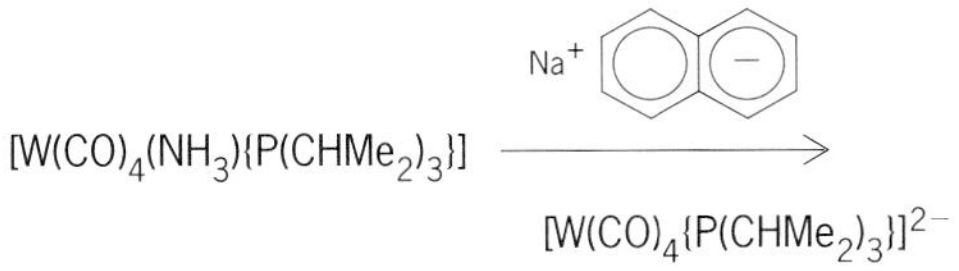

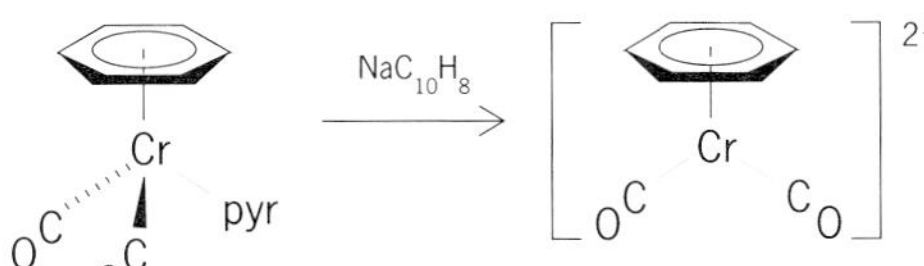

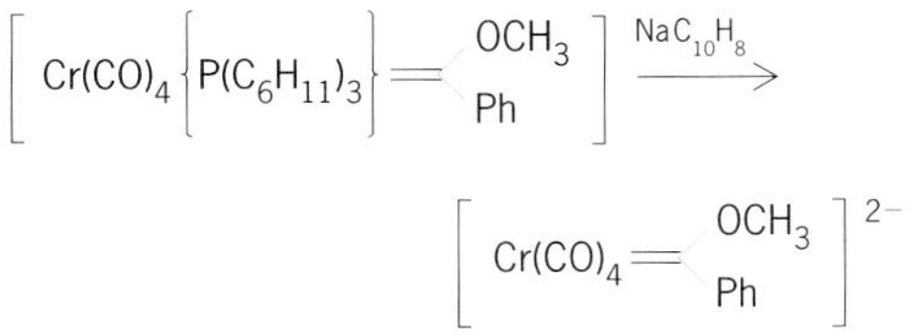

turally characterized; and although the complex has stabilizing interactions between the lithium counterion (Li^+) and Fe, it is chemically reasonable to formulate the complex as containing $[Fe(C_2H_4)_4]^{2-}$.

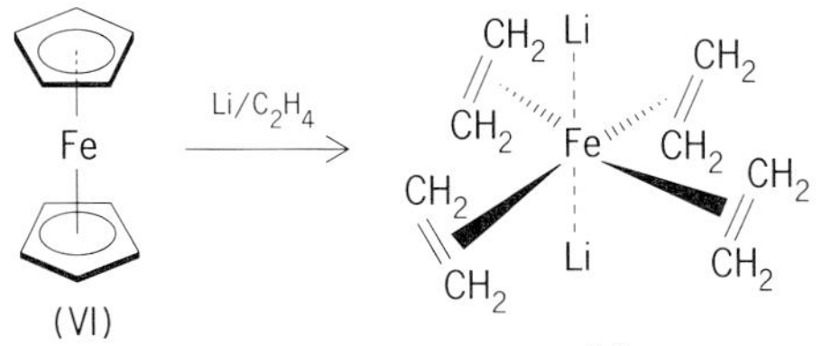

The most recent advances in the chemistry of highly reduced complexes formed by sets of identical ligands have involved the synthesis of the first isonitriles (polyisonitrile complexes of metals in negative oxidation states), for example, structure (VII) in reaction (8),

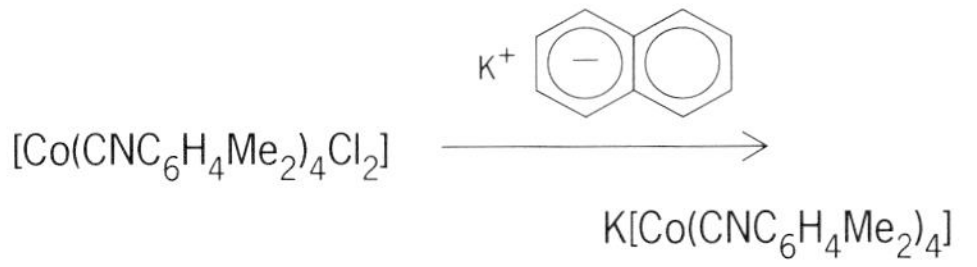

where Me = CH_3; and of arene sandwich compounds of metals in negative oxidation states, such as $[V(\eta^6\text{-}C_6H_6)_2]^-$, where V = vanadium.

Highly reduced complexes of σ-donors.

The stability of complexes of metals in negative oxidation states is ascribed to the π-accepting ability of ligands such as CO and alkenes, but there are a few highly reduced complexes in which the ligands do not have significant π-acceptor character. Examples include $[Co(PMe_3)_4]^-$ and $[Co(dmpe)_2]^-$, where Me = CH_3 and dmpe = bisdimethylphosphinoethane. The alkyl phosphane ligands should have little π-acceptor character, and although the molecules are very powerful reducing agents (for example, they reduce the fluoropolymers), their existence suggests that the factors that stabilize highly reduced complexes are not completely understood.

Reductively induced hapticity shifts.

Direct reduction of 18-electron precursors in which all of the ligands are π-acceptors and one is ligand bonded side-on through several carbon atoms can result in a change in the number of coordinated

Superreduced carbonylmetalates.

It has been determined that the conventional carbonylmetalates do not mark the limits of low-oxidation-state synthesis, as established by the preparation of the chromium group tetracarbonyl tetra-anions, as shown in reaction (6), where M = Cr, molybdenum (Mo), or W.

$$[M(CO)_4(Me_2NCH_2CH_2NMe_2)] \xrightarrow[\text{ammonia}]{\text{Na in liquid}} Na_4[M(CO)_4] \quad (6)$$

The structures of these materials are unknown, and they may contain coupled carbonyl ligands, but they unambiguously contain transition metals in the 4− oxidation state and represent the lowest-valent transition-metal complexes reported to date. Knowledge of superreduced carbonylmetalates (carbonylmetalates containing metals in oxidation states of 3− or 4−) has rapidly expanded, and examples are known for metals from the vanadium group through the cobalt group.

Highly reduced complexes with other π-acceptor ligands.

Early success in carbonylmetalate chemistry led to an assumption that CO was unique in its ability to stabilize low and ultralow oxidation states of transition metals, and the atoms that force the ligand to adopt an unusual co-ordination mode. Arene ligands, for example, can be forced to coordinate through 4 carbons only [reaction (9), where $KC_{10}H_8$ = potassium naphthalenide], and

π-cyclopentadienyl ligands can be forced to coordinate through 3 carbons only.

Reactivity of highly reduced complexes. This reactivity can involve electrophilic addition either to the metal or to the ligand.

Electrophilic addition to the metal. The most important use of highly reduced complexes to date is as intermediates in organic synthesis. In the case of the iron carbonyl complex $[Fe(CO)_4]^{2-}$, oxidative addition of a carbon-centered electrophile can be used to prepare highly reactive iron alkyls; and subsequent carbonyl insertion can modify the addend to give an acyl group, which can be removed to give a carbonylated organic product containing a carboxylic acid group (VIII), an acyl halide (IX), an unsymmetrical ketone (X), or an aldehyde (XI), depending on the conditions as shown in reactions (10), where R, R$'$ = alkyl group;

X = halogen; L = donor ligand; O_2 = dioxygen; and H^+ = hydrogen ion.

Carbonylmetalates are also inherently important in organometallic synthesis, and an elegant example involves sequential addition of an acyl halide (XII) and an electrophilic alkylating agent (XIII) to give a carbene complex (XIV), as shown in reaction (11), where Me = CH_3 and Ph = C_6H_5.

Electrophilic addition to the ligand. The extensive back-donation of electron density to ligand-based π-acceptor orbitals can induce unprecedented reactivity modes such as the addition of electrophiles

to η^4-arene ligands under mild conditions as shown in reaction (12), where CO_2 = carbon dioxide and Me_3SiCl = trimethylsilylchloride.

One of the most remarkable modifications of ligand reactivity in a highly reduced complex is the reversal of the polarity of the carbene ligand in dianionic carbene complexes. The unsaturated carbon in low-valent carbene complexes stabilized by heteroatomic substituents is typically electrophilic, but in carbene complexes of chromium in the 2− oxidation state [Cr(−2)] the unsaturated carbon is nucleophilic, as demonstrated by facile biscarboxylation, as shown in reaction (13),

where Ph = C_6H_5, CO_2 = carbon dioxide, H^+ = hydrogen ion, and CH_2N_2 = diazomethane.

For background information SEE CHEMICAL BOND-ING; COORDINATION CHEMISTRY; COORDINATION COM-PLEXES; INORGANIC CHEMISTRY; METALLOCENES in the McGraw-Hill Encyclopedia of Science & Technology.

N. John Cooper

Bibliography. H. Behrens, Four decades of metal carbonyl chemistry in liquid ammonia: Aspects and prospects, *Adv. Organomet. Chem.*, 18:1–53, 1980; J. P. Collman, Disodium tetracarbonylferrate—a transition metal analog of a Grignard reagent, *Acc. Chem. Res.*, 8:342–347, 1975; J. E. Ellis, Highly reduced metal carbonylanions: Synthesis, characterization, and chemical properties, *Adv. Organomet. Chem.*, 31:1–51, 1990; K. Jonas, Alkali metal derivatives of metal carbonyls, *Adv. Organomet. Chem.*, 18:97–152, 1981.

Particle accelerator

Particle accelerators are the largest tools that are constructed to learn about the world. A full mathematical theory (called for historical reasons the standard model of particle physics) describes the basic laws of nature. These laws have been tested in a variety of ways, and are known to describe phenomena well from distances of about 10^{-18} m (one-thousandth the diameter of a proton) out to the edge of the observable universe, an extraordinary range of about 10^{42}. In spite of this success, a number of clues indicate that this theory is only part of the story.

The Superconducting Supercollider (SSC), presently under construction, is the newest tool that is needed to search for a deeper understanding of nature. Until the end of the nineteenth century, scientists could use only light from the Sun or from lamps to look as deeply as possible into matter with microscopes. One result of quantum theory is that every particle behaves like a wave as well as a pointlike object, and the size of structure that can be probed by a wave is no smaller than its wavelength, about 10^{-7} m for light. Another result is that the wavelength of a particle decreases as particle energy increases, so that it is possible to probe more deeply with a more energetic probe. In 1897, electrons were accelerated to energies that did not occur in the natural world, and scientists began to study matter with artificial probes; they found atoms (10^{-10} m), nuclei (10^{-14} m), protons and neutrons (10^{-15} m), and, most recently, quarks and leptons. The Supercollider can be thought of as the world's largest microscope. It is essential to extending the present understanding of nature; no alternative facility can be guaranteed to provide the needed measurements.

SSC design. Einstein's equation, $E = mc^2$ (where E is energy, m is mass, and c is the speed of light), implies that energy can be converted into matter. Thus, if protons are accelerated until they have a very large energy and are then made to collide head-on, new particles with masses up to the limit allowed by this equation can be created and studied. That process is, in essence, what is done at any accelerator.

At the SSC two beams of protons will be accelerated in opposite directions, until they acquire an energy of 20 TeV, and then collided. The protons will attain about 10^{13} times the energy that an electron acquires from a flashlight battery, and about 20 times the energy of the next largest facility, the Tevatron collider now taking data at Fermi National Accelerator Laboratory (Fermilab) near Chicago. In order to understand how an accelerator works, two results from electromagnetism are need: (1) an electrically charged particle is accelerated by an electric field and acquires energy from it; and (2) the path of an electrically charged particle will curve in a magnetic field. In principle it would be possible to make do with only electric fields, but then the length of the accelerator would make it impossible to build. The solution was to use magnets to steer the particles around so that they traversed the same path repeatedly, each time with more energy. The larger the magnetic field that can be achieved, the greater the curvature, so the magnetic-field size determines the accelerator size.

The acceleration process starts by ionizing some hydrogen gas by using electrical discharges or some other method. An applied electric field then accelerates the positively charged protons one way and the electrons the other, so that they are separated and their energy is increased. The SSC will use the protons. In order to reduce costs and increase efficiency, a series of accelerators will be used; there will be five of increasing size for the SSC, each steering the beam into the next (**Fig. 1**). Finally, the two beams are made to collide in order to study what emerges.

The largest of the rings, where the protons will circle through two pipes in opposite directions, will be slightly oval-shaped (because of considerations concerning experimentation), with a circumference of about 54 mi (87 km). It will be in a tunnel typically about 100 ft (30 m) below ground and about 10 ft (3 m) in diameter. Every few miles around the ring will be access holes to power and refrigeration units and for service. Detectors will be placed in the experimental halls, where the beams will cross and the protons will collide. The SSC campus will have the office buildings, library, computing center, and so forth.

The SSC will use superconducting magnets to steer the protons around the rings for two main reasons. First, superconducting magnets can provide magnetic fields several times stronger than can ordinary iron-core magnets, so the size of the ring can be decreased and construction costs reduced. Second, they require considerably less power to do the same job than does a regular electromagnet, so operating costs are reduced. Since two previous accelerators (the Tevatron and HERA, which is being completed near Hamburg, Germany) have similar superconducting magnets, the technology is not likely to present any major problems. However, because the SSC will require about 20 times more magnets then these lower-energy colliders, the main engineering challenge will be to manufacture reliably this many magnets, each about 50 ft (15 m) long.

Physics motivation. All the extraordinary diversity of nature—including protons and neutrons, nuclei, atoms, molecules and all of chemistry and biol-

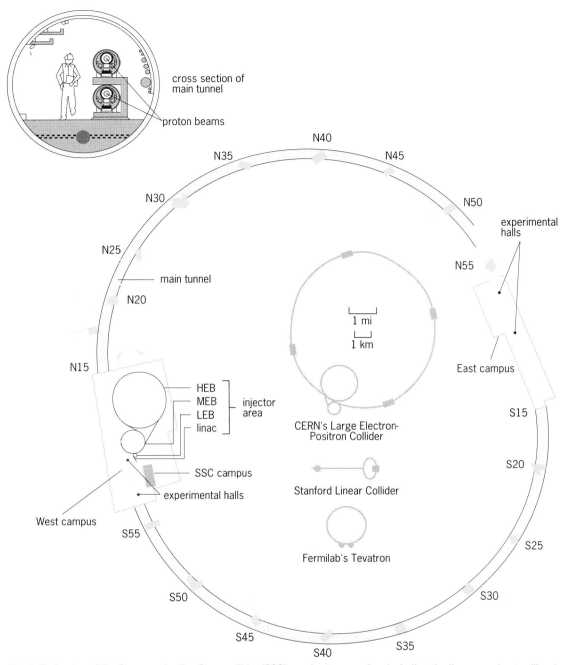

Fig. 1. Scale plan of the Superconducting Supercollider (SSC) accelerator complex, including the linear accelerator (linac) where the acceleration begins; the low-, medium-, and high-energy boosters (LEB, MEB, and HEB); and the main tunnel. Also shown to scale are the existing accelerators at CERN, Stanford, and Fermilab. Even numbers around the main tunnel designate exit and vent facilities, and odd numbers designate refrigeration and power-supply facilities. A cross section of the main tunnel, with the two rings of magnets carrying proton beams, is shown.

ogy, planets, stars, and galaxies, out to the edge of the observable universe, and back in time to within 10^{-15} s after the beginning of the universe—is now believed to be described by only four particles, which interact via three forces transmitted by the exchange of force quanta. The particles are two quarks, called up and down quarks, and two leptons, called the electron and the electron neutrino. The forces are the gravitational, strong, and electroweak forces. The carriers (quanta) of the forces are the gravitons (gravity), eight

gluons (the strong force), and four electroweak bosons (W^+, W^-, Z^0, and γ). All of these particles have been observed and studied in detail, except for individual gravitons whose interactions are too weak to allow the gravitons to be observed with present technology. One great achievement of the 1960s and 1970s was to show that the weak force and the electromagnetic force really were one unified electroweak force.

In addition, four more quarks and four more leptons have been produced at accelerators. (At present, the evidence for the top quark and for the tau neutrino is indirect but compelling.) However, they either live such a short time (10^{-6} s for the longest of the unstable ones) or interact so little that they do not play any direct role in the observable universe. It is not known why these additional quarks and leptons exist. One reason to build the SSC is to find out if there are any more quarks or leptons, and to gain insight into the puzzling role of the extra quarks and leptons in nature by probing matter more deeply. All these particles and the ways they interact to form nuclei, atoms, and matter are summarized in **Fig. 2**.

Because of the large cost of the SSC and the consequent need to curtail other particle physics because of limited funding, the United States particle physics community has devoted extensive efforts, with individual research and with five large 3-week workshops (held from 1982 to 1990), to deciding what facilities are needed to ensure scientific progress. Hundreds of worker-years have gone into thinking about these physics goals. The results can be divided into two areas, hypothetical new physics and Higgs-sector physics.

Hypothetical new physics. Often in the past the most important discovery at a new accelerator has been unanticipated, even when the accelerator was built to answer definite questions. However, at the SSC it is less likely that entirely unanticipated phenomena will occur, because never before has such a good description of nature been available. Many hypothetical new kinds of particles and forces have been considered. New heavier quarks, leptons, or force quanta could appear as experiments are performed at the SSC. In general, the SSC will extend the knowledge of the existence of such objects by about a factor of 20 over the Tevatron, and by a factor of at least $2\frac{1}{2}$ over the Large Hadron Collider, whose construction is being considered as a collaborative effort of European countries. All such searches are open-ended. Perhaps no new objects of these kinds will be found, so it will be much more likely that the known ones really are the only building blocks of matter.

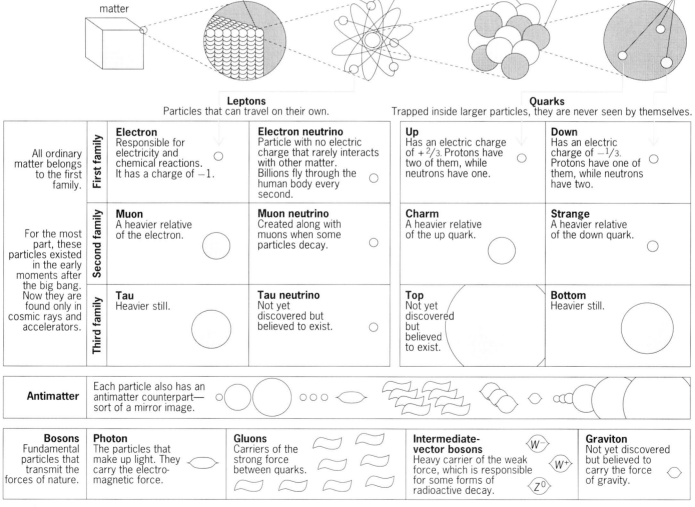

Fig. 2. Summary of the particles and forces that make up the current description of nature. The way in which matter is built up is also shown schematically.

There is one kind of hypothetical new physics, called supersymmetry, whose existence most experts feel is particularly well motivated. In addition, something is known about the masses that must be produced to decide whether it exists. It basically states that the underlying theory has a new symmetry requiring that a partner exist for each known quark, lepton, and force quantum. There are some very tentative indications that the present theory works a little better if these partners exist. The motivations for the existence of supersymmetry disapper if the partners are too heavy, and the SSC is powerful enough either to find the partners or to prove they do not exist. An important aspect is that many of the leading ideas to extend the present theory require that supersymmetry be valid in nature. If it is not, new approaches must be invented. Thus, this is one area where the SSC must produce important consequences, no matter what form the natural laws take.

Higgs-sector physics. This area is another where it is known that the SSC will be able to produce important results, without yet knowing what they will be. The mathematical theory of quarks and leptons interacting by the exchange of force quanta (the standard model) was originally formulated as a theory where all the particles are massless, but in reality, particles do have mass. In the 1960s, a method that kept the theory consistent and allowed the masses to be introduced was invented by using the so-called Higgs mechanism. Ba-

sically, the entire universe is imagined to be pervaded by a Higgs field (analogous to an electromagnetic field). Any particle that interacts with the Higgs field acquires mass. While this mechanism works mathematically, it is not known if the mechanism operates in reality. Two kinds of tests can be carried out at the SSC to answer this question. First, if the Higgs field exists, it will have quanta just as any field does, and they can be detected. If they are not detected, there is no Higgs field. Second, candidates for some of those quanta are already known to exist as parts of the W and Z bosons, but for technical reasons these bosons cannot be studied to learn Higgs-sector physics unless they can be collided at the energies planned for the SSC; then it is possible to find out if they behave as quanta of a Higgs field or something else.

Detectors. In order to achieve the physics goals of the SSC, it is necessary to observe and study the many particles that emerge from the collisions. Because the experiments are building on decades of study, only about one collision in 10^8 will be likely to reveal new physics. The search for these few significant collisions must be carried out electronically by the detector in a fraction of a second. Each quark, lepton, or force quanta either is long-lived enough to enter the detector or decays into particles that are. By building a detector with several components, each of which senses a different property of the particles, it is possible to get enough information to reconstruct what

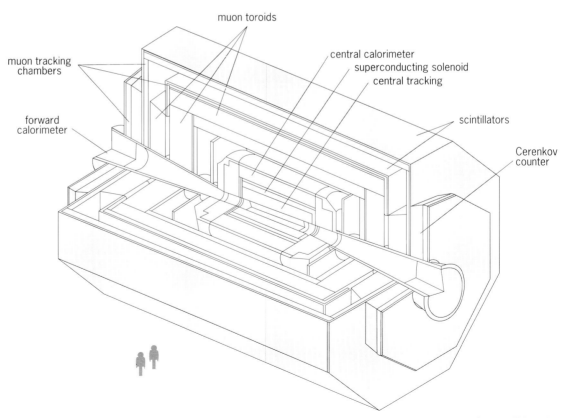

Fig. 3. Cutaway schematic view of a particle detector of the type that will be used at the Superconducting Supercollider. The beams enter from opposite directions along the cylindrical axis and collide. Different elements are combined to ensure that all the information needed to interpret a complicated event is collected.

occurred in the main collision. This ability is based on years of learning at several accelerators how each particle will interact. **Figure 3** shows schematically how a detector is constructed from components, each designed to detect certain particles and measure their energy and path.

Eventually the SSC will probably have six detectors, each emphasizing different goals. When the SSC begins operation, probably two somewhat different and complementary large detectors will be in place, and perhaps a third special-purpose one. Each of the two large detectors will require about 500 physicists to design and build it, because of the many complicated and subtle things it must do, at the limit of modern capabilities. The groups to build the detectors have begun work; it will take at least until 1998 to finish them. The detectors are expensive because there is only one of each kind, with almost all components containing new and sophisticated electronics or materials.

For background information SEE ELECTROWEAK IN-TERACTION; FUNDAMENTAL INTERACTIONS; HIGGS BOSON; LEPTON; PARTICLE ACCELERATOR; PARTICLE DETECTOR; QUANTUM MECHANICS; QUARKS; STANDARD MODEL; SU-PERCONDUCTING DEVICES; SUPERSYMMETRY in the McGraw-Hill Encyclopedia of Science & Technology.

Gordon L. Kane

Bibliography. P. C. W. Davies, *The Forces of Nature*, 2d ed., 1986; G. L. Kane, *Modern Elementary Particle Physics*, 1987; G. 't Hooft, Gauge theories of the forces between elementary particles, *Sci. Amer.*, 242(6):104–138, June 1980; Universities Research Association, *To the Heart of Matter: The SuperConducting SuperCollider*, 1989.

Patent

The first section of this article discusses plant variety protection through certificates and patents, and the second discusses microbiological patents. Since genetic engineering has added new dimensions to the ability to make genetically stable changes within and among species, the need has increased for plant breeders and microbiologists to protect their inventions and for companies to profit from their investments. Patents are legal documents that grant inventors exclusive commercial rights to their inventions for a fixed period of time. The patent system was designed to allow publication of technical advancements but to protect the discoverer's exploitation of the development without competition from others.

Plant Variety Protection

In 1790 the U.S. Congress passed the first General (Utility) Patent Law. This law did not originally include protection of plants, because of perceived inability to describe life-forms in sufficient detail to meet the disclosure requirement of the patent law, and because of the belief that plants were simply "products of nature." Later rulings made protection possible. In 1930, Congress enacted the Plant Patent Act, which covered varieties of asexually reproduced plant cuttings, bulbs,

and spores, but not tubers. Seeds were excluded because of the concern that varieties reproduced by seed were not stable because of their heterozygosity and would not produce uniform offspring.

In 1970 Congress passed the Plant Variety Protection Act to provide protection of sexually produced varieties. Most of the food-producing crops fall under this act. Passage of this act was supported by the seed trade because of its interest in a mechanism to protect intellectual property.

Protecting intellectual property. The three patent acts have allowed developers of varieties of asexually and sexually propagated varieties to seek protection of intellectual property. However, the documentation required under each legislative act is different. To obtain a Utility or General Patent, documentation is required to prove that the variety is useful (it must serve some specific purpose), novel (it must not be previously known to exist in that form in nature), and unobvious (it must be sufficiently different from previous discoveries so as not to be obvious at the time of the invention to someone having ordinary skill in the art of plant selection and evaluation). A complete written description that enables one skilled in the practice of plant selection and evaluation to reproduce the invention is also required. A Utility Patent excludes others from making, using, or selling the variety without compensating the developer.

To obtain a Plant Patent under the Plant Patent Act requires a description of the new variety as complete as possible, documenting that it is different from other, previously described varieties. The developer is exempt from the requirements that the variety be useful, novel, and unobvious. Since passage of the Plant Patent Act, over 6000 plant patents have been issued by the U.S. Department of Commerce.

The Plant Variety Protection Act provides patentlike protection through the issuance of a certificate. The certificate holder has the right to market the variety and collect royalties for 18 years. A variety may be protected if it is novel. To meet the novelty requirement, it must be distinct, uniform, and stable. If a variety differs from all previously developed varieties by one or more morphological, physiological, or other characteristics, it is considered distinct. Hybrids, which are protected through control of inbred lines, are excluded. Inbred lines can be protected under the act. Under the Plant Variety Protection Act, a protected variety may be used in research to develop other varieties without permission of the certificate holder. This act also permits a farmer to save seed for personal use and to sell seed to another farmer; this provision is commonly called the farmer saved seed provision. There are no similar exemptions in patent laws. However, there is a court-decreed exemption in patent law referred to as experimental use that is similar in concept to the research exemption. By 1991 about 2400 certificates had been issued, with about 600 pending by the Plant Variety Protection Office of the U.S. Department of Agriculture.

Issues. Since the passage of the Plant Variety Protection Act of 1970 and the utilization of genetic engineering techniques in plant improvement, a number of issues have arisen. With the ability to protect varieties under the Plant Variety Protection Act, the Patent Office interpreted the intent of Congress that life-forms not be provided protection under the Utility Patent provision. However, the Supreme Court held that the fact that the claimed invention encompassed living material did not preclude Utility Patent protection. The Patent and Trademark Office ruled in 1985 that it is permissible to patent any seed that meets legal requirements. These two decisions provided the framework for granting Utility Patents on varieties. To date, Utility Patents have not been used extensively to protect varieties, primarily because of the difficulty of meeting the requirement that a variety be unobvious.

The primary source of variety protection has been through the Plant Variety Protection Act. The requirements of distinctness, stability, and uniformity are not difficult to meet for most crop species. Also, the plant breeding community has not seriously objected to the research exemption. However, the flagrant abuse of the farmer saved seed exemption and the difficulty in prosecuting offenders have caused concern by private companies. Large volumes of seed of protected varieties have been sold in the seed trade as "variety not stated," so that the ability of a company to detect infringement, and hence to obtain a return on its investment, is limited. If this loophole is not corrected through an amendment to the act, the companies will reduce their investments in plant breeding or use the Utility Patent option. The use of the certified seed option by the applicant provides some protection in that the variety can be sold only by variety name as a class of certified seed.

The emergence of genetic engineering as a tool in plant breeding has raised issues relating to the novel and unobvious criteria of the Utility Patent. Does the rearrangement of genetic material within a crop species qualify for a patent? How much change is required; that is, what "minimum distance" from another variety is necessary for a variety to be considered distinct? Does the isolation and identification of a genetic sequence meet the requirements for a patent? If so, what impact does that have on others working on the same trait but a different sequence? If a trait or genetic sequence is already present in nature, is it unobvious and novel? In cases where genetic material is moved among unrelated organisms, it seems more apparent that something novel has been created, and the requirements under the Utility Patent are thus more easily met. Many of these questions will be answered by the legal process.

Ethical concerns. Several ethical concerns that were raised during consideration of the Plant Variety Protection Act still exist. Fears were originally raised that the act would create monopolies that would control germplasm and the food crops. These monopolies would then profit by escalating food prices. It was also feared that the seed industry would be dominated by multinational corporations, small companies would cease to exist, and the exchange of germplasm would be adversely affected. Since the passage of the Plant Variety Protection Act, there has been an increase in industry-developed varieties that are equal to or superior to varieties that have been released by the public sector. For example, private soybean research has increased fivefold. Some restriction of germplasm exchange has occurred, but companies have also contributed seeds of protected varieties to national germplasm collection centers. Multinational corporations have become dominant in the development of varieties. However, competition has moderated price increases.

Concerns have also been raised about the use of genetic engineering concepts in variety development; about the introduction of pest resistance that might be transferred to weeds, or biological products that might mutate and become pests; and about the insertion into crops of pesticide-resistant genes that link a herbicide to a specific variety (many feel that this step would require pesticide use in crop production). The ultimate impact of the Plant Variety Protection Act is yet to be determined.

Variety protection and the emergence of genetic engineering have had a dramatic impact on the release of varieties by universities. Drastic reductions have occurred in the practical, field-related breeding programs in universities. These programs have been replaced by genetic engineering research. Many university geneticists are supported in part by industry, and proprietary rights are assigned to the company providing support. Where universities have the option, many are obtaining some form of protection for their varieties and germplasm. The reduction of breeding efforts has also impacted the evaluation and maintenance of germplasm collections and the training of plant breeders. At the same time, the industry has expanded breeding programs to incorporate genetically engineered traits into new varieties. Therefore, farmers will become more dependent upon varieties produced by private enterprise. This outlook supports the need for revising the farmer saved seed section of the Plant Variety Protection Act to ensure a strong private sector in plant breeding.

International protection. A number of countries have come together to form the International Union for the Protection of Plant Varieties (UPOV). There are currently 17 member countries, including most western European nations, Japan, the Union of South Africa, Israel, New Zealand, Canada, the United States, and Australia. The UPOV Convention works toward producing uniform plant variety protection laws among member nations and protecting the rights of plant breeders. The United States is represented at meetings by an official of the Patent and Trademark Office. At a conference in March 1991, the rules under which the UPOV Convention operates were revised substantially. The revision included broadening the scope of protection to lengthen the term of protection and to require national treatment of convention decisions. For the United States to comply with provisions of the convention and thus remain a member, the Plant Variety Protection Act will have to be amended. One of the key amendments will be to the section that deals with the farmer saved seed option, since under the rules adopted by the UPOV Convention selling the progeny of a protected variety with permission of the owner is not permitted.

In countries where there are no intellectual property rights, there is no protection, or protection is gained through the control of inbred or breeding lines or through trade secrets. The goal of UPOV member countries is to encourage countries without plant protection to develop variety protection laws and become members.   *B. E. Caldwell*

Microbiological Patents

As with any patent, microbiological patents must comply with four major criteria; novelty, inventiveness, utility, and disclosure.

1. Novelty. In order to be patented, a microorganism or its products must teach something new with no prior disclosure of the product or process to anyone "skilled in the art." A detailed literature search for prior knowledge is essential to satisfy this requirement. One type of patent application that must be carefully examined is that disclosing the discovery of a new organism in nature. This discovery may sometimes be considered merely a natural phenomenon or a product of nature. The United States courts have ruled that a microorganism isolated as a pure culture is a "manufactured" product and therefore patentable under certain circumstances. Another concern in obtaining a patent is when to disclose the discovery. The United States allows an inventor to publish the discovery up to 1 year before a patent application is filed, but, after this, the published material is considered "prior knowledge" and a patent cannot be obtained. In many other countries, publication invalidates the patent application.

2. Inventiveness. A degree of invention or original concept must be shown, and the discovery must not be obvious to a worker in that area of science.

3. Utility. The discovery must have a practical use, and some countries require that it be industrially important.

4. Disclosure. This written report gives anyone skilled in the area of discovery the ability to reproduce the stated results.

U.S. Patent Law. Microbiological inventions are a unique type of invention. This was underscored in 1949, when the U.S. Patent and Trademark Office implemented a requirement that cultures be deposited in conjuction with patent applications concerning microbiological inventions. The reasoning was that for chemical, electrical, or mechanical patents, a diagram or formula can sufficiently describe the invention, whereas in a microbiological patent, illustrations and narrative descriptions are generally inadequate to define the microorganism used and inadequate to comply with the requirement for a full and complete disclosure. In keeping with U.S. Patent Law, the subject strain of the patent application must be deposited with a culture collection that is to maintain this strain for at least the life of the patent (17 years). The depositor has the option of making the patent culture freely available from the date of deposit or requesting that it not be

distributed until issuance of the patent. Other industrial countries have implemented similar requirements, and many have also established a national patent culture depositary.

Budapest Treaty. Signed in 1977 and ratified in 1980, the treaty is a significant document for biotechnology because it allows a single deposit to be made in an approved culture collection called an International Depositary Authority, even if patent applications are being filed in more than one country. Before the treaty was in effect, strains of microorganisms had to be deposited in all countries for which patent applications were made. Ratification of the Budapest Treaty has had a major impact on simplifying the requirements for culture deposits. The ratifying countries are Belgium, Spain, the former Soviet Union, Bulgaria, France, Germany, Hungary, Japan, the United Kingdom, Switzerland, Liechtenstein, Republic of the Philippines, Sweden, Austria, Denmark, Finland, Norway, Italy, the Netherlands, Australia, Korea, Czechoslovakia, and the United States.

In order for a culture collection to be recognized as an International Depositary Authority under the Budapest Treaty, an application of request must be made to the World Intellectual Property Office in Geneva, Switzerland. The culture collection is required to maintain high standards of expertise in culture handling and preservation, as well as to maintain accurate records and to give the depositor the assurance that all records are kept secret. Some collections may elect to accept only certain types of microorganisms. A culture collection is granted International Depositary Authority status when it meets the necessary requirements and when the country in which it is located recommends such recognition and assures that the culture collection will comply with all requirements.

Under the Budapest Treaty, both the depositor and the International Depositary Authority have certain obligations. The first obligation is the shipment of the microorganism to the International Depositary Authority. Shipments of plant and animal pathogens are regulated by the U.S. Department of Agriculture.

When an International Depositary Authority receives a strain, it has the responsibility to complete the deposit form and designate the strain with a date of deposit and an accession number. Sufficient time is needed for the International Depositary Authority to determine viability of the patent culture following receipt. If the strain is not viable, the date of deposit will be delayed until a viable culture is received. The curator then determines the best way to preserve the culture. Following this step, the International Depositary Authority issues a receipt and performs periodic viability testing.

The depositor is required to provide the International Depositary Authority only with limited information concerning the nature of the deposit, such as name of the organism, strain number, and growth requirements. At no time is the depositor required or requested to give the nature of the activity for which the patent is sought or the country in which the application is being made. The International Depositary Authority contracts to maintain the strain for a minimum of 30 years. The depositor must also maintain the strain for the full term of the agreement, in case the International Depositary Authority cannot provide a sample (for example, the strain has died). Microorganisms deposited in accordance with the Budapest Treaty can be distributed only under the conditions of the treaty.

The deposit status of old strains, even those deposited before the Budapest Treaty became effective, can be changed to come under the jurisdiction of the treaty without redepositing the microorganism.

For background information SEE BREEDING (PLANT); GENETIC ENGINEERING; PATENT in the McGraw-Hill Encyclopedia of Science & Technology.

J. L. Swezey; C. P. Kurtzman

Bibliography. I. J. Bousfield, *Living Resources for Biotechnology*, 1988; B. E. Caldwell et al., *Intellectual Property Rights Associated with Plants*, Amer. Soc. Agron. Spec. Publ. 53, 1989; R. S. Crespi, *Patenting in the Biological Sciences*, 1982; I. P. Cooper, *Biotechnology and the Law*, 1982; W. Lesser, Patenting seeds in the United States of America: What to expect, *Ind. Prop.*, pp. 360–367, September 1986; R. Saliwanchik, *Legal Protection for Microbiological and Genetic Engineering Inventions*, 1982; U.S. Plant Variety Protection Act, 84 Stat 1542, 7USC 2321 et seq., December 24, 1970; S. B. Williams, Jr., Protection of plant varieties and parts as intellectual property, *Science*, 225:18–23, 1984.

Pesticide

Many uses of pesticides permit them to move rapidly through the soil. The major example of routine pesticide application to soils is the use of herbicides to control weeds in gardens, lawns, and agricultural fields. Insecticides to control insect pests, either applied directly to the soil or washed into it after application to plants or structures, also enter the soil. Other examples include nematicides added to soil for nematode (worm) control, acaricides for mite control, and fungicides for combatting infestations of fungi such as molds, mushrooms, and liverworts. Beyond these intentional applications, there are many cases in which pesticides are added to soil through improper disposal by farmers or homeowners, by accidental spills, or through the wash-up and cleaning of application equipment. Much research is being conducted into these issues, particularly the movement of pesticides in soil and the basic processes that influence it.

The relative importance of the various dissipation pathways of a pesticide is usually assessed from (1) its solubility, (2) vapor density, (3) sorption (usually as a distribution coefficient), and (4) persistence (usually in terms of half-life). Solubility and vapor density are chemical properties that can be measured easily and accurately; the others depend upon soil and environmental conditions and are less certain.

Once introduced to the soil system, a pesticide is subjected to physical, chemical, and microbiological processes that determine how far it moves and how long it persists. These processes occur at different rates in different soils, and they always vary according to the specific pesticide. The characteristics of the processes can change in response to temperature, rainfall, sunlight, and the amount and type of organic materials in the soil. The unique molecular structure of each pesticide also confounds the situation; each pesticide exhibits a unique responsiveness to environmental conditions. Thus prediction of the movement of a pesticide through soil is a challenging exercise. In each of the areas outlined below, numerous research efforts are under way.

Water transport. The more water that is added to the soil, the higher the probability that a pesticide will be moved beyond the point of application. Downward displacement of the chemical is known as leaching. Initially this vertical redistribution results in the chemical being moved deeper into the root zone; if water is applied beyond that needed for plant growth, it will leach the pesticide to deeper depth and, perhaps, eventually to groundwater. Development of methods to predict this movement will require more information concerning the relationship among several variables.

The concentration of soil-applied pesticide in the water that infiltrates the soil depends upon the solubility of the pesticide, its rate of dissolution, and the infiltration rate. Thus a very soluble pesticide subjected to a very low rainfall rate enters the profile as a narrow band of concentrated solution. As the infiltration rate increases and solubility or rate of dissolution decreases (both of which may be influenced by product formulation), the concentration in the infiltrating water may be lower. High intensities of rainfall may give rise to immediate transport of pesticide to deeper depths through larger cracks and apertures (termed macropores). Several rain events may be required to leach all of the pesticide into the profile. Surface runoff and erosion may remove some of the pesticide from the field and cause contamination of surface waters. Leaching in macropores is less if the chemical is physically incorporated into the soil before water is applied. The relationship between the physical structure of the soil, the occurrence and distribution of macropores, and the effect upon pesticide movement constitute one of the most active areas of pesticide–soil research.

Sorption. Once in the soil, the pesticide can react with organic or mineral surfaces in a process known as sorption. This process decreases the concentration of the pesticide in the soil water. It is not necessarily an instantaneous process, since some of the sorption sites may not be easily accessible. The partitioning of pesticide between sorbed phase and solution phase is often assumed to be at equilibrium in a soil and to be characterized by a distribution coefficient. However, it is not necessarily correct to assume that at equilibrium the ratio of the amount sorbed to the amount remaining in solution equals the distribution coefficient, since it is doubtful that equilibrium is ever attained. It is now believed that a combination of kinetic and equilibrium processes determines the distribution of pesticide between sorbed and solution phases for at least a few pesticides, and possibly many of them.

Additionally, the contents of soil water are changing continuously, owing to infiltration, evaporation, and plant transpiration, with cycles of wetting and drying that are particularly extreme near the soil surface. The pesticide is subjected to corresponding cycles of adsorption (increased interaction with soil surfaces) and desorption (decreased interaction with soil surfaces). Given these processes, the distribution coefficient is now recognized to be only an approximate index of the relative sorption of different chemicals.

After each infiltration event, there is some downward movement of water that may transport dissolved pesticide to greater depths. Movement of soil water within a field varies in rate and direction and with time and depth. This variation in movement tends to disperse a well-defined band of chemical; some moves through larger-than-average soil pores in advance of the maximum pesticide concentration, and some is retarded through movement in smaller, more tortuous pores. Although sorption attenuates the movement of some of the pesticide relative to that of water, it does not prevent the movement completely; even low pesticide concentrations in moving water may present a potential environmental hazard. If water flows predominantly through larger pores, it may not transport pesticide residing in stagnant regions, so that the pesticide in larger pores may move farther than expected. Usually, such convective transport exceeds that by molecular diffusion, but movement into and out of soil aggregates may be controlled by diffusion processes. Additionally, a small proportion of the chemical in the gaseous phase can increase overall diffusive transport, because gaseous-phase diffusion is more rapid than diffusion in solution. Some of the chemical may also be absorbed by plants or adsorbed onto root surfaces.

Microbial action. Pesticides are also subject to transformation and degradation by soil bacteria. Pesticide persistence is historically characterized by a half-life, which is the time it takes for an initial amount of pesticide to be degraded to one-half its original mass. However, it is now well recognized that bacteria do not degrade a given chemical at a steady rate. The persistence of a pesticide depends not only on how fast it moves but also on how quickly it degrades. There is an optimum range of temperature and water content above and below which degradation rates decrease. In addition, growth of a bacterial population capable of degrading a particular chemical may be slow. This distribution of bacteria varies with depth, with fewer active cells being available in dry soil at the surface or at depths that have lower contents of organic matter. There may be aggregates into which bacteria cannot penetrate but into which the chemical may diffuse. Beyond that, the applied pesticide may transform to a degraded product that may be more toxic and more mobile than the parent compound. Further definition of the relationship between pesticide type, environmental variables, and the half-life is needed.

Volatilization. Volatilization (pesticide evaporation) is at a maximum immediately after application, especially if the chemical is applied to the soil or plant surfaces. Three factors determine how much

will be lost; the vapor density of the substance, the atmospheric conditions, and the time that elapses before incorporation into the profile by infiltration or cultivation. At no other time after application will the opportunity for volatilization to the atmosphere be shorter or the vapor density gradients be higher. Once the bulk of the chemical is in the soil, volatilization slows dramatically.

Process interaction. The net result of these interacting processes is that the distribution of chemical in the profile becomes more diffuse and solution concentrations decrease with time. Sporadic events of high rainfall or the cumulative effects of frequent events of rainfall may leach chemicals below the root zone, after which degradation is much slower and sorption is less. The major components of mass balance involving pesticide fate are thus determined in the biologically and chemically active root zone, which fortunately is also the region most amenable to control by good management. Biodegradation of the parent chemical and of any toxic degradation products is the ideal way in which a pesticide is removed from soil. Less desirable pathways are leaching and volatilization.

Areas of research. The complex and interactive nature of pesticide reactions and movements in soil presents soil scientists with the challenge of predicting potential environmental hazards by using data that may be variable, estimated, or measured on an inappropriate scale. Further refinements in the methods of integrating all these processes into a logical assessment of pesticide movement constitute an important area of research activity.

One of the most promising areas of pesticide research is the development of simulation models that can integrate the above suite of processes into a comprehensive assessment of pesticide movement. Current versions of these models range from complex mathematical representations of basic processes to much more simplified conceptualizations of soil–pesticide systems. Yet a variety of unknown factors remain. For example, it is presently unclear how to represent macropore transport in these models. It is also unclear to what degree basic soil and pesticide properties need to be known in order to use a model for accurate prediction of pesticide movement. Such issues will remain a focus of research efforts for years to come.

For background information SEE AGRICULTURAL CHEMISTRY; PESTICIDE; SOIL in the McGraw-Hill Encyclopedia of Science & Technology.

R. J. Wagenet

Bibliography. H. H. Cheng et al. (eds.), *Pesticides in the Soil Environment: Processes, Impacts, and Modeling*, Soil Science Society of America, 1990; W. J. Lyman et al. (eds.), *Handbook of Chemical Property Estimation Methods: Environmental Behavior of Organic Compounds*, 1982; W. A. Jury, D. D. Focht, and W. J. Farmer, Evaluation of pesticide groundwater pollution potential from standard indices of soil-chemical adsorption and biodegradation, *J. Environ. Qual.*, 16:422–428, 1987; M. R. Overcash and J. M. Davidson (eds.), *Environmental Impact of Nonpoint Source Pollution*, 1980.

Petroleum processing

Coking has been the traditional upgrading process available to refiners for converting heavy petroleum residues to lighter liquid products and salable petroleum coke. Recent advances in coking have involved improvements in yields, mechanical reliability, safety, and environmental aspects. In specialty cokes, there have also been improvements in quality. In the coking process, petroleum residues are heated to approximately 900–950°F (482–510°C), the range in which they undergo thermal decomposition to materials boiling below 1100°F (593°C) and to petroleum coke, similar in many respects to coal.

Fuel-grade coke. The production of fuel-grade coke has increased worldwide. This trend is driven by price differentials between light and heavy crude oils as well as by international commercial agreements. Much fuel-grade coke is shot coke, which naturally forms in small spheres and is less desirable than standard grades of coke because it is harder to grind. The market for fuel coke is reasonably stable, and includes cement manufacturers, power companies, and, increasingly, facilities using cogeneration. Fuel coke is classified predominantly by sulfur content and the Hardgrove grindability index.

Feedstock pretreatment. Feedstock pretreatment for coker operations is receiving closer scrutiny by refiners for a number of reasons. Since sodium is an accelerant of furnace coking, good desalter operations are necessary to hold the concentration of sodium ions in the coker feed below threshold levels that would cause excessive coking in the heater; these threshold values may start as low as 15 parts per million by weight. Caustic injection downstream of the desalter is normally not desirable. Because gas oils that could be distilled under vacuum will convert to produce coke, pretreatment by distillation operations at a high cut point under vacuum is employed to reduce coke production. In general, these cut points are greater than 1050°F (566°C). A new process that takes this concept one step further by integrating solvent deasphalting and delayed coking has been developed. The high effective cut point of the asphalt–solvent mix used as coker feed results in a dramatically reduced coke yield.

Hydrotreating residues for coker feed will result in increased yields of liquid products in the coker and lower sulfur levels for both liquids and coke. When used to produce anode-quality coke from otherwise unsuitable feeds, proper hydrotreating technology is critical, as is the selection of coker operating conditions. Hydrotreating that yields large quantities of light distillate products from petroleum residues may produce marginally stable coker feeds, which would have a deleterious effect on the length of the heater run when 100% hydrotreated feed is being processed.

Yield improvement. New cokers that will yield maximum amounts of liquid are being developed. They use low-pressure coking—about 15 lb/in.^{2} (103 kilopascals) gage pressure in the top of the coke drum. When downstream processing conditions permit, the design also includes provision for operation at ultralow recycle ratios. Several of these cokers now in operation use these features separately, and one coker uses them combined. These units are operated for maximum yield of liquids and minimum yield of coke. They produce increased yields of heavier distillate products at the expense of the unwanted coke yield and, to a lesser extent, gas and lighter liquids. Normally, this operation benefits the refinery by increasing the production of cracking feedstock. However, the heavy coker gas oil produced has a higher end point than would be the case in an operation that produces a higher yield of coke.

Effective operation of an ultralow-recycle coker requires control of the recycle, a condition produced by quenching and washing the product vapors in the upper portion of the coke drum. Traditional internal design of the fractionator that uses trayed wash zones is not effective because of low liquid rates that occur when operations are adjusted to achieve low recycle rates. The new design uses a wash zone spray chamber, which is effective at the low liquid rates.

Increased furnace-run length. Increases in the lengths of the coker furnace run are being achieved as a result of better design and some innovations, including the double-fired furnace. A technology known as on-line spalling to remove coke that builds up inside furnace tubes over a period of time has also improved operating-run lengths, with some refiners reporting more than a 2-year run on overfired furnaces. On-line spalling does not require that a heater be shut down, so there is minimal loss of unit production. The procedure provides for one furnace coil at a time to be decoked with steam or condensate while the other coils remain in service. The operation is accomplished by replacing the oil in the coil to be decoked with the spalling fluid and following specific operating guidelines.

Short-cycle operations. Cokers are inherently flexible in operation and able to achieve higher capacity by operating at short cycles. Most older cokers are now being operated on short cycles. In general, a moderate project to remove bottlenecks will permit a coker to operate on coking cycles as low as 16–18 h. Such short cycles are essential if the refiner needs to process heavier feeds or operate at higher capacity.

However, there generally is a loss of operating efficiency associated with the short-cycle operation. Typically the higher loads of vapor require that the unit be operated at higher pressure, and this results in a lower-percentage yield of liquid. This can be somewhat offset by reducing recycle, increasing the amount of steam injected into the heater, and raising the heater outlet temperature. Increased heater firing necessary to maintain levels of coke volatile content will shorten run lengths; on-line spalling solves this problem. Operations around the coke drum are adversely affected. Drum maintenance is increased and drum life is decreased. Higher vapor velocities in the drum will require increased frequency of overhead-line cleaning.

The decoking half of the overall cycle required to remove the coke product from the coke drum is more difficult to shorten. The time required for a typical decoking cycle is 17 h. Most short-cycle schedules are custom-fitted to the unique design features and operational requirements of a specific unit. In developing short cycles, particular care must be given to the direct water quench and heating operations in order to minimize the impact of stresses on the coke drum.

Automation of manual operations. Automation of traditionally manual coker operations is increasing. Techniques and devices include automated unbolting for removing the coke drum bottom flange; remotely operated switches; motor operators and interlocks located in piping necessary to isolate the coke drums; advanced instrumentation for automated quenching of the coke drums; top and bottom unheading devices that eliminate arduous efforts; and computer control of operations that are used to remove coke product regularly from the coke drums.

Environmental aspects. Concerns about the environment are important in refinery units. Two significant developments are the coker blowdown system and the totally enclosed coke handling and dewatering systems. In addition, environmentally desirable systems have been developed for disposal of refinery sludge.

Coker blowdown systems. These are used for recovery of unconverted materials known as wax tailings and of steam generated during the cooling of the coke drum. Modern blowdown systems are totally enclosed and vent noncondensible substances to flare (burning) or to systems using a compressor. Heavy hydrocarbons are recovered in a hot circulated scrubber (blowdown drum), with the steam passing overhead to a condenser and being recovered in a large settling drum.

Environmentally, the scheduled flaring of noncondensible materials by venting during the blowdown may not be desirable from the standpoint of overall refinery emissions. Since these materials are a mixture of hydrogen, methane, and some heavier material, they may have an economic value when recovered. The chief means of recovering these gases are the refinery flare-gas-compression system (a special compressor associated with the refinery combustible vent that recovers the vented gas to minimize losses due to flaring), recycling to the coke-product gas compressor, or a dedicated scavenging type of vent-gas compressor.

Totally enclosed coke handling. In response to stringent environmental regulations in some locations, totally enclosed systems for handling coke that minimize discharges of particulates have been developed. These designs employ enclosed dewatering bins to collect coke following the cutting. Two variations are in operation: the slurry pump system and the gravity discharge system.

Figure 1 depicts the slurry pump type of a totally enclosed dewatering-bin system. During coke cutting, the coke and water drop through a crusher into an enclosed sluice. There they are slurried with additional water; the slurry then flows to a slurry pump that transports the mixture to a dewatering bin. The coke settles in the dewatering bin, and the water is decanted and drained. The dewatered coke is then discharged to a conveyor.

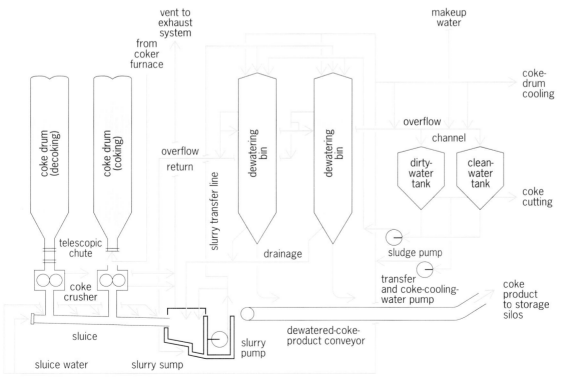

Fig. 1. Schematic diagram of a totally enclosed slurry dewatering-bin system. (*Foster Wheeler USA Corp.*)

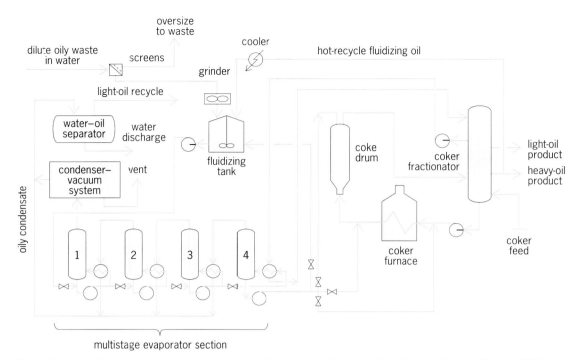

Fig. 2. Schematic diagram of a system for disposal of refinery sludge. (*Dehydro-Tech Corp. and Foster Wheeler USA Corp.*)

An alternative gravity-flow variation of the totally enclosed dewatering-bin system has been developed. The operation is similar to the slurry pump operation except that the coke drums are elevated to allow the coke and water from the cutting operation to drop directly from the crusher into the dewatering bin.

Refinery sludge disposal. Watery refinery sludges can be disposed of in the coker by injection during the initial stage of the coke-drum quench. However, this step does not convert the sludge. In response to the possible need to convert watery refinery sludges, two methods for disposing of and converting this material in a delayed coker have been developed.

The first system (**Fig. 2**) integrates a commercial system with a coker to dewater the sludge and then coke it with the normal coker feed. This process mixes the sludge with a fluidizing oil and evaporates the water under vacuum by using multiple stages. The oil–sludge mixture is converted by the coker to coke and clean products, and the fluidizing oil is recovered.

The second system provides dewatering of wet sludges in the coker blowdown system by utilizing the heat from cooling the coke. The dewatered sludge is mixed with the recovered heavy hydrocarbons in the blowdown drum. An auxiliary heater is provided for periods when there is little or no heat available from coke cooling. The dewatered sludge mixture is converted to coke, gas, and clean liquids by injecting it in the normal coker feed.

For background information *SEE AIR-POLLUTION CONTROL; COKING (PETROLEUM); PETROLEUM PROCESSING* in the McGraw-Hill Encyclopedia of Science & Technology.

Michael J. McGrath; John D. Elliot

Bibliography. R. DeBiase and J. D. Elliott, Recent trends in delayed coking, *Oil Gas J.*, April 19, 1982, or *Hydrocarb. Proc.*, May 1982; J. D. Elliott, Improve coker performance, *Hydrocarb. Proc.*, November 1991; J. D. Elliott, Modern delayed coking: Design and operating considerations for maximum liquid yields, *Oil Gas J.*, February 4, 1991; R. Meyers (ed.), *Petroleum Processing Handbook*, 1985; Research Association for Residual Oil Processing (Japan), *Heavy Oil Processing Handbook*, 1991.

Phloem

Phloem unloading is the removal of photosynthetic assimilates from the phloem (the long-distance food-transport tissue in plants) and their deposition into sinks. A sink is any plant area where there is extensive and immediate use of the assimilates, such as a growing root or flower, or where the assimilates are stored for later use, such as a developing fruit or tuber. Sources are areas where the assimilates are loaded into the phloem, such as mature leaves and storage organs that mobilize assimilates for export to sinks. Thus, assimilates are loaded into the phloem at a source and unloaded at a sink. Phloem loading results in the growth and accumulation of plant tissue or food substances.

Phloem anatomy. Phloem consists primarily of sieve elements and parenchyma cells, and sometimes includes supportive fibers. The two types of sieve elements are sieve-tube members and sieve cells. Sieve-tube members, which occur only in flowering plants, have larger pores in their end walls than in their lateral walls. In sieve cells, which occur in non-flowering plants such as ferns and conifers, all of the pores are small. Dissolved substances flow through the interconnecting pores from one sieve element to the next over relatively long distances. Connections to parenchymatic elements have pores on the sieve-element side and smaller openings called plasmodesmata on the parenchyma-cell side (**Fig. 1**). The parenchyma cells help regulate, load, and unload the sieve elements. In flowering plants, some of these cells attain a high degree of specialization (dense cytoplasm, prominent nuclei, and numerous mitochondria) and are termed companion cells. These cells are very closely associated with the sieve-tube members, which lack nuclei and most other cell components. The two are often considered as a physiological unit, the sieve tube–companion cell complex.

Composition of phloem sap. The primary food substance carried in the phloem of most plants is sucrose, which is the first nonphosphorylated product of photosynthesis in the leaf. Sucrose is electroneutral, relatively inert, and highly soluble in water, and thus ideal for transport without being metabolized along the way. It also provides the carbon skeleton for the synthesis of all organic compounds in the plant. A small percentage of flowering plants transport other sugars as a primary food source (for example, raffinose and stachyose), but these sugars generally have the same properties as sucrose. Amino acids, mineral ions, hormones, and water are also unloaded at sinks.

Unloading pathway. The actual removal of assimilates from the long-distance transport system occurs at the boundary of the sieve elements, but the concept of phloem unloading also includes short-distance transport and deposit into sink tissues. Assimilates may cover this short distance by moving symplastically, that is, from cell to cell via interconnecting plasmodesmata; or they may leave the symplast and move apoplastically, that is, in the space outside the collective plasmalemma of the interconnected cells (Fig. 1). This space, the apoplast, is made up of cell walls and intercellular spaces. Assimilates usually move from the

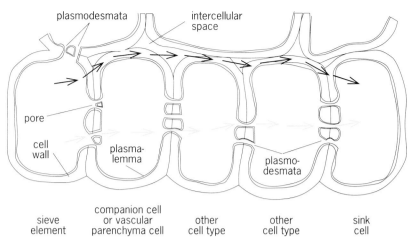

Fig. 1. Transverse section through cells involved in phloem unloading. Black arrows indicate direction of flow in the apoplastic pathway, and colored arrows indicate direction of flow in the symplastic pathway.

apoplast into the symplast at the sink where they will be metabolized or stored. Unloaded substances can sometimes move directly into an adjacent sink cell, such as occurs in potato. However, there are usually several cell layers to traverse (see **table**); and even in potatoes, assimilates may move through companion cells or vascular parenchyma cells before deposit into a storage cell.

Assimilates moving in seeds and grains must cross cell layers (see "Corn grain" in the table) and a symplastic barrier. The unloading maternal tissues, consisting of the seed coat and nucellus (and in the case of grains, also the fruit wall), lack plasmodesmatal connections to the endosperm. Therefore, assimilates must move into the apoplast at the interface between the nucellus and the endosperm, whereupon they are absorbed into the symplast of the endosperm (**Fig. 2**). Plasmodesmata are also absent between the endosperm and the embryo, so assimilates approaching this inter-

Pathways and events in phloem unloading in various types of sink tissues

Sink organ	Cell layers between sieve elements and sink tissue	Apoplastic versus symplastic pathway	Hydrolysis of sucrose	Principal sink tissue	Sink metabolic result
Sugarcane stem	Vascular parenchyma or companion cell, sclerenchyma sheath cells	Unloading into apoplast, transport to sink in symplast	Some hydrolyzed, most unaltered	Storge parenchyma	Storage as sucrose
Potato tuber	None, or vascular parenchyma or companion cells	Symplastic unloading and transport to sink	Unaltered	Storage parenchyma	Storage as starch
Abutilon flower at time of pollination	Vascular parenchyma or companion cells, nectariferous cells, neck cells	Unloading into apoplast, transport to sink in symplast	One-half unaltered, one-half hydrolyzed	Stalk cells of nectary hairs	Secretion of nectar containing about equal parts of sucrose, fructose, and glucose
Corn grain	Vascular parenchyma or companion cells, placenta-chalaza, endosperm transfer cells	Unloading via symplast, release into apoplast, uptake into symplast	Hydrolyzed	Endosperm	Storage of starch
Sugarbeet young leaves	Companion cells, parenchyma cells	Symplastic unloading and transport to sink	Hydrolyzed	Young leaf tissues	Metabolized for energy and structural components needed for growth

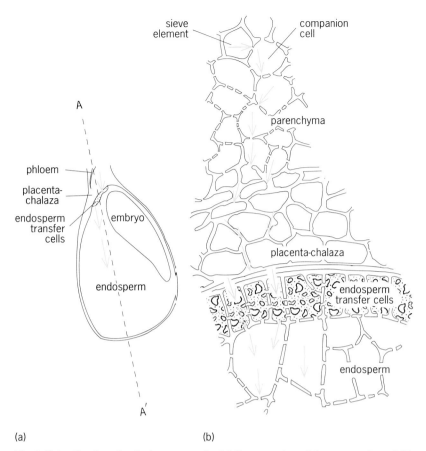

(a)   (b)

Fig. 2. Unloading in a developing corn grain. (*a*) Cross section of the corn grain and (*b*) an enlarged section through plane AA'. Arrows show direction of unloading from sieve elements to endosperm. (*After J. H. Thorne, Phloem unloading of C and N assimilates in developing seeds, Annu. Rev. Plant Physiol., 36:317–343, 1985*)

face must again enter the apoplast before absorption by the developing embryo.

Apoplastic barriers also occur in some sink regions. In mature surgarcane stems, apoplastic flow from vascular bundles is prevented by thick-walled, lignified sclerenchyma cells, the outermost of which are also suberized (**Fig. 3**). Sucrose may be unloaded into the apoplast of the phloem and then taken up by vascular parenchyma cells lining the inner side of the sheath. It must then pass symplastically through the sheath via plasmodesmata to the storage tissue, where some of it escapes to the cell walls and intercellular spaces that constitute a second apoplastic compartment.

Chemical change during unloading. Unloaded sucrose and other substances may cross the distance to the principal sink tissues unchanged, or they may be chemically altered along the way. Sucrose, for instance, is sometimes hydrolyzed, that is, split into glucose and fructose (see table). These two hexoses may remain separate, or they may be resynthesized into sucrose before or during storage. Hydrolysis is thought to occur in the apoplast by action of the enzyme invertase, but the site of this event is difficult to pinpoint because in many cases it is not known precisely where sucrose first enters the apoplast. Indeed, much of the evidence relating to apoplastic or symplas-

tic pathways and hydrolysis is currently circumstantial or indirect.

Models of apoplastic unloading. Virtually any living plant cell can actively absorb sucrose across its plasmalemma from the apoplast. Therefore, any model of unloading from the sieve element–companion cell complex has to be reconciled with immediate reloading. The simplest model is the pump–leak system, in which passive diffusion of sucrose down a concentration gradient from the sieve element–companion cell complex to the apoplast is the leak and active reloading of the complex is the pump. This system would work only if the sucrose concentration in the complex significantly exceeds that of the apoplast and reloading is inhibited. A second model is facilitated unloading, in which a sucrose-proton carrier molecule moves sucrose to the apoplast through inhibition of active transport of protons. A third model is active unloading, once favored but now regarded as unlikely. As indicated earlier, any of these models may operate at a distance of several cells from the apoplast in the immediate vicinity of the sieve element–companion cell complex.

Models of symplastic unloading. Unloading via the symplast could be by either diffusion or

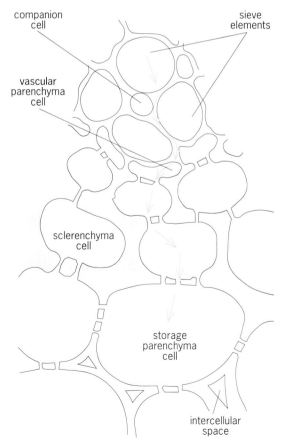

Fig. 3. Cross section of a portion of a vascular bundle and surrounding tissue in a sugarcane stem showing unloading. Arrows show pathway of unloading from sieve elements to storage parenchyma. Shaded walls are impregnated with lignin and suberin, which restrict apoplastic flow.

mass flow, and passive or active. Transport through plasmodesmata has been considered a strictly passive process. However, the presence of an osmotic potential that is lower in some sink companion cells than in adjacent sieve elements suggests active symplastic transport to the companion cells. In general, though, symplastic flow out of sieve elements is probably passive, since the sucrose concentration in sink sieve elements probably exceeds that in the surrounding cell types.

Control of phloem unloading. Different sinks of a plant receive different amounts of assimilate from the sources. For instance, a rapidly growing cucumber receives 40% of the assimilate being fixed in mature leaves, while the roots and young leaves receive considerably less than in plants with no growing fruit. The main factors that determine how much assimilate is unloaded into a given sink are (1) the membrane unloading area and plasmodesmatal frequency in that area, (2) the rate of unloading per unit area, and (3) the spatial or biochemical isolation of assimilates once they are unloaded. In addition, sinks must be able to effectively lower the solute concentration in their sieve elements and thus establish a favorable concentration gradient between source and sink. With apoplastic models, the rate of unloading is controlled by the solute concentration in the sink apoplast, which in turn may be regulated by invertase hydrolysis of unloaded sucrose. A relatively high apoplastic solute concentration characterizes strong sinks, probably because it promotes both assimilate transport into the sink region and uptake into storage tissues. For symplastic models, the rate of unloading would depend on plasmodesmatal regulation of flow and specific mechanisms to prevent backflow into the sieve elements.

For background information SEE PHLOEM; PLANT TRANSPORT OF SOLUTES; PRIMARY VASCULAR SYSTEM (PLANT) in the McGraw-Hill Encyclopedia of Science & Technology.

David G. Fisher

Bibliography. K. J. Oparka, What is phloem unloading?, *Plant Physiol.*, 94:393–396, 1990; J. W. Patrick, Sieve element unloading: Cellular pathway, mechanism and control, *Physiologia Plantarum*, 78: 298–308, 1990; I. F. Wardlaw, The control of carbon partitioning in plants, *New Phytol.*, 116:341–381, 1990.

Photochemistry

Photoinduced intramolecular proton transfer is light-initiated transfer of a proton from one site in a molecule to another site. The driving force for photoinduced proton transfer stems from the strong connection between acid–base properties and electronic distribution, which in turn is dependent on electronic state. For example, the acidity constant (pK_a) of aromatic alcohols can increase by as much as five units on going from the ground electronic state to the first excited state. Photoinduced proton transfer imparts certain photophysical properties that make compounds participating in

this process commercially valuable as polymer photostabilizers and organic dyes. From a fundamental standpoint, compounds that are subject to photoinduced proton transfer can serve as excellent systems for accurately modeling the kinetics of ultrafast proton transfer reactions. This modeling is leading to new insights into the underlying mechanism of proton transfer, a reaction that is found in inorganic, organic, and biological chemistry.

One of the best-known photoinduced proton transfer systems is 3-hydroxyflavone. The stable ground-state (S_0) isomer (I) of this compound is the normal form. Photoexcitation of this form produces an electronically excited normal form that rapidly ($< 10^{-12}$ s) is converted to a tautomer form (II) of the excited electronic

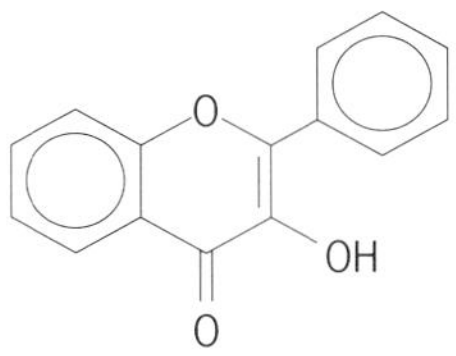

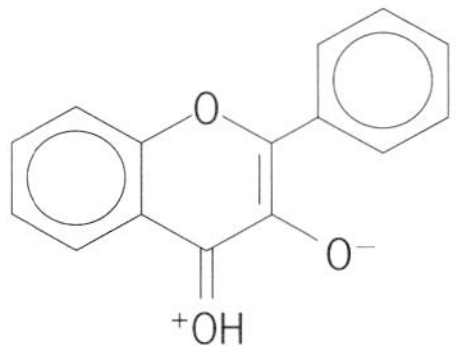

state (S_1) by intramolecular proton transfer. Due to the energetics of the proton transfer process in S_0 and S_1, the fluorescence ($S_1 \rightarrow S_0$) of the tautomer form is extraordinarily shifted to a longer wavelength from the normal absorption and emission band (**Fig. 1**). This large shift in energy between the absorption and emission spectra for excited-state intramolecular proton transfer (ESIPT) compounds is a hallmark of compounds that exhibit this type of reaction. Indeed, this large energy shift is a potentially useful property.

Practical utility. Commercially, the most important compound exhibiting excited-state intramolecular proton transfer is tinuvin [2-(2H-benzotriazole-2-yl)-p-cresol]. Tinuvin is added to polymers to protect them from photodamage, such as yellowing. A colorless compound, tinuvin absorbs strongly in the ultraviolet owing to the $S_0 \rightarrow S_1$ absorption of the normal form. A key process in the stabilizing mechanism of tinuvin is the ultrafast S_1 normal-to-tautomer interconversion, which rapidly and efficiently converts a substantial portion of the original ultraviolet photon energy to harmless heat. Thus, proton transfer plays a role in the important ability of this compound to protect polymers from ultraviolet photochemical damage.

Another potential practical use of excited-state intramolecular proton transfer is in the so-called proton transfer laser. Many laser media employ a four-level system to produce the necessary population inversion

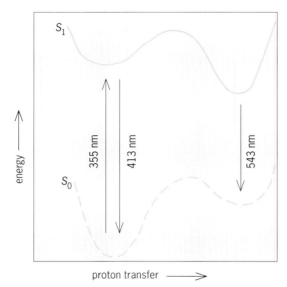

Fig. 1. Schematic representation of the energy dependence versus proton transfer reaction coordinate for 3-hydroxyflavone in the ground (S_0) and excited (S_1) electronic states. The wavelengths (in nanometers) are for the peaks of the absorption and emission bands.

for the optical amplification that is at the heart of laser action. The proton transfer scheme of Fig. 1 has been demonstrated to allow for efficient laser action in a conventional dye laser system over a broad range of wavelengths. The proton transfer laser is an extension of the more conventional dye laser, which involves two electronic states and a single isomer of the dye. The proton transfer offers a broader gain spectrum. The utility of proton transfer dyes in lasers is somewhat limited by deleterious photodegradation of the dyes, but this problem may be ultimately solved by synthesizing more stable derivatives of the most promising laser dyes that exhibit excited-state intramolecular proton transfer.

These compounds are also being used in research on collection of solar energy involving absorption by polymer sheets doped with fluorescent laser dyes. By a combination of photon absorption/reemission and total internal reflection, appropriately prepared polymer sheets can effectively collect photon energy all along the sheet by utilizing absorption by the dye used as a doping agent. The energy is then transmitted to the edge of the sheet by energy transfer within the sheet. The compounds exhibiting excited-state intramolecular proton transfer are particularly promising, because the wavelength of the tautomer emission is so strongly shifted from the normal absorption band that the photon energy can be transmitted over long distances without reabsorption in the sheet. A related technology is being developed for use in sophisticated imaging particle detectors for experiments using the Supercollider. Particles would be detected by scintillation in a large bundle of optical fibers made from polystyrene doped with dyes exhibiting excited-state intramolecular proton transfer. The polystyrene acts as a scintillator, converting the energy of the particles to an excitation in the fiber. The trapping of the excitation

by the molecular dopant produces a normal excited state and subsequently a tautomer excited state, which emits a photon in the wavelength region of tautomer emission. This emission can travel over long distances in the fiber, since the fiber absorbs only at much shorter wavelengths. Ultimately, the photon is converted into an electrical signal when it reaches the light detector at the end of the fiber. A system of many fibers monitored by an array of detectors and a computer system will be used to construct a multidimensional (dimensions of space, energy, time, and so forth) image of collision events in the Supercollider. *See* Particle accelerator.

Basic mechanism. Much of the research on photoinduced proton transfer has been concerned with fundamental aspects of this type of reaction. Two types of experiments on excited-state intramolecular proton transfer in the last decade have been particularly revealing. One of these, ultrafast time-resolved fluorescence spectroscopy, has led to important information on both the kinetics of the excited-state proton transfer and the dynamics of associated relaxation processes, such as vibrational relaxation. The other technique is vibrationally resolved electronic spectroscopy of these compounds in media where the spectra are resolved; the media include seeded supercooled beams, cryogenic matrices (solid samples) of noble gases, and cryogenic doped crystals. Spectroscopic measurements in these media have led to detailed information on the vibrational structure of S_1 and S_0 of the normal and tautomer forms.

A severe complication in studying the dynamic and static spectroscopy of compounds exhibiting excited-state intramolecular transfer is their enormous sensitivity to intermolecular interactions, for example, solute–solvent interactions in solution. Furthermore, some of these compounds are strong hydrogen-bond donors or acceptors, so that they are extremely sensitive to hydrogen-bonding impurities. Nevertheless, careful experimentation in many laboratories has characterized the photophysics of some of these compounds in isolated environments, in ordinary solution, involving weakly and strongly interacting solvents. Also, well-defined complexes of such molecules with a controllable number of solvent molecules have been studied by matrix isolation and in supercooled molecular beams.

For compounds such as methylsalicylate, 2-(2-hydroxyphenyl)benzothiazole, 3-hydroxyflavone, and others in the gas phase and in weakly interacting solvents, such as alkanes, the proton transfer (tautomer) emission appears within a very short time ($< 10^{-12}$ s) after excitation of the normal form. The excited-state intramolecular proton transfer reaction is this rapid even at cryogenic temperatures (4 K or $-453°$F). This behavior has been taken as evidence that the excited-state proton transfer process is not thermally activated, since lowering the temperature does not slow down the proton transfer rate. Some researchers have interpreted the rapid proton transfer rate as evidence that the S_1 interconversion of the normal form to the tautomer form does not have a significant energy barrier. However, this view is in contrast to quantum chemistry calculations on a number of these compounds, which predict nonzero barriers on the order of tens of kilojoules per mole. The apparent contradiction may be simply resolved by invoking quantum-mechanical proton tunneling through a small energy barrier to account for the rapid, low-temperature proton transfer rate of these compounds.

Proton tunneling in compounds that exhibit excited-state intramolecular proton transfer is manifested in a number of ways. The insight gained by these studies is beginning to have an impact on the general understanding of nuclear tunneling in proton transfer.

Frequency-resolved measurements. The strongest evidence for proton tunneling in compounds that exhibit intramolecular proton transfer and related molecules is found in frequency-resolved spectroscopy of the $S_0 \rightarrow S_1$ absorption and $S_1 \rightarrow S_0$ fluorescence bands of symmetric (zero driving force) and nearly symmetric intramolecular proton transfer compounds, such as the large-molecule aromatics tropolone and 9-hydroxyphenalenone and the intermediate-size system malonaldehyde (**Fig. 2a**). These compounds exhibit certain spectral features indicating that the vibrationless levels of S_0 and S_1 are split (by tens of wavenumbers) by proton tunneling between two isoenergic isomers separated by a higher-energy form. In other words, the splitting is due to proton tunneling in a symmetric double minimum potential.

For the symmetric intramolecular proton transfer compounds such as malonaldehyde, proton tunneling effectively delocalizes the proton between the two isomers. If a small amount of asymmetry is introduced, by the environment as with chemical derivatization,

the proton becomes localized in the more stable well, and the spectra no longer give direct information on the tunneling process. Thus, frequency-resolved measurements on compounds like 3-hydroxyflavone do not give information on the proton tunneling.

Theoretical studies on malonaldehyde demonstrate that intramolecular tunneling is a multidimensional process involving tunneling motion of all the atoms in the molecule. However, the motion of the proton that is transferring is especially favored because of its light mass. It is interesting that the calculated tunneling path differs significantly from the usual calculated reaction coordinate, which is a classical prediction for the proton transfer. Theory also shows that the tunneling splitting is increased greatly by exciting vibrational modes of the molecule that are coupled to the proton transfer process.

1-(Acylamino)anthraquinones. The class of compounds shown in Fig. 2b was recently observed to exhibit excited-state intramolecular proton transfer. These compounds offer a near-ideal case for studying the phenomenon. They exhibit dual emission, that is, both normal and tautomer emission bands, because there is only a slight energy difference between the normal and tautomer forms in S_1. Furthermore, the energy gap between the normal and tautomer forms can be adjusted by varying the functional group (R) and the solvent.

The pattern of the emission and absorption spectra strongly indicates that a small energy barrier separates the normal and tautomer forms in S_1. At cryogenic temperatures, some of the compounds in this class exhibit dual emission, which apparently is due to proton delocalization resulting from proton tunneling between the nearly isoenergetic S_1 normal and tautomer forms.

On the other hand, by varying the R group the proton can be localized in the normal or tautomer form. The amount of the tautomer emission is clearly correlated with the acidity of the donor nitrogen-hydrogen (N-H) group. Empirically, a plot of the log of the ratio of tautomer emission intensity to the normal emission intensity versus the pK_a of the corresponding anilide is linear. This is clear evidence that the normal tautomer equilibrium constant is sensitive to the driving force of the proton transfer.

Investigation of ultrafast processes occurring in the compounds shown in Fig. 2b reveals that the tautomer form is produced so rapidly in S_1 that the solvent motion cannot keep up with the proton transfer process. Apparently, the proton tunneling occurs faster than solvent motion for these compounds. Thus, proton tunneling is implicated in the fundamental and practical aspects of intramolecular photoinduced proton transfer.

For background information SEE LASER; MATRIX ISOLATION; MOLECULAR BEAMS; PHOTOCHEMISTRY; TAUTOMERISM; ULTRAFAST MOLECULAR PROCESSES in the McGraw-Hill Encyclopedia of Science & Technology.

Paul F. Barbara

Fig. 2. Intramolecular proton transfer reaction. (*a*) Malonaldehyde; r_1 represents the length of the bond between H_1 and O_2; r_2 represents the distance of the hydrogen bond between H_1 and O_3; r_3 represents the distance between the atoms O_2 and O_3. The subscripts on the atoms indicate position. (*b*) 1-(Acylamino)anthraquinones.

Bibliography. P. F. Barbara and H. P. Trommsdorff (eds.), Special Issue: Spectroscopy and Dynamics of Elementary Proton Transfer in Polyatomic Systems, *Chem. Phys.*, 136:153–360, 1989; P. F. Barbara, P. K. Walsh, and L. E. Brus, Picosecond kinetic and vibrationally resolved spectroscopic studies of intramolecular excited-state hydrogen atom transfer, *J. Phys. Chem.*, 93:29–34, 1989; E. M. Kosower and D. Huppert, Excited state electron and proton transfers, *Annu. Rev. Phys. Chem.*, 37:127–156, 1986; T. P. Smith et al., Excited state intramolecular proton transfer in 1-(acylamino)anthraquinones, *J. Amer. Chem. Soc.*, 113:4035-4036, 1991.

Planetary physics

Planetary rings have fascinated astronomers and physicists since Galileo first observed Saturn's rings in 1610. The interpretation of Galileo's observations raised many questions about the nature of planetary rings, as did the much more recent observations of rings around Uranus, Jupiter, and Neptune. Ring systems exhibit a diverse array of phenomena, and physicists from Galileo onward have met the challenge of understanding how fundamental physical processes can sculpt the often bizarre forms. The study of rings also provides information about processes that have formed the planets and satellites of the solar system.

Four unique ring systems. Although Saturn's rings were initially observed by their reflected sunlight with small telescopes, the present knowledge of ring systems has been gleaned through a much broader use of the electromagnetic spectrum, from ultraviolet to radio wavelengths. Also, spacecraft fly-bys, notably the *Voyager* missions, have provided information on the structure of these systems at over 10,000 times the spatial resolution possible with ground-based telescopes. The occultation technique, practiced both from spacecraft and from the Earth, provides extremely high spatial resolution. An occultation observation involves recording the intensity of the light from a star or the level of a radio signal from a spacecraft as a ring system passes between the source and the observer. The occultation technique is so potent, even when practiced from Earth, that two of the four ring systems were discovered with it.

Although the particles in all planetary ring systems obey the same physical laws, the four known ring systems appear strikingly different from one another.

Saturn. Visible from Earth with high-power binoculars, Saturn's rings are the most visually spectacular in the solar system. Broad and brilliant, the system is composed of numerous ringlets of various breadths and brightnesses, and is the most complex of the four systems (**Fig. 1**). The main components are denoted the A, B, and C rings, with A and B separated by a band visible from Earth and known as the Cassini division. These rings are extremely flat, with a vertical thickness about 10^{-6} of their diameter. The particles composing these rings are mainly centimeter-sized ice balls.

Uranus. Discovered by accident during a 1977 occultation of a star, these rings are narrow with well-defined boundaries and exhibit small deviations from perfect circles (**Fig. 2**). The average Uranian ring par-

Fig. 1. *Voyager 2* photograph of the lit face of Saturn's B ring, obtained on August 25, 1981, from a distance of 743,000 km (461,000 mi). It covers a range of about 6000 km (3700 mi) and shows the ring structure broken up into about 10 times more ringlets than was previously suspected. The narrowest features are due to a combination of differences in ring-particle number density and light-scattering properties. (*NASA*)

ticle is larger and much darker than that of Saturnian rings. Interspersed among the narrow rings is a tenuous sheet of small particles.

Jupiter. The rings of Jupiter, discovered during the *Voyager 2* fly-by in 1979, are faint and tenuous, in contrast to Saturn's bountiful rings. Broad ringlets with sharply defined boundaries characterize Jupiter's system. These rings are composed mainly of micrometer-sized particles, and *Voyager* images show three distinct bands.

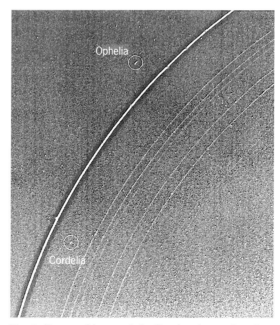

Fig. 2. *Voyager 2* image of the Uranian rings, taken January 21, 1986, at a distance of 4.1×10^6 km (2.5×10^6 mi) and resolution of about 36 km (22 mi). Two shepherd satellites associated with the rings, designated Cordelia and Ophelia, are visible on either side of the bright epsilon ring; all nine of the known Uranian rings are visible. (*NASA*)

Neptune. Neptune's rings were discovered with occultation observations in 1984. With Earth-based occultations, however, only sections of Neptune's rings are dense enough to be detectable, and are called ring-arcs. The *Voyager* spacecraft found that the rings completely encircle the planet but some sections are much denser than others (**Fig. 3**). These dense sections are the only parts of the rings that were detected from Earth. The rings appear bright in Fig. 3 because microscopic ring particles scatter sunlight toward the camera. The particle-size distribution in Neptune's rings is thus quite different from that of Uranus's rings, which contain few dust-size grains.

Early work on Saturn's rings. When Galileo observed Saturn's rings through his small and imperfect telescope in 1610, he interpreted them to be two large satellites of Saturn. In 1612, when the rings were edge-on to the Earth and therefore not visible, Galileo had no explanation for their apparent disappearance. The idea that the Saturnian rings were a set of satellites persisted until 1655, when C. Huygens applied a working theory of how satellites behave to his own observations of Saturn and determined that an inclined, rotating, solid, and relatively thick ring was orbiting Saturn.

Although there had been speculation that Saturn's ring was composed of particles, it was not until 1857 that J. C. Maxwell demonstrated that under newtonian physics even minutely subdivided rings could not exist: "The only system of rings which can exist is one composed of an indefinite number of unconnected particles revolving around the planet with different velocities according to their respective distances." Subsequently, spectroscopic data were obtained and indicated that the orbital velocities of the rings decreased with increasing distance from the planet, as expected for particulate rings. The opposite trend of velocity would have been observed for solid rings.

The extreme flatness of Saturn's rings can be understood in terms of what would be expected if a large collection of particles was orbiting a planet. If initially there were a spherical cloud of particles, collisions would occur, since the individual particle orbits would precess at different rates. These collisions would cause the ring to become flat. Next, spreading of rings would occur because particles closer to the planet would move faster than particles farther from the planet. The difference in particle speed, called keplerian shear, would cause collisions between the particles. These collisions would cause the faster particles to lose angular momentum and therefore to drop closer to the planet. The slower particles would tend to be bumped farther out from the planet. Thus, the rings would spread out in a thin sheet.

Another concept in the historical development of the understanding of Saturn's rings is the Roche limit. Within a radius $2.46\,R(\rho_p/\rho_o)$, where R is the radius of the planet, ρ_p the density of the planet, and ρ_o the density of the orbiting body, a fluid body would break up because of the tidal force from the planet. This tidal force increases with decreasing distance to

Fig. 3. *Voyager 2* image of Neptune's two main rings, about 53,000 km (33,000 mi) and 63,000 km (39,000 mi) from the center of the planet, seen backlit by the Sun. The image of the planet was greatly overexposed to capture detail in the rings. At the right in the outer ring is the main clumpy arc, composed of three features, each about 6–8° long, that were first detected from Earth-based occultations. (*NASA*)

the planet, so that even a solid body, if it was large enough, would break into pieces. Collisions between these pieces would carry the disintegration further, so that a ring of orbiting particles would be formed.

The physical effects just discussed are sufficient to understand what can be seen of Saturn's rings through an Earth-based telescope. However, in order to understand the results of the high-resolution observations of all the ring systems made from spacecraft and with Earth-based occultations, more subtle physical effects must be considered.

Gravitational forces from satellites. In analyzing the dynamics of satellites and ring particles around a planet, a useful approach is to think in terms of gains and losses of angular momentum by the satellites and particles. A force in the direction of motion will cause a particle or satellite to gain angular momentum: the semimajor axis of orbit will be increased, but in this larger orbit the velocity of the body will be less. The opposite happens when a force is applied that opposes a body's motion: the orbit becomes smaller and the body speeds up. The total angular momentum of a planet's rotation and the orbital motion of its rings and satellites is conserved, but angular momentum can be transferred between bodies through their gravitational interaction. An example of angular momentum transfer is seen in the tides raised on the Earth by the Moon: the Moon's orbit is expanding through its gain of angular momentum, while the Earth's rotation is gradually slowing down.

The synchronous orbit is that in which a satellite's angular motion exactly matches that of the surface of the planet. For orbits lying within the synchronous orbit, the satellite is moving faster than the tidal bulge it raises on the surface of the planet; hence its orbit evolves inward, until it gets far enough inside the Roche limit to break up or to impact the planet. The distance for this occurrence depends on the strength of the material composing the satellite. For orbits outside the synchronous orbit, the satellite's orbit will evolve outward until the planetary rotation is slowed enough to match the satellite's orbital period.

Gravitational forces from small satellites can affect the configuration of ring particles because, although satellite mass is small, the distances between the satellites and particles are small also. The most dramatic effect of small satellites on rings is the formation of narrow rings by so-called shepherding satellites. The first known narrow rings, those of Uranus, initially defied explanation. After all, ring particles should spread out through collisions, not clump into narrow rings. However, two nearby small satellites, one orbiting inside the ring and the other orbiting outside, can inhibit the spreading and produce narrow rings. The outer satellite gains angular momentum from the ring particles, while the inner satellite loses angular momentum to the ring particles; the net result is an effective repulsion of the ring particles by both satellites, and prevention of the spreading of the ring. (Two shepherd satellites are shown in Fig. 2.)

Ring particles whose orbital period is in an even-integer ratio with a satellite are said to be in resonance with the satellite. The persistent, periodic tugs from such a satellite can introduce eccentricities into the ring particle orbits, which lead to collisions with other particles until all particle orbits near the resonance are depleted.

Waves in a sheet of ring particles. A large disk of particles, such as Saturn's rings, exhibits collective behavior, analogous to the wave motions that can be induced in a thin sheet of elastic material. Two kinds of waves occur. First, there are wave motions within the plane of the ring particles. These are known as spiral density-waves, since they have a spiral structure when viewed from above, analogous to the spiral structure of some galaxies. A second type of particle wave produces motions perpendicular to the plane of the particles, known as bending waves, since they resemble waves induced in a flapping bed sheet. Both types of waves are excited at locations of resonances with satellites outside the rings. The spiral density-wave propagates outward from the resonance location, and the bending wave propagates toward the planet.

Behavior of small particles. Because of their much larger surface area in proportion to their mass, micrometer-sized particles are significantly affected by weak forces that depend on the particle area. Poynting-Robertson drag describes the loss of particle angular momentum in response to collisions with photons from the planet and the Sun. Plasma drag describes the loss of angular momentum from collisions with the charged particles of the plasma trapped by the planetary magnetic field. If a small ring particle acquires a charge, the influence of the planetary magnetic field can significantly affect its motion. The transient, dark "spokes" discovered by *Voyager* in Saturn's rings are believed to be primarily a charged-particle phenomenon.

Prospects. Since the mid-1970s there have been great strides in the understanding of the physics of planetary rings, but several important questions remain. For example, there is no generally accepted explanation for the sharp inner edges of Saturn's A, B, and C rings and the clumpiness of Neptune's ring-arcs. Also, the age of the rings remains unknown. It is not known whether the rings were formed concurrently with the planets themselves from particles inside the Roche limit that could not gravitationally bind into a satellite, or whether the ring systems have formed and dissipated several times during the 4.5×10^9-year history of the solar system through the breakup of bodies that orbited within the Roche limit.

At least one ring system can be expected to form around Mars within the next 5×10^7 years or so, when the orbit of its moon Phobos decays sufficiently for this satellite to break up from tidal forces. Neptune's ring system may become much more extensive when the orbit of Triton decays and this massive satellite meets the same fate. The time scale for this, however, will be much longer than for Phobos and is uncertain.

Two new spacecraft missions have been designed to study the ring systems of Jupiter and Saturn. The *Galileo* spacecraft will tour the Jovian system within a few years; and the *Cassini* mission, a joint venture for the National Aeronautics and Space Administration (NASA) and the European Space Agency (ESA), is scheduled to be launched for a tour of the Saturn system after the year 2000. Meanwhile, Earth-based occultation observations will continue to provide high-resolution data for the ring systems of Uranus, Saturn, and Neptune. These data alone may lead to answers to several of the dynamical questions about the rings.

For background information *SEE JUPITER; ROCHE LIMIT; SATURN; SPACE PROBE; URANUS* in the McGraw-Hill Encyclopedia of Science & Technology.

James L. Elliot; Lyn E. Elliot

Bibliography. J. K. Beatty and A. Chaiken (eds.), *The New Solar System*, 3d ed., 1990; J. Elliot and R. Kerr, *Rings: Discoveries from Galileo to Voyager*, 1984; R. Greenberg and A. Brahic (eds.), *Planetary Rings*, 1984.

Plant growth

For over a decade, scientists have been preoccupied by the observed increase in global carbon dioxide (CO_2) concentration. If such injection of CO_2 into the atmosphere continues, the CO_2 concentration will double by the middle of the twenty-first century. So far, the efforts of plant physiologists have largely focused on the direct responses of plants to CO_2 enrichment, and it is generally accepted that high CO_2 acts as a fertilizer for plant growth. The projected climate change associated with the global CO_2 increase, known as the greenhouse effect, has received less attention from plant physiologists. However, the existing information on plant responses to CO_2 and to temperature is now being used to focus on physiological phenomena likely to vary as a result of global climatic change.

Direct effects of CO_2. As early as the beginning of the twentieth century, agronomists became aware that plants in an atmosphere elevated in CO_2 grew faster and often had superior yields. Today there is ample evidence to support those findings, and the physiological mechanisms involved are well understood. In most plants, the first acceptor of CO_2 is an enzyme called ribulose bisphosphate carboxylase/oxygenase (RUBISCO). RUBISCO has a dual function: the carboxylase function, which incorporates CO_2 in the part of the Calvin cycle where carbon is transformed into carbohydrates (carbon fixation); and the oxygenase function, which is responsible for the release of carbon from the plant through photorespiration. A loss of as much as 30% of the carbon originally fixed by plants can be attributed to photorespiration. The two functions of RUBISCO are competing in the leaves for the available carbon. If the external CO_2 concentration is high, the carboxylase function is favored; therefore the loss of carbon through photorespiration is reduced. Furthermore, under enriched atmospheric CO_2 concentrations, more CO_2 is available for fixation by RUBISCO, so that photosynthetic rates are enhanced. When photosynthesis is measured at various concentrations of CO_2, the stimulation due to high CO_2 levels is on the order of 20–50%. This enhanced capacity to fix carbon increases growth rates, at least in the early stages of plant growth. The compilation of data obtained on the CO_2 response of 120 different species indicates that the average stimulation of vegetative growth due to elevated CO_2 is 33%. High CO_2 also has a strong impact on flowering and reproduction. Under enriched conditions, plants generally mature more rapidly than at ambient CO_2 levels. A review of observations on the yield of 56 species, most of them agricultural crops, suggests that doubling the atmospheric CO_2 concentration should result in a 32% increase in agricultural yield.

Effects of other environmental factors. These responses of plants to high CO_2 are closely linked with other environmental factors. For example, if nutrients such as nitrogen are limiting, the effect of high atmospheric CO_2 concentration becomes negligible. However, of all the environmental factors that can influence plant responses to elevated CO_2, temperature is the key one. It is believed that the rise of atmospheric CO_2 concentration will cause a global climatic change. The greenhouse effect is generally described as an increase in the global annual mean temperature of the order of 5°F (3°C). An examination of different climatic models shows that the projected temperature change will not spread uniformly over the globe. The predicted temperature variation is minimum in the tropics and increases with latitude. Maximum warming is expected in winter—climatic models predict as much as 29°F (16°C) increase in winter temperature in the Arctic compared with 3.6°F (2°C) near the Equator. Local scenarios of climate change at three points in the Northern Hemisphere—Mexico City, Mexico; Indianapolis, Indiana, United States; and White Horse, Yukon Territory, Canada (respectively located near 20°, 40°, and 60° latitude north)—have been used to translate the climatic predictions into the daily reality of weather fluctuations. For each of the sites, the monthly temperature increment was extracted from the zonal mean temperature of a general-circulation meteorological model that was published in 1986 and assumes a doubling in atmospheric CO_2. The derived temperature increments were then added to the 5-year average mean temperature from the meteorological stations at each location (**Fig. 1**). For each latitude, it is possible to observe a change in the length of the growing season. For example, in White Horse (Fig. 1a) the predicted monthly mean temperature in March (37°F or 3°C) is higher than the actual temperature for April (32°F or 0°C). The frost-free period is likely to increase by nearly 2 months at high latitude. In Indianapolis (Fig. 1b) the projected temperature never drops below 32°F (0°C). Climatologists also predict an increase in the frequency of extremely warm temperature events. In summary, with a doubling in atmospheric CO_2, plants are likely to be exposed to a longer growing season, milder winter, and more frequent heat stress.

Physiological responses of plants. In order to understand the physiological responses of plants to temperature under a CO_2-enriched atmosphere, both instantaneous and long-term photosynthetic carbon uptake patterns should be analyzed. Experimental temperature curves of photosynthesis and simulations using mathematical models have been used to compare the photosynthetic response to temperature under different atmospheric concentrations of CO_2. **Figure 2** shows the temperature response curves of the garden pea, *Pisum sativum*, measured under various CO_2 concentrations.

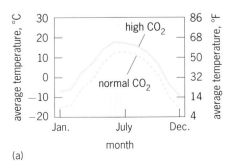
(a)

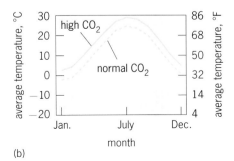
(b)

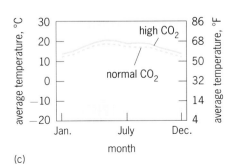
(c)

Fig. 1. Current and projected monthly mean temperature at high and normal CO_2 levels for three locations in North America: (*a*) White Horse, Yukon Territory, Canada (60°N), (*b*) Indianapolis, Indiana, United States (40°N), and (*c*) Mexico City, Mexico (20°N).

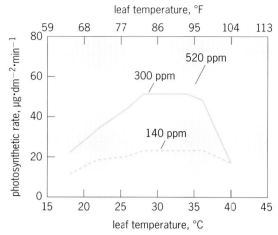

Fig. 2. Temperature response curves of *Pisum sativum* for photosynthesis at three atmospheric CO₂ concentrations. (*After E. O. Hellmuth, The effect of varying air-CO₂ level, leaf temperature, and illuminance on the CO₂ exchange of the dwarf pea, Pisum sativum L. var. Meteor, Photosynthetica, 5:190–194, 1971*)

Photosynthetic rate increases with increasing CO_2, while the curve sharpens. The change in temperature response curve can be interpreted in terms of the recognized limitations of photosynthesis. At low CO_2 concentration, temperature curves are broad and flat: photosynthesis is limited mainly by the activity of RUBISCO. In other words, because the supply of CO_2 is a limiting factor, photosynthesis is almost insensitive to the effect of temperature. At high CO_2 concentrations, the curves are sharp and steep, and reflect a much stronger photosynthetic response to temperature than at low CO_2 concentrations.

The plants used in the experiment described above were grown at ambient CO_2 levels. However, plants can acclimate to environmental conditions. Therefore, it is important to also analyze their physiological response when they are grown in a CO_2-enriched atmosphere. A theoretical model for the photosynthetic response of plants that have been grown in a CO_2-enriched atmosphere suggests that the optimal temperature for photosynthesis should be higher for plants grown at high CO_2 compared to that of plants grown at ambient CO_2. The model also predicts that the range of optimal temperature for photosynthesis should be narrower for plants grown at high CO_2. Only a few studies have tested this model.

The seeming contradictions in experimental evidence accumulated thus far suggest that the photosynthetic response to CO_2 and temperature may also depend upon genetic factors as well as other environmental factors. For radish, carrot, water fern, and water hyacinth grown under natural lighting, it appears that under low-temperature conditions (below 64°F or 18°C), growth is reduced by enriched CO_2. Conversely, it was shown that growth and survival of the tropical crop okra at low temperature is improved by enriched CO_2. Under normal CO_2 concentration, okra plants can grow at day/night air temperatures of 68/52°F (20/14°C). However, growth is satisfac-

tory if the atmospheric CO_2 concentration is doubled. Likewise, growth and yield of soybeans at low temperature are enhanced by a high CO_2 concentration.

Another species that has been shown to respond positively to the interaction of CO_2 and temperature is the weed barnyard grass, *Echinochloa crus-galli*. Barnyard grass is the only species for which physiological parameters have been scrutinized in an attempt to understand the mechanisms by which temperature and high CO_2 interact in the process of plant growth. In that study, plants were subjected to night chilling at a temperature of 45°F (7°C). The experiments were carried out at high and ambient CO_2 levels. Under ambient CO_2 concentration, night chilling reduces photosynthetic carbon uptake. The reduction of the photosynthetic rate is associated with a reduction in the activity of the photosynthetic enzymes. Once carbon is photosynthetically fixed by a plant, it is translocated from the leaves to the rest of the plant. In barnyard grass, as in many other plant species, carbon translocation is impaired by chilling. Following one night of

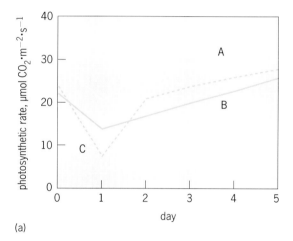

(a)

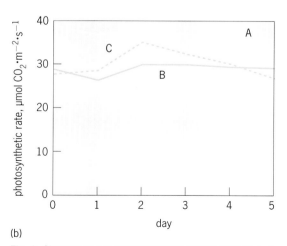

(b)

Fig. 3. Changes in net photosynthesis before (day 0) and after one night of chilling at 45°F (7°C) for plants of *Echinochloa crus-galli* from Quebec, Canada (A), North Carolina (B), and Mississippi (C). Plants were grown and measured under (*a*) ambient (350 microliters/liter) and (*b*) enriched (675 μl/liter) CO₂ concentrations.

chilling, yellow or white streaks appear on the leaves of barnyard grass. These streaks are believed to be due to a disruption of the chloroplasts themselves. Chilling of barnyard grass under an enriched atmospheric CO_2 concentration altered the responses of all the physiological parameters analyzed. The most striking result was for photosynthetic rates (**Fig. 3**): whereas at ambient CO_2 concentration photosynthesis was reduced by as much as 70%, rates remained unchanged after one night of chilling under enriched CO_2 conditions. As in the case of photosynthesis, the low-temperature damages were lessened if the chilling took place in an atmosphere high in CO_2. Following a low-temperature stress, the activity of two of the primary enzymes involved in CO_2 fixation (phosphoenolpyruvate carboxylase and the NADH-malate dehydrogense) was less reduced under high CO_2 concentrations. In the case of barnyard grass, it was apparent that high CO_2 had a beneficial effect on various physiological traits linked to the metabolism of carbon.

Heat stress. A phenomenon correlated with the global climate warming is an increase in the frequency of high-temperature events. In higher plants, heat stress is due to a disruption in the functional integrity of the photosynthetic apparatus at the chloroplast level that leads to low photosynthetic efficiency. Many plants respond to heat stress by triggering physiological mechanisms, such as evaporative cooling, which result in a decrease in temperature. A large body of evidence shows that high atmospheric CO_2 generally results in stomatal closure. If stomates close, transpiration will decrease and, with it, the potential for evaporative cooling. The consequences of reduced transpiration on plant responses to heat stress under enriched CO_2 have not been fully documented. A comparative study of two vine species was conducted for an extended time at two temperature regimes, 77 and 104°F (25 and 40°C), and two CO_2 concentrations, high and ambient. The results indicated that CO_2-enriched plants become bigger and that transpiration at high CO_2 decreases dramatically. Preliminary results from another experiment suggest that the weed *Abutilon* is more sensitive to heat shock when it is grown under high CO_2.

Because temperature is important in determining species distribution, there is little doubt that if the rise of atmospheric CO_2 is associated with a new climatic pattern, major changes in community composition are likely to occur. These changes will depend on the interactive response of plants to CO_2 and temperature. A survey addressing the joint effect of CO_2 and temperature indicates that for plants grown under artificial conditions, CO_2 generally acts as a remedy to both high- and low-temperature damages. Conversely, reports of plants grown under natural conditions suggest that some species might become more sensitive to low-temperature damages at high CO_2. Likewise, the potential remains for plants to be more susceptible to heat stress under enriched CO_2 because of lower transpiration rates. The interplay of these responses will determine how natural and managed communities will cope with the correlated increase in CO_2 and in temperature.

For background information *SEE CLIMATIC CHANGE; GREENHOUSE EFFECT; PHOTORESPIRATION; PHOTOSYNTHESIS; PLANT GROWTH; PLANT RESPIRATION* in the McGraw-Hill Encyclopedia of Science & Technology.

Catherine Potvin

Bibliography. F. A. Bazzaz, The response of natural ecosystems to the rising global CO_2 levels, *Annu. Rev. Ecol. Syst.*, 21:167–196, 1990; S. B. Idso et al., Effects of atmospheric CO_2 enrichment on plant growth: The interactive role of air temperature, *Agr. Ecosys. Environ.*, 20:1–10, 1987; C. Potvin, Amelioration of chilling by CO_2 enrichment, *Physiol. Veg.*, 23:345–352, 1985; C. Potvin et al., Effect of low temperature on the photosynthetic metabolism of the C_4 grass *Echinochloa crus-galli*, *Oecologia*, 69:499–506, 1986.

Plant hormones

In higher plants, a complex system of hormones plays a critical role in controlling growth and development. Abscisic acid (ABA), gibberellins (GA), auxins, ethylene, and cytokinins are the five major classes of low-molecular-weight compounds that are collectively referred to as plant hormones or phytohormones. Exogenous application of plant hormones, either in their naturally occurring forms or as synthetic analogs, has been shown to influence a diverse set of growth processes in plants.

Recent research has focused on the use of plant hormone mutants in investigations, and on how conjugates of plant hormones control the hormone levels within plant tissues.

Plant Hormone Mutants

Physiological and biochemical studies have clearly established the importance of hormones in plant growth and development. However, such approaches have failed to elucidate the molecular mechanism of hormone action. Several problems are encountered in attempting to correlate an observed response in isolated tissues or intact plants with the levels of endogenous or applied hormones: It is technically difficult to measure trace amounts of active hormones in target tissues, and exogenously applied hormones may differ in their uptake, transport, or compartmentation. In addition, nearly all developmental processes in plants appear to involve two or more hormones interacting cooperatively or antagonistically.

An exciting alternative approach to the study of phytohormones is the use of single-gene mutants that are blocked in some step of hormone response or biosynthesis. A mutational approach to the complexities of plant hormone action offers several advantages, including the ability, in some plant species, to clone and characterize genes required for hormone processes.

Identifying mutants. One method of identifying phytohormone mutants has been to examine breeding populations of various species for individuals that display variation in growth and development. Such individuals may be mutants and may have defects that can be correlated with altered hormone biosynthesis or response. Another approach to the identification of hormone mutants is to directly screen mutagenized populations. A variety of mutagenic agents, including chemical mutagens and ionizing radiation, have been used to induce mutations. An alternative strategy for generating mutagenized populations is via insertional mutagenesis. Insertional mutagenesis may be carried out in some plant species by using transposable elements such as transposons. Transposons are mobile segments of deoxyribonucleic acid (DNA) that may cause a mutation by inserting into a particular gene. Another agent used for insertional mutagenesis in plants is the transferred DNA (T-DNA) of the bacterium *Agrobacterium tumefaciens*. When this bacterium infects certain plants, it transfers T-DNA, a segment of its own DNA, into the plant cells. The random insertion of T-DNA into the plant's nuclear DNA may cause a mutation when it integrates into a specific gene. Both transposons and T-DNA offer a means of tagging genes of interest, facilitating the molecular isolation of such genes.

Mutant groups. In general, phytohormone mutants can be divided into two groups. The first group consists of mutants that are defective in hormone biosynthesis and degradation. Such mutants have been used to study hormone metabolism and to examine the developmental consequences of altered hormone levels. For example, mutants of maize and pea that are deficient in gibberellins are characterized by a dwarf phenotype due to a reduction in stem elongation. These dwarf mutants can be restored to normal height by treatment with active gibberellins. Careful analyses of these mutants resulted in the characterization of the gibberellin biosynthetic pathway.

The second group of mutants consists of those that are altered in one or more hormone responses. One method of identifying such hormone-response mutants is to screen mutagenized populations for mutants that exhibit increased resistance to growth-inhibiting levels of applied hormone. Hormone-insensitive or -resistant mutants might be altered with respect to a hormone-binding protein, such as a receptor. Alternatively, these mutants could be defective at another step in the pathway that leads from hormone perception to developmental response (signal transduction pathway). A number of interesting hormone-response mutants are currently being analyzed.

Abscisic acid mutants. Physiological and genetic evidence in a variety of species suggests that abscisic acid is required for induction of seed dormancy. Abscisic acid also appears to play a role in several stress responses such as stomatal control during drought stress. Mutants that are insensitive to abscisic acid have been isolated in a number of species, including *Arabidopsis* and maize. In *Arabidopsis*, mutations in three different genes, *Abi1*, *Abi2*, and *Abi3*, result in plants that do not contain reduced endogenous abscisic acid levels but are much less responsive to applied abscisic acid than wild-type plants. The recessive mutant *abi3* is altered with respect to seed dormancy only, while the dominant mutant *abi1* and the reces-

sive mutant *abi2* are abnormal for all abscisic acid responses examined, including reduced seed dormancy and extreme wilting during water stress. Seeds of the *vp1* mutants of maize exhibit precocious germination (that is, germination that occurs before final seed maturation). This aspect of the *vp1* mutant phenotype is probably due to reduced sensitivity to abscisic acid in *vp1* embryos. The *Vp1* gene has been cloned through transposon tagging, so that the relationship between the *Vp1* gene product and abscisic acid sensitivity could be investigated.

Gibberellin mutants. A large number of gibberellin mutants have been identified in various species, including maize, pea, wheat, barley, tomato, and *Arabidopsis*. Loss of gibberellin function usually produces a dwarf phenotype, making it relatively easy to identify mutants in which gibberellin biosynthesis or response is disrupted. Unlike the majority of gibberellin mutants, dominant dwarf mutants of maize and *Arabidopsis* that do not respond to the application of exogenous gibberellin have been isolated. The dominant mutant *gai* of *Arabidopsis* and the dominant mutant *D8* of maize appear to contain levels of biologically active gibberellins that are similar to or higher than those found in wild-type plants. This result combined with a gibberellin-insensitive phenotype indicates that these mutants may be defective in perceiving or transducing the gibberellin signal.

Auxin mutants. Because high concentrations of auxin are toxic to seedlings and cultured cells, it has been possible to screen for mutants that are capable of normal development in the presence of growth-inhibiting levels of the hormone. All the auxin-resistant mutants that have been identified in several plant species exhibit a variety of developmental abnormalities consistent with defects in auxin-related processes.

In *Arabidopsis*, recessive mutations in either the *Axr1* or *Aux1* genes produce plants that display increased auxin resistance. However, the phenotype of the recessive mutants *axr1* and *aux1* differs significantly. The *axr1* mutants exhibit morphological alterations throughout the plant, including decreases in plant height, hypocotyl elongation, root gravitropism, and fertility. Differences in leaf morphology and an increase in shoot lateral branching are also observed in the *axr1* mutants (**Fig. 1**). In contrast, *aux1* mutants are morphologically normal except for a defect in root gravitropism. Physiological evidence suggests that auxin plays a central role in the response of plant organs to gravity. Other auxin-resistant mutants such as the dominant mutant *axr2* in *Arabidopsis* and the recessive mutant *dgt* in tomato possess defects in both shoot and root gravitropism. Biochemical studies have shown that the shoots of *dgt* mutants are deficient in an auxin-binding protein that may function as a receptor.

The dominant mutant *rac⁻* of tobacco was isolated by selecting cultured cell lines that were auxin-resistant. Plants regenerated from the *rac⁻* cell line were rootless and had to be maintained by grafting into normal plants. One of the earliest measurable responses of plant cells to auxin treatment is a rapid change in the movement of ions across cellular membranes. Because the *rac*⁻ mutant is altered in this early auxin response, the *rac*⁻ mutation may affect one of the first steps in auxin signal transduction.

Ethylene mutants. Many growth processes in plants are affected by increases in the level of the gaseous hormone ethylene. For example, hypocotyl elongation in wild-type seedlings grown in darkness is inhibited by exogenous ethylene. Several ethylene-resistant mutants in *Arabidopsis* have been isolated by screening mutagenized populations for seedlings that form an elongated hypocotyl in the presence of high concentrations of ethylene. Plants that are heterozygous or homozygous for the dominant *etr1* mutation are abnormal in a wide range of ethylene responses that are normally observed in wild-type plants. The lack of sensitivity to exogenous ethylene combined with reduced ethylene binding suggests that the *etr1* mutation may directly affect an ethylene receptor.

The recessive ethylene-resistant mutant, *ein2*, was identified in a similar screen. The *ein2* mutants are insensitive to ethylene in both hypocotyls and roots and, like *etr1* mutants, do not exhibit major differences in vegetative growth or flower development. Ethylene is also thought to be involved in the response of plants to environmental stresses such as wounding. Ethylene-resistant mutants should be useful in determining the role of ethylene during stress responses as well as during normal growth and development.

Interaction. Phytohormone interactions during plant development are well documented. A single hormone may regulate the level of a second hormone, or one hormone may alter the sensitivity of tissues to another hormone. In the case of auxin and cytokinin, their ratio appears to be more critical than the concentration of each hormone. Several hormone-resistant mutants of *Arabidopsis* and *Nicotiana* display insensitivity to more than one hormone. The phenotype of multiple hormone resistance indicates that one mutation can interfere with the ability of plants to respond to several hormones. Thus, it seems likely that some hormone signal transduction pathways may partially overlap. Further study of both hormone-deficient and hormone-response mutants will help elucidate the nature of hormone interactions as well as expand understanding of plant hormone action in general.   *Cynthia Lincoln*

Conjugates of Plant Hormones

The conjugation of plant hormones is a common phenomenon in the plant kingdom. A conjugated hormone is a hormone that is covalently linked to other compounds, commonly to sugars or amino acids. Conjugated hormones have been found in all plants examined, so they represent a significant part of hormone metabolism. The conjugation of plant hormones is believed to play a fundamental role in the regulation of the levels of active free hormones in the plant tissue. The plant can regulate the rate of its growth either by inactivating hormones through conjugation or by releasing them from conjugates. There are indications that the conjugation of hormones and their subsequent hydrolysis may be important in maintaining a steady-state concentration of the hormones in plant tissues that are responsive to environmental conditions. Alterations of the environmental conditions would affect the hormonal balance, which in turn would cause changes in the rate of plant growth.

The conjugates are also believed to be storage forms of the hormones from which the hormones are released at the proper time during plant development. Other functions proposed for conjugates include a role in the transport of hormones within the plant and the protection of hormones against enzymatic degradation. While it has also been suggested that the process of hormone conjugation results only in hormone deactivation, several universal features and the physiological behavior of conjugates of various groups of hormones suggest an important role for these compounds in the control of plant growth and development.

Fig. 1. Morphology of mutant *Arabidopsis* plants. (*a*) Wild-type (left) and *axr1* (right) rosettes. (*b*) Wild-type (left) and *axr1* (right) mature plants.

Auxin conjugates. All conjugates of the major auxin, indole-3-acetic acid (IAA), are derivatives of the IAA carboxyl group. In amide conjugates the carboxylic group of IAA is linked to amino acids by a peptide bond, and in ester conjugates the carboxylic group of IAA is linked either to a sugar or to a sugar alcohol, *myo*-inositol, by an ester bond.

It is well established that the IAA conjugates are most abundant in the seed storage tissue, endosperm or cotyledons, where they accumulate during seed maturation while the level of free IAA drops. At full maturity, they may constitute more than 80% of the total IAA pool in the seed.

The ester conjugates of IAA, especially IAA-*myo*-inositol, are transported in *Zea mays* (corn) from germinating kernels to the shoots, where they are hydrolyzed to provide the IAA required for the growth of the seedling. Thus, the ester conjugates of corn kernels play the role of the IAA storage in the seed, while serving as a transport form of IAA from the endosperm to the shoots.

IAA conjugates are also present in vegetative organs. Thus their biological role is not limited to the storage of the free hormone in the seed. They seem also to be involved in hormonal homeostasis.

It has also been suggested that IAA conjugates are involved in the protection of IAA against attack by degradative enzymes. Experiments with both IAA-*myo*-inositol and some amino acid IAA conjugates have shown these IAA conjugates to be stable in the presence of degradation peroxidases that would destroy free IAA.

Gibberellin conjugates. Most common gibberellin conjugates have a glucose linked to hydroxyl (gibberellin-*O*-glucosides) or carboxyl (gibberellin-*O*-glucosyl esters) groups of the gibberellin molecule (**Fig. 2**). Besides the glucosyl conjugates, acetyl derivatives of gibberellin as well as alkyl esters have been isolated from plants.

It has been frequently proposed that gibberellin conjugates serve as a pool from which free gibberellins can be liberated as needed. The formation of gibberellin conjugates would take place at times when the biologically active free gibberellins have to be removed, for example, during environmental stress. Again, seed development and germination are the most widely studied systems involving the possible reversible conjugation of gibberellins. During seed maturation, free-gibberellin levels decrease and gibberellin conjugates are formed. The release of free gibberellins from conjugates during germination has also been reported. That the ratio of free to conjugated gibberellins during germination of corn kernels can be influenced by light provides evidence of the involvement of plant hormone conjugates in the homeostatic regulation of hormone levels.

The significance of gibberellin conjugation in the regulation of gibberellin pools in plants seems to depend on the taxonomic identity of the plant. For example, in *Vicia faba*, where 2-*beta*-hydroxylation represents the main route of gibberellin degradation, the glucosylating activity is usually very low. In contrast, in *Phaseolus coccineus* the glucosylation of gibberellins is a main route of gibberellin deactivation, and 2-*beta*-hydroxylation seems to be unimportant. Thus, although the presence of gibberellin conjugates appears to be widespread in the plant kingdom, in some plants conjugation seems to be only a minor route of gibberellin deactivation.

The presence of gibberellin conjugates in the bleeding sap of some trees at the time preceding the bud break suggests the involvement of these compounds in ending bud dormancy and regulating the early growth of shoots. Their presence in the translocation ducts of trees may also indicate more general function of these compounds in the long-distance transport of gibberellins. The involvement of gibberellin conjugates in ending seed dormancy has also been postulated.

Cytokinin conjugates. There are a great variety of cytokinin conjugates. However, their physiological role is even less well defined than that of auxin and gibberellin conjugates. Sugar derivatives predominate among naturally occurring cytokinin conjugates. As shown in **Fig. 3**, sugar moieties can be attached at different positions on the cytokinin molecule. All cytokinin glucosides have their sugar linked to a nitrogen of the purine ring at position 3, 7, or 9 (cytokinin-*N*-glucosides) or to the oxygen of the isoprenoid side chain of the cytokinin molecule (cytokinin-*O*-glucosides). Cytokinin ribosides or ribotides (riboside-5'-phosphates) are also frequently found in many plants. In this case, the linkage with the cytokinin base occurs at the 9 position of the purine nitrogen. Additionally, a unique cytokinin conjugate in which alanine is attached to the purine-ring nitrogen of the free base at the 9 position has been isolated.

The biological activity of various cytokinin conjugates differs markedly. For example, cytokinin-*O*-glucosides exhibit a high activity in tissue culture and leaf senescence bioassays in contrast to the much more stable cytokinin-*N*-glucosides, which show very low activity in cytokinin bioassays. Metabolically stable cytokinin conjugates such as cytokinin-*N*-glucosides are considered to be deactivation products formed to lower the free-hormone level. Whether these very stable cytokinin conjugates are storage forms from which free hormone can be released at the proper time or the final products of hormone deactivation is not known. Better candidates for storage functions seem to be the

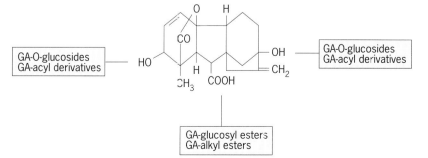

Fig. 2. Conjugation sites for gibberellin A₃ (GA₃) and types of conjugates formed.

cytokinin-*O*-glucosides, which, although much more stable than a free base, can readily liberate free cytokinin when they are applied to a plant tissue. The level of cytokinin-*O*-glucosides has also been observed to drop rapidly during the germination of *Z. mays* seeds, the lateral bud development in bean seedlings, and the breaking of dormancy in potato tubers. Ribosides and ribotides of cytokinins are postulated to be involved in the transport of these hormones within the plant tissue. Ribotide formation appears to be connected with cytokinin uptake and transport across cell membranes, while ribosides are proposed to be important in the translocation of cytokinins in the xylem. Free-base, riboside, and ribotide forms of cytokinins are interconvertible in the plant tissue. The enzymatic reactions interconverting these compounds may play a significant role in maintaining the proper level of active cytokinin in plants.

Abscisic acid conjugates. Abscisic acid (ABA), a major plant growth inhibitor, undergoes conjugation primarily with glucose, which results in the formation of glucosyl esters (carboxyl linkage) and *O*-glucosides (hydroxyl linkage) [**Fig. 4**]. The major metabolites of abscisic acid, phaseic acid and dihydrophaseic acid, were also found to form these two types of glucosyl conjugates.

Early data on the physiological functions of abscisic acid conjugates suggested that these conjugates could also serve as a storage or transport form of the free acid. However, as more data accumulated, these roles seemed to be less likely. For example, it was proposed that a large increase in the level of abscisic acid in plants under water stress was a result of the hydrolysis of abscisic acid conjugates. However, it was demonstrated that this increase in the free-hormone level was accompanied by an increase in the level of abscisic acid conjugates as well. Both free and conjugated abscisic acid decreased when the plants were watered again. Therefore, the role of abscisic acid conjugates in the regulation of free-hormone level, at least in

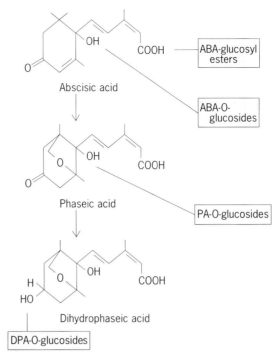

Fig. 4. Conjugation sites of abscisic acid (ABA) and its major metabolites, phaseic acid (PA), and dihydrophaseic acid (DPA); and types of conjugates formed.

this case, was improbable. Further evidence that the abscisic acid conjugates are in fact final products of abscisic acid metabolism comes from their localization in the vacuoles of the plant cells and their inability to pass through the cell wall. Thus, abscisic acid conjugates have not been shown to play a comparable role to that of the conjugates of growth promoters, such as auxins, gibberellins, and cytokinins. Although in many cases they are formed through the same chemical reaction of glucosylation, abscisic acid conjugates appear to be irreversible end products of abscisic acid metabolism.

Prospects. The characterization of the enzymes that catalyze the synthesis and hydrolysis of hormone conjugates as well as genes that encode these enzymes in the plant tissue will make possible the manipulation of plant growth and development by modifying the relationships between plant hormones.

For background information SEE ABSCISIC ACID; AUXIN; CYTOKININS; ETHYLENE; GIBBERELLIN; PLANT HORMONES in the McGraw-Hill Encyclopedia of Science & Technology.

Krystyna Bialek

Bibliography. A. B. Bleecker et al., Insensitivity to ethylene conferred by a dominant mutation in *Arabidopsis thaliana, Science*, 241:1086–1089, 1988; J. D. Cohen and R. S. Bandurski, Chemistry and physiology of the bound auxins, *Annu. Rev. Plant Physiol.*, 33:403–430, 1982; R. R. Finkelstein and C. R. Somerville, Three classes of abscisic acid (ABA)-insensitive mutations of *Arabidopsis* define genes which control overlapping subsets of ABA responses, *Plant Physiol.*, 94:1172–1179, 1990; H. Klee and M. Estelle, Molecular genetic approaches to plant hor-

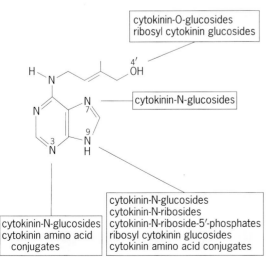

Fig. 3. Conjugation sites for *trans*-zeatin, a major cytokinin, and types of conjugates formed.

mone biology, *Annu. Rev. Plant Physiol. Plant Mol. Biol.*, 42:529–551, 1991; D. S. Letham and L. M. Palni, The biosynthesis and metabolism of cytokinins, *Annu. Rev. Plant Physiol.*, 34:163–197, 1983; C. Lincoln et al., Growth and development of the *axr1* mutants of *Arabidopsis, Plant Cell*, 2:1071–1080, 1990; J. MacMillan (ed.), Hormonal regulation of development. I: Molecular aspects of plant hormones, *Encyclopedia of Plant Physiology*, vol. 9, 1980; K. Schreiber, H. R. Schutte, and G. Sembdner (eds.), *Conjugated Plant Hormones: Structure, Metabolism and Function*, 1987.

Plant morphogenesis

There is a tremendous diversity in form and growth habit in higher plants. With the exception of some algae, plants are multicellular organisms composed of a variety of cell, tissue, and organ types. A long-standing question in developmental biology has been to what extent cells and tissues act autonomously or interact to regulate the development of a multicellular organism. Can one cell or a group of cells induce developmental changes in another cell or group of cells? In animals, cells can migrate to a site of action. In plants, cells are fixed in position. Yet, passageways connect plant cells and allow chemical messages to move from cell to cell. Plants also have vascular tissue that assists in the rapid movement of many compounds.

In plants possessing cells that are all genetically identical, it is difficult to determine whether adjacent cell layers interact. The utilization of plants composed of cells or tissues that are genetically dissimilar (that is, genetic mosaics) has yielded useful information on the regulation of plant development.

Although genetic mosaics can arise spontaneously, the mutations that give rise to them are random, uncharacterized, and frequently of no scientific importance. Two systems can be utilized to create fully characterized genetic mosaics: chimeral synthesis or radiation treatments.

Chimeras. Chimeral synthesis allows the creation of genetic mosaics composed of diverse but characterized genotypes. Unlike radiation treatment techniques, which require the use of plants in which extensive genetic maps have been developed, graft chimeras can be made between relatively uncharacter-ized genetic stocks. The technique does require graft compatibility between the components and the ability to regenerate adventitious shoots from the graft union. These requirements are met in several plant families, but the technique is most easily performed in the Solanaceae (representatives of this family include tomato, tobacco, pepper, and potato). This technique is possible because of the anatomy of the shoot apex. The shoot tip of most higher plants contains independent cell layers. When an entire cell layer is genetically distinct from adjacent layers, a chimera composed of normal and mutant cells is formed. Chimeras can perpetuate the genetically distinct cell line indefinitely, because the distinct cells lie within the shoot apex and continue to produce daughter cells as the plant grows. Unlike most radiation-induced genetic mosaics, chimeras can be perpetuated by vegetatively propagating the plant from stem cuttings that contain a preformed axillary vegetative bud. Since the layered apex is the ultimate source of organs that have cell layers descended from specific apical layers, the relative position of normal and mutant cells is somewhat predictable. While there has been limited success in creating chimeras by regenerating plants from mixed cell cultures, the most successful technique is to make graft chimeras as shown in **Fig. 1**. In this figure, two plants are reciprocally grafted (I). After the graft union has healed and the scion is growing, all but a thin layer of scion cells are removed and discarded (II). If all shoots that form in leaf axils are removed from the rootstock, the plant may eventually form adventitious shoots from the cut surface (II). A blowup of this formation of adventitious shoots on the cut surface is shown in (III). A small fraction of these shoots will be chimeras in that they originate from at least one scion cell and one rootstock cell. Generally, a complex chimera is formed with sectors of both components growing within a single shoot (IV). If this complex shoot is pruned, the dormant buds present in the axil of each leaf will grow out. A closeup of a cross section of the shoot tip (viewed from above) at the cut made in (IV) is shown in (V). Surrounding the main shoot are five smaller shoots, each of which arose in a leaf axil and is now growing out after pruning. Shoots 1 and 2 are not chimeral. Shoots 3 and 4 are still complex, unstable chimeras. Shoot 5, however, has apical layers that are pure. This shoot can be maintained by rooting cuttings because all of the buds in the leaf axils of this

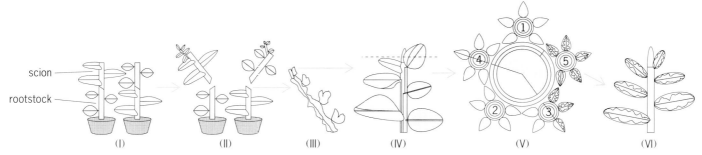

Fig. 1. Grafting technique to generate chimeral plants. Explanation is given in text.

Fig. 2. Genetic mosaics. (*a*) Leaves of tobacco species and chimera. From left to right, the leaves are *Nicotiana tabacum, N. glauca*, and a chimeral leaf possessing cells of both species. The chimeral leaf is from an interspecific graft chimera that was synthesized by using the procedure outlined in Fig. 1. (*b*) Corn leaf showing a mutant clone (the white stripe) composed of cells derived from a single cell that had chromosome damage caused by the irradiation procedure outlined in Fig. 3 (*photograph courtesy of P. Becraft*).

shoot will have the same arrangement of tissues as the shoot apex that gave rise to them (VI).

Most of the graft chimeras existing today were generated accidentally when chimeral shoots arose from the graft unions of plants that were grafted for horticultural purposes. Some of these graft chimeras have been developmentally analyzed. The most interesting is a *Camellia* chimera composed of two species. Traits such as anthocyanin (a red pigment) synthesis, cell size, and fragrance showed complete independence at the cell and tissue level. Most organ or whole-plant traits showed interactions. These traits included floral morphology, leaf size, time of bloom, and plant architecture. In a carnation chimera composed of genotypes for different flower color, the chimera had a flower color different from either of the components. The color was not the result of seeing one cell color overlying another cell color. Instead, there is evidence that a biochemical interaction that altered the pathways to pigment formation occurred between petal layers.

Most recently, graft chimeras have been synthesized with prior planning to study interactions between specific genotypes. A complete set of chimeras has been synthesized between two species of tobacco. In gen-

eral, the plants show independence for many cellular traits (such as flower pigment, epidermal hair morphology, and chlorophyll content), but in many organs (such as leaves and flowers) the chimeras display intermediate phenotypes (**Fig. 2***a*). In tomato, several graft chimeras have been generated to study traits such as fruit structure and abscission, petal formation, and insect resistance. Since there are numerous useful mutants and an extensive genetic map in tomato, the future outlook for utilizing this species as the model to study the interaction of cell layers in plants is promising.

Radiation treatments. Monocots (a major division of flowering plants that includes grasses and palms) cannot be permanently grafted together because of the anatomy of their vascular system. However, obtaining genetic mosaics is still possible. By utilizing genetically characterized stocks, plants can be bred to have a desired genetic constitution. For the past few decades, geneticists have accumulated genetic stocks by inducing mutations or discovering spontaneous mutations. These stocks have been used to map the position of genes on specific chromosomes. For major crop plants such as corn, soybean, and tomato, genetic maps are quite extensive.

Marker genes are those responsible for creating phenotypically obvious traits. A genetic stock that has both a characterized marker gene and a gene of interest can be created through conventional plant breeding. When plants are bred so that both of these genes are on the same arm of the same chromosome, they can be used to create genetic mosaics that have adjacent cells or tissues composed of normal and mutant cells. The procedure (**Fig. 3**) utilizes radiation to break chromosomes. When a chromosome is broken, genetic material is lost. Since this breakage does not occur in all cells, a genetic mosaic is created. Because radiation damage does not generally target specific chromosomes and regions of chromosomes, phenotypic markers must be used so that the plant's appearance will assist in determining the plant's genotype. For example, if corn seed heterozygous for albinism (the marker gene) and a gene of interest are irradiated,

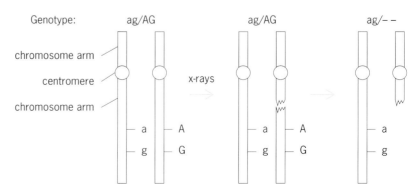

Fig. 3. Diagram of a pair of chromosomes containing a marker gene (a = albinism) and a gene of interest (g). Because both the gene for albinism and the gene of interest are recessive, the heterozygous stock will not express these genes. However, if x-ray exposure breaks the arm of the chromosome possessing the dominant allele, the remaining hemizygous stock will express the recessive traits.

a fraction of the seedlings will be genetic mosaics. When germinated and grown, these plants will possess stems, leaves, and floral organs with albino sectors composed of cells that are descendants of the cells with chromosome damage (Fig. 2b). Each sector is called a clone because it is derived from one cell. If the gene of interest is located farther away from the centromere (the structure connecting the two arms of a chromosome), it too must have been lost when the chromosome arm broke away. Thus, it is assumed that any albino tissue is also expressing the gene of interest. The albino sectors are traceable within the plant and will form certain regions of plant tissues and organs. By observing the morphology of albino sectors that encompass whole organs or parts of organs, it is possible to determine whether the gene of interest behaves autonomously or interacts with adjacent non-mutant tissue.

Recently, the above technique has been used to study several developmental mutations in corn. The most studied mutant is *Knotted*. This mutant perturbs normal leaf development by causing extra cell divisions in localized regions, giving a bumpy or knotted texture to the leaves. This abnormality apparently originates in the epidermis of the leaf where cell divisions parallel to the leaf surface cause a localized thickening. In time, all cell layers are affected. Interestingly, when genetic mosaics that had a normal epidermis covering *Knotted* mesophyll cells were generated, the epidermis still divided abnormally. The indication is that the mesophyll had induced these divisions and that the epidermis was not acting independently. By using molecular biology, the *Knotted* gene has been cloned and may be used in genetic engineering experiments aimed at studying its expression in other plant systems.

The mutant *Teopod 1 (Tp1)* prolongs the expression of juvenile traits, so that reproductive organs are transformed into leaflike structures. Genetic mosaics that possess sectored organs composed of mutant and *Tp1* tissues were created. Even the "normal" sectors of the mosaic organs possessed the mutant phenotype. As in *Knotted*, the indication is that this mutation is nonautonomous in nature and that intercellular communication must be occurring for organs to appear mutant even though some of their cells do not possess the mutant gene. Because most of the known mutants are cell-autonomous, the above examples are exceptional. Some mutants appear to behave autonomously in certain tissues, yet in others communication between cells affects gene expression. For example, the *liguleless-1* (*lg1*) gene eliminates the auricle and ligule of the corn leaf. The auricle is a wedge of pale-green tissue that acts like a hinge between the leaf blade and leaf sheath. The ligule is a fringe of tissue located where the auricle and sheath join. The wild-type (*Lg1+*) allele appears to have a dual function: a ligule-initiating function that is autonomous and affects the epidermis, and an auricle-initiating function in internal tissue. Wild-type internal tissue that contacts *lg1* epidermis appears able to induce the mutant epidermis to form a rudimentary ligule. Apparently, some communication occurs

between cells over short distances, possibly owing to the diffusion of a gene product.

For background information *SEE DEVELOPMENTAL BIOLOGY; DEVELOPMENTAL GENETICS; EMBRYONIC INDUCTION* in the McGraw-Hill Encyclopedia of Science & Technology.

Michael Marcotrigiano

Bibliography. Y. P. S. Bajaj (ed.), *Biotechnology in Agriculture and Forestry*, vol. 11: *Somaclonal Variation in Crop Improvement I*, 1990; P. W. Becraft and M. Freeling, Sectors of *liguleless-1* tissue interrupt an inductive signal during maize leaf development, *Plant Cell*, 3:801–807, 1991; B. J. Niflin (ed.), *Oxford Surveys of Plant Molecular and Cell Biology*, vol. 7, 1991; S. Poethig, A non-cell-autonomous mutation regulating juvenility in maize, *Nature*, 336(6194):82–83, 1988.

Plant pathology

The understanding of plant pathology has grown as new information has become available. This article discusses recent studies that have elucidated the mechanisms involved in induced resistance to infection in plants, and commonly occurring plant diseases.

Induced Resistance Against Pathogens

The idea that plants can be immunized against pathogens has been recognized for a long time and has been documented in the scientific literature for over 80 years. Early descriptive studies showed that pretreatment with a pathogen resulted in protection against subsequent treatment with related pathogens. At least three phenomena can account for such protection: viral cross-protection, biological control, and induced or acquired resistance.

In viral cross-protection, pretreatment of plants with an attenuated virus (that is, one that systemically infects the plant but causes only mild symptoms) can decrease the severity of a subsequent infection by more virulent forms of the same virus.

Biological control involves the colonization of the plant with nonpathogenic microorganisms such as bacteria or fungi that can protect the plant against infection by pathogens. Such protection results from competition for nutrients, production of inhibitory compounds such as antibiotics by the colonizing organism, or the parasitic exploitation of the pathogen by the protecting microbe.

In contrast, induced or acquired resistance, also known as immunization, is a systemic response of the plant itself to pathogen infection. Thus, to a certain extent, plant immunization is analogous to immunization in animals and humans. A first inoculation with virtually any pathogen capable of causing a necrotic lesion results in resistance toward subsequent infections, even in tissue untouched by the inducing infection. In addition, this resistance acts against infections caused by a wide variety of pathogens, including pathogens that are not related to the inducing organism. For example, in cucumber or tobacco a first inoculation with a fungus, bacterium, or virus protects the plant against subsequent infections by other fungal, bacterial, or viral pathogens in both the infected and uninfected parts of the plant. **Table 1** details the wide range of plants in which induced resistance has been found. Understanding the way that induced resistance works may eventually lead to methods of improving the health of agricultural crops.

Phenomenology of induced resistance. In all reported cases the induction of resistance is dependent on the production of necrotic lesions by the inducing pathogen. The level of protection is related to the size and number of lesions produced during the first infection. Thus, induced resistance is not an all-or-none phenomenon, as is observed in genetically determined resistance to a specific pathogen. Despite this variation in expression, induced resistance can be extremely effective and long-lasting. For instance, in cucumber plants infected on the first leaf with tobacco necrosis virus or the fungus *Colletotrichum lagenarium*, induced resistance lasts for several weeks. A second, booster inoculation gives protection up to the time of flowering. The effectiveness of systemic induced resistance has been confirmed in field trials for bean, cucumber, and tobacco. In tobacco plants immunized by subepidermal stem injections with spores of *Peronospora tabacina*, the growth of the plants can be enhanced even in the absence of the challenge pathogen.

A variety of chemicals can also induce resistance, but their effects remain confined to the treated areas. These chemicals include acetylsalicylic acid (aspirin), salicylic acid, polyacrylic acid, salts of heavy metals, oxalate, and phosphate.

Mechanisms of induced resistance. Because the inducing infection results in broad protection against multiple pathogens, it is likely that diverse resistance mechanisms are activated in the plant. Biochemical evidence for at least some of these mechanisms has recently been obtained. In cucumber, lignification of epidermal cells occurs more rapidly and to a larger extent in systemically induced plants than in uninduced controls. This increased synthesis of lignin, a phenolic polymer usually associated with woody tissues, might prevent pathogen invasion in any of several ways, including mechanical reinforcement of the cell wall; protection of other cell wall polymers from the action of hydrolytic enzymes produced by the pathogen; formation of a hydrophobic barrier preventing the leakage of cellular solutes from the plant cell to the pathogen or toxins from the pathogen to the plant cell; direct antimicrobial action of unpolymer-

Table 1. Plants, inducing agents, and the disease organisms affected

Plant	Inducing agent	Protects against	Disease
Tobacco	Tobacco mosaic virus	*Thielaviopsis basicola*	Black root rot
		Phytophthora parasitica	Black shank
		Pseudomonas tabaci	Wildfire
		Peronospora tabacina	Blue mold
		Pseudomonas syringae	Canker
		Tobacco mosaic virus	Local lesions
	Tobacco necrosis virus	*Thielaviopsis basicola*	Black root rot
		Phytophthora parasitica	Black shank
		Pseudomonas tabaci	Wildfire
		Tobacco mosaic virus	Local lesions
		Tobacco necrosis virus	Local lesions
	Thielaviopsis basicola	*Thielaviopsis basicola*	Black root rot
		Tobacco necrosis virus	Local lesions
		Tobacco mosaic virus	Local lesions
	Peronospora tabacina	*Peronospora tabacina*	Blue mold
	Pseudomonas syringae	Tobacco mosaic virus	Local lesions
Cucumber	*Colletotrichum lagenarium*	*Colletotrichum lagenarium*	Anthracnose
	Pseudoperonospora cubensis		
	Pseudomonas lachrymans		
	Tobacco necrosis virus		
	Colletotrichum lagenarium	*Cladosporium cucumerinum*	Scab
		Fusarium oxysporum f.sp. *cucumerinum*	Wilt
		Pseudomonas lachrymans	Angular leaf spot
	Pseudomonas lachrymans	*Colletotrichum lagenarium*	Anthracnose
	Tobacco necrosis virus	*Sphaerotheca fuliginea*	Powdery mildew
		Cladosporium cucumerinum	Scab
		Fusarium oxysporum f.sp. *cucumerinum*	Wilt
		Pseudomonas lachrymans	Angular leaf spot
		Tobacco necrosis virus	Local lesions
Watermelon	*Colletotrichum lagenarium*	*Colletotrichum lagenarium*	Anthracnose
	Fusarium oxysporum f.sp. *cucumerinum*		
Muskmelon	*Colletotrichum lagenarium*	*Colletotrichum lagenarium*	Anthracnose
Tomato	*Phytophthora infestans*	*Phytophthora infestans*	Late blight
Bean	*Colletotrichum lindemuthianum*	*Colletotrichum lindemuthianum*	Anthracnose
	Colletotrichum lindemuthianum		
Potato	Hyphal wall components of *Phytophthora infestans*	*Phytophthora infestans*	Late blight
Rice	*Pseudomonas syringae*	*Pyricularia oryzae*	Blast
Barley	*Erysiphe graminis* f.sp. *hordei*	*Erysiphe graminis* f.sp. *hordei*	Powdery mildew

ized lignin precursors; or growth inhibition of invading fungi by lignification of their hyphal tips.

The most thoroughly documented biochemical change associated with the onset of induced resistance is the synthesis of new proteins. A set of so-called pathogenesis-related (PR) proteins is produced locally and systemically in response to inducing infections. Pathogenesis-related proteins were described initially as acid-extractable, low-molecular-weight, protease-resistant proteins that accumulate in the intercellular fluid of leaves. Recently basic homologs of acid pathogenesis-related proteins have been found; their compartmentalization is typically within the vacuole of the plant cells. These proteins have been observed in a multitude of dicotyledonous and monocotyledonous plant species, including bean, celery, cucumber, cowpea, pea, potato, tobacco, tomato, barley, maize, rice, and wheat.

Complementary deoxyribonucleic acids (cDNAs) encoding all of the pathogenesis-related proteins from tobacco have now been cloned. Several additional cDNAs that are associated with the induction of resistance have also been isolated. In all cases examined, the induction of pathogenesis-related protein expression occurs at the level of transcription of messenger ribonucleic acid (mRNA). The timing and amount of accumulation of the mRNAs transcribed by all these genes correlates well with the magnitude and timing of the induced resistance response. Based on the physical properties of the proteins and sequences of the cDNA clones, the pathogenesis-related proteins have been assigned to several distinct classes. The function of most of the pathogenesis-related proteins is still unknown. Two classes, however, have been shown in cell culture to have chitinase and β-1,3-glucanase enzymatic activity. Interestingly, the substrates for these enzymes, chitin and β-1,3-glucan, occur in fungal cell walls; indeed, mixtures of chitinase and glucanase are antimicrobial in fungal cultures. Another case of pathogenesis-related proteins is structurally similar to proteinase inhibitors, which are thought to play a role in resistance to insect attack. Thus, the existing data are consistent with the hypothesis that accumulation of pathogenesis-related and other proteins is a major component of induced resistance. It also appears that different groups of proteins are directed against different types of pathogens. However, proof that these proteins play a causal role in resistance must await further experiments, in which individual proteins are expressed at high levels in transgenic plants.

Nature of the systemic signal. Since resistance can be induced systemically by localized infections, the production of a signal that activates the resistance mechanisms has been hypothesized. Evidence from stem girdling experiments and grafting experiments has indicated that the putative signal moves through the phloem tissue of the plant's vascular system. Analyses of cucumber phloem collected at various times after an inducing infection showed an increase in salicylic acid. Similar results were obtained in tobacco, where the level of salicylic acid increased in the uninfected leaves of immunized plants and pre-

ceded the accumulation of the mRNA encoding one of the major pathogenesis-related proteins. Exogenous application of salicylic acid can induce resistance in cucumber and tobacco. This induction is not due to a direct fungicidal effect of this compound or of its metabolites. Salicylic acid cannot induce systemic resistance after localized application because of its rapid metabolism or difficulties of gaining access to the vascular system. Salicylic acid induces the same set of pathogenesis-related genes as that expressed in response to a resistance-inducing pathogen infection. The level of salicylic acid in induced tobacco leaves is sufficient to account for the induction of accumulation of a major pathogenesis-related protein. Salicylic acid is therefore a promising candidate for the signal that activates induced resistance.

In a first step toward understanding the link between salicylic acid and activation of pathogenesis-related gene expression, control sequences required for induction of the gene encoding the major pathogenesis-related protein of tobacco have been studied. Varying lengths of DNA upstream of the transcribed region of the gene have been assayed in transgenic plants for their ability to potentiate induction of a heterologous gene. These experiments have shown that sequences sufficient for induction of gene expression by pathogen infection, salicylic acid, and other chemical inducers are found within the first 660 base pairs of DNA upstream of the transcription start site. Information on such cis-acting regions (genetic regions affecting the activity of genes on that same DNA molecule) is essential for identifying trans-acting factors (diffusible product encoded on another DNA molecule) that regulate gene expression in response to biological or chemical induction. In addition, a putative receptor for salicylic acid has been identified. Further biochemical studies will eventually elucidate the signal transduction pathway by which the inducing stimulus is perceived and by which it is ultimately translated into gene expression.

Prospects. The broad protection offered by plant immunization might be of practical value for crop protection. One approach could be to pretreat crops with microorganisms. However, this approach is problematic on a large scale because inoculation should be performed only at discrete sites on the plant. Alternatively, chemicals that act like the endogenous signal by triggering induced-resistance mechanisms could be sought. Application of such compounds could be suitable for conventional large-scale farming. Such treatments could even be complementary to existing disease control practices, which typically involve treatments designed to kill invading pathogens directly. Initial efforts to find resistance-inducing compounds have yielded some results. Oxalate, phosphate, and 2,6-dichloroisonicotinic acid and its ester derivative have been found to induce resistance and associated biochemical changes in much the same way as preinoculations with pathogens. Another approach is to engineer plants for increased resistance by overexpression of certain genes that are important for resistance and its induction. The potentially synergistic combination of treating a transgenic crop with an environmentally safe chemical also merits further consideration. Such plants would be engineered to contain chimeric genes, the promoters of which respond to a particular chemical. Upon chemical treatment, agriculturally useful traits such as resistance to pathogens, insects, herbicides, or even environmental stress such as drought, heat, or cold could be induced. Regulated expression of a phenotype might be desirable to decrease the selective pressure that invariably acts on that trait and causes its loss of efficacy.   *J. P. Métraux; J. Ryals; E. Ward*

Common Problems of Indoor Foliage Plants

Some of the most common problems of indoor plants can be tied to environmental conditions such as low light, too much or too little water, or excess minor elements found in irrigation water or the potting medium. Infectious diseases, caused by plant pathogens such as fungi, bacteria, and viruses, are influenced by these environmental factors. Other problems are caused by insects and mites.

Diseases of indoor plants. Diseases can affect roots, stems, and leaves, depending upon which pathogen is involved. Probably the most common type of disease of indoor plants is root rot, caused by such fungi as *Cylindrocladium, Fusarium, Phytophthora, Pythium,* and *Rhizoctonia* spp. Phomopsis dieback is a devastating disease that occurs on indoor weeping figs (*Ficus benjamina*); it causes a slow decline typified by leaf loss, twig death, stem cankers, and death of the tree. One of the more easily diagnosed diseases, powdery mildew, is typified by a powdery white coating on leaves of grape ivy (*Cissus*), begonia, rose, or African violet (*Saintpaulia*). The final type of indoor disease is caused by bacteria that are often systemic and can cause leaf spot, stem rot, and wilt.

Root rots and overwatering. Symptoms of both overwatering and root rot disease include yellowing of lower leaves (chlorosis), wilting, and stunting. Overwatering can exacerbate existing problems such as bacterial stem rots, fertilizer deficiency, and root rot diseases. Since symptoms of overwatering and root rot are similar, it is advisable to check the root systems of plants prior to purchasing and installation. One of the most serious root diseases is caused by *Cylindrocladium spathiphylli* on *Spathiphyllum* spp. (Peace lily).

Cylindrocladium root and petiole rot of *Spathiphyllum* spp. has been a serious problem for foliage producers since the late 1970s. The earliest symptoms are slight wilting and chlorosis of lower leaves. These leaves gradually turn necrotic, and petiole bases become rotted and detach from the main plant. When moisture is abundant, spores splash onto petioles and leaves and cause elongated, dark-brown to black lesions with bright-yellow halos. Lesions may be pinpoint size or reach 0.4 in. (1 cm) in diameter. The root system of plants showing slight loss of lower leaves may appear unaffected. Occasionally, petioles bend sharply downward, to about 0.5 in. (1.3 cm) from the soil line. During the winter, petiole epinasty and a general chlorosis may be common symptoms. In advanced stages of disease development, the entire top of the plant becomes completely separated from the roots. Development of symptoms is associated with the site of the initial infection. The fungicide benomyl currently provides the best control of Cylindrocladium root and petiole rot.

Indoor plants exhibiting overwatering symptoms can be examined periodically to determine if root rot is involved. If roots appear white and healthy, the watering schedule should be modified until symptoms disappear. If root disease is a problem (roots appear sparse, brown, and rotted), watering frequency should be reduced and a fungicide applied to the potting medium.

Phomopsis dieback of Ficus. Since 1980, the use of *Ficus* spp. in the interior landscape industry has increased dramatically. If the correct conditions for production of these plants are not maintained, trees may become diseased after installation. The most common cause of loss has been Phomopsis dieback. This disease appears initially on small twigs, which wilt and drop their leaves. The pathogen progresses into larger twigs or stems and causes girdling, and then death of the tissue above the girdling site.

The most commonly isolated organism is *Phomopsis cinerescens*. Dieback organisms apparently enter the plant through wounds created by pruning, transport, or adverse weather conditions such as freezing. During commercial production of the plants, these wound sites frequently remain latent. Infection may not be apparent until the plants have been installed indoors and are exposed to environmental stresses such as imbalances in water or light. The only effective control appears to be use of *Ficus* plants free of obvious wounds and twig death, and maintenance under optimum conditions after installation. The *Ficus* must be fully acclimatized for the interior environment in order to minimize the stress of transition.

Powdery mildew. This fungal disease, caused by *Oidium* spp., is quite common on the flowers, petioles, and leaves of African violet, grape ivy, and begonias. A powdery white mass of fungal mycelium and spores can cover an entire flower or leaf or can remain in circular lesions of up to 0.5-in. (1.25-cm) diameter on either leaf suface. The flowers and the leaves of cultivars exhibit differential resistance to the powdery mildew fungus. Benomyl is a safe and effective treatment for powdery mildew.

Bacterial infections. Bacterial infections on indoor foliage plants are less common than viral and fungal diseases. Certain bacterial diseases can be latent in an apparently healthy plant, whose leaves and stems may develop symptoms when the plant is placed indoors. Symptoms include marginal or interveinal spots on leaves. Stem infections usually occur at the soil line and often result in cankers or soft spots that can cause the stem to lodge (break off at the soil line). Before this stage, the leaves on the affected stem can turn yellow and wilt.

There are no effective means to control systemic bacterial infections. Leaf spots may be partially controlled by removing affected leaves and spraying the remaining plant with a copper fungicide. Stem bases on new plants should be checked for signs of infection prior to purchasing or installation to avoid this problem completely.

Environmental disorders. Indoor foliage plants are produced in greenhouses or shadehouses with tropical or subtropical conditions and then transferred to interior environments with cool temperatures and relatively low light and humidity. This change in the environment can result in symptoms that may be similar to those caused by biotic agents (fungi, bacteria, and viruses). A symptom such as leaf chlorosis can be caused by insufficient fertilizer, pesticide phytotoxicity, or cold soil temperatures (**Table 2**).

Light level. Light affects the growth of all plants. Although there is little published research concerning diseases of foliage plants caused by inappropriate light levels, light remains one of the most important aspects of maintaining a healthy foliage plant indoors. Over time, low light causes severe damage to some foliage plants. Symptoms of low light include oversized or elongated leaves, leaf abscission, thin weak stems, and stunting. If the light level remains too low for a plant, growth will eventually cease and the plant may die.

Chilling and cold damage. Temperature extremes, particularly low temperatures, can cause problems in indoor foliage plants. This sensitivity to low temperatures is probably due to the tropical or subtropical origins of most foliage plants. The Silver Queen cultivar of *Aglaonema commutatum* is especially sensitive to chilling temperatures below 50°F (10°C). A greasy appearance develops on leaves within 48 h of exposure to cold, with older leaves most severely affected. Chilling injury can also occur on plants during shipment. For example, weeping fig (*Ficus benjamina*) is fairly cold-tolerant but exhibits heavy leaf abscission and death of some terminal shoots following chilling damage.

Cold-water damage on African violets. Leaf spot of African violet (*Saintpaulia ionantha*) has been a serious problem for violet producers. Symptoms include areas of yellowed or whitened tissue on the upper surfaces of leaves. Lesions appear on the leaves first as water-soaked areas, which rapidly turn white or brown. African violet leaf spot occurs when cold water is used. Damage increases as the difference between the air temperature and water temperature increases. Water temperatures below 50°F (10°C) cause damage

Table 2. Symptoms and possible causes of indoor plant disorders

Symptoms	Possible causes	Symptoms	Possible causes
Leaves		Stems	
Powdery coating	Powdery mildew fungus	Rot at soil line	High soil salinity
Speckling	Mite or insect feeding		Slow-release fertilizer placed against stems
Young leaves chlorotic	Iron too low		Overwatering
	Poor soil drainage		Poor soil drainage
Older leaves chlorotic	Low nitrogen or potassium		Stem disease (caused by bacteria or fungi)
	High soil salinity	Lesions or cankers	Sun scald
	Overwatering		Mechanical injury
	Poor soil drainage		Stem disease (caused by bacteria or fungi)
	Root disease caused by pathogens	Cracks	Mechanical injury
Chlorotic spots	Cold-water injury		Stem disease (caused by bacteria or fungi)
Water-soaked spots	Temperature extremes	Thin and weak stems	Fertilizer extremes
	Bacterial infections		Low light
Necrotic margins or tips	Boron or fluoride toxicity	Roots	
	High soil salinity	Poor development	High soil salinity
	Temperature extremes		Soil temperature extremes
	Drying		Plant potted too deep
Necrotic spots	Cold-water injury	Rots	Root disease caused by pathogens
	Fertilizer toxicity		High soil salinity
	Sun scorch		Poor soil drainage
	Cold injury	Entire plant	
	Bacterial infections	Plant stunted	Root disease caused by pathogens
Leaves too large	Fertilizer excess		Fertilizer extremes
	Low light		Low light
Leaves too small	Fertilizer deficiency		Temperature extremes
	Copper deficiency		Poor soil drainage
	High soil salinity		Pot-bound root system
	Root-bound plants	General chlorosis	High light intensity
Leaves long and narrow	Low light		High soil salinity
Leaves very thin	Excess nitrogen		High leaf temperature and cold soils
	Low light	Wilting	Insufficient soil moisture
Holes	Mechanical injury		Low humidity
	Insect feeding		Poor root system
Defoliation (leaf drop)	High soil salinity		
	Abrupt light reduction		
	Chilling injury		
	Drying		
	Poor soil drainage		
	Low soil moisture		

to most African violets, regardless of air temperatures. A small differential between water temperature and air temperature when both factors are low is sufficient to cause damage to African violet leaves.

Fluoride toxicity. The most common micronutrient toxicity is caused by fluoride. The list of sensitive plants is long, with most members of the genus *Dracaena* exhibiting some damage. The first plant found to be sensitive to fluoride was Ti plant (*Cordyline terminalis*). Symptoms on Ti plant are typically marginal chlorosis and necrosis. The most widely recognized fluoride damage on an indoor foliage plant is found on *Dracaena deremensis*. Elliptical necrotic lesions in interveinal chains form in the white band of tissue on leaves, and are usually less than 0.5 in. (1.25 cm) long. *Dracaena fragrans* can also develop symptoms of marginal chlorosis and necrosis or sometimes dark-green ring spots or mottling. Parlor palm (*Chamaedorea elegans*) and areca palm (*Chrysalidocarpus lutescens*) are also sensitive to fluoride. Foliar tip burn and necrotic lesions are common on parlor palm, while lesions on areca palm are elliptical and may form in interveinal chains. Toxic levels of fluoride can develop from a variety of sources, including superphosphate fertilizer, perlite, water, and some peats. Increasing the pH of the potting medium by additions of dolomite or calcium hydroxide reduces fluoride damage in some plants. Fluoride-sensitive plants should be grown in a potting medium with a pH above 6.0 to reduce solubility of fluoride.

Mercury toxicity. One of the most recently discovered sources of air pollution for plants is interior latex paints used for walls and ceilings. The most sensitive plants show symptoms of marginal chlorosis and necrosis followed by leaf abscission and sometimes plant death. Research has identified the air-pollution source as the mercury that is used as a mildewcide in some latex paints. When a wide variety of foliage plants were subjected to airborne mercury from such paints, only *Ficus* and *Dieffenbachia* spp. were found sensitive. Levels of mercury that are high enough to be toxic to weeping figs continue to be released for at least 6 months after interior application of paint.

For background information SEE BREEDING (PLANT); GENETIC ENGINEERING; PLANT PATHOLOGY in the McGraw-Hill Encyclopedia of Science & Technology.

Ann R. Chase

Bibliography. A. R. Chase, *Compendium of Ornamental Foliage Plant Diseases*, 1987; G. T. Cole and H. C. Hoch (eds.), *The Fungal Spore and Disease Initiation in Plants and Animals*, 1991; J. N. Joiner, *Foliage Plant Production*, 1981; J. Malamy et al., Salicylic acid: A likely endogenous signal in the resistance response of tobacco to viral infections, *Science*, 250:1002–1004, 1990; J. P. Métraux et al., Increase in salicylic acid at the onset of systemic acquired resistance in cucumber, *Science*, 250:1004–1005, 1990; S. S. Woltz, Nonparasitic plant pathogens, *Annu. Rev. Phytopathol.*, 16:403–430, 1978; J. L. Wray (ed.), *The Biochemistry and Molecular Biology of Inducible Enzymes and Proteins in Higher Plants*, 1991.

Plant physiology

Inadequate water supply is one of the major factors limiting plant growth. Many of the effects are not caused directly by water stress but rather are due to the plant's ability to recognize processes that precede stressful limitations of water supply. Plants respond to environmental cues and information associated with limited water supply in order to optimize growth, maturation, and reproduction. The conversion of environmental information into plant responses requires a muiltistage process that may occur slowly or very rapidly (in less than 15 s).

Studies in experimental environments have demonstrated that plants respond to temperature, direction of gravity, length of the day, spectrum of light, direction of light, chemical composition of the soil, intensity of the wind, gas composition in the atmosphere, and water content of both the soil and the atmosphere. Further studies have been initiated on the mechanisms through which plants sense and respond to the environment in a coordinated manner. Plant regulation of responses to water availability is primarily by means of transpiration, which is the evaporation from aerial plant parts.

Exchange of water. Terrestrial plants form a living conduit through which soil water passes into the atmosphere. For a plant to grow, it must maintain its intracellular water as a dilute solution, draw carbon dioxide from the atmosphere, and use the energy of the light to fix the carbon dioxide into more complex plant metabolites. Rapid evaporation of plant water is prevented by a waxy layer called cuticle that covers most of the exterior of the plant (the cuticle is virtually impermeable to liquids, water vapor, carbon dioxide, and many other gases). In addition, tiny apertures called stomata penetrate through the cuticle of the leaves and some other aerial parts of the plant. Stomata open to let in atmospheric carbon dioxide or close to prevent the loss of plant water. The stomata of many plants can respond to a decrease in water supply by closing within seconds. The stomatal behavior is dictated through genetically controlled strategies that allow plant water to escape in order to trap atmospheric carbon dioxide. Thus the plant's ability to conserve water is compromised by the necessity of obtaining carbon dioxide as a nutrient from the atmosphere. The efficiency of the exchange of water for carbon dioxide is determined largely by the concentration gradients of both water and carbon dioxide between the inside of the plant and the atmosphere. The concentration of carbon dioxide is usually higher in the atmosphere, while the water concentration is higher in the plant.

The plant's strategy for water use is further complicated by the requirement of light energy to combine the carbon dioxide with sugars and other plant metabolites. If carbon dioxide is obtained from the atmosphere at the time when the light energy is most intense, the rate of evaporation will increase. The plant's strategy entails both rapid and longer-term responses. A sudden increase in the demand for water is rapidly countered by closing the stomata, so that the rate of water loss is substantially decreased. However, in many plant species, longer-term exposure to a restricted water supply increases tolerance to a subsequent but more severe drought through a number of water-conserving mechanisms.

The efficiency of the plant's combined strategy is expressed as the ratio of soil water that is expended, to the dry matter that the plant accumulates. There are large species variations in this efficiency. For example, the strategy employed by rice plants results in the expenditure of 0.2 gal (1 liter) of water to obtain 0.05 oz (1.5 g) of dry matter, of which only 0.035 oz (1 g) is contributed from carbon dioxide; corn plants have developed a second strategy that concentrates atmospheric carbon dioxide and results in a recovery of 0.11 oz of dry matter per quart (3.3 g per liter) of soil water expended. Some succulent plants, such as pineapple, have incorporated metabolic strategies that allow them to absorb carbon dioxide at night and fix it into sugars during the daylight hours. Pineapple can obtain 1.2 oz of dry matter per quart (35 g per liter) of water expended. Unfortunately this strategy, though efficient, results in slow growth.

Chemical signals. Often the first symptom of limited water supply is a marked decrease in the elongation rate of leaf cells. The slowed growth is not necessarily related to decreased water supply from the soil or decreased water content in the aerial portions of the plant. This phenomenon is associated with root signals that are transmitted to the aboveground portions of the plant through the xylem. If a small portion of the roots are dried while the remaining roots are given adequate moisture, the roots in dried soil generate increased levels of the plant hormone abscisic acid. This hormone is carried in water from the dried roots to the stomata of aerial plant parts and signals the closure of stomata and a decrease in the rate of growth.

Increased rates of abscisic acid synthesis in the roots or aerial plant parts are generated in response to environmental cues. But the nature of these cues and the mechanism that converts them into a regulatory signal like abscisic acid have not been explained. It is known that ribonucleic acid (RNA) and protein synthesis must precede changes in the rate of abscisic acid synthesis. Once abscisic acid is formed, it may be transported to cells that are sensitive to it. It is not known if abscisic acid transport mechanisms are passive or directed, but abscisic acid does move through the xylem of plants in drying soils. Abscisic acid is then perceived by abscisic acid–sensitive cells, probably through an abscisic acid–sensitive molecule or molecular complex (receptor). Although little is known of the steps leading from the environmental cue to the reception of abscisic acid as a signal, a great deal is known of cell response to this hormone.

Although abscisic acid appears to be the major plant hormone controlling water loss, other molecules perhaps have similar effects on stomata. Cytokinins, another class of plant hormone, are known to have the opposite effect of abscisic acid on stomatal closure.

Rapid stomatal response. Stomata respond rapidly to environmental stimuli that are either sensed directly or sensed and transduced into a secondary signal such as abscisic acid. These rapid responses have two functions: they protect the plant from sudden changes in the rate of evaporation (for example, the effect of transition from shade to full sun), and they protect the plant from wasteful evaporation (for example, the stomata will close if light is not available for photosynthesis).

Stomata close if the substomatal cellular level of carbon dioxide rises, the level of light suddenly decreases, or the water supply is restricted. Abscisic acid can substitute for all of these stimuli and induce stomata closure. Environmental events that increase abscisic acid concentration may also close the stomata. Studies have not yet shown how these environmental changes are sensed by the plant and if they are transformed into a signal that increases the concentration of abscisic acid. In fact, stomata often close before any measurable change in abscisic acid levels can be detected. Despite the lack of correlation between abscisic acid levels and rapid stomatal responses, abscisic acid may still control stomatal responses by being rapidly redistributed during plant stress. Recently developed techniques that allow the measurement of abscisic acid in single cells will determine if this hormone indeed controls rapid responses.

Hormone model of plant responses. The study of plant hormones has lagged far behind the study of animal hormones. Animal hormone studies, nevertheless, indicate possible mechanisms through which plant hormones may act. Most hormones act by binding a receptor complex associated with the hormone-sensitive cell and modifying the receptor properties. The modification is then amplified by a series of biochemical events until changes, predetermined by the plant's genetic material, are enacted. Evidence indicates that some abscisic acid–induced responses are a direct effect of abscisic acid on lipid bilayers that do not require any specific receptors. For example, abscisic acid changes the growth pattern of blue-green algae and increases the twitch strength of rat vas deferens smooth muscles through mechanisms that are not believed to involve abscisic acid receptors. Further evidence demonstrates that abscisic acid can change the properties of lipid bilayer preparations that lack proteins. This evidence raises the intriguing possibility that some abscisic acid action on plants may not require a specific receptor. Nevertheless, the potential for abscisic acid responses that are not mediated by receptors does not preclude the existence of specific receptors for this hormone. In fact, a major thrust of current abscisic acid research is directed at identifying both receptors for abscisic acid and amplification mechanisms for converting abscisic acid into plant responses. Considerable progress has been achieved by researchers studying the molecular biology of proteins whose synthesis is induced by abscisic acid.

Many researchers have demonstrated that abscisic acid induces synthesis of specific, and often unusual, plant proteins. These studies have led to the isolation of messenger RNAs (mRNA) that are synthesized in response to abscisic acid and code for these proteins. Molecular biological techniques have made possible the isolation of the deoxyribonucleic acid (DNA) coding for the mRNA and protein sequence of a wheat embryo protein that is synthesized in response to abscisic acid. The isolated genetic material is composed of both coding sequences, which are transcribed to produce mRNA and protein, and regulatory DNA sequences, which determine when the plant produces the protein that is induced by abscisic acid. The regulatory sequence of the gene has a binding site for a class of proteins, known as leucine zipper proteins, that regulate the expression of DNA. In addition, the regulatory sequence possesses a sequence that is similar to those known for proteins that are responsive to cyclic adenosinemonophosphate (cAMP). This DNA sequence information suggests that abscisic acid could exert its control of this genetic material and similar genes through its effects on cellular cyclic adenosinemonophosphate.

Regulation of abscisic acid. For adenosinemonophosphate to be an effective signal, its fate within the plant must be finely regulated and the cells must be sensitive to it. When environmental conditions change rapidly, stored abscisic acid may be released, so that a rapid response is possible. As mentioned previously, when environmental conditions are varied more gradually, mRNA and protein synthesis are prerequisites for increasing abscisic acid levels. Presumably the proteins synthesized are enzymes required for the synthetic pathway of abscisic acid. In response to high levels of abscisic acid, the plant inhibits further synthesis of abscisic acid and synthesizes enzymes that rapidly degrade the abscisic acid into inactive products. A continued high level of abscisic acid also decreases the plant's sensitivity to it.

Syndrome of responses to limited water. Plant responses to limited water and the resultant production (and redistribution) of abscisic acid may involve the activation of hundreds of responsive genes. The responses occur as a complex alteration of growth characteristics that occur at all growth stages and affect all plant parts. Thus the combined responses of plants to limited water supply can be described as a syndrome. The initial responses are reduced evaporation and decreased elongation rates of growing parts. Next a period of specific protein synthesis occurs. The plant becomes increasingly tolerant of several stresses, including freezing, heating, and salinity. Changes occur at all levels of organization: solutes that are believed to protect against desiccation, such as proline, glycyl betaine, and sucrose, accumulate; vacuoles contract, possibly because of the accumulation of the protective solutes; cell-wall thickness increases, as do deposits of waxy materials on the plant surface, so that the resistance of the plant to evaporation is increased. In addition, the plant's morphology may change with a decrease in leaf size and a decreased length of stem occurring between each leaf. The decreased rate of

growth of aerial parts is often accompanied by an increase or maintenance of the root growth rate. Perennial and biennial plants may be driven into winter dormancy by limited water availability. Limiting water conditions often lower plant fertility by decreasing pollen formation, causing the abortion of flowers, and prompting premature loss of some fruit. Even though the responses to limited water are dictated by the plant's genetic code, this syndrome is seen in plants with widely differing genetic backgrounds.

Prospects. Over the last century, agricultural production has increased during each decade and current production exceeds all previous levels. Demand for both food and renewable industrial products will continue to increase. As there is no reserve of underutilized land, this demand must be met with increased production from the same land base. Fortunately, there are water-efficient plants that have strategies that might be incorporated into crop species. Once more is known of the mechanisms of efficient responses to conditions of limited water, molecular biologists will be able to insert relevant genes that enhance efficiency and performance and plant breeders will be able to select for plants with desired responses.

For background information SEE ABSCISIC ACID in the McGraw-Hill Encyclopedia of Science & Technology.

Ping Fu; Martin J. T. Reaney

Bibliography. W. J. Davies and J.-H. Zhang, Root signals and the regulation of growth and development of plants in drying soil, *Annu. Rev. Plant Physiol. Plant Mol. Biol.*, 42:55–76, 1991; L. V. Gusta et al., Abscisic acid analogs: Biological activity and interactions with the plant growth regulator abscisic acid, *Comments Agr. Food Chem.*, 2:143–170, 1990; H. G. Jones, *Plant and Microclimate: A Quantitative Approach to Environmental Plant Physiology*, 1983; T. A. Mansfield, A. M. Hetherington, and C. J. Atkinson, Some current aspects of stomatal physiology, and plant molecular biology, *Annu. Rev. Plant Physiol. Plant Mol. Biol.*, 41:55–75, 1990.

Plant respiration

Plant respiration, unlike animal respiration, is only partially sensitive to cyanide, a potent inhibitor of cytochrome-mediated oxygen uptake in mitochondria. In plants, a cyanide-insensitive alternative oxidase located on the inner mitochondrial membrane functions in the terminal electron transfer to oxygen. This oxidase is inhibited by a class of compounds known as arylhydroxamic acids. The plant respiratory electron-transport system is branched at the ubiquinone step, so that electrons can pass to oxygen via either the cytochrome oxidase or the alternative oxidase (**Fig. 1**). Flow through the alternative pathway occurs primarily when the cytochrome pathway is overloaded or impaired. The alternative pathway presumably functions as an overflow mechanism for transferring electrons to oxygen when they are produced in excess by some physiological or metabolic process. Examples of such

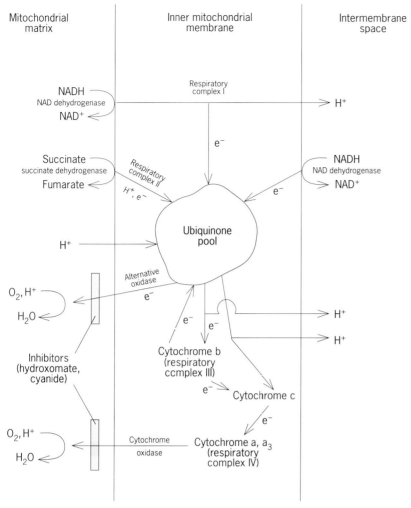

Fig. 1. Combined linear and two-dimensional representation of the plant electron-transport system localized on the inner mitochondrial membrane. Respiratory complexes I through IV consist of two or more proteins that function in electron transport.

processes are heat production in thermogenic tissues, mobilization of highly reduced storage compounds such as lipids, and the maintenance of certain biosynthetic processes requiring production of carbon skeletons. After the ubiquinone branch step, the alternative pathway does not conserve energy. The pathway is probably present in all plants but in widely varying capacities. The pathway is also present in algae, fungi, and some lower animals. The capacity of the pathway is known to be regulated by various environmental, developmental, and chemical factors.

Biochemistry and molecular biology. Respiratory oxygen uptake is the terminal step in which electrons are consumed after transfer through numerous compounds starting with intermediates of respiratory carbon metabolism. The biochemical process of electron transport from the carbon intermediate donor (via the reduced form of nicotinamide adenine dinucleotide, or NADH) to oxygen is shown in Fig. 1. This scheme shows two important differences in the electron-transport system between plant mitochondria and animal mitochondria: the presence of the external

NADH dehydrogenase and the alternative oxidase. The presence of the alternative oxidase constitutes a branch in the pathway at which electrons can be transferred to oxygen via either the cytochrome or the alternative pathways. The branch point is at the ubiquinone pool. As can be seen in Fig. 1, the alternative pathway bypasses the two energy-conserving, proton-translocating steps involving cytochromes. In fact, the only possibility for energy conservation during electron transport via the alternative pathway is when NADH produced within the mitochondria is the substrate. With isolated mitochondria, the utilization of a Krebs cycle acid as a substrate that produces NADH is required. Since succinate dehydrogenase donates electrons directly to ubiquinone, no energy conservation is observed when succinate is the substrate. The primary source of electrons for the electron-transport system, and thus both oxidases, is Krebs cycle intermediates that are in the inner mitochondrial matrix. Differences in the ability of substrates to support alternative oxidase activity in isolated mitochondria probably reflect the relative maximal activities of the corresponding dehydrogenase and cytochrome oxidase.

Antibodies to the alternative oxidase recognize 35-, 36-, and 37-kilodalton proteins. A complementary DNA (cDNA) clone has been isolated that codes for a 38.9-kDa protein containing a 63-amino-acid transit peptide. This clone represents a nuclear gene encoding for a precursor protein of one or more of the alternative oxidase proteins that are transported into the mitochondria and are assembled there into a functional complex. From the sequence of the cDNA clone, it is predicted that the protein contains three regions of amino acids that are likely to form α-helices. These regions would also be expected to traverse a membrane. This prediction is consistent with the alternative oxidase being a membrane-based enzyme.

Measurements. A discussion of the function of and environmental influences on the pathway requires an understanding of how the pathway is measured. Only recently has the catalytic activity of the enzyme been measured in solubilized preparations. Thus most research on the pathway has used inhibitors of the cytochrome and of the alternative pathways in intact plant tissues and isolated mitochondria. To estimate the alternative pathway capacity, the rate of oxygen uptake is first measured in the absence of any inhibitor, then in the presence of cyanide, followed by a measurement in the presence of both cyanide and hydroxamate. The sites of action of these inhibitors are shown in Fig. 1. The capacity of the pathway is determined by the difference between the rate in the presence of both inhibitors and the rate in the presence of cyanide alone. This measurement is referred to as the capacity because the presence of cyanide ensures that all electrons are forced through the alternative pathway. Rates are frequently expressed as a percent of the control rate. The activity of the pathway is determined from the difference in the rate of oxygen uptake in the absence of inhibitors and the rate in the presence of hydroxamate alone. Measured values of the activity are lower than the capacity because some of the electrons flow through the cytochrome pathway during measurements of the activity.

In general, the interpretation is that the activity represents flow through the pathway in the experimental conditions under which the measurement was made. The capacity represents an estimate of maximal potential activity. Measurements from intact tissues presumably reflect the physiological status of the tissue. Capacity measurements from isolated mitochondria correlate well with capacity measurements from intact tissues. Activity measurements made with isolated mitochondria are interpreted differently. Those measurements are normally made in the presence of substrate levels that saturate the particular dehydrogenase that catalyzes the transfer of electrons to the electron-transport system. Under those conditions, the flow of electrons through the alternative oxidase is determined by the extent to which the particular dehydrogenase activity exceeds the capacity of the cytochrome pathway, so that spillover into the alternative pathway occurs.

In recent years, it has become possible to measure the amount of alternative oxidase protein based on the availability of monoclonal antibodies to it. These antibodies were produced against the alternative oxidase of the voodoo lily (*Sauromatum guttatum*), but they cross-react with the oxidase from several species. Immunological estimates of the amount of alternative oxidase protein made after electrophoresis and electrotransfer of mitochondrial proteins have confirmed capacity estimates made from inhibitor experiments.

Alternative oxidase capacity. It is now recognized that alternative-pathway respiration must be present in all plant tissues but in widely varying capacities or amounts. In some cases, the capacity may be large relative to the total respiratory capacity, but the activity may be very low or essentially zero. If the capacity of the alternative pathway exceeds the activity of the cytochrome pathway, cyanide may stimulate the rate of respiration because of the diversion of electrons through the uncontrolled alternative pathway. At the other extreme, when the alternative pathway capacity is low, the respiration of plant tissues is almost completely inhibited by cyanide.

Generalizing about what plant tissues might contain the greatest alternative-pathway capacity is difficult. Older tissues, such as mature leaves, might be expected to have a large capacity. Leaves have a rapid respiration rate when they are young and developing. When the leaves become photosynthetic, respiration rates decline. At this stage of development, some leaves have a large alternative respiratory capacity relative to the respiration rate. During fruit ripening, respiration rates increase during a development stage called the climacteric. Alternative-pathway respiration is a component of the respiration that is induced through the action of the hormone ethylene. In the thermogenic (heat-producing) tissue of the voodoo lily, the capacity of the alternative pathway increases at a specific time during flower development, and its level is induced by salicylic acid.

Functions. The general function for alternative respiration is to provide a mechanism for the utilization or disposition of electrons during a period in which metabolic processes require electrons to be disposed of without energy conservation in the form of proton motive forces or adenosinetriphosphate (ATP) synthesis. Some examples of proposed specific functions, including how they fit this general statement of the function, follow.

In thermogenesis, heat is produced from the energy released when electrons are transferred from compounds with a high reducing potential to oxygen. In order to generate sufficient heat to warm the tissue, energy must not be used for other purposes, and electron transport must be fast enough to produce adequate amounts of energy. Therefore, electron transport must be uncoupled from ATP production. In the breakdown of energy-rich storage compounds such as lipids, more electrons are stored than are needed to supply the energy requirements for growth of the plant. The alternative pathway provides a mechanism to uncouple electron flow from the energy needs and to allow the carbon to be mobilized to supply the carbon require-

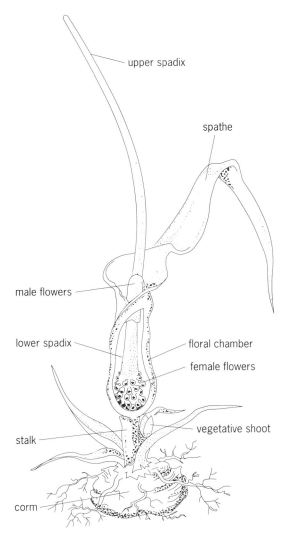

Fig. 2. Inflorescence of *Sauromatum guttatum* (voodoo lily). *(After L. Taiz and E. Zeiger, Plant Physiology, Benjamin/ Cummings, 1991)*

ments of growth. There may be situations in which carbon skeletons are needed for growth in excess of the electrons and energy provided when these carbon compounds are produced from stored substances.

The most extensively studied example of the alternative pathway is the development of the inflorescence of the voodoo lily, shown in **Fig. 2**. At day 5 of the development of this inflorescence, the spadix heats to temperatures as much as $16°F$ ($9°C$) above the ambient air temperature. This heating causes the volatilization of amines that emit an odor of rotting meat and attract insects for the pollination of the flowers contained within the spathe. Early in the growth of the spadix, the cytochrome pathway is actively providing energy for growth. At day 4, the alternative pathway becomes very active, and on day 5 the activity of the cytochrome pathway drops dramatically while the alternative pathway continues to be very active. Removal of the tip of the spadix at about day 3 prevents these changes, but if the salicylic acid is applied to the cut stump the respiratory changes are restored. This example illustrates a function of the alternative pathway in thermogenesis, dramatic changes in its capacity during a sequence of developmental events during flowering, and control by a chemically defined endogenous substance.

For background information *SEE BIOLOGICAL OXIDATION; CYTOCHROME; KREBS CYCLE; MITOCHONDRIA; PLANT RESPIRATION* in the McGraw-Hill Encyclopedia of Science & Technology.

Cecil Stewart

Bibliography. D. Davies (ed.), *The Biochemistry of Plants*, vol. 2, 1980; D. A. Day et al., Regulation of alternative pathway activity in plant mitochondria, *Plant Physiol.*, 95:948–953, 1991; R. Douce and D. A. Day (eds.), *Encyclopedia of Plant Physiology*, vol. 18, 1985; D. M. Rhoads and L. McIntosh, Isolation and characterization of a cDNA clone encoding an alternative oxidase protein of *Sauromatum guttatum* (Schott), *Proc. Nat. Acad. Sci.*, 88:2122–2126, 1991.

Polymer

Recent advances in the field of polymers include the development of a series of degradable plastics produced from polylactic acid, and the preparation of dendritic polymers with controlled molecular architecture.

Degradable Plastics

Interest in bio- and photodegradable plastics is rapidly expanding. Major environmental concerns such as land litter and waste materials on beaches and in the ocean are directly affected by plastics that can remain intact for up to hundreds of years. Stimulated by public demand, strong action by the state legislatures and the Congress are forcing industry to explore degradable and recyclable plastics. These plastics should completely degrade to innocuous compounds, have suitable physical and mechanical properties, and be cost-effective. Few, if any, of the degradable plastics currently available, including those containing

starch, meet these criteria. However, polymers and copolymers (for plastic coating and films) that can be made with lactic acid have very useful mechanical properties and degradability characteristics, are approved by the U.S. Food and Drug Administration (FDA), and break down completely into harmless products. Lactic acid is a chemical compound with a small molecule; it is found in organisms from microbes to humans. The availability of low-cost, high-carbohydrate food by-products (that is, starch from potato waste streams and lactose from whey permeate in cheese manufacture) and cost-effective bioconversion and recovery processes should stimulate the development of lactic acid as a commodity chemical and its use in making biodegradable and photodegradable plastics. (A commodity chemical is one that is produced in the hundreds of millions of pounds per year.)

Bioconversion. Agricultural and industrial waste streams have become an economic burden as well as a serious environmental problem. In the United States, billions of pounds of cheese whey permeate and approximately 10^{10} lb (4.5×10^9 kg) of potato waste each year are typically discarded or sold as cattle feed at a low price, but sometimes require expensive transportation. Some of the large processors of french-fried potatoes are purchasing hundreds of acres for spreading their potato waste on land. As a potential solution to the problem, the Argonne National Laboratory is developing bioconversion technology that uses enzymes or microbial organisms to convert substances in waste streams to lactic acid, which is then used to make environmentally safe, degradable plastics. The process not only will help to solve a waste problem for the food industry but will also save energy and be economically attractive.

The bioconversion of food wastes into higher-priced products of broader utility is very promising. For example, two of the three major wastes from the production of french-fried potatoes are highly enriched with carbohydrates, quite clean, and available in large volumes. About half of the potato is by-product. Other foods that are 70–75% starch (dry weight), such as corn, sorghum, and wheat, also are sources of by-products that are excellent candidates for direct enzymatic or microbial conversions.

Lactic acid has a number of advantages over other products (such as methane and ethanol) that could be made from such waste streams. Lactic acid is approved by the Food and Drug Administration and is biocompatible and nonallergenic. It is made by fermentation of inexpensive, high-carbohydrate food by-products (**Fig. 1**); it can also be made from petrochemicals, but the cost of production is higher. Promising technology is being developed for the recovery of lactic acid, and will probably result in major reductions in the cost of producing it.

Among the many industrial applications of lactic acid, an important use is as a major precursor [polylactic acid or PLA; structure (I)] for the expanding degradable plastics market. Because the physical properties and degradation rate of polylactic acid can be

$$\left[\begin{array}{c} O \\ \| \\ O \quad C \quad CH_3 \\ C \diagup \quad \diagdown \\ \| \quad CH_3 \quad O \\ O \end{array} \right]_n$$

(I)

adjusted and because polylactic acid is 100% degradable, it is generally superior to the starch-filled plastics, which are typically degradable only to the extent of the starch content (5–15%).

Prototype systems. Although the initial substrate for the Argonne process is potato waste, the process (**Fig. 2**) will be extended to convert other food wastes to lactic acid and other products. Proprietary technology for bioconverting more that 90% of the starch in potato wastes to glucose has been developed; the process requires less than 8 h compared to conventional technology requiring more that 100 h. The three steps (gelatinization, processes of liquefaction and enzymatic saccharification) typically required to process starch to glucose have been combined into two steps, and the doses of the thermally stable carbohydrases, alpha-amylase and glucoamylase, together with the pH and temperature parameters have been adjusted. Normally, higher glucoamylase concentrations during saccharification produce elevated amounts of maltose. Fortunately, this does not occur in the Argonne process. Unfermented maltose and other residual sugars contaminating the final product, lactic acid, could interfere with polylactic acid synthesis by functioning as

branch points, that is, as extra chains attached to the primary polymer chain of the lactic acid.

The glucose and other products of starch hydrolysis are subsequently fermented by strains of homofermentative lactic acid bacteria. The lactic acid is recovered, concentrated, and then further purified to a polymer-grade product. Batch fermentation of saccharified potato syrup, yielding lactic acid and glucose at 15 and 0.007 oz/gal (110 and 0.05 g/liter) respectively, can be done in less than 20 h, as compared to the 4 days required to make the so-called natural or fermentation lactic acid with the conventional technology employed in Europe and South America. To minimize cell biomass as a waste product, efficient continuous fermentation systems are being developed. Continuous fermentations (lasting more than 300 h) with high yields of sodium lactate have been achieved. A continuous system involves a cell recycle bioreactor, microfiltration for cell separation, and electrodialysis, a stage of downstream processing that results in the formation of lactic acid and sodium hydroxide. Sodium hydroxide is then recycled to the bioreactor to save chemical costs, and the lactic acid is further purified. Ion-exchange chromatography and the production of methyl lactate are being examined as final stages of purification.

Excellent plastics based on lactic acid, mostly for premium-priced medical applications, have been developed by various research groups. Polymers and copolymers that will be cost-effective and will have improved characteristics involving biodegradability and photodegradability are being constructed and test-

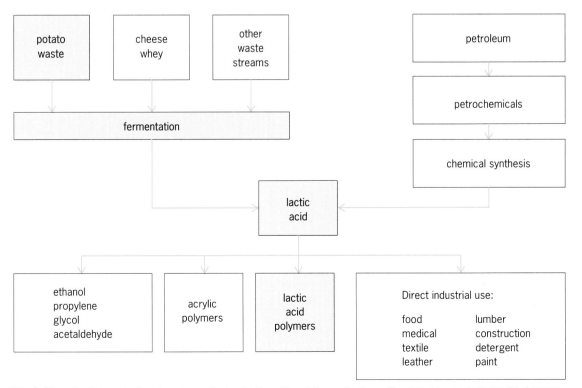

Fig. 1. Flow chart showing the steps in producing lactic acid and its applications. The tinted boxes indicate the Argonne process.

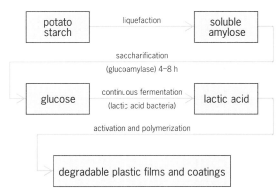

Fig. 2. Mechanism of the Argonne process, in which potato starch is converted to degradable plastics and coatings.

ed for a wide range of applications. A novel route of polylactic acid synthesis that promises to lower unit cost of production has been developed. In addition, modified polylactic acid films and coatings that possess photodegradability and increased sensitivity to water hydrolysis are being developed. The economical synthesis of polymers and copolymers constructed primarily from lactic acid and materials having so-called programmed degradability should stimulate agricultural applications for polylactic acid films and coatings.

Economic assessment. A lactic acid production plant would process two waste streams from a facility that produces french-fried potatoes with a total volume of 90,000 gal (340 kiloliters) of waste containing 10–12% starch per day. There would be a very favorable return on investment. The starch in solid potato waste and primary peel effluent would be enzymatically hydrolyzed to glucose, the glucose continuously fermented to lactic acid, and the lactic acid recovered and purified. With 90% bioconversion and 90% product recovery efficiencies, the projected capacity of the plant would be 16.2×10^6 lb (7.36×10^6 kg) of high-quality lactic acid per year. Although heat-stable (high-purity) lactic acid is required to make high-molecular-weight polylactic acid (for self-supporting films), a less pure lactic acid might be usable to prepare coatings from low-molecular-weight polylactic acid.

Market for degradable plastics. Public and legislative action has prompted a demand for degradable products such as bags for grocery, trash, and composting; six-pack beverage O-rings; packaging for the fast food industry; films for agricultural mulch; and backings for disposable diapers. This demand has created a need for millions of pounds of new materials. Some of these products are already mandated by law in many states. It has been projected that the North American market for degradable plastics will increase from 850×10^6 lb (386×10^6 kg) in 1992 to 1950×10^6 lb (886×10^6 kg) by 2000.

One promising end use for degradable plastics such as polylactic acid is in programmable systems for delivering fertilizers and pesticides, that is, systems for sustained release of the chemicals during the course of a growing season. A time-released pesticide eluting from a degradable plastic matrix could considerably improve the efficiency of pesticide use (with less waste and more cost effectiveness) and minimize pollution of groundwaters. A collaborative study between the Argonne National Laboratory and the Tennessee Valley Authority's National Fertilizer and Environmental Research Center is examining the efficacy of using polylactic acid for sustained release of urea-based fertilizers. Other potential markets for degradable polylactic acid plastics and coatings include 150×10^6 lb (68×10^6 kg) for agricultural mulch films, tens of millions of pounds for specialty bags or sacks such as for compost waste, 500×10^6 lb (227×10^6 kg) for marine plastics, and more than 200×10^6 lb (91×10^6 kg) for degradable conditioner coatings for paperboard stock. None of these applications, with the exception of degradable packaging materials, should impact recycling efforts or affect the composition of landfills; they are targeted primarily toward one-way degradable items, especially for agricultural uses.

One intriguing aspect of the agricultural use of polylactic acid is the potential for indirectly stimulating plant growth. A recent study determined that oligomers of lactic acid residues (dimers through decamers) may act as plant growth regulators but only the L isomer is effective. Although the mechanism of action has yet to be characterized, it is known that oligomers in the parts-per-million range can substantially increase plant size. Since polylactic acid degrades by water hydrolysis, it is reasonable to expect that oligomers (dimers through decamers) will be produced and, in turn, these degradation products may function as plant growth regulators or nutrient uptake facilitators. This hypothesis is being tested.

Prospects. Widespread use of polylactic acid other than for biomedical applications will hinge largely on its unit cost of production rather than on the cost of the monomer lactic acid. A number of research projects are engaged in developing cost-effective technologies. The bioconversion of inexpensive carbon sources such as potato by-product and cheese whey permeate to higher-value products is an attractive alternative to petroleum-based feedstocks. The environmental dilemma of by-product disposal can, in part, be alleviated via conversion of by-products to environmentally acceptable, degradable materials.

Robert D. Coleman

Dendritic Macromolecules

Dendritic macromolecules, or dendrimers, are a new class of polymers that have gained importance recently because of the rising demand for specialty polymers with novel and improved properties. They are also referred to as starburst, hyperbranched, arborol, or fractal polymers. Dendritic macromolecules are geometrically and structurally similar to fractals and dendrites. They are characterized by a high degree of branching that originates from a single focal point or core, by branches at almost every repeating unit, and by a large number of reactive functional groups at the extremities of the branches. As the molecular weights of the macromolecules increase, the molecules assume a spherical, three-dimensional shape, reminiscent of many important globular biopolymers such as hemoglobin; and their surfaces become progressively more congested until, at a certain point, they take on ball-like forms with densely packed, nonporous surfaces.

Divergent growth approach. The synthesis of dendritic macromolecules has been accomplished with two distinct methodologies. The initial approach was a starburst or divergent strategy first reported in 1985. In this approach the dendrimer grows by diverging from a small core molecule; this method is similar to the protect-deprotect strategy that protein and nucleic acid chemists use to make biopolymers.

Polyamidoamines are the best-known example of this strategy. As shown in **Fig. 3**, they are grown from ammonia as the initiating core; reaction with methyl acrylate (H_2C=$CHCOOCH_3$) gives a triester (II), which can be further reacted with an excess of ethylenediamine ($H_2NCH_2CH_2NH_2$) to regenerate the reactive amino groups (NH_2) at the end of each branch. This two-step sequence is called a generation growth (highlighted area in Fig. 3). At this stage, there are three amino groups at the periphery of the growing macromolecule (III); repetition of this two-step sequence leads to larger dendritic macromolecules in which the molecular weight and number of reactive groups at the chain ends double at each gener-

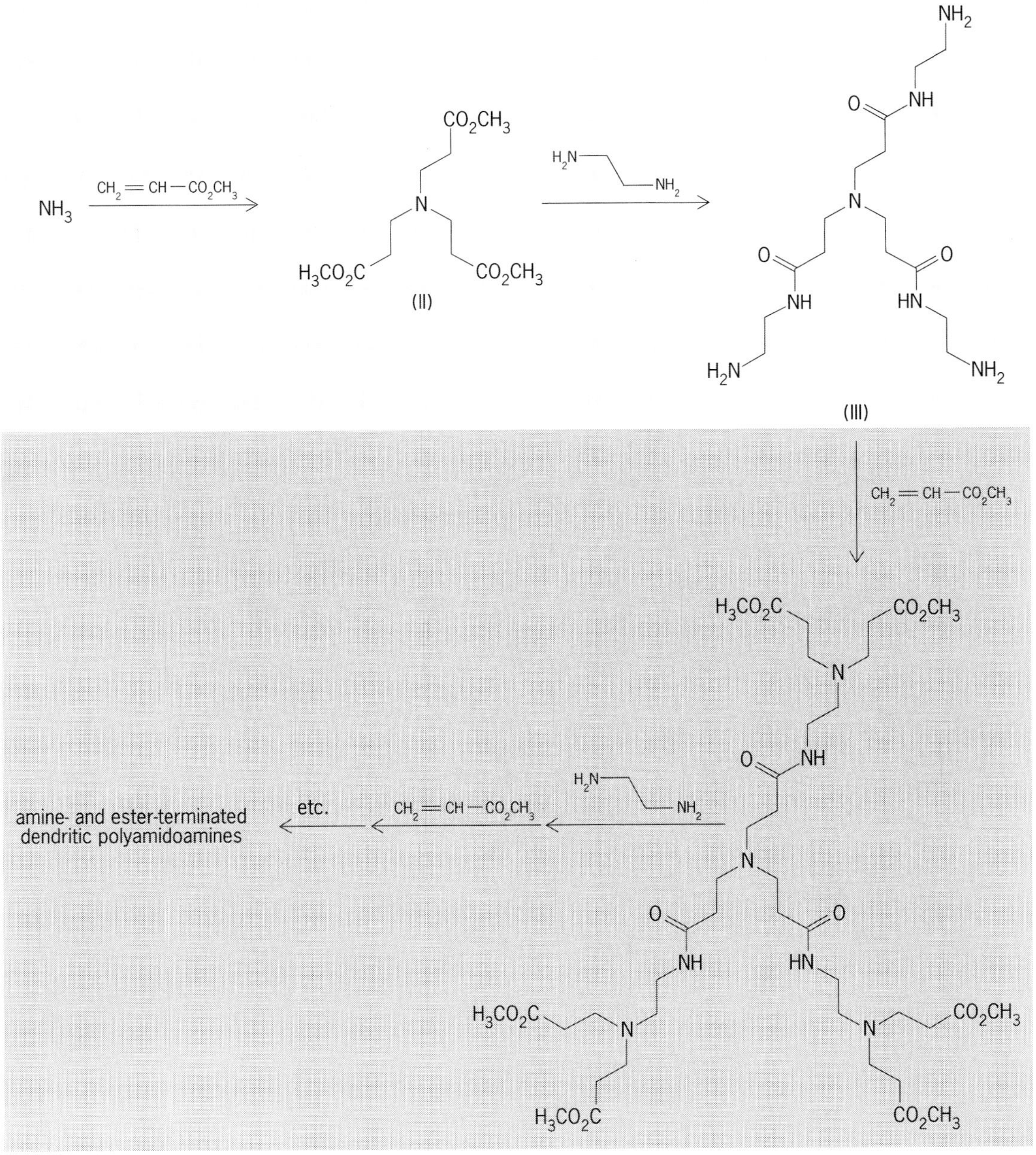

Fig. 3. Reactions for the synthesis of dendritic polyamidoamines by the starburst or divergent approach. The multiple arrows indicate a series of steps.

ation growth. By using this synthetic strategy, dendritic polyamidoamines approaching monodispersity were prepared with molecular weights up to 350,000, 768 reactive terminal groups, and a diameter of 9.8–10.5 nanometers, depending on the measurement technique. Several other novel applications of the divergent approach have appeared, leading to dendritic macromolecules based on repeating units of ether, amine, siloxane, or phosphonium cation.

A significant feature of the divergent approach is the rapid increase in the number of reactive groups at the periphery of the growing macromolecule. This leads to major difficulties, as the number of reactions required for generation growth increases dramatically. Incomplete reactions lead to structural imperfections, while ever-increasing excesses of reagents are required to force reactions toward completion. Despite these shortcomings, the divergent approach provides valuable access to a most unusual family of polymers, the first of which was commercialized in early 1990.

Convergent growth approach. To overcome these difficulties encountered in the divergent growth approach, the convergent growth approach was developed in 1989. The basic concept utilizes the symmetrical nature of these macromolecules to advantage, and is in essence the reverse of the divergent strategy.

Construction of the macromolecule is started at what will ultimately become its periphery and, at each step, growth is designed to occur via reaction of only a very limited number of reactive sites. In the basic approach discussed here, coupling of a single site of the growing molecule with two sites of the monomer is involved in every generation growth (**Fig. 4**). This is in sharp contrast to the divergent approach, where growth must involve simultaneous additions at a progressively larger number of sites.

The starting material contains what will eventually constitute the surface functionalities of the dendritic macromolecule as well as a reactive functional group (f_r). This material is then condensed with the monomer unit. The monomer itself has at least two coupling sites (c) and a protected functional group (f_p). After coupling, f_p is activated to f_r and the two-step generation growth process, coupling and activation, is continued by successive iterations, for example, until the dendritic "wedge" (IV) is obtained. The dendrimer (IV) has a single reactive group (f_r) at its focal point, and the final reaction is attachment to a polyfunctional core to provide the spherical, dendritic macromolecule (V), which has 64 surface functional groups (**Fig. 4**). The structure also demonstrates the fractal nature of these macromolecules. The convergent growth approach has been realized in practice with the synthesis of macromolecules containing repeating units of ether, amide, ester, and phenylene, with molecular weights up to 85,000.

While the divergent approach is superior in its ability to reach higher molecular weights, the convergent

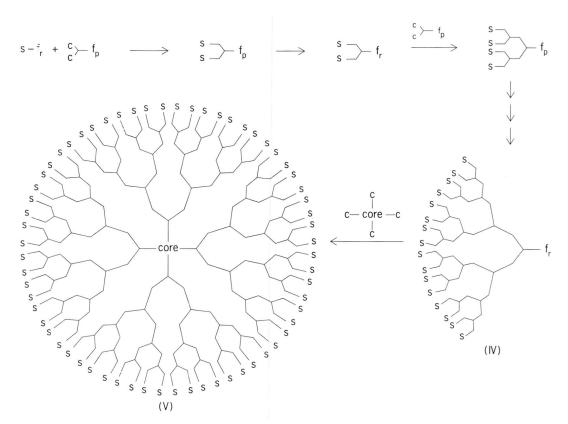

Fig. 4. Schematic representation of the convergent growth approach to dendritic macromolecules. s = surface functionality, f_r = reactive functional group, c = coupling site, f_p = protected functional group. The multiple arrows indicate a series of steps.

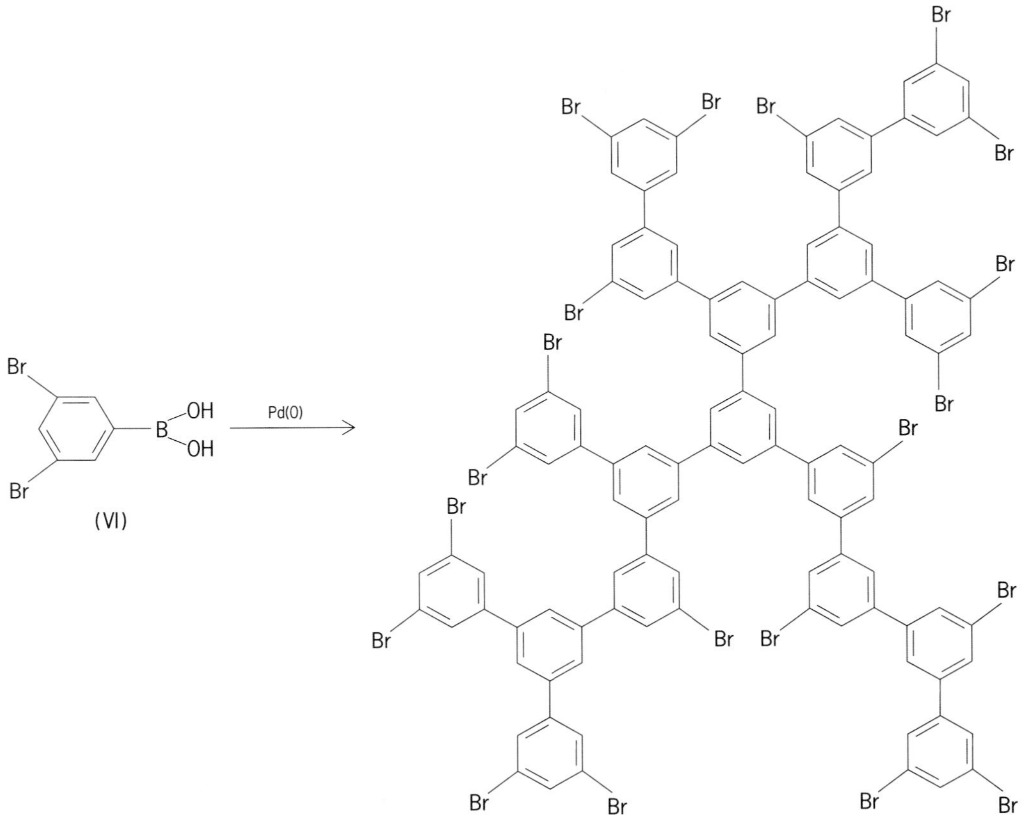

Fig. 5. One-step synthesis of a bromo-terminated hyperbranched polyphenylene, a boronic acid derivative; Pd(0) = catalyst with a central palladium (0) atom.

growth approach has a number of advantages at lower molecular weights (less than 50,000). The smaller number of coupling reactions and the resultant minimization of failure sequences (incomplete reactions at the peripheries of the dendritic macromolecules) allow a greater degree of control over the architecture of the final polymers.

For example, dendritic macromolecules that have different functional groups covering well-defined regions of the periphery of the macromolecule have been synthesized by using different starting materials and different dendritic wedges at specific points in the synthesis. Therefore, dendritic macromolecules having one, two, or n functional groups in specific areas of the periphery can be synthesized. The only requirement is that the functional group at the periphery is compatible with the reaction conditions used for generation growth.

The presence of the unique functional group at the focal point of the dendritic wedges can also be employed to control the macromolecular architecture. Macromonomers are large macromolecules or polymers that have a single polymerizable group attached. A novel class of dendritic macromonomers has been synthesized and polymerized under various conditions to give linear polymer chains to which spherical den-

dritic fragments are attached. Linear polystyrene, prepared anionically, has been terminated by dendritic fragments to give the polymeric analogs of structures shaped like barbells and lollipops.

Properties. The determination of the physical properties of dendritic macromolecules prepared by either the divergent or convergent approaches is still in its infancy. However, a number of interesting results have already been reported. The three-dimensional, globular shape of the dendrimers renders them up to 100 million times more soluble than their linear analogs, and they are soluble in a greater variety of solvents. In traditional linear polymers, the viscosity increases as the molecular weight increases. However, a plot of viscosity versus molecular weight for dendrimers follows a bell-shaped curve; above molecular weights of about 5000 the viscosity actually decreases as molecular weight increases. The polyphenylene dendrimers also possess a remarkable compatibility with several vinyl polymers, and they have made a significant contribution to the mechanical properties of the blended polymers. Dendritic macromolecules have also been investigated as carriers for drug-delivery systems, contrasting agents for nuclear magnetic resonance imaging (where they are several orders of magnitude stronger than conventional contrast agents), and models for electrochemical and photochemical processes; in addition, they recently became commercially available as calibrating agents for sieves, filtration devices for either air or solutions. The reason for the enhanced ability of dendritic macromolecules to act as drug-delivery systems or as contrasting agents is the extremely large number of functionalities at the periphery. For example, 1000 or more paramagnetic metal atoms such as gadolinium can be bound to the surface of a dendrimer so that the number of metal atoms per unit volume when compared to normal systems is greatly increased.

More recently, researchers have synthesized a covalently bound dendritic analog of micelles where the dendrimer has a completely hydrocarbon interior and a large number of ionic carboxylic groups at the surface. These materials have great promise since they show the same properties as traditional micelles but do not suffer disadvantages such as concentration-dependent structures.

Developing methodologies. Both the divergent and convergent approaches to dendritic macromolecules have a serious deficiency: they are multistep syntheses with a least two steps per generation growth. Therefore, while both approaches, especially the convergent, give highly defined and novel macromolecular architectures that are excellent for investigating the new and improved properties resulting from hyperbranching, they nevertheless are time-consuming and tedious. A number of workers have investigated the one-step synthesis of these macromolecules via the polymerization of AB_x monomers, where A and B represent reactive functional groups and x is 2 or greater. Initial interest in the theoretical aspects of the polymerization of AB_x dates back to the early 1950s; however, few examples of the purposeful attempts to control molecular architecture through the use of AB_x monomers were investigated until recently. **Figure 5** shows the one-step polymerization of an AB_2 monomer (VI) to give a hyperbranched polyphenylene macromolecule (VII); the reaction between aryl boronic acids with aryl halides to give phenylenes is a standard Suzuki coupling catalyzed by a palladium (0) species, normally $Pd(PPh_3)_4$. While a high-molecular-weight polymer (about 10,000) is obtained in only one step, growth is irregular and contains many failure sequences. These polymers are not strictly dendritic in nature, but they possess a very high degree of branching and are hence termed hyperbranched macromolecules. It is envisaged that the properties of these hyperbranched polymers obtained from the one-step polymerization of AB_x monomers will approach those obtained for the "perfect" dendritic macromolecules prepared by either the divergent of convergent approach.

For background information SEE POLYMER; POLYMERIZATION in the McGraw-Hill Encyclopedia of Science & Technology.

C. J. Hawker

Bibliography. C. J. Hawker and J. M. J. Frechet, Preparation of polymers with controlled molecular architecture: A new convergent approach to dendritic macromolecules, *J. Amer. Chem. Soc.*, 112:7638–7647, 1990; A. Kinnersley, Promotion of plant growth polymers of lactic acid, *Plant Growth Regulat.*, 9:137–146, 1990; D. A. Tomalia, A. D. Naylor, and W. A. Goddard III, Starburst dendrimers: Molecular-level control of size, shape, surface chemistry, topology, and flexibility from atoms to macroscopic matter, *Angewandte Chemie, International Edition in English*, 29:138–175, 1990; Y. H. Kim and O. W. Webster, Water-soluble hyperbranched polyphenylene, "a unimolecular micelle," *J. Amer. Chem. Soc.*, 112:4592–4594, 1990; K. L. Wooley, C. J. Hawker, and J. M. J. Frechet, Hyperbranched macromolecules via a novel double-stage convergent growth approach, *J. Amer. Chem. Soc.*, 113:4252–4261, 1991.

Printing

Recent advances in printing include development of digital printing systems, desktop color publishing systems, and keyless offset printing.

Digital Printing Systems

The explosion of desktop color publishing systems in prepress, the development of high-speed printing plates, and the introduction of fast-makeready, high-quality, small-format presses have combined to make it possible for lithography to enter the short-run color printing market. The development of digital printing systems has been an important factor.

There are three main approaches to digital printing systems: (1) The digital information is used to make plates or cylinders that produce printed images with conventional inks and presses and that apply pressure to transfer the images to the paper. (2) High-speed digital office copier systems, such as digital color laser copiers or digital document printers, are used. (3) The digital data are used to modulate electrostatic, magnetic, or ionic charges on a charged, sensitive, flat or cylindrical surface on which images are produced with special toners and transferred to the substrate.

Digital cylinder-making processes. Most gravure printing cylinders are made on electromechanical engravers. A number of commercial systems are designed to have links from their prepress systems to the electromechanical engraver. This engraver comprises two machines with associated electronics. For conventional gravure cylinder production, prints of the images are mounted on a cylinder in the first machine in the order of the printing layout. Special reading heads, usually one for each row of images across the cylinder, are used to scan the images and record them as photographic densities that are converted by a computer into electrical impulses. These impulses are fed to engraving heads with diamond styli that are mounted on the second machine. Here the engraving heads are arranged as on the input machine and are positioned so that the diamond styli engrave variable-area, variable-depth cells in the copperplated printing cylinder as it rotates in the machine. Each engraved cell corresponds to a scanned area of the image on the first machine.

For digital imaging, the prepress links replace the images on the first machine and send the image information as digital data directly to the computer of the electromechanical engraver, which converts them to electrical impulses that drive the engraving heads on the second machine. The electromechanical engraver engraves 4000 cells per second. It takes eight engraving heads on an 80-in. (200-cm) cylinder to produce the engraved cylinder in 1 h. Electron-beam engraving has been developed for making gravure cylinders, and is much faster than electromechanical engraving. One head engraves 120,000–150,000 cells per second, so an 80-in. (200-cm) cylinder can be produced in 20 min.

Digital plate processes. There are four types of digital plates. Three use conventional lithographic principles, which depend on the mutual repellence of ink and water and need a critical ink–water balance in printing. The fourth uses a lithographic press, but the process uses dry platemaking and printing principles. The three digital lithographic plates use high-speed processes that can be driven directly from digital data by using lasers, but they require liquid processing and ink and dampening solution for printing. They are silver halide, electrographic, and high-speed-photopolymer systems. The fourth type uses a spark discharge system that requires no processing in making the plates and no water for printing them.

Silver halide plates. These plates use photosensitive coatings similar to photographic film, except that the silver halide emulsions are slower and, for color reproduction, are coated on anodized aluminum. They are somewhat light-sensitive, and thus must be handled in yellow filtered light. The laser-exposed coatings are processed in developers that render the coatings insoluble in the image areas, and the unexposed coatings are dissolved by special chemical treatments. The processing solutions contain heavy-metal (silver) pollutants, which must be either carted away to special treating plants or treated in-plant with silver recovery chemicals before being drained into municipal sewers. The finished plate consists of hardened oleophilic (ink-receptive) coating in the image areas and lithographically treated and gummed hydrophilic (water-receptive) surfaces in the nonprinting areas. The plates are printed on the press with ink and water.

Electrophotographic plates. These plates are made of anodized aluminum coated with organic photoconductors that are charged by a corona discharge, exposed by lasers, and developed by liquid toners, which adhere to the exposed (image) areas and are repelled by the unexposed (nonprinting) areas. The coating in the unexposed areas is dissolved, and the areas are treated with a lithographic "etch" and gum. The toning process must be vented, as most toners are dispersed in isopar, a liquid immiscible with water and considered a toxic material by the U.S. Occupational Safety and Health Administration (OSHA). The process of removing the coating leaves slightly ragged edges on images and a heavily polluted solution that

must be carted away for treatment. Like the silver halide plates, these plates are run on the press with ink and water.

High-speed photopolymer plates. These plates use special high-speed dye-sensitized photopolymers coated on anodized aluminum. They are so light-sensitive that they must be handled in filtered light. On exposure to lasers, the polymers in the image areas increase in molecular size and become water-insoluble and oleophilic. Following the exposure, the unexposed areas are dissolved, and the anodized aluminum in the nonimage areas is treated with lithographic etch and gummed. The processing solutions are water-based and do not carry many pollutants, so they can be drained into most municipal sewer systems.

Spark discharge system. This system is significantly different from all the other platemaking processes. It is nonphotosensitive, so it can be handled in ordinary daylight without danger of exposure or fogging. It is a completely dry system, so there are no polluted solutions that require disposal. Most importantly, it is a dry printing system that does not depend on ink and water to separate the image and nonimage areas, so much less skill is required for producing and printing these plates.

The spark discharge system is based on proprietary technology. The plate consists of a dimensionally stable polyester base that is ink-receptive and is coated with vacuum-deposited aluminum, which serves as a ground for the spark discharge, and silicone rubber, which resists wetting by ink. The images are produced by digitally operated electrodes consisting of very fine wires that burn through the silicone rubber and aluminum coatings in the image areas corresponding to digital signals from the press computer or imagesetter. No processing is required beyond wiping the burnt material from the surface of the plate. A feature of one press is that the four color plates are exposed on the press simultaneously in register, so that practically no makeready is required prior to printing.

Digital office copier systems. For short-run digital color printing, two special color copiers have been developed; they use a so-called intelligent processing unit to output digitally composed color pages at the rate of five per minute. It is anticipated that a number of manufacturers of color copiers will develop similar capabilities. These systems will satisfy the short-run color printing market in the run-length categories of less than 100 to about 200. Above 200 copies, digital plate systems can produce color prints at lower cost.

Digital printing from charged surfaces. Three types of charged surfaces are used for digital printing processes: electrophotographic, magnetographic, and ion-deposition.

Electrophotographic. Electrophotographic printing is similar to high-speed office copier systems. This pressureless printing process uses an electrostatic photoconductor that is charged by a corona discharge, is imaged by lasers modulated by digital signals, and is developed by dry or liquid toners, and a system for transferring the toned image on the photoconductor to a substrate. Digital electrophotographic systems have worked fairly well for single-color or spot-color printing, but they are very slow: 100–300 ft/min (30–90 m/min) as compared with web offset at 1000–2000 ft/min (300–600 m/min).

An example is a web press for printing business forms and checks that is linked to an electronic page layout system and a database of variable information. Paper is imaged directly with liquid toner at 300 dots/in. resolution on two sides or in two colors on one side at 300 ft/min (90 m/min). One-quarter of each page can be different on successive impressions. About 20 presses are in use worldwide for various types of printing requiring variable information.

Magnetographic. Magnetographics is a plate or cylinder pressureless printing process. It is similar in principle to electrophotographic printing, except the photoconductor is magnetic instead of electrostatic, the toners are magnetic, and a device similar to a printing press is used for transferring the image to the substrate. Magnetographics is a short-run process that has a breakeven point with lithography at about 1500 copies. Its limitations are slow speed, toner cost, and lack of light-colored or transparent toners, so it is not suitable at present for process color printing. The image requires toner thicknesses up to 30 micrometers, whereas ink films can be as thin as 2 μm, so that toner costs are very high in comparison with ink costs.

Ion-deposition. This technology is similar to other electronic printing systems. The process consists of four steps: (1) An electrostatic image is generated by directing an array of charged particles (ions) from a patented ion cartridge toward a rotating drum, which consists of very hard anodized aluminum maintained at a temperature of about 54°C (130°F). (2) As the drum rotates, a single-component magnetic toner is attracted to the latent electrostatic image on the drum. (3) The toned image on the drum is transfixed to plain paper through cold pressure fusing. (4) Most of the toner (99.7%) is transferred to the paper and the remainder is scraped off the drum with a reverse-angle doctor blade, and then the drum is ready for reimaging.

Over 3000 ion printing systems are currently in use for volume and variable printing of invoices, reports, manuals, forms, letters, and proposals as well as specialty printing of tags, tickets, and checks. This system has not been used for color printing because of the paper distortion that can be caused by the cold pressure fusing during the printing cycle.

A modified ion-deposition printing system is under development. The process uses two unique materials that overcome the deficiencies of previous processes. The process is claimed to have ink-density rather than area modulation (continuous-tone effect), no image moiré, 100% transfer of ink, and instant drying of ink on contact with the substrate. Such a process could have print-on-demand as well as short-run and long-run color capability.　　　　*Michael Bruno*

Desktop Color Systems

Desktop color systems are different things to different people. A basic desktop color publishing system is a personal computer with a color monitor and a color printer, with reproductions made on a color copier. Better-quality systems use desktop scanners and imagesetters, and professional systems utilize an interface (link) that connects to sophisticated prepress equipment and produces originals that can be used in the printing process to make multiple prints. Generally, the cost of the desktop color publishing systems increases in direct proportion to the quality and productivity required.

Desktop publishing systems are used to produce pages of information with text, artwork, and photographs that are ready for duplication on a color photocopier or printing on a multicolor press. Before desktop publishing, prepress services were provided by professionals known as typesetters and color separators and required expensive equipment, training, and support. Modern desktop publishing equipment offers similar functionality for less money. Three types of equipment are required: workstations, scanners, and output devices.

Workstations. The central component of a desktop publishing system is the computer, or workstation. Most widely used are personal computers. More sophisticated workstations are available when higher-quality images or greater productivity is required. Computers and their operating systems differ in their capabilities and requirements, and advantages and disadvantages. Considerations in choosing a computer include the available software (for example, programs for page layout and for retouching color photographs); the operating system and its features (for example, how easily it can be customized); the availability of inexpensive clones for the particular computer; and the capability to perform more than one task at a time (multitasking).

Scanners. A scanner converts a mechanical image, such as a photograph, into an electronic image and stores it either on magnetic tape or on an electronic disk. The two basic scanner technologies are charge-coupled-device and photomultiplier-tube.

Charge-coupled-device scanners are less costly than the photomultiplier-tube scanners. This affordability has made it possible for many companies to use them to publish newsletters and magazines in-house, and for individuals to practice desktop publishing as a hobby or a part-time job. The problem with charge-coupled-device scanners is that they lack the dynamic range needed for high-quality work. Dynamic range refers to a scanner's ability to capture the gradations of an image from the lightest highlight to the darkest shadow.

The photomultiplier-tube scanner is more sensitive. It utilizes unsharp masking information to control the exposure of the film, ensuring high quality. Information from a photomultiplier-tube scanner is transmitted to proprietary prepress equipment that can handle the large amounts of data generated from this scanner. The advantage of photomultiplier-tube scanners is the dynamic range of color captured; the disadvantage is the large files created, which are cumbersome for all but the most powerful desktop publishing computers to manipulate.

Output devices. The laser printer is the common denominator of all desktop publishing. Laser printers are used to proof pages or to serve as the final output devices. For low-cost color work, color laser printers can be coupled with color copiers. When higher quality is required, more sophisticated output devices known as imagesetters are used. The first imagesetters were modified typesetting devices, with a hardware translator known as a raster image processor (RIP). Collectively the raster image processor and output engine were termed imagesetters, because they could print images. These devices output onto a photographic medium, either resin-coated paper or film, which is developed with a process based on photographic techniques.

Until recently, imagesetters suffered from quality problems involving misaligned registration and moiré. Film that has been printed from an imagesetter can exhibit obvious registration problems when the crop marks or the color separation targets do not align. Moiré problems are distortions in the patterns of dots, known as rosette patterns, which are essential for the accurate appearance of color.

Another disadvantage of imagesetters is the time required to output a page. For example, a page with color scans may take an hour or more to print. In production environments, it is often necessary to output 40 or more pages overnight, and so slower imagesetters are not viable.

Newer imagesetters have resolved some of the quality issues, and are either roll-based (capstan) devices or drum-based devices. In roll-based devices, both the imaging laser beam and the film move at the same time, resulting in artifacts. In contrast, the drum-based equipment moves the medium after the imaging is completed. Since all the mechanical problems of supplying and moving the medium are handled when no imaging is occurring, fewer artifacts result.

Interfacing. Among the most recent advantages in color desktop publishing is the development of interfaces (links). These interfaces link traditional prepress equipment to the personal computers used with desktop publishing. These links are offered by companies known as service bureaus.

These firms represent a new industry that has developed with desktop publishing. They offer services to individuals or companies that utilize desktop publishing technology but cannot purchase the expensive input or output devices. Typical services include scanning with photomultiplier-tube devices, printing from an imagesetter, printing 35-mm slides, and linking files to traditional prepress equipment.

One advantage of the prepress links is that the color input and output are done on high-quality equipment. Another advantage is that the production stages requiring the greatest expertise (that is, color scanning and separations) remain in the hands of experienced technicians.

There is an advantage in linking the various desktop publishing functions to combine typesetting, tints, and line work from desktop publishing computers with high-quality scans. Users can accomplish publishing tasks on affordable equipment. The disadvantage of these systems is the array of problems that have to be solved in order to make the equipment work reliably. All the linking solutions are plagued by problems, both major and minor. However desktop color publishing is defined, the advantage is that smaller, less expensive computers allow more people access to technology that produces color published materials.   *Howard Fenton*

Keyless Offset Printing

Offset lithographic printing is the most widely practiced printing process in the world. A significant offset printing process change such as keyless inking is of major importance to the graphic arts world.

Conventional offset. Offset printing is synonymous with lithographic printing. In lithographic printing, it is essential not only to supply ink to a printing plate that carries the image to be printed, but also to supply a dampening water solution to the printing plate. The dampening solution forms a water layer in the nonimage regions of the plate. The water prohibits transfer of ink from the inking rollers to all portions of the printing plate except the intended image areas. Dampening solution must be applied continuously to the printing plate, because water is lost by misting off, and by evaporation from, the press rollers. Some of the dampening water is forced into the ink and is carried away as part of the image being printed. Water may also be lost to the paper from the nonimage areas on the printing plate.

In conventional, high-speed offset lithographic printing systems, ink is continuously made available to the inking rollers by means of numerous adjustable column or zone input keys. These keys are mechanisms located across the width of the printing press and supply to each zone an amount of ink corresponding to the image content in the same zone of the printing plate. Each inking zone corresponds typically to about 2 in. (5 cm) of press width. A newspaper offset press may have up to 32 ink input control zones for each side of the paper web being printed. An 80-page newspaper produced only with black ink requires setting and adjusting 640 inking keys while printing. Hundreds of other keys need manipulation if significant portions of the newspaper include color graphic advertising or color pictorial content. Color printing requires application of three additional inks to the page: cyan, magenta, and yellow. The inking keys corresponding to three additional printing couples are required for each side of each web of paper on which color will be printed.

Zoned inking keys vary from manually adjusted screws or levers that open and close an ink input slit at each zone across a printing couple, to sophisticated semiautomatic zoned ink pumps that precisely inject required amounts of ink through fixed slits. The latter systems can be remotely controlled at a console by the press operator. These semiautomatic inking systems are more expensive, but provide ongoing savings because they involve less waste and fewer operator errors than manual inking control systems. In either case, zoned ink input control is complicated and requires highly skilled operators or a considerable increase in

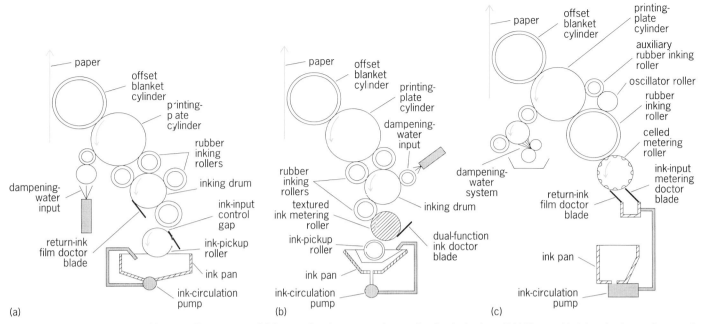

Keyless offset systems. (*a*) Conventional press couple retrofitted to be keyless. (*b*) Lithographic ink-train-dampening couple retrofitted to be keyless. (*c*) Lithographic press couple designed to be keyless.

the number of mechanical, electronic, and software control components. Accordingly, the concept of simplified inking in offset printing systems has considerable appeal to the printers.

Simplified offset inking. At the present time, simplified offset inking also means keyless offset inking. The appeal of simplified inking systems has resulted in a considerable number of keyless offset press installations in three of the world's major printing market regions: North America, Japan, and Europe. The reasons for purchase of keyless offset systems are (1) retention of the inherently advantageous offset printing qualities, (2) simplification of press control and of press maintenance, and (3) expectation of more consistent printed product output. These factors yield lower production costs and equivalent or better product quality.

In keyless systems, ink input is self-regulating. Press simplifications include elimination of the mechanical components or pumps, and of the electronics required to adjust and control the inking pumps or keys. Certain keyless offset press designs also have fewer inking rollers. Keyless inkers do not require the multiple inking rollers considered necessary in conventional, keyed press systems to assure smooth, consistent ink input to the printing plate. Having fewer rollers also contributes to lower maintenance and lower inventory costs. With keyless inking, hard-to-find skilled printing technicians are no longer required.

Improved and simplified control of the printing process is obtained, because the ink input is mechanically controlled to be uniform in time and in location across the press. No incorrect ink input adjustments are made, because there are no inking keys or pumps to adjust. User experience with the most successful keyless offset presses indicates that these can also be operated

without need for setting changes for dampener input. With conventional offset presses, each change in ink input might require a dampener input adjustment to regain the correct balance of ink and water inputs.

Keyless offset trend. Keyless offset printing presses have appeared all over the world virtually simultaneously since the mid-1980s. There may be a technology battle, for keyless offset may involve patent protection that could affect the positioning of press manufacturers in the marketplace. Conventional keyed offset presses from various manufacturers are virtual duplicates of each other from a process viewpoint; this is not the case with keyless offset presses.

Six press manufacturers with headquarters in three different countries have provided nearly 900 keyless offset printing couples to newspaper printers and publishers. Each set of two couples can continuously print one color on both sides of a 16-page newspaper section at up to 70,000 sections per hour. The number of couples in a complete printing press depends upon the number of sections and the number of color pages required for the specific newspaper.

Keyless press components. Regardless of differing physical details among the various components for keyless printing, a successful keyless offset press must accomplish virtually all functions of a conventional keyed offset press, as well as perform three additional functions: (1) it must continuously provide a uniform ink input film to all regions of the printing plate without operator assistance to adjust for the plate's nonuniform ink requirement, independent of where on the printing plate the images are located; (2) it must continuously remove the nonuniform residual portion of the input ink film remaining on the inking rollers that was not picked up by the printing plate image regions for transfer to the paper, to avoid buildup of ink on the inking rollers due to the continuous uniform supply of input ink; and (3) it must avoid inking interference due to possible buildup of the water that is required at the printing plate for image–nonimage differentiation, caused in part by the fact that some water is continuously being forced by the press into the ink on the inking rollers.

Six keyless offset printing press configurations are available commercially. One of the earliest configurations was developed and patented by the American Newspaper Publishers Association. Two systems use microscopically celled rollers to meter fixed quantities of ink into the press. Accurate and uniform ink input is accomplished by overfilling the metering roller cells with ink and then scraping (doctoring) the overfill ink from the roller's surface. Two systems use a conventional slow-speed ink input roller and adjustable roller-to-roller gaps to meter a uniform ink amount into the press system of rollers. Two systems can use a textured, rather than celled, metering roller plus doctor blade to meter the input ink. The three most widely used keyless offset systems are shown in the **illustration**. The lithographic press couple (illus. *a*) has conventional gap control of ink input, with a return ink-film scraping blade and recirculation of the scraped-off ink. The lithographic ink-train-dampening newspaper press couple (illus. *b*) is made keyless by fitting the press with a textured ink-metering roller plus doctor blade to provide uniform ink input and recirculation of the scraped-off ink. The lithographic press couple designed to be keyless (illus. *c*) has a minimum number of rollers; the keyless requirements are met by using two doctor blades operating on a celled ink-metering roller and an ink-recirculation system.

All keyless inking presses utilize a return-ink film doctor blade to remove the nonuniform ink film due to partial ink removal by the printing plate. The ink metering and return-film doctoring functions can be accomplished with one or two doctor blades. All keyless offset press systems have an integral means for recirculating the scraped-off ink for reuse by the press, since much more ink is being scraped off rollers than is being printed out. All these systems print black reasonably well. As yet, none print full-color copy acceptably.

Operational systems. Differences in keyless printing press systems center primarily on overcoming problems that customarily are not present in keyed offset printing systems. In conventional keyed offset printing, there is no scraped-off return ink. In keyless offset, removal of return ink is a requirement, and that ink contains dampening water. Since the return ink must be reused, excessive water can accumulate either in the ink or as separate layers on inking components. Free water prevents smooth quantitative transfer of ink to the printing plate, and the printed product fails to meet quality standards. This problem associated with keyless offset is particularly apparent in attempting to print the more demanding three- and four-color formats. Solving this ink–water interaction barrier remains the only hurdle to a general evolution of the whole industry into keyless offset printing.

For background information *SEE COMPUTER VISION; DIGITAL COMPUTER; MICROCOMPUTER; PHOTOCOPYING PROCESSES; PRINTING; REPROGRAPHICS* in the McGraw-Hill Encyclopedia of Science & Technology.

T. A. Fadner

Bibliography. M. H. Bruno, *Status of Printing in the U.S.A. 1989–90*, GAMA Communications, Salem, New Hampshire, 1991; M. H. Bruno, *What's New(s) in Graphic Communications*, no. 93, July-August 1991; T. A. Fadner and L. J. Bain, A perspective in keyless inking, *TAGA J.*, pp. 443–470, 1987; H. Fenton, State of desktop color publishing, *Color Publ.*, 1(1):8–17, Spring 1991; S. Hanaford, *An Introduction to Digital Color Prepress*, AGFA Compugraphic Division, Wilmington, Massachusetts, 1990; H. D. Matalia and M. Navi, Method of enhancing inking in offset presses, U.S. Patent 4,407,196, October 4, 1983.

Programming languages

A constraint describes relationships that must hold among a number of variables. Because real-world problems can often be naturally modeled by means of constraints, it is desirable that constraints be usable directly as programming constructs. Constraint programming languages have been designed to meet this goal, and their programs are often more expressive, simpler, and more declarative than traditional programs.

Drawbacks of traditional programs. An example is provided by Ohm's law, Eq. (1), which

$$V = IR \tag{1}$$

describes the relationship between voltage V, current I, and resistance R in a resistor. In a procedural programming language, this relationship can be used only to calculate the value of one of the variables, given previously computed values for the other two. For example, in ALGOL, one of the statements in the programming code (2) would have to be used to calculate either V, I, or R, respectively.

```
V := I * R
I := V / R          (2)
R := V / I
```

This requirement is undesirable for a number of reasons. While a single constraint describes the behavior of the physical object, programs require multiple descriptions depending on the way the relationship is to be used. Furthermore, the statements must be ordered so that the variables on the right-hand sides of the assignments have already been given values. Finally, values are assigned to variables only by propagating through directed forms of constraints, one at a time.

Early constraint languages. To avoid these restrictions, there have been efforts since the late 1960s to implement the notion of constraints as a programming construct, for example, in the systems SKETCH-PAD and THINGLAB. The earliest efforts, which were partially successful, revolved around augmenting a procedural language with constraints that are solved by ad hoc techniques. One primary technique is local propagation, based on a generalization of dataflow systems. Informally, a system of constraints is said to be solved by local propagation if all the variables in the system have a single value associated with them after a finite number of local propagation steps. A local propagation step occurs when a constraint has a sufficient number of determined variables for some of its other variables to be determined. These newly determined variables may then precipitate further local propagation steps in other constraints.

For example, a simple electric circuit consisting of a voltage cell and two resistors R_1 and R_2, all connected in series, is typically represented as a constraint network built of simpler units such as adders and multipliers (see **illus**.). The two multipliers are used to represent Ohm's law applied to the two resistors. By Kirchhoff's law, the sum of voltage drops is equal to the cell voltage, and this is represented by using the adders. Now, given I, and given both resistances R_1 and R_2, the cell voltage V can be calculated by local propagation of the values upward through the network to give V.

However, although there is enough information to determine V_1 and V_2 if V, R_1, and R_2 are given, they cannot be calculated by local propagation. Any attempt to direct the equations into a dataflow graph will result in a cyclic dependency. Thus, the solving of simultaneous linear equations is what is really needed. In short, local propagation is a general but weak technique, and it has a number of drawbacks. In particular, it cannot deal with constraints involving cyclic interdependencies.

As a separate issue, the two instances of Ohm's law could have been expressed as a single constraint (template), of which each resistor is a dynamic instance. This approach would make it much easier to reason about complex systems where many different instances of one component type occur.

The example illustrates two of the major weaknesses of early constraint languages. The constraint-solving facilities were generally not powerful enough, and the languages were not expressive enough. Consequently, the constraint-handling mechanisms in these early systems fell short of what was needed to obtain a general-purpose constraint language. Instead, the languages were primarily applications-oriented, for example, in the analysis of electric circuits, geometric layout, or graphical user interfaces.

Features of constraint languages. A constraint programming language can be characterized by the problem domain over which the constraints are defined, how the constraints are interpreted by the program's execution model, how the solvability of constraints affects the control of program execution, and how and to what extent the constraints are solved.

The most common problem domains involve arithmetic. Real-number arithmetic constraints are useful in applications such as the modeling of physical systems. Integer arithmetic constraints are useful in combinato-

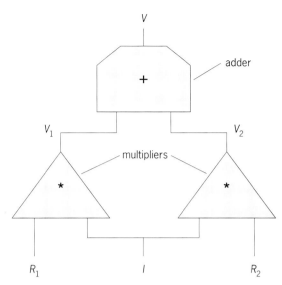

Constraint network of a simple electric circuit.

rial problems. Other examples are finite domains (often user-defined), domains of tree structures, and domains of sets.

Constraints appearing in a program may represent constraints as such or may be templates for constructing constraints at run time. The former are static constraints, the latter dynamic constraints, and run-time copies of such constraints are constraint instances. As mentioned in the circuit example, if only static constraints are available, constraints will have to be written for each component. However, with dynamic constraints, a constraint template is defined for each type of component, and then for each component instance a copy (or instance) of the template is constructed, possibly with extra values filled in. This procedure leads to the notion of collecting instances of constraints at run time.

The control issue involves how the solvability of constraints collected at some point in a computation may influence the future collection of more constraints. In most early constraint languages, the constraints themselves had no effect on program control. Hence these languages sacrificed a key advantage of using constraints. In the more recent languages that are based on logic programming, the unsolvability of constraints is used to limit the goal-directed search to useful directions.

Many important problem domains give rise to constraints that are infeasible to solve in general. Thus, to obtain a practical constraint language over such a domain, it is necessary to select a subclass of constraints that can be solved feasibly. Executing a constraint program involves both collecting constraints and determining if the constraints are solvable. There are many approaches to constraint solving, and, as mentioned above, a main one used in many early constraint languages is local propagation. In general, the solving of constraints requires specialized techniques specific to the problem domain at hand. For example, more powerful algorithms, like the simplex algorithm for linear arithmetic constraints and symbolic algebra techniques for nonlinear arithmetic constraints, are used in newer systems. Many systems also use custom-made algorithms for solving classes of the constraints that can be encountered; those constraints that are unsolved can be deferred from consideration until the time when they become simpler (as the result of variables taking known values, for example). *SEE SYMBOLIC COMPUTING.*

Newer constraint languages. While any language can be augmented with constraint facilities, an important issue is how the underlying language interacts and integrates with those constraint facilities. In the case of the older systems, which are based on extending a procedural language, this interaction was often clumsy and required the user to give a great deal of information about how constraints were to be collected and solved.

Most recent efforts have concentrated on extending the declarative languages: the functional and logic-based languages. The overwhelming majority of this work is based on the logic-based languages, and in particular, the constraint logic programming (CLP) languages. These are a generalization of logic programming languages in which unification is replaced by constraint solving over some domain of computation. A key feature is the use of dynamically created constraints and integration of the constraints within a clear semantic framework.

Constraint logic programming. A constraint logic programming language is defined by specifying the domain of computation over which the constraints in programs are defined. A number of languages based on this scheme have been designed and implemented. In the language CLP($\mathcal{R}$), there are real arithmetic constraints as well as the constraints found in PROLOG, equations over trees. In PROLOG III, there are PROLOG-like constraints, rational arithmetic constraints, boolean equations, and equations over strings. In CHIP, there are constraints over finite sets as well as rational arithmetic constraints.

Example. CLP($\mathcal{R}$) is used here for illustration. The syntax of CLP($\mathcal{R}$) is essentially that of PROLOG, but terms can now include arithmetic subterms, and explicit constraints using = on terms and > or $\geq$ on arithmetic terms may appear in the body of a rule or goal. The language is illustrated by the example program (3). The "mortgage" predicate relates the five

```
mortgage(P, Time, IR, MP, Bal) :-
      Time = 1,
      Bal = P + (P * IR/1200) - MP.
mortgage(P, Time, IR, MP, Bal) :-    (3)
      Time > 1,
      NewP = P + (P * IR/1200) - MP,
      mortgage(NewP, Time - 1, IR,
            MP, Bal).
```

primary parameters in mortgage: the principal (P); the life of the mortgage in months (Time); the annual interest rate (IR), which is compounded monthly; the monthly payment (MP); and finally the outstanding balance (Bal). The recursive definition above states that a mortgage for one month is given by the simple interest calculation: the monthly interest to be paid is given by the annual interest rate/1200 times the principal; the balance is simply the principal plus accrued interest minus payment. Mortgages for greater than a month are determined by calculating the new principal by adding interest and subtracting payment, and then proceeding as for a mortgage of one less month with this new principal.

To ask how much are the monthly repayments of a $100,000 mortgage at 12% for 30 years, a query of the form (4) is run. The answer obtained from running

```
?- mortgage(100000, 360, 12, MP, 0)  (4)
```

query (4) is MP = 1028.61. The main point of this example, however, is that it is possible to ask, not for the values of, but for the relationship between P, MP, and B. For example, running the query (5) results in

```
?- mortgage(P, 120, 12, MP, Bal).  (5)
```

the answer Bal = 0.302995*P − 69.700522*MP.

Implementation issues. The primary implementation issue in constraint programming languages is, of course, the cost of constraint solving. Often, very sophisticated algorithms are required in order to make the language practical. Where constraints are dynamic, as in the case of constraint logic programming languages, there is the added problem that solving is repeatedly done on constraint collections that are progressively larger. Thus, it is important that the constraint solver be incremental; that is, adding new constraints to a solvable collection of constraints must not necessitate the cost of entirely resolving the original collection. In the case of logic-programming-based languages, which again include constraint programming languages, there is a third issue caused by backtracking: the state of run-time structures must be saved in case computation needs to be rolled back to a particular point.

Status. So far, constraint logic programming languages have shown promise, and there have been major applications in diverse fields such as as architectural design, graphics, user interfaces, protocol testing, resource allocation, financial analysis, electrical engineering, and music. However, the area is still maturing and much research work remains to be done.

For background information *SEE DATAFLOW SYSTEMS; LINEAR PROGRAMMING; PROGRAMMING LANGUAGES* in the McGraw-Hill Encyclopedia of Science & Technology.

Joxan Jaffar; Spiro Michaylov; Peter J. Stuckey; Roland H. C. Yap

Bibliography. A. Colmerauer, An introduction to PROLOG-III, *Commun. ACM*, 33(7):69–90, July 1990; J. Jaffar et al., The CLP($\mathcal{R}$) language and system: An overview, *Spring COMPCON 91: IEEE Computer Society International Conference*, 36:376–381, 1991; J. Jaffar and J.-L. Lassez, Constraint Logic Programming, *Proceedings of ACM Symposium on Principles of Programming Languages*, 14:111–119, 1987; P. van Hentenryck, *Constraint Satisfaction in Logic Programming*, 1989.

Prospecting

Biogeochemical prospecting is one of several chemical methods for the exploration and discovery of natural resources, including metal deposits, petroleum deposits, and geothermal energy. Recent prospecting in arid and semiarid terrains, covered by thick unconsolidated deposits, has shown that accumulations of trace metals in plants is a more effective guide to ore than traditional methods based on soil and sediment geochemistry. The success of this method is attributed to dispersion of trace metals by groundwater–ore interaction and the ability to recognize favorable patterns in the data.

Interest in biogeochemical exploration methods began in the early 1900s in Russia and Scandinavia. It was not until the general availability of atomic absorption spectrophotometry, capable of measuring concentrations as low as parts per million (ppm), that biogeochemical prospecting came into wide use. The bulk of literature has been written since 1960. The research reflected the mining industry's interest in particular commodities, starting with uranium (1950s and 1960s), base metals (1970s), and gold (1980s). Primary interest in gold and other noble metals (platinum and palladium in particular) continues, and they have been the focus of recent progress in biogeochemical prospecting.

Plant sampling. Plant sampling is commonly used where other geochemical materials are either unavailable or unlikely to relate to mineralization. Permafrost landscapes and dense coastal forests are often covered with thick carpets of mosses and lichens, which make soil and rock sampling difficult or impossible. Continental and intermountain glacial deposits can be several hundred meters thick. Alluvium and pediment gravel deposits can also be thick and extensive. For example, they obscure nearly half of all of the Basin and Range terrains in Nevada. Deeply leached and chemically depleted laterites and saprolites of the Sierra Nevada and temperate regions of the southeastern United States leave little except plant material for geochemical sampling. Wherever overburden conditions limit the accessibility of the more conventional geochemical materials (rock, sediment, soil), biogeochemistry becomes the method of choice.

Recent work in arid regions has shown concentrations of trace metals in plants to be more directly related to concentrations of metals in groundwater than in the soils in which the plants are rooted. This distinction was not apparent until terrains covered by hundreds of meters of barren overburden and deep groundwater were tested. What was once regarded as only a substitute soil sample is now regarded as a surrogate soil and groundwater sample. Consequently, biogeochemistry is now applied in all terrains and as a complement to other geochemical methods, with an understanding that it can provide information about mobile elements far beneath the surface.

Species. Early investigators realized that some plant species accumulate more of certain trace metals than others. *Astragalus* was identified early in the 1950s as a selenium accumulator; therefore, it is an indicator plant for uranium deposits rich in selenium. Now, identification of accumulators is less important. Analytical methods such as inductively coupled plasma spectrometry (ICP), instrumental neutron activation (INA), and graphite furnace/atomic absorption spectrophotometry (GF/AAS) permit detection of background concentrations for most elements down to low part-per-billion levels. Consequently, biogeochemical sampling includes almost every species of perennial shrubs and trees. For example, in the desert region of the southwestern United States, biogeochemical surveys are composed of homogeneous and mixed populations of sagebrush (*Artemisia*), greasewood (*Sarcobatus*), shadscale (*Atriplex*), mountain mahogany (*Cercocarpus*), creosote bush (*Larrea*), mesquite (*Prosopis*), acacia (*Acacia*), pinyon pine (*Pinus*), and juniper (*Juniperus*). In the early 1980s, many of these species would not have been suitable for gold exploration because gold concentrations would not have been detectable. Advances in analytical chemistry have made it possible to survey previously unsuitable terrains, including desert terrains, terrains that do not have accumulator plants, and terrains that do not support deeply rooted phreatophytes.

Trace-metal dispersion. The interaction of groundwater and areas of mineralization tends to mobilize the component metals to varying degrees based on several factors, the most important of which are pH, equilibrium oxidation potential (Eh), and organic complexing. The metals disperse from the source down the hydrologic gradient, and in arid environments they move vertically to the surface. The ascent of metals from deep bedrock sources to relatively shallow root zones is a mechanism that is still poorly understood, but it seems to be a function of several forces, including mass-gradient, concentration-gradient, capillary-action, cation-exchange, and electromotive forces. Several indirect areas of research, including electrogeochemical methods and enzyme-controlled soil extraction methods, may soon provide answers.

Metal concentrations in a plant are limited by the intensive oxidative character of the groundwater, the extensive contact of groundwater with areas of mineralization, the vertical and lateral distance of the plant to the source, the occurrence of structures as pathways for water migration, and the mobility of metal cations through the substrate. Estimation of the presence of an orebody as well as its location and depth is complicated by these factors. Consequently, pattern recognition rather than criteria based on absolute concentrations is used for the successful interpretation of biogeochemical data.

Data interpretation. The Galaxy deposit near the Alligator Ridge Gold Mine, White Pine County,

Nevada, provides a good example of biogeochemistry applied to mineral exploration. Gold concentrations hosted in the carbonaceous Pilot Shale average 0.02–0.1 oz/ton (0.5–3.1 g/metric ton) in a structurally prepared and epithermally altered body that lies 82 ft (25 m) below the surface (**Fig. 1**). This body is overlain by barren shale (A–G in Fig. 1) and variably thick alluvium (H–K). Soil geochemistry identified a weakly mineralized jasperoid (E) but did not indicate the orebody. The biogeochemical survey was done after systematic drilling around the jasperoid and after the deposit had been discovered.

Pinyon pine (*Pinus monophylla*; A–F in Fig. 1) and big sagebrush (*Artemisia tridentata*; G–K) are respectively dominant in the high- and low-relief areas. Needles and leaves were sampled at 100-ft (30-m) intervals along two parallel lines that tested background and mineralized terrains (**Fig. 2**). The samples were analyzed for gold and several pathfinder elements (that is, arsenic and antimony) by instrumental neutron activation. Lower limits of detection were 0.1 ppb gold, 0.01 ppm arsenic, and 0.01 ppm antimony.

The fact that pinyon accumulated more gold than sagebrush (**Fig. 3***a*) is known as a species-effects principle. However, the abilities of pinyon and sagebrush to accumulate arsenic and antimony were about equal (Fig. 3*b* and *c*). Had different parts of the same plants (such as twigs) been sampled, different concentrations would have been reported. For instance, the values for gold probably would have been lower and those for arsenic and antimony higher.

Geochemists use the contrast between background and anomalous concentrations to determine the likelihood of a mineral occurrence. In this example, the contrast in values of gold from pinyon is exceptional, but not from sagebrush. The results of this survey based only on gold in sagebrush would not have led to discovery. However, the values for arsenic and antimony are highly contrasted in both species and effectively define the limits of the hidden orebody. More important, the concentration profiles reveal a systematic increase through several sample stations toward a common point. In other words, the pattern, even more than the difference between the highest and lowest values in the pattern, determines the likelihood of discovery. In this example, as in most others, the use of several elements to develop an exploration target is essential to the success of the interpretation.

The mining industry is studying various biogeochemical patterns to understand how the signatures vary with depth, lateral distance, and contact with groundwater. For instance, high-contrast anomalies are obtained when groundwater is in contact with an area of mineralization but not when groundwater is beneath it. Consequently, in arid environments where groundwater may be as much as 195 ft (60 m) from the surface, deep deposits are readily identified, while exposed deposits are not even weakly detected.

Biogeochemists are also studying the patterns of related elements such as rare earths, alkaline earths, and transition metals to learn how they can be used to locate bedrock contacts, faults, and other major

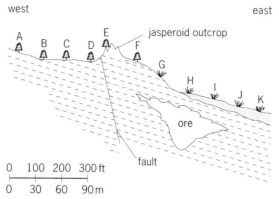

Fig. 1. Generalized cross section of the Galaxy deposit (Nevada) in Pilot Shale showing the barren shale (sample locations A–G) and alluvial overburden (H–K), fault, mineralized jasperoid outcrop, and distribution of pinyon (A–F) and sagebrush (G–K).

structures hidden beneath an obscuring cover. Recent multielement biogeochemical surveys have been used to map bedrock in buried terrains with an accuracy that competes with the more frequently applied geophysical methods (such as magnetic, gravity, or resistivity methods). The distinct advantage is the chemical character of the lithologies and structures imparted by the biogeochemical data.

Petroleum and geothermal exploration. Known associations of metals with petroleum and geothermal systems make biogeochemistry a viable exploration tool for these resources. Petroleum sources are associated with enriched concentrations of iron, manganese, nickel, uranium, and vanadium; and geothermal systems often have the same metal associates as the paleogeothermal analogs that created the various epithermal metal deposits common in the western United States.

Though more actively used for metals exploration, biogeochemistry linked with remote sensing is increasingly applied in other resource areas. One attribute

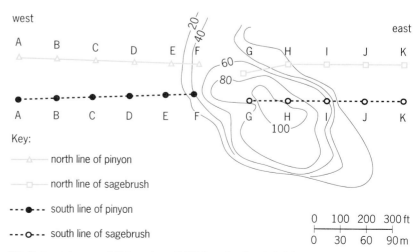

Fig. 2. Isopach map of Galaxy deposit (20-ft or 6-m isopleths) in plan view showing the relation of survey lines to the deposit thickness, the barren shale (A–G), and the alluvial overburden (H–K). Contours are in feet (1 ft = 0.3 m).

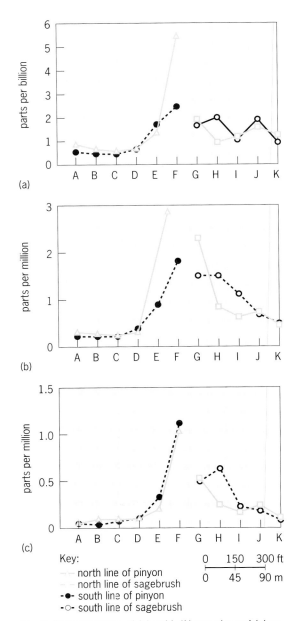

Key:

— north line of pinyon
— north line of sagebrush
-●- south line of pinyon
-○- south line of sagebrush

```
0    150   300 ft
0    45    90 m
```

Fig. 3. Concentrations of (*a*) gold, (*b*) arsenic, and (*c*) antimony in pinyon needles and sagebrush leaves over the Galaxy deposit.

that lends itself to satellite and airborne methods is an affiliated science known as geobotany. Geobotanical anomalies are manifested by (1) anomalous distribution of species and plant communities, (2) altered phenological cycles such as early senescence and late blooms, (3) stunted growth and reduced coverage, and (4) altered leaf pigmentation and transpiration rates.

For background information SEE ACTIVATION ANALYSIS; GEOBOTANICAL INDICATORS; GEOCHEMICAL PROSPECTING; PROSPECTING; REMOTE SENSING in the McGraw-Hill Encyclopedia of Science & Technology.

Shea Clark Smith

Bibliography. D. Carlisle et al. (eds.), *Mineral Exploration: Biological Systems and Organic Matter* (Rubey Vol. V), 1986; D. R. Cohen, E. L. Hoffman, and I. Nichol, Biogeochemistry: A geochemical method for gold exploration in the Canadian Shield, *J. Geochem. Explor.*, 29:49–73, 1987; J. A. Erdman and J. C. Olson, The use of plants in prospecting for gold: A brief overview with selected bibliographic and topic index, *J. Geochem. Explor.*, 24:281–304, 1985; A. W. Rose, H. E. Hawkes, and J. S. Webb (eds.), *Geochemistry in Mineral Exploration*, 2d ed., 1979.

Quinoa

Quinoa (*Chenopodium quinoa*) is grown in the highlands of South America. Because it is resistant to drought and frost and can survive on poor soils with minimal care, it is an important crop there. Quinoa is the staple of inhabitants of the upper Andes, where it is boiled, toasted, or ground into tortilla flour. Over the past decade, interest has focused on quinoa because of the high nutritional quality of its grain. Quinoa is also grown in Colorado in the United States, and in Finland, and is presently under investigation in a number of countries, including Great Britain, Japan, and Canada.

Quinoa is closely related to two other chenopods native to South America, hauazontle (*C. nuttalliae*) and caniha (*C. pallidicaule*). These grain crops belong to the family Chenopodiaceae, which also includes sugarbeet, mangel, and beetroot (all *Beta* species), and spinach (*Spinacia oleracea*), all of which have been domesticated in the Andes.

Quinoa is an annual broad-leaved plant, 4–6 ft (1–2 m) high, with large clusters of small white or pink seeds. The grain is conical, cylindrical, or ellipsoidal with a diameter of 0.07–0.1 in. (1.8–2.6 mm). Quinoa is sown at the high altitudes of the Andes from September through November and harvested around March. At maturity the plants are cut, bundled, dried, and flailed, and the grain is winnowed.

Geographical distribution. In South America, the largest producer of quinoa is Peru, where it is grown mainly in the Mantaro Valley around Huancayo and the altiplano near Puno. A smaller crop is grown in Bolivia close to Lake Titicaca, as well as in Columbia, Chile, Ecuador, and Argentina.

Agronomic practices. Little has changed in the cultivation of quinoa since the Spanish conquest, and it remains a subsistence crop. Some mechanization is found in the altiplano in Peru, where quinoa usually follows potatoes in crop rotation. This rotation is important since potatoes usually provide residual fertilizer, although some additional nitrogen may be required. A well-drained level seed bed is necessary so that the small seeds are not subjected to waterlogging.

A dry-season cultivation method was developed in the altiplano, where moisture is limited. In this method a fallow field is plowed to catch the rains between December and March. The soil is then plowed again in early April to destroy weeds and to conserve water for the final plowing and sowing of the quinoa crop in September. The optimum seed density for quinoa is reported to be 3.6–5.4 lb/acre (4–6 kg/hectare) in the Bolivian altiplano. To minimize waterlogging and dehydration of seeds, a sowing depth of 0.4–0.6 in. (1.0–1.5 cm) is recommended.

Breeding. The objective of breeding is to produce a higher-yielding quinoa seed that is resistant to pests and diseases as well as low in saponin content. The first scientific attempt to improve quinoa was made in 1965 by the Food and Agriculture Organization (FAO). Subsequent breeding programs included that in Colorado, where quinoa is grown at elevations of 6560–11,480 ft (2000–3500 m). In Jokoinen, Finland, around 50 quinoa varieties have been grown to maturity.

Quinoa germplasm was first collected in 1964 in Peru, and later by the Bolivian Institute of Andean Crops. The major collection of quinoa germplasm is located at the National Technical University of Altiplano in Peru. In order to retain viability, the germplasm should be stored at low temperature and humidity. A smaller collection of quinoa germplasm is located in Cambridge, England.

Composition. Proximate composition of quinoa varieties shows protein at 10–18%, crude fat 4.5–8.8%, carbohydrates 54.1–64.2%, crude fiber 2.1–4.9%, and ash 2.4–3.7%. Quinoa has an amino acid balance that is far superior to the major cereal crops (see **table**). Of particular note are the substantially higher lysine levels, which make quinoa grain an excellent protein source.

Nutritional quality. The nutritional quality of quinoa proteins results in excellent growth rates in experimental animals. Early work showed quinoa protein fed to rats was equivalent in quality to dried milk protein. Later research found that mixing 20% quinoa flour with 80% wheat flour improved PER (protein efficiency ratio) in rats equivalent to a 0.2% supplementation of lysine. The separation of quinoa proteins by electrophoresis showed that quinoa seed polypeptides can be classified as either albumin or globulin, the more soluble proteins. These soluble proteins are high in lysine and methionine and contribute to the nutritional quality of quinoa seed.

Bitter factors. One major reason that quinoa is not accepted more generally is the presence of bitter compounds known as saponins. These are triterpene glycosides located predominantly in the outer layers of the quinoa seed, the perianth and pericarp, the seed coat, and the cuticlelike covering. The major saponin aglycones or sapogenols (saponins in which the sugar moiety has been removed) in quinoa are phytolaccagenic acid (>40%), hederagenin (25%), and oleanolic acid (30%). The structures of these three sapogenols are shown in the **illustration**.

Saponins afford protection to the plant against attack by birds, insects, and fungi but must be removed from the seed prior to consumption. In addition to having a bitter and soapy taste, saponins are known to hemolyze red blood cell membranes. The saponins must be removed because of the higher red blood cell requirement of the Andean population living at high altitudes. More recent studies, however, suggest that saponins may have a beneficial effect in the human diet by lowering plasma cholesterol. A recent study of 17 quinoa cultivars reported that saponin content ranged from 0.14 to 0.73%.

Once saponin is removed, quinoa seeds offer a wide variety of nutrients that are essential for normal growth and development: protein, minerals (potassium, calcium, phosphorus, magnesium, and iron), and B vitamins (thiamin, riboflavin, niacin, and tocopherols). To eliminate the bitter taste of quinoa, saponins were traditionally removed by continuous washing of the grain, and the ancient Incas were reported to soak bags of quinoa in river water to detoxify it. Commercial processes developed to reduce or remove saponins from quinoa include abrasive dehulling, alkali washing or dry scouring, and roller milling. Washing or abrasive dehulling was found to be extremely effective in reducing the saponin content in quinoa by about 80%. The relatively higher proportion of phytolaccagenic acid remaining in the residual saponins, however, suggests that the three major saponins are located in different parts of the cell.

Amino acid content of quinoa and five cereal grains, in grams per 16 grams nitrogen*

Amino acid	Quinoa	Wheat	Rice	Corn	Barley	Oats
Isoleucine	3.6	3.3	3.8	3.7	3.6	3.8
Leucine	6.0	6.7	8.2	12.5	6.7	7.3
Lysine	5.6	2.9	3.8	2.7	3.5	3.7
Methionine	2.0	1.5	2.3	1.9	1.7	1.7
Cystine	—	2.5	1.1	1.6	2.3	2.7
Phenylalanine	4.1	4.5	5.2	4.9	5.1	5.0
Tyrosine	2.8	3.0	3.5	3.8	3.1	3.3
Threonine	3.5	2.9	3.9	3.6	3.3	3.3
Arginine	7.0	4.6	8.3	4.2	4.7	6.3
Histidine	2.4	2.3	2.5	2.7	2.1	2.1
Alanine	4.7	3.6	6.0	7.5	4.0	4.5
Aspartic acid	7.3	4.9	10.3	6.3	5.7	7.7
Glutamic acid	11.9	29.9	20.6	18.9	23.6	20.9
Glycine	5.2	3.9	5.0	3.7	3.9	4.7
Proline	3.1	9.9	4.7	8.9	10.9	5.2
Serine	3.7	4.6	5.4	5.0	4.0	4.7

*After L. Coulter and K. Lorenz, Quinoa: Composition, nutritional value, food applications, *Lebensmittel-Wissenschaft und Technologie*, 23:203–207, 1990.

Oleanolic acid: R = CH$_3$, R' = CH$_3$

Hederagenin: R = CH$_2$CH, R' = CH$_3$

Phytolaccagenic acid: R = CH$_2$OH, R' = COOCH$_3$

Structure of sapogenols of quinoa. (*After C. L. Ridout et al., Quinoa saponins: Analysis and preliminary investigations into the effects of reduction by processing, J. Sci. Food Agri., 54:165–176, 1991*)

Food applications. The primary use of quinoa is for tortillas. Quinoa seeds can also be fermented into a beverage known as *chicha* or used to prepare a coarse bread known as *Kispina*. A variety of quinoa-based products, including pasta, are manufactured in the United States and marketed as gluten-free products. Many health food stores in North America sell desaponized quinoa seeds or flour from Ecuador, Bolivia, or Peru. The grain can also be used in soups and cooked and served like rice.

For background information SEE GRAIN CROPS in the McGraw-Hill Encyclopedia of Science & Technology.

N. A. Michael Eskin

Bibliography. G. S. Chauhan et al., Nutrients and antinutrients in quinoa seed, *Cereal Chem.*, 69:85–88, 1992; L. Coulter and K. Lorenz, Quinoa: Composition, nutritional value, food application, *Lebensmittel-Wissenschaft und Technologie*, 23:203–207, 1990; J. A. Gonzalez et al., Quantitative determination of chemical compounds with nutritional value from Inca crops: *Chenopodium quinoa (quinoa), Plant Foods Human Nutr.*, 39:331–337, 1989; R. Inness, Quinoa's comeback, *IDRC Rep.*, 18(2):22–23, 1989; C. L. Ridout et al., Quinoa saponins: Analysis and preliminary investigations into the effects of reduction by processing, *J. Sci. Food Agri.*, 54:165–176, 1991; J. C. Risi and N. W. Galwey, The *Chenopodium* grains of the Andes: Inca crops for modern agriculture, *Adv. Appl. Biol.*, 10:145–216, 1984; N. W. Simmonds, The grain chenopods of the tropical American highlands, *Econ. Bot.*, 19:223–235, 1964.

Radiation

Spontaneous emission of light or radiation occurs when isolated atoms or molecules in an excited state make a transition to a state of lower energy. The excitation may be a result of thermal, mechanical, electrical, or optical excitation of the atoms or molecules. Familiar examples of spontaneous emission are the light from a hot object, the glow from a phosphorescent coating, and the triboluminescent emission of light when a piece of hard candy is crunched in the dark. Other examples of spontaneous emission occur in laboratory situations such as laser excitation of solids, liquids, and gases, and in lasers themselves.

Physics of spontaneous emission. For many years physicists believed they had a firm understanding of how wave disturbances, in particular electromagnetic waves, interacted with macroscopic media, which absorb, reflect, or refract light. The basis was the concept of the electromagnetic field and its governing laws, the Maxwell equations. Spectroscopic measurements, however, showed that matter exhibits sharp and discrete responses to electromagnetic waves of different frequencies. This discrete atomic response was explained by using quantum-mechanical solutions for the allowed energies of the atomic systems. These solutions led to the concept of discrete atomic states with well-defined energy levels between which an atom can make transitions. These concepts were joined to the physics of how electromagnetic waves interact with atoms to form a theory of absorption and stimulated emission of radiation by an atom or molecule. This physical description explained how an electromagnetic wave of the right frequency (1) can induce an atom to absorb electromagnetic energy and go to a higher energy level; and (2) can induce an excited atom to make a transition to a state of lower energy, with energy being added to the electromagnetic wave (a process called stimulated emission).

This description of the interaction of a classical electromagnetic field and a quantum-mechanical atomic system explains absorption and even basic laser action, but it cannot explain spontaneous emission. The reason is that if there is no wave to start with, there is nothing to induce the atomic transition giving off the light. In such a model, an excited atom in vacuum cannot lose its energy by radiation.

The solution to the problem came when it was realized that quantum mechanics must apply to everything, including the electromagnetic field. One important aspect of the quantization of any system—mechanical, electromagnetic, or otherwise—is that it is never possible for the energy to be zero. This consequence of the Heisenberg uncertainty principle states that, by virtue of their quantization, all systems will have a so-called zero-point energy. The implication is that a mass on a spring can never be completely still when the quantized spring system is in its lowest-energy state; it is still vibrating with some zero-point energy.

The quantization of the electromagnetic field leads to a zero-point electromagnetic field energy just as it does for the mass on a spring and the atoms and molecules discussed above. The discrete states of the electromagnetic field are described by photons whose energy is proportional to their frequency, and the state where no photons are present is called the electromagnetic vacuum state. This state has no average electric field E, but does have a zero-point energy of intensity (proportional to E^2) within a given enclosure. The presence of a zero-point electromagnetic vacuum makes it possible to understand spontaneous emission. The process of spontaneous emission can in some sense be viewed as stimulated emission by the zero-point electromagnetic field energy, which results in a fully quantum-mechanical treatment of both the atom and the field.

Inhibition of spontaneous emission. Although the interaction between atoms and the electromagnetic vacuum leading to spontaneous emission is a fundamental process, it is not an unchangeable fact of nature. In 1946, E. Purcell first realized that the spontaneous emission rate of atoms and molecules could be altered. He showed that, by altering the strength of the vacuum zero-point fluctuations, it is possible to enhance spontaneous emission. In particular, it can be shown that the application of quantum mechanics to the elecromagnetic field is associated with satisfying certain boundary conditions. These boundary conditions are imposed by the material properties and dimensions of the enclosure in which the atom is placed. In the case of conducting cavities, the Maxwell equations require the electric field to be zero at the cavity walls. This requirement in turn manifests itself in the condition that only certain wavelengths (and hence frequencies) of electromagnetic radiation can exist within the enclosure. In the simplest rectangular geometry, no radiation whose wavelength is larger than half the box dimension can satisfy the boundary conditions. These solutions, called modes, determine the types of photons that can exist in the enclosure. In addition, the mode characteristics are preserved even for the vacuum state, when the box is void of all but the zero-point energy of that mode.

The properties of radiation in enclosures and structures have measurable effects on excited atoms or molecules. An atom that radiates at a wavelength λ in free space, where all wavelengths are allowed, will not emit spontaneously inside a box where modes of that wavelength are not present. This effect, which has been experimentally confirmed, occurs because no zero-point fluctuation energy exists to stimulate spontaneous emission. Furthermore, should the atom emit, the photon it generates is not a physically acceptable solution of radiation in the box.

Experimental observations. An important experiment demonstrating that the electromagnetic mode structures of metallic cavities could inhibit as well as enhance spontaneous emission used atoms in high Rydberg states. These atoms are excited to states near the ionization limit, where the energy spacings between the states are small. The small energy separations between levels result in radiation with wavelengths of several tenths of a millimeter. These long wavelengths allow the construction of waveguides and cavities whose dimensions are smaller than the wavelength, so that those modes are eliminated from the structure. Kleppner's experiments used an atomic cesium beam with the atoms optically excited to the $n = 22$ state, from which they would spontaneously emit a photon with a wavelength of $\lambda = 450$ micrometers in free space. The inhibition was observed by sending a beam of atoms through two gold-plated parallel plates with a separation of less than $\lambda/2$. The plates were separated by using six disk-shaped quartz spacers with a thickness of 231.1 μm, resulting in a separation of $d = 1.027(\lambda/2)$. By switching the atomic resonance in and out of the inhibition condition ($\lambda/2 > d$) by using the Stark effect in a static electric field, inhibition of spontaneous emission could be observed. The maximum tuning of the transition wavelength using the applied field was limited to a range of about 4% by the onset of field ionization of the cesium atoms. Even with such a small tuning range, an inhibition factor of at least 20 was measured. In other words, the atoms took at least 20 times longer to spontaneously emit in this structure than in a vacuum.

Decay of cyclotron motions. In another experiment, the radiative decay of the orbital cyclotron motion of a single electron was inhibited when the electron was located within a metallic cavity formed by the electrodes of a Penning trap that confined the electron. The experiment utilized the energy levels that result when an electron is in an applied magnetic field. The resulting cyclotron motion produces transitions whose wavelength can be tuned to be several millimeters long. A strong magnetic field was applied to produce a wavelength of radiation at $\lambda_c = 2$ mm. These experiments had to be performed at a temperature of 4 K ($-452°$F) so that the blackbody thermal radiation did not produce any sizable stimulated-emission effects on the orbiting electron. The results of tuning the radiation wavelength with the magnetic field showed that the spontaneous-emission rate changed from the free-space value when inhibition condition was approached.

Emission in the visible spectrum. The extension of these experiments to the visible spectral region, where the wavelength of spontaneously emitted radiation would be in the range from 0.4 to 0.7 μm, has posed an experimental challenge. The reason is that it is a difficult task to produce submicrometer cavities to exclude the appropriate modes. A new approach to this problem was the use of periodic structures. Since periodic structures do not allow electromagnetic waves whose wavelength is twice the structure period to propagate owing to Bragg scattering, they should be effective in inhibiting spontaneous emission as well.

The realization of inhibited spontaneous emission in the visible spectrum using a periodic dielectric structure has been recently demonstrated. Since a perfectly periodic, three-dimensional structure would be nearly impossible to fabricate, the experiments were performed by using colloidal crystals. These structures are composed of a lattice of charged polystyrene spheres in water. The spheres are about 0.1 μm in diameter and carry a charge of about a thousand electrons on their surface. The strong electrostatic repulsion between the spheres at high densities (10^{14}/cm^3) causes them to form an ordered structure with a face-centered cubic (fcc) [111] plane-to-plane spacing that can be varied from 0.2 to 0.5 μm.

These colloidal crystals were doped with dyes that served as the spontaneously emitting molecules. The measurements had to be performed by using picosecond lasers and pump-probe techniques, since the natural free-space spontaneous lifetime is of the order of nanoseconds for these dyes. The pump-probe technique employs two lasers capable of producing light pulses 4 ps long. A pulse from the pump laser excites the dye-doped, colloidal crystal, while a pulse from the second laser probes the response at a time determined by an optical path delay. Since the response at different times directly measures the number of molecules still excited, the spontaneous-emission rate could be measured in this way. When the crystal-plane spacing was adjusted to the correct value for the emitting (sulfurhodamine) dye, spontaneous emission was weakly inhibited.

Potential applications. Spontaneous emission is a random process that mirrors the fluctuating nature of the zero-point energy. This randomness results in photons being emitted with arbitrary phases in any system that has excited atoms or molecules, such as a laser. The spontaneous emission present while stimulated emission is occurring in a laser results in a noise component in the output. The inhibition of spontaneous emission from lasers could result in oscillators with extremely small linewidth-to-frequency ratios. An optical device with these properties would be of tremendous use for communications applications as well as fundamental physical measurements such as gravity-wave detection.

For background information SEE ATOMIC STRUCTURE AND SPECTRA; CAVITY RESONATOR; COLLOIDAL CRYSTALS; NONRELATIVISTIC QUANTUM THEORY; PARTICLE TRAP; PHOTON; RYDBERG ATOM; ULTRAFAST MOLECULAR PROCESSES; WAVEGUIDE in the McGraw-Hill Encyclopedia of Science & Technology.

Nabil M. Lawandy

Bibliography. G. Gabrielse and H. Dehmelt, Observation of spontaneous emission, *Phys. Rev. Lett.*, 55:67–70, 1985; R. G. Hulet, E. S. Hilfer, and D. Kleppner, Inhibited spontaneous emission by a Rydberg atom, *Phys. Rev. Lett.*, 55:2137–2140, 1985; J. Martorell and N. M. Lawandy, Observation of spontaneous emission in a periodic dielectric structure, *Phys. Rev. Lett.*, 65:1877–1880, 1990; E. M. Purcell, Spontaneous emission probability at radio frequencies, *Phys. Rev.*, 69:681, 1946; E. Yablonovitch, Inhibited spontaneous emission in solid state physics and electronics, *Phys. Rev. Lett.*, 58:2059–2062, 1987.

Radio spectrum allocations

Recent developments in the area of radio spectrum allocations include the study of the allocation process by the International Telecommunication Union (ITU), the convening of a World Administrative Radio Conference (WARC) in early 1992, and spectrum allocation changes in the United States.

Study of allocation process. The process of international allocation of the radio spectrum is under study by the International Telecommunication Union and its 162 member administrations. ITU radio spectrum allocations and associated rules for radio services are contained in the international Radio Regulations. By the 1980s, it became evident that the pace

of technological change was outstripping the ability of the ITU to keep the international Radio Regulations current. The Radio Regulations can be changed only by a World Administrative Radio Conference, which is held as needed. The product of a WARC is an international treaty that, to become law in the United States, must be ratified by the Senate.

Standards for radio services are developed by the International Radio Consultative Committee (CCIR), an agency of the ITU. The CCIR organizes its work into 4-year study periods and has study groups to deal with standards of the various radio services. Traditionally, this deliberative body has required one or two study periods to develop, by consensus, new recommendations. In addition, publication of documents in its three working languages (English, French, and Spanish) has taken about a year, and translations into other languages have required even longer. Clearly, these delays have diminished the influence of CCIR recommendations on new radio systems.

High Level Committee report. At its 1989 Plenipotentiary Conference in Nice, France, the ITU decided to establish a High Level Committee to study how the ITU can keep abreast of technological change. The High Level Committee was asked to review the structure and methods of the ITU and to make recommendations that would bring about the necessary changes. The High Level Committee's final report was submitted to the ITU Administrative Council at its May-June 1991 meeting.

The High Level Committee report called for changes to streamline the organization, more frequent opportunities to change the Radio Regulations, more timely preparation of new standards, and greater transfer of telecommunications technology to developing countries. Some of these changes are administrative in nature and are being implemented. Others will require approval of the Plenipotentiary Conference scheduled for December 1992. Nevertheless, implementation of the High Level Committee's recommendations seems likely, possibly with some minor modifications.

New ITU structure. The new structure of the ITU is to include (1) a radiocommunication organ, consisting of the present CCIR and the International Frequency Registration Board (IFRB); (2) a standards organ, having the functions of the present International Telegraph and Telephone Consultative Committee (CCITT), and possibly having responsibility for radio systems that would connect to the global public telecommunications network; and (3) a development organ to serve the needs of developing countries. The full scope of the standards organ is currently under discussion. One viewpoint is that the standards organ should carry on the traditional role of the CCITT and concern itself with the public telephone and data network; another is that it should expand into development of standards for radio systems that may become part of, or connected to, the public network. This uncertainty leaves the mission of the radiocommunication organ somewhat unclear, particularly with regard to radio systems that may be associated with the global network.

More frequent WARCs. The High Level Committee has recommended that WARCs be held every 2 years instead of aperiodically in order to accelerate changes to the Radio Regulations, which include spectrum allocations. Having WARCs every 2 years would undoubtedly speed up the consideration of changes to the Radio Regulations and the international frequency allocation table. Since it takes a minimum of 2 years to get ready for a WARC, preparation for future conferences would become a continuous effort for both member administrations and other interested organizations.

Allocation alternatives. A Voluntary Group of Experts has been formed to determine how the Radio Regulations can be simplified and made more flexible. Similarly, CCIR Task Group 1/1 has been established to study ways of apportioning the radio spectrum as alternatives to the present system of allocating blocks of frequencies by radio service. Several alternatives are under active consideration: allocation by technical characteristics, such as modulation, bandwidth, transmitted power, and antenna gain; allocation by functional categories, namely, type of service (communications or radio determination), coverage (point-to-point or area), and use (terrestrial, aeronautical, or space); allocation by electromagnetic compatibility, which would require an interference study of the current and proposed uses of a block; and allocation by a generic approach in making broader categories of radio services and making allocations globally or regionally.

While it is too early to predict the outcome of this study of allocation alternatives, agreement on a radical departure from the service allocation system seems unlikely. Nevertheless, some change toward greater flexibility is likely, to facilitate new uses of the radio spectrum.

WARC-92. A World Administrative Radio Conference was held in Spain in February-March 1992. It decided on a variety of spectrum allocation matters, including the possible extension of the high-frequency broadcasting spectrum. The United States had proposed expanding the high-frequency broadcasting allocations by 1325 kHz. European broadcasters also wanted to see these allocations expanded, but many developing countries were opposed. They saw no need for more foreign broadcasts targeted at their countries; and expansion of the broadcasting spectrum would come at the expense of fixed (point-to-point) service users. Most industrialized countries do not have many fixed users in the high-frequency bands, since such users have moved to microwaves or, more recently, to optical-fiber transmission.

The WARC-92 agenda also included allocations for satellite sound broadcasting in the frequency range 500–3000 MHz. This subject has been vigorously debated by many member administrations but particularly in the United States. In its Notice of Inquiry concerning WARC-92, the U.S. Federal Communications Commission proposed three band options: 728–788 MHz, 1493–1525 MHz, and a band around 2400 MHz. However, ultrahigh-frequency (UHF) television interests were opposed to the first option because they wanted to use the band at 728–788 MHz for high-definition television. Aeronautical interests strongly objected to the possible loss of spectrum in the band at 1493–1525 MHz used for flight-test telemetry; and the band around 2400 MHz would take additional spectrum from aeronautical telemetry and possibly the amateur service. Proponents of satellite sound broadcasting favored a band around 1500 MHz.

Other items on the WARC-92 agenda were the microwave spectrum for high-definition television via satellite; frequencies for future public land mobile telecommunications systems (FPLMTS), that is, third-generation cellular telephones; allocations for low-Earth-orbit data satellites; and additional allocations for mobile satellites. Most of the attention centered on new allocations in the 1–3-GHz band, with regard to new services and their compatibility with services having existing allocations therein. As new services enter the 1–3-GHz band, pressure will increase on the fixed service to move upward in frequency or to use optical-fiber communications where possible. Possible displacement of the fixed service has been a matter of particular concern to developing countries, some of which have recently established new fixed-service operations in the 1–3-GHz band.

Spectrum allocation changes in U.S. In response to criticism that entrenched technologies have the advantage over new technologies in spectrum allocation, the Federal Communications Commission has decided to give preference to pioneers (that is, people with new and innovative uses of the radio spectrum that are not covered under the present rules). To minimize possible abuses of the preference, the Commission ruled that spectrum allocations to pioneers would still require a petition, a notice of proposed rule making, and a report and order.

In a separate proceeding, the Commission reallocated the 220–222-MHz band from the amateur service to the land mobile service for narrow-band technology using 5-kHz-wide channels. The Commission also issued notices of inquiry concerning personal communications systems and digital audio broadcasting. Because of the need to find additional radio spectrum for private and public uses, the U.S. Congress has held hearings on a possible release of 200 MHz of spectrum from the federal government to the civilian sector.

The National Telecommunications and Information Administration (NTIA) issued a Notice of Inquiry asking the public for information on how the spectrum should be managed. Numerous respondents said that data concerning federal government use of the radio spectrum should be more readily available and that the decision-making process should be more open.

The NTIA also conducted a study to determine possible frequencies for wind profiler radars, which presently operate at 404 MHz and are a source of interference to COSPAS-SARSAT (search-and-rescue satellites). Wind profilers are special radars designed to sense clear-air turbulence of winds aloft. Unlike radiosondes, which are released periodically, wind profilers can provide virtually continuous readings of upper-air conditions and contribute to the safety of air travel. For a given radio frequency, the wind profiler senses best the turbulence over a particular range of altitudes; at a frequency around 50 MHz, winds at the highest altitudes are sensed, while a profiler operating around 1000 MHz senses winds near the ground. The assignment of a band around 400 MHz is a practical compromise for sensing turbulence at various altitudes. There are 31 wind profilers now operating in the United States. Plans call for expansion to around 300, which may be contingent on allocation of a suitable frequency band. The ITU intends to take up the question of an allocation for wind profilers at a radio conference subsequent to WARC-92.

For background information SEE RADIO SPECTRUM ALLOCATIONS in the McGraw-Hill Encyclopedia of Science & Technology.

Paul L. Rinaldo

Bibliography. High Level Committee To Review the Structure and Function of the International Telecommunication Union, *Tomorrow's ITU: The Challenges of Change*, 1991; International Telecommunication Union, *Radio Regulations*, 1990; National Telecommunications and Information Administration, U.S. Department of Commerce, *U.S. Spectrum Management Policy: Agenda for the Future*, 1991.

Recombination (genetics)

Site-specific recombination systems occur in a variety of organisms ranging from bacteria to humans. These systems typically consist of a limited number of enzymes known as recombinases that perform genetic recombination between short defined deoxyribonucleic acid (DNA) sequences. Several such systems have been extensively studied in order to understand why these organisms possess site-specific recombination machinery. In addition, these systems have been studied as simple models for the more complex general recombination that occurs during meiosis. Recently, biologists have begun to exploit these systems as tools that can be transferred to other species in order to perform certain types of genetic manipulations with an ease and precision not previously possible.

Site-specific recombination. The simplest site-specific recombination mechanisms rely on a single protein recombinase that recognizes and catalyzes recombination between specific DNA sequences. This recombination consists of a controlled breakage and rejoining of two different segments of DNA with a consequent rearrangement of the DNA on either side of the breakpoint. One well-studied example occurs in bakers' yeast, *Saccharomyces cerevisiae*. In this system the recombinase is called FLP, and the sequence on which it acts is called FRT, for FLP Recombination Target. An FRT consists of 34 base pairs of DNA, with two inverted repeats of a 13-base-pair sequence flanking a central spacer region of 8 base pairs. The spacer possesses an inherent asymmetry that confers directionality to the recombination reaction; that is, FRTs must be aligned in a parallel fashion for recombination to occur.

The outcomes of FLP-catalyzed recombination between FRTs on the same strand of DNA depends on the relative orientation of the two FRTs. If the FRTs are in the same direction, FLP-mediated recombination results in the excision of a circle of DNA carrying all the material between FRTs as well as one FRT (**Fig. 1***a*). The other FRT remains behind. FLP-mediated recombination between inverted repeats causes all the DNA between FRTs to invert with respect to the flanking sequences (Fig. 1*b*).

Transfer to other species. Recombination between directly repeated FRTs provides a simple demonstration that this site-specific recombination system can function when it is transferred to other organisms. When the FLP-FRT system was transferred from bakers' yeast to the fruit fly, *Drosophila melanogaster*, two constructs were made. The first consisted of an *FLP* gene modified so that it would remain off until the fly's body temperature was raised higher than normal. In order to visualize FLP-mediated recombination, a second construct was made. The *white* gene, a gene that is required to give *Drosophila* eyes their normal red pigmentation, was placed between direct repeats of the FRT sequence. (In *Drosophila*, genes are named for the phenotype they produce when mutated; in this

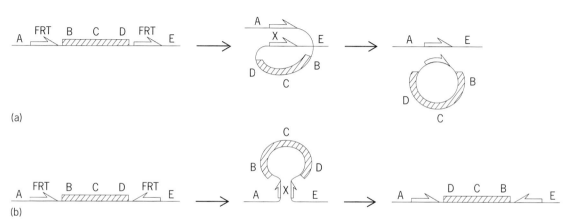

(a)

(b)

Fig. 1. Consequences of exchange between FRTs (*a*) in the same orientation and (*b*) in inverse orientation. The X indicates the recombination reaction catalyzed by FLP, and A–E indicate arbitrary sites in the DNA strand.

case a gene that is required to make red eye pigment is called the *white* gene.) Both constructs were then integrated into the *Drosophila* genome. This integration is performed by placing the DNA sequence to be integrated within a transposable-element vector. This transposable element has the ability to jump from place to place within the chromosomes of *Drosophila*. When the purified DNA of this vector is injected into young embryos, the element can jump into chromosomes and carries the inserted DNA with it. These flies can then be mated to establish a strain of flies carrying that DNA.

Flies were produced that carried both constructs and in which the normal copy of *white* was nonfunctional because of a mutation it carried, so that the only source of functional white product was the gene flanked by FRTs. When the body temperature of one of these flies was raised to a high level for a short time early in development (before eye pigment is synthesized), FLP was made. The FLP catalyzed exchange between FRTs in a portion of the cells of the flies, so that the *white* gene was excised from the chromosome. Because *Drosophila* cells have no mechanism for maintaining an extrachromosomal circle of DNA, the *white* gene is lost from the cells as they continue to divide. Cells in the eye that are mitotic descendants of a cell that has lost the *white* gene show up as a white patch; those cells that retain the *white* gene are pigmented. (Such a group of cells that are related by descent is known as a clone.) **Figure 2** shows the compound eye from such a fly.

Another site-specific recombination system known as the Cre-*lox* system derived from a bacterial virus has been shown to work in both plant and mammalian cells. The FLP-FRT system has been shown to work in mammalian cells as well. These two systems, although recognizing different DNA sequences, are quite similar in their functional aspects.

Mosaic analysis. Animals that are composed of cells with different genotypes, such as the *Drosophila* in Fig. 2, are known as genetic mosaics. The study of mosaic animals is one of the most useful methods for learning about animal development and the genes that govern it. For instance, Fig. 2 illustrates two of the purposes of mosaic analysis. First the obvious white patches show that the *white* gene acts in a cell-autonomous fashion. That is, for this gene, the phenotype of a cell depends on its own genotype and is not influenced by the genotype of neighboring cells. Thus, mosaic analysis can elucidate information about the nature of a gene product. Second, the distinct boundaries between pigmented and nonpigmented patches demonstrate a key feature of development in this organism—that mitotically related cells tend to stay grouped together with little mixing. The loss events seen in the eye of Fig. 2 are too numerous to determine with certainty which patches represent single clones and which represent the confluence of more than one clone. However, a milder heat treatment can be used to produce clones at a much lower frequency. It is then possible to be confident that a marked patch of cells represents

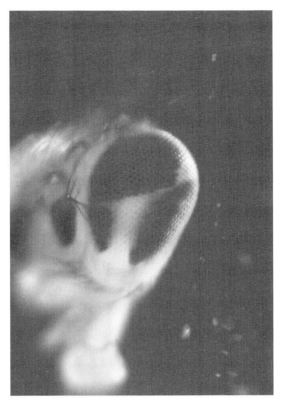

Fig. 2. Compound eye from the fruit fly, *Drosophila melanogaster*, showing eye color mosaicism resulting from the action of the FLP site-specific recombinase.

a single clone. Such a mosaic provides an arbitrary marking of a cell and its descendants, so that the fate of an individual cell at a specific time in development can be determined. This type of study has contributed to construction of a detailed fate map of the *Drosophila* embryo that shows which adult structures are derived from particular parts of the embryo.

Another use of mosaics is to determine when a gene acts during development. Site-specific recombination makes it possible to turn a gene on or off at different times. A gene's DNA sequence can be altered so that its protein-coding region is separated from its regulatory region by an FRT. If another FRT in inverted orientation is placed on the opposite side of the coding region, FLP can be used to invert a portion of the gene. This recombination can be effectively used to turn the gene on and off. By determining the earliest stage at which a gene can be turned off and still provide its full function, and conversely the latest stage at which it can be turned on and still function normally, it is possible to precisely time the stage at which that gene is necessary.

Mosaics may also be used to determine in which tissues a gene is required. In *Drosophila*, for instance, there are many mutations that, when homozygous, make females sterile. There are many possible causes for such sterility: the females may have a behavioral defect that prevents them from mating, they may have a defect in their genitalia that prevents mating or laying

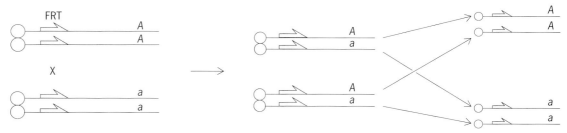

Fig. 3. Mechanism of mosaic production by mitotic recombination. A pair of homologous chromosomes from a cell in the second gap (G2) of the cell cycle (that is, after DNA replication) is shown at left. This cell is heterozygous for a normal gene (*A*) and a recessive mutation in that gene (*a*). The X indicates FLP-mediated recombination between FRTs on homologous chromosomes. The arrangement of chromosomes within a cell after this recombination event is shown in the middle panel. Mitosis can sort out the recombinant chromosomes to the two daughter cells as diagrammed, so that the two homozygous cells shown at the right result.

eggs, or they may have more subtle defects in their germline cells or in the somatic cells that contribute to the production of the egg. There are molecular assays that can determine in which cells the gene is expressed, but a mosaic analysis is needed to define those cells in which the expression is functionally significant.

Mitotic recombination. Site-specific recombination has also been used in *Drosophila* to generate mosaics by a method known as mitotic recombination. When an insertion of FRTs is made homozygous, FLP is able to catalyze recombination between the FRTs on homologous chromosomes. If such recombination occurs during the second gap (G2) of the cell cycle, the arrangement of chromatids indicated in **Fig. 3** will be produced. If the parent cell is heterozygous for a normal gene and a recessive mutation in that gene, then mitotic segregation produces daughter cells that are homozygous for genes that are farther away from the centromere than is the FRT. This method allows the production of animals that are mosaic for function in virtually every gene that has been identified by mutation, irrespective of whether that gene has been cloned or not.

X-rays have traditionally been used to induce mitotic recombination by causing chromosome breakage and rejoining events, but site-specific recombination represents an improvement in many respects. Mitotic recombination catalyzed by FLP at FRTs is much more efficient than x-rays and occurs at defined sites in the genome. X-rays, on the other hand, induce breaks at random sites and cause substantial cell death.

Further applications. Because site-specific recombination can be used to rearrange DNA sequences in a precise and designed fashion within a developing animal, it promises to be a very useful technique for novel types of genetic manipulations. It has been used for integrating additional DNA into the genome of an organism at a specific site where a recombination target has already been inserted. It might also find application in generating specific chromosome rearrangements for genetic and developmental studies.

For background information *SEE GENETIC ENGINEERING; MITOSIS; MOSAICISM; RECOMBINATION (GENETICS)* in the McGraw-Hill Encyclopedia of Science & Technology.

Kent Golic

Bibliography. N. L. Craig, The mechanism of conservative site-specific recombination, in A. Campbell, B. Baker, and I. Herskowitz (eds.), *Annu. Rev. Genet.*, 22:77-105, 1988; K. G. Golic, Site-specific recombination between homologous chromosomes in *Drosophila, Science*, 252:958–961, May 17, 1991; A. S. Wilkins, *Genetic Analysis of Animal Development*, 1986.

Recycling technology

Recent research directed at improvements in waste management has produced methods for recycling plastics and a number of advances in recycling of aluminum.

Plastics

The end of the twentieth century will bring a sharp change in waste management alternatives that are available for some of the more populated areas of the United States. Today, approximately 75% of municipal solid waste is disposed of in landfills. Within the next 20 years, the Environmental Protection Agency (EPA) predicts that 80% of the landfills will be closed.

The generation of municipal solid waste will continue to increase, while the number of options for managing waste will decrease because of environmental or public health risks or public opposition to various current or proposed practices. State and local governments will be encouraged to develop recycling programs to decrease the amount of municipal solid waste. The Office of Technology Assessment (OTA) of the U.S. Congress has recently recommended for state and local governments a guiding framework that includes both waste prevention and materials management, with recycling as the preferred materials management scheme.

A major barrier to recycling of plastics is lack of separation of plastic from the solid waste stream. Plastic discards represent an estimated 8% by weight and up to 18% by volume of the municipal solid waste generated in the United States. At present, only 1.1% of this postconsumer plastic (plastic waste generated by the final user) is recycled. This recovery rate, as a percent of waste material generated, trails recovery of ferrous metals (5.8%), glass (12.0%), paper (25.6%), and aluminum (31.7%).

Most of the effort to recycle postconsumer plastic is on high-density polyethylene (HDPE) and polyethylene terephthalate (PET) resins. Plastic waste made of these two resins is more easily identified and separated from other plastic waste as a result of industry standardization of materials for soft drink bottles (polyethylene terephthalate), base cups from these soft drink bottles (high-density polyethylene), and milk bottles (high-density polyethylene).

Plastic available for recycling. Approximately half of plastic waste consists of single-use convenience packaging and containers. Many manufacturers prefer plastic for packaging because it is lightweight, resists breakage and environmental deterioration, and can be processed to suit specific needs. Once plastics are discarded, these attractive physical properties become detriments.

Six resins are used for most commodity plastic products: low-density polyethylene (LDPE), high-density polyethylene, polyethylene terephthalate, polyvinyl chloride (PVC), polypropylene (PP), and polystyrene (PS). These single-resin plastics are considered suitable for recycling because they become pliable when heated and can be remolded.

Plastics are also used with paper and metal as multicomponent films, mostly for packaging. They are more difficult to recycle than single-resin plastic material discards, and can be converted only to mixed-plastic products.

Recycled plastic products. Secondary materials are products recovered from the waste stream and reformulated into new physical forms serving end uses other than those of the original materials. A variety of secondary materials are produced from plastic waste. The outer plastic layer of disposable diapers has been recycled into garbage bags, flower pots, and plastic lumber. Polyethylene terephthalate is currently recycled into scouring pads, fiber fill for jackets, paint brushes, carpet fibers, and other products. Polyethylene terephthalate is also recycled to produce polyol for use in urethane foam and furniture, and unsaturated polyester for boat hulls, pools, auto body parts, and parts for appliances. High-density polyethylene is recycled into base cups for soft drink bottles, flower pots, toys, and pallets. Projects to recycle bottles made from polyvinyl chloride resins have focused on products such as bottles for shampoo and vegetable oil.

Recycling postconsumer polystyrene foam has been targeted in certain pilot projects by industry for reuse as coat hangers, building insulation, office accessories, trash receptacles, and flower pots. As much as 95% of clean preconsumer industrial plastic waste (plastic scrap from an industrial or manufacturing process) is recycled. Companies have also recycled unclean industrial plastic waste for use in inner layers, with virgin resins as outer layers, for products such as multilayer detergent bottles; however, the actual amount of unclean plastic industrial scrap that is recycled is unknown.

Research in recycling of plastics has expanded to include the development of lumber from a mix of all types of postconsumer plastic. Such a mix, commonly called commingled plastics, comprises various plastic resins, pigments, additives used in manufacturing, and nonplastic contaminants. Plastic lumber has a number of advantages over wood, such as resistance to rot, chemicals, water, and insects. Plastic lumber is being used for fence posts, poles, marine pilings and bulkheading, dock surfaces, park benches, landscape timbers, retaining walls, palettes, and parking space bumpers.

Methods and technologies. Postconsumer plastic may be collected in a curbside recycling program or at designated drop-off centers. The Borough of the Bronx in New York City has implemented a buy-back system for plastic containers. Nine states have successful bottle deposit programs for the collection of polyethylene terephthalate soda bottles. Some of these states use a reverse vending machine for the collection of these soda bottles; the machine shreds the plastic for later processing.

There are a number of problems associated with the collection of plastic waste: its high volume can burden an existing collection program; the material does not crush easily; and plastics made from different resins may be mixed together.

To assist the public and laborers working in material recovery facilities with identifying different plastics, the Society of the Plastics, Industry sponsors a voluntary coding system. The coding system consists of a triangular arrow stamp with a number in the center and letters underneath to identify the resin used in the container. The codes are as follows: 1, PET; 2, HDPE; 3, V (vinyl); 4, LDPE; 5, PP; 6, PS; and 7, other (including multilayer resins).

The collected plastic waste is usually separated manually from the waste stream, and often it is cleaned to remove adhesives or other contaminants. It is sorted further, based on different resins. Mechanical techniques to separate plastics are available, and they can be used to sort plastics based on unique physical or chemical properties. Several research projects are being pursued at several universities. One of these groups is developing automated systems for separating plastic containers made from different resins. The separation technology is based on the use of photoelectric beams and infrared technology to identify different resins. Another group is developing a chemical separation technique utilizing selective solvent dissolution and flash devolatilization to recover individual resins. After separation, the resin usually undergoes shredding or grinding to form pellets.

The technologies used to manufacture recycled plastic are virtually the same as those used to manufacture products from virgin plastic resins. The recycled products are melted; for some products, additives or virgin resins are used to improve the properties. The plastic is then extruded into specific products or pellets.

For secondary lumber-type materials made from mixed plastic waste, the technology generally used is known as Extruder Technology 1 (ET/1). ET/1 has three main components: an extruder, a molding unit, and an extraction unit. The extruder consists of a short adiabatic screw rotating at a high angular velocity (number of revolutions per minute); it is used for melting the plastic at temperatures of 360–400°F (182–204°C). Plastics with higher melt temperatures, such as polyethylene terephthalate, become encapsulated within the melted plastic. The molding unit consists of linear molds mounted on a turret that rotates through a water cooling tank. The product shrinks within the mold during cooling, and is ejected from the mold by compressed air.

Technologies employed in Europe recycle plastic with the addition of 30% industrial preconsumer plastic materials to manufacture substitutes for lumber and concrete products, and they remove the low-density and high-density polyethylenes from commingled plastic for recycling. Other technologies are being developed there to recover particular plastic resins from mixed plastic waste.

Engineering and environmental issues. A significant problem in plastic recycling is the presence of contaminants such as dirt, glass, metals, chemicals from previous usage, toxics from metallic-based pigments, and other nonplastic materials that are part of or have adhered to the plastic products. The U.S. Food and Drug Administration (FDA) has expressed concern over potential risks in using recycled plastic products in contact with food. Other constraints involve inconsistencies in the amount of different plastic resins in commingled plastic wastes used for recycling, and engineering aspects of recycled plastic products, such as lessened chemical and impact resistance, strength, and stiffness, and the need for additional chemicals to counteract other types of degradation from reprocessing. There may be limitations to the number of times a particular plastic product can be effectively recycled as compared to steel, glass, or aluminum, which can be recycled many times with no loss of their properties and virtually no contamination.

The long-term engineering properties of recycled products are still largely unknown. The many products currently made from recycled plastic include gimmick toys, penholders, and paper trays; these may appear in the waste stream in a short time.

The recycling of plastics creates a waste stream of its own—contaminated wastewater and air emissions. Many additives used in processing and manufacturing plastics, such as colorants, flame retardants, lubricants, and ultraviolet stabilizers, are toxic and may be present in the waste stream.

Water is commonly used as a coolant in the manufacture of recycled plastic products, and it may become contaminated with residues associated with recycled plastic. Recycled high-density polyethylene can produce odors during processing and severe smoke conditions from blowmolding. The reason may be incomplete cleaning and processing of the plastic resin. Virgin resins exhibit neither odor nor smoke.

Economic and social acceptability. A number of important economic issues must be resolved in order for plastic recycling to be successful. Collection for recycling is taking place on a small scale, compared to the amount of plastic waste being generated. Limited supplies of plastic discards are resulting

in the recycling businesses operating at lower capacity, thus hampering profitability. It has been estimated that in 1989 recycling rates for high-density polyethylene (milk containers) and polyethylene terephthalate (soda bottles) equaled approximately 1.5% and 10% respectively of the amount of virgin resins used; the recycling of these resins provided only about one-third of their potential market. For polyvinyl chloride, less than 0.1% used in 1989 was recovered. Potential markets could absorb a hundredfold increase of recycled polyvinyl chloride. The total amount of recycled plastics contributes only 4% of the actual plastics market.

To date, markets for recycled single-resin plastics have had more success than mixed-plastic products. The reason may be that bottle return laws lead to better collection of single-resin polyethylene terephthalate beverage containers and to focusing by the plastics industry on recycling technologies for single-resin plastics. Marketing problems that arise with products made from commingled plastics include inconsistency in feedstock, lack of suitable engineering specifications, and unpredictable performance.

An important issue related to the marketability of recycled products is whether such products are, in fact, cheaper than the products they will be replacing. Plastic lumber, for example, is expensive compared to wooden lumber, concrete, and other materials for which it can be substituted. However, product lifetime must also be taken into consideration with the initial cost of a material; the average lifetime of plastic lumber and other products made from commingled plastics is undetermined. Other factors, such as plastic lumber's low maintenance costs, are also significant considerations.

For products made from commingled plastics, marketing problems due to a reduced esthetic appeal may arise: there is a lack of black and very dark colors, and impurities such as bits of paper or metal can be seen within the material.

A notable marketing advantage for recycled plastic products over items made from virgin plastics or other material is that they are appealing from an environmental standpoint to many consumers. Some in the industry argue, however, that this advantage is limited: consumers may not buy recycled products that are more costly than nonrecycled products.

Hopefully, recycling plastics will become a long-term endeavor. It is an essential part of a national waste prevention and waste reduction strategy. Recycling plastic that may otherwise be diverted to landfills, incinerators, or roadside litter is a positive step in reducing the waste stream and should be incorporated in all programs for solid-waste management. Research, engineering specifications, government regulations, and continued interest are necessary for future success of products made from recycled postconsumer plastic. However, plastic recycling is only in its infancy and has been largely driven by the plastics industry's concern that consumers would boycott their product. Approximately 16 states or local government jurisdictions have legislation pending, and a number of states have already placed restrictions on the use of certain plastic products. More opposition is likely to follow unless high-quality, well-engineered materials are available. Industry must foster engineering specifications, content labeling, and environmental measures to assure the future success of products made from recycled-plastics waste material. The credibility and future of the plastics industry as a whole may be at stake.

R. L. Swanson; Vincent T. Breslin; Marci L. Bortman

Aluminum

Aluminum recycling is practically as old as the commercial application of the metal itself. The recycling developed rapidly for two reasons: the metal was much more expensive than the traditional metals with which it competed, and remelting aluminum was relatively easy and required only about 5% of the energy it took to make the same amount of metal from ore. While the price of the metal has come down considerably, the energy issue is still valid. Furthermore, concerns about excessive litter caused by discarded cans, hazardous waste products, and lack of landfill space give impetus to the drive to increase recycling of all materials, with aluminum leading the way. Presently, recycled aluminum represents more than one-third of total aluminum shipments in the United States, which amounted in 1990 to about 790,000 metric tons (870,000 tons).

Some of this metal passed through the molten phase several times, first as primary metal, then as production scrap, and then as secondary or customer scrap; this processing resulted in an estimated 550,000 metric tons (610,000 tons) of dross and, ultimately, 240,000 metric tons (260,000 tons) of lost metal. These losses represent a cost of approximately $350 million. It is understandable that a sizable portion of the industry's research and development efforts is aimed at reduction of metal loss during processing in the molten phase (the so-called melt losses).

Dross is a mixture of oxides, contaminants, trapped metal, and gas that floats on top of the molten metal bath in the processing furnaces; it is skimmed off and usually shipped out for recovery of the metal components. Lately, generation of dross has increased significantly, because expansion of recycling efforts has forced the processors to use the more challenging, more dross-generating scrap forms, such as oily machining chips, metal/polymer composites, and used beverage containers. Over the years a specialized industry dedicated to treatment of dross has evolved.

Early developments. Aluminum recycling efforts started almost immediately after the invention of the primary metal production process by C. M. Hall and P. L. Heroult in 1888. The first oil-fired furnaces had the size and shape of a 55-gallon drum with the axis horizontal. They could be rotated to allow the molten metal to be tapped by way of an open pouring spout. The rapidly growing primary industry soon needed larger and more efficient remelting technology for scrap. In the early 1920s came the development of the reverberatory or open-hearth furnace, which consists of a single chamber with a relatively shallow hearth, a large door, and usually a set of two burners firing in a special pattern to maximize initially the convective heat transfer to the pile of cold scrap, pushed in place through the door by way of mobile charging machines. In the later stages of the melting process, radiation becomes the predominant mode of heat transfer. The continued popularity of this type of stationary furnace is due to its simplicity, made possible by the application of the submerged, resealable tap hole. The furnace is designed to handle mostly bulky scrap forms with a high volume–to–surface area ratio, typically the heavy-gage processing scrap generated by the primary producers. With the exception of some baling of lighter-gage scrap forms and scalping chips (produced by smoothing the cast ingots), no special scrap preparation steps are performed.

When the melting is completed, the molten metal bath is skimmed off (to remove dross), gently stirred (to equalize the temperature), and transferred to a holding furnace where the quality of the metal is brought up to specifications prior to casting. Historically, the melting of dirty, low-grade scrap and the reclamation of dross were relegated to secondary processors, which typically use rotary salt furnaces and perform no scrap preparation steps. The rotary salt furnace is essentially a short rotary kiln, gas- or oil-fired from one end with the flue located on the other end. Salt (sodium chloride or a mixture of 50–65% sodium chloride and 50–35% potassium chloride) is charged with the scrap or dross and heated simultaneously. When the mixture starts to melt, there is formed a viscous mass in which the molten metal coalesces and migrates to the lower portion of the fluid bath. The motion created by the rotating drum enhances heat transfer as well as coalescence of the smaller, isolated metal droplets in the salt–oxide mixture.

The metal is tapped through a resealable hole in the drum wall, or through the charge opening by tilting the furnace and decanting first the metal and then the salt cake. The metal is either poured into hot metal crucibles and transported to its final destination or cast in sow molds as remelt scrap ingot (RSI). The salt cake, which contains up to 10% aluminum that is difficult to recover, is usually discarded, although environmental concerns are encouraging salt cake treatment for salt recycling and generation of an inert by-product.

Trends. Modern recycling started in the early 1960s, when the traditional distribution of roles between primary and secondary industry began to shift. The primary industries started to buy what was considered typical secondary material, namely discarded consumer products (old scrap) such as cookware, electrical cable, litho scrap and, since 1970, used beverage containers. Presently, the primary industry recycles more old scrap than the secondary industry. The first consequence of this trend was that the recycling activity grew from an opportunistic, volatile, low-capital endeavor into a focused, solidly rooted, sophisticated industry. There are strong interdependent relationships between primary and secondary producers, the latter often delivering molten metal of acceptable chemical composition "just in time," which requires reliable, advanced equipment and control instrumentation.

The second consequence was that more and more scrap went back into the product stream from which it originated, for example, cans into can sheet and pans into cooking utensils, because that is the logical destination for well-segregated, well-prepared scrap. From a simple reuse of metal in a degraded form, the process evolved into a true recycling of metal into its own product. This closed-loop recycling requires strict quality and process control from scrap preparation to postmelt cleaning.

Can recycling technology. Figure 1 depicts the can manufacturing process for an integrated, state-of-the-art plant for recycling used beverage cans. It demonstrates the closed-loop nature of the process and the level of sophistication required for including scrap forms that are badly contaminated in such a high-quality product loop.

Used beverage containers are received from the collection centers as bales of approximately 1.5 m³ (53 ft³) and 400 kg (880 lb; the average density is 270 kg/m³ or 17 lb/ft³, the lowest density that will allow filling a closed railroad car to capacity). Alternatively, the containers are received as briquettes of a maximum density of 500 kg/m³ (35 lb/ft³), which can be stacked on skids (offering storage advantages to the supplier, but some hardship on the equipment of the receiver). In the shredding operation, bales and briquettes are broken apart, and the cans are shredded to ensure that no trapped liquid or extraneous material reaches the melting furnaces and causes damage or injuries from molten metal explosions. From the shredder the material passes over a magnetic separator to remove ferrous contaminants and over a set of screens to remove the dirt; and then it passes through an air knife in which the heavy nonferrous and nonmagnetic materials, such as lead, zinc, and stainless steel scrap, drop out while the shredded cans pass on to the delacquering units.

There are two basic thermal delacquering methods. One is based on a relatively long exposure time at a safe temperature, and the other is based on staged temperature increases to just below melting for as short an exposure time as possible. The first method uses a pan

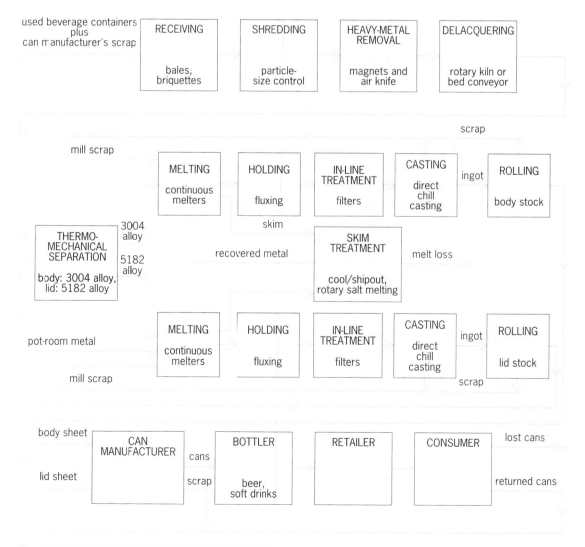

Fig. 1. Flow diagram of closed-loop can manufacturing and recycling.

conveyor on which a bed of used beverage containers approximately 20 cm (8 in.) deep moves through a chamber held at about 520°C (968°F), with gases that are produced by combustion being diluted with air to provide the proper atmosphere as well as temperature for the delacquering process (part pyrolysis, part combustion). The second method uses a rotary kiln with a sophisticated recirculating system at various entry points for the gases produced by combustion. The temperature in the last stage is near 615°C (1140°F), very close to the temperature at which incipient melting occurs in 5XXX series alloys, the 3–4% magnesium-containing alloys used for the can lids.

Both systems have inherent problems of control resulting in nonuniform delacquering. Too low a temperature or too short exposure times at proper temperatures will leave a tarlike coating that causes increased losses upon melting, due to premelt burning. Too high a temperature or too lengthy exposure times at proper temperatures will cause considerable oxidation of the scrap, also resulting in increased melt losses. In the pan delacquering system, beds that are very deep cause temperature gradients that result in the above-mentioned problems; while in the kiln, gas flows are high and the particles are physically agitated, so that nonuniform residence times lead to the same problems. Proper operating controls for these delacquering units, which treat about 20 metric tons (22 tons; about 1,250,000 used beverage containers) of scrap per hour, are vital for producing low-melt-loss feedstock. A new development, based on creating a fluid bed on a shaking conveyor, using the hot delacquering gases in the fluidizing medium, promises to reduce the control problems plaguing both existing methods.

The hot, delacquered used beverage containers may then move into the thermomechanical separation chamber, held at a specific temperature and neutral atmosphere, in which a gentle mechanical action breaks up the 5182 alloy lids into small fragments along grain boundaries weakened by the onset of incipient melting. An integrated screening action removes the fragments as soon as they can pass the screen to avoid over-fragmentation. This process requires a very narrow operating control capability to avoid melting of entire 5182 alloy particles, which would then cluster with the still-solid 3004 alloy particles from the can body. The screened-out particles of 5182 alloy are transported to the lid stock melters, and the large particles of 3004 alloy continue directly into the body stock melting furnaces.

Nowadays most melting facilities for used beverage containers throughout the industry are dedicated units designed to handle the enormous volumes and to minimize the melt losses inherent in melting thin-walled material. Larger companies have developed their own proprietary processes. The still-significant amounts of dross are removed and treated for metal recovery. The dross weight may typically be 15% of the original charge. The recovered metal (5–7% of the original charge) from this dross is used only in manufacturing body stock because of its high manganese and contaminant level.

The metal from these dedicated melters is often transferred to on-line melting furnaces in which additional bulky scrap is remelted and primary metal is charged to create the desired volume of the proper alloy composition. From there, metal is transferred to the holding furnaces, where minor composition adjustments are made and metal quality treatments are performed, for example, gas fluxing to remove hydrogen. Some metal treatment, such as inclusion removal, may occur in in-line treatment units; and in this aspect most major companies have also developed their own preferred methods and technology. The now-clean metal with the correct composition is cast into ingots of up to 18 metric tons (20 tons). During casting and rolling of the ingot to sheet, about 42% of the weight of the original melt may be shaved (scalped), cropped, or slit off. This so-called in-house or runaround scrap is returned directly to the remelters. The body and lid sheet are shipped to a can manufacturer. As a result of can fabrication, about 20% of the sheet (or 13% of the original melt) is returned to the aluminum manufacturer as skeleton scrap. On a global basis (that is, based on the national recycling rate), 55% of a melt consists of "new" (production-related) scrap. If all cans were returned as used beverage containers and total melt losses were 5% of the melt, this 5% would be the only makeup metal required from the pot room (that is, primary metal) to close the loop (provided the market remained constant). In order to drive the process losses that low, a special melt technology was developed.

Continuous melting concept. It is undesirable for low-density scrap to be exposed to the corrosive furnace atmosphere; thus, either the scrap has to be submerged quickly in a heel of superheated molten metal inside the furnace, or molten metal has to be taken out of the furnace for external mixing with the scrap. The advantage of the latter option is that the inevitable dross (fortunately in much smaller amounts) can be captured outside the furnace as well. The resulting metal stream, cooled down but still molten, can be returned to the furnace for reheating by means of a molten metal pump. In this manner a steady-state condition can be maintained if the heat required for melting a constant mass flow of scrap particles (plus makeup for heat losses) is equal to the net heat input into the furnace, as shown in the equation below.

$$\dot{m}_s C_p(T_m - T_s) + \dot{m}_s C_p(T_h - T_m) + \dot{m}_s H_{s-l}$$
$$= \dot{m}_l C_p(T_h - T_l)$$

Here $\dot{m}_s$ is the scrap charge rate, C_p is the alloy's specific heat, T_m is the melting temperature of the alloy, T_s is the temperature of the charge, T_h is the high temperature (and also the tapping temperature) of the melt, H_{s-l} is the heat of fusion of the alloy, $\dot{m}_l$ is the circulation rate of the molten metal, and T_l is the low temperature of the melt.

It takes about 10^6 J/kg to heat aluminum from 25°C to 730°C (77°F to 1346°F), and the specific heat of molten aluminum is 10^3 J/kg C. Thus, 10^3 J per kilogram of charge is needed to maintain steady state. This is supplied by the cooling of the superheated, circulating molten metal: $\dot{m}_l(T_h - T_l)$. Therefore, if the molten metal circulation rate is tied to the scrap feed rate, the temperature span is constant. For instance, if a circulation rate of 20 times the charge rate is selected, the temperature difference across the charge bay will be 50°C (90°F). Obviously, the metal will regain 50°C (90°F) during the return trip through the furnace for steady state to be maintained. Control of the steady state is maintained by matching the scrap charge rate with a selected furnace firing rate, which is based on the heat transfer efficiency of the entire melt complex. Steady state is reached when the temperature at any given location in the melting complex remains constant. Hence, process control can be based on monitoring $T_h - T_l$.

The mass balance is equally simple. The output (metal overflow) is equal to the charge rate minus the dross generation rate times the oxidation constant (an empirically determined correction for weight gain due to oxidation).

A few events can cause deviation from steady state; for example, oxide buildup on the metal in the furnace will change the heat transfer efficiency, or sludge buildup in the passageways will affect the circulation rate. Diagnostic sensors monitoring the process detect the changes in the early stages and alert the operator.

The key to successful operation of a continuous melt facility is the charge system. The hot delacquered cans need to arrive just in time at a predetermined rate—for example, 10 metric tons (11 tons) per hour—and be submerged in the superheated molten metal stream. Several vortex-inducing methods have been developed to achieve the high rates of ingestion to match the large thermal potential of the more efficient, modern systems. In principle, the circulating molten metal stream is forced to enter a bowl-shaped charge bay tangentially and leave through a duct in the center of the bottom so that the otherwise floating scrap particles are dragged down and submerged in the swirl that is created (**Fig. 2**).

The particles melt in the turbulent stream before it enters the next confined area, the skim bay, where the dross floats to the surface, while the clean molten metal returns through an underpass to the furnace for reheating. Continuous overflow or periodic tapping of limited amounts of metal keeps the metal level at or near optimum for the process.

Continuous dross treatment. Dross floats on molten metal because a considerable amount of gas is trapped in the product. The proportions of gas, metal, and oxide depend on the alloy, the atmosphere, and the mass-flow conditions during the formation. In the case of continuous melting of used beverage containers, the apparent density of freshly formed dross is usually less than 1.3 g/cm³ (0.75 oz/in.³), while the density of the molten alloy is about 2.3 g/cm³ (1.3 oz/in.³).

Advantage was taken of this density ratio by designing an in-line continuous dross treatment process in which the dross is continuously manipulated mechanically in a confined zone at the metal bath level. The newly formed dross coming from the charge bay and floating up into the treatment zone constantly lifts the treated, dry, sandlike material to the exit (Fig. 2).

The treatment is based on the principle that a certain shear force will tear oxide skins of approximately the same strength simultaneously, so that small neighboring metal droplets can coalesce at the surface of mechanically driven manipulating blades. The blades move the dross mass gradually to the periphery of the cylindrical treatment bay, where the blades have a higher linear velocity and thus increased shear forces. The combined rotary and upward movements result in a steady-state treatment condition, which yields about half of the metal (trapped in the original dross) by letting it rejoin the metal stream, while the metal confined in the fine particles exiting the unit can be recovered in part by conventional salt melting. The 50% direct recovery makes this process economically attractive.

Secondary dross treatment. Traditionally, the primary industry left the metal recovery and disposal of the dross treatment by-product (salt cake) to secondary processors, but environmental considerations have encouraged an industry-wide review of the practices that affect dross formation, handling, and treatment. Essentially, there are two basic approaches for solving the problems associated with hazards in waste streams. The first one eliminates the use of

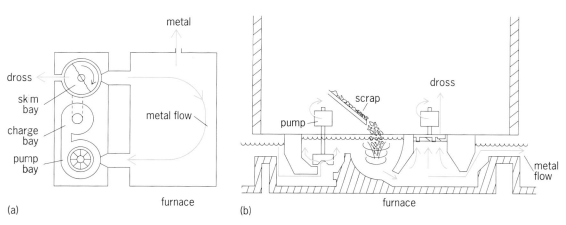

(a)

(b)

Fig. 2. Diagram of the operation of the advanced continuous scrap melting system. (*a*) Plan view. (*b*) Side view.

salt in the dross treatment process and instead applies plasma technology using inert gases to achieve high recovery of metals and minimal formation of oxides and other reaction products in essentially the same rotary furnaces that are used for salt-based dross treatment. The process claims a 90% recovery of available metal and a by-product that can be converted into a value-added product by means of a calcining step. The economics of this process have not been entirely elucidated.

The other approach continues the use of salt to promote coalescence of metal droplets in the dross treatment step and adds a treatment process to recover salt from the salt cake. This process consists of a series of crushing and screening steps followed by leaching, filtering, drying, and crystallizing. An evaluation of this process found that with proper material preparation and careful process control, (1) an essentially inert by-product can be obtained, (2) sufficiently improved metal recovery can be achieved through increased salt use in the furnace, and (3) the process is not competitive at today's landfill costs unless a productive use of the by-product is included.

Both processes will most likely not be economically attractive until landfill costs rise significantly above present levels, especially if the metal values left in the dross by the primary producers are reduced as a consequence of improved melting and skimming practices.

For background information SEE ALUMINUM; FLUIDIZATION; PLASTICS PROCESSING in the McGraw-Hill Encyclopedia of Science & Technology.

Jan H. L. van Linden

Bibliography. Environmental Protection Agency, *Characterization of Municipal Solid Waste in the United States: 1990 Update*, PB90-215112, 1990; G. J. Kulik and J. C. Daley, Aluminum dross processing in the 90s, *2d International Symposium: Recycling of Metals and Engineered Materials*, 1990; T. J. Nosker et al., Commingled plastics recycling and environmental concerns, *EPA Meeting on Solid Waste*, San Diego, pp. 1–15, 1989; Office of Technology Assessment, *Facing America's Trash*, OTA-O-424, 1989; J. H. L. van Linden, Aluminum recycling: Everybody's business, technological challenges and opportunities, *Light Metals 1990: Proceedings of the Annual Meeting of the Metallurgical Society of AIME*, 1990; J. S. Viland, A secondary's view of recycling, *Light Metals 1990: Proceedings of the Annual Meeting of the Metallurgical Society of AIME*, 1990.

Relativity

Albert Einstein's theory of special relativity (1905) plays a fundamental role in the understanding of nature. Historically, optical experiments have provided tests for the validity of special relativity. Since 1970, lasers with highly stabilized output frequencies have been developed. Combined with high-resolution, nonlinear spectroscopic techniques and frequency metrology, these lasers are now being used in a new generation of optical tests of special relativity.

The central postulates of special relativity are that the laws of nature are the same in all uniformly moving frames of reference, and that the speed of light in vacuum is a constant, independent of the motion of either the source emitting the light or of an observer receiving the light. These postulates lead to interesting results. For example, an observer measuring distances with a moving meter stick would find the length to be shortened, while an observer measuring time with a moving clock would find the time interval to be lengthened. Laser spectroscopy has been used to test both the constancy of the speed of light and the length-contraction and time-dilation effects to a high degree of accuracy.

Michelson-Morley experiments. One type of experiment tests the constancy of the speed of light with propagation direction. It tests whether the speed of light remains unchanged to an observer in a certain reference frame who looks out in different directions in space; that is, it tests whether space is isotropic with respect to the propagation of light. This type of experiment is known as the Michelson-Morley experiment, after A. A. Michelson and E. W. Morley, who in 1887 developed an interferometric method to test the ether theory of light propagation.

In a Michelson interferometer (**Fig. 1**), a light beam from a monochromatic source was split into two paths that were oriented at right angles to each other. The two beams were reflected by mirrors and were recombined to form an interference pattern. A shift in the interference pattern would result if there were a change in the path lengths of the two arms. The apparatus was mounted on a turntable and was rotated at a slow rate so that the two arms scanned out different directions in space. A constant speed of light would produce no change in the fringe pattern, but if the speed of light changed with respect to directions, then the difference in the travel time of light in the two arms would be equivalent to a difference in the two optical path lengths, and there would be a variation

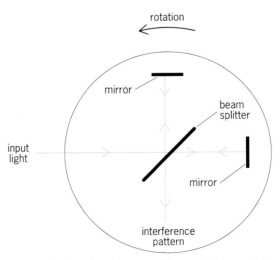

Fig. 1. Michelson interferometer. Change in the speed of light with direction will cause a shift in the interference pattern as the apparatus is rotated.

in the interference pattern correlated to the rotation of the table.

The Michelson-Morley experiment achieved a limit of 3×10^{-10} in the isotropy of space with the propagation of light. It was recognized early in the development of lasers that the sensitivity of the Michelson-Morley experiment could be greatly enhanced if the optical path-length measurements were carried out with frequency-counting techniques using highly stabilized lasers.

In a laser experiment to test the isotropy of space (**Fig. 2**), the arms of the Michelson interferometer were replaced by a Fabry-Pérot cavity and a helium-neon laser mounted on a rotating table. The Fabry-Pérot cavity was an optical resonance cavity formed by a pair of mirrors separated by a spacer of fixed length. When the wavelength of the laser satisfied the standing-wave boundary condition of the cavity, a sharp resonance occurred in the transmission of the cavity owing to constructive interference of the multiply reflected beams of light inside the cavity. Thus, by using a feedback scheme (servo control), the laser wavelength could be locked to the center of a transmission fringe of the cavity. Any variation in the cavity length would be manifested as a variation in the laser wavelength. A small part of the laser beam was diverted up along the table rotation axis in order to be compared with a reference laser, which was kept fixed in the laboratory. The reference laser was also a helium-neon laser, but its frequency was stabilized to a molecular transition in methane.

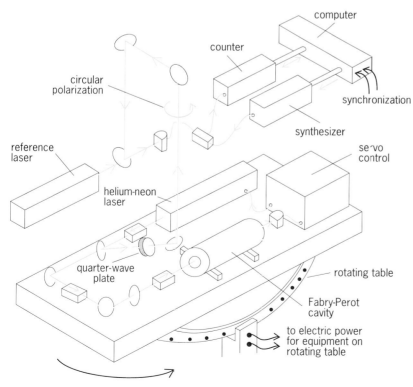

Fig. 2. Laser isotropy-of-space experiment. (*After A Brillet and J. L. Hall, Improved laser tests of the isotropy of space, Phys. Rev. Lett., 42:549–552, 1979*)

If space were not isotropic with direction, the length of the Fabry-Pérot cavity would vary as the turntable was rotated, and there would be a change in the frequency of the first helium-neon laser. However, the frequency of the reference helium-neon laser should not be affected. By comparing the output frequency of the two lasers, the variation (or the lack of it) with respect to spatial direction could be obtained. With this arrangement, the experiment was sensitive to length changes as small as 6×10^{-16} m. The experiment sets a limit of $\pm 2 \times 10^{-15}$ in the fractional variation, $\Delta c / c$, in the speed of light with direction.

Kennedy-Thorndike experiments. While the question of isotropy of space is settled for a reference frame, a second issue remains: whether the speed of light has the same value in different reference frames moving relative to one another. The classical experiment for testing this question was performed by R. J. Kennedy and E. M. Thorndike in 1932, using a Michelson interferometer with arms of unequal length. They used the Earth as a moving reference frame. As the Earth moved around the Sun, the velocity of the Earth changed in magnitude and in direction. The different locations of the Earth in its orbit were taken to be the different moving reference frames for carrying out the test. The experiment looked for any variation in the round-trip speed of light as the Earth traveled around the Sun. Because of the difficulties associated with keeping the system in operation for long periods of time, this experiment was not attempted again until very recent times. Again, a highly stable Fabry-Pérot cavity was used as the length standard that controlled the frequency of a helium-neon laser, and a second helium-neon laser, stabilized to a molecular transition (this time in iodine), was used as a reference. With about 90 days of data, spanning a period of more than 2 years, the experiment found no variation at the level of 2×10^{-13}. The result represents a 300-fold improvement over the 1932 result and established that, to the level of 2×10^{-13}, the speed of light has the same numerical value in different reference frames.

The above two experiments test the constancy in the round-trip speed of light. They do not yet constitute a complete set of tests for special relativity, since they do not address the time-dilation question in an unambiguous way. Thus a third type of experiment is needed.

Tests of relativistic Doppler effect. One consequence of the time-dilation effect is the relativistic Doppler shift. Classical Doppler effects are familiar: an example is the change in pitch of a train whistle when the train is approaching or departing. Likewise, an observer will detect a shift in the ticking rate (that is, the frequency) of a moving clock. In special relativity, the Doppler frequency shift is modified, owing to the time-dilation effect, by an extra factor over the classical result. The presence of this time-dilation term implies that a moving clock will always have a Doppler shift, even when the source and observer are moving in perpendicular directions.

Since atoms have precise transition frequencies, they can be used as clocks to measure time intervals. Modern tunable dye lasers and spectroscopic techniques are particularly well suited to excite a specific atomic transition and to measure the transition frequency. In a typical laser experiment, the frequency shift of a beam of fast-moving atoms is measured against atoms at rest and is compared to the prediction of special relativity. The atoms may be accelerated in particle accelerators to speeds approaching the speed of light, although the experiments with the most sensitivity at the present time are performed with much smaller velocities, a few thousandths of the speed of light.

This class of experiments is known as Ives-Stilwell experiments, after H. E. Ives and G. R. Stilwell, who first observed the existence of a transverse Doppler shift in a beam of 30-keV molecular hydrogen ions in 1938. Recent laser experiments have utilized atoms ranging from a 120-keV neon beam to an 800-MeV hydrogen beam. In the neon experiment, which has the best accuracy so far, a technique known as two-photon spectroscopy (**Fig. 3**) was used. Two light beams of the same frequency (from a single laser) but traveling in opposite directions are used to excite an atom that is moving with velocity v in the laboratory. Because of the Doppler effect, the laser beam that is traveling in the same direction as the atom will have its frequency down-shifted, in the atomic frame, while the laser beam that is traveling in the opposite direction has its frequency up-shifted. These shifts contain both the classical part, $\pm v/c$, and the special-relativity part, $[1 - (v^2/c^2)]^{-1/2}$, where c is the speed of light. For small values of v/c, the small relativistic term is completely masked by the large classical terms. However, there are cases in which an atom can make a transition by simultaneously absorbing two photons, one from each of the two counter-propagating beams. The total transition frequency is then the sum of the two photons; the classical Doppler shifts cancel, so that only the relativistic contribution remains.

In the experiment, a tunable dye laser was used to excite the transition in a fast neon beam with well-defined velocity, while a second dye laser was used to excite the same transition in slow thermal atoms. The thermal atoms were considered to be essentially clocks at rest. The frequencies of the two lasers were stabilized to the centers of the two-photon transitions of their respective atoms, and the frequency difference between them was measured by counting the beat frequency produced by the two lasers in a photodiode. The result was found to agree with the predictions of special relativity to better than 7 parts per million. The accuracy of the laser experiments has surpassed the best results from lifetime studies of elementary particles by a factor of more than 100.

Constancy of one-way speed of light. The observed anisotropy in the 3-K cosmic background radiation may be explained to be due to the motion of the Earth relative to the rest frame of the cosmic radiation. Thus, it is interesting to see if there is also a preferred-frame effect in the one-way speed of light. The Michelson-Morley and Kennedy-Thorndike experiments do not check for this, since, in an interferometer or in an optical cavity, the light always travels in two directions because it is reflected by the mirrors. A relativistic Doppler-shift experiment can be used as a test. If there is an anisotropy in the one-way propagation speed of light, then the Doppler shift of a beam of fast atomic clocks will vary as a function of the direction of the atomic beam in space. If the atomic beam is fixed on Earth, then the rotation of the Earth on its axis would impose a sidereal variation on the measured Doppler shift. At present, results from laser experiments indicate that the one-way effect must be smaller than 3×10^{-9}.

Space experiment. All the experiments described so far are ground-based. A rocket in free fall represents a moving reference frame, and the first space experiment was carried out in 1976. An atomic clock (a hydrogen maser) was carried aboard a rocket and launched to an altitude of 10,000 km (6000 mi). The frequency of the hydrogen maser in flight was compared to that of a maser clock on the ground. The large (classical) Doppler shifts were canceled by a separate go–return link, which also removed atmospheric and ionospheric effects. The objective of the experiment was to measure the gravitational redshift on the atomic clock, but the results also set a limit on the constancy of the one-way speed of light and on the validity of the relativistic Doppler effect. Although the accuracy of this experiment is now surpassed by ground-based experiments, it was the first direct test of the one-way speed of light, and marked the beginning of the use of high-accuracy clocks in space to test relativity.

Prospects. The laser Michelson-Morley and Kennedy-Thorndike experiments already utilize the most advanced laser stabilization and frequency metrology techniques, and elaborate control of the systematic effects that may affect the accuracy of the experiments. It does not appear that major improvements in these experiments will be forthcoming in the near future. On the other hand, the experiments concerning relativistic Doppler shift and the one-way speed of light can be improved with the use of faster atoms, narrower

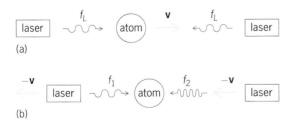

Fig. 3. Arrangement for two-photon spectroscopy. (*a*) Laboratory frame. The atom is moving with velocity v, and the two laser beams have the same frequency f_L. (*b*) Atom frame. The laser beams are Doppler-shifted to frequencies $f_1 = \gamma f_L[1 - (v/c)]$ and $f_2 = \gamma f_L[1 + (v/c)]$, where $\gamma = [1 - (v^2/c^2)]^{-1/2}$ is the time-dilation factor. When the atom makes a transition by absorbing a photon from each beam, the classical Doppler shifts are canceled, but the relativistic shift remains: $f_1 + f_2 = 2\gamma f_L$.

atomic transitions, and better frequency-stable lasers. Experiments being considered involve using heavy-ion storage rings to deliver beams of high-energy and laser-cooled atoms. Accuracy on the order of 10^{-9} appears to be feasible for the relativistic Doppler effect. This level of accuracy is important for experiments proposed to test the equivalence principle or metric theories of gravitation.

For background information SEE CLOCK PARADOX; DOPPLER EFFECT; INTERFEROMETRY; LASER SPECTROS-COPY; LIGHT; RELATIVITY in the McGraw-Hill Encyclopedia of Science & Technology.

Siu Au Lee

Bibliography. M. P. Haugan and C. M. Will, Modern tests of special relativity, *Phys. Today*, 40(5):69–76, May 1987; D. Hils and J. L. Hall, Improved Kennedy-Thorndike experiment to test special relativity, *Phys. Rev. Lett.*, 64:1697–1700, 1990; E. Riis et al., Test of the isotropy of the speed of light using fast-beam laser spectroscopy, *Phys. Rev. Lett.*, 60:81–84, 1988; E. F. Taylor and J. A. Wheeler, *Spacetime Physics*, 1966.

Reptilia

The Therapsida, a group long known informally as the mammallike reptiles, gave rise evolutionarily to all mammals, including humans. For that reason, therapsids are of substantial interest to vertebrate paleontologists and zoologists. Their anatomy and diversity are documented by a rich fossil record, especially from rocks of Triassic age in southern Africa. Important therapsid fossils have also been collected in North and South America, Eurasia, and even in Antarctica. Work by paleontologists over the past century had seemingly established that therapsids first appeared geologically in the upper Permian and became extinct in the Middle Jurassic, an interval of about 100 million years (m.y.). However, two recent discoveries, if confirmed by further work, will force an important revision of this stratigraphic range as well as a major reevaluation of certain other aspects of therapsid history.

Pelycosaurs to therapsids. Paleontologists traditionally have classified therapsids as the more advanced of two orders included in the reptilian subclass Synapsida: The more archaic group in this subclass, the Pelycosauria, represents the first stage of evolution from ancestral stem reptiles (the earliest amniotes) to mammals. Unfortunately, the next stage, the transition from primitive pelycosaurs to therapsids, is poorly documented by fossils but is crucial to a thorough understanding of the phylogenetic sources of Mammalia. The available evidence has suggested that the transition took place during the middle Permian, among animals that retained a persistently primitive (pelycosaurlike) anatomy of the skull and jaws while they evolved a more progressive organization of the limbs and limb girdles. These latter locomotory adaptations reflect an increasingly active mode of life that seemed to mark the origin of the Therapsida. Nonetheless,

a significant morphological gap between pelycosaurs and therapsids has persisted, with fossils that document the actual steps in this transition being virtually unknown.

Restudy of Tetraceratops insignis. The first of the two discoveries modifying this outline of therapsid history comes from restudy of *Tetraceratops insignis*, a reptile from the lower Permian of Texas that was described near the turn of the century as a pelycosaur. *Tetraceratops* is known only from a single specimen that includes the skull and lower jaw. Restudy of the specimen following removal of the enclosing rock matrix revealed additional anatomical characters. They indicate that *Tetraceratops* is not a pelycosaur at all but a therapsid, and one that appears to fill the morphological gap between the two groups. Seven characters of the dentition, jaw mechanism, and palate are more derived than in any known pelycosaur and occur otherwise only in therapsids; six additional skull and dental features are pelycosaurlike, and are more primitive than in any previously known therapsid. While this mosaic of primitive and derived characters thus places *Tetraceratops* in the gap between pelycosaurs and therapsids, *Tetraceratops* is classified as a therapsid because it possesses the seven uniquely therapsid characters.

It will be of great interest to see whether the postcranial skeleton, if discovered, will be consistent with this estimate of the position of *Tetraceratops* within the Synapsida, and whether modifications of the skull and the dentition actually preceded those of the postcranial skeleton in primitive therapsids that remained at a pelycosaur level of locomotory specialization.

Discovery of new therapsids. The second discovery concerns the collection of therapsid fossils from rocks of late Paleocene age at a locality near Cochrane in Alberta, Canada. The Cochrane fossils are 100 m.y. younger than the youngest previously discovered therapsids (from the Middle Jurassic of Europe, China, and Mexico), and they furnish the first record of therapsids from the Cenozoic Era, the so-called Age of Mammals. The fossils are from a single new species (new at the generic, family, and perhaps infraordinal level). The fossils (see **illus.**) include a fragment of a tiny lower jaw made up of a dentary bone containing three teeth, plus four isolated teeth from other individuals. Surfaces on the posterior parts of the dentary indicate that two or more small bones were loosely attached there to produce a compound lower jaw. A compound lower jaw is found in all nonmammalian tetrapods, including therapsids, and in certain primitive Mesozoic mammals, but in no living or extinct mammals of Paleocene age or younger. Additionally, the Cochrane dentary, although incomplete, shows several features that are explicitly therapsidlike but that are not found in even the most primitive mammals. For example, although broken, the dentary appears to have originally lacked the rounded process or condyle for attachment to the skull found in all mammals. Therefore the attachment must have been provided by one or more of the missing bones, as in therapsids. In addition, the specialized dentary surfaces for insertion of the major lateral jaw-closing muscle, which are unique to therapsids and mammals, are limited in their extent. This limiting indicates that the muscle was still relatively small, at a premammalian stage of evolution.

Dentition of new species. Evidence from the dentition of the species also indicates that they are therapsids: the postcanine teeth fit into deep dentary sockets by means of only a single root with a single pulp cavity. In contrast, comparable teeth in mammals are socketed but two-rooted, with each root having its own pulp cavity. The crowns of the teeth from the new species are strongly developed, with three major cusps arranged in a simple pattern to form a broadly open triangle in some specimens and a nearly straight anteroposterior line in others, depending upon the position of the tooth in the jaw. There is no trace of a posterior heel, as on the lower molars of tribosphenic mammals, including marsupials and placentals characteristic of the Late Cretaceous and Cenozoic.

While the crowns of some of these teeth, especially those having a more triangular pattern of the cusps, resemble those of certain Mesozoic pretribosphenic mammals termed symmetrodonts, the resemblance is only superficial. For example, the lower postcanines of symmetrodonts are two-rooted, have a ridge or cingulum that encircles the base of the crown, and have a thick enamel covering; none of these characteristics are true of the Cochrane fossils. The closest morphological similarities of the Cochrane teeth are to the teeth of certain poorly known Triassic therapsids, including *Microconodon* and *Therioherpeton*. Further, preliminary work indicates that the ultrastructure of the enamel of the Cochrane teeth does not resemble that known for symmetrodonts or other mammals, and is otherwise therapsidlike in lacking prisms (nonprismatic enamel is widely distributed among therapsids, but it has been reported in only a few, primitive mammals).

Apparently, the new therapsid was a small, probably insectivorous form that either was a relict member of the important therapsid subgroup, the Cynodontia (which includes the immediate ancestors of mammals), or was a collateral relative of cynodonts. Support for the latter alternative is provided by the lack of a groove along the inner side of the Cochrane jaw (found in all cynodonts but lost independently within several lineages of early mammals) and the absence of evidence of teeth replacement (the teeth are replaced throughout life in most, but not all, cynodonts, while in most mammals they are replaced only once). However, these appear to be minor specializations that could have readily evolved during the more than 100 m.y. that separate the Cochrane species from any possible antecedents among Mesozoic therapsids.

For background information *SEE MAMMALIA; PELYCOSAURIA; REPTILIA; SYNAPSIDA; THERAPSIDA* in the McGraw-Hill Encyclopedia of Science & Technology.
Richard C. Fox

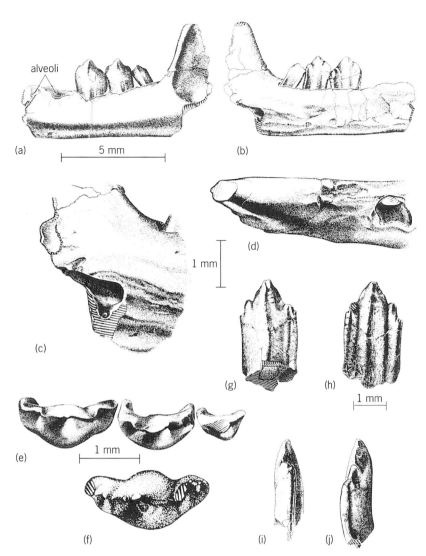

Jaw and teeth of a new therapsid (Cochrane fossil) from the late Paleocene. (*a*) Labial view and (*b*) lingual view of the incomplete left dentary with three teeth; (*c*) posteromedial view of smooth articulation surfaces for the postdentary bones; (*d*) dorsal view of the scar for the coronoid bone behind the posteriormost tooth; (*e*) enlargement of tooth crowns in occlusal view. (*f*) Occlusal view, (*g*) labial view, (*h*) lingual view, (*i*) anterior view, and (*j*) posterior view of an isolated lower left molariform tooth. (*After University of Alberta Laboratory for Vertebrate Paleontology Specimen Cat. No. 32358 and 32359*)

Bibliography. T. M. Bown and K. D. Rose (eds.), *Dawn of the Age of Mammals in the Northern Part of the Rocky Mountain Interior, North America*, 1990; R. L. Carroll, *Vertebrate Paleontology and Evolution*, 1988; N. Hotton et al. (eds.), *The Ecology and Biology of Mammal-like Reptiles*, 1986; M. Laurin and R. Reisz, *Tetraceratops* is the oldest known therapsid, *Nature*, 345:249–250, 1990.

Rhizosphere

Living roots of grass plants exert beneficial effects on the soil structure, and they generally improve the stability of aggregates. In the bulk soil, several agents and factors—for example, texture, organic matter, microbial activity, climatic conditions, and soil management—control the state of the soil structure.

At the soil–root interface (the rhizosphere), organic materials released from the roots during growth (root exudates) also play an important role in soil aggregation. Root exudates have various molecular weights, and the amount released varies from 1 to 20% of the net photosynthesis of the plant. They cause substantial modifications in the chemical and microbiological properties of the rhizosphere, resulting in shifts in availability of plant nutrients and sharp increases in the microbial activity at the root surface.

Most of the obvious improvements in soil aggregation around plant roots have been attributed to the presence of root exudates. However, because of the complexity of the mechanisms of the reactions at the soil–root interface, the specific effects of root exudates on soil aggregation can hardly be observed in experiments involving growing plants. For that reason, the role of exudates has not been clearly demonstrated until recently. The use, as a model, of intact root mucilage collected from maize (corn, *Zea mays*) plants has shown that the exudates with high molecular weights that are secreted by active roots play a predominant role in building up and stabilizing the aggregates in the rhizosphere. Evidence of the firm adsorption of exudates on clay minerals has been produced, and observations made with scanning electron microscopy have revealed the formation of microaggregates between clay particles and root mucilage. Mucilaginous exudates produced at the root tip cement the soil particles, so that aggregates are protected against destruction by water. While growing roots may cause some detrimental effects, such as mechanical compaction of the soil aggregates, root exudates remain the most important factor in increasing the stability of the aggregates in the rhizosphere.

Root exudates. Plant roots release a large number of different compounds of various molecular weights. High-molecular-weight exudates (mucilage) are the dominant material exuded by root tips, while exudates with low molecular weights are produced mostly behind the tip.

Mucilage is made primarily of polysaccharides (95.0%), and it also contains some proteins and mineral ash. Two weak acidities, with average values for the reciprocal of the equilibrium constant (pK_a) of 4.4 and 10.1, representing 2.4 and 0.6 milliequivalents (meq) per gram of dry matter, have been identified. Exudates with low molecular weights exhibit stronger acidic properties, which provide them with the ability to form complexes with metal ions; and they play a major role in the solubilization of nutrients in the rhizosphere.

The determination of the specific properties of root exudates requires intact material that is not contaminated by soil particles or microorganisms. Collection of intact exudates may involve using sterile plant cultures in a nutrient solution under controlled environmental conditions. The soluble compounds are collected by filtration of the nutrient solution, and the mucilage, firmly adhering to the root surface, is collected by pumping under vacuum. Amounts are gener-

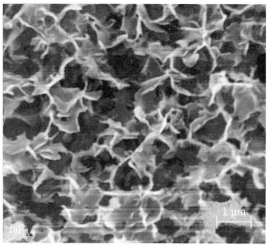

Fig. 1. Microorganization of clay and clay–mucilage complex. Root mucilage was mixed with a suspension of a calcium-montmorillonite. (*a*) Calcium-montmorillonite. (*b*) Calcium-montmorillonite–mucilage. (*From L. Habib et al., Adsorption de mucilages racinaires de maïs sur des argiles homoioniques: Conséquences sur la microorganisation des complexes formés, Compt. Rend. Acad. Sci., 310:1541–1546, 1990*)

ally sufficient for analysis of the released compounds, but too low for studying their fate and role in the rhizosphere.

Larger amounts of root exudates, mostly mucilaginous material, can easily be obtained from maize plants grown in the field. Shoots of intact 2-month-old plants are dipped into distilled water to ensure rehydration. Apical exudates are then collected by vacuum filtration. This method allows for the collection of relatively large amounts of material, and makes it possible to carry out classical studies on the fate of organic material in soils, that is, studies of adsorption, biodegradation, and the effect on the physical properties of the soil.

Exudate adsorption on clay minerals. Soil aggregation and stabilization is primarily the result of the formation of firm bonds between minerals and organic materials. Mucilage released by roots is readily adsorbed on clay minerals; the rate of adsorption is higher on montmorillonite than on kaolinite, owing to higher charge density in montmorillonite. As with most anionic polymers, adsorption is stronger in the presence of calcium (Ca) as the compensating cation than in the presence of monovalent sodium (Na).

Adsorption of mucilages, due to the formation of cation bridges, is restricted to the external surface of the clay. Observations made with scanning electron microscopy have revealed the fibrillar structure of mucilages, whose strands extend through the clay pores and bind the particles together (**Fig. 1***a*).

A small amount of mucilage will flocculate a calcium-montmorillonite clay and form a microaggregated structure up to 3 micrometers thick. This action may represent one of the early steps in soil aggregation in the vicinity of root tips (Fig. 1*b*). Other interactions or associations can be identified at higher concentrations of exudates. These are similar to the interactions described with microbial polysaccharides, where an increase in water-holding capacity of the clay is generally observed.

Exudates and aggregate stability. Incorporating freshly collected exudate into a biologically active soil is a means of simulating rhizosphere conditions. It may help to understand the fate of exudates and their specific role on aggregation around growing roots. Following their mixing with the soil, exudates serve as a source of carbon and energy for the heterotrophic soil microflora, as assessed from the measurement of evolved carbon dioxide (CO_2; **Fig. 2**). While low-molecular-weight compounds are rapidly degraded, mucilage, which consists mostly of polysaccharide molecules, is slowly broken down. Biodegradation occurs only 2–3 days after incorporation into the soil; however, later on, except where it is protected by adsorption on clays, the material is quickly consumed by microorganisms, and visible zones of lysis appear under the electron microscope.

Immediately after its release in the soil, fresh mucilage sticks to soil particles very strongly, especially to clay minerals. This sticking causes a spectacular increase in aggregate stability, as revealed with tests involving sieving of aggregates in water (**Fig. 3**). This process, which occurs very quickly, is independent of

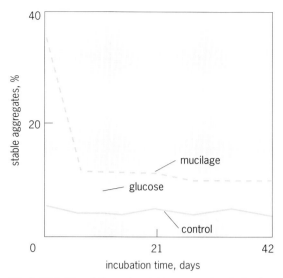

Fig. 3. Evolution of the water-stable aggregates during incubation of unamended, glucose-amended, and mucilage-amended silty clay soil. (*After J. L. Morel et al., Influence of maize root mucilage on soil aggregate stability, Plant and Soil, 136:111–119, 1991*)

microbial activity and is due in part to the reduction of the wettability of the soil aggregates. After a short period of time, the protection of the aggregates from destruction by water decreases proportionally to the biodegradation of the exudates. Where sufficient clay is present, however, a longer sticking effect due to mucilage is measured. Mucilage is then replaced by newly synthesized microbial polysaccharides, which establish new bonds with minerals. In older root parts, soil particles are aggregated by a complex gelatinous material (mucigel) composed of root mucilage and debris, bacterial slime, and colloidal mineral and organic matter. Because of mucigel, aggregate stability of clayey soils remains significantly higher in soil receiving mucilage than in unamended soil.

Aggregate stability around roots. In general, roots grow in soil pores that are wider than their own diameter. Under such favorable conditions, tips of living roots locally release amounts of mucilage that strongly adhere to soil particles, covering the pore wall, filling the interstices, and embedding soil particles. The root somehow slides into this protecting glove, and the surrounding soil aggregates become more resistant to destruction by water. Later, colonization of mucilage by microorganisms partially destroys the organo-mineral associations to an extent that depends on the abundance of clay minerals. The disappearance of mucilage is effectively replaced by microbial extracellular polysaccharides. This replacement ensures the persistence of the organo-mineral associations and long-lasting aggregate stability.

An examination of grass roots often reveals soil material firmly attached to the root and forming strong sheaths stabilized by microorganisms and root mucilages. Such a soil–root continuum, resistant even to washing with water, maintains bridges that form a buffer, keeping the root moist; and this moistness

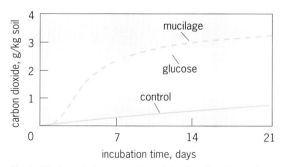

Fig. 2. Biodegradation of root exudates in silty clay soil as measured with cumulative CO_2 evolution from unamended, glucose-amended, and mucilage-amended soil. (*After J. L. Morel et al., Influence of maize root mucilage on soil aggregate stability, Plant and Soil, 136:111–119, 1991*)

contributes to the transfer of nutrients to the root surface. Such a scheme requires sufficient clay content to establish firm and durable bonds with the mucigel.

When growing in soil pores smaller than their diameter, roots exert a force of mechanical compaction on the soil and crush the aggregates, so that the organic matter is exposed to decomposition by microorganisms. The presence of highly metabolizable exudates enhances the biodegradation of the existing organic matter. In addition, the mechanical effects of root growth may weaken the aggregates.

Another detrimental effect of root growth is caused by some soluble complexing exudates, which may destroy the organo-mineral associations of the soil. This destruction occurs when cations forming bridges between organic molecules and clay surfaces are removed by complexing exudates. However, for the most part, these negative effects induced by growing roots seem to be compensated by the effects of high-molecular-weight exudates on the aggregates. Evidence provided by visual observations of roots growing in soil and the greater aggregate stability of rhizosphere soil associated with grass roots confirm the predominance of root exudates in the mechanisms controlling soil aggregation in the vicinity of roots.

For background information SEE RHIZOSPHERE; ROOT (BOTANY); SOIL CHEMISTRY in the McGraw-Hill Encyclopedia of Science & Technology.

Jean Louis Morel

Bibliography. L. Habib et al., Adsorption de mucilages racinaires de maïs sur des argiles homoioniques: Conséquences sur la microorganisation des complexes formés, *Compt. Rend. Acad. Sci.*, 310: 1541–1546, 1990; H. M. Helal and D. R. Sauerbeck, Carbon turnover in the rhizosphere, *Z. Pflanzenernähr. Bodenk.*, 152:211–216, 1989; J. L. Morel et al., Influence of maize root mucilage on soil aggregate stability, *Plant and Soil*, 136:111–119, 1991; J. B. Reid, M. J. Goss, and P. D. Robertson, Relationship between the decrease in soil stability effected by the growth of maize roots and changes in organically bound iron and aluminium, *J. Soil Sci.*, 33:397–410, 1982.

Satellite astronomy

In December 1990, the Astro-1 payload, attached to the space shuttle *Columbia* in low Earth orbit, flew a 9-day mission. Four sophisticated telescopes made up the astronomy-dedicated payload that was interactively controlled to observe many astronomical sources. Each instrument was uniquely developed to measure properties of ultraviolet or x-ray emission from celestial objects. Thus, astronomers could gain greater insight into the physical mechanisms of the hot, sometimes violent universe glimpsed through these energetic windows of the electromagnetic spectrum.

Ultraviolet and x-ray observations. Ultraviolet light and x-radiation are generated by very hot or energetic events. In the neighborhood of the Sun, extremely hot (25,000–100,000 K) stars emit most of their radiation in the ultraviolet. Massive hot

stars throw off their outer envelopes at the prodigious rates of a solar mass every 10^5 years at thousands of kilometers per second. Interacting binary star systems, through collisions of the extended stellar envelopes, generate x-rays and ultraviolet radiation. Material spiraling into black holes is detectable by the resultant x-ray emissions. The mysterious central region, or nucleus, of an active galaxy emits prodigious amounts of radiation in the radio, infrared, visible, ultraviolet, and even x-ray portions of the spectrum. Astrophysical objects are observed throughout the electromagnetic spectrum to gain insight into the physical processes that power these sources.

X-radiation and ultraviolet radiation must be observed from above the Earth's atmosphere, since atmospheric gases absorb the energetic photons (and thereby protect life). Instruments tailored to respond to ultraviolet radiation or x-radiation are built and then launched above the atmosphere. Early studies, in the 1950s, were done by small rockets that could lift a compact instrument above the absorbing atmosphere for several hundred seconds, long enough to point the instrument at a bright astronomical source and record some spectral information. By the 1960s and 1970s, spacecraft such as the *Orbiting Astronomical Observatory* satellites—*Copernicus*, *Uhuru*, and *Einstein*—provided very useful information on astronomical sources in the ultraviolet and x-ray portions of the electromagnetic spectrum. Many astronomical sources of fundamental interest could not be studied simply because they were too faint and the instruments not sufficiently sensitive.

Newer techniques, more efficient optics, and more sophisticated detectors led to greatly improved instruments that were prepared for space flight by short sorties on small rockets. An alternative method of flying multiple instruments on a shuttle mission was thought to be quicker and less costly than building a satellite for each instrument.

Astro-1 instruments. In 1978, the National Aeronautics and Space Administration (NASA) provided an opportunity for the space community to propose experiments for the Spacelab, or reusable payload hardware. Several hundred teams responded, with approximately 50 experiments being selected for study. Four concepts came together as the dedicated payload that flew as Astro-1.

Ultraviolet Imaging Telescope (UIT). This 15-in. (38-cm) telescope recorded ultraviolet radiation over a field of view 40 arc-minutes in diameter (slightly larger than the lunar diameter), in selected wavelength bands from 120 to 310 nanometers. A Ritchey-Chrétien two-mirror system was used to image the field onto a two-stage image intensifier, with film as the recording medium. Much can be learned from comparison of ultraviolet and visible images (see **illus.**). A nearby galaxy, composed of 10^{10}–10^{11} stars, appears as a very diffuse, extended structure at visible wavelengths. Stars like the Sun and cooler stars dominate the field. In the ultraviolet, the very hot, massive, but short-lived stars dominate the field, even though less than 1% of the stars are so massive. As these stars

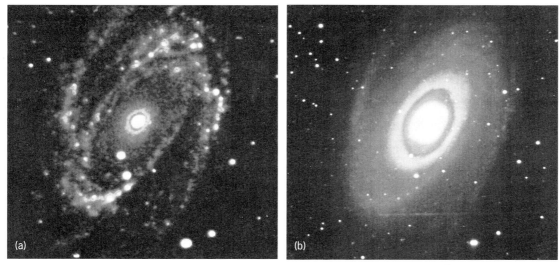

Comparison of images of the spiral galaxy M81 in the constellation Ursa Major, about 1.2×10^7 light-years from Earth. (*a*) Galaxy as photographed by the Ultraviolet Imaging Telescope on the Astro-1 mission. The bright spots in the curved spiral arms of the galaxy are concentrations of very young, hot stars, revealing regions where new stars are forming at a rapid rate. (*b*) Galaxy photographed in red light with a 36-in. (0.9-m) telescope at Kitt Peak National Observatory near Tucson, Arizona. (*NASA, Goddard Space Flight Center*)

are short-lived in stellar time scales, they serve as signposts pointing out regions of recent star formation. Studies of nearby galaxies in the ultraviolet and visible regions provide much information on star formation, star deaths, nonthermal emission regions, and galactic structure in general.

Hopkins Ultraviolet Telescope (HUT). This instrument recorded far-ultraviolet spectra of selected objects. Since the telescope operated at wavelengths from 42.5 to 175 nm, where no mirror is very efficient (10% reflectivity from each surface), a spectrograph was placed at the primary focus on the single, fast mirror. Knowledge that an object emits much ultraviolet radiation may not be sufficient to understand why it does so. Dispersing the radiation into a spectrum provides many clues as to the causes of the emission. Lines attributable to specific ions not only help identify what elements are present but also give insight as to their abundances, degree of ionization, temperature, and even densities. In the far-ultraviolet, gases at 10^5–10^6 K are studied. The Hopkins Ultraviolet Telescope recorded much important information on the blast waves from supernovae, the atmospheres of flare stars, the nuclei of active galaxies, and distant quasars. One key result places a strong upper limit on the mass of neutrinos, important because if neutrinos had significant mass, they could "close" the universe, bringing it to the density at which it would eventually collapse. SEE NEUTRINO.

Wisconsin Ultraviolet Photopolarimetry Experiment (WUPPE). In this instrument, a Cassegrain mirror system fed light into a spectrometer that measured the polarization of ultraviolet radiation. Light from a uniform, spherical source (for example, a normal star) is not polarized. Light that passes through the interstellar medium and is preferentially scattered by interstellar dust grains becomes linearly polarized because of the preferential orientation of the dust grains

by the galactic magnetic field. Light emitted by a rotationally distorted star, a tidally distended member of a binary star system, or a magnetic white dwarf can be circularly polarized. Indeed, the polarization measurements provide key information on the geometry of many types of stars. The WUPPE obtained much information on hot stars within the Milky Way Galaxy and information on a few extragalactic objects. Already these measurements are demonstrating that some prevailing theories of stellar atmospheres must be modified.

Broad-Band X-ray Telescope (BBXRT). This instrument utilized a tightly nested set of very lightweight, gold-coated mirrors that by glancing incidence focused x-rays onto a silicon diode detector. The solid-state detector responded to x-rays with energies between 500 eV and 12 keV, with 100-eV spectral resolution. The Broad-Band X-ray Telescope was able to conclusively identify x-ray emissions from many heavy elements formed in the interior of massive stars and thrown into the interstellar medium. Ranging from oxygen to iron, these elements are the building blocks of the solid Earth and of life. Supernovae, active galaxies, interacting binary systems, neutron stars, and black-hole candidates were measured by the Broad-Band X-ray Telescope, and the data are providing much new insight on these objects.

Astro-1 mission. Each of the four instruments was developed and tested as a payload for small rockets. Flight on the space shuttle meant that a larger instrument could be built and that ground and flight crews could interactively control the instruments for rapid acquisition, target identification, and real-time evaluation of the data. Moreover, instead of observation of one object for a few hundred seconds on a small rocket flight, the instrument could observe objects for as much as several thousand seconds, and the opportunity to observe well over 100 objects was possible during a 10-day shuttle flight. As operation of the ultraviolet instruments is complex and requires highly trained skills, real-time interaction with the instruments enhanced the scientific data greatly. Two payload specialists, selected from the instrument teams, flew on board the shuttle. Both the mission specialists and the payload specialists were professional astronomers with considerable observational training.

Planning. The shuttle mission had to be planned before flight with preselected astronomical sources and preplanned observing sequences. Provision for real-time replanning that would respond to launch slips and changes in resources was essential to maximize the scientific data to be obtained. The planning activities were carried out primarily by on-line computer systems and databases.

Response to malfunctions. Launch occurred on December 2, 1990, at 6:40 Greenwich Time. As on any shuttle flight, some hardware malfunctioned. The high-data-rate recorder failed, and some problems with pointing and acquisition occurred. Both on-board computer terminals used by the crew to control the payload also failed. However, the crew and the ground support teams knew alternative solutions to each problem, and after a short replanning period the collection of scientific data began. In the case of the failed computer terminals, the ground teams quickly responded by interactively controlling the instruments from the ground, with the on-board crew continuing to control the shuttle pointing and to do target acquisition plus guiding. All four instruments were controlled in parallel for the rest of the mission.

Results. Nearly 300 observations of over 120 astronomical objects were recorded. Within the next few years, an estimated 100 refereed papers based on these observations will be published in scientific journals. Several special sessions of scientific meetings have been set aside to present the first results of Astro-1. The data and their interpretation are already influencing theories of star formation in nearby galaxies, the activity of star formation and destruction in the nuclei of active galaxies, the synthesis of elements in supernovae, velocities in the shock fronts of interstellar blast waves, and the atmospheres of massive hot stars. Astro-1 is indeed scientifically very successful.

Reflight of Astro instruments. The three ultraviolet instruments are scheduled to be carried aboard the shuttle as Astro-2 in late 1994. A newer version of the Broad-Band X-ray Telescope will fly as part of a Japanese satellite, Astro-D, scheduled for launch in early 1993.

For background information SEE ASTRONOMICAL SPECTROSCOPY; ASTROPHYSICS, HIGH-ENERGY; GALAXY, EXTERNAL; GAMMA-RAY ASTRONOMY; SATELLITE ASTRONOMY; STELLAR EVOLUTION; X-RAY ASTRONOMY; X-RAY TELESCOPE in the McGraw-Hill Encyclopedia of Science & Technology.

Theodore Gull

Bibliography. W. P. Blair and T. R. Gull, Astro: Observatory in a shuttle, *Sky Telesc.*, 79:591–595, June 1990; S. P. Maran, Astro: Science in the fast lane, *Sky Telesc.*, 81:591–596, June 1991.

Satellite launch system

On April 5, 1990, the world's first privately developed and operated space transportation system launched two satellites into orbit. The flight of the Pegasus vehicle (**Fig. 1**) was the culmination of a commercial joint-venture development effort of Orbital Sciences Corporation (OSC) and Hercules Aerospace Company. Combining advances in aerodynamics, guidance and control, propulsion, and materials technology, Pegasus exploits the inherent advantages of air launch from a carrier aircraft to provide a flexible and inexpensive Earth-to-space transportation system for satellites. These advances have created new opportunities for the use of space by making access to orbit affordable and relatively free of constraints for defense, science, and commercial customers.

Benefits of air launch. Prior to Pegasus, all orbital transportation systems relied on complex and expensive ground-based launch pads. These facilities require extensive ground support equipment for erecting, stabilizing, and servicing the launch vehicle; must be refurbished after each lift-off; and are limited in number, so that utilization conflicts occur. Ground launches are additionally constrained by weather. Most significantly, because ground-launched rockets must fly through the densest part of the Earth's atmosphere immediately after lift-off, performance is reduced through aerodynamic resistance and lower propellant efficiency, and the rocket is subjected to high aerodynamic pressures.

However, with recent advances in lightweight-composite structures and solid-rocket propulsion and an improved understanding of hypersonic aerodynamic flight, it became possible to build a winged lift-assisted launch vehicle that was large enough to put a significant payload mass into orbit but still small enough to be carried aloft and air-launched by a conventional aircraft. These advances allowed development of a three-stage rocket that is carried aloft by a transport-bomber-class aircraft and can launch payloads of up to 500 kg (1100 lb) into low Earth orbits. With this breakthrough, Pegasus is able to begin a more efficient trajectory above the densest layers of the Earth's atmosphere, utilize the additional lift provided by its wing, and break free of many constraints associated with ground launch pads.

Fig. 1. First flight of Pegasus on April 5, 1990.

Fig. 2. Pegasus just after release for launch by NASA B-52 carrier aircraft.

Pegasus vehicle. The Pegasus vehicle is a winged, three-stage, solid-rocket booster weighing approximately 18,600 kg (41,000 lb) and measuring 15.5 m (51 ft) in length and 1.27 m (4.2 ft) in diameter. It is lifted by a carrier aircraft to a level flight condition of about 12,500 m (41,000 ft) and Mach 0.8. After release from the aircraft and stage-1 motor ignition, the vehicle's autonomous guidance and flight-control system provides all the guidance necessary to insert small payloads into a wide range of suborbital and orbital trajectories. **Figure 2** shows Pegasus after release from the carrier aircraft and just prior to stage-1 ignition.

The Pegasus vehicle design combines advanced technologies and conservative design margins to achieve high performance and reliability at reduced cost. The vehicle incorporates seven major elements: three solid-rocket motors, a payload fairing, a lifting wing, an avionics assembly, and an aft skirt assembly including three movable control fins (**Fig. 3**).

The three solid-rocket motors and payload fairing were designed and optimized specifically for the Pegasus program. Advanced composite materials are used extensively in the carbon-composite delta wing and the three foam-core graphite-composite fins. These fins are driven by electromechanical fin actuators supported in the cylindrical aluminum aft skirt. A graphite-composite avionics structure and aluminum-honeycomb deck support the payload and most vehicle avionics. The two-piece graphite-composite payload fairing of 1.3-m (4.2-ft) outside diameter is pyrotechnically separated and encloses 18.6 m^3 (656 ft^3) of payload space, as well as the avionics subsystem and stage-3 motor.

For separable payloads, an optional marmon clamp-band payload separation system is available. In this system, a metal band is wrapped around the circumference of a special cylindrical joint between the payload and the Pegasus vehicle and holds the joint and the two structures together. Once orbit has been attained, the metal band is cut so that the payload can separate from the launch vehicle. Another optional system, the hydrazine auxiliary propulsion system (HAPS), can be mounted inside the avionics structure, and provides up to 73 kg (161 lb) of hydrazine (N_2H_4) for orbit raising and adjustment as a fourth stage. When combined with the Pegasus standard on-board Global Positioning System (GPS) receiver, the HAPS provides autonomous precision orbit-injection capability.

The optional PegaStar integrated spacecraft bus can provide extended (5–10-year) attitude control, propulsion for makeup and adjustment, data storage, elec-

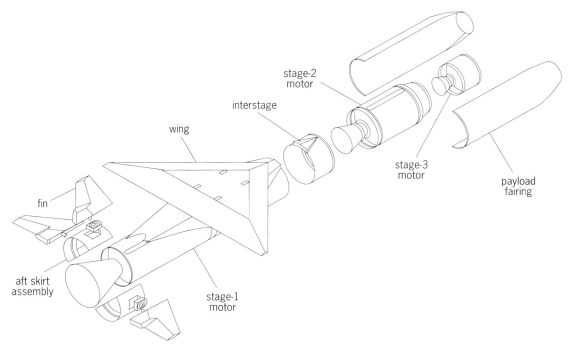

Fig. 3. Expanded view of the Pegasus vehicle configuration.

trical power, and telemetry support for a wide range of mission applications. PegaStar replaces the standard avionics section and can be configured to provide precision three-axis, nadir-pointing, and spin-stabilized attitude control.

The Pegasus avionics system is simple, robust, and reliable. A microprocessor-based flight computer controls all flight events and executes the vehicle's autopilot program. An additional microprocessor in the flight computer supports the autopilot processor and processes vehicle telemetry. Vehicle attitude, velocity, and navigation information is provided by a strap-down inertial measurement unit in conjunction with the Global Positioning System receiver.

Integration and launch operations. Pegasus launch missions are managed by a commercial launch support team, which provides complete advance mission planning, coordination of interfaces between the spacecraft and the launch vehicle, prelaunch systems integration and checkout, processing of documentation, launch-site operations dedicated exclusively to the user's requirements, and postflight analysis.

Pegasus field integration is straightforward and requires a minimum of ground-support equipment and facilities. The horizontal integration of the vehicle allows easy access for component installation, test and inspection, and payload integration.

After integration with the spacecraft payload, Pegasus is mated to the carrier aircraft, and preflight testing is conducted. The aircraft then departs and proceeds to the launch point over an open ocean area. The typical launch sequence begins with the release of Pegasus from the carrier aircraft at an altitude of approximately 12,500 m (41,000 ft) and a speed of Mach 0.8. Approximately 5 s after release from the aircraft, stage-1 ignition occurs. The vehicle quickly accelerates to supersonic speed while beginning an aerodynamic pull-up maneuver. At approximately 35 s, a maneuver is initiated to depress the trajectory, and the vehicle angle of attack quickly approaches zero. Attitude control during the stage-1 burn is provided by the three active control fins.

Stage-2 ignition occurs shortly after stage-1 burnout, and the payload fairing is jettisoned as quickly as payload dynamic pressure limitations will allow, normally at approximately 112,500 m (369.000 ft). Stage-2 burnout is followed by a long coast, during which the payload and stage 3 nearly achieve orbital altitude. Stage 3 then provides the additional velocity necessary to circularize the orbit. Stage-3 burnout typically occurs approximately 10 min after launch and 2200 km (1400 mi) downrange of the launch point. Attitude control during stage-2 and stage-3 powered flight is provided by the motor thrust-vector controllers and the nitrogen cold-gas reaction-control system.

The optional HAPS stage operates after stage-3 motor separation. The subsystem's hydrazine thrusters perform final orbit injection and adjustment burns prior to spacecraft separation.

Applications. Pegasus has opened a new era of opportunities for greater involvement in space activities. The low launch cost of Pegasus and its flexible approach to operations, together with dramatic advances in small-satellite mission capabilities, have stimulated a greater diversity of customers interested in using space to enhance national defense, science, and commerce. The U.S. Defense Advanced Research Projects Agency (DARPA), the National Aeronautics and Space Administration (NASA), and the U.S. Air Force, which all contributed to the commercial development effort, became the first organizations to benefit from the new technology; by 1991 they had selected Pegasus for a total of over 55 firm launches and options. The U.S. Department of Defense is now developing and deploying on Pegasus small space systems to conduct missions at a fraction of the cost required in the early 1980s. Similarly, NASA is conducting space science and exploration missions involving Pegasus more quickly and less expensively than before. Many private companies are moving ahead with the space-based communication and remote-sensing business ventures that would have been economically infeasible prior to Pegasus. Much as the advent of the personal computer broadly expanded people's access to powerful information technology, Pegasus and many of the small space systems it has generated have dramatically increased access to the benefits of space.

For background information SEE COMPOSITE MATERIAL; LAUNCH COMPLEX; ROCKET PROPULSION; SATELLITE (SPACECRAFT); SATELLITE NAVIGATION SYSTEMS in the McGraw-Hill Encyclopedia of Science & Technology.

David W. Thompson

Bibliography. *International Symposium of the Society of Photo-Optical Instrumentation Engineers*, April 1, 1991; Orbital Sciences Corp., *Pegasus Payload User's Guide Release 2.00*, 1991; A. Wilson (ed.), *Interavia Space Directory* 1990–1991; *World Guide to Commercial Launch Vehicles*, 1991.

Seed

The period from planting to seedling establishment is a critical phase of crop production that determines the potential plant population and the uniformity among plants. Both of these factors, in turn, influence the marketable yield and quality of a crop. A wide range of cultural practices and seed treatments have been developed to improve the success of seed germination and seedling emergence from the soil. Seed priming is a physiologically based method to increase the speed and uniformity of seed germination and seedling emergence. Seed priming is a process of controlled hydration of seeds that allows pregerminative metabolism to occur but delays actual germination. Seeds treated in this way are redried, stored, marketed, and planted just like untreated seeds. However, because of prehydration treatment, metabolic processes are advanced, and germination occurs rapidly and uniformly after the primed seeds are planted.

Seed germination and priming. As shown in **Fig. 1**, the water content of dry seeds imbibed on

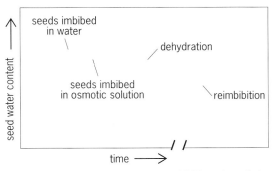

Fig. 1. Generalized time courses of imbibition of seeds in water and in a priming treatment.

water increases rapidly, then reaches a plateau where water content either does not change or increases very slowly. The initial increase is a purely physical phenomenon due to the hydration of the proteins, carbohydrates, and other constituents of the seed, and will occur in both living and dead seeds. During the plateau phase, metabolism is activated, respiration increases, and molecular damage sustained by the seeds during their dry storage period is repaired. Seeds also switch their molecular machinery from a developmental mode, responsible for the accumulation of storage reserves in the developing seed and the preparation for desiccation, to a germinative mode, in which the proteins and enzymes that are needed to mobilize the stored reserves and initiate growth are synthesized. Following this period of activation, seeds enter the third phase, during which embryo growth begins and germination occurs. Plant tissues grow by absorbing water that expands their cells, so the water content increases further during this phase.

Seed priming alters this imbibition pattern by extending the plateau (activation) phase and preventing the seeds from entering the growth phase. This alteration is done by controlling the maximum seed water content (Fig. 1). If the water content of the plateau phase of imbibition is reduced slightly, the seeds will not be able to germinate, as they will be unable to absorb sufficient water for cell expansion. If the water content is not too low, however, the pregerminative metabolism of the plateau phase can proceed. Different types of seeds may differ in the specific mechanisms initiated during priming; in general, however, seed priming allows the physiological processes normally associated with germination to occur but prevents the final phase of cell expansion and growth. Seeds can be held in this state for extended periods to allow all the seeds to prepare for germination. The seeds can be redried after the priming treatment while the metabolic advancement accomplished during the hydrated period is retained. These seeds can then be packaged, marketed, and planted just like untreated seeds. When the seeds are reimbibed, the length of the plateau phase is markedly shortened, and the seeds progress almost directly from imbibition to growth (Fig. 1).

Most of the benefits of seed priming for crop establishment, growth, and yield can be traced back to the more rapid and uniform germination and emergence of seedlings (**Fig. 2**). In addition, seed priming can overcome dormancy that limits germination under some conditions, as in the case of lettuce seeds planted in hot weather. Primed seeds are commercially available for a number of vegetable crops. Increasingly, flower seeds for the bedding plant industry are also being primed. For both vegetables and flowers, seed priming offers considerable advantages for greenhouse production of seedlings. Primed turfgrass seeds are also starting to appear on the market. The inherently slow germination and emergence of many grass seeds and the desire for quick establishment of seeded areas on golf courses, parks, sports fields, and lawns make this potentially a major application of priming technology.

Seed priming methods. The methods used to prime seeds must provide the environmental conditions required for pregerminative metabolism to proceed, including proper temperature, oxygen supply, and light intensity, while they also control seed moisture content. One way is to imbibe the seeds in an osmotic solution containing salts or organic osmotic agents. By reducing the osmotic potential of the solution (that is, by increasing the concentration of solutes) in the range of -5 to -15 atm (-0.5 to -1.5 megapascals), the seed moisture content can be controlled to the desired level. Various salts, including potassium nitrate (KNO_3), potassium phosphate (K_3PO_4), and sodium chloride ($NaCl$), have been used as osmotic solutes, but the most commonly used solute is polyethylene glycol (PEG), an inert polymer with molecules too large to enter the cells of the seeds. PEG is very effective osmotically, but it also increases the viscosity of solutions and reduces their ability to supply oxygen to the seed. Seed priming systems, therefore, bubble air through the solution and agitate the seeds to keep them suspended. After the desired time in the priming solution, the seeds are quickly rinsed and dried.

The osmotic-solution approach to priming has been successful for vegetable and flower seeds and is the basis of most current commercial priming techniques. However, this approach has disadvantages for the large-scale applications that would be required for

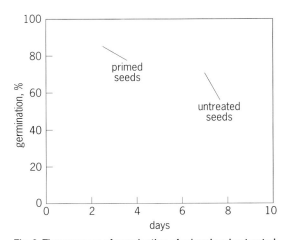

Fig. 2. Time courses of germination of primed and untreated seeds when they are imbibed on water.

priming field seeds such as corn, cotton, or beans. Large volumes of solution and substantial quantities of osmotic agents would be needed, and the engineering aspects of scaling up the solution process have only recently been addressed. An alternative approach is to control the seed water content by simply limiting the amount of water supplied to the seeds. In one method, called drum priming, the seeds are placed in a rotating drum, and controlled quantities of water are sprayed onto them to bring the seed water content to the desired level. The seeds can then remain in the rotating drum while priming, with the tumbling action providing good aeration. After the desired priming period, air can be flushed through the drum to dry the seeds. This approach requires no solutions or osmotic agents and minimal handling of the seeds, and could presumably be scaled up by using larger drums. A modification of this approach is called solid matrix priming. This method controls the seed water content by mixing together measured amounts of seeds, water, and a granular matrix, such as clay, coal, or shale particles. The absorbent matrix acts as a water reservoir to determine the final water content of the seeds. By selecting a material with the proper water-holding properties and varying the proportions of seeds, matrix, and water, the desired level of seed water content can be achieved during priming. The matrix maintains a porous structure in the seed–matrix mixture, allowing aeration of the seeds, and may also absorb germination inhibitors that are released from the seeds. After drying, the matrix particles can be removed by screening.

With any of the priming methods, the process can be further modified by the addition of plant growth regulators, plant protection chemicals, or beneficial microorganisms to the priming solution or imbibitional water.

Prospects. Whether priming will be utilized in major agronomic crops depends upon the success of current efforts to scale up the process for the volumes of seeds required, and upon the potential economic benefits to be obtained from those crops. Increasingly, the seed industry is applying multiple technological approaches to improve seed quality and cultivar performance, including hybridization, recombinant deoxyribonucleic acid (DNA) technology, tissue-cultured embryos, beneficial microorganisms, improved seed cleaning and separation, seed treatments, film coating, pelleting, and modified planting techniques. Seed priming is one component of this overall effort to develop more efficient, environmentally safe, and profitable crops and production practices.

For background information *SEE DORMANCY; PLANT GROWTH; SEED* in the McGraw-Hill Encyclopedia of Science & Technology.

Kent J. Bradford

Bibliography. K. J. Bradford, Manipulation of seed water relations via osmotic priming to improve germination under stress conditions, *HortScience*, 21:1105–1112. 1986; J. A. Eastin, Solid matrix priming of seeds, U.S. Patent No. 4,912,874, April 3, 1990; D. Gray. H. R. Rowse, and R. L. K. Drew, A comparison of two large-scale seed priming techniques, *Ann.* *Appl. Biol.*, 116:611–616, 1990; R. B. Taylorson (ed.), *Recent Advances in the Development and Germination of Seeds*, 1989.

Sewage

During the last 20 years, the use of composting to process municipal wastewater sludge into a beneficial material has become a widely accepted practice. Sludge composting is symbolic of the change in waste management practices. A scientific, technological, and conceptual framework exists, and an expansion of composting to aid in the recycling of organic residual materials is anticipated.

At the end of 1990, a survey indicated that 276 municipalities in the United States were using or planning to use composting in their sludge management. Of that total, 133 projects were operational, 44 under construction, 35 in planning and design, 20 in pilot operations, and 44 under consideration. Compared to data from a 1989 survey, there was a gain of 50 sludge composting projects.

Sludge composting. Various methods are used to compost sewage sludge, including windrow, static pile, aerated static pile, aerated windrow, and in-vessel. Of the operational facilities, the composting methods being used in 1990 comprised 76 aerated-static-pile systems, 24 in-vessel systems, 25 windrow systems, 6 aerated-windrow systems, and 2 static-pile systems. All of these methods involve mixing of the sewage sludge with a bulking material to provide porosity, adjust moisture content, and create an optimum carbon-to-nitrogen ratio for composting.

Windrow systems. The windrow method entails formation of a composting pile, which is then turned on a regular basis throughout the period necessary to produce a finished (stable) compost. In open-windrow systems, the material is stacked into elongated piles (windrows) directly after it has been reduced in size. "Open" refers to the fact that the material is not placed in a vessel or container (reactor) specifically constructed for composting. With the turned-windrow system, the piles are aerated by tearing down and rebuilding the pile. The frequency of this step is determined by the aeration requirements of the material, principally involving structural strength, moisture content, temperature, and biological nature of the material.

Static-pile systems. The static-pile and aerated-static-pile methods do not entail turning; the key to the latter approach is blowing air through the pile. The static-pile system differs from the turned-pile system in that it is not periodically torn down and rebuilt; and as mentioned, it may receive forced aeration. The static pile functions best with material that is relatively homogeneous and sufficiently granular to impart a porous texture to the mass of composting material.

Other systems. The aerated-windrow method is a combination of aeration and turning. Mechanical in-vessel options are those that are below a roof and mostly enclosed. There are numerous system designs, from vertical silos to horizontal troughs. Enclosed systems have certain characteristics in common. The systems are more or less closely controlled and involve a two-stage operation: an active stage followed by a maturation stage. Regardless of the type of reactor (enclosure unit), aeration is accomplished by tumbling, stirring, and forced aeration.

Pressure for adoption. The driving forces for the increased adoption of composting for sludge are the same forces that have led to the surge in recycling as much of the solid-waste stream as possible. In the United States, a series of state and federal legislative acts have banned traditional practices of sludge disposal such as ocean dumping and, in a few cases, landfilling. Stricter air-pollution controls and public opposition have made incineration of sludge far less acceptable as well as costly. The lack of landfill space and higher charges for waste disposal (tipping fees) and hauling costs have decreased the use of burial as a management practice. Development of land for residential and industrial use has made it more difficult to find suitable sites to practice direct land application of sludge. Also, there has been growing public demand for elected officials to implement policies that would convert the waste products of society to beneficial resources. All these factors have contributed to the view that composting is an economically and environmentally viable method for municipal sludge management.

Research and development. One of the most significant efforts to advance the knowledge of sludge composting and to develop sound processing methods was a joint project in the 1970s at the Biological Waste Management Laboratory of the Agricultural Research Service at Beltsville, Maryland. This research project, cosponsored by the U.S. Department of Agriculture and the Environmental Protection Agency (EPA), provided valuable data on aeration, temperature control, use of bulking agents such as wood chips and sawdust, mixing and materials-handling equipment, pathogen destruction, and end-product utilization. In addition, this work led to the development of a sludge composting technique known as the Beltsville aerated-static-pile method.

In the years since, as the number of research projects has steadily increased, the methods have steadily improved owing to experience and data collected. There is a better understanding of aspects of the subject known as composting realities; these include the need for odor control methods, the parameters for heavy metals, and the amount of physical area for processing and storing material. There also is a greater understanding of the composting process and its management requirements.

An industry now provides a commercial infrastructure for sludge composting. A number of consulting engineering firms have staff that are trained in the planning, designing, and permitting of sludge composting projects. Equipment manfacturers now offer windrow turners, screens, and mixers as well as totally enclosed systems for processing dewatered sludges into usable products for the horticultural and agricultural industry. In some cases, fertilizer and soil-conditioner firms have taken over the responsibility for marketing the finished compost, while at many facilities the municipalities themselves have established markets for their own material.

Concerning market development, some states have already established policies to specify that sludge-based composts be used in public landscaping projects. For example, Maine's Department of Agriculture is developing such compost-standard guidelines for its Department of Transportation. Legislation has been enacted in Maine that requires state agencies to give preference to recycled products such as compost. The U.S. Congress has considered legislation that would mandate use of compost by the Department of Transportation. Research projects at many state universities have shown the value of sludge compost in preparation of potting mixtures, turf growing, and other applications. Currently the EPA is in the final stages of issuing its beneficial-use regulations for sludge that will also provide a firmer base for sludge compost marketing programs.

Another driving force for project development in the United States is the interest of many cities and counties in composting of yard waste. Many states have now banned the landfilling of such vegetative materials as leaves, grass clippings, and woody debris. As a result, it often makes economic sense to cocompost yard waste with sewage sludge. In the 1990 survey cited above, about 70 projects were using or planning to use yard waste in the sludge composting mix. The yard waste serves as a bulking agent, helping to improve porosity of the material; it also reduces or eliminates the need to purchase wood chips to serve as a bulking agent.

A great deal of research on a broad range of issues relating to sludge composting is being carried out. Current projects include studies of the long-term effects of the application of sludge to dryland wheat, evaluation of the impacts of nitrogen and heavy metals from compost on groundwater quality, development of methods to utilize sludge from paper manufacturing, chemical analysis and characterization of sewage sludge, use of sludge for reclamation of mined land, and identification of the variables involved in producing a compost that would be useful in horticulture.

Odor control. Recognizing that composting has become a very important method for stabilizing and processing municipal sewage sludge in the United States, the EPA issued in 1989 a report on in-vessel composting. The report reviewed considerations for project planning, design, and operation. Case studies cover projects at eight locations in the United States. Information was collected on system performance related to odor generation, mix ratios, moisture removal, detention times, materials handling, production of quality product for a market, and costs.

Much attention was given to odor control. Based on data collected in 1988, of the eight facilities surveyed, six had received odor complaints from the local community. At five, the complaints were serious enough to cause the plant owners to retrofit the facilities with additional odor-treatment equipment. Two plants ceased operations while the new equipment was being installed. A third plant was operated at reduced capacity to limit odor production while the new treatment equipment was piloted, designed, and installed. The other two plants continued to operate, but they were proceeding with retrofitting the odor-treatment system as quickly as possible.

Almost all of these problems can be attributed to incomplete odor-control plans. To be effective, an odor-control plan must include five elements: control of the composting process; an inventory of potential odor sources; an odor collection and containment system; an odor-treatment system; and effective dispersal of residual odors.

Compost markets. As with all recycling programs, the managers of a sludge composting project must have strategies for marketing and utilization in mind at the outset. Potential users of sludge compost include personnel responsible for landscaping public works projects, nursery operators, turf growers, homeowners, and farmers. Potential applications for the material include topsoil blending, landfill cover, and reclamation of strip-mined land.

The oldest marketing effort for sludge compost was begun in California. Since the mid-1920s, a commercial firm has had a contract with the Los Angeles County Sanitation District to purchase sludge from the drying beds; it markets a complete line of garden soil conditioners utilizing composted sludge.

For background information SEE SEWAGE; SEWAGE SOLIDS in the McGraw-Hill Encyclopedia of Science & Technology.

Jerome Goldstein

Bibliography. BioCycle Staff (eds.), *The Art and Science of Composting*, 1991; EPA Center for Environmental Research Information, *In-Vessel Composting of Municipal Wastewater Sludge*, 1989; N. Goldstein and D. Riggle, Sludge composting maintains momentum, *BioCycle*, pp. 26–31, December 1990.

Ship powering and steering

The concept of magnetohydrodynamic (MHD) ship propulsion has been discussed for many years, but always hypothetically. Now it is becoming a reality as superconducting magnets become a viable commercial technology. With the discovery of high-temperature superconductors, there is a real possibility that magnetohydrodynamic propulsion can be developed for practical use.

Basic principle and characteristics. The basic principle of magnetohydrodynamic ship propulsion is based on Fleming's left-hand law of electromagnetism. If a magnetic field perpendicular to the motion of a ship is generated in seawater by a magnet placed aboard ship, and an electric current is generated in seawater perpendicular to the magnetic flux and the axis of the ship, then seawater will be pushed rearward by an electromagnetic force (the Lorentz force) and the ship will be propelled by the reaction. This principle is exactly the same as that of a magnetohydrodynamic pump. Since the reaction of the

Lorentz force acts directly on the magnet (not through hydrodynamic pressure as in conventional propelling devices such as screw propellers), no rotating part outboard a ship is necessary; therefore, magnetohydrodynamic propulsion will be less affected by cavitation and will be relatively quiet in operation. The absence of cavitation may be advantageous when magnetohydrodynamic propulsion is applied to high-speed ships, provided that the thruster is lightweight and highly efficient. The quiet operation may be advantageous for ships that specifically require silent operation, such as submarines. However, bubbles generated by electrolysis may cause underwater acoustic problems.

Types of MHD thrusters. Since the Lorentz force is a vector product of the magnetic flux and the

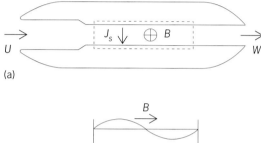

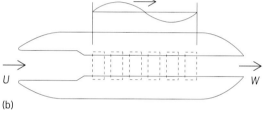

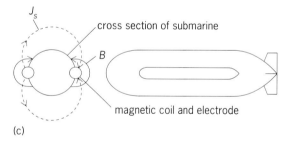

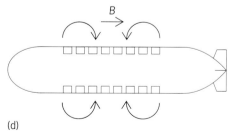

Fig. 1. Methods of magnetohydrodynamic propulsion, applied to a submarine. U = inlet speed = ship speed; W = outlet speed; J_s = electric current; B = magnetic flux. (a) Internal flow, direct current. The symbol $\oplus$ indicates that the magnetic flux B is pointing into the paper. (b) Internal flow, induced current. The graph above the submarine shows the variation of the value of the magnetic flux B along the duct. (c) External flow, direct current. (d) External flow, induced current. (*After S. Way, Electromagnetic propulsion for cargo submarines, J. Hydronaut., 2(2):49–57, 1968*)

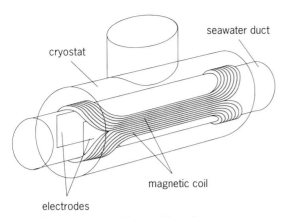

Fig. 2. Basic structure of internal-flow, direct-current magnetohydrodynamic thruster.

electric current, it will be maximized when the magnetic flux and the electric current intersect at a right angle. Four types of magnetohydrodynamic thrusters that meet this condition can be considered (**Fig. 1**), based on combinations of either the internal-flow system or the external-flow system with either a direct current or an induced current. However, at the present state of superconducting magnet technology, the induction-type magnetohydrodynamic thruster, requiring alternating-current superconducting magnets, is not practical to use. An external-flow, cross-field-type magnetohydrodynamic thruster was tested in 1984 in Japan and proposed for an icebreaker. However, present research and development pertain mainly to internal-flow-type magnetohydrodynamic thrusters. Consequently, only internal-flow, direct-current, cross-field-type magnetohydrodynamic ship propulsion will be described.

MHD thruster structure and efficiency. An internal-flow-type magnetohydrodynamic thruster consists of a seawater duct surrounded by a pair of dipole superconducting magnet coils that provide magnetic flux perpendicular to the duct's axis (**Fig. 2**). A pair of electrodes is provided in the duct to generate direct electric current perpendicular to the magnetic flux and the duct's axis. The whole system is contained in a cryostat and kept at 4 K ($-452°$F) by liquid helium to maintain the superconductivity of the coils.

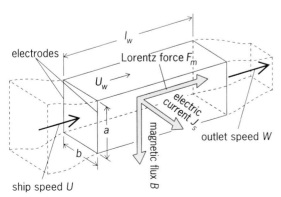

Fig. 3. Relationship of major parameters in producing magnetohydrodynamic propulsion. Symbols are explained in text.

The following symbols are used in **Fig. 3** and in the text for the major parameters involved in producing magnetohydrodynamic propulsion. (The units in which the quantities are expressed are given in parentheses.)

B (T) = magnetic flux density
J_s (A) = electric current
b (m) = duct width
a (m) = duct height and electrode height
l_w (m) = length of the parallel part of the magnetic coil [same as length of the electrode]
U (m/s) = ship speed
U_w (m/s) = flow speed
W (m/s) = outlet speed
σ (S/m) = conductivity of seawater
ρ (kg/m^3) = density of seawater
F_m (N) = total Lorentz force
T (N) = thrust to propel ship at velocity U

Using Fig. 3, Eq. (1) can be formulated for the

$$F_m = J_s B b \ \text{(N)} \qquad (1)$$

Lorentz force F_m. The electric potential between electrodes that is necessary to generate the electric current J_s in seawater is given by Eq. (2). The electric ef-

$$V_s = B U_w b + \frac{b J_s}{\sigma a l_w} \qquad (2)$$

ficiency of the thruster (the ratio of the work done by the Lorentz force to the electrical power input) is given by Eq. (3). From Eqs. (1), (2), and (3), it

$$\eta_E = \frac{F_m U_w}{V_s J_s} \qquad (3)$$

follows that this efficiency is given by Eq. (4), where

$$\eta_E = \left(1 + \frac{F_m}{\sigma a b l_w B^2 U_w}\right)^{-1} \qquad (4)$$

the loss due to viscous drag in the duct is neglected. Since the efficiency η_E is fairly small, the second term of the denominator is dominant. From Eq. (4), for a constant Lorentz force, the efficiency η_E increases almost proportionately to the square of the magnetic flux density B. Therefore the use of a strong magnet with minimum power loss (that is, a superconducting magnet) is the key to making magnetohydrodynamic propulsion technically feasible. Since the product abl_w is the seawater volume in the duct, using a duct of large volume is also effective in improving the magnetohydrodynamic efficiency. Improving the conductivity of seawater σ is also effective because the rather low efficiency of magnetohydrodynamic thrusters is mainly due to the joulean heat loss. Increasing the flow speed U_w is also effective. However, the viscous drag in the duct should also be considered.

Total MHD propulsive efficiency. The total propulsive efficiency is the ratio of the work done by the thrust to the electric power input, and is therefore given by Eq. (5). If the flow speed in the duct U_w

$$\eta = \frac{TU}{J_s V_s} \qquad (5)$$

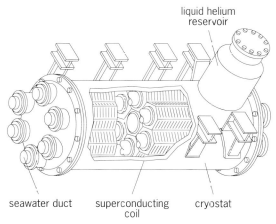

Fig. 4. Cutaway view of superconducting magnetohydrodynamic thruster.

is assumed to be equal to the outlet speed W (which is equivalent to assuming that there is no contraction at the outlet) and if the jet fraction W/U is denoted as in Eq. (6), then η can be expressed as in Eq. (7).

$$\frac{W}{U} = \gamma \qquad (6)$$

$$\eta = \frac{2}{\gamma + 1} \left[1 + \frac{\rho}{\sigma B^2} \left(\frac{\gamma^2 - 1}{\gamma} \right) \frac{U}{2l_w} \right]^{-1} \qquad (7)$$

From Eq. (7), it follows that the efficiency η can be improved if (1) the magnetic flux density B is increased; (2) the conductivity of seawater σ is improved; (3) the jet fraction γ is reduced; or (4) the length of the magnetic field l_w is increased. The effect of the sectional area of the duct ab does not appear explicitly in this formula. However, if the sectional area of the duct is increased, γ decreases accordingly, and consequently the efficiency η will be improved. Therefore, an increased duct volume makes a significant contribution to improving η.

Equation (7) also shows that the efficiency η decreases as the ship speed U is increased. Therefore, while magnetohydrodynamic propulsion may appear to be an advantageous means for propelling high-speed ships because of the diminished effect of cavitation, it may not necessarily be suitable in view of the loss of propulsive efficiency.

A trial calculation on a middle-size submersible using Eq. (7) showed that overall efficiencies greater than 50% are feasible at speeds of 40 knots (20 m/s) and higher, provided a magnetic flux density greater than 5 teslas can be maintained over a volume of the order 3500 ft^3 (100 m^3). If the viscous drag in the duct is taken into account, the efficiency will be a little less than this amount. Nevertheless, this figure appears to be very promising. However, the assumed power of the magnet is far beyond the present state of the art.

Recent progress of MHD propulsion. Extensive research projects on magnetohydrodynamic propulsion have been conducted recently, mainly in the United States. In the 1980s, fundamental and theoretical advances were achieved, concerning chiefly the application to naval surface ships and submarines. Experimental tests (for example, using a large superconducting magnet whose magnetic flux density was 6 T and whose inner diameter was 3.3 ft or 1 m) were also performed. The practical application of magnetohydrodynamic propulsion to surface merchant ships has also been studied in Japan since 1985, and the experimental *Yamato-1* has been constructed. The ship is the first prototype magnetohydrodynamic ship ever built.

Experimental MHD ship. The *Yamato-1* is 98 ft (30 m) in length, has a displacement of 204 tons (185 metric tons), and is equipped with two magnetohydrodynamic thrusters. The specifications of the ship are shown in the **table**. Each magnetohydrodynamic thruster consists of six unit coils that form a lotuslike ring so as to reduce leakage of magnetic flux outside the thruster (**Fig. 4**). Leakage of magnetic flux is thus reduced so that magnetic shielding is not necessary at any part of the ship. This is extremely helpful in reducing the ship's weight. Specifications of each unit coil and a set of magnetohydrodynamic thrusters are shown in the table.

The ship (**Fig. 5**) was completed in 1991, and sea trials were to be completed in 1992. The attained propulsive efficiency was very low as predicted. The efficiency might be higher if a duct of large volume was adopted. However, at the present state of superconducting magnet technology, the magnet, whose size determines that of the duct, is almost the largest that can be manufactured. Nevertheless, the construction and trials of *Yamato-1* demonstrated that a ship can be propelled by magnetohydrodynamic thrusters with all the necessary equipment aboard the ship. It also helped identify problems to be resolved in the future.

Prospects. At the present stage of technology, the efficiency of magnetohydrodynamic propulsion is significantly lower than that of a conventional propulsion system under basically the same conditions. This low efficiency is mainly due to the joulean heat loss associated with the passage of an electric current through seawater. However, the efficiency may be improved in the future by (1) developing much stronger superconducting magnets within reasonable weight limits; (2) developing very large superconducting magnets

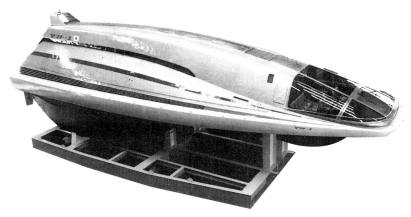

Fig. 5. Experimental magnetohydrodynamic ship *Yamato-1*.

Specifications of the experimental ship *Yamato-1*

Characteristic	Specification
Overall length	98 ft (30 m)
Length between perpendiculars	86.6 ft (26.4 m)
Breadth (molded)	34.1 ft (10.39 m)
Depth (maximum)	12.1 ft (3.69 m)
Draft (maximum)	8.8 ft (2.69 m)
Displacement (including water in the ducts)	204 tons (185 metric tons)
Speed (at Lorentz force of 3600 lbf or 16,000 N)	About 8 knots (4 m/s)
Hull material	Aluminum
Number of persons	10, including a crew of 3
Unit coil (6 per thruster)	
Inner diameter of coil	1.18 ft (0.360 m)
Inner diameter of duct	0.79 ft (0.240 m)
Coil length (total)	12.14 ft (3.70 m)
Coil length (parallel part)	12.11 ft (3.69 m)
Number of turns	220 × 2 layers × 2 poles
Normal electric current	4600 A
Magnetic flux density at center (for single unit)	3.5 T
Magnetohydrodynamic thruster (2)	
Compound magnetic flux density at center	4 T
Electrode current (normal)	2000 A
Lorentz force	1800 lbf/set (8000 N/set)
Thrust	900 lbf/set (4000 N/set)
Weight	20 tons/set (18 metric tons/set)

(magnets with large inner diameter and length) within reasonable weight limits; (3) using high-temperature superconductors to reduce the cryostat's weight and volume; (4) developing much stronger, lightweight coil collars; and (5) significantly improving the conductivity of seawater by chemical or physical means.

If the efficiency of magnetohydrodynamic thrusters is reasonably improved, they should be a suitable propelling device for high-speed ships and for ships specifically requiring silent operation. The latter feature will be emphasized if fuel cells are used as a main power source in the future.

For background information *SEE CAVITATION; ELECTROMAGNETIC PUMPS; MAGNETIC FIELD; SHIP POWERING AND STEERING; SUPERCONDUCTING DEVICES* in the McGraw-Hill Encyclopedia of Science & Technology.

Seizo Motora

Bibliography. A. Iwata, E. Tada, and Y. Saji, Experimental and theoretical study of superconducting electromagnetic ship propulsion, *5th LIPS Propeller Symposium*, 1983; D. L. Mitchell and D. U. Gubser, Magneto-hydrodynamic ship propulsion and superconducting magnets, *J. Superconduct. Sci. Technol.*, 1:349–364, 1988; Ship and Ocean Foundation, *Proceedings of the International Symposium on Magneto-hydrodynamic Ship Propulsion (MHDS 91)*, Kobe, Japan, 1991; S. Way, Electromagnetic propulsion for cargo submarines, *J. Hydronaut.*, 2(2):49–57, 1968.

Snail

The distribution of marine benthic organisms is the result of several biological and nonbiological parameters interacting during all stages of the life cycle. Biological parameters include the occurrence of prey, predators, and parasites. Nonbiological parameters include environmental factors such as temperature, siltation, and water depth, as well as historical factors such as tectonics and paleoclimates.

Considerable progress in terrestrial biogeography has resulted from the discovery of plate tectonics. However, the frontiers of marine zoogeographic provinces differ markedly from those of terrestrial provinces. This difference may be a consequence of marine biota differing from terrestrial biota in the considerable dispersal capacities of their larval stages.

Larval life histories. There are different ways of classifying the larval development of marine invertebrates. However, from an ecological perspective, perhaps the planktotrophic/nonplanktotrophic dichotomy best describes the larval life histories of marine snails.

In the planktotrophic type of development, a veliger larva (**Fig. 1**) of the snail hatches from an egg cap-

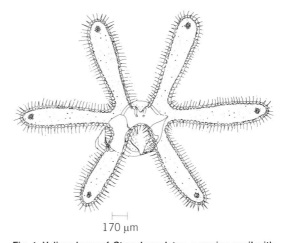

170 μm

Fig. 1. Veliger larva of *Strombus alatus*, a marine snail with planktotrophic development. (*After C. Thiriot-Quiévreux, Summer meroplanktonic prosobranch larvae occurring off Beaufort, North Carolina, Estuaries, 6:387–398, 1983*)

sule and spends from several weeks to months in the plankton. It feeds on phytoplankton until it metamorphoses on the bottom and becomes a crawling benthic snail. Planktotrophic larvae are also referred to as meroplanktonic larvae, as opposed to holoplanktonic for animals, which spend their entire life cycle in the plankton.

In the nonplanktotrophic type of development, there is no long-lived feeding larva in the plankton. The developing embryo gets its nourishment from the yolk-rich egg or other intracapsular reserves; this mode of development can also be termed lecithotrophic. A short (from several hours to several days) nonfeeding planktonic stage occurs in some species, whereas others undergo metamorphosis within the egg capsule and hatch as crawling juveniles (**Fig. 2**).

The simultaneous occurrence of the two modes of development in the same species is termed poecilogony. However, to date no species of marine snail has been proven to exhibit both kinds of development. Past cases that had suggested poecilogony have been reinterpreted as evidence of two closely related species, each with its own mode of development. Mode of development is therefore a species-specific character.

The planktotrophic/nonplanktotrophic dichotomy occurs in most phyla of marine invertebrates. Shelled gastropods and bivalves, however, are unique in that their mode of development can be inferred from the morphology of the larval shell, or protoconch. The protoconch is the aragonitic shell formed by the veliger or the embryo before metamorphosis. It is preserved on the apex of the adult, postmetamorphic shell, or

teleonconch, unless it is broken, corroded, or dissolved. Snails with different modes of development differ in the morphology of their protoconchs. Both protoconchs and teleoconchs can be fossilized, and gastropods and bivalves are probably the only animals in which such clues to their larval biology can be preserved in the fossil record.

Modes of development. The planktotrophic and nonplanktotrophic life-history strategies have long been assumed to draw on contrasting dispersal capabilities. There is a spectrum in which species with veligers spending long periods of time in the plankton are expected to have higher dispersal capacities than species lacking a planktonic larval stage. At one end of the spectrum are the long-lived planktotrophic larvae (termed teleplanic larvae), which spend periods of up to 3–6 months in the surface plankton and are passively transported by currents over thousands of kilometers. The adults of these species may have amphi-atlantic distributions; that is, they occur on both sides of the Atlantic. Among marine snails, a positive correlation between teleplanic larvae and extensive adult distribution is encountered in the families Tonnidae (tun shells), Ranellidae (also known as Cymatiidae; tritons and trumpet shells), and Architectonicidae (sundial shells). Results obtained in the Pacific archipelagoes have shown that isolated, oceanic islands tend to have high percentages of "good dispersers" or of species derived from them.

At the other end of the spectrum are those species whose life cycle is spent entirely in the benthos. Such species, referred to as holobenthic, undergo metamorphosis inside the egg capsule and hatch as heavy crawling snails, with very restricted or no passive dispersal transport. Offshore islands can be predicted to be out of reach for continental populations of such holobenthic species, a situation parallel to that encountered in terrestrial biota. Among marine snails, a correlation between holobenthic development and restricted adult distribution is encountered in the families Volutidae (volutes) and Marginellidae (marginellids).

The larval life histories of many species of marine gastropods, however, fall in the middle of this spectrum—they neither have teleplanic larvae nor hatch as heavy crawling snails. Their dispersal capacity can be directly related to the duration of their larval planktonic phase. At 1.5 knots (1.7 mi/h or 2.75 km/h), a water mass is capable of rafting larvae over a distance of 41 mi (66 km) per day. Although the speed of many marine surface currents is on the order of only 6 mi (10 km) per day, the Gulf Stream is traveling at up to 136 mi (220 km) per day.

As a consequence of different dispersal capacities, the planktotrophic/nonplanktotrophic dichotomy can also be used to predict the genetic structure of species and their evolutionary tempo. Populations of species with long-lived planktonic larvae are expected to exchange genetic material more freely than populations of species with a holobenthic life cycle. In the fossil record, species with nonplanktotrophic larval development are expected to have narrower geographical ranges and shorter temporal distributions than species

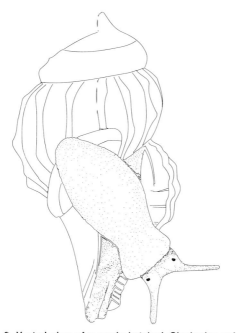

Fig. 2. Ventral view of a newly hatched *Glaphyrina vulpicolor,* a marine snail with nonplanktotrophic development and intracapsular metamorphosis. Shell length = 0.17 in. (3.25 mm). (*After M. C. Pilkington, The eggs and hatching stages of some New Zealand prosobranch molluscs, J. Roy. Soc. New Zeal., 4:411–431, 1974*)

with nonplanktotrophic development. Scattered pieces of evidence support these predictions, but much remains speculative and requires research.

Development and distribution. The Vitoria-Trindade seamount chain (**Fig. 3**), in the western South Atlantic, offers a unique situation where the correlation between mode of development and distribution patterns has been tested. A group of isolated, oceanic islands is at the eastern end of the chain (Trindade Island, Martin Vaz Archipelago), at 684 mi (1100 km) off the Brazilian coast, and six major, linearly arranged seamounts occur between the coast of Brazil and these islands, between latitudes 20° and 21°S. The relative chronology of the formation of the seamount chain is that of older structures at the western end, and more recent structures at the eastern extremity. Trindade and Martin Vaz, the youngest structures, are upper Pliocene (2.3–2.9 million years) in age. The predominantly flattened tops of the seamounts are the result of growth and erosion of calcareous algal deposits over a volcanic pedestal, associated with eustatic processes during the Pleistocene.

Summits of seamounts rise relatively close to the surface (27–82 fathoms or 50–150 m), but shallower, subtidal depths are present only in the oceanic islands. Distances of 31–155 mi (50–250 km) separate each seamount from its closest neighbors. The water depth between the seamounts is 1087–2717 fathoms (2000–5000 m). Geological evidence indicates that elements forming the Vitoria-Trindade seamount chain were never connected to any land mass. Consequently, the distribution of benthic marine animals on the summit plateaus of the seamounts can be accounted for only by passive dispersal.

A total of 244 shelled gastropod species has been recorded from the seamount chain and nearby continental shelf. The filtering effect of distance is evidenced by the significant decrease of the number of species per seamount or island with increasing distance from the shelf to the oceanic islands: from 99 species near the continent to 33 on Trindade.

The most drastic species decline concerns those species with intracapsular metamorphosis. These species experience a rapid decrease in percentages and absolute numbers away from the coast. For example, the volutes do not disperse farther east than Vitoria seamount, and the marginellids drop from a maximum of eight species on Vitoria to a single species on Martin Vaz. This pattern can be related to the difficulty of passive dispersal by rafting of species with large nonplanktonic eggs, juveniles, or adults.

However, mean percentages of planktotrophs and lecithotrophs in the seamounts of the chain are not significantly different. The widespread distribution of lecithotrophic types indicates that oceanic barriers of the order of magnitude found between seamount plateaus of the Vitoria-Trindade chain do not hamper regular transport of short-lived, lecithotrophic larvae. It may also indicate that factors other than larval dispersal are important, including dispersal of adults or egg masses by rafting.

The gastropod fauna of oceanic structures such as the Vitoria-Trindade seamount chain or Northern Mariana Islands demonstrate that the lack of a planktonic larval stage is correlated with a reduced capacity for crossing oceanic barriers. In contrast to terrestrial biogeographic patterns, passive dispersal is an efficient dispersal mechanism in the ocean. Larval dispersal,

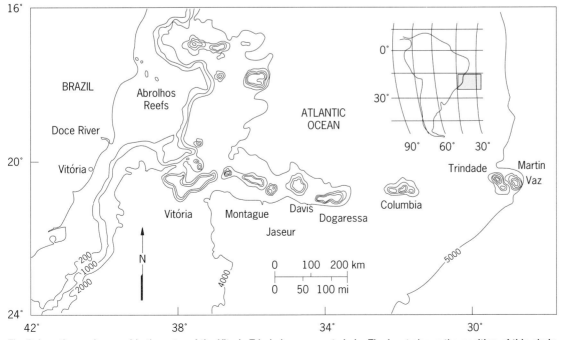

Fig. 3. Location and general bathymetry of the Vitoria-Trindade seamount chain. The inset shows the position of this chain in the southwestern Atlantic Ocean. (*After J. Leal and P. Bouchet, Distribution patterns and dispersal of prosobranch gastropods along a seamount chain in the Atlantic Ocean, J. Mar. Biol. Ass. U.K., 71:11–25, 1991*)

which appears instantaneous on the geological time scale and can account for dispersal over hundreds or thousands of kilometers. has no equivalent in terrestrial biota. This is certainly one of the major differences between marine and terrestrial zoogeography, and may explain why models based on plate tectonics have proved to be of limited value in marine zoogeography.

For background information SEE GASTROPODA; MARINE ECOLOGY; SNAIL; ZOOGEOGRAPHY in the McGraw-Hill Encyclopedia of Science & Technology.

P. Bouchet

Bibliography. D. Jablonski and R. Lutz, Larval ecology of marine benthic invertebrates: Paleobiological implications, *Biol. Rev.*, 58:21–89, 1983; J. Leal and P. Bouchet, Distribution patterns and dispersal of prosobranch gastropods along a seamount chain in the Atlantic Ocean, *J. Mar. Biol. Ass. U.K.*, 71:11–25, 1991; G. Vermeij, E. A. Kay, and L. Eldredge, Molluscs of the Northern Mariana Islands, with special reference to the selectivity of oceanic dispersal barriers, *Micronesica*, 19:27–55, 1985.

Software engineering

Although computers have been programmed for over 40 years, programming remains an error-prone activity. There are fundamentally two ways to attempt to ease this difficulty: by using better languages and by using better environments. This article focuses on software development environments.

A software development environment is intended to ease the burden of software development. Its essential role is to act as a clerical assistant for the programmer, providing information that makes it easier to understand what is already in the program, how the program works, and what is yet to be done. Very few development environments realize these goals completely.

Difficulties of programming. First, it is necessary to understand why many people find programming very difficult.

Abstractions problem. When a person is writing software, it is necessary both to devote attention to the low-level, intricate details of the system being implemented and to maintain a clear vision of the "big picture" and structure of what is being built. A style of programming that constructs a layered hierarchy of abstractions can ease this process. Unfortunately, maintaining this structure can be difficult and, in many situations, unfeasible since access to low-level details is needed. Further, as software evolves, software is often patched to provide the needed additional functionality, even if such a technique upsets the existing abstraction structures.

Precision problem. Computers require programs that are written in extremely precise computer languages instead of the rather flexible natural languages used in everyday life. It is extremely difficult for people to express themselves in a programming language so that contradictions do not creep into the design and implementation of a system.

Nonphysicality problem. While software is engineered, programming is different from more traditional forms of engineering. Until now, engineering has largely been concerned with physical materials, which are amenable to certain laws that govern, for example, their elasticity. This relationship affects the design space within which an engineer works by implicitly shaping the available design options and providing metrics against which to argue the merits of the design. Whether programs are amenable to analogous rules is not yet known. Thus, much of software development is still a matter of trial and error.

Basic environment tools. The basic tools that go into a development environment are an editor, a language translator, and a debugger.

An editor allows the programmer to enter and modify programs. Editors fall into two major categories. Text-oriented editors treat a program as a long stream of text, and the user manipulates the program as if it were any combination of characters, just as in using a word processor. Language-oriented editors have some knowledge of a particular language and can constrain the user to produce only programs that are correct according to some criterion, usually the absence of syntax errors. The language-oriented approach gives more direct support to the user, but significant constraints are often placed on the freedom with which users can manipulate the program structure. The converse is generally true for text-oriented editors.

A language translator checks various syntactic and semantic constraints and then translates the program so that it may be executed. A debugger allows step-by-step execution of the program so that errors can be identified and understood.

Integrated environments. A basic consideration in different environments is whether or not they are integrated. A nonintegrated environment provides a loosely coupled set of tools that the user can deploy in various combinations to assist with program development. A well-known example is the UNIX Programmers WorkBench, which provides a set of independent editors, compilers, and debuggers, as well as some of the more advanced tools discussed below. The biggest problem with this approach is knowing how to fit these tools together, or even what they provide. The advantages of using families of independent tools is that it is possible to employ a mix-and-match approach to selecting the tools that make up the environment, for example, choosing to use a particular editor, compiler, and debugger.

An integrated environment provides the user with a single tool that combines the various facilities discussed above with a single interface, so it is much easier to use. For example, the user could invoke the compiler or debugger with a single key press from within the editor when editing. Similarly, when compiling or debugging, the user can be shown the place in the source code where problems have been found, and can be taken directly to that spot to make program changes.

The major advantages of an integrated environment are that it is easier to learn and use and that it usually provides a window-based interface. If an environment is simple and powerful, programmers are more productive. The major problem with integrated environments is that they are usually closed: they support a fixed set of tools, and it is not possible to add new tools or substitute for existing tools.

To get around this problem, some recent environments research has focused on open integrated systems. These systems attempt to provide a framework into which existing tools can be connected to provide an integrated environment while addition and deletion of tools are allowed. The Field system is probably the best research example. SEE INFORMATION TECHNOLOGY.

Advanced environment functions. Editors, compilers, and debuggers are not the only tools needed in an environment. These tools provide no support for group activities or software evolution.

Group support. Software is usually developed by teams of programmers. Significant development environment support is needed to keep the programmers from tripping over each other. The basic problem is that if two programmers modify the same program file simultaneously, one set of changes will be lost when the files are written to disk; and if they work on separate copies of the file, the versions can diverge and need to be merged to get a copy that contains the changes of both workers. This problem is enhanced when the changes conflict; a simple syntactic merge of the code does not work; and a deeper, semantic merge is needed. A second problem is that programmers affect one another as they work; if one programmer modifies a file, another programmer's file may need to be changed as well to deal with the first programmer's changes (for example, changing the number, order, or type of parameters to a procedure).

Support for the multiprogrammer versioning problem is still an open issue, but most software development environments should provide some clerical means of dealing with multiple versions of the same file.

Support for the ways in which the work of one programmer can affect the work of others is also an open research problem. Research in the area of computer-supported cooperative work (CSCW) attempts to provide theories and tools that explain and support the ways in which members of groups interact. Software process modeling is investigating how the processes of software development can be understood, modeled, analyzed, and supported, so that ways of producing more reliable software in a more efficient fashion can be discovered.

Software evolution. Software evolves. Over time, an organization can release many different versions of a piece of software, with differing features and capabilities. At any time, one of these releases can contain older versions of parts of the system and newer versions of others. For example, many UNIX tools are being taken from the vendors' priority user interfaces and incorporated in the common X11 windowing environment. Thus the tool may use older versions of the database and computation components but new user interface components. It is quite possible that after several years software producers will be releasing systems that have new computational components but that use the old proprietary user interface in one version (for those who prefer that interface) and a new user interface component in another version.

Configuration management tools are a component of many software development environments, and support the ability to specify a configuration as a family of different versions of the files making up the system. This approach allows the development team to easily control the makeup of different releases of an evolving system.

Both integrated and nonintegrated environments can provide version and configuration management tools. Very little support for software process modeling or computer-supported cooperative work can be found in currently available environments outside the research laboratory.

Program analysis. Support for program analysis is another kind of functionality that environments, integrated or nonintegrated, can provide. Much of the research to support program analysis has been focused on parallel or object-oriented languages. Parallel language analysis is oriented toward providing the user a view of the program that helps in understanding where the bottlenecks in the parallel execution of a program are occurring, such as data dependency graphs. Object-oriented languages, such as Smalltalk or C++, use an inheritance hierarchy, a way of sharing code among abstract data types to promote reuse and the evolution of common interfaces to related data types. Browsers allow programmers to understand the inheritance hierarchy and manipulate it and the program fragments constituting it in a simple and intuitive way. Such tools are usually found in newer, research environments but are also becoming more used in commercially available systems. SEE CONCURRENT PROCESSING.

For background information SEE PROGRAMMING LANGUAGES; SOFTWARE ENGINEERING in the McGraw-Hill Encyclopedia of Science & Technology.

Simon M. Kaplan

Bibliography. G. E. Kaiser, S. M. Kaplan, and J. Micallef, *Multiuser, Distributed Language-Based Environments*, 1988; *Proceedings of the ACM Conference on Organization Computing Systems*, Atlanta, November 1991; S. P. Reiss, *Connecting Tools Using Message Passing in the FIELD Environment*, 1990; T. W. Reps and T. Teitelbaum, *The Synthesizer Generator*, 1989.

Soil

A saline seep results from a soil salinization process, often accelerated by dryland farming, that allows water to move through salt-laden substrata below the root zone (**Fig. 1**). Saline seep refers to intermittent or continuous saline water discharge at or near the surface of the soil, downslope from recharge areas under dryland (rain-fed) conditions. This process reduces or eliminates the growth of crops in the discharge area because of increased soluble concentrations of salt in

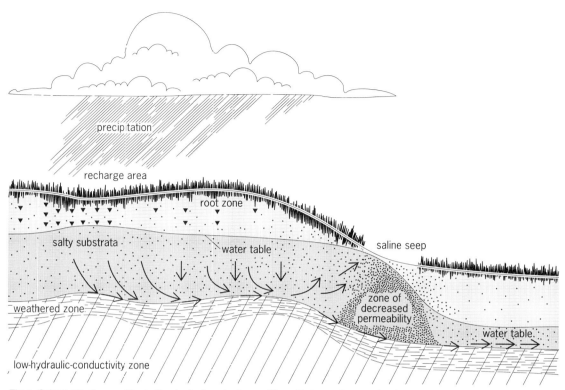

Fig. 1. Schematic diagram illustrating typical geologic conditions that contribute to development of saline seep. The triangles indicate downward movement of water through the soil profile. (*After P. L. Brown et al., Saline-Seep Diagnosis, Control and Reclamation, USDA Cons. Res. Rep. 30, 1983*)

the root zone (**Fig. 2**). Saline seeps can be differentiated from other saline soil conditions by their recent and local origin, saturated root-zone profile, shallow water table, and sensitivity (short-term response) to precipitation and cropping systems.

Occurrence of saline seeps. Saline seeps occur frequently in dryland farming areas throughout the North American Great Plains, with an estimate of nearly 10^6 hectares (2.5×10^6 acres) of productive cropland salinized. Saline seep problems are present in Australia, India, Iran, Turkey, and Latin America. Saline seeps result from a combination of geologic, climatic, hydrologic, and cultural (land-use) conditions. The primary cause is a change in vegetation from grassland or forest to a cropping system that is less

efficient in water use, such as a crop–summer fallow rotation, which allows precipitation in the recharge areas to move below the root zone and provide seepage water. The characteristics, hydrology, and causes of most saline seeps are similar regardless of geographic location.

In the United States, the crop–summer fallow system of dryland farming has contributed significantly to the development of the saline seep problem in the Great Plains but is not the only cause. Seep development is encouraged by periods of above-normal precipitation; restricted surface and subsurface drainage due to construction of roads or pipelines; large snow drifts at windbreaks, roadways, and such; gravelly and sandy soils; obstructions (such as roads) across natural drainageways; unplugged or poorly cased artesian water wells; leaky ponds and dugouts; and crop failures. Water conservation practices, such as forming level bench terraces, have contributed to saline seep development.

Seep development generally occurs on sidehills or toe slopes (bottom part of a sidehill) of rolling-to-undulating topography, where permeable material is underlain by less permeable strata, a circumstance conducive to development of perched water tables. An understanding of the geology and circumstances that cause a particular saline seep to form will help in designing effective control or prevention measures. In general, while agronomic practices work well to control most seeps, some may require additional drainage and land leveling to achieve hydrologic control.

Fig. 2. Typical saline seep discharge area in Montana.

Table 1. Chemical composition of waters associated with saline seeps in the Great Plains*

| Location | pH | Electrical conductivity, dS/m[†] | Ion concentration, mmol/liter | | | | | | |
			Calcium (Ca^{2+})	Magnesium (Mg^{2+})	Sodium (Na^+)	Bicarbonate (HCO_3^-)	Nitrate (NO_3^-)	Chloride (Cl^-)	Sulfalte (SO_4^{2-})
Montana recharge	8.4	5	7	11	18	3.8	4.3	0.7	21
Montana seep	8.2	9	8	21	66	9.8	0.4	0.8	52
Montana seep	7.9	14	10	37	109	8.1	29.5	2.6	80
Montana seep	8.4	26	1	108	211	4.0	5.4	7.6	225
Montana recharge	8.2	7	3	21	39	2.4	6.2	11.2	44
North Dakota seep	3.7	10	9	36	59	—	5.7	2.1	70
North Dakota seep	4.6	8	9	30	40	—	4.7	2.5	55
Oklahoma seep	8.1	5	15	16	26	—	0.6	12.3	27
Oklahoma seep	8.2	3	3	17	13	—	—	16.0	15

*From A. D. Halvorson, Management of dryland saline seeps, in K. K. Tanji (ed.), *Agricultural Salinity Assessment and Management*, ASCE Man. Rep. Eng. Prac. 71, American Society of Civil Engineers, 1990.

[†] Decisiemens per meter.

Water quality and saline seeps. As the water passes through the soil profile toward the perched or permanent water table, salts are dissolved and moved downward. Often the shallow groundwater associated with saline seeps is unsuitable for human or livestock consumption because of high levels of nitrate (NO_3; >0.7 mmol/liter) and other salts, and for irrigation because of total salt concentration. Calcium (Ca^{2+}), magnesium (Mg^{2+}), and sodium (Na^+) are the dominant cations and sulfate (SO_4^{2-}) is the dominant anion in most of the shallow groundwater associated with saline seeps. Sulfates are the dominant anion in the water and soil system in the Great Plains, while chlorides are generally low in the northern Great Plains but tend to be slightly higher in the southern Great Plains (**Table 1**). Little, if any, of the nitrate in the water originated from nitrogen fertilization practices, because little, if any, nitrogen fertilizer was used by dryland farmers in the Great Plains prior to the early 1970s, when the saline seep problem was first researched. Much of the nitrate was of geologic origin.

Identification. Early detection and diagnosis of a saline seep problem is important in designing and implementing control and reclamation practices in order to prevent further damage. Early detection may allow a farmer to minimize the damage by changing current cropping systems.

Visual symptoms of impending saline seep development are (1) vigorous growth of kochia (*Kochia scoparia*) or other weeds after grain harvest in areas where normally the soil should be too dry to support weed growth; (2) presence of salt crystals on soil surface; (3) prolonged wetness in small areas of the soil surface following rain; (4) tractor wheel slippage or equipment bog-down in isolated areas of a particular field or water seepage into wheel tracks, with salt crystals visible as soil dries; (5) rank crop growth accompanied by lodging (stem breaking) in localized areas that previously produced normal crop growth, which may indicate a rising water table where soil salinity is not yet high enough to reduce crop growth and yield; (6) increased infestations of salt-tolerant weeds; (7) stunted or dying trees in a shelterbelt or windbreak; and (8) poor seed germination.

Methods for measuring soil salinity based on the electrical conductivity of the soil have been developed for identifying potential saline seep areas. Four-electrode resistivity and electromagnetic inductive techniques have been used to characterize soil-profile salinity levels of saline seep areas and to identify recharge areas. These electrical conductivity methods can be used for detecting and delineating saline seeps, for measuring and mapping field soil salinity, and for verifying areas of high and low salinity in the field without need for laboratory analyses. Thus, salinity in suspected saline seep areas can be monitored in comparison to surrounding nonseep areas. Existing saline seeps generally have high levels of salinity at the soil surface, and these levels decrease with soil depth. Developing seep areas generally have low-to-medium levels of salinity at the soil surface, with higher salinity at shallow (1–2 m or 3–6 ft) soil depths and lower salinity at greater depths. Soil salinity generally increases gradually with increasing soil depth in the recharge area (**Fig. 3**).

Delineating the location and approximate size of recharge areas is essential to designing successful control treatments. Generally, recharge areas are located a

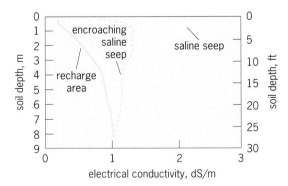

Fig. 3. Typical four-probe electrical conductivity readings as a function of soil depth in a saline seep recharge area, encroaching saline seep area, and saline seep area. (*After A. D. Halvorson and J. D. Rhoades, Assessing soil salinity and identifying potential saline-seep areas with field soil resistance measurements, Soil Sci. Soc. Amer. J., 38:576–581, 1974*)

Table 2. Yields, in metric tons/hectare,* of several crops grown in two reclaimed saline seeps in 1978 and 1979 compared to average county yields in northeastern Montana[†]

Crop	1978	1979	Average	1978	1979	Average
		Seep A			Richland County	
Spring wheat	2.5	1.6	2.1	2.2	1.4	1.8
Barley	4.5	2.1	3.3	2.4	1.3	1.9
Oats	3.4	1.6	2.5	2.0	1.2	1.6
Alfalfa	5.7	9.8	7.8	4.3	3.4	3.9
		Seep B			Roosevelt County	
Spring wheat	2.4	1.8	2.1	1.8	1.3	1.6
Barley	3.9	3.3	3.6	2.1	1.4	1.8
Oats	5.3	2.2	3.8	1.8	1.2	1.5
Corn (silage)	16.9	3.5	10.2	17.9	11.2	14.6

* 1 metric ton/ha = 0.45 ton/acre.
[†] From A. D. Halvorson, Saline-seep reclamation in northern Great Plains, *Trans. ASAE*, 27:773–778, 1984.

short distance upslope from the discharge or seep area. Information from test holes, water table levels, salinity measurements, visual observations, and topography can be used to delineate the approximate recharge-area location. A combination of probing, mapping field salinity levels, and drilling test holes is an effective way to locate recharge areas.

Control. Since seeps are caused by water moving below the root zone in the recharge area, there will be no permanent solution to the saline seep problem unless control measures are applied to the recharge area. There are two general procedures for managing seeps. In the first, ponded surface water is drained mechanically before it infiltrates, and the lateral flow of subsurface water is intercepted with drains before it reaches the discharge area. The second method is to let crops use the water before it percolates below the root zone.

Hydraulic control can be quickly and effectively accomplished with subsurface interceptor drains located on the upslope side of the seep area. However, a suitable outlet for disposal of the saline water needs to be available. Outlet considerations must include not only an easement for transport of drainage water across intervening lands but also the effect of drainage waters on the quality of the receiving streams or reservoirs. The water is saline, usually high in both sulfate and nitrate, and disposal is difficult because of environmental, physical, and legal constraints. Therefore, subsurface drainage is generally not satisfactory because of disposal problems and the economics of dryland crop production. The best approach is to use the soil water for crop growth before the water becomes saline.

Hydraulic control of saline seep areas can be accomplished by planting crops and employing cropping systems that will effectively use soil water in the recharge area. This approach requires identification of the recharge area, followed by adoption of appropriate cultural practices to minimize deep percolation. Any delay in implementing control practices can lead to a larger problem that is more difficult to manage. Alfalfa (*Medicago sativa*), seeded in recharge areas, effec-

tively controls or stops excessive percolation. Saline seep areas have dried sufficiently with alfalfa to obtain normal grain and forage crop yields (**Table 2**). Once a saline seep area has been controlled, reclaimed, and returned to normal crop production, soil water in the recharge area must be continually managed to prevent recurrence.

Flexible cropping systems, which involve planting a crop in years when stored soil water and expected growing-season precipitation are sufficient to produce an economic crop yield, have been used to control saline seeps. Using flexible, small-grain cropping systems to gain hydraulic control of seep discharge areas is a slower process than using alfalfa. Inclusion of safflower (*Carthamus tinctorius*) or sunflower (*Halianthus annuus*), which are normally deeper-rooted than small grains, will help deplete the stored soil water to greater depths.

Reclamation. Before reclamation of a saline seep area can proceed, the flow of water from the recharge area must be reduced to the extent that the water table depth in the saline seep has been lowered sufficiently (to more than 150 cm or 59 in.) to prevent movement of salts by capillary action from the water table into the root zone. Both research and farmer experiences show that reclamation occurs quite rapidly. With a water table depth in the seep area that is more than 150 cm (59 in.), reclamation procedures to remove salts from the root zone can proceed. Rate of reclamation depends on the amount of precipitation received to leach the salts. Therefore, practices such as snow trapping or fallowing in the summer in the salt-affected area will enhance water movement through the profile and hasten the reclamation process.

Socioeconomic concerns. Saline seeps do not respect property lines. A recharge area on one farmer's property can supply water to a discharge area on a neighbor's farm, or the seep discharge can contaminate a stream, natural drainageway, or farm pond. Except for small, uncomplicated seeps, most farmers need help in diagnosing their saline seep problem and in developing cropping systems or other control measures. When a recharge area is on an adjacent farm, the cooperation of landowners is needed. Knowledgeable individuals or agencies can assist by characterizing the problem and recommending control measures. Legislation may provide procedures for farmers to form salinity control districts to achieve collectively what cannot be done individually.

Saline seep is not just a farming problem. Any loss of farmland decreases a nation's food and tax base. Salty water from seeps can pollute fresh surface waters and add to the salinity of groundwater. The saline seep problem has political implications, involving such questions as subsidies, crop acreage allotments, and landowner rights. In the United States, federal farm programs have sometimes adversely affected progress in controlling saline seeps by restricting the acreage that can be planted to small grains or other seep control crops.

For background information SEE AGRICULTURAL SOIL AND CROP PRACTICES; GROUNDWATER HYDROLOGY in the McGraw-Hill Encyclopedia of Science & Technology.

Ardell D. Halvorson

Bibliography. W. W. Berg, J. W. Naney, and S. J. Smith, Salinity, nitrate, and water in rangeland and terraced wheatland above saline seeps, *J. Environ. Qual.*, 20:8–11, 1991; A. L. Black et al., Dryland cropping strategies for efficient water-use to control saline seeps in the northern Great Plains, U.S.A., *Agr. Water Manag.*, 4:295–311, 1981; W. L. Hargrove (ed.), *Cropping Strategies for Efficient Use of Water and Nitrogen*, Spec. Publ. 51, ASA-CSSA-SSSA, Madison, Wisconsin, 1988; K. K. Tanji (ed.), *Agricultural Salinity Assessment and Management*, ASCE Man. Rep. Eng. Prac. 71, American Society of Civil Engineers, 1990.

Soil ecology

The traditional view of ecosystems suggests that the importance of animals derives from their consumption of energy and materials. In contrast, recent studies suggest that terrestrial animals do not consume a great percentage of the material produced by green plants in most ecosystems. Instead, the importance of animals to the structuring and functioning of ecosystems is related to activities other than consumption. For example, both pollination and disease can control the structure and functioning of ecosystems, yet the importance of animals as pollinators of plants is not related to the quantity of energy or materials that the animals consume; nor does their important role as agents in the transmission of disease relate directly to their involvement in the energy dynamics of ecosystems.

Another animal activity that significantly affects the composition and functioning of ecosystems is burrowing by fossorial species. The term fossorial (adapted to a burrowing mode of life) is often restricted to burrowing organisms that rely on the resources that they extract belowground.

Only the direct effects of an animal's digging activities on soil properties and the subsequent influences on the ecosystem will be considered in this article.

Such an approach ignores effects of fossorial animals that are related to removing whole plants or their parts during feeding, and effects of fertilization due to the production of feces.

Fossorial species. Numerous animals dig in soil. Often this digging involves merely scratching through the soil surface to obtain food. Many large mammals obtain consumable plant parts in this manner. Obviously, this digging can have some effect on ecosystem properties; however, the more intense, subterranean burrowing activities of the fossorial organisms have a more pronounced effect. Burrowing is a common activity in taxonomic groups, ranging from minute insects to large mammals. Species that are especially familiar as burrowers include moles, prairie dogs, a variety of rodents, badgers, foxes, some rabbits, armadillos, a host of insect species (especially ants, wasps, beetles, termites, and cicadas), some tortoises, a few birds, and frogs. Fossorial species occur worldwide in most ecosystem types. However, they are most common in areas that are not heavily forested, and they are generally small. Among the largest is the aardvark (*Orycteropus afer*) of Africa.

While some burrowers exhibit few adaptations to a subterranean life and spend only part of their time belowground, others are highly specialized. A prime example of specialization is a family of rodents confined to the New World—the Geomyidae, usually called pocket gophers. This family consists of 5 genera and about 35 species that occur from British Columbia, Canada, to Colombia, South America. Pocket gophers are familiar in many areas of North America because of characteristic mounds that are created during the burrowing process and linear casts of soil that appear on the ground surface following the melting of winter snow. All members of the family spend their lives in burrows, where they forage, mate, give birth, and die; they seldom come to the surface. Their feet, teeth, muscles, sense organs, and respiratory characteristics are adapted to a subterranean life.

Effects of burrowing animals. As animals burrow or dig, they can alter the physical, chemical, or biological properties of the soils. These changes affect the soil's properties and its inhabitants. This observation was described by Charles Darwin in 1888 in a work that detailed the effects of burrowing by earthworms.

Physical effects in soils result from merely turning the material over—changing its bulk density, mixing different combinations of particle sizes in soil, or aerating soils by burrow creation. The changes in bulk density (how compacted the soil is) are similar to those caused by a person digging in a yard. When a hole is dug and then refilled with the removed material, there will be an excess of soil that forms a mound because the soil volume has increased. As burrowers dig, the increased volume of soil must be recompacted, used to fill an existing tunnel, or moved to the surface to keep the burrow system clear. Soil moved to the surface covers the existing soil and plants; because of its lesser bulk density, this excavated soil is more permeable to water (has a higher rate of infiltration). Since the burrowing activities of groups of animals, such as pocket gophers, might disturb more than 10% of the soil surface of an area each year, the effect on the bulk density of surface soils is significant. Even in situations where much smaller animals, such as insects, are the agents of soil disturbance, significant quantities of soil can be carried to the surface. Ants in Australia bring nearly 420 kg of soil per hectare (375 lb per acre) per year to the surface. In the pinelands of Florida as much as 580 kg of soil per hectare (517 lb per acre) is deposited on the surface in a year.

Soil brought to the surface often has a different texture (mixture of soil particle sizes) from that of surface soil. The texture variation causes many changes in soil properties, affecting the capture of seeds, aeration, infiltration of water, and ease of penetration by roots and burrowers. Some animals move different particle sizes around to suit some purpose; for example, many ants differentially cover their mounds with small pebbles, and so alter the particle size distribution of surface soils.

Certain landscapes of the Pacific Northwest of North America and lowland and highland areas of Africa and southern South America are dominated by soil mounds, known as Mima mounds, that reach 2 m (6.6 ft) in height and 20 m (66 ft) in diameter. These structures are attributed to a combination of geomorphic processes and digging by several rodents, which include pocket gophers (family Geomyidae) in North America, mole rats (Bathyergidae) in Africa, and tuco tucos (Ctenomyidae) in South America.

Often chemical changes in soils result from the same mixing processes that cause changes in soil particle size and bulk density. Soils brought to the surface may contain higher quantities of certain chemicals than those possessed by surface soils already affected by leaching and by the extraction of materials by green plants. Beetles in Florida bring material to the surface that is higher in extractable phosphorus, calcium, and magnesium compared to the surface soil. Soils moved to the surface by pocket gophers contain increased quantities of some nitrogen compounds, phosphorus, and potassium.

One significant chemical change in soils is caused by animals incorporating organic matter into their burrows, for example, by animals dragging plant material down from the surface. The movement of organic matter deeper into the soil layer can increase the rate of decomposition, and it extends the organic horizon farther into the soil, so that the soil profile and its properties are dramatically changed.

Among the biological changes in the soil in response to burrowing is the transport of seeds and spores from below the soil surface, where they cannot germinate, to the surface, where they can germinate and grow. This activity may also bring germinating plants into contact with the spores of fungi that develop beneficial relationships with the plants. For example, the formation of a mutually beneficial structure termed a mycorrhiza results from the interaction of a fungus mycelium and a plant root. In fact, over 90% of the world's green plants form this association, often in an obligatory way.

Some ecosystem consequences. The effects on ecosystems by the digging of fossorial species are numerous and complex. The species composition of the ecosystem can be altered by burrowing.

For example, plants can be covered during excavation, certain species can be favored by the digging, plants can be infected by mycorrhizal fungi, and obligate relationships can be established.

Covered plants. Plants that are covered with soil during the excavation process may not survive. Plants with belowground storage structures (such as bulbs or corms) probably survive better than annuals or perennials with little storage tissue. In such cases, the species composition of the ecosystem may change. In an extreme example, pocket gophers digging in subalpine meadows of Utah may prevent the succession of communities to the forest stage for as long as 1000 years.

Favored species. Digging itself may favor some species over others. In a desert–grassland transition area of Arizona, the digging activities of kangaroo rats (*Dipodomys*) maintain desert shrub conditions, while removal of the rats enhances the establishment of grasses.

Mycorrhizal fungi. Plants establishing an excavated material may be infected with mycorrhizal fungi more often than plants establishing elsewhere, so those on animal-handled soils gain an advantage. Plants growing on mounds benefit from a soil that has better drainage and more favorable mixes of particle size compared to nonmound soil. These conditions favor germination as well as growth. Following the Mount Saint Helens volcanic eruption in 1980, more plant species and more individual plants became established on the mounds of the few pocket gophers that survived in some of the devastated areas than on adjacent volcanic materials that were not mixed by surviving pocket gophers. Additionally, plants on mounds developed more mycorrhizae and had higher rates of survival.

Obligate associates. Many animals use burrows of other species for refuge, some in an obligate relationship. Tortoise burrows in the southeastern United States harbor over 267 species of animals. In one area, 22% were obligate associates of these burrows.

For background information SEE ECOSYSTEM; MYCORRHIZAE; SOIL ECOLOGY in the McGraw-Hill Encyclopedia of Science & Technology.

James A. MacMahon

Bibliography. D. C. Anderson, Belowground herbivory in natural communities: A review emphasizing fossorial animals, *Quart. Rev. Biol.*, 62:261–286, 1987; N. Huntley and R. Inouye, Pocket gophers in ecosystems: Patterns and mechanisms, *BioScience*, 38:786–793, 1988; O. J. Reichman and S. C. Smith, Burrows and burrowing behavior by mammals, *Curr. Mammal.*, 2:197–244, 1990; A. D. Whicker and J. K. Detling, Ecological consequences of prairie dog disturbances, *BioScience*, 38:778–785, 1988.

Solar energy

Traditionally, applications of concentrated solar thermal energy have centered on the production of electricity and process heat, with the goal of providing a cheap, renewable energy source. Tighter regulations on waste disposal and increased concerns for the environment, however, have provided the impetus for development of a new application for solar energy: destruction of hazardous wastes. Research, development, and demonstration activities are under way to bring these new solar technologies for cleanup of contaminated water and soil into everyday use.

Purifying polluted water. Of the estimated 280×10^6 tons (252×10^6 metric tons) of hazardous waste generated in the United States each year, nearly three-quarters is in the form of wastewater contaminated with solvents, corrosive substances, metals, and other toxic material. Perhaps as many as one-quarter of the supplies of potable water are likewise contaminated. Existing treatment methods either are inefficient and costly or merely transfer the contaminants from one environmental medium to another. For example, air stripping is in common use; this is a technique in which air is bubbled through the water, but it simply moves the organic contaminants into the air, where they are diluted to so-called acceptable concentrations. Granular activated carbon can be used to remove organic compounds from water or air, but the problems and cost of disposal or regeneration of the carbon remain. One of the new solar technologies being developed to help solve these problems is a photocatalytic process capable of cleaning up large volumes of groundwater or industrial wastewater that are contaminated with low levels of toxic organic chemicals.

In this solar detoxification process for water purification, sunlight is focused by mirrored parabolic troughs onto a glass reactor through which the contaminated water is flowing (**Fig. 1**). Ultraviolet energy in the concentrated beam is absorbed by a semiconductor catalyst and activates it in the waste stream. Electron-hole pairs generated at these activated sites react with water, dissolved oxygen, or added oxidants (such as hydrogen peroxide or ozone); very reactive oxidizers (principally hydroxyl radicals) are formed, and attack the organic molecules. All reactions occur at warm-water temperatures (30–85°C or 86–185°F). The organic compounds are completely mineralized (converted to simpler inorganic compounds) in reacting with the oxidizers to form only carbon dioxide (CO_2), water (H_2O), and (if the organic compounds are halogenated) dilute, easily neutralized, simple mineral acids such as hydrochloric acid (HCl). The destruction of trichloroethylene (C_2Cl_3H), a common pollutant, is summarized by the overall reaction shown below.

$$2C_2Cl_3H + 3O_2 + 2H_2O \rightarrow 4CO_2 + 6HCl$$

The catalyst for this reaction, titanium dioxide (TiO_2; commonly used as a paint pigment or as an ingredient in sunscreen, cosmetics, and toothpaste), is nontoxic and can be reused many times.

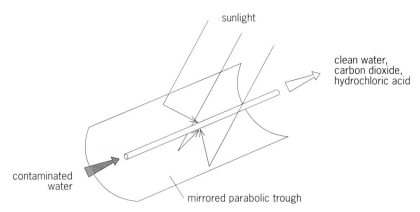

Fig. 1. Diagram of the solar detoxification process for water purification.

The solar detoxification process is capable of destroying most organic compounds, including industrial solvents, pesticides, fuel components, wood preservatives, and dyes such as those found in textile-mill effluents. Unlike many current technologies in which the toxic chemicals in the water are released to the air or absorbed by materials that must subsequently be treated to remove the toxic compounds, the solar detoxification process destroys the organic compounds in a single step. It does not generate hazardous intermediate compounds, so no additional cleanup of the water is required.

Recent laboratory and engineering-scale experimental results conducted for the U.S. Department of Energy have graphically demonstrated the effectiveness of this process. In the laboratory, this work has expanded on the pioneering photochemical work of recent decades, which demonstrated the potential of ultraviolet light and a catalyst to destroy complex organic molecules, by assessing the effects of light intensity, catalyst and oxidant concentration, and reactor geometry on destruction rates in solar-powered systems. Possible improvements to the catalyst have also been evaluated. At a larger scale, by using parabolic trough systems similar to those used to produce over 350 MW of electric power in the Mojave Desert, the destruction of chlorinated solvents (such as trichloroethylene) and organic dyes at low concentrations in water has been successfully demonstrated. For example, at water flow rates up to 100 liters (25 gal) per minute, the destruction of 0.1–5 parts per million (ppm) trichloroethylene to levels below the 0.005-ppm drinking-water standard has been accomplished in 2–4 min by using the parabolic trough test facility shown in **Fig. 2**. This facility, over 210 m (700 ft) long with 465 m² (5000 ft²) of reflective mirrors, concentrates the incoming sunlight by a factor of 30 and increases the reaction rate of the process to commercially attractive levels. Generally, no intermediate products are observed, and the organic compounds are completely mineralized. These successes have led to the recent installation of a field test system to demonstrate the process on contaminated groundwater in California. This year-long test will give researchers valuable feedback on how effectively the process performs at a Superfund reme-

Fig. 2. Experimental parabolic trough system for treatment of contaminated groundwater.

diation site. It will be an important step in bringing this innovative technology into commercial use.

The solar detoxification process is convenient, because it can be done on site at near-ambient temperatures. It limits the site owner's liability to on-site operations, and eliminates both the need to handle the same wastes several times and the possibility of further damage to the environment. A complete facility would include several thousand square meters of mirrored troughs, and would be controlled by a simple computer system to adjust the rate of processing as the sunlight intensity changes throughout the day. In sunny parts of the country, such a system would fit on less than an acre and could clean tens of thousands of gallons of contaminated groundwater or industrial wastewater per day throughout the year. Smaller systems might be used for community wells or for effluents from industrial processes. To improve the cost competitiveness of solar detoxification systems, work is under way to increase catalyst and reactor efficiencies and to decrease solar collector costs; systems that use the sunlight directly (without concentrating mirrors) are also being investigated.

As results from the ongoing test and demonstration programs become available and the process is refined, this technology should emerge as a viable option for purifying polluted water.

Cleaning contaminated soils. Although not as well developed as the technology for purifying water, a second set of solar processes is emerging to help in the cleanup of soils contaminated with organic substances. Depending on the application and the specific wastes, these high-temperature, solar-driven processes for destroying concentrated organic wastes use either photolytic or thermal/catalytic steam reforming chemistry to convert the wastes to harmless by-products.

In both processes, a reflective parabolic dish or heliostat (Sun-tracking mirror) field is used to concentrate sunlight up to 1000 times through a quartz window into an insulated reactor vessel. In a photolytic process, the ultraviolet portion of the solar spectrum is absorbed at a high temperature (700–1000°C or

1290–1830°F) directly by the organic molecules and provides the energy necessary to break chemical bonds and oxidize the waste to carbon dioxide, water, and halogen acids (which are easily neutralized to simple salts). In the steam reforming process, the sunlight is absorbed by a rhodium catalyst; the hot catalytic surface then initiates reactions with steam to convert the organic compounds to products similar to those in the photolytic process plus significant amounts of hydrogen and carbon monoxide, which could potentially be used for their fuel value or as chemical feedstocks. In both cases, the waste reacts completely in fractions of a second. Unlike incineration, the solar-driven reactions do not produce an array of products of incomplete combustion, which can potentially be more hazardous than the original wastes. This type of reactor exhibits excellent characteristics of heat and mass transfer so that the system can be compact and highly efficient. Laboratory and pilot-scale tests using concentrated sunlight have demonstrated the potential of solar energy to effectively destroy a range of concentrated organic compounds, including dioxins and industrial solvents. **Figure 3** shows a scaled-up version of a solar reforming system (in this case, reforming methane as a means of storing solar energy as chemical energy) as it might be used commercially.

The primary potential application of these systems is the decontamination of soils. The organic contaminants are first desorbed from the soil (perhaps by utilizing heat recovered from the high-temperature solar reactor) and then routed to the solar reactor. Clean soil is returned to the environment. Other potential applications include the destruction of organic compounds during on-site regeneration of the granular activated carbon used in many waste treatment operations, and the destruction of contaminated industrial solvent waste streams and stored inventories of wastes and environmentally hazardous materials such as chlorofluo-

Fig. 3. Solar parabolic dish used for solar chemistry experimentation; this 220-m² (2500-ft²) solar concentrator is powering a solar reforming reactor essentially identical to one that would destroy concentrated organic chemicals desorbed from soils.

rocarbons. A single solar parabolic dish (10-m or 30-ft diameter; identical to those being developed to power Stirling engines for generation of electricity) could destroy hundreds of kilograms of chlorinated hydrocarbons per day. Because these solar energy systems are relatively small and inexpensive, they can be sized to a particular waste stream and operated on the site where the waste is generated. Thus waste hauling and the concomitant costs, liabilities, and public opposition will be avoided.

Before the soil treatment technologies are ready for commercial implementation, however, experimental and systems research programs are needed to verify that high destruction rates and efficiencies are achievable over a range of processing conditions and waste materials and to establish the economic practicality of the systems. These activities are under way, and it is anticipated that on-site field demonstrations will begin soon.

Prospects. Solar-energy-driven technologies for cleaning the environment are not operational. However, ongoing research programs have brought them very close to reality. Within a couple of years, solar-powered water cleanup technology should be commonplace, and continued development should make solar energy a practical tool for soil cleanup by the late 1990s.

For background information *SEE PHOTOCHEMISTRY; SOLAR ENERGY; WATER POLLUTION* in the McGraw-Hill Encyclopedia of Science & Technology.

Craig E. Tyner

Bibliography. D. F. Ollis, Contaminant degradation in water: Heterogeneous photocatalysis degrades halogenated hydrocarbon contaminants, *Environ. Sci. Technol.*, 19:480–484, 1985; J. E. Pacheco, Photocatalytic destruction of chlorinated solvents with solar energy, *ASME International Solar Energy Conference Proceedings*, 1991; C. E. Tyner, Application of solar thermal technology to the destruction of hazardous wastes, *Solar Energy Mater.*, 21:113–129, 1990.

Solar neutrinos

The search for neutrinos emitted by the Sun was stimulated by the desire to experimentally verify that the Sun generates energy by nuclear fusion reactions, the fusion of hydrogen into helium. Prior to the twentieth century, the prevailing scientific opinion was that solar energy is generated by the gradual release of gravitational energy as the Sun slowly shrinks. The gravitational energy released by the Sun to date is only sufficient to supply power at the present rate for about 10^7 years. With the development of radioactive dating in the early part of the twentieth century, it rapidly became clear that the Earth was at least several billion years old and that biological activity, which indicated a warm Earth, had been going on for at least 5×10^8 years. Based on this discrepancy, the gradual-gravitational-collapse source of solar energy was replaced in the 1920s by the idea that solar energy is generated by nuclear fusion.

Neutrino production in the Sun. The Sun emits energy at the rate of 4×10^{26} joules per second. It is assumed that this energy is generated by the fusion of protons, the nuclei of hydrogen atoms, into helium nuclei. Since the helium nucleus consists of two protons and two neutrons, it is necessary that two of the protons be converted into neutrons. There is a sequence of reactions beginning with reactions (1a) and (2). In 85% of the reaction sequences, reaction (2)

$$P + P \rightarrow D + e^+ + \nu_e \qquad (1a)$$

$$P + D \rightarrow {}^3\text{He} + \gamma \qquad (2)$$

is followed by reaction (3). The other 15% of the time,

$${}^3\text{He} + {}^3\text{He} \rightarrow {}^4\text{He} + 2P \qquad (3)$$

the next step is reaction (4), which is followed 99.9%

$${}^3\text{He} + {}^4\text{He} \rightarrow {}^7\text{Be} + \gamma \qquad (4)$$

of the time by an electron capture by ${}^7\text{Be}$, reaction (5), and 0.1% of the time by a proton interaction with

$$e^- + {}^7\text{Be} \rightarrow {}^7\text{Li} + \nu_e \qquad (5)$$

${}^7\text{Be}$, reaction (6). Since each fusion of four hydrogen

$$P + {}^7\text{Be} \rightarrow {}^8\text{B} + \gamma$$
$${}^8\text{B} \rightarrow {}^8\text{Be} + e^+ + \nu_e \qquad (6)$$

nuclei into one helium nucleus releases 4.2×10^{-12} J, there must be about 10^{38} fusion reactions per second. In each complete fusion reaction, two protons are converted into neutrons, and thus two neutrinos are released, leading to a flux of 6×10^{10} neutrinos per square centimeter per second at the Earth. It is the detection of these neutrinos that is the goal of solar neutrino observations.

There are three basic reactions occurring in the Sun—$P + P$, $e^- + {}^7\text{Be}$, and the decay of ${}^8\text{B}$—that produce neutrino secondaries. These reactions, their predicted neutrino fluxes, and the energy ranges of the emitted neutrinos are given in the **table**. The rate for reaction (1a) can be calculated approximately from the energetics discussion above. In order to calculate the rates for reactions (5) and (6), it is necessary to know the temperature profile in the core of the Sun and the cross sections for various nuclear reactions at the energies that these nuclei have in the core of the Sun.

Although the neutrinos carry away only about 2% of the energy released in the fusion process, their production, via the weak nuclear force, plays a crucial role in determining the rate of energy generation in the Sun. If the weak nuclear force were slightly stronger, permitting a higher fusion rate, the Sun would emit power at a much higher rate and its lifetime would be drastically shortened. The conditions for biological evolution on the Earth would be drastically altered. Similarly, if the weak nuclear force restricted nuclear fusion to a lower rate, the Earth would be a very cold place with permanently frozen oceans.

Neutrino detection reactions. Because neutrinos have no electric charge, they can be observed only by their interaction with particles in the material

Neutrino-producing reactions in the Sun

Reaction		Predicted neutrino flux, $\nu_e/(cm^2 \cdot s)$	Neutrino energy range
$P + P \rightarrow D + e^+ + \nu_e$	(1a)	6×10^{10}	0–0.44 MeV
$P + P + e^- \rightarrow D + \nu_e$	(1b)	4×10^8	1.44 MeV
$e^- + {}^7Be \rightarrow {}^7Li + \nu_e$	(5)	7×10^9	0.86 MeV, 90% 0.38 MeV, 10%
$P + {}^7Be \rightarrow {}^8B + \gamma$ ${}^8B \rightarrow {}^8Be + e^+ + \nu_e$	(6)	6×10^6	0–14 MeV

of a detector. There are three basic reactions available for neutrino detection: reaction (7), which involves

$$\nu_e + N \rightarrow P + e^- \qquad (7)$$

only the electron neutrino ν_e; reaction (8), which

$$\nu + e^- \rightarrow \nu + e^- \qquad (8)$$

involves all neutrino species, with the reaction rate for ν_e seven times higher than that for other neutrino species; and neutrino distintegration of nuclei, a neutral-current reaction that has the same rate for all neutrino species.

Solar neutrino detection. In 1946, B. Pontecorvo proposed using the reaction, $\nu_e + {}^{37}Cl \rightarrow {}^{37}Ar + e^-$, an interaction of type (7), in which one of the neutrons in chlorine-37 (${}^{37}Cl$) is converted into a proton, to detect solar neutrinos. In 1965, R. Davis, in a pioneering effort, began the construction of the first solar neutrino detector, using the chlorine in 672 tons (610 metric tons) of perchloroethylene (C_2Cl_4). Because neutrinos have a very weak interaction with matter, it is necessary to utilize very massive detectors. The solar-neutrino-induced argon-37 (${}^{37}Ar$) production rate in 672 tons (610 metric tons) of perchloroethylene predicted by the standard solar model is only about 1.5 atoms per day. The cosmic-ray flux at the surface of the Earth, however, would produce millions of argon-37 atoms per day in a detector of this size. In order to avoid this cosmic-ray-induced background, the detector was placed 4850 ft (1478 m) beneath the surface of the Earth in the Homestake Gold Mine in Lead, South Dakota (**Fig. 1**). Almost all of the cosmic rays are absorbed by the rock above the detector. The cosmic rays that penetrate to the depth of the detector produce less than 0.1 argon-37 atom per day.

Because the reaction that converts chlorine-37 to argon-37 requires a neutrino with energy greater than 0.81 MeV, it is sensitive only to reactions (1b), (5), and (6), but not reaction (1a). The observations of the past two decades with this detector give an average production rate of one argon-37 atom in 2 days, about 27% of the rate predicted by the standard solar model.

A second unusual and unexpected feature of the solar neutrino flux measurements with the Homestake detector is that the solar neutrino flux appears to vary with the solar activity cycle, the detected neutrino flux being lower during high solar activity and higher during solar quiet periods. The detected neutrino flux during the solar maximum is about half that during the solar minimum.

Since 1987, a second solar neutrino detector, the Kamiokande water Cerenkov detector in Kamioka, Japan, has been in operation (**Fig. 2**). This detector uses the electrons in a 660-ton (600-metric-ton) fiducial volume of water to look for neutrino-electron elastic scattering, reaction (8). This detector has an electron energy threshold of 7.5 MeV, and thus is sensitive only to solar neutrinos from process (6). The results from the Kamioka detector are that the observed signal is about 46% of that predicted by the standard solar model.

In 1989, a third solar neutrino detector, in the Baksan Neutrino Observatory in the northern Caucasus in the former Soviet Union, that employs a target of 66

Fig. 1. Solar neutrino detector in Homestake Gold Mine in Lead, South Dakota. The 98,000-gallon (370,000-liter) tank containing the perchloroethylene target material is in a cavity approximately 30 ft wide, 32 ft high, and 60 ft long (9 m by 10 m by 18 m) that is 4850 ft (1478 m) below the surface of the Earth.

tons (60 metric tons) of metallic gallium, became operational (**Fig. 3**). Because the neutrino energy required to convert gallium-71 (^{71}Ga) to germanium-71 (^{71}Ge), a detection reaction of type (7), is only 0.23 MeV, this detector is the first one to be sensitive to the prime fusion reaction in the Sun, the fusion of two protons into a deuteron, reaction (1a). Preliminary results from the Baksan gallium solar neutrino detector are that the very low neutrino flux detected is also less than half of that predicted.

Explanations of the results. The original purpose in looking for neutrinos emitted by the Sun was to experimentally verify that the Sun's energy is generated by nuclear fusion. This experimental verification was established by the Homestake detector in the mid-1970s and confirmed by the Kamiokande detector in the late 1980s. Two new problems now arise: why is the observed flux less than that predicted, and why does the emitted flux vary with time? There are two possible causes for the reduced neutrino flux. One is that the present rate of fusion reactions in the Sun is less than that predicted from the power emitted by the Sun's surface. There is no presently compelling argument to support this possibility.

The second possible explanation for the reduced observed neutrino flux is that the electron neutrinos produced in the solar core convert to another, undetectable form as they transit the Sun. One such mechanism, the MSW mechanism suggested by S. P. Mikhe'ev, A. Yu. Smirnov, and L. Wolfenstein, involves the conversion of electron neutrinos into other neutrino species, either muon or tau neutrinos. This process would require a new particle interaction, one that has not yet been observed on Earth. If this conversion indeed occurs, it will be the first example of a particle-physics interaction that requires an environment—in this case, very high density over path lengths comparable to or larger than the Earth's diameter—not available on Earth. The MSW mechanism would result in the missing neutrino flux leaving the Sun as nonelectronic neutrinos. These could be detected only by neutral-current reactions, that is, reaction (8) and neutrino disintegration of nuclei. The difference between the neutrino flux measured by the Kamiokande detector, 46% of the predicted rate, and the lower flux observed at Homestake might result from the fact that Kamiokande is sensitive to nonelectron neutrinos while the Homestake detector is not. Additional experiments are required to verify this possibility.

Another neutrino conversion process involves a magnetic interaction between an assumed magnetic moment of the neutrino and an extended large-magnitude magnetic field in the Sun. The effect of this interaction is to rotate the magnetic moment of the electron neutrino so that its magnetic moment, instead of pointing along the neutrino direction of motion, points opposite to the direction of motion. Another way of stating this effect is that it converts left-handed neutrinos into right-handed neutrinos. Right-handed neutrinos will not interact with matter in the conventional way and thus may pass through the detectors without being observed. Since the magnetic fields inside the Sun may

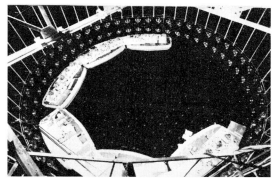

Fig. 2. Kamiokande water Cerenkov detector in Kamioka, Japan. Photomultiplier tubes that line the sides of the detector and several rafts that were used during the installation of the detector are visible.

vary during the solar cycle, such a mechanism might be responsible for the time-varying signal observed by the Homestake detector.

New solar neutrino detectors. Given the limited amount of data that is presently available, neither of the above suggested explanations has yet been critically tested. What is now needed is a database obtained by several complementary, high-detection-rate detectors operating for one or more solar cycles. All three of the presently operating detectors have a final detected signal rate of about one count per 3–4 days. New detectors should have neutrino observation rates of 5–10 counts per day. They should permit determination of the fluxes from each of the neutrino generating processes (1a), (1b), (5), and (6). They should also permit the detection of nonelectron neutrinos. Finally, this set of new detectors must remain in continuous operation for one or more decades so that the flux over one or more solar cycles can be observed.

Several new detectors that satisfy the above specifications are now under construction, and several others are being designed or contemplated. Among these are a 33-ton (30-metric-ton) gallium chloride detector

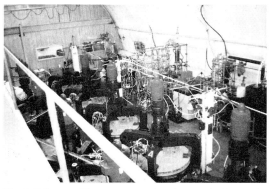

Fig. 3. Baksan gallium solar neutrino detector in the northern Caucasus in the former Soviet Union. The tops of the reactor vessels that hold the gallium are visible, as well as the stirring motors that mix the gallium, a liquid metal, with hydrogen peroxide and dilute hydrochloric acid during the procedure that extracts the germanium-71 (^{71}Ge) from the gallium.

(GALLEX) at an underground laboratory, Gran Sasso, in Italy; an enlarged version of the Kamiokande detector (Superkamiokande); a heavy-water (D_2O) Cerenkov detector at a nickel mine in Sudbury, Ontario, Canada; a 3300-ton (3000-metric-ton) perchloroethylene detector at Baksan in the former Soviet Union; an 1100-ton (1000-metric-ton) iodine detector at Homestake. and boron- and liquid-argon-based detectors at Gran Sasso. The GALLEX detector began taking data in 1991; the others will begin to operate in the mid-1990s.

In addition to the neutrino questions posed above, the new generation of solar neutrino detectors will permit exploration of a number of other possible oddities in the behavior of neutrinos. One unusual consequence of the MSW effect is that the detected solar neutrino rate at night, when the Sun is on the other side of the Earth and the neutrinos must traverse the Earth to reach the detector, may be higher than it is during the day, when the Sun is overhead. This effect results from the reconversion of nonelectron neutrinos into electron neutrinos as these neutrinos pass through the Earth. It has also been suggested that there may be short-term variations in the solar neutrino detection rate, possibly associated with gigantic solar flares or extended magnetic disturbances in the Sun. Again, the higher detection rates and the correlations possible with data from multiple simultaneous detectors in the second half of the 1990s will permit these questions to be addressed.

Thus, the measurement of neutrinos emitted by the Sun, which began a quarter century ago as an attempt to demonstrate that the Sun is powered by nuclear fusion reactions, has evolved into the use of the Sun as a laboratory in which a new type of weak interaction can be studied, and the use of the neutrino as a probe of density and magnetic field structure within the Sun.

For background information *SEE CERENKOV RADIATION; NEUTRAL CURRENTS; NEUTRINO; PROTON-PROTON CHAIN; SOLAR NEUTRINOS* in the McGraw-Hill Encyclopedia of Science & Technology.

Kenneth Lande

Bibliography. J. N. Bahcall, *Neutrino Astrophysics*, 1989; M. L. Cherry, K. Lande, and W. A. Fowler (eds.), *Solar Neutrinos and Neutrino Astronomy*, AIP Conf. Proc. 126, 1985.

Solid-state chemistry

Low-temperature synthesis of solid-state compounds, in the range 200–700°C (400–1300°F), has emerged as an attractive alternative to traditional preparative routes for new materials at higher temperatures. Techniques designed to circumvent the high reaction temperatures of traditional solid-state syntheses (sometimes known as chemie douce) must decrease the time needed for reactants to interdiffuse; this can be accomplished by enhancing the coefficient of interdiffusion, or by decreasing the distance over which reactants must diffuse. Synthesis of nonmolecular solid-state compounds typically takes place through direct reaction of the elements or binary oxides, with high reaction temperatures necessary to obtain effective solid-solid diffusion. These syntheses require elevated temperatures: about 450°C (850°F) for compounds containing the low-melting-point elements sulfur (S), selenium (Se), or phosphorus (P); near 1000°C (1800°F) for metal oxides; and almost 2500°C (4500°F) for metal carbides, borides, and silicides. These high temperatures are necessary because the kinetics of solid-solid interdiffusion are very slow, and only at elevated temperatures can reactions take place on time scales convenient for the synthetic scientist.

The principal interest in low-temperature syntheses of solid-state compounds is the opportunity to examine previously unexplored regions of simple phase diagrams. For example, materials that might be stable only at temperatures below 1000°C (1800°F) can be isolated if appropriate synthetic conditions are found for reaction at lower temperatures. Some of the materials isolated from low-temperature syntheses can be described as metastable, because decomposition to more thermodynamically stable products is kinetically very slow at these reduced temperatures. Examples of interesting chemical and physical features that can be kinetically stabilized by use of lower reaction temperatures include transition-metal oxides containing the metal in a high formal oxidation state, low-dimensional materials such as one-dimensional chainlike compounds or two-dimensional sheetlike solids, and materials containing extensive homoatomic bonding, either metal-metal or nonmetal-nonmetal.

Three different approaches can be used to effect the low-temperature synthesis of extended solids: (1) dissolving reactants in some solution medium or solvent (sometimes known as a flux) to provide the means for solution reactivity, (2) choosing suitable precursor materials that might be easily decomposed at low temperatures to form either new materials or finely divided solids that might undergo rapid solid-solid interdiffusion at low temperatures, and (3) appropriate deintercalation of intercalated compounds. The solution approach has long been utilized for the growth of large single crystals and in studies involving electrosynthesis. However, the use of this method for the synthesis of new phases has only recently been fully appreciated, and this research area is growing rapidly.

Solution techniques. Perhaps the simplest means to decrease temperatures necessary for the production of extended solids is to increase the rate of interdiffusion of reactants by dissolving them in a suitable solvent medium. This approach is the same procedure favored by most molecular chemists in their synthetic efforts. The utility of the solution techniques can be demonstrated by comparing typical diffusion coefficients (D) in solutions at room temperature, $D \sim 10^{-5}$–10^{-6} cm^2/s, versus diffusion coefficients for solid-solid interdiffusion of metal oxides at 1500°C (2700°F), $D \sim 10^{-8}$–10^{-10} cm^2/s. However, nonmolecular solids themselves often need elevated temperatures to form and crystallize, thereby requiring that the solvent of choice be capable of withstanding the temperatures necessary for product formation.

Although this requirement eliminates most common organic solvents, it does include many inorganic compounds that have sufficiently low melting points, even though they may be solids at room temperature. Such inorganic solvents include highly ionic compounds (molten salts) such as potassium chloride (KCl; melting point $770°C$ or $1400°F$) and sodium hydroxide (NaOH; melting point $360°C$ or $680°F$), more covalent species such as sodium borate ($Na_2B_4O_7$; melting point $741°C$ or $1366°F$) or mixtures of sodium tungstate and tungsten trioxide (Na_2WO_4/WO_3; melting point $\sim 550°C$ or $1022°F$), and even low-melting-point elements such as tin (melting point $232°C$ or $450°F$).

The increased diffusivity and lower reaction temperatures possible when preparing extended solids by solution techniques provide several important advantages over more conventional solid-state synthetic routes. Solution synthesis allows for selectivity in precipitation of the product from solution, and thus solids can be prepared at temperatures much lower than would be possible by direct reactions in the solid state. Impurity phases can be eliminated through proper control of solution conditions. In addition, crystallinity, particle size, morphology, and microstructure can be changed by adjusting the reaction conditions. Most importantly, with low-temperature syntheses it is possible to prepare entirely new materials that are not stable at higher temperatures.

Several new compounds have been prepared that demonstrate the versatility of using low-temperature syntheses in molten salt media. The low temperatures and high oxygen activity possible in molten hydroxides offer an attractive route for the synthesis of metal oxides. Furthermore, the water solubility of alkali hydroxides facilitates their removal from the desired product, as they can be readily washed away with cold water and the product isolated by filtration. Reaction of lanthanum oxide (La_2O_3) and copper(II) oxide (CuO) dissolved in the molten sodium hydroxide–potassium hydroxide (NaOH-KOH) eutectic at $300°C$ ($572°F$) in air yields material that is structurally identical to lathanum copper(II) oxide (La_2CuO_4) typically prepared at $1000°C$ ($1800°F$) by solid-state reaction. The material prepared under the low-temperature conditions is, however, different from the high-temperature material because it displays superconducting behavior below 35 K ($-396°F$), while the product of the solid-state reaction is not superconducting. All attempts to prepare potassium-doped barium bismuthate ($Ba_{1-x}K_xBiO_3$) by solid-state techniques have been unsuccessful. However, when barium carbonate ($BaCO_3$) and bismuth oxide (Bi_2O_3) are reacted in KOH at $400°C$ ($750°F$), a new material of composition $Ba_{1-x}K_xBiO_3$ is found. When solution conditions of the KOH are properly adjusted, $Ba_{0.60}K_{0.40}BiO_3$ is formed; when isolated, this solid is observed to superconduct with a critical temperature (T_c) of 30 K ($-342°F$). Similarities between reaction conditions in molten hydroxides and high-pressure dioxygen (O_2) solid-state syntheses have been utilized to prepare materials such as strontium nickel oxide ($Sr_5Ni_4O_{11}$) and

barium nickel oxide ($BaNiO_3$), which contain nickel in formal oxidation states of III and IV, oxidation states previously achieved through high-pressure techniques only.

Like the hydroxide solutions, polysulfide solutions, for example, potassium sulfide and sulfur mixtures (K_2S/S) at $375°C$ ($700°F$), have been useful as media for the preparation of new metal-sulfide/polysulfide systems such as $K_4Ti_3S_{14}$, containing infinite polyanionic chains of $[Ti_3(S_2)_6(S)_2]_n^{4n-}$ (Ti = titanium). Reactivity in polyselenide melts has yielded a variety of new compounds, including a novel two-dimensional-type structure containing sodium (Na), gold (Au), and selenium (Se): ($NaAuSe_2$). The low-melting-point metal solvents, for example, tin, have been particularly useful in the preparation of metal phosphides. This is due to the relative instability of tin-metal and tin-phosphorus phases with respect to the binary metal phosphides, for example, molybdenum phosphide (MoP_4) and the ternary metal phosphides, for example, europium nickel phosphide ($EuNi_5P_3$). The tin solvent is most commonly separated from the desired product by dissolution of the excess tin with dilute hydrochloric acid.

Decomposition techniques. The decomposition, or precursor, route to low-temperature synthesis of solids closely resembles the traditional high-temperature reactivity in that product is formed by direct combination of, for example, the component oxides. However, high reaction temperatures are avoided by the preparation of very finely divided and well-intermixed reactants. This effect leads to smaller particle sizes and consequently smaller diffusion distances, where equilibrium might be achieved rapidly even at low reaction temperatures.

The synthetic technique is typically accomplished according to some variation of the following procedure, given for metal oxides: (1) preparation of aqueous solutions of the component metal ions in the presence of some complexing agent (nitrate, citrate, and oxalate ions are most commonly used); (2) mixing of the individual solutions, taking care to adjust solution conditions to prevent precipitation of any single component; (3) gentle evaporation of the water to form a precipitate composed of an intimate mixture of the complexed metal ions, for example, a mixture of neodymium nitrate and nickel nitrate [$Nd(NO_3)_3/Ni(NO_3)_2$]; (4) thermal decomposition of this mixture under mild ($\sim 400-600°C$ or $750-1100°F$) conditions to produce finely divided metal oxides; and (5) continued heating under low-temperature conditions to yield the desired product. This approach has proven a suitable low-temperature technique for the preparation of compounds containing metals in unusually high oxidation states, for example, copper(III) ion (Cu^{3+}) in $YBa_2Cu_3O_{7-x}$ (Y = yttrium; Ba = barium) and nickel(III) ion (Ni^{3+}) in $NdNiO_3$, two materials that in the high-temperature, solid-state approach require preparation under $1-1000$ atm (10^5-10^8 pascals) O_2. Although the decomposition approach is well suited for preparation of ceramic samples, its principal limitation is that it is not a suitable means for growth of crystals large enough for single-crystal structural and physical studies.

Intercalation and deintercalation. Intercalation chemistry, important in solid-state battery technology, involves the insertion of one species, for example, alkali metal cations, between the layers of an otherwise two-dimensional material, for example, titanium sulfide (TiS_2). Deintercalation, under rather mild conditions, has been used as a means for the isolation of several new materials and also of other known materials that have proven difficult to prepare using standard high-temperature techniques. For instance, oxidative deintercalation of the thiospinel copper titanium sulfide ($CuTi_2S_4$) with bromine (Br_2) in acetonitrile at room temperature has yielded a new cubic modification of TiS_2 in a defect spinel structure, which converts to the layered structure above 400°C (750°F). Similar procedures have been used to isolate a new form of titanium dioxide (TiO_2) different from the previously known anatase, rutile, and brookite modifications. Low-temperature deintercalation of lithium niobium oxide ($LiNbO_2$) with nitrosonium tetrafluoroborate ($NOBF_4$) in organic solvents has been used to produce a novel superconducting (T_c = 5.0 K or −450°F) ion modification with composition $Li_{0.45}NbO_2$.

For background information SEE SOLID-STATE CHEMISTRY; SUPERCONDUCTIVITY in the McGraw-Hill Encyclopedia of Science & Technology.

Gary F. Holland

Bibliography. A. K. Cheetham and P. Day (eds.), *Solid State Techniques*, 1987; A. R. West, *Solid State Chemistry and Its Applications*, 1987.

Space flight

The year 1991 was dominated by news of the collapse of the Soviet Union, an event certain to have both immediate and long-term effects on its space program. Confusion, even chaos, in Russia and the other republics was reflected in markedly reduced activity in space; for example, Soviet launches were down 20% from the figure for 1990. Perhaps the most extraordinary development was seen in efforts by the Soviets to open up hitherto-secret space-related production, research, and other facilities for commercial purposes.

The United States space program was marked by advances in several areas of space science. In April 1991, the Compton Gamma-Ray Observatory became the second great observatory sponsored by the National Aeronautics and Space Administration (NASA), as it joined the Hubble Space Telescope in a major effort to explore the cosmos. In September, deployment of the *Upper Atmosphere Research Satellite (UARS)* from the space shuttle initiated NASA's Mission to Planet Earth, a two-decade-long coordinated research program to study the Earth as a complete environmental system. A Total Ozone Mapping Spectrometer launched aboard a Soviet *Meteor* satellite in August ensures that ozone data will continue to be available for several years.

By the end of 1991, the *Magellan* spacecraft had completed radar mapping of virtually the entire surface of Venus, and the *Ulysses* spacecraft had set its trajectory for Jupiter on its way to study the poles of the Sun.

On its way to Jupiter, the *Galileo* planetary probe passed by the asteroid Gaspra and returned the first closeup image ever taken of an asteroid. The future success of the *Galileo* mission was cast in doubt, however, when the spacecraft's main antenna failed to deploy upon command in April. Four attempts, including one on December 17, led scientists to consider the possibility that only a fraction of the data and images expected from *Galileo* may ever be received.

Preliminary design of Space Station *Freedom*'s astronaut-tended configuration was completed in 1991 following a congressionally mandated restructuring of the *Freedom* program.

Significant space launches in 1991 are listed in **Table 1**. The total numbers of payloads launched by various countries are given in **Table 2**.

United States Space Activity

Seven NASA space shuttle flights were planned for 1991, but because of investigation of a faulty external tank door mechanism, one flight (STS 42) was rescheduled for January 1992. The six flights accomplished on schedule demonstrated the unique qualities and versatility of the space shuttle.

NASA's space shuttle program augmented its fleet of reusable space planes with the delivery of *Endeavor* on April 25 in a ceremony at the Rockwell facility in Palmdale, California. This newest NASA orbiter displays significant safety improvements over earlier models, including redundant nose-wheel steering and a drag-chute system. *Endeavor*'s first flight was launched on May 9, 1992, and returned on May 16.

The first shuttle mission flown in 1991 (April 5, STS 37/Gamma-Ray Observatory) involved two extra-vehicular-activity missions, one planned, the other unplanned. The planned extravehicular activity, the first in 5 years for NASA, demonstrated aids to mobility for accomplishing work in space that have been especially designed for use on Space Station *Freedom* (**Fig. 1**). The unplanned extravehicular activity was necessary to achieve full deployment of the Gamma-Ray Observatory's high-gain antenna.

The shuttle *Discovery* successfully completed, for the Department of Defense, one of the most complicated missions ever flown (April 28, STS 39/Air Force Payload). Dozens of maneuvers were performed, deploying canisters from the cargo bay, and permitting the Strategic Defense Initiative Organization to gain important data and observations on rocket plumes.

In the first mission since *Skylab* to do intensive research into the effects of weightlessness on humans (June 5, STS 40/Spacelab Life Sciences), essential data were obtained for use by NASA in planning for long-duration shuttle missions and for Space Station *Freedom*.

A *Tracking and Data Relay Satellite (TDRS)* was launched on *Atlantis* (STS 43/TDRS-E), which set a record as the heaviest mission flown so far, with a

Table 1. Some significant space launches in 1991

Payload or vehicle	Date	Country	Purpose or outcome
Cosmos 2133	Feb. 14	Soviet Union	Geosynchronous communications satellite to talk to *Mir* crews
Almaz (Diamond)	Mar. 31	Soviet Union	Largest remote-sensing spacecraft
Cosmos 2139–2141	Apr. 4	Soviet Union	Three GLONASS navigation spacecraft
STS 37/Gamma-Ray Observatory	Apr. 5	United States	Launch Gamma-Ray Observatory; test personal mobility aids
STS 39/Air Force Payload	Apr. 28	United States	Gather data for Strategic Defense Initiative
Soyuz TM-12	May 18	Soviet Union	Three-person crew for science aboard *Mir*
STS 40/Spacelab Life Sciences	June 5	United States	Study effects of weightlessness on humans
STS 43/*Tracking and Data Relay Satellite (TDRS)*	Aug. 2	United States	Deploy *TDRS*; heaviest mission flown so far (253,000 lb or 115,000 kg at liftoff)
STS 48/*Upper Atmosphere Research Satellite (UARS)*	Sept. 12	United States	Mission to Plane: Earth initiated with *UARS*; perform atmosphere research
Soyuz TM-13	Oct. 2	Soviet Union	Three-person crew for science aboard *Mir*
STS 44/Defense Support Program	Nov. 24	United States	Gather data for Department of Defense; continue NASA research on effects of weightlessness on humans

liftoff weight of 253,000 lb (115,000 kg). Deployment of the *TDRS* was necessary to maintain full operational capability for the network supporting shuttle missions.

A research satellite was placed in orbit to study the upper atmosphere (September 12, STS 48/*Upper Atmosphere Research Satellite*). The final shuttle launch of 1991 (STS 44/Defense Support Program), in addition to its classified defense mission, yielded research data on the effects of weightlessness on humans in preparation for the Extended-Duration Orbiter missions planned for 1992. STS 44, originally planned for a 10-day mission, was terminated safely on the seventh day, following the failure of a redundant inertial measurement unit aboard the spacecraft.

Mission to Planet Earth. Deployment of the *Upper Atmosphere Research Satellite* (**Fig. 2**) dramatically expanded NASA's research in ozone depletion. Data from this satellite are used to create three-dimensional maps of ozone and chemicals important in ozone depletion. Instruments aboard the spacecraft also provide scientists with comprehensive data on upper-atmosphere winds and on energy inputs from the Sun into Earth's upper atmosphere.

Preliminary data from the *Upper Atmosphere Satellite* has clearly illustrated the link between low levels of ozone and high levels of chlorine monoxide, a key intermediate compound in the chemical chain reaction that leads to ozone depletion. Other ongoing studies sponsored by NASA maintained the agency's lead in international efforts to understand ozone depletion. Data from the Total Ozone Mapping Spectrometer (TOMS) aboard the *Nimbus 7* satellite indicated that the ozone depletion problem continues to be serious.

Astrophysics. The Compton Gamma-Ray Observatory (**Fig. 3**) discovered bursts of gamma radiation coming from outside the narrow plane of stars that make up the Milky Way Galaxy. The implication is that there are unknown sources of gamma rays either relatively near the solar system or from objects well outside the Milky Way Galaxy. In July 1991, the observatory also detected the most distant and most luminous source of gamma rays ever observed, Quasar 3C279, approximately 7×10^9 light-years from Earth.

Table 2. Successful launchings in 1991*

Country or organization	Number of launches
Soviet Union	59
United States	18[†]
European Space Agency	8
Japan	2
People's Republic of China	1
Total	88

* Achieved Earth orbit or beyond.
[†] Includes 12 NASA launches and 6 commercial launches.

Fig. 1. Astronaut J. L. Ross attaches a tether to a guide wire in a lengthy extravehicular activity in the cargo bay of the orbiter *Atlantis* on April 8, 1991. He and astronaut J. Apt evaluated translation devices that could be predecessors of devices to be used on Space Station *Freedom*. (*NASA*)

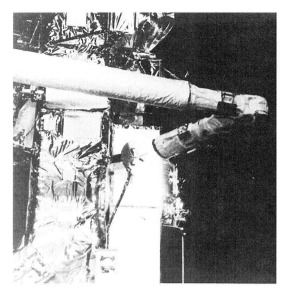

Fig. 2. Deployment of the *Upper Atmosphere Research Satellite* on September 15, 1991. The satellite is separating from the Remote Manipulator System and beginning to move away from the payload bay of the orbiter *Discovery*. (*NASA*)

The quasar emits about 10^7 times the energy of the entire Milky Way Galaxy.

In other missions, the NASA Soft X-Ray Telescope was launched in August aboard the Japanese *Solar-A* satellite. Data from the *Cosmic Background Explorer (COBE)* were used to create galactic-scale maps of the distributions of nitrogen, carbon, and interstellar dust, enabling astronomers to better understand the heating and cooling processes that take place in the Milky Way Galaxy.

Solar system exploration. The *Magellan* mission to Venus completed its primary objective of mapping 70% of the planet's surface ahead of schedule. By the end of 1991, *Magellan* had mapped 93.5% of the planet. The geological data returned by *Magellan* included evidence of volcanic activity, some of which appears to be quite recent, and images of the longest channel in the solar system, a 4200-mi-long (6800-km) chasm across the planet's surface. *SEE VENUS.*

NASA's other planetary probes continued on their trajectories. *Ulysses*, on its way to study the Sun's poles in a joint mission with the European Space Agency, set its trajectory for Jupiter, where it will investigate the planet's magnetic field and interaction with the solar wind. In August, when *Ulysses* passed behind the Sun (relative to the Earth), scientists used radio signals from the spacecraft to investigate the outer atmosphere of the Sun.

But the excitement surrounding the *Galileo* mission, encouraged by receipt of the first closeup image of an asteroid early in the year, turned to disappointment and then concern when the spacecraft's main antenna failed to deploy in April. NASA engineers believe that 3 of the antenna's 18 ribs are stuck as a result of unexpected friction on the 3-in. (8-cm) titanium pins that keep the ribs properly seated against the antenna's main shaft. Following a fourth attempt to free the antenna, on December 17, ground controllers began to face the possibility that they may have to rely on the spacecraft's alternate, smaller, lower-data-rate antenna when *Galileo* arrives in the vicinity of Jupiter in 1995. If the main antenna cannot be deployed, the full value of the *Galileo* mission will not be realized.

Space Station Freedom. Preliminary design of *Freedom*'s human-tended configuration was completed in 1991, and construction and testing of hardware proceeded on schedule toward a launch of the station's first elements in November 1995.

A budget shortfall in fiscal year 1991 and a congressional directive to significantly reduce out-year spending prompted NASA to begin restructuring the *Freedom* program in late 1990. The restructuring was completed in the spring of 1991. *Freedom*'s new design is less expensive, smaller, and easier to assemble in orbit, and it requires fewer shuttle flights to build. Major new features of the redesigned station include shorter modules that can be launched fully outfitted, and a preintegrated truss structure that is assembled and verified on the ground, so that the need for in-orbit extravehicular activity is significantly reduced. *SEE SPACE STATION.*

Commercial space activities. United States commercial launch firms made six successful launches during 1991, all devoted to placing communications satellites in orbit.

General Dynamics Space Systems suffered a setback in April with the launch failure of an Atlas

Fig. 3. Deployment of the Compton Gamma-Ray Observatory on April 7, 1991. The satellite is at the end of the Remote Manipulator System of the orbiter *Atlantis*. (*NASA*)

Centaur carrying a Japanese *BS-3H* communications relay satellite. The Centaur stage and payload were destroyed 245 mi (394 km) downrange at a height of 109 mi (175 km). Investigation indicated that the probable cause of the mishap was contamination of turbomachinery of one of the two Centaur engines by solid particulates and ice.

This loss was offset by General Dynamics' return to space on December 7, with the successful launch of the firm's first Atlas 2, which placed a 4138-lb (1877-kg) *Eutelsat 2 F3* telecommunications satellite in orbit (in time to telecast the February 1992 Winter Olympic Games from France). This success was regarded as critical to General Dynamics' effort to become a major participant in the commercial launch industry. The *Eutelsat* spacecraft, built by a European consortium, is expected to have an operational lifetime of 7 years.

Soviet Space Activity

Perestroika, the ongoing effort to restructure Soviet society and the Soviet economy, resulted in some notable developments in 1991. The huge Yuzhny Machine Building Factory located at Dnepropetrovsk, which produces the space program's Energia booster as well as intercontinental ballistic missiles and spy satellites for the Soviet military, began conversion of military assembly lines to production of trolley buses and aircraft parts. Plant managers began openly to solicit investors from the United States and other countries to participate in joint ventures to produce a wide variety of civilian machine products.

Commercial space services. Although United States policy makers remain hesitant toward easing restrictions that prevent Soviet rockets from launching United States or western European satellites, the Soviets lobbied to overcome such restrictions and even suggested that Energia, the world's most powerful rocket, be used in building Space Station *Freedom*.

Even more remarkably, the Soviet Ministry of Defense offered its hitherto-top-secret Military Satellite Control Center for lease to any group, including foreign entities, for use in controlling scientific or commercial space missions. The offer for use of the Moscow-based facility was designed to generate much-needed hard currency for the Soviet government.

United States technicians were permitted to visit the control center for the first time on August 15, when the center was used for controlling the launch of a NASA ozone mapper aboard a Soviet *Meteor 3* spacecraft. The Defense Ministry published an advertising prospectus in English, French, and Russian, outlining the capabilities of the facility and the terms for its lease.

Space program reorganization. Turmoil in the Soviet Union accelerated during 1991 with an abortive military coup d'état in August, and culminated in the dissolution of the union in December, with the formation of a Commonwealth of Independent States, including the Russian republic and 11 other now-independent states. These events raised profound questions concerning the future of the Soviet space program. The Commonwealth of Independent States did agree, however, to maintain the space program

on a joint or collaborative basis, whose exact terms remain ill-defined.

In November, five of the Soviet Union's major aerospace institutes announced their intention to form an independent research agency, modeled along the lines of NASA. In early 1992, an agreement between Russia and Kazakhstan gave Russia possession and operational responsibility for the Baikonur launch complex, located in Kazakhstan. Since the other two launch sites are in Russia, this agreement in effect placed the space program under the control of the Russian republic.

The five institutes that announced their intention to join forces are the Central Aerohydrodynamics Research Institute, the Central Institute of Aviation Motors, the All-Union Institute of Aviation Materials, the Central Aviation Systems Institute, and the Flight Research Institute. The reorganization is an effort to counter the threat to continued research funding, resulting from the collapse of the central government. Although aircraft and rocket design bureaus can look forward to privatization and commercial work in order to maintain their viability, aerospace research must continue to rely on government funding.

Mir space station. A Progress M7 resupply vehicle nearly collided with the *Mir* space station on March 23, 1991, during an automatic docking attempt. The incident was described as a near catastrophe, which might have resulted in the deaths of the two cosmonauts aboard *Mir*.

Investigation indicated that the problem was most likely caused by one of the cosmonauts who had been working outside the space station during an earlier extravehicular activity and may have accidentally misaligned a critical *Mir* docking-system antenna, so that the Progress missed its proper approach. The Progress, which was launched March 19, was approaching the *Kvant* module's aft docking port when a ground controller realized the difficulty and transmitted commands causing the craft to veer away, missing *Mir* by only about 40 ft (12 m).

On March 26, the *Mir* cosmonauts undocked their *Soyuz TM-11* spacecraft from *Mir*'s forward multiple docking adapter and, using manual controls, succeeded in docking it at the faulty aft dock location. On March 28, the Progress, which had remained in orbit near *Mir*, made a successful automatic docking maneuver at the forward port, which had been vacated by the *Soyuz*.

The Soviets conducted scientific research aboard *Mir* during 1991, notably in the area of space biology, and they met with success in efforts to interest foreign companies in utilizing *Mir* for research in space on a commercial basis. On December 5, the Japanese electronics firm, Fujitsu, Ltd. of Tokyo, announced that it had contracted for a series of biological and materials experiments to be conducted aboard *Mir* in 1992.

As part of a reevaluation of the Soviet space program, necessitated by collapse of the central government and uncertain out-year budgets, design work was halted on a new *Mir* space station, *Mir 2*, which would have featured a truss structure similar to NASA's Space Station *Freedom*. As an alternative, the Soviets have begun fabricating a modernized, updated model of the core vehicle that is part of the existing *Mir*. The vehicle, colloquially referred to as *Mir 1.5*, is less costly than an entirely new design, and is expected to be launched as a replacement for the existing space station in 1994 or 1995.

Buran. A second *Buran* shuttle orbiter was completed in 1991 at the Baikonur Cosmodrome. Although this vehicle could be flown on a crewless mission to the *Mir* space station some time in 1992, it is believed to be as much as 2–4 years away from being fully prepared for crewed flight. As a result of budget stringencies and other uncertainties clouding the Soviet space program, the overall future of the *Buran* shuttle program is uncertain, and high-level decisions on the vehicle's future role and mission have yet to be made.

The Soviet flight-safety-equipment manufacturer Zvezda validated its K-36RB ejection seat system for speeds up to Mach 4, so that the system qualified for use aboard *Buran*. The ejection seat and associated equipment, including the cosmonaut's flight suit, were tested as piggyback payloads on *Soyuz* boosters that were used to launch Progress resupply craft to the *Mir* space station. A series of tests with the K-36RB released at high altitudes and at speeds up to Mach 4.1, led to refinements in the safety system. The system was adapted from standard ejection equipment used on Soviet high-performance combat aircraft.

European Space Activity

The European Space Agency's *ERS 1* satellite was lofted into orbit on July 16, 1991, aboard an Ariane 4 booster. This event signaled a major expansion in Europe's earth sciences program, in cooperation with NASA's Mission to Planet Earth. The spacecraft features six instrument packages that effectively establish the European Space Agency's primacy in investigations of ocean-area climatology and meteorology. After some initial difficulty, ground controllers managed to fully deploy the spacecraft's wind scatterometer, which measures sea-surface wind speeds and direction.

Construction of a second research satellite (*ERS 2*) is proceeding on schedule for launch in 1994; this second spacecraft, also to be launched under the aegis of Mission to Planet Earth, will house a Global Ozone Monitoring Experiment. The ERS program has required a major expansion of the European Space Agency's ground data-handling facilities, including an office for data users at the European Space Research Institute at Frascati, Italy. Data from *ERS 1* are available to users within 3 h of transmission from the satellite.

In a related effort, a United States–French satellite, the *Topex/Poseidon*, remained on schedule for launch in July 1992. The spacecraft is designed to help scientists compile a detailed picture of ocean currents, eddies, and other circulation features on a worldwide basis every 10 days.

Arianespace began its initial marketing efforts for the Ariane 5 heavy-lift launcher with publication in October of the first edition of its Ariane 5 users' manual. The manual includes technical operational documentation for satellites to be carried aboard the vehicle. Designers of Ariane 5 expect to reduce launch costs from the current Ariane 4 series by 30–40% so that a competitive position with respect to future United States and Japanese boosters will be maintained. Ariane 5's primary mission will be to carry single or multiple satellite payloads into geostationary transfer orbit; other planned missions include placing the European Space Agency's *Hermes* crewed spaceplane in low Earth orbit.

In November 1991, a casting problem with the lower segment of an Ariane 5 solid-propellant booster resulted in what is expected to be a 6-month delay in start-up of full-scale static firings of Ariane 5 boosters, previously targeted for March 1992. The problem occurred as a batch of propellant that was poured into the booster segment failed to harden properly. The entire segment was rendered useless, and another had to be cast. The Ariane 5's solid booster is made up of three segments and features a total of 518,000 lb (235 metric tons) of composite propellant. The lower segment, which developed the problem, is designed to contain 229,000 lb (104 metric tons) of solid propellant.

During 1991, Arianespace booked two Mexican telecommunications satellites, *Solidaridad 1* and *2*, for launch in 1993–1994. This scheduling brought to 34 the total of pending orders for commercial launch services now claimed by Arianespace.

Asian Space Activities

Japan and China continued their space activity in 1991.

Japan. A fatal accident on August 9 resulted in delay of Japan's development of its large H-2 booster. A pressure-generated explosion during engine component testing, which killed a Mitsubishi Heavy Industries worker, was attributed to incorrect closure of a steel test-chamber door. The accident could push back deployment of Japan's space-based *GMS-5* weather satellite, now scheduled to be launched into orbit in 1994 with the first production-model H-2.

Japan formally entered the commercial launch business in late 1991 with a bid by Rocket Systems Corporation of Japan to launch *Inmarsat 3* satellites aboard the H-2 booster, when it becomes operational. Rocket systems Corporation is a consortium of 75 Japanese companies, including the H-2's prime contractor, Mitsubishi Heavy Industries. The firm could launch its first commercial payload as early as 1995, assuming that planned H-2 test flights in 1993–1994 are not delayed and that they prove successful. The two-stage H-2 rocket is designed to carry 8800 lb (4.0 metric tons) to geostationary transfer orbit, or 9300 lb (4.2 metric tons) in an augmented-power version.

The *Superbird-A* spacecraft, which had only recently begun to provide Japan's Space Communications Corporation with relay satellite capability, failed in geosynchronous orbit on December 23 when propellant in its attitude-control system developed a leak. The *Superbird-A* began to lose attitude control on December 18 when a valve malfunctioned or was accidentally left open, so that oxidizer was lost. The mishap had a major effect on television broadcasts in Japan.

Japan continued development of a hypersonic single-stage-to-orbit horizontal-takeoff-and-landing spaceplane. A new ramjet facility, which neared completion in 1991 at the Kakuda Research Center, will be used to test a scramjet engine under development for the spaceplane.

In addition to the scramjet, work in Japanese propulsion technology is focusing on liquefied-air-cycle-engine (LACE) power plants. These units eliminate the need for carrying on-board oxidizers because they liquefy air at low altitudes and concentrate oxygen to be used later in rocket flight. LACE is also intended for use in an H-2-based vehicle that will launch Japan's *HOPE* shuttle orbiter.

China. The People's Republic of China lost a major selling point for its commercial launch services, a heretofore-perfect launch record, with the December 28 failure of a Long March 3 rocket to place a Chinese communications satellite in its intended orbit. The accident was attributed to malfunction of the rocket's third stage. The *DFH-2* telecommunications satellite cannot perform its function, and is effectively lost, because of its position in a geosynchronous transfer orbit.

The mishap occurred as the United States and European governments were applying diplomatic pressure on China to raise the price of its launch services to reflect their true costs. Its low price schedules have made China a strong contender for the commercial launch market. The accident reportedly will not affect agreements for China to launch three satellites in 1992, including a research satellite for the Swedish Space Corporation and two communications satellites for Aussat, Australia's communications satellite company.

For background information *SEE APPLICATIONS SATELLITES; COMMUNICATIONS SATELLITE; GAMMA-RAY ASTRONOMY; SATELLITE (ASTRONOMY); SPACE BIOLOGY; SPACE FLIGHT; SPACE PROBE; SPACE SHUTTLE; SPACE STATION* in the McGraw-Hill Encyclopedia of Science & Technology.

Robert J. Griffin, Jr.

Bibliography. Five space agencies emerge from Soviet chaos, *Space News*, vol. 2, no. 44, December 16–22, 1991; $1.4 billion *Galileo* mission appears crippled, *Washington Post*, p. A3, December 18, 1991; Soviet ejection seat for *Buran* Shuttle qualified for deployment at up to Mach 4, *Aviat. Week Space Technol.*, 134(23):44–48, June 10, 1991.

Space probe

President Bush's call for a crewed mission to Mars has given new importance to the September 1992 scheduled launch of the *Mars Observer*, the first of a series of relatively low-cost spacecraft to explore the solar system. Further exploration of the Martian surface, whether by robots or by astronauts, will rely on information returned by the *Mars Observer*.

Past missions to Mars have left major gaps in knowledge of the planet. The *Viking* and *Mariner* missions looked at only a part of the Martian surface at a resolution of about 250 m (820 ft). As a comparison, the *Mars Observer* camera will have a resolution of 3 m (10 ft). The earlier missions did not detect a magnetic field, and study of such a field (if it exists) is important to understanding the interior composition and thermal qualities of Mars. More detailed information concerning the surface features, elements, minerals, and atmosphere of Mars is needed.

Mission scenario. After a journey of 333 days, the spacecraft will enter an elliptical orbit around Mars (**Fig. 1**). The orbit will be carefully adjusted through a few intermediate steps over a duration of several weeks until the spacecraft is in its nearly circular, nearly polar mapping orbit. In this orbit the spacecraft will cross the Martian equator almost at right angles, and pass over each pole about every 2 h. The mapping orbit is Sun-synchronous, so that the sunlight will always be at approximately the same angle at the planet surface below the spacecraft. In other words, the spacecraft will always see the surface of Mars as it appears at 2:00 P.M., Mars time, when the spacecraft crosses the equator on the sunlit side. This position of the Sun represents the best compromise between the mapping requirements of the science instruments and the projection of shadows to reveal ground relief. The amount of reflected light (albedo) can also be detected.

An exact mapping orbit is crucial for the *Mars Observer* scientific objectives. Therefore the spacecraft will circle Mars at a nearly constant altitude of approximately 389 km (242 mi), with an orbital period of 117 min. It will be traveling southward on the daylight side of the planet and northward on the night side. At this altitude the spacecraft will be orbiting at the edge of the Martian atmosphere. This orbit will afford maximum uniformity of instrument coverage across the planet surface.

Science and innovation. The seven scientific experiments aboard the *Mars Observer* (see **table**), with a total science payload mass of 166 kg (366 lb), are expected to observe the surface features and the gravitational field, to determine the abundance of elements and composition of minerals on the surface, to determine the nature of a magnetic field (if it exists), to explore the makeup and circulation of the atmosphere, and to obtain a picture of the climate over the four seasons of one Martian year (687 Earth days). The Mars balloon relay, while not a scientific experiment, is included as an instrument in the table.

In an innovative approach to remote data gathering, scientists will receive information from the spacecraft at their home institutions, rather than having to be at the Mission Control Center. Principal investigators will control their space-borne experiments through a computer network linking them to the *Mars Observer* operations center at the Jet Propulsion Laboratory in Pasadena, California, where the Deep Space Network of the National Aeronautics and Space Administration (NASA) has its headquarters. In the same way, data returned from each experiment will be sent electronically to investigators at their home institutions. This cost-effective method of controlling individual instruments

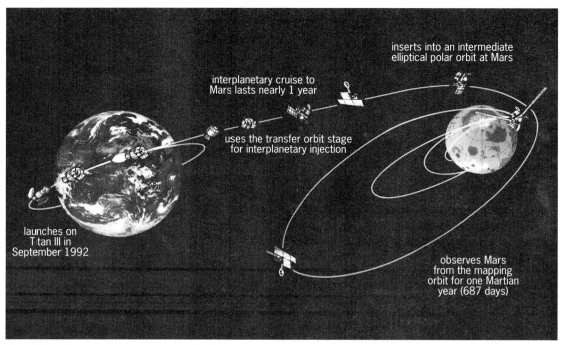

Fig. 1. *Mars Observer* mission scenario. (*Jet Propulsion Laboratory*)

will allow scientists rapid access to data and permit quick changes in science observations.

Another important aspect of this mission involves international cooperation. Portions of the scientific instruments on the spacecraft are being provided by European suppliers. Oxford University, England, is providing the cells and some detectors for the pressure modulator infrared radiometer instrument, and the electron reflectometer is being provided by the Center for Cosmic Ray Studies in Toulouse, France.

The French space agency will provide equipment to relay scientific information from planned instrument packages that will descend to Mars from Russian spacecraft to be launched in 1994 and 1996. The 1994 Russian mission is to include surface penetrators and a meteorological station on Mars; the 1996 Russian mission includes a near-surface balloon-borne instrument package, as well as a possible surface rover. Data from all of these experiments will be stored in the large solid-state memory of the *Mars Observer* camera, where they will be encoded and processed for return to Earth.

Spacecraft. The *Mars Observer* spacecraft is the first of a series of planetary explorers to be launched by NASA. It is being constructed for the Jet Propulsion Laboratory and NASA by the Astro Space Division of the General Electric Company. The contract, which includes spare spacecraft components and mission support operations, provides for profits to be based on the amount of scientific data returned over the useful life of the mission. The *Mars Observer* (**Fig. 2**) consists of two basic parts: the spacecraft itself, called the bus, and the scientific instruments, called the payload.

The bus is box-shaped and approximately 2.9 m high by 3.2 m wide by 2.9 m deep (9.5 ft by 10.5 ft by 9.5 ft). The spacecraft's antenna, instrument booms, and solar arrays will be folded close to the bus and will unfurl from the spacecraft bus during the 11-month

Instruments and objectives of *Mars Observer* mission

Instrument	Objectives
Gamma-ray spectrometer (boom-mounted)	Surface elemental and mineralogical composition determination
Laser altimeter	Topographical study
Line-scan camera	Study of climate, surface geology, and surface and atmospheric interactions
Magnetometer and electron reflectometer (boom-mounted)	Global study of intrinsic magnetic field
Pressure modulator infrared radiometer	Atmospheric temperature, water vapor, dust, and pressure profiles
Thermal-emission spectrometer	Mapping of mineral content of rocks, frosts, and clouds
Radio science (ultrastable oscillator)	Study of gravitational field, atmosphere refractive indices, and temperature profiles
Mars balloon relay	Data relay for Russian Mars 1994 and 1996 penetrators, balloon, and Mars rover experiments, and for high-resolution pictures and surface and boundary-layer characteristics

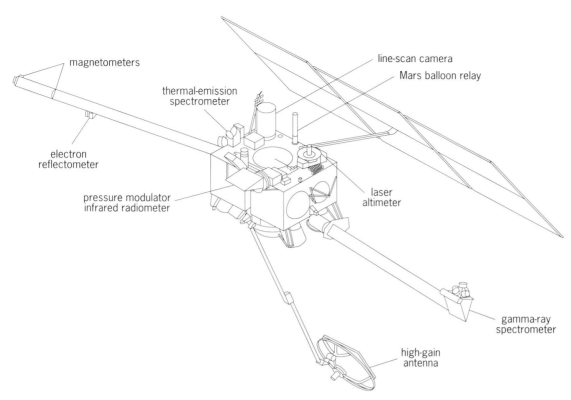

Fig. 2. Locations of scientific instruments on *Mars Observer* (shown in mapping configuration).

cruise to Mars. The main communication antenna will be raised on a 5.5-m (18-ft) boom to clear the 3.7 by 6.7 m (12 by 22 ft) solar array, which will not be fully unfolded until the spacecraft reaches Mars. Once at Mars, the spacecraft will power its seven instruments to conduct the mission experiments.

Mapping Mars. The spacecraft will be nadir-oriented; that is, most of its instruments will always point toward the surface of Mars directly below the spacecraft. The scientific instruments have fields of view ranging from 100 m (328 ft) to hundreds of kilometers on the planet surface. Thus, the instruments will see the entire planet every 26 days. During the Martian year the *Mars Observer* will be able to go through this cycle of observation about 26 times. The spacecraft will be able to see the seasonal changes, which will be useful in monitoring weather, winds, atmospheric variations, and other changing patterns on Mars.

While the scientific instruments on the spacecraft will be pointed at Mars, the spacecraft solar array will be constantly adjusted to face the Sun, and the spacecraft's antenna will also be changing position so that it will point toward the Earth, since it will be sending data back to the Earth and receiving commands from the Earth. While the spacecraft is orbiting Mars, the positions of the Sun and the Earth also will be changing gradually as the Earth and Mars orbit the Sun. Orbital information periodically sent to the spacecraft by the mission team will be used by the on-board computers to determine the proper pointing directions for the solar array and antenna.

Mapping data collected by the scientific instruments will be recorded continuously during the 687-day mapping period. Once each day the spacecraft transmitter will relay the science data back to Earth. The NASA Deep Space Network will receive these transmissions and relay the information to the Mission Control Center at the Jet Propulsion Laboratory.

Mission phases. Four mission phases have been defined for the *Mars Observer*: launch, cruise, orbit insertion, and mapping. **Figure 3** shows the planned time line for these phases.

The launch phase extends from the start of the launch countdown to the separation of the spacecraft from the transfer orbit stage, after interplanetary injection. The cruise phase begins when the spacecraft starts its interplanetary flight toward Mars, following separation from the transfer orbit stage. The orbit insertion phase begins with the initial elliptical orbit of Mars and ends when the spacecraft has reached the proper mapping orbit. The spacecraft will remain in this mapping orbit for 10 days while scientists check out the equipment and make certain that everything is ready.

The mapping phase is the period of gathering scientific data with the instruments and returning the data to Earth. This phase extends for one Mars year after the end of the orbit insertion phase. Overlapping this phase, beginning in September 1995, will be the additional Mars balloon relay activity. This activity was scheduled to end in February 1996, but the mission may be extended to accommodate the Russian Mars 1996 mission.

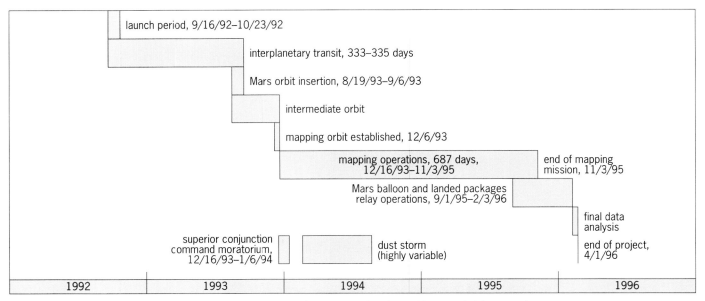

Fig. 3. Nominal time line for *Mars Observer* mission. It could be extended for several years.

For background information *SEE MARS; SPACE PROBE* in the McGraw-Hill Encyclopedia of Science & Technology.

David D. Evans

Bibliography. A. L. Albee and F. D. Palluconi, *Mars Observer*'s global mapping mission, *EOS Transac.*, 71(39):1099, 1107, September 25, 1990; E. C. Ezell and L. N. Ezell, *On Mars*, NASA SP-4212, 1984; C. R. Spitzer, *Viking Orbiter Views of Mars*, NASA SP-441, 1980.

Space station

Space Station *Freedom*, which will become permanently occupied in the year 2000, will be a notable achievement in space exploration and technology. It marks a new era in which humans will no longer be limited to short, periodic trips into space. The space station will continue the development of the space infrastructure that will enable humankind to continue its quest to explore the universe, understand and preserve the Earth, and extend human presence into the solar system. On *Freedom*, technologies that make possible missions to achieve these goals will be developed, and future astronauts will be qualified for long-duration space voyages. The low-gravity environment in which the space station operates will permit scientific studies not possible on Earth, particularly in the areas of materials science and life science. Ultimately, the space station may serve as a transportation node in Earth orbit, a place where future trips to the Moon, Mars, and beyond could be assembled, tested, and launched.

The roots of Space Station *Freedom* extend back to the beginning of the civilian space program. Ever since the late 1950s, the United States has made plans for some type of space station. The first space station launched by the National Aeronautics and Space Ad-

ministration (NASA) was *Skylab*. *Skylab 2*, *3*, and *4* were launched in 1973 and occupied by three different astronaut crews for a total of 171 days.

Following *Skylab*, plans for a permanent space station continued to develop, and in 1984 President Reagan announced that the United States would develop a permanently occupied space station and invited other countries to participate. Canada, Japan, and 10 European nations have joined with the United States to build Space Station *Freedom*, the largest international cooperative science and technology program ever undertaken.

Design. Space Station *Freedom* will be composed of four pressurized modules mounted on a 353-ft-long (108-m) truss. It will operate at an altitude ranging 150–240 nautical miles (278–444 km), in a 28.5°-inclination orbit. The four modules will be linked by tunnellike structures called resource nodes. These nodes will serve as passageways and as command-and-control centers for space station systems and communications. One of the nodes will be outfitted with a viewing tower, called a cupola, to allow the crew to direct and oversee external operations.

The four partners on the project are the United States, Canada, the European Space Agency (ESA), and Japan. Each is providing elements for the overall space station structure. The United States is responsible for the overall assembly and verification of Space Station *Freedom*. It is providing many of the basic space station components, including two of the pressurized modules (a laboratory module and a habitation module to house the station crew), logistics elements to resupply the station, and a major portion of the space station infrastructure, including the truss, the power and communication systems, the environmental life-support system, the data system, and the guidance, navigation, and control system. Canada is also providing elements of the space station infrastructure. It is responsible for providing the robotic elements that will be employed to assemble, service, and operate Space Station *Freedom*. The Canadian robotic elements are collectively known as the Mobile Servicing System. The European Space Agency is providing the attached Columbus laboratory module and a Columbus Free Flyer, which will coorbit with Space Station *Freedom*. Japan is providing a third laboratory module called the Japanese Experiment Module. This module will include an externally attached facility to accommodate experiments exposed to space and an experiment logistics module that will carry payloads from Earth to the station on board the shuttle.

Space station supplies will be carried back and forth on the space shuttle in logistics carriers that will remain attached to one of the resource nodes between shuttle flights. In order to maintain the correct altitude and attitude over the 30-year life of the space station, the facility will be outfitted with hydrazine propulsion modules that will be attached to the truss. Each propulsion module will be outfitted with both reboost and attitude-control thrusters, and fuel tanks.

The space station power system will consist of three photovoltaic solar arrays, 34.4 ft (10.5 m) wide by 89 ft (27 m) long, located along the truss. These panels will provide sun-generated power to the station. When the station is in the Earth's shadow, power will be provided by nickel-hydrogen storage batteries.

Assembly. Assembly of the space station is scheduled to begin in late 1995 or early 1996. The station will be assembled in stages by using the space shuttle as the launch vehicle to bring hardware elements from the Earth to be assembled in orbit. The space shuttle is capable of launching in excess of 36,000 lb (16,000 kg) of payload into the space station orbit, in addition to the astronaut crew necessary for space station assembly and utilization and space station operations.

Assembly of the space station will be conducted over a 5-year period by using 17 shuttle flights. The space station will be assembled by members of the shuttle crew with the assistance of Remote Manipulator System robots. The space shuttle Remote Manipulator System is a robotic arm located in the shuttle cargo bay and is used to launch and retrieve cargo payloads. The space station Remote Manipulator System, a component of the Mobile Servicing Center provided by Canada, will be used for assembly, operation, and maintenance activities.

The space station truss will be launched in eight different segments and joined together by the crew with bolt fixtures as the Remote Manipulator System holds the segments together. The truss segments will be preintegrated on Earth; that is, the subsystems, such as power and communications, will be attached to the structure before launch. The truss segments will be connected by the crew as the truss is assembled on-orbit. The space station core elements, including the pressurized modules, solar arrays, propulsion modules, and robotic elements, will be mounted to the truss structure. The truss structure will also include utility ports to accommodate small external payloads.

The first milestone in the assembly will be the first-element launch (FEL). The second milestone is man-tended-capacity (MTC; **Fig. 1**), which will be attained on flight six of the assembly sequence. The MTC configuration includes the United States laboratory module and 15 user payload racks, which will allow scientists to begin utilization of the space station. MTC will allow at least two members of the space shuttle crew to conduct research and tend to experiments while the shuttle is docked at the station during assembly flights. In addition to the assembly flights, three utilization flights per year will be flown during this utilization phase in order to support use of the space station. These flights will bring payload racks and user supplies to the station, and the crew will be dedicated to operating and tending experiments. The utilization phase will last for approximately 3 years, during which other key elements of the space station will be assembled.

The next assembly milestone is permanently manned capability (PMC; **Fig. 2**). PMC will include all three pressurized laboratories and an assured-crew-return vehicle, which will provide the space station astronauts with an emergency return-to-Earth capability. In addition, three photovoltaic arrays will be assembled, which will provide over 56 kW of power. The habitation module will be attached, and will provide living quarters for the four-person crew. NASA and the other partners plan to continue assembly of the space station following PMC. The follow-on phase includes provisions for a fourth solar array, which will increase total power to 75 kW, and a second habitation module to support an eight-person crew, as well as possibly enhanced hardware and systems, as program requirements become better defined.

Fig. 1. Concept of Space Station *Freedom* in the man-tended capability (MTC) configuration. The space shuttle is bringing the United States laboratory module to be attached to the station.

Fig. 2. Concept of Space Station *Freedom* in the permanently manned capability (PMC) configuration, with the space shuttle approaching the station.

The astronaut crew that will work and live on board Space Station *Freedom* will be international and varied in fields of expertise. The space station will initially operate with a four-person crew. During the follow-on phase the space station manned base (SSMB) will operate with an eight-person crew composed of two four-person teams that stay on board for 6 months each. There will be two station operators, one of which is the space station commander, who will be responsible for the space station operations and functions and the overall safety of the crew. Station scientists will be responsible for operating and maintaining all station scientific payloads. Payload scientists are a class of astronauts whose unique skills allow them to work on special scientific experiments. Each space station astronaut will be required to undergo a minimum of 1 year of basic training, as well as several years of advanced space station systems and payload-specific training. In addition to specialized individual training, the space station crew will participate in integrated training to develop an international team that will be compatible for the long-duration stays on the space station.

Applications. Space Station *Freedom* for the first time will provide the opportunity to learn how to assemble and operate a permanent platform in space. It will provide the capability to maintain a continuous presence in space so that scientific microgravity and life sciences experiments can be performed, and will also provide a base to attach external payloads. Space Station *Freedom* will also serve as a test bed for technology to be applied for future missions. Propulsion systems, communications, materials, and assembly techniques are examples of technologies currently under study. Long-duration exposure to an extremely low gravity environment (one-millionth

the gravity on Earth) cannot be duplicated on Earth. It will enable all cooperating nations to conduct in-depth, space-based research and analysis to a degree not previously possible. Further, Space Station *Freedom* will provide the opportunity for payloads to be placed in low Earth orbit for potential commercial exploitation.

The space station will provide unique capabilities to study the nature of physical, chemical, and biological processes in the low-gravity environment. The results from these studies can be applied to advance knowledge and capabilities in such fields as fluid physics, materials science, combustion science, automation and robotics, space structures, energy systems and thermal management, biotechnology, plant and animal physiology, exobiology and biomedical research, astrophysics, astronomy, and Earth sciences.

The research focus of Space Station *Freedom* will be on progressive science investigations, many requiring hands-on scientist involvement using sophisticated experiment hardware. With this in mind, the primary users of the space station will be from the fields of science, technology, and commerce.

One of the most important applications for the space station is to study the ability of humans to adapt to long-duration weightlessness. For human flight, the space station will provide the crucial life sciences experience necessary to understand how long-term exposure to microgravity and solar radiation affects the human body. These effects must be well understood before any long-term planetary exploration missions can be attempted.

Prospects. In the future the space station can play a number of roles. The space station is being designed so that it can evolve and provide new capabilities as necessary. For example, the station could evolve to serve as a transportation node where spacecraft could be assembled, fueled, and verified before launch. The space station could act as a staging ground for both human and robotic space flight.

Space Station *Freedom* can play a critical role for the United States and its partners for future space exploration. Space Station *Freedom* is the cornerstone of the United States civil space program that will make future missions possible. The United States can learn invaluable lessons as the station is constructed and assembled. These lessons can be carried forward in this and other space programs.

The benefits to be gained by the United States and its international partners in the space station program are likely to go well beyond scientific discoveries. The technical and political challenges that must be overcome with the integration of such a complex project can lay the foundation from which even greater international cooperative projects may be launched.

For background information SEE SPACE BIOLOGY; SPACE FLIGHT; SPACE PROCESSING; SPACE STATION; SPACE TECHNOLOGY; WEIGHTLESSNESS in the McGraw-Hill Encyclopedia of Science & Technology.

Richard H. Kohrs

Bibliography. *Report to Congress on the Restructured Space Station Program*, March 20, 1991.

Space Telescope

On April 25, 1990, the deployment of the Hubble Space Telescope from the cargo bay of the space shuttle *Discovery* inaugurated an era of large, international observatories in space. The telescope now orbits Earth every 96 min at an altitude of 380 mi (610 km). As the first orbiting observatory of the National Aeronautics and Space Administration's Great Observatories Program, Hubble is intended to observe cosmic sources in the optical, ultraviolet, and near-infrared parts of the electromagnetic spectrum with a sensitivity about 50 times better and a resolution about 10 times better than can normally be achieved with ground-based telescopes. In astronomers' terms, the design goals are to see 28th-magnitude cosmic objects with 0.1-arcsecond resolution.

Throughout the first year of operation, Hubble was put through an extensive series of systems checks and calibrations in an attempt to test its thousands of different operating modes among its six on-board scientific instruments—two cameras, two spectrographs, a photometer, and an astrometric device. In the course of this commissioning process, a number of unexpected problems were discovered. Foremost are an inability to focus well the telescope optics and an inability to keep the spacecraft steady during all parts of its orbital motion around the Earth. Despite these operational problems, the telescope is providing unprecedented rich and detailed views of a variety of cosmic objects that can be obtained only with a telescope above Earth's atmosphere.

Problems. The first problem, known technically as spherical aberration, concerns the shape of the main (8-ft or 2.4-m diameter) mirror. The mirror was polished to the wrong prescription and is 2 micrometers flatter at the edges than called for by design, so that the telescope cannot direct all of its collected light to a single focus. The resulting raw image is blurred, with much light scattered in a halo around any celestial target. With computer enhancement algorithms, however, much of the degraded image can be restored; thus the resolution expected for the Space Telescope is in fact being achieved. The trade-off is that the scattered light in a source's halo must be discarded, which is why the greatest loss arising from Hubble's poor optics is in sensitivity. The telescope can see well only to the 24th or 25th magnitude, and cannot work on the faintest, cosmologically significant targets.

The second major problem concerns pointing instabilities during part of each orbit. As the spacecraft crosses the terminator separating orbital day and night (thus twice each orbit), the solar arrays that collect light from the Sun to power the telescope tend to oscillate. This oscillation causes the main hull of the spacecraft to wobble, so that the delicate pointing system is upset and the camera optics are degraded.

Both these problems will probably be addressed during a shuttle servicing mission in the next few years. Astronauts will install a new set of solar arrays less subject to contraction and expansion, so that the telescope will be more steady throughout its orbit.

Fig. 1. Image of Saturn taken with the Hubble Space Telescope. This image is computer-enhanced and reveals structure as small as 420 mi (670 km) across, much as the ringed planet would appear to the unaided eye if it were only twice as far away as the Moon. (*National Aeronautics and Space Administration*)

They are expected to install a new wide-field camera (already in development) with corrective optics that will compensate for the aberration of the main, faulty mirror, and perhaps additional corrective optics for additional instruments. (The main mirror cannot be replaced on Hubble unless the entire spacecraft is brought back to the Earth, an unlikely prospect.)

Observations. In the meantime, science data are being transmitted to Earth stations at a rapid rate. The Hubble telescope has proved that it can do first-class astronomical science. Among many science observations made thus far, the telescope has been able to track with unprecedented clarity a rare storm on Saturn, discover a luminescent ring around Supernova 1987A, resolve the inner structures of several previously mysterious star clusters, uncover detailed structure in the morphology of extragalactic jets, and probe the chemical history of the early universe by taking a spectrum of a distant quasar.

Planets. Regarding planets, Hubble observations of Saturn have emphasized details never before photographed from Earth (**Fig. 1**). When a massive storm erupted in September 1990, Space Telescope images revealed a rapidly changing scalloped structure, resembling the cloud bands of Jupiter. The bright, white cirrus-cloud-like features, probably moist matter upwelling from deeper in Saturn's atmosphere and bubbling up to form ammonia ice crystals, provide a rare opportunity to trace the dynamics of Saturn's otherwise bland meteorology.

Mars has been observed extensively by the Space Telescope. By comparing images made in ultraviolet light that does not reach the Earth's surface, astronomers have found evidence for absorption by ozone in the Martian atmosphere, especially over the dry polar regions. Because water vapor initiates chemical reactions that remove ozone from the Martian air, ultraviolet data acquired over many years permit monitoring of the amount of water, as well as ozone, in the planet's atmosphere.

Stars and stellar systems. The Space Telescope has also examined a variety of stars and stellar systems, and many of these observations are contributing to an improved understanding of stellar evolution. For example, the Orion Nebula, a region of star formation, has been shown by Hubble to be rich in arcs, filaments, and other intricate structures. A new level of detail has been uncovered, including resolved features as small as the solar system, despite the distance to the Orion Nebula of about 1500 light-years.

In a star system close to "death," the Hubble telescope has resolved the inner core of the novalike object R Aquarii. Camera images directly reveal rapidly moving geysers of plasma with a length of 2.5×10^{11} mi (4×10^{11} km or 2500 times the distance between the Sun and the Earth). The luminescent streams are twisted by the force of the nova explosion and channeled upward and outward by strong magnetic fields. This stellar volcano erupts from two very bright knots that probably harbor the double star system, one member of which is a cool red giant star. The red giant apparently spills some of its mass onto its companion star (a white dwarf), which periodically becomes overloaded with infalling matter, after which it explodes and blasts matter back into space.

Supernova 1987A. The Hubble telescope has directly observed the aftermath of an exploded star, SN1987A, the southern-hemisphere supernova whose bright flash reached the Earth in February 1987. The Hubble image (**Fig. 2**) shows clearly an elliptical, luminescent ring of gas surrounding the expanding cloud of debris; at 0.1-arc-second resolution, detailed structure within the ring can also be observed. By comparing the present diameter and its expansion velocity, researchers have inferred that about 5000 years ago the progenitor star developed a stellar wind, and in the process evolved from an old red giant star into a blue giant star. Such direct measurements of evolutionary events for an individual star have never been obtained before on time scales so diverse.

An unexpected bonus from the SN1987A observation is a precise measurement of 169,000 light-years to the supernova's host galaxy, the Large Magellanic Cloud. (This measurement is accurate to 5%, compared to the previously best known value of 15%.) The improved distance scale was made possible by Hubble's accurate measurement of the ring's angular diameter, and, combined with the ring's physical size (known from other spacecraft studies), astronomers used the conventional method of trigonometry to yield the new distance. This kind of work is providing a new basis for the determination of the size and age of the universe.

Globular clusters. The Space Telescope has observed the centers of several globular clusters, including M14, M15, and 47 Tucanae. Globular clusters are tightly bound aggregates comprising on the order of 10^5 stars held together by gravity in a region only a few light-years across. Initial Hubble observations have presented a very different picture from what was presumed from ground-based observations. In particular, the hearts of these systems lack what many astronomers thought would be present: a large black hole. With the extremely high resolution of the Hubble telescope, tens of thousands of stars can be clearly resolved within the innermost light-year, but there seems to be no indirect signature (a great cusp of light) at the heart of the globular, and thus apparently no black hole.

External galaxies. By contrast, in a normal, quiescent, elliptical galaxy where astronomers had not expected to find the signature of a black hole, they have. In a remarkably detailed view of NGC 7457, which lies about 10^7 light-years away, the Hubble image reveals that stars are much more tightly concentrated at the center of the galaxy than was previously known, at least 30,000 times more densely than the stars that are observed within the galactic neighborhood of the Sun. Apparently, the Hubble data provide new evidence that dense cores are common to most galaxies, and so, perhaps, are black holes.

The Hubble Space Telescope has begun to probe the distant universe. Here, among many novel observations, the telescope has examined closely the jets of fast-moving matter emanating from active galaxies. When seen with Hubble's new level of resolution, jets in galaxies known as 3C66B and PKS0521-36 display remarkably detailed structures, including knots of emission and numerous instabilities along the edge of the plasma flow. The jets' braided appearance, in fact, suggests that electrons spiral along channels or hollow tubes that extend for tens of thousands of light-years from the center of the galaxies.

For background information *SEE GALAXY, EXTERNAL; MARS; NOVA; ORION NEBULA; SATELLITE ASTRONOMY; SATURN; STAR CLUSTERS; STELLAR EVOLUTION; SUPERNOVA* in the McGraw-Hill Encyclopedia of Science & Technology.

Eric J. Chaisson

Bibliography. E. Chaisson and R. Villard, The science mission of the Hubble Space Telescope, *Vistas Astron.*, 33:105–141, 1990; Special issue on initial Space Telescope results, *Astrophys. J.*, vol. 369, no. 2, March 10, 1991.

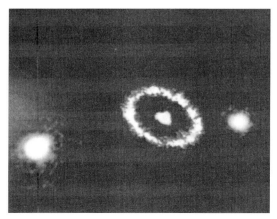

Fig. 2. Hubble Space Telescope's view of a mysterious ring of matter surrounding the remnants of Supernova 1987A. (*National Aeronautics and Space Administration and European Space Agency*)

Spread spectrum communication

Spread spectrum communication is a technique that has been widely used for military applications. In the last few years there has been a rapid increase in nonmilitary applications of spread spectrum, especially in wireless communication systems such as cellular car phone systems, personal communication networks, and wireless local-area networks (LANs).

Advantages. The military community has been interested in spread spectrum waveforms primarily because of their interference rejection properties and their characteristically low probability of intercept. These attributes allow communications to take place in the presence of intentional jamming, and make it difficult for an intelligent adversary to discover the transmissions of the desired signal, since the spectral density of the transmitted waveform can be hidden below the thermal noise level. Both properties are attributable to the fundamental nature of a spread spectrum waveform, namely its bandwidth being greater than (and typically much greater than) the bandwidth of the underlying information.

This increase in bandwidth can be accomplished in various ways, but the most common are direct sequence and frequency hopping. In direct sequence, the information sequence is multiplied by a high-bit-rate (or, in standard terminology, by a high-chip-rate) sequence. The resulting product has a bandwidth comparable to that of the high-chip-rate spreading sequence. In frequency hopping, the information signal has its carrier "hopped" over a wide range in a seemingly random manner. In either case, the properties discussed above are the direct result of the spectrum spreading.

These same properties have also made spread spectrum techniques of interest to the civilian sector. For example, the interference rejection properties of these waveforms make them of interest both in situations where multipath must be combatted and in situations where multiple-accessing capability is needed. In the first case, the interference that the spread spectrum waveforms must overcome is the reflected signals due to the multipath nature of the channel. In the second case, a technique called code-division multiple access (CDMA) is employed, in which each user in the system is given its own spreading sequence, and a receiver desiring to listen to the message of one of the users must rely on the spread spectrum interference suppression capability to attenuate sufficiently the energy received from all the unwanted signals on the channel.

Early nonmilitary applications. The National Aeronautics and Space Administration (NASA) has used code-division multiple access for many years on a communication link of its Tracking and Data Relay Satellite System. In this system, low-altitude satellites transmit information to synchronous satellites, which then relay the information to fixed ground stations. Because some of the energy from the low-altitude satellites bounces off the Earth and is reflected to the synchronous satellite, a multipath situation exists. This situation, combined with the need for multiple-accessing capability (because there are many low-altitude satellites), made spread spectrum signaling an appropriate choice.

Another nonmilitary application of spread spectrum technology has been in the orbital spacing of satellites; the closer the satellites are, the more satellites can be accommodated in a given orbit, but the more intersatellite interference is produced. Spread spectrum techniques have been used to allow the receivers located in closely spaced satellites to withstand higher levels of interference.

A different application has been to situations in which a regulatory agency has imposed a maximum on the power spectral density allowed for a signal. With conventional communications, such a constraint would result in a corresponding constraint on transmitted power. However, since by its very nature spread spectrum takes the energy of the information-bearing portion of the waveform and disperses it over a much larger bandwidth, the resulting power spectral density is decreased. Thus, if the signal is first spread, the constraint on spectral density can be met, and yet the total power can be made as large as is necessary, if the spreading is done over a sufficiently wide bandwidth.

Wireless applications. While these nonmilitary applications indicate that spread spectrum can be used advantageously in a wide set of circumstances, in fact up until a few years ago, spread spectrum was used primarily by the military. However, this limited use has dramatically changed because of the desire of consumers to have wireless and, in particular, mobile communications.

Cellular car phones. An example is the cellular car phone. A given geographical area (say, a city) is divided into regions called cells and in each cell is a base station. All mobile users in a cell transmit directly to the base station controlling that cell (that is, they do not transmit directly to other mobile users, either in their own cell or in any other cell). The base station then transmits the mobile user's message to the intended party, typically through the wired telephone system. When the return message is generated, it is sent back to the base station and then transmitted to the original mobile user on a different frequency band.

With the current cellular car phones, the modulation format is analog frequency modulation, with each signal occupying a bandwidth of 30 kHz. Multiple users simultaneously access the channel through a frequency-division multiple-access (FDMA) scheme. Because of the tremendous demand from consumers, the capacity of such systems is essentially saturated in many major cities (for example, Los Angeles). Thus, the industry has made a decision that the next generation of cellular car phones will employ digital, rather than analog, modulation. This modification allows a broader choice in the multiple-accessing technique; there is still the option of employing frequency-division multiple access, but both time-division multiple access (TDMA) and code-division multiple access can also be considered.

The industry has now (seemingly) ruled out frequency-division multiple access, and is concentrating its efforts on deciding between time-division multiple access and code-division multiple access. Among the motivations for using code-division multiple access are, again, its ability to function well in a multipath environment. The multipath-producing nature of the mobile radio channel is evident from consideration of a typical environment in which a car phone operates, namely, a congested urban area with numerous obstructions to radio waves such as buildings and other vehicles. Also, privacy is inherent in the use of spread spectrum because of its characteristically low probability of intercept. In other words, with the current analog FM system, any passerby who happens to be tuned to the frequency band over which a given subscriber is communicating can eavesdrop on that conversation; with spread spectrum, such simple eavesdropping will no longer be possible.

Personal communication networks. A related application to cellular car phones is personal communication networks. It is envisioned that each subscriber in a personal communications network will have a pocket-sized handset to carry about. An identification number will be dedicated to the subscriber, so that instead of needing one telephone number for the office, one for the home, and one for the car, only a single number will be required. *SEE MOBILE RADIO.*

In part for the same reasons that spread spectrum is attractive to the cellular car phone industry, it is an attractive option for personal communications networks. However, another motivation for the use of spread spectrum in this latter application is the possibility of sharing common spectrum with narrow-band (that is, conventional) users. Specifically, the frequency band from 1850 to 1990 MHz is currently occupied by a variety of microwave users, from power and utility companies to state and local agencies such as police and hospitals. These microwave users employ conventional communication techniques, which use signals having bandwidths that are typically no larger than 10 MHz. Because the microwave usage in this frequency range is somewhat sparse in any given location of any city (that is, only a fraction of the band from 1850 to 1990 MHz is occupied at a specific location), there is the possibility of sharing the overall spectrum and thus increasing the spectral efficiency of that band. One proposal for such spectral sharing is to overlay on that band a direct-sequence, code-division multiple-access, personal communications network operating with a chip rate of 24 megachips per second. The result is a transmitted spread spectrum signal occupying about 48 MHz from the base to the subscribers and from the subscribers back to the base. Thus, 96 of the 140 MHz are used, and the remaining 44 MHz serves as guard bands.

Wireless local-area networks. While mobile communications is perhaps the ultimate form of wireless, or untethered, communications, there are situations where just the untethered aspect is sufficient to generate strong interest. One example is indoor wireless local-area networks, and it is the ability of spread spectrum waveforms both to combat multipath and to allow multiple accessing capability that is the driving force. The motivation for wireless local-area networks is twofold: avoiding the initial cost of building wiring, and avoiding the continually recurring costs of rewiring as systems are upgraded.

Other applications. Many other applications for spread spectrum communications have been receiving serious consideration. One application is communication between a coach and, say, the quarterback of a football team. While such communication using conventional narrow-band waveforms is now available, it is vulnerable in that someone from the opposite team can listen to the transmission. As indicated above, spread spectrum has an inherent aspect of privacy.

Another potential application, which makes use of the property of low probability of intercept, is to police radars. If such radars were to employ spread spectrum signaling, they would suddenly render useless most of the radar detectors used by motorists.

For background information *SEE LOCAL-AREA NETWORKS; MOBILE RADIO; MULTIPLEXING; SPACE COMMUNICATIONS; SPREAD SPECTRUM COMMUNICATION* in the McGraw-Hill Encyclopedia of Science & Technology.

Laurence B. Milstein

Bibliography. M. K. Simon et al., *Spread Spectrum Communications*, vols. 1–3, 1985; R. E. Ziemer and R. L. Peterson, *Digital Communications and Spread Spectrum Systems*, 1985.

Steel manufacture

The technology of steel-sheet production systems is undergoing significant changes. Thin-slab casting and other continuous casting processes are being planned to change the way that molten steel is converted into sheet steel for use in automotive components, appliances, structural applications, welded pipe, metal grating, roofing, siding, and so forth. The thin-slab continuous casting process is an efficient alternative to the conventional production of sheet steel.

By the early 1990s, over 82% of all crude steel was continuous-cast, while in 1965 this method represented only 1% of the production. This change was driven on a global scale by both technical and economic forces. The primary economic forces were energy conservation and reduction of capital investment and labor costs by producing steel castings closer to net shape. Costs are reduced by eliminating slab handling, batch-reheating furnaces, and the breakdown rolling-mill operations (heavy mechanical reductions required to break thick ingots down to thin slabs). Also, an important technical incentive is the rapid solidification achieved by casting thinner sections.

Current conventional practice is to cast steel continuously in thick slabs that are 200–250 mm (8–10 in.) thick and 2000 mm (80 in.) wide. A thick slab (strand) is cast at a rate of about 275 tons (250 metric tons) per hour at a withdrawal speed of roughly 1.2 m (4.0 ft) per minute. Casters producing thick slabs operate primarily in integrated steel mills.

Mini-mill versus integrated steel mill. An integrated steel mill accepts an input of iron ore and produces so-called hot metal (pig iron) by reducing this ore to metallic iron. The molten pig iron is transformed within the same plant to steel in reaction vessels known as oxygen converters. An integrated steel mill is characterized by the production of hot metal, high annual throughput, and high capital investment.

Mini-mills do not produce hot metal by reducing iron ore. They produce molten metal by remelting purchased scrap and purchased direct-reduced iron. This molten metal is cast continuously into shapes for later fabrication into rods, billets, and other products. Mini-mills may lead the way toward the production of sheet steel by continuous casting of molten steel into thin slabs.

Thin-slab flat-rolling steel mill. The first major commercial production mill of this kind is now operational in Crawfordsville, Indiana. A mini-mill capable of 800,000 tons (730,000 metric tons) per year started up in 1989 and reached design operating capacity during 1991. This plant combines thin-slab casting and hot-strip rolling in a continuous process. An innovative process developed in Germany is being used. The plant layout is shown in **Fig. 1**.

Melting furnaces. Steel is produced in two eccentric bottom-tap electric-arc furnaces of split-shell design. The furnaces have a shell with a diameter of 670 cm (270 in.), use electrodes that are 61 cm (24 in.) in diameter, and have water-cooled roofs and side walls. The nominal capacity is 136 tons (123 metric tons) with tap-to-tap time of 80–90 min; 65-megavolt-ampere transformers are used.

Ladle treatment. The plant has a ladle-treatment facility that includes a vacuum degassing unit to produce steels with exceptionally low carbon or nitrogen content. The ladle-treatment station removes inclusions, improves homogeneity, and distributes alloy additions evenly. Ladles leave this area with a homogenized steel temperature of 1575–1595°C (2867–2903°F), according to steel grade, and a free-oxygen content of less than 20–40 parts per million (ppm) after calcium-silicon treatment, addition of aluminum, and ladle stirring.

Casting. After ladle treatment, ladles are taken directly to the casting tower (**Fig. 2**). Molten steel is poured into a tundish. The ladle contains 113 tons (103

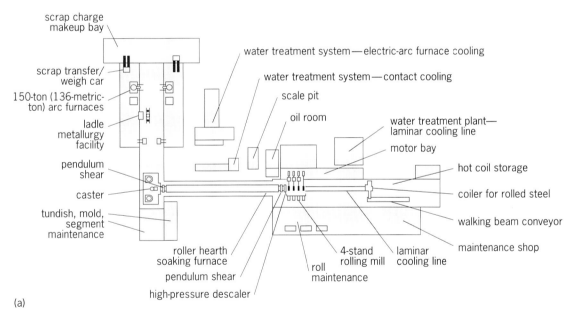

(a)

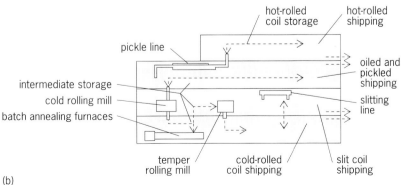

(b)

Fig. 1. Plant layout of a facility for thin-slab casting and sheet-steel fabrication. (*a*) Continuous casting and rolling operations. (*b*) Cold-rolling, finishing, and shipping operations.

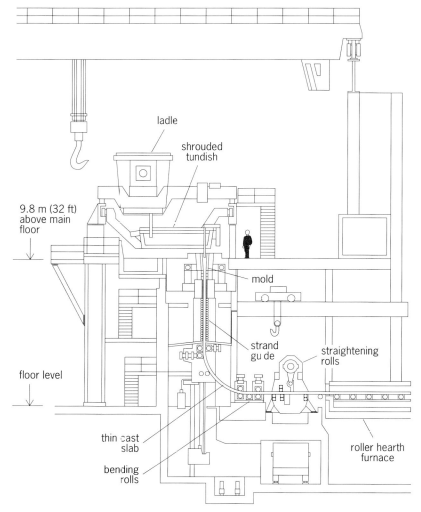

Fig. 2. Thin-slab caster. (*Nucor Steel, Crawfordsville, Indiana*)

ment section. The slab is bent on a 3-m (10-ft) radius from the vertical to the horizontal. At this point the slab, at a temperature of about 980°C (1796°F), is traveling at a speed of 4 m (13 ft) per minute.

Shearing and reheating. The bent thin slab is sheared before entering the reheating furnace. It is sheared into lengths of 42–46 m (138–151 ft), weighing 18–22 tons (16–20 metric tons).

The solidified thin slab moves continuously through a gas-fired tunnel soaking furnace that is 162 m (531 ft) long; it is shown schematically in Fig. 1. This roller hearth furnace receives a slab that is 45 m (148 ft) long. The residence time of the slab in the furnace is 25 min. The thin slab has an entry temperature of more than 900°C (1650°F) and an exit temperature of 1100°C (2012°F). This furnace equalizes the temperature, prevents further heat loss, and brings the average thin-slab temperature up to 1100°C (2012°F) for rolling. The length of the furnace allows for transient storage of up to three lengths of thin slab in the event that minor problems develop in the production line.

Descaling. As a result of the 25-min dwell time of a typical slab in the reheating furnace, a thin but toughly adhering scale forms on the surface. This formation requires a high-pressure descaling operation as soon as the slab exits the furnace.

Hot-strip rolling. Next, the thin slab goes directly into a four-stand finishing mill that reduces the thickness from 50 mm (2 in.) to 2.5 mm (0.1 in.) The thin slab enters the first of the four high hot-strip rolling mills at a speed of 0.25 m (0.82 ft) per second and exits the last mill at 5 m (16 ft) per second. The temperature of the strip falls only 120°C (216°F) in this process. This hot-strip mill has an automatic gage-and-profile-control system. Each strand is driven by a 7000-kW motor.

The hot-strip output ranges from 2.5 mm (0.1 in.) to 16 mm (0.64 in.), depending on customer require-

metric tons), and a heel of 18–23 tons (16–21 metric tons) is left in the furnace. The molten steel flows into a funnel-shaped copper mold through a submerged nozzle (**Fig. 3**). The flow of molten steel from the tundish into this mold is controlled automatically by a stopper rod and a radioactive device for measuring molten metal. The submerged entry nozzle with the funnel-shaped mold is the most crucial part of the thin-slab casting process.

The molten-metal meniscus in the mold is controlled within 1–1.5 mm (0.04–0.06 in.). The mold oscillates with a stroke of 3 mm (0.12 in.; 6 mm or 0.24 in. peak to peak) at a frequency of 60 strokes per minute. Mold powders are placed in the mold chamber to improve withdrawal by minimizing sticking of the slab to the mold.

As heat is extracted by water flowing at high velocity within the copper mold, the steel is solidified. The thin slab is withdrawn at a rate of about 4 m (13 ft) per minute. As it is removed from the mold cavity, the casting deforms into a rectangular slab 50 mm (2 in.) thick and 1320 mm (53 in.) wide.

Figure 2 shows this slab coming from the contain-

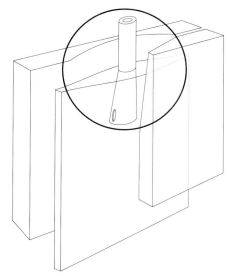

Fig. 3. Diagram showing funnel-shaped mold and submerged nozzle. (*SMS Schloemann-Siemag Aktiengesellschaft, Dusseldorf, Germany*)

ments. The thin-gage products of 2.5–4 mm (0.1–0.16 in.) are generally low in carbon content (0.04–0.08%). The heavy gages, used for structural and welded-pipe applications, generally have a carbon content of 0.18–0.26% and a manganese content of 0.9%.

Coiling. The rolled sheet passes through a laminar cooling section, where it is sprayed from both top and bottom by 16 banks of water jets. The strip is made into coils at a temperature of 530–730°C (986–1346°F) by a hydraulically operated coiler; each coil weighs 18–22 metric tons (20–24 tons) and has an inside diameter of 760 mm (30 in.) and an outside diameter of up to 1930 mm (77.2 in.). Known as hot bands, these coils are either sold to service centers or stored on site for later cold-rolling into sheet product.

Finishing. Finishing operations include rolling to meet specifications of thickness, width, and temper (strength). These operations are carried out in a separate building (Fig. 1) housing equipment for cold rolling, slitting, pickling, and annealing, and a temper mill. Finishing operations include any packaging or protective oiling operations before shipping the so-called semifinished steel-sheet product.

Personnel and productivity. Mini-mills use minimal personnel as compared with large integrated steel mills. Generally, incentives are used to motivate mini-mill workers to achieve high productivity. The Crawfordsville operation utilizes only 1.5 worker-hours of labor per ton of shipment versus 3–6 worker-hours per ton for a typical conventional flat-products plant. The Crawfordsville plant is operated with 386 people. The production plant is operated 24 h every day of the week, with four crews working 12-h shifts. Their work schedule is 4 days on duty followed by 4 days off. Many aspects of personnel practice in mini-mills differ from those in integrated mills. Personnel practice is one of the important factors that have led to increased competition and progress in this critical segment of the economy.

Product quality. A high level of internal metal cleanliness is claimed for the thin-slab casting process. This is obtained by using careful scrap control and good melting practice. Phosphorus, sulfur, aluminum, and residual elements are maintained at levels that produce acceptable quality of surfaces and edges. The level of the grain size at the surface and midthickness is an ASTM 10 Plus, which is rated as very fine. This quality feature is a direct result of thin-slab casting. The as-cast size of the microstructural features is dictated by the solidification rate of the steel. Faster solidification rates lead to finer features, with less segregation of alloying elements. Many properties of such rapidly solidified steels, including the responsiveness to heat treatment, are enhanced because of the reduced time and lower temperature required for homogenization of alloying elements within the steel microstructure.

Thin-slab versus thick-slab casting. The economics of thin-slab casting, practiced by mini-mills, versus conventional thick-slab casting, used by large integrated steel mills, are a major force in changing how sheet steel will be produced in the future. Conventional thick-slab casting generally requires large-tonnage production to be economical and competitive. The total unit cost of about $90/ton ($100/metric ton) for a product of hot-rolled strip produced from thick slab compares favorably with about $68 ($75) for hot-rolled strip produced from thin-slab casting at an annual capacity of 880,000 tons (800,000 metric tons). This capacity is similar to that for the Crawfordsville facility. These economics illustrate the reason for the current competitive advantage of this new sheet-steel production system. A mini-mill plant of lower (that is, 400,000 tons or 360,000 metric tons per year) capacity would lose much of its competitive advantage as compared with a thick-slab casting process now considered conventional in a large integrated steel mill. At a plant capacity of about 800,000 tons (730,000 metric tons) per year, thin-slab casting has a significant competitive advantage of $24/ton ($26/metric ton) over conventional practice.

Other thin-slab processes. An alternative process to the thin-slab casting process used in Crawfordsville is a continuous casting process using parallel-sided water-cooled molds; it is under development in Germany, with a production plant being installed in Cremona, Italy. The main difference is that the newer process does not require a concave-cavity mold to accommodate a nozzle for delivering molten steel into the cavity. Instead, it uses a flat, wide flounder-shaped nozzle for entry of the metal. A thin slab 60–80 mm (2.4–3.2 in.) thick emerges from the mold and then is squeeze-rolled to 20 mm (0.8 in.) while hot. The slab still has molten steel in the core during the squeeze-rolling. This plant became operational in early 1992; it represents another important step in the evolution of these thin-slab casting processes.

Other thin-slab processes include a continuous process based on casting with a water-cooled mold and processes based on twin-belt casting.

Direct strip casting. Another advance involves the development of direct strip-casting processes. A sufficiently thin cross section is cast, so that the still-hot metal requires only final rolling. Such processes are under active development and hold promise for bringing further revolutionary changes in systems for producing sheet steel. Extensive research and development under way in Japan, Germany, and the United States promises to eliminate the hot-rolling of steel slabs. Examples of advanced strip-production systems include the melt-drag, the twin-belt, and the Osprey processes. In the melt-drag continuous casting process, a cooled roll is spun in contact with molten metal, which is dragged while heat is extracted so that the metal solidifies. Twin-belt is a continuous process in which two moving, water-cooled steel belts extract heat from molten metal so that a continuously cast slab of metal is produced. In the Osprey process, molten metal is sprayed into a chamber in which the pressure has been reduced, and a fine spray of droplets is solidified onto a slab or other substrate. A number of patents record many approaches to these technological innovations.

Prospects. Thin-slab casting is a viable alternative for the efficient production of sheet steel. The production of this semifinished commodity will be more widely distributed geographically both to serve end users better and to improve access to steel scrap. Production of direct reduction iron will be increasingly important as a source of iron units to these strip producers. Such technological innovations will create job opportunities in new steel-strip-producing facilities. The ultimate beneficiary will be the consumer, who will receive a quality product at lower costs.

For background information SEE ARC HEATING; STEEL; STEEL MANUFACTURE in the McGraw-Hill Encyclopedia of Science & Technology.

J. Keverian

Bibliography. J. C. Argarwal and M. J. Loreth, Economic analysis of thin section casting, *Iron and Steel Maker*, 17(11):39–42, November 1990; W. D. Huskonen, Nucor starts up thin-slab mill, *Metal Producing*, 27(8):33–46, 1989; F. K. Iverson and K. Busse, *A Review of First Year CSP Operations at Nucor Steel's New Thin Slab Casting Facility*, Technical Report—Continuous Casting and Rolling, SMS Schloemann-Siemag Aktiengesellschaft, Dusseldorf, Germany, 1990.

Stress (psychology)

The first section of this article reviews the effects of stress on immune system functioning; the second section discusses posttraumatic stress disorder.

Immune System Functioning

The field of psychoneuroimmunology deals with the relationships among the central nervous system, the endocrine system, and the immune system, and the ways in which psychological and physical stressors can affect immune system functioning and health.

Indicators. Components of the immune system that can be used as indicators of immune functioning include granulocytic leukocytes, lymphocytes, the products of B lymphocytes (antibodies), and macrophages and the monocytes from which they are derived. Techniques used to determine immune function include measuring absolute levels of these components, assessing relative levels (for example, the ratio of T helper cells to T suppressor cells), and examining the functional status of the component such as the ability to reproduce when stimulated by a mitogen or the ability to kill invading cells.

Stress effects on immune response. It was thought that the central nervous system could not influence the immune system. However, links between psychological stressors and immunological response have recently been demonstrated.

Neuroendocrine system. The most likely way that stress affects immune components is through the neuroendocrine system. The effects of stress on endocrine functioning are well established: stress activates the sympathetic adrenal-medullary system and the hypothalamic-pituitary-adrenocortical system, so that circulation of catecholamines (epinephrine and norepinephrine), adrenocorticotropic hormone (ACTH), and corticosteroids (cortisol) increases. In addition, the release of endogenous opioids is associated with stress. These factors are known to affect immune components. For example, catecholamines increase the release of lymphocytes and reduce their functional effectiveness. Cortisol has a depressive effect on immune function by reducing lymphocyte numbers and suppressing natural-killer-cell activity. Opioids reduce lymphocyte response to mitogens and reduce natural-killer-cell activity.

Acute stressors. In animals, exposure to loud noise and electric shock are related to lymphocyte suppression. In humans, acute stressors such as taking examinations are associated with increased antibodies to herpes virus, indicating immunocompromise that allowed the recurrence of a dormant herpes virus. Medical students also show lower percentage of T cells and natural killer cells and decreased effectiveness of lymphocytes and cytotoxic activity of natural killer cells during examination periods. An unusual acute stressor that has been studied is the strain associated with splashdown for astronauts. Following the stress of the reentry period, astronauts showed decreased white blood counts (which include both T and B cells) compared to preflight and inflight levels.

Chronic stressors. The chronic stressors that have been linked to immune system functioning include bereavement, problems in relationships, being a caregiver, and living with an environmental disaster. The bereaved have lower natural-killer-cell activity and suppressed T-cell response to mitogen challenge relative to a control group of nonbereaved who are matched on characteristics such as age, sex, and race. Recently separated or divorced individuals showed lower levels of helper T cells and natural-killer cells. This was particularly true for those individuals who were still attached emotionally to their ex-spouses, and who had not initiated the separation. The lack of satisfactory relationships is another type of relationship stress: loneliness is associated with poorer antibody response and lower natural-killer-cell activity. Providing care for a chronically ill family member can be a source of great stress and has implications for immune functioning. Caregivers of Alzheimer's patients show higher Epstein-Barr virus antibody titers and have a lower percentage of T cells than matched noncaregivers. These effects occur even when differences in nutrition, alcohol use, and caffeine consumption are controlled for. Finally, in a recent study some volunteers were intentionally exposed to a respiratory virus. Those who reported more psychological stress in their lives just prior to being exposed to the virus were more likely to develop a clinically documented viral infection.

Problems. Although a fair number of studies indicate that stress is related to a variety of decrements in immune system functioning, there are some difficulties in reaching clear conclusions. Not all the results support the hypothesis that stress reduces immune functioning; in some studies, stress was associated with an increase in immune functioning rather than an impairment. Both increases and decreases in immune functioning attributable to stress are sometimes found in the same study. A vexing problem is that studies measuring more than one immune system parameter sometimes find conflicting effects of stress on the separate immune components. There is no one measure that gives the best overall assessment of immune system status. Because many of the studies are correlational, it is possible that uncontrolled variables such as amount of sleep or caffeine consumption may have been changed by the stressor and may be responsible for the immune impairment attributed to stress. Despite these problems, there is an established pathway (the neuroendocrine system) by which stress can affect the immune system; and there is strong, but not conclusive, evidence that stress adversely affects immune functioning.

Implications for health. Another important question concerns the health implications of stress-related changes in the immune system. Are such changes in immune system components likely to leave an organism vulnerable to disease or impair its ability to recover from a disease? At least three problems face researchers in this area. First, it is not known what degree of change in immune components is likely to leave an organism vulnerable to disease. Thus, it is difficult to determine whether the stress-related changes in immune functioning are large enough to be meaningful. Second, the time delay between the experience of stress and immune system impairment that is large enough to have disease consequences has not been established. It is not known if stress is likely to have immediate or delayed consequences for health. For example, one would expect the infection with upper respiratory disease or herpes virus to occur within a week after the stress, but health outcomes such as cancer or acquired immune deficiency syndrome (AIDS) are not likely to be evident for months or years. Third, there are difficulties in establishing that a stressor caused an immune system decrement and that the immune impairment was responsible for decreased health. For example, the bereaved have poorer immune functioning and have more illnesses than nonbereaved controls. However, the poorer health of the bereaved may not be due to the loss in immune functioning but may be caused by other factors associated with the loss of a spouse such as poor nutrition or lack of sleep. Despite these problems, research programs are beginning to address the issue of health implications, especially in the areas of cancer, genital herpes, and AIDS. The results are complex, but generally indicate that emotional reactions to stress such as hopelessness, negative affect, or a lack of positive affect are associated with impairment of at least one immune component and also with a health outcome—for example, length of survival in cancer, recurrence in herpes, or progression of disease in AIDS.

Interventions. A final question is whether interventions to reduce stress can enhance immune functioning or control the damage done by exposure to stressful circumstances. The promising first studies in this area have shown that a group of college students who were instructed to express their emotions (which presumably reduces stress) through writing about a traumatic experience for 4 consecutive days had increased mitogen response relative to a control group who wrote about neutral topics. In another study, older adults who received relaxation training to reduce stress had significantly higher levels of natural killer cells and lower antibody titers to the herpes virus than those randomly assigned to a control group that received no relaxation training. In another study, cancer patients who received training in relaxation and coping showed enhanced immune functioning relative to the control group who received no instruction. Although the problem of determining the health implications of changes in immune functioning affects these studies, the findings are important because they indicate that the immune system can be strengthened through techniques that reduce stress.   *Suzanne C. Thompson*

Posttraumatic Stress Disorder

Posttraumatic stress disorder is a psychiatric disorder that became a formal diagnostic entity in 1980. Conditions similar to this disorder, however, have been described for a long time. Terms such as traumatic neurosis, war neurosis, and shell shock had been used to loosely identify a poorly understood condition in which soldiers became psychologically dysfunctional during or following combat. Posttraumatic stress disorder, on the other hand, has been carefully described, and a large body of research continues to accumulate on the validity of the diagnosis, causative and biological factors, and methods of treatment.

Diagnosis. Posttraumatic stress disorder is currently considered to be one of the anxiety disorders. The characteristic features of anxiety disorders are fear, particularly in the absence of a real-life threat to safety, and avoidance behavior. For a diagnosis of posttraumatic stress disorder, four criteria must be met. First, the person must have been exposed to an extremely stressful and traumatic event beyond the range of normal human experience. Examples include combat, a direct threat to one's life such as armed robbery or rape, and natural disasters such as earthquakes. Second, the individual must periodically and persistently reexperience the event. This reexperiencing can take different forms, such as recurrent dreams and nightmares, an inability to stop thinking about the event, flashbacks during which the individual experiences reliving the trauma, and auditory hallucinations. The third criterion involves persistent avoidance of events related to the trauma, and psychological numbing that was not present prior to the trauma. Individuals with posttraumatic stress disorder often withdraw from others, feel emotionally deadened, lose interest in activities, and avoid contact with anything even remotely similar to the traumatic events. Children who develop posttraumatic stress disorder will sometimes regress to an earlier stage of development. They may become mute or lose their toilet skills. The fourth criterion for a diagnosis of posttraumatic stress disorder is the presence of enduring symptoms of anxiety and arousal. These symptoms can be manifested in different forms, including anger, irritability, a very sensitive startle response, an inability to sleep well, and physiological evidence of fear when reexposed to events that remind the person of the trauma.

Posttraumatic stress disorder can occur at any age, and the symptoms typically begin soon after the cessation of the traumatic event. Most knowledge of posttraumatic stress disorder comes from research on combat soldiers and other victims of war. Perhaps the most widely studied group of soldiers are from the Vietnam War. Research on this group has provided valuable information on the development of posttraumatic stress disorder, the course of the illness, and how it can best be treated.

Validity. When posttraumatic stress disorder was introduced as a formal psychiatric diagnosis, there was controversy about its validity. Some researchers and clinicians thought that posttraumatic stress disorder was not really a disorder but a "pseudocondition" or, perhaps, a conscious faking of symptoms for sympathy or for financial compensation from the government. Others worried that posttraumatic stress disorder would be overdiagnosed in lawsuits that involved some form of distressing event. After approximately 10 years of research, however, there now appears to be a consensus that posttraumatic stress disorder does exist and that it can be an extremely disabling condition.

Efforts have been made to identify physiological markers of the disorder that are unlikely to be under conscious control. For example, data from research comparing the biological eye-blink startle reflex of combat veterans indicated that veterans with posttraumatic stress disorder did indeed have a more sensitive startle response than veterans without the disorder.

Causative factors. Research on the incidence of posttraumatic stress disorder has resulted in some interesting findings. Estimates of occurrence of the disorder in large groups of Vietnam veterans have ranged from 15 to 50%. This range is probably influenced by the subjective nature of perceived stress. What is mildly stressful for some individuals may be moderately stressful or highly stressful for others. Posttraumatic stress disorder symptoms appear to range over a continuum of severity, and it is unlikely that the disorder is an all-or-none phenomenon. After a traumatic event, the person's life is changed, and the memories of the traumatic event cannot be "surgically" removed. The degree of the posttraumatic stress response is likely to be influenced by a complex interaction of personality, nature of the trauma, and posttraumatic events.

Based on research with Vietnam veterans, the best predictor of the onset of posttraumatic stress disorder is the severity of the trauma. A high frequency of combat exposure, long length of tour of duty, and exposure to gruesome events are characteristics that are frequently shared by veterans who developed posttraumatic stress disorder. Premilitary variables, such as level of psychological disturbance and childhood behavior problems, do not seem to be predisposing factors. Behav-

iors during and after the stressful events, however, may play a role in the unfolding of the symptoms. At least one study has suggested that alcohol and other substance abuse behaviors that occurred during the course of traumatic events were associated with a greater likelihood of chronic posttraumatic stress disorder. One possible explanation is that chronic substance abuse negates or dampens a natural adjustment process that involves gradual diminishing of the memories. Indeed, one theory maintains that the high levels of avoidance behavior that characterize the illness actually potentiate and prolong the symptoms. It is well established that failure to expose oneself to stimuli that evoke fear allows the fear response to remain strong. Increased exposure, however, diminishes the fear response. In short, many symptoms of posttraumatic stress disorder may be self-perpetuating.

Events that occur after the cessation of the traumatic event also seem to be important in whether symptoms become severe and intractable. Vietnam veterans who reported poor self-care, substance abuse, and few social support systems during the first 6 months after returning to the United States were more likely to manifest significant chronic posttraumatic stress disorder symptoms.

Physiological changes. In addition to psychological factors, there are important biological events that accompany stress. Physiological arousal responses in patients with posttraumatic stress disorder have been well studied. Compared to combat veterans who do not exhibit posttraumatic stress disorder, veterans with the disorder show greater increases in heart rate, respiration rate, and skin conductivity when reexposed to traumatic stimuli. Posttraumatic stress disorder may also be associated with structural and physiological changes in the brain. Researchers have demonstrated that neuronal death can occur under conditions consistent with heightened and prolonged stress. Other studies have documented decreases in the alpha-frequency activity in the electroencephalographs of combat veterans diagnosed with posttraumatic stress disorder. When individuals are placed under high levels of stress, the activity levels of two brain neurotransmitters, norepinephrine and serotonin, are reduced. Reductions in these chemical transmitters are not typically found under mild or moderate stress conditions. Stressful events also affect the activity levels of the pituitary and adrenal glands. These fluctuations in adrenergic hormonal activity are likely to be associated with periodic alterations in the brain's endogenous opiates. All of these physiological changes are probably complexly related to the persistence, waxing, and waning of symptoms in posttraumatic stress disorder. Researchers are only beginning to unravel the interactions between physiological, cognitive, and behavioral factors involved in the disorder. Another biological issue that is likely to be pertinent relates to the effects of chronic stress on organs other than the brain. Extreme and prolonged stress has been associated with a variety of physical ailments, including heart attacks, ulcers, colitis, and decreases in immunological functioning. Thus, chronic posttraumatic stress disorder may be associated with characteristic physical diseases and impaired biological functions that are secondary to prolonged stress.

Associated disorders. Posttraumatic stress disorder is rarely the only psychological problem that an affected person manifests. When an individual is diagnosed as having posttraumatic stress disorder, particularly after the disorder has been present for a number of years, it is not uncommon to also find significant depression, substance abuse and dependence, marital problems, and intense, almost debilitating anger. A history of problems associated with substance abuse and anger, such as driving-while-intoxicated and assault charges, is frequently present. These problems are best conceptualized as secondary to the posttraumatic stress disorder and seem to reflect abortive efforts to cope with the disorder. Initial findings indicate that the primary symptoms of posttraumatic stress disorder—anxiety, fear, and avoidance—are actually quite amenable to psychological treatment efforts. The secondary problems commonly associated with the chronic disorder are much more difficult to treat. Again, an accurate, prompt diagnosis is a key feature in the effective treatment of posttraumatic stress disorder.

Treatment. Most of the psychological treatments for posttraumatic stress disorder involve reexposure to the traumatic event. This reexposure is typically imaginal and can range from simply talking with the individual about the trauma to having the person vividly imagine reliving the traumatic event. This latter behavioral procedure is called implosion therapy or flooding. While flooding is not appropriate for all posttraumatic stress disorder patients, the procedure can dramatically decrease anxiety and arousal, intrusive thoughts, avoidance behavior, and emotional numbing. Unfortunately, other problems that may accompany the disorder, such as substance abuse and interpersonal difficulties, are less sensitive to the beneficial effects of flooding. Along with specific behavior interventions, posttraumatic stress disorder patients should be involved in psychotherapeutic treatment for secondary problems.

For background information SEE CELLULAR IMMUNOLOGY; ENDOCRINE MECHANISMS; IMMUNOLOGY; NORADRENERGIC SYSTEM; STRESS (PSYCHOLOGY) in the McGraw-Hill Encyclopedia of Science & Technology.

Robert W. Butler

Bibliography. R. W. Butler et al., Combat-related posttraumatic stress disorder in a nonpsychiatric population, *J. Anxiety Dis.*, 2:111–120, 1988; *Diagnostic and Statistical Manual of Mental Disorders*, 3d ed. rev., American Psychiatric Association, 1987; R. Glaser et al., Stress-related immune suppression: Health implications, *Brain Behav. Immun.*, 1:7–20, 1987; L. C. Kolb, A neuropsychological hypothesis explaining posttraumatic stress disorders, *Amer. J. Psych.*, 144:989–995, 1987; L. Michelson and L. M. Ascher, eds., *Anxiety and Stress Disorders*, 1987; A. O'Leary, Stress, emotion, and human immune function, *Psych. Bull.*, 108(3):363–382, 1990; G. F. Solomon, Psychoneuroimmunology and human immunodeficiency virus infection, *Psych. Med.*, 7:47–57, 1989.

Supercomputer

The parallelism characteristic of supercomputers can be exploited by writing parallel programs or alternatively by using a parallelizing compiler (usually called a restructurer) to translate existing or newly written sequential code into a parallel form tailored to the organization of the supercomputer. Explicit parallel code is usually more difficult to write and understand than sequential code. By allowing the exploitation of parallelism from sequential code, restructurers can facilitate the task of the programmer and help decrease the cost of developing and maintaining parallel programs. Also, restructurers make possible the efficient execution of old sequential code (dusty decks) on modern supercomputers.

Restructurers have traditionally been part of the standard software of practically all supercomputers. FORTRAN, sometimes with vector or concurrent extensions or both, is the source language accepted by most compilers. FORTRAN predominates because most of the time supercomputers are used for numerical applications, which have traditionally been programmed in FORTRAN. Recently, parallelizing C compilers have become more common even though present translators cannot deal effectively with pointers, which are used in the C language.

Practically all existing restructurers are targeted at vector computers or shared-memory multiprocessors. There are no commercially available translators for distributed-memory multiprocessors. The study of techniques for such translators is the subject of active research. Restructuring for distributed-memory machines is more difficult than restructuring for shared-memory multiprocessors. Part of the reason is that in distributed-memory machines, the restructurer not only has to detect parallelism to achieve good performance but also has to distribute the data properly across the local memories of the system. SEE CONCURRENT PROCESSING.

Existing restructurers for shared-memory multiprocessors and vector machines are only partially successful. For some programs they produce very efficient parallel code, but for many others they fail to detect much of the parallelism present in the program. Further improvement of the compiler technology is clearly needed. Fortunately, recent experiments indicate that much better results can be obtained by applying modified versions of current techniques.

Dependence relations and graphs. A discussion of the methodology used by restructurers may begin by considering a sequence of assignment statements. In a sequential program, the statements in a sequence execute in the order in which they appear. However, many assignment statement sequences can be automatically translated into parallel form, because two assignment statements can execute in parallel without affecting the final outcome if they are not dependent on each other.

Given two statements, S and T, such that S precedes T in the sequence, there are four cases in which T is said to be dependent on S:

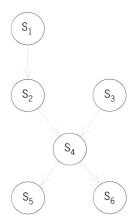

Fig. 1. Dependence graph of the sequence of assignment statements in example (1) of the text.

1. S writes a variable that is read by T.
2. S reads a variable that is written by T.
3. Both S and T write the same variable.
4. There exists a statement U such that U is dependent on S and T is dependent on U.

It is easy to see why these situations preclude the fully parallel execution of S and T. In the first case, T uses a value computed by S. In the second case, S should fetch the variable before T modifies it. To understand why the third case precludes complete parallelism, it is helpful to consider a statement W that follows S and T in the sequence and that reads the variable, say v, written by S and T. Clearly, W should read the value written by T. However, this cannot be guaranteed if S and T execute in the parallel and write v in an unpredictable order.

The dependence relation is usually represented as a directed graph, known as a dependence graph, where the nodes represent statements, and an arc from node N_1 to N_2 indicates that the statement represented by

N_2 is directly dependent on the statement represented by N_1. This dependence graph is used to determine which parallelization transformations are valid. Two statements can execute in parallel if there is no path connecting them in the dependence graph. For example, the sequence of assignment statements in example (1) has the dependence graph of **Fig. 1**.

$$
\begin{array}{lll}
S_1: & a = b + c & \\
S_2: & d = a + c & \\
S_3: & e = f + m & \\
S_4: & n = e + d & (1) \\
S_5: & p = n + 1 & \\
S_6: & q = n + b &
\end{array}
$$

From the information in this graph, a restructurer could transform the previous sequence into a maximal parallel form where the sequence $S_1; S_2$ executes in parallel with S_3, and, after S_4 completes, S_5 executes in parallel with S_6. This maximally parallel code may not be faster than the serial code if the overhead associated with the creation and coordination of the parallel threads is too high. A restructurer has to be aware of the overhead and should generate parallel code only when it will improve performance over the serial version. Otherwise, the code should be left unmodified.

Parallelization of DO loops. In many numerical programs, much of the execution time, and often most of the parallelism, is inside DO loops. A DO loop is a compound statement whose body (a sequence of statements) is executed once for each value assumed by the index variable. The loop header specifies the index variable and its range of values. For example, in the header (2), the index variable is i and it assumes

$$
\text{DO } i = 1, n \qquad (2)
$$

the values $1, 2, \ldots, n$, in that order.

Before describing DO-loop transformations, it is necessary to describe their dependence graph because DO-loop transformations are constrained by the dependence relation.

Construction of dependence graphs. If the number of iterations of a DO loop were known at compile time, the loop could be transformed into a sequence of assignment statements by unrolling it, that is, replacing the loop with one copy of the body for each value assumed by the index variable. Given a statement S in the body of a loop, S in each iteration is called an instance of S. After unrolling the loop, each statement instance becomes a separate statement. Hereafter, the instance of S corresponding to iteration j will be denoted $S(j)$.

To describe the dependence graph of a DO loop it is convenient to consider the unrolled form of the loop. Given two statements, S and T, T is said to be dependent on S if, in the unrolled form of the loop, there is an instance of T that is dependent on an instance of S. In the dependence graph of a DO loop, there is one and only one node for each statement in the loop body, and the arcs represent the dependence relation.

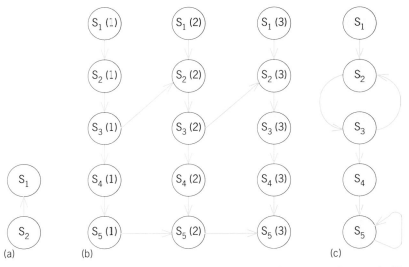

Fig. 2. Dependence graphs of DO loops. (*a*) Dependence graph of the loop in example (3) of the text. (*b*) Dependence relation between the 15 statement instances of the loop in example (5) of the text. (*c*) Dependence graph of this loop.

For example, it can be seen that when the loop in example (3) is unrolled, as shown in (4), the in-

$$
\begin{aligned}
&\text{DO } i = 1,\ 3\\
&\qquad S_1: \qquad a(i) = b(i) + c(i)\\
&\qquad S_2: \qquad c(i+1) = d(i) + 1\\
&\text{END DO}
\end{aligned}
\tag{3}
$$

$$
\begin{aligned}
S_1(1): &\qquad a(1) = b(1) + c(1)\\
S_2(1): &\qquad c(2) = d(1) + 1\\
S_1(2): &\qquad a(2) = b(2) + c(2)\\
S_2(2): &\qquad c(3) = d(2) + 1\\
S_1(3): &\qquad a(3) = b(3) + c(3)\\
S_2(3): &\qquad c(4) = d(3) + 1
\end{aligned}
\tag{4}
$$

stances of S_1 corresponding to iterations 2 and 3 [that is, S_1 (2) and S_1 (3)] are dependent on the instances of S_2 corresponding to iterations 1 and 2 [that is, S_2 (1) and S_2 (2)], respectively. The dependence graph of the loop has the form given in **Fig. 2***a*.

Cycles are possible in the dependence graphs of DO loops. For example, the dependence relation between the 15 statement instances of the loop in example (5)

$$
\begin{aligned}
&\text{DO } i = 1,\ 3\\
&\qquad S_1: \qquad a(i) = b(i) + c(i)\\
&\qquad S_2: \qquad d(i) = e(i-1) + a(i)\\
&\qquad S_3: \qquad e(i) = d(i) + 1\\
&\qquad S_4: \qquad f(i) = e(i) * 2\\
&\qquad S_5: \qquad s = f(i) + s\\
&\text{END DO}
\end{aligned}
\tag{5}
$$

is given in Fig. 2*b*. Then it can be seen that the dependence graph of the loop has the form given in Fig. 2*c*.

In practice, restructurers do not build the dependence graph by unrolling loops but by solving systems of equations. For example, to determine whether there is a dependence from S_3 to S_2 caused by the array e, it is necessary to determine whether the subscripts of the array e in both statements coincide. This can be answered by determining whether the equation $k = j - 1$ has a solution under some constraints. In this equation, k represents the subscript of e in S_3, and $j - 1$ represents the subscript of e in S_2. The constraints are that both k and j are within the loop limits (1 and 3 in this case) and that $k < j$. In general, the equation for computing the dependence relation is more complex than the one in this example because a statement can be surrounded by several DO loops, there can be several subscripts, and the subscript expressions can be more elaborate. Several techniques have been developed to determine when no dependence exists. Because of the desirability of decreasing compilation times and also because of mathematical limitations, these techniques are conservative in that sometimes they fail to detect that there are no solutions to an equation and therefore produce a dependence graph with more arcs than are necessary. Fortunately, these extra arcs seem to be introduced quite infrequently, and, in any case, the extra dependence arcs do not affect the correctness of the transformations applied by restructurers.

The discussion below contains the term π-block, which is defined as the set of statements in a strongly connected component of the dependence graph. Two nodes, v and w, belong to the same strongly connected component if there is a path in the dependence graph from v to w and from w to v. A node that does not belong to any cycle is also a strongly connected component consisting of one node. For example, in the previous loops there are four π-blocks: $\{S_1\}$, $\{S_2, S_3\}$, $\{S_4\}$, and $\{S_5\}$. One immediate observation is that all iterations of the nodes that do not belong to any cycles in a loop dependence graph can be executed in parallel because the statement instances corresponding to such nodes do not depend on each other.

Parallel code for vector machines. Several strategies can be followed to generate parallel code based on a dependence graph. When the target of the restructurer is a vector machine, one simple strategy starts by distributing the loop, that is, by breaking the original loop in such a way that each π-block ends up in a separate loop with the same index variable, initial value, last value, and increment as the original loop. After distribution, each loop surrounding a node that does not belong to any cycles in the dependence graph can be easily transformed into a vector operation.

These ideas can be illustrated by considering again the DO loop of example (5). After distribution, it takes form (6). The loops corresponding to statements S_1 and

$$
\begin{aligned}
&\text{DO } i = 1,\ 3\\
&\qquad S_1: \qquad a(i) = b(i) + c(i)\\
&\text{END DO}\\
&\text{DO } i = 1,\ 3\\
&\qquad S_2: \qquad d(i) = e(i-1) + a(i)\\
&\qquad S_3: \qquad e(i) = d(i) + 1\\
&\text{END DO}\\
&\text{DO } i = 1,\ 3\\
&\qquad S_4: \qquad f(i) = e(i) * 2\\
&\text{END DO}\\
&\text{DO } i = 1,\ 3\\
&\qquad S_5: \qquad s = f(i) + s\\
&\text{END DO}
\end{aligned}
\tag{6}
$$

S_4 can be transformed into the vector statements in (7).

$$
\begin{aligned}
S_1: &\qquad a(1:n) = b(1:n) + c(1:n)\\
S_4: &\qquad f(1:n) = e(1:n) * 2
\end{aligned}
\tag{7}
$$

After this transformation, there are still two loops that remain sequential because of the dependence cycles. One way in which a restructurer can deal with cycles in the dependence graph is by using pattern matching and replacing the loop containing the dependence graph cycle by a call to a function or subroutine that executes in parallel. For example, most restructurers that generate vector statements can recognize loop (8)

$$
\begin{aligned}
&\text{DO } i = 1,\ n\\
&\qquad S_5: \qquad s = f(i) + s\\
&\text{END DO}
\end{aligned}
\tag{8}
$$

and generate code (9).

$$
s = s + \text{SUM}\,(f(1:n))
\tag{9}
$$

The function $\text{SUM}\,(f(1:n))$ produces the sum of all the elements of the vector f and can be evaluated

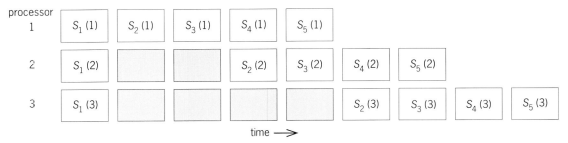

processor 1 | $S_1(1)$ | $S_2(1)$ | $S_3(1)$ | $S_4(1)$ | $S_5(1)$

2 | $S_1(2)$ | | | $S_2(2)$ | $S_3(2)$ | $S_4(2)$ | $S_5(2)$

3 | $S_1(3)$ | | | | $S_2(3)$ | $S_3(3)$ | $S_4(3)$ | $S_5(3)$

time $\longrightarrow$

Fig. 3. Execution of the loop in example (5) of the text on a multiprocessor. Color areas indicate idle processors.

in parallel. If sufficient resources are available, the number of steps needed to complete SUM is only the smallest integer equal to or greater than $log_2 n$, in contrast to the n steps required by the sequential version.

Parallel code for multiprocessors. There are also several possible strategies for loop parallelization when the target computer is a multiprocessor. If the dependence graph is acyclic, the loop can be transformed into a parallel DO loop whose iterations can be executed in parallel with each other. For example, the loop of example (3) can be transformed into code (10).

$$
\begin{aligned}
&a(1) = b(1) + c(1) \\
&\text{PARALLEL DO } i = 1, 2 \\
&\quad c(i + 1) = d(i) + 1 \\
&\quad a(i + 1) = b(i + 1) + c(i + 1) \\
&\text{END PARALLEL DO} \\
&c(4) = d(3) + 1
\end{aligned} \tag{10}
$$

An alternative approach to parallelizing this loop is to distribute the loop and then create two parallel loops, one surrounding each statement. However, a single parallel loop is usually better because it allows saving in overhead.

Several strategies are also possible when there are cycles in the dependence graph and the target machine is a multiprocessor. One possibility consists in allowing the different iterations to execute in parallel but inserting synchronization instructions to guarantee that successive iterations of a given π-block execute serially with each other. In the case of the loop of example (5), the execution could, for example, follow the order of the time lines shown in **Fig. 3**.

For background information SEE CONCURRENT PRO-CESSING; DIGITAL COMPUTER PROGRAMMING; MULTIPRO-CESSING; PROGRAMMING LANGUAGES; SUPERCOMPUTER in the McGraw-Hill Encyclopedia of Science & Technology.

David A. Padua

Bibliography. U. Banerjee, *Dependence Analysis for Supercomputing*, 1988; D. A. Padua, D. J. Kuck, and D. H. Lawrie, High-speed multiprocessors and compilation techniques, *IEEE Trans. Comput.*, C-29: 763–776, 1980; D. A Padua and M. J. Wolfe, Advanced compiler optimizations for supercomputers, *Commun. ACM*, 29:1184–1201, 1986; M. J. Wolfe, *Optimizing Supercompilers for Supercomputers*, 1989.

Superconducting devices

Although various types of superconducting devices have been proposed, almost all are still in early stages of research. Only the Josephson junction, a device that utilizes the tunneling of superconducting electrons, has found practical application. The Josephson junction has been applied to many areas of electronics, including high-sensitivity magnetic sensors, electromagnetic detectors, and voltage standards. Another application of the Josephson junction is in the construction of an ultrahigh-speed computer. Not yet commercially available, such a computer is rapidly being developed.

Ultrafast computer requirements. The construction of an ultrafast computer requires a switching element that is fast and consumes little power. The reason for the second requirement is that fast switching in itself is not sufficient for ultrahigh calculation speeds, and the propagation delay time between individual elements must also be minimized. To reduce the delay, the device spacing must be reduced, and low power consumption is required to prevent overheating of the switching elements. For example, at present, a large-scale computer having a central processing unit of 20×20 in. (50×50 cm) might consume a few kilowatts of power. To increase the operation speed of this unit by a factor of 10, it is necessary both to increase the switching speed by a factor of 10 and to decrease the board size to 2×2 in. (5×5 cm). At present levels of power consumption, a computer of these dimensions would burn out as soon as it was turned on. Therefore, the power consumption must be three orders of magnitude smaller than that of conventional elements. The Josephson junction is unique in its ability to fill this requirement.

Figure 1 shows the relationship between switching time and power consumption for various high-speed devices. The switching speed of the Josephson junction is one order of magnitude faster than that of conventional bipolar transistor logic or emitter-coupled logic (ECL). In addition, the Josephson junction consumes three orders of magnitude less power than emitter-coupled logic. Thus, the faster Josephson junction can be more densely integrated, so that an ultrahigh-speed computer is possible.

Josephson junction development. Until 1983, lead alloy was mainly used as the superconducting material for Josephson junctions. However, the lead-alloy junction was extremely unstable. The superconducting current flowing through the junction increased with the storage time at room temperature, and the junctions broke down after a few years. Thus, the lead alloy junction could not be reliably applied either to high-speed circuits or to small-scale applications such as magnetic sensors, which require only a few junctions. It was thought that the Josephson computer would never be realized.

In 1983, a Josephson junction was fabricated from niobium instead of lead alloy. The junction consists of niobium electrodes sandwiching an ultrathin aluminum oxide layer a few nanometers thick. This is the most reliable junction at present. The introduction of a high-quality junction made from niobium has opened the door to developing ultrahigh-speed Josephson computers. A Josephson microprocessor, data processor, digital signal processor, and memory are already feasible. These circuits operate much faster and consume less power than semiconductor circuits. It is now feasible to fabricate all components necessary to construct a Josephson computer, although the present integration density of the circuits is lower than that of semiconductor devices.

High-speed circuits. Logic gates with functions such as OR and AND can be constructed by combining a few Josephson junctions, although the junction itself cannot isolate an output signal from an input signal because it is a two-terminal device. More than 10 types of Josephson logic gates have been proposed. Some gates have been replaced by new improved gates. Josephson logic gates now being used include Josephson interferometer logic (JIL), 4-junction logic (4JL), resistor-coupled Josephson logic (RCJL), and modified variable-threshold logic (MVTL). These gates exhibit switching times of a few picoseconds per gate with power consumption of several microwatts per gate (Fig. 1). Among them, the fastest logic gate is MVTL, whose switching time is 1.5 ps, with a power consumption of 12 μW. This switching speed is faster than that of any other logic gate, including gates based on semiconductor devices. *See* Transistor.

Various high-speed large-scale integrated circuits have been fabricated with niobium Josephson junctions. They include a 4-bit microprocessor, an 8-bit digital signal processor, and a 4-kilobit memory. The **table** compares typical speeds and power consumptions of AM2901-type 4-bit microprocessors for silicon, gallium arsenide (GaAs), and niobium Josephson junctions. (The AM2901 is considered to be the world's standard 4-bit microprocessor.) Clearly, the Josephson junction is superior to the semiconductor devices. Other microprocessors have also been developed. Almost all microprocessors operate at a clock frequency on the order of 1 GHz, which is one order of magnitude faster than that of semiconductor processors. The largest Josephson circuit fabricated is an 8-bit digital processor having 6300 gates (**Fig. 2**). This chip functions in a manner similar to commercially available complementary metal-oxide semiconductor (CMOS) chips, and can be operated nearly 100 times faster than a CMOS digital signal processor.

Josephson memory circuits can also be manufac-

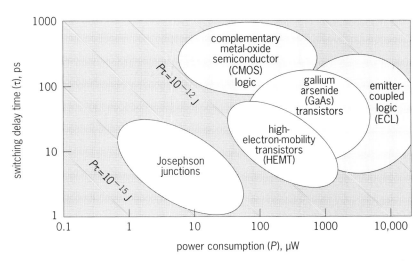

Fig. 1. Relationship between switching time and power consumption for various high-speed devices.

tured, although the circuit density is not as high as that of semiconductor circuits. The integration level of current Josephson memory is 1 to 4 kilobits. The typical access time is 500 ps, with a power consumption of 5 to 20 mW per chip.

An important circuit needed to realize Josephson computers is the interface circuit between the Josephson circuits and the semiconductor circuits. Josephson logic circuits operate with a logic swing voltage of less than 3 mV, so that a driver circuit is necessary to amplify the output voltage to 1 V to drive room-temperature semiconductor circuits. The driver circuit has been fabricated to demonstrate that the semiconductor circuits can be driven in this manner. An output voltage from a Josephson gate was amplified to 150 mV with a Josephson driver, and further amplified to 1.7 V by a gallium arsenide (GaAs) comparator immersed in liquid helium. An output voltage of 1.7 V is sufficient to drive semiconductor circuits at room temperature. The output interface circuit could operate up to a clock frequency of 800 MHz.

Development of Josephson computer. At present, several large-scale integrated circuits have been fabricated for the Josephson computer. They include logic having a few thousand gates, a few kilobits of memory, and an interface to amplify the signal to the semiconductor circuits. Although their integration density is not high, they operate at much faster

Comparison of performance of 4-bit microprocessors (type AM2901)

Device	Clock frequency, MHz	Power dissipation, W
Silicon	30	1.4
Gallium arsenide (GaAs)	72	2.2
Niobium Josephson junction	770	0.005

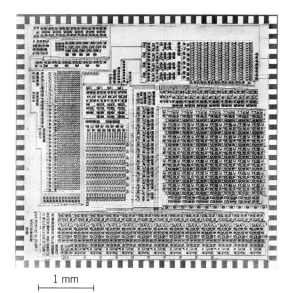

1 mm

Fig. 2. Eight-bit digital signal processor (DSP), a Josephson circuit with 6300 gates.

speeds and lower power consumptions than semiconductor circuits. The next step in development is to demonstrate the high-speed performance of a prototype Josephson computer.

A large-scale Josephson computer would be very useful in a variety of applications that require high-speed computation. However, since the technology has not yet been established, the first Josephson computer will be built on a small scale. Most likely, it will be a special-purpose computer, such as those needed by radio astronomers or medical researchers. To manufacture a special-purpose computer, a hybrid system consisting of Josephson and semiconductor devices would be most practical. A hybrid computer would relieve bottlenecks in high-speed computation by having them performed by Josephson devices and having other calculations done by semiconductor devices.

In this system, it is important to reduce the distance between the liquid-helium and room-temperature devices, because this distance determines the transmission time between Josephson devices and semiconductor devices. A shorter distance results in higher speeds. The distance was an order of 3 ft (1 m) for conventional cooling systems. However, a cooling system that shortens the distance to 0.8 in. (2 cm) has recently been developed. Thus, it may be possible to operate a hybrid computer at clock frequencies as high as 1 GHz.

After the usefulness of a hybrid computer system has been demonstrated, a large-scale high-speed Josephson computer could be developed.

For background information SEE *INTEGRATED CIRCUITS; JOSEPHSON EFFECT; SUPERCOMPUTER; SUPERCONDUCTING DEVICES* in the McGraw-Hill Encyclopedia of Science & Technology.

Shinya Hasuo

Bibliography. M. Gurvitch, M. A. Washington, and H. A. Huggins, High quality refractory Josephson tunnel junctions utilizing thin aluminum layers, *Appl. Phys. Lett.*, 42:472–474, 1983; S. Hasuo and T. Imamura, Digital logic circuits, *Proc. IEEE*, 77:1177–1193, 1989; S. Kotani et al., An 8-b Josephson digital signal processor, *IEEE J. Solid State Circ.*, 25:1518–1525, 1990; S. Kotani et al., A sub-ns clock cryogenic system for Josephson computers, *Digest of Technical Papers of the IEEE International Solid-State Circuitry Conference*, pp. 32–33, February 1991.

Symbolic computing

Symbolic systems are a special class of computer programs that perform computations with constants and variables according to the familiar rules of algebra, calculus, and other branches of mathematics. They are also called algebraic computation systems, computer algebra systems, or symbolic computation systems. They are used in science and engineering teaching and research to perform nonnumerical calculations that would be done less effectively by hand or numerical means.

Characteristics of symbolic systems.
Symbolic systems manipulate symbols and equations, unlike conventional scientific systems that manipulate numbers. For example, if a symbolic system is given the expression $x + x$, it will normally return the answer $2x$ and not question the absence of a value for x. Although most provide a language for programming in the conventional sense, they are normally used in the style of a calculator, in which an expression is given as input and the system responds with the answer. As a result, they are well suited for use on personal computers and workstations. Some of their capabilities have also recently appeared in calculators.

Another important characteristic of symbolic systems is that they produce exact results for calculations that, when done by hand, are often tedious, error-prone, and difficult. Even the arithmetic is normally exact; numbers are manipulated with as many digits as are necessary for a given computation, unlike numerical languages that usually impose a fixed precision on their manipulations. Systems can now perform calculations that could not be done in a lifetime by hand. In other words, they can solve problems that are incapable of realistic hand solution, such as advanced calculations in fields such as celestial mechanics, general relativity, and elementary particle physics.

The long-term goal of symbolic systems is to automate as much of the mathematical problem-solving process as possible. However, there still remains much of mathematics for which the theory is insufficient for computer implementation. In the meantime, symbolic systems provide a tool whose purpose is to increase scientific and engineering productivity. In most cases, this tool is used in the initial steps of the problem-solving process, with numerical computation as the final step. Of course, there are problems in which numerical steps are interleaved with symbolic steps, but most calculations involve transforming the problem by various symbolic steps to a form in which the final numerical computation is relatively easy.

The field began seriously in the 1960s. The first systems were mostly designed to solve particular problems in various application areas such as those mentioned above. Systems have also been developed for mathematical disciplines such as group theory and number theory. Although such special-purpose systems are still widely used, there has been increasing emphasis on general-purpose systems, which can solve problems in a wide range of areas, including those served by the special-purpose systems. In fact, some of the general-purpose systems now in use were originally designed for a specific application, and evolved with time into a more general problem-solving tool. One result of this evolution, however, is reduced efficiency; general-purpose systems tend to be slower than special-purpose systems. However, as computers have increased in power and the general-purpose software has become more sophisticated, this has been less of a concern except for the most difficult problems in a given application area.

Facilities offered. There are now many general-purpose symbolic systems available. The best known of these still under active development are Axiom (also known as Scratchpad), Derive, MACSYMA, Maple, Mathematica, and REDUCE. With the exception of Derive, which is designed for IBM PC–compatible machines, these systems normally require a computer with several megabytes of memory to operate effectively. Although each system has its own strengths and weaknesses, they all share a common basic set of capabilities. These include:

1. Exact integer arithmetic with any number of digits, and floating-point arithmetic to whatever precision is desired.

2. Multivariate polynomial manipulation, including the expansion of polynomials and collection of like terms, factorization, and reduction of rational expressions to lowest terms by the elimination of greatest common divisors.

3. Support for elementary functions, such as sines, cosines, logarithms, and exponentials, including the basic formulas for their manipulation.

4. Symbolic differentiation and integration of expressions. All systems can differentiate almost any expression, but the facilities for integration vary. Limited facilities for the solution of differential equations are also provided by most systems, again with wide variation in capabilities between systems.

5. Matrix manipulation, including the computation of determinants, inverses, and eigenvalues and eigenvectors.

6. Power series manipulation, including the addition, multiplication, and division of such series and division of such series and their reversion.

7. Solution of one or more algebraic equations. Most systems can solve nonlinear as well as linear systems, although the capabilities vary.

8. Production of output in a variety of forms, including expressions specifically optimized for numerical computation in the syntax of a numerical programming language, and output usable by typesetting systems.

9. A variety of additional algebraic capabilities

such as limits, and closed-form sums and products of sequences. These capabilities are usually limited in scope.

Most systems also provide for the graphing of algebraic expressions in either two or three dimensions, and some have a notebook-style user interface that includes text and mathematical output of book quality. In this area, too, there is considerable variation among systems, with Mathematica advertising these capabilities as among its most important.

In addition, each system provides specific facilities for various application areas. It is therefore important for potential users to understand the extent to which a given symbolic system is tailored to their needs before making a choice of which to use.

Of all these areas, the two where the most progress recently has been made have been symbolic integration and the solution of systems of nonlinear polynomial equations. In symbolic integration, considerable advances have occurred during the past few years in the algorithmic determination of integrals of algebraic functions where they exist in closed form. Axiom has the most complete implementation of these algorithms at present. For example, Axiom can show that expression (1) is equal to expression (2), which would

$$\int \tan \left(\frac{\text{atan } (x)}{3} \right) dx \qquad (1)$$

$$\frac{4}{9} \log \left(3 \tan^2 \left(\frac{\text{atan } (x)}{3} \right) - 1 \right) - \frac{1}{6} \tan^2 \left(\frac{\text{atan } (x)}{3} \right)$$

$$+ x \tan \left(\frac{\text{atan } (x)}{3} \right) + \frac{8}{9} \qquad (2)$$

have been impossible for any algebra system to do as recently as 1990. [Here, atan (x) is the arc tangent of x.]

As far as nonlinear polynomial equations are concerned, the last few years have seen the development of a method known as Groebner bases, which has enabled researchers to find closed-form analytic solutions to many sets of coupled nonlinear equations for which such solutions were previously unknown. These analytic solutions have given researchers increased insight into problems for which the numeric methods, because of the approximations involved, sometimes gave wrong answers.

Applications. There are now several thousand published papers acknowledging the use of symbolic systems. Nearly every scientific and engineering discipline has examples of applications. In physics, there have been important applications in quantum chromodynamics and quantum electrodynamics, celestial mechanics, general relativity, astrophysics, plasma physics, fluid mechanics, optics, and quantum mechanics. In chemistry, symbolic systems have been used in molecular electronic structure calculations, solutions to reaction rate equations, finding steady-state solutions of complex reacting systems, and kinetic energy operator calculations. In mathematics, these methods have been applied to essentially every area, including numerical analysis, coding theory, probability theory, cryptography, number theory, group theory, finite fields, geometrical theorem proving, algebraic geometry, homology, and topology. In engineering, applications can be found in electrical network analysis, circuit design, turbine design, ship hull design, helicopter rotor design, control theory, image and signal processing, and antenna design.

Educational uses. For a long time, symbolic systems were limited in use to researchers with access to large mainframe computers. However, the personal computer and workstation have changed this and have encouraged the development of new products, such as Mathematica, designed for a mass market. Such systems are now being used in the teaching of mathematics, with this use expected to grow significantly.

As the audience for symbolic systems changes from sophisticated researchers to students, different characteristics of these systems become important. One common source of concern is that different systems often give different (but equivalent) answers to the same problem. This arises because the concept of evaluation of algebraic expressions is not as well-defined as for numerical expressions. Sometimes described as simplification, such evaluation is normally designed to reduce expressions to as simple a form as possible. However, in doing so, several subjective decisions must be made by the system designer. For example, expressions often look simpler if partial factorizations are done. On the other hand, such factorizations may prevent various algorithms from working properly, so some systems limit such factorization. Consequently, the results of a given calculation, although equivalent, do not always look the same. However, most systems do offer options for formatting results in different ways, which give users some control over the form of output.

Even more important, systems often carry out transformations that are correct most of the time but can fail in certain regions. For example, essentially all systems will perform transformation (3), even though this fails

$$\frac{x}{x} \rightarrow 1 \qquad (3)$$

if x happens to be zero. As a result, even though symbolic systems are almost always correct, they sometimes compromise in the interest of efficiency. The alternative would be to carry results in the above example for the two cases $x = 0$ and $x \neq 0$. In nontrivial applications, such an approach rapidly leads to results containing a long chain of conditional clauses, making the answers incomprehensible. Some systems also make more assumptions than others and can easily produce logical inconsistencies if they are pushed too far. Such concerns are obviously very important in teaching context, but the complete solution of this problem remains a research issue at the present time.

Another issue that teaching use has raised is the provision of an audit trail for calculations—in other words, a trace that shows the steps the system took to get to the displayed result. This enables users to understand how the calculation was performed, and also gives them an opportunity to determine where things

went wrong with their formulation of the problem if they believe the answer to be incorrect. This again is a difficult issue since in many cases the methods used by symbolic systems are quite different from those used in hand calculations. The latter often rely on heuristics in finding an answer, whereas the former depend much more on algorithms developed specifically for automatic computation. A trace of such algorithms for a case such as factorization or integration is often useless to all but those completely knowledgeable about the structure of the algorithm.

Prospects. The advent of inexpensive computer hardware, plus commercially supported software, is bringing symbolic systems into mainstream use. With increased access to such techniques, considerable progress can be expected in this field during the next few years. The expected gains in computing power plus improved software will enable problems that are now seen as research challenges to be solved by high school students. There remain a number of problems, however. The challenge for the algorithm developer is to bring the goal of automated mathematics nearer to reality. The challenge for the system designer is to build systems that are easier to use and to extend to new problem domains. The challenge for the educator is to adapt the relevant learning materials to incorporate use of these techniques. The challenge for the researcher is to determine the proper place for this tool in his or her work.

For background information SEE ALGORITHM; INTEGRATION; POLYNOMIAL SYSTEMS OF EQUATIONS in the McGraw-Hill Encyclopedia of Science & Technology.

Anthony C. Hearn

Bibliography. A. Boyle and B. F. Caviness (eds.), *Future Directions for Research in Symbolic Computation*, SIAM Reports on Issues in the Mathematical Sciences, 1990; J. H. Davenport, Y. Siret, and E. Tournier, *Computer Algebra, Systems and Algorithms for Algebraic Computation*, 1989; D. Harper, C. D. Wooff, and D. E. Hodgkinson, *A Guide to Computer Algebra Systems*, 1991.

Systems engineering

Systems integration is a logical, objective procedure for applying in an efficient, timely manner new or expanded performance requirements to the design, procurement, installation, and operation of an operational configuration consisting of distinct modules (or subsystems), each of which may embody inherent constraints or limitations. Typically, systems integration requires the coordination of preexisting and coexisting system components with newly developed system components. Thus, systems integration may be the result of significant growth of an existing system; it may occur from the merging of two or more systems; it may come about as the result of combining existing system fragments with commercial off-the-shelf components; or it may involve combinations of these.

Systems integration methodologies. Implementation of a systems integration process combines systems engineering and systems management, and applies these methodologies to the complex problems associated with melding existing systems and new technologies to form a more capable system. The systems integration process applies this combination of skills to large-scale programs that integrate hardware, software, facilities, personnel, procedures, and training to achieve specific goals and objectives. Achievement of the goals of systems integration requires personnel who possess sound engineering skills combined with the ability to integrate technology and operations to provide technical and managerial direction. Systems engineering represents an approach that is essential for the design, development, integration, and operation of large-scale systems. Systems management covers the tasks and related activities associated with planning, control, and operations to achieve the goals and objectives of a specific systems integration engineering program. Systems management provides the necessary supervision skills to move complex designs to a successful conclusion for both the customer and the contracting company. Systems integration programs cover the entire range from procurement of major weapons systems to business management systems.

Systems integration life cycle. Systems integration programs generally are managed through application of a systems development life cycle. This life cycle usually comprises seven phases that range from the identification of requirements and the statement of specifications to operational deployment of the system. Minimally, the system life-cycle phases are requirements definition, design and development, and operation and maintenance. The seven-phase life cycle that is most commonly used comprises requirements definition and specifications; feasibility analysis; program and project planning; logical and physical design; design implementation and system tests; operational deployment; and review, evaluation, and retirement.

The seven-phase life-cycle approach utilizes the tools of systems engineering to develop, analyze, and implement large-scale systems integration programs that are intended to enhance productivity, improve quality of the product, and increase the functionality of the product in operation. A principal advantage of this approach is that it divides large, complex issues and problems into well-defined sequences of simpler problems and issues that are easier to understand, manage, and build.

Systems management. Systems management activities require an in-depth knowledge of the methodologies used for systems integration programs, knowledge of the management techniques related to implementation and management of large-scale integration programs, and the ability to manage the integration of existing system details and constraints together with new requirements. A primary systems management concern is the ability to understand and work with existing systems and resources that have probably been in place for a number of years and have undergone a number of modifications, and to determine how the new integration program should be managed to assure that the new needs of the user are met.

Strategic plan. A major requirement to help ensure the effectiveness of a systems integration program is a strategic plan that serves as a guideline and roadmap for the systems integration project. A systems integration acquisition strategy consists of two parallel paths: the systems specification component itself, and the auditing component. The key goals are to establish the client's functional needs and to establish the general technical capabilities of the system.

The seven principal components of the systems integration strategic plan are to (1) assemble the best available talent to form the technical and management team; (2) provide a guiding plan in the assessment and management of risk; (3) seek out the best contractors and subcontractors to assist in the systems integration project; (4) place exacting contractual obligations on vendors who are successful winners of equipment and software contracts for the systems integration project; (5) develop the best working relationships with the contracting agency without compromising the independence of the systems integration contractor; (6) provide an audit trail that anticipates challenges and protests and implement this for all contractors, subcontractors, suppliers, and other vendors; and (7) follow the strategic plan, carry out the processes, follow the procedures, and verify and validate the outcomes.

In administering the tenets of this strategic plan to ensure success of the systems integration program, a suitable systems engineering methodology that encompasses the entire integration program—from requirements, through design and construction, to test, and finally to deployment and maintenance—is required. Systems management methodologies must be in place to support problem understanding and communication between all parties at all stages of development. The systems integration engineering approach must enable capture of design and implementation needs early in the process, especially interface and interactive needs associated with bringing together new and existing equipment and software. This approach must support both a top-down and a bottom-up design philosophy as well as provide full compliance with system-level quality assurance, risk assessment, and evaluation. The engineering approach must support definition and documentation of all aspects of the program to assure that proper audit-trail information is collected and managed over the life of the program. These activities provide a framework that allows appropriate systems management techniques to be applied to all aspects of the program.

The successful application of a systems integration strategic plan requires a commitment to the project by both the contracting agency and the contractor. It includes the identification of so-called champions in both groups to administer the project and involves both top-down and bottom-up approaches to management and design. A plan to manage subcontractors and provide interfaces to other major players in the project is necessary when a number of suppliers and subcontractors are involved. Adequate preparation and training of personnel responsible for all aspects of the systems integration project is essential for success. A wel_-defined cost management and control procedure for cost awareness and cost control that has been tested and is operational is needed. The ability to understand the objectives of the program, measure and assess progress, and determine and assign appropriate metrics to ascertain systems integration project status is essential. A plan for handling exigencies determines the way in which consideration is given to issues raised by constraints and alterables related to requirements that necessitate an interface with existing systems. A risk-assessment and protest-avoidance plan (designed to avoid protests that might arise because of breaches of the terms and conditions in contracts, failure to provide for free and open competition for procurements, or bidder unhappiness over loss of a contract) coupled with the audit-trail plan is required to assure the management team of adequate information to supervise the systems integration program. Finally, a performance evaluation plan for verification and validation of the system is necessary.

Systems integration program steps. The first step in a systems integration program is preparation of a system implementation and integration plan that will identify, analyze, and prioritize technical and functional integration issues and propose solutions to problems. This plan will identify probable areas of risk, initiate the audit trail for tracking all activities, and provide for appropriate verification and validation methods. The next step is to conduct an impact assessment of existing constraints on the proposed system. This step should allow for refinement of parameters within the architectural structure and for evaluation of alternatives. These alternative systems and processes should be capable of being interfaced and interoperable with existing systems, integrated with other systems, and compatible with other system requirements.

Following these two initial steps, the systems integration project will generally follow the system life cycle presented above. Key activities to be conducted during this process include identification of new technological approaches that will enhance operational functionality of the new system; identification of significant cost-drivers that represent a high percentage of total costs; and identification of methods that will reduce costs while retaining benefits. Following these steps will improve benefit/cost ratios and assist in the clarification of requirements to be placed on subcontractors. Another step is to make certain all subsystem interfaces work and to ensure proper size, scale, speed, and so forth, of all elements. Proper life-cycle management will lead to minimization of life-cycle costs and assure that the system undergoes graceful degradation rather than catastrophic collapse as components fail. In turn, retrofitting or expansion will be simplified. Complete system documentation must be accomplished. Finally, implementation of the strategic plan will serve to assist in the management of reduction of risk due to unforeseen events by providing requirements and design traceability and by maintenance of the audit trail.

Establishing an audit trail and embedding this information in the systems integration process begins immediately upon receipt of the systems integration contract. The way to embed an audit trail in a systems integration program, either by the agency or by the systems integration contractor, is to (1) utilize the system-level requirements and document these in a database; (2) establish whether or not issues are present and, if so, resolve them; (3) assign appropriate validation and verification metrics to requirements; (4) establish guidelines for subcontractors; (5) provide solid, unambiguous procurement instruments (detailed contract specifications for hardware and software procurement, usually acquired from suppliers); and (6) track these activities from specification through installation and operation.

For background information SEE SYSTEMS ANALYSIS; SYSTEMS ENGINEERING in the McGraw-Hill Encyclopedia of Science & Technology.

James D. Palmer

Bibliography. H. Eisner, *Computer Aided Systems Engineering*, 1988; G. M. Hall, Know-how versus know-when in systems integration, *Inform. Strategy*, 6(2):22–27, 1990; A. P. Sage and J. D. Palmer, *Software Systems Engineering*, 1990; R. H. Thayer (ed.), *Software Engineering Project Management*, 1987.

Telescope

Since 1980 there has been a strong resurgence in the design and construction of large optical and infrared telescopes. A significant advance in telescope design came with the Palomar 5-m (200-in.) Hale telescope, completed in 1949. For the next 30 years, telescopes were designed around the Palomar model and, because of the weight and complexity of the primary mirror, were size-limited. New concepts in telescope design have developed since 1975 and permit the construction of appreciably larger telescopes. While these ideas have had the greatest impact on the design of the primary mirror, they also influence the optical design as a whole and the mechanical schema of the protective enclosure. Primaries consisting of a mosaic of mirror segments 10 m (400 in.) in diameter are already under construction, and "rigid" honeycomb single mirrors up to 8 m (320 in.) in diameter and thin, flexible single mirrors up to 8 m (320 in.) in diameter have been proposed.

Increasing the size of telescopes has great scientific value, both because most scientific studies are limited by the amount of available light and because larger telescopes can often provide greater angular resolution.

Fundamental limitations. The fundamental limitations on the power of telescopes involve diffraction, seeing, detectors, gravity, and temperature.

Diffraction. The angular resolution achievable with a telescope is fundamentally limited by diffraction, an effect caused by the finite wavelength of the light being used. The limiting angular resolution θ set by diffraction is $\theta = \lambda/D$, where λ is the wavelength of light being used and D is the diameter of the telescope. For a 10-m (400-in.) telescope at wavelength $\lambda = 500$ nanometers, an angular resolution of 0.01 arc-second is possible.

Seeing. Thermal turbulence in the atmosphere limits the image quality for most large ground-based telescopes. Atmospheric turbulence theory characterizes the atmosphere by a scale length r_0 that varies nearly linearly with wavelength. For visible light, the best astronomical sites on Earth have $r_0 = 0.20$ m (8 in.) and a corresponding angular resolution of $\theta = \lambda/r_0 = 0.5$ arc-second. However, as the wavelength increases, r_0 increases, and at some wavelength it equals the diameter of the telescope at which point the seeing limit and the diffraction limit become equal. Beyond this wavelength, image sizes are set by the diffraction limit. The local thermal environment around the telescope can also substantially contribute to the seeing. For example, if the primary mirror is $1.0°$C ($1.8°$F) hotter than the ambient air, image degradation from thermal turbulence will be about 0.5 arc-second. Therefore, great efforts are now made to avoid trapping heat in the telescope dome.

Detectors. For most of the twentieth century, the photographic plate was the light detector of choice for astronomical observations. Since light is quantized into photons, the quantum efficiency is a central assessment of the efficiency of a detector. Photographic plates have quantum efficiencies of roughly 1%; hence, improving the detector quantum efficiency has been a very effective means of increasing a telescope's light-gathering power. With modern charge-coupled devices (CCDs) as detectors, quantum efficiencies of 80% are now achieved, and additional improvements will be slight. At infrared wavelengths, area detectors with similarly high quantum efficiencies are also now available. Thus, further improvements in telescope power can come only from building larger telescopes.

Gravity. For large mirrors, the force of gravity will naturally lead to distortions of the mirror shape that will significantly degrade the image quality, unless exquisite care is used in designing and building the support system. For a given support system, the gravity-induced deformations δ will vary according to the equation below, where c_1 is geometrical

$$\delta = \frac{c_1 \rho D^4}{E h^2}$$

constant, ρ is the material density, D is the mirror diameter, E is the material modulus of elasticity, and h is the mirror thickness. Hence a 10-m (400-in.) mirror will have 16-times-larger gravity deformations than a 5-m (200-in.) mirror for a given support system and mirror thickness. This simple fact is central to the design considerations for large telescopes. Segmented-mirror telescopes minimize this problem by using relatively small segments and an active control system to position the segments. Large honeycomb mirror designs minimize this problem by making the mirror very thick and stiff. Large, thin monolithic mirrors handle this problem by using an active control system to sense the mirror shape and then adjust forces on the back of the flexible mirror to modify or control the shape.

Temperature. Telescopes operate in the ambient air. Because the air temperature varies throughout the day and night, it is virtually impossible to maintain a constant mirror temperature. In addition, it is desirable that the temperature of the surface of the mirror match that of the air in order to avoid the image-degrading effects of thermal turbulence. Large mirrors in contact with the ambient air will rarely be in thermal equilibrium because of their large thermal mass; hence temperature variations will arise within the mirror. Since all materials expand or contract when their temperature changes, such temperature variations can cause significant distortions of the mirror shape. The effect grows more significant with larger mirrors and with materials having a larger coefficient of thermal expansion. Different telescope designs attempt to eliminate this problem in different ways. For segmented-mirror telescopes (the Keck telescope in particular), this problem is handled by selecting a low-expansion mirror material (for the Keck telescope, Zerodur, with a coefficient of thermal expansion $\alpha = 5 \times 10^{-9}/°C = 3 \times 10^{-9}/°F$), a small segment diameter, and a small mirror thickness (0.07 m or 3 in.), and by precooling the mirror during the day. Honeycomb monolithic mirrors are currently designed to use Pyrex-like materials, because of their ease of casting. Since these materials have a large thermal coefficient ($\alpha = 3 \times 10^{-6}/°C = 2 \times 10^{-6}/°F$), a sophisticated thermal control system involving actively flushing ambient air through the mirror interior is planned. Large, thin monolithic mirrors are designed to handle thermal problems by actively controlling the mirror shape and by precooling the mirror so that it is never hotter than the ambient air.

Segmented-mirror telescopes.

In segmented-mirror telescopes the primary mirror is made up of a mosaic of mirror segments. For large telescopes this design has a number of advantages: high-quality mirror materials are readily available in these sizes; handling, polishing, and aluminizing equipment is compact and relatively inexpensive; the mirror is very lightweight; the segments are relatively easy to support against gravitational forces; the mirror segments can be made with the benefits of mass production because of their numbers; and the risk of catastrophic breakage is greatly reduced. For ground-based applications there are alternatives, but large space telescopes will almost certainly be segmented because of limitations on the size and mass of payloads carried by launch vehicles. Although not as sophisticated as optical segmented telescopes, radio telescopes for some time have had their primaries made up of surface panels (segments) that are attached to a steel support structure. These primaries typically are not actively controlled, since the tolerances are thousands of times looser for them than for optical telescopes.

Keck telescope. The Keck telescope, a joint project of the University of California and the California Institute of Technology, is a 10-m (400-in.) telescope (**Fig. 1**). Its primary mirror consists of 36 hexagonal segments, each 1.8 m (72 in.) at its widest diameter (**Fig. 2**). Completed in early 1992, it has four times the collecting area of the Hale 5-m (200-in.) telescope on Palomar Mountain. However, because of its unique

Fig. 1. Keck telescope. (*a*) Observatory on Mauna Kea, Hawaii (*California Association for Research in Astronomy*). (*b*) Model of telescope. Efficient space-frame design minimizes the weight and costs and maximizes the stiffness.

use of electronics and its segmented array, the Keck weighs less than half as much as the Hale telescope.

Although there are several advantages to building segmented-mirror telescopes, there are unique difficulties as well. Polishing the mirror segments poses unusual problems because they all must be part of the identical parent surface and because they are off-axis sections of the parent. Stressed-mirror polishing, deliberately deforming the mirror elastically during polishing, was developed to solve this problem. With suitable applied forces, the mirror can be deformed so that a sphere can be polished, and then the sphere elastically

relaxes into the desired off-axis shape when the forces are removed. For the Keck telescope the distortions extend up to 100 micrometers. Large segmented mirrors also require an active control system to maintain the correct relative positions of the mirror segments; this requirement is not found in conventional single-mirror telescopes.

The Keck telescope has a relatively conventional Ritchey-Chrétien optical design, with an $f/1.75$ hyperbolic primary mirror and a 1.4-m-diameter (56-in.) convex hyperbolic secondary used to provide an $f/15$ focus that is free of comatic aberrations. The focal

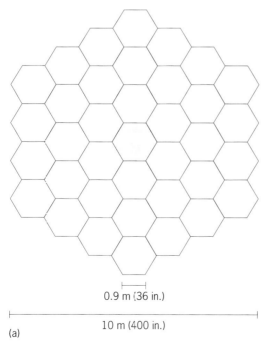

0.9 m (36 in.)

10 m (400 in.)

(a)

Fig. 2. Keck telescope mirror. (*a*) Arrangement of 36 hexagonal segments that form the primary mirror. The central segment is omitted so that light, reflected from a secondary mirror, can reach the Cassegrain focus underneath the primary mirror. (*b*) View of mirror during construction, with 18 of its 36 segments in place. (*Courtesy of Peter French*)

length is unusually short by recent standards, although current large telescopes usually have very short focal lengths in order to reduce costs associated with the length of the telescope and the resulting size of the protective enclosure. The length of the Keck telescope is about the same as that of the smaller Hale telescope (Fig. 1).

The 36 segments of the Keck telescope are actively controlled, since the steel structure that supports them deforms up to 0.6 mm under the forces of gravity, a much larger amount than the allowed tolerances of roughly 50 nm. This control system is a central ingredient for a segmented-mirror telescope, and here it consists of sensors to detect segment motion, a mathematical algorithm to calculate the error in the segment positions based on the sensor readings, and actuators or pistons to move the segments back to their desired positions. There are 168 displacement sensors located in pairs on every intersegment edge. These can detect intersegment motions normal to the surface as small as 4 nm. Commands to the actuators (three per segment) correct any measured position errors, and the actuators can make moves as small as 4 nm as well. Thus the control system keeps the segments aligned to tolerances much smaller than the wavelength of visible light.

To ensure the best image quality, the Keck telescope is located on Mauna Kea, Hawaii, probably the best astronomical site on Earth. This site has superb seeing; a high probability of clear skies; low latitude, giving better sky coverage; and low humidity, improving infrared observing conditions.

Other projects. Another segmented-mirror telescope, the Segmented Spectroscopic Telescope (SST), is being planned by the University of Texas and Pennsylvania State University. This telescope avoids many of the complexities of the Keck telescope, albeit at some loss of versatility. In this design, the 8-m (320-in.) primary is spherical, leading to a very small field of view, and the telescope does not move to follow stars, so that sky coverage is limited. These changes greatly simplify the engineering problems and reduce the cost of the telescope.

The radio astronomy community is also taking advantage of the capabilities of actively controlled mirror segments. The National Radio Astronomical Observatory (NRAO) is currently building a 100-m (328-ft) radio telescope in Green Bank, West Virginia, with actively controlled mirror segments. The Green Bank telescope will have greatly improved performance over any passively controlled telescope of comparable size.

The National Aeronautics and Space Administration (NASA) has a program to study segmented-mirror optical systems for space astronomy. The major issues relate to segment fabrication and control and the unique problems of erection and assembly in space. A variety of NASA projects will rely on this technology development, ranging from optical to short-wavelength radio telescopes, and including telescopes up to 30 m (100 ft) in diameter.

Telescope arrays. The technological innovations used for segmented telescopes make much larger telescopes practical. However, a related idea has many merits as well. Arrays of telescopes can provide large collecting areas, and superior angular resolution over a single telescope with the same collecting area. The improvement is possible since the limiting angular resolution is λ/D, where D can be either the diameter of the telescope or the size of the baseline of an array of telescopes. The major difficulty with arrays is the challenge of combining the light from the individual telescopes.

There are two types of telescope arrays, those with all telescopes attached to a common mechanical mount and those with the telescopes mechanically independent. The Multiple Mirror Telescope (MMT), an optical telescope built in the late 1970s in Arizona, consists of six 1.8-m (72-in.) telescopes held in a common mount. By virtue of the common mount, the optical combining of the six images is relatively straightforward. This telescope has an effective baseline of 7 m (280 in.).

Arrays are easier to build, but the light from the independent telescopes must be coherently combined to achieve the diffraction-limited resolution implied by the baseline.

In radio astronomy, this technique is used successfully in a number of circumstances. The Very Large Array (VLA) in New Mexico consists of 27 telescopes, each 25 m (82 ft) in diameter, that can be moved to provide baselines up to about 50 km (30 mi). There are arrays that have baselines roughly equal to the diameter of the Earth. For these extremely large arrays, the data are collected on magnetic tape by using atomic clocks for time references and are combined at a later time by using computers. For astronomical work in the visible and infrared regions, such data storage technology is not yet available, and the light from the separate telescopes must be combined optically in order to achieve the highest angular resolution.

Optical interferometer arrays have been used spo-

Fig. 3. Model of the four 8-m (320-in.) telescopes planned by the European Southern Observatory (ESO) for the Very Large Telescope (VLT). These telescopes will be used both in independent fashion and in combined fashion, where the light from all telescopes will be brought together for interferometric measurements with very high angular resolution. (*European Southern Observatory*)

radically throughout the twentieth century, but only now are arrays with large mirrors being planned. A twin of the Keck telescope will be constructed by 1996, and the two 10-m (400-in.) telescopes will then have a baseline of 85 m (279 ft) that can be used for interferometry. The European Southern Observatory (ESO) is planning to build four 8-m (320-in.) telescopes (thin monoliths) in Chile to be used both as four separate telescopes and as an interferometer array with a baseline of about 100 m (328 ft; **Fig. 3**).

For background information *SEE OPTICAL TELESCOPE; RESOLVING POWER (OPTICS); TELESCOPE; X-RAY TELESCOPE* in the McGraw-Hill Encyclopedia of Science & Technology.

Jerry E. Nelson

Bibliography. L. D. Barr (ed.), *Advanced Technology Optical Telescopes IV*, Proc. SPIE 1236, 1990; J. R. Gustafson and W. Sargent, The Keck Observatory: 36 mirrors are better than one, *Mercury*, 43(2):43–53, March–April 1988; B. Martin, J. M. Hill, and R. Angel, The new ground-based optical telescopes, *Phys. Today*, 44(3):22–30, March 1991; J. Nelson, The Keck telescope, *Amer. Sci.*, 77:170–176, 1989.

Tent caterpillar

Populations of several species of forest caterpillars rise and fall with amazing regularity. Some well-known examples of caterpillars that undergo population fluctuations are the Douglas-fir tussock moth, which has a 10-year cycle in western North America; the forest, western and eastern tent caterpillars (**Fig. 1**), which cycle with a periodicity of 10–13 years in deciduous forests from British Columbia to Ontario; and the Eastern spruce budworm, with a 30-year cycle in eastern North America.

Population outbreaks of caterpillars often result in defoliation of trees, reduction of tree growth, and sometimes tree death. Therefore, population ecologists have studied forest caterpillars in search of vulnerabilities that might allow future outbreaks to be controlled. Population studies involve counting caterpillars to estimate their abundances and changes in abundance over time. Techniques include counting caterpillars on a predetermined number of tree branches, counting

Fig. 1. Eastern tent caterpillars on their silk tent.

eggs laid on trees, collecting adult moths at traps, and surveying forests from aircraft to determine the extent of defoliation each year. In addition, researchers attempt to discover causes of caterpillar death, including predators, parasitoids, poor weather, disease, and insufficient or poor-quality food. Many hypotheses have been proposed to explain population cycles of forest caterpillars, including hypotheses that incorporate the effects of food limitation, increases in predation, increases in parasitization, and impacts of weather.

Weather hypothesis. One popular explanation is that weather drives the caterpillar population cycles. For example, mild springs could lead to population outbreaks if warm temperatures improved conditions for feeding, survival, and growth of caterpillars. Poor weather, such as cool and rainy spring conditions, could reduce caterpillar survival and cause population decline. Extremely cold winters could kill eggs and trigger population decline.

An intensive study of weather and the patterns of defoliation caused by forest tent caterpillars in Ontario over a 40-year period (**Fig. 2**) has been used to test the hypothesis. An analysis of this study shows no consistent patterns. Increases in defoliation occurred sometimes following cool springs and sometimes following warm springs. Similarly, decreases in defoliation occurred in both warm and cool years. Therefore, the data do not support this weather hypothesis.

While temperatures below $-40°F$ ($-40°C$) can kill eggs, four province-wide population declines of the forest tent caterpillar in Ontario were not preceded by severely cold winters. The northern distribution of this species may be influenced by low temperatures, but year-to-year changes in populations do not seem to be.

For British Columbia populations of tent caterpillars, population cycles occur in synchrony even though some populations are at sea level, where caterpillars develop in April and May, and others are at elevations of 1640–2300 ft (500–700 m), where caterpillars develop in June and July. The summer and winter weather conditions experienced by the populations over this range of elevations can be quite different. Even so, their population dynamics remain very similar, and the populations increase and decline in synchrony. These observations indicate that population fluctuations are not associated with particular patterns of warm or cold weather.

Induced plant defenses. A recent hypothesis to explain population cycles is that trees respond to caterpillar feeding by increasing the levels of defensive compounds in their leaves. This hypothesis predicts that as caterpillar populations increase, the quality of their food plants declines. Leaves with more defensive chemicals are predicted to be less edible for the caterpillars, and caterpillars feeding on leaves from previously damaged trees should therefore grow more slowly, pupate at a smaller size, and survive poorly. Moths emerging from smaller pupae should lay fewer eggs. Thus, this hypothesis predicts that the decline of a population should be related to the history of tree damage caused by insect feeding; sustained defoliation should precede population decline.

Observations of field populations of tent caterpillars do not support the predictions of the induced-defense hypothesis. In fact, just the opposite was found: as populations increase, so do the average number of eggs produced by females (**Fig. 3**a and c). This observation suggests that food quality might initially improve following damage from caterpillars. Once the population of tent caterpillars reaches peak density and begins to decline, moths produce fewer eggs. For several years after the populations begin to decline, moths lay fewer eggs.

The caterpillar survival rate begins to drop as the population density increases (Fig. 3b). Laboratory observations show that the survival of caterpillars that are fed leaves from previously damaged trees is similar to that of caterpillars that are fed leaves from undamaged trees. In the laboratory, most caterpillars from declining populations die regardless of the history of trees from which leaves are gathered to provide food.

The phase of decline for cyclical populations begins with poor survival of late stages of caterpillars and pupae. The summer of peak population density characteristically begins with many large egg masses that give rise to abundant caterpillars. The moths emerging from pupae at the end of the summer of peak density produce fewer eggs, and the population declines with far fewer eggs being laid to initiate the next generation.

Introduction experiments. A characteristic of peak populations of western tent caterpillars is that trees previously uninfested one year are suddenly infested with many tents the next year. When density is high, moths must fly to new areas to lay their eggs. It is interesting that these newly initiated populations decline in synchrony with surrounding populations despite the variation in their history. To determine the relationship between the amount and duration of feeding damage by caterpillars and the dynamics of the populations, egg masses were introduced to alder trees in areas either with no tent caterpillars or with low densities of caterpillars. Introduced populations usually declined at the same time as control populations. In several cases in which eggs were introduced several years before the peak density would be reached in the control populations, the population decline was delayed by one year. But populations initiated with eggs from peak populations declined at the same time as control populations. The introduced egg masses appear to carry the elements of future demise. Eggs from a declining population do not initiate a population outbreak even if introduced to an area with undamaged food plants.

An experiment in which caterpillars are removed from trees in an area where an outbreak is about to occur has also been carried out. By keeping a population from reaching outbreak densities for several years, it is possible to determine if a population eruption will occur when caterpillars are no longer removed. This experiment tests whether preventing damage to food plants will prevent the cyclical population decline. The results showed that reducing a population of tent caterpillars for several years would not result in a delayed population outbreak. Protecting the food plants from

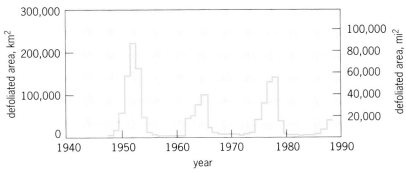

Fig. 2. Population cycles of forest tent caterpillars in Ontario as shown by the extent of forest defoliation. (*After C. Daniel, Climate and outbreaks of the forest tent caterpillar in Ontario, M.Sc. thesis, Department of Zoology, University of British Columbia, 1990***)**

caterpillar damage did not change the dynamics of the population cycle.

Both sets of experiments show that population cycles of tent caterpillars are very resilient and nearly impossible to modify through experimental introduction or removal of insects. The many attempts at controlling outbreaks of forest caterpillars by using insecticides agree with these experimental results as well. Although trees can be protected on a yearly basis with insecticide spraying, caterpillar populations cannot be permanently suppressed with insecticides, and the population outbreak continues unaltered on a regional scale.

Viral disease. There are a variety of wasp and fly parasitoids that attack tent caterpillars. In addition, an externally transmitted disease, nuclear polyhedral virus, is characteristic of declining populations. Caterpillars contract the disease by consuming leaves that are contaminated with polyhedra. The protein coat of this virus is digested in the caterpillar gut, and the virus attacks the nuclei of cells lining the gut. The infection spreads into other tissue, and after about a week the caterpillar dies and millions of polyhedra are released onto the foliage. Polyhedra are spread by rain, wind, and insect activity.

While viral disease has long been associated with declining populations of tent caterpillars and other forest lepidopterans, its role has not been understood. Several characteristics of viral disease are compatible with characteristics of population cycles of tent caterpillars. For example, experiments have shown that disease is transmitted in the environment and is also carried on egg masses. Therefore, the synchronous declines of both natural populations and experimental populations from introduced eggs could be explained by disease being carried on the eggs. In addition, disease could kill insects not killed by other agents late in the population outbreak. This possibility would explain the synchronous declines of populations that vary in the degree of parasitism from flies and wasps.

Theoretical analyses show that population cycles will occur if there is a delay in the recovery of populations for several generations after each population decline. Studies of tent caterpillars have found that both fecundity of moths and survival of the early caterpillar stages remain low after the initial decline. Currently

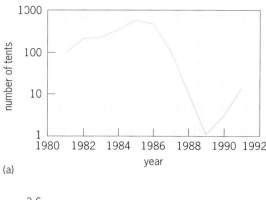

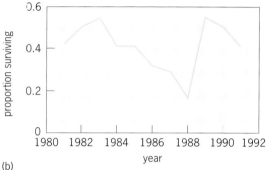

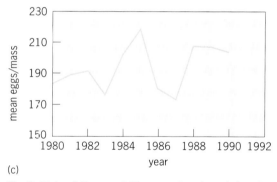

Fig. 3. Plots of three variables as a function of time for western tent caterpillars on a 15-acre (6-hectare) island in British Columbia over a population cycle. (*a*) Number of tents; (*b*) caterpillar survival rate; (*c*) moth fecundity. The number of eggs per mass is for female moths at the end of the generation for which density data are given.

the interactions between tent caterpillars and their nuclear polyhedral virus are being studied to discover if disease can reduce moth fecundity and how disease is maintained in low-density populations. Persistence of virus in the population could delay the recovery of the population.

Studies of other species of cyclic lepidopterans show that although populations in some areas might not reach sufficient densities to cause defoliation in each cycle, they may still remain in synchrony with other populations in subsequent outbreaks. The reasons why populations remain in synchrony are not fully understood, but dispersal of moths among populations may be sufficient to maintain regional patterns. If the moths that disperse from populations in the late increase phase are carrying disease, and parasitoids move to

areas of high host density, cyclic dynamics may be maintained. Not only does survival change as disease and parasitoids build up in increasing populations, but the fecundity of moths, as shown in the number of eggs they produce, also changes.

Although population outbreaks have not yet been stopped, spraying populations with polyhedral virus has potential. In British Columbia a population outbreak of tussock moths was terminated a year early by spraying with nuclear polyhedral virus. Because this virus is very specific, it is much safer than chemical sprays and has no effect on humans and other animals. It is still expensive to obtain the virus in sufficiently large quantities, since virus can be produced only by rearing infected caterpillars and extracting polyhedra after they die. Research into better production and spraying procedures may make polyhedral viruses the preferred biological insecticide of the future.

For background information SEE CATERPILLAR; INSECT CONTROL (BIOLOGICAL); LEPIDOPTERA; POPULATION ECOLOGY in the McGraw-Hill Encyclopedia of Science & Technology.

Judith H. Myers

Bibliography. J. R. Gould, J. S. Elkinton, and W. E. Wallner, Density-dependent suppression of experimentally created gypsy moth, *Lymantria dispar* (Lepidoptera: Lymantriidae), populations by natural enemies, *J. Anim. Ecol.*, 59:213–233, 1990; K. D. Murray and J. S. Elkinton, Environmental contamination of egg masses as a major component of transgenerational transmission of gypsy moth nuclear polyhedrosis virus (LdMNPV), *J. Invert. Path.*, 53:324–334, 1989; J. H. Myers, Can a general hypothesis explain population cycles of forest lepidoptera?, *Adv. Ecol. Res.*, 18:179–242, 1988; J. H. Myers, Population cycles of western tent caterpillars: Experimental introductions and synchrony of fluctuations, *Ecology*, 71:986–995, 1990.

Tooling

Tribology, the science of wear and friction, is an area in which applications of diamond-film technology will be increasing. Diamond, with its unique combination of properties, is an ideal industrial cutting-tool material for the turning, milling, and drilling of nonferrous alloys, composites, and other abrasive work materials.

Physical properties of diamond. Although graphite, soot, and carbon black have the same chemical composition as diamond (all are forms of carbon), this trio and diamond have very different crystal structures. Graphite consists of layers of condensed, six-numbered aromatic rings of sp^2-hybridized carbon atoms. These rings are strongly linked in a single plane and weakly held between the planes by van der Waals bonds. These layers can slide over each other, so that graphite is a soft material used as a lubricant. Soot and carbon black are microcrystalline forms of graphite.

The crystallographic network of diamond consists completely of covalently bonded, aliphatic sp^3-hybridized carbon atoms arranged tetrahedrally with a uniform distance between atoms. The tetrahedrons are connected at their tips to form the crystal lattice. Diamond's structure accounts for many of its properties, such as extreme hardness, outstanding wear resistance, and low coefficient of friction, that make it useful as a free-standing piece that has been brazed to a device or as a coating on cutting-tool inserts, end mills, and drills.

Mechanical properties of diamond film. The properties of chemical-vapor-deposited diamond film that render it important as a cutting-tool material include hardness, abrasion resistance, desirable friction characteristics, high thermal conductivity, low coefficient of thermal expansion, and chemical inertness.

Single-crystal natural diamond exhibits an indentation hardness (Knoop scale) in the range 5700–10,400 kg/mm^2 (8.1–14.8 $\times$ 10^6 lb/in.^{2}). Chemical-vapor-deposited diamond that exhibits 100% sp^3 bonding has measured hardness in the range 8000–10,000 kg/mm^2 (11.4–14.2 $\times$ 10^6 lb/in.^{2}). A large number of these diamond films have measurable quantities of sp^2 bonding, particularly in the grain boundary area. The presence of these weaker graphitic bonds results in hardnesses of these diamond films that are below those of natural crystals.

The observed abrasion resistance of both free-standing chemical-vapor-deposited diamond and the diamond films applied as a coating exceeds that of tungsten carbide by one to greater than two orders of magnitude. Chemical-vapor-deposited diamond also has an abrasion resistance that is two to ten times greater than conventional high-pressure–high-temperature sintered polycrystalline diamond materials.

When chemical-vapor-deposited diamond is polished, it exhibits friction properties equal to those of the fluoropolymers at ambient temperatures. The friction properties of diamond are believed to be the result of the surface chemistry of the materials. Free dangling bonds on the surface of diamond have great affinity for hydrogen. The atomic surface layer of hydrocarbon thus formed creates an ultralow-friction, low-adhesion surface when in contact with a wide variety of materials, including mating diamond surfaces. This low-friction, chemically inert surface results in a cutting-tool material that functions efficiently at a low cutting temperature without galling. The material is suitable for nonferrous and composite applications.

The thermal conductivity of diamond is the highest of all materials. Chemical-vapor-deposited diamond has values of 8-20 W/(cm K), and type 2A single-crystal natural diamond has a value of 22 W/(cm K). The latter type is a variety that is effectively free of nitrogen as an impurity and has enhanced optical and thermal properties. The next-highest thermal conductivity is 4.29 W/(cm K) for silver. The high thermal conductivity of chemical-vapor-deposited diamond results in a cutting tools that can conduct heat away from the cutting edge, so that tool life is increased and possible heat damage and distortion of the workpiece being machined are reduced. Diamond also has a low specific heat; therefore it conducts the heat generated in cutting to another heat sink, such as the substrate brazed or coated with diamond.

Diamond has a comparatively low thermal expansion coefficient. Its percent of thermal expansion between ambient temperature and 750°C (1382°F) is about 0.2%. The value is nonlinear, and the coefficient for diamond increases rapidly with temperature; therefore, the operating range of interest must be defined before the extent of thermal expansion mismatch can be determined. In order to take full advantage of the properties of chemical-vapor-deposited diamond either as a thick plate brazed to a substrate or as a coating, the matching of the thermal expansion rates of the substrate and diamond is critical for the diamond to perform well and not crack, fracture, or spall off.

Thermal expansion rates are particularly important when the diamond film is deposited directly as a coating onto a substrate. The deposition process is typically carried out between 700 and 1000°C (1290 and 1830°F). This high temperature limits the materials onto which the film can be coated. The low thermal expansion of diamond further limits the available substrates. The engineering materials onto which the diamond can be nucleated and deposited for cutting tools are silicon nitride (Si_3N_4) ceramic and tungsten carbide with a low content of cobalt ($\leq$6%). Silicon nitride has the closest match of thermal expansion coefficient for these substrate materials, and it is a better chemical match for diamond nucleation and growth than tungsten carbide. Tungsten carbide presents greater difficulty because of cobalt in the grain boundaries introduced as a sintering aid. The cobalt at the tool surface causes graphite to form preferentially instead of diamond, and the result is a weakened adhesion of the coating as well as of the diamond film itself.

The mismatching due to the thermal expansion coefficient is one of the key limitations in commercial development of chemical-vapor-deposited diamond films for cutting tools and other high-force applications, wherein the diamond film is subjected to very high tensile, compressive, and shear forces as well as high temperatures and large fluctuations in temperature.

Another important factor in cutting-tool performance is the chemical or thermal stability of the tool material either as a monolith or as a coating. Diamond is one of the most chemically inert substances. However, it cannot be used in machining ferrous-, nickel-, or titanium-based alloys without elaborate cooling methods (liquid carbon dioxide or nitrogen) because of its reactivity and chemical wear (carbon diffusion out of the diamond) at the high contact pressures and temperatures that are generated during the machining process. On the other hand, for machining aluminum and other ductile nonferrous alloys, plastics, and abrasive composite materials, diamond has no peer.

The wear environment at the cutting-tool tip involves extreme temperature. A temperature of up to 1000°C (1830°F) can be present locally during continuous turning, and the temperature may fluctuate between 700 and 200°C (1300 and 390°F) within microseconds in the interrupted cut of milling. Extreme force at the cutting-tool tip results in contact pressures of 200–500 kg/mm^2 (285,000–710,000 lb/in.^{2}). Thus chemical-vapor-deposited diamond tool material must have sufficient bulk strength and fracture toughness to be utilized as a brazed or free-standing piece, and the diamond coating must have high-temperature hardness relative to the substrate and workpiece material, high-temperature chemical stability, good adhesion to substrate, and good microfracture toughness.

The chemical stability of chemical-vapor-deposited diamond film is superior to polycrystalline diamond materials produced at high pressures. The latter begin to oxidize at 600°C (1100°F), and at 700°C (1300°F) they degrade very rapidly in either air or vacuum because of the cobalt grain boundary phase needed as a sintering aid. The chemical-vapor-deposited diamond begins to degrade slowly in air at 700°C (1300°F), but in vacuum or in an inert atmosphere it is stable at 1200°C (2200°F). This increased chemical and thermal stability over polycrystalline diamond material results in significant improvements in performance when machining is performed in very corrosive environments or at the high temperatures generated in processing very abrasive advanced composite materials.

The characteristics of diamond film coating that provide good adhesion to substrates and good microfracture toughness are strong functions of film thickness, microstructure, and conditions of deposition. The deposition conditions must be optimized to produce a diamond film with good adherence, which is a function of low intrinsic stress (stress in the film due to growth conditions), low extrinsic stress (stress due to mismatch of thermal expansion with substrate), and maximal mechanical and chemical bonding with the substrate surface. Coating thickness must also be optimized; if it is too thick, the diamond coating begins to exhibit inherently brittle bulk behavior due to increased stress, which can induce premature microfracture or disbonding with the substrate. This residual stress in the coating also affects film microhardness, which is important because at the high cutting temperatures most materials lose their microhardness (this applies also to other hard coatings such as titanium carbide, titanium nitride, and aluminum oxide).

High-temperature microhardness is important in two modes of tool wear. Tool wear at the hot crater zone on the rake face (top surface) of the tool depends on the chemical inertness of the coating as well as the microhardness at high temperature. The abrasive wear resistance is directly related to the microhardness at the cutting temperature of the flank face of the tool. Inevitably the mechanical properties of the coating must be related in terms of its microstructure (grain size, crystallinity, defects, and so forth), which is a function of the parameters of the coating process.

Machining. Diamond film is the ultimate cutting-tool material for machining aluminum, aluminum composites, and advanced composites.

For aluminum and aluminum composites, either brazed-on "thick-film" free-standing diamond film or complex geometries coated with diamond film are beginning to be applied in production environments. Higher machining speeds are achievable with diamond film than with tungsten carbide and even poly-

crystalline diamond materials. Diamond-film tools exhibit a tool-life increase of 15–300 times over tungsten carbide tools and 2–10 times over high-pressure polycrystalline-diamond-material tools.

Diamond film also has many advantages in machining advanced composites. Materials such as graphite, carbon-carbon, carbon-filled phenolics, fiberglass, and honeycomb materials quickly wear out tungsten carbide tooling. Polycrystalline diamond is often inadequate for machining these materials because it cannot be used in applications requiring complex geometries. Also, polycrystalline diamond becomes corroded chemically in machining corrosive plastic-based materials; because it wears faster, it is uneconomical in such applications. Diamond-film cutting tools are unsurpassed in tool life and performance when these advanced composites are machined, and their use is growing rapidly.

Prospects. Diamond film is a very active area of research and development in Europe, Japan, and the United States. Besides the applications in tribology as related to cutting tools, areas under development include optics (infrared windows or radomes) and electronics (heat sinks, substrates, packaging, and even semiconductors). Diamond-film technology has immense potential for numerous applications and industries. SEE DIAMOND.

For background information SEE CHEMICAL BONDING; DIAMOND; MACHINE TOOLS; TOOLING; VAPOR DEPOSITION in the McGraw-Hill Encyclopedia of Science & Technology.

Robert A. Hay

Bibliography. Diamond and diamond-like materials, *J. Mater. Res.*, 5(11):2273–2609, November 1990; R. Messier et al., *New Diamond Science and Technology*, 1991; Y. Tzeng et al., *Applications of Diamond Films and Related Materials*, Mater. Sci. Monogr. 73, 1991.

Toxic shock–like syndrome

In the past few decades, dramatic declines in the prevalence of severe infections from the bacterium group A streptococcus (*Streptococcus pyogenes*) have been observed in developed countries. However, since the late 1980s, outbreaks of severe group A streptococcal infections have been reported in North America, Europe, New Zealand, and Australia. One such outbreak is the newly recognized toxic shock–like syndrome (TSLS), also known as streptococcal toxic shock syndrome, a serious illness that can affect individuals of any age.

First described in 1987, toxic shock–like syndrome is related to toxic shock syndrome (TSS). The latter, caused by infection with *Staphylococcus aureus*, received media attention during the early 1980s because of its high, though not exclusive, association with menstrual periods and the use of certain tampons. Toxic shock–like syndrome is caused by infection with toxin-producing group A streptococci, which also cause pharyngitis ("strep" sore throat), scarlet fever, impetigo, erysipelas, necrotizing fasciitis, myositis, puerperal fever, septicemia, and poststreptococcal sequelae, including rheumatic fever, acute glomerulonephritis, and erythema nodosum. Some investigators believe that toxic shock–like syndrome is a new illness resulting from the appearance of more virulent strains of group A streptococci. However, toxic shock–like syndrome more likely represents a severe course of scarlet fever, a disease that has existed for centuries.

Major outbreaks of severe scarlet fever have occurred in association with wars throughout history, and epidemics in the United States at the turn of the century necessitated quarantine procedures. Scarlet fever usually begins with throat or skin streptococcal infection and is typified by extremely high fevers (the patients appear to radiate heat) and a diffuse sunburn rash.

Symptoms. The clinical symptoms of toxic shock–like syndrome resemble closely those of toxic shock syndrome. Patients often suffer from high-grade fever, hypotension, rash, and multiple-organ-system involvement. In severe cases patients may develop hypotension and shock, renal dysfunction, respiratory failure, and disseminated intravascular coagulation, leading to amputations. Hypotension and shock are the major causes of death in toxic shock–like syndrome. Patients also suffer from massive nonpitting edema, especially in the extremities, and from respiratory failure due to fluid leakage into the lungs. Case fatality rates of up to 50% have been reported.

Route of transmission. The most common route of transmission of the organism in toxic shock–like syndrome appears to be through breaks in the skin. This mode is in contrast to typical scarlet fever, where a major route of transmission appears to be by droplets from the throat. Although group A streptococci rarely reside on the skin of healthy individuals, breakage in the skin may allow the organism to enter from some, as-yet-unidentified, human source (group A streptococci are pathogens of humans only).

A particularly troubling route of transmission appears to be through lesions in children infected with chickenpox. This disease causes a transient immunosuppression and provides skin breaks where streptococci may establish infection. Chickenpox lesions often scab over and may show some signs of inflammation. Since streptococcal infections tend to be more common in children than in adults, it is important that these lesions be monitored. In the majority of children thus far identified with toxic shock–like syndrome associated with chickenpox, one or more limbs have had to be amputated.

The organism may occasionally be transmitted orally. *Streptococcus pyogenes* has been isolated from the throat of patients with toxic shock–like syndrome who do not have histories of skin breakage. However, the oral route has not been proven and is being debated among researchers. Indeed, toxic shock–like syndrome cases occurring after pharyngeal infection will likely be referred to as probable toxic shock–like syndrome.

Streptococcal pyrogenic exotoxins. Like toxic shock syndrome, toxic shock–like syndrome is a toxin-mediated illness. Streptococci that cause it produce streptococcal pyrogenic exotoxins (SPEs; also

known as scarlet fever toxins, erythrogenic toxins, and lymphocyte mitogens), of which there are three types, A, B, and C, based upon differential antibody reactivity. A streptococcus that causes toxic shock–like syndrome can produce one or more SPEs, depending on the presence and expression of the SPE gene (*spe*). Interestingly, the SPE-B gene (*spe B*) is present in the chromosomes of all group A streptococci, but SPE-B protein is made by only half of these streptococci. The *spe A* and *spe C* are variable traits in that they are present in some organisms but not in others, and these two toxins are encoded by bacteriophages.

The SPEs belong to a larger family of pyrogenic toxins made by *Staphylococcus aureus* as well as group A streptococcus. These toxins have several common biological activities (**Fig. 1**), including pyrogenicity (ability to induce fever), enhancement of host susceptibility to lethal endotoxin shock, nonspecific stimulation of T lymphocytes (leading to B-cell immunosuppression, release of many lymphokines such as interleukin 2 and gamma interferon, and enhancement of delayed hypersensitivity and rash), and induction of release of monokines such as tumor necrosis factor alpha and interleukin 1 from macrophages. The pyrogenic toxins share many physicochemical properties. They are all relatively low-molecular-weight (21,000–44,000), single-polypeptide-chain globular proteins that are true exotoxins in that they are secreted immediately upon synthesis and have a defined signal peptide. Many of the pyrogenic toxins exhibit extensive similarity in their amino acid sequences (see **table**). For example, there are high percentages of similarity among SPE-A and staphylococcal enterotoxin (SE) types B, C_1, C_2, and C_3 (made by *S. aureus*). SPE-B and toxic shock syndrome toxin 1 (TSST-1) made by *S. aureus* show the least similarity to other pyrogenic toxins.

The pyrogenicity of SPEs depends on both release of tumor necrosis factor alpha and interleukin 1 from macrophages and direct toxin effects on the central nervous system (Fig. 1). These latter effects may depend on toxin interaction with macrophagelike cells of the central nervous system or direct stimulation of the fever response control center in the hypothalamus. In addition, tumor necrosis factor alpha is capable of inducing hypotension and shock and is thought to mediate many effects of endotoxin shock, and thus probably contributes to the severe manifestations of toxic shock–like syndrome.

Perhaps the most significant biological activity of the toxins is their ability to amplify the lethal effects of endotoxin by a factor of up to 100,000. Thus, experimental animals challenged with a sublethal dose of SPE show a greatly enhanced susceptibility to otherwise nontoxic doses of gram-negative bacterial endotoxin. The animals show many of the severe manifestations of toxic shock–like syndrome, and investigators hypothesize a role for this phenomenon in hypotension and shock associated with toxic shock–like syndrome. Animals that survive the combination of SPE and endotoxin show significant signs of heart damage, as do many patients with toxic shock–like syndrome.

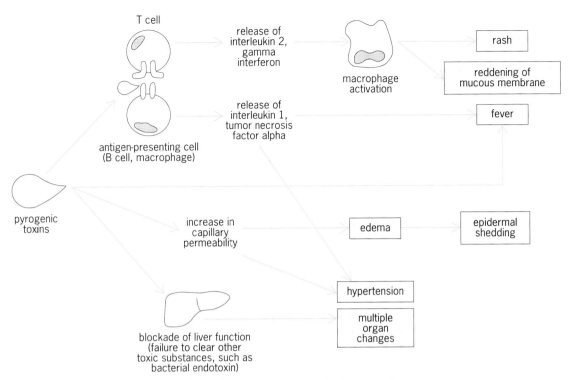

Fig 1. Model for the development of toxic shock syndrome and toxic shock–like syndrome.

T-cell stimulation.   Recently, the pyrogenic toxins have served as important tools for understanding mechanisms of T-cell stimulation by antigens. Probably, the ability of SPEs to stimulate T cells is important for the development of some symptoms of toxic shock–like syndrome. As with other antigens, T-cell proliferation by the pyrogenic toxins first depends on the toxin's binding to surface glycoproteins, referred to as class II major histocompatibility complex (MHC) molecules, on macrophages or B cells. However, unlike most antigens, which require specific types of class II MHC, the pyrogenic toxins can bind to any class II MHC; this ability suggests that the need for class II MHC is nonspecific.

After binding to the class II MHC molecules, most antigen–MHC complex interacts with both alpha (α) and beta (β) chains of the T-cell receptor complex that is specific for the particular antigen (**Fig. 2**). The

alpha and beta chains form an antigen recognition groove in the T-cell receptor complex that is specific for the antigen–MHC complex. In the case of pyrogenic toxins, this specific stimulation is bypassed, as the toxin–MHC complex interacts outside the antigen recognition groove and only with the beta chain of the T-cell receptor complex. Therefore, the SPEs are able to nonspecifically stimulate T cells, regardless of antigen–MHC complex specificity of the T cells.

Emergence of the syndrome.   Prior to the description of toxic shock–like syndrome in 1987, severe illness from group A streptococcal infection was uncommon in developed countries. One reason for the seemingly sudden increase in number and severity of streptococcal illnesses appears to be the change in the SPE production profile of the organism. Several reports suggest that the increased virulence of group A streptococcus was due to the increased numbers

Percentage of similarity between aligned pyrogenic toxin amino acid sequences*

	SPE-B	SPE-C	SE-A	SE-B	SE-C$_1$	SE-C$_2$	SE-C$_3$	SE-D	SE-E	TSST-1
SPE-A	22	28	33	48	46	45	46	35	34	25
SPE-B	.	22	22	18	20	21	21	22	21	26
SPE-C	.	.	27	26	30	31	30	30	27	20
SE-A	.	.	.	34	31	32	33	50	70	23
SE-B	.	.	.	.	69	68	70	36	35	23
SE-C$_1$	.	.	.	.	.	97	94	33	31	22
SE-C$_2$	.	.	.	.	.	.	96	33	31	22
SE-C$_3$	.	.	.	.	.	.	.	33	33	22
SE-D	.	.	.	.	.	.	.	.	52	24
SE-E	.	.	.	.	.	.	.	.	.	26

* SFEs are produced by group A streptococci; SEs and TSST-1 are produced by *Staphylococcus aureus*.

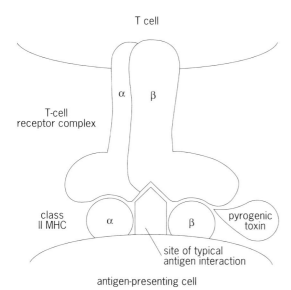

Fig. 2. Schematic diagram of interaction of pyrogenic toxin with class II MHC and T-cell receptor complex.

Labels in figure: T cell; T-cell receptor complex; α; β; class II MHC; α; β; pyrogenic toxin; site of typical antigen interaction; antigen-presenting cell

of SPE-A producers isolated. Evidence indicates that strains that produce SPE-A are more virulent since SPE-A is made in greater quantity than SPE-B or SPE-C; in addition, SPE-A is more toxic than other SPEs in animal models of toxic shock–like syndrome.

Ten years prior to the description of toxic shock–like syndrome, the occurrence of producers of SPE-A was rare. In fact, during the periods 1976–1979 and 1983–1986, no SPE-A-producing group A streptococci were isolated, even from patients with severe streptococcal illness. In contrast, several isolates of group A streptococci from scarlet fever cases in the early twentieth century are maintained in laboratories, and consistently produce SPE-A. Currently, 87% of toxic shock–like syndrome isolates are producers of SPE-A. No strains from uncomplicated pharyngitis have been reported to be SPE-A-positive, and the overall presence of SPE-A producers in a general collection of group A streptococci is about 15%. The increase in SPE-A production by recent strains could be explained by infectiouslike transmission of the bacteriophage that encodes SPE-A, from strains resident in the community to newly emerging strains.

In addition to SPE-A association, streptococci that cause toxic shock–like syndrome appear to have an association with specific M protein types, which are cell-surface proteins that allow the organism to resist phagocytosis in the host. The host must develop antibodies against M protein in order to develop immunity to the microbe. It is known that various microorganisms undergo a cyclic evolution, with particularly virulent strains appearing and disappearing. In the case of group A streptococci, this may be due to the appearance of strains with specific M protein types that are more resistant to phagocytosis and thus are able to escape the host immunity and confer greater virulence. For example, approximately 60% of toxic shock–like syndrome isolates belong to M protein types 1 and 3, with M1 predominating. In the late 1980s there was

a significant resurgence of these two M protein types, and genetic analyses showed that M1 and M3 isolates are closely related. Possibly, there has been a resurgence of toxic shock–like syndrome as a consequence of a dramatic increase in M1 and M3 streptococcal isolates.

Other, as-yet-unidentified factors probably contribute to the increased virulence of toxic shock–like syndrome isolates. However, SPEs, notably SPE-A, present in such isolates are able to induce symptoms of toxic shock–like syndrome in animals, and share biological activities and sequence similarities with other toxins known to cause toxic shock syndrome. In addition, recent analyses indicate that antibody levels against SPE-A are low in the general population; therefore, susceptibility to the toxin effects is high. Thus, it is likely that SPEs are important contributors to this newly emergent and very severe illness.

For background information *SEE CELLULAR IMMUNOLOGY; RHEUMATIC FEVER; SCARLET FEVER; STREPTOCOCCUS; TOXIC SHOCK SYNDROME* in the McGraw-Hill Encyclopedia of Science & Technology.

Peter K. Lee; Patrick M. Schlievert

Bibliography. K. Belani et al., Association of exotoxin producing group A streptococci and severe disease in children, *Ped. Infect. Dis. J.*, 10:351–354, 1991; G. A. Bohach et al., Staphylococcal and streptococcal pyrogenic toxins involved in toxic shock syndrome and related illnesses, *Crit. Rev. Microbiol.*, 17:251–272, 1990; J. M. Musser et al., *Streptococcus pyogenes* causing toxic-shock-like syndrome and other invasive diseases: Clonal diversity and pyrogenic exotoxin expression, *Proc. Nat. Acad. Sci. USA*, 88:2668–2672, 1991; D. L. Stevens et al., Severe group A streptococcal infections associated with a toxic shock-like syndrome and scarlet fever toxin A, *N. Engl. J. Med.*, 321:1–7, 1989.

Toxicology

Recent advances in toxicology involve elucidation of the mechanism of cancer introduction by chromium(VI) and by carcinogen activation of transcription factors.

Chromium(VI) Carcinogenesis

Chromium as a substance capable of causing human cancer was first clearly recognized in November 1889, when a 47-year-old workman who had been exposed to chromate was admitted to Glasgow Royal Infirmary in Scotland with a tumor, an adenocarcinoma, in the left turbinate body in his nose. Since that time, many epidemiological studies have established that inhalation of chromium trioxide and chromates can cause lung cancer in workers in the chromate, ferrochromium, and chromium electroplating industries and also in welders working with stainless steel. In these workers, cancer of the lung is the predominant form, with a much lower frequency of cancers of the nasal cavities and larynx.

An increased incidence of cancer in humans appears to be associated only with the inhalation of hexavalent chromium [Cr(VI)] compounds; trivalent chromium [Cr(III)] appears to be noncarcinogenic. The reason for the inability of the generally very reactive Cr(III) species to cause cancer in contrast to the rather less reactive Cr(VI) species raises a number of interesting chemical and biochemical questions.

Biochemistry of chromium. Trivalent chromium is generally the most stable state of the element in nature. The trivalent chromium cation (Cr^{3+}) is highly reactive; it readily forms complexes with proteins, deoxyribonucleic acid (DNA), ribonucleic acid (RNA), and other complexing ligands in biological systems. Because of its high ionic charge and small ionic radius, 0.0615 nanometer, the Cr^{3+} ion readily hydrolyzes to form insoluble hydroxy species. Hydrolysis and complex formation combine to make Cr(III) poorly absorbable from the gastrointestinal tract ($\sim 1\%$) and almost unable to pass through cell membranes to reach the internal environment of cells. In contrast, the Cr(VI) compounds, which are frequently much more soluble, can pass more readily through cell membranes and are absorbed to a greater extent (about 10% ingested) from the gastrointestinal tract.

One of the most important factors controlling the activity of Cr(VI) in the human or animal body is the relative ease with which it can be reduced to Cr(III) in many body fluids and cells, as shown in reaction (1).

$$CrO_4^{2-} + 4H_2O + 3e^- \rightarrow Cr(OH)_3 + 5OH^- \qquad (1)$$

The standard reduction potential for this reduction of Cr(VI) to Cr(III) is -0.12 V. The mechanism of this three-electron reduction process is not yet fully understood, but it may be a stepwise reaction involving Cr(V) and Cr(IV) intermediates.

The reduction of Cr(VI) can occur in blood plasma; in the red blood cells; in the fluids of the gastrointestinal tract and the lungs; in lung macrophages; and in liver cells, kidney cells, and many other types of cell within the body. At the cellular level the reduction appears to occur on the cell membrane and in the mitochondria, the endoplasmic reticulum, and the nuclear membrane within the cells. It appears to involve a number of enzymatic and nonenzymatic systems, for example, glutathione, cytochrome P-450, and oxidoreductases. Two cell types with a poor ability to reduce Cr(VI) are the parenchymal cells in the human lung and muscle cells. This may be of special relevance in relation to cancer induction, since in humans the parenchymal cells of the lung may become transformed into the cancer cells; similarly, in rats, following intramuscular injection of Cr(VI), tumors arise at the injection site.

In general, soluble Cr(VI) entering the bloodstream following absorption from the gastrointestinal tract or the lungs will rapidly be reduced to Cr(III) in either the plasma, erythrocytes, or tissue cells. The intracellularly formed Cr(III) appears to react readily with proteins and other cell structures to form chemical species that, except when present in large amounts, cause few harmful effects. Within the lungs, the macrophages (the phagocytic cells present in the alveoli or air sacs and in the walls of the lung airways) are especially active

in the reduction of inhaled Cr(VI) and probably play a very important role in reducing the harmful effects that could result from the inhalation of chromates. However, some fraction of the solubilized Cr(VI) that enters the lungs following inhalation will enter into the parenchymal cells lining the airways (the bronchi and bronchioles), and it is these cells that not only have a limited capacity to reduce Cr(VI) to Cr(III) but also can be transformed into cancer cells.

In industrial exposures to chromium compounds the lungs are the most important entry route, and it seems likely that chromium will be transported from the lung only when the capacity of the lung fluids and the bronchial cells to reduce Cr(VI) to the noncarcinogenic Cr(III) is exceeded.

Mechanisms of cancer induction. Despite quite intensive study since 1970, the actual mechanism of cancer induction by chromium remains unclear. The generally accepted theory of cancer induction, or carcinogenesis, is that the inducing agent, the carcinogen, causes a nonlethal change in a cell that, after cell division, becomes fixed as an inheritable mutation, and a mutated cell may then, after one or more further steps, develop into a malignant tumor.

For chromium the nature of the chemical species that causes the actual chemical change in the DNA and leads to the inheritable mutation is not yet known. Studies with experimental animals and with cells in culture have shown that treatment with Cr(VI) can lead to damage to the genetic material in the cell nucleus. The genetic material in the cell nucleus is found in the chromatin, which is composed of proteins and the double-stranded, helical DNA. Chromium-induced damage to DNA may include strand breaks, cross links between the two strands, and, in the case of chromatin, DNA–protein crosslinks. Some of this damage will be so severe that the cell will become incapable of undergoing cell division (lethal damage), and some will be repaired by the special repair enzymes present in most cells. However, a small fraction of this nonlethal damage may escape repair or be irreparable, and may subsequently become an inheritable mutation that could later develop into a cancer.

Biochemical studies that have been carried out in liver and other mammalian cells in the laboratory suggest that the reduction of the Cr(VI) to Cr(III) is essential for the reaction with DNA. However, studies involving living organisms (rats) indicate that if Cr(III) does not reach the genetic material it does not appear to damage the DNA. This apparent contradiction could be explained if the chromium species causing the carcinogenic change was not in fact Cr(III) itself but some other reactive species formed during the reduction of Cr(VI).

The reduction of Cr(VI) can be seen as a stepwise three-electron reaction (2). Cr(V) is known to be suf-

$$Cr(VI) + e^- \rightarrow Cr(V) + e^- \rightarrow Cr(IV) + e^- \rightarrow Cr(III) \quad (2)$$

ficiently stable both in the organism and in the test tube to be detectable by electron-spin trapping methods, and this species is known to damage DNA by formation of free radicals. However, it seems likely that Cr(V) would need to be formed very close to the DNA in order to cause damage; thus the site of Cr(VI) reduction inside the cell may be a critical factor.

The hypothesis that Cr(VI) reduction corresponds to Cr activation is attractive, but the cells most at risk for cancer induction, the lung parenchyma in humans and muscle cells in experimental animals, are the two cell types known to have the lowest ability to reduce Cr(VI). This difficulty could be explained if only a fraction of the Cr(VI) reduction products are carcinogenic, for example Cr(V), and if they can actually cause DNA damage only if they are formed close to the genetic targets in the cell nucleus, for example by reduction in or on the nuclear membrane. In cells with strong Cr(VI)-reducing activity, the Cr(III) may be produced very rapidly and then sequestered equally rapidly as stable and relatively noncytotoxic complexes. In cells with low Cr(VI)-reducing activity, the location of the reducing centers within the cells may also be different, with the result that Cr(VI) can penetrate nearer to the cell nucleus before being reduced, thus giving the active species a better chance of reaching the DNA in sufficient concentration to cause a carcinogenic transformation of the cell. Although this explanation of Cr(VI) carcinogenesis is supported by good experimental evidence, it remains speculative, and much more evidence is needed to prove it.

The ability of Cr(V) to produce potentially carcinogenic changes in lung cells needs to be demonstrated experimentally in order to confirm this species as the actual, proximal carcinogen. Further studies of the reduction of Cr(VI) in cells, with special relevance to the intermediate species formed, their lifetimes, and their ability to diffuse into the nuclear material within the cell, could yield valuable information. Also valuable would be a detailed investigation of the intracellular location of the sites of Cr(VI)-reducing activity within cells with high and low reducing abilities. In addition, in order to fully understand chromium carcinogenesis in humans it is necessary to know more not only about the carcinogenic mechanisms themselves but also about the possible effects of concomitant exposures such as to other metals, to other types of environmental or industrial toxic agents, and to tobacco smoke. *David M. Taylor*

Transcription Factors

In contrast to many chemical carcinogens, both dioxin and peroxisome proliferators (or metabolites derived from them) do not appear to be genotoxic; that is, they do not appear to bind to DNA or cause mutations that are thought to be initiating events in carcinogenesis. However, dioxin is one of the most carcinogenic chemicals in certain sensitive species when they are fed extremely low amounts. Dioxin tends to accumulate in tissues, because it is poorly metabolized and very soluble in cellular lipids. This persistence together with the potent toxicity and carcinogenicity seen in some species has led to fears that even very low levels of exposure to dioxin in the environment can lead to adverse affects in humans. As a result, the U.S. Environmental Protection Agency (EPA) has set very low exposure limits to this and related chemicals based on mathematical models employed for genotoxic carcinogens. It may be very costly to attain these low exposure limits, and, in fact, these limits may be unnecessarily stringent. Although a report published in 1991 indicates that workers chronically exposed to high levels of dioxin show an increased incidence of some cancers, epidemiological data generally indicate that accidental exposures to dioxin that exceed current safeguard limits have not produced significant adverse effects. Moreover, the model used to derive these exposure limits may not be consistent with the mechanism by which dioxin is toxic and carcinogenic.

Regulatory systems. Dioxin and related compounds activate a cellular protein, the dioxin receptor, by binding to it; and it is this dioxin–protein complex that is thought to mediate the toxic effects of dioxin and related polychlorinated hydrocarbons. The receptor–dioxin complex is known, in turn, to regulate gene expression selectively. In a similar fashion, peroxisome proliferators, a second group of nongenotoxic carcinogens in rodents, can activate a recently identified protein, the peroxisome proliferator–activated receptor, that also selectively regulates gene transcription. This mechanism of action is similar to that of hormones such as estrogens and glucocorticoids. The changes that occur when the expression of responsive genes is altered by the chronic activation of these transcription factors by environmental chemicals is thought to lead to toxic or carcinogenic sequelae.

The molecular details of this process are best understood for the stimulation of the expression of the cytochrome P-450 1A1 gene by dioxin. The receptor–dioxin complex binds to short, specific DNA sequences (enhancers) upstream of the gene's promoter, and it is this binding that is thought to stimulate transcription of the P-450 1A1 gene. In the absence of dioxin, the receptor is found in the cytoplasm of the cell in association with a heat-shock protein, HSP-90, and possibly a second unidentified protein. HSP-90 is also seen to associate with other ligand-activated transcription factors, such as the glucocorticoid receptor; and HSP-90 appears to block the binding of either receptor to its DNA recognition sequence. The association of either the dioxin or glucocorticoid receptors with their respective activating ligands appears to facilitate the dissociation of the receptors from HSP-90.

In the case of the dioxin receptor, a second, recently discovered protein, the aryl hydrocarbon receptor nuclear translocator (arnt), is required in order for the receptor–dioxin complex to accumulate in the nucleus and bind to regulatory sites of the P-450 1A1 gene. Mutant cells that do not express a competent arnt protein accumulate the receptor–dioxin complex in the cytosol, and no induction of P-450 1A1 is seen. The arnt protein exhibits a basic helix-loop-helix structure, which is seen for a variety of transcription factors that form homodimers and heterodimers. Cross-linking studies suggest that two proteins of unequal sizes bind to the DNA recognition sites of the P-450 1A1 gene, and only one of these appears to bind dioxin. The arnt protein may correspond to the other component of the oligomeric dioxin receptor. However, this has not been established experimentally.

The peroxisome proliferator–activated receptor is likely to function in a similar manner. This receptor is related to a large family of transcription factors that includes the glucocorticoid receptor, which shares many functional characteristics with the dioxin receptor. A recent study indicates that the peroxisome proliferator–activated receptor controls the induction of the P-450 4A6 gene in response to peroxisome proliferators, and that this is mediated by an upstream regulatory element of the P-450 4A6 gene. It is unclear, however, whether or not these two regulatory systems that govern the induction of P-450 genes by environmental chemicals are related. It is not yet known whether the peroxisome proliferator–activated receptor binds to the regulatory element of the P-450 4A6 gene as a homodimer or heterodimer. Receptors such as the glucocorticoid receptor that are structurally related to the peroxisome proliferator–activated receptor generally form homodimers, and they have not been found to associate with basic helix-loop-helix proteins. In addition, the peroxisome proliferator–activated receptor does not exhibit the basic helix-loop-helix motif of the arnt protein. This motif is thought to be important for both DNA binding and dimer formation. Thus, the dioxin receptor may represent a new class of ligand-activated transcription factors that is distinct from the peroxisome proliferator–activated receptor.

Mechanisms. Studies with genetically defined strains of mice suggest that the dioxin receptor mediates the carcinogenic and toxic effects of dioxin. Some strains exhibit a variant form of the dioxin receptor that is expressed at lower concentrations and that exhibits a lower apparent affinity for dioxin. Thus, the concentrations of dioxin required to activate the receptor are higher in strains of mice that express the variant of low affinity and capacity, and the toxic and carcinogenic dose of dioxin differs in a corresponding manner. In addition, the relative toxicity of congeners of dioxin roughly parallels their relative affinity for the dioxin receptor. Moreover, the concentration of dioxin must be sufficiently high in the cell in order to activate the receptor in accordance with the equilibrium constant, which characterizes the binding affinity of dioxin to the receptor. Thus, at very low concentrations the receptor would not be effectively activated, and gene expression would not be altered. These implications have prompted the EPA to review the basis for promulgating safe exposure limits to dioxin, because the incorporation of a mathematical model for receptor activation might lead to estimates of safe exposure limits that are different from those for genotoxic carcinogens that assume that even a single molecule of dioxin could be carcinogenic.

The intermediacy of the dioxin receptor implies that persistent alterations in the expression of a gene effected by the dioxin receptor are likely to lead to initiation or promotional events in carcinogenesis as well as to toxicity. The induction of the P-450 1A1 gene by dioxin could alter the relative risk of initiation events from other chemical carcinogens. Cytochrome P-450 1A1 and other P-450 enzymes often catalyze the activation of chemical carcinogens to genotoxic forms as well as the detoxication of these carcinogens. Thus, alterations in the relative expression of these enzymes can affect the balance between activation and detoxication and influence the risk of initiation. In a similar manner, peroxisome proliferators lead to increases in the cellular content of enzymes that yield reactive products from the partial reduction of oxygen that could lead to genetic damage and tumor initiation. However, it is not clear that these effects are sufficient to explain the carcinogenicity of these chemicals.

On the other hand, both dioxin and peroxisome proliferators are also promoters of carcinogenicity. Promoters can lead to experimentally detectable tumors when applied following exposure to initiators at concentrations where tumors are not detected for either the initiator or promoter alone. By analogy to the phorbol ester promoters (potent carcinogens), which activate the transcription factor AP-1, the promotional activity of dioxin and peroxisome proliferators could reflect the role of the transcription factors activated by these chemicals in regulating the expression of genes that govern cellular proliferation. Such a mechanism might also underlie initiation as well as promotion in carcinogenesis in much the same way that alterations in the expression or function of proto-oncogenes that are transcription factors can lead to cellular transformation. In this case, environmental chemicals would alter the function of their target transcription factors over a prolonged period of time, leading to progressive cellular changes that culminate in transformation.

Significance. Identification of the genes that lead to carcinogenesis and toxicity is necessary to clearly establish the underlying mechanism of how these toxic chemicals act through their receptors. In addition, the interactions between both the dioxin receptor and the peroxisome proliferator–activated receptor with other proteins to form oligomeric complexes need further delineation. Different oligomers may regulate different genes, and competition between transcription factors for limited amounts of a partner may lead to pathological consequences. Moreover, there may be additional receptors with a similar function for other P-450 genes that are induced by different chemicals occurring in the environment. Several of these inducing chemicals, such as phenobarbital, have also been shown to be tumor promoters. In addition to providing a more complete picture of how environmental chemicals can alter gene expression by nongenotoxic mechanisms, the characterization of these receptors could lead to the design of inexpensive assays for identifying chemicals that can activate these receptors and that could have pathological effects.

For background information *SEE CHROMIUM; DIOXIN; FREE RADICAL; ONCOLOGY; REACTIVE INTERMEDIATES; TOXICOLOGY* in the McGraw-Hill Encyclopedia of Science & Technology.

Eric F. Johnson

Bibliography. J. C. Bailar III, How dangerous is dioxin, *N. Engl. J. Med.*, 324:260–262, 1991; B. J.

Culliton, U.S. government orders new look at dioxin, *Nature*, 352:753, 1991; E. C. Hoffman et al., Cloning of a factor required for activity of the Ah (dioxin) receptor, *Science*, 252:954–958, 1991; I. Issemann and S. Green, Activation of a member of the steroid hormone receptor superfamily by peroxisome proliferators, *Nature*, 347:645–650, 1990; J. P. Landers and N. J. Bunce, The Ah receptor and the mechanism of dioxin toxicity, *Biochem. J.*, 276:273–287, 1991; S. Langard, Basic mechanisms of the carcinogenic action of chromium: Animal and human data, *Toxicol. Environ. Chem.*, 24:1–7, 1989; S. Langard, One hundred years of chromium and cancer: A review of epidemiological evidence and selected case reports, *Amer. J. Ind. Med.*, 17:189–215, 1990; E. Merian (ed.), *Metals and Their Compounds in the Environment*, 1991.

Transistor

Transistors are key elements of modern electronics. Their speed (or their maximum frequency of operation) determines the ultimate speed and capabilities of all electronic systems, from computers to communications, and from consumer and industrial electronics to defense systems. This article will consider transistors operating at the highest possible frequencies. Some of these devices have already demonstrated record speeds of operation. Others illustrate new concepts that may achieve an ultimate transistor speed in the near future.

Transistors are semiconductor devices that control larger output voltages and currents by smaller input voltages or currents. The vast majority of transistors used in all kinds of systems, from digital watches to supercomputers, are field-effect transistors (FETs). A field-effect transistor can be visualized as a parallel-plate capacitor where one of the "plates," the gate,

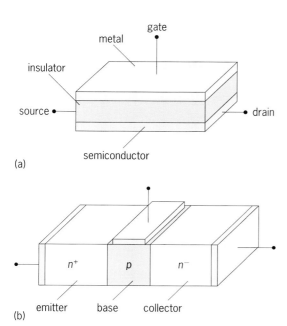

(a)

(b)

Fig. 1. Basic transistor types. (*a*) Field-effect transistor. (*b*) Bipolar junction transistor.

controls the amount of charge induced into the other "plate," the semiconductor channel connecting the drain and source contacts (**Fig. 1***a*). Another type of transistor is the bipolar junction transistor (BJT), which consists of semiconductor layers of opposite polarities, *p* and *n*, forming an *npn* structure (Fig. 1*b*). The contacts of the bipolar junction transistor are called the emitter, base, and collector. In the typical operating regime of a bipolar junction transistor, a forward-biased emitter-base *pn* junction injects carriers through a narrow base region into a reverse-biased collector-base *pn* junction.

A never-ending effort to increase the transistor's speed of operation started immediately after the transistor was invented. Higher speeds have been achieved through three different approaches (or their combination): making devices smaller so that electrons (or holes) have to travel shorter distances to respond to a signal, using new high-speed materials, and using new transistor designs or even new principles of transistor operation.

Electron-beam lithography. Usually, transistors are fabricated by using a photolithography process. In this process, a semiconductor wafer is covered with a special material called a photoresist. The patterns defining transistor contacts and other features are printed on the photoresist in a procedure similar to the exposure of a photographic film. In a standard photolithography process, the photoresist is exposed by an ultraviolet light, and the minimum transistor feature size is determined by the wavelength of this light. Typically, this size is on the order of a micrometer. The use of electron beams to expose the photoresist can reduce this minimum size by a factor of 100, down to 10 nanometers or so, with a commensurate increase in transistor frequency of operation.

Compound-semiconductor transistors. Conventional transistors are usually made of silicon, an element belonging to the fourth column of the periodic table (an element with four valence electrons). Elements from columns 3 and 5 of the periodic table (such as gallium and arsenic) can be combined to create compound semiconductors with the same number of valence electrons per atom as in silicon (four). Compounds such as gallium arsenide (GaAs), indium phosphide (InP), indium arsenide (InAs), indium antimonide (InSb), and aluminum arsenide (AlAs) have semiconducting properties and band structures somewhat similar to classic elemental semiconductors, such as silicon or germanium. At the same time, they provide a wide variety of materials with different band gaps (direct and indirect), different lattice constants, and other physical properties for use in semiconductor devices. Some of these materials can be further combined to form solid-state solutions, such as $Al_xGa_{1-x}As$, where the composition, x, may vary continuously from 0 to 1, with a corresponding change in physical properties from those of gallium arsenide into those of aluminum arsenide.

Semiconductor materials. Gallium arsenide is the most studied and best-understood compound-semiconductor material. It has proven indispensable for many device applications, from ultrahigh-speed transistors to lasers and solar cells. Its room-temperature lattice constant (0.5653 nm) is very close to that of aluminum arsenide (0.5661 nm), and the heterointerface between the two materials has a very small density of interface states. New technologies such as molecular-beam epitaxy (MBE) and metal-organic chemical vapor deposition (MOCVD) make it possible to grow these materials with very sharp and clean heterointerfaces and to have very precise control over doping and composition profiles, literally (in the case of molecular-beam epitaxy) changing these parameters within an atomic distance. Other compound semiconductors, important for applications in ultrahigh-speed submicrometer devices, include $In_xGa_{1-x}As$, $Al_xIn_{1-x}As$, $In_xGa_{1-x}P$, gallium phosphide (GaP), indium phosphide, and aluminum nitride (AlN).

Advantages. There are several advantages of gallium arsenide and related compound semiconductors for applications in submicrometer devices. The effective mass of electrons in gallium arsenide is much smaller than in silicon ($0.067\,m$ in gallium arsenide compared to $0.98\,m$ longitudinal effective mass and $0.19\,m$ transverse effective mass in silicon, where m is the free electron mass). The result is a much higher electron mobility in gallium arsenide: approximately 8500 $cm^2/(V{\cdot}s)$ in pure gallium arsenide at room temperature compared to 1500 $cm^2/(V{\cdot}s)$ in silicon. Moreover, in high electric fields, the electron velocity in gallium arsenide is also larger than in silicon; this characteristic is very important in small (submicrometer) devices where the electric fields are high. The light electrons in gallium arsenide are much more likely to experience ballistic transport, that is, to move through the short active region of a high-speed device without having any collisions (with lattice imperfections or lattice vibration quanta). As a result, their velocity may be boosted far beyond the values expected for long devices. Such ballistic transport may become important in very short devices with sizes on the order of 0.1 μm or less. In somewhat longer gallium arsenide devices (with dimensions between 0.1 and 1.5 μm), overshoot effects are important. These effects are related to the finite time that it takes for an electron to relax its energy. They may also result in boosting the electron velocity to considerably higher levels than the stationary values. Overall, in many compound semiconductors, such as gallium arsenide, electrons move faster than in silicon so that the speed of operation is higher. For this reason, the next generations of supercomputers, which demand an ultimate speed, will probably utilize compound-semiconductor transistors. Both heterostructure field-effect transistors (HFETs), with gate lengths as short as 80 nm, and heterojunction bipolar transistors (HBTs) have already demonstrated very high frequencies of operation.

Cutoff frequency. A transistor speed can be characterized by a cutoff frequency, which is more or less the maximum frequency of transistor operation. Compound-semiconductor field-effect transistors with ultrasubmicrometer gates, and heterojunction bipolar transistors with highly doped ultrasubmicrometer base regions, have achieved cutoff frequencies approaching 200 GHz. These frequencies of operation may be further increased in the near future by utilizing new designs where the electric field profile in the active region is optimized by using variable doping or variable composition.

Applications. Compound-semiconductor transistors are used mostly for microwave and ultrahigh-speed applications, where their high-speed properties are important. Other possible uses include optoelectronics, radiation-hard electronics, high-temperature electronics, power devices, and electronics operating in harsh environments.

Alternative materials. The ability to produce high-quality heterojunctions by growing layers of different compound semiconductors on top of each other has led to the development of many novel transistors promising even higher speeds of operation. Both heterojunction field-effect transistors and heterojunction bipolar transistors have been implemented by utilizing different material systems, such as AlGaAs/GaAs, AlGaAs/InGaAs/GaAs, InGaAs/InP, and AlInAs/InP. The first heterojunction bipolar transistors were implemented by using the AlGaAs/GaAs material system. However, the use of indium gallium arsenide ($In_xGa_{1-x}As$) as the base material has several advantages over gallium arsenide. The low-field mobility of indium gallium arsenide is higher than the mobility of gallium arsenide. Electrons in indium gallium arsenide achieve higher velocities without being transferred to satellite valleys with heavy electron mass. This characteristic makes ballistic and overshoot transport in the base much more efficient. Also, in AlGaAs/InGaAs/GaAs heterojunction bipolar transistors, the composition of indium gallium arsenide can be graded, with the percentage of indium increasing from the emitter toward the base. This grading creates an additional built-in field assisting electron transport across the base. The potential advantages of InGaAs/InP and AlInAs/InP heterojunction bipolar transistors are related: a smaller built-in emitter junction potential, more pronounced ballistic and overshoot effects, a high electron drift velocity in indium phosphide, and low surface recombination velocity. In addition, the quaternary InGaAsP emitter can be used with a band gap corresponding to the wavelength $\lambda = 1.3$ μm, compatible with fiber-optics communication systems. Experimentally, the best heterojunction bipolar transistors were fabricated by using InP/$In_{0.53}Ga_{0.47}As$/InP or AlInAs/$In_{0.53}Ga_{0.47}As$/InP material systems. (This particular composition of indium gallium arsenide has a lattice constant that is very close to that of indium phosphide.) These devices exhibit both high gain and exceptionally high-speed operation.

It has been suggested that materials such as silicon carbide or diamond could, in principle, reach even higher speeds of operation than more conventional compound semiconductors such as gallium arsenide, especially at elevated temperatures of operation. Silicon carbide transistors have already demonstrated frequencies of operation in the gigahertz range. However, silicon carbide technology has not yet matured, and diamond technology is still in its infancy, so that it is difficult to determine the full potential of these materials for applications in ultrahigh-speed transistors.

Ballistic electron transport. Many of the new transistors proposed in recent years, such as the hot-electron transistor (HET) or vertical ballistic transistor (VBT), utilize ballistic (or collisionless) electron transport, which should, in principle, lead to a higher speed of operation. In a hot-electron transistor (**Fig. 2**), the emitter region is separated from the base region by a thin barrier. Electrons tunnel through the emitter–base barrier into the very narrow base region, which they traverse ballistically. The fraction of the electrons coming to the collector depends on the collector voltage, which controls the height of the barrier at the base–collector interface. At reverse collector voltages, the device acts as a hot-electron spectrometer, which allows the electron energy distribution to be determined through measurement of the collector current. At forward collector biases, this transistor exhibits an amplification action.

In the vertical ballistic transistor (**Fig. 3**a), electrons are injected from the cathode electrode into a very short region separating it from the anode. Ideally, they move through this region ballistically (that is, without experiencing any scattering events) because the region is much shorter than the mean free path between collisions. The effective cross section of this region is modulated by the potential of the gate contacts, so that there is a transistor action. This device is quite similar to a more conventional static induction transistor (SIT). So far, technological difficulties have prevented these transistors from surpassing the speed of more conventional heterojunction field-effect transistors and heterojunction bipolar transistors. However, hot-electron transistors have been already used for obtaining important information about semiconductor parameters, and as a result the physics of electron transport in high electric fields is better understood.

Real-space-transfer transistors. Real-space-transfer (RST) transistors utilize the effect of the electron energy increase in high electric fields. (This

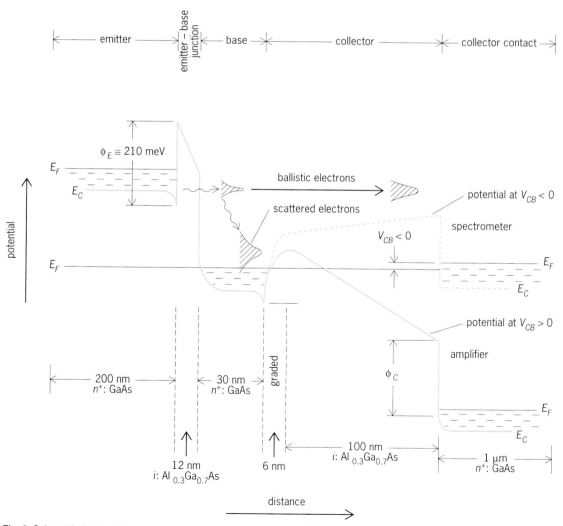

Fig. 2. Schematic band diagram of a hot-electron transistor for two different values of the collector-base voltage V_{CB}. E_C = energy of bottom of conduction band; E_F = Fermi energy; ϕ_E = emitter work function; ϕ_C = collector work function. (*After M. Heiblum et al., Direct observation of ballistic transport in GaAs, Phys. Rev. Lett., 55:2200–2203, 1985*)

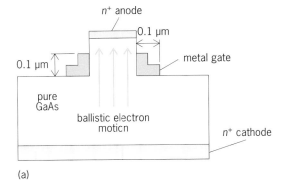

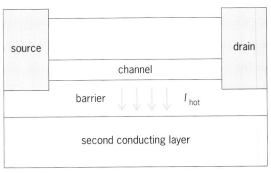

Fig. 3. Diagrams of ultrahigh-frequency transistors. (*a*) Vertical ballistic transistor (*after L. F. Eastman et al., Ballistic electron motion in GaAs at room temperature, Electr. Lett., 16:524–525, 1980*). (*b*) Real-space-transfer transistor (*after S. M. Sze, ed., High Speed Devices, John Wiley and Sons, 1990*).

increase is usually characterized by introducing a notion of an increasing electron temperature.) Figure 3*b* illustrates the principle of operation of a typical real-space-transfer transistor, called a negative-resistance field-effect transistor (NERFET). In this device, the increase in electron temperature with the increase in the drain-to-source voltage leads to an increase in the hot-electron current I_{hot} flowing across a barrier into a second conducting layer; as a consequence, there is a drop in the drain-to-source current, that is, to a negative differential resistance (which is governed by the electron temperature in the channel). This device should be able to achieve a very high speed of operation because the modulation of the drain-to-source current in this regime is determined by the energy relaxation time (of the order of 1 picosecond or so for gallium arsenide) and by the time constant of the electric field variation. The latter time is determined by the electron transit time across the high-field region near the drain and may be several times shorter than the transit time of electrons across the gate, so that a limit is set for the intrinsic speed of conventional field-effect transistors.

Prospects. Novel ultrahigh-speed transistors utilizing compound semiconductor materials are expected to reach subpicosecond speeds of operation, speeds that are now achieved only by two-terminal devices, which already operate in a terahertz frequency range. (At the present time, the ultrahigh-frequency transis-

tors operate at about 10 times lower frequencies.) Such devices will make it possible to build notebook supercomputers, foot-size satellite dishes, video telephones, and intelligent robots.

For background information SEE INTEGRATED CIRCUITS; SEMICONDUCTOR; SEMICONDUCTOR HETEROSTRUCTURES; TRANSISTOR in the McGraw-Hill Encyclopedia of Science & Technology.

Michael Shur

Bibliography. M. Shur, *GaAs Devices and Circuits*, 1987; S. M. Sze (ed.), *High Speed Devices*, 1990; C. T. Wang (ed.), *Introduction to Semiconductor Technology: GaAs and Related Compounds*, 1990.

Tubeworm

The marine worm *Riftia pachyptila* attains a length of about 5 ft (1.5 m) and a diameter of nearly 2 in. (5 cm) even though it has no mouth and no digestive system. This animal is found in warm springs at ocean depths greater than 1 mi (1.6 km). There is no photosynthetically derived food in that region, and animals living there must rely upon a certain type of bacteria for nutrition. These bacteria utilize an inorganic chemical source for energy requirements rather than the energy of sunlight, as is the case in green plants. *Riftia* incorporates these bacteria into a structure inside its body and harvests them as food. It is an exquisite symbiotic relationship.

Riftia pachyptila was first collected in 1977 during geological investigations of an area of sea-floor spreading on the Galápagos Rift. The rift is an east-west extension from the East Pacific Rise, a linear spreading center adjacent to South and Central America. Using the submersible *Alvin*, at depths of about 8200 ft (2500 m) researchers found communities of animals living in the flow of heated water issuing from fissures on the ocean floor, so-called hydrothermal vents.

Vent environment. The hydrothermal vent environment, as well as the animal communities of the vents, is anomalous relative to that of the nonvent sea floor at the same depth. Water is heated deep in the crust of the Earth and, as it moves upward, leaches various materials from the rocky material through which it passes. When the heated water reaches the surface of the sea floor on the Galápagos Rift, it may have a temperature as high as 73°F (23°C), whereas seawater over the sea floor at nonvent sites is about 36°F (2°C). Among the minerals and chemicals suspended or dissolved in the vent water is hydrogen sulfide. Ordinarily this inorganic chemical acts as a poison by interfering with an enzyme that is crucial for the metabolism of animals.

It was found that, among the bacteria associated with hydrothermal vents, certain types are at the base of the food chain of the vent community. These are the sulfide-oxidizing bacteria, which use the chemical energy bound in sulfide to carry on their metabolism. In converting sulfide to elemental sulfur and to thiosulfate, the bacteria obtain energy to convert carbon dioxide to organic materials necessary for their life.

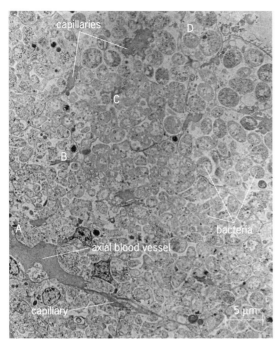

Fig. 1. *Riftia pachyptila.* Cross section of a trophosomal lobule, one quadrant. Zones A, B, C, and D of bacterial development are discussed in the text. (*Transmission electron micrograph by D. S. Rowan, Bryn Mawr College*)

merous lobules (**Fig. 1**), with superficial blood vessels connected to a single axial vessel within the lobule via capillaries that pass between bacteria-containing cells, the bacteriocytes (**Fig. 2**).

The circulatory system of *R. pachyptila* (**Fig. 3**) is closed; that is, the blood is confined to blood vessels and capillaries. Blood flowing from the branchial plume enters the single ventral vessel through a valve that prevents backflow, and is moved posteriorly through the vestimentum to the trunk mainly by the contractions of a thin layer of muscles in the vessel wall. In the trunk, vessels carry blood to the trophosomal lobules and ramify over the surface of each trophosomal lobule. Blood passes from the extralobular network to a single central axial vessel through capillar-

Bacterial growth is so prolific that, directly or indirectly, the bacteria feed all of the animals present at the vent and there is no reliance on photosynthetically derived food.

Classification. *Riftia pachyptila* is a vestimentiferan worm, a small group of marine animals. At present there is disagreement as to whether *Riftia* and the other vestimentiferans constitute a phylum, the Vestimentifera, or whether they are a subphylum of the Pogonophora. There are 12 species distributed among seven genera; all are restricted to similar sulfide environments.

Adult morphology and anatomy. *Riftia pachyptila* lives in a tough, flexible, white tube attached to rocks, shells, or the tubes of other *Riftia*. The body of the worm comprises four regions. The most anterior region is a red plume of branchial filaments; the color is due to hemoglobin carried in blood passing through the blood vessels of the filaments, the site for the uptake of oxygen and other materials. As many as 227,800 branchial filaments are fused into thin lamellae that are arranged perpendicular to the plume axis on the right and left sides. A second region, the vestimentum, serves to hold the worm at the open end of the tube with the plume extended into the surrounding water. The trunk is the third and longest region, and the last region is quite short and segmented.

Internally, running the length of the trunk are longitudinal muscles, two major blood vessels, the nerve cord, and either male or female reproductive organs. Also present is the trophosome, a central structure provided with peripheral and internal blood vessels and composed of cells containing sulfide-oxidizing bacteria. The substructure of the trophosome consists of nu-

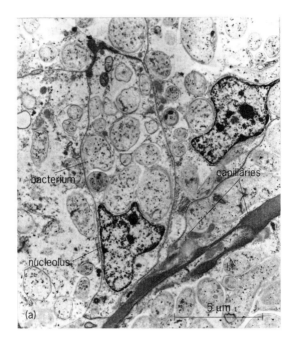

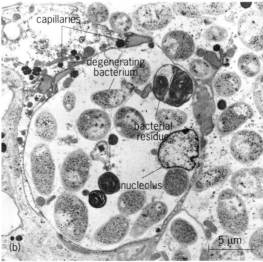

Fig. 2. Bacteriocytes of (*a*) zone C of trophosomal lobule and (*b*) zone D of trophosomal lobule. (*Transmission electron micrographs by S. L. Gardiner, Bryn Mawr College*)

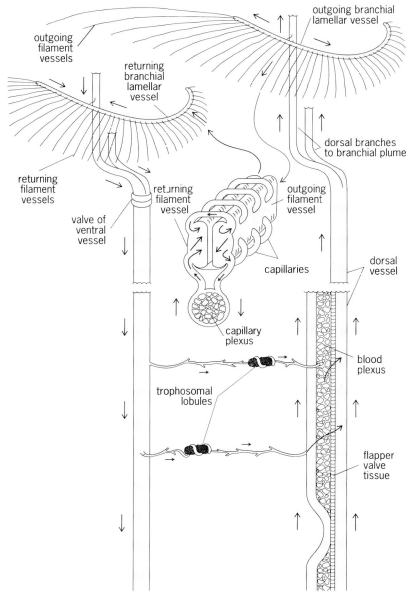

Fig. 3. Summary of certain elements of the circulatory system of *Riftia pachyptila*; anterior at top. Arrows indicate direction of blood flow. Branchial vessels of only one branchial lamella are shown. Areas in color represent blood. (*Adapted from M. L. Jones, The Vestimentifera, their biology, systematic and evolutionary patterns, in L. Laubier, ed., Actes du Colloque Hydrothermalisme, Biologie et Ecologie, Oceanol. Acta, Special Volume no. 8, 1988*)

ies passing between bacteriocytes. Axial vessels unite to form larger and larger vessels that carry blood to the dorsal vessel via a smaller secondary longitudinal vessel or via a blood plexus between the two. The dorsal vessel, in essence a heart, has a thick layer of muscles whose contractions serve as the primary force in moving blood forward in that vessel and throughout the blood system; backflow here is prevented by longitudinal, "flapper-valve" tissue over the openings of vessels entering the dorsal vessel.

Blood moves anteriorly to the two halves of the branchial plume via two branches of the dorsal vessel. These branches give rise to outgoing blood vessels at the base of each branchial lamella, and these, in turn, send a vessel into each of the filaments in the

lamella. At intervals of 5 micrometers along the length of a filament, capillaries pass between epithelial cells of the filament wall and meet with a returning vessel running to the base of the filament. At every third or fourth capillary there is a capillary plexus within a flattened lobe on one side of the filament. Blood flow along the filament in both directions is facilitated by a series of ring muscles that apparently "milk" the blood through the vessels of the filaments and through the capillary system, thus acting as auxiliary "hearts" in each filament. Blood is collected from the filaments by returning branchial lamellar vessels and finally passes from the plume in two vessels that unite just anterior to the valve in the ventral vessel.

Hemoglobin molecules of *Riftia* transport both sulfide and oxygen at the same time, the sulfide to the endosymbiotic bacteria in the trophosome and the oxygen both to tissues of the worm and to the bacteria. The two are bound so tightly to the hemoglobin molecule that the sulfide is not oxidized by the oxygen and is not able to poison any critical enzyme in the worm.

Endosymbionts.   The endosymbionts of *R. pachyptila* are gram-negative, sulfide-oxidizing bacteria that are located in bacteriocytes (Fig. 2). They have not yet been cultured in the laboratory and have not been identified. Generally they are about 3–5 μm in diameter. The raw materials that they need for chemosynthesis—sulfide, oxygen, and carbon dioxide—are brought to them in the trophosome via the blood system of the worm.

In the fully developed trophosome there appears to be a gradient of bacterial development. Cells lining the axial vessel of lobules have no bacteria (Fig. 1, zone A). Adjacent to these is a zone of bacteriocytes with relatively small bacteria (Fig. 1, zone B), and farther away from the axial vessel is a zone of bacteriocytes that contain larger bacteria (Fig. 1, zone C; Fig. 2*a*). At the outer margin of the lobule there are flattened peritoneal cells that lack bacteria. Just internal to these, there is a zone of bacteriocytes in which bacteria appear to be undergoing a chemical breakdown (Fig. 1, zone D; Fig. 2*b*). It is thought that this last bacterial zone is the site of digestion of bacteria, the products of which are probably taken up by the peripheral layer of the lobule. Here, then, is the source of nutrition for adult *Riftia*.

Larval development and association.   In *Riftia* and other vestimentiferans, the sexes are separate; that is, there are female worms and male worms. Based on histological observations of sperm in the oviduct of females, fertilization appears to be internal. There are no indications of early development of fertilized eggs in the oviduct, so it is assumed that such development occurs in the surrounding seawater. Ultimately a trochophore-type larval stage is reached. The larva has a mouth and gut, and it is capable of dispersal in the plankton. Eventually the trochophore larva settles at a hydrothermal vent site, and the next part of its life cycle proceeds. If no vent site is encountered, the larva probably lives on its supply of yolk or on whatever food it may find, and its prospects are not promising.

In later larval development a ventral medial process arises near the anterior end. The process has a ciliated opening (mouth) that leads to a gut and then to a posterior anal opening. At this stage there is a complete gut. When the larva, settled at a hydrothermal vent, is about 0.25 mm (0.01 in.) in length, bacteria are found in the cells of the gut wall. There is disagreement as to whether these bacteria are ingested by the larva from the surrounding environment or whether bacteria are passed on by the parent via sperm or egg, being present in yolk granules from the beginning of larval development. Soon the mouth and anus are closed off; the isolated gut with its bacteriocytes (cells of the former gut lining, filled with bacteria) undergoes growth, obliterating the gut lumen; the mass of bacteriocytes becomes well vascularized; the trophosome is formed, and gutless symbiotic life begins.

For background information SEE *DEEP-SEA BACTERIA; DEEP-SEA FAUNA; HYDROTHERMAL VENT; MARINE MICROBIOLOGY; TUBEWORMS* in the McGraw-Hill Encyclopedia of Science & Technology.

Meredith L. Jones

Bibliography. C. R. Fisher, Chemoautotrophic and methanotrophic symbioses in marine invertebrates, *Rev. Aquat. Sci.*, 2(3,4):339–436, 1990; M. L. Jones and S. L. Gardiner, On the early development of the vestimentiferan tube worm *Ridgeia* sp. and observations on the nervous system and trophosome of *Ridgeia* sp. and *Riftia pachyptila, Biol. Bull.*, 177:254–276, 1989; L. Laubier (ed.), *Actes du Colloque Hydrothermalisme, Biologie et Ecologie, Oceanol. Acta*, Special Volume no. 8, 1988.

Underground mining

When facing a decision on a complex problem, experts rarely have complete information. Often, the information that is available appears overwhelming in sheer volume, and perhaps some is contradictory. A human assimilates this information; filters it by using a blend of analytical, judgmental, and experiential factors; and arrives at a decision. The exact nature of this decision-making process is not well understood and may be different for various experts considering the same problem. In some complex situations, the expert may approach a final solution by constantly fine-tuning the decision.

Expert systems. These is a class of computer programs that attempt to solve problems much as a human expert would. They collect information and process it by using rules of thumb and more formal mathematical logical algorithms to arrive at a decision. A knowledge base of relevant information is constructed from insights gained through interviews with one or more human experts to establish the data and the procedures that they weigh when solving a complex problem.

Expert-system technology has just begun to affect the mining industry. Many complex mining problems lend themselves to the power of expert-system solutions. Once human expert knowledge is accurately captured and recorded, it survives indefinitely and provides continuity of expertise. Also, an expert system allows true expert knowledge to be available for many more problems than the single human would have time to consider.

In mining, many problems are quite complex, requiring the best judgment of human experts. The U.S. Department of the Interior's Bureau of Mines has applied expert-system technology to complex problems in equipment maintenance, methane control, dust control, and roof bolting.

Hydraulic diagnostic expert system. Machine breakdowns due to hydraulic-system failures can contribute to prolonged delays for repairs that result in lost production time and increased operating expenses. The Bureau of Mines has developed a sensor-based hydraulic diagnostic expert system for maintenance of a continuous-mining machine. This system will decrease the occurrence of breakdowns.

The hydraulic diagnostic expert system can pinpoint problems occurring within the hydraulic system by applying the same reasoning techniques used by expert maintenance mechanics in troubleshooting. It can also make diagnoses by reasoning about the theoretical aspects of the problem and by knowing how the machine reacts to certain problems. Diagnoses are based on information received from 38 machine-mounted sensors that measure system pressures, flows, temperatures, fluid levels, and the amount of ferrous debris in the hydraulic oil. A distributed hardware–software interface between the sensors and the expert system handles data acquisition, storage, preprocessing, and transfer of sensor status information to the expert system. The operator need not input data to the system in normal operation.

The hydraulic diagnostic expert system can be used on existing mining machines or on the computer-assisted mining machines of the future. It was developed by using the expert-system shell Goldworks II to accommodate an on-line sensor interface.

Expert system for methane control. Methane gas occurs naturally within coalbeds and is held in equilibrium within the coal's micropore structure. Mining disrupts this equilibrium so that the gas dissociates from the coal and flows through the natural fractures of the coalbed to the mine opening. This emission of methane gas must be carefully controlled to prevent explosive accumulations.

A number of methods can control methane emissions, ranging from forced ventilation to methane drainage by boreholes. The selection of the most efficient method depends on a number of conditions, such as mining-system design, mine plans, and coalbed characteristics. METHPRO was developed to assist mining companies in selecting the appropriate methane-control technique for their particular conditions. METHPRO is a knowledge-based expert system that incorporates the research results from the Bureau of Mines' Methane Control Program.

METHPRO employs four subsystems: a natural-language interface, the knowledge base, an inference engine, and a working memory. The natural-language-interface subsystem provides the user with the interactions expected in a face-to-face consultation with an expert. METHPRO uses 10 different knowledge bases and over 600 rules to represent the expert-level decision-making process for methane control. The inference engine is used to draw logical conclusions and to control the reasoning process. The working memory provides a mechanism for entering data for a particular problem and keeping track of the status of the solution. The expert-system shell used for building METHPRO is known as INSIGHT 2+.

The key issues with the development of any computer model or program are the validity and application of the program to real-world problems. METHPRO was evaluated at a number of mining sites to determine the accuracy of the program's findings. These sites included a wide range of conditions such as both longwall and room-and-pillar mines, and high- and low-gas-content coalbeds. The results of the evaluations compared favorably to realistic conditions, and METHPRO was released to the mining community.

METHPRO has greatest application to those sites where methane emissions are highest. METHPRO is an excellent tool for determining the suitability of various technologies for methane drainage versus ventilation design. These sites are in Alabama, Illinois, Pennsylvania, Virginia, and West Virginia. Underground coal mines in these states account for more than 250×10^6 ft^3 (7×10^6 m^3) of methane emissions per day. Methane drainage before and during mining is often necesary to prevent interruptions to mining. METHPRO has been successfully used in these cases to recommend solutions to methane control problems.

Expert system for dust control. DUST-PRO, another expert system developed by the Bureau of Mines, analyzes dust-control problems on longwall shearers and continuous-miner sections. The system for longwall shearers comprises (1) advice for new users on the basic dust-control measures every longwall should use; (2) advice on several more refined approaches if compliance cannot be achieved with the basic measures; and (3) a special menu that provides choices for users with specific interests.

The expert system for dust control on continuous miners first determines the source of the respirable dust by querying the user about the mine ventilation system, the mining sequence, and compliance with federal standards. If the problem is determined to be the roof bolter (a machine used to install large screwed-in rods that act as permanent supports for the roof) rather than the mining machine, the expert system focuses on roof bolter maintenance. DUSTPRO was written by using the INSIGHT 2+ expert-system shell.

Roof-bolting expert system. The Bureau of Mines is developing an expert system to improve the ability of mining engineers to evaluate the effectiveness of roof support for ground control in the coal mine. This expert system compares the capacities of roof bolts to the predicted support required for entry stability. Results of tests on various types of roof bolts are evaluated, and the maximum allowable loading based on the anchorage capacity and yield strength of the support is estimated.

The user manually enters (keyboards) geological information, including rock properties, geometry of the opening, in-place stresses, bolt pattern parameters, and rock types. The expert system uses this information to compare the predicted loading to the capacity of the roof support and advises the operator regarding the adequacy of the design and offers suggestions for improvements, if needed. As ground conditions change, a reassessment to change the support design is required. One possible approach is to collect real-time sensor information regarding changing roof conditions from the roof bolter itself and to feed it into the expert system to dynamically update the data concerning the current roof conditions. Researchers have collected data on drill-bit position, penetration rate, thrust, torque, and rotation rate by using an instrumented roof bolter. These data formed the input pattern for a neural-network model that identifies different types of strata and roof features. The user interfaces with the expert system by using the information from the model and the manually entered geologic data. The mining engineer now has a tool to refine continually the roof support design using information that is regularly updated as mining progresses.

For background information SEE DISTRIBUTED SYSTEMS (CONTROL SYSTEMS); EXPERT SYSTEMS; NEURAL NETWORK; SIMULATION; UNDERGROUND MINING in the McGraw-Hill Encyclopedia of Science & Technology.

U.S. Bureau of Mines

Bibliography. F. N. Kissell and R. L. King, Three expert systems which give advice on coal mine methane and dust control, *Proceedings of the 4th International Mine Ventilation Congress*, Brisbane, Queensland, Australia, 1988; J. Mitchell, A knowledge-based system for hydraulic maintenance of a continuous miner, *Proceedings of the 9th West Virginia University International Mining Electrotechnology Conference*, 1988.

Underwater sound

Since the 1920s, when sonic depth finders gained widespread acceptance, sound waves have been the most effective means of remotely sensing the sea floor. The reason is that other forms of radiation, such as electromagnetic waves commonly used in radar, are attenuated over a few tens of meters in the ocean, whose average depth throughout the world is about 4 km (2.4 mi). Improvements in electronics, computers, and devices for mass storage of digital data have permitted evolution of the technology from broad-beam sonars yielding a single depth point per transmission cycle to multi-narrow-beam sonars mapping a swath of soundings for each cycle. Likewise, acoustic images of the sea floor provided by side-looking sonars have evolved from mere scan-line displays of the echoes received as a function of time, one scan line per ping, to range-corrected acoustic images for bathymetric purposes. Recently, there have been a number of advances in acoustic imaging of the sea floor and swath bathymetry for deep-water

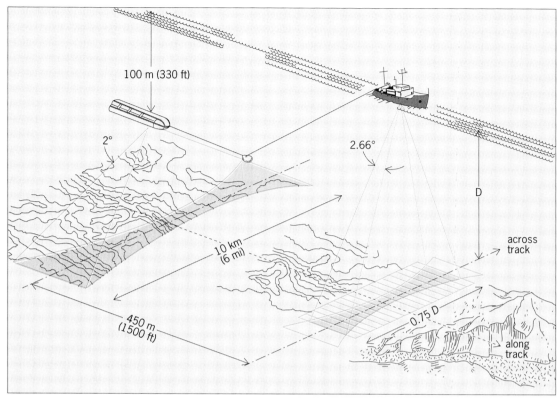

Fig. 1. Diagram showing the geometry of swath mapping sonars: multibeam echo sounder (right) and side-looking sonar (left). In the multibeam geometry, the shaded area represents the intersection of the transmit and receive patterns and corresponds to the elemental sea-floor areas from which echoes are received. In the side-looking sonar case, the same beam is used to transmit and receive. D = water depth. (*From C. de Moustier, P. L. Lonsdale, and A. N. Shor, Simultaneous operation of the SeaBeam multibeam echo sounder and the SeaMARC II bathymetric sidescan sonar system, IEEE J. Oceanic Eng., 15(2):84–94, 1990*)

sonars. **Figure 1** shows two common types of sonars used for sea-floor swath mapping: multibeam echo sounders and phase-measuring side-looking sonars.

Multibeam echo sounders. In most multibeam echo sounders, transducer arrays are permanently mounted on the underside of a ship hull. This type of sonar implements the cross-fan-beam technology, in which the transmitting and receiving acoustic arrays are mounted with their long axes perpendicular. The transmitting array is usually mounted with its longest dimension parallel to the vessel's fore–aft axis so as to form an acoustic beam that is narrow in that direction and broad in the athwartships dimension. The receiving array, being perpendicular to the fore–aft axis of the vessel, forms a beam that is narrow athwartships and broad fore–aft.

These relative beam widths stem from a physical relationship between beam width, array size, and acoustic wavelength such that the width of the acoustic beam in a given direction is inversely proportional to the number of acoustic wavelengths across the effective aperture of the array in that direction. For example, a line array must be about 25 wavelengths long to form a $2°$ beam in the direction, also called broadside, perpendicular to its face. At 12 kHz, a widely used acoustic frequency for oceanic echo sounders, the wavelength is roughly 0.12 m (0.4 ft), yielding arrays 3 m (10 ft) long.

The intersection of the transmit and receive beams

patterns is a small angular sector delimited by their respective narrow dimensions ($2.66°$ in Fig. 1). Constraint of echoes to small angular sectors is the salient characteristic of narrow-beam echo sounders, as it affords good spatial discrimination when determining the bearing of sea-floor echoes. In order to obtain a swath of such soundings, several contiguous receive beams are formed simultaneously by a process called steering. This entails adjusting for the time or phase delays sensed by the array elements as an acoustic wavefront impinges on them from a given direction. By bringing all the echoes from this wavefront to a common phase reference, a beam is "steered" in that direction. Today, this operation is usually performed by computers on the digital representation of the echoes received at each array element.

SeaBeam, the first commercial multibeam echo sounder, became operational in the late 1970s and has been installed aboard a score of research vessels throughout the world. This system forms 16 receive beams and provides soundings with an angular resolution of about $2.66°$ over a swath whose width is roughly three-fourths of the water depth (D) in the survey area. Although this system is still operational aboard many vessels, it is being supplanted by more capable sonars introduced since the mid-1980s. Notable improvements include more beams, greater swath coverage, and greater sounding accuracy.

The Hydrosweep system, manufactured in Germany, became operational in 1985. It produces 59 beams over a 90° athwartships sector for a swath width of twice the water depth down to 9000 m (30,000 ft). The advantage of this system is its ability to switch transmit and receive functions between its hull-mounted arrays. As the athwartships survey pattern is flipped along track, soundings that were previously measured at nadir (the point on the sea floor that lies vertically below the sonar) without being affected by refraction are measured over angles extending up to 45° away from nadir and are subject to refraction. Refraction refers to the bending of sound rays traveling at oblique incidence through the water column because of changes in the speed of sound encountered as a function of depth and horizontal distance. Availability of both types of soundings allows for a refraction correction to be computed and then applied to the next set of soundings to be collected in the conventional athwartships survey mode.

Today's most advanced multibeam echo sounder, the EM12, manufactured in Norway, became operational in early 1990. This sonar produces up to 151 beams over a 150° athwartships sector for a swath coverage of more than seven times the water depth down to 2000 m (6600 ft). Beyond a depth of 2000 m (6600 ft) the number of beams and the corresponding athwartships angular sector are reduced to account for transmission losses incurred as sound waves travel through the water column. The 150° coverage is obtained by mounting one pair of transmit and receive arrays on each side of the vessel's center line, creating what amounts to a double system; this configuration became operational in 1991. Sounding accuracies reported from surveys performed with the EM12 show sounding standard deviations averaging 0.2–0.3% of the water depth over a 140° athwartships sector. This level of accuracy is achieved by combining the formation of narrow beams with differential phase measurements as is done for bathymetric side-looking sonars to refine the bearing estimate. By comparison, soundings made with the 16-beam SeaBeam system have a standard deviation of about 1% of the water depth.

Recent upgrades of the SeaBeam system, manufactured in the United States, have yielded the SeaBeam 2000, producing 121 beams over a 120° sector down to a depth of 4500 m (15,000 ft), and reducing to a 90° sector with 91 beams at a depth of 10,000 m (33,000 ft). The manufacturer has predicted that sounding accuracies will give standard deviations of less than 0.5% of water depth.

Side-looking sonars. By comparison with the multibeam sonars, long-range side-looking sonars such as the Gloria system, or phase-measuring side-looking sonars, are generally simpler to design and build. Acoustic arrays are usually mounted on a tow body that is deployed from ships of opportunity, so that flexibility is added to the survey operation. An obvious drawback of any towed system is the risk of losing the tow body and its sonars when the tow cable breaks.

Side-looking sonars use the same acoustic array for the transmit and receive functions. This array is usually long enough to provide a 2° beam width in azimuth (angle measured in the fore–aft horizontal plane of the tow body) and narrow enough for a beam width in excess of 70° in the athwartships direction. In order to obtain maximum coverage on either side of the tow body, one such array is mounted on each side. To

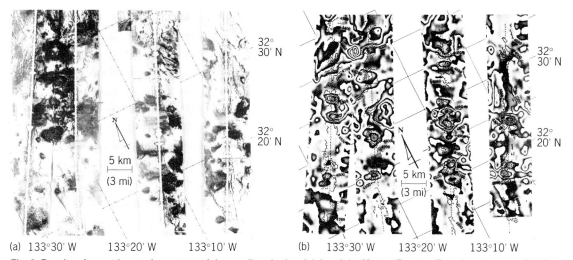

Fig. 2. Results of a swath mapping survey of the sea floor in the vicinity of the Murray Fracture Zone (northeastern Pacific), using simultaneously the SeaBeam and SeaMARC II sonar systems. (a) The SeaMARC II acoustic image appears as four parallel swaths of black-and-white sonograms in which several small volcanoes can be seen (black dots of various sizes). Dark areas represent strong sea-floor echoes, light tones represent weak echoes (light gray) or shadows (white). (b) For bathymetry, the corresponding swaths are shaded contours overlaid with SeaBeam bathymetry (black contour lines at 100-m or 330-ft interval, with tick marks pointing downhill). The SeaMARC II bathymetry is displayed by using eight cycles of a gray scale. Each cycle, running from black to white, represents a depth interval of about 170 m (560 ft). Over the entire display, depth ranges from about 3900 to 5200 m (13,000 to 17,000 ft). (*Modified from C. de Moustier, P. L. Lonsdale, and A. N. Shor, Simultaneous operation of the SeaBeam multibeam echo sounder and the SeaMARC II bathymetric sidescan sonar system, IEEE J. Oceanic Eng., 15(2):84–94, 1990*)

avoid interference between the two sides, the arrays are operated at different acoustic frequencies. As an example, the system known as SeaMARC II operates at 11 and 12 kHz. With broad athwartships beams, these sonars cannot determine accurately the bearing of sea-floor echoes; they rely on the time of arrival of these echoes, plus knowledge of the altitude of the tow body above the sea floor and a flat-bottom assumption, to infer echo arrival angles and hence the lateral position of the corresponding sea-floor features. An acoustic image of the sea-floor can thus be formed by displaying the intensity of the echoes as a function of horizontal range athwartships. This image has the characteristics of a reversed shadowgram, in which features oriented toward the arrays return relatively strong echoes, appearing dark in the image, whereas features oriented away from the ensonification field (the sea floor illuminated by the sound waves) return little or no echoes, so that light impressions (acoustic shadows) are left in the image (**Fig. 2**).

In the early 1980s, SeaMARC II, a long-range side-looking sonar system, was fitted with arrays consisting of two parallel transducer rows roughly a half wavelength apart, on each side of the tow body. The intention was to improve the bearing resolution of the system by allowing measurement of the phase difference between echoes received at each row. By analogy to the beam-forming technique mentioned above, this phase difference corresponds to the difference in travel time from one row to the next, and it is therefore related to the angle of arrival of the echo. Such phase measurements are then converted to bearing angles that, along with the times of arrival of the corresponding echoes, serve to compute depths and horizontal distances. As a result, this sonar system is capable of outputting both a bathymetric swath map of the sea floor and an acoustic image of the same area (Fig. 2). The bathymetry delineates the morphology of the sea floor, and the acoustic image provides information on its texture or roughness at a scale that is orders-of-magnitude finer than the bathymetry. Compared to the multibeam echo sounders, side-looking sonar systems suffer from an inherent ambiguity problem whenever echoes from different features arrive at the same time, as is commonly the case in rough terrains. The accuracy of soundings obtained with such systems is usually on the order of 3% of the altitude of the tow body above the sea floor. Since 1990, newer bathymetric side-looking sonars have been built with improved echo processors and more carefully designed transducer arrays, some of them including a third row for better bearing accuracy.

Joint use. Over the past few years, a number of research expeditions have used multibeam echo sounders and side-looking sonars in conjunction to take advantage of the high bathymetric resolution (a few meters) of multibeam systems and the imaging abilities of the side-looking sonars. An example of joint use of the SeaBeam and SeaMARC II systems is shown in Fig. 2. However, since the mid-1980s it has been shown that multibeam echo sounders could also

provide acoustic images of the sea floor by displaying the echoes received by each beam in their proper spatial frame across track. Recently, such acoustic images have been made an integral function of the EM12 multibeam sonar. Nonetheless, the combined use of towed and hull-mounted sonars is likely to continue as researchers strive to collect swath maps of the sea floor at different scales of resolution to gain a greater understanding of the morphology and of the acoustic properties that characterize specific substrates (for example, volcanic rocks or different types of sediments), to elaborate geoacoustic models of the sea floor.

For background information SEE ECHO SOUNDER; SONAR; UNDERWATER SOUND in the McGraw-Hill Encyclopedia of Science & Technology.

Christian de Moustier

Bibliography. *Int. Hydrog. Rev.*, vol. 65, no. 2, July 1988; *Mar. Technol. Soc. J.*, December 1986; *Proceedings of the IEEE Oceans' 91 Conference*, Honolulu, October 1991; Special issue on bathymetry and sea-floor acoustic remote sensing, *IEEE J. Oceanic Eng.*, vol 14, no. 4, October 1989.

Vaccination

Vaccination, the immunization of individuals to prevent infectious disease, is a highly cost-effective public health measure. The first modern vaccine to prevent smallpox was developed nearly 200 years ago. Its use decreased the incidence of smallpox in several countries, and its worldwide application during 1967–1977 led to the eradication of smallpox. There are now about 30 viral and bacterial vaccines for medical use, a similar number close to registration, and many more under development.

Types of vaccines. Some current viral vaccines and some in late-stage clinical trials are listed in the **table**. Almost all of these vaccines are against

Viral vaccines

| Type | Agents | |
	Current	Under trial
Live, attenuated	Vaccinia	Cytomegalo
	Measles	Hepatitis A
	Yellow fever	Influenza
	Mumps	Dengue
	Polio*	Rota
	Adeno*	Parainfluenza
	Rubella	Japanese en-
	Varicella	cephalitis
	zoster	Polio*
Inactivated	Polio	Hepatitis A
	Influenza	
	Rabies	
	Japanese	
	encephalitis	
Subunit	Hepatitis B	
	Influenza	

* Administered orally.

viruses that cause acute infections (for example, influenza or smallpox) in most people, as distinct from viruses that cause chronic, persisting infections [for example, human immunodeficiency virus (HIV), leading, in time, to acquired immune deficiency syndrome (AIDS)].

Attenuated live vaccines. Most vaccines are live (that is, infectious), attenuated preparations; the process of attenuation reduces the virulence of the organism while retaining the power to induce a strong immune response. Some of these vaccines are very effective, for example, those against measles and yellow fever; they may give long-lived, perhaps lifelong, immunity after one or two doses. They are usually safe except in those people whose immune system is compromised.

Noninfectious vaccines. Some vaccines are composed of inactivated, whole virus. Because such preparations are noninfectious, they are generally safe for all. However, some vaccines cause ill effects in a small number of people or in certain populations, and this aspect is discovered only when the vaccine is administered to perhaps several hundred thousand people. For example, immunization with a preparation composed of whole inactivated respiratory syncytial virus actually sensitized some babies, so that later exposure to the natural infection resulted in more severe pathology than was experienced by unvaccinated babies. Those vaccines listed in the table are safe and generally effective.

Noninfectious vaccines may be composed of only part (a subunit) of the virus. For example, only the surface antigen of the hepatitis B virus is used in the hepatitis B vaccine. This surface antigen is present in the blood of infected people and so can easily be isolated for use in vaccines.

Bacterial vaccines. It has been difficult to make attenuated bacterial vaccines. Inactivated whole bacteria, such as the *Bordetella pertussis* vaccine for whooping cough, or subunit vaccines, such as the *Streptococcus pneumoniae* vaccine for pneumonia, are more common.

Stimulation of immune response. A vaccine must be both safe to use and capable of protecting vaccinees against disease when they are later exposed to the natural infection. An effective vaccine for prophylactic use stimulates the adaptive components of the immune response, that is, the lymphocytes that show immunological memory, namely B (bone marrow–derived) and T (thymus-derived) lymphocytes. B cells make antibodies; if some of these antibodies bind to the virus (for example, in **Fig. 1**, to the envelope protein component, gp120, of the HIV), the virus may be prevented from infecting a cell. There are two types of T cells: one type, cytotoxic T cells, may kill cells once they are infected with viruses or bacteria; the other type, helper T cells, helps new B cells to become antibody-secreting cells, and other new T cells to become cytotoxic T cells. Helper T cells play a central role in controlling immune responses.

Vaccines are occasionally used for therapy, for ex-

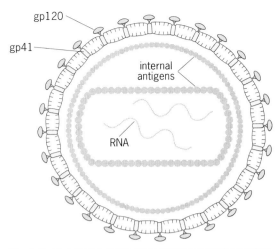

Fig. 1. Some features of the human immunodeficiency virus (HIV). The envelope protein, gp160, has two protein components, gp120 and gp41, the latter spanning the viral lipid membrane.

ample, in the treatment of rabies victims. In this case, the generation of additional responses, such as natural-killer-cell activity and production of some interferons, may be useful. Vaccines for the treatment of people already infected with HIV are currently under development.

Immune response to live virus and to inactivated virus. Noninfectious antigen, such as inactivated virus, is taken up by an antigen-presenting cell, such as a macrophage (**Fig. 2**). It is broken down to peptides, some of which may bind inside the cell to certain host-cell proteins, called major histocompatibility (MH) proteins, so that self protein–foreign peptide complexes are formed. After these complexes are transported to the cell surface, they may be recognized as foreign antigen by specific receptors on resting T cells. If the antigen presenting cell also secretes special factors called interleukins, the resting T cell becomes activated, produces additional interleukins, and differentiates to become a helper T cell.

In contrast, an infectious agent, such as a live virus, may infect the antigen presenting cell (Fig. 2). Since live viral preparations always contain some noninfectious particles, their processing will quickly lead to helper-T-cell formation. As part of the infectious process, however, some of the newly synthesized viral proteins are degraded to different peptides, some of which may bind to other MH molecules. These complexes may be recognized by specific receptors on another class of resting T cells. However, to differentiate further, this class of cell needs help (in the form of interleukins) from the already-formed helper T cell. With this help, the resting T cell becomes a cytotoxic cell, capable of killing virus-infected cells, so that the formation and release of new viral progeny is prevented. For this reason, cytotoxic T cells appear after the formation of helper T cells during an infection.

Some infectious agents cause chronic persisting infections. The reason may lie in the generation of suppressor T cells, which suppress other immune responses. To be most effective, a vaccine should not induce the formation of these suppressor cells.

MH proteins are so diverse (polymorphic) that they may differ even between individuals. Particular peptides may or may not bind to an individual's MH proteins, and an individual protein–peptide complex may or may not be recognized by a T cell. These factors determine an individual's immune response to an infection or a vaccine. Whole-organism vaccines are generally very effective because there are many proteins that can contribute peptides, so that every vaccinee should respond to some of them. A potential flaw with subunit vaccines is that the number of peptides is more limited.

Role of different immune responses. After the initial infection by a microorganism, helper T cells are first formed, followed by cytotoxic T cells and then by secretion of antibody by B cells. Since infection can be prevented only by the presence of a preexisting specific antibody, a prime requirement of prophylactic vaccines is to stimulate B-cell responses so that antibody is continually present months or even years after the vaccination. This antibody should reduce the size of the challenge infection by over 99% so that the host's immune system can comfortably cope with the remaining infectious agent. If, however, the infectious agent eludes this antibody barrier, the presence of a large pool of memory helper T cells and cytotoxic T cells generated by the vaccine should rapidly be activated. Infected cells can then be destroyed before viral progeny accumulates. Ideally, a vaccine should induce specific antibody formation for a long period after vaccination, generate large pools of memory lymphocytes, and be a rich source of different peptides so that some T cells of all recipients will respond.

Eluding the host's immune response. In order to coexist with the host, infectious agents have developed various strategies, such as antigenic diversity, infectious sanctuaries, infection of immune cells, and poor antibody responses.

Antigenic diversity. Great antigenic variation can occur, especially with surface antigens of viruses containing a ribonucleic acid (RNA) genome. Protective antibody is usually directed to these antigens; if the antigens vary, antibody produced following vaccination may not prevent infection. The classic example of antigenic diversity is influenza virus; diversity is even more marked with HIV. As the internal antigens of a virus, usually a rich source of peptides recognized by cytotoxic T cells, are less antigenically variable than surface antigens, cytotoxic T cells should still be able to help in the recovery from infection. A vaccine that contains a "cocktail" of surface antigens with different antigenic specificities, such as the influenza virus and bacterial pneumococcal vaccines, may overcome this problem by generating antibodies with a range of specificities.

Infectious sanctuaries. Such sanctuaries are sites (cells, organs) where the immune process fails

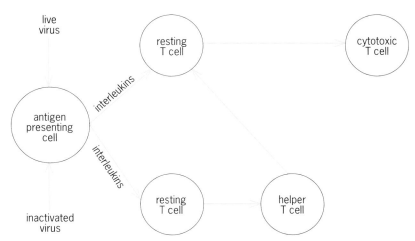

Fig. 2. Model of immune response to live virus and to inactivated virus.

to prevent the continuing formation of infectious virus progeny. Thus, some viruses, such as HIV, may reach and infect parts of the body, such as the brain or the epididymus, which are often inaccessible to components of the immune system. In another situation, such as adenovirus infection, the infected cell produces much less MH protein than is normally the case. As a result, fewer MH protein–peptide complexes are expressed at the cell surface, so that the chances for the cell to be recognized by a cytotoxic T cell are greatly reduced. Therefore, infectious adenovirus continues to be produced in such cells. A vaccine should aim to stop or limit an infection before these events happen.

Infection of immune cells. Some viruses infect, and thus affect the function of, important cells of the immune system; for example, HIV infects helper T cells, and Epstein-Barr virus infects B cells.

Poor antibody responses. HIV and the agent of scrapie, to name two, induce poor or delayed antibody responses. The cause may be suppressor effects, which new vaccines are aiming to overcome.

Vaccine development. There are still many diseases for which vaccines are needed. For various reasons, such as the inability to grow enough of the agent, three new approaches to vaccine development, all based on the subunit principle, are being evaluated. (A fourth approach, anti-idiotype vaccines, is less well established.)

Peptide synthesis. As only parts of antigens (perhaps only one or two peptides in the whole protein) are recognized by T and B cells, the synthesis and use of only these parts to develop a vaccine is being avidly pursued, especially for the development of a malaria vaccine.

Transfection of cells with DNA coding for antigens. Deoxyribonucleic acid (DNA) coding for foreign proteins can be introduced into different cells and can be expressed so that a large amount of the foreign protein is made. Thus hepatitis B virus surface antigen is now made in transfected yeast cells, and is the first genetically engineered vaccine registered for human use. The HIV gp120 protein component (Fig. 1) has been made in transfected mammalian cells, and is being assessed for safety and immunogenicity in clinical trials.

Chimeric live vectors as vaccines. With the concept of exploiting the advantages of live-agent vaccines, DNA coding for foreign proteins has been inserted into the DNA of some existing vaccines, including viruses (vaccinia, adeno) and bacteria (*Salmonella*; Bacillus Calmette-Guérin, or BCG). The foreign proteins are sythesized and induce an immune response when the viruses or bacteria infect host cells. A chimeric live vaccinia virus containing DNA that codes for the HIV gp120 protein is also now in clinical trials. This approach is already working well for veterinary vaccines, particularly in controlling two important viral infections, rabies and rinderpest.

Such live vectors can be programmed to express DNA coding for several antigens, from either the same or different infectious agents, so that they can potentially be regarded as multivalent vaccines, protecting against several diseases.

Concerns about new approaches. A vaccine consisting of only one or a few proteins may not contain sufficient T-cell or B-cell epitopes to elicit a response from all people. Even though the hepatitis B surface antigen contains several proteins, up to 14% of people may respond poorly or not at all. It may become necessary to add another viral protein component, known to contain peptide sequences recognized by helper T cells, to the vaccine to increase efficacy.

Recently, immunization of mice with a chimeric vaccinia virus containing a subunit preparation of the lymphocytic choriomeningitis (LCM) virus resulted in the production of little protective antibody and low levels of cytotoxic T cells. When the mice were challenged with infectious LCM virus, there were insufficient cytotoxic T cells to kill the cells infected with virus. By the time the cytotoxic T-cell response had substantially increased, there were many more infected cells, and their killing resulted in severe immunopathology.

Advance knowledge of these potential difficulties will help in the development of safe vaccines by these new techniques. There are still many infectious diseases of global medical importance, including HIV infection leading to AIDS, for which vaccines are urgently needed.

For background information *SEE ANIMAL VIRUS; CELLULAR IMMUNOLOGY; IMMUNOLOGY; VACCINATION* in the McGraw-Hill Encyclopedia of Science & Technology.

Gordon Ada

Bibliography. G. L. Ada, The immune response to antigens: The immunological principles of vaccination, *Lancet*, 335:523–526, 1990; E. Egea et al., The cellular basis for lack of antibody response to hepatitis B vaccine in humans, *J. Exp. Med.*, 173:531–538, 1991; C. Flexner, New approaches to vaccination, *Adv. Pharmacol.*, 21:51–99, 1990; F. A. Murphy et al., Vaccinia-vectored vaccines: Risks and benefits, *Res. Virol.*, 140: 463–491, 1989; S. Oehen et al., Vaccination for disease, *Science*, 251:195–198, 1991.

Venus

The *Magellan* mission to map Venus is the first new United States planetary mission in 12 years. This mission completes the reconnaissance exploration of the inner planets of the solar system, providing vital new information about the geological processes of Earth-size planets, and begins a new phase of highly focused missions of scientific exploration.

Magellan mission. The *Magellan* mission science objectives were established in 1983, when the project was approved by the National Aeronautics and Space Administration (NASA). The objectives were to improve the knowledge of the geological history of Venus by analysis of the surface morphology and electrical properties and the processes that control them. Included in the objectives are tectonic and volcanic histories and evidence for past climates.

The *Magellan* spacecraft was launched from Cape Canaveral, Florida, aboard the space shuttle on May 4, 1989, and was placed in orbit around Venus on August 10, 1990. The orbit is elliptical and near-polar, with a period of 3.26 h. The periapsis altitude (closest point) is 182 mi (293 km) at 9.9° north latitude. The high point in the orbit, apoapsis, is nearly 5000 mi (8000 km). After a period of testing and setting up the radar and navigation for mapping, *Magellan* began mapping on September 15, 1990. The primary mission lasted 243 days, the number of Earth days it takes for Venus to turn once on its axis, and ended on May 15, 1991. This period is a Venus sidereal day. During this 8-month period, *Magellan* mapped 83% of the planet. Each day *Magellan* maps 7.3 orbits of image data in a strip 12 mi (19 km) wide and about 10,000 mi (16,000 km) long, extending from the North Pole to below 70° south latitude.

The long, narrow strips of image data are mosaicked by computer into a set of large images that can be more easily studied. The amount of image data returned by *Magellan* is more than double the images returned by the United States planetary program since the first *Mariner* missions to Mars in 1964 through the present, a period including more than a dozen planetary missions.

Magellan radar system. *Magellan* uses synthetic aperture radar. Synthetic aperture radars "look" at the surface several times as the spacecraft moves over the surface, so that an antenna that is much longer than the actual physical antenna is built up. Although radar uses a part of the electromagnetic spectrum that includes much longer wavelengths than visible-light cameras (12.6 cm versus 0.5 micrometer), the processed images look much like conventional digital television images. There are some important differences in the geometry of camera systems and radars, but the same kinds of measurements and stereo can be done with both systems. The resolution of the images is 400–984 ft (120–300 m), and the images are processed with 246-ft (75-m) picture elements (pixels).

Radar images generally provide information about the roughness of the surfaces at a scale of about a tenth of a wavelength to several wavelengths. In the case of *Magellan*, surfaces that are smooth at a scale of

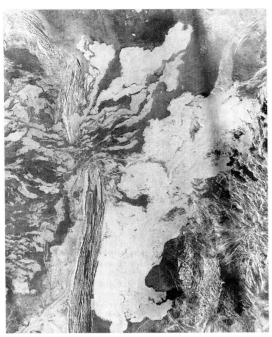

Fig. 1. *Magellan* image of a portion of the Lada region, showing a system of radar-bright and -dark lava flows encountering and breaching a ridge belt and then pooling in a vast, radar-bright deposit on the right side of the image. The lava flows and ridge belts are evidence of the volcanism and tectonism that is present everywhere on Venus. (*Jet Propulsion Laboratory; NASA*)

about 1 in. (2.5 cm) or less appear darker, and surfaces that have roughness heights of a few inches to several feet or more appear brighter in the images. The radar system is also used as an altimeter to map the surface topography by sending the signal straight down to reflect off the surface beneath the path of the spacecraft. The altimeter measures relative height, and approaches 16-ft (5-m) accuracy in smooth areas and 290-ft (88-m) accuracy in extremely rough areas. A third operating mode of the radar is as a radiometer. Between the bursts of radar and altimetry signals are quiet periods when the passive radiation of the surface at 12.6-cm wavelength is recorded. In this way the microwave brightness temperature of the hot surface of the planet is measured. The comparison of this emission with the actual physical temperature of the surface, which varies only with elevation, provides information about the electrical properties of the surface.

Volcanism and tectonism. The surface of Venus is complex and variable, with many types of volcanic and tectonic features. Venus has been revealed as a planet with at least as complex a geologic record as Earth, with many of the same geologic processes revealed on its surface. The dominant process is volcanism, seen in many forms on the plains, which make up about 85% of the surface. On Venus, volcanism, the eruption of molten rock onto the surface, and tectonism, the faulting and folding of crustal rocks, act much like erosion by running water on Earth to modify the landscape, and eventually erase the highest mountains. There is abundant evidence of continuing volcanism and tectonism everywhere on the planet,

in the vast lava floods and the fractures, faults, and ridge belts of the volcanic plains (**Fig. 1**). Pervasively fractured and faulted older terrains are also seen. The tessera, a comprehensive term for highly fractured terrains (**Fig. 2**), probably represent the oldest rocks, but the deformation that has so completely distorted them appears to be continuing.

Wind-generated landforms. The upper atmosphere of Venus, at 30–40 mi (50–65 km) above the surface, has high-speed winds blowing at speeds above 200 mi/h (90 m/s) from east to west. At the surface, winds are variable with typical speeds of 2–3 mi/h (1–1.5 m/s) in various directions. This surface wind speed is slow but, in the dense Venusian atmosphere, fast enough to move small sand and dust particles. In many regions of Venus, the *Magellan* images show streaks that appear to be in the lee of topographic obstacles (**Fig. 3**). In some cases the streaks are bright and may indicate removal of fine material, revealing a rougher underlying surface. In other areas the opposite is seen, with a dark streak in the lee of small volcanic domes and other topographic obstacles. Similar streaks are seen on Earth in radar images of desert regions, and record the direction of the maximum winds capable of transporting sand.

Volcanic landforms. A wide range of volcanic landforms of quite varied sizes and shapes is seen in the *Magellan* images. The most frequently occurring volcanic form is the small dome or lava shield. It generally has a circular outline and a height of 325–650 ft (100–200 m) and diameter of 1.2–5 mi (2–8 km). There may be more than 100,000 small volcanic domes on Venus. They resemble the smaller shield volcanoes on Iceland or Hawaii. Many of these domes occur in clusters of 50–100 individual volcanoes. Larger volcanic structures are fewer but probably represent far more total volume. There are on the order of 3000 intermediate-sized (12–62 mi or 20–100 km

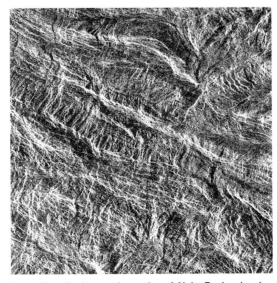

Fig. 2. *Magellan* image of a portion of Alpha Regio, showing the complex pattern of intersecting ridges and valleys called tessera. (*Jet Propulsion Laboratory; NASA*)

Fig. 3. *Magellan* image of a portion of the northwestern Ovda region, showing wind streaks 300 mi (500 km) northeast of Mead, the largest impact crater known to exist on Venus, with a diameter of 170 mi (275 km). The streaks most likely represent debris from the impact that has been modified by surface winds. (*Jet Propulsion Laboratory; NASA*)

Steep-sided domes observed on Venus suggest thicker, more viscous lavas. On Earth such lavas generally have a higher silica content and are often associated with explosive volcanism. One curious type of dome is a circular form 15–43 mi (25–70 km) in diameter and 300–2000 ft (100–600 m) high. In general, these domes resemble giant pancakes (Fig. 4*d*).

Plains. A wide variety of thin, areally extensive lava flows are seen on the plains. This type of volcanism may be responsible for forming the plains on Venus. Many sinuous channels have been discovered on the plains. The channels are 1600 ft to 1 mi (0.5 to 1.5 km) wide and remarkably constant in width (**Fig. 5**). They are often leveed and may terminate in broad plains deposits. The longest of these channels, located in Rusalka Planitia, is more than 3700 mi (6000 km) long. The channels may have formed by cutting and melting of hot lavas that were extruded at high volume rates. Some channels appear to have been produced by erosion, with carving of streamlined islands, as is seen in channels cut by running water on Earth and Mars. Candidates for the liquid that cut the channels, other than water (which cannot exist on Venus), are sulfur, carbonates, and high-temperature lavas.

Current activity. Many volcanic regions on Venus appear to be young and may be currently active. The only way to determine whether a region is currently active is to observe a change between visits, which can occur at 8-month intervals as the planet slowly turns beneath the orbit plane of the spacecraft.

Tectonic features. Venus and Earth have many features in common. Among these are global systems of fractures and, as evidence that the crust has been shortened by compressive forces, elevated mountain ranges that are much longer than they are wide. However, the tectonic style of Earth appears to be simpler than that of Venus. The Earth's crust is divided into a few rigid plates. New crust on Earth is produced at spreading centers along mid-ocean rises, and crust is destroyed at collision boundaries, where it sinks back into the mantle. Along such boundaries on Earth are found the most violent earthquakes, the

in diameter) volcanoes (**Fig. 4***a*). Often, individual lava flows are associated with these larger structures. Approximately 100–150 volcanoes on Venus are larger than 62 mi (100 km) in diameter (Fig. 4*b*). The larger volcanic structures, like the smaller ones, are built up of multiple individual lava flows, much like the island of Hawaii. And, like Hawaii, they are generally situated on a broad domical crustal rise. Large complex rift systems, where the crust has been broken and pulled apart, often cut through the larger volcanoes.

Calderas and domes. Large depressions, called calderas on Earth, are also seen on Venus. The largest calderalike structure on Venus, Sacajawea Patera (Fig. 4*c*), a depression in the high plains of western Ishtar Terra, is 120 mi (200 km) in diameter and 0.6–1.2 mi (1–2 km) deep. This depression is probably caused by subsurface migration of magma and collapse of the surface.

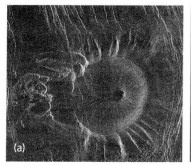
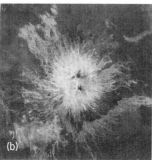
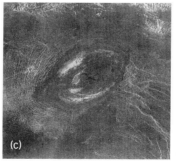
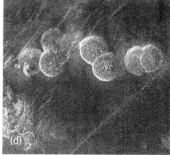

Fig. 4. *Magellan* images of volcanic landforms. (*a*) Intermediate-sized volcano in the Eistla region, approximately 41 mi (66 km) across the base. To the left, the rim of the structure appears to have been breached by dark lava flows that emanated from a shallow summit pit and traveled to the left along a channel. (*b*) The volcano Sapas Mons, about 250 mi (400 km) across and 0.9 mi (1.5 km) high. The flanks are composed of numerous overlapping lava flows, and the summit area consists of a pair of flat-topped mesas. (*c*) Sacajawea Patera, the largest calderalike structure on Venus. (*d*) Portion of the eastern edge of Alpha Regio, showing seven circular, domical hills averaging 15 mi (25 km) in diameter with maximum heights of 2500 ft (750 m). These features can be interpreted as eruptions of viscous lava from vents on the relatively level ground. (*Jet Propulsion Laboratory; NASA*)

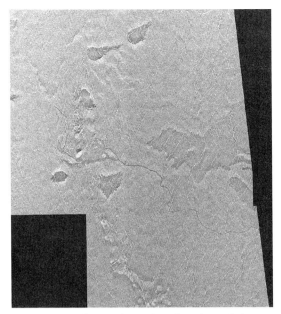

Fig. 5. *Magellan* image of a portion of southeast Guinevere Planitia, showing a sinuous channel, which is about 300 mi (500 km) long in this image and has been traced for approximately an additional 300 mi (500 km). The channel maintains a nearly constant width between 2500 ft (750 m) and 1.2 mi (2 km) throughout its length. (*Jet Propulsion Laboratory; NASA*)

highest mountains, the deepest trenches, and the most explosive and abundant volcanoes. The crust of Venus also seems to deform in similar ways, but a global system is not evident. Locally, and perhaps on a smaller scale, many of the kinds of tectonic deformation that are observed on Earth may be occurring on Venus. On both planets, the surface expression of crustal deformation results from motions within the underlying mantle.

Impact craters. Nearly 1000 impact craters have been mapped on Venus from the *Magellan* images (**Fig. 6**). Observations of asteroids and comets that cross the orbit of Venus suggest that this number of objects would collide with Venus in several hundred million years. Their average velocity is about 45,000 mi/h (20 km/s). This view is consistent with the number of impact craters found on Earth, which formed within the same period of time. On Venus, the impact craters appear to be fresh and unmodified, while those on Earth are generally highly eroded by rain and the runoff of water. The dense atmosphere causes smaller objects to break up completely before they reach the surface, and this process has completely prevented the formation of impact craters smaller than about 2 mi (3 km) in diameter, and has caused the formation of a smaller number of craters less than about 15 mi (25 km) in diameter.

The ejecta deposits of crushed and melted rock thrown from the craters appear to have flowed out over the surface to produce petallike lobes of radar-bright ejecta. That many of the impact craters are surrounded by radar-dark zones possibly indicates smooth surfaces. These may be smoothed regions where the surface was modified by the impact, perhaps by the shock wave associated with passage of the projectile through the atmosphere.

Impact craters are used to determine the age of the surface on the Moon and on other planets and satellites. Since impacts are believed to form randomly in time and space, the number of impacts on a surface gives a measure of how long it has been exposed to impact bombardment. Some highly cratered surfaces on the Moon date back nearly 4×10^9 years. The number of craters on Venus may vary from place to place, and so a possible range of ages from 0 to 800×10^6 years is indicated.

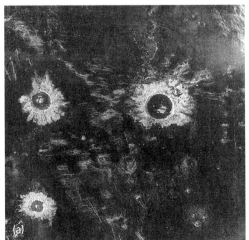
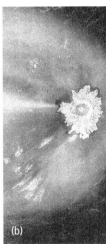

Fig. 6. *Magellan* images of impact craters. (*a*) Three impact craters, with diameters of 23–30 mi (37–50 km), in a region of fractured plains in the Lavinia region. Small volcanic domes are visible at lower right. (*b*) Adivar crater, about 19 mi (30 km) in diameter. Streaks extending over a broad area, particularly to the left of the crater, are believed to result from the interaction of crater materials with high-speed winds in the upper atmosphere. (*c*) Aurelia crater, about 20 mi (32 km) in diameter. (*d*) Stuart crater, 41.5 mi (67 km) in diameter, which is classified as a peak-ring crater because of the concentric ring of hills on its floor. (*Jet Propulsion Laboratory; NASA*)

Mission prospects. *Magellan* will continue to monitor and observe Venus for as long as the spacecraft continues to function. It is not possible to predict when some vital part will fail, but the mission is expected to continue well into the 1990s. Most of the planet will be imaged more than once to provide stereo images and to search for recent volcanism, eruptions that have occurred since *Magellan* first passed over the region. The search by *Magellan* continues for evidence that Earth's sister planet remains as violent today as the images suggest it has been in the recent past.

For background information SEE *RADAR; SPACE PROBE; VENUS* in the McGraw-Hill Encyclopedia of Science & Technology.

R. Stephen Saunders

Bibliography. J. W. Head et al., Venus volcanism: Initial analysis from *Magellan* data, *Science*, 252:276–287, 1991; R. S. Saunders et al., The *Magellan* Venus radar mapping mission, *J. Geophys. Res.*, 95:8339-8355, 1990; R. S. Saunders et al., An overview of Venus geology, *Science*, 252:249–252, 1991; R. S. Saunders and G. H. Pettengill, *Magellan*: Mission summary, *Science*, 252:247–249, 1991.

Virus

Viruses use various mechanisms to evade the defenses of the host's immune system. The first section of this article reviews viral molecular defense mechanisms; the second section discusses in detail antigenic variation.

Viral Molecular Defense Mechanisms

In order to replicate and spread in an animal, viruses must contend with defenses designed to combat foreign invaders. Viruses have developed a wide variety of counterdefenses to enhance their survival in the host. An understanding of these viral defense mechanisms might enable researchers to develop safer and more effective live vaccines, better antiviral therapies, and better comprehension of the host immune response to infection.

Host defenses. Host responses to invading viruses can be classified as specific or nonspecific. Nonspecific defenses include complement and cytokines such as interferon (IFN) and tumor necrosis factor (TNF), and involve a variety of cells that contribute to the inflammatory response. The serum proteins that make up the complement system can directly attach to the surface of some viruses, which then lose infectivity; this process is known as virus neutralization. Attachment of complement components to the virus can also result in the efficient engulfment and destruction of the virus by macrophages. Interferon production by infected cells signals neighboring cells to activate mechanisms to prevent their own infection. Interferon acts by inducing the expression of a protein kinase that, in the presence of double-stranded ribonucleic acid (RNA), phosphorylates the small subunit of a protein synthesis initiation factor (eIF2-α). Phosphorylation of this factor inhibits further protein synthesis in infected cells. Interferon can also induce the synthesis of a second enzyme, $2',5'$-oligoA synthetase, which, in the presence of double-stranded RNA, causes the activation of a ribonuclease and results in the destruction of both viral and cellular RNA. Both mechanisms prevent infection by inhibiting viral protein synthesis. Tumor necrosis factor produced by macrophages and lymphocytes can suppress virus replication and mediate the lysis of cells infected by some viruses. The mechanism of tumor necrosis factor–mediated cytolysis is not known.

Specific host immune responses against viruses include humoral (antibody) and cellular (T-cell) mediated defenses. Viruses induce B cells to produce antibodies that recognize specific regions within proteins (epitopes) expressed on the virus particle or infected cell. These circulating antibodies can attach to the virus particle and lead to virus neutralization or efficient engulfment and destruction of the virus by macrophages. Complement can enhance antibody-dependent viral neutralization. Antibody along with complement can also cause the lysis of infected cells.

Cellular immunity is a complex host-defense system that allows the host to recognize and specifically eliminate cells infected with virus. This host-defense system relies on the presentation of viral antigens by the host major histocompatibility complex (MHC) molecule. Almost all cells express class I MHC molecules. In infected cells, viral proteins are processed into small portions (peptides) that associate with class I MHC molecules, and are transported to the plasma membrane. Recognition of the viral antigen–MHC complex by the T-cell receptor on cytotoxic T cells leads to attack on and killing of virus-infected cells.

Viral evasion of host defenses. Before entry into the intracellular environment, virus particles may be attacked by complement and antibody. For herpes simplex viruses type 1 and 2, their envelope glycoprotein C protects them from complement-mediated virus neutralization. This protein is also expressed on the membrane of herpes-infected cells and protects them from complement-mediated cell lysis. Glycoprotein C alters the complement cascade by causing the accelerated decay of the enzyme complex (C3 convertase) of the alternative pathway that cleaves a complement protein (C3), and by interfering with the interaction of a major product of C3 cleavage (C3b) with the fifth component of complement (C5).

Cells infected by vaccinia virus, a member of the poxvirus family, secrete a viral protein that regulates the complement cascade by accelerating the decay of the C3 convertase of the classical pathway, and assists the protease Factor I in causing the inactivation of C3b and C4b (a major product of cleavage of the fourth component of complement). The secretion of this viral protein in the local environment by infected cells may protect newly released virus particles from direct complement-mediated attack and from antibody-dependent, complement-enhanced virus neutralization. In addition, the secreted viral protein may protect infected cells from complement-mediated cell lysis.

Interferons are a first-line defense against viral infections. Adenovirus and vaccinia virus are relatively insensitive to the effects of interferons. RNA, transcribed from the adenovirus VA$_1$ gene, binds to the cellular protein kinase, so that the protein kinase is prevented from phosphorylating eIF2-α and interfering with protein synthesis. Vaccinia virus also encodes an inhibitor of the protein kinase, as well as an inhibitor of the ribonuclease activated by the product of the 2′,5′-oligoA synthetase. Thus, by preventing interferon-mediated protein synthesis inhibition and viral RNA degradation, vaccinia virus is able to overcome the effects of interferon. Recently another unique mechanism for regulating interferon production was proposed for the Epstein-Barr virus (EBV), a member of the herpes virus family. An Epstein-Barr virus gene, BCRFI, was found to have significant homology to a T-cell cytokine synthesis inhibitory factor that inhibits the synthesis of an interferon. Thus, it is possible that Epstein-Barr virus has captured a cytokine gene to inhibit interferon production and therefore enhance the virus's survival in the host.

Tumor necrosis factor can cause the lysis of virus-infected cells by a yet-unknown mechanism. An adenovirus protein inhibits the tumor necrosis factor–mediated lysis of adenovirus-infected cells. A protein secreted by the Shope fibroma virus (SFV), another member of the poxvirus family, has been shown to have remarkable similarity to the cellular receptor for tumor necrosis factor. It has been hypothesized that the Shope protein is secreted in a soluble form that competes for tumor necrosis factor. By binding tumor necrosis factor, this viral product might inhibit the cytokine's antiviral effect.

The antibody response is a major host defense against viral infection. An effective means of escaping this response is by changing the sequence of the outer viral proteins, a mechanism known as antigenic variation. The best example of this viral molecular defense mechanism is seen with influenza virus and retroviruses. Influenza virus and retroviruses may also avoid neutralizing antibody by displaying carbohydrate side chains near important antigenic sites that block antibody binding. Another potential viral counterdefense mechanism against the host humoral response is the secretion of excessive amounts of a target viral protein in a soluble form to bind antibodies. This role is postulated for gp120 shed from HIV-infected cells, for secreted G protein from rhabdovirus-infected cells, and for secreted noninfectious viral envelope particles from hepatocytes infected with hepatitis B virus. These released viral proteins might act as decoys and interfere with antibody interaction with the virus and virus-infected cells. In the presence of virus-specific antibody, measles virus proteins on the infected cell membrane appear to undergo capping toward one pole of the cell and are shed as antigen–antibody complexes. The result is an infected cell surface void of viral proteins. This antibody-induced modulation of viral antigens allows the infected cell to evade immune surveillance.

Viruses have evolved ways of escaping cell-mediated immune responses. Adenovirus down-regulates the expression of class I MHC on infected cells during the course of the infection by two different mechanisms: one virus encoded protein turns off the synthesis of class I MHC molecules, while another protein binds to class I MHC molecules and prevents their proper transport to the cell surface. Because of the decreased expression of viral antigen–MHC complexes on adenovirus-infected cells, the infected cells become less efficient targets for killing by cytotoxic T cells. A similar strategy is used by Epstein-Barr virus in Burkitt's lymphoma cells. However, instead of down-regulating MHC, Epstein-Barr virus appears to diminish the amount of surface adhesion molecules, which are important in the interaction of infected cells with cytotoxic T cells. Cytotoxic T cells are again rendered less effective in killing infected cells.

Many viruses cause a generalized immunosuppression in the host. By directly infecting important cells involved in the effective activation and implementation of the immune response, some viruses enhance their survival. An example of a local immunosuppressive molecular counterdefense mechanism has been studied in poxviruses. Cowpox virus encodes several proteins with similarity to the superfamily of serine protein inhibitors. When studied on the chorioallantoic membrane of a developing chick embryo, a mutant containing a deletion of a gene encoding one of these proteins results in a pock lesion infiltrated with inflammatory cells. The wild-type cowpox virus forms a lesion with a relative paucity of neutrophils. Thus, the viral encoded protein inhibits the host's normal inflammatory response to infection.   *Stuart N. Isaacs; Bernard Moss*

Antigenic Variation

Antigenic variation is a change in a virus sufficient to functionally abrogate recognition by an immunoglobulin molecule. Viruses with an RNA genome include many of the most variable human pathogens, for example, the picornaviruses (including poliovirus and human rhino virus), influenza virus, and the retroviruses. All of these viruses seem to have roughly similar and very high rates of mutation (approximately 10^{-3} substitution per nucleotide per year is fixed in the population). Furthermore, within a given virus population there are many genetically distinct viruses that together are thought of as forming a quasi-species.

Given the extraordinarily accelerated evolution of viruses compared to their hosts, it is not surprising that many viruses can achieve extreme antigenic diversity, as manifested in the 100 or more antigenically distinct serotypes of human rhinovirus, nor that the antigenic variation occurs relatively quickly. This variation can occur by point mutations, in which case antigenic drift (minor change in a surface antigen) results, or in the case of viruses with a segmented genome (such as influenza virus) gene reassortment that leads to antigenic shift (abrupt major change in a surface antigen) can occur. Antigenic shift has occurred several times during the twentieth century with influenza virus, and arises when genes for the principal viral antigens are picked up from a reservoir of antigenically diverse but rather static avian viruses; the result is a major pandemic in the human population.

In addition to producing viruses with improved fitness, which then become dominant via natural selection, antigenic variation may lead to changes that affect other properties of the virus; one example is the point mutation in influenza virus hemagglutinin that caused in 1983 a normally harmless avian influenza virus to wreak havoc among chickens in the United States. The virus was rapidly eradicated, but at the cost of some 17 million birds. A further complexity is that the way antigenic variants become fixed in the population and transmitted is likely to vary with the life-style of the virus. Thus the selective pressures will be different for, say, a retrovirus that goes through many hundreds of cycles of replication in each host and a picornavirus with far more frequent transmission.

Methods. A number of methods in molecular biology and molecular biophysics are available to study antigenic variations. The ability to sequence the viral genome allowed antigenic variation to be understood in terms of the nucleotide sequence. Recently, the very powerful technique known as the polymerase chain reaction has begun to be used in amplifying a tiny number of copies of genetic information. Of particular importance in understanding antigenicity has been the use of monoclonal antibodies (MAbs) as reagents and the various enzyme-linked immunosorbent assay (ELISA) methods. The ELISA techniques work by attaching antigen or antibody to a solid matrix and then monitoring antibody–antigen interactions by an enzyme linked to a reporter antibody. Such results must, of course, be interpreted with caution (distortion of the antigen may result from binding to the matrix), and are complemented by methods such as neutralization assays. Recently, crystallography has allowed the direct visualization not only of the antigen and antibody components of the system but also of the complete antigen–antibody complex.

Recognition. The recognition of antigen by antibody occurs between the amino end of the heterodimeric antibody arm and the surface of the antigen. Crystal structures of several of these recognition complexes reveal some common features: the interaction is via the same weak noncovalent forces (ionic) that hold protein oligomers together. These forces are not sufficient to produce much distortion of a stable protein structure, but achieve a strong specific interaction by spreading the energetic load over a considerable area of complementary surfaces (usually 7–8 square nanometers). As a result, many residues are subsumed within the footprint of the antibody on the antigen. This footprint is equated with an epitope. Overlapping footprints can be grouped to form an antigenic site. A simple model of antigenic variation relates changes in surface residues with interference in the specific antigen–antibody interaction. Some questions remain, however. Would such a change suffice to prevent all antibodies from recognizing the site? Does this model represent what viruses actually do?

Fortunately it is often easy to artificially induce antigenic variation by selection of viruses resistant to neutralization by particular monoclonal antibodies. Extensive studies have mapped the amino acid changes in these escape mutants and also in naturally occurring antigenic variants onto the three-dimensional structure of several viral antigens. The picture is surprisingly clear: the crucial changes almost invariably occur on the surface of the antigen (see **illus.**), usually (in the case of the monoclonal antibody studies) involving residues in the epitope. The surfaces of the viral antigens are not smooth, so the antigenic sites generally occur on the outcrops accessible to the blunt-ended antibody. These projections are often separated by depressions or canyons. The restriction of antigenic residues to certain regions of the viral surface allows some of the important biological functions of the virus to be concealed from immunological surveillance. A mechanism of particular interest occurs in the human rhinoviruses (and probably others) where the site at which the virus attaches to its rather slender cell surface receptor lies in a deep canyon too narrow for antibody penetration. This mechanism has been termed the canyon hypothesis.

Mechanisms. If attention is focused on the epitopes themselves, a question arises as to whether the observed changes leading to escape are scattered evenly over the epitopic residues. Surprisingly, they are not; they tend to be concentrated in a handful of residues. The natural variation observed shows that many other surface residues can change without compromising the biological properties of the virus. For a given monoclonal antibody, there is in fact a subset of residues in the epitope that is of key importance in the recognition process and provides "good value" in terms of antigenic variation. It is not known whether there are general rules defining this property. In the field, the mix of immune selection and other evolutionary processes will be different than in the laboratory situation, and evasion of a single antibody will provide little advantage to a virus.

Are the results obtained with monoclonal antibodies relevant to this situation? Do viruses invoke more subtle mechanisms? Recently structures have become available for a number of escape mutants and natural

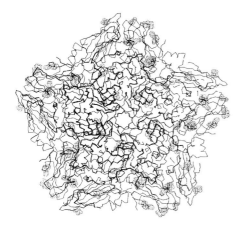

Monoclonal antibody–escape mutants mapped onto the three-dimensional structure of foot-and-mouth disease virus. Two orthogonal views of a pentamer of the virus capsid are shown. Sites of amino acid substitutions in monoclonal antibody–mutants are represented as dotted spheres. In the right view, the outer surface of the capsid is on the left. Note the segregation of the escape changes on this surface.

antigenic variants, in particular, for poliovirus, foot-and-mouth disease virus, and influenza virus hemagglutinin and neuraminidase. These results fall into two classes, simple and more complex. The simple results comprise the majority of the monoclonal antibody–escape mutants and show that escape is achieved by the minimal mechanism of substitution at a single amino acid. The other class of results shows a more interesting mechanism of amplification of conformational variation by using the particular chemical or conformational properties of the antigen. In the case of poliovirus, an example was a substitution at a proline residue that appears to be responsible for a significant rearrangement of an antigenic loop between two different serotypes of the virus. Another example comes from two monoclonal antibody–selected variants of foot-and-mouth disease virus, where it was found that a major antigenic loop was conformationally switched by an apparently innocuous change in a residue proposed to underlie this loop in the parent virus. In this case, the switch was primed by tension placed on the appropriate loop with a disulfide bond.

For background information SEE ANTIBODY; ANTIGEN; ANTIGEN-ANTIBODY REACTION; CELLULAR IMMUNOLOGY; COMPLEMENT; IMMUNOLOGY; MONOCLONAL ANTIBODIES; VIRUS in the McGraw-Hill Encyclopedia of Science & Technology.

D. Stuart; E. Fry; D. Logan

Bibliography. R. M. L. Buller and G. J. Palumbo, Poxvirus pathogenesis, *Microbiol. Rev.,* 55:80–122, March 1991; B. N. Fields and D. M. Knipe (eds.), *Virology,* 1990; D. J. Filman et al., Structural factors that control conformational transitions and serotype specificity in type 3 poliovirus, *EMBO,* 8:1567–1579, 1989; S. L. Harris et al., Glycoprotein C of herpes simplex virus type 1 prevents complement-mediated cell lysis and virus neutralization, *J. Infect. Dis.,* 162:331–337, August 1990; K. W. Moore et al., Homology of cytokine synthesis inhibitory factor (IL-10) to the Epstein-Barr virus gene BCRFI, *Science,* 248:1230–1234, June 1990; N. P. Parry et al., Structural and serological evidence for a novel mechanism of antigen variation in foot-and-mouth disease virus, *Nature,* 347:569–572, 1990; R. E. Phillips et al., Human immunodeficiency virus genetic variation that can escape cytotoxic T cell recognition, *Nature,* 354:453–459, 1991; M. G. Rossmann et al., Structure of a human common cold virus and functional relationship to other picornaviruses, *Nature,* 317:145–153, 1985; C. A. Smith et al., A receptor for tumor necrosis factor defines an unusual family of cellular and viral proteins, *Science,* 248:1019–1023, May 1990; W. R. Tulip et al., Crystal structures of neuraminidase-antibody complexes: Immunological recognition, *Cold Spring Harbor Symposium on Quantitative Biology,* vol. 54, 1989.

Vision

This article focuses on three areas in vision research: the relationship between genes, photopigments, and vision; color vision; and visual fatigue.

Genes, Photopigments, and Vision

From examination of pedigrees of visual disorders, it has long been appreciated that intimate relationships exist between genes, photopigments, and vision. In fact, an analysis of color blindness carried out more than 80 years ago provided the first entry on the map of the human genome (an X-chromosome location for color blindness). In the past few years, technical advances in molecular biology have been used in conjunction with the more traditional approaches of electrophysiology and psychophysics to substantially enhance understanding of the biological mechanisms underlying normal and abnormal vision.

Photopigment structure and operation. The complex process of sight is initiated when incoming light activates photopigment molecules arrayed in the membranes of photoreceptors. The amino acid sequence for the most ubiquitous photopigment rhodopsin, which is found in rod receptors, was first inferred from gene nucleotide sequences in 1982. It has provided a good general model for a superfamily of protein molecules typically located on cell surfaces. These molecules include not only other photopigments but also receptors for biogenic amines, receptors for peptides and nonpeptides, odorant receptors, and dozens of other receptor types. Like photopigments, all of these receptors have seven transmembrane helical segments. The transmembrane portions of these molecules show considerable sequence similarities. The connecting loops and side chains, which are the sites for posttranslational modifications, are more divergent in sequence among these molecules.

Photopigment molecules contain a small chromophore embedded in a pocket in the seventh transmembrane segment. Photoisomerization of the chromophore (which takes only 20 picoseconds) initiates a chain of events that results in change in the electrical potential across the membrane of the photoreceptor. It was inferred from psychophysical experiments conducted more than 50 years ago that a rod can be activated by a single photon. Recent electrophysiological measurements have provided direct confirmation of this prediction. The remarkable sensitivity of the photoreceptor is due in part to the fact that multiple events occurring between isomerization and membrane voltage change include a series of highly regulated stages that yield an aggregate amplification of about 100,000.

The flow of information during these initial stages involves interactions of activated photopigment with proteins and guanyl mono-, di-, and triphosphates (cyclic GMP, GDP, and GTP). These interactions modulate the flow of sodium and calcium ions (Na^+ and Ca^{2+}) across the photoreceptor membrane. The essence of operation is that in a dark environment Na^+ and Ca^{2+} enter the photoreceptor through channels held open by cGMP. Channel closure, which leads to a change in membrane potential, comes about by depletion of cGMP. This closure is triggered from the interactions of photoisomerized pigment with a G protein, transducin, which in turn activates an enzyme that hydrolyzes cGMP. The process is halted by deactivation of transducin and the operation of at least two mechanisms that quench the stimulating effects of photoisomerized rhodopsin. The recovery cycle is initiated by a drop in the level of Ca^{2+} in the cell that is detected by a calcium-sensitive protein. This stimulates guanylate cyclase so that again the cGMP level is elevated, membrane channels are opened, and the membrane voltage is restored to its dark state. There is much less direct information about how these events are played out in the other class of retinal photoreceptors, the cones, but evidence suggests that the similarity of operation is considerable.

Spectral tuning of photopigments. The initial transduction stage just described embodies a weighting of the effectiveness of light as a function of its wavelength. The spectral sensitivity curves for photopigments are tuned in the sense that, for any pigment type, sensitivity to light declines smoothly in both directions from the most efficient wavelength value. Most retinas contain more than one pigment type, a feature that allows subsequent neural interactions to compare and contrast signals originating in different pigment classes. It has long been known that a multiplicity of pigment types can be found among various species; that variety is usually believed to reflect the demands imposed and the opportunities offered by different visual environments. These natural variations in photopigments raise questions about the mechanisms controlling the number of photopigment types and the means for their spectral tuning.

One hint about the mechanisms controlling the spectral positioning of photopigments comes from measurements of photopigments in different species. An early survey of the spectral positioning of rod pigments in many different species of fishes suggested that these photopigments were restricted to some set of preferred spectral locations. A more recent examination of the spectral positioning of cone pigments in a variety of different mammals yielded the same conclusion. The implication in both cases is that pigment positioning is quantized in the sense that pigments are limited to one or another of several spectral locations. It appears that the spectral peak locations for photopigments are staggered along the wavelength scale at spaced intervals of about 6 nm. Two implications follow: evolutionary change in photopigments involves discrete jumps in spectral positioning, and the mechanisms controlling photopigment tuning are likely to involve discrete changes in the photopigment molecule.

Two procedures are currently being employed to study the molecular basis of spectral tuning. One involves site-specific mutagenesis, a molecular biological procedure that makes it possible to substitute amino acids at various locations in the photopigment molecule and then examine the effects of these substitutions on the spectral sensitivity of the mutated pigment. At present this technique has been used successfully only with rod photopigments. A second approach is to use gene sequence information to derive the structures of naturally occurring cone photopigments, thus exploiting the fact that evolution has already provided a range of spectral variation.

An instance of this second approach involved comparisons of eight primate cone pigments, two human pigments and six pigments obtained from neotropical monkeys. These eight pigments represent five distinct spectral locations, covering a range of peak positions from 530 to 560 nm. Amino acid sequences were inferred from the nucleotide sequences of the X-chromosome genes specifying these pigments. The sequences of photopigment genes are all highly similar, so it was not surprising that only a small number of amino acid differences were found among the eight pigments. Pairwise comparisons of the sequences suggested that substitutions at only three amino acid locations (positions 180, 227, and 285) are associated with variations in spectral tuning among these pigments. Each of these substitutions was associated with a spectral shift of a different magnitude. For example, substitution of serine for alanine at position 180 yielded a 6-nm shift in peak position toward the longer wavelengths. The analysis further implied that the effects of spectral change engendered by these substitutions are summative; that is, the spectral shift produced by changes at two locations is the linear sum of the spectral shifts produced by substitutions at either of the locations alone. These experiments provide a working hypothesis about the mechanisms underlying spectral tuning that may now be investigated further, both by direct alteration of photopigments at these crucial locations and by additional investigation of structure in naturally occurring photopigments that vary in their spectral tuning.

Genes and visual disorders. Emerging knowledge about the relationships between genes and photopigment is providing a deeper understanding of how genetic variations influence both normal and abnormal vision. A prominent case involves the polymorphisms of human color vision. The most common color vision defects, those degrading red/green discriminations, arise from changes in the X-chromosome genes. These changes result either from unequal meiotic recombinations that yield additions and deletions of genes from the normal complement, or from the production of new genes. The latter (fusion genes) result from intragenic recombinations and so derive part of their sequences from different parent genes. Depending on where the crossover occurs, this recombination may shift the spectrum of the produced photopigment and thus alter the color vision of an individual. Some progress has also been made in understanding the molecular basis of a form of complete color blindness called blue cone monochromacy, a condition leaving only one normal cone pigment type instead of the usual three. This condition apparently can arise in two ways: either from a loss of sequence in the vicinity of two pigment genes that perhaps compromises their normal operation, or from an unequal recombination in conjunction with a point mutation.

Progress has also been made in understanding the mechanisms involved in retinitis pigmentosa, a group of inherited retinal degenerative diseases that typically culminate in blindness. Sequence examination of the rhodopsin gene in affected individuals suggests that at least three different point mutations may be associated with one form of this disease. Definition of the genetic basis for retinitis pigmentosa may provide a first step toward the eventual treatment of the disease.

Gerald H. Jacobs

Color Vision

All visual sensation involves regions that possess shape or spatial extent, duration or temporal extent, and color. Predicting the perceived color, or color appearance, of regions in the visual environment and understanding the visual mechanisms that give rise to it are fundamental objectives of color science. The trichromatic theory of color vision, proposed by T. Young in the seventeenth century and elaborated by H. von Helmholtz and J. C. Maxwell in the late nineteenth century, is able to predict many aspects of color appearance. Photoreception by cones on the human retina and the spectral responsivities of the pigments they contain provide the mechanism of trichromatic vision. Much research in recent years has examined aspects of color appearance that cannot be easily explained within the trichromatic theory. A variety of different models have been proposed, some that are straightforward extensions and generalizations of the trichromatic theory, and others that are a major reworking of it. This section provides a brief overview of the relationship of trichromatic theory to color appearance and of anomalies of current interest, followed by examples of new concepts designed to accommodate the anomalies within extended models of color vision.

The principle of univariance, a cornerstone of the trichromatic theory, states that the only signals available to the brain are three excitations that are proportional to the number of light quanta absorbed by each of three visual pigments. This principle provides a detailed explanation of the existence of equivalence classes of spectra, all members of which match in color, regardless of viewing conditions. Through the opponent-channels model of the trichromatic theory, these three excitations provide a rough account of color appearance. This model hypothesizes the rearrangement of the excitations into three bipolar channels: one signaling whiteness versus blackness, a second signaling redness versus greenness, and a third signaling blueness versus yellowness. Any color appearance should be expressible as the conjunction of three quantities, one from each of the pairs. For example, a light pink is the conjunction of a large amount of whiteness, a small amount of redness, and a very small amount of blueness. An indigo color consists of a large amount of blueness combined with small amounts of redness and blackness. In each case the exact quantities of the three attributes determine the exact color appearance.

Color appearance anomalies. When applied to isolated patches of color, the above description of color appearance works surprisingly well. When it is applied to more typical visual stimuli containing many objects and sources of illumination, however, a variety of anomalies emerge.

First class of anomalies. One class of anomalies appears in stimuli that combine areas of color with strong achromatic contours. An example of this effect may be seen by means of the construction of a stimulus like **Fig. 1**, with the light-gray area replaced by pale yellow. From a distance it appears as a black curve filled with a yellow color: there are points inside the curve that trichromatic theory predicts to be white but appear yellow, and there are points outside the curve that trichromatic theory predicts to be yellow but appear white. Similar effects appear in much pictorial art, such as comic books, in which black inking is used to provide colored shapes with fine detail even though the color printing has very low resolution; figurative art, in which dark underpainting is used to define the outline of facial features that are overpainted with realistic skin colors; and high-quality printing, in which unsharp masking is used to improve the subjective quality of the image by emphasizing edges.

Second class of anomalies. A second class of anomalies appears in stimuli like **Fig. 2**. In this version, the gray area appears darker when surrounded by white than it does when surrounded by black. If the gray area were to be replaced by a dark-yellow color, it would appear yellow when surrounded by black, brown when surrounded by white. (Careful choice of the yellow color is needed to make this illustration effective, and it will probably be necessary to restrict the field of view to the center and surround only.) In both cases, simple trichromatic theory predicts the same color appearance, but visual inspection proves the reverse. In fact, brown, which is a distinct color sensation, seems not to occur in isolated patches of color, but only when at least two colors appear together. Attempts to encompass brown in the opponent-channels model make it a conjunction of yellowness and blackness, emphasizing that the appearance of blackness arises most easily when a visual stimulus contains more than one color. Thus, the white-black axis of the opponent-channels model is seen to be problematic since its various appearances (white, gray, black) arise, not from specific trichromatic excitations, but from contrasts between regions of different lightness within a single stimulus. The fundamental role of contrast in mediating the appearance of white, gray, and black has been observed in many experiments.

Third class of anomalies. A fundamental role for contrast in the appearance of white and black is connected to a third class of anomalies associated with surfaces of objects. The three excitations of trichromatic theory are produced by the spectral power distribution of light entering the eye, a quantity that is the product of the spectral power distributions of the sur-

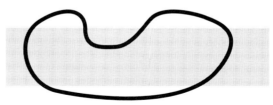

Fig. 1. Model of a stimulus that combines areas of color with strong achromatic contours.

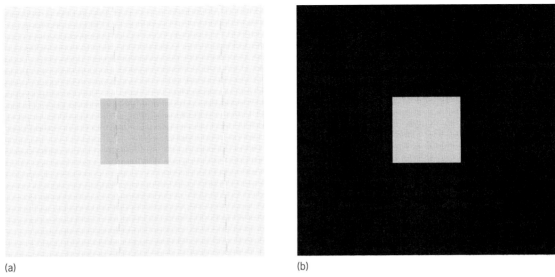

(a)   (b)

Fig. 2. Central color is the same in *a* and *b* but appears different because of the surrounding area.

face of a perceived object and of the light with which it is illuminated. Within trichromatic theory, these two quantities cannot be disentangled: a white surface illuminated by red light cannot be distinguished from a red surface illuminated by white light. Yet color appearance in complex stimuli is usually associated with the surfaces of objects, and the ability of color to inform the observer about useful properties of objects—for example, red/green means ripe/unripe when applied to tomatoes—is effective only if the observer can perceive the color of the surface. Such observations are often generalized into the concept of color constancy, which hypothesizes that the appearance of surfaces is independent of the spectral character of the source of illumination. Color constancy, if it exists, then allows the observer direct access to surface properties. The observation that lightness perception depends on contrast is evidence in favor of color constancy, because contrast is explicitly formulated to remain constant when the illumination level changes.

New models and theories. Several families of models and theories that help to explain some of these anomalies have been elaborated in the past few years.

First group of theories. This group may be related to the structure of the receptive fields of retinal ganglion cells. Most red-green-sensitive ganglion cells have centers and surrounds that are opposed in polarity [for example, the center may be excitatory (+), while the surround is inhibitory (−)], and respond to cones of different responsivities (for example, center responds to red, surround responds to green). Since the spectral responsivities of red and green cones are very similar, red-green-sensitive cells have two characteristic modes of response. For stimuli that vary spatially on scales comparable to the dimension of the center of the receptive field, which can be as small as a single cone, these cells respond strongly to black-white stimuli. For example, a stimulus with a bright area the size of the cell center surrounded by dark gives a strong response

when the bright area is aligned with the center of the cell. On the other hand, for stimuli that vary spatially on scales larger than the surround area this cell signals the red-green balance of the stimulus: positive if the stimulus is red, and negative if the stimulus is green. This description leads to two important conclusions. First, the fact that color and spatial information is multiplexed in the optic nerve suggests that there ought to be aspects of color appearance that are inseparable from the spatial character of the stimulus. This observation can be elaborated into an explanation of the second set of anomalies. Second, stimuli defined by color are capable of supporting less spatial detail than ones that are defined by lightness. This observation is supported by measurements of modulation transfer functions for color stimuli, and provides a theoretical framework for understanding the first set of anomalies, in which the shape defined by lightness contours dominates the shape defined by color.

The notion that different attributes of the visual stimulus may have different properties and be processed separately accords well with recent developments in the physiology of the visual cortex. Recording techniques reveal many different stimulus representations encoding such attributes as motion, depth, and color. Psychophysical investigation of the interaction of these stimulus attributes is an area of active research.

Second group of theories. This group of theories is based on the idea that color appearance is determined, not by cone responses within a region, but by differences of cone responses at edges in the vicinity of the region. A recent influential example of such a theory is the Land's retinex theory. Theories that depend on edges easily explain the second group of anomalies, since regions that differ in appearance are defined by edges that are quite different. They also provide a basis for color constancy, because in the presence of uniform illumination, changes at edges factor out illumination effects. Thus, edge-based theories can produce color appearance that is independent of illumination. Unfortunately, human vision seems not to possess good color constancy, but is intermediate between no color constancy and perfect color constancy, so that edge-based theories tend to overpredict illumination independence when they are used to simulate color appearance.

Third group of theories. This group of theories has its antecedents in computer vision, and attempts to handle the third group of anomalies by suggesting that visual stimuli have many fewer degrees of freedom than physical analysis suggests. Some theories of this type demonstrate that simple lighting geometries create extensive redundancy in physical scene descriptions. Once these redundancies are taken into account, solving for surface properties is much simpler. Because of their emphasis on simple lighting and internal consistency, these theories produce results that are qualitatively similar to edge-based theories. Other theories consider the surfaces themselves, and find that the surfaces have many fewer degrees of freedom than are mathematically possible. Linear models, for example, analyze groups of surfaces, and show that linear combinations of four to eight components are adequate to produce satisfactory matches to the spectral reflectance of large classes of surfaces. All theories in this third group suggest that the input bandwidth of the cones is almost sufficient to provide a unique description of the visual stimulus. If this is the case, the third set of anomalies is cleared up very elegantly, since the observer directly perceives the appearance of the illumination and of the surface simultaneously. If this is so, many color constancy experiments are misconceived and must be reinterpreted.

William Cowan

Visual Fatigue

The issue of visual fatigue has received attention lately because of the increasing number of visual display units in the workplace. However, symptoms of visual fatigue are not confined to this type of display. Although there have been numerous attempts to establish that visual-display-unit–related complaints are more frequent than complaints from workers performing comparable tasks, when controlled experiments have been performed (that is, with equivalent working conditions), visual-display-unit operators report symptoms at about the same rate as workers not using the units. The study of visual fatigue encompasses prolonged exposure to many types of near-work visual displays such as books or radar screens. Consequently, the etiology and prevention of visual display assumes considerable practical importance throughout the general population.

Symptoms. Visual fatigue includes a constellation of symptoms that should not be confused with the induction mechanisms. That is, visual fatigue describes an impairment of function, not the cause of the impairment. Subjective symptoms of visual fatigue may be divided into two components: pain-fatigue and impaired focal response. Pain-fatigue includes pains, headaches, burning eyes, red eyes, and general fatigue.

Impaired focal response includes blurring of either near or far sight, double images, and increased impressions of flicker from visual display units, television screens, or fluorescent lighting.

Objectively recorded symptoms include findings of pupillary constriction and prolonged shifts of accommodation (focal power of the eye lens) and vergence (disjunctive movement of the eyes in which the fixation axes are not parallel, as in convergence or divergence) in the direction of the visual task. For example, prolonged near-viewing followed by a switch of gaze to a distant object often results in accommodative and convergent responses being temporarily biased inward.

Theories. There is no consensus on what causes visual fatigue. Classical explanations cite ocular muscle fatigue. Convergence and accommodation to near-displays or prolonged dynamic activity of the eye muscles is hypothesized as causing an accumulation of metabolic by-products resulting in a loss of accurate muscular control. Several lines of evidence argue against muscle fatigue as the sole induction process for visual fatigue. For example, the frequently observed inward movements of the near-points of accommodation and convergence after sustained near-viewing are problematic because fatigued muscles are presumably unable to maintain contractive effort. Such fatigue should result in an outward movement of the near-points rather than inward movements. In addition, the muscle fibers of the extraocular muscles (that is, the muscles that control vergence) are smaller and more numerous than any in the body. In these muscles only a few fibers at at time contract; thus each fiber would presumably be highly resistant to fatigue since the fibers essentially take turns at maintaining the muscular load.

Given the high degree of reserve capacity of the extraocular muscles, it has been argued that visual fatigue may be confined to the ciliary muscles (the muscles that control accommodation). If fatigue of the ciliary muscles is related to the pain in eyestrain, the pain should be modified if the ciliary muscles are selectively paralyzed. However, temporary paralysis of the ciliary muscles does not significantly alter the pain of eyestrain.

Theories based on adaptation rather than on ocular muscle fatigue explain visual fatigue symptoms as manifestations of the normal operation of the visual system's response to different kinds of visual strain. Adaptive explanations emphasize the role of visual stressors in challenging normal visual function. The visual system has several well-known adaptative abilities. For example, if an eye muscle loses contractive capacity through aging or injury, the oculomotor control system modifies its innervational support to compensate for the loss of contractive ability. Adaptive phenomena are often accompanied by feelings of discomfort.

Visual stressors. There are a vast array of potential visual stressors, but they may be divided into three categories: degraded viewing conditions, task-specific stressors, and degraded visual function.

Degraded viewing conditions include displays with inherent distortion, with low contrast or luminance, with noticeable flicker, with glare, with poor color contrast, or with inherent blurring. Another degraded viewing condition occurs when an individual performs visual work with an inappropriate corrective aid.

Task-specific stressors affect visual work. Most near or fine-detailed work falls into this category. One recently discovered type of task-specific stressor is text-induced pattern contrast adaptation. This stressor occurs during prolonged viewing of grating patterns (**Fig. 3**) resulting in reduced contrast sensitivity to patterns of similar spacing and orientation. Pattern contrast adaptation is frequently accompanied by symptoms synonymous with visual fatigue. Text displays, with their repetitive line structure, act similarly. There is evidence that relates pattern contrast adaptation with losses of accommodative and vergence accuracy in some individuals. Text-induced pattern contrast adaptation has even been reported to induce seizures in individuals prone to them. These findings suggest that the spatial layout of information could be relevant in the induction of some types of visual fatigue.

Degraded visual function refers to inherent ocular problems such as myopia, or convergence insufficiency. Each individual's visual system has its own preferred operational range of clear vision. The primary function of a corrective aid such as reading glasses is to shift the operational range of the visual system to be consistent with the required visual work. Reduction in visual capacity is a normal accompaniment to the aging process. For example, the eye's lens becomes stiffer and requires greater muscular effort to change shape. Loss of lens elasticity is important because the eye's ability to change focal distance depends on a pliable lens.

Stressors and visual fatigue. Visual fatigue occurs when a stressor or combinations of stressors overwhelm the ability of the visual system to adapt to the strain. Visual fatigue is characterized by the gradual onset and intensification of symptoms over time, reflecting the visual system's response, or lack of it, to prolonged visual work. Immediately apparent visual difficulty, with no prior visual work experience, suggests the presence of significant sources of visual strain. Sustained visual effort under these conditions is likely to induce symptoms consistent with visual fatigue.

Stressors that place abnormal levels of load on the extraocular or ciliary muscles are frequently implicated as a cause of muscle fatigue. The issue is what load level is considered abnormal. Designing a work environment based on the average individual's visual capacity does not help the 40-year-old myopic person with convergence insufficiency. Each person's visual system is unique, so that what is normal for one person may not be suitable for another. Work environments need to be designed with flexibility.

In adaptative theories, symptoms of visual fatigue reflect the operation of several active control processes that are attempting to maximize performance to prevailing conditions. Feelings of stress signal that a control process is under strain. If different combinations of visual stressors are present, it is possible that multiple adaptative processes could work against each other.

Treatment. Treatment of visual fatigue consistent with both explanations involves frequent work breaks, appropriate corrective aids to ensure that an individual is working well within his or her accommodative and convergence ranges, and use of visual displays that conform to ergonomic standards. Effects of pattern contrast adaptation can be reduced by avoiding reverse video text displays (that is, light letters on dark backgrounds), shortening line lengths, and working at a line separation greater than double space.

For background information SEE COLOR; COLOR VISION; EYE (VERTEBRATE); PERCEPTION; PHOTORECEPTION; VISION in the McGraw-Hill Encyclopedia of Science & Technology.

Robert Lunn

Bibliography. K. R. Boff, L. Kaufman, and J. P. Thomas (eds.), *Handbook of Perception and Human Performance*, vol. 1: *Sensory Processes and Perception*, 1986; D. H. Brainard and B. A. Wandell, An analysis of the retinex theory of color vision, *J. Opt. Soc. Amer.*, A3:1651–1661, 1986; T. P. Dryja et al., Mutations within the rhodopsin gene in patients with autosomal dominant retinitis pigmentosa, *New Engl. J. Med.*, 332:1302–1307, 1990; R. Lunn and W. P. Banks, Visual fatigue and spatial frequency adaptation to video displays of text, *Hum. Factors*, 28:457–464, 1986; L. T. Maloney and B. A. Wandell, Color constancy: A method for recovering surface spectral reflectance, *J. Opt. Soc. Amer.*, A3:29–33, 1986; F. V. Malmstrom et al., Visual fatigue: The need for an integrated model, *Bull. Psychonom. Soc.*, 17(4):183–186, 1981; M. Neitz, J. Neitz, and G. H. Jacobs, Spectral tuning of pigments underlying red-green color vision, *Science*, 252:971–974, 1991; B. Ramazzini, *Diseases of Workers* (transl. by W. C. Wright from *De Morbis Artificum*, 1713), 1964; E. J. Rinalducci (ed.), *Video Displays, Work, and Vision*, 1983; R. Shapley, Visual sensitivity and parallel retinocortical channels, *Annual Rev. Psych.*, 41:635–658, 1990; L. Stryer, Visual excitation and recovery, *J. Biol. Chem.*, 266:10711–10714, 1991; J. Westphal, *Colour: A Philosophical Introduction*, 2d ed., 1991.

One of the major problems in the study of VF is the commonsense appeal of muscle fatigue explanations. The concept of VF provides a classic example of everybody "knowing" what the term represents, but most everybody disagreeing on a specific definition.

Fig. 3. The horizontal grating and the text display are potent sources of pattern contrast adaptation.

Volcanology

The volcanic eruption of Mount Saint Helens (Washington) on May 18, 1980, was the largest eruption in the twentieth century in the conterminous United States, and the fourth largest worldwide in that period. Its impact on a modern, technological society was enormous. The effects were felt in all walks of life throughout the Pacific Northwest—by the volcanologist studying the eruption, by the truck driver coping with increased vehicle maintenance due to ash contamination, by the farmer who saw increased crop yields from the mulching effect of the ash, by the people who lost homes, loved ones, and livelihood. Around 230 mi^2 (600 km^2) of land was destroyed, and 57 people lost their lives. The damage and cleanup has been conservatively estimated at $1 billion. The eruption was the most costly natural disaster in the history of the United States.

Mount Saint Helens is one of a series of volcanoes that extend from northern California (Mount Lassen) into southern British Columbia, Canada (**Fig. 1**). These volcanoes originate through a process known as subduction. Off the coast of North America, a piece of oceanic crust is being subducted or pushed under the edge of the continent. As the oceanic crust is driven farther under the continental crust, heat and pressure drive fluids off the subducted ocean crust. These fluids help melt the lowermost parts of the continental crust to produce liquid rock (magma). The magma finds its way to the surface to erupt and form volcanoes. In this manner the Cascade chain of volcanoes was formed, of which Mount Saint Helens is one.

Mount Saint Helens eruption patterns. Mount Saint Helens is geologically one of the youngest volcanoes of the Cascade chain. Its development began 40,000 years ago, but the peak as it appeared on May 18, 1980, was built only 3000 years ago, during the Spirit Lake Eruptive Stage. When a volcano is growing, it goes through many eruptive stages, made up of a series of eruptive periods that feature explosions, extrusion of lava, or other volcanic phenomena. The stage is characterized by the extrusion of magma of a distinctive chemical type. Eruptive stages can be separated by long periods of quiet, sometimes lasting thousands of years. Since Mount Saint Helens started growing, it has gone through four eruptive stages, each consisting of several eruptive periods.

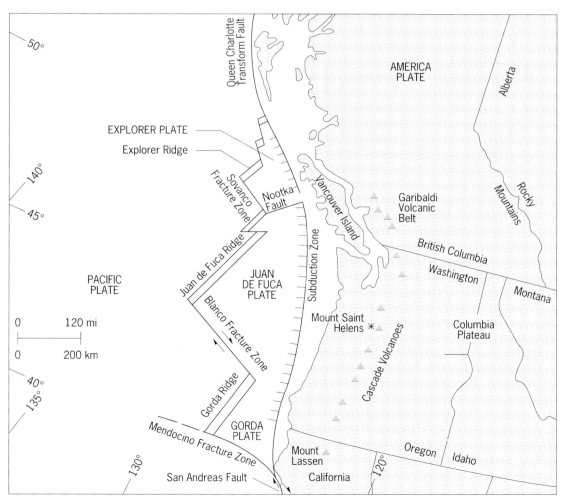

Fig. 1. Map showing the location of the Cascade volcanoes along the west coast of North America.

The eruptive periods at Mount Saint Helens start with an explosive phase that produces avalanches of hot, broken fragments of cooling lava (termed pyroclastic flows) and thick layers of ash. This violent explosive phase is followed by emission of lava to produce short flows (a few kilometers or less) and domes (the buildup of lava around the vent area). The most recent of these periods (prior to 1980) was the Goat Rocks Eruptive Period, part of the Spirit Lake Eruptive Stage. This period lasted from 1800 to 1857; it produced ash, a short lava flow, and a lava dome known as Goat Rocks. The present eruptive period followed a pattern similar to those of the earlier eruptive periods.

Recent Mount Saint Helens eruption. The start of the present eruptive period was signaled by an earthquake on March 20, 1980, that measured 4.2 on the Richter scale. As magma begins its ascent to the Earth's surface, the first clues of an impending eruption are earthquakes created by the movement of the magma. On March 27, the first small explosions occurred. These explosions were phreatic in nature: they occurred as groundwater became superheated by the rising magma; the flashing to steam created small explosions that sent steam and particles of preexisting rock into the air. Phreatic activity continued intermittently until the morning of May 18. As the ash and steam explosions were occurring, the north flank of the mountain became noticeably distended. A zone of deformation 1 mi (1.5 km) across and 1.25 mi (2 km) long was identified high on the volcano's north flank. This region was deforming at a rate of 4.6 ft (1.4 m) a day. The deformation was caused by magma rising within the mountain to form what became known as the cryptodome. The composition of the magma was dacite, a volcanic rock that is made up of about 62%

Fig. 2. Sequential photographs of the initial 2 min of Mount Saint Helens eruption on May 18, 1980: (*a*) 08:32, (*b*) 08:32.83, (*c*) 08:33.0, (*d*) 8:33.3, (*e*) 08:33.5, and (*f*) 08:33.7 (all Pacific Daylight Time). They show the progress of the first two avalanche blocks down the flank of the mountain, and the rapid spreading of the ash cloud from the paroxysmal lateral blast both northward (to the right in the photographs) and eastward (toward the viewer). (*Courtesy of P. Hickson*)

silica and is potentially highly explosive when it is still in a liquid state.

On the morning of May 18, 1980, an earthquake with a magnitude of 5.1 on the Richter scale occurred at 08:32 Pacific Daylight Time; it dislodged the distended and oversteepened north flank of the volcano. The resultant avalanche depressurized the dacitic magma in the cryptodome, and the hot-water (or hydrothermal) system built up around it to trigger a complex sequence of events that has been subdivided into six phases. The first phase, begun by the almost instantaneous removal of the pressure cap, caused a gigantic explosion that sent fractured rock and steam coursing down the volcano's slopes in all directions. The exact nature of this event, termed the paroxysmal lateral blast (**Fig. 2**), is still a matter of debate. This blast of relatively hot (570°F or 300°C) particles, steam, and entrained air rushed down the mountainside at speeds in excess of 330 ft/s (100 m/s). It was largely directed northward, and destroyed 244 mi^2 (632 km^2) of land. At its furthest point it traveled 20 mi (32 km) from the mountain and fanned out over a 180° sector. The blast can be separated into two parts: An initial explosion began 30 s after the earthquake and involved the magma of the cryptodome. The second explosion, 90 s later, was larger and involved the surrounding hydrothermal system.

It is now known that the avalanche that triggered the explosions was not a single event but involved three separate failures. Two of these failures can be seen in Fig. 2; the third is obscured by the growing clouds of ash and steam. The total volume of all three avalanches was 1.8 mi^3 (2.5 km^3). The avalanche traveled 14 mi (23 km) from the vent area before coming to rest and blocking the Toutle River drainage at an average thickness of 148 ft (45 m).

The eruption that followed could be subdivided into five separate phases, with the avalanche as the first phase. Phase two of the eruption was a Plinian eruption column, a towering column of ash up to 19 mi (30 km) in height. This rising column of fragmented rock particles was blown eastward in a complex pattern dictated by the winds in the jet stream. Blowing ash was observed across Montana and North Dakota as the ash cloud below 10,000 ft (3000 m) moved east. Fast-moving winds between 8000 and 40,000 ft (2400 and 12,000 m) carried the cloud east across the continent in about 3 days. Ash from this cloud at about 30,000–40,000 ft (9000–12,000 m) circled the Earth in 17 days. Above 53,000 ft (16,000 m), the ash cloud moved slowly southeast over Texas and the Gulf coast. Ash was deposited in appreciable amounts throughout much of eastern Washington, Idaho, and southern Canada. The intensity of the eruption built during the third phase, and the most distal tephra units were erupted during the fourth phase of the eruption, which lasted from 15:00 to 17:50 PDT. The deposited ash covered over 98,000 mi^2 (250,000 km^2) and had the widest-reaching effect of all the events at Mount Saint Helens. During phases five and six the eruption waned and stopped.

During all phases of the eruption, hot avalanches of ash and blocks (pyroclastic flows) spilled from the eruption crater. These eruptions and the earlier landslide melted copious quantities of snow and ice on the volcano's flanks and generated lahars (or debris flows) that sent mud and debris down all of the surrounding rivers. The most devastating was down the north fork of the Toutle River. The Toutle drains into the Cowlitz River and then into the Columbia River. The mud and debris that swept down the Toutle flowed all the way to the shipping lanes of the Columbia and silted them and blocked the channels.

Posteruption events. The eruption lasted into May 19. When the ash cleared, there was a 1970-ft-deep (600-m), 1.2 mi by 2.5 mi (2 km by 4 km) horseshoe-shaped crater with a rim 1300 ft (400 m) lower than the previous symmetrical summit. Since May 1980, the volcano has been intermittently active. Following the catastrophic events of May 18, there have been 21 distinctive eruptive episodes, preceded by seismicity and ground deformation. Of these, five had small columns that spread ash a few tens of kilometers from the vent. The last of these events, during October 16–18, 1980, was followed by the building of a dome in the crater. The dome continued to grow until October 1986. As of May 1990, the dome base was 3478 ft by 2820 ft (1060 m by 860 m), rose 876 ft (267 m) above the vent area, and had a volume of 97×10^6 yd^3 (64×10^6 m^3).

The dome was built in several stages over the 6-year period. It grew by intrusion of magma (endogenous growth) and by breakouts of lava onto its surface (exogenous growth). These lava breakouts formed short (660–1300 ft or 200–400 m), thick flows (lava lobes). The viscosity of the lava is very high, so the flows move very slowly, generally less than 2 in./min (5 cm/min). Since October 1986, there have been numerous small phreatic explosions. These explosions are worrisome, because they have no clear precursor warnings; as more people visit the crater, the potential for their being injured or even killed becomes greater.

Mount Saint Helens has been a remarkable laboratory for studying volcanoes. The work is invaluable in understanding volcanic eruptions elsewhere in the world, as well as those of the Cascade volcanoes. Though in historical terms the eruption was relatively small, its attributes give it a prominent place in scientific circles. The event occurred in a developed country with the expertise to monitor the volcano with the latest technological equipment. The cataclysmic eruption occurred in daylight hours during clear weather. Therefore a number of ground-based, airborne, and satellite observations could be made and correlated with later ground studies of the deposits from the eruption. This eruption was the best-documented eruption anywhere in the world. Activity subsequent to May 18 was at a subdued level so that there could be detailed monitoring and observations that enabled the prediction of many of the later events, days to weeks ahead. This work continues to be of paramount importance and allows the lessons learned at Mount Saint Helens to be used on volcanoes elsewhere to mitigate the effects of an eruption and ultimately to save lives.

For background information SEE MAGMA: PLATE TEC-TONICS; SEISMOLOGY; VOLCANOLOGY in the McGraw-Hill Encyclopedia of Science & Technology.

Catherine J. Hickson

Bibliography. C. J. Hickson and D. W. Peterson (eds.), Special Symposium Commemorating the 10th Anniversary of the Eruption of Mount Saint Helens, *Geosci. Can.*, 17:125–187, 1990; S. A. C. Keller (ed.), *Mount St. Helens: Five Years Later*, 1986; P. W. Lipman and D. L. Mullineaux (eds.), *The 1980 Eruptions of Mount St. Helens, Washington*, USGS Prof. Pap. 1250, 1981.

Water pollution

Although vast amounts of oil are shipped throughout the world and the numbers of companies involved in oil handling continues to grow, the technology for cleaning up oil spills has not grown proportionately. There was little research in spill response in the United States throughout the 1980s. European countries remained active during that time and joined to build an integrated research and development program. The participating European countries have united not only because of their focus on the commercially lucrative drilling in the North Sea but also because of political environmental organizations such as the Paris Commission of the United Nations, concerned with implementation of the International Law of the Sea.

Oil spill response factors. In addition to the problems of old technology, spill response is hampered by weather and by limitations inherent in people and equipment. Four major factors contribute to the quality and effectiveness of a response: geographic location; weather and sea conditions; the availability of the proper spill-fighting equipment; and the experience, readiness, and ability to make decisions on the part of the people responding.

Weather and sea conditions. Favorable weather enables a response to begin and to continue in as safe an environment as possible. Good weather also contributes to calm seas, which are critical to successful oil removal. Stiff winds and high seas make spill-response equipment difficult to control and sometimes render it ineffective. Winds and currents spread oil and increase the area that must be cleaned. As the oil spreads, lighter fractions evaporate so that it becomes more dense. The oil slick becomes more difficult to treat as it breaks up and the area affected widens. In turn, mechanical retrieval becomes more difficult.

Logistics and communications. The technological advances in the operational equipment are only one component of effective response. The response community has learned the importance of logistics and communications, two areas which have not been emphasized enough in traditional concepts of operations. Catastrophic spills, such as the Valdez, Alaska, spill, require enormous amounts of heavy equipment. Given the nature of spills, it is essential to minimize the time lapse between the first report of a spill and the actual response. Locating the correct amount and mix of oil-spill-response equipment and moving it to the scene is often a formidable task.

Communications are important. Command posts need to be quickly established to support complex and often remote land, sea, and air operations. Valuable decision-making time may be lost if information is not available to the hundreds of people in the spill-response and cleanup chain of command. More attention has been paid in recent years to the whole of the spill-fighting operation, with the realization that lack of teamwork at a spill leads to duplication of efforts, or worse, no effort at all.

Organization and training. The ability to work together and make decisions quickly is one key to a successful cleanup operation. The nature of a spill and the jurisdictional boundaries crossed by the oil mean that conflicting ideas, values, and legal statutes may come into play. Recent experience has pointed to the need for one person to be in charge who has the authority to make decisions. The need for a well-trained response team has also become apparent, and there is growing interest in cooperation and training.

Remote sensing. Knowing precisely where the spilled oil is located, particularly its largest concentrations, is vital. One promising area of logistical and communications research is remote sensing. It is hoped that eventually remote sensing technology will enable responders to locate oil at night and under conditions of reduced visibility, so that they can function both day and night.

Oil containment and recovery. Quick arrival of responders on the scene and then quick decision making about fighting the spill will help lead to an effective response. After stabilizing the spill source and ensuring personnel safety, the next step is to contain the oil at the spill site or as close to it as possible. This action will minimize both the work required to later recover the oil and the damage the oil can do to the marine and shore environments. Then, the most effective means to remove the oil from the water must be decided.

Booms. These are long sections of floating barriers that are deployed immediately to form a fence to contain, control, or channel the spilled oil in the water. Depending on location and conditions, the boom may have an underwater skirt and may also have above-water splash guards attached. The calmer the water, the more effective the booms will be in containing oil prior to its removal.

Mechanical removal. Once the oil is controlled, skimmers can be placed to mechanically remove the oil from the water surface. Although skimmers take a variety of forms, they all pull oil from the water and, in the process, bring some of the water with it. Skimmers require a storage vessel at the scene, usually a barge into which the oil can be pumped as it is skimmed. When the mixture separates, the water is decanted so that oil storage capacity is maximized. The recovered oil can sometimes be used for its original purpose or can be processed as a lesser-grade product.

Crude oils and crude-oil products have differing fates and effects in water. Environmental factors such as air and water temperature, winds, waves, and currents affect oil and the ability to recover it when it is spilled. Since many of the lighter and more volatile components of the oil evaporate over time, the total volume spilled cannot be recovered. Also, the effects of the wind and current break up and spread the slick over a large area and agitate the oil sufficiently to cause emulsification and increased volume. Emulsified oil, often called mousse, is particularly difficult to skim from the water surface because of its high viscosity. Many skimmers work along the oil–water interface. When that interface is irregular because of wave action, the machine is particularly inefficient. Often what is recovered by mechanical means is largely oily water. The problems of disposing of large volumes of oily water slow the recovery process.

Because of all these difficulties, the amount of oil that can be removed following a major oil spill generally has been relatively small, although mechanical removal at a site is usually the first option considered, and a good effort is often made. Optimum work in the best of conditions might yield a removal of about 15% of the spilled oil, and in many cases the amount has been much lower.

In addition to mechanical recovery of oil, there are other options at an oil spill, each with its own special set of considerations.

Dispersants. Chemical dispersants applied to oil on the water surface cause it to break apart and become more susceptible to further breakdown by natural actions. As a result, the oil no longer floats but is dispersed into the water column. A decision on whether to use dispersants involves weighing the effects of untreated oil left in the water against the effects of the oil dispersed into the water column.

On-site burning. Oil controlled inside fireproof booms can be burned to remove it from the water surface. More research is needed on this countermeasure to fully understand its ramifications. However, this method can quickly reduce large volumes of oil. The burning process transforms the waste in the water (and the evaporative hydrocarbons) into atmospheric waste (that is, particulates and the products of combustion) and also creates a carbon residue in the water. Thus, aside from operational constraints, a policy decision has to be reached as to whether the resulting air pollution is more acceptable than the water or shoreline pollution that may ensue if the oil is not recovered. Depending on the type of oil, responders may have difficulty igniting the oil and sustaining the burning process. On-site burning is still a technology in early stages of development.

Shoreline cleanup. Regardless of the measures taken to counter the spill, some of the oil will inevitably reach the shore. Workers can clean up by hand, using buckets, sorbent materials, and any other means they can devise. Larger-scale cleanup of the shoreline can be carried out by pressurized steam and hot water, but this approach is controversial because of its impact on the shore environment.

At times, the best approach to shoreline cleanup may be to leave the oil alone. Mechanical recovery and other human intervention could further disturb the fragile ecosystems of marshes and other sensitive environments. Flushing oil off a beach and back into the water where it can be removed using sorbent materials or skimmers can be a useful technique. There is, however, much controversy over whether the use of pressurized or hot water increases damage to the shoreline ecosystem.

Bioremediation. Bioremediation is the use of microbes or nutrients to enhance the degradation and oxidation of the hydrocarbon molecules that compose oil. The microbes can be naturally occurring marine bacteria, nonindigenous natural bacteria, or genetically engineered microbes. Bioremediation is a proven and commercially successful technique for cleaning up wastes on land, for example, in sewage treatment facilities or in refineries. The technology is less understood in a marine environment, and warrants more study as an offshore and shoreline measure. Bioremediation has the advantage of being noninvasive. Manually removing oil from a salt marsh, for example, might be too harsh a measure for the delicate ecology. Bioremediation would take longer than manual cleanup but would not result in additional harm to the marsh.

In the marine environment, dilution and the general movement of water masses make it more difficult to maintain an effective proportion of bioremediant to oil than in a land-based bioremediation effort. Oil on land stays in one place, so that remediation techniques can be more effective. Time, in the sense of a long enough contact between bioremediant and oil, is far more critical in the marine environment.

Prospects for improved response. There is no one best means of removing spilled oil from water. Rather, there are several techniques that can be used separately or together, depending on location, conditions, and the type and amount of oil spilled. Advances in technology will bring incremental improvements in the equipment used to clean up oil. More promising is a focus on improved teamwork, more realistic and more frequent training, and an evaluation of existing countermeasures to gain consensus in their appropriate use. Controlled spills, in which a regulatory agency introduces a small amount of oil into the water for the purpose of testing response measures, need more serious consideration as experimental platforms.

Effect of spills on marine environment. The release of large volumes of oil into a marine environment is certainly a potential disaster. Marine and shore life that cannot survive the contact with the oil or the cleanup measures will die. At the same time, researchers have learned that the marine environment is resilient and has a vast capacity to withstand the effects of oil. Scientists investigating a marine environment decades after a spill still observe the unmistakable effects of the damage, but also report that the environment has significant capacity for coping with the oil over the long term. The extent of that natural recovery capability is an ongoing subject of study.

Studies on the effects of untreated oil on the environment versus the effects of the best efforts to remove that oil are also important. The question of how to strike a balance between artificial and natural cleanup lies at the heart of the future spill-response research. A related question centers on whether society will be patient enough to allow long-term bioremediation approaches to replace the more immediate mechanical methods when appropriate. There needs to be a better understanding of the dividing line between acute toxic effects of oil and the longer-term effects, which are slower to be noted. In the first instance, mechanical or other conventional methods are probably the first defense; bioremediants used on shorelines may be useful to minimize the effects of longer-term chronic effects.

For background information SEE PETROLEUM MICRO-BIOLOGY; WATER CONSERVATION; WATER PURIFICATION in the McGraw-Hill Encyclopedia of Science & Technology.

John D. Costello

Bibliography. U.S. Congress, Office of Technology Assessment, *Coping with an Oiled Sea: An Analysis of Oil Spill Response Technologies*, 1990.

Weather

A singularity in meteorology is a condition that does not follow the trend indicated by statistical features of the broad climatological record. The climate of a place is usually described in terms of monthly and annual averages of temperature, precipitation, and wind; these averages do not indicate fluctuations at periods much shorter than a month. But detailed examinations of daily records show episodes of significant weather tending to recur on particular calendar dates.

Particular singularities. As databases and the means of examining them improve, more singularities are found; knowledge of them should aid the scheduling of weather-sensitive events. However, a science of singularities is not being pursued today as much as during the years right after World War II, and

the reality of singularities is arguable.

January thaw. Perhaps the most recognized singularity of temperature in the United States is the January thaw, frequently apparent in New England. It is manifested as a rather sudden rise in temperature about January 21, followed by a sharp decline. A singularity observed in the Washington, D.C., area appears as a sudden decline of temperature during the period November 21–25. Effects of both singularities are felt over considerable areas, but the singularities do not occur every year.

There have been partially successful efforts to tie such singularities to large features of the atmospheric general circulation. Thus, the January thaw was shown to be more likely during seasons when the zonal index, a measure of the strength of circumpolar flow, was relatively low. During such seasons an additional warm period tended to be prominent during the period January 6–9.

Occurrence in California. A remarkable maximum of precipitation occurring in mid-November at Point Mugu, California, has been related to a more diffuse maximum of precipitation over much of the western United States. It has been theorized that this episode results from planetary-scale wave adjustments induced by the topography of Asia and North America, excited by the zonal wind changes characteristic of the season. The circumpolar westerlies intensify, sometimes rather abruptly, as the winter season approaches.

Indian summer. Another singularity, Indian summer, which appears as halcyon days in September in the northeastern United States, was shown to be more likely and more pronounced during Septembers with high zonal index, that is, with relatively strong westerlies aloft.

Occurrence in Oklahoma. A recent study of precipitation in central Oklahoma has revealed an interesting singularity. **Figure 1** shows monthly average precipitation at Oklahoma City; the principal maximum occurs in May, and there is a secondary maximum in September. The peaks occurring in May and September are probably caused by critical positioning and strength of the axis of westerly winds aloft, coupled with moisture from the Gulf of Mexico that is brought northward by disturbances in that westerly flow. **Figure 2** presents daily average values of precipitation on all days and the number of days with measurable precipitation during the period of record after smoothing a series of data points by calculation of running averages. Appreciable oscillations now appear throughout, but only a few of these are significant in that similar oscillations have tended to occur near the same date in many individual years. These oscillations are typical of such records, and most will not be well duplicated in the future. The oscillation marked "singularity" in Fig. 2 seems particularly robust, since it appears in five of six subsets of the record shown, in earlier data, and at adjacent stations to a range of 50 mi (80 km). Figure 2 shows that average daily precipitation increases during late winter and spring to a maximum about May 30, and then it declines rather sharply to a minimum about June

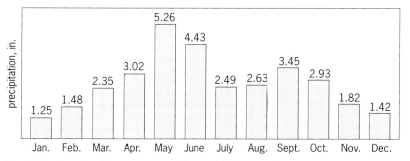

Fig. 1. Bar chart showing average monthly precipitation in inches at Oklahoma City during the period 1931–1990. 1 in. = 2.54 cm.

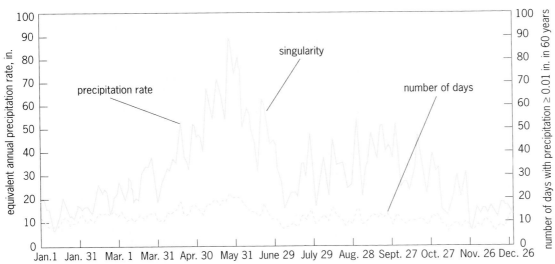

Fig. 2. Graph showing 7-day binomially weighted running averages of daily precipitation at Oklahoma City during 1931–1990 and the number of days with 0.01 in. or more precipitation during that period. 1 in. = 2.54 cm.

18. Subsequently the average daily precipitation and the number of days with precipitation rise to a secondary maximum about June 23 before falling precipitously to a droughty condition particularly common during the first half of July. These changes in a period of 1 month are large—they encompass a factor of about 6 in average daily values of precipitation rate.

The prominent maximum over a period of 5 days around June 23 has been studied intensively. It appears in the Oklahoma City records during five of the six decades from 1931 through 1990, and it also appears prominently in the overall averages from 1901 to 1930. It occurs in much the same way in records of stations within 50 mi (80 km) of Oklahoma City, but not at stations farther than 100 mi (160 km). Since the phenomenon is persistent throughout the record and there can be no indication of it in monthly averages, it is properly denoted as a singularity. The increase in average daily rainfall is formed by an increase and then a decrease in both the percentage of days with rain and the amount of rain on such days.

The broad trend of precipitation depicted in Fig. 2 is well explained. Precipitation in Oklahoma, as elsewhere poleward of the tropics, is largely associated with disturbances in the meandering westerly flow of winds whose latitudinal center expands and contracts with the seasons; as the warm season advances, air from the Gulf of Mexico is drawn farther northward, enters Oklahoma more often, and has higher moisture content. Both the fraction of days with precipitation and the amount of precipitation on rainy days increase. In late spring and summer, however, the core of the westerlies is usually north of Oklahoma, and this stimulus to disturbances is less active in this region. As the frequency of this stimulus lessens, both the fraction of days with rain and the average amount of rain on such days decline.

Such a simple plausible explanation is lacking for the singularity of June 23, and it is conceivable that the singularities cited may represent only random irregularities in the records. It should be borne in mind that a record displaying many random irregularities usually displays a few that are persistent over the period of that record, also by chance. An irregularity that corresponds to a previously hypothesized phenomenon, then, is properly given greater weight. It is to be hoped that both discovery and understanding of true singularities will be enhanced as detailed analyses of data interplay with advanced numerical and mathematical studies of the enormously complex Earth–Sun–atmosphere system.

Potential application. The mere discovery of such phenomena has practical value, since it serves to focus attention; and singularities persistent into the future can be a basis for improved forecasts. Forecasts are beneficially weighted with knowledge of detailed climatology as well as current observations on atmospheric flow. For example, a farmer in central Oklahoma needing to gather hay is probably well advised to schedule this activity around June 28 or later, and thereby to dodge the much larger chance of rain that seems to occur only a few days earlier.

For background information SEE ATMOSPHERIC GENERAL CIRCULATION; PRECIPITATION (METEOROLOGY); WEATHER; WEATHER FORECASTING AND PREDICTION in the McGraw-Hill Encyclopedia of Science & Technology.

Edwin Kessler

Bibliography. R. A. Bryson (ed.), *Climates of North America*, 1973; E. Kessler and S. Stadler, Some warm-season singularities in central Oklahoma precipitation, *Rep. Okla. Climatol. Surv.*, 1992; S. W. Lyons, Recurrent mid-November precipitation frequency maxima associated with planetary-scale wave adjustments, *Geophys. Sci. Tech. Note*, no. 113, Pacific Missile Test Center, Point Mugu, California, 1986; E. W. Wahl, Singularities and the general circulation, *J. Meteorol.*, 10:42–45, 1952.

Weather forecasting and prediction

The ability to foresee severe weather events is important to society. Meteorologists can determine with reasonable accuracy those general areas threatened by thunderstorms and severe weather, so that thunderstorm and tornado watches are sometimes issued. However, short-term, detailed prediction of thunderstorms based on computer models of atmospheric motion has not been attempted because of the absence of reliable, high-spatial- and high-temporal-resolution observations of storms and the near environments. This situation is rapidly changing because of a network of scanning Doppler radars now being installed across the United States. The national Doppler network, known as NEXRAD or Next Generation Radar (formally WSR-88D), will provide nearly continuous single-Doppler coverage across the contiguous United States and, in many regions, coarse-resolution dual-Doppler observations from which the three-dimensional wind field can be estimated. SEE AERONAUTICAL METEOROLOGY; METEOROLOGICAL RADAR.

Atmospheric predictability. Because the atmosphere is essentially a thin layer of fluid on a rotating sphere, its motion and evolution are governed by the same physical laws that apply to planetary motion or a baseball trajectory. Although astronomers can predict very accurately the occurrence of eclipses or the passing of comets several decades or centuries in advance, meteorologists have difficulty determining the weather several days or weeks ahead. The reason is the multiplicity of and interaction among temporal and spatial scales of atmospheric motion. For example, large-scale planetary waves that persist for weeks generally determine the position and intensity of midlatitude cyclonic storms and associated fronts, most of which last for several days and produce convective thunderstorms that last for 1–2 h and generate short-lived turbulent eddies both aloft and near the ground. The intimate linkage of these events and scales through complex transfers of moisture, heat, momentum, and energy explains the difficulty of atmospheric prediction and ultimately defines a fundamental limit to predictability. That is, because all scales of motion cannot be completely observed continuously, some degree of uncertainty or error will always exist in the initial state provided to a numerical forecast model. Once in the model, this error can grow in amplitude and move from one spatial scale to the next; eventually the entire forecast is contaminated and the model solution is rendered indistinguishable from a randomly chosen atmospheric state, even if the model is perfect and computer resources are unlimited. The time required for such growth to occur is deemed the predictability limit of the flow; for large-scale motion this is approximately 2 weeks.

In general, the large-scale planetary waves of the atmosphere behave as if they were two-dimensional, or uniform in their vertical extent. Theory states that the energy associated with disturbances and waves within such a framework tends to remain in the large

scales, and thereby the longevity and predictability of these phenomena is enhanced. Conversely, individual thunderstorms and their larger aggregates are highly three-dimensional; thus they transfer large amounts of energy from large to small scales, where dissipation by turbulence occurs. As a result, thunderstorms tend to be shorter-lived and hence less predictable than their larger-scale atmospheric counterparts. A notable exception occurs for thunderstorms that exhibit strong rotation about a vertical axis, a feature which appears to impede the down-scale transfer of energy and thus to enhance the lifetime of the storm. These supercell thunderstorms are common in the Great Plains of the United States during the spring and often produce devastating tornadoes.

The accuracy of a numerical prediction is strongly influenced by the complexity of physical processes and the ability to correctly account for their effects, in some instances by using empirical laws that are difficult or impossible to validate with observations. Examples include condensation and evaporation; the creation and depletion of various types, sizes, and concentrations of particles of frozen or liquid precipitation; and infrared and ultraviolet heating and cooling.

Storm-scale prediction models. The first computer models designed to simulate the evolution of isolated dry convective thermals, principally using two-dimensional plane and axisymmetric geometry, were constructed in the late 1950s and early 1960s. Advances in numerical techniques, representation of physical processes, and computing capability during the subsequent decade led to more advanced models, culminating in the early to mid 1970s with the first fully three-dimensional models capable of simulating precipitating convection. The introduction of the Cray-

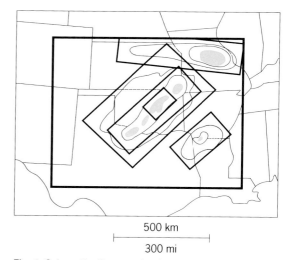

500 km

300 mi

Fig. 1. Schematic diagram showing the computational domain (bold rectangle) of a storm-scale prediction model having a horizontal dimension of 1000 × 1000 km (600 × 600 mi). The smaller boxes represent subdomains of successively higher spatial resolution used to better represent the thunderstorms, which are shown by hatching and solid-and-white areas. (*After K. K. Droegemeier, Toward a science of storm-scale prediction, 16th Conference on Severe Local Storms, American Meteorological Society, Kananaskis Provincial Park, Alberta, Canada, October 22–26, 1990*)

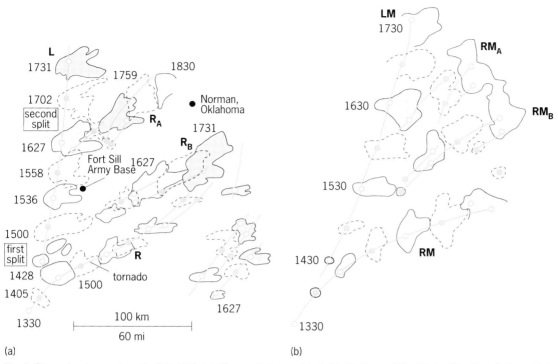

Fig. 2. Storm development on April 3, 1964. (*a***)** Observed storms, labeled L, R, R$_A$, and R$_B$. First split = time during which the initial thunderstorm split into right and left moving storms. Second split = time during which the storm from the first split once again splits to form two more storms. (***b***) Numerically modeled storms, corresponding to observed storms (LM, RM, RM$_A$, and RM$_B$). Observed reflectivities with radar reflectivity factors greater than 12 dBz at 0° elevation angle and modeled rainwater contents greater than 0.5 g/kg (0.08 oz/lb) at 0.4 km (0.24 mi) altitude are enclosed by solid and broken-line contours at approximately 30 min intervals, Central Standard Time. Maxima in these fields are connected by solid lines. (***After R. B. Wilhelmson and J. B. Klemp, A three-dimensional numerical simulation of splitting severe storms on 3 April, 1964, J. Atm. Sci., 38:1581–1600, 1981***)

1 supercomputer in 1976 revolutionized this development, and today the meteorological community has at its disposal a number of very sophisticated models that run at speeds approaching 10^9 floating operations per second on the fastest supercomputers available. When cast in a form appropriate for representation by a computer, the governing equations yield several million dependent variables and produce billions of output data points during the course of a calculation.

Because thunderstorms often occur in relative isolation, only a small fraction of the storm prediction model domain may be occupied by active convection at any given time. As a result, it is desirable from a computational standpoint to use high spatial resolution only in regions of substantial weather activity. Although there are a number of methods for accomplishing this task, one of the most widely used involves grid nesting or adaptive grid refinement, in which multiple finer-scale and possibly overlapping subgrids of arbitrary orientation are placed within a coarser base grid in response to the evolving flow (**Fig. 1**). For various situations this strategy can reduce the requirements for computer memory by one to three orders of magnitude relative to a single grid of uniformly high resolution, so that the use of existing supercomputers for real-time storm-scale weather prediction becomes practical.

Numerical storm prediction. Short- and medium-range weather forecasts (out to several days) have improved considerably in the past few years,

largely because of advances in the prediction models and the computers on which they run. Horizontal resolutions of 50 km (30 mi) will soon be a reality in operational global models, and thus further improvements in prediction on this scale are anticipated. The practical prediction of weather on the scale of individual thunderstorms and their environments has been much slower to evolve. The reasons are the unavailability of data (apart from those generated by special observing programs operating over limited regions and for limited periods of time); the absence of computers having sufficient speed and capacity to resolve relevant events on, and interactions among, these scales, particularly for short-lived phenomena whose forecasts must be generated and transmitted to the public quickly enough to be of practical use; and insufficient understanding of processes relating to predictability on these scales.

Despite these limitations, three-dimensional numerical simulation models, in conjunction with field observations by storm intercept teams and data obtained from research Doppler radars, have been used with great success since the mid-1970s to gain insight into the fundamental structure of severe storms. It is important in this context to distinguish between simulation, which involves using a computer model to recreate a previously observed event through repeated trial-and-error experiments, and prediction, which involves the use of current observations to numerically forecast the future state of the atmosphere. **Figure 2***a* illustrates notable recent successes in numerical storm simulation. It shows a time sequence of radar echoes for a complex of severe storms that occurred on April 3, 1964. A numerical simulation of this same event (Fig. 2*b*) exhibits remarkable similarity to the observations in both the placement and timing of the individual storms. Such simulations suggest that the practical prediction of thunderstorms using computer models is indeed plausible, given accurate initial observations and a suitable model.

The challenges of storm-scale prediction are varied and difficult. For example, it is not yet clear which phenomena on the storm scale are predictable, nor are those factors that influence or control predictability well understood. In addition, the sensitive dependence of a numerical forecast to uncertainties or errors in the specified initial state, first recognized by E. Lorenz as a principal impediment to large-scale atmospheric predictability, remains largely unexplored for storm-scale motions. Finally, correct representations of and interactions among complex physical processes in numerical models must be improved if accurate forecasts are to be achieved. *SEE CHAOS.*

Data assimilation. One of the most important aspects of numerical weather prediction is the quality of the initial state provided to the forecast model. Without an accurate representation of current conditions, the prediction of future events is difficult to impossible. Data assimilation is the process by which a forecast model and observational data are made compatible in an effort to arrive at the best initial state possible. This task is generally quite arduous, because the atmosphere contains considerably more complexity and information than can be represented by the model equations. Combined with a variety of possible errors in the collection and transmission of observational data, along with gaps in data coverage and the inability to directly gather information for each state variable predicted by the model, the assimilation process is as computationally intensive and scientifically challenging as the forecast problem itself.

Since it is impossible to directly observe all parameters and data fields necessary to initialize a numerical forecast model, techniques have been developed to retrieve such information from observed quantities in a manner consistent with the equations of the prediction model and the associated laws governing atmospheric motion. In this context, data assimilation and parameter retrieval differ significantly for global-scale and thunderstorm-scale motions. On the global scale, all necessary fields (temperature, horizontal wind, humidity, and pressure) except for vertical velocity can be obtained with varying degrees of accuracy from in-place observations by virtue of upper-air balloons and other remote sensing devices. The associated mass (temperature and pressure) and wind fields satisfy balances such as geostrophic and hydrostatic equilibrium and, with proper filtering of unwanted, nonmeteorological modes such as sound and gravity waves, are appropriate for use as model initial conditions.

On the thunderstorm scale, Doppler radar is the principal source of detailed observations within the storm volume; however, in sharp contrast to global-scale observations, only the time history of one horizontal (radial or along the radar beam) wind component and the three-dimensional radar reflectivity (related to rainfall intensity) are available from a single Doppler radar; all other physical variables must be deduced from these two quantities. To further complicate this process, no simple balances apparently exist between the mass and wind fields within thunderstorms; rather, all relationships are determined by the complicated dynamics governing storm evolution.

Exploiting this information, two broad classes of methods have been developed for retrieving unobserved quantities from single-Doppler radar data. The first, shown schematically in **Fig. 3***a*, is known as forward retrieval; it involves repeated insertions of single-Doppler-radar wind data and perhaps reflectivity into a forecast model. With each insertion, the model is constrained to accept the externally supplied information, while it uses its own equations to determine all other fields subject to statistical and dynamical constraints. After some number of insertions, the assimilation process ceases, and a forward prediction is begun. Of critical importance to this method is the time resolution of the observations, nominally 5 min for the new national network of Doppler radars.

The second technique, known as adjoint data assimilation (Fig. 3*b*), is a more elegant approach based on control theory; it seeks to optimize, in an objective manner, all observed data. In contrast to the prediction model, whose dependent variables are physical quantities such as wind, temperature, and pressure, the dependent variables of the adjoint model are related to the sensitivity of the forecast to its initial state.

In the adjoint method, the prediction model is provided with an initial state from, for example, its own previous forecast or a larger-scale model. A prediction is then initiated, with output from the model saved periodically for some length of time, perhaps 20 or 30 min. At the end of the prediction, the adjoint model is run backward in time in an attempt to minimize the discrepancy between the output from the forward model and any observations (in this case, a single-Doppler wind component) that were taken during the forward prediction step. As a consequence, all unobserved quantities are retrieved in a manner consistent with the observations and the model's dynamical equations.

The adjoint portion of the assimilation cycle provides information on how to adjust the initial state of the forward prediction model so that the observations and forecast are in optimal agreement. The assimilation process shown in Fig. 3*b* repeats in an iterative fashion until a converged initial state is reached. At this point, a forward prediction is initiated.

A principal concern of both data assimilation techniques, particularly the adjoint method, is the uniqueness of the retrieved fields. It is plausible that more than one set of retrieved mass and wind fields exists for a given time history of single-Doppler wind data,

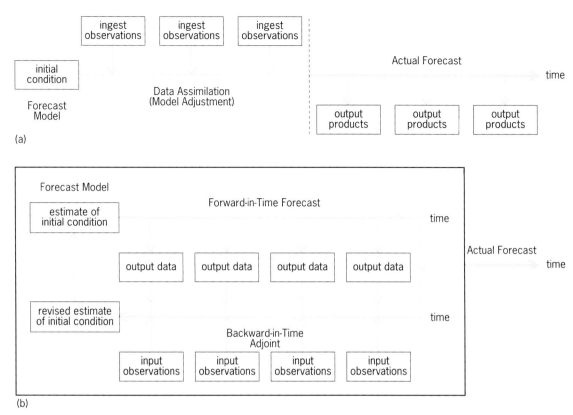

(a)

(b)

Fig. 3. Schematic diagram of (*a*) forward and (*b*) adjoint data assimilation.

and considerable effort is being expended to address this issue by using both theoretical and practical approaches.

Computational requirements. Historically, meteorology has been one of the principal driving forces behind the development and utilization of digital computers. The success of storm-scale weather prediction will depend upon three key elements of computing: (1) fast computers, probably of the massively parallel class supporting tens of thousands of processors and operating at billions to trillions of operations per second; (2) fast digital networks, capable of sending billions of pieces of information per second over distances of thousands of miles with great reliability; and (3) powerful tools for converting digital data into three- and four-dimensional color images that can be displayed interactively, perhaps in holographic form.

Prospects. In contrast to the present methodology of operational numerical weather prediction, in which one national center (for example, the National Meteorological Center in the United States) is responsible for gathering global observations and running a suite of global and regional forecast models, storm-scale prediction will likely take the form of local centers operated by federal agencies (for example, the Department of Transportation), state governments, municipalities, or the private sector. This setup would involve data from nearby Doppler radars and other observing systems being collected by individual centers and assimilated into short-term forecast models. The associated predictions might range in duration from one to several hours, and perhaps extend to only 30 min over airports where transient low-level wind shear poses a threat to both commercial and general aviation.

With storm-scale weather prediction will come an entirely new methodology of operational forecasting in which staff meteorologists will watch the evolution of the model forecast as it is being computed, examine model output and observations interactively by using animation and three-dimensional color images, and create customized, computer-based data-analysis tools appropriate for local climatologies and influences. In order to deal with the massive amounts of data and greatly enhanced capabilities associated with storm-scale prediction, high-performance computer workstations and graphical display systems will be used, most of which will exceed the power of the fastest mainframe computers available today. In addition, the time required for disseminating vital weather information to the public will be greatly reduced so that the short-duration forecasts will retain their predictive value.

For background information *SEE ATMOSPHERIC GENERAL CIRCULATION; CHAOS; DIGITAL COMPUTER; RADAR METEOROLOGY; SIMULATION; SUPERCOMPUTER; WEATHER FORECASTING AND PREDICTION* in the McGraw-Hill Encyclopedia of Science & Technology.

Kelvin K. Droegemeier

Bibliography. B. W. Atkinson, *Meso-Scale Atmospheric Circulations*, 1986; W. R. Cotton and R. A. Anthes, *Storm and Cloud Dynamics*, International Geophysics Series, vol. 44, 1989; J. B. Klemp, Dynamics of tornadic thunderstorms, *Annu. Rev. Fluid Mech.*,

19:369–402, 1987; D. K. Lilly, Numerical prediction of thunderstorms—Has its time come?, *Quart. J. Roy. Meteorol. Soc.*, 116:779–798, 1990.

Wetlands management

The management of fresh-water wetlands is determined largely by the characteristics and functions of individual wetland ecosystems. However, emotions and politics have also influenced the management and regulation of wetlands. Questions about the production and allocation of wetland resources and benefits—wildlife, flood control, timber, agriculture, biodiversity, and recreation—must be addressed. Generally, a wetland can be managed to produce multiple benefits, but sometimes the uses conflict. Frequently, the prerogatives of private landowners are limited so that society's needs can be met.

In the United States, wetlands are defined by legislation. Section 404 of the Clean Water Act (1977) defines wetlands as areas that are inundated or saturated by surface water or groundwater at a frequency and duration sufficient to support, and under normal circumstances do support, a prevalence of vegetation typically adapted for life in saturated soil conditions. This legislation also states that wetlands generally include swamps, marshes, bogs, and similar areas.

Delineation of wetlands. The surface soils of all wetlands are saturated with water for periods that are long enough for the soil oxygen to be depleted for a part of the growing season. Hence, wetlands occur only where geological formations or variations in topography impede drainage or permit surface flooding for extended periods. Saturation with water usually gives the soils chemical and visual characteristics that are indicative of wetland conditions. The presence of wetlands soils is also indicated by the occurrence of plants that can live in anaerobic soil by transporting oxygen to their roots internally, by respiring anaerobically, or by adapting to seasonal soil saturation.

Generalizing about wetlands hydrology is difficult, because the hydrology of each wetland is unique. Hydrology is especially complex where soil saturation is maintained by subterranean flow of water. This condition is common in lakes in the northern Midwest and in the sandy flatwoods of Florida. A flatwood is a nearly level zone in a forest, not clearly defined, that contains imperfectly drained acid soils.

Wetland types. These types include swamps, riverbottom terraces, coastal plain wetlands, and constructed wetlands.

Swamps. These are flooded for long periods and frequently support stands of cypress and water tupelo. This type of swamp forest is shown in **Fig. 1**; it has been logged repeatedly and regenerates naturally from the residual stand of trees. Black willow and certain other species can reproduce vegetatively in standing water under some conditions, but not even the most hydric forest species can regenerate from seed in standing water. The characteristic vegetation of deep swamps arises by regeneration from seed, which can occur only when mineral soil is exposed for a 2–3-month period during the growing season. Thus, natural regeneration in swamps depends on the occurrence of periodic droughts.

As sediment gradually fills swamps, they are invaded by mesic species, with hydric species being replaced. Filling and replacement of species can take from less than 10 years to several hundred years and is a normal geologic succession. Artificial alteration of hydrology can accelerate or retard the filling of swamps with sediment. Road construction and water control projects often alter the hydrology of swamps. When such projects result in deeper flooding, tree stands are harmed and natural regeneration does not occur. Improved drainage promotes invasion by species less tolerant to flooding. Forestry activities that do not alter hydrology have no adverse effect on the function of swamps. It is not feasible to convert deep swamps to agriculture.

Riverbottom terraces. These are better drained than cypress-tupelo swamps; they support ashes, oaks, hickories, maples, sweetgum, persimmon, and other mesic species. Generally, ecological succession is slower on riverbottom terraces than in swamps, and is not substantially impacted by sedimentation. Fire has little impact on terrace vegetation because the terraces are moist and terrace forest communities do not carry fire well. Riverbottom terraces have enormous value for wildlife and species diversity as well as for productive agriculture and forestry.

Coastal plain wetlands. These wetlands occupy large areas in the South and Southeast. They vary in function and vegetation according to hydrology and nutrient supply. In pre-Columbian times, much of the flatwoods portion of this area was in loblolly, slash, and longleaf pine forests that were maintained in a subclimax state by periodic fire. Sites that did not burn supported swamp tupelo, cypress, and many of the broadleafed hardwood species common on moist river bottoms.

Constructed wetlands. These are created by intentionally restricting drainage to develop hydric or partially hydric environments. Constructed wetlands include greentree reservoirs (mesic woodland in which

Fig. 1. Swamp forest with cypress and swamp tupelo that is flooded much of the year.

normal drainage is blocked during the dormant season) and wetlands that function as water purification systems.

Greentree reservoirs attract waterfowl and are used as a substitute for altered waterfowl habitat. Unless the hydrology of greentree reservoirs is managed carefully, however, tree productivity declines and other functions of the wetland are lost.

The use of constructed wetlands to remove contaminants from wastewater is relatively new. The hydrology of a constructed wetland that removes nitrates from sewage-treatment-plant effluents or heavy metals from mining wastewater must prevent both escape of contaminants from the wetland and accumulation of excessive levels of toxic materials. Constructed wetlands may be stocked with plants that will absorb toxic materials, combine them with organic matter, and render the chemicals inert, or may be created on soils that will hold the pollutants. SEE LAND RECLAMATION.

Introduction of nutrient-rich effluents into wetlands systems can result in the rapid breakdown of peat under extreme reducing conditions and in the production of large quantities of methane, a greenhouse gas. Technology for dealing with this problem is not yet available.

Loss of wetlands. Many fresh-water wetlands are transitory. Sedimentation tends to raise the level of the soil surface and to drain wetlands, especially in river floodplains. The accumulation of organic matter can alter the hydrology of wetlands; this process is common in peat bogs in the north-central states. Ground fire can reverse the accumulation of organic matter; this process occurs often in swamps and on cypress domes on the Atlantic Coastal Plain. Hence, the ecological functions and value of wetland areas can change over time.

Fresh-water wetlands are also lost as a result of subsidence of the land elevation of sea level. Subsidence and salt-water intrusion has caused significant losses of wetlands near the Mississippi River delta in recent years.

About 55–60% of the pre-Columbian wetlands in the United States have been altered by agriculture or destroyed by development. This loss has had a severely adverse impact on migratory waterfowl and other wildlife. Most prairie wetlands of the Midwest and northern Midwest have been converted to agriculture. Previously, these sites consisted of glacial till covered with tall-grass prairie that was maintained by fire. Except in the northern plains states, almost all of this wetland type has been converted to extremely productive agricultural land. The surviving prairie wetland sites are important to migratory waterfowl and for groundwater recharge.

The construction of dams, levees, and other water-control structures has had a profound impact on wetlands, particularly in river bottoms. The impact of public-works projects on wetlands has come under increased scrutiny in recent years. Agriculture and the development of tillable land have affected very large areas of wetlands. Most convertible wetland areas have already been brought under the plow, however, and

Fig. 2. Managed forest wetland with planted pine trees on a southeastern flatwoods site. Note the minor drainage in the foreground.

conversion to agriculture has almost ceased. In some instances, agricultural land has reverted to forested wetlands or has been replanted to forests.

Many riverbottom wetland areas have been converted to agriculture during the last 75–100 years. This process has affected very large acreages in the Mississippi, Ohio, and Missouri river valleys.

Coastal plain wetlands have been less affected by conversion to agriculture. They are small, scattered, and hard to drain, and their soils are infertile. Many of these areas have been altered, however, by the construction of roads, water-control structures, and utility corridors and, to a lesser extent, by intensive forestry practices. In pre-Columbian times, much of the coastal plain flatwoods was in southern pine forests that were maintained by periodic fire. Development of the coastal plain has led to the control of fire, and hardwoods have invaded much of the original pine-grass savanna. Many of the remaining wetlands are privately owned and are managed primarily for timber production (**Fig. 2**). Some of the silvicultural practices employed in flatwoods areas are controversial. However, these practices are normally exempt from regulation under the Clean Water Act. Urban encroachment, which is quickly replacing agriculture as a threat to wetlands, is regulated by this act.

Legislation in the United States. Concern over loss of fresh-water wetlands has prompted states and the federal government to pass legislation to protect and manage these areas for the benefit of society. The most significant federal laws are the Clean Water Act of 1977 and the Food Security Act of 1985. Congress is currently considering updating the Clean Water Act and other legislation in order to prevent further net loss of fresh-water wetlands. States have passed or are in the process of adopting legislation to address wetlands protection issues of local concern.

For background information SEE ECOLOGICAL SUCCESSION; ECOSYSTEM; HYDROLOGY; SWAMP, MARSH, AND BOG in the McGraw-Hill Encyclopedia of Science & Technology.

William H. McKee, Jr.

Bibliography. Federal Interagency Committee for Wetland Delineation, *Federal Manual for Identifying and Delineating Jurisdictional Wetlands*, 1989; D. D.

Hook et al. (eds), *The Ecology and Management of Wetlands*, vols. 1 and 2, 1988; D. D. Hook and R. Lee, *Proceedings: Symposium on the Forested Wetlands of the Southern United States*, U.S. Department of Agriculture, Forest Service, Gen. Tech. Rep. SE-50, 1988; W. J. Mitsch and J. G. Gosselink, *Wetlands*, 1986.

Whale

Many hundreds of fossil species of cetaceans (whales, dolphins, and porpoises) have been named in the past 150 years, and since about 1970 there has been a worldwide surge in fossil whale work. The fossil record includes many exceptionally well-preserved skulls and skeletons that serve to demonstrate most of the major steps in the origin of whales, the subsequent development of an impressive diversity of fossil groups, and a slight decrease in diversity as the living types evolved.

Important fossil features. The fossil study of cetaceans permits documentation of changes and adaptations in their skeletal anatomy that provide clues to their ancestry and trace their evolution. Important features in this study include overall body size; stage of maturity; numbers of vertebrae, ribs, teeth, and fingers within the forelimb (pectoral flipper); extent of fusion of the neck vertebrae; shape of teeth; distribution of air sinuses extending from the middle ear into other parts of the skull; position of foramina (holes) in bones, especially those of the skull, that mark the course of nerves and blood vessels; and shapes of individual bones, with emphasis on those of the head, and on muscle scars and articulation surfaces. From the bones can be deduced information about soft anatomical structures, physiology, and behavior of the fossil species, using modern animals as analogs.

The anatomy of living whales demonstrates their ancestry. For example, the fact that whales share important features of physiology and anatomy (sometimes only vestigial) with all other mammals proves that they belong to the great lineage of placental mammals that evolved from reptiles during the Mesozoic Era. In modern cetaceans, such features include small hindlimb bones deep within their bodies, rudimentary facial hairs (whiskers), and reduced olfactory nerves (once used for smell).

Archaeocetes. The earliest fossil whales are the extinct cetacean suborder Archaeoceti (archaeocetes). Whether or not the archaeocetes gave rise to the later toothed whales (Odontoceti) and the baleen whales (Mysticeti) has been a matter of controversy for many years. A monophyletic origin for baleen and toothed whales is supported by recent studies of whale chromosomes. The studies have shown that, considering the similarities in both the number and morphology of chromosomes and even in the structures of deoxyribonucleic acid (DNA) and ribonucleic acid (RNA) molecules within the chromosomes, it is highly unlikely that the two main living groups of whales could have evolved from separate groups of terrestrial mammals.

Morphological characters indicating that archaeocetes are whales are a characteristic ear anatomy, a long palate, a bony shelf projecting over the eye from the top of the skull, nostrils located on top of the snout, an elongate body with long tail and short neck, reduced hindlimbs, paddle-shaped front limbs, and a point of flexion in the tail vertebrae that allows up and down movement of horizontal tail flukes. These features constitute a suite found nowhere else in the animal world except in the archaeocetes and the modern baleen and toothed whales. It is highly unlikely that this unique group of characters would appear independently in three unrelated groups of mammals, and the suite is highly suggestive of close evolutionary relationships.

Archaeocete fossils. The first archaeocete bones found in the United States were originally believed to represent a large fossil reptile subsequently named *Basilosaurus*, meaning king of the reptiles. Five years later the animal was correctly reinterpreted as a whalelike mammal and renamed *Zeuglodon*. The first name has scientific validity, however, so today *Basilosaurus* is still used for the largest known archaeocete.

German and British scientists searching for Egyptian antiquities also discovered abundant archaeocete fossils in the Fayum near Cairo. Around the turn of the century, several pioneering studies of these fossils appeared. Fossil archaeocetes have also been discovered in Great Britain, Russia, Nigeria, India, British Columbia, Australia, and Antarctica. Sites in Egypt, India, and Pakistan are in what was once part of the ancient Tethys Sea, extending east-west through the Middle East to Southeast Asia. The Mediterranean Sea is a vestige of the former Tethys.

Primitive archaeocetes. In 1982, the most primitive known braincase of any archaeocete was named *Pakicetus inachus*. The fossil, collected in the Himalayan region of Pakistan from sediments that were laid down about 50 million years ago (m.y.a.) near the northern margin of the Tethys Sea, seems to have many of the requisite structures to be a predecessor of the fully aquatic late Eocene archaeocetes. Nothing more than this braincase and possibly some jaw parts are known of *Pakicetus*, but it has been speculated that the animal was amphibious and had hindlegs.

The Protocetidae are the oldest and most primitive archaeocetes and include the middle Eocene fossils from Pakistan. Known fossils are limited to those from Tethys, *Pappocetus lugardi* from Nigeria, and a few interesting and problematic bones from the southeastern United States. All protocetids were small-bodied animals, less than 10 ft (3 m) long, with nontelescoped skulls, and the nostrils were on the anterior part of the snout. They had the normal mammalian dental formula, with two roots on the first and second premolars, and three roots on all the following teeth. That *Protocetus atavus* had facets on the sacral vertebrae for articulation with large innominate (pelvic) bones suggests that it and the still-earlier *Pakicetus* were both amphibious and had hindlimbs.

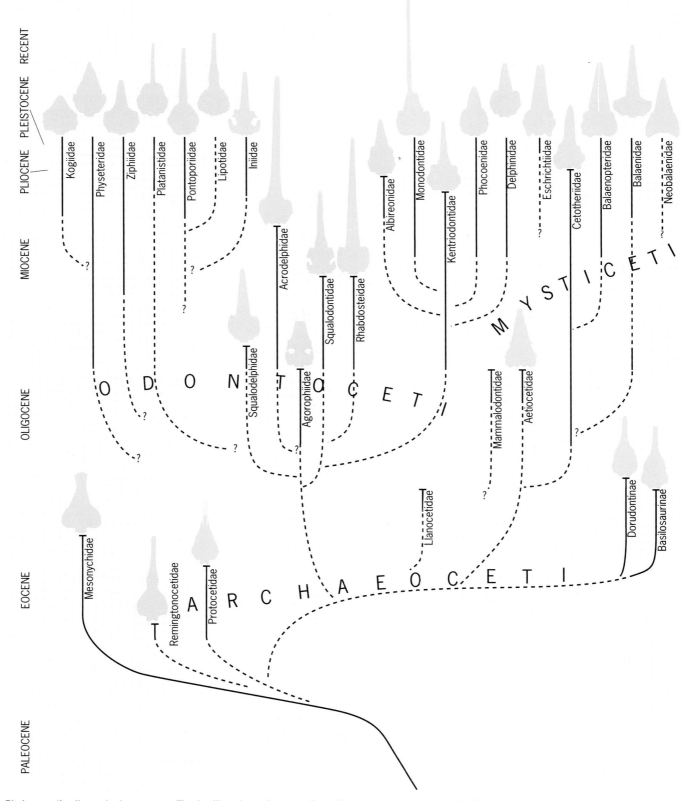

Phylogeny (family tree) of cetaceans. The families shown here are those that are now widely recognized by a majority of researchers. In general, the groups toward the sides of the illustration are the most divergent, and those toward the center are the more generalized. The solid lines indicate the time periods for which there is a relatively complete fossil record that documents an uncontrovertible evolutionary lineage. The broken lines indicate suspected lineages and relationships between families. Question marks indicate problem areas for which there are few or no known fossils. (*Copyright © 1992 by P. Folkers and L. Barnes; used with permission*)

In 1986, fossils of a radically divergent group of archaeocetes, the family Remingtonocetidae, were reported. Known from fossils in India, they seem to have evolved almost crocodilelike skulls and teeth, and could not possibly have been ancestral to the later groups of more highly evolved whales.

The later and more highly evolved Basilosauridae, discovered virtually around the world, include the medium-sized, generalized Dorutontinae and the more derived, giant Basilosaurinae. Like those of the Protocetidae, their skulls were not telescoped, but their molars and premolars were more specialized because none were three-rooted, they had accessory denticles on the anterior and posterior edges, and the third molar was lost. *Basilosaurus* is known to have had hindlegs. Generalized dorudontines have been suggested as possible ancestors of the later groups of true cretaceans, the mysticetes and odontocetes.

Derivation of cetaceans. Through the years, scientists have postulated markedly different possible derivations for cetaceans, some founded on sound fossil and modern anatomical evidence, others purely conjectural. Fossil jaw parts and teeth of several species of early archaeocetes have been found in Eocene-age sediments in Pakistan and India. Two of these, *Gandakasia potens* and *Ichthyolestes pinfoldi*, were believed by the scientists who originally named them to be terrestrial "proto-ungulate" mammals of the condylarth family Mesonychidae. Recently these animals have been reinterpreted as archaeocete whales. Mesonychids are the extinct terrestrial group that some researchers have suggested were the ancestors of whales. This switch in the interpretation of the affinities of *Gandakasia* and *Ichthyolestes* serves to emphasize both the degree of similarity between the archaeocetes and mesonychids and the types of radically different conclusions that can sometimes be reached by different workers studying the same fossils.

Mesonychids were large-bodied mammals that during the Paleocene and Eocene developed a variety of adaptations. As judged by their dentitions, some were carnivorous, some herbivorous, and some omnivorous. They share some skull and dental characters with archaeocetes, and a close relationship between the two groups is quite probable. Some are inferred to have been fish eaters, and the fact that their fossils are sometimes found in estuarine or lagoonal sediments suggests that a variety of mesonychids might have been evolving toward an aquatic life. Mesonychid fossils are also found in some of the same rock units that have produced early archaeocetes in Pakistan and India.

It is also significant that some paleontologists have believed mesonychids to be ancestors of the modern ungulate or hoofed mammals. That whales and ungulates share a common ancestry through the extinct early mesonychids is supported by data from living animals that indicate that fetal blood sugar, blood composition, chromosomes, insulin, uterine morphology, and tooth enamel microstructure of whales are most like those of the living artiodactyls, the cloven-hoofed ungulates. This relationship was proposed on the basis of anatomical evidence as early as 1883.

Summary of cetacean evolution. In the phylogeny shown in the **illustration**, the broad relationships of the cetacean families appear against a scale of geologic time. Details of the interrelationships vary among different published references, and new specimens are always being discovered and announced, so the gaps are gradually being filled. Examples are the relatively recently recognized families Albireonidae, Remingtonocetidae, Llanocetidae, and Mammalodontidae. Except for the Mammalodontidae and the Llanocetidae, all family lineages are represented by an outline of the skull of a typical genus of the family. Two subfamilies, Dorudontinae and Basilosaurinae, of the archaeocete family Basilosauridae are shown to indicate two different lineages in that group.

The major steps in the evolution of cetaceans can be summarized as follows: The earliest archaeocetes are good intermediates between terrestrial mesonychids and later archaeocetes, and thereby directly or indirectly ancestral to the earliest Odontoceti and Mysticeti that appeared in the Oligocene. The most primitive baleen whales (Aetiocetidae) had teeth, and their cranial structure closely resembled that of archaeocetes. By the middle Oligocene, true baleen-bearing mysticetes appeared. Most modern mysticete families had appeared by the late Miocene, about 12 m.y.a. Most of the Oligocene odontocetes were small or medium-sized animals. Most of the more recent groups of odontocetes could have evolved from them, but none of the families that were alive during the Oligocene survived to the present. Other groups of dolphinlike families diversified in the Miocene Epoch, approximately 23–5 m.y.a., most becoming extinct. The sperm whales (Physeteridae) and the beaked whales (Ziphiidae) are among the few odontocete groups that originated in the early or middle Miocene and survived to the present. The other typical Miocene groups were succeeded by, and in some cases ancestral to, such modern families as the Delphinidae (true dolphins), Phocoenidae (porpoises), and Monodontidae (belugas, narwhals), all of which were relative latecomers, not appearing in the fossil record until about 12 m.y.a. Modern diversity, both in numbers of species and in variety of morphology, of toothed whales had appeared by the Miocene, but it was a subsequent wave of new late Miocene and Pliocene groups that appeared as the forerunners of the living types.

For background information SEE CETACEA; MAMMALIA in the McGraw-Hill Encyclopedia of Science & Technology.

Lawrence G. Barnes

Bibliography. L. G. Barnes, D. P. Domning, and C. E. Ray, Status of studies on fossil marine mammals, *Mar. Mammal Sci.*, 1(1):15–53, 1985; T. Bishop, C. Craig, and C. Creagh (eds.), *Whales, Dolphins and Porpoises*, 1988; P. D. Gingerich et al., Origin of whales in epicontinental remnant seas: New evidence from the Early Eocene of Pakistan, *Science*, no. 220, pp. 403–406, 1983; A. R. Kellogg, *A Review of the Archaeoceti*, Carnegie Inst. Wash. Publ. 482, 1936.

Wine

Winemaking is the sequence of unit operations that transforms juices of grapes and other fruits into wines. It is a traditional process, known for more than 4000 years. Present operations range from the ethanol fermentation by yeast and the optional malolactic fermentation by bacteria to the enzymatic, physical, and chemical reactions associated with process treatments and aging reactions.

Maturity assessment. The discovery that the terpene family of grape volatiles, responsible for the characteristic aromas of cultivars such as Riesling, Gewurztraminer, and Muscats, are present in juices mostly as nonvolatile glycosylated forms has led to several developments related to the assessment of the maturity of these cultivars. The developments include analytical methods to determine both the free and potential terpenes, studies of their distribution within berries, evaluations of their enzymatic and chemical hydrolysis, and evaluations of the relationship between viticultural practices and the levels of terpenes in grapes, juices, and wines. However, because these white cultivars account for only a minor fraction of the grapes used for the production of white wines today, the assessment of maturity based on terpene analysis has had only limited application.

The more general application of spectrophotometric assessment of maturity by measuring the accumulation of major biochemicals such as total phenol content (estimated at 280 nanometers wavelength) and the red pigment formation (estimated at 520 nm) due to the anthocyanins has yet to be widely adopted. By far the most common method of maturity assessment continues to be the measurement of sugar and soluble solids based on solution density or refractive index.

Juice and must treatments. The practice of allowing extended contact between the skins and the juice of white grapes prior to their separation and fermentation, generally referred to as skin contact, has become less common with the introduction of membrane-type presses and more practical draining devices. In contrast, there has been some interest in the corresponding treatment (cold maceration) of red musts prior to fermentation on the skins; this treatment permits manipulation of the color and the phenolic and flavor components to yield characteristics other than those normally obtained during the fermentation of the skins. However, no published data support the claim that more distinctive wines are made by this treatment.

Ethanol fermentation. An ongoing search for improved or more suitable yeast strains of *Saccharomyces cerevisiae* for ethanol fermentation continues. The major criteria are related to the minimizing formation of traces of undesirable by-products (such as hydrogen sulfide, sulfite, acetic acid, and urea) rather than any further enhancement of the specific growth rate or of the speed and extent of the fermentation. The seasonal variation in the levels of all nutrients and growth factors continues to be an overriding aspect of the poor reproducibility and the inconsistency of such evaluations of yeast strains.

The concern about trace levels of the potential human carcinogen ethyl carbamate in wines has led to numerous investigations of its precursors, the role of yeast strains and nutrition in urea formation, and the subsequent chemical behavior of the yeast in wines and wine distillates. Urea has been identified as the yeast by-product that can lead to formation of carbamate. Studies with a radioactive-labeled substrate have shown that the major precursor is arginine in grape juice, with secondary effects due to other carbamyl compounds. While the conditions that influence the synthesis, excretion, and utilization of urea during fermentations are not completely understood, there are effects due to the yeast strain and the nutritional status of the juice. Typical levels of 10–20 milligrams of urea per liter can be formed, and can lead to 1–10 micrograms of carbamate per liter during the life of typical table wines.

Some species of yeast secrete a toxic protein to which they are immune but which is lethal to other, sensitive strains. This feature is referred to as the killer factor, and the yeast is referred to as killer yeast. The selection and breeding of this killer phenomenon in wine yeast has met with only limited success. Apparently, the natural course of ethanol development during the fermentation is a much more important factor in the selection when several competitive strains are present.

More practical evaluation and selection relates to low-foaming yeast strains. The goal is the reduction of the headspace required for foam breakdown in white-juice fermentations and more effective use of fermenter volumes during the limited fermentation season.

There has been only limited introduction of a technique known as proton flux for the evaluation of yeast during propagation and fermentation. This method seems to hold more promise for measurement of viability during adaptation of cultures for the secondary fermentation of sparkling wines, rather than for the general production of table wines.

Malolactic fermentation. Concurrent ethanol and malolactic fermentation is becoming a more widely accepted commercial practice. The initiation and extent of the conversion of malic acid to lactic acid by strains of *Leuconostoc* and *Lactobacillus* is far more certain and complete during the juice stage than in young wines. The practice of permitting the malolactic fermentation to occur in young Chardonnay wines as a flavor treatment is declining in extent and incidence in favor of other treatments.

Although commercially available freeze-dried bacterial cultures now are widely accepted, the development of a more practical propagation medium for wineries to use with these cultures is still needed.

Wine treatments. Among the treatments now commonly used to modify wine flavors is the practice of allowing white wines to remain in contact with the yeast cells for weeks or months after the fermentation; the practice is called sur-lie. It is generally performed on a wine that has undergone ethanol fermentation in a partly filled barrel; rather than separating the suspended matter, similar wine is added to fill the barrel. The contents may be either mixed periodically or left undisturbed. It has been documented that cellular components are released by leakage in the months that follow, although the true autolysis of the cells generally takes a year or more.

A second treatment is that of aging dry wines in new oak barrels in order to obtain oak extractives under conditions in which both oxygen and vapor permeate the container. It has received considerable attention with respect to the type of oak, the region of production, and the thermal treatments during the coopering process. The nature and level of extractives can be varied by altering the extent of charring during barrel manufacture. However, the overall differences observed vary with the wine and the barrels used in such studies, and there appear to be seasonal inconsistencies.

The adsorption of wine components to subthreshold sensory levels or to concentrations that ensure some aspect of physical stability is receiving renewed attention as the interest in the reduction of solid waste develops. At present a number of colloidal adsorptive agents such as ion-exchange clays (bentonites) and natural proteins (casein, gelatin, albumen, and isinglass) require the use of filtration with diatomaceous-earth cakes or pads for subsequent clarification. Efforts directed at developing methods that would generate minimum amounts of waste involve immobilized agents, regenerative agents, and packed and fluidized beds that would eliminate the need for diatomaceous earth.

Crossflow and membrane applications. Recently a number of applications of crossflow and membrane filtration to wines have been studied; in addition, gas separation membranes for the on-site production of process nitrogen have been introduced. The wine applications include high-end ultrafiltration (100,000 molecular-weight cutoff) and crossflow microfiltration (nominal pore sizes of 0.2 and 0.45 micrometer) for the clarification of suspended and colloidal matter, the limited application of ultrafilters with cutoffs values of 10,000 and 20,000 for heat stability, and the application of reverse-osmosis membranes for the production of nonalcohol and low-alcohol wines.

Crossflow microfiltration is of interest because of its ability to handle higher loads of suspended solids and because of the possibility of eliminating diatomaceous earth as a processing aid, which would achieve reduction of solid waste. The earliest trials using submicrometer back-flushed filters of porous stainless steel for juice and wine took place in 1976; there have been several recent tests with suspended yeast and the sediments from bentonite and other additions of fining agents using synthetic polymeric membranes. This approach has been shown to be technically feasible.

Use of ultrafilters for removing heat-unstable, haze-forming proteins with molecular-weight cutoff values of 10,000, 20,000, and 30,000 has had limited success because of the concomitant reduction of some polymeric phenols, which seems to be caused by solute interactions with the membrane materials under the pH and ethanol conditions occurring in the wine. Further, heat-unstable peptide fractions and peptide–tannin complexes pass through the membranes in some

wines, and the purpose of the treatment is defeated. Ultrafilters with molecular-weight cutoff values less than 10,000 significantly reduce color and flavor, and they have limited application as wine treatments other than in the recovery of overextracted wines or fractions obtained from pressing the skins for blending purposes. These ultrafilters have not yet received governmental approval in the United States, as they are deemed to cause major changes in the chemical composition of the wine.

The reverse-osmosis process for the production of ethanol-free wines essentially involves the two-step separation of ethanol and water from other wine components and the subsequent addition of process water to restore the original concentration of the remaining wine components. Only traces of other low-molecular-weight organics are removed, primarily acetic acid and some acetate esters. There is almost complete retention of inorganic ions due to the rejection of charged species by the membranes. The use of these membranes to concentrate color in red table wines or to reduce the ethanol content, while technically feasible, has not been adopted, since most countries have fairly strict wine regulations governing such compositional changes. There is little merit in using currently available membranes for the concentration of grape juices since the juices are already close to the practical osmotic pressure limit of 240–250 g of sugar per liter.

The use of commercially available gas-separation membranes for the on-site generation of process nitrogen from air has begun at some locations, primarily because of the costs and difficulties associated with delivery by tanker. The elimination of more than 95% of atmospheric oxygen is required in order to keep oxygen pickup under storage and bottling operations below 0.4 mg/liter at equilibrium. There is presently little interest in the recovery and purification of fermentation carbon dioxide, either as a process gas or in recovery schemes to remove organic volatiles.

Physical stability testing. The accelerated, conductivity-based crystallization tests for the potassium bitartrate stability in wines are gaining acceptance and superseding the concentration-product concept and the freezing-test approach. The accelerated test involves bringing the wine sample to the desired temperature, noting its conductivity, and then adding a determined amount of fine potassium bitartrate crystals to speed up any crystallization that might take place. After 30 min or so, a final conductivity reading is taken, and the results can be interpreted to determine the extent of the instability. An analogous accelerated test for the calcium tartrate stability, in which changes in calcium content rather than conductivity are followed, has now been developed.

There is also an increasing awareness of colloidal stability in general, with the corresponding introduction of the ethanol-addition test for quantification of the colloidal haze. In this test, the precipitation of insoluble and marginally soluble components is induced and quantified by nephelometry. The test also provides a basis for the evaluation of alternative stability treatments.

Process computer applications. The most common application of computers to winemaking involves the use of databases for tracking the source and composition of wine lots that are consolidated and blended. In the United States, record keeping and reporting for both state and federal agencies cover not only ethanol content and volume for excise tax purposes but also the proportions and grape sources for the varietal and appellation aspects of label approval.

Many wineries use computers to monitor temperatures in the fermentation and storage tanks, but none at this time routinely monitor the progress of the fermentations or perform actual process diagnosis or control. Difficulties in sampling and on-line analysis exist because of the concentration gradients occurring in unagitated tanks, samples supersaturated with dissolved carbon dioxide, and the presence of particulate materials and yeast that need to be removed before measurements can be taken. More generalized methods, such as juice density, derived from differential pressure measurements at two heights in the fermenter, suffer from errors associated with the changing contribution of suspended materials in the measurement zone and necessary temperature corrections. The application of a very accurate pressure transducer combined with an inert-gas bubbling system for following fermentation rates and extents by weight-loss measurements is only now at the point of commercial trials. Advantages are the capability of monitoring as many as 120 tanks with a single transducer, freedom from any internal fittings or sampling device, the constant or predictable contribution of suspended solids, and the temperature independence of the measurement. The corresponding benchtop procedure of fermentation monitoring by weighing has been used in research studies for some 50 years. There are now fermentation models and routines for estimation of parameters that can be used for the early diagnosis of problem fermentations and for the prediction of individual and collective cooling loads, but these are not yet being applied to wine fermentations.

The application of process computers to filtration, centrifugation, and distillation (either batch or continuous) is presently limited to algorithms developed by equipment manufacturers rather than by winery process engineers. Therefore operation of the equipment has generally been optimized for applications other than wine, and the results are not usually part of an integrated process-monitoring scheme.

An event-driven winery simulation model of harvest operations has been developed recently; it has applications in both daily harvesting and production planning as well as in evaluation of alternative winery designs. This model can evaluate the interaction of the arrival of many separate loads of grapes and the type, number, and size of the various pieces of winery equipment. The impact of arrival rate and grape mix as well as the consequence of certain production decisions can be evaluated, and the points of process limitation can be identified.

Antimicrobial chemical additives. In the United States, the federal authorities recently approved the use of dimethyl dicarbonate (commercially available under the name Delcorin) as a transient antimicrobial additive. This application provides a chemical alternative to sterile filtration prior to bottling. In wine, this compound has a half-life of approximately 30 min at bottling temperatures. It is effective with most wine yeast strains; however, at levels active for these yeasts, it has little if any effect on the strains of bacteria found in wines.

Prospects. Current developments in biotechnology, chemical instrumentation, and computers provide exciting opportunities for engineered microorganisms, enzyme-based manipulations and assays, the identification of trace components, and the measurement and rapid simulation and optimization of process conditions. However, applications are delayed by limits in the understanding of many aspects of winemaking. For example, the development of improved yeast strains for wine fermentations can be achieved only with advances in the knowledge of the genetics, biochemistry, and transport and energy systems of the current strains. The introduction and adoption of technology is also governed by financial considerations, governmental regulations, and collective attitudes.

For background information SEE FERMENTATION; FILTRATION; INDUSTRIAL MICROBIOLOGY; TERPENE; ULTRAFILTRATION; WINE; YEAST, INDUSTRIAL in the McGraw-Hill Encyclopedia of Science & Technology.

Roger Boulton

Bibliography. E. Dimitiriadis and P. Williams, The development and use of a rapid analytical technique for the estimation of free and potentially volatile monoterpene flavorants of grapes, *Amer. J. Enol. Vitic.*, 35:66–71, 1984; M. Malfieto-Ferreira, J. P. Miller-Guerra, and V. Loureiro, Proton extrusion as an indicator of the adaptive state of yeast starters for the continuous production of sparkling wines, *Amer. J. Enol. Vitic.*, 41:219–222, 1990; F. F. Monteiro, E. K. Trousdale, and L. F. Bisson, Ethyl carbamate formation in wine: Use of radioactively labelled precursors to demonstrate the development of urea, *Amer. J. Enol. Vitic.*, 40:1–8, 1989; C. S. Ough et al., Factors contributing to urea formation in commercially fermented wines, *Amer. J. Enol. Vitic.*, 41:68–73, 1990; C. Peri, M. Riva, and P. Decio, Crossflow membrane filtration of wines: Ultrafiltration, microfiltration and intermediate cut-off membranes, *Amer. J. Enol. Vitic.*, 39:162–168, 1988.

X-ray optics

New optics make it possible to focus and control x-rays and neutrons over broad angular and energy ranges. Now x-rays can be focused for medical therapy and materials analysis, intense quasiparallel x-ray and neutron beams can be produced from divergent sources for medical diagnosis and x-ray lithography, and x-ray telescopes that detect x-rays at higher energies and are much lighter and less expensive than previous models can be built. These optics show promise for improved efficiency and effectiveness in almost all applications using x-rays or slow neutrons, and make possible many new applications.

Principles of Kumakhov optics. In 1986, new optics for x-rays and neutrons were proposed by M. A. Kumakhov. Now called Kumakhov optics, they have been developed to the point that important applications in science, medicine, and industry have been demonstrated. Their advantage is the ability to operate over a much larger angular range (up to 30°) and energy range (0.5–100 keV) than had been possible with the optics previously available.

Kumakhov optics are based on total external reflection, except that instead of only one or two reflections, as is currently the practice in x-ray microscopy or x-ray astronomy, multiple reflections from specially shaped surfaces are used. Conventional x-ray optics is strongly constrained by the fact that x-rays are reflected from smooth solid surfaces only at very small angles of incidence (**Fig. 1a** and *b*). The maximum reflection angle, the so-called critical angle for total external reflection (also called the Fresnel angle), is typically only 0.2° (4 milliradians) for 10-keV x-rays, and is even smaller at higher x-ray energies. Kumakhov optics (Fig. 1*c*) allow the x-rays to be deflected through angles tens or even hundreds of times larger than the Fresnel angle. The necessary condition under which deflection of the type shown in Fig. 1*c* can be carried

out efficiently (ensuring that the x-rays are incident on the channel walls at an angle less than the Fresnel angle for each scattering) is given by inequality (1),

$$R_c > \frac{2d}{\theta_c^2} \qquad (1)$$

where R_c is the radius of curvature of the channel, d is the spacing between the channel walls, and θ_c is the Fresnel angle.

Many different forms for the reflecting surfaces can in principle be used, but a particularly useful form is small-diameter hollow capillary tubes. These capillaries can be arranged to meet a wide variety of needs, just as glass fibers can be used to control visible light.

Configurations of capillary tubes designed to capture x-rays from a divergent source and then to guide them to form a small focal spot or to form a quasiparallel beam are shown in **Fig. 2**. Except for very special capillary designs, the x-ray beam emerging from each of the capillary tubes has a divergence close to the critical angle for total external reflection corresponding to the x-ray energy and the material from which the capillary is constructed. The same parameters also define the angular range over which each capillary tube captures x-rays. For the glass materials from which most of the Kumakhov lenses have been constructed so far, this angle is given approximately by Eq. (2),

$$\theta_c = \frac{30}{E} \qquad (2)$$

where the photon energy E is given in kiloelectronvolts and the critical angle θ_c in milliradians (1 milliradian $\approx 0.06°$).

Kumakhov lens construction. The first Kumakhov lens (**Fig. 3**) was constructed from 2107 hollow tubes made from a borosilicate glass similar to Pyrex. The inside diameter of each tube was 0.36 mm (0.0142 in.) and the outside diameter was 0.42 mm (0.0165 in.). First-generation x-ray lenses such as this were quite large (0.5–1 m or 20–40 in. in length) because single-capillary glass tubes of less than about 0.3 mm (0.0012 in.) outside diameter were too difficult to make and use because of their flexibility and fragility. Thus, single capillaries with the smaller inside diameter needed for smaller lenses or for higher-energy x-rays would necessarily have such thick walls that their efficiency would be very limited.

A geometric parameter γ that controls the efficiency with which a capillary tube will capture and transmit x-rays is given by Eq. (3), where R is the bending

$$\gamma = R \frac{\theta_c^2}{2d} \qquad (3)$$

radius of the capillary tube, θ_c is the angle for total external reflection, and d is the internal diameter of the capillary. It was clear that to reduce the lens size, or to increase x-ray energy, the inside diameter d would need to be decreased substantially. To do this, a new technology was developed in which a single glass fiber contains many parallel open channels separated by thin glass walls. It is now possible to fabricate such

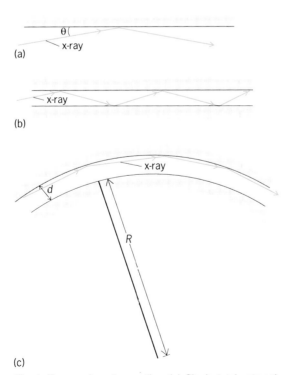

(a)

(b)

(c)

Fig. 1. X-ray and neutron optics. (*a*) Single total external reflection, which must be at an incident angle *θ* less than the critical or Fresnel angle. (*b*) Guiding of x-rays or neutrons along a straight hollow channel by multiple reflections between smooth surfaces. (*c*) Deflection of x-rays and neutrons through a bent hollow tube by multiple reflections from the channel walls. *R* = channel radius of curvature; *d* = spacing between channel walls.

polycapillary fibers with channel widths as small as 1 micrometer and wall thicknesses of less than 0.1 μm. This advance has allowed the range of x-ray energies that can be controlled to be extended to 100 keV, and has also allowed such optics to be used for control and focusing of slow neutron beams.

Lenses made from polycapillaries are called second-generation Kumakhov lenses. For both first- and second-generation lenses, the cross section (diameter) of each capillary channel is uniform along its length. It is also possible to construct lenses in which the cross section is not constant. In this case, the motion of x-rays through the lens is more complicated but for some applications offers distinct advantages such as simpler and automated construction, smaller size, and reduction of the focal spot size from about 300 μm to a few micrometers (potentially to a few tenths of a micrometer). Such third-generation Kumakhov lenses are being developed and investigated.

Application of Kumakhov optics.
Because of their broad angular range (up to 30°), broad energy range (0.5–100 keV), and high efficiency (typically, 10–50% depending on the x-ray energy and the maximum deflection angle), Kumakhov optics for x-rays and neutrons have the potential to be used in virtually all applications that involve x-rays or slow neutrons.

Medical imaging. Although the use of x-rays in medicine is extensive, the inability to control and focus them has restricted applications. The potential offered by Kumakhov optics has stimulated a broad review of the use of x-rays in medicine, and probably everything

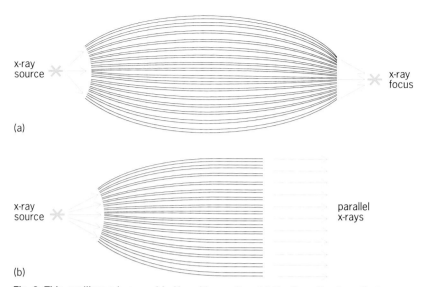

Fig. 2. Thin capillary tubes used in Kumakhov optics. (*a*) Configuration to collect x-rays from a divergent source and focus them onto a small spot. (*b*) Configuration to collect x-rays from a divergent source and form a quasiparallel beam. In the reverse direction, a parallel beam of x-rays can be focused onto a small spot.

from dental x-ray imaging to computed tomography will be affected. The medical imaging applications under most active development are angiography and mammography.

Angiography is the imaging of the coronary system with x-rays. It typically involves the introduction of a contrast agent into the heart cavity by use of a catheter, followed by the transmission of x-rays through the

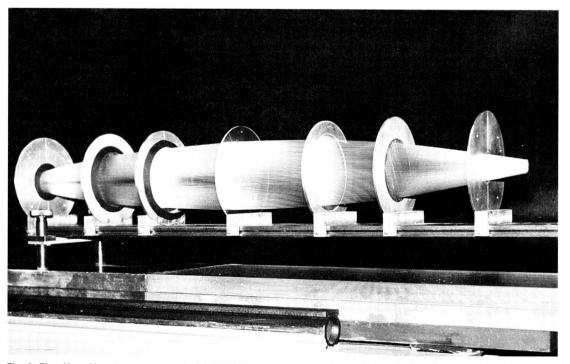

Fig. 3. First Kumakhov lens, composed of 2107 hollow glass capillary tubes. Each tube has 0.36 mm (0.0142 in.) inside diameter and 0.42 mm (0.0165 in.) outside diameter. The length of the lens is 96 cm (38 in.), and the diameter is 11 cm (4.3 in.).

chest to expose sensitivity-enhanced x-ray film. This procedure produces a picture of the coronary arteries and of the heart, so that blocking of blood flow can be detected.

Kumakhov optics offer the possibility of increasing the efficiency and resolution of synchrotron-based angiography by more effectively collecting, shaping, and collimating the beam. Even more important is the chance to make such measurements with less expensive hospital- and clinic-based x-ray sources.

Current radiographic imaging of the human breast is subject to degradation from point-source distortion and interference by scattered x-rays, so its resolution and sensitivity for detection of lesions (scarring), growths, or changes in tissue density are limited. A point source followed by a normal grid or collimator has a relatively small transmission efficiency; in contrast, a parallel beam (produced by a Kumakhov lens and an ordinary point source) followed by a Kumakhov array designed to select only those transmitted photons that are still on a parallel path can have a transmission efficiency of over 50%. The results are a substantially reduced dose to the patient and a more portable source. Most important, however, is the possibility for a large improvement in resolution (50 μm seems to be straightforward) for detection of calcification or other changes in tissue density. Such an improvement could result in detection of breast cancer at an earlier stage before metastasis (spreading through the lymph system) begins.

Medical therapy. Radiation therapy of tumors has been hampered by the inability to focus the radiation, as healthy tissue receives more intense exposure than does the buried tumor. To reduce such exposure, it has become common practice to use very high energy radiation (10–20 MeV) and complex motion of the patient and the source during the exposure. In turn, the expense is greatly increased and the general usefulness of radiation therapy, especially at early stages when the tumors are small, is reduced. The ability to sharply focus x-ray onto a very small spot offers the possibility of designing an "x-ray scalpel" for the noninvasive treatment of inoperable localized tumors.

The use of capillary optics to control beams of low-energy neutrons, while gamma rays and higher-energy neutrons are filtered out, may also make boron-neutron capture therapy practical for treatment of malignancies such as melanomas. It may even be possible to use x-ray and neutron "light guides" to introduce low-energy x-rays and neutrons through an endoscope to tumors inside the body without any exposure of healthy tissue.

Materials analysis. Materials analysis and study can be enhanced in many ways by the use of Kumakhov optics. Many techniques that previously required the intense and parallel x-rays from a synchrotron may now be carried out with laboratory-based x-ray sources used in combination with Kumakhov optics. These sources include extended x-ray absorption fine structure (EXAFS), small-angle x-ray scattering (SAXS), x-ray photoelectron spectroscopy (XPS), and microtomography.

As with x-rays, the ability to focus and control neutron beams opens new possibilities for neutron diffraction analysis and neutron impurity depth profiling, and may make neutron microscopy possible.

X-ray lithography. X-ray lithography is a process with the potential for making high-density computer memory chips. Problems with getting sufficient x-ray beam intensity and the need for a parallel beam of radiation have caused most companies to concentrate on very expensive synchrotron sources rather than the conventional x-ray sources. Kumakhov optics may make x-ray lithography possible with high-intensity standard sources. Such sources make x-ray lithography systems small enough to be incorporated into existing production facilities, could be compatible with existing orientation and vacuum requirements, and could be easier and much less expensive to implement.

X-ray astronomy. Potential applications of Kumakhov optics to x-ray astronomy are receiving considerable attention. The first such application will be a concentrator of radiation for spectroscopic studies of known astrophysical x-ray sources; it is important in the study of the properties of supernovae, black holes, and colliding galaxies. Kumakhov optics make it possible to extend the available x-ray energy range by more than a factor of 10 (to 100 keV) and at the same time to substantially reduce the cost and weight of the instruments relative to other current or proposed techniques.

For background information SEE FIBER-OPTIC IMAGING; INTEGRATED CIRCUITS; MAMMOGRAPHY; MEDICAL IMAGING; NEUTRON DIFFRACTION; RADIOGRAPHY; RADIOLOGY; SYNCHROTRON RADIATION; X-RAY ASTRONOMY; X-RAY DIFFRACTION; X-RAY MICROSCOPE; X-RAY OPTICS; X-RAY TELESCOPE; X-RAYS in the McGraw-Hill Encyclopedia of Science & Technology.

Walter M. Gibson; Muridan A. Kumakhov

Bibliography. M. A. Kumakhov, Channeling of photons and new x-ray optics, *Nucl. Instrum. Meth.*, B48: 283–286, 1990; M. A. Kumakhov and F. F. Komarov, Multiple reflections from surface x-ray optics, *Phys. Rep.*, 191(5):289–350, 1990.

Xylem

Water in plants is pulled from the soil through a network of xylem conduits to replace that evaporated from leaves. The pull is transmitted by hydrogen bonding between water molecules, and places the water under subatmospheric pressure usually less than that of the vapor pressure of water. Liquid water under these conditions is under tension and vulnerable to rapid vaporization, or cavitation, which leads to a vapor- and air-filled conduit termed an embolism. An embolism reduces the hydraulic conductivity of the xylem and potentially restricts gas exchange in leaves, so that their productivity is reduced. Vulnerability of the xylem to cavitation in response to water stress, freezing stress, and developmental events varies considerably between species, and is an important factor in the adaptation of plants to the environment.

Structure of xylem conduits. Xylem conduits are well designed for transport of water under tension. They are derived from elongated cells that have thick cellulosic walls impregnated with lignin for rigidity. Thus inward collapse such as occurs when water is sucked up a soggy paper straw is prevented. Xylem conduits are dead at maturity and, because they lack a protoplast and cell membrane, form a tube for water flow. There are two basic types of conduits: tracheids, derived from single elongated cells and less than 0.1 in. (3 mm) in length; and vessels, derived from several cells stacked in longitudinal files forming tubes up to several meters long. Gymnosperm xylem has only tracheids, and angiosperm xylem has tracheids plus vessels. Interconduit pits provide a channel between adjacent conduits for water flow (**Fig. 1**a). These pits are thin areas in the common wall spanned by the pit membrane, a derivation of the primary cell walls of the conduits. The pit membrane is a porous mesh of cellulose microfibrils and pectic matrix materials; its structure varies in conifer tracheids and angiosperm vessels. Intertracheid membranes have a central thickened region, the torus, that is held in place by radial strands composing the thin and porous margo (Fig. 1b). Intervessel pits lack a torus and are uniformly thin and porous (Fig. 1c).

In addition to facilitating flow between conduits, pits are valves that prevent the spread of embolism in the event of cavitation or damage to the xylem. If a branch is cut from a tree, air rushes into the severed conduits, and water pulls back into the tree. If air intake is not checked by pits between conduits, pruning would be lethal because it would embolize the entire conducting system.

Pits of tracheids and vessels confine embolism differently. In conifer tracheids, as the air-water interface is pulled into the relatively large pores (about 0.3 micrometer in diameter) of the margo region of the pit membrane (Fig. 1b), capillary forces generated by surface tension of water and adhesion of water to the hydrophilic cellulose of the margo momentarily prevent passage of air into the next tracheid. The resulting pressure difference across the membrane pushes the membrane against the chamber wall, and the central torus region that lacks pores seals off the pit aperture to prevent air passage (**Fig. 2**a). In most vessels the pores in the membrane are generally smaller than those in the margo of conifers, and a torus is absent. The spread of embolism is confined by the capillary forces at these pores (Fig. 2b); because these pores are smaller in diameter (to less than 0.05 μm) they can sustain correspondingly higher pressure differences than can the larger pores of the conifer margo.

Xylem structure exhibits several compromises to accommodate the conflicting requirements for efficiency of water conduction and safety from embolism. The presence of multiple conduits is one example: the most efficient system for water conduction would be a single branched pipe connecting all parts. However, this system would provide the least margin of safety because a

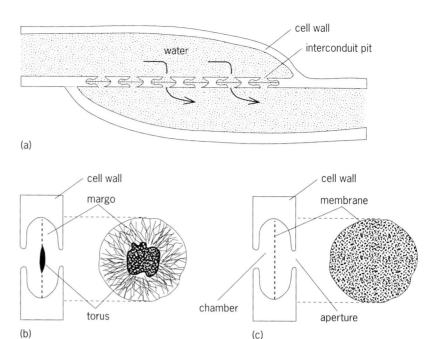

(a)

(b)   (c)

Fig. 1. Structure and function of xylem conduits. (*a*) Overlapping xylem conduits with interconduit pits for water flow. (*b*) Interconduit pit structure in tracheids of conifers and (*c*) in vessels of angiosperms.

single cavitation would kill the plant. The compromise is a multiple-conduit system with pits acting as check valves to confine embolism.

Embolism and water stress. The relationship between xylem tension and embolism can be determined by measuring the reduction in hydraulic conductivity of xylem due to embolism as tensions increase in water-stressed stems. This relationship varies considerably between species and shows no correlation with conduit type: gymnosperm tracheids and angiosperm vessels show comparable variation in cavitation resistance. There is good correlation with water availability in the habitat. For example, the arid-land chaparral shrub *Ceanothus megacarpus* resists cavitation, whereas the water birch, *Betula occidentalis*, that grows adjacent to streams does not. The xylem is resistant enough in most species to prevent significant embolism by normal water-stress levels. However, the correlation between vulnerability and habitat suggests there are trade-offs between cavitation resistance and hydraulic efficiency.

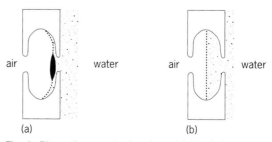

(a)   (b)

Fig. 2. Pits acting as check valves. (*a*) Tracheid pit and sealing action of the torus. (*b*) Vessel pit and sealing action of the air–water meniscus at membrane pores.

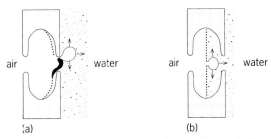

Fig. 3. Failure of check valves and the resulting cavitation. (a) Displacement of the torus allows air to enter through large pores of the margo membrane. **(b)** Displacement of the air–water meniscus in pores of the vessel pit membrane allows air to enter an adjacent vessel, so that cavitation and embolism ensue.

The mechanism of cavitation by water stress is simply the failure of the check valve at the pit membrane. In conifers, this failure is due to displacement of the torus from its sealing position over the aperture (**Fig. 3**a). Vulnerable conifers such as fir, *Abies lasiocarpa*, have weak pit membranes that cannot hold the torus in place as well as the strong membranes of resistant species, such as juniper, *Juniperus virginiana*. In vessels, vulnerable species such as the water birch have large pores in their pit membranes that unseal at relatively low tensions (Fig. 3b), as opposed to the small pores of more resistant species such as the mangrove *Rhizophora mangle* that retain a meniscus against higher tensions.

This mechanism suggests the nature of the trade-off between conducting efficiency and safety: in pit membranes, maximum permeability to water confers hydraulic efficiency but reduces safety from embolism because it increases permeability at air–water interfaces. The correlation of cavitation vulnerability and habitat may result from an evolutionary compromise between competing selective pressures for hydraulic efficiency and safety.

Embolism and freezing. Woody plants growing in temperate climates are subject to embolism that is caused by freezing and thawing. The xylem water is essentially pure at all times of the year and freezes during winter. Gases dissolved in the water freeze out as bubbles because they are insoluble in ice. If large enough, these bubbles serve as nucleating sites for cavitation when water is placed under tension after thawing.

Many vessel-bearing trees embolize almost completely during the winter because of repeated freezing and thawing. Coniferous trees that have tracheids show much lower levels of embolism in winter. Laboratory studies have shown that vulnerability to freezing-induced embolism is a function of the volume of the conduit: small-volume tracheids of conifers do not embolize in response to freeze-thaw cycles whereas vessels do, with larger vessels being the most vulnerable. Large conduits are vulnerable because the bubbles that freeze out are larger and more likely to serve as nucleating sites for cavitation after thawing. Again, a trade-off exists between the greater efficiency of large-volume conduits and their greater vulnerability

to freezing-induced cavitation.

Mechanism of embolism repair. Many vessel-bearing trees of the temperate zone can repair embolism developed during winter. The simplest mechanism involves production of new xylem before leafing-out that accommodates the tree's conducting requirements. This mechanism occurs in ring-porous hardwoods, such as oak, ash, and locust, whose large-diameter vessels (up to 500 μm) have sufficient conducting capacity to carry enough water in one annual ring. Other species refill embolized vessels by above-atmospheric xylem pressures generated in spring before leafing out. These include maples (*Acer* spp.), grapevines (*Vitis* spp.), and birches (*Betula* spp.).

The ability of xylem to avoid or repair freezing-induced embolism is an important factor for the success of woody plants in cold climates. Certainly the resistance of conifers to freezing-induced cavitation relates to their dominance in boreal and high-altitude forests.

Embolism and aging. In at least one species, quaking aspen, vessels embolize with age. This process is not due to environmental stress but rather to a loss of the check-valve function of pits. The pores enlarge from less than 0.1 μm to over 0.5 μm through removal of matrix materials in the pit membranes. It is not known if this degradation is mediated by enzymes. The larger pores cause cavitation at moderate tensions, which leads to a uniform deactivation of the functional xylem (sapwood). The regulation of sapwood area by this means may be a general phenomenon. The development of the inner core of nonfunctional xylem, or heartwood, may begin with developmentally programmed cavitation of sapwood by loss of check-valve function in pits.

Plant productivity. Although it is important to understand the functional significance of xylem structure and its performance under stress and through development, to appreciate its broader significance, its influence on plant productivity via restrictions on gas exchange must be quantified. Because of the necessary compromise between conducting efficiency and safety, xylem structure is closely tailored to the demands of a species' habitat. Therefore, xylem structure may limit the ability of leaves to transpire and access carbon dioxide for photosynthesis. Modeling studies have corroborated this relationship, but future research must provide empirical evaluation.

For background information SEE CELL WALLS (PLANTS); PLANT-WATER RELATIONS; XYLEM in the McGraw-Hill Encyclopedia of Science & Technology.

J. Sperry

Bibliography. J. S. Sperry and M. T. Tyree, Water-stress-induced xylem embolism in three species of conifers, *Plant Cell Environ.*, 13:427–436, 1990; M. T. Tyree and F. W. Ewers, The hydraulic architecture of trees and other woody plants, *New Phytol.* (Tansley review), 119:345–360, 1991; M. T. Tyree and J. S. Sperry, Vulnerability of xylem to cavitation and embolism, *Annu. Rev. Plant Physiol.*, 40:19–38, 1989; M. H. Zimmerman, *Xylem Structure and the Ascent of Sap*, 1983.

Zener diode

A Zener diode is a *pn* junction that is doped to have a safe reverse-voltage breakdown, where current begins to flow at some voltage and increases greatly for a small further voltage rise. A Zener is often used to define a voltage level by simply supplying the diode with a roughly constant current, which most simply can be drawn through a resistor from a higher voltage supply.

Zeners are available in selected voltages from 2 to 200 V and with safe power dissipations up to 50 W. Since their performance is stable and predictable, and they are small and usually inexpensive, they are much used in circuits as regulators, voltage limiters, and coupling elements. This article describes a particular refinement to allow their use as a voltage standard.

Principle of operation. When a voltage is applied to a diode that is the reverse of that needed for forward conduction (**Fig. 1***a*), a normal rectifying diode begins to pass a current only when the applied voltage reaches a few hundred volts. This so-called reverse avalanche current rises sharply with voltage, and the diode is likely to be destroyed without a protective resistor R in the circuit.

The mechanism that causes this reverse current is shown in Fig. 1*b*. In the *n* region of the diode, electrons (negative carriers) are present in high concentrations, due to the doping of the semiconductor during manufacturing. Similarly, in the *p* region, many holes (positive carriers) are present. High concentrations of electrons and holes cannot exist next to each other since some electrons diffuse from the *n* region into the *p* region, and some holes diffuse in the opposite direction. A so-called depletion region, containing few, if any, mobile carriers, therefore forms around the junction.

Under the applied potential shown in Fig. 1*b*, the electrons in the *n* region are pulled away from the junction, and the depletion region widens. Conduction will occur only due to the presence of minority carriers, from impurities in the material. Since as pure a material as possible is used, this so-called leakage current is normally very small (**Fig. 2**).

When the applied voltage becomes large, a large voltage gradient occurs at the depletion region. Carriers in the leakage current are accelerated to high velocities and ionize semiconductor atoms in the region they enter. The result is additional holes and electrons, which are now also accelerated in opposite directions and collide and ionize repeatedly, and a so-called avalanche results.

The leakage current is multiplied by a factor that increases sharply with applied voltage, so that reverse breakdown occurs (Fig. 2). In diodes used for high-voltage rectification, one side of the junction is left very lightly doped, and up to 2 kV can be withstood; whereas breakdown voltages down to 6 V can be obtained, because of the avalanche mechanism, if only one side of the junction is highly doped. A temperature rise increases the breakdown voltage.

If the doping density on both sides of the junction

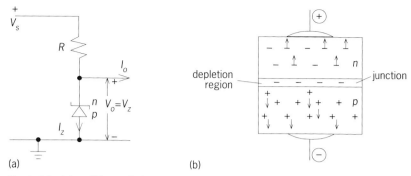

Fig. 1. Principles of Zener diode operations. (*a*) **Basic Zener regulator circuit.** (*b*) **Diagram of current carriers in material around a** *pn* **junction.**

is high, the depletion region becomes very thin, and a different breakdown mechanism known as Zener or field breakdown occurs. The thin depletion region allows little space for avalanche ionization, but it raises the junction field to high values. A combination of a high field, of the order 10^7 V/m, and a narrow depletion region, below 10 nanometers, allows electrons to tunnel directly from the valence band in the *p* region to the conduction band in the *n* region, a quantum-mechanical effect. The high electric field supplies the energy for electrons to traverse the gap in the energy bands. This gap decreases slowly with increasing temperature, and so the temperature coefficient of the Zener voltage is negative.

Zener breakdown fields are reached at voltages below 5 V. When the breakdown voltage is between 4 and 6 V, both avalanche and Zener mechanisms operate, and the breakdown voltage may become nearly independent of temperature.

Voltage–current characteristic. The voltage–current characteristic of a Zener diode (Fig. 2) has a normal forward characteristic (to the right), where a current flows once a forward voltage V_F has been applied (about 0.6 V for silicon). To the left, when V_{pn} is negative, that is, the *n* region is more positive, as in Fig. 1*a*, a small leakage current rises sharply at the reverse breakdown. The slope of the characteristic

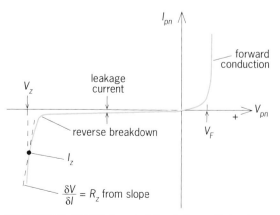

Fig. 2. Voltage–current characteristic of a Zener diode, showing the usual operating point at voltage V_z for a current I_z in the reverse direction.

there can be used to derive an internal or slope resistance R_z for the diode for small current changes in the region of I_z.

Once in breakdown, the voltage across the diode, V_z, remains nearly constant, so long as the permitted power dissipation, $P = I_z \cdot V_z$, as given by the manufacturer, is not exceeded. The Zener diode is most used in this mode, providing an essentially constant voltage at some level less than the supply voltage V_s. In Fig. 1a, R is chosen as given by Eq. (1) with the

$$R = \frac{V_s - V_o}{I_z} = \frac{(V_s - V_z) \cdot V_z}{P'} \qquad (1)$$

expression for power dissipation given above and with P' being some safe factor below P, the maximum power dissipation (for $I_o = 0$). As the current output I_o is increased, the source resistance of the circuit is not R, but rather that of R in parallel with R_z. The diode resistance R_z is of the order of 20 Ω for $I_z = 10$ mA. This low source resistance allows the voltage change ΔV to be calculated for some current change ΔI from Eq. (2). When large output currents I_o are desired, a

$$\Delta V = \frac{\Delta I \cdot R \cdot R_z}{R + R_z} \qquad (2)$$

power transistor is added as a shunt or series regulator.

Compensated Zeners. These are avalanche Zeners, with a breakdown voltage of about 5.6 V, in series with a forward pn junction. The positive temperature coefficient of the avalanche Zener offsets the -2 mV/$°$C temperature coefficient of the 0.6 V dropped across a normal forward-biased silicon junction. At a Zener current of 7.5 mA and an output voltage of 6.2 V, compensated Zeners have a lower slope resistance (about 10 Ω) than a single Zener junction. Temperature coefficients (change of Zener voltage for change of temperature) range from 0.01%/$°$C down to 0.0005%/$°$C for the most expensive devices.

To provide the reference voltage for modern five- and six-digit voltmeters, the compensated Zener's defects, which are chiefly its voltage changes with current and temperature, require improvement.

To ensure a constant Zener current, the circuit of **Fig. 3**a is often used. Here the Zener current, $(V_o - $

$V_z)/R_1$, is drawn from the higher output voltage V_o, which is itself derived from the stable Zener voltage. In a design to provide, say, $V_o = 10$ V to calibrate a digital voltmeter, R_2 and R_3 are preferably packed together, so that they track each other well in temperature. They should be related by Eq. (3) to give the

$$V_z = \frac{V_o \cdot R_3}{R_2 + R_3} \qquad (3)$$

correct feedback to the operational amplifier A. Alternatively, separate low-temperature-coefficient resistors can be used if a less stable output is acceptable.

To ensure a constant Zener temperature, a four-lead integrated circuit is available in which a complete temperature-control circuit with heater surrounds the Zener junction, so that the chip is maintained at about 70$°$C (150$°$F). Unfortunately, this elevated temperature results in the Zener voltage changing more rapidly with time.

Only a few manufacturers guarantee Zener voltage stability over time. Noise in the Zener voltage and an erratic or large rate of change with time are often related to effects at the semiconductor surface, where impurities reside. A buried Zener junction made by ion implantation reduces both these effects.

Voltage standards. The international standard of voltage is defined by the Josephson junction constant. Since the voltage steps of a Josephson junction are very small, typically 70 microvolts, some ingenuity is needed to relate them to the voltage levels of 1.018 V of the Weston standard cell or 10 V of a Zener-stabilized supply, which are used for routine measurements in national laboratories.

Standard cell banks, which hold a group of Weston cells at constant temperature, have been used since 1911 as international voltage standards. Carefully kept Weston cells age by 1 ppm per year or less and have, until now, outrivaled alternatives using Zener diodes. By 1974, a 10-V dc standard called a reference amplifier was in production, using a custom-made Zener diode and an npn transistor in an integrated circuit as the basic internal reference. The forward-biased base-to-emitter junction gave excellent compensation against the temperature coefficient of the Zener diode

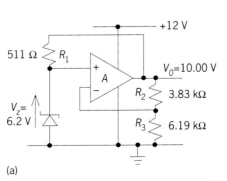

(a)

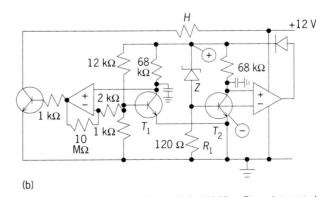

(b)

Fig. 3. Zener diode circuits. (*a*) Circuit to provide a constant current to a precision Zener diode. (*b*) Ultra-Zener integrated circuit, with Zener *Z*, transistors T_1 and T_2, and heater *H*. The left-hand circuit controls temperature and the right-hand one controls the Zener current.

by a suitable choice of resistors in the circuit. The transistor also amplified directly the error between the Zener and the feedback to voltages and so obtained a very low noise circuit.

In 1987, a subsurface Zener reference was produced in an eight-lead integrated circuit called the ultra-Zener (Fig. 3b). As well as the Zener Z, the circuit has a heater resistor H, a temperature-sensing transistor T_1, and a second transistor T_2, whose base-emitter junction would largely compensate the effect of temperature on the Zener voltage. When used with a dual operational amplifier and a few further transistors and resistors, it provides an output of 7 V from the points + to −. The expected effect of temperature on voltage is only 0.05 ppm/°C, and the expected change in the 7-V output is only 2 μV per month. The stability-with-time figure aroused great interest, especially since the change rate is expected to vary as the square root of time; thus, only double the quoted change would be expected in 4 months, and so on.

The advantages of a solid-state standard over a Weston cell bank are robustness and good ability to withstand electrical damage. With a 10-V output level, rather than 1.018 V, it is easier to protect against errors due to thermal emf's in the circuit wiring.

The standard would use the Zener circuit of Fig. 3b to replace the Zener in Fig. 3a. A further refinement is desirable to control the drift that occurs with time in the R_2-R_3 divider chain in Fig. 3a. A time-division voltage divider can be added to produce a second feedback voltage derived from the 10-V output V_o. This derived voltage is made an exact fraction of the supply by switching the supply voltage into a filter for a fixed fraction of the time and putting in zero volts for the remainder. The difference between this signal and the voltage at the tap between R_2 and R_3 is applied to a high-gain amplifier that controls a field-effect transistor shunting R_3, so that any drift is slowly compensated. As a result, the voltage output of the standard must essentially rely on the performance of the Zener only.

For background information SEE ELECTRICAL UNITS AND STANDARDS; ION IMPLANTATION; JOSEPHSON EFFECT; JUNCTION DIODE; SEMICONDUCTOR DIODE; VOLTAGE MEASUREMENT; WET CELL; ZENER DIODE in the McGraw-Hill Encyclopedia of Science & Technology.

Peter J. Spreadbury

Bibliography. P. Horowitz and W. Hill, *The Art of Electronics*, 2d ed., 1989; P. J. Spreadbury, The ultra-Zener: A portable replacement for the Weston cell, *IEEE Trans. Instrum. Meas.*, 40:343–346, 1991; C. D. Todd, *Zener and Avalanche Diodes*, 1970; C. Zener, A theory of electrical breakdown of solid dielectrics, *Proc. Roy. Soc. (London)*, 145:523–529, 1934.

Contributors

The affiliation of each Yearbook contributor is given, followed by the title of his or her article. An article title with the notation "in part" indicates that the author independently prepared a section of an article; "coauthored" indicates that two or more authors jointly prepared an article or section.

A

Abraham, Dr. Frederick D. *Blueberry Brain Institute, Waterbury Center, Vermont.* CHAOS—in part.

Ada, Dr. Gordon L. *John Curtin School of Medical Research, Australian National University, Canberra City.* VACCINATION.

Aguilera, Dr. Roberto. *President, Servipetrol Ltd., Calgary, Alberta, Canada.* OIL AND GAS WELL DRILLING—in part.

Anderson, Dr. J. D. *Plant Hormone and Weed Science Laboratories, Beltsville Agricultural Research Center (West), Agricultural Research Service, U.S. Department of Agriculture, Beltsville, Maryland.* ALLELOPATHY—coauthored.

Anslyn, Dr. Eric. *Department of Chemistry, University of Texas, Austin.* ENZYME—in part.

Appelbaum, Prof. Frederick R. *Clinical Research Division, Fred Hutchinson Cancer Research Center, University of Washington School of Medicine, Seattle.* GROWTH FACTOR.

Armstrong, Elizabeth. *Department of Biological Sciences, University of Arkansas.* LEAF—coauthored.

Athale, Dr. Ravindra A. *Department of Electrical and Computer Engineering, George Mason University, Fairfax, Virginia.* OPTICAL INFORMATION SYSTEMS.

Atwood, Prof. Jerry L. *Department of Chemistry, University of Alabama.* MOLECULAR RECOGNITION—coauthored.

B

Babbitt, Captain J. Randolph. *President, Air Line Pilots Association, Washington, D.C.* AIRCRAFT INSTRUMENTATION.

Baer, Prof. Norbert S. *Conservation Center of the Institute of Fine Arts, New York University.* BOOK PRESERVATION.

Bailey, Dr. B. A. *Plant Hormone and Weed Science Laboratories, Beltsville Agricultural Research Center (West), Agricultural Research Service, U.S. Department of Agriculture, Beltsville, Maryland.* ALLELOPATHY—coauthored.

Bamber, Dr. Roger N. *Marine and Freshwater Biology Unit, Fawley Aquatic Research Laboratories, Ltd., Hampshire, England.* MYSIDACEA.

Barbara, Prof. Paul F. *Department of Chemistry, University of Minnesota.* PHOTOCHEMISTRY.

Barnes, Dr. Lawrence G. *Curator and Section Head, Vertebrate Paleontology Section, Natural History Museum, Los Angeles, California.* WHALE.

Bartsch, Dr. James A. *Department of Agricultural and Biological Engineering, Cornell University.* FOOD ENGINEERING—in part.

Bass, Prof. George E. *Department of Anthropology, Nautical Archaeology Program, Texas A&M University.* MARINE ARCHAEOLOGY.

Beldman, Dr. Gerrit. *Department of Food Science, Agricultural University, Wageningen, The Netherlands.* FOOD MANUFACTURING—coauthored.

Belohlavek, Dr. Marek. *Biodynamics Research Unit, Department of Physiology and Biophysics, Mayo Clinic/ Foundation, Rochester, Minnesota.* MEDICAL IMAGING—coauthored.

Benkovic, Prof. Stephen J. *Department of Chemistry, Pennsylvania State University.* ANTIBODY—coauthored.

Bialek, Dr. Krystyna. *Plant Hormone Laboratory, Agricultural Research Service, U.S. Department of Agriculture, Beltsville, Maryland.* PLANT HORMONES—in part.

Blackwell, Dr. James A., Jr. *Deputy Director, Political Military Studies, Center for Strategic & International Studies, Washington, D.C.* INTELLIGENT WEAPONS.

Bortman, Dr. Marci L. *Waste Management Institute, Marine Sciences Research Center, State University of New York, Stony Brook.* RECYCLING TECHNOLOGY—coauthored.

Bouchet, Dr. Philippe. *Museum National d'Histoire Naturelle, Laboratoire de Biologie des Invertebres Marins et Malacologie, Paris, France.* SNAIL.

Boulton, Prof. Roger. *Department of Viticulture and Enology, University of California, Davis.* WINE.

Bradford, Prof. Kent J. *Department of Vegetable Crops, University of California, Davis.* SEED.

Brandler, Philip. *Acting Director, Food Engineering Directorate, U.S. Army Natick Research, Development and Engineering Center, Natick, Massachusetts.* FOOD—coauthored.

Breslin, Dr. Vincent T. *Waste Management Institute, Marine Sciences Research Center, State University of New York, Stony Brook.* RECYCLING TECHNOLOGY—coauthored.

Bruno, Michael H. *Graphic Arts Consultant, Bradenton, Florida.* PRINTING—in part.

Bruschi, Howard J. *Director, Nuclear Plant Programs, Nuclear and Advanced Technology Division, Westinghouse Electric Corporation, Pittsburgh, Pennsylvania.* NUCLEAR REACTOR—in part.

Bryant, Dr. Harold N. *Department of Biological Sciences, University of Calgary, Alberta, Canada.* CAT.

Bushnell, Dennis M. *Associate Chief, Fluid Mechanics Division, Langley Research Center, National Aeronautics and Space Administration, Hampton, Virginia.* DRAG REDUCTION.

Butler, Dr. Robert W. *Department of Pediatrics, Memorial Sloan-Kettering Cancer Center, New York, New York.* STRESS (PSYCHOLOGY)—in part.

Byram, Kenneth V. *U.S. Federal Aviation Administration, Washington, D.C.* GUIDANCE SYSTEMS.

C

Caldwell, Prof. B. E. *Assistant Director, ANR/CRD, North Carolina Extension Service, North Carolina State University.* PATENT—in part.

Cantrell, Dr. John H. *Langley Research Center, National Aeronautics and Space Administration, Hampton, Virginia.* MEDICAL ULTRASONIC TOMOGRAPHY.

Carlton, James T. *Director, Program in American Maritime Studies, Williams College-Mystic Seaport, Mystic, Connecticut.* EXTINCTION (BIOLOGY).

Caskey, C. Thomas. *Institute for Molecular Genetics, Human Genome Program Center, Howard Hughes Medical institute, Baylor College of Medicine, Houston, Texas.* HUMAN GENETICS—coauthored.

Chaisson, Dr. Eric J. *Scientist, Head of Educational and Public Affairs, Space Telescope Science Institute, Baltimore, Maryland.* SPACE TELESCOPE.

Chase, Prof. A. R. *Institute of Food and Agricultural Sciences, University of Florida, Apopka.* PLANT PATHOLOGY—in part.

Close, Dr. Frank. *Rutherford Appleton Laboratory, Science and Engineering Research Council, Oxon, England.* NUCLEAR FUSION.

Coey, Prof. J. M. D. *Head, Department of Pure and Applied Physics, Trinity College, Dublin, Ireland.* MAGNETIC MATERIALS.

Coleman, Dr. Robert D. *Environmental Research Division, Argonne National Laboratory, Argonne, Illinois.* POLYMER—in part.

Cooper, Prof. N. John. *Chair, Department of Chemistry, University of Pittsburgh.* ORGANOMETALLIC COMPOUNDS.

Costello, Vice Admiral John D. *Marine Spill Response Corporation, Washington, D.C.* WATER POLLUTION.

Cowan, Dr. William. *Director, Computer Science Laboratory, University of Waterloo, Ontario, Canada.* VISION—in part.

Currie, Dr. Philip J. *Head, Dinosaur Research, Tyrrell Museum of Palaeontology, Drumheller, Alberta, Canada.* DINOSAUR.

D

Damian, Prof. Raymond T. *Department of Zoology, University of Georgia.* MOLECULAR MIMICRY.

Darney, Dr. Philip D. *Department of Obstetrics, Gynecology and Reproductive Sciences, University of California, San Francisco.* BIRTH CONTROL.

Darsch, Gerald. *Chief, Food Technology Division, Food Engineering Directorate, U.S. Army Natick Research, Development and Engineering Center, Natick, Massachusetts.* FOOD—coauthored.

de Moustier, Dr. Christian. *Marine Physical Laboratory, Scripps Institution of Oceanography, University of California, La Jolla.* UNDERWATER SOUND.

DeSanctis, Dr. Gerardine. *Information & Decision Sciences Department, Curtis L. Carlson School of Management, University of Minnesota.* DECISION THEORY.

Dhir, Prof. V. K. *Mechanical, Aerospace and Nuclear Engineering Department, University of California, Los Angeles.* HEAT TRANSFER.

Dick, Dr. James D. *Associate Director, Microbiology Division, Laboratory Medicine, Johns Hopkins Medical Institutions, Baltimore, Maryland.* GASTROINTESTINAL TRACT DISORDERS.

Doll, Dr. Gary. *Physics Department, General Motors Research Laboratories, Warren, Michigan.* LASER DEPOSITION.

Dougherty, Prof. C. T. *Department of Agronomy, University of Kentucky.* GRASSLAND ECOSYSTEM.

Droegemeier, Dr. Kelvin K. *School of Meteorology and Center for Analysis and Prediction of Storms, University of Oklahoma.* WEATHER FORECASTING AND PREDICTION.

Durante, Raymond W. *Executive Consultant, AECL Technologies, Bethesda, Maryland.* NUCLEAR POWER.

E

Edwards, Prof. R. Lawrence. *Department of Geology and Geophysics, University of Minnesota.* DATING METHODS—in part.

Elliot, Dr. James L. *Department of Earth, Atmospheric, and Planetary Sciences, Massachusetts Institute of Technology.* PLANETARY PHYSICS—coauthored.

Elliot, Lyn E. *Department of Earth, Atmospheric, and Planetary Sciences, Massachusetts Institute of Technology.* PLANETARY PHYSICS—coauthored.

Elliott, J. D. *Manager, Petroleum Processing, Foster Wheeler USA Corporation, Clinton, New Jersey.* PETROLEUM PROCESSING—coauthored.

Erickson, Dennis C. *Senior Electronics Engineer, Bonneville Power Administration, Vancouver, Washington.* OPTICAL FIBERS.

Eskin, Prof. N. A. Michael. *Department of Foods and Nutrition, University of Manitoba, Winnipeg, Manitoba, Canada.* QUINOA.

Evans, David D. *Manager, Mars Observer Project, Jet Propulsion Laboratory, California Institute of Technology, Pasadena.* SPACE PROBE.

F

Fadner, Dr. T. A. *Staff Scientist, Graphic Systems Division, Rockwell International Corporation, Westmont, Illinois.* PRINTING—in part.

Fanta, Dr. Christopher H. *Longwood Medical Area Adult Asthma Center, Brigham and Women's and Beth Israel Hospitals and Harvard Medical School, Boston, Massachusetts.* ASTHMA—coauthored.

Fenton, Howard. *President, DTP Ink, Oradell, New Jersey.* PRINTING—in part.

Fertel, Marvin S. *Vice President, Technical Programs, U.S. Council for Energy Awareness, Washington, D.C.* NUCLEAR REACTOR—in part.

Fisher, Dr. David G. *Department of Physiological and Biological Sciences, Maharishi International University, Fairfield, Iowa.* PHLOEM.

Fleming, Dr. Ian. *University Chemical Laboratory, University of Cambridge, England.* ASYMMETRIC SYNTHESIS.

Foley, James M. *Manager, Business Development, Satellite Communications, Motorola Inc., Chandler, Arizona.* MOBILE RADIO.

Fox, Prof. Richard C. *Laboratory for Vertebrate Paleontology of the Department of Geology and Zoology, University of Alberta, Edmonton, Alberta, Canada.* REPTILIA.

Fry, Dr. Elizabeth. *Department of Zoology, Laboratory of Molecular Biophysics, University of Oxford, England.* VIRUS—coauthored.

Fu, Dr. Ping. *Crop Science Department, University of Saskatchewan, Saskatoon, Saskatchewan, Canada.* PLANT PHYSIOLOGY—coauthored.

G

Gad-el-Hak, Prof. Mohamed. *Department of Aerospace & Mechanical Engineering, University of Notre Dame.* CONTROL SYSTEMS.

Gallino, Dr. R. *Istituto di Fisica Generale dell'Universita, Turin, Italy.* METEORITE—coauthored.

Gibson, Dr. Walter M. *Director, Center for X-ray Optics, University at Albany, State University of New York.* X-RAY OPTICS—coauthored.

Gies, Robert M. *President, Rocky Mountain Geological Engineering, Limited, Calgary, Alberta, Canada.* GEOLOGICAL ENGINEERING.

Glantz, Prof. Stanton A. *Member, Cardiovascular Research Institute, University of California, San Francisco.* ENVIRONMENTAL HEALTH.

Gokel, Prof. George W. *Associate Chairman for Research and Graduate Studies, University of Miami.* MOLECULAR RECOGNITION—coauthored.

Goldstein, Jerome. *Publisher, BioCycle, Journal of Waste Recycling, Emmaus, Pennsylvania.* SEWAGE.

Golic, Dr. Kent. *Department of Biology, University of Utah.* RECOMBINATION (GENETICS).

Greenleaf, Dr. James F. *Biodynamics Research Unit, Department of Physiology and Biophysics, Mayo Clinic/Foundation, Rochester, Minnesota.* MEDICAL IMAGING—coauthored.

Gregory, Dr. Christopher. *Department of Immunology, Medical School, University of Birmingham, England.* CELLULAR IMMUNOLOGY—in part.

Griffin, Robert J., Jr. *Department of Energy, Washington, D.C.* SPACE FLIGHT.

Guckel, Prof. Henry. *Department of Electrical Engineering, University of Wisconsin, Madison.* MECHANISM.

Gull, Dr. Theodore. *Goddard Space Flight Center, Greenbelt, Maryland.* SATELLITE ASTRONOMY.

Gust, Douglas A. *Drilling Engineering Manager, Sperry-Sun Drilling Services of Canada, Calgary, Alberta, Canada.* OIL MINING.

H

Halvorson, Dr. Ardell D. *Research Leader, Agricultural Research Service, U.S. Department of Agriculture, Akron, Colorado.* SOIL.

Hamelink, Jane. *Group Leader, Automatic Dependent Surveillance, Mitre Corporation, McLean, Virginia.* AIR-TRAFFIC CONTROL.

Hansen, Dr. Anthony R. *Center for Atmospheric and Space Sciences, Augsburg College, Minneapolis, Minnesota.* CHAOS—in part.

Harley, Dr. Calvin B. *Department of Biochemistry, McMaster University, Hamilton, Ontario, Canada.* CELL SENESCENCE.

Harris, Dr. William M. *Department of Biological Sciences, University of Arkansas.* LEAF—coauthored.

Hasuo, Dr. Shinya. *Fujitsu Laboratories, Ltd., Atsugi, Japan.* SUPERCONDUCTING DEVICES.

Hawker, Dr. Craig J. *Department of Chemistry, University of Queensland, Brisbane, Queensland, Australia.* POLYMER—in part.

Hay, Robert A. *Diamond Film Division, Norton Company, Northboro, Massachusetts.* TOOLING.

Hayes, William C. *Editor, Electrical World, McGraw-Hill, Inc., New York, New York.* ELECTRICAL UTILITY INDUSTRY.

Haynes, Dr. Benjamin W. *Division of Environmental Technology, Bureau of Mines, U.S. Department of the Interior, Washington, D.C.* LAND RECLAMATION.

Hearn, Dr. Anthony C. *Resident Scholar, Rand Corporation, Santa Monica, California.* SYMBOLIC COMPUTING.

Heller, Prof. Joseph. *Department of Zoology, Hebrew University of Jerusalem, Israel.* MOLLUSCA.

Hernandez, Dr. Ernesto. *South Atlantic Area Citrus & Subtropical Products Laboratory, Agricultural Research Service, U.S. Department of Agriculture, Winter Haven, Florida.* FOOD MANUFACTURING—in part.

Hickson, Dr. Catherine J. *Research Scientist, Geological Survey of Canada, Vancouver, British Columbia, Canada.* VOLCANOLOGY.

Hirschberg, Prof. Daniel S. *Information & Computer Science, University of California, Irvine.* DATA COMPRESSION.

Holland, Dr. Gary F. *Staff Scientist, Advanced Development, Rocket Research Company, Redmond, Washington.* SOLID-STATE CHEMISTRY.

Huang, Dr. Thomas T. *Ship Hydromechanics Department, David Taylor Research Center, Bethesda, Maryland.* HYDRODYNAMICS—coauthored.

Huner, Dr. Jay V. *Director, Crawfish Research Center, University of Southwestern Louisiana, Lafayette.* AQUACULTURE.

I

Isaacs, Dr. Stuart N. *Laboratory of Viral Diseases, National Institute of Allergy and Infectious Diseases, National Institutes of Health, Bethesda, Maryland.* VIRUS—coauthored.

Israel, Dr. Elliot. *Longwood Medical Area Adult Asthma Center, Brigham and Women's and Beth Israel Hospitals and Harvard Medical School, Boston, Massachusetts.* ASTHMA—coauthored.

Iverson, Dr. Brent. *Department of Chemistry and Biochemistry, University of Texas, Austin.* ANTIBODY—in part.

J

Jacobs, Dr. Gerald H. *Department of Psychology, University of California, Santa Barbara.* VISION—in part.

Jaffar, Dr. Joxan. *IBM T.J. Watson Research Center, Yorktown Heights, New York.* PROGRAMMING LANGUAGES—coauthored.

Jameson, Dr. Donald A. *Administrative Division, U.S. Department of Agriculture Forest Service, Washington, D.C.* FOREST AND FORESTRY.

Jensen, Prof. Farrell E. *Department of Economics, Brigham Young University, Provo, Utah.* DAIRY CATTLE PRODUCTION.

Jewell, Dr. Jack. *Photonics Research Inc., Broomfield, Colorado.* LASER.

Jochens, Dr. Ann E. *Project Scientist, U.S. World Ocean Circulation Experiment, Department of Oceanography, Texas A&M University.* OCEAN CIRCULATION.

Johnson, Dr. Eric F. *Department of Molecular and Experimental Medicine, Research Institute of Scripps Clinic, La Jolla, California.* TOXICOLOGY—in part.

Jones, Dr. Meredith L. *Zoologist Emeritus, Department of Invertebrate Zoology, National Museum of Natural History, Smithsonian Institution, Washington, D.C.* TUBEWORM.

K

Kane, Dr. Gordon L. *Randall Laboratory of Physics, University of Michigan.* PARTICLE ACCELERATOR.

Kaplan, Prof. Simon M. *Department of Computer Science, University of Illinois, Urbana.* SOFTWARE ENGINEERING.

Karp, Dr. Alan H. *Senior Staff Member, IBM, Palo Alto Scientific Center, Palo Alto, California.* CONCURRENT PROCESSING.

Kendig, Dr. James W. *Department of Pediatrics, University of Rochester Medical Center, Rochester, New York.* INFANT RESPIRATORY DISTRESS SYNDROME.

Kessler, Prof. Edwin. *Retired; formerly, Director, National Severe Storms Laboratory, Norman, Oklahoma.* WEATHER.

Keverian, Prof. J. *Materials Engineering Department, Drexel University, Philadelphia, Pennsylvania.* STEEL MANUFACTURE.

Kohrs, Richard H. *Director, Space Station Freedom, Office of Space Flight, National Aeronautics and Space Administration, Washington, D.C.* SPACE STATION.

Kozak, Dr. Mark. *Nora Eccles Harrison Cardiovascular Research and Training Institute, University of Utah.* BLOOD—coauthored.

Kravitz, Prof. Edward A. *Department of Neurobiology, Harvard Medical School, Boston, Massachusetts.* HORMONE.

Kroto, Dr. Harold W. *School of Chemistry and Molecular Sciences, University of Sussex, Brighton, England.* CARBON.

Kumakhov, Dr. Muridan A. *Institute for Roentgen Optical Systems, World Laboratory, Moscow, Russia.* X-RAY OPTICS—coauthored.

Kurtzman, Dr. C. P. *Research Leader, Microbial Properties Research, National Center for Agricultural Utilization Research, Agricultural Research Service, U.S. Department of Agriculture, Peoria, Illinois.* PATENT—coauthored.

L

Lande, Prof. Kenneth. *Department of Astronomy, David Rittenhouse Laboratory, University of Pennsylvania.* SOLAR NEUTRINOS.

Lawandy, Prof. Nabil M. *Department of Physics, Brown University.* RADIATION.

Lee, Dr. Peter K. *Department of Microbiology, University of Minnesota Medical School, Minneapolis.* TOXIC SHOCK-LIKE SYNDROME—coauthored.

Lee, Dr. Siu Au. *Department of Physics, Colorado State University.* RELATIVITY.

Levy, Prof. J. C. *Secretary General, British National Committee for FEANI, London, England.* ENGINEERING.

Lincoln, Dr. Cynthia. *Agricultural Research Service, U.S. Department of Agriculture, Albany, California.* PLANT HORMONES—in part.

Lippard, Prof. Stephen J. *Department of Chemistry, Massachusetts Institute of Technology.* BIOINORGANIC CHEMISTRY—in part.

Logan, Dr. Derek. *Department of Zoology, Laboratory of Molecular Biophysics, University of Oxford, England.* VIRUS—coauthored.

Loik, Dr. Michael E. *Department of Biology and Laboratory of Biomedical and Environmental Sciences, University of California, Los Angeles.* COLD HARDINESS (PLANT).

Lunn, Dr. Robert. *School of Optometry, University of California, Berkeley.* VISION—in part.

M

McCain, Dr. Douglas C. *Department of Chemistry, University of Southern Mississippi, Hattiesburg.* LEAF—in part.

McGrath, Michael J. *Manager, Process Design Operations, Foster Wheeler USA Corporation, Clinton, New Jersey.* PETROLEUM PROCESSING—coauthored.

McIntyre, Dr. Thomas M. *Nora Eccles Harrison Cardiovas-* *cular Research and Training Institute, University of Utah.* BLOOD—coauthored.

McKee, Dr. William H., Jr. *Research Soil Scientist, Southeastern Forest Experiment Station, U.S. Department of Agriculture, Charleston, South Carolina.* WETLANDS MANAGEMENT.

MacMahon, Dr. James A. *Dean, College of Sciences, Utah State University.* SOIL ECOLOGY.

Mallinson, John. *Mallinson Magnetics, Inc., Carlsbad, California.* MAGNETIC RECORDING—coauthored.

Marcotrigiano, Dr. Michael. *Department of Plant and Soil Sciences, University of Massachusetts.* PLANT MORPHOGENESIS.

Margulis, Dr. Lynn. *Sciencewriters, Amherst, Massachusetts.* GAIA HYPOTHESIS—coauthored.

Metraux, Prof. Jean-Pierre. *Institut de Biologie Vegetale et Phytochimie, Université de Fribourg, Switzerland.* PLANT PATHOLOGY—coauthored.

Michaylov, Dr. Spiro. *School of Computer Science, Carnegie Mellon University, Pittsburgh, Pennsylvania.* PROGRAMMING LANGUAGES—coauthored.

Micozzi, Dr. Marc S. *Director, National Museum of Health & Science, Armed Forces Institute of Pathology, Washington, D.C.* FORENSIC ANTHROPOLOGY.

Miller, Prof. Martin W. *Department of Food Science and Technology, University of California, Davis.* FOOD SPOILAGE.

Miller, Michael. *President, Safety Boss, Ltd., Calgary, Alberta, Canada.* OIL-FIELD FIRES—coauthored.

Milstein, Prof. Laurence. *Department of Electrical & Computer Engineering, University of California, San Diego.* SPREAD SPECTRUM COMMUNICATION.

Moore, Prof. Randy. *Chairman, Department of Biological Science, Wright State University, Dayton, Ohio.* LEAF—in part.

Morel, Prof. Jean Louis. *Laboratoire Associé Agronomie et Environnement, Ecole Nationale Superieure d'Agronomie et des Industries Alimentaires, Vandoeuvre les Nancy, France.* RHIZOSPHERE.

Moss, Dr. Bernard. *Laboratory of Viral Diseases, National Institute of Allergy and Infectious Diseases, National Institutes of Health, Bethesda, Maryland.* VIRUS—coauthored.

Motora, Prof. Seizo. *Director, Ship & Ocean Foundation, Tokyo, Japan.* SHIP POWERING AND STEERING.

Myers, Dr. Judith H. *Departments of Zoology and Plant Science, University of British Columbia, Vancouver, British Columbia, Canada.* TENT CATERPILLAR.

N

Nair, Prof. P. K. Ramachandran. *Department of Forestry, Institute of Food and Agricultural Sciences, University of Florida.* AGROFORESTRY.

Nelson, Dr. Jerry E. *Project Scientist, California Association for Research in Astronomy, Kamuela, Hawaii.* TELESCOPE.

Norman, Dr. Eric B. *Nuclear Science Division, Lawrence Berkeley Laboratory, University of California, Berkeley.* NEUTRINO.

Nussinovitch, Dr. Amos. *Department of Biochemistry and Human Nutrition, Hebrew University of Jerusalem, Rehovot, Israel.* FOOD MANUFACTURING—in part.

P

Padua, Dr. David A. *Associate Director, Center for Supercomputing Research and Development, University of Illinois, Urbana.* SUPERCOMPUTER.

Palmer, Prof. James D. *Associate Dean, School of Informa-*

tion Technology and Engineering, George Mason University, Fairfax, Virginia. SYSTEMS ENGINEERING.

Pegg, Prof. David J. *Department of Physics, University of Tennessee.* ATOMIC PHYSICS.

Pegram, Dr. William J. *Department of Geology and Geophysics, Yale University.* DATING METHODS—coauthored.

Peterson, Dr. David L. *Ecosystem Science and Technology Branch, NASA-Ames Research Center, Moffet Field, California.* FOREST RESOURCES.

Peterson, Dr. Per A. *Chairman, Department of Immunology, Scripps Research Institute, La Jolla, California.* ANTIGEN—in part.

Plant, Prof. Richard E. *Vice Chair, Department of Agronomy and Range Science, University of California, Davis.* AGRICULTURE.

Pomeroy, Prof. Lawrence R. *Institute of Ecology, University of Georgia.* MARINE ECOSYSTEM.

Potvin, Dr. Catherine. *Department of Biology, McGill University, Montreal, Quebec, Canada.* PLANT GROWTH.

Prescott, Prof. Stephen M. *Nora Eccles Harrison Cardiovascular Research and Training Institute, University of Utah.* BLOOD—coauthored.

R

Raiteri, Dr. C. M. *Osservatorio Astronomico di Torino, Turin, Italy.* METEORITE—coauthored.

Rana, Dr. Tariq M. *Department of Chemistry and Laboratory of Chemical Biodynamics, University of California, Berkeley.* BIOINORGANIC CHEMISTRY—in part.

Ray, Prof. Peter. *Associate Chairman, Department of Meteorology, Florida State University.* METEOROLOGICAL RADAR.

Reaney, Dr. Martin J. T. *Crop Science Department, University of Saskatchewan, Saskatoon, Saskatchewan, Canada.* PLANT PHYSIOLOGY—coauthored.

Riley, Dr. M. A. *Department of Physics, Florida State University.* NUCLEAR STRUCTURE.

Rinaldo, Paul L. *Manager, Technical Development, American Radio Relay League, Inc., Newington, Connecticut.* RADIO SPECTRUM ALLOCATIONS.

Robertus, Prof. Jon D. *Director, Center for Structural Biology, University of Texas, Austin.* IMMUNOTHERAPY.

Rossiter, Dr. Belinda J. F. *Institute for Molecular Genetics, Human Genome Program Center, Baylor College of Medicine, Houston, Texas.* HUMAN GENETICS—coauthored.

Rothblatt, Martin A. *President and CEO, Multi-Technology Analysis & Research Corporation, Washington, D.C.* DIRECT BROADCAST RADIO SATELLITES.

Ruben, Prof. John. *Department of Zoology, Oregon State University.* AVES.

Ryals, Dr. J. *Agricultural Biotechnology Research Unit, CIBA-GEIGY Corporation, Research Triangle Park, North Carolina.* PLANT PATHOLOGY—coauthored.

S

Sagan, Dr. Dorion. *Sciencewriters, Amherst, Massachusetts.* GAIA HYPOTHESIS—coauthored.

Sage, Dr. Andrew P. *First American Bank Professor and Dean, School of Information Technology and Engineering, George Mason University, Fairfax, Virginia.* INFORMATION TECHNOLOGY.

Sand, Dr. Wayne. *Research Applications Program, National Center for Atmospheric Research, Boulder, Colorado.* AERONAUTICAL METEOROLOGY.

Saunders, Dr. R. Stephen. *Jet Propulsion Laboratory, Pasadena, California.* VENUS.

Savanick, Dr. George A. *Group Supervisor, Novel Fragmentation, Bureau of Mines, U.S. Department of the Interior, Minneapolis, Minnesota.* BOREHOLE MINING.

Savitz, Dr. David A. *Department of Epidemiology, School of Public Health, University of North Carolina.* ELECTRIC POWER SYSTEMS.

Schertler, Dr. Ronald J. *Chief, ACTS Experiments Office, Lewis Research Center, National Aeronautics and Space Administration, Cleveland, Ohio.* COMMUNICATIONS SATELLITE.

Schlievert, Prof. Patrick M. *Department of Microbiology, University of Minnesota Medical School, Minneapolis.* TOXIC SHOCK-LIKE SYNDROME—coauthored.

Schwartz, Dr. Jeffrey. *Department of Psychiatry and Biobehavioral Sciences, University of California School of Medicine, Los Angeles.* AFFECTIVE DISORDERS.

Schwartzberg, Prof. Henry G. *Department of Food Science, University of Massachusetts.* FOOD MANUFACTURING—in part.

Sedjo, Dr. Roger A. *Senior Fellow and Director, Forest Economics and Policy Program, Resources for the Future, Washington, D.C.* FOREST ECOSYSTEM.

Sen, Prof. Ayusman. *Department of Chemistry, Pennsylvania State University.* METHANE.

Shaw, William E. *Director, Development Engineering, Key Technology, Inc., Walla Walla, Washington.* FOOD ENGINEERING—in part.

Sheffer, Dr. Albert L. *Longwood Medical Area Adult Asthma Center, Brigham and Women's and Beth Israel Hospitals and Harvard Medical School, Boston, Massachusetts.* ASTHMA—coauthored.

Shimizu, Dr. Yoji. *Department of Microbiology and Immunology, University of Michigan Medical School, Ann Arbor.* CELLULAR IMMUNOLOGY—in part.

Shur, Dr. Michael. *Department of Electrical Engineering, University of Virginia.* TRANSISTOR.

Sims, Dr. Judith L. *Utah Water Research Laboratory, Utah State University.* BIOREMEDIATION.

Smith, Dr. S. Clark. *President, Minerals Exploration & Environmental Geochemistry, Reno, Nevada.* PROSPECTING.

Smith, Dr. William E. *Ship Hydromechanics Department, David Taylor Research Center, Bethesda, Maryland.* HYDRODYNAMICS—coauthored.

Sperry, Dr. John. *Department of Biology, University of Utah.* XYLEM.

Spreadbury, Dr. Peter J. *Department of Engineering, University of Cambridge, England.* ZENER DIODE.

Stanton, Dr. Timothy K. *Woods Hole Oceanographic Institution, Woods Hole, Massachusetts.* FOOD WEB.

Stephenson, H. G. *President, Norwest Mine Services Ltd., Calgary, Alberta, Canada.* OIL MINING.

Stewart, Prof. Cecil. *Department of Botany, Iowa State University of Science and Technology.* PLANT RESPIRATION.

Stewart, Dr. Jon D. *Department of Chemistry, Pennsylvania State University.* ANTIBODY—coauthored.

Stroh, Dr. Patricia. *School of Oceanography, University of Washington.* MID-OCEANIC RIDGE.

Stuart, Dr. David. *Department of Zoology, Laboratory of Molecular Biophysics, University of Oxford, England.* VIRUS—coauthored.

Stuckey, Dr. Peter J. *Department of Computer Science, University of Melbourne, Victoria, Australia.* PROGRAMMING LANGUAGES—coauthored.

Swanson, Dr. R. Lawrence. *Director, Waste Management Institute, Marine Sciences Research Center, State University of New York, Stony Brook.* RECYCLING TECHNOLOGY—coauthored.

Swezey, J. L. *Microbial Properties Research, National Center for Agricultural Utilization Research, Agricultural Research Service, U.S. Department of Agriculture, Peoria, Illinois.* PATENT—coauthored.

Swift-Hook, Prof. Donald T. *Department of Electronic and Electrical Engineering, King's College, London, England.* ENERGY SOURCES.

T

Tanzi, Dr. Rudolph E. *Molecular Neurogenetics Laboratory, Massachusetts General Hospital, Charlestown.* ALZHEIMER'S DISEASE.

Taylor, Prof. David M. *Director, Institut für Genetik und für Toxikologie von Spaltstoffen, Kernforschungszentrum Karlsruhe, Karlsruhe, Germany.* TOXICOLOGY—in part.

Terrazas, Dr. Teresa. *Department of Biology, University of North Carolina.* LATERAL MERISTEM.

Thompson, David W. *Orbital Sciences Corporation, Fairfax, Virginia.* SATELLITE LAUNCH SYSTEM.

Thompson, Dr. Suzanne. *Department of Psychology, Pomona College, Claremont, California.* STRESS (PSYCHOLOGY)—in part.

Thornton, Prof. Richard D. *Department of Electrical Engineering and Computer Science, Massachusetts Institute of Technology.* INSTRUMENTATION.

Tomonari, Dr. Kyuhei. *Transplantation Biology Section, Medical Research Council, Clinical Research Centre, Harrow, England.* ANTIGEN—in part.

Turekian, Dr. Karl K. *Department of Geology and Geophysics, Yale University.* DATING METHODS—coauthored.

Tyner, Dr. Craig E. *Sandia National Laboratories, Albuquerque, New Mexico.* SOLAR ENERGY.

V

Vallee, Dr. Richard B. *Principal Scientist, Worcester Foundation for Experimental Biology, Shrewsbury, Massachusetts.* MICROTUBULES.

van Linden, Dr. J. H. L. *Alcoa Laboratories, Aluminum Company of America, Alcoa Center, Pennsylvania.* RECYCLING TECHNOLOGY—in part.

Van Valkenburgh, Dr. Blaire. *Department of Biology, University of California, Los Angeles.* MAMMALIA.

Voragen, Prof. Fons. *Department of Food Science, Agricultural University, Wageningen, The Netherlands.* FOOD MANUFACTURING—coauthored.

W

Wabl, Prof. Matthias. *Department of Microbiology and Immunology, University of California, San Francisco.* IMMUNOGLOBULIN.

Wagenet, Prof. R. J. *Chair, Department of Soil, Crop and Atmospheric Sciences, Cornell University.* PESTICIDE.

Wahlen, Prof. Martin. *Geological Research Division, Scripps Institution of Oceanography, University of California, La Jolla.* CLIMATOLOGY.

Ward, Dr. Eric R. *Senior Scientist, Agricultural Biotechnology Research Unit, CIBA-GEIGY Corporation, Research Triangle Park, North Carolina.* PLANT PATHOLOGY—coauthored.

Whittaker, Dr. James W. *Department of Chemistry, Carnegie Mellon University, Pittsburgh, Pennsylvania.* ENZYME—in part.

Wood, Roger. *IBM Magnetic Recording Institute, San Jose, California.* MAGNETIC RECORDING—coauthored.

Worthington, Dr. Paul F. *Senior Research Associate, Reservoir Description, BP International Limited, Sunbury Research Centre, Middlesex, England.* BOREHOLE LOGGING.

Wyman, Richard E. *Vice President, Research, Canadian Hunter Exploration, Ltd., Calgary, Alberta, Canada.* OILFIELD FIRES—coauthored.

Y

Yap, Dr. Roland H. C. *IBM T.J. Watson Research Center, Yorktown Heights, New York.* PROGRAMMING LANGUAGES—coauthored.

Yarbrough, Dr. Walter A. *Materials Research Laboratory, Pennsylvania State University.* DIAMOND.

Z

Zimmerman, Dr. Guy A. *Nora Eccles Harrison Cardiovascular Research and Training Institute, University of Utah.* BLOOD—in part.

Index

Index

Asterisks indicate page references to article titles.